Prentice-Hall International, Inc., *London*
Prentice-Hall of Australia, Pty. Ltd., *Sydney*
Prentice-Hall of Canada, Ltd., *Toronto*
Prentice-Hall of India Private Limited, *New Delhi*
Prentice-Hall of Japan, Inc., Tokyo

Radiation and Scattering of Waves

Radiation and Scattering of Waves

LEOPOLD B. FELSEN

Professor of Electrophysics
Polytechnic Institute of Brooklyn

NATHAN MARCUVITZ

Professor of Applied Physics
New York University

Prentice-Hall, Inc., Englewood Cliffs, New Jersey

To

Sima (F.) and Muriel (M.)

for their patience and forebearance

Preface

Classical field theory is concerned with the space-time behavior of physical variables describing field phenomena excited by prescribed sources. In the linear regime the methodology of description is to a large extent independent of the nature of the field and equally applicable to acoustic, electromagnetic, plasma and other fields. Within a stated space-time domain, the general linear field requires a specification of the field variables and prescribed sources, usually in terms of partial differential equations, with uniqueness of solution following from a statement of boundary and initial conditions. Solution of the so specified field problem can be effected by formal field representations whose reduction to rapidly convergent forms in appropriate space-time domains poses problems of special interest.

The general field problem is a scattering or diffraction problem distinguished by excitation from sources located either at finite distances or at infinity, and by spatial and (or) temporal complexities in the scattering region. Equivalence concepts permit replacement of the scatterers by (initially unknown) "induced currents;" they reduce the overall problem to that of finding fields radiated by prescribed and induced sources in domains of relatively simple geometrical shape. It is this latter radiation problem of determining fields excited in relatively simple regions by arbitrary sources, and the concomitant propagation of these fields, with which this book is primarily concerned. The determination of induced currents is regarded as a distinctly separate problem; it frequently poses analytical questions of considerable difficulty and usually requires integral equation techniques or the treatment of infinite sets of simultaneous equations.†

For linear fields, wherein the superposition principle is applicable, the basic radiation problem is that of determining the field excited by a point source. This is the so-called Green's function problem. Green's functions are scalar for the simple acoustic field, dyadic functions for the vector electromagnetic field, and $N \times N$ matrix functions for more complex fields. For a general linear field the components

---

†For an account of original pioneering waveguide applications, see Julian Schwinger and David Saxon, *Discontinuities in Waveguides*, Gordon and Breach, New York (1969); L. A. Weinstein, *The Theory of Diffraction and the Factorization Method*, Golem Press, Boulder, Colorado (1969).

of a dyadic or matrix Green's function are not usually independent, but for "separable" regions the overall Green's function may be decomposed (scalarized) into a number of independent scalar Green's functions. Thus, in the case of "separable" regions, dyadic electromagnetic Green's functions are reducible to scalar acoustic-type Green's functions, an observation that implies the direct applicability of results from one field to that of another. The central theme of this book revolves essentially about the evaluation of Green's functions in homogeneous and inhomogeneous regions of planar, cylindrical, spherical, etc., symmetry.

A Green's function may be represented in various ways as a superposition of wave functions that display the symmetries of a field region. Thus, in a linear, homogeneous, stationary, unbounded region, the plane wave functions $\exp[i(\mathbf{k}\cdot\mathbf{r} - \omega t)]$ constitute a convenient set capable of representing completely a relatively arbitrary space-time dependent field; $\mathbf{k}$ and ω denote, respectively, the wave vector and radian frequency, with $(\mathbf{k}, \omega) = (k_x, k_y, k_z, \omega)$ spanning an appropriate spectrum. In the so-called Fourier–Laplace representation the $(\mathbf{k}, \omega)$ spectrum is continuous over (almost) all real values from $-\infty$ to $+\infty$, and the resulting plane waves comprise a complete orthogonal set in space-time; a characteristic feature of such a representation is that the plane-wave field amplitudes are determined by a simple algebraic analysis. In what shall be called space- or time-guided wave representations, only three of the $(\mathbf{k}, \omega)$ periodicities are employed to define the wave spectrum, the remaining one being determined by a so-called dispersion equation; the resulting plane wave set has orthogonality and completeness properties on a three-dimensional hypersurface of space-time, and the field amplitudes are determined by solutions of ordinary differential equations. These alternative plane wave representations are typical of similar representations that obtain for bounded and anisotropic regions; each representation is characterized by convergence properties and ranges of applicability that are useful in the solution of different types of field problems.

Although the stated field representations are formally exact field solutions, the integrations occurring therein must be performed explicitly so as to yield rapidly convergent or closed form expressions for the field. This process can be effected exactly in a number of cases but generally requires approximations. A powerful and physically significant approximation procedure is provided by the function-theoretic method of "saddle point integration." This method yields asymptotic "quasi-optic" field approximations in regions illuminated by field sources, and can be modified to apply as well to "shadow" and "transition" (penumbra) regions. Such methods are related intimately to ray theories,† wavepacket propagation, WKB procedures, etc., employed in the solution of systems of partial differential equations.

With the knowledge of a proper set of modes or waves as a base, modal representation of solutions to field problems requires a two-fold procedure: (1) an analysis or transform process to determine the dependence of modal amplitudes on sources, and (2) a modal synthesis or inverse transform for the evaluation of the space-time dependence of the desired fields. The various chapters of this book develop and illustrate these modal analysis and synthesis procedures in a wide range of radiation and wave-scattering applications. Although the table of contents provides a detailed indication of subject organization, it may be desirable to elucidate and place in

---

†J. B. Keller, "A Geometric Theory of Diffraction," Calculus of Variations and its Applications, *Symposia Appl. Math.*, McGraw-Hill, New York, **8** (1958), pp. 27–52.

proper perspective interrelationships among the topics covered. With this intent, we sketch below some of the guiding themes that underly the organization of the chapters.

Chapter 1 is devoted both to the formulation of linear field problems and to an indication of methods of their solution. Features, properties, and methodology common to linear acoustic, electromagnetic, and plasma fields are emphasized within the context of a first order (partial differential equation) field theory; "reduced" (second and higher order) field formulations are considered subsequently. Green's functions for the above fields are introduced in comparative form highlighting their similarities and interrelationships. Exact modal representations of these Green's functions are presented in alternative ways and evaluated in closed form for simple unbounded homogeneous regions. Approximate evaluations, equally valid for inhomogeneous, anisotropic, and dispersive regions, are considered in some detail, firstly via saddle point integration and then by ray-optic and transport equation techniques. This first chapter ought not to be neglected in a first reading; it is intended to knit together with a unified viewpoint, and to anticipate with simple illustrations, many of the applications in subsequent chapters. The introductory comments to the various sections should be helpful in providing a quick overall perspective.

Chapters 2 and 3 are concerned with the modal analysis of fields in regions that generally are bounded and inhomogeneous. In Chapter 2, eigenvalue problems that provide a modal basis for transformation of vector electromagnetic problems into transmission-line (ordinary differential equation) problems are deduced for electromagentic fields in uniform and spherical waveguide regions. Techniques for solution of transmission-line equations, to which field equations are reduced in waveguide regions, are reviewed via network-theoretic and one-dimensional scalar Green's function methods. Chapter 3 contains explicit expressions for vector and scalar mode functions and their orthogonality properties for a variety of waveguide cross-sections. By classical methods, characteristic Green's function (resolvent) methods, and the method of comparison equations, these results are derived from both exact and approximate treatment of one-dimensional Sturm–Liouville type problems appropriate to homogeneous or inhomogeneously filled cross-sections. Apart from their relevance to applications treated subsequently in this book, Chapters 2 and 3 can form the basis for a course dealing with transmission-line and related eigenvalue problems.

Chapter 4 contains an extensive discussion of saddle point methods of integration necessary for approximate closed form synthesis of modal representations. An account of steepest descent integration is included, with particular attention to mathematically uniform descriptions of effects arising from the presence of different types of singularities near saddle points and from the confluence of several saddle points. These effects relate physically to field descriptions within so-called transition or penumbra regions separating "light" and "shadow" areas—or more generally, different propagation modes—in a field. Although the physical significance of various integral representations is emphasized, this chapter is self-contained and may serve as a reference to the theory of asymptotic evaluation of integrals.

Applications of the preceding theory to the explicit determination of fields radiated by sources in isotropically stratified planar, cylindrical and spherical regions are presented in Chapters 5 and 6. Although examples relate primarily to the electromagnetic fields, scalarization is frequently permissible (either directly or by decomposition), in which event the results then apply as well to acoustic and other scalar problems. Because of the complexity of several of the calculations, an attempt has been made to standardize the format for presentation of many of the results. After statement of the problem, a summary and physical interpretation of the calculated results are first presented in their various ranges of applicability; this is followed, under the heading of *Discussion*, by a more detailed indication of the function-theoretic analysis and limitations, if any. This separation of theory and results is intended to appeal both to the application- and theory-oriented reader; it should provide a type of handbook listing of the problems solved, as is evident from the table of contents for these chapters.

Chapters 7 and 8 are concerned with extensions and applications to fields in anisotropic regions. Uniaxial media are considered in Chapter 7, while gyrotropic and somewhat more general anisotropic media are treated in Chapter 8. The anisotropic regions under consideration are intended to apply to crystalline, plasma and ferromagnetic type media and, in a "reduced" electromagnetic formulation, are characterized by dyadic (tensor) permittivity and permeability parameters. This view of such media ignores certain non-electromagnetic effects, but when applicable, does provide a quantitative indication of many of the dispersive wave phenomena to be expected.

Concerning overall philosophy of subject presentation, much effort in this book has been expended on developing and applying a unified formalism for systematized eigenmode and transmission-line (network) analysis of linear field problems. Whether such systematization is justified in the solution of one or two individual problems is debatable. However, for analysis of classes of field problems having similar but not identical features, elimination of redundant aspects becomes almost essential. While the treatment thus emphasizes techniques applicable to broad classes of problems, an attempt has been made, by self-contained problem statement and frequent cross referencing, also to serve the reader interested in only a particular case. The guided-wave approach alluded to above has been found successful for many electromagnetic and acoustic field problems, and it may prove to be equally valid in similar applications for plasma, solid state, and other fields.

A note of apology ought to be sounded because of a possible unevenness in portions of this book resulting from the chronology of its preparation. Much of the material has been presented in lectures by the authors over the past 15–20 years, mostly at the Polytechnic Institute of Brooklyn and partially, by one of the authors, at New York University; in fact, a series of widely distributed reports issued some time ago by the Microwave Research Institute of the Polytechnic under the title "Modal Analysis and Synthesis of Electromagnetic Fields" constituted a first draft of portions of this manuscript. For initial support in the preparation of these reports, the authors express their appreciation to the Air Force Cambridge Research Laboratories, Bedford, Massachusetts; they also gratefully acknowledge the sponsorship by the Air Force Cambridge Research Laboratories and by the Joint Services Electronics Program of research, the results of which are included in this book. Although a good deal of effort has been expended in continually revising the text material, the authors do not feel that the presentation has been optimized in all respects. One feature of subject treatment should be mentioned in this context. The imaginary unit $\sqrt{-1}$ is designated throughout the book as i or $-j$, depending on whether the subject matter relates primarily to mathematical physicists or engineers. Usage of $\exp(-i\omega t)$ and $\exp(+j\omega t)$ in these respective disciplines has been fairly customary in the treatment of time-harmonic fields. Evidently, the division is not unambiguous, but engineers have traditionally been concerned more with transmission-line and network aspects of the overall field problem and less with scattering and diffraction. To minimize confusion, the time dependence is stated whenever relevant, and the facility in switching from one dependence to another is often useful when comparing various results published in the technical literature. Finally, it should be mentioned that no attempt has been made to include a comprehensive bibliography; however, the references cited provide adequate background information.

A number of individuals have contributed to the preparation of this book. We have benefited from comments and criticisms by colleagues and students. With special gratitude, we would like to acknowledge the efforts of Mrs. Margaret Bartoli who did much of the painstaking work in the typing and organization of the final manuscript. For providing necessary services and facilities, thanks in large measure are due to the Electrophysics Department of the Polytechnic, and in the final stages also to the School of Engineering and Science, New York University. Finally, we gratefully note the continued encouragement and patience of our respective families, which made completion of this effort possible.

L. B. FELSEN
N. MARCUVITZ

New York, N.Y.

Contents

Contents

Contents **xxi**

Contents

xxvii

8. FIELDS IN ANISOTROPIC REGIONS   **821**

Radiation and Scattering
of Waves

1. Space- and Time-Dependent Linear Fields

1.1 FORMULATION OF VECTOR FIELD AND SCALAR POTENTIAL PROBLEMS

Field equations describe implicitly the space- and time-dependent response of a linear field to a known excitation and pose a problem of determining field solutions subject to initial and boundary conditions. Since knowledge of symmetry (reciprocity) properties of a field frequently facilitates the determination of explicit field solutions, it is desirable to infer such properties from the general form of the field equations. For this purpose one considers certain auxiliary or adjoint problems, related to the original field problem in such a way as to reveal the space–time symmetry of the original field.

If field problems are phrased in terms of Green's functions, which describe the field response to "point-source excitation," the desired properties appear most succinctly as symmetries in these Green's functions. Relevant Green's functions for a number of classical linear fields are defined herein so as to emphasize common features in the description of any linear field and to expedite the application of generic mathematical solutions to several different fields. Since scalar, vector, or n-component fields are to be considered, the corresponding Green's functions assume scalar, dyadic, or matrix forms. Scalar Green's functions are most convenient for analysis, and hence wherever possible, dyadic and matrix Green's functions will be decomposed into their independent scalar components. Such a scalarization cannot always be effected.

Field equations may be phrased either as a system of first-order partial differential equations, or on elimination of some of the field variables, as higher-order equations. One advantage of the first-order formulation is its applicability in unchanged form to both homogeneous or inhomogeneous media (i.e., media whose constitutive parameters are respectively independent of, or dependent on, position). The second-order or "reduced" formulations are usually dependent upon whether they relate to homogeneous or inhomogeneous media. Both first- and second-order formulations will be presented for a number of linear fields.

In reduced formulations of a composite field, effects of one type of field are exhibited as a modification of the constitutive parameters of another type. For example, in an electromagnetic-plasma field, dynamical effects of charged particles in ionized media appear in a reduced electromagnetic formulation as a modification of the equivalent permittivity of the electromagnetic medium. Whereas the constitutive parameters in a first-order formulation are non-dispersive, the constitutive parameters in a reduced formulation exhibit spatial and temporal dispersion. Spatial and temporal dispersion in constitutive parameters obtains when the parameters depend, respectively, on spatial and temporal derivative operators, in addition to a possible dependence on space and time variables. It should be noted that it is also conventional to employ the term "dispersion" in connection with the frequency dependence of plane-wave constituents of a field. These two uses of the term are not necessarily equivalent and should be distinguished.

1.1a The Scalar Acoustic Field

General properties

A non-viscous uncharged fluid of mass density $n_0 m$ and static isotropic pressure p_0 provides a homogeneous or inhomogeneous background capable of supporting acoustic phenomena. In the linear regime the acoustic field is describable by small-amplitude variations of pressure $p = p(\mathbf{r}, t)$ and velocity $\mathbf{v} = \mathbf{v}(\mathbf{r}, t)$ obeying the Euler field equations[1]:

$$\frac{1}{\gamma p_0}\frac{\partial p}{\partial t} + \nabla \cdot \mathbf{v} = -s,$$

$$\nabla p + n_0 m \frac{\partial \mathbf{v}}{\partial t} = -\mathbf{f}. \tag{1}$$

The excitation terms $s = s(\mathbf{r}, t)$ and $\mathbf{f} = \mathbf{f}(\mathbf{r}, t)$ represent the scalar particle source and the impressed vector force densities, respectively; γ is the specific heat ratio for the fluid, and ∇ is the spatial gradient operator. The pressure and velocity fields are uniquely defined by Eqs. (1) if one imposes on the excitation the requirement $s = 0 = \mathbf{f}$ for $t \leq t_1$, and on the fields both the initial condition,

$$p = 0 = \mathbf{v} \qquad \text{for } t \leq t_1, \tag{1a}$$

and the boundary condition,

$$p = \alpha \mathbf{v} \cdot \mathbf{n} \qquad \text{on } S. \tag{1b}$$

The unit vector $\mathbf{n}$ is normal to the surface S (if any) bounding the volume within which the field equations (1) are applicable, and α is a parameter characteristic of the boundary.

On multiplication of the two field equations (1) by p and $\mathbf{v}$, respectively, and addition, one derives the conservation-of-energy statement:

$$\nabla\cdot p\mathbf{v} = -\frac{\partial}{\partial t}\left(\frac{1}{2\gamma p_0}p^2 + \frac{n_0 m}{2}\mathbf{v}^2\right) - sp - \mathbf{f}\cdot\mathbf{v}. \tag{2}$$

The vector $p\mathbf{v}$ is identified as the instantaneous acoustic power flow per unit area at $\mathbf{r}, t$ and the terms in parentheses as the total stored energy density. The quantities $-sp$ and $-\mathbf{f}\cdot\mathbf{v}$ represent the power supplied per unit volume at $\mathbf{r}, t$ by the impressed source and force densities, respectively.

For homogeneous media one derives a second-order set of field equations that follow readily from Eq. (1) on elimination of $\mathbf{v}$ or p. For simplicity we omit the excitation terms and thereby obtain the homogeneous (wave) equations

$$\nabla^2 p - \frac{1}{a^2}\frac{\partial^2}{\partial t^2}p = 0, \tag{3a}$$

$$\nabla\nabla\cdot\mathbf{v} - \frac{1}{a^2}\frac{\partial^2}{\partial t^2}\mathbf{v} = 0, \tag{3b}$$

where $a = (\gamma p_0/n_0 m)^{1/2} = (\gamma\kappa T_0/m)^{1/2}$ is the acoustic (thermal) speed, T_0 the background temperature, and κ the Boltzmann constant. The presence of source terms leads to an evident modification of Eqs. (3). The "longitudinal" character of the acoustic field becomes manifest on taking the divergence $(\nabla\cdot)$ and curl $(\nabla\times)$ of Eq. (3b). It is thereby apparent that in the source-free case $\nabla\cdot\mathbf{v}$ obeys a wave equation of the same form as Eq. (3a); the vorticity $\nabla\times\mathbf{v}$ remains constant in time and is, in fact, zero because of Eq. (1a), if the velocity field is bounded. As will be indicated in Sec. 1.2a, these statements imply that for acoustic propagation of a single plane wave, the only non-vanishing component of the velocity field is along the direction of propagation (i.e., is "longitudinal").

The solution of spatially homogeneous acoustic field problems can be based on either first-order or second-order field equations. A virtue of the latter is that since the acoustic field is basically scalar, the field description may be expressed in terms of a single scalar variable, which may be a pressure, a velocity potential, etc. Nevertheless, we shall employ a procedure based on the first-order field equations (1), since this has the advantages of (a) displaying in simple terms the field dependence on the excitations s and $\mathbf{f}$, (b) applying to both homogeneous and inhomogeneous media, and (c) permitting the use of a mathematical format common to all linear field problems.

Linearity of the field equations (1) implies that the pressure and velocity at any space–time point $\mathbf{r}, t$ can be expressed in terms of the excitations s and $\mathbf{f}$ as

$$p(\mathbf{r}, t) = -\int G_{11}(\mathbf{r}, \mathbf{r}'; t, t')s(\mathbf{r}', t')\,d\mathbf{r}'\,dt' - \int \mathbf{G}_{12}(\mathbf{r}, \mathbf{r}'; t, t')\cdot\mathbf{f}(\mathbf{r}', t')\,d\mathbf{r}'\,dt'$$

$$\mathbf{v}(\mathbf{r}, t) = -\int \mathbf{G}_{21}(\mathbf{r}, \mathbf{r}'; t, t')s(\mathbf{r}', t')\,d\mathbf{r}'\,dt' - \int \mathscr{G}_{22}(\mathbf{r}, \mathbf{r}'; t, t')\cdot\mathbf{f}(\mathbf{r}', t')\,d\mathbf{r}'\,dt',$$

$$\tag{4}$$

where $d\mathbf{r}'\, dt'$ is the differential space–time volume element and the integration extends over regions wherein the excitations are non-vanishing. From Eq. (4) one identifies $G_{11}(\mathbf{r}, \mathbf{r}'; t, t')$ and $\mathbf{G}_{21}(\mathbf{r}, \mathbf{r}'; t, t')$ as a scalar and a vector Green's function representing, respectively, the negative of the pressure and velocity at $\mathbf{r}, t$ due to a "unit" source[†] of fluid particles at $\mathbf{r}', t'$. Similarly, $\mathbf{G}_{12}(\mathbf{r}, \mathbf{r}'; t, t')$ and $\mathscr{G}_{22}(\mathbf{r}, \mathbf{r}'; t, t')$ are identifiable as vector and dyadic Green's functions representing (when dot product multiplied from the right by a unit vector $\mathbf{e}'$) the pressure and velocity, respectively, at $\mathbf{r}, t$ arising from a "unit" vector force density[‡] acting in the direction $\mathbf{e}'$ at $\mathbf{r}', t'$.

The field representation in Eq. (4) reduces the problem of solving the field equations (1) to the determination of the four acoustic Green's functions G_{11}, $\mathbf{G}_{12}$, $\mathbf{G}_{21}$, and $\mathscr{G}_{22}$. The primary virtue of this reduction is that, in solution of the field problem for the Green's functions, complexities associated with the functional form of the excitations $s(\mathbf{r}, t)$ and $\mathbf{f}(\mathbf{r}, t)$ are eliminated. To ascertain the defining equations for the Green's functions, one substitutes the representation (4) into Eqs. (1), whence, in view of the arbitrariness of the excitations s and $\mathbf{f}$, one obtains[§]

$$\frac{1}{\gamma p_0}\frac{\partial}{\partial t}G_{11} + \mathbf{\nabla}\cdot\mathbf{G}_{21} = \delta(\mathbf{r} - \mathbf{r}')\delta(t - t'), \qquad \frac{1}{\gamma p_0}\frac{\partial}{\partial t}\mathbf{G}_{12} + \mathbf{\nabla}\cdot\mathscr{G}_{22} = 0,$$

$$\mathbf{\nabla}G_{11} + n_0 m\frac{\partial}{\partial t}\mathbf{G}_{21} = 0, \qquad \mathbf{\nabla}\mathbf{G}_{12} + n_0 m\frac{\partial}{\partial t}\mathscr{G}_{22} = \mathbf{1}\delta(\mathbf{r} - \mathbf{r}')\delta(t - t'),$$

$$(5)$$

subject to boundary and initial conditions

$$G_{11} = \alpha\mathbf{n}\cdot\mathbf{G}_{21} \qquad \text{on } S, \qquad\qquad \mathbf{G}_{12} = \alpha\mathbf{n}\cdot\mathscr{G}_{22} \qquad \text{on } S, \qquad (5a)$$

$$G_{11} \equiv 0 \equiv \mathbf{G}_{21} \qquad \text{for } t \leq t', \qquad \mathbf{G}_{12} \equiv 0 \equiv \mathscr{G}_{22} \qquad \text{for } t \leq t', \qquad (5b)$$

where $\mathbf{n}$ is the outward normal vector on the surfaces S (if any) bounding the region within which the field is defined.

Since the form of the boundary conditions (5a) may complicate the determination of the acoustic Green's functions, it is frequently desirable to introduce other Green's functions, defined by simpler boundary conditions such as $G_{11} = 0 = \mathbf{G}_{12}$ on S. In this event, occurring in the case of bounded regions, the representation (4) must be generalized by addition of surface integral terms arising from singular source distributions on the boundary surface S.

On elimination of $\mathbf{G}_{21}$ and $\mathbf{G}_{12}$ from Eqs. (5) one obtains, for a homogeneous medium, second-order partial differential equations defining the scalar

---

[†] A unit source has the space–time form $\delta(\mathbf{r} - \mathbf{r}')\delta(t - t')$, where $\delta(\mathbf{r} - \mathbf{r}')$ and $\delta(t - t')$ are, respectively, the three- and one-dimensional Dirac delta functions. The three-dimensional delta function is defined by $\delta(\mathbf{r}) = 0$ or ∞, depending on whether $\mathbf{r} \neq 0$ or $\mathbf{r} = 0$, and its volume integral is unity for $\mathbf{r}$ within the integration volume.

[‡] A unit vector force density at $\mathbf{r}', t'$ has the space–time form $\mathbf{e}'\delta(\mathbf{r} - \mathbf{r}')\delta(t - t')$, where $\mathbf{e}'$ is a unit vector.

[§] The unit dyadic $\mathbf{1}$ is defined by $\mathbf{1}\cdot\mathbf{A} = \mathbf{A}\cdot\mathbf{1} = \mathbf{A}$, where $\mathbf{A}$ is an arbitrary vector.

Green's function $G_{11}(\mathbf{r}, \mathbf{r}'; t, t')$ and the dyadic Green's function $\mathscr{G}_{22}(\mathbf{r}, \mathbf{r}'; t, t')$:

$$\left(\nabla^2 - \frac{1}{a^2}\frac{\partial^2}{\partial t^2}\right)G_{11} = -n_0 m \frac{\partial}{\partial t}\delta(\mathbf{r} - \mathbf{r}')\delta(t - t'), \tag{6a}$$

$$\left(\nabla\nabla - \frac{1}{a^2}\frac{\partial^2}{\partial t^2}\mathbf{1}\right)\cdot\mathscr{G}_{22} = -\frac{1}{\gamma p_0}\frac{\partial}{\partial t}\delta(\mathbf{r} - \mathbf{r}')\delta(t - t')\mathbf{1}. \tag{6b}$$

Knowledge of the Green's functions defined in Eqs. (6) permits, on recourse to Eqs. (5), determination of $\mathbf{G}_{21}$ and $\mathbf{G}_{12}$ by time quadrature.

It is of interest to explore properties of acoustic Green's functions that can be inferred prior to the explicit solution of Eqs. (5). For example, the invariance of the form of Eqs. (5) to arbitrary linear (coordinate) displacements in space and time implies that, in an unbounded homogeneous region, the solutions of (5) are functions of the differences $\mathbf{r} - \mathbf{r}'$ and $t - t'$, i.e., that

$$G_{ij}(\mathbf{r}, \mathbf{r}'; t, t') = G_{ij}(\mathbf{r} - \mathbf{r}', t - t'), \tag{6c}$$

where $i, j = 1, 2$ distinguish the various scalar, vector, and dyadic Green's functions.

Additional properties of acoustic Green's functions can be inferred by relating the solutions of the field equations (1) to those of an "adjoint" problem. To this end, adjoint equations, as well as boundary and initial conditions, are chosen so as to evolve a reciprocity relation, Eq. (9), between the original and adjoint fields. Equations for the adjoint pressure and velocity fields $p^+ = p^+(\mathbf{r}, t)$ and $\mathbf{v}^+ = \mathbf{v}^+(\mathbf{r}, t)$, respectively, follow from the original field equations (1) on applying the reflection transformation $\partial/\partial t \to -\partial/\partial t$ and $\nabla \to -\nabla$:

$$-\frac{1}{\gamma p_0}\frac{\partial}{\partial t}p^+ - \nabla\cdot\mathbf{v}^+ = -s^+,$$

$$\tag{7}$$

$$-\nabla p^+ - n_0 m \frac{\partial\mathbf{v}^+}{\partial t} = -\mathbf{f}^+.$$

They are subject to boundary and initial conditions

$$p^+ = -\alpha\mathbf{n}\cdot\mathbf{v}^+ \qquad \text{on } S, \tag{7a}$$

$$p^+ = 0 = \mathbf{v}^+ \qquad \text{for } t \geq t_2, \tag{7b}$$

which display a similar spatial and temporal reflection, and to excitations $s^+ = s^+(\mathbf{r}, t)$ and $\mathbf{f}^+ = \mathbf{f}^+(\mathbf{r}, t)$, which vanish for $t > t_2$. The adjoint equations (7) differ from the original acoustic equations (1) in that they are "time reversed," have different excitations, and yield ingoing rather than outgoing wave solutions. Ingoing wave solutions have a functional form $F(|\mathbf{r}| + at)$ and may vanish for $t > t_2$, whereas outgoing solutions have the form $F(|\mathbf{r}| - at)$ and may vanish for $t < t_1$.

By appropriate multiplication of Eqs. (1) by p^+ and $\mathbf{v}^+$, of Eqs. (7) by p and $\mathbf{v}$, and addition of the resulting equations, one deduces the "reciprocity relation"

$$\nabla\cdot(p\mathbf{v}^+ + p^+\mathbf{v}) + \frac{1}{\gamma p_0}\frac{\partial}{\partial t}(pp^+) + n_0 m\frac{\partial}{\partial t}(\mathbf{v}\cdot\mathbf{v}^+)$$

$$= +ps^+ - p^+s - \mathbf{f}\cdot\mathbf{v}^+ + \mathbf{f}^+\cdot\mathbf{v} \qquad (8)$$

between solutions of the original and adjoint equations. On integration of Eq. (8) over the space–time region bounded by the spatial surface S and the times $t_1, t_2 > t_1$, and on use of the divergence theorem as well as the boundary and initial conditions (1a) and (1b) and (7a) and (7b), one derives the integral form of the reciprocity relation:

$$0 = \iiint d\mathbf{r}\int_{t_1}^{t_2} dt[ps^+ - p^+s - \mathbf{f}\cdot\mathbf{v}^+ + \mathbf{f}^+\cdot\mathbf{v}]. \qquad (9)$$

Linearity of the adjoint equations (7) implies that the adjoint field can be expressed in terms of adjoint Green's functions in a manner similar to Eqs. (4):

$$p^+(\mathbf{r}, t) = -\int G_{11}^+(\mathbf{r}, \mathbf{r}'; t, t')s^+(\mathbf{r}', t')\,d\mathbf{r}'\,dt' - \int \mathbf{G}_{12}^+(\mathbf{r}, \mathbf{r}'; t, t')\cdot\mathbf{f}^+(\mathbf{r}', t')\,d\mathbf{r}'\,dt',$$

$$\mathbf{v}^+(\mathbf{r}, t) = -\int \mathbf{G}_{21}^+(\mathbf{r}, \mathbf{r}'; t, t')s^+(\mathbf{r}', t')\,d\mathbf{r}'\,dt' - \int \mathscr{G}_{22}^+(\mathbf{r}, \mathbf{r}'; t, t')\cdot\mathbf{f}^+(\mathbf{r}', t')\,d\mathbf{r}'\,dt',$$

$$(10)$$

where the adjoint Green's functions, G_{ij}^+, distinguished by the superscript $^+$, play the same role in the adjoint field as the Green's functions in the original field. The defining equations for the adjoint Green's functions follow from Eqs. (7) by reflection of the original Green's function equations (5) and (6), together with boundary and initial conditions, in the manner indicated in Eqs. (7).

The reciprocity condition, Eq. (9), provides a relation between the fields p, $\mathbf{v}$ excited by s, $\mathbf{f}$ and the adjoint fields p^+, $\mathbf{v}^+$ excited by s^+, $\mathbf{f}^+$. If the various excitations are chosen to be of the "point-source" form,

$$s = \delta(\mathbf{r} - \mathbf{r}')\delta(t - t'), \qquad s^+ = \delta(\mathbf{r} - \mathbf{r}'')\delta(t - t''),$$

$$\mathbf{f} = 0, \qquad\qquad\qquad \mathbf{f}^+ = 0,$$

then, from Eqs. (4) and (10), the corresponding field solutions are

$$p = -G_{11}(\mathbf{r}, \mathbf{r}'; t, t'), \qquad p^+ = -G_{11}^+(\mathbf{r}, \mathbf{r}''; t, t''),$$

$$\mathbf{v} = -\mathbf{G}_{21}(\mathbf{r}, \mathbf{r}'; t, t'), \qquad \mathbf{v}^+ = -\mathbf{G}_{21}^+(\mathbf{r}, \mathbf{r}''; t, t'\,),$$

and hence one infers from Eq. (9) that

$$G_{11}(\mathbf{r}'', \mathbf{r}'; t'', t') = G_{11}^+(\mathbf{r}', \mathbf{r}''; t', t''). \qquad (11a)$$

Similarly, the point-source excitations may be chosen as

$$s = 0, \qquad\qquad\qquad s^+ = 0,$$

$$\mathbf{f} = \mathbf{e}'\delta(\mathbf{r} - \mathbf{r}')\delta(t - t'), \qquad \mathbf{f}^+ = \mathbf{e}''\delta(\mathbf{r} - \mathbf{r}'')\delta(t - t''),$$

where $\mathbf{e}'$ and $\mathbf{e}''$ are unit vectors characterizing the directions of the point force densities at $\mathbf{r}'$ and $\mathbf{r}''$. Since, from Eqs. (4) and (10),

$$p = -\mathbf{G}_{12}(\mathbf{r}, \mathbf{r}'; t, t') \cdot \mathbf{e}', \qquad p^+ = -\mathbf{G}_{12}^+(\mathbf{r}, \mathbf{r}''; t, t'') \cdot \mathbf{e}'',$$

$$\mathbf{v} = -\mathscr{G}_{22}(\mathbf{r}, \mathbf{r}'; t, t') \cdot \mathbf{e}', \qquad \mathbf{v}^+ = -\mathscr{G}_{22}^+(\mathbf{r}, \mathbf{r}''; t, t'') \cdot \mathbf{e}'',$$

one deduces from Eq. (9) that

$$\mathbf{e}' \cdot \mathscr{G}_{22}(\mathbf{r}', \mathbf{r}''; t', t'') \cdot \mathbf{e}'' = \mathbf{e}'' \cdot \mathscr{G}_{22}^+(\mathbf{r}'', \mathbf{r}'; t'', t') \cdot \mathbf{e}',$$
$$\tilde{\mathscr{G}}_{22}(\mathbf{r}', \mathbf{r}''; t', t'') = \mathscr{G}_{22}^+(\mathbf{r}'', \mathbf{r}'; t'', t'), \tag{11b}$$

where $\tilde{\mathscr{G}}_{22}$ is the transposed dyadic to $\mathscr{G}_{22}$. The second equation in Eq. (11b) follows from the first on recognizing that a scalar is equal to its transpose and that $\mathbf{a} \cdot \mathscr{D} \cdot \mathbf{b} = \mathbf{b} \cdot \tilde{\mathscr{D}} \cdot \mathbf{a}$ where $\mathbf{a}$ and $\mathbf{b}$ are vectors and $\mathscr{D}$ is a dyadic. Correspondingly, with non-vanishing excitations

$$s = \delta(\mathbf{r} - \mathbf{r}')\delta(t - t'), \qquad \mathbf{f}^+ = \mathbf{e}''\delta(\mathbf{r} - \mathbf{r}'')\delta(t - t''),$$

or

$$\mathbf{f} = \mathbf{e}'\delta(\mathbf{r} - \mathbf{r}')\delta(t - t'), \qquad s^+ = \delta(\mathbf{r} - \mathbf{r}'')\delta(t - t''),$$

one infers, respectively,

$$\mathbf{G}_{21}(\mathbf{r}'', \mathbf{r}'; t'', t') = \mathbf{G}_{12}^+(\mathbf{r}', \mathbf{r}''; t', t''), \tag{11c}$$

$$\mathbf{G}_{12}(\mathbf{r}'', \mathbf{r}'; t'', t') = \mathbf{G}_{21}^+(\mathbf{r}', \mathbf{r}''; t', t''). \tag{11d}$$

Other reciprocity properties of the acoustic Green's functions follow from the observations, manifest from a comparison of Eqs. (1) and (7), that the adjoint Green's functions are time reversed, and also, as a consequence of spatial reflection, have reversed velocity components:

$$G_{ij}^+(\mathbf{r}, \mathbf{r}'; t, t') = (-1)^{i+j} G_{ij}(\mathbf{r}, \mathbf{r}'; -t, -t'), \tag{12a}$$

whence, from relations (11a)–(11d), one finds

$$\tilde{G}_{ij}(\mathbf{r}', \mathbf{r}; t', t) = (-1)^{i+j} G_{ji}(\mathbf{r}, \mathbf{r}'; -t, -t'), \tag{12b}$$

with the same notational comments as in Eq. (6c). The reciprocity relations, Eqs. (12), applicable to a general class of acoustic field problems in both homogeneous and inhomogeneous media, frequently simplify the explicit determination of the acoustic Green's functions.

As can be inferred from Eqs. (6) and (5), or as can be verified by direct substitution into Eqs. (5), one can express all the desired acoustic Green's functions in a homogeneous medium in terms of a single scalar Green's function $g(\mathbf{r}, \mathbf{r}'; t, t')$ as follows:†

$$G_{11}(\mathbf{r}, \mathbf{r}'; t, t') = n_0 m \frac{\partial}{\partial t} g(\mathbf{r}, \mathbf{r}'; t, t'),$$

$$\mathscr{G}_{22}(\mathbf{r}, \mathbf{r}'; t, t') = \left(\frac{1}{\gamma p_0} \frac{\partial}{\partial t} + \frac{\nabla \times \nabla \times \mathbf{1}}{n_0 m (\partial/\partial t)} \right) g(\mathbf{r}, \mathbf{r}'; t, t'), \tag{13a}$$

$$G_{21}(\mathbf{r}, \mathbf{r}'; t, t') = -\nabla g(\mathbf{r}, \mathbf{r}'; t, t') = G_{12}(\mathbf{r}, \mathbf{r}'; t, t')$$

---

†Note that

$$\frac{1}{\partial/\partial t} f(t) \equiv \int_{-\infty}^{t} f(t)\, dt.$$

where $g(\mathbf{r}, \mathbf{r}'; t, t')$ satisfies the wave equation

$$\left(\nabla^2 - \frac{1}{a^2}\frac{\partial^2}{\partial t^2}\right) g(\mathbf{r}, \mathbf{r}'; t, t') = -\delta(\mathbf{r} - \mathbf{r}')\delta(t - t') \tag{13b}$$

plus boundary and initial conditions in keeping with (5a) and (5b). The ability to express all four acoustic Green's functions in terms of one scalar function is a general consequence of the scalar nature of the acoustic field but is a property not shared by general vector fields. In view of the reciprocity properties (12b) of the acoustic Green's functions, for a homogeneous and time-invariant medium, it follows from Eq. (13a) that

$$g(\mathbf{r}, \mathbf{r}'; t, t') = g(\mathbf{r}', \mathbf{r}; -t', -t). \tag{13c}$$

Conversely, it is evident that if Eq. (13c) obtains, relations (12b) follow.

Scalar Green's function for unbounded space

For an unbounded homogeneous region the scalar Green's function g displays the property (6c) that $g(\mathbf{r}, \mathbf{r}'; t, t') = g(\mathbf{r} - \mathbf{r}', t - t')$, and hence its defining equation (13b) reduces to

$$\left(\nabla^2 - \frac{1}{a^2}\frac{\partial^2}{\partial t^2}\right) g(\mathbf{r}, t) = -\delta(\mathbf{r})\delta(t), \tag{14}$$

wherein for simplicity we have chosen $\mathbf{r}' = 0 = t'$. A solution of Eq. (14) satisfying the initial (causality) condition of vanishing for $t \leq 0$ (or equivalently the outgoing wave condition) is

$$g(\mathbf{r}, t) = \frac{\delta[t - (r/a)]}{4\pi r}, \qquad r = |\mathbf{r}|. \tag{15a}$$

This solution can be inferred by considering a sphere of radius r, volume V, and surface S, centered at $r = 0$, whence the divergence theorem yields

$$\iiint_V \nabla \cdot \nabla g \, dV = \iint_S \frac{\partial g}{\partial r} \, dS = 4\pi r^2 \frac{\partial g}{\partial r},$$

and also

$$\iiint_V g \, dV \to 0 \qquad \text{as } \mathbf{r} \to 0.$$

From the spherically symmetric equation (14), on volume integration about $\mathbf{r} = 0$, one then finds that

$$4\pi r^2 \frac{\partial g}{\partial r}\bigg|_{r\to 0} = -\delta(t).$$

Since the general (spherically symmetric) solution of Eq. (14) satisfying the outgoing radiation condition has the functional form $g(\mathbf{r}, t) = F[t - (r/a)]/r$, one infers Eq. (15a), which we now write as

$$g(\mathbf{r}, \mathbf{r}'; t, t') = \frac{\delta[t - t' - (|\mathbf{r} - \mathbf{r}'|/a)]}{4\pi|\mathbf{r} - \mathbf{r}'|}, \tag{15b}$$

whence, on substitution into Eq. (13a), one obtains the various Green's functions for an unbounded homogeneous fluid.

If a point source $s(\mathbf{r}, t)$ at $\mathbf{r}' = 0$ has a time-dependent amplitude specified by $s(t)$ for $t \geq 0$, the corresponding (potential) field $\varphi(\mathbf{r}, t)$ may be obtained from Eq. (15b) on multiplication by $s(t')$ and integration over t' between the limits $t' = 0$ and $t' = t - r/a$. From the result

$$\varphi(\mathbf{r}, t) = \frac{s(t - r/a)}{4\pi r} \, U\!\left(t - \frac{r}{a}\right), \tag{15c}$$

where $U(\tau)$ is the unit step function, which equals 1 for $\tau > 0$ and 0 for $\tau < 0$, one observes that the time dependence of the field at $\mathbf{r}$ is the same as that of the source, but retarded by the time $\tau = r/a$ required for the field to propagate from the source to $\mathbf{r}$. Furthermore, the causality requirement $\varphi \equiv 0$ for $t < 0$ is seen to be sharpened to the condition $\varphi \equiv 0$ for $t < r/a$ (i.e., the first response at $\mathbf{r}$ is observed at a time τ after initiation of the excitation).

For a bounded region the scalar Green's function g cannot be expressed as simply as in Eq. (15b) but can be represented in terms of appropriate eigenfunctions of the region, as will be discussed in Sec. 1.4 and Chapter 2.

1.1b   The Vector Electromagnetic Field

General properties

In a vacuum whose constitutive parameters are the dielectric constant ϵ_0 and the permeability μ_0, the vector electric field intensity $\mathbf{E}(\mathbf{r}, t)$ and magnetic field intensity $\mathbf{H}(\mathbf{r}, t)$ satisfy at any point the Maxwell equations,[†]

$$\epsilon_0 \frac{\partial \mathbf{E}}{\partial t} - \nabla \times \mathbf{H} = -\mathbf{J},$$

$$\nabla \times \mathbf{E} + \mu_0 \frac{\partial \mathbf{H}}{\partial t} = -\mathbf{M}, \tag{16}$$

where the source excitations $\mathbf{J}(\mathbf{r}, t)$ and $\mathbf{M}(\mathbf{r}, t)$ are, respectively, the vector electric current density and magnetic current density.[2a] One associates electric and magnetic charge densities $\rho(\mathbf{r}, t)$ and $\rho_m(\mathbf{r}, t)$ with the above current densities via the continuity equations:

$$\nabla \cdot \mathbf{J} = -\frac{\partial \rho}{\partial t},$$

$$\nabla \cdot \mathbf{M} = -\frac{\partial \rho_m}{\partial t}. \tag{16a}$$

At any time t, the field equations (16) are supplemented by the auxiliary equations

---

[†] The first-order Maxwell equations (16) are also valid for inhomogeneous media, wherein ϵ_0 and μ_0 are replaced by $\mathbf{r}$ dependent parameters.

$$\nabla \cdot \epsilon_0 \mathbf{E} = \rho,$$
$$\nabla \cdot \mu_0 \mathbf{H} = \rho_m, \tag{16b}$$

which make explicit the field dependence on the charge densities and which follow from Eq. (16) in view of Eqs. (16a) and (16c) after a divergence and time-quadrature operation. For uniqueness of the fields $\mathbf{E}$ and $\mathbf{H}$, one requires that the excitations $\mathbf{J}$ and $\mathbf{M}$ vanish for $t < t_1$, and that the fields satisfy the initial (causality) condition

$$\mathbf{E} \equiv 0 \equiv \mathbf{H} \qquad \text{for } t \leq t_1, \tag{16c}$$

and the boundary condition

$$\mathbf{n} \times \mathbf{E} = \mathscr{L} \cdot \mathbf{H} \tag{16d}$$

on the surface S (if any) bounding the field region. The unit vector $\mathbf{n}$ is the outward normal to the surface S, and $\mathscr{L}$ is an appropriate "impedance" dyadic having components transverse to $\mathbf{n}$ and characteristic of the boundary surface.

On multiplication of the two Maxwell equations (16) by $\mathbf{E}$ and $\mathbf{H}$, respectively, one readily derives the conservation-of-energy theorem:

$$\nabla \cdot (\mathbf{E} \times \mathbf{H}) = -\frac{\partial}{\partial t} \left(\frac{\epsilon_0 \mathbf{E}^2}{2} + \frac{\mu_0 \mathbf{H}^2}{2} \right) - \mathbf{J} \cdot \mathbf{E} - \mathbf{M} \cdot \mathbf{H}. \tag{17}$$

One identifies $\mathbf{E} \times \mathbf{H}$ as the instantaneous electromagnetic power flow per unit area at $\mathbf{r}, t$; the term in parentheses as the total stored electromagnetic energy density; $-\mathbf{J} \cdot \mathbf{E}$ as the power per unit volume supplied by the electric current excitation; and $-\mathbf{M} \cdot \mathbf{H}$ as the power per unit volume supplied by the magnetic current excitation.

In the absence of excitation, and for spatially homogeneous media, one obtains on elimination of $\mathbf{H}$ or $\mathbf{E}$ from Eqs. (16) the second-order vector-field equations

$$\nabla \times \nabla \times \mathbf{E} + \frac{1}{c^2} \frac{\partial^2}{\partial t^2} \mathbf{E} = 0,$$
$$\nabla \times \nabla \times \mathbf{H} + \frac{1}{c^2} \frac{\partial^2}{\partial t^2} \mathbf{H} = 0, \tag{18}$$

where $c = (\mu_0 \epsilon_0)^{-1/2}$ is the speed of light in vacuum. The source-excited form of the second-order field equations follows readily from Eqs. (16) by the addition of equivalent source currents to the right-hand side of Eqs. (18), as does their counterpart for inhomogeneous media. In the absence of sources one observes from Eqs. (18) that $\nabla \cdot \mathbf{E} = 0 = \nabla \cdot \mathbf{H}$; this condition is later shown to imply that, in a uniform medium, there are no longitudinal components of $\mathbf{E}$ and $\mathbf{H}$ in the direction of propagation of a single plane-wave field.

Linearity of the field equations (16) implies that the fields can be expressed in terms of the excitation currents by the integral representations

$$\mathbf{E}(\mathbf{r}, t) = -\int \mathscr{G}_{11}(\mathbf{r}, \mathbf{r}'; t, t') \cdot \mathbf{J}(\mathbf{r}', t') \, d\mathbf{r}' \, dt' - \int \mathscr{G}_{12}(\mathbf{r}, \mathbf{r}'; t, t') \cdot \mathbf{M}(\mathbf{r}', t') \, d\mathbf{r}' \, dt',$$

$$\tag{19a}$$

$$\mathbf{H}(\mathbf{r}, t) = -\int \mathscr{G}_{21}(\mathbf{r}, \mathbf{r}'; t, t')\cdot\mathbf{J}(\mathbf{r}', t')\, d\mathbf{r}'\, dt' - \int \mathscr{G}_{22}(\mathbf{r}, \mathbf{r}'; t, t')\cdot\mathbf{M}(\mathbf{r}', t')\, d\mathbf{r}'\, dt',$$

$$(19b)$$

where the integrals are extended over four-dimensional space–time volume elements $d\mathbf{r}'\, dt'$ within which the currents $\mathbf{J}$ and $\mathbf{M}$ are non-vanishing. If current sources are present at infinity, their contributions $\mathbf{E}_{\text{inc}}$ and $\mathbf{H}_{\text{inc}}$ to Eqs. (19a) and (19b), respectively, must be indicated explicitly. From Eqs. (19), one readily identifies the dyadic components $\mathscr{G}_{11}(\mathbf{r}, \mathbf{r}'; t, t')\cdot\mathbf{e}'$ and $\mathscr{G}_{21}(\mathbf{r}, \mathbf{r}'; t, t')\cdot\mathbf{e}'$ as negatives of the vector electric and magnetic fields, respectively, at $\mathbf{r}, t$ produced by a unit electric current density† at $\mathbf{r}', t'$ in the direction $\mathbf{e}'$. Correspondingly, $\mathscr{G}_{12}(\mathbf{r}, \mathbf{r}'; t, t')\cdot\mathbf{e}'$ and $\mathscr{G}_{22}(\mathbf{r}, \mathbf{r}'; t, t')\cdot\mathbf{e}'$ are negatives of the electric and magnetic fields produced at $\mathbf{r}, t$ by a unit magnetic current density at $\mathbf{r}', t'$ in the direction $\mathbf{e}'$.

The four dyadic functions $\mathscr{G}_{ij}$ play the fundamental role of Green's functions for the electromagnetic field. Their defining equations are readily obtained on substitution of Eqs. (19) into the field equations (16), whence, on noting the arbitrariness of $\mathbf{J}$ and $\mathbf{M}$,

$$\epsilon_0 \frac{\partial \mathscr{G}_{11}}{\partial t} - \nabla \times \mathscr{G}_{21} = \mathbf{1}\delta(\mathbf{r} - \mathbf{r}')\delta(t - t'),$$

$$\nabla \times \mathscr{G}_{11} + \mu_0 \frac{\partial \mathscr{G}_{21}}{\partial t} = 0,$$

and $\hfill (20)$

$$\epsilon_0 \frac{\partial \mathscr{G}_{12}}{\partial t} - \nabla \times \mathscr{G}_{22} = 0,$$

$$\nabla \times \mathscr{G}_{12} + \mu_0 \frac{\partial \mathscr{G}_{22}}{\partial t} = \mathbf{1}\delta(\mathbf{r} - \mathbf{r}')\delta(t - t').$$

Equations (20) are to be subject to the initial and boundary conditions

$$\mathscr{G}_{11} = 0 = \mathscr{G}_{21} \quad \text{for } t \leq t', \qquad \mathbf{n} \times \mathscr{G}_{11} = \mathscr{L}\cdot\mathscr{G}_{21} \quad \text{on } S, \qquad (20a)$$

$$\mathscr{G}_{12} = 0 = \mathscr{G}_{22} \quad \text{for } t \leq t', \qquad \mathbf{n} \times \mathscr{G}_{12} = \mathscr{L}\cdot\mathscr{G}_{22} \quad \text{on } S, \qquad (20b)$$

where $\mathbf{n}$ and $\mathscr{L}$ are defined as in Eq. (16d). It is often desirable to simplify the Green's function problem by specification of simpler boundary conditions, for example,

$$\mathbf{n} \times \mathscr{G}_{11} = 0 \quad \text{on } S, \qquad \mathbf{n} \times \mathscr{G}_{12} = 0 = \mathbf{n} \times (\nabla \times \mathscr{G}_{22}) \quad \text{on } S. \qquad (20c)$$

In this case the representation (19) must be generalized so as to include surface integrals representing the effect of "induced currents" arising from the difference between the field boundary conditions (16d) and the Green's function boundary conditions (20c).

---

†A unit current density at $\mathbf{r}', t'$ in the direction $\mathbf{e}'$ has the space–time form
$$\mathbf{e}'\delta(\mathbf{r} - \mathbf{r}')\delta(t - t').$$

On elimination of the Green's functions $\mathcal{G}_{12}$ and $\mathcal{G}_{21}$ from the defining equations (20), one obtains for homogeneous media, or vacuum, the second-order partial differential equations

$$\nabla \times \nabla \times \mathcal{G}_{11} + \frac{1}{c^2}\frac{\partial^2}{\partial t^2}\mathcal{G}_{11} = \mu_0 \frac{\partial}{\partial t}\mathbf{1}\delta(\mathbf{r} - \mathbf{r}')\delta(t - t'), \tag{21a}$$

$$\nabla \times \nabla \times \mathcal{G}_{22} + \frac{1}{c^2}\frac{\partial^2}{\partial t^2}\mathcal{G}_{22} = \epsilon_0 \frac{\partial}{\partial t}\mathbf{1}\delta(\mathbf{r} - \mathbf{r}')\delta(t - t') \tag{21b}$$

for the electric and magnetic types of Green's functions $\mathcal{G}_{11}(\mathbf{r}, \mathbf{r}'; t, t')$ and $\mathcal{G}_{22}(\mathbf{r}, \mathbf{r}'; t, t')$. Knowledge of the Green's functions $\mathcal{G}_{11}$ and $\mathcal{G}_{22}$ permits determination of the remaining Green's functions $\mathcal{G}_{12}$ and $\mathcal{G}_{21}$ on use of Eqs. (20) and time quadrature.

If field symmetries exist, properties of the electromagnetic Green's functions can be inferred prior to their explicit determination. For example, in an unbounded, homogeneous, stationary region in which the field equations (20) are invariant under arbitrary linear space–time displacements, one readily infers that solutions of Eqs. (20) are functions of the differences $\mathbf{r} - \mathbf{r}'$ and $t - t'$, i.e.,

$$\mathcal{G}_{ij}(\mathbf{r}, \mathbf{r}'; t, t') = \mathcal{G}_{ij}(\mathbf{r} - \mathbf{r}'; t - t') \tag{22}$$

for $i, j = 1, 2$.

Additional symmetry properties follow from consideration of an adjoint-field problem. The adjoint field is so defined as to permit the derivation of the reciprocity relation (24). With this intent, equations for adjoint electric and magnetic fields $\mathbf{E}^+ = \mathbf{E}^+(\mathbf{r}, t)$ and $\mathbf{H}^+ = \mathbf{H}^+(\mathbf{r}, t)$ are derived from the original field equations (16) by a temporal and spatial reflection transformation $\partial/\partial t \to -\partial/\partial t$ and $\nabla \to -\nabla$; the former effects a time reversal and the latter a reversal of the $\mathbf{E}/\mathbf{H}$ ratio for source-free fields, as can be inferred from Eqs. (23), (23a) and (23b). One thus obtains

$$-\epsilon_0 \frac{\partial \mathbf{E}^+}{\partial t} + \nabla \times \mathbf{H}^+ = -\mathbf{J}^+,$$

$$-\nabla \times \mathbf{E}^+ - \mu_0 \frac{\partial \mathbf{H}^+}{\partial t} = -\mathbf{M}^+, \tag{23}$$

which are subject to "reflected" initial and boundary conditions

$$\mathbf{E}^+ = 0 = \mathbf{H}^+ \qquad \text{for } t \geq t_2, \tag{23a}$$

$$\mathbf{n} \times \mathbf{E}^+ = -\tilde{\mathcal{Z}} \cdot \mathbf{H}^+ \qquad \text{on } S. \tag{23b}$$

and correspond to excitations $\mathbf{J}^+$ and $\mathbf{M}^+$ that vanish for $t > t_2$. The adjoint field differs from the original field defined by equations (16) in that it is "time reversed," the $\mathbf{E}/\mathbf{H}$ ratio is reversed, and the boundary is characterized by the transpose impedance dyadic $-\tilde{\mathcal{Z}}$. The adjoint field admits ingoing (advanced) wave solutions of the functional form $F(t + r/c)$ rather than the characteristic outgoing (retarded) wave solutions $F(t - r/c)$ of the original field, an inference from causality in Eq. (16c) and from its time-reversed analogue in Eq. (23a).

The adjoint and original fields are connected by a reciprocity relation,

$$\nabla\cdot(\mathbf{E}\times\mathbf{H}^+ + \mathbf{E}^+\times\mathbf{H}) + \epsilon_0\frac{\partial}{\partial t}(\mathbf{E}\cdot\mathbf{E}^+) + \mu_0\frac{\partial}{\partial t}(\mathbf{H}\cdot\mathbf{H}^+)$$

$$= \mathbf{J}^+\cdot\mathbf{E} - \mathbf{J}\cdot\mathbf{E}^+ + \mathbf{M}^+\cdot\mathbf{H} - \mathbf{M}\cdot\mathbf{H}^+, \tag{24}$$

derivable on suitable multiplication of Eqs. (16) by $\mathbf{E}^+$, $\mathbf{H}^+$, and of Eqs. (23) by $\mathbf{E}$, $\mathbf{H}$. On integration of Eq. (24) over the space–time domain bounded by the surface S and the times $t_1, t_2 > t_1$, and on use of the divergence theorem together with the initial and boundary conditions (16c) and (16d), and (23a) and (23b), one infers an integral form of the reciprocity relation:

$$0 = \iiint d\mathbf{r}\int_{t_1}^{t_2} dt(\mathbf{J}^+\cdot\mathbf{E} - \mathbf{J}\cdot\mathbf{E}^+ + \mathbf{M}^+\cdot\mathbf{H} - \mathbf{M}\cdot\mathbf{H}^+). \tag{25}$$

To display succinctly the relationship between the original and adjoint fields implicit in the reciprocity relation (25), it is first necessary to introduce adjoint dyadic Green's functions. Linearity of the adjoint equations (23) indicates that the adjoint fields are representable in a form similar to that in Eqs. (19):

$$\mathbf{E}^+(\mathbf{r}, t) = -\int \mathscr{G}_{11}^+(\mathbf{r}, \mathbf{r}'; t, t')\cdot\mathbf{J}^+(\mathbf{r}', t')\, d\mathbf{r}'\, dt'$$

$$- \int \mathscr{G}_{12}^+(\mathbf{r}, \mathbf{r}'; t, t')\cdot\mathbf{M}^+(\mathbf{r}', t')\, d\mathbf{r}'\, dt', \tag{26a}$$

$$\mathbf{H}^+(\mathbf{r}, t) = -\int \mathscr{G}_{21}^+(\mathbf{r}, \mathbf{r}'; t, t')\cdot\mathbf{J}^+(\mathbf{r}', t')\, d\mathbf{r}'\, dt'$$

$$- \int \mathscr{G}_{22}^+(\mathbf{r}, \mathbf{r}'; t, t')\cdot\mathbf{M}^+(\mathbf{r}', t')\, d\mathbf{r}'\, dt', \tag{26b}$$

where the adjoint Green's functions, distinguished by the superscript $^+$, have the same significance for the adjoint field as the Green's functions discussed under Eqs. (20) have for the original field. Defining equations for the adjoint Green's functions may be obtained by employing the same reflection transformations as in Eqs. (23), but we shall omit this step and proceed to infer properties of the adjoint Green's functions from the reciprocity relation (25).

To utilize the reciprocity relation (25) we consider a number of different choices of excitation for the original and adjoint fields. For example, if

$$\mathbf{J} = \mathbf{e}'\delta(\mathbf{r} - \mathbf{r}')\delta(t - t'), \qquad \mathbf{J}^+ = \mathbf{e}''\delta(\mathbf{r} - \mathbf{r}'')\delta(t - t''),$$

$$\mathbf{M} = 0, \qquad\qquad\qquad \mathbf{M}^+ = 0,$$

there follows from Eqs. (19) and (26) that

$$\mathbf{E}(\mathbf{r}, t) = -\mathscr{G}_{11}(\mathbf{r}, \mathbf{r}'; t, t')\cdot\mathbf{e}', \qquad \mathbf{E}^+(\mathbf{r}, t) = -\mathscr{G}_{11}^+(\mathbf{r}, \mathbf{r}''; t, t'')\cdot\mathbf{e}'',$$

$$\mathbf{H}(\mathbf{r}, t) = -\mathscr{G}_{21}(\mathbf{r}, \mathbf{r}'; t, t')\cdot\mathbf{e}', \qquad \mathbf{H}^+(\mathbf{r}, t) = -\mathscr{G}_{21}^+(\mathbf{r}, \mathbf{r}''; t, t'')\cdot\mathbf{e}'', \tag{27}$$

whence on substitution of the above fields and excitations into Eq. (25),

$$\mathbf{e}'\cdot\mathscr{G}_{11}^+(\mathbf{r}', \mathbf{r}''; t', t'')\cdot\mathbf{e}'' = \mathbf{e}''\cdot\mathscr{G}_{11}(\mathbf{r}'', \mathbf{r}'; t'', t')\cdot\mathbf{e}'$$

or

$$\mathcal{G}_{11}^{+}(\mathbf{r}', \mathbf{r}''; t', t'') = \tilde{\mathcal{G}}_{11}(\mathbf{r}'', \mathbf{r}'; t'', t'),$$

where $\tilde{\mathcal{G}}_{11}$ is the transpose of the dyadic $\mathcal{G}_{11}$. In a similar manner, from the point excitations

$$\mathbf{J} = 0, \qquad\qquad\qquad \mathbf{J}^{+} = 0,$$
$$\mathbf{M} = \mathbf{e}'\delta(\mathbf{r} - \mathbf{r}')\delta(t - t'), \qquad \mathbf{M}^{+} = \mathbf{e}''\delta(\mathbf{r} - \mathbf{r}'')\delta(t - t''),$$

one infers

$$\mathcal{G}_{22}^{+}(\mathbf{r}', \mathbf{r}''; t', t'') = \tilde{\mathcal{G}}_{22}(\mathbf{r}'', \mathbf{r}'; t'', t'), \tag{28b}$$

and from the excitations

$$\mathbf{J} = \mathbf{e}'\delta(\mathbf{r} - \mathbf{r}')\delta(t - t'), \qquad \mathbf{J}^{+} = 0,$$
$$\mathbf{M} = 0, \qquad\qquad\qquad \mathbf{M}^{+} = \mathbf{e}''\delta(\mathbf{r} - \mathbf{r}'')\delta(t - t''),$$

one obtains

$$\mathcal{G}_{12}^{+}(\mathbf{r}', \mathbf{r}''; t', t'') = \tilde{\mathcal{G}}_{21}(\mathbf{r}'', \mathbf{r}'; t'', t'). \tag{28c}$$

For the special case wherein $\mathcal{L} = \tilde{\mathcal{L}}$, one infers from Eqs. (23) that the adjoint Green's functions are identical to the original Green's functions defined in Eqs. (20) but with time, $\mathcal{G}_{11}/\mathcal{G}_{21}$ and $\mathcal{G}_{12}/\mathcal{G}_{22}$ ratios reversed. One thus obtains from Eqs. (28) as the reciprocity relations for electromagnetic Green's functions in a vacuum bounded by a surface on which $\mathcal{L} = \tilde{\mathcal{L}}$:

$$\mathcal{G}_{ij}(\mathbf{r}, \mathbf{r}'; t, t') = (-1)^{i+j}\tilde{\mathcal{G}}_{ji}(\mathbf{r}', \mathbf{r}; -t', -t), \tag{29}$$

where $i, j = 1, 2$. The relations (28) and (29) frequently facilitate explicit evaluation of the electromagnetic Green's functions. (See also Sec. 1.5b.)

Dyadic Green's functions in free space (**invariant evaluation**)

Electric and magnetic fields excited by prescribed sources may be expressed via Eqs. (19) in terms of the electromagnetic Green's functions for the regions containing these sources. This technique, employing Green's functions that are independent of the form and distribution of sources, is to be contrasted with the related classical method of scalar and vector potentials (both electric and magnetic) that are dependent on the form and distribution of sources. In the following we shall represent explicitly the dyadic Green's functions for a number of different regions, the form of representation depending on the symmetry properties of the region. From Eqs. (21) it is seen that explicit determination of the Green's functions $\mathcal{G}_{ij}$ is essentially concerned with inversion of the dyadic operator $[\mathbf{\nabla} \times (\mathbf{\nabla} \times \mathbf{1}) + \mathbf{1}(\partial^2/c^2 \, \partial t^2)]$. In free space, the inversion is simple and may be effected by an analytical method equivalent to that of finding the classic scalar and vector potentials of a point source; in bounded spaces this method no longer applies and must be rephrased. Alternatively, as shown in this section, the inversion may be accomplished by an operator method that is applicable to both bounded and unbounded regions and constitutes a succinct, invariant procedure. The term "invariant" implies that the method and result are independent of the choice of coordinate system for the space in question.

Classical method

For free space (unbounded and homogeneous), the four electromagnetic dyadic Green's functions $\mathscr{G}_{ij}$ can be expressed more or less classically in terms of a single scalar Green's function. The result for $\mathscr{G}_{11}$ follows from the second-order wave equation (21a) with $\mathscr{G}_{11}(\mathbf{r}, \mathbf{r}'; t, t')$ subject to an initial (causality) condition of vanishing for $t < t'$ (with a thereby implied outgoing wave condition). On taking the divergence of the first of Eqs. (20), one observes that

$$\epsilon_0 \frac{\partial}{\partial t} \nabla \cdot \mathscr{G}_{11} = \nabla \delta(\mathbf{r} - \mathbf{r}')\delta(t - t').$$

On substitution of this relation into Eq. (21a) and noting that $\nabla \times (\nabla \times \mathscr{G}_{11}) = \nabla\nabla \cdot \mathscr{G}_{11} - \nabla^2 \mathscr{G}_{11}$, one finds [see footnote on p. 7]

$$\left(\nabla^2 - \frac{1}{c^2}\frac{\partial^2}{\partial t^2}\right)\mathscr{G}_{11} = -\left(\mu_0 \frac{\partial}{\partial t}\mathbf{1} - \frac{\nabla\nabla}{\epsilon_0(\partial/\partial t)}\right)\delta(\mathbf{r} - \mathbf{r}')\delta(t - t'). \tag{30a}$$

Introduction of a scalar Green's function $g(\mathbf{r}, \mathbf{r}'; t, t')$ via

$$\mathscr{G}_{11}(\mathbf{r}, \mathbf{r}'; t, t') = \left(\mu_0 \frac{\partial}{\partial t}\mathbf{1} - \frac{\nabla\nabla}{\epsilon_0(\partial/\partial t)}\right)g(\mathbf{r}, \mathbf{r}'; t, t') \tag{30b}$$

indicates that the expression (30b) satisfies Eq. (30a) provided $g(\mathbf{r}, \mathbf{r}'; t, t')$ vanishes for $t < t'$ (thus obeying an outgoing wave condition) and is defined by the scalar wave equation

$$\left(\nabla^2 - \frac{1}{c^2}\frac{\partial^2}{\partial t^2}\right)g(\mathbf{r}, \mathbf{r}'; t, t') = -\delta(\mathbf{r} - \mathbf{r}')\delta(t - t'). \tag{30c}$$

This scalar Green's function g, except for the presence of the light speed c rather than the sound speed a, is identical to that discussed in connection with Eq. (13b) and hence has the solution

$$g(\mathbf{r}, \mathbf{r}'; t, t') = \frac{\delta[t - t' - (|\mathbf{r} - \mathbf{r}'|/c)]}{4\pi|\mathbf{r} - \mathbf{r}'|}. \tag{31}$$

By duality to Eq. (30b), the dyadic Green's function $\mathscr{G}_{22}$ is expressible in terms of g in Eq. (31) as

$$\mathscr{G}_{22}(\mathbf{r}, \mathbf{r}'; t, t') = \left(\epsilon_0 \frac{\partial}{\partial t}\mathbf{1} - \frac{\nabla\nabla}{\mu_0(\partial/\partial t)}\right)g(\mathbf{r}, \mathbf{r}'; t, t'), \tag{32a}$$

and hence to satisfy Eqs. (20), the remaining electromagnetic Green's functions may be chosen as

$$\mathscr{G}_{12} = \nabla \times \mathbf{1}g = -\mathscr{G}_{21}. \tag{32b}$$

Operator method

An operator procedure for the determination of the Green's functions $\mathscr{G}_{ij}$ may be based on the dyadic operator identity $\nabla \times (\nabla \times \mathbf{1}) = \nabla\nabla - \nabla^2\mathbf{1}$, expressed in the suggestive operator form

$$\mathbf{1} = \mathbf{1}_L + \mathbf{1}_T = \frac{\boldsymbol{\nabla}\boldsymbol{\nabla}}{\nabla^2} - \frac{\boldsymbol{\nabla} \times (\boldsymbol{\nabla} \times \mathbf{1})}{\nabla^2}, \tag{33a}$$

where $\mathbf{1}_L$ and $\mathbf{1}_T$ are orthogonal unit dyads, "longitudinal" and "transverse" to the vector operator $\boldsymbol{\nabla}$, with the property $\mathbf{1}_L \cdot \mathbf{1}_T = 0$. The latter property implies that in a basis oriented along $\boldsymbol{\nabla}$, the matrix representative of Eq. (33a) is diagonal. In view of the unit dyads defined in Eq. (33a), the basic operator in Eqs. (21) may be represented as

$$\boldsymbol{\nabla} \times (\boldsymbol{\nabla} \times \mathbf{1}) + \frac{1}{c^2}\frac{\partial^2}{\partial t^2}\mathbf{1} = \frac{1}{c^2}\frac{\partial^2}{\partial t^2}\frac{\boldsymbol{\nabla}\boldsymbol{\nabla}}{\nabla^2} + \left(\nabla^2 - \frac{1}{c^2}\frac{\partial^2}{\partial t^2}\right)\frac{\boldsymbol{\nabla} \times (\boldsymbol{\nabla} \times \mathbf{1})}{\nabla^2}.$$

$$\tag{33b}$$

Because of the orthogonality property of the unit dyads, one verifies, by direct (dot product) multiplication, the general inverse relation

$$\left[A\frac{\boldsymbol{\nabla}\boldsymbol{\nabla}}{\nabla^2} - B\frac{\boldsymbol{\nabla} \times (\boldsymbol{\nabla} \times \mathbf{1})}{\nabla^2}\right]^{-1} = \left[\frac{1}{A}\frac{\boldsymbol{\nabla}\boldsymbol{\nabla}}{\nabla^2} - \frac{1}{B}\frac{\boldsymbol{\nabla} \times (\boldsymbol{\nabla} \times \mathbf{1})}{\nabla^2}\right], \tag{33c}$$

where A and B are scalars, or scalar operators, that commute with $\boldsymbol{\nabla}$. Accordingly, one obtains from Eqs. (33a)–(33c) the desired inverse operator:

$$\left[\boldsymbol{\nabla} \times (\boldsymbol{\nabla} \times \mathbf{1}) + \frac{1}{c^2}\frac{\partial^2}{\partial t^2}\mathbf{1}\right]^{-1} = \left(\frac{1}{c^2}\frac{\partial^2}{\partial t^2}\right)^{-1}\frac{\boldsymbol{\nabla}\boldsymbol{\nabla}}{\nabla^2}$$

$$+ \left(\nabla^2 - \frac{1}{c^2}\frac{\partial^2}{\partial t^2}\right)^{-1}\frac{\boldsymbol{\nabla} \times (\boldsymbol{\nabla} \times \mathbf{1})}{\nabla^2}$$

$$= \left[\mathbf{1} - \left(\frac{1}{c^2}\frac{\partial^2}{\partial t^2}\right)^{-1}\boldsymbol{\nabla}\boldsymbol{\nabla}\right]\left[-\left(\nabla^2 - \frac{1}{c^2}\frac{\partial^2}{\partial t^2}\right)^{-1}\right].$$

$$\tag{33d}$$

The operator identities (33) are valid† for operation on any space–time function and, in particular, on the point-source function $\delta(\mathbf{r} - \mathbf{r}')\delta(t - t')$. In view of Eq. (30c), the inverse operator in the rightmost set of brackets of Eq. (33d), when operating on $\delta(\mathbf{r} - \mathbf{r}')\delta(t - t')$, is recognized as the scalar Green's function $g(\mathbf{r}, \mathbf{r}'; t, t')$. The dyadic Green's functions $\mathscr{G}_{11}$ and $\mathscr{G}_{22}$ then follow from Eqs. (21) and (33d) in the form already displayed in Eqs. (30b) and (32a), with the related Green's functions $\mathscr{G}_{21}$ and $\mathscr{G}_{12}$ given in Eq. (32b).

Field of an electric dipole current

As an application of the above free-space Green's functions, we consider the fields excited by the sudden creation at $\mathbf{r} = 0$ and $t = 0$ of an electric charge dipole of moment $\mathbf{p}(\mathbf{r}, t) = \mathbf{p}\delta(\mathbf{r})U(t)$, where $U(\alpha)$ is the Heaviside step function, which equals 1 for $\alpha > 0$ and 0 for $\alpha < 0$. This excitation gives rise to an impulsive electric current density $\mathbf{J}(\mathbf{r}, t) = \partial\mathbf{p}(\mathbf{r}, t)/\partial t = \mathbf{p}\,\delta(\mathbf{r})\,\delta(t)$ and

---

†The range and/or domain[5] of the relevant operators are determined by the boundary and initial conditions concomitant to Eqs. (21).

to space- and time-dependent electric and magnetic fields given by Eqs. (19)
as

$$\mathbf{E}(\mathbf{r}, t) = -\mathscr{G}_{11}(\mathbf{r}, 0; t, 0)\cdot\mathbf{p},$$
$$\mathbf{H}(\mathbf{r}, t) = -\mathscr{G}_{21}(\mathbf{r}, 0; t, 0)\cdot\mathbf{p}, \tag{34}$$

where the relevant free-space Green's functions $\mathscr{G}_{11}$ and $\mathscr{G}_{21}$ follow from Eqs.
(30b), (32b), and (31). To cast the latter in a form that more clearly dis-
tinguishes the nature of the various field contributions, we rewrite $\mathscr{G}_{11}$ of Eqs.
(30b) and (31) as [note that $\int_{-\infty}^{t} \delta(\tau - r/c)\, d\tau = U(t - r/c)$]

$$\mathscr{G}_{11}(r, 0; t, 0) = \mu_0 \frac{\delta'(t - r/c)}{4\pi r}\mathbf{1} - \nabla\nabla\frac{U(t - r/c)}{4\pi\epsilon_0 r}$$

$$= \sqrt{\frac{\mu_0}{\epsilon_0}}\frac{1}{4\pi r}\left[(\mathbf{1} - \mathbf{r}_0\mathbf{r}_0)\frac{\delta'(t - r/c)}{c} + (\mathbf{1} - 3\mathbf{r}_0\mathbf{r}_0)\frac{\delta(t - r/c)}{r}\right]$$

$$+ (\mathbf{1} - 3\mathbf{r}_0\mathbf{r}_0)\frac{U(t - r/c)}{4\pi\epsilon_0 r^3}, \tag{34a}$$

and $\mathscr{G}_{21}$ of Eqs. (32b) and (31) as

$$\mathscr{G}_{21}(\mathbf{r}, 0; t, 0) = -\nabla \times \mathbf{1}\frac{\delta(t - r/c)}{4\pi r}$$

$$= \left[\frac{\delta(t - r/c)}{r} + \frac{\delta'(t - r/c)}{c}\right]\frac{\mathbf{r}_0 \times \mathbf{1}}{4\pi r}, \tag{34b}$$

where $\delta'(\alpha) = d\delta(\alpha)/d\alpha$. The first two terms in brackets in Eq. (34a) represent
"radiation" and "near-field" contributions in the form of impulsive spherical
fronts traveling outward at the speed of light; the third term, which exists only
in the region $r < ct$ following the spherical front, is the static electric field of
the charge dipole. The radiative term is transverse to $\mathbf{r}_0$ and decays like $1/r$,
whereas the other terms have the familiar $1/r^2$ and $1/r^3$ decay. Correspondingly, in Eq. (34b) the magnetic field is seen to be non-static and in the form
of impulsive spherical fronts moving outward with the speed of light. The $\mathbf{q}$
component of field produced by a dipole in the direction of $\mathbf{p}$ is readily obtained from the dyadics in Eqs. (34a) and (34b) by pre- and post-multiplication by
$\mathbf{q}$ and $\mathbf{p}$, respectively; note that $\mathbf{r}_0 \times \mathbf{1}\cdot\mathbf{p} = \mathbf{r}_0 \times \mathbf{p}$.

Free space, dyadic Green's functions (transversely invariant)—Hertz potentials

The free-space, dyadic Green's functions in Eqs. (30b), (32a), and (32b)
were evaluated in a ∇-oriented basis that is invariant to choice of a spatial
vector coordinate basis. It is of interest to carry out a similar evaluation in a
basic oriented along a constant unit vector $\mathbf{a}$ but invariant in the surface transverse to $\mathbf{a}$. Representations of Green's functions in such a basis are particularly
convenient for the analysis of field problems containing planar-stratified scattering structures with axis of stratification along the preferred $\mathbf{a}$ direction.
Although the spatial cross section transverse to $\mathbf{a}$ is of infinite extent in this
consideration, the method is readily generalizable to cylindrical regions of

arbitrary but finite cross section. As in the previous invariant representations, the dyadic Green's functions are still representable in terms of a single scalar Green's function, which, however, is different from the scalar Green's function defined in Eq. (30c). An additional feature of these dyadic Green's function representations is that they provide a simple Hertz potential representation of the electric and magnetic fields produced by current sources, and they lead to a clarification of certain singularities characteristic of such representations.

As has been shown, via both classical and operator methods, the dyadic Green's function $\mathcal{G}_{11}$ can be expressed as in Eq. (30b) by an invariant dyadic operator and a single scalar Green's function g. We shall utilize that invariant result and reexpress the dyadic operators $\mathbf{1}$ and $\nabla\nabla$ of Eq. (30b) in a basis oriented along an axial direction defined by the constant unit vector $\mathbf{a}$.

On decomposition of the gradient as $\nabla = \nabla_t + (\mathbf{a}\cdot\nabla)\mathbf{a}$, one reexpresses the unit dyadic operator on the right of Eq. (30b) by use of the identity

$$\nabla_t^2 \mathbf{1} = \nabla_t^2 \mathbf{aa} + \nabla_t\nabla_t + (\nabla \times \mathbf{a})(\nabla \times \mathbf{a}), \tag{35a}$$

where ∇_t is the invariant component of ∇ transverse to $\mathbf{a}$ and $\nabla_t^2 = \nabla^2 - (\mathbf{a}\cdot\nabla)^2$.† Equation (35a) may be verified by noting that $\mathbf{a}, \nabla_t, \nabla \times \mathbf{a} = \nabla_t \times \mathbf{a}$, are orthogonal vectors with magnitudes $1, \sqrt{\nabla_t^2}, \sqrt{\nabla_t^2}$, respectively, whence formally

$$\mathbf{1} = \mathbf{aa} + \frac{\nabla_t\nabla_t}{\nabla_t^2} + \frac{(\nabla \times \mathbf{a})(\nabla \times \mathbf{a})}{\nabla_t^2}. \tag{35b}$$

To reexpress the dyadic $\nabla\nabla$, one employs the identities

$$\nabla\nabla = \nabla_t\nabla_t + (\nabla_t\mathbf{a} + \mathbf{a}\nabla_t)(\mathbf{a}\cdot\nabla) + \mathbf{aa}(\mathbf{a}\cdot\nabla)^2$$

$$[\nabla \times (\nabla \times \mathbf{a})][\nabla \times (\nabla \times \mathbf{a})] = \nabla_t\nabla_t(\mathbf{a}\cdot\nabla)^2 - (\nabla_t\mathbf{a} + \mathbf{a}\nabla_t)(\mathbf{a}\cdot\nabla)\nabla_t^2$$
$$+ \mathbf{aa}\nabla_t^2\nabla_t^2,$$

which, on multiplication by ∇_t^2 and addition, yield

$$\nabla_t^2\nabla\nabla = (\mathbf{aa}\nabla_t^2 + \nabla_t\nabla_t)\nabla^2 - [\nabla \times (\nabla \times \mathbf{a})][\nabla \times (\nabla \times \mathbf{a})]. \tag{36}$$

From Eqs. (35a) and (36) one thus derives for application to the dyadic operator of Eq. (30b):

$$\nabla_t^2\left(\mu_0\frac{\partial}{\partial t}\mathbf{1} - \frac{\nabla\nabla}{\epsilon_0(\partial/\partial t)}\right) = -\frac{1}{\epsilon_0(\partial/\partial t)}(\mathbf{aa}\nabla_t^2 + \nabla_t\nabla_t)\left(\nabla^2 - \frac{1}{c^2}\frac{\partial^2}{\partial t^2}\right)$$

$$+ \mu_0\frac{\partial}{\partial t}(\nabla \times \mathbf{a})(\nabla \times \mathbf{a})$$

$$+ \frac{[\nabla \times (\nabla \times \mathbf{a})][\nabla \times (\nabla \times \mathbf{a})]}{\epsilon_0(\partial/\partial t)}. \tag{37}$$

From Eq. (30b) one obtains, on introducing a scalar function $\mathscr{S}(\mathbf{r}, \mathbf{r}'; t, t')$

---

†Note that dot-product multiplication of Eq. (35a) from the right by a vector $\mathbf{A}$ yields an expression for $\nabla_t^2\mathbf{A}$, with $(\nabla \times \mathbf{a})\cdot\mathbf{A} \equiv \nabla\cdot(\mathbf{a} \times \mathbf{A})$.

defined by $\nabla_t^2 \mathscr{S} = g$, the alternative representation of the free-space dyadic Green's function $\mathscr{G}_{11}$:

$$\mathscr{G}_{11}(\mathbf{r}, \mathbf{r}'; t, t') = \nabla_t^2 \left(\mu_0 \frac{\partial}{\partial t}\, \mathbf{1} - \frac{\nabla\nabla}{\epsilon_0(\partial/\partial t)} \right) \mathscr{S}(\mathbf{r}, \mathbf{r}'; t, t'). \tag{38}$$

Substituting the operator identity (37) into Eq. (38), and noting that one may put $\nabla = -\nabla'$ when operating on a function of $\mathbf{r} - \mathbf{r}'$ such as $\mathscr{S}(\mathbf{r}, \mathbf{r}'; t, t')$, one finds

$$\mathscr{G}_{11}(\mathbf{r}, \mathbf{r}'; t, t') = \frac{1}{\epsilon_0(\partial/\partial t)} \left(\mathbf{aa} + \frac{\nabla_t \nabla_t}{\nabla_t^2} \right) \delta(\mathbf{r} - \mathbf{r}')\, \delta(t - t')$$

$$+ \left[\mu_0 \frac{\partial}{\partial t} (\nabla \times \mathbf{a})(\nabla' \times \mathbf{a}) \right.$$

$$\left. - \frac{[\nabla \times (\nabla \times \mathbf{a})][\nabla' \times (\nabla' \times \mathbf{a})]}{\epsilon_0(\partial/\partial t)} \right] \mathscr{S}(\mathbf{r}, \mathbf{r}'; t, t'), \tag{38a}$$

where $\mathscr{S}$ is a free-space scalar Green's function defined by

$$\nabla_t^2 \left(\nabla^2 - \frac{1}{c^2} \frac{\partial^2}{\partial t^2} \right) \mathscr{S}(\mathbf{r}, \mathbf{r}'; t, t') = \delta(\mathbf{r} - \mathbf{r}')\, \delta(t - t') \tag{38b}$$

and satisfies an initial condition of vanishing for $t < t'$. It is of interest to note that the first term on the right of Eq. (38a) contains a singular Green's function term $\bar{G}$ which, when evaluated in a ρ, φ, z coordinate system, has the form

$$\bar{G} = \frac{\delta(\mathbf{r} - \mathbf{r}')\, \delta(t - t')}{\nabla_t^2},$$

which for $\rho' = 0$ may be written more conventionally as

$$\nabla_t^2 \bar{G} = \frac{1}{\rho} \frac{\partial}{\partial \rho}\, \rho \frac{\partial}{\partial \rho}\, \bar{G} = \delta(\boldsymbol{\rho})\, \delta(z - z')\, \delta(t - t'),$$

whence the ρ-dependent part is the static-line-source Green's function $(1/2\pi) \ln(\alpha\rho)$. On restoring $\boldsymbol{\rho}' \neq 0$, one has, explicitly,

$$\bar{G} = \left[\frac{1}{2\pi} \ln \alpha|\boldsymbol{\rho} - \boldsymbol{\rho}'| \right] \delta(z - z')\, \delta(t - t'), \tag{38c}$$

where α is a constant; this term is non-vanishing (and singular) only on the plane $z = z'$ and at $t = t'$.

Just as in Eq. (30b), the expression for $\mathscr{G}_{11}$ in Eq. (38a) requires only one scalar Green's function $\mathscr{S}$ (if $z \neq z', t \neq t'$). From Eqs. (31) and (38b) one has

$$\nabla_t^2 \mathscr{S} = -\frac{\delta[\tau - (|\mathbf{r} - \mathbf{r}'|/c)]}{4\pi|\mathbf{r} - \mathbf{r}'|}, \tag{39a}$$

where $\tau = t - t'$. In a cylindrical ρ, φ, z coordinate system with z in the direction $\mathbf{a}$, one finds, if $\mathbf{r}' = 0$ and hence $|\mathbf{r} - \mathbf{r}'| = \sqrt{\rho^2 + z^2}$, that

$$\frac{1}{\rho} \frac{\partial}{\partial \rho}\, \rho \frac{\partial}{\partial \rho}\, \mathscr{S} = -\frac{\delta[\tau - (\sqrt{\rho^2 + z^2}/c)]}{4\pi\sqrt{\rho^2 + z^2}}.$$

Multiplication by ρ and integration from 0 to ρ yields

$$\frac{\partial}{\partial \rho}\,\mathscr{S}(r, 0; \tau) = \frac{-c}{4\pi\rho}\, U\!\left(\tau - \frac{|z|}{c}\right) U\!\left(\frac{r}{c} - \tau\right), \tag{39b}$$

where the Heaviside unit function $U(\alpha)$ equals 1 for $\alpha > 0$ and 0 for $\alpha < 0$. In many applications (see Sec. 5.2), only $\nabla_t^2 \mathscr{S}$ or $\nabla_t \mathscr{S}$, rather than $\mathscr{S}$ itself, needs to be known; in such cases, only the results in Eqs. (39) are necessary to determine the field.

A similar development leads to representation of $\mathscr{G}_{21}$, $\mathscr{G}_{12}$, and $\mathscr{G}_{22}$ and also provides an alternative to the free-space representation of Eqs. (32a), (32b), and (31). By Eq. (35a) one first reexpresses the dyadic $\nabla \times \mathbf{1}$ of Eq. (32b) by use of the identity

$$\nabla \times \mathbf{1}\nabla_t^2 = \nabla \times \mathbf{aa}\nabla_t^2 + \nabla \times \nabla_t\nabla_t + (\nabla \times \nabla \times \mathbf{a})(\nabla \times \mathbf{a})$$

$$= -(\nabla \times \mathbf{a})(\nabla \times \nabla \times \mathbf{a}) + (\nabla \times \nabla \times \mathbf{a})(\nabla \times \mathbf{a}) \tag{40}$$

where in the second line we have employed the vector relations $\nabla \times (\nabla \times \mathbf{a}) = \nabla_t(\mathbf{a}\cdot\nabla) - \nabla_t^2\mathbf{a}$, $\nabla = \nabla_t + \mathbf{a}(\mathbf{a}\cdot\nabla)$, and $\mathbf{a} \times \nabla_t = -\nabla \times \mathbf{a}$, with $\mathbf{a}$ the unit vector along the symmetry direction. From Eqs. (32b) and (30c) one finds that

$$\nabla_t^2\!\left(\nabla^2 - \frac{1}{c^2}\frac{\partial^2}{\partial t^2}\right)\mathscr{G}_{21}(\mathbf{r}, \mathbf{r}'; t, t') = \nabla_t^2\nabla \times \mathbf{1}\delta(\mathbf{r} - \mathbf{r}')\,\delta(t - t')$$

and hence, by Eqs. (40) and (38b),

$$\mathscr{G}_{21}(\mathbf{r}, \mathbf{r}'; t, t') = -[(\nabla \times \mathbf{a})(\nabla' \times \nabla' \times \mathbf{a})$$

$$+ (\nabla \times \nabla \times \mathbf{a})(\nabla' \times \mathbf{a})]\mathscr{S}(\mathbf{r}, \mathbf{r}'; t, t') \tag{41}$$

is the desired alternative representation of the magnetic-type dyadic Green's function (with $\nabla' = -\nabla$) in unbounded free space. As before, the Green's functions $\mathscr{G}_{12} = -\mathscr{G}_{21}$ of Eq. (41) and $\mathscr{G}_{22}$ may be obtained from $\mathscr{G}_{11}$ in Eq. (38a) on the duality replacements $\epsilon_0 \longleftrightarrow \mu_0$.

To exhibit explicitly the electric and magnetic fields produced by an electric current density $\mathbf{J}(\mathbf{r}, t)$, one utilizes Eqs. (19), (20), (38), and (41) to obtain (with restrictions noted below)

$$\mathbf{E}(\mathbf{r}, t) = -\mu_0\frac{\partial}{\partial t}\nabla \times \mathbf{a}\,\Pi''(\mathbf{r}, t) + \nabla \times (\nabla \times \mathbf{a})\,\Pi'(\mathbf{r}, t),$$

$$\tag{42a}$$

$$\mathbf{H}(\mathbf{r}, t) = \nabla \times (\nabla \times \mathbf{a})\,\Pi''(\mathbf{r}, t) + \epsilon_0\frac{\partial}{\partial t}\nabla \times \mathbf{a}\,\Pi'(\mathbf{r}, t),$$

where the scalar functions

$$\Pi''(\mathbf{r}, t) = \int \nabla' \times \mathbf{a}\mathscr{S}(\mathbf{r}, \mathbf{r}'; t, t')\cdot\mathbf{J}(\mathbf{r}', t')\,d\mathbf{r}'\,dt',$$

$$\tag{42b}$$

$$\Pi'(\mathbf{r}, t) = \frac{1}{\epsilon_0(\partial/\partial t)}\int \nabla' \times (\nabla' \times \mathbf{a})\mathscr{S}(\mathbf{r}, \mathbf{r}'; t, t')\cdot\mathbf{J}(\mathbf{r}', t')\,d\mathbf{r}'\,dt'$$

are Hertz potentials associated with the electric current density $\mathbf{J}$ [recall that

$\nabla' \times (\nabla' \times \mathbf{a})\mathscr{S} \equiv \nabla' \times [\nabla' \times (\mathbf{a}\mathscr{S})]$, etc.]. From Eqs. (42a) one observes that since $\nabla \times \mathbf{a}\,\Pi'$ is a vector transverse to $\mathbf{a}$, the Hertz potential Π' does not contribute an $\mathbf{a}$ component of magnetic field and Π'' does not contribute an $\mathbf{a}$ component of electric field; for this reason Π' and Π'' are frequently termed *E*-mode and *H*-mode potentials, respectively, with respect to $\mathbf{a}$. The *electric-field* representation in Eq. (42a) is not valid in planes transverse to $\mathbf{a}$ containing the source $\mathbf{J}(\mathbf{r}, t)$, since in the transition from Eqs. (38a) to (42a), singular contributions in the indicated planar region from the first term of Eq. (38a) have been omitted. This omission and consequent restriction on the region of applicability of the Hertzian potential representation of the electric field in Eq. (42a) should be recognized. Note also that for the magnetic-field representation in Eq. (42a), which is applicable even in source regions, $\epsilon_0(\partial/\partial t)\,\Pi'$ and not the potential Π' need be calculated.

The significance of the somewhat peculiar potential function in Eq. (39b) becomes evident on calculation of the fields radiated by an impulsive point electric current element $\mathbf{J}(\mathbf{r}, t) = \mathbf{y}_0 p\, \delta(\mathbf{r})\, \delta(t)$, where p is the electric dipole moment strength and $\mathbf{y}_0$ is perpendicular to $\mathbf{a} \equiv \mathbf{z}_0$. The portion of the x component of magnetic field, contributed via Eq. (42a) by the potential Π' in Eq. (42b), is found to be [note that $\delta'(\alpha) = (d/d\alpha)\delta(\alpha)$, and sgn $\alpha = \pm 1$ for $\alpha \gtrless 0$]

$$H'_x = p\,\frac{x^2 - y^2}{4\pi\rho^4}\left[(\text{sgn } z)\,\delta\!\left(t - \frac{|z|}{c}\right) - \frac{z}{r}\,\delta\!\left(t - \frac{r}{c}\right)\right]$$

$$+ p\,\frac{y^2 z}{4\pi\rho^2 r^2}\left[\frac{\delta(t - r/c)}{r} + \frac{1}{c}\,\delta'\!\left(t - \frac{r}{c}\right)\right], \tag{43a}$$

and the portion contributed by Π'' is

$$H''_x = p\,\frac{y^2 - x^2}{4\pi\rho^4}\left[(\text{sgn } z)\,\delta\!\left(t - \frac{|z|}{c}\right) - \frac{z}{r}\,\delta\!\left(t - \frac{r}{c}\right)\right]$$

$$+ p\,\frac{x^2 z}{4\pi\rho^2 r^2}\left[\frac{\delta(t - r/c)}{r} + \frac{1}{c}\,\delta'\!\left(t - \frac{r}{c}\right)\right], \tag{43b}$$

whence the total x component becomes

$$H_x = H'_x + H''_x = p\,\frac{z}{4\pi r^2}\left[\frac{\delta(t - r/c)}{r} + \frac{1}{c}\,\delta'\!\left(t - \frac{r}{c}\right)\right], \tag{43c}$$

where

$$r^2 = x^2 + y^2 + z^2 = \rho^2 + z^2.$$

As shown in Fig. 1.1.1, the constituent fields H'_x and H''_x are seen to comprise a combination of plane and spherical wavefronts, whereas the composite field H_x exhibits only the spherical wavefronts expected for a localized source. The decomposition of H_x into H'_x and H''_x corresponds to a similar decomposition of the source function $\mathbf{J}(\mathbf{r}, t)$ into portions that radiate only *E*-mode fields (Π') with respect to z, and only *H*-mode fields (Π''), respectively. These equivalent

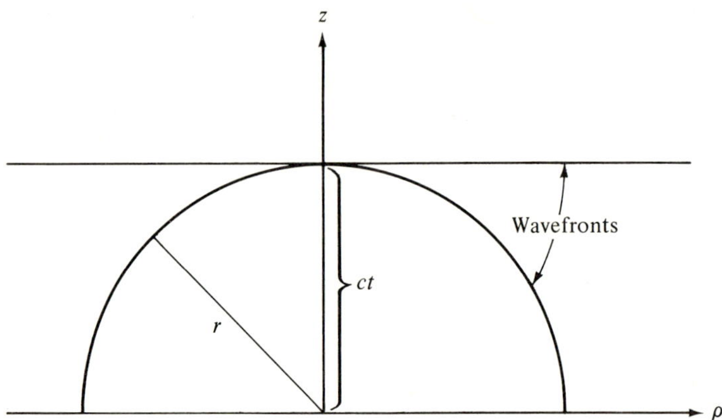

FIG. 1.1.1   Wavefronts generated by *E*-mode or *H*-mode potentials.

source distributions occupy the whole plane $z = 0$ and each gives rise to a plane-wave response in addition to the spherical wave, with the former canceled when the composite field is calculated. The *E*- and *H*-mode contributions to the y component of magnetic field behave in a similar manner:

$$H_y' = -H_y'' = \frac{pxy}{4\pi\rho^2}\left[\frac{\operatorname{sgn} z}{\rho^2}\delta\left(t - \frac{|z|}{c}\right) - \frac{z}{r}\left(\frac{1}{r^2} + \frac{1}{\rho^2}\right)\delta\left(t - \frac{r}{c}\right)\right.$$

$$\left. - \frac{z}{cr^2}\delta'\left(t - \frac{r}{c}\right)\right], \tag{44a}$$

but the total y component vanishes:

$$H_y = H_y' + H_y'' = 0. \tag{44b}$$

The z component of magnetic field is contributed entirely by the *H*-mode potential Π'' in Eqs. (42),

$$H_z \equiv H_z'' = -\frac{px}{4\pi r^2}\left[\frac{\delta(t - r/c)}{r} + \frac{1}{c}\delta'\left(t - \frac{r}{c}\right)\right], \tag{45}$$

and in contrast to the components H_x'' and H_y'' transverse to the guide (symmetry) axis z, does not contain the plane-wave constituent. The results in Eqs. (43c), (44b), and (45) are, respectively, the same as the x, y, and z components of $\mathbf{H} = -p\mathscr{G}_{21}\cdot\mathbf{y}_0$ obtained from the formulation in Eq. (34b) (noting that $\mathbf{r}_0 r = \mathbf{x}_0 x + \mathbf{y}_0 y + \mathbf{z}_0 z$).

In contrast to the magnetic-field expressions, the transverse electric field, calculated on substitution of Eq. (39b) into Eqs. (42), exhibits a singular behavior in the source plane $z = 0$. For the above y-directed electric current element, the contributions E_x' and E_x'' from the *E*- and *H*-mode potentials Π' and Π'', respectively, are found to be given by

$$E'_x = \frac{pxy}{4\pi\rho^2}\sqrt{\frac{\mu_0}{\epsilon_0}}\left\{\frac{2\delta(t-|z|/c)}{\rho^2} + \frac{\delta(t-r/c)}{\rho^2 r^4}\left[\rho^4 - 2z^2(2\rho^2+z^2)\right]\right.$$

$$\left. - \frac{z^2}{r^3 c}\,\delta'\!\left(t-\frac{r}{c}\right)\right\} + \frac{3pxy}{4\pi\epsilon_0 r^5}\,U\!\left(t-\frac{r}{c}\right) - \frac{pxy}{\pi\epsilon_0 \rho^4}\,\delta(z), \qquad (46a)$$

$$E''_x = \frac{pxy}{4\pi\rho^2}\sqrt{\frac{\mu_0}{\epsilon_0}}\left\{-\frac{2\delta(t-|z|/c)}{\rho^2} + \frac{2\delta(t-r/c)}{\rho^2} + \frac{1}{rc}\,\delta'\!\left(t-\frac{r}{c}\right)\right\},$$

$$(46b)$$

whence $E_x = E'_x + E''_x$ becomes

$$E_x = \frac{pxy}{4\pi r^3}\sqrt{\frac{\mu_0}{\epsilon_0}}\left[\frac{3\delta(t-r/c)}{r} + \frac{1}{c}\,\delta'\!\left(t-\frac{r}{c}\right)\right] + \frac{3pxy}{4\pi\epsilon_0 r^5}\,U\!\left(t-\frac{r}{c}\right)$$

$$- \frac{pxy}{\pi\epsilon_0 \rho^4}\,\delta(z). \qquad (47)$$

One observes that both the E-mode and H-mode contributions contain the plane and spherical wavefronts depicted in Fig. 1.1.1. The E-mode portion has in addition the static dipole field given by the second term in Eq. (46a) and in the source plane $z = 0$, the last term in Eq. (46a) behaves singularly. This singularity in the total field expression (47) is spurious since it destroys the continuity of E_x across the $z = 0$ plane, and its presence represents a slight deficiency in the Hertz potential formulation. Cancellation of the singularity is assured by retention of the operator term $(\nabla_t \nabla_t / \nabla_t^2)$ contained in the exact Eq. (38a) but omitted in the formulation of Eqs. (42). [See the result obtained on pre- and postmultiplication of this term by $-px_0$ and y_0, respectively, noting the footnote to Eq. (38a).] The above calculation demonstrates that care must be exercised in the use of the Hertz potential formulas (42) in planes containing sources and oriented perpendicular to $\mathbf{a}$. However, the fields computed exterior to the source planes can be employed on these planes if necessary continuity requirements are satisfied [i.e., if the singularity can be isolated and ignored, as in Eq. (47)]. One verifies readily that with the last term omitted, E_x in Eq. (47) agrees with the expression obtained from the alternative formula, $-px_0 \cdot \mathscr{G}_{11} \cdot y_0$, in Eq. (34a).

Dyadic Green's functions for bounded cylindrical regions

For the case of a uniform waveguide of arbitrary (but non-varying) cross section transverse to the guide axis $\mathbf{a}$, and bounded by perfectly conducting walls s, the electric-field Green's function $\mathscr{G}_{11}$ implicitly defined by Eqs. (21a) must satisfy a boundary condition

$$\mathbf{v} \times \mathscr{G}_{11} = 0 \qquad \text{on } s, \qquad (48)$$

$\mathbf{v}$ being the outward normal vector at the walls. The representation in Eq. (38a) of the free-space dyadic $\mathscr{G}_{11}$ satisfies an outgoing wave condition in all directions but is inadequate to satisfy the requirement in Eq. (48). However, as derivable from the inverse-operator derivation, the representations in Eqs. (38a) and (41) are still valid inverses if two different scalar Green's functions $\mathscr{S}'$ and $\mathscr{S}''$ are inserted into these expressions:

$$\mathscr{G}_{11} = \frac{1}{\epsilon_0(\partial/\partial t)}\left(\mathbf{aa} + \frac{\nabla_t \nabla_t}{\nabla_t^2}\right)\delta(\mathbf{r} - \mathbf{r}')\,\delta(t - t')$$

$$+ \mu_0\frac{\partial}{\partial t}(\nabla \times \mathbf{a})(\nabla' \times \mathbf{a})\mathscr{S}''(\mathbf{r}, \mathbf{r}'; t, t')$$

$$- \frac{[\nabla \times (\nabla \times \mathbf{a})][\nabla' \times (\nabla' \times \mathbf{a})]\mathscr{S}'(\mathbf{r}, \mathbf{r}'; t, t')}{\epsilon_0(\partial/\partial t)} \tag{49a}$$

$$\mathscr{G}_{21} = -[(\nabla \times \mathbf{a})(\nabla' \times \nabla' \times \mathbf{a})\mathscr{S}'(\mathbf{r}, \mathbf{r}'; t, t')$$

$$+ (\nabla \times \nabla \times \mathbf{a})(\nabla' \times \mathbf{a})\mathscr{S}''(\mathbf{r}, \mathbf{r}'; t, t')], \tag{49b}$$

where both $\mathscr{S}'$ and $\mathscr{S}''$ still satisfy Eq. (38b) but are distinguished by their boundary conditions, deducible from the requirement (48):

$$\frac{\partial}{\partial v}\mathscr{S}'' = 0 \qquad\qquad \text{on } s,$$

$$\nabla_t^2\mathscr{S}' = 0 = \mathscr{S}' \qquad \text{on } s. \tag{49c}$$

Compatibility of the conditions $\nabla_t^2\mathscr{S}' = 0$ and $\mathscr{S}' = 0$ on s can be inferred from the homogeneous form of Eq. (38b). Initial conditions of vanishing for $t \leq t'$ and hence outgoing wave conditions at $\pm\infty$ on the guide axis are also required.

The Green's functions in Eqs. (49) can be utilized for the representation of the fields excited by an electric current density $\mathbf{J}$ in a uniform waveguide with perfectly conducting walls. The field representations in Eqs. (42) are still applicable but, now, instead of the expressions in Eqs. (42b), one finds for the electric-type Hertz potentials,

$$\mathbf{\Pi}''(\mathbf{r}, t) = \int \nabla' \times \mathbf{a}\mathscr{S}''(\mathbf{r}, \mathbf{r}'; t, t')\cdot\mathbf{J}(\mathbf{r}', t')\,d\mathbf{r}'\,dt',$$

$$\mathbf{\Pi}'(\mathbf{r}, t) = \frac{1}{\epsilon_0(\partial/\partial t)}\int \nabla' \times (\nabla' \times \mathbf{a})\,\mathscr{S}'(\mathbf{r}, \mathbf{r}'; t, t')\cdot\mathbf{J}(\mathbf{r}', t')\,d\mathbf{r}'\,dt'. \tag{50}$$

It should be noted that for linear stationary systems, the scalar Green's functions $\mathscr{S}'$ and $\mathscr{S}''$ both have the form

$$\mathscr{S}(\mathbf{r}, \mathbf{r}'; t, t') = \mathscr{S}(\mathbf{r}, \mathbf{r}'; t - t'), \tag{51a}$$

and since, in view of the reciprocity property (29) of the Green's function $\mathscr{G}_{11}$, $\mathscr{S}(\mathbf{r}, \mathbf{r}'; t, t') = \mathscr{S}(\mathbf{r}', \mathbf{r}; -t', -t)$, one has

$$\mathscr{S}(\mathbf{r}, \mathbf{r}'; t - t') = \mathscr{S}(\mathbf{r}', \mathbf{r}; t - t'). \tag{51b}$$

The above reciprocity properties permit one to anticipate the general forms that $\mathscr{S}'$ and $\mathscr{S}''$ must assume:

$$\mathscr{S}'(\mathbf{r}, \mathbf{r}'; t, t') = \sum_\alpha A_\alpha \Phi_\alpha(\mathbf{r}, t)\Phi_\alpha(\mathbf{r}', -t'),$$

$$\mathscr{S}''(\mathbf{r}, \mathbf{r}'; t, t') = \sum_\beta B_\beta \psi_\beta(\mathbf{r}, t)\psi_\beta(\mathbf{r}', -t'), \tag{51c}$$

and Eq. (38b) implies that for $\mathbf{r} \neq \mathbf{r}', t \neq t'$ the mode functions Φ_α and ψ_β must satisfy

$$\left(\nabla^2 - \frac{1}{c^2}\frac{\partial^2}{\partial t^2}\right)\Phi_\alpha = 0, \qquad \Phi_\alpha = 0 \text{ on } s,$$

$$\left(\nabla^2 - \frac{1}{c^2}\frac{\partial^2}{\partial t^2}\right)\psi_\beta = 0, \qquad \frac{\partial\psi_\beta}{\partial v} = 0 \text{ on } s, \tag{51d}$$

with the amplitude constants A_α and B_β being determined by the singularity of Eq. (38b) at $\mathbf{r} = \mathbf{r}', t = t'$. In the steady state, representations of this type will be exhibited in Secs. 2.3 and 5.2.

In a manner similar to the derivations in Eqs. (38a) and (41), or by duality, one infers that the general solutions of Eqs. (20) and (21b) for $\mathscr{G}_{22}$ and $\mathscr{G}_{12}$ in a uniform waveguide of perfectly conducting walls have the form

$$\mathscr{G}_{22} = +\frac{1}{\mu_0(\partial/\partial t)}\left(\mathbf{aa} + \frac{\nabla_t\nabla_t}{\nabla_t^2}\right)\delta(\mathbf{r} - \mathbf{r}')\,\delta(t - t')$$

$$+ \epsilon_0\frac{\partial}{\partial t}(\nabla \times \mathbf{a})(\nabla' \times \mathbf{a})\mathscr{S}'(\mathbf{r}, \mathbf{r}'; t, t')$$

$$- \frac{[\nabla \times (\nabla \times \mathbf{a})][\nabla' \times (\nabla' \times \mathbf{a})]\mathscr{S}''(\mathbf{r}, \mathbf{r}'; t, t')}{\mu_0(\partial/\partial t)}, \tag{52a}$$

$$\mathscr{G}_{12} = (\nabla \times \mathbf{a})[\nabla' \times (\nabla' \times \mathbf{a})]\mathscr{S}''(\mathbf{r}, \mathbf{r}'; t, t')$$

$$+ [\nabla \times (\nabla \times \mathbf{a})](\nabla' \times \mathbf{a})\mathscr{S}'(\mathbf{r}, \mathbf{r}'; t, t'), \tag{52b}$$

whence, by Eq. (19b), the electric and magnetic fields produced by a magnetic current density $\mathbf{M}(\mathbf{r}, t)$ are

$$\mathbf{E}(\mathbf{r}, t) = \nabla \times (\nabla \times \mathbf{a})\,\Pi'(\mathbf{r}, t) - \mu_0\frac{\partial}{\partial t}\nabla \times \mathbf{a}\,\Pi''(\mathbf{r}, t),$$

$$\mathbf{H}(\mathbf{r}, t) = \epsilon_0\frac{\partial}{\partial t}\nabla \times \mathbf{a}\,\Pi'(\mathbf{r}, t) + \nabla \times (\nabla \times \mathbf{a})\,\Pi''(\mathbf{r}, t), \tag{53a}$$

where

$$\Pi'(\mathbf{r}, t) = -\int \nabla' \times \mathbf{a}\,\mathscr{S}'(\mathbf{r}, \mathbf{r}'; t, t')\cdot\mathbf{M}(\mathbf{r}', t')\,d\mathbf{r}'\,dt',$$

$$\Pi''(\mathbf{r}, t) = \frac{1}{\mu_0(\partial/\partial t)}\int \nabla' \times (\nabla' \times \mathbf{a})\,\mathscr{S}''(\mathbf{r}, \mathbf{r}'; t, t')\cdot\mathbf{M}(\mathbf{r}', t')\,d\mathbf{r}'\,dt'. \tag{53b}$$

Since the field expressions in Eqs. (53a) and (42a) are identical, it is manifest that the Hertz potentials in the presence of both electric and magnetic current densities $\mathbf{J}$ and $\mathbf{M}$ are given by superposition of Eqs. (50) and (53b). There is a restriction in the range of applicability of the magnetic-field representation in Eq. (53a) to planes transverse to $\mathbf{a}$ not containing the source $\mathbf{M}(\mathbf{r}, t)$; as observed previously, this results from a singularity in $\mathscr{G}_{22}$ noted in Eq. (52a) but not included in Eq. (53a).

1.1c   The Plasma Field (One-Component Fluid Model)

General properties

A plasma, comprising a system of charged particles in thermal motion, exhibits collective wave phenomena that in certain ranges may be approximately described in terms of a homogeneous, collisionless, fluid model. We restrict our considerations to the case wherein only plasma electrons are assumed mobile. The associated plasma field, linearized and suitably averaged, is describable by an electric field $\mathbf{E}(\mathbf{r}, t)$, a magnetic field $\mathbf{H}(\mathbf{r}, t)$, an electron pressure $p(\mathbf{r}, t)$, and an average electron velocity $\mathbf{v}(\mathbf{r}, t)$ obeying the field equations[3]

$$
\begin{aligned}
\epsilon_0 \frac{\partial \mathbf{E}}{\partial t} - \nabla \times \mathbf{H} \qquad\qquad &-n_0 q \mathbf{v} \qquad\qquad &= -\mathbf{J}, \\[6pt]
\nabla \times \mathbf{E} + \mu_0 \frac{\partial \mathbf{H}}{\partial t} \qquad\qquad & &= -\mathbf{M}, \\[6pt]
\frac{1}{\gamma p_0} \frac{\partial p}{\partial t} + \qquad &\nabla \cdot \mathbf{v} \qquad\qquad &= -s, \\[6pt]
n_0 q \mathbf{E} \qquad + \quad \nabla p \ + n_0 m &\left(\frac{\partial \mathbf{v}}{\partial t} - \omega_c \mathbf{b}_0 \times \mathbf{v} \right) &= -\mathbf{f}.
\end{aligned}
\tag{54}
$$

As before, ϵ_0 is the vacuum dielectric constant, μ_0 the vacuum permeability, $-n_0 q$ and $n_0 m$ are the background electron charge and mass densities, p_0 is the background electron pressure (in a cold plasma $p_0 \sim n_0 T_0 = 0$), and γ is the specific-heat ratio for electrons. The plasma is polarized by a static magnetic field $\mathbf{B}_0 = B_0 \mathbf{b}_0$, thereby introducing an electron "cyclotron" (gyro) frequency $\omega_c = q B_0 / m$ into the field description. The field is assumed to be excited externally by an electric current density $\mathbf{J}(\mathbf{r}, t)$, a magnetic current density $\mathbf{M}(\mathbf{r}, t)$, an electron source density $s(\mathbf{r}, t)$, and a force density $\mathbf{f}(\mathbf{r}, t)$. It is to be observed that the first two rows of Eqs. (54) comprise the conventional Maxwell equations appropriate to a charged fluid of electric current density $-n_0 q \mathbf{v}$ with applied current densities $\mathbf{J}$ and $\mathbf{M}$. The latter two rows constitute the Euler equations for a charged inviscid fluid, including a Lorentz force density $-n_0 q (\mathbf{E} + \mathbf{v} \times \mathbf{B}_0)$ and applied excitations s and $\mathbf{f}$ [see Eqs. (1) and (16)]. The presence of $\mathbf{v}$ dependent and $\mathbf{E}, \mathbf{H}$ dependent terms in the Maxwell and Euler equations, respectively, produces coupling between the electromagnetic and dynamical fields considered separately in Secs. 1.1a and 1.1b. The arrangement of Eq. (54) anticipates transition to the matrix equation (57) and to the abstract formulation in Sec. 1.1d.

To assure a unique description of the fields $\mathbf{E}, \mathbf{H}, p,$ and $\mathbf{v}$ defined by Eqs. (54), one imposes the additional requirements that the excitations $\mathbf{J}, \mathbf{M}, s,$ and $\mathbf{f}$ vanish for $t \leq t_1$ and that the fields satisfy the initial conditions

$$
\mathbf{E} \equiv \mathbf{H} \equiv p \equiv \mathbf{v} \equiv 0 \qquad \text{for } t \leq t_1
\tag{54a}
$$

and the boundary conditions

$$\left.\begin{array}{r} \mathbf{n} \times \mathbf{E} = \mathscr{L} \cdot \mathbf{H} \\[2mm] p = \alpha \mathbf{v} \cdot \mathbf{n} \end{array}\right\} \text{ on } S, \tag{54b}$$

where $\mathbf{n}$ is the outward unit vector normal to the surface S (if any) bounding the region within which the field is defined, while α and $\mathscr{L}$ are boundary ("impedance") parameters for the dynamical and electromagnetic fields, respectively.

An energy-conservation theorem is readily inferred on suitably multiplying Eqs. (54) by $\mathbf{E}$, $\mathbf{H}$, p, and $\mathbf{v}$, respectively, and adding, whence one obtains

$$\nabla \cdot (\mathbf{E} \times \mathbf{H} + p\mathbf{v}) = -\frac{\partial}{\partial t}\left(\frac{\epsilon_0 \mathbf{E}^2}{2} + \frac{\mu_0 \mathbf{H}^2}{2} + \frac{p^2}{2\gamma p_0} + \frac{n_0 m \mathbf{v}^2}{2}\right)$$
$$- \mathbf{J} \cdot \mathbf{E} - \mathbf{M} \cdot \mathbf{H} - sp - \mathbf{f} \cdot \mathbf{v} \tag{55}$$

as a natural generalization of Eqs. (2) and (17). The vector $(\mathbf{E} \times \mathbf{H} + p\mathbf{v})$ is identified as the total instantaneous plasma power flow per unit area at any space–time point $\mathbf{r}, t$ in the plasma field. Correspondingly, the time derivative on the right of Eq. (55) reveals the total instantaneous plasma field energy density, while the remaining terms display the instantaneous power per unit volume supplied to the field by the excitations $\mathbf{J}, \mathbf{M}, s$, and $\mathbf{f}$. The latter are evident combinations of previously defined electromagnetic and acoustic power and energy densities.

The linearity of the plasma field Eqs. (54) implies a corresponding linear dependence of the fields $\mathbf{E}$, $\mathbf{H}$, p, and $\mathbf{v}$ on the excitations $\mathbf{J}, \mathbf{M}, s$, and $\mathbf{f}$. Thus, in an evident matrix generalization of Eqs. (4) and (19), the fields at any space–time point $\mathbf{r}, t$ can be represented as

$$\begin{pmatrix} \mathbf{E}(\mathbf{r}, t) \\ \mathbf{H}(\mathbf{r}, t) \\ p(\mathbf{r}, t) \\ \mathbf{v}(\mathbf{r}, t) \end{pmatrix} = -\iiiint \begin{pmatrix} \mathscr{G}_{11} & \mathscr{G}_{12} & \mathbf{G}_{13} & \mathscr{G}_{14} \\ \mathscr{G}_{21} & \mathscr{G}_{22} & \mathbf{G}_{23} & \mathscr{G}_{24} \\ \mathbf{G}_{31} & \mathbf{G}_{32} & \mathbf{G}_{33} & \mathbf{G}_{34} \\ \mathscr{G}_{41} & \mathscr{G}_{42} & \mathbf{G}_{43} & \mathscr{G}_{44} \end{pmatrix} \cdot \begin{pmatrix} \mathbf{J}(\mathbf{r}', t') \\ \mathbf{M}(\mathbf{r}', t') \\ s(\mathbf{r}', t') \\ \mathbf{f}(\mathbf{r}', t') \end{pmatrix} d\mathbf{r}' \, dt', \tag{56}$$

where the integrals are to be extended over all space–time volume elements $d\mathbf{r}' \, dt'$ within which the excitations are finite. The scalar, vector, or dyadic nature of the Green's function matrix elements $G_{\alpha\beta} \equiv G_{\alpha\beta}(\mathbf{r}, \mathbf{r}'; t, t')$ and the appropriate definition of the matrix product in Eq. (56) is determined by the scalar or vector character of the field components on the left side of the equation. The Green's function $G_{\alpha\beta}$ represents the negative of a field produced at $\mathbf{r}, t$ by a unit current applied at $\mathbf{r}', t'$. For example, $-\mathscr{G}_{11} \cdot \mathbf{e}$ is the vector electric field produced at $\mathbf{r}, t$ by a unit electric current density at $\mathbf{r}', t'$ in the direction $\mathbf{e}$, etc.

The field representation in Eq. (56) reduces the plasma field problem to one of determining the Green's functions $G_{\alpha\beta}$. The defining equations for these Green's functions are obtained on substitution of Eq. (56) into Eqs. (54). For example, the Green's functions $G_{\alpha 1}$ excited by a point dyadic electric current density $\mathscr{J} = \mathbf{1}\,\delta(\mathbf{r} - \mathbf{r}')\,\delta(t - t')$ are defined by

$$\begin{pmatrix} \epsilon_0 \dfrac{\partial}{\partial t}\mathbf{1} & -\nabla \times \mathbf{1} & 0 & -n_0 q\mathbf{1} \\[8pt] \nabla \times \mathbf{1} & \mu_0 \dfrac{\partial}{\partial t}\mathbf{1} & 0 & 0 \\[8pt] 0 & 0 & \dfrac{1}{\gamma p_0}\dfrac{\partial}{\partial t} & \nabla \\[8pt] n_0 q\mathbf{1} & 0 & \nabla & n_0 m\left(\dfrac{\partial}{\partial t}\mathbf{1} - \omega_c \mathbf{b}_0 \times \mathbf{1}\right) \end{pmatrix} \cdot \begin{pmatrix} \mathscr{G}_{11} \\[8pt] \mathscr{G}_{21} \\[8pt] \mathbf{G}_{31} \\[8pt] \mathscr{G}_{41} \end{pmatrix} = \begin{pmatrix} \mathscr{I} \\[8pt] 0 \\[8pt] 0 \\[8pt] 0 \end{pmatrix} \tag{57}$$

where in conformity with Eq. (56) we have employed a matrix notation for the basic field operator of Eqs. (54). For uniqueness, Eqs. (57) are to be subject to initial and boundary conditions which typically are of the form [see Eqs. (54a) and (54b)]

$$G_{\alpha 1} \equiv 0 \qquad \text{for } t \leq t' \tag{57a}$$

and

$$\left. \begin{aligned} \mathbf{n} \times \mathscr{G}_{11} &= \mathscr{L} \cdot \mathscr{G}_{21} \\ \mathbf{G}_{31} &= \alpha \mathbf{n} \cdot \mathscr{G}_{41} \end{aligned} \right\} \qquad \text{on } S. \tag{57b}$$

Note that, despite the dot-product multiplication on the left of Eq. (57), the product of the 43 vector matrix element and the vector $\mathbf{G}_{31}$ is simple (i.e., ∇G_{31}).

Symmetry and reciprocity properties of the plasma Green's functions are derivable from an adjoint problem in the manner illustrated in Secs. 1.1a and 1.1b. To avoid repetitiousness we merely state the result of such a derivation:

$$G_{\alpha\beta}(\mathbf{r}, \mathbf{r}'; t, t') = (-1)^{\alpha+\beta}\widetilde{G}_{\beta\alpha}(\mathbf{r}', \mathbf{r}; -t', -t), \tag{58}$$

where $\sim$ denotes the transpose operation for dyadics . The reciprocity property 58 generalizes and contains the previously obtained properties (12) and (29) for the acoustic and electromagnetic fields.

Dyadic Green's functions for an unbounded, isotropic, electron plasma

Closed-form expressions for the Green's functions $G_{\alpha\beta}$ of an unbounded plasma with no magnetic field ($\omega_c = 0$) can be derived by generalization of the inverse-operator procedure employed in Eqs. (33) et seq. for free-space Green's functions. In this case one finds that the overall plasma field (i.e., the $G_{\alpha\beta}$) can be expressed or scalarized in terms of two distinctive scalar Green's functions characteristic of wave propagation at light and acoustic speeds. One first solves Eq. (57) for $\mathscr{G}_{11}(\mathbf{r}, \mathbf{r}'; t, t')$ by elimination of $\mathscr{G}_{21}$, $\mathbf{G}_{31}$, and $\mathscr{G}_{41}$, whence one obtains

$$\left\{ \epsilon_0 \frac{\partial}{\partial t}\left[1 + \frac{\omega_p^2}{(\partial^2/\partial t^2) - a^2\nabla^2} \right]\frac{\nabla\nabla}{\nabla^2} - \frac{(\partial^2/\partial t^2) + \omega_p^2 - c^2\nabla^2}{\mu_0(\partial/\partial t)}\frac{\nabla \times (\nabla \times \mathbf{1})}{c^2\nabla^2} \right\} \cdot \mathscr{G}_{11}$$
$$= \delta(\mathbf{r} - \mathbf{r}')\,\delta(t - t'), \tag{59a}$$

where an operational form† has been employed for simplicity of expression. Utilizing the inverse relation (33c), one can invert the bracketed operator in Eq. (59a) by writing in successive steps:

$$\mathscr{G}_{11} = -\frac{\nabla^2 - (1/a^2)(\partial^2/\partial t^2)}{\epsilon_0(\partial/\partial t)} \frac{\nabla\nabla}{\nabla^2} g_a - \mu_0 \frac{\partial}{\partial t} \frac{\nabla \times (\nabla \times \mathbf{1})}{\nabla^2} g_c, \qquad (59b)$$

$$\mathscr{G}_{11} = \mu_0 \frac{\partial}{\partial t} \mathbf{1} g_c - \frac{\nabla\nabla}{\epsilon_0(\partial/\partial t)} \left[g_a + \frac{\partial^2/\partial t^2}{\nabla^2} \left(\frac{g_c}{c^2} - \frac{g_a}{a^2} \right) \right], \qquad (59c)$$

$$\mathscr{G}_{11} = \mu_0 \frac{\partial}{\partial t} \mathbf{1} g_c - \frac{\nabla\nabla}{\epsilon_0(\partial/\partial t)} \left[g_a + \frac{\partial^2/\partial t^2}{(\partial^2/\partial t^2) + \omega_p^2} (g_c - g_a) \right], \qquad (59d)$$

where g_a and g_c are two different scalar Green's functions defined in Eq. (60b). Finally, on rearranging the g_a and g_c terms, one rewrites Eq. (59d) as

$$\mathscr{G}_{11} = \left(\mu_0 \frac{\partial}{\partial t} \mathbf{1} - \frac{\nabla\nabla}{\epsilon_0} \frac{\partial/\partial t}{(\partial^2/\partial t^2) + \omega_p^2} \right) g_c - \frac{\nabla\nabla}{\epsilon_0} \frac{\omega_p^2}{(\partial/\partial t)[(\partial^2/\partial t^2) + \omega_p^2]} g_a, \tag{60a}$$

where

$$g_u = \frac{-1}{\nabla^2 - (1/u^2)[(\partial^2/\partial t^2) + \omega_p^2]} \delta(\mathbf{r} - \mathbf{r}')\delta(t - t'), \qquad u = a, c, \qquad (60b)$$

with $c = 1/\sqrt{\mu_0\epsilon_0}$, $a = \sqrt{\gamma p_0/n_0 m}$, and $\omega_p = \sqrt{n_0 q^2/m\epsilon_0}$. Equations (60b) define operationally the two scalar Green's functions, in terms of which an isotropic one-component plasma field can be represented. It should be noted that when the plasma frequency $\omega_p \to 0$, the expression for $\mathscr{G}_{11}$ in Eq. (60a) reduces to that for the electromagnetic Green's function $\mathscr{G}_{11}$ given in Eq. (30b).

In an unbounded medium the outgoing solution $g_u(\mathbf{r}, \mathbf{r}'; t, t')$ of Eq. (60b) can be determined in the form (writing $\mathbf{r} - \mathbf{r}' = \mathbf{r}$ and $t - t' = t$)

$$g_u(\mathbf{r}, t) = \int_{-\infty}^{+\infty} \frac{\exp\{-i[\omega t - \sqrt{\omega^2 - \omega_p^2}(r/u)]\}}{4\pi r} \frac{d\omega}{2\pi} . \qquad (61a)$$

This generalization of the $\omega_p = 0$ case,

$$\frac{\delta[t - (r/u)]}{4\pi r} = \int_{-\infty}^{+\infty} \frac{\exp\{-i[\omega t - \omega(r/u)]\}}{4\pi r} \frac{d\omega}{2\pi},$$

follows from the recognition that in a temporal Fourier representation of Eq. (60b) (in which $\partial/\partial t$ is replaced by $-i\omega$), the cases $\omega_p = 0$ and $\omega_p \neq 0$ differ only by the replacement $\omega \to \sqrt{\omega^2 - \omega_p^2}$ (see Sec. 1.5e). The integral in Eq. (61a) is tabulated as[4]

$$g_u(\mathbf{r}, t) = \frac{\delta[t - (r/u)]}{4\pi r} - \frac{\omega_p}{u} \frac{J_1[\omega_p\sqrt{t^2 - (r^2/u^2)}]}{\sqrt{t^2 - (r^2/u^2)}} \frac{U[t - (r/u)]}{4\pi}, \qquad (61b)$$

where $J_1(x)$ is the first-order Bessel function, and the step function $U(x)$ equals 1 or 0 depending on whether $x > 0$ or < 0, respectively.

---

†One can multiply this equation by $[(\partial^2/\partial t^2) - a^2 \nabla^2]\nabla^2$ to remove the inverse operators.

The scalar field g_u of Eq. (61b) is seen to be formed of a spherical wave front identical with that in a medium having $\omega_p = 0$, followed by a dispersion-dominated trailing "wake" that persists indefinitely. Dispersion implies that the various harmonic wave constituents in Eq. (61a), required to synthesize the field, travel at different speeds. At observation times sufficiently long to permit replacement of the Bessel function by its large-argument approximation, $J_1(\alpha) \sim (2/\pi\alpha)^{1/2} \cos[\alpha - (3\pi/4)]$, the field may be interpreted in terms of wave packets, or bundles of plane waves, of appropriate wavenumber and frequency; this aspect is discussed in detail in Sec. 1.6.

On observing from Eq. (61b) that since $\mathcal{G}_{11}$ must vanish for $t < (r/u) = \tau$, one evaluates†

$$\frac{1}{(\partial^2/\partial t^2) + \omega_p^2}\, \delta(\iota - \tau) = \frac{\sin \omega_p(t - \tau)}{\omega_p}\, U(t - \tau) \tag{62a}$$

and obtains, on multiplication by $g_u(\mathbf{r}, t)$ and integration,

$$\frac{1}{(\partial^2/\partial t^2) + \omega_p^2}\, g_u(\mathbf{r}, t) = \int_{-\infty}^{t} \frac{\sin \omega_p(t - \tau)}{\omega_p}\, g_u(\mathbf{r}, \tau)\, d\tau. \tag{62b}$$

On substitution of Eqs. (61b) and (62) into the operational solution in Eq. (59b), one then finds that

$$\mathcal{G}_{11}(\mathbf{r}, \mathbf{r}'; t, t') = \frac{\mu_0}{4\pi}\, \mathbf{1}\, \frac{\partial}{\partial t}\left[\frac{\delta[t - (r/c)]}{r} - \frac{\omega_p}{c}\, \frac{J_1[\omega_p\sqrt{t^2 - (r^2/c^2)}]}{\sqrt{t^2 - (r^2/c^2)}}\, U\!\left(t - \frac{r}{c}\right)\right]$$

$$- \frac{1}{4\pi\epsilon_0}\, \nabla\nabla \left\{\left[\frac{\cos \omega_p[t - (r/c)]}{r}\right.\right.$$

$$\left. - \frac{\omega_p}{c} \int_{r/c}^{t} \cos \omega_p(t - \tau)\, \frac{J_1(\omega_p\sqrt{\tau^2 - (r^2/c^2)})}{\sqrt{\tau^2 - (r^2/c^2)}}\, d\tau\right] U\!\left(t - \frac{r}{c}\right)\right\}$$

$$- \frac{1}{4\pi\epsilon_0}\, \nabla\nabla \left\{\left[\frac{1 - \cos \omega_p[t - (r/a)]}{r}\right.\right.$$

$$\left. - \frac{\omega_p}{a} \int_{r/a}^{t} [1 - \cos \omega_p(t - \tau)]\, \frac{J_1(\omega_p\sqrt{\tau^2 - (r^2/a^2)})}{\sqrt{\tau^2 - (r^2/a^2)}}\, d\tau\right] U\!\left(t - \frac{r}{a}\right)\right\}. \tag{63}$$

Equation (63) displays explicitly the electric field produced by a suddenly created electric dipole at $t = 0 = \mathbf{r}$ as a superposition of two spherical wavefields traveling at speeds c and a, each associated with a dispersive wake.

Reduced formulation of plasma field

It is of interest to note that the first-order form (54) of the plasma field equations can be recast into a higher-order form on elimination of the field variables p and $\mathbf{v}$. One obtains by this reduction an electromagnetic phrasing of the plasma field equations as

---

†As is verified by the operation of $(\partial^2/\partial t^2) + \omega_p^2$ on both sides of Eq. (62a).

$$\epsilon\left(\nabla, \frac{\partial}{\partial t}\right)\frac{\partial}{\partial t}\cdot\mathbf{E} - \nabla\times\mathbf{H} = -\mathbf{J}',$$

$$\nabla\times\mathbf{E} + \mu_0\frac{\partial\mathbf{H}}{\partial t} = -\mathbf{M},$$

$$(64)$$

where

$$\epsilon\left(\nabla, \frac{\partial}{\partial t}\right) = \epsilon_0\left(1 + \frac{\omega_p^2}{(\partial^2/\partial t^2)\mathbf{1} + a^2\nabla\nabla}\right)$$

is an equivalent dielectric constant operator for the isotropic ($\omega_c = 0$) plasma and $\mathbf{J}'$ represents an equivalent electric current density involving a somewhat complicated operator expression in $\mathbf{J}, \mathbf{f}$, and s. The dependence of the equivalent dielectric constant on ∇ and $\partial/\partial t$ is frequently referred to as a spatial and temporal dispersion property; this is to be contrasted with the non-dispersive character of the parameters in the first-order formulation. (See Sec. 1.5).

An alternative fluid dynamic rephrasing of the plasma field equations in terms of p and $\mathbf{v}$ can be obtained on elimination of the field variables $\mathbf{E}$ and $\mathbf{H}$ from Eqs. (54).

1.1d   General Linear Field (Abstract Formulation)

The commonality in analytical procedures employed to describe the acoustic, electromagnetic, and the one-component plasma fields in Secs, 1.1a–1.1c, suggests their applicability to any linear field that is invariant under spatial and temporal displacements. This observation finds its most succinct expression when the field equations for a linear field are represented in the operator form

$$L\Psi = -\Phi \tag{65}$$

where $L = L(\nabla, \partial/\partial t)$ is a linear operator descriptive of the field equations, $\Psi = \Psi(\mathbf{r}, t)$ is a wavevector characterizing the field variables, and $\Phi = \Phi(\mathbf{r}, t)$ is a wavevector describing the excitation. Since L is a (unbounded) derivative operator, it is necessary for uniqueness to state that Ψ lies in a prescribed domain $\mathcal{D}_L$ of the operator L—a remark that is equivalent to the statement of initial and boundary conditions on the elements of Ψ. We list below the matrix forms taken by the operator L and the wavevectors Ψ and Φ for the fields considered in Secs. 1.1a–1.1c.

Acoustic field

$$L \to \begin{pmatrix} \dfrac{1}{\gamma p_0}\dfrac{\partial}{\partial t} & \nabla \\[2ex] \nabla & n_0 m\dfrac{\partial}{\partial t}\mathbf{1} \end{pmatrix},$$

$$\Psi \to \begin{pmatrix} p \\ \mathbf{v} \end{pmatrix}, \qquad \Phi \to \begin{pmatrix} s \\ \mathbf{f} \end{pmatrix}. \tag{66}$$

Electromagnetic field

$$
L \to
\begin{pmatrix}
\epsilon_0 \dfrac{\partial}{\partial t}\mathbf{1} & -\nabla \times \mathbf{1} \\[2ex]
\nabla \times \mathbf{1} & \mu_0 \dfrac{\partial}{\partial t}\mathbf{1}
\end{pmatrix},
\tag{67}
$$

$$
\Psi \to \begin{pmatrix} \mathbf{E} \\ \mathbf{H} \end{pmatrix}, \qquad
\Phi \to \begin{pmatrix} \mathbf{J} \\ \mathbf{M} \end{pmatrix}.
$$

One-component plasma field

$$
L \to
\begin{pmatrix}
\epsilon_0 \dfrac{\partial}{\partial t}\mathbf{1} & -\nabla \times \mathbf{1} & 0 & -n_0 q\mathbf{1} \\[2ex]
\nabla \times \mathbf{1} & \mu_0 \dfrac{\partial}{\partial t}\mathbf{1} & 0 & 0 \\[2ex]
0 & 0 & \dfrac{1}{\gamma p_0}\dfrac{\partial}{\partial t} & \nabla \\[2ex]
n_0 q\mathbf{1} & 0 & \nabla & n_0 m\left(\dfrac{\partial}{\partial t}\mathbf{1} - \omega_c \mathbf{b}_0 \times \mathbf{1}\right)
\end{pmatrix},
\tag{68}
$$

$$
\Psi \to \begin{pmatrix} \mathbf{E} \\ \mathbf{H} \\ p \\ \mathbf{v} \end{pmatrix}, \qquad
\Phi \to \begin{pmatrix} \mathbf{J} \\ \mathbf{M} \\ s \\ \mathbf{f} \end{pmatrix}.
$$

The matrix–wavevector product $L\Psi$ in Eq. (65) requires modest care in evaluation, as the matrix elements of the operator L are scalars, vectors, or dyadics, while the wavevector elements are scalars or vectors. Proper identification of the product is evident on comparison of Eq. (66) with Eq. (1), of Eq. (67) with Eq. (16), and of Eq. (68) with Eq. (54).

To effect combinatorial operations on Eq. (65) one defines the inner product of two wavevectors and is led thereby to the concept of an adjoint operator.[5] The inner product of two wavevectors Ψ^+ and Ψ is defined as the four-dimensional space–time integral

$$
(\Psi^+, \Psi) \equiv \int \Psi^+(\mathbf{r}, t) \cdot \Psi(\mathbf{r}, t)\, d\mathbf{r}\, dt;
\tag{69a}
$$

for example, in the case of the real electromagnetic field described by the wavevector in Eq. (67), this relation takes the more explicit form

$$
(\Psi^+, \Psi) = \int [\mathbf{E}^+ \cdot \mathbf{E} + \mathbf{H}^+ \cdot \mathbf{H}]\, d\mathbf{r}\, dt.
\tag{69b}
$$

If the inner product of the wavevectors Ψ^+ and $L\Psi$ can be related to the inner product of the wavevectors $L^+\Psi^+$ and Ψ as follows,

$$
(\Psi^+, L\Psi) = (L^+\Psi^+, \Psi),
\tag{70}
$$

then the operator L^+ is said to be the adjoint of the operator L for Ψ and Ψ^+ lying in suitable domains $\mathscr{D}_L$ and $\mathscr{D}_{L^+}$, respectively. The operator L is to be identified as one of the matrix derivative operators shown in Eqs. (66–68); Eq. (70) then constitutes an integration-by-parts theorem in a four-dimensional space–time volume wherein the boundary contributions vanish by restricting Ψ and Ψ^+ to suitable domains.

The definition of the adjoint operator L^+ in Eq. (70) permits the introduction of an adjoint problem

$$L^+\Psi^+ = -\Phi^+, \tag{71}$$

where for uniqueness Ψ^+ is subject to appropriate boundary and initial conditions (i.e., Ψ^+ lies in a prescribed domain $\mathscr{D}_{L^+}$) and where the wavevector Φ^+ is arbitrary. The wavevector Ψ of the original problem in Eq. (65) is not unrelated to the wavevector Ψ^+ of the adjoint problem in Eq. (71). In fact, in view of Eq. (70), it is evident that

$$(\Psi^+, \Phi) = (\Phi^+, \Psi), \tag{72}$$

an adjointness relation that is equivalent to and generalizes the previously encountered adjointness relations in Eqs. (9) and (25).

One introduces a Green's function operator G for the linear field Ψ by

$$\Psi = -G\Phi, \qquad \text{whence } LG = 1, \tag{73a}$$

and, correspondingly, an adjoint Green's function G^+ for the adjoint field Ψ^+ by

$$\Psi^+ = -G^+\Phi^+, \qquad \text{whence } L^+G^+ = 1. \tag{73b}$$

Equations (73) constitute succinct defining equations for the Green's functions of a general linear field. Matrix representatives of G and G^+ are denoted by

$$G \to (G_{ij}(\mathbf{r}, \mathbf{r}'; t, t')) \qquad \text{and} \qquad G^+ \to (G_{ij}^+(\mathbf{r}, \mathbf{r}'; t, t')) \tag{73c}$$

with the matrix elements G_{ij} and G_{ij}^+ identifiable as scalars, vectors, or dyadics; see, for example, the representations in Eqs. (4), (19), and (56). In view of the adjointness relation (72), the adjoint Green's function G^+ is related to the original Green's function G by

$$(G^+\Phi^+, \Phi) = (\Phi^+, G\Phi), \tag{74a}$$

whence one infers from Eq. (69a) and the arbitrariness of the excitation wavevectors Φ and Φ^+ the matrix element relations

$$G_{ij}^+(\mathbf{r}, \mathbf{r}'; t, t') = \tilde{G}_{ji}(\mathbf{r}', \mathbf{r}; t', t), \tag{74b}$$

which generalizes previous results derived in Eqs. (11) and (28); note in Eq. (74b) that the transpose symbol is necessary only if G_{ij} is a dyadic element.

The relation (74b) can be cast as a reciprocity relation involving only the original Green's function G if the adjoint G^+ is expressible in terms of G. Such a possibility exists if the original field operator L possesses symmetry properties under a time-reversal transformation in an appropriate domain. For example, if the time-reversed field operator $L(\nabla, -\partial/\partial t)$ is denoted by $\hat{L}$, then for all the operators L^+ considered above, $L^+ = T\hat{L}T$, where T is a diagonal operator with ± 1 matrix elements such that $T^{-1} = T$. It follows that

$$G^+ = T\hat{G}T, \tag{75}$$

where $\hat{G}$ is defined in a suitable domain by $\hat{L}\hat{G} = 1$ and has a matrix representative

$$\hat{G}_{ij}(\mathbf{r}, \mathbf{r}'; t, t') = G_{ij}(\mathbf{r}, \mathbf{r}'; -t, -t'). \tag{76}$$

From Eqs. (74b), (75), and (76) one then infers in an appropriate domain the reciprocity property

$$G_{ij}(\mathbf{r}, \mathbf{r}'; t, t') = (-1)^{i+j}\tilde{G}_{ji}(\mathbf{r}, \mathbf{r}'; -t', -t) \tag{77}$$

for the original field, a result that is a generalization of the previously noted relations (12), (29), and (58).

1.2   PLANE-WAVE FIELD REPRESENTATIONS

Closed-form solutions for the space- and time-dependent fields excited by arbitrary, space–time distributed sources are not generally possible. Although a number of formal solutions for such problems have been obtained in Sec. 1.1 via operator or equivalent techniques, their explicit evaluation frequently requires a complicated integration process, depending on the space–time distribution of the sources. The determination of field solutions for free-space sources of harmonic plane-wave form is, however, much simpler to effect, because the operator analysis becomes essentially algebraic. Thus, in suitable regions, if the source distributions can be analyzed into their plane-wave constituents, the corresponding field response can generally be ascertained by algebraic techniques and the desired space–time fields evaluated by synthesis (integration) of the constituent plane-wave responses. This analysis and synthesis procedure provides a very effective methodology for the calculation of power flow, asymptotic evaluation of far fields, etc., for appropriate linear fields. In the following, attention will be concentrated on plane-wave analyses of fields, and their Green's functions; the corresponding space- and time-dependent fields are derivable therefrom via transform relations to be stated.

Fields in linear, homogeneous, stationary, unbounded regions are invariant under space–time translations and hence are representable as superpositions of plane waves of the form $\exp[i(\mathbf{k}\cdot\mathbf{r} - \omega t)]$. The vector wavenumber $\mathbf{k}$ and angular frequency ω, which characterize the wave periodicities along the spatial coordinate $\mathbf{r}$ and time t, can be so selected that such plane waves constitute a complete set capable of representing relatively arbitrary space–time functions. The mathematical basis for such a representation is provided by the four-dimensional Fourier-integral theorem,[6] whereby an integrable space–time function $F(\mathbf{r}, t)$ is representable as

$$F(\mathbf{r}, t) = \int F(\mathbf{k}, \omega)e^{i(\mathbf{k}\cdot\mathbf{r}-\omega t)}\frac{d\mathbf{k}\,d\omega}{(2\pi)^4} \qquad (1a)$$

with the regular transform amplitude $F(\mathbf{k}, \omega)$ given by

$$F(\mathbf{k}, \omega) = \int F(\mathbf{r}, t)e^{-i(\mathbf{k}\cdot\mathbf{r}-\omega t)}\,d\mathbf{r}\,dt. \qquad (1b)$$

Each of the fourfold integrals in Eqs. (1a) and (1b) extends from $-\infty$ to $+\infty$, with $d\mathbf{k}$ and $d\mathbf{r}$ denoting volume elements in $\mathbf{k}$ and $\mathbf{r}$ space, respectively. For notational simplicity the same symbol, but with different arguments, designates the function $F(\mathbf{r}, t)$ and its transform $F(\mathbf{k}, \omega)$.

The Fourier transforms (1a) and (1b) can be combined into the more succinct form of a "completeness relation," which constitutes a plane-wave representation of the four-dimensional space–time delta function:

$$\delta(\mathbf{r} - \mathbf{r}')\,\delta(t - t') = \int e^{i[\mathbf{k}\cdot(\mathbf{r}-\mathbf{r}')-\omega(t-t')]}\frac{d\mathbf{k}\,d\omega}{(2\pi)^4}. \qquad (2a)$$

The transform relations (1) are recoverable from Eq. (2a), as is evident on multiplication of the latter by $F(\mathbf{r}', t')$ and integration over all space–time volume elements $d\mathbf{r}'\,dt'$. The transform relations (1) also imply an "orthogonality" property

$$(2\pi)^4\,\delta(\mathbf{k} - \mathbf{k}')\,\delta(\omega - \omega') = \int e^{-i[(\mathbf{k}-\mathbf{k}')\cdot\mathbf{r}-(\omega-\omega')t]}\,d\mathbf{r}\,dt. \qquad (2b)$$

In the domain of integrable functions, the range of integration in Eqs. (1a) and (2a) spans all real frequencies ω and real wavenumbers $\mathbf{k}$. In many physical problems there appear non-Fourier-integrable, but bounded, "causal" functions that vanish for t less than some finite time and that may be finite as t approaches infinity. To assure the existence of the transform (1b) of such a causal function, it is sufficient to shift slightly the ω contour of integration from the real ω axis into the Im $\omega > 0$ region. Equations (1) and (2) so modified constitute the Fourier–Laplace integral theorem. That Eqs. (1) provide a complete representation of a causal function $F(\mathbf{r}, t)$, vanishing for $t < 0$, may be verified by contour deformation of the ω integral path in Eq. (1a) into the upper half-plane Im $\omega > 0$ where, in view of the regularity property of $F(\mathbf{k}, \omega)$ in this region, the integral vanishes by Cauchy's theorem.

In a linear, homogeneous, stationary, unbounded medium, Green's functions possess an $\mathbf{r} - \mathbf{r}'$, $t - t'$ space–time dependence and hence, by Eq. (2), are representable in the form

$$g(\mathbf{r}, \mathbf{r}'; t, t') = \int g(\mathbf{k}, \omega)e^{i[\mathbf{k}\cdot(\mathbf{r}-\mathbf{r}')-\omega(t-t')]}\frac{d\mathbf{k}\,d\omega}{(2\pi)^4}. \qquad (3)$$

A characteristic (eigen) feature of such a plane-wave representation (basis) is that it algebraizes (diagonalizes) the derivative operators ∇ and $\partial/\partial t$ in field equations that are invariant to space–time translations. In such a basis

$$\nabla \equiv i\mathbf{k} \qquad \text{and} \qquad \frac{\partial}{\partial t} \equiv -i\omega. \qquad (4)$$

This algebraic property implies that the representation (3) reduces space- and time-dependent Green's function problems in linear, homogeneous, stationary, unbounded regions to simple algebraic problems of determining $g(\mathbf{k}, \omega)$. In passive lossless systems it can be anticipated that $g(\mathbf{k}, \omega)$ will possess singularities for real values of both ω and $\mathbf{k}$, since impulsive excitation of such systems generates waves that are propagated to infinity. Alternatively stated, since natural plane-wave solutions of the source-free field equations exist for certain real ω and $\mathbf{k}$, the Green's function $g(\mathbf{k}, \omega)$ must exhibit "resonances" or singularities at the corresponding ω and $\mathbf{k}$. Accordingly, to define uniquely integrals of the form (3) it will be desirable to distinguish between the analytic representation of $g(\mathbf{k}, \omega)$ in the half-plane Im $\omega > 0$ (to assure causality) and the singular, but integrable, representations for Im $\omega = 0$, with $\mathbf{k}$ real in both instances.

1.2a   The Acoustic Field

As noted in Sec. 1.1a, acoustic fields excited in linear homogeneous regions are expressible in terms of a scalar Green's function defined, as in Eq. (1.1.13b), by

$$\left(\nabla^2 - \frac{1}{a^2}\frac{\partial^2}{\partial t^2}\right)g(\mathbf{r}, \mathbf{r}'; t, t') = -\delta(\mathbf{r} - \mathbf{r}')\,\delta(t - t'), \tag{5}$$

subject to the initial condition $g \equiv 0$ for $t \leq t'$. In the case of an unbounded region this implies for given t, because of the finite propagation speed a, the boundary condition $g \equiv 0$ as $|\mathbf{r} - \mathbf{r}'| \to \infty$ [see Eq. (1.1.15c)]. In view of the algebraic property (4) and the representation in Eq. (3) one infers from Eq. (5) the simple transform solution

$$g(\mathbf{k}, \omega) = \frac{1}{k^2 - (\omega^2/a^2)}, \qquad \text{Im } \omega > 0, \tag{6a}$$

where the constraint Im $\omega > 0$ assures in this domain the analyticity of $g(\mathbf{k}, \omega)$, in keeping with the causal nature of the Green's function. To continue $g(\mathbf{k}, \omega)$ onto the real ω axis (in the limit Im $\omega \to 0+$), one first observes that a real singular function such as $1/x$, undefined at $x = 0$, can be made unique by the following limiting process:

$$\frac{1}{x} = \lim_{\epsilon \to 0+} \frac{1}{x - i\epsilon} = \lim_{\epsilon \to 0}\left(\frac{x}{x^2 + \epsilon^2} + i\,\frac{\epsilon}{x^2 + \epsilon^2}\right) = P\frac{1}{x} + \pi i \delta(x),\dagger \tag{6b}$$

where P denotes the "principal part" and serves in the sense of Cauchy to exclude the singular point $x = 0$. Equation (6b) is to be interpreted as a "distribution" which is given conventional meaning on multiplication by a suitable function of x and integration over x.[7] In accordance with Eq. (6b), one continues $g(\mathbf{k}, \omega)$ as

---

$\dagger$The relation $\pi\delta(x) = \lim\limits_{\epsilon \to 0} [\epsilon/(x^2 + \epsilon^2)]$ can be verified on integrating both sides of Eq. (6b) about $x = 0$.

$$g(\mathbf{k}, \omega) = P\frac{1}{k^2 - (\omega^2/a^2)} + \pi i \delta\!\left(k^2 - \frac{\omega^2}{a^2}\right), \qquad \text{Im } \omega = 0. \qquad (6c)$$

Knowledge of the amplitude $g(\mathbf{k}, \omega)$ permits, via Eq. (3), the evaluation of the space- and time-dependent Green's function†

$$g(\mathbf{r}, t) = \begin{cases} \displaystyle\iint \frac{e^{i(\mathbf{k}\cdot\mathbf{r} - \omega t)}}{k^2 - (\omega^2/a^2)} \frac{d\mathbf{k}\, d\omega}{(2\pi)^4} = \frac{\delta[t - (r/a)]}{4\pi r}, & t > 0, \\ \\ 0, & t \leq 0, \end{cases} \qquad (7)$$

where for simplicity $\mathbf{r}, t$ in Eqs. (7) denote $\mathbf{r} - \mathbf{r}'$, $t - t'$, and where the result of the integration is known from the alternative evaluation in Eq. (1.1.15a). It should be noted that pole singularities of the integrand in Eq. (7), or equivalently of $g(\mathbf{k}, \omega)$, occur at those values of $\mathbf{k}, \omega$ that satisfy a "dispersion equation" $k^2 - (\omega^2/a^2) = 0$, a "resonance" relation that will be considered below in more detail.

The various acoustic Green's functions $G_{11}(\mathbf{r}, \mathbf{r}'; t, t'), \mathscr{G}_{22}(\mathbf{r}, \mathbf{r}'; t, t')$, etc., are derivable from $g(\mathbf{r}, \mathbf{r}'; t, t')$, as previously noted in Eqs. (1.1.13a). It readily follows from these equations together with Eqs. (4) and (6c) that in the $\mathbf{k}, \omega$ transform space, the acoustic Green's functions have the form (for Im $\omega = 0$),

$$G_{11}(\mathbf{k}, \omega) = \pi \omega n_0 m\, \delta\!\left(k^2 - \frac{\omega^2}{a^2}\right) + P\frac{1}{j(\omega/\gamma p_0) + (k^2/j\omega n_0 m)}$$

$$\mathscr{G}_{22}(\mathbf{k}, \omega) = \left[\pi \frac{\omega}{\gamma p_0} \delta\!\left(k^2 - \frac{\omega^2}{a^2}\right) + P\frac{1}{j\omega n_0 m + [k^2/(j\omega/\gamma p_0)]}\right]\mathbf{1}_L + \frac{1}{j\omega n_0 m}\mathbf{1}_T$$

$$\mathbf{G}_{12}(\mathbf{k}, \omega) = \mathbf{G}_{21}(\mathbf{k}, \omega) = j\mathbf{k}\left[-\pi j \delta\!\left(k^2 - \frac{\omega^2}{a^2}\right) + P\frac{1}{k^2 - (\omega^2/a^2)}\right], \qquad (8)$$

where the delta functions and the principal value symbol P are to be omitted when Im $\omega \neq 0$.‡ The unit dyadics $\mathbf{1}_L$ and $\mathbf{1}_T$ are defined by

$$\mathbf{1}_L = \frac{\mathbf{k}\mathbf{k}}{k^2}, \qquad \mathbf{1}_T = \frac{-\mathbf{k} \times (\mathbf{k} \times \mathbf{1})}{k^2}, \qquad \mathbf{1} = \mathbf{1}_L + \mathbf{1}_T, \qquad (8a)$$

which are longitudinal and transverse, respectively, to the direction $\mathbf{k}_0 = \mathbf{k}/k$ of plane-wave propagation. The significance of these transformed Green's functions is enhanced if one first rewrites the acoustic field equations (1.1.1) in transform space as

$$j\frac{\omega}{\gamma p_0} p(\mathbf{k}, \omega) - j\mathbf{k}\cdot\mathbf{v}(\mathbf{k}, \omega) = -s(\mathbf{k}, \omega),$$

$$-j\mathbf{k}p(\mathbf{k}, \omega) + j\omega n_0 m\mathbf{v}(\mathbf{k}, \omega) = -\mathbf{f}(\mathbf{k}, \omega), \qquad (9)$$

---

†It should be remarked that the ω integration in Eqs. (7) yields an additional term $\delta[t + (r/a)]/4\pi r$, which vanishes for $t > 0$ [see Eq. (1.3.25)].

‡The notation $j = -i$ will be employed occasionally to emphasize network interpretations and corresponds for any harmonic constituent to a time dependence $\exp(j\omega t)$; the half-plane Im $\omega > 0$ in the i notation becomes Im $\omega < 0$ in the j notation. Also note that $G_{11}(\mathbf{k}, \omega) = G_{11}(-\mathbf{k}, \omega), \mathscr{G}_{22}(\mathbf{k}, \omega) = \mathscr{G}_{22}(-\mathbf{k}, \omega)$, etc.

whence their algebraic solution can be expressed in terms of the transformed acoustic Green's functions in Eqs. (8) as

$$p(\mathbf{k}, \omega) = -G_{11}(\mathbf{k}, \omega)s(\mathbf{k}, \omega) - \mathbf{G}_{12}(\mathbf{k}, \omega)\cdot\mathbf{f}(\mathbf{k}, \omega),$$

$$\mathbf{v}(\mathbf{k}, \omega) = -\mathbf{G}_{21}(\mathbf{k}, \omega)s(\mathbf{k}, \omega) - \mathscr{G}_{22}(\mathbf{k}, \omega)\cdot\mathbf{f}(\mathbf{k}, \omega). \tag{10}$$

That Eqs. (10) are indeed the solution of Eqs. (9) in an unbounded space can be verified by direct inversion of Eqs. (9), a fact that is already evident from Eqs. (1.1.4) and (1.1.6).

A network schematization of the acoustic field equations in $\mathbf{k}, \omega$ transform space provides a pictorial view of interrelations among the field variables and also may be used to calculate dispersion properties (see p. 2) and power radiation. If one introduces the characteristic $\mathbf{k}_0, \mathbf{T}_0', \mathbf{T}_0'' = \mathbf{k}_0 \times \mathbf{T}_0'$ unit vector coordinate system shown in Fig. 1.2.1, vector fields can be resolved into longitudinal (L) and transverse (T) components as

$$\mathbf{v}(\mathbf{k}, \omega) = v_L\mathbf{k}_0 + v_{T'}\mathbf{T}_0' + v_{T''}\mathbf{T}_0'',$$

$$\mathbf{f}(\mathbf{k}, \omega) = f_L\mathbf{k}_0 + f_{T'}\mathbf{T}_0' + f_{T''}\mathbf{T}_0'',$$

whence Eqs. (9) can be separated into longitudinal equations,

$$j\frac{\omega}{\gamma p_0}\hat{p} - kv_L = -\hat{s},$$

$$k\hat{p} + j\omega n_0 m v_L = -f_L \tag{11a}$$

and transverse equations

$$j\omega n_0 m v_{T'} = -f_{T'},$$

$$j\omega n_0 m v_{T''} = -f_{T''}, \tag{11b}$$

where $\hat{p} = -jp(\mathbf{k}, \omega)$ and $\hat{s} = -js(\mathbf{k}, \omega)$. As depicted in Fig. 1.2.2, Eqs. (11) can be schematized as a steady-state network whose elements comprise a "capacitance" $1/\gamma p_0$, an "inductance" $n_0 m$, and an ideal transformer of turns ratio k. The source terms $\hat{s}$ and f_α play the role of an "applied current" generator with infinite shunt impedance and "applied voltage" generators with zero series impedance, respectively, while $\hat{p}$ and v_α act as the "voltage" across the capacitance $1/\gamma p_0$ and the "currents" through the inductances $n_0 m$. The inde-

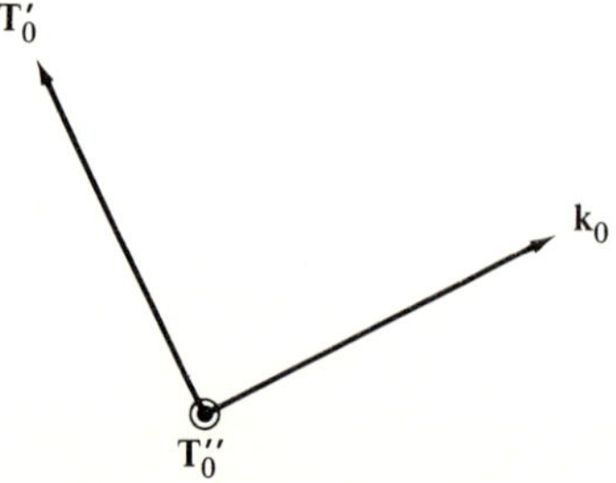

FIG. 1.2.1 $\mathbf{k}_0, \mathbf{T}_0', \mathbf{T}_0''$ coordinate system.

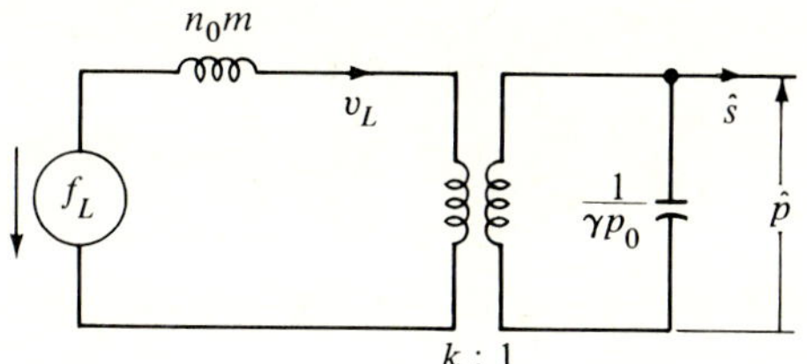

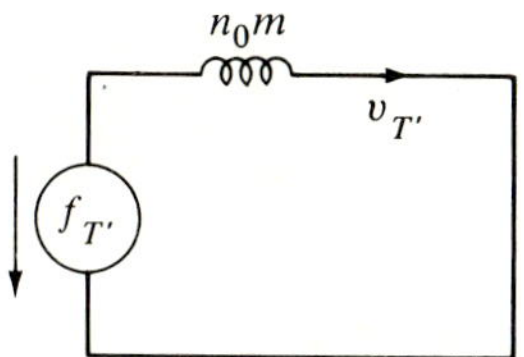

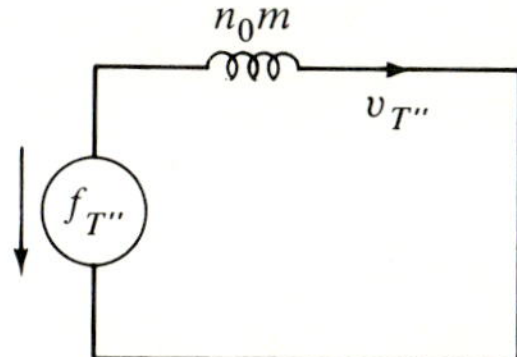

FIG. 1.2.2 Acoustic network.

pendence or uncoupling of the longitudinal (L) and transverse (T) circuits is manifest.

In the absence of excitation (i.e., $\hat{s} = 0 = f_\alpha$), it is apparent from Eqs. (11), or the longitudinal network picture, that non-vanishing source-free acoustic fields are possible for those $\mathbf{k}, \omega$ that satisfy the "resonance condition" (total mesh impedance $= 0$)

$$j\omega n_0 m + \frac{k^2}{j\omega/\gamma p_0} = 0, \tag{12a}$$

or, equivalently, the "dispersion equation" [secular determinant $= 0$ as in Eq. (44)]

$$\left(k^2 - \frac{\omega^2}{a^2}\right) = \left(k + \frac{\omega}{a}\right)\left(k - \frac{\omega}{a}\right) = 0, \qquad a = \sqrt{\frac{\gamma p_0}{n_0 m}}. \tag{12b}$$

These permitted real values of $\mathbf{k}, \omega$ evidently characterize two plane-wave fields traveling in $\pm \mathbf{k}_0$ directions with acoustic speed a. The pressure p and longitudinal velocity v_L are the only non-vanishing field components associated with these source-free waves. From Eqs. (11), or the network, one infers for the wave structure (characteristic impedance) of these longitudinal waves,

$$\frac{p}{v_L} = \pm\sqrt{\frac{n_0 m}{1/\gamma p_0}} \tag{12c}$$

with the $\pm$ signs referring to waves traveling in the $\pm \mathbf{k}_0$ directions, respectively.

Steady-state power radiated by acoustic source

At each point $\mathbf{r}, t$ the total power density supplied by a source $s(\mathbf{r}, t)$ to an acoustic field is equal to $-p(\mathbf{r}, t)s(\mathbf{r}, t)$, as derived in Eq. (1.1.2). Thus, in a non-dissipative unbounded medium the total acoustic *energy* supplied to the field by a distributed source $s(\mathbf{r}, t)$ is

$$-\int p(\mathbf{r}, t)s(\mathbf{r}, t) \, d\mathbf{r} \, dt = -\int p(\mathbf{k}, \omega)s^*(\mathbf{k}, \omega) \frac{d\mathbf{k} \, d\omega}{(2\pi)^4}, \qquad (13a)$$

where the right-hand member follows by use of Eqs. (1a) and (2), plus the observation that $s(-\mathbf{k}, -\omega) = s^*(\mathbf{k}, \omega)$ for $\mathbf{k}, \omega$ real. The total *power* radiated at frequency ω_0 by a harmonic source of the form $s(\mathbf{r})e^{j\omega_0 t}$ is the real part of the energy (13a) delivered to the field provided the source transform $s(\mathbf{k}, \omega) = s(\mathbf{k})2\pi\delta(\omega - \omega_0)$, obtained from Eq. (1b), is substituted into Eq. (13a),

$$P_{\text{rad}}(\omega) = -\text{Re} \int p(\mathbf{k}, \omega)s^*(\mathbf{k}) \frac{d\mathbf{k}}{(2\pi)^3} = \text{Re} \int G_{11}(\mathbf{k}, \omega)|s(\mathbf{k})|^2 \frac{d\mathbf{k}}{(2\pi)^3}, \qquad (13b)$$

where $\omega = \omega_0$ in Eq. (13b). It is apparent that only the real (resistive) part of the Green's function $G_{11}(\mathbf{k}, \omega)$, given in Eqs. (8), is necessary for calculation of the radiated power in an unbounded space.

For the special case of a complex harmonic point source of particles $s(\mathbf{r}) = S\delta(\mathbf{r})$ [i.e., $s(\mathbf{k}) = S$], the radiated acoustic power at frequency ω is, by Eqs. (13b) and (8),

$$P_{\text{rad}}(\omega) = \int \omega n_0 m\pi\delta\left(k^2 - \frac{\omega^2}{a^2}\right)|S|^2 \frac{d\mathbf{k}}{(2\pi)^3}$$

$$= \frac{1}{4\pi} \sqrt{\frac{n_0 m}{1/\gamma p_0}} |Sk_a|^2, \qquad k_a = \frac{\omega}{a}, \qquad (14)$$

where the integration over $\mathbf{k}$ has been effected in a spherical coordinate system wherein $d\mathbf{k} = 4\pi k^2 \, dk$ and it has been noted that

$$\delta\left(k^2 - \frac{\omega^2}{a^2}\right) = \frac{\delta(k - \omega/a)}{2k} \qquad \text{for } 0 < k < \infty.$$

[Note that $\delta(ax) = |a|^{-1}\delta(x)$].

In a similar manner, an applied force density $\mathbf{f}(\mathbf{r}, t)$ of harmonic form $\mathbf{f}(\mathbf{r})e^{j\omega_0 t}$ will radiate at frequency $\omega = \omega_0$ the acoustic power

$$P_{\text{rad}}(\omega) = \text{Re} \int \mathbf{f}^*(\mathbf{k}) \cdot \mathscr{G}_{22}(\mathbf{k}, \omega) \cdot \mathbf{f}(\mathbf{k}) \frac{d\mathbf{k}}{(2\pi)^3}, \qquad (15)$$

where $\mathbf{f}(\mathbf{k})$ is the spatial Fourier transform of $\mathbf{f}(\mathbf{r})$. For an unbounded medium, wherein only longitudinal waves can be radiated, Eq. (15) reduces by Eqs. (8) to

$$P_{\text{rad}}(\omega) = \text{Re} \int G_{22L}(\mathbf{k}, \omega)|f_L(\mathbf{k})|^2 \frac{d\mathbf{k}}{(2\pi)^3}, \qquad (16)$$

where the subscript L denotes the longitudinal component. The example of a point harmonic force density $\mathbf{f}(\mathbf{r}) = \mathbf{F}\delta(\mathbf{r})$, of complex vector amplitude $\mathbf{F}$, leads by Eqs. (16) and (8) to a radiated power of the form

$$
\begin{aligned}
P_{\text{rad}}(\omega) &= \int \pi \frac{\omega}{\gamma p_0} \delta\!\left(k^2 - \frac{\omega^2}{a^2}\right) |\mathbf{F}\cdot\mathbf{k}_0|^2 \frac{d\mathbf{k}}{(2\pi)^3} \\
&= \pi \frac{\omega}{\gamma p_0} \int_0^\infty \frac{k^2\,dk}{(2\pi)^2} \int_0^\pi \sin\theta\,d\theta \left[|F|^2 \cos^2\theta \frac{\delta(k - \omega/a)}{2k}\right] \\
&= \frac{1}{12\pi} \sqrt{\frac{1/\gamma p_0}{n_0 m}}\, |Fk_a|^2, \qquad k_a = \frac{\omega}{a},
\end{aligned}
\tag{17}
$$

where for the integration a spherical coordinate system is employed in which the polar angle θ is the angle between $\mathbf{F}$ and $\mathbf{k}_0$.

1.2b   The Electromagnetic Field

As indicated in Eqs. (1.1.31) and (1.1.32), the electromagnetic fields in an unbounded region are derivable from a scalar Green's function defined by

$$
\left(\nabla^2 - \frac{1}{c^2}\frac{\partial^2}{\partial t^2}\right) g(\mathbf{r}, \mathbf{r}'; t, t') = -\delta(\mathbf{r} - \mathbf{r}')\delta(t - t')
\tag{18}
$$

subject to the boundary condition $g \equiv 0$ as $|\mathbf{r} - \mathbf{r}'| \to \infty$ at any finite time t, and to the initial condition $g \equiv 0$ for $t \le t'$. Paralleling the discussion in Sec. 1.2a of the almost identical equation for the acoustic field, one observes that, in the $\mathbf{k}, \omega$ transform space, the solution of Eq. (18) is

$$
g(\mathbf{k}, \omega) = \begin{cases} \dfrac{1}{k^2 - (\omega^2/c^2)}, & \text{Im } \omega > 0, \\[2ex] P\dfrac{1}{k^2 - (\omega^2/c^2)} + \pi i \delta\!\left(k^2 - \dfrac{\omega^2}{c^2}\right), & \text{Im } \omega = 0, \end{cases}
\tag{19}
$$

whence in a space–time representation [see note to Eq. (7)],

$$
g(\mathbf{r}, t) = \begin{cases} \displaystyle\iint \dfrac{e^{i(\mathbf{k}\cdot\mathbf{r} - \omega t)}}{k^2 - (\omega^2/c^2)} \dfrac{d\mathbf{k}\,d\omega}{(2\pi)^4} = \dfrac{\delta[t - (r/c)]}{4\pi r}, & t > 0, \\[2ex] 0, & t < 0, \end{cases}
\tag{20}
$$

where $\mathbf{r}, t$ hereafter denote $\mathbf{r} - \mathbf{r}', t - t'$, respectively. The electromagnetic Green's functions $\mathscr{G}_{11}(\mathbf{r}, t), \mathscr{G}_{22}(\mathbf{r}, t)$, etc. are expressible, as shown in Eqs. (1.1.30)–(1.1.32), in terms of $g(\mathbf{r}, t)$. In $\mathbf{k}, \omega$ space these relations become, for Im $\omega = 0$,

$$
\mathscr{G}_{11}(\mathbf{k}, \omega) = \frac{\mathbf{1}_L}{j\omega\epsilon_0} + \left[\pi\omega\mu_0\delta\!\left(k^2 - \frac{\omega^2}{c^2}\right) + P\frac{1}{j\omega\epsilon_0 + (k^2/j\omega\mu_0)}\right]\mathbf{1}_T,
$$

$$
\mathscr{G}_{22}(\mathbf{k}, \omega) = \frac{\mathbf{1}_L}{j\omega\mu_0} + \left[\pi\omega\epsilon_0\delta\!\left(k^2 - \frac{\omega^2}{c^2}\right) + P\frac{1}{j\omega\mu_0 + (k^2/j\omega\epsilon_0)}\right]\mathbf{1}_T,
\tag{21}
$$

$$
\mathscr{G}_{12}(\mathbf{k}, \omega) = -\mathscr{G}_{21}(\mathbf{k}, \omega) = -j\mathbf{k} \times \mathbf{1}\left[-\pi j\delta\!\left(k^2 - \frac{\omega^2}{c^2}\right) + P\frac{1}{k^2 - (\omega^2/c^2)}\right],
$$

where the ("resistive and conductive") delta functions and P symbols are to be omitted for Im $\omega \neq 0$, where the longitudinal and transverse dyads $\mathbf{1}_L$ and $\mathbf{1}_T$, respectively, are defined in Eq. (8a), and where $j = -i$.[†]

The role played by the transformed Green's functions of Eqs. (21) in the description of the electromagnetic field becomes evident if one rewrites the Maxwell equations (1.1.16) in $\mathbf{k}, \omega$ space as

$$j\omega\epsilon_0 \mathbf{E}(\mathbf{k}, \omega) + j\mathbf{k} \times \mathbf{H}(\mathbf{k}, \omega) = -\mathbf{J}(\mathbf{k}, \omega),$$
$$-j\mathbf{k} \times \mathbf{E}(\mathbf{k}, \omega) + j\omega\mu_0 \mathbf{H}(\mathbf{k}, \omega) = -\mathbf{M}(\mathbf{k}, \omega). \tag{22}$$

The solution of these transformed Maxwell equations is evidently the transform of the space- and time-dependent solution given by Eqs. (1.1.19). In view of the $\mathbf{r} - \mathbf{r}'$ and $t - t'$ dependence of the Green's function for a homogeneous, stationary, unbounded medium, the transform of Eqs. (1.1.19) yields, either by direct transformation or by the Fourier convolution theorem[6], the simple relation

$$\mathbf{E}(\mathbf{k}, \omega) = -\mathscr{G}_{11}(\mathbf{k}, \omega) \cdot \mathbf{J}(\mathbf{k}, \omega) - \mathscr{G}_{12}(\mathbf{k}, \omega) \cdot \mathbf{M}(\mathbf{k}, \omega),$$
$$\mathbf{H}(\mathbf{k}, \omega) = -\mathscr{G}_{21}(\mathbf{k}, \omega) \cdot \mathbf{J}(\mathbf{k}, \omega) - \mathscr{G}_{22}(\mathbf{k}, \omega) \cdot \mathbf{M}(\mathbf{k}, \omega). \tag{23}$$

On decomposition of the vectors $\mathbf{E}, \mathbf{H}, \mathbf{J}$, and $\mathbf{M}$ into components along the characteristic coordinates $\mathbf{k}_0, \mathbf{T}_0', \mathbf{T}_0''$ shown in Fig. 1.2.1,

$$\mathbf{E}(\mathbf{k}, \omega) = E_L \mathbf{k}_0 + E_{T'} \mathbf{T}_0' + E_{T''} \mathbf{T}_0'', \text{ etc.,}$$

Eqs. (22) separate into the longitudinal equations

$$j\omega\epsilon_0 E_L = -J_L, \qquad j\omega\mu_0 H_L = -M_L \tag{24a}$$

and the transverse equations

$$j\omega\epsilon_0 E_{T'} - k\hat{H}_{T''} = -J_{T'}, \qquad j\omega\epsilon_0 E_{T''} - k\hat{H}_{T'} = -J_{T''},$$
$$kE_{T'} + j\omega\mu_0 \hat{H}_{T''} = -\hat{M}_{T''}, \qquad kE_{T''} + j\omega\mu_0 \hat{H}_{T'} = -\hat{M}_{T'}, \tag{24b}$$

where

$$\hat{H}_{T''} \equiv +jH_{T''}, \quad \hat{M}_{T''} \equiv +jM_{T''}, \quad \hat{H}_{T'} \equiv -jH_{T'}, \quad \hat{M}_{T'} \equiv -jM_{T'}. \tag{24c}$$

The transformed electromagnetic equations (24) can be schematized by the network shown in Fig. 1.2.3, whose elements comprise "inductances" μ_0, "capacitances" ϵ_0, and ideal transformers with turns ratio $k : 1$. The electric field components E_α ($\alpha = L, T', T''$) appear as "voltages" across the capacitances ϵ_0, the magnetic field components $\hat{H}_\alpha$ are "currents" through the inductances μ_0, and J_α and $\hat{M}_\alpha$ act as "applied currents" and "applied voltages." The resonant nature of the transverse circuits, and their independence or decoupling from the longitudinal circuits, pictorializes the transverse nature of the electromagnetic field.

Source-free electromagnetic fields in the form of plane waves $\exp\left[-j(\mathbf{k}\cdot\mathbf{r} - \omega t)\right]$ are possible only for those values of $\mathbf{k}, \omega$ that admit non-vanishing solutions of the homogeneous ($\mathbf{J} = 0 = \mathbf{M}$) field equations (24). Thus, transverse-wave solutions of Eqs. (24b) are possible whenever

---

[†] Note that in the j notation the relevant half-plane Im $\omega > 0$ in Eqs. (19) becomes Im $\omega < 0$.

$$j\omega\epsilon_0 + \frac{k^2}{j\omega\mu_0} = 0 \tag{25a}$$

or

$$k^2 - \frac{\omega^2}{c^2} = 0, \tag{25b}$$

and longitudinal solutions only for $\omega = 0$. Equations (25) constitute a dispersion relation [determinant of Eqs. (24b) $= 0$] for electromagnetic waves, or equivalently a condition for resonance (total "admittance" at P equal 0) of the transverse networks illustrated in Fig. 1.2.3. From Eqs. (24b), or the network, one ascertains that the wave fields have the ratio (characteristic impedance)

$$\frac{E_{T'}}{H_{T''}} = \pm\sqrt{\frac{\mu_0}{\epsilon_0}} = \frac{-E_{T''}}{H_{T'}}, \tag{26}$$

with $\pm$ signs distinguishing waves traveling in $\pm\mathbf{k}_0$ directions, respectively.

Steady-state power radiated by electric and magnetic currents in free space

Calculation of the electromagnetic energy radiated by electric or magnetic current sources in a non-dissipative unbounded region can be readily effected in a $\mathbf{k}, \omega$ basis. Since the power per unit volume supplied by a real electric

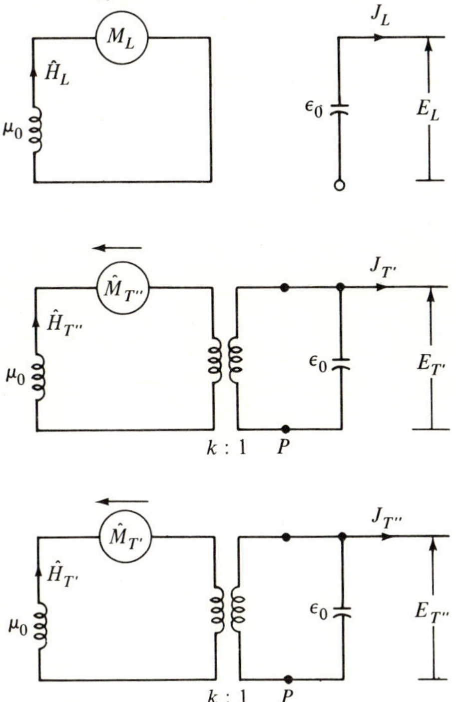

FIG. 1.2.3   Electromagnetic network.

current density $\mathbf{J}(\mathbf{r}, t)$ to the field at $\mathbf{r}, t$ is $-\mathbf{E}(\mathbf{r}, t) \cdot \mathbf{J}(\mathbf{r}, t)$, the total *energy* delivered to the field over an infinite time interval is

$$-\int \mathbf{E}(\mathbf{r}, t) \cdot \mathbf{J}(\mathbf{r}, t)\, d\mathbf{r}\, dt = -\int \mathbf{E}(\mathbf{k}, \omega) \cdot \mathbf{J}^*(\mathbf{k}, \omega) \frac{d\mathbf{k}\, d\omega}{(2\pi)^4}, \tag{27}$$

where, as in the analogous acoustic relation (13a), the right-hand member follows from the left by Parseval's theorem (Sec. 5.2d), or directly on use of Eqs. (1a), the analogue of Eqs. (2), and the observation that $\mathbf{J}(-\mathbf{k}, -\omega) = \mathbf{J}^*(\mathbf{k}, \omega)$ for real $\omega, \mathbf{k}$. In the case of a harmonic current source of the form $\mathbf{J}(\mathbf{r})e^{j\omega_0 t}$, for which correspondingly $\mathbf{E}(\mathbf{r}, t) = \mathbf{E}(\mathbf{r})e^{j\omega_0 t}$, the average real *power* delivered to the field, and hence radiated at the frequency $\omega = \omega_0$, is [see Eq. (13a) at seq.]

$$P_{\text{rad}} = -\text{Re} \int \mathbf{J}^*(\mathbf{k}) \cdot \mathbf{E}(\mathbf{k}) \frac{d\mathbf{k}}{(2\pi)^3}, \tag{28a}$$

$$= \text{Re} \int \mathbf{J}^*(\mathbf{k}) \cdot \mathscr{G}_{11}(\mathbf{k}, \omega) \cdot \mathbf{J}(\mathbf{k}) \frac{d\mathbf{k}}{(2\pi)^3}, \tag{28b}$$

$$= \int R(\mathbf{k}, \omega) |J_T(\mathbf{k})|^2 \frac{d\mathbf{k}}{(2\pi)^3}. \tag{28c}$$

$\mathbf{J}(\mathbf{k})$ and $\mathbf{E}(\mathbf{k})$ are the (root mean square) complex Fourier spatial transforms of $\mathbf{E}(\mathbf{r})$ and $\mathbf{J}(\mathbf{r})$. Equation (28b) follows by Eq. (23), and, as indicated in Eqs. (21), $\text{Re}\, \mathscr{G}_{11}(\mathbf{k}, \omega) = R(\mathbf{k}, \omega)\mathbf{1}_T$ defines the real (resistive) component $R(\mathbf{k}, \omega)$ of the impedance dyadic.

For a point harmonic source of electric current density $\mathbf{J}(\mathbf{r}) = \mathbf{J}^0 \delta(\mathbf{r}) = \mathbf{I}l\delta(\mathbf{r})$, where $\mathbf{J}^0 = \mathbf{I}l$ is the "vector current moment," the power radiated in the form of transverse waves is, by Eqs. (28c) and (21) [note comments under Eq. (14)],

$$P_{\text{rad}} = \frac{\pi}{2}\sqrt{\frac{\mu_0}{\epsilon_0}} \int_0^\pi 2\pi \sin\theta\, d\theta \int_0^\infty \frac{k^2\, dk}{(2\pi)^3} |Il \sin\theta|^2 \delta\left(k - \frac{\omega}{c}\right)$$

$$= \frac{1}{6\pi}\sqrt{\frac{\mu_0}{\epsilon_0}} |Ik_0 l|^2, \qquad k_0 = \frac{\omega}{c}, \qquad l = |\mathbf{l}|, \tag{29}$$

where the integration has been performed in a spherical coordinate system with volume element $d\mathbf{k} = 2\pi \sin\theta k^2\, d\theta\, dk$, θ being the angle between $\mathbf{k}$ and the source direction $\mathbf{l}$. For a point magnetic current source $\mathbf{M}(\mathbf{r}) = \mathbf{M}^0 \delta(\mathbf{r}) = \hat{I}l\delta(\mathbf{r})$, the average radiated power leads in a similar manner to the dual results

$$P_{\text{rad}} = \text{Re} \int \mathbf{M}^*(\mathbf{k}) \cdot \mathscr{G}_{22}(\mathbf{k}, \omega) \cdot \mathbf{M}(\mathbf{k}) \frac{d\mathbf{k}}{(2\pi)^3}$$

$$= \int G(\mathbf{k}, \omega) |\mathbf{M}_T(\mathbf{k})^2| \frac{d\mathbf{k}}{(2\pi)^3}$$

$$= \frac{1}{6\pi}\sqrt{\frac{\epsilon_0}{\mu_0}} |\hat{I}k_0 l|^2, \qquad k_0 = \frac{\omega}{c}, \tag{30}$$

where $G(\mathbf{k}, \omega)\mathbf{1}_T = \mathrm{Re}\,\mathscr{G}_{22}(\mathbf{k}, \omega)$ is the real (conductive) part of the admittance dyadic $\mathscr{G}_{22}$ given in Eq. (21).

1.2c   The Plasma Field

For the one-component plasma field described in Sec. 1.1c, the overall Green's function problem, as defined in Eqs. (1.1.56) and (1.1.57), requires the evaluation of a large number of subsidiary Green's functions $G_{\alpha\beta}(\mathbf{r}, \mathbf{r}'; t, t')$. Although for an isotropic ($\omega_c = 0$) plasma it is possible to relate all such Green's functions to two scalar Green's functions of the form given implicitly by Eqs. (5) and (18), we shall instead discuss the direct solution of the Green's function in Eqs. (1.1.57) for the special case $\omega_c = 0$. An identical but more complicated procedure exists for $\omega_c \neq 0$.

For an unbounded plasma medium, one introduces the $\mathbf{k}, \omega$ Green's function transform $G_{\alpha\beta}(\mathbf{k}, \omega)$ by

$$G_{\alpha\beta}(\mathbf{r}, \mathbf{r}'; t, t') = \int G_{\alpha\beta}(\mathbf{k}, \omega) e^{i(\mathbf{k}\cdot\mathbf{r}'' - \omega t'')} \frac{d\mathbf{k}\,d\omega}{(2\pi)^4}, \tag{31}$$

where $\mathbf{r}'' = \mathbf{r} - \mathbf{r}'$, $t'' = t - t'$. Hence in $\mathbf{k}, \omega$ space the defining equation (1.1.57) for the $G_{\alpha 1}$ becomes, when $\omega_c = 0$,

$$\begin{pmatrix} -i\omega\epsilon_0\mathbf{1} & -i\mathbf{k}\times\mathbf{1} & 0 & -n_0 q\mathbf{1} \\ +i\mathbf{k}\times\mathbf{1} & -i\omega\mu_0\mathbf{1} & 0 & 0 \\ 0 & 0 & -i\dfrac{\omega}{\gamma p_0} & i\mathbf{k} \\ n_0 q\mathbf{1} & 0 & i\mathbf{k} & -i\omega n_0 m\mathbf{1} \end{pmatrix} \cdot \begin{pmatrix} \mathscr{G}_{11} \\ \mathscr{G}_{21} \\ \mathbf{G}_{31} \\ \mathscr{G}_{41} \end{pmatrix} = \begin{pmatrix} \mathbf{1} \\ 0 \\ 0 \\ 0 \end{pmatrix}, \tag{32}$$

with similar equations for $G_{\alpha 2}$, $G_{\alpha 3}$, and $G_{\alpha 4}$, except that the unit source term on the right is in the second, third, and fourth row, respectively. Inverting Eq. (32), one can ascertain the various $G_{\alpha\beta}(\mathbf{k}, \omega)$. For instance (with $j = -i$ and $\mathrm{Im}\,\omega \neq 0$),

$$\mathscr{G}_{11}(\mathbf{k}, \omega) = \frac{\mathbf{1}_L}{j\omega\epsilon_0 + \{n_0^2 q^2 / [j\omega n_0 m + k^2/(j\omega/\gamma p_0)]\}}$$

$$+ \frac{\mathbf{1}_T}{j\omega\epsilon_0 + [k^2/(j\omega\mu_0)] + [n_0^2 q^2/(j\omega n_0 m)]}, \tag{33}$$

where the longitudinal and transverse dyads $\mathbf{1}_L$ and $\mathbf{1}_T$ are defined in Eq. (8a). The physical significance of the transformed Green's functions $G_{\alpha\beta}(\mathbf{k}, \omega)$ is manifest on transformation of Eq. (1.1.56), whereby, for the case of a linear, homogeneous, stationary, unbounded medium, applicability of Eq. (1.1.22) and the convolution integral permits simple interpretations. In particular, it is apparent that $\mathscr{G}_{11}(\mathbf{k}, \omega)$ yields the electric field $\mathbf{E}(\mathbf{k}, \omega)$ excited only by an electric current density $\mathbf{J}(\mathbf{k}, \omega)$,

$$\mathbf{E}(\mathbf{k}, \omega) = -\mathscr{G}_{11}(\mathbf{k}, \omega)\cdot\mathbf{J}(\mathbf{k}, \omega), \tag{34}$$

whence one observes that $\mathscr{G}_{11}$ is a generalization of the electric Green's function $\mathscr{G}_{11}$ of Sec. 1.2b.

One can express Eq. (33) in an alternative form that makes explicit the dependence of the dyadic Green's function $\mathscr{G}_{11}$ on two scalar Green's functions. Thus, in the notation of Sec. 1.1c,

$$\mathscr{G}_{11}(\mathbf{k}, \omega) = j\omega\mu_0\left[\left(1 - \frac{k^2 c^2}{\omega^2 - \omega_p^2}\,\mathbf{1}_L\right)G_{eo}(\mathbf{k}, \omega) + \frac{\omega_p^2}{\omega^2}\frac{k^2 c^2}{\omega^2 - \omega_p^2}G_{ea}(\mathbf{k}, \omega)\mathbf{1}_L\right],$$

$$(35)$$

where the "optic-type" scalar Green's function transform G_{eo} is defined in the j notation by

$$G_{eo}(\mathbf{k}, \omega) = \begin{cases} \dfrac{1}{k^2 - [(\omega^2 - \omega_p^2)/c^2]}, & \text{Im } \omega < 0, \\[3mm] P\,\dfrac{1}{k^2 - [(\omega^2 - \omega_p^2)/c^2]} - \pi j\delta\left(k^2 - \dfrac{\omega^2 - \omega_p^2}{c^2}\right), & \text{Im } \omega = 0. \end{cases}$$

$$(35a)$$

The "acoustic-type" scalar Green's function G_{ea} is similarly defined but with c, the speed of light, replaced by a, the acoustic speed. The representation in Eq. (35a) of G_{eo} implies that the corresponding representation of the real (resistive) part of the dyadic Green's function $\mathscr{G}_{11}(\mathbf{k}, \omega)$ for Im $\omega = 0$ is

$$\text{Re}\,\mathscr{G}_{11}(\mathbf{k}, \omega) = \pi\omega\mu_0\left[\frac{\omega_p^2}{\omega^2}\frac{c^2}{a^2}\delta\left(k^2 - \frac{\omega^2 - \omega_p^2}{a^2}\right)\mathbf{1}_L + \delta\left(k^2 - \frac{\omega^2 - \omega_p^2}{c^2}\right)\mathbf{1}_T\right],$$

$$(36)$$

while the imaginary part is the principal part of that shown in Eq. (33).

The transformed plasma field equations (1.1.54) in $\mathbf{k}, \omega$ space, when expressed in terms of the components of $\mathbf{E}, \mathbf{H}, p, \mathbf{v}$ and $\mathbf{J}, \mathbf{M}, s, \mathbf{f}$ along the coordinate axes $\mathbf{k}_0, \mathbf{T}_0', \mathbf{T}_0''$ shown in Fig. 1.2.1, constitute 10 scalar equations for the 10 unknown field components of $\mathbf{E}, \mathbf{H}, p,$ and $\mathbf{v}$. These equations are readily inferred from Eqs. (1.1.54) on use of the operator algebraizations $\partial/\partial t = j\omega$ and $\nabla = -j\mathbf{k}$. A network schematization of the resulting equations is shown in Fig. 1.2.4. This isotropic one-component plasma network, which represents a coupling of the electromagnetic and acoustic networks of Figs. 1.2.2 and 1.2.3, respectively, contains the same circuit elements as the latter networks with the addition of an ideal coupling transformer of turns ratio $n_0 q$; the "voltages" and "currents" are similarly defined. The characteristic independence of the longitudinal and transverse networks of Fig. 1.2.4, together with their evident resonant nature, implies that both longitudinal and transverse waves can be propagated independently in an isotropic plasma.

Source-free plasma fields having the form of plane waves $\exp\left[-j(\mathbf{k}\cdot\mathbf{r} - \omega t)\right]$ are possible for those values of $\mathbf{k}, \omega$ for which there exist finite solutions of the $\mathbf{k}, \omega$ transformed equations (1.1.54) with $\mathbf{J} = \mathbf{M} = s = \mathbf{f} = 0$. The various possibilities are displayed by the zeros of the determinant of these equations or,

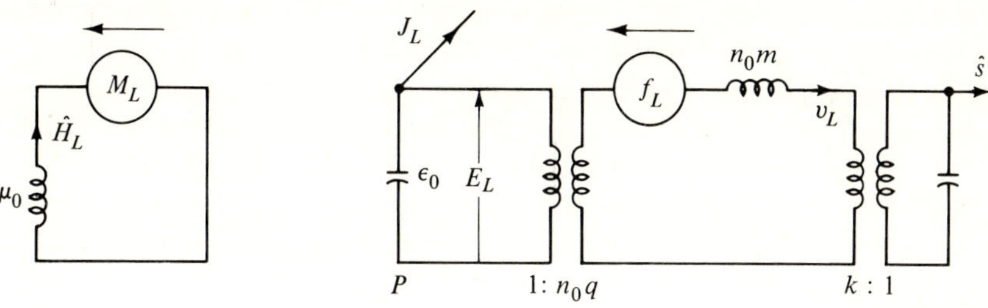

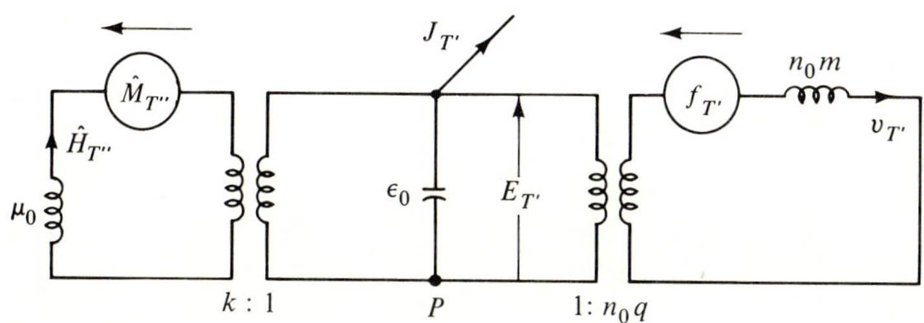

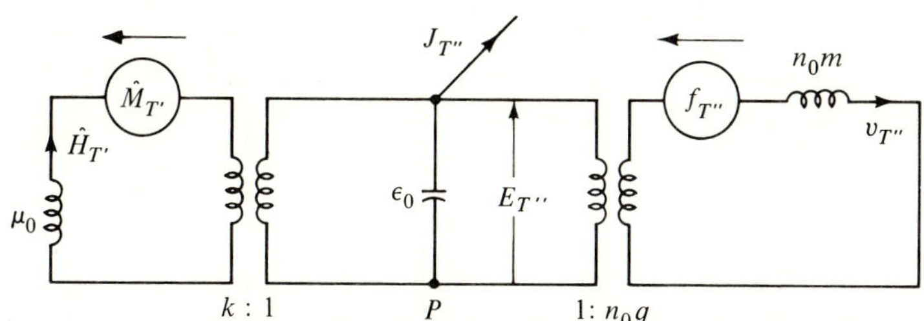

FIG. 1.2.4   One-component-plasma network.

equivalently, of the determinant of the matrix in Eq. (32). These yield the following dispersion relations:

$$j\omega\epsilon_0 + \frac{n_0^2 q^2}{j\omega n_0 m + [k^2/(j\omega/\gamma p_0)]} = 0$$

or

$$k^2 - \frac{\omega^2 - \omega_p^2}{a^2} = 0 \qquad (37a)$$

for the longitudinal waves, and

$$j\omega\epsilon_0 + \frac{k^2}{j\omega\mu_0} + \frac{n_0^2 q^2}{j\omega n_0 m} = 0$$

or

$$k^2 - \frac{\omega^2 - \omega_p^2}{c^2} = 0 \qquad (37b)$$

for the transverse waves. The dispersion relations (37) also characterize the pole singularities of the Green's function $\mathscr{G}_{11}$ of Eq. (33) and the zeros (resonances) of the total admittances at terminal plane P of the longitudinal and transverse networks of Fig. 1.2.4. The field-component ratios that characterize the field structure of these waves can be inferred from the transformed Eqs. (1.1.54), or from the resonant voltages and currents of the plasma network of Fig. 1.2.4, but will not be given explicitly at this point (see Chapter 8).

Steady-state power radiated by electric currents in an unbounded plasma

Calculation of the plasma field energy [i.e., the electromagnetic plus acoustic energy defined in Eq. (1.1.55)] radiated by sources in a plasma medium is relatively simple in $\mathbf{k}, \omega$ space. For example, in the non-dissipative case, the average real power delivered in electromagnetic form to the plasma by an electric current density $\mathbf{J}(\mathbf{r}) \exp(j\omega t)$ and thereby radiated as plasma field energy per unit time is [see Eqs. (28) and (35)]

$$P_{\text{rad}} = -\text{Re} \int \mathbf{J}^*(\mathbf{k}) \cdot \mathbf{E}(\mathbf{k}) \frac{d\mathbf{k}}{(2\pi)^3}$$

$$= \text{Re} \int \mathbf{J}^*(\mathbf{k}) \cdot \mathscr{G}_{11}(\mathbf{k}, \omega) \cdot \mathbf{J}(\mathbf{k}) \frac{d\mathbf{k}}{(2\pi)^3}, \tag{38}$$

wherein $\mathbf{J}(\mathbf{k})$ and $\mathbf{E}(\mathbf{k})$ are (root mean square) spatial Fourier transforms of the harmonic electric current density and electric field, respectively. This result generalizes the purely electromagnetic relation (28), but it should be noted that the radiated plasma field power is both electromagnetic and acoustic [see Eq. (1.1.55)]. For a one-component isotropic plasma, $\text{Re}\,\mathscr{G}_{11}(\mathbf{k}, \omega)$ is given in Eq. (36).

The plasma power radiated into a one-component isotropic medium by a point electric current dipole of harmonic current density $\mathbf{J}(\mathbf{r}, t) = \mathit{Il}\delta(\mathbf{r}) \exp(j\omega t)$ is calculated via Eqs. (38) and (36). Employing a spherical coordinate system in $\mathbf{k}$ space with $d\mathbf{k} = 2\pi k^2 \sin\theta\, d\theta\, dk$, and with θ as the polar angle between the vector $\mathbf{k}$ and the dipole direction $\mathbf{l}$, one obtains from Eqs. (38) and (36),

$$\begin{aligned}
P_{\text{rad}} &= \pi\omega\mu_0 \int \left[\frac{\omega_p^2}{a^2} \frac{c^2}{a^2} \delta\!\left(k^2 - \frac{\omega^2 - \omega_p^2}{a^2}\right) |Il \cos\theta|^2 \right. \\
&\qquad \left. + \delta\!\left(k^2 - \frac{\omega^2 - \omega_p^2}{c^2}\right) |Il \sin\theta|^2 \right] 2\pi \sin\theta \frac{k^2\, dk\, d\theta}{(2\pi)^3} \\
&= \frac{1}{6\pi}\sqrt{\frac{\mu_0}{\epsilon_0}} |Ik_0 l|^2 \sqrt{1 - \frac{\omega_p^2}{\omega^2}} \left[\frac{1}{2}\left(\frac{c}{a}\right)^3 \frac{\omega_p^2}{\omega^2} + 1 \right], \qquad \omega > \omega_p
\end{aligned} \tag{39}$$

$$P_{\text{rad}} = 0, \qquad\qquad\qquad\qquad\qquad\qquad\qquad\qquad \omega < \omega_p$$

where $k_0 = \omega/c$. The large first term in the brackets represents the portion of the power radiated in the form of longitudinal waves; the second term represents the transverse-wave contribution. For $\omega_p \to 0$, the expression in Eqs. (39) reduces to similar ones derived previously in Eq. (29) for the electromagnetic field.

The result in Eq. (39) is not realistic because of the assumed infinitesimal length of the radiating element. Results for a finite radiator can be obtained in a similar manner from Eq. (38) on use of an appropriate $\mathbf{J}(\mathbf{k})$.[8]

1.2d General Linear Field

Plane-wave representations exist for a general class of fields in a linear. homogeneous, stationary (time independent), and unbounded medium. As discussed in Sec. 1.1d, such a field is describable by a general linear field equation (1.1.65) and leads to a Green's function $G(\mathbf{r}, \mathbf{r}'; t, t')$ defined by

$$L\left(\nabla, \frac{\partial}{\partial t}\right)G(\mathbf{r}, \mathbf{r}'; t, t') = \delta(\mathbf{r} - \mathbf{r}')\delta(t - t'), \tag{40}$$

with G subject to a causality condition (outward propagation). Since, in general, L is representable by a square matrix whose elements are either scalar, vector, or dyadic operators, the Green's function G is likewise represented as a square matrix whose elements are subsidiary Green's functions of $\mathbf{r}, \mathbf{r}'$ and t, t'. A unit matrix is implicit in the right-hand delta-function term of Eq. (40). The invariance of Eq. (40) under arbitrary space–time displacements, manifest in the independence of L on the coordinates $\mathbf{r}, t$, implies the existence of plane-wave representations as in Eqs. (1) and leads via the property (4) to the transformed equation

$$L(\mathbf{k}, \omega)G(\mathbf{k}, \omega) = 1 \tag{41}$$

in $\mathbf{k}, \omega$ space. As in Eq. (3), $G(\mathbf{k}, \omega)$ is the Fourier–Laplace transform of the Green's function, and for simplicity we denote $L(\mathbf{k}, \omega) \equiv L(i\mathbf{k}, -i\omega)$.

The significance of $G(\mathbf{k}, \omega)$ is evident from the transform of the Green's function field relationship (1.1.73), which constitutes the solution of the general linear field equation (1.1.65). Thus, on use of the convolution theorem, one obtains

$$\mathbf{\Psi}(\mathbf{k}, \omega) = -G(\mathbf{k}, \omega)\mathbf{\Phi}(\mathbf{k}, \omega), \tag{42}$$

where $\mathbf{\Psi}(\mathbf{k}, \omega)$ and $\mathbf{\Phi}(\mathbf{k}, \omega)$, as in Eq. (1b), are the Fourier–Laplace transforms of $\mathbf{\Psi}(\mathbf{r}, t)$ and $\mathbf{\Phi}(\mathbf{r}, t)$. It is to be noted that $G(\mathbf{k}, \omega)$ is representable by a square matrix, with algebraic elements, given explicitly via Eq. (41) as the inverse

$$G(\mathbf{k}, \omega) = L^{-1}(\mathbf{k}, \omega) \tag{43}$$

of the known matrix $L(\mathbf{k}, \omega)$. The singularity properties of $G(\mathbf{k}, \omega)$ in the complex $\mathbf{k}, \omega$ planes determine the dispersion properties of plane waves characteristic of the source-free fields within the given medium. For example, singularities of $G(\mathbf{k}, \omega)$ occur at those values of $\mathbf{k}, \omega$ for which

$$\det L(\mathbf{k}, \omega) = 0. \tag{44}$$

Explicit knowledge of the transformed Green's function $G(\mathbf{k}, \omega)$ from Eq. (43) permits by Eq. (1a) the determination of the space- and time-dependent Green's function of Eq. (40) as

$$G(\mathbf{r}, \mathbf{r}'; t, t') = \int L^{-1}(\mathbf{k}, \omega)e^{i[\mathbf{k}\cdot(\mathbf{r}-\mathbf{r}')-\omega(t-t')]} \frac{d\mathbf{k}\, d\omega}{(2\pi)^4}. \tag{45}$$

By Eq. (42) the corresponding solution of the general linear field problem in Eq. (1.1.65) becomes

$$\Psi(\mathbf{r}, t) = -\int L^{-1}(\mathbf{k}, \omega)\Phi(\mathbf{k}, \omega)e^{i(\mathbf{k}\cdot\mathbf{r}-\omega t)} \frac{d\mathbf{k}\, d\omega}{(2\pi)^4}. \tag{46}$$

Although Eq. (46) provides a formal solution of the field problem in Eq. (1.1.65), it requires explicit evaluation of the product $L^{-1}\Phi$ and a fourfold integration over $\mathbf{k}$ and ω.

In many instances the above-indicated procedure can be simplified considerably by utilizing, ab initio, symmetry properties, with respect to a time or space direction, of the unbounded region under consideration. These simplifications, discussed in Secs. 1.3 and 1.4, are also applicable to (stationary, homogeneous) bounded regions.

1.3 GUIDED-WAVE (OSCILLATORY) REPRESENTATIONS IN TIME

As described in Sec. 1.2, the plane waves $\exp(i\mathbf{k}\cdot\mathbf{r} - i\omega t)$, $-\infty < (k_x, k_y, k_z, \omega) < \infty$, possess four-dimensional space–time orthogonality properties characteristic of the spatial and temporal symmetry properties of an unbounded, linear, homogeneous, stationary medium. However, since ω and $\mathbf{k}$ are as yet unrelated, these plane waves do not represent solutions of the source-free field equations, nor are they characteristic for the polarization structure of the field. Such plane waves permit a complete algebraization of the space–time operators in a linear field and thereby reduce a field problem in an unbounded region to one of simple algebraic matrix inversion and evaluation of four-dimensional Fourier transforms. The time-guided fields, $\Psi_\alpha(\mathbf{k}) \exp[i\mathbf{k}\cdot\mathbf{r} - \omega_\alpha(\mathbf{k})t]$, $-\infty < k_x, k_y, k_z < \infty$, in this section have orthogonality properties characteristic both of the three-dimensional spatial volume containing the field and of the polarization structure (i.e., they are characteristic solutions of the source-free field equations, but they are not orthogonal along the time coordinate). Guided waves in time permit a representation of a linear field in terms of characteristic oscillations, require evaluation of only three-dimensional Fourier transforms, and are useful for solving radiation and initial-value problems. Such representations may also be generalized to accommodate media whose constitutive parameters vary with time.

In an unbounded region the mathematical basis for these guided-wave representations is provided by two transform theorems. The three-dimensional Fourier integral theorem yields a representation of an integrable wave function $\Psi(\mathbf{r}, t)$ in the form†

---

† Note that functions and their transforms are represented by the same symbols and distinguished, where necessary, by their arguments.

$$\Psi(\mathbf{r}, t) = \int \Psi(\mathbf{k}, t)e^{i\mathbf{k}\cdot\mathbf{r}} \frac{d\mathbf{k}}{(2\pi)^3}, \tag{1a}$$

where

$$\Psi(\mathbf{k}, t) = \int \Psi(\mathbf{r}, t)e^{-i\mathbf{k}\cdot\mathbf{r}} \, d\mathbf{r}, \tag{1b}$$

and, as in the previous section, $d\mathbf{k}$ and $d\mathbf{r}$ denote coordinate-independent volume elements in the $\mathbf{k}$ and $\mathbf{r}$ spaces, respectively. In a homogeneous bounded space, eigenfunctions $\Phi(\mathbf{r})$ rather than $\exp(i\mathbf{k}\cdot\mathbf{r})$ would appear in Eqs. (1). Independently of the function Ψ, the Fourier transforms in Eqs. (1a) and (1b) can be combined into a "completeness relation," or three-dimensional delta-function representation, as

$$\delta(\mathbf{r} - \mathbf{r}') = \int e^{i\mathbf{k}\cdot(\mathbf{r}-\mathbf{r}')} \frac{d\mathbf{k}}{(2\pi)^3}. \tag{2a}$$

To assure completeness, the integration is extended over all real points in the three-dimensional $\mathbf{k}$ space. Correspondingly, any integrable function of $\mathbf{r} - \mathbf{r}'$, such as a Green's function $G(\mathbf{r}, \mathbf{r}'; t, t')$ in an unbounded homogeneous medium, is representable as

$$G(\mathbf{r}, \mathbf{r}'; t, t') = \int G(\mathbf{k}; t, t')e^{i\mathbf{k}\cdot(\mathbf{r}-\mathbf{r}')} \frac{d\mathbf{k}}{(2\pi)^3}. \tag{2b}$$

One also infers from the transform relations (1) the "orthogonality" property,

$$(2\pi)^3 \delta(\mathbf{k} - \mathbf{k}') = \int e^{-i(\mathbf{k}-\mathbf{k}')\cdot\mathbf{r}} \, d\mathbf{r}. \tag{2c}$$

In Eqs. (1), the transformed wavefunction $\Psi(\mathbf{k}, t)$, which is a one-column matrix or vector in a finite (n)-dimensional polarization space, can be further represented for given $\mathbf{k}$ as a superposition of eigenvectors $\Psi_\alpha(\mathbf{k})$:

$$\Psi(\mathbf{k}, t) = \sum_\alpha \Psi_\alpha(\mathbf{k})a_\alpha(\mathbf{k}, t), \qquad \alpha = 1, 2, \ldots, n, \tag{3a}$$

where the coefficients a_x are given in general by the weighted, Hermitean (i.e., complex-conjugate) scalar product of the vector Ψ and an adjoint vector Ψ_α^+ as

$$a_\alpha(\mathbf{k}, t) = \frac{(W^+ \Psi_\alpha^+(\mathbf{k}), \Psi(\mathbf{k}, t))}{2N_\alpha(\mathbf{k})}, \tag{3b}$$

with W^+ as the "weighting" operator. For given $\mathbf{k}$ the Ψ_α and Ψ_α^+ comprise a complete biorthogonal set of base vectors (in polarization space) with the properties

$$(W^+ \Psi_\alpha^+, \Psi_\beta) = 2N_\alpha \delta_{\alpha\beta} = \begin{cases} 2N_\alpha, & \alpha = \beta, \\ 0, & \alpha \neq \beta, \end{cases} \tag{3c}$$

N_α being a normalization constant. The transforms (3a) and (3b) may be synthesized, independently of $\Psi(\mathbf{k}, t)$, into a completeness relation for the polarization space,

$$1 = \sum_{\alpha} \frac{\mathbf{\Psi}_{\alpha}(\mathbf{k}) \, W^{+} \mathbf{\Psi}_{\alpha}^{+}(\mathbf{k})}{2N_{\alpha}(\mathbf{k})}, \tag{4}$$

where the unit operator 1 on the left is represented by a unit matrix in polarization space.

The transforms in Eqs. (1) and (3) provide a representation of the wavefunction $\mathbf{\Psi}(\mathbf{r}, t)$ in both geometrical and polarization spaces as

$$\mathbf{\Psi}(\mathbf{r}, t) = \int \sum_{\alpha} a_{\alpha}(\mathbf{k}, t) \mathbf{\Psi}_{\alpha}(\mathbf{k}) e^{i\mathbf{k}\cdot\mathbf{r}} \frac{d\mathbf{k}}{(2\pi)^3}, \tag{5a}$$

where

$$a_{\alpha}(\mathbf{k}, t) = \int \frac{(W^{+}\mathbf{\Psi}^{+}(\mathbf{k}), \mathbf{\Psi}(\mathbf{r}, t))}{2N_{\alpha}(\mathbf{k})} e^{-i\mathbf{k}\cdot\mathbf{r}} \, d\mathbf{r}. \tag{5b}$$

Alternative to these transforms is the completeness relation,

$$1\delta(\mathbf{r} - \mathbf{r}') = \int \sum_{\alpha} \frac{\mathbf{\Psi}_{\alpha}(\mathbf{k}) \, W^{+} \mathbf{\Psi}_{\alpha}^{+}(\mathbf{k})}{2N_{\alpha}(\mathbf{k})} e^{i\mathbf{k}\cdot(\mathbf{r}-\mathbf{r}')} \frac{d\mathbf{k}}{(2\pi)^3}, \tag{6}$$

from which one observes that the $\mathbf{\Psi}_{\alpha}(\mathbf{k})e^{i\mathbf{k}\cdot\mathbf{r}}$ constitute a complete set of characteristic fields for the representation of solutions to a general linear field problem (see Secs. 1.1d and 1.2d) in an unbounded medium. In particular, one infers from Eq. (6) that a general (matrix) Green's function for a linear field is representable as

$$G(\mathbf{r}, \mathbf{r}'; t, t') = \int \sum_{\alpha} G_{\alpha}(\mathbf{k}; t, t') \frac{\mathbf{\Psi}_{\alpha}(\mathbf{k}) \, \mathbf{\Psi}_{\alpha}^{+}(\mathbf{k})}{2N_{\alpha}(\mathbf{k})} e^{i\mathbf{k}\cdot(\mathbf{r}-\mathbf{r}')} \frac{d\mathbf{k}}{(2\pi)^3}, \tag{7}$$

where the coefficients G_{α} are to be determined from the equation defining G, as we shall illustrate below.

1.3a General Linear Field

A general field in a linear, homogeneous, and stationary medium is described, as in Sec. 1.1d, by a linear operator $L(\nabla, \partial/\partial t)$. If the operator L is decomposed into two components, one depending only on $\partial/\partial t$, the other only on ∇, the defining equation of the Green's function in Eq. (1.1.73a) can be put into the form

$$i\left[M(\nabla) + \frac{W}{i}\frac{\partial}{\partial t}\right]G(\mathbf{r}, \mathbf{r}'; t, t') = \delta(\mathbf{r} - \mathbf{r}')\delta(t - t'), \tag{8}$$

subject to the initial condition $G = 0$ for $t \leq t'$, and to appropriate boundary conditions. As noted in Sec. 1.1d, the Green's function G is in general representable as a matrix whose elements may be scalar, vector, or dyadic functions. Examples of the component operators M and W can be inferred from the $L = iM + W(\partial/\partial t)$ operators listed in Sec. 1.1d.

A representation of the Green's function G in the form shown in Eq. (7) requires the knowledge of characteristic vectors $\mathbf{\Psi}_{\alpha}$. To ascertain the latter one seeks solutions of the source-free field equation $L\mathbf{\Psi} = 0$ with a time dependence

$\exp(-i\omega_\alpha t)$. In view of the representation of the operator L in Eq. (8), the eigenvectors $\boldsymbol{\Psi}_\alpha$ and the associated eigenfrequencies ω_α are evidently described in coordinate space by

$$M(\nabla)\boldsymbol{\Psi}_\alpha(\mathbf{r}) = \omega_\alpha W\boldsymbol{\Psi}_\alpha(\mathbf{r}). \tag{9}$$

For a translationally invariant, unbounded medium, one can choose $\boldsymbol{\Psi}_\alpha(\mathbf{r}) = \boldsymbol{\Psi}_\alpha(\mathbf{k})e^{i\mathbf{k}\cdot\mathbf{r}}$,† and hence the eigenvalue problem in Eq. (9) takes the algebraic form

$$M(\mathbf{k})\boldsymbol{\Psi}_\alpha(\mathbf{k}) = \omega_\alpha(\mathbf{k})W\boldsymbol{\Psi}_\alpha(\mathbf{k}), \tag{10a}$$

where $M(\mathbf{k}) \equiv M(i\mathbf{k})$ is a matrix operator with algebraic elements and $\boldsymbol{\Psi}_\alpha(\mathbf{k})$ is a multicomponent vector in polarization space. The corresponding adjoint eigenvalue problem,‡

$$M^+(\mathbf{k})\boldsymbol{\Psi}_\alpha^+(\mathbf{k}) = \omega_\alpha^*(\mathbf{k})W^+\boldsymbol{\Psi}_\alpha^+(\mathbf{k}) \tag{10b}$$

is associated with the conjugate eigenvalue ω_α^* and with adjoint operators M^+ and W^+, whose off-diagonal elements are transposed conjugates to those of M and W. By means of Hermitean products of vectors in polarization space, one forms from Eqs. (10) a relation connecting the solutions $\boldsymbol{\Psi}_\alpha^+$ of Eq. (10b) and $\boldsymbol{\Psi}_\beta$ of Eq. (10a),

$$(M^+\boldsymbol{\Psi}_\alpha^+, \boldsymbol{\Psi}_\beta) - (\boldsymbol{\Psi}_\alpha^+, M\boldsymbol{\Psi}_\beta) = \omega_\alpha(W^+\boldsymbol{\Psi}_\alpha^+, \boldsymbol{\Psi}_\beta) - \omega_\beta(\boldsymbol{\Psi}_\alpha^+, W\boldsymbol{\Psi}_\beta). \tag{10c}$$

Since by definition of the adjoint operators [see Eqs. (1.1.69)] the Hermitean products on both the left and right are equal, one derives the biorthogonality properties,

$$(\boldsymbol{\Psi}_\alpha^+, W\boldsymbol{\Psi}_\beta) = 2N_\alpha \delta_{\alpha\beta}, \tag{11}$$

which relation has already been noted in Eq. (3c). The associated completeness relation in the polarization space is

$$1 = \sum_\alpha \frac{\boldsymbol{\Psi}_\alpha(\mathbf{k})\, W^+\boldsymbol{\Psi}_\alpha^+(\mathbf{k})}{2N_\alpha(\mathbf{k})}, \tag{12}$$

as anticipated in Eq. (4). For the special case $M^+ = M$ and $W^+ = W$, one observes that $\boldsymbol{\Psi}_\alpha^+ = \boldsymbol{\Psi}_{\alpha*}$, where $\boldsymbol{\Psi}_{\alpha*}$ is a solution of Eq. (10a) corresponding to the eigenvalue $\omega_\alpha = \omega_\alpha^*$.

In the **k**-transform space the Green's function equation (8) becomes, via the representation in Eq. (2b), the ordinary matrix differential equation,

$$\left[M(\mathbf{k}) + \frac{W}{i}\frac{d}{dt}\right]G(\mathbf{k}; t, t') = \frac{\delta(t - t')}{i}. \tag{13}$$

In view of the completeness relation (12), the transformed matrix Green's function $G(\mathbf{k}; t, t')$ can be represented as

---

†For notational simplicity the same symbol is employed for $\boldsymbol{\Psi}_\alpha(\mathbf{r})$ and $\boldsymbol{\Psi}_\alpha(\mathbf{k})$. The latter occurs most frequently and will usually be abbreviated as $\boldsymbol{\Psi}_\alpha$.

‡From Eq. (10c) one infers that the adjoint eigenvalue $\omega_\alpha^+ = \omega_\alpha^*$ provided there exists a $\boldsymbol{\Psi}_\alpha$ and $\boldsymbol{\Psi}_\alpha^+$ such that $(W^+\boldsymbol{\Psi}_\alpha^+, \boldsymbol{\Psi}_\alpha) \neq 0$.

$$G(\mathbf{k}; t, t') = \sum_\alpha \frac{\boldsymbol{\Psi}_\alpha(\mathbf{k})}{2N_\alpha} (\boldsymbol{\Psi}_\alpha^+(\mathbf{k}), WG(\mathbf{k}; t, t'))$$

$$= \sum_\alpha \frac{\boldsymbol{\Psi}_\alpha(\mathbf{k})\boldsymbol{\Psi}_\alpha^+(\mathbf{k})}{2N_\alpha} G_\alpha(\mathbf{k}; t, t'), \tag{14}$$

where the G_α are scalar coefficients defined by $\boldsymbol{\Psi}_\alpha^+ G_\alpha = (\boldsymbol{\Psi}_\alpha^+, WG)$. Substitution of Eq. (14) into Eq. (2b) then provides a representation of the space- and time-dependent Green's function

$$G(\mathbf{r}, \mathbf{r}'; t, t') = \int \sum_\alpha G_\alpha(\mathbf{k}; t, t') \frac{\boldsymbol{\Psi}_\alpha(\mathbf{k})\boldsymbol{\Psi}_\alpha^+(\mathbf{k})}{2N_\alpha} e^{i\mathbf{k}\cdot(\mathbf{r}-\mathbf{r}')} \frac{d\mathbf{k}}{(2\pi)^3}, \tag{15}$$

as in Eq. (7). To determine the coefficients G_α, one forms the Hermitean product of both sides of Eq. (13) with $\boldsymbol{\Psi}_\alpha^+(\mathbf{k})$, and using Eq. (14) obtains as the defining equation for G_α,

$$\left[\frac{d}{dt} + i\omega_\alpha(\mathbf{k})\right]G_\alpha(\mathbf{k}; t, t') = \delta(t - t')$$

subject to the initial condition $G_\alpha = 0$ for $t \le t'$. Since $(d/dt)U(t - t') = \delta(t - t')$, where U is the Heaviside unit function, its solution is

$$G_\alpha(\mathbf{k}; t, t') = \begin{cases} e^{-i\omega_\alpha(\mathbf{k})(t-t')} & \text{for } t > t', \\ 0 & \text{for } t < t'. \end{cases} \tag{16}$$

Hence, from Eq. (15), one infers that the oscillatory space- and time-dependent Green's function has the form

$$G(\mathbf{r}, \mathbf{r}'; t, t') = \begin{cases} \displaystyle\int\!\!\int \sum_\alpha \frac{\boldsymbol{\Psi}_\alpha(\mathbf{k})\boldsymbol{\Psi}_\alpha^+(\mathbf{k})}{2N_\alpha(\mathbf{k})} e^{i[\mathbf{k}\cdot(\mathbf{r}-\mathbf{r}')-\omega_\alpha(\mathbf{k})(t-t')]} \frac{d\mathbf{k}}{(2\pi)^3}, & t > t', \\ 0, & t < t'. \end{cases}$$
$$\tag{17}$$

It should be noted that Eq. (17) also may be obtained from Eq. (1.2.45) by contour integration in the ω plane.

Equation (17) is the desired oscillatory representation of the Green's function of a linear field and satisfies the causality requirement of vanishing before a certain time and being outward going thereafter. Each of the component oscillations in the representation is a plane wave that can be viewed as either outward or inward going, depending on the point $\mathbf{r}$ of observation. The presence of both outgoing and ingoing plane waves in the representation of an overall outward radiating field, although correct, is somewhat redundant and raises the question as to whether it is possible to identify and eliminate the ingoing contributions.

A plane-wave oscillation of wavenumber $\mathbf{k}$ has the form $\exp[i(\mathbf{k}\cdot\mathbf{r} - \omega(\mathbf{k})t)]$, with the frequency $\omega(\mathbf{k})$ either positive or negative because of the time reversibility of the plane-wave fields under consideration. At a fixed observation point $\mathbf{r}$, a positive $\omega(\mathbf{k})$ implies the plane-wave oscillation is outgoing in the $\mathbf{r}$ direction if $\mathbf{k}$ is such that $\mathbf{k}\cdot\mathbf{r} > 0$ and ingoing if $\mathbf{k}\cdot\mathbf{r} < 0$; a converse statement is made if $\omega(\mathbf{k})$ is negative. These observations permit one to identify an outgoing wave at $\mathbf{r}$ by a frequency $\omega_+(\mathbf{k})$ which is positive if $\mathbf{k} \cdot \mathbf{r} > 0$ and negative if $\mathbf{k} \cdot \mathbf{r} < 0$; a frequency $\omega_-(\mathbf{k})$, defined as negative if $\mathbf{k} \cdot \mathbf{r} > 0$ and positive if $\mathbf{k} \cdot \mathbf{r} < 0$, distinguishes an ingoing wave at $\mathbf{r}$. Such frequency definitions are characterized by the property $\omega_\pm(\mathbf{k}) = -\omega_\pm(-\mathbf{k})$.

With the $\omega_\pm(\mathbf{k})$ identification of outgoing and ingoing wave contributions at $\mathbf{r}$ it is possible to obtain a representation alternative to that in Eq. (17) by subtracting therefrom the ingoing wave contributions. Distinguishing the $\omega_\pm(\mathbf{k})$ oscillations in the α sum of Eq. (17) by corresponding subscripts $\alpha_\pm$, one subtracts the ingoing α_- oscillations which satisfy the source-free equation (9). This subtraction must be carried out for all t in order not to introduce a singularity into G and thereby violate Eq. (8) [i.e., the subtracted part must satisfy the homogeneous equation (9)]. In this subtraction process, contributions from modes with $\omega_\alpha = 0$ (if they exist) are weighted by a factor of $\frac{1}{2}$, since these modes are, so to speak, both inward and outward going at $\mathbf{r}$. The resulting Green's function is no longer causal, since it is now finite for $t < t'$ in contrast to that in Eq. (17). For $t > t'$, however, it is identical with the causal Green's function in Eq. (17) and moreover contains only outgoing α_+ contributions. Performing the subtraction, one obtains as an alternative (outgoing) oscillatory representation of the Green's function of the linear field

$$G(\mathbf{r}, \mathbf{r}'; t, t') = \int \sum_{\alpha_+} \frac{\Psi_\alpha(\mathbf{k})\Psi_\alpha^+(\mathbf{k})}{2\epsilon_\alpha N_\alpha(\mathbf{k})}\, e^{i[\mathbf{k}\cdot(\mathbf{r}-\mathbf{r}')-\omega_\alpha(\mathbf{k})(t-t')]}\, \frac{d\mathbf{k}}{(2\pi)^3}, \qquad t > t', \qquad (18)$$

where

$$\epsilon_\alpha = \begin{cases} 2 & \text{for } \omega_\alpha = 0, \\ 1 & \text{for } \omega_\alpha \neq 0. \end{cases}$$

The representation in Eq. (18) is somewhat simpler to evaluate than that in Eq. (17) because of the omission of α_- contributions. Although Eqs. (17) and (18) yield identical results for $t > t'$, the representation in Eq. (18) suffers from the presence of a singularity at $t = t'$ [see the example of Eq. (25b)] which arises from the discontinuous $\mathbf{k}$ dependence of the $\omega_\pm(\mathbf{k})$ frequency definitions. This singularity is not troublesome if the Green's function representation in Eq. (18) is employed to represent fields produced by sources that vanish before the time t of observation of the field; a spurious contribution is introduced if the source acts beyond the observation time, and this may be difficult to identify. A somewhat similar singularity at certain z arises in connection with Hertz potential representations, as noted in Sec. 1.1c and at the end of Sec. 1.3c.

The representations in Eqs. (17) and (18) are in a more useful form than that in (1.2.45), not only because they require one less integration but also because the $\Psi_\alpha(\mathbf{k})$ are simpler to obtain than the elements of the inverse matrix $L^{-1}(\mathbf{k}, \omega)$. In the following we shall determine the oscillatory (plane-wave) eigenvectors $\Psi_\alpha(\mathbf{k})$ and eigenfrequencies $\omega_\alpha(\mathbf{k})$ for a number of linear fields. The resulting information permits one to obtain oscillatory representations of

Green's functions for these fields, in the form of Eqs. (17) or (18), with the explicit space–time dependence determined by integration over $\mathbf{k}$. We shall normally define the eigenfrequencies $\omega(\mathbf{k})$ as continuously dependent on $\mathbf{k}$; the discontinuously defined $\omega_{\pm}(\mathbf{k})$ frequencies shall only be employed in connection with the representation in Eq. (18).

1.3b The Acoustic Field

For the homogeneous linear acoustic field of Sec. 1.1, the oscillatory eigenvalue problem in Eq. (9), which defines source-free solutions with harmonic time dependence $\exp(-i\omega t)$, has the form

$$\nabla \cdot \mathbf{v} = \frac{i\omega}{\gamma p_0}\, p,$$
$$\nabla p = i\omega n_0 m \mathbf{v} \tag{19}$$

where $p = p(\mathbf{r}, \omega)$ and $\mathbf{v} = \mathbf{v}(\mathbf{r}, \omega)$ denote the complex amplitudes of pressure and velocity at the point $\mathbf{r}$, and as in Eq. (1.1.1), $n_0 m$ and $1/\gamma p_0$ are the acoustic parameters. In an unbounded region the steady-state oscillatory solutions of Eqs. (19) are representable as

$$\begin{pmatrix} p(\mathbf{r}, \omega) \\ \mathbf{v}(\mathbf{r}, \omega) \end{pmatrix} = \begin{pmatrix} p(\mathbf{k}, \omega) \\ \mathbf{v}(\mathbf{k}, \omega) \end{pmatrix} e^{i\mathbf{k}\cdot\mathbf{r}}, \tag{20a}$$

where for notational simplicity, the same symbols p and $\mathbf{v}$ are employed for the $\mathbf{r}$- and $\mathbf{k}$-dependent fields. In the wavevector notation of the previous section, Eq. (20a) becomes

$$\boldsymbol{\Psi}(\mathbf{r}, \omega) = \boldsymbol{\Psi}(\mathbf{k}, \omega)e^{i\mathbf{k}\cdot\mathbf{r}} \tag{20b}$$

with the associated operators $M = M^+$ and $W = W^+$ of Eqs. (9) and (10) following from Eqs. (19) as

$$iM \rightarrow \begin{pmatrix} 0 & \nabla \\ \nabla & 0 \end{pmatrix}, \qquad W \rightarrow \begin{pmatrix} \dfrac{1}{\gamma p_0} & 0 \\ 0 & n_0 m\mathbf{1} \end{pmatrix}. \tag{21}$$

In Eq. (21) the 11 elements of M and W are recognized as scalars, the 12 and 21 elements as vectors, and the 22 elements are dyadics.

One infers from Eqs. (19), or Eqs. (10a) and (21), that eigenoscillations of the form (20) exist only for frequencies $\omega_\alpha(\mathbf{k})$ corresponding to the vanishing of $\det[M(\mathbf{k}) - \omega_\alpha(\mathbf{k})W]$, wherein one notes that $\nabla \equiv i\mathbf{k}$. There are four such frequencies, ordered as follows:

$$\omega_{\pm 1} = \pm ka, \qquad \omega_2 = 0, \qquad \omega_3 = 0, \tag{22}$$

where $k = |\mathbf{k}|$ and $a = (\gamma p_0/n_0 m)^{1/2}$ is the acoustic speed. The frequencies (22) are seen to be the source-free resonant frequencies of the acoustic network depicted in Fig. 1.2.2. In the $\mathbf{k}_0$, $\mathbf{T}_0'$, $\mathbf{T}_0''$ basis shown in Fig. 1.2.1, the eigenvector solutions $\boldsymbol{\Psi}_\alpha(\mathbf{k}, \omega) \equiv \boldsymbol{\Psi}_\alpha = \boldsymbol{\Psi}_\alpha^+$ of Eqs. (19), corresponding to the eigenfrequencies ω_α noted in Eq. (22), are

$$\Psi_{\pm1} \rightarrow \begin{pmatrix} \sqrt{\gamma p_0} \\ \pm\dfrac{\mathbf{k}_0}{\sqrt{n_0 m}} \end{pmatrix}, \qquad \Psi_2 \rightarrow \begin{pmatrix} 0 \\ \dfrac{\mathbf{T}_0'}{\sqrt{n_0 m}} \end{pmatrix}, \qquad \Psi_3 \rightarrow \begin{pmatrix} 0 \\ \dfrac{\mathbf{T}_0''}{\sqrt{n_0 m}} \end{pmatrix}, \qquad (23a)$$

with orthonormality properties as in Eq. (11) defined in terms of the normalization constants

$$N_{\pm1} = 1, \qquad 2N_2 = 1, \qquad 2N_3 = 1. \tag{23b}$$

Oscillatory representation of acoustic Green's function

The time-dependent (matrix) Green's function for the acoustic field may be evaluated from the oscillatory representation in either Eq. (17) or (18). Equation (17) includes both outgoing and ingoing waves and yields, with $\mathbf{r} - \mathbf{r}'$ and $t - t'$ denoted for simplicity by $\mathbf{r}$ and t, for $t > 0$:

$$G(\mathbf{r}; t) = \int \left(\frac{\Psi_1 \Psi_1}{2} e^{-ikat} + \frac{\Psi_{-1} \Psi_{-1}}{2} e^{+ikat} + \Psi_2 \Psi_2 + \Psi_3 \Psi_3 \right) e^{i\mathbf{k}\cdot\mathbf{r}} \frac{d\mathbf{k}}{(2\pi)^3},$$

$$(24a)$$

whereas Eq. (18), with only outgoing waves, yields for $t > 0$,

$$G(\mathbf{r}; t) = \int \frac{1}{2} (\Psi_1 \Psi_1 e^{-i\omega_{+1}t} + \Psi_2 \Psi_2 + \Psi_3 \Psi_3) e^{i\mathbf{k}\cdot\mathbf{r}} \frac{d\mathbf{k}}{(2\pi)^3}, \tag{24b}$$

with $\omega_\pm$ definitions noted in the discussion preceding Eq. (18). On substitution of the appropriate eigenfrequencies and eigenvectors from Eqs. (22) and (23) into Eq. (24a), one obtains for the 11 matrix element in successive steps

$$G_{11}(\mathbf{r}; t) = \int \frac{\gamma p_0}{2} (e^{-i\omega_{+1}t} + e^{-i\omega_{-1}t}) e^{i\mathbf{k}\cdot\mathbf{r}} \frac{d\mathbf{k}}{(2\pi)^3}$$

$$= n_0 m \frac{\partial}{\partial t} \int (e^{-ikat} - e^{ikat}) e^{i\mathbf{k}\cdot\mathbf{r}} \frac{d\mathbf{k}\,a}{-2ik(2\pi)^3}.$$

On doing the θ and φ integrations in a polar coordinate system wherein $\mathbf{k}\cdot\mathbf{r} = kr\cos\theta$ and $d\mathbf{k} = k^2 \sin\theta\, d\theta\, d\varphi\, dk$, one finds

$$G_{11}(\mathbf{r}; t) = n_0 m \frac{\partial}{\partial t} \int_0^\infty [e^{ik(r-at)} + e^{-ik(r-at)} - e^{ik(r+at)} - e^{-ik(r+at)}] \frac{d(ka)}{8\pi^2 r}$$

$$= n_0 m \frac{\partial}{\partial t} \left[\frac{\delta[t - (r/a)]}{4\pi r} - \frac{\delta[t + (r/a)]}{4\pi r} \right]$$

$$= n_0 m \frac{\partial}{\partial t} \frac{\delta[t - (r/a)]}{4\pi r} \qquad \text{for } t > 0. \tag{25a}$$

Note in Eq. (25a) that the $\delta[t + (r/a)]$ term, contributed by the ingoing waves in the representation, vanishes for $t > 0$. For the alternative representation in Eq. (24b), involving only outgoing waves, one employs a similar procedure in the integration over the polar angle θ, but with the $\omega_\pm(\mathbf{k})$ definitions discussed in connection with Eq. (18); one notes that the frequency $\omega_{+1}(\mathbf{k})$ becomes $\omega_{+1} = ka$ for $0 < \theta < \pi/2$ and $\omega_{+1} = -ka$ for $\pi/2 < \theta < \pi$. Thus, one deduces from Eq. (24b) the same result as in Eq. (25a):

$$
\begin{aligned}
G_{11}(\mathbf{r};\,t) &= \int \frac{\gamma p_0}{2}\, e^{i(\mathbf{k}\cdot\mathbf{r}-\omega_+t)}\,\frac{d\mathbf{k}}{(2\pi)^3} \\[4pt]
&= \frac{\gamma p_0}{2}\left[\int_0^{\pi/2} e^{i(kr\cos\theta - kat)}\,\frac{k^2\sin\theta\, d\theta\, dk}{(2\pi)^2} + \int_{\pi/2}^{\pi} e^{i(kr\cos\theta + kat)}\,\frac{k^2\sin\theta\, d\theta\, dk}{(2\pi)^2}\right] \\[4pt]
&= n_0 m\,\frac{\partial}{\partial t}\int_0^{\infty}\left[e^{ik(r-at)} + e^{-ik(r-at)} - e^{-ikat} - e^{ikat}\right]\frac{d(ka)}{8\pi^2 r} \\[4pt]
&= n_0 m\,\frac{\partial}{\partial t}\left[\frac{\delta[t-(r/a)]}{4\pi r} - \frac{\delta(t)}{4\pi r}\right] \\[4pt]
&= n_0 m\,\frac{\partial}{\partial t}\,\frac{\delta[t-(r/a)]}{4\pi r} \qquad \text{for } t>0;
\end{aligned}
\tag{25b}
$$

the $\delta(t)$ contribution in the next-to-last equation is spurious and arises from our ingoing–outgoing decomposition. The results in Eqs. (25a) and (25b) are in the form anticipated in the acoustic G_{11} Green's function result in Eq. (1.1.13a); the other acoustic Green's functions can be similarly evaluated from the remaining matrix elements of G in Eqs. (24).

1.3c The Electromagnetic Field

Source-free electromagnetic oscillations with time dependence $\exp(-i\omega t)$ are determined by the steady-state electromagnetic field equations

$$
\begin{aligned}
\nabla \times \mathbf{H} &= -i\omega\epsilon_0\mathbf{E}, \\
\nabla \times \mathbf{E} &= i\omega\mu_0\mathbf{H},
\end{aligned}
\tag{26}
$$

subject to appropriate conditions at the boundary (if any) of the field region. Equations (26) pose an eigenvalue problem in the form of Eq. (9), wherein $\mathbf{E} = \mathbf{E}(\mathbf{r}, \omega)$ and $\mathbf{H} = \mathbf{H}(\mathbf{r}, \omega)$ denote the complex electric- and magnetic-field amplitudes, and ϵ_0 and μ_0 are the customary electromagnetic vacuum parameters. For free space the steady-state solutions of Eq. (26) are spatially representable as

$$
\begin{pmatrix} \mathbf{E}(\mathbf{r}, \omega) \\ \mathbf{H}(\mathbf{r}, \omega) \end{pmatrix} = \begin{pmatrix} \mathbf{E}(\mathbf{k}, \omega) \\ \mathbf{H}(\mathbf{k}, \omega) \end{pmatrix} e^{i\mathbf{k}\cdot\mathbf{r}},
\tag{27a}
$$

or in the eigenvector notation of Sec. 1.3a as

$$
\Psi(\mathbf{r}, \omega) = \Psi(\mathbf{k}, \omega)e^{i\mathbf{k}\cdot\mathbf{r}},
\tag{27b}
$$

where in view of Eq. (26), the associated operators $M = M^+$ and $W = W^+$ are†

$$
iM \rightarrow \begin{pmatrix} 0 & -\nabla \times \mathbf{1} \\ \nabla \times \mathbf{1} & 0 \end{pmatrix}, \qquad W \rightarrow \begin{pmatrix} \epsilon_0 & 0 \\ 0 & \mu_0 \end{pmatrix}\mathbf{1}.
\tag{28}
$$

Equations (26) have eigensolutions of the spatial form (27), whereby $\nabla \equiv i\mathbf{k}$, only for the six eigenfrequencies

---

†Note that $M = M^+$, since ∇/i is Hermitian and $\mathbf{a} \times \mathbf{1} = -\widetilde{\mathbf{a} \times \mathbf{1}}$ is antisymmetric.

$$\omega_{\pm 1} = \omega_{\pm 2} = \pm\, kc,$$
$$\omega_3 = \omega_4 = 0, \tag{29}$$

where $k = |\mathbf{k}|$ and $c = (\mu_0\epsilon_0)^{-1/2}$ is the speed of light in vacuum. The frequencies (29), some of which are doubly degenerate, are recognizable as the source-free resonances of the electromagnetic network depicted in Fig. 1.2.3. The associated resonant fields, which also follow from Eq. (26), are characterized by the eigenvectors $\mathbf{\Psi}_\alpha \equiv \mathbf{\Psi}_\alpha(\mathbf{k}, \omega)$, where in the $\mathbf{k}_0$, $\mathbf{T}_0'$, $\mathbf{T}_0''$ basis of Fig. 1.2.1,

$$\mathbf{\Psi}_{\pm 1} \rightarrow \begin{pmatrix} \mathbf{T}_0'/\sqrt{\epsilon_0} \\ \pm\mathbf{T}_0''/\sqrt{\mu_0} \end{pmatrix}, \quad \mathbf{\Psi}_{\pm 2} \rightarrow \begin{pmatrix} \mathbf{T}_0''/\sqrt{\epsilon_0} \\ \mp\mathbf{T}_0'/\sqrt{\mu_0} \end{pmatrix},$$

$$\mathbf{\Psi}_3 \rightarrow \begin{pmatrix} \mathbf{k}_0/\sqrt{\epsilon_0} \\ 0 \end{pmatrix}, \quad \mathbf{\Psi}_4 \rightarrow \begin{pmatrix} 0 \\ \mathbf{k}_0/\sqrt{\mu_0} \end{pmatrix}. \tag{30a}$$

These possess the $\mathbf{k}$ independent orthonormality properties in Eq. (11) with the normalization constants

$$N_{\pm 1} = 1 = N_{\pm 2}, \qquad 2N_3 = 1 = 2N_4. \tag{30b}$$

Oscillatory representation of electromagnetic Green's function

An oscillatory representation of the time-dependent (matrix) Green's function of the electromagnetic field is provided by Eq. (18). On use therein of the known eigenfrequencies (29) and eigenvectors (30) one obtains, since $\mathbf{\Psi}_\alpha^+ = \mathbf{\Psi}_\alpha$ in free space, for $t > t'$ (i.e., omitting a singularity at $t = t'$):

$$G(\mathbf{r}, \mathbf{r}'; t, t') = \int \frac{1}{2}\Big[(\mathbf{\Psi}_1\mathbf{\Psi}_1 + \mathbf{\Psi}_2\mathbf{\Psi}_2)e^{-i\omega_{+1}(t-t')}$$
$$+ \mathbf{\Psi}_3\mathbf{\Psi}_3 + \mathbf{\Psi}_4\mathbf{\Psi}_4\Big]e^{i\mathbf{k}\cdot(\mathbf{r}-\mathbf{r}')}\frac{d\mathbf{k}}{(2\pi)^3}, \tag{31}$$

whose 11 matrix element is identically the impedance dyadic Green's function $\mathscr{G}_{11}(\mathbf{r}, \mathbf{r}'; t, t')$ of Sec. 1.1b (with $\mathbf{r} - \mathbf{r}'$, $t - t'$ written as $\mathbf{r}$, t):

$$\mathscr{G}_{11}(\mathbf{r}, t) = \int\Big[\Big(\frac{\mathbf{T}_0'\mathbf{T}_0'}{2\epsilon_0} + \frac{\mathbf{T}_0''\mathbf{T}_0''}{2\epsilon_0}\Big)e^{-i\omega_{+1}t} + \frac{\mathbf{k}_0\mathbf{k}_0}{2\epsilon_0}\Big]e^{i\mathbf{k}\cdot\mathbf{r}}\frac{d\mathbf{k}}{(2\pi)^3}$$

$$= \mathbf{1}\int\frac{e^{i(\mathbf{k}\cdot\mathbf{r}-\omega_{+1}t)}}{2\epsilon_0}\frac{d\mathbf{k}}{(2\pi)^3} - \int\mathbf{k}_0\mathbf{k}_0\frac{e^{i\mathbf{k}\cdot\mathbf{r}}}{2\epsilon_0}(e^{-i\omega_{+1}t} - 1)\frac{d\mathbf{k}}{(2\pi)^3} \tag{32a}$$

$$= \Big(\mathbf{1}\mu_0\frac{\partial}{\partial t} - \frac{\nabla\nabla}{\epsilon_0(\partial/\partial t)}\Big)\int\frac{e^{i(\mathbf{k}\cdot\mathbf{r}-\omega_{+1}t)}}{-2ik/c}\frac{d\mathbf{k}}{(2\pi)^3}, \tag{32b}$$

$$\mathscr{G}_{11}(\mathbf{r}, t) = \Big(\mathbf{1}\mu_0\frac{\partial}{\partial t} - \frac{\nabla\nabla}{\epsilon_0(\partial/\partial t)}\Big)\frac{\delta(t - r/c)}{4\pi r}, \qquad t > 0. \tag{32c}$$

In successive steps we have utilized in Eqs. (32) the identities $\mathbf{1} - \mathbf{k}_0\mathbf{k}_0 = \mathbf{T}_0'\mathbf{T}_0' + \mathbf{T}_0''\mathbf{T}_0''$, $\nabla\nabla = -k^2\mathbf{k}_0\mathbf{k}_0$, $\partial/\partial t = -ikc$, $(\partial/\partial t)^{-1} = \int_0^t dt$, and the ω integrated result of the integral representation in Eq. (1.2.20) [see Eq. (1.3.25)].

The representation in Eq. (32c) is seen to be the previously obtained dyadic Green's function of Eqs. (1.1.30b) and (1.1.31). The other electromagnetic dyadic Green's functions are determined from the remaining matrix elements of Eq. (31) in a similar manner.

The noted singularity of oscillatory (time-guided) representations (31) of space- and time-dependent Green's functions should be contrasted with analogous properties of space-guided representations discussed in Secs. 1.1b and 1.4. In the space-guided formulation, a z-guided mode is excited by a distribution of sources with appropriate temporal and transverse spatial periodicities in a plane transverse to z; modal excitation by sources spatially confined to a plane $z = z'$ is determined on replacing such sources by equivalent distributions occupying the entire $z = z'$ plane [see Eqs. (1.1.43) and (1.4.15)]. A time-guided mode, on the other hand, is excited by sources distributed throughout space with appropriate spatial periodicity $\mathbf{k}$; modal excitation by spatially and temporally confined sources is determined on replacing such sources by equivalent volume distributions. When a simplified representation of either the space-guided [see Eqs. (1.1.42)] or time-guided [see Eq. (18)] Green's functions is used, spurious contributions are introduced into the former on the source plane $z = z' = 0$ [Eq. (1.1.47)] and into the latter at $t = t' = 0$ [see Eq. (25b)]. In both instances, these anomalies are evidently confined to hyperplanes containing the source and oriented perpendicular to the guiding axis in the four-dimensional $\mathbf{r}, t$ space.

1.3d   The Plasma Field

As can be inferred from the time-dependent field equations of Sec. 1.1c, source-free harmonic oscillations of an isotropic one-component fluid model of a plasma are defined by

$$
\begin{aligned}
\nabla \times \mathbf{H} \qquad\qquad + n_0 q \mathbf{v} &= -i\omega\epsilon_0 \mathbf{E}, \\
\nabla \times \mathbf{E} \qquad\qquad\qquad &= i\omega\mu_0 \mathbf{H}, \\
-\nabla \cdot \mathbf{v} &= -i\frac{\omega}{\gamma p_0}p, \\
n_0 q \mathbf{E} \qquad\qquad + \nabla p &= i\omega n_0 m \mathbf{v},
\end{aligned}
\tag{33}
$$

subject to appropriate boundary conditions. Equations (33) constitute an eigenvalue problem of the type shown in Eq. (9), wherein ω plays the role of an eigenvalue and the associated eigenvector components are the electric-field intensity $\mathbf{E} = \mathbf{E}(\mathbf{r}, \omega)$, the magnetic-field intensity $\mathbf{H} = \mathbf{H}(\mathbf{r}, \omega)$, the linearized pressure $p = p(\mathbf{r}, \omega)$, and the linearized velocity $\mathbf{v} = \mathbf{v}(\mathbf{r}, \omega)$. In an unbounded, stationary, homogeneous plasma, steady-state solutions of Eqs. (33) are representable in the form of a 10 (scalar)-component wavevector as

$$
\begin{pmatrix}
E(\mathbf{r}, \omega) \\
H(\mathbf{r}, \omega) \\
p(\mathbf{r}, \omega) \\
v(\mathbf{r}, \omega)
\end{pmatrix}
=
\begin{pmatrix}
E(\mathbf{k}, \omega) \\
H(\mathbf{k}, \omega) \\
p(\mathbf{k}, \omega) \\
v(\mathbf{k}, \omega)
\end{pmatrix}
e^{i\mathbf{k}\cdot\mathbf{r}},
\tag{34a}
$$

or, in wavevector notation, as

$$
\Psi(\mathbf{r}, \omega) = \Psi(\mathbf{k}, \omega)e^{i\mathbf{k}\cdot\mathbf{r}}.
\tag{34b}
$$

Substitution of Eqs. (34) into (33) permits one to rewrite the latter in the abstract form of Eq. (10), wherein the operators $M(\mathbf{k}) = M^{+}(\mathbf{k})$ and $W = W^{+}$ are

$$
M \rightarrow
\begin{pmatrix}
0 & -\mathbf{k} \times \mathbf{1} & 0 & in_0 q\mathbf{1} \\
\mathbf{k} \times \mathbf{1} & 0 & 0 & 0 \\
0 & 0 & 0 & \mathbf{k} \\
-in_0 q\mathbf{1} & 0 & \mathbf{k} & 0
\end{pmatrix},
$$

$$
W \rightarrow
\begin{pmatrix}
\epsilon_0 \mathbf{1} & 0 & 0 & 0 \\
0 & \mu_0 \mathbf{1} & 0 & 0 \\
0 & 0 & \dfrac{1}{\gamma p_0} & 0 \\
0 & 0 & 0 & n_0 m\mathbf{1}
\end{pmatrix}.
\tag{35}
$$

Eigenvector solutions of Eqs. (10) or (33) exist for eigenfrequencies $\omega = \omega_\alpha(\mathbf{k})$ defined by the following dispersion relations:

$$
\omega_{\pm 1} = \omega_{\pm 2} = \pm \sqrt{\omega_p^2 + k^2 c^2},
$$
$$
\omega_{\pm 3} = \pm \sqrt{\omega_p^2 + k^2 a^2},
\tag{36}
$$
$$
\omega_4 = \omega_5 = \omega_6 = \omega_7 = 0, \quad \text{where } c = \frac{1}{\sqrt{\mu_0 \epsilon_0}} \text{ and } a = \sqrt{\frac{\gamma p_0}{n_0 m}}.
$$

The first four frequencies define optic-type oscillations, the $\omega_{\pm 3}$ are the plasma oscillations, and the latter four are static zero-frequency oscillations† of longitudinal and transverse types. These 10 frequencies characterize the free resonances of the source-free plasma network displayed in Fig. 1.2.4. The corresponding resonant fields are described by eigenvectors $\Psi(\mathbf{k}, \omega_\alpha(\mathbf{k})) = \Psi_\alpha(\mathbf{k}) \equiv \Psi_\alpha$ of the type shown in Eq. (34). In the $\mathbf{k}_0, \mathbf{T}_0', \mathbf{T}_0''$ basis illustrated in Fig. 1.2.1, the Ψ_α have the form:

---

†It is assumed that $k \neq 0$ for the zero-frequency oscillation ω_5, which incidentally need not satisfy Poisson's equation. The case $k = 0$ necessitates a special consideration of the constant positive background required for charge neutrality.

$$\Psi_{\pm 1} \rightarrow \begin{pmatrix} \dfrac{\mathbf{T}'_0}{\sqrt{\epsilon_0}} \\[2ex] \dfrac{kc}{\omega_{\pm 1}}\dfrac{\mathbf{T}''_0}{\sqrt{\mu_0}} \\[2ex] 0 \\[2ex] \dfrac{-i\omega_p}{\omega_{\pm 1}}\dfrac{\mathbf{T}'_0}{\sqrt{n_0 m}} \end{pmatrix}, \quad \Psi_{\pm 2} \rightarrow \begin{pmatrix} \dfrac{\mathbf{T}''_0}{\sqrt{\epsilon_0}} \\[2ex] \dfrac{-kc}{\omega_{\pm 2}}\dfrac{\mathbf{T}'_0}{\sqrt{\mu_0}} \\[2ex] 0 \\[2ex] \dfrac{-i\omega_p}{\omega_{\pm 2}}\dfrac{\mathbf{T}''_0}{\sqrt{n_0 m}} \end{pmatrix},$$

$$\Psi_{\pm 3} \rightarrow \begin{pmatrix} \dfrac{\omega_p}{\omega_{\pm 3}}\dfrac{\mathbf{k}_0}{\sqrt{\epsilon_0}} \\[2ex] 0 \\[2ex] \dfrac{-ika}{\omega_{\pm 3}}\sqrt{\gamma p_0} \\[2ex] \dfrac{-i\mathbf{k}_0}{\sqrt{n_0 m}} \end{pmatrix}, \quad \Psi_4 \rightarrow \begin{pmatrix} 0 \\[2ex] \dfrac{\mathbf{k}_0}{\sqrt{\mu_0}} \\[2ex] 0 \\[2ex] 0 \end{pmatrix}, \quad \Psi_5 \rightarrow \begin{pmatrix} \dfrac{\mathbf{k}_0}{\sqrt{\epsilon_0}} \\[2ex] 0 \\[2ex] i\dfrac{\omega_p}{ka}\sqrt{\gamma p_0} \\[2ex] 0 \end{pmatrix}, \qquad (37a)$$

$$\Psi_6 \rightarrow \begin{pmatrix} 0 \\[2ex] \dfrac{i\mathbf{T}'_0}{\sqrt{\mu_0}} \\[2ex] 0 \\[2ex] \dfrac{kc}{\omega_p}\dfrac{\mathbf{T}''_0}{\sqrt{n_0 m}} \end{pmatrix}, \quad \Psi_7 \rightarrow \begin{pmatrix} 0 \\[2ex] \dfrac{i\mathbf{T}''_0}{\sqrt{\mu_0}} \\[2ex] 0 \\[2ex] -\dfrac{kc}{\omega_p}\dfrac{\mathbf{T}'_0}{\sqrt{n_0 m}} \end{pmatrix}.$$

These eigenvectors possess the orthogonality properties in Eq. (11) with normalization constants

$$N_{\pm 1} = N_{\pm 2} = N_{\pm 3} = 1, \quad 2N_4 = 1, \quad 2N_5 = 1 + \frac{\omega_p^2}{k^2 a^2},$$

$$2N_6 = 2N_7 = 1 + \frac{k^2 c^2}{\omega_p^2}. \qquad (37b)$$

Oscillatory representation of plasma Green's function

As shown in Eqs. (18), the (matrix) Green's function $G(\mathbf{r}, \mathbf{r}'; t, t')$ of the plasma field is expressible in terms of the oscillatory eigenvectors in Eq. (37a). As a particular example, the dyadic element $\mathscr{G}_{11}(\mathbf{r}, \mathbf{r}'; t, t')$, which represents the negative of the vector electric field at $\mathbf{r}, t$ produced by a point electric current source at $\mathbf{r}', t'$ is given by [we denote $\mathbf{r} - \mathbf{r}', t - t'$ by $\mathbf{r}, t$ and we employ the $\omega_{\pm}(\mathbf{k})$ frequencies defined in connection with the discussion of Eq. (18)]

$$\mathcal{G}_{11}(\mathbf{r}, t) = \frac{1}{2}\int\left[\boldsymbol{\Psi}_1\boldsymbol{\Psi}_1 e^{-i\omega_1 t} + \boldsymbol{\Psi}_2\boldsymbol{\Psi}_2 e^{-i\omega_2 t} + \boldsymbol{\Psi}_3\boldsymbol{\Psi}_3 e^{-i\omega_3 t}\right.$$

$$\left. + \frac{\boldsymbol{\Psi}_5\boldsymbol{\Psi}_5}{1 + (\omega_p^2/k^2 a^2)}\right]e^{i\mathbf{k}\cdot\mathbf{r}}\frac{d\mathbf{k}}{(2\pi)^3}$$

$$= \frac{1}{2\epsilon_0}\int\left[(\mathbf{T}_0'\mathbf{T}_0' + \mathbf{T}_0''\mathbf{T}_0'')e^{-i\omega_1 t}\right.$$

$$\left. + \mathbf{k}_0\mathbf{k}_0\frac{\omega_p^2}{\omega_p^2 + k^2 a^2}(e^{-i\omega_3 t} - 1) + \mathbf{k}_0\mathbf{k}_0\right]e^{i\mathbf{k}\cdot\mathbf{r}}\frac{d\mathbf{k}}{(2\pi)^3}$$

$$= \frac{1}{2}\int\left[\mathbf{1}\mu_0\frac{\partial}{\partial t} + \frac{\boldsymbol{\nabla}\boldsymbol{\nabla}}{\epsilon_0}\frac{\partial}{\partial t}\frac{1}{k^2 c^2}\right]\frac{e^{i(\mathbf{k}\cdot\mathbf{r} - \omega_1 t)}}{-i\omega_1}\frac{c^2 d\mathbf{k}}{(2\pi)^3}$$

$$- \frac{\boldsymbol{\nabla}\boldsymbol{\nabla}}{\epsilon_0}\frac{\omega_p^2}{\partial/\partial t}\int\frac{e^{i(\mathbf{k}\cdot\mathbf{r} - \omega_3 t)}}{ik^2\omega_3}\frac{d\mathbf{k}}{(2\pi)^3}. \tag{38}$$

On integration over polar and azimuthal angles in $\mathbf{k}$ space and on change of variable from k to $\omega = \sqrt{\omega_p^2 + k^2 u^2}$, as required, with $u = a$ or c, one obtains

$$\mathcal{G}_{11}(\mathbf{r}, t) = \left(\mathbf{1}\mu_0\frac{\partial}{\partial t} - \frac{\boldsymbol{\nabla}\boldsymbol{\nabla}}{\epsilon_0}\frac{\partial/\partial t}{(\partial^2/\partial t^2) + \omega_p^2}\right)g_c(\mathbf{r}, t)$$

$$- \frac{\boldsymbol{\nabla}\boldsymbol{\nabla}}{\epsilon_0}\frac{\omega_p^2}{(\partial/\partial t)[(\partial^2/\partial t^2) + \omega_p^2]}g_a(\mathbf{r}, t), \tag{39a}$$

where

$$g_u(\mathbf{r}, t) = \int_{-\infty}^{+\infty}\frac{\exp\left[-i(\omega t - \sqrt{\omega^2 - \omega_p^2}\, r/u)\right]}{4\pi r}\frac{d\omega}{2\pi}, \qquad \text{Im}\ \omega > 0, \tag{39b}$$

with $u = c$ or a. A known Fourier transformation [see Eqs. (1.1.61)] indicates that the expressions (39) for $\mathcal{G}_{11}(\mathbf{r}, t)$ are identical for $t \neq t'$ to those previously derived in Eqs. (1.1.59)–(1.1.61).

1.4 GUIDED-WAVE REPRESENTATIONS IN SPACE

As we have seen, fields in linear, homogeneous, and stationary media can be represented in alternative ways. The $\exp\left[i(\mathbf{k}\cdot\mathbf{r} - \omega t)\right]$ plane-wave representations of Sec. 1.2 reduce a field problem in an unbounded medium to a simple algebraic problem for the $\mathbf{k}$, ω-dependent field amplitudes. Similarly, the oscillatory $\exp(i\mathbf{k}\cdot\mathbf{r})$ representations of Sec. 1.3 reduce a field problem to an ordinary differential equation problem for the time- and k-dependent field amplitudes. In this section we shall introduce another representation well suited to the solution of field problems in media with a symmetry axis along some spatial direction, say z. The guided waves $\exp(i\mathbf{k}_t\cdot\boldsymbol{\rho} - i\omega t)\exp(i\kappa z)$ considered below are plane waves with real transverse (to z) wavenumber $\mathbf{k}_t$ and frequency ω. They possess orthogonality properties in time and in the cross section defined by the radius vector $\boldsymbol{\rho}$ transverse to z; they are field solutions in the medium and are distinguished by wavenumbers $\kappa = \kappa(\mathbf{k}_t, \omega)$ that are characteristic both of the cross-sectional shape and the polarization structure of the field. Such wave fields permit a representation of a linear field in terms of characteristic guided waves with $\exp(i\kappa z)$ dependence, require evaluations of three-dimensional transforms (two dimensional in space and one dimensional in time), and are well adapted to the solution of boundary-value problems in uniform stratified regions.

In a uniform region whose cross section transverse to z is unbounded and described by a radial vector $\boldsymbol{\rho}$, an integrable wave function $\Psi(\mathbf{r}, t)$ is representable by means of the three-dimensional Fourier integral theorem as†

$$\Psi(\boldsymbol{\rho}, t; z) = \iiint \Psi(\mathbf{k}_t, \omega; z)e^{i(\mathbf{k}_t \cdot \boldsymbol{\rho} - \omega t)}\frac{d\mathbf{k}_t d\omega}{(2\pi)^3}, \tag{1a}$$

where

$$\Psi(\mathbf{k}_t, \omega; z) = \iiint \Psi(\boldsymbol{\rho}, t; z)e^{-i(\mathbf{k}_t \cdot \boldsymbol{\rho} - \omega t)}d\boldsymbol{\rho}\, dt \tag{1b}$$

and $d\mathbf{k}_t d\omega$ denotes the "volume" element in $\mathbf{k}_t, \omega$ space while $d\boldsymbol{\rho}\, dt$ is the "volume" element in $\boldsymbol{\rho}, t$ space; $\mathbf{k}_t$ is the wavevector component of $\mathbf{k}$ transverse to z. In a homogeneous, transversely *bounded* waveguide, $\exp(i\mathbf{k}_t \cdot \boldsymbol{\rho})$ in Eqs. (1) would be replaced by a suitable transverse eigenfunction $\Phi_\alpha(\boldsymbol{\rho})$ (see Chapter 3). The completeness relation equivalent to the transform relations (1a) and (1b) is

$$\delta(\boldsymbol{\rho} - \boldsymbol{\rho}')\delta(t - t') = \iiint e^{i\mathbf{k}_t \cdot (\boldsymbol{\rho} - \boldsymbol{\rho}')}e^{-i\omega(t - t')}\frac{d\mathbf{k}_t d\omega}{(2\pi)^3}, \tag{2a}$$

where for completeness the integration is to be extended over all real wavenumbers $\mathbf{k}_t$ and frequencies ω from $-\infty$ to $+\infty$. From the transform relations (1), one also infers the orthogonality property

$$(2\pi)^3\delta(\mathbf{k}_t - \mathbf{k}_t')\delta(\omega - \omega') = \int e^{-i(\mathbf{k}_t - \mathbf{k}_t')\cdot\boldsymbol{\rho} + i(\omega - \omega')t}d\boldsymbol{\rho}\, dt. \tag{2b}$$

A Green's function in the basis of Eq. (2a) would be represented as

$$G(\mathbf{r}, \mathbf{r}'; t, t') = \iiint G(\mathbf{k}_t, \omega; z, z')e^{i\mathbf{k}_t \cdot (\boldsymbol{\rho} - \boldsymbol{\rho}')}e^{-i\omega(t - t')}\frac{d\mathbf{k}_t d\omega}{(2\pi)^3}. \tag{3}$$

Singularities of $G(\mathbf{k}_t, \omega; z, z')$ are frequently encountered on the real ω-integration path in Eq. (3); to satisfy causality (i.e., $G \equiv 0$ for $t < t'$), one may deform this path by analytic continuation into the region $\mathrm{Im}\,\omega > 0$ of the complex ω plane. As in the oscillatory representation of Sec. 1.3, the transformed wavefunction $\Psi(\mathbf{k}_t, \omega; z)$ of Eqs. (1) is a one-column matrix (vector) in an n-dimensional "polarization" space and is representable in terms of a set of modes or eigenvectors $\Psi_\alpha(\mathbf{k}_t, \omega)$ as

$$\Psi(\mathbf{k}_t, \omega; z) = \sum_\alpha \Psi_\alpha(\mathbf{k}_t, \omega)a_\alpha(\mathbf{k}_t, \omega; z), \tag{4a}$$

---

†Note that functions and their transforms are represented by the same symbol and distinguished, where necessary, only by their arguments.

where α is a multicomponent summation index distinguishing the αth mode. The amplitude a_α is determined as the weighted Hermitian inner product of the wave vector Ψ and an adjoint eigenvector $\Psi_\alpha^+(\mathbf{k}_t, \omega)$ as

$$a_\alpha(\mathbf{k}_t, \omega; z) = \frac{(\Gamma^+\Psi_\alpha^+(\mathbf{k}_t, \omega), \Psi(\mathbf{k}_t, \omega; z))}{2N_\alpha(\mathbf{k}_t, \omega)}, \tag{4b}$$

Γ being a "weight operator. The eigenfunctions Ψ_α and Ψ_α^+ constitute a complete biorthogonal set in polarization space and possess the orthogonality properties

$$(\Gamma^+\Psi_\alpha^+, \Psi_\beta) = 2N_\alpha\delta_{\alpha\beta}, \tag{4c}$$

where N_α is the normalization constant and $\delta_{\alpha\beta}$ is the Kronecker delta, which vanishes when the mode indices α and β distinguish modes with different κ_α and κ_β. If the κ_α are n-fold degenerate, the various wavevectors $\Psi_{\alpha i}$ and their adjoints $\Psi_{\alpha j}^+$ with $i, j = 1, 2, \cdots, n$ should be chosen to satisfy the orthogonality properties

$$(\Gamma^+\Psi_{\alpha i}^+, \Psi_{\alpha j}) = 2N_{\alpha i}\delta_{ij}. \tag{4d}$$

In polarization space the completeness relation equivalent to the transforms in Eqs. (4a) and (4b) is provided by the unit-operator representation

$$1 = \sum_\alpha \frac{\Psi_\alpha(\mathbf{k}_t, \omega)\Gamma^+\Psi_\alpha^+(\mathbf{k}_t, \omega)}{2N_\alpha(\mathbf{k}_t, \omega)}. \tag{5}$$

The function and polarization space transforms in Eqs. (1) and (4) permit a representation of the overall space–time field $\Psi(\mathbf{r}, t)$ as

$$\Psi(\boldsymbol{\rho}, t; z) = \int \sum_\alpha \Psi_\alpha(\mathbf{k}_t, \omega)a_\alpha(\mathbf{k}_t, \omega; z)e^{i(\mathbf{k}_t\cdot\boldsymbol{\rho} - \omega t)}\frac{d\mathbf{k}_t\,d\omega}{(2\pi)^3} \tag{6a}$$

with

$$a_\alpha(\mathbf{k}_t, \omega; z) = \int\frac{(\Gamma^+\Psi_\alpha^+(\mathbf{k}_t, \omega), \Psi(\boldsymbol{\rho}, t; z))}{2N_\alpha(\mathbf{k}_t, \omega)}e^{-i(\mathbf{k}_t\cdot\boldsymbol{\rho} - \omega t)}d\boldsymbol{\rho}\,dt. \tag{6b}$$

The corresponding completeness relation in function and polarization space is

$$1\delta(\boldsymbol{\rho} - \boldsymbol{\rho}')\delta(t - t') = \int \sum_\alpha \frac{\Psi_\alpha(\mathbf{k}_t, \omega)\Gamma^+\Psi_\alpha^+(\mathbf{k}_t, \omega)}{2N_\alpha(\mathbf{k}_t, \omega)}e^{i\mathbf{k}_t\cdot(\boldsymbol{\rho} - \boldsymbol{\rho}')}e^{-i\omega(t - t')}\frac{d\mathbf{k}_t\,d\omega}{(2\pi)^3},$$

$$\tag{7}$$

which states that the set of guided mode fields $\Psi_\alpha(\mathbf{k}_t, \omega)e^{i(\mathbf{k}_t\cdot\boldsymbol{\rho} - \omega t)}$ is capable of completely representing the solution of a general linear field problem in a transversely unbounded medium. In particular, as we shall demonstrate, the (matrix) Green's function in a transversely unbounded medium is representable as

$$G(\mathbf{r}, \mathbf{r}'; t, t') = \int \sum_\alpha G(\mathbf{k}_t, \omega; z, z')\frac{\Psi_\alpha(\mathbf{k}_t, \omega)\Psi_\alpha^+(\mathbf{k}_t, \omega)}{2N_\alpha(\mathbf{k}_t, \omega)}e^{i\mathbf{k}_t\cdot(\boldsymbol{\rho} - \boldsymbol{\rho}')}e^{-i\omega(t - t')}\frac{d\mathbf{k}_t\,d\omega}{(2\pi)^3}.$$

$$\tag{8}$$

1.4a General Linear Field

As noted in Sec. 1.1d, a general linear field in a homogeneous, stationary medium is describable by a linear operator $L(\nabla, \partial/\partial t)$. In a medium displaying symmetry along the z direction, the operator L may be decomposed into two components, one depending only on $\partial/\partial z$ and the other on ∇_t, $\partial/\partial t$, where $\nabla_t = \nabla - \mathbf{z}_0(\partial/\partial z)$ is the component of the vector derivative ∇ transverse to z. The defining equation (1.1.73a) for the Green's function $G(\mathbf{r}, \mathbf{r}'; t, t')$ of a general linear field is thus expressible in the form

$$-i\left[K\left(\nabla_t, \frac{\partial}{\partial t}\right) - \frac{\Gamma}{i}\frac{\partial}{\partial z}\right]G(\boldsymbol{\rho}, \boldsymbol{\rho}', t, t'; z, z') = 1\delta(\boldsymbol{\rho} - \boldsymbol{\rho}')\delta(t - t')\delta(z - z') \quad (9)$$

and in an unbounded region is subject to causality (or equivalently, an outgoing wave condition as $|z - z'| \to \infty$). The component operators K and Γ may be inferred from the L operators noted in Sec. 1.1d and will be listed below for a number of fields. The elements of the matrix representatives of these operators are scalars, vectors, or dyadics, and so, correspondingly, are those of the Green's function G.

To represent in the form (8) the Green's function G defined by Eq. (9), one must first ascertain the characteristic field vectors $\boldsymbol{\Psi}_\alpha$. The latter are defined as solutions of the source-free field equations $L\boldsymbol{\Psi}_\alpha = 0$ for fields $\boldsymbol{\Psi}_\alpha$ with a z dependence $\exp(i\kappa_\alpha z)$. Introducing this z dependence, and noting the decomposition of the operator L in Eq. (9), one obtains from $L\boldsymbol{\Psi}_\alpha = 0$ the eigenvalue equation

$$K\left(\nabla_t, \frac{\partial}{\partial t}\right)\boldsymbol{\Psi}_\alpha(\boldsymbol{\rho}, t) = \kappa_\alpha\Gamma\boldsymbol{\Psi}_\alpha(\boldsymbol{\rho}, t) \tag{10}$$

defining the eigenvectors $\boldsymbol{\Psi}_\alpha$ and eigenvalues κ_α. For the case of an unbounded homogeneous (translationally invariant) cross section transverse to z, $\boldsymbol{\Psi}_\alpha(\boldsymbol{\rho}, t) = \boldsymbol{\Psi}_\alpha(\mathbf{k}_t, \omega)\exp[i(\mathbf{k}_t\cdot\boldsymbol{\rho} - \omega t)]$ and hence in $\mathbf{k}_t$, ω space the eigenvalue problem in Eq. (10) takes the form

$$K(\mathbf{k}_t, \omega)\boldsymbol{\Psi}_\alpha(\mathbf{k}_t, \omega) = \kappa_\alpha(\mathbf{k}_t, \omega)\Gamma\boldsymbol{\Psi}_\alpha(\mathbf{k}_t, \omega), \tag{11a}$$

where $K(\mathbf{k}_t, \omega)$ is the algebraic matrix operator $K(\nabla_t, \partial/\partial t)$ with the substitutions $\nabla_t = i\mathbf{k}_t$ and $\partial/\partial t = -i\omega$, and $\boldsymbol{\Psi}_\alpha(\mathbf{k}_t, \omega)$ is a multicomponent vector in an appropriate "polarization" space. The corresponding adjoint eigenvalue problem† is

$$K(\mathbf{k}_t, \omega)^+\boldsymbol{\Psi}_\alpha^+(\mathbf{k}_t, \omega) = \kappa_\alpha(\mathbf{k}_t, \omega)^*\Gamma^+\boldsymbol{\Psi}_\alpha^+(\mathbf{k}_t, \omega) \tag{11b}$$

and possesses complex-conjugate eigenvalues κ_α^* and adjoint eigenvectors $\boldsymbol{\Psi}_\alpha^+$ characteristic of adjoint operators K^+ and Γ^+, whose off-diagonal matrix elements are the transposed conjugates of the elements of K and Γ. From Eqs.

---

†The adjoint eigenvalue $\kappa_\alpha^+ = \kappa_\alpha^*$, provided there exists a $\boldsymbol{\Psi}_\alpha$ and $\boldsymbol{\Psi}_\alpha^+$ such that the inner product $(\Gamma^+\boldsymbol{\Psi}_\alpha^+, \boldsymbol{\Psi}_\alpha) \neq 0$. In dissipative systems it is sometimes more convenient to define the inner product without the conjugate operation.

(11), on forming appropriate Hermitian inner products of the vectors $\mathbf{\Psi}_\alpha$ and $\mathbf{\Psi}_\beta^+$, one finds that

$$(\kappa_\beta - \kappa_\alpha)(\mathbf{\Psi}_\beta^+, \Gamma\mathbf{\Psi}_\alpha) = (K^+\mathbf{\Psi}_\beta^+, \mathbf{\Psi}_\alpha) - (\mathbf{\Psi}_\beta^+, K\mathbf{\Psi}_\alpha) = 0,$$

from which one infers in the $\mathbf{k}_t, \omega$ polarization space the previously noted biorthogonality property in Eq. (4c) and the abstract completeness relation (5),

$$1 = \sum_\alpha \frac{\mathbf{\Psi}_\alpha(\mathbf{k}_t, \omega)\Gamma^+\mathbf{\Psi}_\alpha^+(\mathbf{k}_t, \omega)}{2N_\alpha(\mathbf{k}_t, \omega)}. \tag{12}$$

For the Hermitian case wherein $K = K^+$ and $\Gamma = \Gamma^+$, one infers from Eqs. (11) that $\mathbf{\Psi}_\alpha^+ = \mathbf{\Psi}_{\alpha^*}$, where $\mathbf{\Psi}_{\alpha^*}$ is the eigenvector with the eigenvalue κ_α^*.

In the $\mathbf{k}_t, \omega$ transform space the Green's function equation (9) becomes, on use of the representation in Eq. (3), the matrix ordinary differential equation

$$-i\left[K(\mathbf{k}_t, \omega) - \frac{\Gamma}{i}\frac{d}{dz}\right]G(\mathbf{k}_t, \omega; z, z') = 1\delta(z - z'), \tag{13}$$

where for a region unbounded in the z direction, the Green's function satisfies the boundary condition $G \to 0$ as $|z - z'| \to \infty$, provided that Im $\omega > 0$. This $\mathbf{k}_t, \omega$ transformed Green's function can be further represented via Eq. (12) as

$$G(\mathbf{k}_t, \omega; z, z') = \sum_\alpha \frac{\mathbf{\Psi}_\alpha(\mathbf{k}_t, \omega)}{2N_\alpha(\mathbf{k}_t, \omega)}(\mathbf{\Psi}_\alpha^+(\mathbf{k}_t, \omega), \Gamma G(\mathbf{k}_t, \omega; z, z'))$$

$$= \sum_\alpha \frac{\mathbf{\Psi}_\alpha(\mathbf{k}_t, \omega)\mathbf{\Psi}_\alpha^+(\mathbf{k}_t, \omega)}{2N_\alpha(\mathbf{k}_t, \omega)}G_\alpha(\mathbf{k}_t, \omega; z, z'), \tag{14}$$

where the scalar amplitudes G_α are defined by $\mathbf{\Psi}_\alpha^+ G_\alpha = (\mathbf{\Psi}_\alpha^+, \Gamma G)$. Equation (14), when employed in Eq. (3), leads to the space- and time-dependent Green's function representation anticipated in Eq. (8). The determination of the amplitude coefficients G_α involves transformation of Eq. (13) with respect to the basis $\mathbf{\Psi}_\alpha$. Thus, Hermitian inner-product multiplication of Eq. (13) by $\mathbf{\Psi}_\alpha^+(\mathbf{k}_t, \omega)$ leads, in view of Eq. (14), to

$$\left[\frac{d}{dz} - i\kappa_\alpha(\mathbf{k}_t, \omega)\right]G_\alpha(\mathbf{k}_t, \omega; z, z') = \delta(z - z'), \tag{15}$$

subject to $G_\alpha \to 0$ as $|z - z'| \to \infty$, provided that Im $\kappa_\alpha \neq 0$, as results from Im $\omega > 0$. Equation (15) admits solutions

$$G_\alpha(\mathbf{k}_t, \omega; z, z') = \begin{cases} e^{i\kappa_\alpha(z-z')} & z > z', \\ 0, & z < z', \end{cases} \tag{16a}$$

for modes κ_α, with Im $\kappa_\alpha > 0$, that characterize waves transporting energy (or decaying) in the $+z$ direction, and

$$G_\alpha(\mathbf{k}_t, \omega; z, z') = \begin{cases} 0, & z > z', \\ -e^{i\kappa_\alpha(z-z')} & z < z', \end{cases} \tag{16b}$$

for modes κ_α, with Im $\kappa_\alpha < 0$, carrying energy (or decaying) in the $-z$ direction. Indices $\alpha > 0$ or $\alpha < 0$ will be employed to identify the proper κ_α to be associated with waves traveling in the $+z$ or $-z$ directions, respectively. These identifications pose difficulties in complex media; in simple passive media, waves carrying power in the $+z$ or $-z$ directions are distinguished by group speeds $(\partial \kappa_\alpha/\partial \omega)^{-1} > 0$ or <0, respectively (see Sec. 1.6c). The proper identification may be effected, as noted above and in Eq. (3), by analytic continuation into the region Im $\omega > 0$ with subsequent selection of Im $\kappa_\alpha \gtrless 0$ for the $\pm z$ traveling waves. Alternatively, for real $\mathbf{k}_t$, ω, the identification may be accomplished on assigning small loss to the medium, thereby removing the singularities of $G(\mathbf{k}_t, \omega; z, z')$ from the integration paths, and then passing to the lossless limit.

Substituting Eqs. (16) into Eqs. (14), one obtains with the aid of Eq. (3) the desired Green's function representation,

$$G(\mathbf{r}, \mathbf{r}'; t, t') = \int \sum_{\alpha \geq 0} \frac{\mathbf{\Psi}_\alpha(\mathbf{k}_t, \omega)\mathbf{\Psi}_\alpha^+(\mathbf{k}_t, \omega)}{\pm 2 N_\alpha(\mathbf{k}_t, \omega)} e^{i[\mathbf{k}_t \cdot (\boldsymbol{\rho} - \boldsymbol{\rho}') - \omega(t - t')]} e^{i\kappa_\alpha(z - z')} \frac{d\mathbf{k}_t \, d\omega}{(2\pi)^3}, \quad (17)$$

where the $\alpha \gtrless 0$ and $\pm$ designations correspond, respectively, to the regions $z \gtrless z'$, and the triple integration ranges over all real values of $\mathbf{k}_t$ and ω from $-\infty$ to $+\infty$. The guided-wave representation in Eq. (17) is to be contrasted with the oscillatory Green's function representation of Eq. (1.3.17). In the latter representation the causality requirement of vanishing for $t < t'$ is evidently satisfied; in the guided-wave case causality requires the proper identification of the $+$ and $-$ (i.e., $\alpha \gtrless 0$) traveling waves.

The guided-wave representation in Eq. (17) becomes explicit with a knowledge of the eigenmode vectors $\mathbf{\Psi}_\alpha$ and $\mathbf{\Psi}_\alpha^+$ together with their normalizations N_α. Their evaluation is exhibited for a number of linear fields in the following subsections (see also Secs. 8.3 and 8.4).

1.4b   The Acoustic Field

In a linear acoustic field, the eigenvalue problem in Eq. (10), which characterizes source-free guided-wave solutions with $\exp(i\kappa z)$ space dependence, can be written, via Eqs. (1.1.1), in the form

$$\frac{1}{\gamma p_0} \frac{\partial p}{\partial t} + \boldsymbol{\nabla}_t \cdot \mathbf{v}_t = -i\kappa \mathbf{v} \cdot \mathbf{z}_0,$$

$$\boldsymbol{\nabla}_t p + n_0 m \frac{\partial \mathbf{v}}{\partial t} = -i\kappa p \mathbf{z}_0, \qquad (18)$$

where $p = p(\boldsymbol{\rho}, t; \kappa)$ and $\mathbf{v} = \mathbf{v}(\boldsymbol{\rho}, t; \kappa)$ are the complex amplitudes of the pressure and velocity fields, and where $\mathbf{v} = \mathbf{v}_t + v_z \mathbf{z}_0$, $\boldsymbol{\nabla} = \boldsymbol{\nabla}_t + (\partial/\partial z)\mathbf{z}_0$, and $\mathbf{r} = \boldsymbol{\rho} + z\mathbf{z}_0$ represent vector decompositions transverse and longitudinal to the chosen symmetry direction $\mathbf{z}_0$. In a homogeneous transversely unbounded medium, guided-wave solutions of Eqs. (18) are distinguished by a subscript α and represented as

$$\begin{pmatrix} p(\mathbf{r}, t) \\ \mathbf{v}(\mathbf{r}, t) \end{pmatrix} = \begin{pmatrix} p_\alpha(\mathbf{k}_t, \omega) \\ \mathbf{v}_\alpha(\mathbf{k}_t, \omega) \end{pmatrix} e^{i(\mathbf{k}_t \cdot \boldsymbol{\rho} - \omega t)} e^{i\kappa_\alpha z}, \qquad (19a)$$

where for simplicity of notation, the same symbol is employed for a field and its $\mathbf{k}_t, \omega$-dependent amplitude. In the wavevector notation of Sec. 1.4a, one writes Eq. (19a) as

$$\Psi(\mathbf{r}, t) = \Psi_\alpha(\mathbf{k}_t, \omega)e^{i(k_t \cdot \rho - \omega t)}e^{i\kappa_\alpha z}, \tag{19b}$$

whence Eqs. (18) can be written in the form

$$L\Psi = -i(K - \kappa_\alpha \Gamma)\Psi_\alpha = 0, \tag{20a}$$

where

$$-iK\left(\nabla_t, \frac{\partial}{\partial t}\right) \to \begin{pmatrix} \dfrac{1}{\gamma p_0}\dfrac{\partial}{\partial t} & \nabla_t \\ \nabla_t & n_0 m \dfrac{\partial}{\partial t}\mathbf{1} \end{pmatrix}, \quad K(\mathbf{k}_t, \omega) \to \begin{pmatrix} \dfrac{\omega}{\gamma p_0} & -\mathbf{k}_t \\ -\mathbf{k}_t & \omega n_0 m\mathbf{1} \end{pmatrix},$$

$$\Gamma \to \begin{pmatrix} 0 & \mathbf{z}_0 \\ \mathbf{z}_0 & 0 \end{pmatrix} \tag{20b}$$

are the operator components of Eqs. (10) and (11a); it should be noted that $K = K^+, \Gamma = \Gamma^+$, and thus $\Psi_\alpha^+ = \Psi_{\alpha*}$. Since in the $\mathbf{k}_t, \omega$ basis the biorthogonality property in Eq. (4c) takes the form

$$(\Psi_{\alpha*}(\mathbf{k}_t, \omega), \Gamma\Psi_\beta(\mathbf{k}_t, \omega)) \equiv p_{\alpha*}^*(\mathbf{k}_t, \omega)v_{z\beta}(\mathbf{k}_t, \omega) + v_{z\alpha*}^*(\mathbf{k}_t, \omega)p_\beta(\mathbf{k}_t, \omega)$$
$$= 2N_\alpha \delta_{\alpha\beta}, \tag{21}$$

and evidently involves only the independent field components p and v_z, it is frequently convenient to elide the dependent field component $\mathbf{v}_t$ and denote the abbreviated acoustic field wavevector $\overline{\Psi}$ as

$$\overline{\Psi}(\mathbf{r}, t) \to \begin{pmatrix} p(\mathbf{r}, t) \\ v_z(\mathbf{r}, t)\mathbf{z}_0 \end{pmatrix}, \tag{22}$$

For eigenvectors of the form (19), one notes that $\nabla_t \equiv i\mathbf{k}_t$ and $\partial/\partial t \equiv -i\omega$. Hence, for prescribed $\mathbf{k}_t, \omega$, non-vanishing solutions of Eqs. (18) or (20a) are found to exist only for the two wavenumbers (eigenvalues) given by the zeros of det L:

$$\kappa_{\pm1} = \pm\sqrt{\frac{\omega^2}{a^2} - k_t^2}, \quad \text{where } a = \sqrt{\frac{\gamma p_0}{n_0 m}}. \tag{23}$$

These are recognizable as the two propagation wavenumbers in the transmission-line network depicted in Fig. 1. 4. 1; the per-unit-length elements of the line comprise a series "inductance" $n_0 m$ and a shunt "capacitance" $1/\gamma p_0$ coupled by an ideal transformer of turns ratio $1 : k_t$ to an "inductance" $n_0 m$. The transmission-line interpretation becomes evident on elimination of the dependent field variable $\mathbf{v}_t$ from Eqs. (18), whence in the $\mathbf{k}_t, \omega$ basis the z-dependent acoustic equations (18), in which $i\kappa = \partial/\partial z$, may be expressed in the form

$$\frac{\partial v_z}{\partial z} = i\left(\frac{\omega}{\gamma p_0} - \frac{k_t^2}{\omega n_0 m}\right)p = i\kappa Yp,$$

$$\frac{\partial p}{\partial z} = i\omega n_0 m v_z \qquad = i\kappa Z v_z, \tag{24a}$$

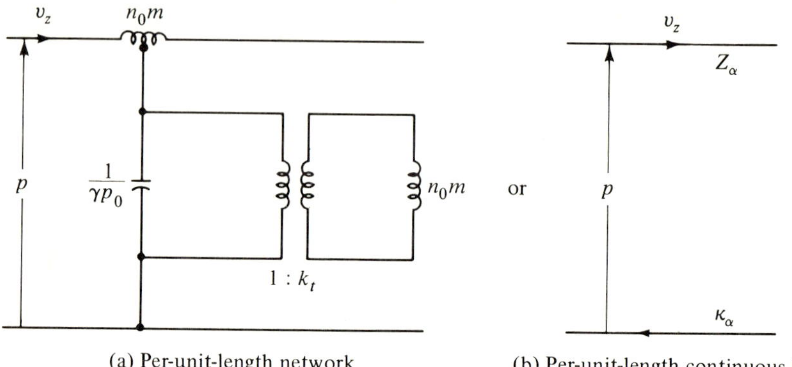

(a) Per-unit-length network   (b) Per-unit-length continuous line

FIG. 1.4.1   Acoustic transmission-line network.

where

$$\kappa = \sqrt{\frac{\omega^2}{a^2} - k_t^2} \quad \text{and} \quad Z = \frac{\omega n_0 m}{\kappa} = \frac{1}{Y}. \tag{24b}$$

κ and Z are the propagation constant and characteristic impedance, and p and v_z play the role of generalized transmission-line "voltage" and "current," respectively. To the eigenvalues $\kappa_{\pm 1}$ in Eq. (23) there correspond eigenvectors $\Psi_{\pm 1}$ of the form (19b), normalized to $2N_{\pm 1}$ in accordance with the biorthogonality property in Eq.(21); these follow from Eqs. (18) or (11a) and (20) as

$$\Psi_{\pm 1} \rightarrow \begin{pmatrix} 1 \\ (\kappa_{\pm 1}\mathbf{z}_0 + \mathbf{k}_t)/\omega n_0 m \end{pmatrix}, \qquad N_{\pm 1} = \frac{\kappa_{\pm 1}}{\omega n_0 m} = \pm Y \tag{25}$$

and display the field structure of the two possible guided eigenwaves for given $\mathbf{k}_t$ and ω.

Knowledge of the guided-wave characteristics in Eqs. (23) and (25) permits an explicit representation of the acoustic Green's functions defined by Eqs. (9) and (20). Thus via Eq. (17), the z-guided-wave representation of the acoustic Green's function G_{11} of Eq. (1.1.4) is given by

$$G_{11}(\mathbf{r}, \mathbf{r}'; t, t') = \int \frac{\omega n_0 m}{2\sqrt{(\omega^2/a^2) - k_t^2}} e^{i(\mathbf{k}_t \cdot \mathbf{\rho} - \omega t)} e^{\pm i \sqrt{(\omega^2/a^2) - k_t^2}\, z} \frac{d\mathbf{k}_t\, d\omega}{(2\pi)^3}, \quad z \gtrless 0,$$

$$\tag{26}$$

where the integration extends over all real $\mathbf{k}_t$, ω from $-\infty$ to $+\infty$, and for simplicity we have set $\mathbf{\rho}' = z' = t' = 0$. Equation (26) provides a representation alternative to that in Eq. (1.3.25a), but when integrated is identical to the closed form result in that equation. It should be noted that Eq. (26) also follows from the $\mathbf{k}, \omega$ representation of G_{11} in Eqs. (1.2.7) and (1.2.8) by integration over the z component κ of the wavevector $\mathbf{k} = \mathbf{k}_t + \kappa \mathbf{z}_0$ (see Sec. 1.5c).

1.4c The Electromagnetic Field

Source-free guided waves in an electromagnetic field have an exp $(i\kappa z)$ dependence and satisfy the homogeneous form of the Maxwell equations (1.1.16) with $\partial/\partial z \equiv i\kappa$:

$$\epsilon_0 \frac{\partial \mathbf{E}}{\partial t} - \nabla_t \times \mathbf{H} = i\kappa \mathbf{z}_0 \times \mathbf{H},$$

$$\nabla_t \times \mathbf{E} + \mu_0 \frac{\partial \mathbf{H}}{\partial t} = -i\kappa \mathbf{z}_0 \times \mathbf{E}. \tag{27}$$

Equations (27) constitute an eigenvalue problem of the type shown in Eq. (10), with $\mathbf{E} = \mathbf{E}(\boldsymbol{\rho}, t; \kappa)$ and $\mathbf{H} = \mathbf{H}(\boldsymbol{\rho}, t; \kappa)$ as the eigenvector amplitudes of the electric and magnetic fields, and κ as the eigenvalue. As before, $\mathbf{V} = \mathbf{V}_t + i\kappa \mathbf{z}_0$ and $\mathbf{r} = \boldsymbol{\rho} + z\mathbf{z}_0$ represent vector decompositions transverse and longitudinal to the guiding direction z. In a transversely unbounded, homogeneous medium, eigenwave solutions of Eq. (27) are representable in the form

$$\begin{pmatrix} \mathbf{E}(\mathbf{r}, t) \\ \mathbf{H}(\mathbf{r}, t) \end{pmatrix} = \begin{pmatrix} \mathbf{E}_\alpha(\mathbf{k}_t, \omega) \\ \mathbf{H}_\alpha(\mathbf{k}_t, \omega) \end{pmatrix} e^{i(\mathbf{k}_t \cdot \boldsymbol{\rho} - \omega t)} e^{i\kappa_\alpha z}, \tag{28a}$$

where the subscript α distinguishes different eigensolutions with a fixed $\mathbf{k}_t, \omega$ dependence. For notational convenience, we employ the same symbol for a field as for its $\mathbf{k}_t, \omega$ amplitude; ambiguity is avoided by specifying, where necessary, the arguments of the field function.

In the $\boldsymbol{\Psi}$ wavevector notation of Sec. 1.4a, one rewrites Eq. (28a) as

$$\boldsymbol{\Psi}(\mathbf{r}, t) = \boldsymbol{\Psi}_\alpha(\mathbf{k}_t, \omega) e^{i(\mathbf{k}_t \cdot \boldsymbol{\rho} - \omega t)} e^{i\kappa_\alpha z}, \tag{28b}$$

whence Eqs. (27) with $\kappa = \kappa_\alpha$ take the general form

$$L\boldsymbol{\Psi}_\alpha = -i(K - \kappa_\alpha \Gamma)\boldsymbol{\Psi}_\alpha = 0 \tag{29a}$$

with the component operators K and Γ of Eqs. (10) and (11a) defined by

$$-iK\left(\nabla_t, \frac{\partial}{\partial t}\right) \to \begin{pmatrix} \epsilon_0 \dfrac{\partial}{\partial t} \mathbf{1} & -\nabla_t \times \mathbf{1} \\ \nabla_t \times \mathbf{1} & \mu_0 \dfrac{\partial}{\partial t} \mathbf{1} \end{pmatrix}, \quad K(\mathbf{k}_t, \omega) \to \begin{pmatrix} \omega\epsilon_0 \mathbf{1} & \mathbf{k}_t \times \mathbf{1} \\ -\mathbf{k}_t \times \mathbf{1} & \omega\mu_0 \mathbf{1} \end{pmatrix},$$

$$\Gamma \to \begin{pmatrix} 0 & -\mathbf{z}_0 \times \mathbf{1} \\ \mathbf{z}_0 \times \mathbf{1} & 0 \end{pmatrix} \tag{29b}$$

The operators $K = K^+$ and $\Gamma = \Gamma^+$ are Hermitian (since $\mathbf{a} \times \mathbf{1} = \mathbf{1} \times \mathbf{a} = -\widehat{\mathbf{a} \times \mathbf{1}}$), and hence $\boldsymbol{\Psi}_\alpha^+ = \boldsymbol{\Psi}_{\alpha^*}$.† In the $\mathbf{k}_t, \omega$ basis, the biorthogonality property (4c) of the eigenvectors $\boldsymbol{\Psi}_\alpha$ takes the specific form

---

†K is Hermitian in the transversely unbounded case and, with suitable transverse boundary conditions, also in the bounded case. In this case the adjoint eigenvector $\boldsymbol{\Psi}_\alpha^+ = \boldsymbol{\Psi}_{\alpha^*}$, where $\boldsymbol{\Psi}_{\alpha^*}$ is the eigenvector with eigenvalue κ_α^*.

$$(\Psi_{\alpha^*}, \Gamma\Psi_\beta) = \mathbf{E}_{\alpha^*}^*(\mathbf{k}_t, \omega)\cdot\mathbf{H}_\beta(\mathbf{k}_t, \omega) \times \mathbf{z}_0 + \mathbf{H}_{\alpha^*}^*(\mathbf{k}_t, \omega)\cdot\mathbf{z}_0 \times \mathbf{E}_\beta(\mathbf{k}_t, \omega)$$

$$= 2N_\alpha \delta_{\alpha\beta} \tag{30}$$

and involves only transverse electric- and magnetic-field components. It is thereby implied that the transverse components $\mathbf{E}_t$, $\mathbf{H}_t$ are the independent components of the electromagnetic field; accordingly, one frequently employs an abbreviated notation for the transverse electromagnetic field wavevector,

$$\overline{\Psi}(\mathbf{r}, t) \rightarrow \begin{pmatrix} \mathbf{E}_t(\mathbf{r}, t) \\ \mathbf{H}_t(\mathbf{r}, t) \end{pmatrix}, \tag{31}$$

whose eigenvectors are represented as in Eq. (28b) and possess the same biorthogonality property as in Eq. (30).

In a transversely unbounded and homogeneous medium, electromagnetic eigenvectors of the form (28) are defined as the non-vanishing resonant solutions of Eqs. (27) with $\nabla_t \equiv i\mathbf{k}_t$ and $\partial/\partial t \equiv -i\omega$. For prescribed $\mathbf{k}_t, \omega$, such solutions exist only for the four degenerate eigenvalues (the zeros of det L)

$$\kappa_{\pm 1} = \pm\sqrt{\frac{\omega^2}{c^2} - k_t^2} = \kappa_{\pm 2}, \qquad \text{where } c = \frac{1}{\sqrt{\mu_0\epsilon_0}}. \tag{32}$$

The existence of four eigenvalues is commensurate with the fact that there are four independent components of the transverse electromagnetic field. In a $\mathbf{k}_t, \omega$ basis, the dependent longitudinal field components E_z, H_z may be eliminated via the relations

$$-\omega\epsilon_0 E_z = \mathbf{k}_t\cdot\mathbf{H}_t \times \mathbf{z}_0 \qquad \text{and} \qquad -\omega\mu_0 H_z = \mathbf{k}_t\cdot\mathbf{z}_0 \times \mathbf{E}_t, \tag{33}$$

derivable from Eqs. (27). On restoring $\partial/\partial z = i\kappa$ in Eqs. (27), one thereby obtains in a $\mathbf{k}_t, \omega$ basis the following defining equations for the z-dependent transverse field amplitudes $\mathbf{E}_t$, $\mathbf{H}_t$:

$$\frac{\partial\mathbf{E}_t}{\partial z} = \left(i\omega\mu_0\mathbf{1} + \frac{\mathbf{k}_t\mathbf{k}_t}{i\omega\epsilon_0}\right)\cdot(\mathbf{H}_t \times \mathbf{z}_0),$$

$$\frac{\partial\mathbf{H}_t}{\partial z} = \left(i\omega\epsilon_0\mathbf{1} + \frac{\mathbf{k}_t\mathbf{k}_t}{i\omega\mu_0}\right)\cdot(\mathbf{z}_0 \times \mathbf{E}_t). \tag{34}$$

The form of the bracketed dyadics in Eq. (34) suggests the following transverse vector decompositions along the directions $\mathbf{k}_{t0}$ and $\mathbf{k}_{t0} \times \mathbf{z}_0$ ($\mathbf{k}_{t0}$ is the unit vector in the direction of $\mathbf{k}_t$):

$$\mathbf{E}_t(\mathbf{k}_t, \omega; z) = E_t'(z)\mathbf{k}_{t0} + E_t''(z)\mathbf{k}_{t0} \times \mathbf{z}_0,$$

$$\mathbf{H}_t(\mathbf{k}_t, \omega; z) = H_t'(z)\mathbf{z}_0 \times \mathbf{k}_{t0} + H_t''(z)\mathbf{k}_{t0}, \tag{35a}$$

which, on substitution into Eqs. (34), uncouple the latter into two independent sets of scalar equations:

$$\frac{\partial E_t'}{\partial z} = \left(i\omega\mu_0 + \frac{k_t^2}{i\omega\epsilon_0}\right)H_t', \qquad \frac{\partial E_t''}{\partial z} = i\omega\mu_0 H_t'',$$

$$\frac{\partial H_t'}{\partial z} = i\omega\epsilon_0 E_t', \qquad \frac{\partial H_t''}{\partial z} = \left(i\omega\epsilon_0 + \frac{k_t^2}{i\omega\mu_0}\right)E_t''. \tag{35b}$$

Equations (35b) are recognized as transmission-line equations for the independent fields specified by the amplitudes E'_t, H'_t and E''_t, H''_t, and can be schematized by the networks shown in Fig. 1.4.2. The amplitudes E'_t, H'_t distinguish the "voltage" and "current" of an E mode (with $H_z = 0$) on a transmission line, depicted in Fig. 1.4.2a, whose series and shunt elements per differential length are expressible in terms of an "inductance" μ_0, a "capacitance" ϵ_0, and an ideal transformer of turns ratio k_t. A corresponding interpretation of the amplitudes E''_t, H''_t is depicted in the H-mode (with $E_z = 0$) network of Fig. 1.4.2b. The equivalent smoothed "two-wire-line" schematizations are also shown in Fig. 1.4.2. and are expressible in terms of characteristic impedances Z', Z'' and propagation wavenumbers κ', κ'' that may be inferred from the networks of Fig. 1.4.2, or Eqs. (35b), as

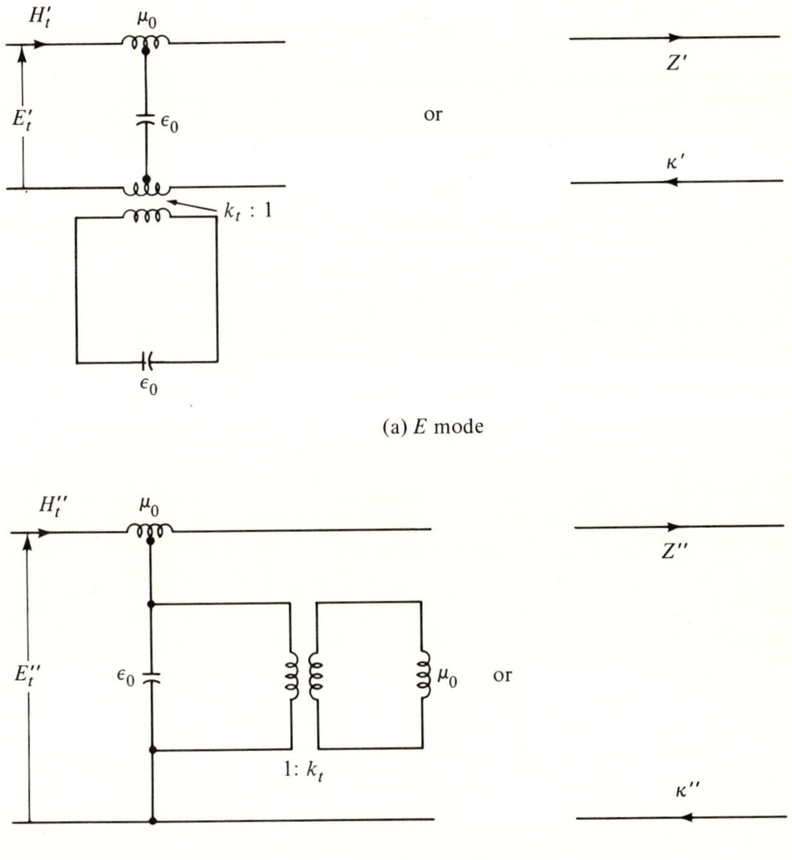

(a) E mode

(b) H mode

FIG. 1.4.2 Electromagnetic transmission-line networks.

$$Z' = \frac{\kappa'}{\omega\epsilon_0} = \frac{1}{Y'}, \qquad Z'' = \frac{\omega\mu_0}{\kappa''} = \frac{1}{Y''},$$

$$\kappa' = \sqrt{\omega^2\mu_0\epsilon_0 - k_t^2} = \kappa''. \tag{36a}$$

The relations (36a) permit the rewriting of Eqs. (35b) in the alternative transmission-line form:

$$\frac{\partial E_t'}{\partial z} = i\kappa' Z' H_t' \qquad \frac{\partial E_t''}{\partial z} = i\kappa'' Z'' H_t'',$$

$$\frac{\partial H_t'}{\partial z} = i\kappa' Y' E_t' \qquad \frac{\partial H_t''}{\partial z} = i\kappa'' Y'' E_t''. \tag{36b}$$

They readily permit the evaluation of the four independent guided-mode wave-vectors $\mathbf{\Psi}_\alpha$ as (note $\mathbf{\Psi}_\alpha^+ = \mathbf{\Psi}_{\alpha*}$)†

$$\mathbf{\Psi}_{\pm 1} \longrightarrow \begin{pmatrix} \pm Z'\mathbf{k}_{t0} - \dfrac{k_t}{\omega\epsilon_0}\mathbf{z}_0 \\[2mm] \mathbf{z}_0 \times \mathbf{k}_{t0} \end{pmatrix}, \qquad \mathbf{\Psi}_{\pm 2} \longrightarrow \begin{pmatrix} \pm Z''\mathbf{k}_{t0} \times \mathbf{z}_0 \\[2mm] \mathbf{k}_{t0} - \dfrac{k_t}{\kappa_{\pm 1}}\mathbf{z}_0 \end{pmatrix}, \tag{37a}$$

with normalization constants

$$N_{\pm 1} = \pm Z' = \frac{\kappa_{\pm 1}}{\omega\epsilon_0}, \qquad N_{\pm 2} = \pm Z'' = \frac{\omega\mu_0}{\kappa_{\pm 2}} \tag{37b}$$

and eigenvalues $\kappa_{\pm 1}$ and $\kappa_{\pm 2}$, as shown in Eq. (32). The wavevectors (37a), which display the field structures of the two independent guided waves for prescribed k_t, ω, satisfy the biorthogonality property in Eq. (30).

With the knowledge of the wavevectors $\mathbf{\Psi}_\alpha$ and eigenvalues κ_α from Eqs. (37) and (32), one can employ Eq. (17) to obtain a guided-wave representation of the electromagnetic-field Green's functions. Thus, by Eq. (17), the matrix electromagnetic Green's function G of Eq. (1.1.19) may be represented as

$$G(\mathbf{r}, \mathbf{r}'; t, t') = \int \left[\frac{\mathbf{\Psi}_{\pm 1}\mathbf{\Psi}_{\pm 1}^*}{2\kappa/\omega\epsilon_0} + \frac{\mathbf{\Psi}_{\pm 2}\mathbf{\Psi}_{\pm 2}^*}{2\omega\mu_0/\kappa} \right] e^{i(\mathbf{k}_t \cdot \mathbf{\rho} - \omega t)} e^{\pm i\kappa z} \frac{d\mathbf{k}_t d\omega}{(2\pi)^3} \tag{38a}$$

where the $\pm$ signs refer to $z \gtrless 0$, respectively, and where, for simplicity, we have set $\mathbf{\rho}' = t' = z' = 0$. To evaluate the dyadic Green's function $\mathscr{G}_{11}$, one substitutes the first-row elements of $\mathbf{\Psi}_\alpha$ from Eq. (37a) into Eq. (38a). Noting that the bracketed term in the integrand of Eq. (38a) yields‡

$$[\quad]_{11} = \frac{1}{2}\left[\frac{\omega\epsilon_0}{\kappa}\mathbf{k}_{t0}\mathbf{k}_{t0} \mp \frac{k_t}{\omega\epsilon_0}(\mathbf{k}_{t0}\mathbf{z}_0 + \mathbf{z}_0\mathbf{k}_{t0}) \right.$$

$$\left. + \frac{k_t^2}{\omega\epsilon_0\kappa}\mathbf{z}_0\mathbf{z}_0 + \frac{\omega\mu_0}{\kappa}(\mathbf{k}_{t0} \times \mathbf{z}_0)(\mathbf{k}_{t0} \times \mathbf{z}_0) \right],$$

---

†See Eq. (8.2.10a).

‡Note that the *ij*th element of the matrix $\mathbf{\Psi}_\alpha\mathbf{\Psi}_{\alpha*}$ is the product of the *i*th-row element of $\mathbf{\Psi}_\alpha$ and the conjugate of the *j*th-row element of $\mathbf{\Psi}_{\alpha*}$; for lossless guides $\mathbf{\Psi}_{\alpha*}^* = \mathbf{\Psi}_\alpha$. The *ij*th element is recognized as $(\varphi_i, \mathbf{\Psi}_\alpha)(\mathbf{\Psi}_{\alpha*}, \varphi_j)$, where φ_i and φ_j are "unit wavevectors" and the Hermitian product is implied.

one obtains

$$\mathscr{G}_{11} = \left(\mu_0 \frac{\partial}{\partial t} \mathbf{1} - \frac{\nabla\nabla}{\epsilon_0(\partial/\partial t)}\right) \int \frac{e^{i\,(\mathbf{k}_t\cdot\mathbf{\rho} - \omega t)}\, e^{\pm i\kappa z}}{-2i\kappa} \frac{d\mathbf{k}_t d\omega}{2(\pi)^3}, \qquad z \gtrless 0, \qquad (38b)$$

where we have taken $\nabla = i k_t \mathbf{k}_{t0} \pm i\kappa \mathbf{z}_0$ and $\partial/\partial t = -i\omega$ terms outside the integral and where $\kappa = \kappa' = \kappa''$ by Eq. (36a) or (32). Equation (38b) provides a guided-wave representation alternative to the oscillatory representation in Eq. (1.3.32b) but is identical to the integrated result shown in Eq. (1.3.32c). One notes that the integral in Eq. (38b) follows from the $\mathbf{k}, \omega$ integral of Eq. (1. 2. 20) by residue integration over the κ components of the wavevector $\mathbf{k} = \mathbf{k}_t + \kappa\mathbf{z}_0$(see Sec 1.5c).

1.5 REDUCED ELECTROMAGNETIC FIELD EQUATIONS

In Secs. 1.1–1.4 the discussion of linear fields has been based primarily on a first-order formulation of the field equations. However, much of the technical literature on electromagnetic wave propagation in homogeneous or inhomogeneous media is based on a "reduced" formulation wherein non-electromagnetic variables have been eliminated from the first-order equations. The reduced field equations are characterized by constitutive parameters (permittivity† and permeability) that depend in general on the derivative operators ∇ and $\partial/\partial t$, and for inhomogeneous but stationary media on the spatial coordinate $\mathbf{r}$ as well. An example is provided by an ionized plasma medium whose electromagnetic properties are described in the cold-fluid approximation by a temporally dispersive (i.e., $(\partial/\partial t)$-dependent) dielectric tensor; inclusion of temperature effects introduces an additional spatial dispersion (∇ dependence), as noted in Eqs. (1.1.64).

The reduced field equations may be solved in terms of Green's functions G_{ij} that are identical to, but fewer in number than, those discussed in Sec. 1.1d, since the $_{ij}$ now refer only to a reduced number of field variables. Thus, in the time-dependent case, the electric field $\mathbf{E}(\mathbf{r}, t)$ excited solely by an electric current density $\mathbf{J}(\mathbf{r}, t)$ in an inhomogeneous medium is given by

$$\mathbf{E}(\mathbf{r}, t) = - \int \mathscr{G}_{11}(\mathbf{r}, \mathbf{r}';t, t')\cdot \mathbf{J}(\mathbf{r}', t')\, d\mathbf{r}' dt',$$

which implies that a reduced field equation for this case is of the form

$$\mathscr{Y}\!\left(\nabla, \frac{\partial}{\partial t};\mathbf{r}\right)\cdot \mathbf{E}(\mathbf{r}, t) = -\mathbf{J}(\mathbf{r}, t),$$

where the derivative (admittance) operator $\mathscr{Y}$ manifestly has the Green's function $\mathscr{G}_{11}$ as its inverse. Alternatively, the reduced field equations can be expressed in terms of both the electric and magnetic-field variables $\mathbf{E}$ and $\mathbf{H}$,

---

†Since medium properties may be spatially variable, we frequently use the term "permittivity" instead of "dielectric constant."

as, for example, in Eqs. (1.1.64), and will involve dispersive permittivity and permeability operators ϵ and μ, respectively, instead of the admittance $\mathscr{Y}$.

Because of the dispersive nature of the constitutive parameters, energy-storage and power-flow relations in a reduced field description are expressed differently than in the first-order formulation of Eq. (1.1.55). To identify energy-density and power-flow expressions, it is necessary to distinguish between harmonic fields of the form $\exp[i(\mathbf{k}\cdot\mathbf{r} - \omega t)]$ and wavepackets that are superpositions of such harmonic fields. A number of energy theorems can be inferred from the dispersive properties (i. e. , from the ω and $\mathbf{k}$ derivatives) of the admittance operator $\mathscr{Y}$ for the harmonic fields. Because time-harmonic fields represent an important class of problems, the presentation emphasizes this regime. Reciprocity properties for reduced time-harmonic fields are similar to those obtained in previous sections and do not require separate treatment.

It has been noted that certain results for spatially inhomogeneous or bounded media can be derived by straightforward modification of those for the homogeneous case. These aspects will be pursued in more detail in the present section. In addition to, or instead of, the radiation (outgoing energy) condition at infinity employed for unbounded regions, electromagnetic fields on boundaries or interfaces must satisfy other boundary conditions. Also, in the presence of medium inhomogeneities, the reciprocity properties of the field are modified from their form in an unbounded homogeneous space. These aspects are considered in Sec. 1.5b.

Propagation in plane-stratified media or in regions bounded by cylindrical, perfectly conducting walls can be analyzed by simple modification of the guided-wave representations of Sec. 1.4 (see Secs. 2.1–2.3 et seq.). For the vector electromagnetic field, this entails a scalarized description carried out with respect to a "waveguide axis" z parallel to the boundary walls or to the direction of stratification. The resulting scalar Green's function problems, stated in Eqs. (1.1.38b) and (1.1.49), are not tied to a special choice of z as the preferred direction and it is often convenient to analyze these scalar problems in alternative "guided" representations with respect to coordinates transverse to z. An introduction to the theory of alternative representations is given in Sec. 1.5c. A more general treatment will be given in Sec 3.3c, and specific applications may be found in Secs. 5.4a and 5.6a.

1.5a Energy Density, Power Flow, and Group Velocity for the Electromagnetic Field

Energy-conservation theorems are important for characterization of energy storage and transport properties in a field. In a first-order field formulation, wherein the constitutive parameters for a medium are non-dispersive, conservation theorems take on a relatively simple form, as is illustrated for a plasma medium in Eq.(1.1.55). In a reduced formulation, wherein non-electromagnetic variables have been eliminated, one is led to vacuum-like electromagnetic field equations but with a permittivity $\epsilon = \epsilon(\nabla, \partial/\partial t; \mathbf{r})$ and permeability $\boldsymbol{\mu} = \boldsymbol{\mu}(\nabla, \partial/\partial t; \mathbf{r})$ which are both anisotropic, spatially and temporally dispersive, and for a spatially inhomogeneous (but stationary) medium, dependent on $\mathbf{r}$ as well. The reduced formulation is exemplified by Eqs. (1.1.64) for the special case of a warm isotropic homogeneous plasma fluid. Because of the dispersive nature of the $\epsilon, \boldsymbol{\mu}$ parameters, energy-conservation theorems for non-monochromatic reduced fields require care in their interpretation. For example, the difficulty with interpreting $\epsilon|E|^2/2$ as a stored (positive) electric energy density in an isotropic but dispersive medium may be appreciated by recalling that, for an isotropic cold plasma under time-harmonic conditions, the equivalent permittivity $\epsilon = \epsilon_0(1 - \omega_p^2/\omega^2)$ may be negative for certain frequency ranges.

Energy-transport properties of a field are investigated most naturally for time-dependent conditions since during a finite interval one may then follow a bundle of energy (wavepacket) excited by a source. As noted in Sec. 1.3, source-excited, time-dependent fields in unbounded regions may be represented rigorously as integral superpositions of oscillatory plane waves with different wavenumber $\mathbf{k}$. The evaluation of these representations by asymptotic techniques will be discussed in Sec. 1.6, where it is found that the field energy is localized in regions wherein the plane waves constituting a wavepacket interfere constructively. The condition for constructive interference leads to the conclusion that, in certain regions, a field behaves *locally* like a plane wave $\Psi \sim \exp(i\psi)$, where $\psi = [\mathbf{k}\cdot\mathbf{r} - \omega(\mathbf{k})t]$, and that its energy is transported with a group velocity $\mathbf{v}_g = \nabla_k \omega(\mathbf{k})$ given by the gradient in $\mathbf{k}$ space of the frequency $\omega = \omega(\mathbf{k})$ determined from the dispersion equation for the medium [see Eq. (1.3.10a)]. The direction of $\mathbf{v}_g$ is called the *ray* direction, and in an anisotropic medium this differs generally from the direction of the wavevector $\mathbf{k}$ along which equiphase surfaces (wavefronts) $\psi = $ constant advance with a phase speed ω/k. Relations governing the detailed behavior of rays, wavepackets, wavefronts, etc., will be considered in Secs. 1.6 and 1.7 both for time-dependent and for time-independent (harmonic) conditions. In the present section, energy-transport properties are inferred not from specific integral representations of the field, but from general conservation theorems derivable from reduced electromagnetic field equations.

In the following we limit our considerations to a reduced electromagnetic field description with only temporally dispersive, space-dependent, dyadic parameters ϵ and $\boldsymbol{\mu}$; the effects of spatial dispersion can be included but for the most part are omitted for simplicity of presentation. Medium anisotropy is retained to emphasize the distinction between phase- and energy-propagation characteristics. Since applications to dispersive media in this book relate primarily to anisotropic plasmas, the permeability μ is usually assumed to be a scalar function, thereby simplifying some of the manipulations and also exhibiting within a single formula expressions appropriate to tensor and scalar media. The extension to media with dyadic permeability is easily accomplished.

The reduced time-dependent Maxwell equations descriptive of the electromagnetic field excited by prescribed electric and magnetic current distributions $\mathbf{J}(\mathbf{r}, t)$ and $\mathbf{M}(\mathbf{r}, t)$ in a temporally dispersive, spatially inhomogeneous, anisotropic medium are given by

$$\nabla \times \mathbf{E}(\mathbf{r}, t) = -\frac{\partial}{\partial t} \mathbf{B}(\mathbf{r}, t) - \mathbf{M}(\mathbf{r}, t).$$

$$\nabla \times \mathbf{H}(\mathbf{r}, t) = \frac{\partial}{\partial t} \mathbf{D}(\mathbf{r}, t) + \mathbf{J}(\mathbf{r}, t). \tag{1a}$$

The constitutive relations are assumed to be of the form

$$\mathbf{B}(\mathbf{r}, t) = \mathbf{\mu}\left(\frac{\partial}{\partial t}, \mathbf{r}\right) \cdot \mathbf{H}(\mathbf{r}, t), \qquad \mathbf{D}(\mathbf{r}, t) = \mathbf{\epsilon}\left(\frac{\partial}{\partial t}, \mathbf{r}\right) \cdot \mathbf{E}(\mathbf{r}, t), \tag{1b}$$

which are applicable in the absence of spatial dispersion and can equally well be phrased in terms of linear integral operators. For only electric current excitation (i.e., $\mathbf{M} = 0$), elimination of $\mathbf{H}$ from Eqs. (1) leads to†

$$\left[\frac{\partial}{\partial t} \mathbf{\epsilon} + \nabla \times \left(\frac{1}{(\partial/\partial t)\mathbf{\mu}} \cdot (\nabla \times \mathbf{1})\right)\right] \cdot \mathbf{E}(\mathbf{r}, t) = -\mathbf{J}(\mathbf{r}, t) \tag{2a}$$

or

$$\mathcal{Y}\left(\nabla, \frac{\partial}{\partial t}; \mathbf{r}\right) \cdot \mathbf{E}(\mathbf{r}, t) = -\mathbf{J}(\mathbf{r}, t) \tag{2b}$$

where the spatially and temporally dispersive dyadic (admittance) operator $\mathcal{Y}$ evidently characterizes the reduced electromagnetic field. As in Sec. 1.1, the solution of Eqs. (2) is expressible as

$$\mathbf{E}(\mathbf{r}, t) = -\int \mathcal{G}_{11}(\mathbf{r}, \mathbf{r}'; t, t') \cdot \mathbf{J}(\mathbf{r}', t') \, d\mathbf{r}' \, dt' \tag{3}$$

in terms of a dyadic Green's function $\mathcal{G}_{11}$ which is the inverse of the admittance operator $\mathcal{Y}$. The desired energy-transport properties of the field are deducible from the dispersive properties of the operator $\mathcal{Y}$.

Energy density and power flow

Before ascertaining the equation of energy transport, we shall first attempt to identify energy-density and power-flow expressions for the field. From the source-free ($\mathbf{J} = 0 = \mathbf{M}$) form of the field equations (1a), one derives the well-known relation

$$\nabla \cdot (\mathbf{E} \times \mathbf{H}) = -\left(\mathbf{E} \cdot \frac{\partial \mathbf{D}}{\partial t} + \mathbf{H} \cdot \frac{\partial \mathbf{B}}{\partial t}\right). \tag{4}$$

$\mathbf{S} = \mathbf{E} \times \mathbf{H}$ is the instantaneous density of electromagnetic power flow at $\mathbf{r}, t$ (i.e., the Poynting vector), but it is difficult to interpret the expression in parentheses on the right of Eq. (4) as the time derivative of an instantaneous

---

†Note that $1/(\partial/\partial t)$ is the integral operator $\int dt$ and $1/\mathbf{\mu}$ is the inverse dyadic $\mathbf{\mu}^{-1}$.

energy density W. This difficulty stems from the temporally dispersive nature of the constitutive parameters in the relations

$$\frac{\partial \mathbf{D}}{\partial t} = \frac{\partial}{\partial t}\left[\epsilon\left(\frac{\partial}{\partial t}, \mathbf{r}\right)\cdot \mathbf{E}\right], \qquad \frac{\partial \mathbf{B}}{\partial t} = \frac{\partial}{\partial t}\left[\mu\left(\frac{\partial}{\partial t}, \mathbf{r}\right)\cdot \mathbf{H}\right], \qquad (5)$$

in consequence of which it is not possible to identify an energy density W for general time-dependent fields. An energy interpretation on a time average basis is, however, possible for monochromatic steady-state fields of the form

$$\mathbf{E}(\mathbf{r}, t) = \sqrt{2}\ \mathrm{Re}\ [\bar{\mathbf{E}}(\mathbf{r})e^{-i\omega t}], \qquad \mathbf{H}(\mathbf{r}, t) = \sqrt{2}\ \mathrm{Re}\ [\bar{\mathbf{H}}(\mathbf{r})e^{-i\omega t}] \qquad (6)$$

where $\bar{\mathbf{E}}(\mathbf{r})$ and $\bar{\mathbf{H}}(\mathbf{r})$ are conventional root-mean-square (rms) complex time-independent field amplitudes, whence the relations (5) become†

$$\frac{\partial \mathbf{D}}{\partial t} = -\sqrt{2}\ \mathrm{Re}\ [i\omega\epsilon(\omega, \mathbf{r})\cdot\bar{\mathbf{E}}e^{-i\omega t}],$$

$$\frac{\partial \mathbf{B}}{\partial t} = -\sqrt{2}\ \mathrm{Re}\ [i\omega\mu(\omega, \mathbf{r})\cdot\bar{\mathbf{H}}e^{-i\omega t}]. \qquad (7)$$

With this complex amplitude notation the time average ‡ of Eq. (4) can be expressed as

$$\nabla\cdot\bar{\mathbf{S}} = -\frac{\overline{\partial W}}{\partial t}, \qquad (8)$$

where

$$\bar{\mathbf{S}} = \mathrm{Re}\ \bar{\mathbf{E}} \times \bar{\mathbf{H}}^*,$$

$$\frac{\overline{\partial W}}{\partial t} = -\frac{i\omega}{2}[\bar{\mathbf{E}}^* \cdot (\epsilon - \epsilon^+) \cdot \bar{\mathbf{E}} + \bar{\mathbf{H}}^* \cdot (\mu - \mu^+) \cdot \bar{\mathbf{H}}] \qquad (8a)$$

and ϵ^+ and μ^+ denote the transposed conjugate (Hermitian adjoint) dyadics to ϵ and μ. Although the expression for $\overline{\partial W}/\partial t$ in Eq. (8a) represents dimensionally the time rate of increase of energy per unit volume, it does not permit an identification of an average energy density $\bar{W}$. For a medium with a real scalar permeability $\mu = \mu^+ = \mu\mathbf{1}$,

$$\frac{\overline{\partial W}}{\partial t} = \bar{\mathbf{E}}^* \cdot \sigma \cdot \bar{\mathbf{E}} \qquad (9)$$

represents conventionally the power dissipation per unit volume in terms of a medium conductivity dyadic§ $\sigma = -i\omega(\epsilon - \epsilon^+)/2$. If the medium is lossy,

---

†Note in Eq. (5) that $\partial/\partial t$ commutes with ϵ and μ in stationary media.

‡In complex notation the time average of the product of harmonic time-dependent quantities AB is $(1/T)\int_0^T AB\,dt = \mathrm{Re}\ \bar{A}\bar{B}^* = \frac{1}{2}(\bar{A}\bar{B}^* + \bar{A}^*\bar{B})$, where $T = 2\pi/\omega$ is the oscillation period.

§Losses due to conduction currents $\bar{\mathbf{J}}_c(\mathbf{r}, \omega) = \sigma(\mathbf{r}, \omega)\cdot\bar{\mathbf{E}}(\mathbf{r}, \omega)$ are expressed in terms of the conductivity dyadic σ included in the dielectric tensor $\epsilon = \epsilon_1 + (\sigma/-i\omega)$. It follows from Eq. (10) that ϵ_1 represents the Hermitian part of ϵ (i.e., $\epsilon_1 = \epsilon_1^+$). Since $\sigma/-i\omega$ is then anti-Hermitian, one concludes that the conductivity dyadic is also Hermitian (i.e., $\sigma = \sigma^+$).

energy is converted into heat and the expression on the right-hand side of Eq. (9) must be positive. This condition applies more generally to Eq. (8a), and for arbitrary fields this is assured only if each of the terms is positive, whence the Hermitian tensors $-i(\boldsymbol{\epsilon} - \boldsymbol{\epsilon}^+)$ and $-i(\boldsymbol{\mu} - \boldsymbol{\mu}^+)$ must be positive definite. For a lossless medium with a real permeability, the average energy density remains constant and the consequent vanishing of the right-hand side of Eq. (8a) requires that (if $\boldsymbol{\mu} = \mu\mathbf{1}$)

$$\boldsymbol{\epsilon} = \boldsymbol{\epsilon}^+, \qquad \mu = \text{real}. \tag{10}$$

Thus, the dyadics representative of the constitutive parameters in a lossless anisotropic medium must be Hermitian. If the medium is isotropic, the constitutive parameters are real but may be negative for certain frequencies.

To calculate the average stored electric energy density in a lossless, dispersive medium, it is necessary to consider non-monochromatic fields. Generalizing Eqs (6), we introduce

$$\mathbf{E}(\mathbf{r}, t) = \sqrt{2}\ \text{Re}\,[\bar{\mathbf{E}}(\mathbf{r}; t)e^{-i\omega t}], \tag{11a}$$

$$\mathbf{D}(\mathbf{r}, t) = \sqrt{2}\ \text{Re}\,[\bar{\mathbf{D}}(\mathbf{r}; t)e^{-i\omega t}], \tag{11b}$$

where $\bar{\mathbf{E}}$ and $\bar{\mathbf{D}}$ are complex rms functions of t that vary slowly during a time interval equal to the period $T = 2\pi/\omega$. Then over the period T, the time average of the first term on the right of Eq. (4) yields

$$\overline{\frac{\partial W_e}{\partial t}} = \frac{1}{2T}\int_0^T (\bar{\mathbf{E}}e^{-i\omega t} + \bar{\mathbf{E}}{}^*e^{i\omega t}) \cdot \frac{\partial}{\partial t}(\bar{\mathbf{D}}e^{-i\omega t} + \bar{\mathbf{D}}{}^*e^{i\omega t})\,dt$$

$$\cong \frac{1}{2T}\int_0^T \left[\bar{\mathbf{E}}{}^*e^{i\omega t} \cdot \frac{\partial}{\partial t}(\bar{\mathbf{D}}e^{-i\omega t}) + \bar{\mathbf{E}}e^{-i\omega t} \cdot \frac{\partial}{\partial t}(\bar{\mathbf{D}}{}^*e^{i\omega t})\right]dt, \tag{12}$$

and since $t = 0$ is an arbitrary reference point, the result in Eq. (12) refers to a typical period. Contributions arising from terms in the integrand containing factors $\exp(\pm i2\omega t)$ have been ignored since they are small compared to the remaining ones. This is verified upon assuming, for example, that during the relevant time interval, $\bar{\mathbf{E}}$ and $\bar{\mathbf{D}}$ may be represented as† [9a]

$$\bar{\mathbf{E}}(\mathbf{r}; t) = \bar{\mathbf{E}}_1(\mathbf{r})e^{-i\omega_1 t}, \qquad \bar{\mathbf{D}}(\mathbf{r}; t) = \bar{\mathbf{D}}_1(\mathbf{r})e^{-i\omega_1 t} \tag{13}$$

with the restriction $\omega_1 \ll \omega$ imposed to ensure the slow variation of these functions over the period T. Integration then shows that the ignored terms are $O(\omega_1/\omega)$ relative to the retained contribution, thereby justifying their omission. From Eq. (13) one has

---

†For a more general derivation utilizing Fourier integral representations of the field, see V. L. Ginzburg, *The Propagation of Electromagnetic Waves in Plasmas*, Pergamon Press, New York, 1964, App. B. Alternatively, in evaluating $(\partial/\partial t)\epsilon(\partial/\partial t)\cdot\bar{\mathbf{E}}(\mathbf{r}; t)\exp(-i\omega t)$ to obtain a result as in Eq. (14), one should also compare the more general procedure in Eq. (27a) below and the footnote thereto.

$$\frac{\partial}{\partial t}(\bar{\mathbf{D}}e^{-i\omega t}) = -i(\omega + \omega_1)\boldsymbol{\epsilon}(\omega + \omega_1) \cdot \bar{\mathbf{E}}e^{-i\omega t} \tag{14}$$

since $\bar{\mathbf{D}}\exp(-i\omega t)$ corresponds to a time-harmonic field with frequency $(\omega + \omega_1)$ which can be expressed in terms of $\bar{\mathbf{E}}\exp(-i\omega t)$ via Eq. (7). Upon expanding the factor $-i(\omega + \omega_1)\boldsymbol{\epsilon}(\omega + \omega_1)$ about ω up to the linear term $O(\omega_1)$, one finds

$$\frac{\partial}{\partial t}(\bar{\mathbf{D}}e^{-i\omega t}) = \left\{-i\omega\boldsymbol{\epsilon}(\omega) \cdot \bar{\mathbf{E}} + \frac{\partial}{\partial\omega}[\omega\boldsymbol{\epsilon}(\omega)] \cdot \frac{\partial\bar{\mathbf{E}}}{\partial t}\right\}e^{-i\omega t} \tag{15}$$

which evidently generalizes the monochromatic result obtained in Eq. (7). Substitution into Eq. (12) then yields for the dominant contribution,

$$\frac{\overline{\partial W_e}}{\partial t} \cong \frac{-i\omega}{2}\bar{\mathbf{E}}^* \cdot [\boldsymbol{\epsilon}(\omega) - \boldsymbol{\epsilon}^+(\omega)] \cdot \bar{\mathbf{E}}$$

$$+ \frac{1}{2}\left\{\bar{\mathbf{E}}^* \cdot \frac{\partial}{\partial\omega}[\omega\boldsymbol{\epsilon}(\omega)] \cdot \frac{\partial\bar{\mathbf{E}}}{\partial t} + \frac{\partial\bar{\mathbf{E}}^*}{\partial t} \cdot \frac{\partial}{\partial\omega}[\omega\boldsymbol{\epsilon}^+(\omega)] \cdot \bar{\mathbf{E}}\right\}. \tag{16}$$

A directly analogous result is obtained for the magnetic energy provided that $\bar{\mathbf{E}}$, $\boldsymbol{\epsilon}$ and $\boldsymbol{\epsilon}^+$ are replaced by $\bar{\mathbf{H}}$, $\boldsymbol{\mu}$ and $\boldsymbol{\mu}^+$, respectively. For the lossless case $\boldsymbol{\epsilon}(\omega) = \boldsymbol{\epsilon}^+(\omega)$, Eq. (16) reduces to

$$\frac{\overline{\partial W_e}}{\partial t} = \frac{\partial}{\partial t}\left\{\frac{1}{2}\bar{\mathbf{E}}^* \cdot \frac{\partial}{\partial\omega}[\omega\boldsymbol{\epsilon}(\omega)] \cdot \bar{\mathbf{E}}\right\}. \tag{17}$$

Thus, the average stored energy density $\bar{W}$ in a lossless, dispersive, electrically anisotropic medium with a scalar permeability μ may be identified as †

$$\bar{W} = \frac{1}{2}\left\{\bar{\mathbf{E}}^* \cdot \frac{\partial}{\partial\omega}[\omega\boldsymbol{\epsilon}(\omega)] \cdot \bar{\mathbf{E}} + \frac{\partial}{\partial\omega}(\omega\mu)|\bar{\mathbf{H}}|^2\right\}, \tag{18}$$

where $\bar{\mathbf{E}}$ and $\bar{\mathbf{H}}$ are to be taken as space- and weakly time-dependent amplitudes of the harmonic field $\bar{\mathbf{E}}(\mathbf{r}, t)\exp(-i\omega t)$ and $\bar{\mathbf{H}}(\mathbf{r}, t)\exp(-i\omega t)$, respectively. Since the stored energy is positive, $\partial(\omega\mu)/\partial\omega$ must be positive and the dyadic $(\partial/\partial\omega)[\omega\boldsymbol{\epsilon}(\omega)]$ must be positive definite. Moreover, the average energy stored by a given electromagnetic field in a physical medium is at least as great as that in vacuum, since energy has to be expended to bring about the polarization effects which are expressed via the constitutive parameters. Thus, $\bar{W} \geq \bar{W}_0$, where $\bar{W}_0$ is the stored energy in vacuum, with $\mu = \mu_0$, $\boldsymbol{\epsilon} = \mathbf{1}\epsilon_0$ (μ_0, ϵ_0 constant). These considerations lead to the sharper condition[9b]

$$\bar{W} - \bar{W}_0 = \frac{1}{2}\left\{\bar{\mathbf{E}}^* \cdot \left[\frac{\partial}{\partial\omega}(\omega\boldsymbol{\epsilon}) - \mathbf{1}\epsilon_0\right] \cdot \bar{\mathbf{E}} + \left[\frac{\partial}{\partial\omega}(\omega\mu) - \mu_0\right]|\bar{\mathbf{H}}|^2\right\} \geq 0. \tag{19}$$

where $\mathbf{1}$ is the unit dyadic.

---

†Since W_e varies slowly over the period T, one has $\bar{W}_e \approx W_e$ in $0 \leq t \leq T$ and $\overline{\partial W_e/\partial t} \approx \partial\bar{W}_e/\partial t$. In the latter relation, the time derivative relates to a time scale long compared to T.

In a lossless, cold electron plasma subjected to a steady external magnetic field $\mathbf{H}_0$ along the z axis of a rectangular coordinate system, one has $\mu = \mu_0$, while the $\boldsymbol{\epsilon}$ tensor has the form

$$\boldsymbol{\epsilon} \rightarrow \begin{pmatrix} \epsilon_1 & -i\epsilon_2 & 0 \\ i\epsilon_2 & \epsilon_1 & 0 \\ 0 & 0 & \epsilon_z \end{pmatrix}, \tag{20}$$

where

$$\frac{\epsilon_1}{\epsilon_0} = 1 - \frac{\omega_p^2}{\omega^2 - \omega_c^2}, \quad \frac{\epsilon_2}{\epsilon_0} = -\frac{\omega_c}{\omega}\frac{\omega_p^2}{\omega^2 - \omega_c^2}, \quad \frac{\epsilon_z}{\epsilon_0} = 1 - \frac{\omega_p^2}{\omega^2} \tag{20a}$$

and

$$\omega_p^2 = \frac{e^2 N}{m\epsilon_0}, \qquad \omega_c = \frac{eB_0}{m},$$

where $-e$ and m represent the electronic charge and mass, respectively. Only the electrons are considered mobile in this simple, zero temperature, collisionless description of a plasma, and ω_p and ω_c denote the plasma and cyclotron frequencies of the electrons, respectively. To derive the form of the permittivity matrix in Eq. (20), one notes for a time-harmonic field $\bar{\mathbf{E}} \exp(-i\omega t)$ that electrons with harmonic velocity $\bar{\mathbf{v}} \exp(-i\omega t)$ in a static magnetic field $\mathbf{B}_0 = B_0 \mathbf{z}_0$ obey the dynamical equation [see Eq. (1.1.54) with $p = 0$ and $\partial/\partial t = -i\omega$]

$$-i\omega m\bar{\mathbf{v}} = -e(\bar{\mathbf{E}} + \bar{\mathbf{v}} \times \mathbf{B}_0), \tag{21}$$

whence

$$(i\omega\mathbf{1} + \omega_c \mathbf{z}_0 \times \mathbf{1}) \cdot m\bar{\mathbf{v}} = e\bar{\mathbf{E}}. \tag{21a}$$

From the dyadic identities

$$(i\omega\mathbf{1} + \omega_c \mathbf{z}_0 \times \mathbf{1}) \cdot (i\omega\mathbf{1} - \omega_c \mathbf{z}_0 \times \mathbf{1}) = -(\omega^2 - \omega_c^2)\mathbf{1}_t - \omega^2 \mathbf{z}_0 \mathbf{z}_0,$$

$$-(\mathbf{z}_0 \times \mathbf{1}) \cdot (\mathbf{z}_0 \times \mathbf{1}) = 1 - \mathbf{z}_0 \mathbf{z}_0 = \mathbf{1}_t,$$

and Eq. (21a), the electron velocity components $\bar{\mathbf{v}}_t$ and $\bar{\mathbf{v}}_z$, transverse and along the magnetic field direction, contribute the current densities

$$-N e\bar{\mathbf{v}}_t = i\omega\epsilon_0 \left[\frac{\omega_p^2 \mathbf{1}_t + i(\omega_c/\omega)\omega_p^2 \mathbf{z}_0 \times \mathbf{1}}{\omega^2 - \omega_c^2}\right] \bar{\mathbf{E}}_t,$$

$$-N e\bar{\mathbf{v}}_z = i\omega\epsilon_0 \frac{\omega_p^2}{\omega^2} \bar{\mathbf{E}}_z. \tag{22}$$

Since the permittivity dyadic for the plasma medium may be defined by the relation

$$-N e\bar{\mathbf{v}} = -i\omega(\boldsymbol{\epsilon} - \epsilon_0 \mathbf{1}) \cdot \bar{\mathbf{E}},$$

one infers from Eqs. (22) and (20a) that

$$\boldsymbol{\epsilon} = \epsilon_1 \mathbf{1}_t + i\epsilon_2 \mathbf{z}_0 \times \mathbf{1}_t + \epsilon_z \mathbf{z}_0 \mathbf{z}_0,$$

which is the dyadic form of the matrix (20). For $[\partial(\omega\boldsymbol{\epsilon})/\partial\omega - \mathbf{1}\epsilon_0]$ to be positive semidefinite [see Eq. (19)], the principal minors of the permittivity matrix in Eq. (20) must be non-negative, whence

$$\frac{\partial}{\partial\omega}\left(\omega\frac{\epsilon_1}{\epsilon_0}\right) \geq 1, \quad \frac{\partial}{\partial\omega}\left(\omega\frac{\epsilon_z}{\epsilon_0}\right) \geq 1, \quad \left[\frac{\partial}{\partial\omega}\left(\omega\frac{\epsilon_1}{\epsilon_0}\right) - 1\right]^2 \geq \left[\frac{\partial}{\partial\omega}\left(\omega\frac{\epsilon_2}{\epsilon_0}\right)\right]^2 .$$

$$(23)$$

It is easily checked that these conditions are satisfied by the constitutive parameters in Eq. (20a), although the elements ϵ_1, ϵ_2, and ϵ_z may themselves be negative.

Average energy transport (group velocity)

For weakly anharmonic fields a kinetic description of the average flow of energy throughout parts of a field can be obtained in terms of wavepackets moving along characteristic trajectories with well-defined group velocity. It will be shown that in an homogeneous, non-spatially-dispersive medium an anharmonic field is characterized by a time-averaged Poynting vector $\bar{\mathbf{S}}$ [Eq. (8a)] and a time-averaged energy density $\bar{W}$ [Eq. (18)] that are related by

$$\bar{\mathbf{S}} = \bar{W}\mathbf{v}_g, \tag{24a}$$

where $\mathbf{v}_g$ is a group velocity characteristic of the energy flow (or of the type of wavepacket) at the point in question. In this case we shall find that one can infer an average energy theorem of the form (8). In view of Eq. (24a), it follows that $\nabla \cdot \bar{\mathbf{S}} = \mathbf{v}_g \cdot \nabla\bar{W}$ provided $\mathbf{v}_g$ is space independent, and Eq. (8) can be written as

$$\frac{\partial}{\partial t}\bar{W} + \mathbf{v}_g \cdot \nabla\bar{W} = 0, \tag{24b}$$

whence

$$\frac{d}{dt}\bar{W} = 0 \qquad \text{if } \mathbf{v}_g = \frac{d\mathbf{r}}{dt}. \tag{24c}$$

Equation (24c) implies that the average energy density $\bar{W}$ is unchanged when referred to a point moving with a wavepacket along a linear trajectory $\mathbf{r} = \mathbf{r}(t)$ with the as-yet-undetermined velocity $\mathbf{v}_g$. Since Eqs. (24) refer to an anharmonic field, $\bar{W}$ and hence $\bar{\mathbf{S}}$ are weakly space- and time-dependent even in a homogeneous medium; we shall show that they are expressible in terms of both the admittance operator $\mathscr{Y}$ of Eqs. (2) and the anharmonic field amplitudes. For inhomogeneous media, Eqs. (24) require generalization, as will be shown in Sec. 1.7.

To explore energy-transport properties under conditions for which Eqs. (24) are applicable, we turn to a consideration of anharmonic fields in lossless dispersive homogeneous media. In accordance with Eq. (2), the electric field $\mathbf{E}$

in an homogeneous medium is described at source-free points by an admittance operator $\mathscr{Y}$ such that

$$\mathscr{Y}\left(\nabla, \frac{\partial}{\partial t}\right) \cdot \mathbf{E}(\mathbf{r}, t) = 0. \tag{25}$$

We seek a slightly anharmonic (wavepacket) solution of Eq. (25) as the real part of

$$\mathbf{E}(\mathbf{r}, t) = \bar{\mathbf{E}}(\mathbf{r}, t)e^{i(\mathbf{k}\cdot\mathbf{r} - \omega t)}, \tag{26}$$

where for prescribed $\mathbf{k}$ both the weakly space- and time-dependent rms amplitude $\bar{\mathbf{E}}(\mathbf{r}, t)$ and the frequency $\omega = \omega(\mathbf{k})$ are to be determined from Eq. (25). In the following, $\bar{\mathbf{E}}$ and $\bar{\mathbf{H}}$ are weakly dependent on both $\mathbf{r}$ and t; they should be distinguished from the fields $\bar{\mathbf{E}} = \bar{\mathbf{E}}(\mathbf{r}; t)$ and $\bar{\mathbf{H}} = \bar{\mathbf{H}}(\mathbf{r}; t)$ in Eqs. (11) et seq., which depend arbitrarily on $\mathbf{r}$. From Eq. (25) and the footnote to Eqs. (8) the total average stored energy density at source-free points does not vary with time [i.e., the average power density Re $(\mathbf{E}^* \cdot \mathscr{Y} \cdot \mathbf{E}) = 0$ by Eq. (9) with Re $\mathscr{Y} = \sigma = 0$], whence

$$\mathbf{E}^*(\mathbf{r}, t) \cdot \mathscr{Y}\left(\nabla, \frac{\partial}{\partial t}\right) \cdot \mathbf{E}(\mathbf{r}, t) + \mathbf{E}(\mathbf{r}, t) \cdot \mathscr{Y}^*\left(\nabla, \frac{\partial}{\partial t}\right) \cdot \mathbf{E}^*(\mathbf{r}, t) = 0,$$

or on substitution of (26) and use of the differentiation rule † ($\omega, \mathbf{k}$ are real),

$$\bar{\mathbf{E}}^* \cdot \mathscr{Y}\left(i\mathbf{k} + \nabla, -i\omega + \frac{\partial}{\partial t}\right) \cdot \bar{\mathbf{E}} + \bar{\mathbf{E}} \cdot \mathscr{Y}^*\left(i\mathbf{k} + \nabla, -i\omega + \frac{\partial}{\partial t}\right) \cdot \bar{\mathbf{E}}^* = 0.$$
$$\tag{27a}$$

For a weakly space- and time-dependent amplitude $\bar{\mathbf{E}} \equiv \bar{\mathbf{E}}(\mathbf{r}, t)$, one can expand the operator $\mathscr{Y}$ of Eq. (27a) in a power series‡ about the point $i\mathbf{k}, -i\omega$, whence, for a lossless medium, wherein $\mathscr{Y}(i\mathbf{k}, -i\omega) = -i\mathscr{B}(\mathbf{k}, \omega)$ and $\mathscr{B} = \tilde{\mathscr{B}}$ is real, one infers from Eq.(27a) that§

$$\frac{\partial}{\partial t}\left[\bar{\mathbf{E}}^* \cdot \frac{\partial}{\partial \omega}\mathscr{B}(\mathbf{k}, \omega) \cdot \bar{\mathbf{E}}\right] - \nabla \cdot \nabla_k[\bar{\mathbf{E}}^* \cdot \mathscr{B}(\mathbf{k}, \omega) \cdot \bar{\mathbf{E}}] = 0, \tag{27b}$$

where ∇_k acts only on $\mathscr{B}(\mathbf{k}, \omega)$ and ∇ only on $\bar{\mathbf{E}}$ and $\bar{\mathbf{E}}^*$. To identify the bracketed terms in Eq. (27b), one observes from Eqs. (2) for a lossless homogeneous medium with a scalar permeability [see Eq. (10)] that

$$\mathscr{B}(\mathbf{k}, \omega) = \omega\boldsymbol{\epsilon} + \frac{\mathbf{k} \times (\mathbf{k} \times \mathbf{1})}{\omega\mu} \equiv \mathscr{B}, \tag{28a}$$

whence noting Eq.(18), one has as a generalization of a familiar network energy theorem:

---

†An operator function $F(\nabla, \partial/\partial t)$ of the derivatives $\nabla, \partial/\partial t$ acting on a product of two space–time functions A, B can be expressed as $F(\nabla, \partial/\partial t)A(\mathbf{r}, t)B(\mathbf{r}, t) = F(\nabla' + \nabla, \partial/\partial t' + \partial/\partial t)A(\mathbf{r}, t)B(\mathbf{r}', t')]_{r=r', t=t'}$, which permits separate differentiation of the two factors.

‡$\mathscr{Y}(i\mathbf{k} + \nabla, -i\omega + \partial/\partial t) = \mathscr{Y}(i\mathbf{k}, -i\omega) + i(\partial/\partial t)(\partial/\partial\omega)\mathscr{Y}(i\mathbf{k}, -i\omega) - i\nabla\cdot\nabla_k\mathscr{Y}(i\mathbf{k}, -i\omega) + \cdots$.

§Note the relation for the transpose of the scalar $\mathbf{a}\cdot\mathscr{F}\cdot\mathbf{b} = \mathbf{b}\cdot\tilde{\mathscr{F}}\cdot\mathbf{a}$.

$$\bar{\mathbf{E}}^* \cdot \frac{\partial \mathscr{B}}{\partial \omega} \cdot \bar{\mathbf{E}} = \bar{\mathbf{E}}^* \cdot \left[\frac{\partial(\omega \epsilon)}{\partial \omega} - \frac{\mathbf{k} \times (\mathbf{k} \times \mathbf{1})}{\omega^2 \mu^2} \frac{\partial}{\partial \omega}(\omega \mu) \right] \cdot \bar{\mathbf{E}}$$

$$= \bar{\mathbf{E}}^* \cdot \frac{\partial}{\partial \omega}(\omega \epsilon) \cdot \bar{\mathbf{E}} + \bar{\mathbf{H}}^* \cdot \frac{\partial}{\partial \omega}(\omega \mu) \bar{\mathbf{H}} = 2\bar{W}, \quad (28b)$$

since $\mathbf{k} \times \bar{\mathbf{E}} = \omega \mu \bar{\mathbf{H}}$. Similarly, by Eq. (28a) one finds in Eq. (27b), if ϵ is non-spatially dispersive, that

$$-\nabla \cdot \nabla_k [\bar{\mathbf{E}}^* \cdot \mathscr{B} \cdot \bar{\mathbf{E}}] = \nabla \cdot \nabla_k \left[\frac{(\mathbf{k} \times \bar{\mathbf{E}}^*) \cdot (\mathbf{k} \times \bar{\mathbf{E}})}{\omega \mu} \right]$$

$$= \nabla \cdot \left[\frac{\bar{\mathbf{E}}^* \times (\mathbf{k} \times \bar{\mathbf{E}}) + \bar{\mathbf{E}} \times (\mathbf{k} \times \bar{\mathbf{E}}^*)}{\omega \mu} \right]$$

$$= \nabla \cdot [\bar{\mathbf{E}}^* \times \bar{\mathbf{H}} + \bar{\mathbf{E}} \times \bar{\mathbf{H}}^*] = 2\nabla \cdot \bar{\mathbf{S}}, \quad (28c)$$

where we have used the identities $\nabla_k \mathbf{k} = \mathbf{1}$ and $\nabla_k(\mathbf{A} \cdot \mathbf{B}) = (\nabla_k \mathbf{A}) \cdot \mathbf{B} + (\nabla_k \mathbf{B} \cdot \mathbf{A})$. In view of the relations (28), Eq. (27b) assumes the form (8):

$$\frac{\partial \bar{W}}{\partial t} + \nabla \cdot \bar{\mathbf{S}} = 0.$$

This conservation of average energy flow for the anharmonic field in Eq. (26) is also applicable more generally to fields other than the electromagnetic.

It now remains to prove the energy-flux relation (24a). We observe from the source-free field equation (25) for a lossless homogeneous medium and the wavepacket solution in Eq. (26) that

$$\mathscr{B}(\mathbf{k}, \omega) \cdot \bar{\mathbf{E}} = 0, \quad (29a)$$

where $\mathscr{B}(\mathbf{k}, \omega)$ is defined in Eq. (28a). Non-vanishing solutions $\bar{\mathbf{E}}$ of Eq. (29a) exist for those frequencies $\omega = \omega_a(\mathbf{k})$ that satisfy the determinental requirement, or dispersion equation, det $\mathscr{B}(\mathbf{k}, \omega_a) = 0$. For an arbitrary incremental change $\delta \mathbf{k}$, it follows from Eq. (29a) that

$$\delta[\bar{\mathbf{E}}^* \cdot \mathscr{B}(\mathbf{k}, \omega_a) \cdot \bar{\mathbf{E}}] = \bar{\mathbf{E}}^* \cdot \delta \mathscr{B}(\mathbf{k}, \omega_a) \cdot \bar{\mathbf{E}} = 0,$$

and since (with $\delta \omega_a = \delta \mathbf{k} \cdot \nabla_k \omega_a$)

$$\delta \mathscr{B}(\mathbf{k}, \omega_a) = \delta \mathbf{k} \cdot \nabla_k \mathscr{B}(\mathbf{k}, \omega_a) + \delta \mathbf{k} \cdot \nabla_k \omega_a \frac{\partial \mathscr{B}}{\partial \omega}, \quad (29b)$$

one infers from the arbitrariness of $\delta \mathbf{k}$ that

$$\nabla_k [\bar{\mathbf{E}}^* \cdot \mathscr{B} \cdot \bar{\mathbf{E}}] + \nabla_k \omega_a \left[\bar{\mathbf{E}}^* \cdot \frac{\partial \mathscr{B}}{\partial \omega} \cdot \bar{\mathbf{E}} \right], \quad (29c)$$

whence from Eqs. (28b) and (28c) one obtains the result asserted in Eq. (24a),

$$\bar{\mathbf{S}} = \bar{W} \mathbf{v}_g, \qquad \mathbf{v}_g = \nabla_k \omega_a \quad (30)$$

where $\mathbf{v}_g$ is the group velocity. Since the vanishing of det $\mathscr{B}(\mathbf{k}, \omega)$ occurs for different dispersion relations $\omega = \omega_a(\mathbf{k})$, the energy flux relation in Eq. (30) and hence the group velocity $\mathbf{v}_g = \nabla_k \omega_a$ are characteristic of different types $(a = 1, 2, \ldots)$ of wavepackets.

1.5b Boundary Conditions, Uniqueness, and Reciprocity Relations for the Electromagnetic Field

Boundary conditions and uniqueness

In a material medium the field equations may be expressed either in the first-order form discussed in Sec. 1.1d or in the reduced electromagnetic form noted, for example, in Eqs. (1.1.64) or (1). In the presence of an externally applied magnetic field, the permittivity and permeability are generally dyadic parameters. For time-harmonic fields, which vary as $\exp(-i\omega t)$, and for a spatially inhomogeneous medium with non-spatially-dispersive parameters, the permittivity and permeability will be expressed as complex dyadics $\boldsymbol{\epsilon}(\mathbf{r}, \omega)$ and $\boldsymbol{\mu}(\mathbf{r}, \omega)$. Under such conditions, one obtains for the steady-state form of the reduced Maxwell equations (1) descriptive of the electromagnetic field excited by prescribed electric and magnetic current distributions $\mathbf{J}$ and $\mathbf{M}$ in a temporally dispersive anisotropic medium:

$$\nabla \times \mathbf{E}(\mathbf{r}) = i\omega\mathbf{B}(\mathbf{r}) - \mathbf{M}(\mathbf{r}), \qquad \nabla \times \mathbf{H}(\mathbf{r}) = -i\omega\mathbf{D}(\mathbf{r}) + \mathbf{J}(\mathbf{r}), \qquad (31a)$$

$$\mathbf{B}(\mathbf{r}) = \boldsymbol{\mu}(\mathbf{r}) \cdot \mathbf{H}(\mathbf{r}), \qquad\qquad \mathbf{D}(\mathbf{r}) = \boldsymbol{\epsilon}(\mathbf{r}) \cdot \mathbf{E}(\mathbf{r}), \qquad (31b)$$

where the ω dependence and the factor $\exp(-i\omega t)$ have been suppressed. Special cases of lossless media, isotropic media with $\boldsymbol{\epsilon} = \mathbf{1}\epsilon, \boldsymbol{\mu} = \mathbf{1}\mu$, etc. are contained within this formulation.

Across a regular surface S of a medium discontinuity, the electromagnetic field vectors must satisfy the conditions

$$\mathbf{v} \times (\mathbf{H}_2 - \mathbf{H}_1) = \mathbf{J}_s, \qquad \mathbf{v} \times (\mathbf{E}_1 - \mathbf{E}_2) = \mathbf{M}_s, \qquad (32a)$$

$$\mathbf{v} \cdot (\mathbf{B}_1 - \mathbf{B}_2) = \eta_m, \qquad \mathbf{v} \cdot (\mathbf{D}_2 - \mathbf{D}_1) = \eta, \qquad (32b)$$

where the subscripts 1 and 2 distinguish quantities in regions 1 and 2 of Fig. 1.5.1. On the interface, $\mathbf{J}_s$ and $\mathbf{M}_s$ are, respectively, electric and magnetic surface current distributions, η and η_m are electric and magnetic surface charge

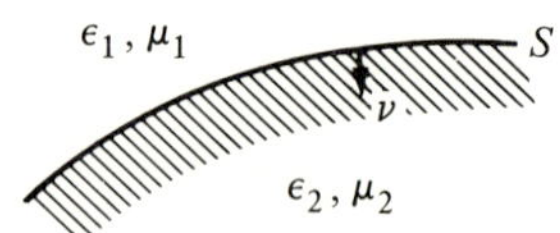

FIG. 1.5.1 Interface between two media.

densities, and $\mathbf{v}$ is a unit vector normal to S and pointing into medium 2. If neither medium is perfectly conducting, no surface currents or charges are induced and $\mathbf{J}_s = \mathbf{M}_s = \eta = \eta_m = 0$; however, if *external* surface current distributions are placed on the interface, these must be taken into account. When medium 2 is a perfect electric conductor, $\mathbf{E}_2$ and $\mathbf{H}_2$ vanish everywhere inside this medium and induced electric currents and charges exist on the surface. Thus, from Eqs. (32),

$$\mathbf{H}_1 \times \mathbf{v} = \mathbf{J}_s, \quad \mathbf{v} \times \mathbf{E}_1 = 0, \quad \mathbf{v} \cdot \mathbf{B}_1 = 0, \quad -\mathbf{D}_1 \cdot \mathbf{v} = \eta. \qquad (33)$$

While the tangential electric field vanishes at the surface of a perfect conductor, one notes from Eq. (32a) that a finite field can be generated next to the surface by *placing* thereon a magnetic surface current distribution. In fact, if the magnetic current distribution is $\mathbf{M}_s$, the tangential electric field generated by it in the immediate vicinity of a perfectly conducting surface is given by

$$\mathbf{v} \times \mathbf{E}_1 = \mathbf{M}_s. \tag{33a}$$

Although physically realizable electromagnetic sources can be described solely in terms of electric charges and currents, the use of equivalent magnetic currents is frequently a convenient artifice; for example, a linear magnetic current element in an isotropic medium is equivalent to a circular electric current flowing around a path of vanishingly small radius in the plane perpendicular to the element. Since Eqs. (32) and (33) apply to any time-harmonic field, the Fourier inversion $f(\mathbf{r}, t) = \int_{-\infty}^{\infty} f(\mathbf{r}, \omega)e^{-i\omega t}\, d\omega$ implies that they are valid also for the time-dependent field.

For an unbounded region it is necessary to know the field behavior on a surface at infinity. If all sources are contained in a finite region, the field behavior at large distances from the sources must meet the physical requirement that energy travel *away* from the source region (i.e., the field solution comprises only "outgoing" waves). This requirement, originated by Sommerfeld, constitutes a boundary condition on a surface at infinity and is referred to as the "radiation condition."[10] More specifically, the transverse fields in a spherically diverging wave in homogeneous medium decay like $1/r$ at large distances r from the source region and behave locally like plane waves traveling outward in the r direction: thus each field component transverse to r behaves like $\exp(ik_n r)/r$. This requirement can be phrased mathematically as[†]

$$\lim_{r \to \infty} r\left(\frac{\partial A}{\partial r} - ik_n A\right) = 0 \tag{34a}$$

where A stands for any field component transverse to r, and $k_n = \omega\sqrt{\mu\epsilon}$ approaches a constant as $r \to \infty$. In view of the local plane-wave character, the relation between the transverse electric and magnetic fields as $r \to \infty$ is simply

$$\mathbf{r}_0 \times \mathbf{E} \to \sqrt{\frac{\mu}{\epsilon}}\, \mathbf{H}, \tag{34b}$$

where $\mathbf{r}_0$ is a unit vector pointing in the $\mathbf{r}$ direction and $\sqrt{\mu/\epsilon}$ is the impedance of unbounded space.

For two-dimensional fields that are independent of a rectilinear coordinate, say x, the outgoing waves excited by source distributions contained in a bounded

---

[†]The present discussion is restricted to the case of isotropic media; the *energy*-radiation condition applies also to anisotropic regions, but its imposition is more complicated since the directions of phase- and energy-propagation are generally different (see Secs. 1.6 and 1.7).

region transverse to x are cylindrical [i.e., each field component behaves like $\exp(ik_n\rho)/\sqrt{\rho}$, where ρ is the radial variable in a plane transverse to the x coordinate]. In this instance, one obtains as the two-dimensional analogue of Eq. (34a),

$$\lim_{\rho \to \infty} \sqrt{\rho}\left(\frac{\partial A}{\partial \rho} - ik_n A\right) = 0, \tag{34c}$$

where A stands for any transverse field component. Equation (34b) obtains also, with $\mathbf{r}_0$ replaced by $\boldsymbol{\rho}_0$.

Equations (34a) and (34c) apply to non-dissipative media, wherein k_n is real. If one assumes the medium to be slightly lossy (as all physical media are), then k_n has a positive imaginary part and the "outgoing wave" condition at infinity can be replaced by the simpler requirement that all fields excited by sources in a finite region vanish at infinity. This simple condition is valid also for anisotropic and inhomogeneous media and forces the choice of a decaying dependence as $r \to \infty$. For non-harmonic time dependence, the radiation condition is subsumed under the causality condition required for time-dependent fields. On analytic continuation into the upper half of the complex ω plane, the causal requirement Im $\omega > 0$ (see Sec. 1.2) may also be used to select the decaying field solution.

In the discussion of the boundary conditions in Eqs. (32) and (33), it was implied that the discontinuity surface S is regular (i.e., it possesses no sharp points, edges, or corners). If geometrical singularities exist on an electrically impenetrable surface, some field components may become infinite in the neighborhood of these singularities. To explore permissible types of divergence in an electromagnetic field, one employs the following energy argument: The energy inside a bounded volume containing any physical set of sources is finite. The singular electromagnetic field behavior near a surface discontinuity on an impenetrable surface such as a perfect conductor is caused by *induced* surface currents whose energy content cannot be larger than that of the true sources. Thus, one imposes the condition that the electric and magnetic energy stored in any volume surrounding the geometrical singularities is finite. In an isotropic non-dispersive medium this requirement can be phrased as[†]

$$\int_\tau [\epsilon|\mathbf{E}|^2 + \mu|\mathbf{H}|^2]\,d\tau = \text{finite}, \tag{35}$$

where τ is a finite volume surrounding the singularity.

Equation (35) puts a definite limitation on the growth of any field component. For example, for the edge singularity in Fig. 1.5.2(a) one chooses a cylindrical volume centered on the edge; if the edge is described by a regular curve, the height of the cylinder is chosen so small that the portion of the edge contained therein is essentially straight. Upon employing cylindrical coordinates, ρ, φ, z, with z directed along the edge, one has $d\tau = \rho\,d\rho\,d\varphi\,dz$. If it is assumed that

---

†See Sec. 1.5a for the effects of dispersion and anisotropy.

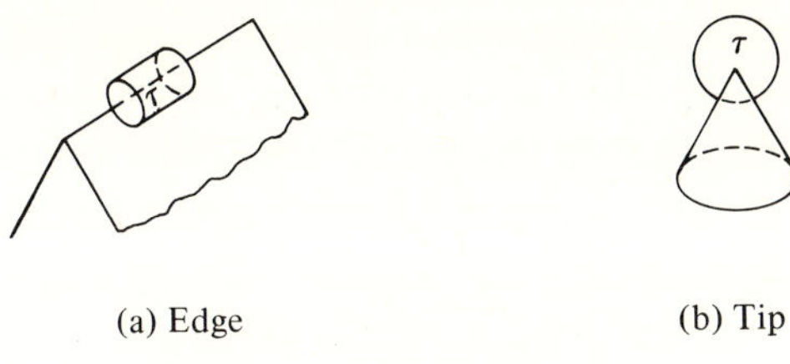

(a) Edge   (b) Tip

FIG. 1.5.2 Surface singularities.

each field component behaves like $f(\rho)\psi(\varphi, z)$ as the edge is approached (i.e., as $\rho \to 0^\dagger$), then condition (35) can be phrased in terms of the ρ behavior alone since the fields are regular functions of φ and z. The pertinent portion of Eq. (35) can be written as (R = radius of cylindrical volume)

$$\int_0^R |f(\rho)|^2 \rho \, d\rho = \text{finite}, \tag{36}$$

from which one infers that $|f(\rho)|^2$ can behave at worst like $\rho^{-2(1-\alpha)}$ as $\rho \to 0$, where α is an arbitrarily small positive quantity. Thus, we infer the "edge condition" that

no component of **E** or **H** can grow more rapidly than $\rho^{-1+\alpha}$, $\alpha > 0$, (37)

where ρ is the distance from the edge.

Similarly, for a tip singularity as in Fig. 1.5.2(b), one surrounds the tip by a spherical volume of radius R. In a spherical (r, θ, φ) coordinate system centered at the tip, the volume element $d\tau = r^2 \sin \theta \, dr \, d\theta \, d\varphi$. If one assumes that the behavior of each field component can be described in the form $g(r)\psi(\theta, \varphi)$ as $r \to 0$, one can again write the pertinent portion of Eq. (35) as

$$\int_0^R |g(r)|^2 r^2 \, dr = \text{finite}, \tag{38}$$

i.e., $|g(r)|^2$ behaves at most like $r^{-3+2\alpha}, \alpha > 0$. Thus, one finds that near a conical tip,

no component of **E** or **H** can grow more rapidly than $r^{-(3/2)+\alpha}$, $\alpha > 0$, (39)

where r is the distance from the tip. The requirements (37) and (39) constitute additional boundary conditions to be imposed on the fields in the presence of surface singularities in order to assure a unique solution of the field problem.

When solving the Maxwell field equations subject to all the above-stated boundary conditions, one is assured that only one (unique) solution is possible. The following uniqueness statement, given without proof, is relevant for a volume of space τ bounded by a surface S (which may be multiply connected) and occupied by a "physical" medium having some dissipation: The steady-state electromagnetic field inside τ is determined uniquely if [2b] (1) the sources (if any) are specified in τ, and (2) either the tangential electric field or the tangential

---

$\dagger$This behavior can be shown to hold rigorously for a straight edge (see Sec. 6.5).

magnetic field is specified on the various parts of S. Included in condition (2) is the case where the tangential electric field on the boundary is linearly related to the tangential magnetic field via [see Eq. (1.1.16d)]

$$\mathbf{v} \times \mathbf{E} = \mathscr{L} \cdot \mathbf{H}, \qquad \mathbf{v} = \text{normal unit vector directed into boundary}, \qquad (40)$$

where $\mathscr{L}$ is an impedance dyadic with non-vanishing components perpendicular to $\mathbf{v}$ (i.e., $\mathbf{v} \cdot \mathscr{L} = \mathscr{L} \cdot \mathbf{v} = 0$). Equation (40) constitutes an "impedance boundary condition" that can be employed to approximate field behavior on the surface of highly lossy or "reactive" regions. The above conditions remain valid when the medium is anisotropic. If a portion of S lies at infinity and all sources are confined to a finite region, uniqueness obtains when the field behavior at infinity is expressed in terms of the previously stated radiation condition. For surfaces with singularities, condition (35) must be imposed, in addition.

If the fields are time dependent, their values at time $t > t_1$ are determined uniquely by their initial values at the reference time t_1. The above-stated conditions on S must also be satisfied for $t \geq t_1$.

Reciprocity relations

Reciprocity relations for the electromagnetic field in vacuum and in a homogeneous plasma have been considered in Secs. 1.1b and 1.1c. Apart from the assumption of medium homogeneity, that derivation involves all field constituents rather than only the electromagnetic field, as in the reduced formulation leading to Eqs.(31). Furthermore, the results have been given for the time-dependent case so that their time-harmonic form, although obtainable without difficulty, is not stated explicitly. We shall now derive various reciprocity conditions for the time-harmonic electromagnetic field in spatially varying anisotropic media directly from Eqs. (31) and compare the results with those given in Sec. 1.1.

In a given region τ bounded by the surface S, consider two sets of field solutions $\mathbf{E}, \mathbf{H}$ and $\hat{\mathbf{E}}, \hat{\mathbf{H}}$, corresponding to excitations by the sources $\mathbf{J}, \mathbf{M}$, and $\hat{\mathbf{J}}, \hat{\mathbf{M}}$, respectively. The medium in the first problem is assumed to be described by the space-dependent dyadic permittivity $\boldsymbol{\epsilon}(\mathbf{r})$ and permeability $\boldsymbol{\mu}(\mathbf{r})$, with $\hat{\boldsymbol{\epsilon}}(\mathbf{r})$ and $\hat{\boldsymbol{\mu}}(\mathbf{r})$ representing the corresponding quantities in the second problem. At the real steady frequency ω, each set of fields is therefore a solution of the Maxwell equations†

$$\nabla \times \mathbf{E} = i\omega\boldsymbol{\mu} \cdot \mathbf{H} - \mathbf{M}, \qquad \nabla \times \mathbf{H} = -i\omega\boldsymbol{\epsilon} \cdot \mathbf{E} + \mathbf{J}, \qquad (41a)$$

$$\nabla \times \hat{\mathbf{E}} = i\omega\hat{\boldsymbol{\mu}} \cdot \hat{\mathbf{H}} - \hat{\mathbf{M}}, \qquad \nabla \times \hat{\mathbf{H}} = -i\omega\hat{\boldsymbol{\epsilon}} \cdot \hat{\mathbf{E}} + \hat{\mathbf{J}}. \qquad (41b)$$

Upon calculating the expressions $\nabla \cdot (\mathbf{E} \times \hat{\mathbf{H}}) = (\hat{\mathbf{H}} \cdot \nabla \times \mathbf{E} - \mathbf{E} \cdot \nabla \times \hat{\mathbf{H}})$ and $\nabla \cdot (\hat{\mathbf{E}} \times \mathbf{H})$, taking their difference, integrating over τ, and invoking the divergence theorem, one finds

---

†Quantities distinguished by $\hat{\ }$ are related to the adjoint field in Sec. 1.1 on identifying $\hat{\mathbf{J}} = -\mathbf{J}^+$ and $\hat{\mathbf{M}} = -\mathbf{M}^+$.

$$\oint_{s} [\mathbf{n} \times \mathbf{E} \cdot \hat{\mathbf{H}} - \mathbf{n} \times \hat{\mathbf{E}} \cdot \mathbf{H}] \, dS$$

$$- i\omega \int_{\tau} [\hat{\mathbf{H}} \cdot \boldsymbol{\mu} \cdot \mathbf{H} - \mathbf{H} \cdot \hat{\boldsymbol{\mu}} \cdot \hat{\mathbf{H}} - \mathbf{E} \cdot \hat{\boldsymbol{\epsilon}} \cdot \hat{\mathbf{E}} + \hat{\mathbf{E}} \cdot \boldsymbol{\epsilon} \cdot \mathbf{E}] \, d\tau$$

$$= \int_{\tau} [\mathbf{H} \cdot \hat{\mathbf{M}} + \hat{\mathbf{E}} \cdot \mathbf{J} - \hat{\mathbf{H}} \cdot \mathbf{M} - \mathbf{E} \cdot \hat{\mathbf{J}}] \, d\tau. \tag{42}$$

The reciprocity statement follows if the right-hand side of Eq. (42) can be equated to zero. Since generally $\hat{\mathbf{H}} \cdot \boldsymbol{\mu} \cdot \mathbf{H} = \mathbf{H} \cdot \tilde{\boldsymbol{\mu}} \cdot \hat{\mathbf{H}}$, etc., the volume integral on the left-hand side vanishes if the medium in the second problem is the "transpose" of that in the first, i.e., if

$$\hat{\boldsymbol{\epsilon}}(\mathbf{r}) = \tilde{\boldsymbol{\epsilon}}(\mathbf{r}), \qquad \hat{\boldsymbol{\mu}}(\mathbf{r}) = \tilde{\boldsymbol{\mu}}(\mathbf{r}). \tag{43a}$$

If the boundary conditions on S are given by the general impedance relation (40) and the corresponding requirement $\mathbf{v} \times \hat{\mathbf{E}} = \hat{\mathscr{L}} \cdot \hat{\mathbf{H}}$ on S for the second problem, then the identification

$$\hat{\mathscr{L}}(\mathbf{r}) = \tilde{\mathscr{L}}(\mathbf{r}) \tag{43b}$$

assures the vanishing of the surface integral.† Thus, if Eqs. (43) apply, the fields $\mathbf{E}, \mathbf{H}$ and $\hat{\mathbf{E}}, \hat{\mathbf{H}}$ are related as follows:

$$\int_{\tau} [\mathbf{H} \cdot \hat{\mathbf{M}} + \hat{\mathbf{E}} \cdot \mathbf{J} - \hat{\mathbf{H}} \cdot \mathbf{M} - \mathbf{E} \cdot \hat{\mathbf{J}}] \, d\tau = 0. \tag{44}$$

The reciprocity relations in Eq. (44) are clarified by special choices of the excitation. If one selects $\mathbf{M} = \hat{\mathbf{M}} = 0$, $\mathbf{J} = \mathbf{J}^{\circ}\delta(\mathbf{r} - \mathbf{r}')$, $\hat{\mathbf{J}} = \hat{\mathbf{J}}^{\circ}\delta(\mathbf{r} - \hat{\mathbf{r}}')$, where $\mathbf{J}^{\circ}$ and $\hat{\mathbf{J}}^{\circ}$ are constant vectors of equal magnitude, then

$$\mathbf{J}^{\circ} \cdot \hat{\mathbf{E}}(\mathbf{r}') = \hat{\mathbf{J}}^{\circ} \cdot \mathbf{E}(\hat{\mathbf{r}}'). \tag{45a}$$

This relation states that in a region described by the parameters $\boldsymbol{\epsilon}, \boldsymbol{\mu}, \mathscr{L}$, the electric-field component excited at $\hat{\mathbf{r}}'$ in the direction of $\hat{\mathbf{J}}^{\circ}$ by an electric current element $\mathbf{J}^{\circ}$ at $\mathbf{r}'$ is identical with the electric-field component excited at $\mathbf{r}'$ in the direction of $\mathbf{J}^{\circ}$ by an electric current element $\hat{\mathbf{J}}^{\circ}$ at $\hat{\mathbf{r}}'$ in a region of the same geometrical configuration described by parameters $\tilde{\boldsymbol{\epsilon}}, \tilde{\boldsymbol{\mu}}, \tilde{\mathscr{L}}$. Similarly, for $\mathbf{J} = \hat{\mathbf{J}} = 0$, $\mathbf{M} = \mathbf{M}^{\circ}\delta(\mathbf{r} - \mathbf{r}')$, $\hat{\mathbf{M}} = \hat{\mathbf{M}}^{\circ}\delta(\mathbf{r} - \hat{\mathbf{r}}')$, where $\mathbf{M}^{\circ}$ and $\hat{\mathbf{M}}^{\circ}$ are constant vectors of equal magnitude,

$$\mathbf{M}^{\circ} \cdot \hat{\mathbf{H}}(\mathbf{r}') = \hat{\mathbf{M}}^{\circ} \cdot \mathbf{H}(\hat{\mathbf{r}}'), \tag{45b}$$

while for $\mathbf{J} = \hat{\mathbf{M}} = 0$, $\quad \hat{\mathbf{J}} = \hat{\mathbf{J}}^{\circ}\delta(\mathbf{r} - \hat{\mathbf{r}}')$, $\quad \mathbf{M} = \mathbf{M}^{\circ}\delta(\mathbf{r} - \mathbf{r}')$,

$$\mathbf{M}^{\circ} \cdot \hat{\mathbf{H}}(\mathbf{r}') = -\hat{\mathbf{J}}^{\circ} \cdot \mathbf{E}(\hat{\mathbf{r}}'). \tag{45c}$$

Although the above reciprocity relations generally involve fields in the "transposed" problem, they pertain to fields in the *same* medium provided the

---

†If a portion of S lies at infinity, the vanishing of the surface integral there is assured by the radiation condition. If S is a perfectly conducting surface containing surface singularities such as edges or tips, one surrounds these singularities by the closed surface S' and applies the edge condition to secure vanishing of the surface-integral contribution from S' as τ' shrinks to zero.

anisotropy, if any, is of the symmetric type [i.e., $\epsilon(\mathbf{r}) = \tilde{\epsilon}(\mathbf{r})$, $\mu(\mathbf{r}) = \tilde{\mu}(\mathbf{r})$, $\mathscr{L}(\mathbf{r}) = \tilde{\mathscr{L}}(\mathbf{r})$]. The latter constraints are, of course, satisfied in isotropic regions. Simplifications occur also in lossless regions wherein $\epsilon = \tilde{\epsilon}^* \equiv \epsilon^+$, $\mu = \mu^+$, $\mathscr{L} = -\mathscr{L}^+$ [see Eq. (10)]. If the complex conjugate of Eqs. (41b) is employed, with $\hat{\epsilon}$ and $\hat{\mu}$ replaced by ϵ and μ, respectively, and the preceding derivation followed with minor modification, one finds in view of the above lossless relation that

$$\hat{\mathbf{J}}^{\circ *} \cdot \mathbf{E}(\hat{\mathbf{r}}') = -\mathbf{J}^{\circ} \cdot \hat{\mathbf{E}}^*(\mathbf{r}'), \qquad \hat{\mathbf{M}}^{\circ *} \cdot \mathbf{H}(\hat{\mathbf{r}}') = -\mathbf{M}^{\circ} \cdot \hat{\mathbf{H}}^*(\mathbf{r}'),$$
$$\hat{\mathbf{J}}^{\circ *} \cdot \mathbf{E}(\hat{\mathbf{r}}') = -\mathbf{M}^{\circ} \cdot \hat{\mathbf{H}}^*(\mathbf{r}'), \tag{45d}$$

with all fields and sources referring to the same lossless region. It should be noted that the results in Eq. (45d) generalize the reciprocity statements in Sec. 1.1b for time-dependent fields, in vacuum.

As in Eqs. (1.1.19) with $dr' \equiv d\tau'$, one may define time-harmonic Green's functions that represent electromagnetic fields arising from arbitrary harmonic current distributions,

$$\mathbf{E}(\mathbf{r}) = -\int_\tau \mathscr{G}_{11}(\mathbf{r}, \mathbf{r}') \cdot \mathbf{J}(\mathbf{r}')\, d\tau' - \int_\tau \mathscr{G}_{12}(\mathbf{r}, \mathbf{r}') \cdot \mathbf{M}(\mathbf{r}')\, d\tau', \tag{46a}$$

$$\mathbf{H}(\mathbf{r}) = -\int_\tau \mathscr{G}_{21}(\mathbf{r}, \mathbf{r}') \cdot \mathbf{J}(\mathbf{r}')\, d\tau' - \int_\tau \mathscr{G}_{22}(\mathbf{r}, \mathbf{r}') \cdot \mathbf{M}(\mathbf{r}')\, d\tau', \tag{46b}$$

where $\mathscr{G}_{11}(\mathbf{r}, \mathbf{r}')$, $\mathscr{G}_{22}(\mathbf{r}, \mathbf{r}')$ and $\mathscr{G}_{12}(\mathbf{r}, \mathbf{r}')$ or $\mathscr{G}_{21}(\mathbf{r}, \mathbf{r}')$ are the dyadic electric, magnetic, and "transfer" Green's functions, respectively. The interpretation of the various $\mathscr{G}_{ij}$ is the same as in Eq. (1.1.19), with evident modifications to accommodate the time-harmonic regime. It follows from Eqs. (31), with $\mathbf{M} = 0$, $\mathbf{J}(\mathbf{r}) = \mathbf{J}^{\circ}\delta(\mathbf{r} - \mathbf{r}')$, that $\mathscr{G}_{11}(\mathbf{r}, \mathbf{r}')$ satisfies the vector differential equation

$$\nabla \times [\mu^{-1}(\mathbf{r}) \cdot \nabla \times \mathscr{G}_{11}(\mathbf{r}, \mathbf{r}')] - \omega^2\epsilon(\mathbf{r}) \cdot \mathscr{G}_{11}(\mathbf{r}, \mathbf{r}') = -i\omega\mathbf{1}\delta(\mathbf{r} - \mathbf{r}'), \tag{47a}$$

were μ^{-1} is the inverse dyadic defined so that $\mu^{-1} \cdot \mu = \mu \cdot \mu^{-1} = \mathbf{1}$, and the constant vector $\mathbf{J}^{\circ}$ has been elided. By dual considerations one finds that

$$\nabla \times [\epsilon^{-1}(\mathbf{r}) \cdot \nabla \times \mathscr{G}_{22}(\mathbf{r}, \mathbf{r}')] - \omega^2\mu(\mathbf{r}) \cdot \mathscr{G}_{22}(\mathbf{r}, \mathbf{r}') = -i\omega\mathbf{1}\delta(\mathbf{r} - \mathbf{r}'). \tag{47b}$$

Equations (47) constitute a generalization of Eqs. (1.1.21), for the harmonic case $\partial/\partial t = -i\omega$, to inhomogeneous and anisotropic media.

As in Sec. 1.1b, the reciprocity conditions in Eqs. (45) can be phrased concisely in terms of the above dyadic Green's functions. In view of the definition of $\mathscr{G}_{11}$ in Eq. (46a) and similarly of $\hat{\mathscr{G}}_{11}$, Eq. (45a) can be written as

$$\mathbf{J}^{\circ} \cdot \hat{\mathscr{G}}_{11}(\mathbf{r}', \hat{\mathbf{r}}') \cdot \hat{\mathbf{J}}^{\circ} = \hat{\mathbf{J}}^{\circ} \cdot \mathscr{G}_{11}(\hat{\mathbf{r}}', \mathbf{r}') \cdot \mathbf{J}^{\circ} = \mathbf{J}^{\circ} \cdot \tilde{\mathscr{G}}_{11}(\hat{\mathbf{r}}', \mathbf{r}') \cdot \hat{\mathbf{J}}^{\circ}, \tag{48}$$

from which

$$\hat{\mathscr{G}}_{11}(\mathbf{r}', \hat{\mathbf{r}}') = \tilde{\mathscr{G}}_{11}(\hat{\mathbf{r}}', \mathbf{r}'). \tag{49a}$$

In a directly analogous manner one notes from Eqs. (45b, c) and the above that

$$\hat{\mathscr{G}}_{22}(\mathbf{r}', \hat{\mathbf{r}}') = \tilde{\mathscr{G}}_{22}(\hat{\mathbf{r}}', \mathbf{r}'), \tag{49b}$$

$$-\hat{\mathscr{G}}_{21}(\mathbf{r}', \hat{\mathbf{r}}') = \tilde{\mathscr{G}}_{12}(\hat{\mathbf{r}}', \mathbf{r}'). \tag{49c}$$

From Maxwell's equations and Eqs. (49) the *total* electromagnetic field in a source-free region can be inferred from the knowledge of *any one* of the dyadic Green's functions. Suppose that we know $\mathscr{G}_{11}(\mathbf{r}, \mathbf{r}')$. From the Maxwell field equations one obtains $\mathscr{G}_{21}(\mathbf{r}, \mathbf{r}')$ in terms of $\nabla \times \mathscr{G}_{11}(\mathbf{r}, \mathbf{r}')$. Equation (49c) then yields $\mathscr{G}_{12}(\mathbf{r}, \mathbf{r}')$, and the Maxwell field equations finally lead to $\mathscr{G}_{22}(\mathbf{r}, \mathbf{r}')$ in terms of $\nabla \times \mathscr{G}_{12}(\mathbf{r}, \mathbf{r}')$.

In a region with symmetric anisotropy $\boldsymbol{\epsilon} = \tilde{\boldsymbol{\epsilon}}, \boldsymbol{\mu} = \tilde{\boldsymbol{\mu}}, \mathscr{L} = \tilde{\mathscr{L}}$, or in an isotropic medium, the Green's functions on the left-hand side of Eqs. (49) are replaced by $\mathscr{G}_{11}(\mathbf{r}', \hat{\mathbf{r}}'), \mathscr{G}_{22}(\mathbf{r}', \hat{\mathbf{r}}'), \mathscr{G}_{21}(\mathbf{r}', \hat{\mathbf{r}}')$, respectively, thereby making the reciprocity statement applicable in one and the same region. The resulting relations are of the same form as in Eqs. (1.1.29) or (1.1.58), on ignoring the time dependence. These considerations apply also to lossless regions with $\boldsymbol{\epsilon} = \boldsymbol{\epsilon}^+, \boldsymbol{\mu} = \boldsymbol{\mu}^+$, and $\mathscr{L} = -\mathscr{L}^+$, wherein one replaces the Green's functions on the left-hand side of Eqs. (49) by $-\mathscr{G}_{11}^*(\mathbf{r}', \hat{\mathbf{r}}'), -\mathscr{G}_{22}^*(\mathbf{r}', \hat{\mathbf{r}}')$, and $\mathscr{G}_{21}^*(\mathbf{r}', \hat{\mathbf{r}}')$, respectively.

Reciprocity conditions analogous to the above are also found to be satisfied by modal Green's functions which represent the voltages or currents excited by unit strength point current or voltage generators on a transmission line. These results are derived in Sec. 2.3c and listed in Eqs. (2.3.15).

1.5c Alternative Representations

When z is a symmetry axis, the guided-wave approach in Sec. 1.4 leads naturally to a field representation in terms of eigenfunctions characteristic of the cross-sectional domain transverse to z. If the cross-sectional region is describable simply in terms of a separable (u, v) coordinate system, it is possible to deduce alternative representations involving eigenfunctions in the (u, z) or (v, z) coordinates. Since alternative field representations usually have different convergence properties, their availability aids in the evaluation of formal field solutions for various parameter regimes. The basic theory underlying the determination of alternative representations is given in Sec. 3.3, and detailed applications to electromagnetic or acoustic fields may be found in Chapter 5. At this time, we make certain observations within the context of problems discussed in this chapter.

Field representations developed in Sec. 1.4 [see also Eqs. (1.1.49) et seq.] for homogeneous unbounded regions can readily be modified to accommodate regions that contain inhomogeneities along the wave-guide axis z, or are bounded in the directions transverse to z. Although the reduction of vector field problems to equivalent scalar problems in such regions usually requires assignment of a special role to the symmetry axis z, the component scalar problems are not necessarily so restricted. For the electromagnetic field, this aspect may be inferred from Eqs. (1.1.38b) and (1.1.49c), wherein the scalar potential functions $\mathscr{S}'$ and $\mathscr{S}''$ satisfy differential equations which, subject to stated boundary conditions, may be solved in any convenient representation. The same statement evidently applies to the acoustic field which is derivable from the scalar Green's function defined in Eq. (1.1.13b), subject to suitable boundary conditions.

A number of alternative field representations have already been given in connection with the development of the plane-wave, time-guided, and space-guided wave representations of Green's functions in Secs. 1.2, 1.3, and 1.4, respectively. It has been noted that the time-guided representations in Eq. (1.3.2b) involve eigenfunctions in the **r** domain; they can be derived from the plane-wave integral representations in Eq. (1.2.3) on performing the ω integration by deforming the integration contour in the complex ω plane about the ω singularities of the Green's function $G(\mathbf{k}, \omega)$, and invoking the residue theorem. Alternatively, the space-guided representations in Eq. (1.4.3) involve eigenfunctions in the (ρ, t)domain and result from those in Eq. (1.2.3) on doing the k_z integration; one deforms the integration contour in the complex k_z plane about the k_z singularities of $G(\mathbf{k}, \omega)$, with subsequent use of the residue theorem. It is evident from these considerations that the plane-wave representation is quite general (although not necessarily most convenient for field evaluations) since either of the guided representations may be derived from it. To pass from one guided representation to another, one first constructs the fourfold integral representation of a field in terms of the Green's function $G(\mathbf{k}, \omega)$, and then eliminates the undesired integration variable in the manner noted. By an alternative procedure, based on "characteristic" Green's functions, guided-wave representations are derived without intervention of an additional integration.

To illustrate these techniques, we consider the one-dimensional version of the scalar potential problem for the isotropic plasma field defined in Eq. (1.1.60b):

$$\left[\frac{\partial^2}{\partial z^2} - \frac{1}{u^2}\left(\frac{\partial^2}{\partial t^2} + \omega_p^2\right)\right]g(z, z'; t, t') = -\delta(z - z')\delta(t - t'), \qquad (50)$$

subject to the causality requirement $g \equiv 0$ for $t < t'$. By the plane-wave integral representation [see Eq. (1.2.31)]

$$g(z, z'; t, t') = \frac{1}{(2\pi)^2}\int\!\!\int_{-\infty}^{\infty} G(k, \omega)e^{ik(z-z')-i\omega(t-t')}\, dk\, d\omega, \qquad (51)$$

where

$$G(k, \omega) = \frac{1}{k^2 - (\omega^2 - \omega_p^2)/u^2}. \qquad (51a)$$

To satisfy causality, the integration path in the complex ω plane runs with Im $\omega > 0$ above the pole singularities at $\omega = \pm\omega_1, \omega_1 = (k^2u^2 + \omega_p^2)^{1/2}$. In the complex k plane, pole singularities lie at $k = \pm k_1 = \pm(\omega^2 - \omega_p^2)^{1/2}/u$, Im $k_1 > 0$, on opposite sides of the integration path Im $k = 0$. On deforming the path into the lower half of the complex ω plane, where $\exp[-i\omega(t - t')]$ decays for $t > t'$, one obtains on use of the residue theorem the following reduction of Eq. (51):†

$$g = \frac{u^2}{2\pi} \int_{-\infty}^{\infty} e^{ik(z-z')} \frac{\sin \omega_1(t - t')}{\omega_1} \, dk, \qquad \omega_1(k) = \sqrt{k^2 u^2 + \omega_p^2}. \tag{52}$$

Alternatively, on deforming the k-plane path into the upper and lower half-planes for $(z - z') > 0$ and $(z - z') < 0$, respectively, one finds

$$g = \frac{1}{2\pi} \int_{-\infty}^{\infty} e^{-i\omega(t-t')} \frac{e^{ik_1|z-z'|}}{-2ik_1} \, d\omega, \qquad k_1(\omega) = \frac{\sqrt{\omega^2 - \omega_p^2}}{u}. \tag{53}$$

The representations in Eqs. (52) and (53) may be obtained directly on use of the time-guided and space-guided formulations of Secs. 1.3 and 1.4, respectively. From the one-dimensional form of Eqs. (1.3.2), utilizing normalized eigenfunctions $\Phi_k(z)$ in the z domain,

$$\delta(z - z') = \int_{-\infty}^{\infty} \Phi_k(z)\Phi_k^*(z') \, dk, \qquad \Phi_k(z) = \frac{e^{ikz}}{\sqrt{2\pi}} \tag{54}$$

$$g(z, z'; t, t') = \int_{-\infty}^{\infty} g_t(t, t'; \lambda_{tk})\Phi_k(z)\Phi_k^*(z') \, dk, \tag{55}$$

where the notation and normalization conforms with the more general discussion in Sec. 3.3. On substitution into Eq. (50), g_t is found to satisfy the equation

$$\left(\frac{d^2}{dt^2} + \lambda_{tk}\right)g_t(t, t'; \lambda_{tk}) = u^2\delta(t - t'), \qquad \lambda_{tk} = \omega_1^2(k), \tag{56}$$

which thus identifies g_t as the one-dimensional Green's function in the time domain. Subject to $g_t = 0$ for $t < t'$, the solution of Eq. (56) is

$$g_t(t, t'; \lambda_{tk}) = u^2 \frac{\sin \sqrt{\lambda_{tk}}\,(t - t')}{\sqrt{\lambda_{tk}}}, \tag{57}$$

whence Eq. (55) yields the same result as Eq. (52). g_t is the analogue of the Green's function G_α defined in Eq. (1.3.16); in the present discussion the difference between them arises from use of the second-order form of the field equations in Eqs. (50).

To obtain the space-guided representation, we employ the one-dimensional form of Eqs. (1.4.2) and (1.4.3),

$$\delta(t - t') = \int_{-\infty}^{\infty} \Phi_\omega(t)\Phi_\omega^*(t') \, d\omega, \qquad \Phi_\omega(t) = \frac{e^{-i\omega t}}{\sqrt{2\pi}}, \tag{58}$$

$$g(z, z'; t, t') = \int_{-\infty}^{\infty} g_z(z, z'; \lambda_{z\omega})\Phi_\omega(t)\Phi_\omega^*(t') \, d\omega. \tag{59}$$

On substitution into Eq. (50), there results the reduced form

$$\left(\frac{d^2}{dz^2} + \lambda_{z\omega}\right)g_z(z, z'; \lambda_{z\omega}) = -\delta(z - z'), \qquad \lambda_{z\omega} = k_1^2(\omega), \tag{60}$$

---

†This integral is tabulated and yields the closed-form solution as[11]

$$g = \frac{u}{2} J_0\left(\frac{\omega_p}{u}\sqrt{u^2(t - t')^2 - (z - z')^2}\right) U[u(t - t') - |z - z'|].$$

whence subject to a radiation condition at $|z - z'| \to \infty$ and in view of $\mathrm{Im}\sqrt{\lambda_{z\omega}} > 0$, the one-dimensional Green's function g_z in the z domain is given by

$$g_z(z, z'; \lambda_{z\omega}) = \frac{\exp(i\sqrt{\lambda_{z\omega}}\,|z - z'|)}{-2i\sqrt{\lambda_{z\omega}}}, \tag{61}$$

so Eqs. (59) and (53) are equivalent. It has thus been demonstrated that appropriate reduction of the plane-wave integral representation in Eq. (51) yields either the time- or space-guided representations, whence, by reconstructing Eq.(51), it is possible to derive one guided representation from the other. Comparison of Eqs. (51), (55), and (59) shows that the reconstruction involves the representation of g_t in Eq. (57), or g_z in Eq. (61), in terms of t- or z-dependent eigenfunctions:

$$-\frac{g_t(t, t'; \lambda_{tk})}{u^2} = \int_{-\infty}^{\infty} \frac{\Phi_\omega(t)\Phi_\omega^*(t')}{\omega^2 - \lambda_{tk}}\,d\omega, \tag{62a}$$

$$g_z(z, z'; \lambda_{z\omega}) = \int_{-\infty}^{\infty} \frac{\Phi_k(z)\Phi_k^*(z')}{k^2 - \lambda_{z\omega}}\,dk, \tag{62b}$$

which forms suggest that there is an intimate relation between Green's functions and spectral representations. This aspect is given further attention in Sec.3.3.

It is significant to observe that either of Eqs. (55) and (59) may also be derived from the other without need of the intermediary relation (51). In the complex ω plane, the space-guided representation in Eqs. (59) and (61) has branch-point singularities at $\omega = \pm\omega_p$. If the corresponding Riemann surface is chosen so that everywhere on the top sheet $\mathrm{Im}\sqrt{\lambda_{z\omega}} > 0$, branch cuts are drawn as shown in Fig. 1.5.3 (see Sec. 5.3b for discussion of Riemann surfaces).

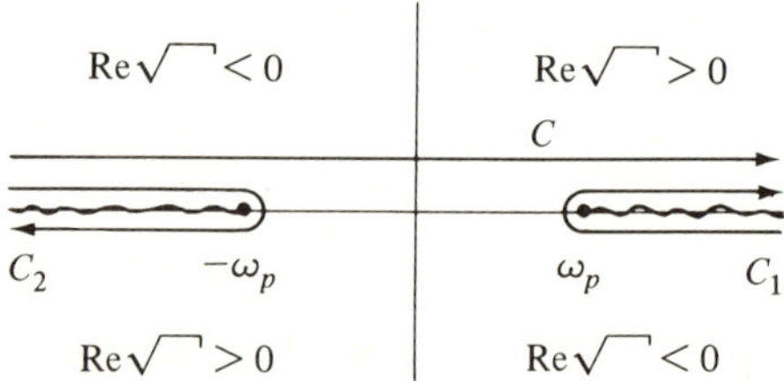

FIG. 1.5.3 Contours in complex ω-plane and behavior of $\mathrm{Re}\sqrt{\omega^2 - \omega_p^2}$ on top sheet of Riemann surface where $\mathrm{Im}\sqrt{\omega^2 - \omega_p^2} > 0$.

For $t > t'$, the exponential behavior of the integrand at infinity in the lower half of the complex ω plane is one of decay, so the contour C can be deformed into the contours C_1 and C_2 around the branch cuts. On changing variables to $-\omega$, one may combine the integral over C_2 with that over C_1; a further change of variable from ω to $\sqrt{\lambda_{tk}}$, with reference to Fig. 1.5.3, leads to Eq. (55).

The preceding considerations will be generalized and phrased succinctly in terms of "characteristic" Green's functions in Sec. 3.3.

1.6 RAY-OPTIC APPROXIMATIONS OF INTEGRAL REPRESENTATIONS

The exact alternative forms for the space- and time-dependent Green's functions developed in Secs. 1.2–1.4 generally lead to integral representations that cannot be evaluated in terms of known functions for arbitrary observation points in space and time. However, they permit useful approximations to be derived when the range of observations is suitably restricted. Concern in this section is with the field behavior at "large" observation distances from the source, in which regions the representation integrals are amenable to asymptotic evaluation by the method of saddle points described in detail in Chapter 4. The principal contributions to an integral will be shown to arise from the vicinity of isolated critical points—stationary (saddle) points, singularities, and endpoints—in the integration interval. One thereby obviates the need for precise evaluation of the integral over the entire integration range. The contribution from each critical point implies a wave process with distinct physical characteristics that are to be examined in detail in Secs. 1.6a–1.6c.

Saddle-point contributions are found to describe wavepackets (narrow bundles of plane waves with limited $\mathbf{k}$, ω values) that move along certain trajectories, called rays, in space–time. The rays are straight in a homogeneous medium but curved when the medium exhibits spatial variation. Energy in a wavepacket is preserved along its space–time trajectory. Both the wavevector $\mathbf{k}$ and the frequency $\omega(\mathbf{k})$ in the plane-wave bundle remain constant in time when the medium is homogeneous, but only ω is invariant in the presence of spatial inhomogeneities. These conclusions may be inferred either from the oscillatory or the guided-wave representations in Secs. 1.6a and 1.6b, respectively; their separate examination reveals distinctive features that illuminate further the propagation mechanism ascribed to each asymptotic solution.

While the saddle-point condition may be stated simply, the location of saddle points by analytical means is generally quite involved when the medium has any but the simplest dispersive properties. However, the $(\mathbf{k}, \omega)$ dispersion surfaces representing the plane-wave dispersion relations are useful for graphically locating the saddle points since the latter may be found thereon by simple geometrical construction. This construction provides the space–time rays determining the spatial and temporal location of a wavepacket. It is performed in four-dimensional $(\mathbf{k}, \omega)$ space when the dispersion surface has no symmetries whatever, but reduces to a plot in three or two dimensions, respectively, when the medium is gyrotropically anisotropic (as in a magnetoplasma) or isotropic. Use of dispersion surfaces also facilitates the charting of space–time rays in an inhomogeneous medium and reveals the intimate connection between propagation and radiation processes under time-dependent and time-harmonic conditions. The conventional rays of geometrical optics emerge as special cases of the space–time rays.

The wavepacket description of field phenomena at distant observation points applies provided that multiple wavepackets, if they exist, are fully formed and well separated in wavenumber and frequency. There are, however, certain regions in space–time, called transition regions, wherein individual wavepackets either are not yet fully developed or interact strongly with adjacent packets. In integral field representations the transition regions are revealed by a confluence of critical points in the integrand. The simple integration procedure applying to isolated critical points is then invalidated; the necessary modifications in the method are indicated in Sec. 1.6c.

1.6a Oscillatory Integral Representations

Homogeneous media

Oscillatory representations of time-dependent Green's functions in an unbounded, stationary, homogeneous, isotropic, or anisotropic medium lead to plane-wave integrals of the form

$$I(\mathbf{r}, t) = \int A(\mathbf{k})e^{i\psi(\mathbf{r},t;\mathbf{k})}\, d\mathbf{k}, \qquad \psi(\mathbf{r}, t; \mathbf{k}) = \mathbf{k} \cdot \mathbf{r} - \omega(\mathbf{k})t, \tag{1}$$

where $d\mathbf{k} = dk_x\, dk_y\, dk_z$ and the integration extends over the infinite volume in $\mathbf{k}$ space. In general, the overall solution may comprise several integrals of the form shown in Eq. (1), one for each mode type, distinguished by different $\omega_\alpha(\mathbf{k})$ and $A_\alpha(\mathbf{k})$ (see Sec. 1.3).

Since the integral in Eq. (1) cannot generally be evaluated in closed form, it is relevant to discuss approximation procedures. For integrals whose integrands contain a large parameter, one of the most useful procedures is the method of saddle points. This method, and its relation to the stationary-phase method, is discussed in detail in Chapter 4 and is applicable if the distance r from the source point $\mathbf{r} = 0$ to the observation point $\mathbf{r}$ is chosen sufficiently large. Under these circumstances, the major contribution to the integral arises from the vicinity of saddle (stationary) points $\mathbf{k}_s$ defined implicitly by $\nabla_k \psi = 0$, where $\nabla_k = \mathbf{x}_0(\partial/\partial k_x) + \mathbf{y}_0(\partial/\partial k_y) + \mathbf{z}_0(\partial/\partial k_z)$, or

$$\mathbf{v}_g(\mathbf{k}) \equiv \nabla_k \omega(\mathbf{k}) = \frac{\mathbf{r}}{t}, \qquad \text{at } \mathbf{k}_s = \mathbf{k}_s(\mathbf{r}, t). \tag{2}$$

Near $\mathbf{k}_s$, the (slowly varying) amplitude function may be approximated by $A(\mathbf{k}) \approx A(\mathbf{k}_s)$ but the phase requires power-series expansion up to the quadratic term in $(\mathbf{k} - \mathbf{k}_s)$(the linear term is absent in view of the saddle-point condition):

$$\psi(\mathbf{r}, t; \mathbf{k}) = \psi(\mathbf{r}, t; \mathbf{k}_s) + \tfrac{1}{2}[(\mathbf{k} - \mathbf{k}_s) \cdot \nabla_{k_s}]^2 \psi(\mathbf{r}, t; \mathbf{k}_s) + \cdots, \tag{3a}$$

$$= \psi(\mathbf{r}, t; \mathbf{k}_s) - \tfrac{1}{2}t[(\mathbf{k} - \mathbf{k}_s) \cdot \nabla_{k_s}]^2 \omega(\mathbf{k}_s) + \cdots, \tag{3b}$$

where it is understood that ∇_{k_s} operates only on ψ or ω. The resulting integral is evaluated on use of the formula [see Eq. (4.7.5)]

$$\int \exp\left\{-i\frac{t}{2}\left[(\mathbf{k} - \mathbf{k}_s) \cdot \nabla_{k_s}\right]^2 \omega\right\} d\mathbf{k} = (2\pi)^{3/2}\frac{e^{-i(\pi/4)\sigma}}{t^{3/2}Q^{1/2}}, \tag{4}$$

where $Q = |\bar{R}_1\bar{R}_2\bar{R}_3|^{-1}$ is the absolute value of the determinant of the matrix $\mathcal{Q}$, with

$$\mathcal{Q} = \begin{bmatrix} \dfrac{\partial^2 \omega}{\partial k_x^2} & \dfrac{\partial^2 \omega}{\partial k_x \partial k_y} & \dfrac{\partial^2 \omega}{\partial k_x \partial k_z} \\[2mm] \dfrac{\partial^2 \omega}{\partial k_y \partial k_x} & \dfrac{\partial^2 \omega}{\partial k_y^2} & \dfrac{\partial^2 \omega}{\partial k_y \partial k_z} \\[2mm] \dfrac{\partial^2 \omega}{\partial k_z \partial k_x} & \dfrac{\partial^2 \omega}{\partial k_z \partial k_y} & \dfrac{\partial^2 \omega}{\partial k_z^2} \end{bmatrix}_{\mathbf{k}=\mathbf{k}_s} = |\nabla_k \nabla_k \omega(\mathbf{k}_s)|. \tag{4a}$$

Here $\sigma = \sum_{j=1}^{3} \operatorname{sgn} \bar{R}_j$, with $\bar{R}_j$ denoting the reciprocal of the elements (eigenvalues) in the diagonalized form of the matrix $\mathcal{Q}$, and $\operatorname{sgn} \bar{R}_j = \pm 1$ for $\bar{R}_j \gtrless 0$. Thus, one finds for the major contribution to the integral in Eq. (1) for large r [and hence, via Eq. (2), also for large t] and real $\psi(\mathbf{r}, t; \mathbf{k}_s)$ (lossless medium):

$$I(\mathbf{r}, t) \sim A(\mathbf{k}_s) e^{i\psi(\mathbf{r},t;\mathbf{k}_s)} \frac{(2\pi)^{3/2}}{(t^3 Q)^{1/2}} e^{-i(\pi/4)\sigma}. \tag{5}$$

$\mathbf{k}_s$ is specified implicitly in Eq. (2) and if this equation has several solutions $\mathbf{k}_{si}$, the single term in Eq. (5) is replaced by a sum over i. Since $I(\mathbf{r}, t)$ is real, there exists for each saddle-point solution $\mathbf{k}_s$, $\omega(\mathbf{k}_s)$ another solution $-\mathbf{k}_s$, $\omega(-\mathbf{k}_s) = -\omega(\mathbf{k}_s)$; the sum over i contains these conjugate pairs as well as different saddle-point species. Additional contributions to the integral may arise from singularities in the integrand (see Sec. 1.6c and Chapter 4) but are ignored for the present.

The approximate asymptotic solution in Eq. (5) has an interesting physical interpretation. Evidently, the field at the space–time point $(\mathbf{r}, t)$ is established by a plane wave $\exp[i\mathbf{k}_s \cdot \mathbf{r} - i\omega(\mathbf{k}_s)t]$ whose amplitude is not $A(\mathbf{k}_s)$, as in the integrand of Eq. (1), but is modified by the last factor in Eq. (5). This modification, noted from the manner of evaluation of the integral, arises from constructive interference of a "packet" or "bundle" of oscillatory plane waves of amplitude $A(\mathbf{k})$ whose wavenumbers lie in a small $\mathbf{k}$-space interval $\delta\mathbf{k} = (\mathbf{k} - \mathbf{k}_s)$ about the saddle-point value $\mathbf{k}_s$. The average phase ψ of the waves in this packet is determined by the central wavenumber $\mathbf{k}_s$, while the composite amplitude changes according to Eq. (5). From Eq. (2), $\mathbf{k}_s$ is constant if $(\mathbf{r}/t)$ is constant so that an observer moving with the wavepacket at the constant velocity $\mathbf{v}_g = (\mathbf{r}/t)$ sees a fixed wavenumber and frequency $\omega(\mathbf{k}_s)$. Stated alternatively, a trajectory on which the frequency ω of a wavepacket remains constant is defined in the $(\mathbf{r}, \mathbf{k})$ phase space by the parametric relations

$$\frac{d\mathbf{r}}{dt} = \nabla_k \omega(\mathbf{k}), \qquad \frac{d\mathbf{k}}{dt} = 0, \qquad \left[\frac{d\omega(\mathbf{k})}{dt} = 0\right]. \tag{6}$$

These trajectory equations for a homogeneous medium are modified, as noted in Eqs. (17a) and (19), in the presence of spatial inhomogeneities. Since the field energy is localized in the moving wavepacket, $\mathbf{v}_g = \nabla_k \omega$ is identified as the group velocity, or energy-transport velocity; the latter is generally different from the phase velocity,

$$\mathbf{v}_p = \frac{\omega}{|\mathbf{k}|} \mathbf{n}_0, \tag{7}$$

with which the equiphase surfaces $\psi = $ constant advance along the normal direction $\mathbf{n}_0$. Only in the dispersionless case $\omega(\mathbf{k}) = kc$ does $\mathbf{v}_g$ equal $\mathbf{v}_p$, with c denoting a $\mathbf{k}$-independent wave-propagation speed. When the observer is at rest at the observation point $\mathbf{r}$, the saddle-point wavenumber $\mathbf{k}_s$ in Eq. (2) changes with time so that the response in Eq. (5) describes a distinct wavepacket at each observation time t.

The amplitude change due to plane-wave interference within a wavepacket centered at $\mathbf{k}_s$ can be deduced in simple physical terms.[12-14] We shall assume that $I(\mathbf{r}, t)$ in Eq. (1) is normalized so that the total energy at any time t is given by

$$W(t) = \int |I(\mathbf{r}, t)|^2 \, d\mathbf{r}, \tag{8a}$$

where the integration is over the entire physical space. By Parseval's theorem [or direct substitution from Eq. (1) with use of Eq. (1.3.2a)], one finds for the energy at $t = 0$,

$$W(0) = (2\pi)^3 \int |A(\mathbf{k})|^2 \, d\mathbf{k}, \tag{8b}$$

so the initial energy represented by waves in the above wavenumber interval $\delta\mathbf{k}$ is $\Delta W = (2\pi)^3 |A(\mathbf{k}_s)|^2 \Delta\mathbf{k}$, where $\Delta\mathbf{k}$ is the volume element corresponding to $\delta\mathbf{k}$. In the absence of dissipation the energy cannot change with time; hence we shall assume that ΔW remains constant and eventually describes the energy in the wavepacket when the latter has formed at $(\mathbf{r}, t)$ values validating Eq. (5). At a particular time thereafter, let the wavepacket occupy the spatial region $\Delta\mathbf{r}$, with corresponding energy $\Delta W = |I|^2 \Delta\mathbf{r}$, whence by equating the above expressions for ΔW,

$$|I| = (2\pi)^{3/2} |A(\mathbf{k}_s)| \sqrt{\frac{\Delta\mathbf{k}}{\Delta\mathbf{r}}}. \tag{9}$$

Although $\Delta\mathbf{k}$ remains constant, $\Delta\mathbf{r}$ changes with time, since individual plane waves within the packet have slightly different group velocities. The mapping from $\mathbf{r}$ to $\mathbf{k}$ space is accomplished via

$$\Delta\mathbf{r} = J\!\left(\frac{x, y, z}{k_x, k_y, k_z}\right) \Delta\mathbf{k}, \tag{10}$$

where J, the Jacobian of the transformation, can be evaluated from $\mathbf{r} = \mathbf{v}_g(\mathbf{k})t$ as given in Eq. (2). Since $x = v_{gx}(\mathbf{k})t = t\partial\omega(\mathbf{k})/\partial k_x$, etc., one finds $|J| = t^3 Q$, with Q defined in connection with Eq. (4), thereby establishing agreement of the results in Eqs. (5) and (9).

The preceding field solution is valid for those values of $(\mathbf{r}, t)$ at which wavepackets are distinct and fully developed. In "transition regions" of $(\mathbf{r}, t)$ space where these conditions are not satisfied, alternative solutions must be derived by the considerations of Sec. 1.6c.

Dispersion surfaces and space–time rays

Calculation of the transient fields from Eq. (5) requires knowledge of the saddle-point wavenumbers $\mathbf{k}_s(\mathbf{r}, t)$ specified implicitly by the condition (2). However, the (plane-wave) dispersion relation $\omega = \omega(\mathbf{k})$ is generally so complicated that Eq. (2) cannot be solved explicitly for the various $\mathbf{k}_s$. Hence, it is useful to employ graphical procedures based (for lossless media) on the real dispersion surface $\omega = \omega(k_x, k_y, k_z)$, or more generally on the implicit relation

$$f(k_x, k_y, k_z, \omega) = 0. \tag{11}$$

In the most general case, f is a hypersurface in the four-dimensional $(\mathbf{k}, \omega)$ space but simplifies when the dispersion relation has certain symmetries. In an isotropic medium, wave-propagation properties are independent of the direction of propagation, whence $\omega(\mathbf{k}) = \omega(k)$, k being the magnitude of the wavevector. Equation (11) then reduces to

$$f(k, \omega) = 0, \tag{11a}$$

which can be plotted in a two-dimensional (k, ω) frame. In a magnetoactive medium, such as a plasma rendered anisotropic by the application of a steady external magnetic field $\mathbf{H}_0 = \mathbf{z}_0 H_0$, plane-wave propagation characteristics do not depend on direction in a plane transverse to $\mathbf{H}_0$. The dispersion surface thus takes the form

$$f(k_\rho, k_z, \omega) = 0, \tag{11b}$$

which, with k_ρ representing the wavenumber component transverse to z, can be plotted in a three-dimensional (k_ρ, k_z, ω) coordinate system.

Graphical determination of the saddle points $\mathbf{k}_s(\mathbf{r}, t)$ satisfying Eq. (2) can be effected with the aid either of the dispersion surface $f(\mathbf{k}, \omega) = 0$ in the four-dimensional $(\mathbf{k}, \omega)$ space, or of its $\omega = \omega(\mathbf{k}) =$ constant projections in three-dimensional $\mathbf{k}$ space. In the former instance, the desired saddle points are located at those points $[\mathbf{k}_s(\mathbf{r}, t), \omega(\mathbf{k}_s)]$ of the dispersion surface whereon the four-vector normal is parallel to the space–time four-vector $(\mathbf{r}/t, 1)$, provided that the coordinate axes k_x, k_y, k_z, ω are chosen parallel to $x, y, z, -t$, respectively. In the latter case, the $\mathbf{k}_s$ distinguish points of the constant $\omega(\mathbf{k})$ projections at which the normal gradient vector is equal to $\mathbf{r}/t$ in both direction and magnitude. The proof of the former statement follows from the observations that the four-vector normal to the dispersion surface $f(\mathbf{k}, \omega) = 0$ is given by the four-dimensional gradient $\square f \equiv (\nabla_k f, \partial f/\partial \omega)$ and is perpendicular to the tangential four-vector $(d\mathbf{k}, d\omega) = (d\mathbf{k}, \nabla_k \omega \cdot d\mathbf{k})$,[†] whence on forming the scalar product of the two vectors, one infers that

$$df = 0 = \nabla_k f + \frac{\partial f}{\partial \omega}\, \nabla_k \omega; \tag{12}$$

---

†For $(\mathbf{k}, \omega)$ and $(\mathbf{k} + d\mathbf{k}, \omega + d\omega)$ on the dispersion surface, $(d\mathbf{k}, d\omega)$ is a tangential vector element.

thus, the four-vector normal $\square f$ is parallel to the four-vector $(\nabla_k \omega, -1)$ which, by Eq. (2) and because of the opposite orientation of the ω and t axes, must be made parallel to $(\mathbf{r}/t, 1)$. The proof of the latter statement above follows from the fact that the normal gradient vector to the surface $\omega = $ constant is the three-vector $\nabla_k \omega$ which, by Eq. (2), must be set equal to $\mathbf{r}/t$. The vector $(\mathbf{r}/t, 1) = (\mathbf{v}_g, 1)$, the group-velocity vector in four-dimensional space, defines a "space–time ray" trajectory $\mathbf{r} = \mathbf{r}(t)$ descriptive of the spatial and temporal location of a moving wavepacket.

The geometrical principles underlying the graphical method for locating the saddle points are relatively simple. However, in a general medium described by the dispersion equation (11), application of this procedure is complicated by the necessity of dealing with normals to four-dimensional surfaces, or with three-dimensional surfaces, when finding the projection of the normal onto hyperplanes $\omega = $ constant. Simplification occurs when the medium anisotropy is of the magnetoactive type since the dispersion surface [Eq. (11b)] can then be plotted in a three-dimensional frame and projections onto planes $\omega = $ constant reduce to curves in the (k_ρ, k_z) wavenumber plane. These considerations are illustrated for electromagnetic wave propagation in a cold electron plasma rendered uniaxially anisotropic by imposition of an infinitely strong external magnetic field H_0 along the z axis. The dispersion equation in this instance (see Sec. 7.2), with c denoting the speed of light in vacuum,

$$ f(k_\rho, k_z, \omega) = k_\rho^2 + \left(1 - \frac{\omega_p^2}{\omega^2}\right)k_z^2 - \frac{\omega^2}{c^2}\left(1 - \frac{\omega_p^2}{\omega^2}\right) = 0, \qquad (13) $$

describes a surface whose intersections with planes $\omega = $ constant yield a family of ellipses when $|\omega|$ exceeds the plasma frequency ω_p, and a family of hyperbolas when $|\omega| < \omega_p$. For simplicity, we consider only the $|\omega| > \omega_p$ portion of the surface, as shown in Fig. 1.6.1(a). It is convenient for dimensional reasons to deal with plots of $(\mathbf{k}c, \omega)$ and $(\mathbf{r}, ct)$, so the normal to the dispersion surface is now defined as $\square f \equiv (c^{-1}\nabla_k f, \partial f/\partial\omega)$. To locate the saddle point corresponding to the group-velocity vector $\mathbf{V}_g = (\mathbf{r}/t, c)$ [see Fig. 1.6.1(b)] we search on the dispersion surface, plotted in the coordinate frame described previously, for normals $\square f$ parallel to $\mathbf{V}_g$ and ascertain thereby the projection $c\mathbf{k}_s$ [Fig. 1.6.1(a)].

Alternatively, one may plot the family of wavenumber curves for various ω as in Fig. 1.6.1(c), and the family of group speed curves $v_g(\mathbf{k}) = v_g(k, \bar{\theta}) = |\nabla_k \omega| = [(\partial\omega/\partial k)^2 + (\partial\omega/k\partial\bar{\theta})^2]^{1/2}$ as in Fig. 1.6.1(d), with k and $\bar{\theta}$ denoting the magnitude and orientation angle of $\mathbf{k}$ in cylindrical polar coordinates. Location of the saddle point is achieved by use of the projected vector $\mathbf{v}_g = \mathbf{r}/t$ in Fig. 1.6.1(b). In Fig. 1.6.1(c), the required codirectionality of the vectors $\nabla_k \omega$ and $\mathbf{r}/t$ is enforced by locating those points on the wavenumber curves having normals parallel to $\mathbf{r}$; one thereby obtains the curve A, described parametrically by $k = k(\bar{\theta})$, on which the saddle-point value $k_s = k(\bar{\theta}_s)$ lies. Note that the vector $\nabla_k \omega$ on a curve $\omega = \omega_i$ points toward the curve described by a value

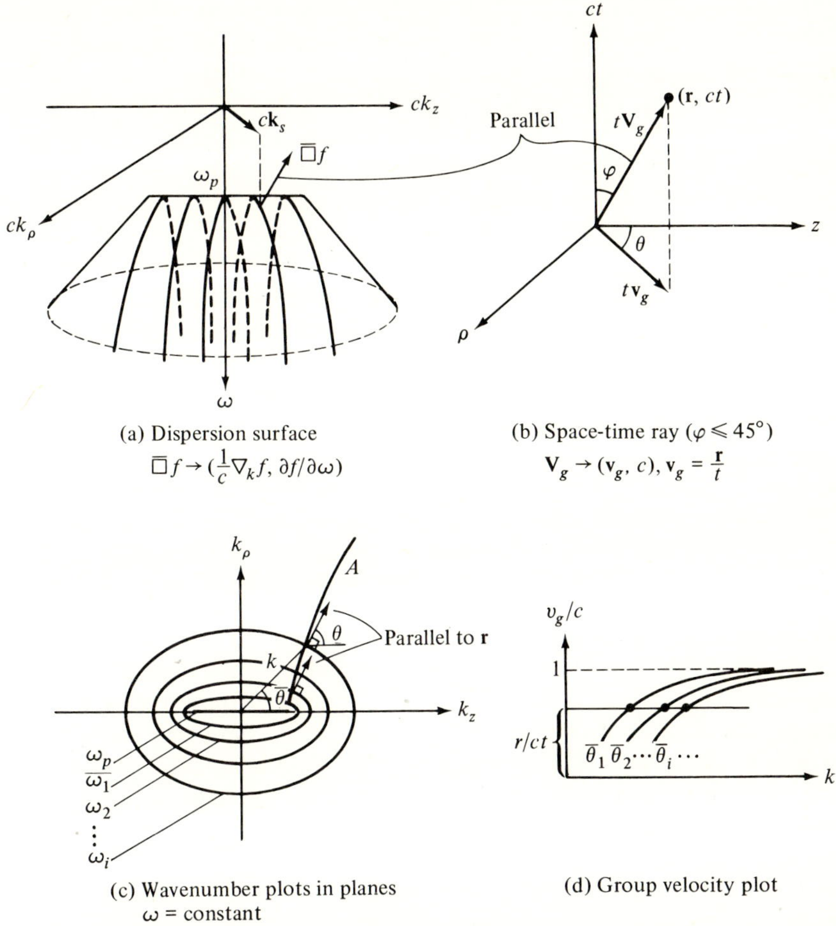

(a) Dispersion surface
$$\overline{\Box}f \rightarrow (\tfrac{1}{c}\nabla_k f,\ \partial f/\partial\omega)$$

(b) Space-time ray ($\varphi \leqslant 45°$)
$$\mathbf{V}_g \rightarrow (\mathbf{v}_g,\ c),\ \mathbf{v}_g = \frac{\mathbf{r}}{t}$$

(c) Wavenumber plots in planes
ω = constant

(d) Group velocity plot

FIG. 1.6.1 Graphical methods for locating the saddle points $\mathbf{k}_s$ on an anisotropic dispersion surface.

$\omega > \omega_i$. The magnitude requirement $v_g = |\nabla_k\omega| = r/t$ is imposed by the construction in Fig. 1.6.1(d), and the saddle point $\mathbf{k}_s$ is that point on A in Fig. 1.6.1(c) which is compatible with the k versus $\overline{\theta}$ values deduced from Fig. 1.6.1(d).

One observes from Fig. 1.6.1(c) that in an anisotropic medium, the wavevector $\mathbf{k}$ and the group velocity vector $\mathbf{v}_g$ generally are nonparallel, so phase and energy propagate along different directions. This distinction is illustrated in Fig. 1.6.2. Let us recall that a wavepacket comprises a bundle of plane waves, whose wave vectors lie within a small cone in $\mathbf{k}$ space, that propagates in the direction of constructive wave interference. Wavefronts advance along the direction of the wavevector $\mathbf{k}$ with a speed given by $v_p = \omega/k$ [see Eq. (7)]; this quantity may be plotted to furnish the phase velocity

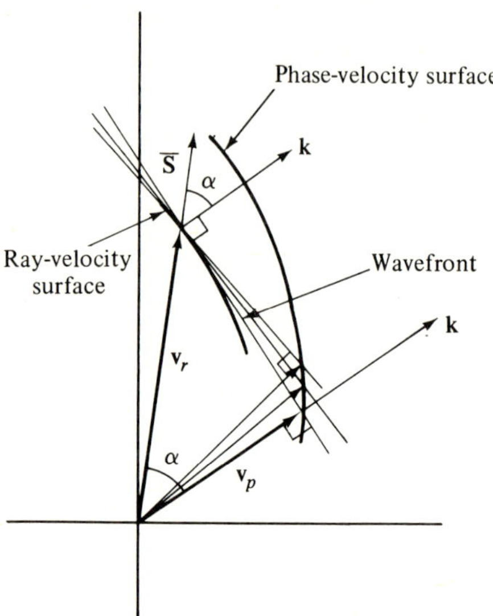

FIG. 1.6.2   Phase- and ray-velocity surfaces.

surfaces, one of which is shown in part in Fig. 1.6.2 for a typical member of the wavenumber diagrams in Fig. 1.6.1(c). Since the vector $\mathbf{v}_p t$ yields the spatial displacement of a wavefront from its location at a reference time $t = 0$, the phase velocity surface may be used to construct the wavefront configuration after a unit time interval.[15] The phase fronts in a wavepacket located at the origin of the $\mathbf{v}_p t$ plot at $t = 0$ will have moved to the locations shown in Fig. 1.6.2 at $t = 1$; their intersections locate the interference maximum and hence the new position of the wavepacket, as shown by the "ray-velocity" vector $\mathbf{v}_r$ in Fig. 1.6.2 along which the energy propagates. Unless the phase-velocity surface is spherical as in an isotropic medium, the phase- and ray-velocity vectors are displaced by an angle α and $v_r = v_p/\cos \alpha$ represents the speed of propagation of the wavefront along the direction of the ray. By carrying out the wavefront construction for all points on the phase-velocity surface, one generates a new surface, the ray-velocity surface or simply "ray surface," which constitutes the envelope of the wavefronts. The energy-transport, or ray, direction corresponding to a given direction of $\mathbf{v}_p$ is inferred by drawing a perpendicular plane at the endpoint of $\mathbf{v}_p$, determining its point of tangency on the ray surface, and drawing a vector from the origin to the contact point, as shown in Fig. 1.6.2. Conversely, to the ray direction specified by a given point on the ray surface corresponds a wavevector direction given by the normal to the surface.

For the special case of an isotropic medium, the two procedures described in connection with Fig. 1.6.1 are equivalent since the relevant dispersion equation (11a) requires only a two-dimensional plot in (k, ω) space. Because of the rotational symmetry of the dispersion surface about the ω axis, the wavenumber plots in Fig. 1.6.1(c) are now circular and therefore the curve A degenerates into a radial straight line. Since $\mathbf{v}_g(\mathbf{k}) = \mathbf{v}_g(k)$, the family of curves in Fig. 1.6.1(d) collapses into a single $\bar{\theta}$-independent curve. For illustration, consider a cold electron plasma having the dispersion relation

$$f(k, \omega) = \omega^2 - (kc)^2 - \omega_p^2 = 0, \tag{14}$$

whence the dispersion surface and space–time ray plots are those in Fig. 1.6.3(a) and (b), with the group-velocity curve given in Fig. 1.6.3(c). The determination of k_s follows as previously. In the present instance, it is also possible (and conventional) to use instead of Fig. 1.6.3(c) the dispersion curve in Fig. 1.6.3(a); by direct imposition of the saddle-point condition $d\omega/dk = r/t$, one locates k_s on drawing a tangent having the slope r/t. The dispersion surface can also be employed to provide a pictorial representation for the amplitude variation noted in Eq. (5). In the discussion following Eqs. (8), it is emphasized that the energy in a wavepacket described by a fixed wavenumber deviation $\Delta\mathbf{k}$ from $\mathbf{k}_s$ remains constant. For the isotropic problem in Fig. 1.6.3, this feature may be illustrated by letting the wavepacket under consideration be characterized by the wave-

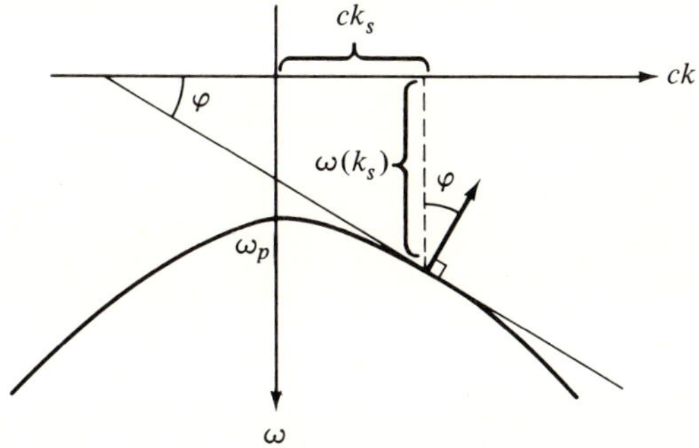

(a) Dispersion surface

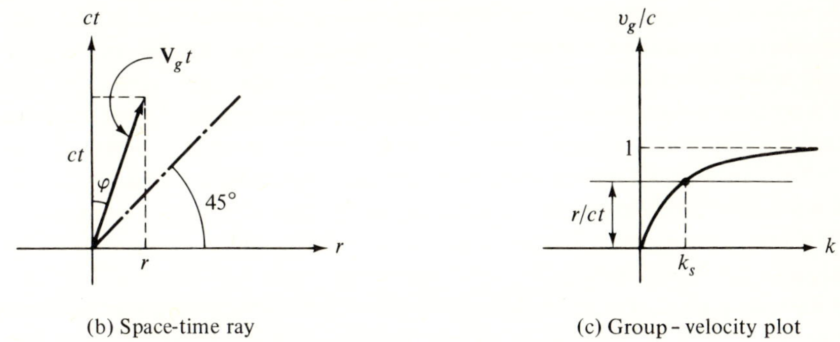

(b) Space-time ray   (c) Group – velocity plot

FIG. 1.6.3 Graphical methods for locating the saddle points $\mathbf{k}_s$ for an isotropic medium.

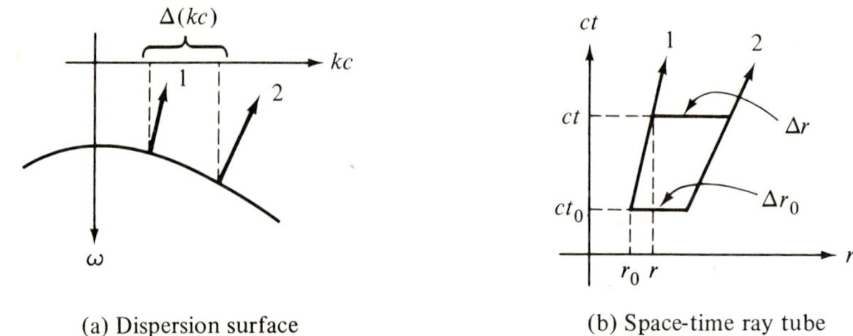

(a) Dispersion surface   (b) Space-time ray tube

FIG. 1.6.4   Conservation of energy in a space–time ray tube.

number interval shown in Fig. 1.6.4(a), with corresponding limiting rays 1 and 2 as indicated. In the configuration space of Fig. 1.6.4(b), these rays form a space–time ray tube, and since this ray tube describes the same wavepacket with fixed wavenumber spread $\Delta(kc)$, the energy in the ray tube remains constant. The constancy of the energy $\Delta W = |I|^2 \Delta r$ implies that the field amplitude $|I(r, t)|$ at t is related to its value $|I(r_0, t_0)|$ at t_0 by the square root of the ratio of the corresponding ray tube cross sections in a plane $t = \text{constant}$:

$$|I(r, t)| = |I(r_0, t_0)| \sqrt{\frac{\Delta(r_0)}{\Delta(r)}}. \tag{15}$$

This relation continues to hold for more general dispersion surfaces characterized by three- or four-dimensional $(\mathbf{k}c, \omega)$ plots. It is equivalent to that given in Eq. (9) since the expression for the energy, $|I(r_0, t_0)|^2 \Delta r_0$, is also proportional to $|A(k_s)|^2 \Delta k$. As is evident from Fig. 1.6.4, for the same wavenumber interval $\Delta(kc)$, a more strongly curved dispersion surface provides a more rapidly diverging ray tube, thereby leading to a more rapidly decreasing amplitude, as expressed by the factor $Q^{-1/2}$ in Eq. (5).

Weakly inhomogeneous media

When the properties of a medium vary with spatial position, it is no longer possible to represent the space- and time-dependent field by the plane-wave superposition in $\mathbf{k}$ space as in Eq. (1), since $A(\mathbf{k}) \exp[i\mathbf{k} \cdot \mathbf{r} - i\omega(\mathbf{k})t]$ is then no longer an eigenfunction in the $\mathbf{r}, t$ domain. Although a valid representation may be achieved through superposition of functions of the form $A(\mathbf{k}, t)$ $\exp(i\mathbf{k} \cdot \mathbf{r})$, the calculation of $A(\mathbf{k}, t)$ is complicated since the functions $\exp(i\mathbf{k} \cdot \mathbf{r})$ do not form an orthogonal set in $\mathbf{r}$ space in the presence of spatial inhomogeneity. However, for "sufficiently slow" medium variations, one may synthesize the field solution approximately by a spectrum of *local* plane waves whose amplitude A and frequency ω are dependent not only on the wavevector $\mathbf{k}$ but also on the average position coordinate $\bar{\mathbf{r}}$, with the latter playing the role of a slow parameter descriptive of the local properties in the medium. For a medium with one-dimensional inhomogeneity, the validity of such an approximate integral representation is substantiated by use of WKB methods (see Sec. 3.5 and 1.6b); for two- and three-dimensional variations, conclusions obtained therefrom are confirmed by the procedure described in Sec. 1.7e.

As suggested by the field representation (1) for a homogeneous medium, it is possible to account for slow variations in the medium properties by first considering integrals appropriate to a piecewise constant medium. One is then led to consider integrals of the form

$$I(\mathbf{r}, t; \bar{\mathbf{r}}) = \int A(\mathbf{k}, \bar{\mathbf{r}}) e^{i[\mathbf{k}\cdot\mathbf{r} - \omega(\mathbf{k},\bar{\mathbf{r}})t]} \, d\mathbf{k} \tag{16}$$

where the range of observation points $\mathbf{r}$ is restricted to a volume τ centered on $\mathbf{r} = \bar{\mathbf{r}}$, τ being small enough to render the coordinate dependent dispersion equation $\omega = \omega(\mathbf{k}, \mathbf{r}) \approx \omega(\mathbf{k}, \bar{\mathbf{r}})$ essentially constant. For fixed $\mathbf{r}, \bar{\mathbf{r}}$, and t, the major contribution to $I(\mathbf{r}, t; \bar{\mathbf{r}})$ in Eq. (16) arises from those points $\mathbf{k} = \mathbf{k}_s(\mathbf{r}, t; \bar{\mathbf{r}})$ that satisfy the saddle-point condition

$$\frac{\mathbf{r}}{t} = \nabla_k \omega(\mathbf{k}, \bar{\mathbf{r}}). \tag{17}$$

Just as for the corresponding condition (5) in a homogeneous medium, there is an interesting physical interpretation of Eq. (17) for $\mathbf{r}, t$ regions wherein wavepackets are fully developed. An observer moving on a trajectory $\mathbf{r} = \mathbf{r}(t)$ in τ with the velocity

$$\mathbf{v}_g = \frac{d\mathbf{r}}{dt} = \frac{\mathbf{r}}{t} = \nabla_k \omega(\mathbf{k}, \bar{\mathbf{r}})$$

sees a wavenumber $\mathbf{k} = \mathbf{k}_s$ dependent on $\bar{\mathbf{r}}$. Since $\bar{\mathbf{r}}$ takes on a different value $\bar{\mathbf{r}}_1$ in a volume τ_1 adjacent to τ, the relevant $\mathbf{k}_s$ values are given by Eq. (17) with $\bar{\mathbf{r}}$ and τ replaced by $\bar{\mathbf{r}}_1$ and τ_1, and similarly for other volume elements τ_i. But since $\omega(\mathbf{k}, \bar{\mathbf{r}})$ depends weakly on $\bar{\mathbf{r}}$ [i.e., $\omega(\mathbf{k}, \bar{\mathbf{r}}_i) \approx \omega(\mathbf{k}, \mathbf{r})$ for $\mathbf{r}$ in τ_i], the wavenumber $\mathbf{k}$ seen by an observer moving with velocity $d\mathbf{r}/dt$ is specified by

$$\frac{d\mathbf{r}}{dt} = \nabla_k \omega(\mathbf{k}, \mathbf{r}). \tag{17a}$$

If the observer moves with a wavepacket on a phase-space trajectory $\mathbf{k} = k[\mathbf{r}(t)] = \mathbf{k}(t), \mathbf{r} = \mathbf{r}(t)$, along which the frequency $\omega(\mathbf{k}, \mathbf{r})$ of the wavepacket is conserved, then one infers from the constancy of the frequency

$$\frac{d}{dt}\omega(\mathbf{k}, \mathbf{r}) = 0 = \nabla_k \omega \cdot \frac{d\mathbf{k}}{dt} + \nabla\omega \cdot \frac{d\mathbf{r}}{dt}, \tag{18}$$

and hence, using Eq. (17a),

$$\frac{d\mathbf{k}}{dt} = -\nabla\omega(\mathbf{k}, \mathbf{r}). \tag{19}$$

For the homogeneous case $\omega = \omega(\mathbf{k})$, Eqs. (17a) and (19) manifestly reduce to Eqs. (6).

The significance of phase-space coordinates and of constant-frequency trajectories in the propagation of a wavepacket in an inhomogeneous medium is discussed further in Sec. 1.7a.

The variability of $\mathbf{k}_s$ on the ray trajectory complicates the use of graphical methods, as discussed in connection with Fig. 1.6.1. For media plane-stratified along z, where $\nabla\omega = \mathbf{z}_0(\partial\omega/\partial z)$, Eq. (19) implies constancy of the component $\mathbf{k}_t$ transverse to z, thereby permitting application of the graphical procedure in Fig. 1.6.9 for charting the ray paths; relevant $\mathbf{k}_s$ values are those corresponding to rays passing through the prescribed point $(\mathbf{r}, t)$. For medium variability in two or three dimensions, Eq. (19) implies constancy of the component $\mathbf{k}_t$ transverse to the $\mathbf{r}$-dependent direction of $\nabla\omega$; for comments concerning the graphical construction in this case, see remarks following Eq. (1.7.27).

1.6b Guided-wave Integral Representations

Field representations in terms of waves guided along a rectilinear spatial coordinate z are useful for stratified media whose properties vary continuously or abruptly along the z direction. Guided-wave integral representations of the field have the form [see Eq. (1.4.3)]

$$I(\mathbf{r}, t) = \int_{-\infty+i\gamma}^{\infty+i\gamma} I(\mathbf{r}, \omega)e^{-i\omega t}\, d\omega, \qquad \gamma > 0, \tag{20}$$

where

$$I(\mathbf{r}, \omega) = \iint_{-\infty}^{\infty} F(\mathbf{k}_t, \omega; z)e^{i\mathbf{k}_t\cdot\boldsymbol{\rho}}\, d\mathbf{k}_t. \tag{21}$$

γ in Eq. (20) is chosen large enough to ensure that the integration path in the complex ω plane lies above all the singularities of $I(\mathbf{r}, \omega)$, as required by causality. The transverse wavenumber $\mathbf{k}_t = \mathbf{x}_0 k_x + \mathbf{y}_0 k_y$ in Eq. (21) ranges over all real values in the plane transverse to z. The separation into $\mathbf{k}_t$ and ω integrations emphasizes the utility of guided-wave field representations for the study of time-harmonic excitation. If ω_0 is the harmonic oscillation frequency, linearity and invariability of the medium properties with time require that the frequency dependence of F has the form† $F(\mathbf{k}_t, \omega; z) = \bar{F}(\mathbf{k}_t, \omega_0; z)\delta(\omega - \omega_0)$, thereby rendering the ω integration in Eq. (20) superfluous and identifying $I(\mathbf{r}, \omega_0)$ as the time-harmonic response function.

Since even in a homogeneous medium, the guided-wave representation provides information that may be compared and contrasted with that deduced from the oscillatory representation in Sec. 1.6a, the homogeneous case is considered first and furnishes the basis for subsequent study of inhomogeneous media.

Homogeneous media (time-harmonic case)

In a homogeneous medium, the z dependence of F is in the form of plane waves, so $I(\mathbf{r}, \omega)$ is written typically as

---

†For this form of F, $\gamma = 0$ in Eq. (20). If $\gamma > 0$, one may use $F(\mathbf{k}_t, \omega; z) = \bar{F}(\mathbf{k}_t, \omega_0; z)/(\omega - \omega_0)$, close the integration contour in the lower half of the complex ω plane when $t > 0$, and evaluate via the residue theorem.

$$I(\mathbf{r}, \omega) = \int A(\mathbf{k}_t, \omega)e^{i\psi(\mathbf{r};\mathbf{k}_t,\omega)}\, d\mathbf{k}_t, \quad \psi(\mathbf{r}; \mathbf{k}_t, \omega) = \mathbf{k}_t \cdot \boldsymbol{\rho} + \kappa(\mathbf{k}_t, \omega)z, \qquad (22)$$

where the longitudinal propagation constant $k_z \equiv \kappa = \kappa(\mathbf{k}_t, \omega)$ follows from a plane-wave dispersion equation $f(\mathbf{k}_t, k_z, \omega) = 0$. For large values of $\mathbf{r} = \boldsymbol{\rho} + z_0 z$, the integral can be approximated by the saddle-point technique discussed in Chapter 4. The saddle points $\mathbf{k}_{ts}$ are defined implicitly by $\nabla_{k_t}\psi = 0$, where $\nabla_{k_t} = \nabla_k - z_0(\partial/\partial k_z) = x_0(\partial/\partial k_x) + y_0(\partial/\partial k_y)$, or on suppression of the ω dependence,

$$\frac{\boldsymbol{\rho}}{z} = -\nabla_{k_t}\kappa(\mathbf{k}_t) \qquad \text{at } \mathbf{k}_{ts} = \mathbf{k}_{ts}(\boldsymbol{\rho}, z). \qquad (23)$$

By proceeding as in Eqs. (3) and (4), one may write for the $\mathbf{k}_{ts}$ contribution to the integral in Eq. (22),

$$I(\mathbf{r}, \omega) \sim A(\mathbf{k}_{ts})e^{i[\mathbf{k}_{ts}\cdot\boldsymbol{\rho}+\kappa(\mathbf{k}_{ts})z]}\frac{2\pi e^{i(\pi/4)[\text{sgn }\hat{R}_1+\text{sgn }\hat{R}_2]}}{z\bar{Q}^{1/2}} \qquad (24)$$

where $\bar{Q} = |\hat{R}_1\hat{R}_2|^{-1}$ is the absolute value of the determinant of the matrix $\bar{\mathcal{D}}$, with

$$\bar{\mathcal{D}} = \begin{bmatrix} \dfrac{\partial^2\kappa}{\partial k_x^2} & \dfrac{\partial^2\kappa}{\partial k_x \partial k_y} \\[3mm] \dfrac{\partial^2\kappa}{\partial k_y \partial k_x} & \dfrac{\partial^2\kappa}{\partial k_y^2} \end{bmatrix}_{\mathbf{k}_t = \mathbf{k}_{ts}}, \qquad (24a)$$

and $(1/\hat{R}_{1,2})$ are the elements in the diagonalized form of the matrix $\bar{\mathcal{D}}$.

It is useful to write the result in Eq. (24) in an invariant form that is more directly descriptive of wave propagation in a homogeneous medium.[13] To this end, one assumes first that the z axis of the coordinate system is rotated so as to coincide with the radius vector $\mathbf{r}$, whence $\boldsymbol{\rho} = 0$ and $z = r$. The saddle-point condition (23) in the rotated $(\bar{\mathbf{k}}_t, \bar{\kappa})$ coordinate system then selects $\bar{\mathbf{k}}_{ts}$ points on the (wavenumber) dispersion surface $\bar{\kappa} = \bar{\kappa}(\bar{\mathbf{k}}_t)$ at which the normal vector is parallel to z (Fig. 1.6.5). At such points P, the diagonalized form of $\bar{\mathcal{D}}$ is equal to $(1/\bar{R}_1\bar{R}_2)$, where $\bar{R}_1$ and $\bar{R}_2$ are the principal radii of curvature of the wavenumber surface at P. The following quantities are invariant under coordinate rotation: the vector $\mathbf{r}$ from the origin in physical space to the observation point, the wave vector $\mathbf{k}_s$ from the origin in wavenumber space to a point P of the wavenumber surface whereon the normal is parallel to $\mathbf{r}$, and the principal radii of curvature at P. Thus, one may write Eq. (24) in the original fixed-coordinate system as

$$I(\mathbf{r}, \omega) \sim A(\mathbf{k}_{ts})e^{i\mathbf{k}_s\cdot\mathbf{r}}\frac{2\pi\, e^{\,i(\pi/4)(\text{sgn }R_1+\text{sgn }R_2)}|\bar{R}_1\bar{R}_2|^{1/2}}{r}. \qquad (25)$$

If several points P on the surface satisfy the above requirements, each contributes an expression of the form (25) to $I(\mathbf{r}, \omega)$ (see subsequent remarks concerning the radiation condition). Singularities in the integrand of Eq. (22) may also contribute to the asymptotic value of the integral but are ignored for the present.

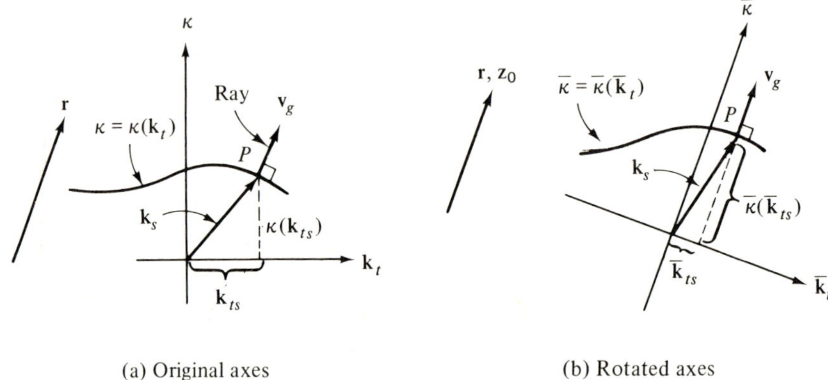

FIG. 1.6.5   Graphical location of saddle point.

It is useful to recall that a wavenumber surface in $\mathbf{k}$ space, corresponding to a fixed oscillation frequency ω, is one of a family of $\omega = \omega(\mathbf{k})$ dispersion surfaces descriptive of the space–time dispersion relation in four dimensions. The normal vector to the surface $\omega = $ constant is given by $\nabla_k \omega(\mathbf{k})$, identified previously in Eq. (2) as the group-velocity vector $\mathbf{v}_g$ in the direction of energy transport, the *ray* direction. By allowing for a slight frequency spread in a time-harmonic signal, one may still assign significance to the group velocity in the present discussion. Thus, Eq. (23) may be interpreted as defining a ray trajectory, $\boldsymbol{\rho}/z = $ constant, along which travels the energy emitted continuously by the source. Since the medium is homogeneous, the rays are straight lines, whence, from Eq. (23),

$$\mathbf{k}_{ts} = \text{const.}, \qquad \kappa(\mathbf{k}_{ts}) = \text{const.} \Rightarrow \mathbf{k}_s = \text{const.} \qquad (26)$$

The graphical location of the saddle points in Fig. 1.6.5 is precisely the same as in Fig. 1.6.1(c). Only those points $\mathbf{k}_s$ on the $\mathbf{k}$ surface may be included in (25) for which the orientation and direction of $\nabla_k \omega$ coincide with that of $\mathbf{r}$. Such points satisfy a radiation condition requiring outflow of energy along the radial direction. For electromagnetic wave propagation in a non-spatially-dispersive, anisotropic dielectric (e.g., a cold magnetoplasma), identification of permitted points is simplified by the recognition that the angle between $\mathbf{k}_s$ and $\mathbf{v}_g$ (codirectional with $\mathbf{r}$) must not exceed $90°$ [see Eq. (1.7.53a)].

As in the analogous result in Eq. (5) for the time-dependent case, the time-harmonic field is established locally by a plane wave $\exp(i\mathbf{k}_s \cdot \mathbf{r})$ whose amplitude $A(\mathbf{k}_{ts})$ in Eq. (22) is modified by the last factor in Eq. (25). The latter arises from the interference of plane waves having wavenumbers within a narrow cone $\Delta\mathbf{k}$ in $\mathbf{k}$ space. Such a cone near $\mathbf{k}_1$ is shown in Fig. 1.6.6(a) and intercepts an area element $\Delta\bar{A}_1$ on the wavenumber surface; the corresponding ray tube, drawn perpendicular to the surface on the boundaries of $\Delta\bar{A}_1$, has its "vertex" at a distance proportional to $\sqrt{|\bar{R}_1 \bar{R}_2|}$, where $\bar{R}_1$ and $\bar{R}_2$ are the principal radii of curvature of the wavenumber surface at $\mathbf{k}_1$. The (constant)

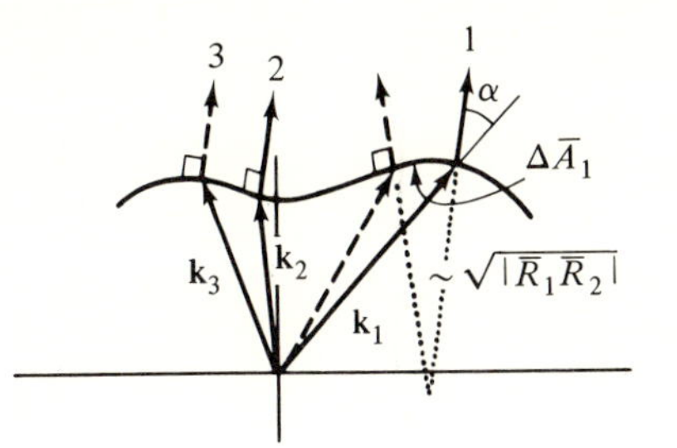
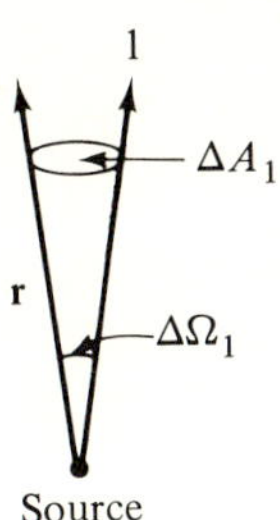

(a) Wavenumber surface   (b) Physical configuration

FIG. 1.6.6 Point source in an anisotropic medium.

energy contained in the plane-wave bundle described by $\Delta\mathbf{k}$ is propagated in space within the ray tube, whence it is evident that the energy density in space varies inversely with the ray-tube cross section, which, in turn, depends on the rate of divergence of the rays (see also Sec. 1.7b). From Fig. 1.6.6(a), a weakly curved surface at $\mathbf{k}_1$ gives rise to a slowly diverging ray cone with weak decay of energy density in the ray tube, whereas the decay is rapid for strong surface curvature. If $\Delta\Omega_1$ denotes the solid angle subtended by the ray cone at the source, then on comparing Figs. 1.6.6(a) and (b), $\Delta\Omega_1 = \Delta A_1/r^2 \propto \Delta\bar{A}_1/\bar{R}_1\bar{R}_2$, so the ray-tube cross section corresponding to a fixed $\Delta\bar{A}_1$ varies like $r^2/\bar{R}_1\bar{R}_2$. Since the field amplitude is proportional to the square root of the energy density, one verifies the behavior exhibited in Eq. (25).[13,14,16]

z-stratified media (time-harmonic case)

In a medium with variability along the z coordinate, the function F in the integrand of Eq. (21) cannot generally be expressed in closed form in terms of known functions. Closed-form expressions for F are obtainable for special z functions descriptive of the medium variation (see Sec. 5.9), but even under this circumstance the integral (21) is usually not calculable exactly. For slow medium variations, however, F may be approximated at almost all z by its local plane-wave (WKB) representation (Sec. 5.8d), whence Eq. (21) becomes

$$I(\mathbf{r}, \omega) = \int A(\mathbf{k}_t, \omega; z) \exp\left\{i\left[\mathbf{k}_t \cdot \boldsymbol{\rho} + \int^z \kappa(\mathbf{k}_t, \omega; \zeta)\, d\zeta\right]\right\} d\mathbf{k}_t, \qquad (27)$$

where both the local longitudinal propagation constant κ, which follows from the z-dependent plane-wave dispersion equation $f(\mathbf{k}_t, \kappa, \omega; z) = 0$, and the local amplitude A depend weakly on z. At large observation distances $\mathbf{r} = \boldsymbol{\rho} + \mathbf{z}_0 z$, or for short wavelengths,† the principal contribution to the integral (27) arises from the vicinity of saddle points $\mathbf{k}_{ts}$ (see Chapter 4); these are defined implicitly

---

†In the time-harmonic problem, the relevant parameter is the normalized distance r/λ, where λ is the local wavelength. Largeness of r/λ can be secured either by sufficiently large r or sufficiently small λ. The latter choice is used in Sec. 1.7a.

on equating to zero the $\mathbf{k}_t$ derivative (∇_{k_t}) of the exponent in the integrand (ω dependence omitted):

$$\boldsymbol{\rho} = -\int^z \nabla_{k_t} \kappa(\mathbf{k}_t, \zeta)\, d\zeta \qquad \text{at } \mathbf{k}_t = \mathbf{k}_{ts}(\boldsymbol{\rho}, z). \tag{28}$$

Following the procedure described previously, one obtains for the $\mathbf{k}_{ts}$ contribution to the integral (27) at the observation point $(\boldsymbol{\rho}, z)$ the result in Eq. (24), provided that the product $\kappa(\mathbf{k}_{ts})z$ in the exponent and denominator of Eq. (24) is replaced by $\int^z \kappa(\mathbf{k}_{ts}, \zeta)\, d\zeta$.

To facilitate location of the saddle points $\mathbf{k}_{ts}$, it is convenient, as in the case of the homogeneous medium [see Eqs. (23) and (26)], to interpret Eq. (28) as defining trajectories in $\mathbf{r}$ space. Of special interest are trajectories on which $\mathbf{k}_{ts}$ = constant. These describe propagation paths of local plane waves since in a homogeneous stratum at level z_i, a plane wave is described by the wavevector $\mathbf{k}(z_i) = \mathbf{k}_t + \mathbf{z}_0 \kappa(\mathbf{k}_t, z_i)$; phase continuity across an interface (Snell's law of refraction) leading to the next stratum z_j requires constancy of the tangential wavevector $\mathbf{k}_t$, whence $\mathbf{k}(z_j) = \mathbf{k}_t + \mathbf{z}_0 \kappa(\mathbf{k}_t, z_j)$. Thus, the differential form of Eq. (28),

$$\frac{d\boldsymbol{\rho}}{dz} = -\nabla_{k_t} \kappa(\mathbf{k}_t, z), \qquad \mathbf{k}_t = \text{constant}, \tag{29}$$

defines a family of curved trajectories along which plane-wave packets described by the parameter $\mathbf{k}_t$ propagate, and the saddle point $\mathbf{k}_{ts}$ selects the trajectory passing through a particular observation point $(\boldsymbol{\rho}, z)$.

Comparison of Eqs. (29) and (23) with Fig. 1.6.5 reveals that for a given value of $\mathbf{k}_t$, the local direction of the trajectory is along the normal to the (wavenumber) dispersion surface. Since the normal direction, determined in accord with the radiation condition, coincides with that of the energy flow, the curves of Eq. (29) characterize ray (energy-flow) trajectories. Their progress through the inhomogeneous medium can be charted by repeating the construction in Fig. 1.6.5(a), with $\mathbf{k}_t$ = constant, for the family of wavenumber plots. This procedure is illustrated in Fig. 1.6.7 for an isotropic medium wherein coincidence of ray and wavevector directions implies that each wavenumber plot is spherical. The medium is divided into thin locally homogeneous layers whose width is so chosen as to yield a good approximation to the given continuous profile [Fig. 1.6.7(a)]; the corresponding family of wavenumber diagrams is shown in Fig. 1.6.7(b). A reference value k_{t0} is assigned to the incident ray 1 in medium n_1, and the trajectory through the medium is plotted as in Fig. 1.6.7(c) by applying the condition $k_t = k_{t0}$ = constant to the ray construction in Fig. 1.6.7(b). One observes that the ray does not penetrate layer n_9 since $k_9 < k_{t0}$; this feature implies total reflection at the upper interface as shown in Fig. 1.6.7.(c) Additional details may be found in Sec. 5.8.

For excitation by a point source, all the rays pass through the source point $(\boldsymbol{\rho}', z') = (0, z')$; for each ray, the parameter k_t is specified in terms of the ray

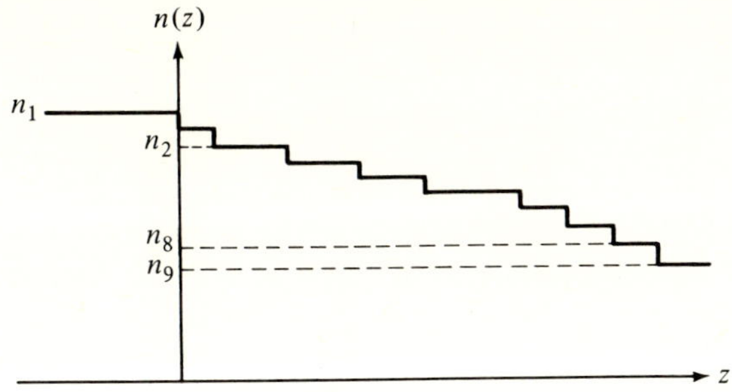

(a) Refractive index profile

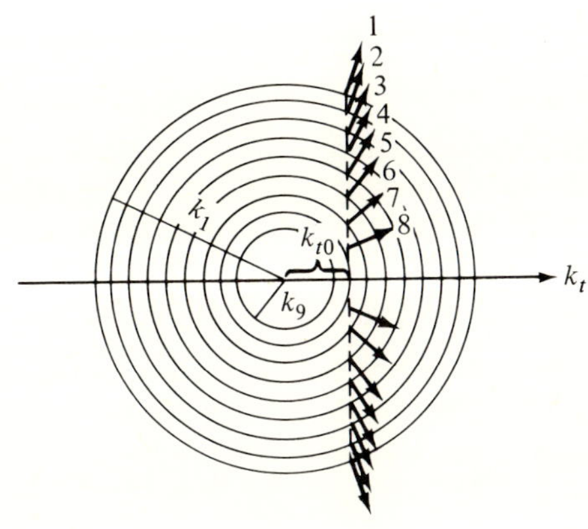

(b) Wavenumber plot
$$(k_i = k_0 n_i, \quad k_0 = \frac{\omega}{c})$$

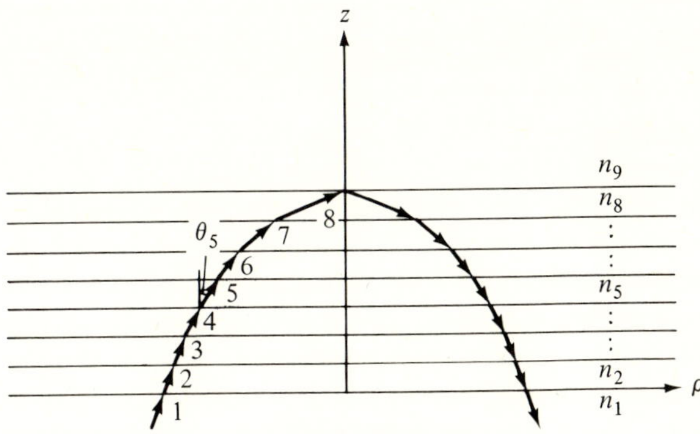

(c) Ray path in space
(constant k_{ts})

FIG. 1.6.7 Finely layered medium.

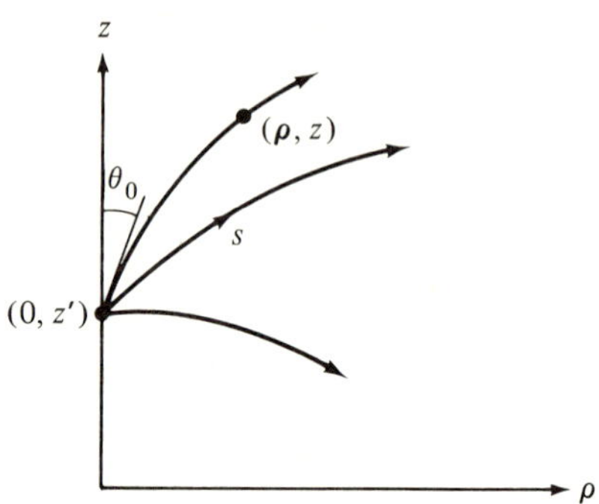

FIG. 1.6.8   Graphical determination of saddle point for an isotropic medium: $k_{ts} = k_0 n(z') \sin \theta_0$.

departure angle θ_0 as $k_t = k(z') \sin \theta_0 = k_0 n(z') \sin \theta_0$, where k_0 is a (constant) reference wavenumber and $n(z)$ is the refractive index. By selecting that ray which passes through the observation point (ρ, z) as in Fig. 1.6.8, one ascertains the saddle point $\mathbf{k}_{ts}$.

z-stratified media (transient case)

For excitation by a transient source, the representation integrals contain in addition to the $\mathbf{k}_t$ integration in Eq. (27) the ω integration in Eq. (20). Relevant saddle points $(\mathbf{k}_{ts}, \omega_s)$ are now determined from the conditions

$$\nabla_{k_t}\psi = \frac{\partial \psi}{\partial \omega} = 0, \qquad \psi = \mathbf{k}_t \cdot \boldsymbol{\rho} + \int^z \kappa(\mathbf{k}_t, \omega; \zeta)\, d\zeta - \omega t. \qquad (30)$$

For propagation of local plane-wave fields, only those saddle points that yield real values of $\mathbf{k}_t$, κ, and ω are of interest. The requirement $\nabla_{k_t}\psi = 0$ has already been discussed in connection with the time-harmonic problem [Eqs. (28) and (29)]. $\partial\psi/\partial\omega = 0$ yields

$$t = \int^z \frac{\partial \kappa(\mathbf{k}_t, \omega; \zeta)}{\partial \omega}\, d\zeta \qquad \text{at } \omega_s = \omega_s(\mathbf{r}, t), \qquad (31)$$

and, in differential form,

$$\frac{dz}{dt} = \frac{1}{\partial\kappa(\mathbf{k}_t, \omega; z)/\partial\omega} \qquad \text{at } \omega_s. \qquad (32)$$

For an interpretation of the simultaneous saddle-point conditions in Eqs. (28) and (31), it is instructive to consider a homogeneous medium with κ independent of z so that the z integrations can be performed trivially. The problem is now the same as in Sec. 1.6a, but the solution here has been obtained by a guided-wave representation of the fields, whereas in Sec. 1.6a the representation involves oscillatory modes. In connection with the latter, it has been shown that the saddle-point condition in Eq. (2) can be satisfied geometrically by locating on the four-dimensional $(\mathbf{k}, \omega)$ dispersion surface those points $(\mathbf{k}_s, \omega_s)$ where the four-vector normal is parallel to the four-vector $(\mathbf{r}/t, 1)$, provided that the coordinate axes k_x, k_y, k_z, ω are chosen parallel to $x, y, z, -t$, respectively. Since the geometrical procedure is independent of the particular form of the dispersion equation [$\omega = \omega(\mathbf{k})$ in Sec. 1.6a and $\kappa = \kappa(\mathbf{k}_t, \omega)$ in the present case], it is to be expected that Eqs. (28) and (31) select the same points on the $(\mathbf{k}, \omega)$ dispersion surface as does Eq. (2). For proof, we observe that at the point $(\mathbf{k}, \omega)$, the slopes with respect to the z axis of projections of the normal vector $\square f = (\nabla_{k_t} f, \partial f/\partial \kappa, \partial f/\partial \omega)$ onto the $(\boldsymbol{\rho}, z)$ and $(-t, z)$ hyperplanes are given by $\nabla_{k_t} f/(\partial f/\partial \kappa)$ and $-(\partial f/\partial \omega)/(\partial f/\partial \kappa)$, respectively. On the other hand, Eqs. (28) and (31), when specialized to a homogeneous medium, define a vector whose corresponding slopes are $-\nabla_{k_t}\kappa$ and $\partial \kappa/\partial \omega$, respectively. In view of the relations

$$\nabla_{k_t} f + \frac{\partial f}{\partial \kappa}\nabla_{k_t}\kappa = 0, \qquad \frac{\partial f}{\partial \omega} + \frac{\partial f}{\partial \kappa}\frac{\partial \kappa}{\partial \omega} = 0, \tag{33}$$

which follow from the dispersion equation $f(\mathbf{k}_t, \omega; \kappa(\mathbf{k}_t, \omega)) = 0$ on forming $df = 0$, the two vectors have the same direction. Since, at a given value of $(\mathbf{r}, t)$, the saddle points for either of the two guided-wave representations are deduced from the same points $(\mathbf{k}_s, \omega_s)$ on the dispersion surface, they describe the same wavepackets. To obtain the saddle points in the oscillatory formulation, one selects the projection $\mathbf{k}_s$, whereas for guided waves along z, the relevant projection is $(\mathbf{k}_{ts}, \omega_s)$.

From the preceding considerations and those in Sec. 1.6a, it is noted that in a homogeneous medium, a wavepacket moves along a straight-line ray path in space–time and preserves constancy of the total wavevector $\mathbf{k} = \mathbf{k}_s$ as well as the frequency $\omega = \omega_s$. When the medium is inhomogeneous, the wavepacket is described by the parameters $(\mathbf{k}_{ts}, \omega_s)$, with $\kappa(\mathbf{k}_{ts}, \omega_s; z)$ variable along z, so only the frequency, not the *total* wavevector $\mathbf{k}$, remains constant [see also Eq. (19)]. As noted from Eqs. (29) and (32), the space–time ray trajectories are now curved. When projected onto a hyperplane perpendicular to the time axis, the space–time rays generate the ray curves in Eq. (29), descriptive of the time-harmonic (i.e., $\omega = $ constant) field (Fig. 1.6.8). As noted previously, this relation permits the time-harmonic ray curves to be identified with paths of continuous energy transport in $\mathbf{r}$ space.

To track the space–time ray curves in an inhomogeneous medium, a graphical procedure analogous to that in Fig. 1.6.7 may be utilized. One constructs an appropriate sequence of dispersion surfaces for successive differential z elements along the path, and imposes the condition $(\mathbf{k}_{ts}, \omega_s) = $ constant. For illustration, we consider a wavepacket moving in the (z, ct) plane such that $\boldsymbol{\rho}$ and therefore $\mathbf{k}_{ts}$ equals zero. If the medium is an isotropic plasma, the dispersion equation for each local value of z_i [Fig. 1.6.9(a)] is given by Eq. (14) with ω_p replaced by $\omega_{pi} = \omega_p(z_i)$; the relevant portions of the corresponding dispersion curves are shown in Fig. 1.6.9(b). A wavepacket at a frequency ω_s is assumed to be incident from region 1 and is therefore characterized by ray segment 1. To effect the transition into region 2, whose properties differ slightly from those in region 1, one imposes the condition $\omega_s = $ constant to construct

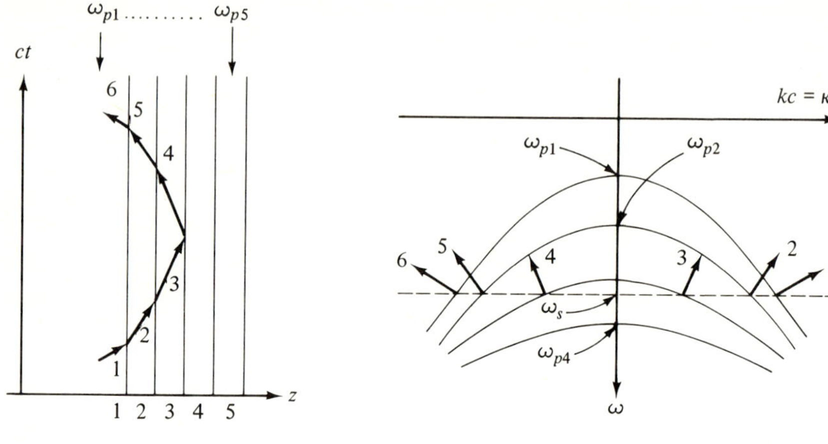

(a) Ray paths in space-time (constant ω_s)   (b) Dispersion surfaces

FIG. 1.6.9   Construction of space–time ray trajectories in an inhomogeneous medium.

ray segment 2, etc. (note that time always increases along a space-time ray path, in accord with causality). Since $\kappa(z)$ decreases with z in the present example, the space–time ray turns and the wavepacket is reflected; no energy at the selected frequency ω_s penetrates beyond layer 3.

With the saddle points determined implicitly from Eqs. (29) and (32), the behavior of the transient field can be ascertained from an asymptotic evaluation of the integrals in Eqs. (27) and (21) by techniques similar to those used for Eq. (1).

Transients in non-dispersive configurations (closed-form inversion of time-harmonic result)

The integration in Eq. (20) generally cannot be performed explicitly, so one must resort to asymptotic or other approximation procedures to evaluate the time-dependent field. An exception occurs for a class of non-dispersive problems in which the time-harmonic solution can be cast as a Laplace integral

$$I(\mathbf{r}, \omega) = \int_0^\infty e^{-s\tau} B(\mathbf{r}, \tau)\, d\tau, \qquad s = -i\omega, \tag{34}$$

where B is independent of s. If Eq. (34) applies, then it follows from the Fourier inversion of Eq. (20), with the causality requirement $I(\mathbf{r}, t) \equiv 0$ for $t < 0$, that

$$I(\mathbf{r}, \omega) = \frac{1}{2\pi} \int_0^\infty I(\mathbf{r}, t) e^{i\omega t}\, dt, \tag{35}$$

whence, from a comparison of Eqs. (34) and (35),

$$I(\mathbf{r}, t) = 2\pi B(\mathbf{r}, t). \tag{36}$$

An example included in the category of integrals (34) is the generic time-harmonic radiation integral of Eq. (5.3.14),

$$I(L, \omega) = \int_{\bar{P}} e^{ikL \cos (w-a)} f(w)\, dw, \tag{37}$$

where $\bar{P}$ is the contour shown in Fig. 5.3.6b. The parameters L and α are assumed to be positive, with α restricted to the range $0 < \alpha < \pi/2$, and the function $f(w)$ is to be independent of $k = \omega/c$. Setting $\omega \to is$ in Eq. (37), with Re s sufficiently large, one may write

$$I(L, \omega) = \int_{i\infty}^{-i\infty} e^{-s(L/c) \cos w}\, f(w + \alpha)\, dw, \tag{38}$$

if it is assumed that the function $f(w)$ has no singularities in the strip $0 < |\text{Re } w| < \pi/2$; if singularities are present, their effect may lead to additional contributions. Since $\exp [-s(L/c) \cos (w - \alpha)]$ decays in the strip $\cos (w_{\text{real}} - \alpha) > 0$, the contour of integration can be shifted to achieve the representation (38) if Re s is large enough. The successive changes of variables $\beta = iw$ and $\tau = (L/c) \cosh \beta$ lead to the formulation

$$I(L, \omega) = -i \int_{L/c}^{\infty} e^{-s\tau} \frac{b(\tau)}{\sqrt{\tau^2 - (L/c)^2}}\, d\tau, \tag{39a}$$

where

$$b(\tau) = f\left[\alpha - i \cosh^{-1}\left(\frac{c\tau}{L}\right)\right] + f\left[\alpha + i \cosh^{-1}\left(\frac{c\tau}{L}\right)\right]. \tag{39b}$$

Equation (39a) is evidently in the form (34), with

$$B(L, \tau) = \begin{cases} 0, & \tau < \dfrac{L}{c}, \\[2mm] \dfrac{-ib(\tau)}{\sqrt{\tau^2 - (L^2/c^2)}}, & \tau > \dfrac{L}{c}. \end{cases} \tag{40}$$

If $v(w) = -if(w)$ is real for real values of w, then $v(w^*) = v^*(w)$ (from the Schwartz reflection principle[17]) and $b(\tau)$ can be written as

$$-ib(\tau) = 2 \operatorname{Re}\left\{-if\left[\alpha - i \cosh^{-1}\left(\frac{c\tau}{L}\right)\right]\right\}. \tag{41}$$

It may be noted that the formulation in Eq. (39a) is useful even when $f(w)$ is dependent on s. Although one cannot then perform the Laplace inversion in closed form, a series expansion of $f(w; s)$ in powers of $1/s$, or of s, permits the derivation of asymptotic results in the time domain, applicable immediately or long after the arrival of the first response, respectively.

1.6c Diffraction and Transition Phenomena

The characterization of the far-zone field in terms of distinct wavepackets as carried out in Secs. 1.6a and 1.6b applies at almost all space–time points $(\mathbf{r}, t)$ but fails in "transition regions" wherein individual wavepackets either are not yet fully developed or are strongly modified by interaction with other wavepackets. The former regime obtains near the time of arrival of the initial disturbance, or wavefront, traveling at the highest propagation speed c in the medium. Since the field vanishes before the first response arrives, the field variables and (or) their derivatives must behave discontinuously across the wavefront. Wavefront field variations are synthesized primarily by very high frequency waves, for which dispersive effects are negligible. In contrast, wavepackets emerge from well-ordered dispersive wavetrains only at sufficiently long observation distances behind the wavefront or, for a stationary observer, at sufficiently long observation times after arrival of the initial response. In the analytical treatment, the transition region between the wavefront and wave-packet regimes is characterized by $\mathbf{k}_s, \omega_s \to \infty$.

Another class of transition phenomena occurs when two or more wavepack-ets have the same local wavenumber, frequency, and group velocity, thereby providing strong interaction that destroys the independent existence of each. These transition regions in $(\mathbf{r}, t)$ space for transient problems, or in $\mathbf{r}$ space for time-harmonic problems, are characterized by a confluence of saddle points and (or) singularities (critical points) in the integral representations of the field. Isolated pole or branch-point singularities near saddle points have been ignored in Secs. 1.6a and 1.6b, but may also provide asymptotic field contributions that can be interpreted in terms of distinct wave processes. These singularities are relevant when intercepted during path deformations required for the asymptotic evaluation of integrals by the saddle-point method (see Chapter 4)

The presence of transition regions can usually be discerned by a divergence in the simple saddle-point calculation of the amplitude of the affected wave constituents. For example, transition regions in $\mathbf{r}$ space for which $\bar{R}_{1,2} \to \infty$ in Eq. (25) correspond to inflection points on the dispersion curve and signify the coalescence of two saddle points. This divergence does not imply unlimited growth of the field but rather the inadequacy of the simple asymptotic formula for a particular wave type. For description of the far field valid at *all* observation points, one employs more complicated uniform asymptotic approximations, given in Chapter 4 for the case of integrals containing adjacent critical points. The somewhat different characterization of transition effects near a wavefront is discussed at the end of this section.

The wave processes mentioned above are classified conveniently as primary and diffraction (secondary) effects, with the former representing dominant contributions and the latter distinguishing corrections thereto. The identification of a particular wave type with a critical point depends on the integral representation employed; in one representation, a primary field may arise from a saddle point, whereas in another, it may be attributed to a singularity. Primary and diffracted asymptotic wave types are characterized conveniently in terms of rays. A list of various ray species, their mechanism of excitation, and their use in constructing the solution of a time-harmonic diffraction problem are illustrated in Sec. 1.7c. Since no detailed discussion of transient propagation in dispersive media is given elsewhere in this book, some aspects concerning the initial formation and subsequent interaction of wavepackets will be considered at this time. The interaction problem is treated first since it relates more directly to the discussion in Secs. 1.6a and 1.6b.

Transient and signal propagation in a magnetoplasma (interaction between wave-packets)

We assume that the time-harmonic field has been approximated asymptotically in the far zone so that $I(\mathbf{r}, \omega)$ in Eq. (20) is known explicitly. In simplified form, the time-dependent field then requires evaluation of the integral $I(\mathbf{r}, t)$:

$$I(\mathbf{r}, t) = \int_{-\infty+i\gamma}^{\infty+i\gamma} f(\omega)e^{iq(\omega)}\, d\omega, \qquad q(\omega) = \xi(\omega)r - \omega t, \qquad (42)$$

where f is an amplitude, r a distance coordinate, and ξ a modified wavenumber equal to k or $k \cos \alpha$ in an isotropic or anisotropic medium, respectively, with α denoting the angle between $\mathbf{k}$ and the ray vector. Although Eq. (42) implies a homogeneous medium, no essential complication arises when slow inhomogeneities, in the form of weak dependence of f and ξ on r, are present. The saddle-point contributions to Eq. (42) at $\omega = \omega_{si}$, treated analogously in Sec. 1.6b, furnish the result [see Eq. (4.2.1) et seq.]

$$I(\mathbf{r}, t) \sim \sum_i \sqrt{\frac{2\pi}{r|\xi''(\omega_{si})|}}\, f(\omega_{si}) \exp\left\{i\left[q(\omega_{si}) + \frac{\pi}{4} \operatorname{sgn} \xi''(\omega_{si})\right]\right\}, \qquad (43)$$

where the saddle points $\omega_{si}(r, t)$ are determined implicitly by $q'(\omega_{si}) = 0 = (r/t) - 1/\xi'(\omega_{si})$, with the prime denoting the derivative with respect to ω.† As noted previously, each i-term in Eq. (43) represents a wavepacket with central frequency ω_{si}, wavenumber $\xi(\omega_{si})$, and group speed $[1/\xi'(\omega_{si})]$. The saddle points ω_{si} may be located graphically as shown in Fig. 1.6.10(a) for a multibranched dispersion surface (representative, for example, of extraordinary wave propagation across the magnetic field in a cold magnetoplasma; see Secs. 8.3a and 8.3b). This should be contrasted with the analogous construction in Fig. 1.6.3, where the saddle-point variable is the wavenumber k rather than the frequency ω. With r fixed, one observes from Fig. 1.6.10(b) that a single saddle point contributes for observation times $r/c < t < t_s$, while three saddle points contribute for $t > t_s$ [see also Fig. 1.6.10(c)]. The time t_s distinguishes the arrival of a wavepacket of frequency ω_{ss} traveling at the maximum group speed corresponding to the inflection point T of the upper branch of the plot in Fig. 1.6.10(a).

One observes from Fig. 1.6.10(a) or 1.6.10(b) that the wavepackets characterized by rays 2 and 3 interact strongly when $t \approx t_s$. In fact, distinct

---

†Since $I(\mathbf{r}, t)$ must be real, saddle points occur in pairs at $(\pm\omega_{si})$, the contribution from $(-\omega_{si})$ being the complex conjugate of that from $(+\omega_{si})$. Only the latter contributions are included in Eq. (43) [see the remarks following Eq. (5)]. Note also, that by expanding $\xi(\omega)$ about ω_s and inverting, one shows readily that $(d\xi(\omega)/d\omega)_{\omega_s} = (d\omega(\xi)/d\xi)_{\xi_s}^{-1}$, with $\omega_s = \omega(\xi_s)$ and $\xi_s = \xi(\omega_s)$, thereby providing alternative expressions for the group velocity.

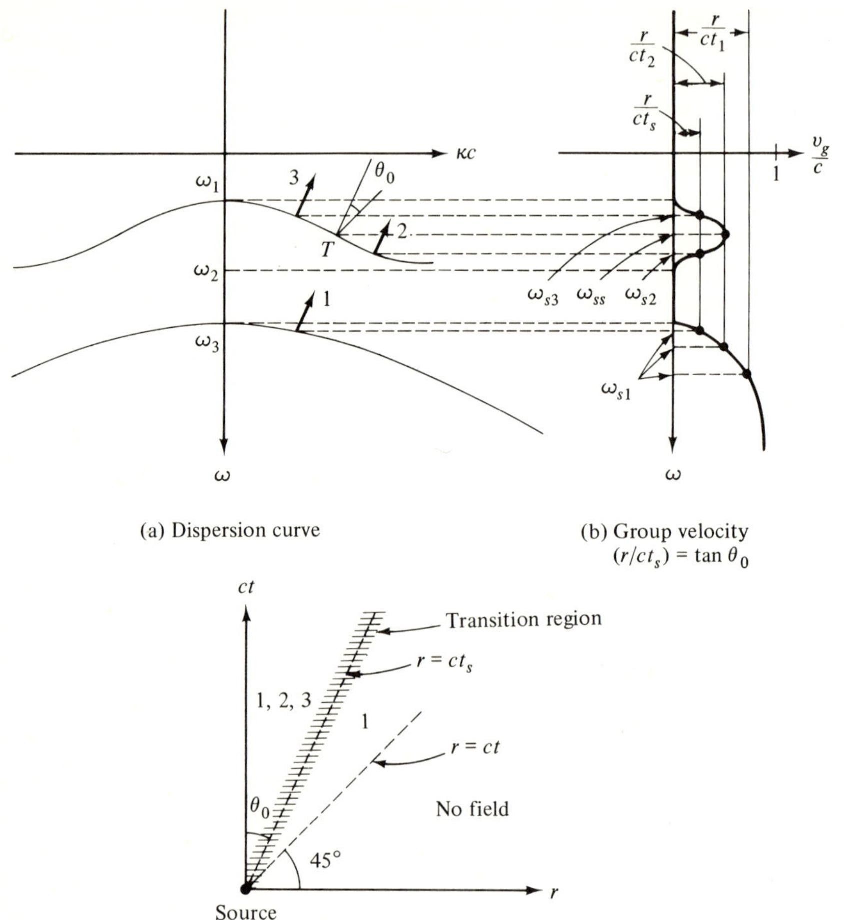

(a) Dispersion curve

(b) Group velocity
$(r/ct_s) = \tan\theta_0$

(c) Space-time regions reached by various ray species

FIG. 1.6.10 Two-branch dispersion curve, group velocity, and space–time ray plots (inflection-point singularity).

wavepackets can be distinguished only at sufficiently long observation times after the arrival at t_s of the cluster of energy with central frequency ω_{ss}. The existence of a transition region near ω_{ss} is manifested by the vanishing of the curvature term $\xi''(\omega_{ss}) = 0$, thereby invalidating Eq. (43). The proper description of the field near ω_{ss} allows for the coalescence of the two adjacent saddle points $\omega_{s2,3}$ and leads to a representation in terms of Airy functions (see Sec. 4.2e), whence this portion of the transient behavior is sometimes called the Airy phase.[18] The enhancement of the transient field observed during the Airy phase is analogous to the enhancement of the time-harmonic field observed near a caustic (see Sec. 5.8d). Both phenomena describe a focusing of energy.

Another interesting transition phenomenon is associated with a pole singularity in the amplitude function $f(\omega)$ in Eq. (42), as occurs for a suddenly switched-on time-harmonic wave source $e^{-i\omega_0 t} U(t)$, where $U(t) = 0$ or 1 for $t < 0$ or $t > 0$, respectively. The Fourier transform of the source function (and hence $f(\omega)$) has a simple pole at $\omega = \omega_0$ which contributes to the integral in Eq. (42) the residue

$$I_r = -2\pi i [(\omega - \omega_0)f(\omega)]_{\omega_0} e^{iq(\omega_0)} U(t - t_0). \tag{44}$$

Equation (44) represents a time-harmonic signal which arrives at the observation point r at a time t_0 specified implicitly by $\omega_s(r, t_0) = \omega_0$. Such a signal appears when, as in Fig. 1.6.11(a) and (b), the original integration path C, on being deformed into the path C' through the saddle point, crosses the pole singularity at ω_0 [the contour C' is drawn for the condition $q''(\omega_s) < 0$ satisfied on the ω_{s1} branch in Fig. 1.6.10(a); see Sec. 4.4a for relevant details on saddle point

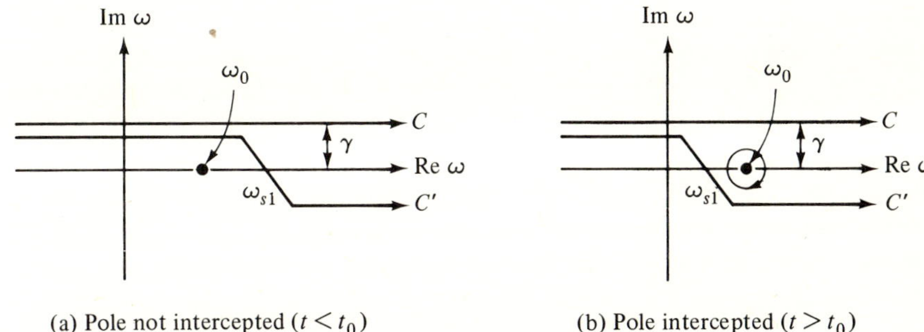

(a) Pole not intercepted ($t < t_0$)   (b) Pole intercepted ($t > t_0$)

FIG. 1.6.11   Original path C in the presence of a pole and deformed path C' through the saddle point.

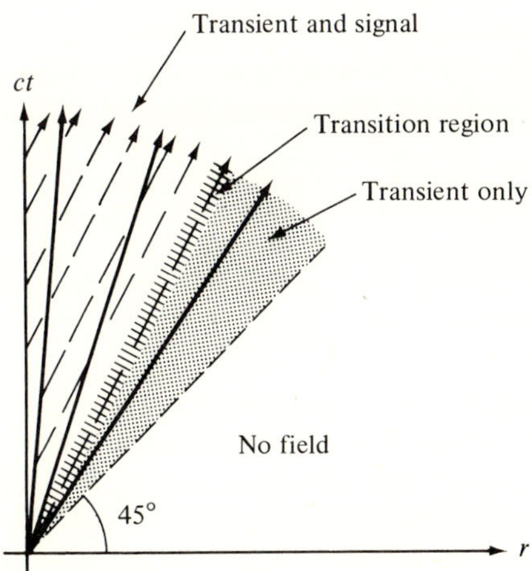

FIG. 1.6.12   Space–time ray plot for transient ($\rightarrow$) and time-harmonic ($--\rightarrow$) signal (pole singularity).

integration]. The time t_0 is required for the harmonic signal to traverse at its group speed $v_{g0} = 1/\xi'(\omega_0)$ the distance from the source to the observation point at r; the wave contribution (44) is maintained at subsequent observation times by the continual arrival of wavepackets at the group speed v_{g0}. The space–time ray diagram in Fig. 1.6.12 depicts both the time-harmonic signal in Eq. (44) and the transient terms in Eq. (43). Near the arrival time t_0, Eq. (43) fails since $f(\omega_0) \to \infty$; in this transition region characterized by adjacent pole and saddle point, the field is represented in terms of a Fresnel integral (Sec. 4.4a). The main signal in Eq. (44) may be regarded as the primary contribution, whereas the transients in Eq. (43) constitute a diffraction effect.[19]

Field behavior near a wavefront prior to formation of a wavepacket

The procedures of Secs. 1.6a and 1.6b fail for saddle-point values $\mathbf{k}_s$, $\omega_s \to \infty$. In view of the limiting form $\omega = kc$ of the dispersion relation as $k, \omega \to \infty$, the surface curvature term proportional to $d^2\omega/dk^2$ or $d^2k/d\omega^2$ in the denominators of Eqs. (5) or (43), respectively, tends to zero in this limit. As noted previously, the medium does not possess dispersive properties at frequencies characterizing the highest propagation speed c of the initial disturbance or wavefront, with a consequent inability for wavepacket formation. Since the fields on and near the wavefront are established by the high frequency wave components, it is to be expected that the high-frequency time-harmonic wave solution characterizes the initial transient response, and conversely. Let us assume that the field $I(\mathbf{r}, t)$ near the wavefront $t = r/c$ behaves as

$$I(\mathbf{r}, t) \sim a\left(t - \frac{r}{c}\right)^{\beta}, \qquad t \geq \frac{r}{c}, \tag{45}$$

where a is time independent and $\beta > -1$. Substitution of this approximate expression into Eq. (35), with the lower limit replaced by r/c since $I \equiv 0$ for $t < r/c$, may be justified for $\omega \to \infty$ [with Im $\omega > 0$; see Eq. (20)] since $\exp(i\omega t)$ then decays rapidly away from $t = r/c$, thereby localizing the effective integration range. Use of the gamma function

$$\Gamma(x) = \int_0^\infty v^{x-1} e^{-v}\, dv, \qquad \mathrm{Re}\, x > 0, \tag{46}$$

in Eq. (35) then yields

$$I(\mathbf{r}, \omega) \sim e^{ikr} \frac{a\Gamma(\beta + 1)}{2\pi(-i\omega)^{\beta+1}}, \qquad \omega \to \infty, \tag{47}$$

where $k \approx \omega/c$ and ω is now allowed to be real. Thus, the field behavior near $t = (r/c)$ is related to the time-harmonic behavior for $\omega \to \infty$ as in Eqs. (47) and (45). If $I(\mathbf{r}, \omega) \sim (2\pi)^{-1} a' \exp(ikr)$, where a' is independent of ω, the corresponding time-dependent function is $I(\mathbf{r}, t) \sim a'\delta[t - (r/c)]$. By including higher-order expansion coefficients in Eq. (45), one may generate higher-order terms in the asymptotic expansion (47), and conversely.[20] This aspect is considered further in Sec. 1.7e.

Even when higher-order terms in inverse powers of ω are included in Eq. (47), the resulting time-dependent field $I(\mathbf{r}, t)$ does not provide a transition to the propagation regime characterizable in terms of wavepackets since the (non-dispersive) dispersion relation $k \sim \omega/c$ is too restrictive. An improved formulation is obtained by retaining the next term in the high-frequency approximation,

$$k(\omega) \sim \frac{\omega}{c} - \frac{\omega_\alpha^2}{c\omega}, \qquad \omega \to \infty, \tag{48}$$

where ω_α is generally a characteristic frequency of the medium. Equation (48) accounts for the onset of dispersion, removes the infinity introduced by the vanishing of $d^2\omega/dk^2$ in Eq. (5) or $d^2k/d\omega^2$ in Eq. (43) (with $\xi \equiv k$), and therefore forms the basis for a uniform approximation which connects the wavefront with the wavepacket regimes. Insertion of Eq. (48) into Eq. (42), with use of the asymptotic behavior of the amplitude function $f \sim B(-i\omega)^\nu, \omega \to \infty$, where $B = $ const., yields via a known integral expression for the Bessel function,[21]

$$I(\mathbf{r}, t) \sim 2\pi B\left(\frac{\tau}{b}\right)^{-(1+\nu)/2} J_{-1-\nu}(2\sqrt{b\tau}), \qquad b = \frac{\omega_\alpha^2 r}{c}, \qquad \tau = t - \frac{r}{c} \tag{49}$$

valid for a small range of observation times at or near r/c; b is a large parameter if $\mathbf{r}$ is large. For very short observation times such that $b\tau \to 0$, the small argument approximation for the Bessel function reduces Eq. (49) to

$$I \sim 2\pi B \frac{\tau^{-(1+\nu)}}{\Gamma(-\nu)}, \tag{50}$$

which agrees with the result in Eq. (45) since in the present instance, $I(r, \omega) \sim B(-i\omega)^\nu \exp(ikr)$. For somewhat longer observation times such that $2\sqrt{b\tau} \gg 1$ (since b is large, the inequalities $b\tau \gg 1$ and $\tau \ll 1$ can be satisfied simultaneously), one finds on use of the large-argument approximation for the Bessel function [see. Eq. (4.2.22b)],

$$I \sim 2B\pi^{1/2} \frac{b^{(\nu+\frac{1}{2})/2}}{\tau^{(\nu+3/2)/2}} \cos\left(2\sqrt{b\tau} + \frac{\nu+1}{2}\pi - \frac{\pi}{4}\right). \tag{51}$$

With $\xi(\omega) \equiv k(\omega)$ given by Eq. (48), one may verify that the resulting expression for I in Eq. (43), added to its complex conjugate, agrees with the expression in Eq. (51). Thus, Eq. (49) provides the desired transition from the wavefront to the wavepacket behavior of the field. For a uniform asymptotic treatment valid for a larger range of τ, see Reference 22.

1.7 RAY-OPTIC APPROXIMATIONS FOR DIFFERENTIAL EQUATIONS

In Sec. 1.6, rigorous integral representations of the time-dependent and time-harmonic field in separable regions were approximated by asymptotic (saddle-point) techniques valid at distant observation points. These asymptotic approximations frequently assumed an invariant ray-optical form that suggests their applicability to classes of problems broader than the separable ones to which they refer. The validity of ray-optical approximations in non-separable inhomogeneous configurations is established in this section by an asymptotic procedure based directly on the differential field equations.

Modal (integral representation), as opposed to direct asymptotic, techniques are based on exact field solutions (although for a limited class of problems), so approximations employed in their evaluation can be validated systematically. Thus, as discussed in Sec. 1.6c, no new representation, but only a modified asymptotic technique, is required for calculating the field in transition regions. While the saddle-point procedure highlights wave interference and thereby wave-packet processes, the introduction of rays and trajectories is not essential to the asymptotic evaluation but does serve to clarify physical propagation mechanisms. In contrast, if one assumes the form of the asymptotic field at the outset as in the direct procedure, one can use the exact differential equations for the field to derive simplified equations for phase and amplitude functions in the assumed field representation. These simpler equations define ray trajectories and energy-conservation theorems descriptive of transport properties that play a direct role in the analysis. For this reason, the direct method is henceforth called the "ray method." Since it describes energy-transport phenomena, the ray method does not furnish initial values of a field constituent; in the vicinity of source or scattering regions, such values must be determined by other (e.g., modal) methods. Like the simple saddle-point method, the ray procedure fails in transition regions. A major modification is required to remedy this failure, the latter being attributable to inadequacy of the initially assumed asymptotic field form. This defect may be removed by recourse to boundary-layer techniques,[23] but these will not be discussed herein. As mentioned, the ray method is not limited to separable configurations and therefore has broad scope. Confidence in its validity is confirmed by comparisons with asymptotic fields derived from exact modal solutions of separable problems, and an effective methodology can be constructed by selective use of both procedures.

The basic features of the ray method have already been described; details of its application depend on the form of the time-dependent or time-harmonic equations defining the field. As may be surmised, the formulation is simplest for scalar fields, with additional complexity arising from polarization, anisotropy, etc., for vector fields. The time-harmonic case is analyzed more easily than the time-dependent case since the presence of dispersion in the latter complicates the structure of the field equations. Although the differences are most pronounced in the calculation of field amplitudes and polarizations, information concerning ray trajectories and phase progression is deduced from first-order partial differential equations having a common structure for all cases. These equations, their formal solution by the method of characteristics, and their trajectory interpretation are discussed in Sec. 1.7a. The presentation then proceeds to time-harmonic propagation of scalar fields in Sec. 1.7b, and to the treatment of time-harmonic vector fields in isotropic and anisotropic media in Sec. 1.7c. Ray constructs developed in Secs. 1.7b and 1.7c for propagation in unbounded media are extended in Sec. 1.7d by introduction of reflected, refracted, and diffracted rays to account for the presence of boundaries and scattering centers. The geometrical theory of diffraction, a ray theory for synthesizing high-frequency fields in the presence of composite objects in terms of simpler constituent ray fields, is also presented in Sec. 1.7d and applied to an illustrative example. In the time-dependent regime, the ray method leads directly from the differential field equations to wavepackets and their trajectories which have been deduced in Secs. 1.6a and 1.6b by asymptotic evaluation of modal integrals. Transient propagation in an isotropic plasma serves to illustrate the ray method for a simple case (Sec. 1.7e). The ray method also permits further elaboration of the relation between high-frequency time-harmonic fields and transient fields near an impinging wavefront.

1.7a   Rays and the Theory of Characteristics

As noted in Sec. 1.6, ray methods are based on asymptotic field representations in terms of assumed local plane waves. In the abstract notation of Sec. 1.1d, one assumes to the lowest order of approximation (see Sec. 1.7e for a complete asymptotic expansion) that for large $\mathbf{r}$, t, a general linear space- and time-dependent field can be represented as (v here should not be confused with the same symbol used on p. 123)

$$\Psi(\mathbf{r}, t) \sim \Psi_0(\mathbf{r}, t)e^{iv\psi(\mathbf{r},t)}, \tag{1}$$

where v denotes the large parameter in the asymptotic development. Although some of the considerations below apply also to lossy media, we shall deal only with the lossless case unless specified otherwise. The wavevector $\Psi(\mathbf{r}, t)$ satisfies the first-order source-free field equations

$$L\left(\nabla, \frac{\partial}{\partial t}; \mathbf{r}, t\right)\Psi(\mathbf{r}, t) = 0, \tag{2}$$

wherein the explicit dependence of the operator L on $\mathbf{r}$ and t signifies the formal applicability of Eq. (2) to media with weak spatial and temporal inhomogeneities. On substitution of Eq. (1) into Eq. (2), one derives on retention of only the dominant term in v the following first-order system of partial differential equations for the phase function ψ and the amplitude function Ψ_0:

$$L\left(iv\nabla\psi, iv\frac{\partial\psi}{\partial t}; \mathbf{r}, t\right)\Psi_0(\mathbf{r}, t) = 0. \tag{3}$$

Retention of $\mathbf{r}$ and t in Eq. (3) and below implies that these quantities are of $O(v)$; moreover, one may regard certain "natural frequencies" in the medium to be $O(v)$ (these aspects, unimportant for the general discussion, are clarified in Sec. 1.7e).

Since in a homogeneous medium, $v\psi = \mathbf{k}\cdot\mathbf{r} - \omega t$, where $\mathbf{k}$ and ω are the (constant) wavevector and wave frequency, respectively, the derivatives $\nabla\psi$ and $-\partial\psi/\partial t$ assume the roles of *local* wavenumber $\bar{\mathbf{k}}$ and frequency $\bar{\omega}$, respectively, normalized with respect to v. From Eq. (3), for non-vanishing Ψ_0, the phase ψ must evidently satisfy the determinantal equation

$$\det L\left(iv\nabla\psi, iv\frac{\partial\psi}{\partial t}; \mathbf{r}, t\right) = 0, \tag{4a}$$

which, in terms of $\bar{\mathbf{k}}$ and $\bar{\omega}$, becomes the local (space- and time-dependent) *dispersion equation* [see Eq. (1.2.44) and the notational statement following Eq. (1.2.41)]

$$\det L(\bar{\mathbf{k}}, \bar{\omega}; \mathbf{r}, t) = 0, \qquad \bar{\mathbf{k}} = \nabla\psi, \qquad \bar{\omega} = -\frac{\partial\psi}{\partial t}, \tag{4b}$$

or, explicitly,

$$\bar{\omega} = \bar{\omega}(\bar{\mathbf{k}}, \mathbf{r}, t), \tag{4c}$$

where the same symbol has been used to denote $\bar{\omega}$ and its functional dependence on $\bar{\mathbf{k}}, \mathbf{r}, t$. The eigenvectors $\Psi_{0\alpha}(\mathbf{r}, t)$ of Eq. (3), corresponding to each solution $\bar{\omega}_\alpha(\bar{\mathbf{k}}, \mathbf{r}, t)$ of the dispersion equation (4b), then yield the polarization of each of the eigenwave constituents of the lowest-order wavevector Ψ_0. The amplitude $\Psi_0(\mathbf{r}, t) = \sum_\alpha A_\alpha(\mathbf{r}, t)\Psi_{0\alpha}(\mathbf{r}, t)$ is synthesized by superposition of the eigenwave solutions; its further determination requires use of a transport equation, the next in a series of equations obtained when $\Psi(\mathbf{r}, t)$ is expanded beyond the first term in Eq. (1) in a series of terms involving decreasing powers of v.[24]

For time-harmonic wave propagation in a time-invariant but spatially inhomogeneous medium, the time dependence $\exp(-i\omega t)$ may be suppressed and the lowest-order approximation of the wavevector assumes the form (see Secs. 1.7b and 1.7c for a complete asymptotic expansion)

$$\Psi(\mathbf{r}) \sim \Psi_0(\mathbf{r})e^{ik_0\psi(\mathbf{r})}, \tag{5}$$

where the reference wavenumber $k_0 = \omega/c$ takes on the role of the large parameter. The time-harmonic field equations follow from Eq. (2) on removal of the time variable t and the replacement $(\partial/\partial t) \to -i\omega$. To a lowest order of approximation, the correspondingly modified Eq. (3) becomes

$$L(\nabla, \mathbf{r}; k_0)\Psi(\mathbf{r}) = 0, \qquad -iL(\nabla, \mathbf{r}; k_0) = M(\nabla) - k_0cW(\mathbf{r}), \tag{6a}$$

where use has been made of the decomposition of the operator L stated in Eq. (1.3.8). Substitution of Ψ from Eq. (5) yields, to a first order of approximation,

$$\bar{L}(\nabla\psi, \mathbf{r})\Psi_0(\mathbf{r}) = 0, \qquad \bar{L}(\nabla\psi, \mathbf{r}) = M(\nabla\psi) + icW(\mathbf{r}). \tag{6b}$$

On introduction of the wavevector $\bar{\mathbf{k}} = \nabla\psi$, normalized with respect to k_0, the determinental equation for ψ becomes

$$\det \bar{L}(\bar{\mathbf{k}}, \mathbf{r}) = 0, \qquad \bar{\mathbf{k}} = \nabla\psi; \tag{6c}$$

this equation is commonly referred to as the *eiconal equation* of geometrical optics. For the special case of an isotropic medium, one has the explicit form [see Eqs. (18a) and (38b)]

$$\bar{\mathbf{k}} = \mathbf{s}\, n(\mathbf{r}), \qquad \mathbf{s} = \frac{\bar{\mathbf{k}}}{\bar{k}}, \tag{6d}$$

where n is the refractive index and $\mathbf{s}$ is a unit vector in the direction of $\nabla\psi$ perpendicular to the wavefront $\psi = $ constant. As in the time-dependent case, calculation from Eq. (6b) of the eigenvectors $\boldsymbol{\Psi}_{0\alpha}(\mathbf{r})$ for each solution of the eiconal equation provides the polarization and amplitudes of the eigenwaves required for the synthesis of the wave vector $\boldsymbol{\Psi}_0(\mathbf{r})$ in Eq. (5). Further properties of $\boldsymbol{\Psi}_0(\mathbf{r})$ are inferred from a transport equation derived on use of an expansion of $\boldsymbol{\Psi}(\mathbf{r})$ in decreasing powers of k_0.

Both the dispersion and eiconal equations (4b) and (6c) are first-order partial differential equations. They are reducible by the method of characteristics to first-order ordinary differential equations that can be integrated formally along special trajectories, the rays defined previously in Sec. 1.6. Consider the generic form

$$G\left(\frac{\partial\varphi}{\partial x_1} \cdots \frac{\partial\varphi}{\partial x_n}, \; x_1 \cdots x_n\right) = 0, \tag{7}$$

defining the function $\varphi(x_1 \cdots x_n)$, where $x_i, i = 1, \ldots, n$, represents either a space or time coordinate. By implicit differentiation, one observes on denoting $(\partial\varphi_i/\partial x)$ by ξ_i that

$$dG = 0 = \sum_{i=1}^{n}\left[\frac{\partial G}{\partial\xi_i}d\xi_i + \frac{\partial G}{\partial x_i}dx_i\right], \tag{7a}$$

which equation can be satisfied if

$$\frac{d\xi_i}{\partial G/\partial x_i} = -\frac{dx_i}{\partial G/\partial\xi_i}, \qquad i = 1, \ldots, n. \tag{7b}$$

By introducing a parameter s and defining the trajectory $x_i = x_i(s)$, $\xi_i = \xi_i(s)$ in ξ, x phase space, one may write Eq. (7b) in a form similar to Hamilton's canonical equations in mechanics,

$$\frac{dx_i}{ds} = \frac{\partial G}{\partial\xi_i}, \quad \frac{d\xi_i}{ds} = -\frac{\partial G}{\partial x_i}, \qquad i = 1, \ldots, n. \tag{8}$$

On these trajectories, the derivative of φ is given by

$$\frac{d\varphi}{ds} = \sum_{i=1}^{n}\frac{\partial\varphi}{\partial x_i}\frac{dx_i}{ds}, \tag{9}$$

whence φ can be determined by integration over s. (For a rigorous derivation, the reader should consult standard references on the theory of partial differential equations.[25,26])

To apply these results to the time-dependent dispersion relation, we let $x_1 = x, x_2 = y, x_3 = z, x_4 = t, \varphi \equiv \psi, G \equiv \bar\omega - \bar\omega(\bar{\mathbf{k}}, \mathbf{r}, t)$. From Eq. (4b) and one of the relations in Eq. (8), $dt/ds = -1$, so the time variable can be used as the parameter descriptive of the ray trajectories $\mathbf{r} = \mathbf{r}(t)$. Then, from Eq. (8),

$$\frac{d\mathbf{r}}{dt} = \nabla_k\bar\omega(\bar{\mathbf{k}}, \mathbf{r}, t), \quad \frac{d\bar{\mathbf{k}}}{dt} = -\nabla\bar\omega(\bar{\mathbf{k}}, \mathbf{r}, t), \tag{10a}$$

whence

$$\frac{d\bar{\omega}(\bar{\mathbf{k}}, \mathbf{r}, t)}{dt} = \frac{\partial\bar{\omega}(\bar{\mathbf{k}}, \mathbf{r}, t)}{\partial t}. \tag{10b}$$

Equation (10b) is not independent but follows from Eqs. (10a) and the derivative formula $(d\bar{\omega}/dt) = (\nabla\bar{\omega}\cdot d\mathbf{r}/dt) + (\nabla_k\bar{\omega}\cdot d\mathbf{k}/dt) + (\partial\bar{\omega}/\partial t)$. As in Sec. 1.6, $\nabla_k\bar{\omega}$ represents the group velocity $\mathbf{v}_g$ of energy transport in the lossless medium. By Eq. (9), on integrating from the space–time point $(\mathbf{r}_1, t_1)$ to $(\mathbf{r}, t)$ along a ray,

$$\psi(\mathbf{r}, t) - \psi(\mathbf{r}_1, t_1) = \int_{(\mathbf{r}_1, t_1)}^{(\mathbf{r}, t)} [\bar{\mathbf{k}}\cdot d\mathbf{r} - \bar{\omega}dt]. \tag{11}$$

For the time-harmonic problem in Eq. (6c), Eqs. (8) yield with $\mathbf{x} \equiv \mathbf{r}$ and $\varphi \equiv \psi$,

$$\frac{d\mathbf{r}}{ds} = \nabla_k G, \quad \frac{d\bar{\mathbf{k}}}{ds} = -\nabla G, \qquad G \equiv \det \bar{L}, \tag{12a}$$

or in an isotropic medium with $G = \bar{\mathbf{k}} - n(\mathbf{r})$ as in Eq. (6d),

$$\frac{d\mathbf{r}}{ds} = \mathbf{s}, \qquad \frac{d}{ds}(n\mathbf{s}) = \nabla n. \tag{12b}$$

By integrating along the ray trajectory $\mathbf{r}(s)$ defined by Eq. (12a) between the points $\mathbf{r}_1$ and $\mathbf{r}$, one solves for the phase function ψ via (9),

$$\psi(\mathbf{r}) - \psi(\mathbf{r}_1) = \int_{\mathbf{r}_1}^{\mathbf{r}} \bar{\mathbf{k}}\cdot\nabla_k G ds = \int_{\mathbf{r}_1}^{\mathbf{r}} \bar{\mathbf{k}}\cdot d\mathbf{r}. \tag{13}$$

The ray equations (10a) and (12), and consequently also Eqs. (11) and (13), simplify for special configurations. If the medium parameters do not vary in space, then $\bar{\omega} = \bar{\omega}(\bar{\mathbf{k}}, t)$, whence $\bar{\mathbf{k}} = $ constant on a space–time ray defined in Eq. (10a). If the medium parameters do not vary with time, then $\bar{\omega} = \bar{\omega}(\bar{\mathbf{k}}, \mathbf{r})$, whence $\bar{\omega} = $ constant on a space–time ray. If as a special case, the medium is merely plane stratified along z so that $\bar{\omega} = \bar{\omega}(\bar{\mathbf{k}}, z)$ then $\bar{\mathbf{k}}_t$, the component of $\bar{\mathbf{k}}$ transverse to z, is also constant; the resulting ray equation (10) is equivalent to Eqs. (1.6.29) and (1.6.32). Finally, if the medium parameters do not vary with either space or time so that $\bar{\omega} = \bar{\omega}(\bar{\mathbf{k}})$, then both $\bar{\mathbf{k}}$ and $\bar{\omega}$ are constant along a ray, as noted in Eq. (1.6.6). These constraints may be used for the graphical construction of ray trajectories in the manner depicted in Figs. 1.6.1, 1.6.7, and 1.6.9.

In the following sections and also in Chapters 5–8, detailed attention is given to the asymptotic evaluation of time-harmonic fields and to their ray-optical interpretation. Time-dependent fields in dispersive media are treated briefly (see Sec. 1.7e), but the Problems at the end of this chapter contain additional results.

1.7b   Scalar Time-Harmonic Fields

Let us consider scalar fields exterior to source regions, for example, the acoustic field treated in Sec. 1.3b. Since the vectorial aspect of the (longitudinal) acoustic field is trivial, as $\mathbf{v}$ is derivable directly from p by differentiation [see

Eq. (1.3.19)], one needs to consider only the scalar pressure p and thereby avoids the use of the $p, \mathbf{v}$ wavevector formalism of Eq. (6a). In an inhomogeneous medium where the background density n_0, the static pressure p_0, and the temperature ($\propto p_0/n_0$) are functions of $\mathbf{r}$, one may verify from Eqs. (1.3.19) that the time-harmonic pressure p satisfies the wave equation

$$\left[\nabla^2 + k_0^2 n^2(\mathbf{r}) - \left(\sqrt{n_0(\mathbf{r})}\nabla^2\frac{1}{\sqrt{n_0(\mathbf{r})}}\right)\right]\frac{p(\mathbf{r})}{\sqrt{n_0(\mathbf{r})}} = 0, \qquad k_0 = \frac{\omega}{a_0}, \qquad (14)$$

where $n(\mathbf{r}) = a_0/a(\mathbf{r})$ is the refractive index (not to be confused with the background density n_0), $a(\mathbf{r}) = [\gamma p_0(\mathbf{r})/mn_0(\mathbf{r})]^{1/2}$ is the local wave-propagation speed, and a_0 is a reference speed in a medium with $n = 1$. If the background density varies sufficiently slowly, the last term in the brackets in Eq. (14) can be neglected, and one obtains the approximate wave equation

$$[\nabla^2 + k_0^2 n^2(\mathbf{r})]u(\mathbf{r}) = 0, \qquad (15)$$

where $u = p/n_0^{1/2}$.

It is desired to construct a high-frequency asymptotic solution of the form (the $\sim$ signifies "asymptotically equal to")

$$u(\mathbf{r}) \sim e^{ik_0\psi(\mathbf{r})} \sum_{m=0}^{\infty} \frac{u_m(\mathbf{r})}{(ik_0)^m}. \qquad (16)$$

In this development, $u_m(\mathbf{r})$ and $\psi(\mathbf{r})$ are assumed to be independent of the wavenumber k_0, which is chosen as the large parameter rather than the observation distance as in Sec. 1.6b. By substituting Eq. (16) into Eq. (15) and assuming that the differentiation can be performed on each term in the sum, one finds with $\bar{\mathbf{k}} \equiv \nabla\psi = n\mathbf{s}$,

$$\{(ik_0)^2[\bar{k}^2 - n^2] + (ik_0)[\nabla\cdot\bar{\mathbf{k}} + 2\bar{\mathbf{k}}\cdot\nabla] + \nabla^2\} \sum_{m=0}^{\infty} \frac{u_m(\mathbf{r})}{(ik_0)^m} \sim 0. \qquad (17)$$

Since this expansion is to hold for arbitrary (though large) values of k_0, the coefficient of each power of k_0 must vanish independently. From the k_0^2 term,

$$\bar{k}^2 = n^2, \qquad (18a)$$

from the k_0^1 term,

$$(\nabla\cdot\bar{\mathbf{k}} + 2\bar{\mathbf{k}}\cdot\nabla)u_0 = 0; \qquad (18b)$$

and from the general term $k_0^{-\nu}$, $\nu = 0, 1, 2, \ldots$, †

$$(\nabla\cdot\bar{\mathbf{k}} + 2\bar{\mathbf{k}}\cdot\nabla)u_m = -\nabla^2 u_{m-1}, \qquad m \geq 1. \qquad (18c)$$

It is recalled that the wavevector $\bar{\mathbf{k}}$ is normalized with respect to k_0. In deriving Eqs. (18b) and (18c), the *transport equations* for the amplitude coefficients in the asymptotic expansion (16), use has been made of Eq. (18a), the *eiconal equation* of geometrical optics.[27] In view of the recursive character of the sys-

---

†In this formal development, Eq. (15) is regarded as exact. If Eq. (15) is only an approximate form derived, for example, from Eq. (14), the higher-order terms in the expansion must be reexamined.

tem of equations described by Eq. (18c), all the coefficients $u_m, m > 1$, can be derived in principle from the lowest-order coefficient u_0.

If the medium is non-dissipative, so that n^2 is positive real, the procedure for solving Eqs. (18) involves the following steps:

1. Determination of the ray trajectories (i.e., the curves parallel to the local time-averaged energy flux density vector, $\bar{\mathbf{S}}$).

2. Evaluation of the phase function ψ by integrating along a ray.

3. Calculation of the lowest-order amplitude u_0 by invoking conservation of energy in a tube of rays.

Steps 1 and 2 have already been formulated in Eqs. (12b) and (13) and are recapitulated below. Concerning step 3 we consider first the time-averaged energy flux density in a scalar harmonic field u, given typically by

$$\bar{\mathbf{S}} = \zeta \, \mathrm{Im}\,(u^*\nabla u), \tag{19}$$

where ζ is a real constant. The lowest-order solution of Eq. (16) in the high-frequency limit is the local plane-wave field

$$u \sim u_0 e^{ik_0\psi}. \tag{20}$$

This "geometric-optical" field† dominates the remaining terms in Eq. (16) if the relative variation in the refractive index, $|\nabla n|/n$, is small compared with the local wavelength $k_0 n$, i.e., if‡

$$\frac{|\nabla n|}{k_0 n^2} \ll 1. \tag{21}$$

Evidently, Eq. (20) is of the same form as Eq. (5). Wherever u_0 is slowly varying [an exception occurs in focal regions; see Eq. (36)], $\nabla u \sim ik_0 u \nabla \psi$, so that

$$\bar{\mathbf{S}} \sim \zeta k_0 |u_0|^2 \nabla\psi = \zeta k_0 |u_0|^2 \bar{\mathbf{k}}. \tag{22}$$

For the acoustic field, $u = p/n_0^{1/2}$, so that in view of Eq. (1.3.19),

$$\bar{\mathbf{S}} = \mathrm{Re}\,(pv^*) = \frac{1}{a_0 m}\,|u_0|^2 \bar{\mathbf{k}}, \tag{22a}$$

which is evidently of the form given in Eq. (22) with $\zeta = (1/\omega m)$. This result for $\bar{\mathbf{S}}$ is used below for integration of the transport equation (18b).

It should be noted that the preceding (also subsequent) considerations are not affected when the series (16) is multiplied by $(ik_0)^{-\beta}$, $0 < \beta < 1$, so fields with fractional power dependence on k_0 may also be accommodated.

Ray trajectories

From Eq. (12b) [see also Eq. (22)], rays are tangent to the "ray vector" $n\mathbf{s}$, where $\mathbf{s}$ is the unit vector defined in Eqs. (6d). The resulting ray equation (12b), repeated for convenience, is

---

†Although the present discussion does not deal with the propagation of light, it is suggested that the term "geometric-optical" be retained since the field so described obeys laws analogous to those in light optics.

‡More precisely, the condition $u_0(\mathbf{r}) \gg k_0^{-1} u_1(\mathbf{r})$ is satisfied if $[k_0 u_0(\mathbf{r})n(\mathbf{r})]^{-1}\nabla^2 u_0(\mathbf{r}) \ll 1$ along the ray trajectory, with $u_0(\mathbf{r})$ given by Eq. (34) (see Problem section).

$$\frac{d}{ds}\, n\frac{d\mathbf{r}}{ds} = \nabla n, \qquad \text{with } \left(\frac{d\mathbf{r}}{ds}\right)^2 = 1. \tag{23}$$

In a homogeneous medium, for which $n = $ constant, Eq. (23) has the solution $\mathbf{r} = \mathbf{A}s + \mathbf{B}$, where $\mathbf{A}$ and $\mathbf{B}$ are constant vectors. The rays in this case are straight lines.

In an inhomogeneous medium where n varies continuously, the rays are smoothly curved. Information about a ray trajectory is obtained by examining the vector derivative $d\mathbf{s}/ds$, which defines a vector perpendicular to the ray curve; its magnitude is the local curvature of the ray,

$$\frac{d\mathbf{s}}{ds} = \frac{\boldsymbol{\tau}}{D}, \tag{24}$$

where $\boldsymbol{\tau}$ is a unit vector and D is the local radius of curvature along the ray. From Eq. (23) and since $\boldsymbol{\tau}\cdot\mathbf{s} = 0$, $\boldsymbol{\tau}\cdot\nabla n = \boldsymbol{\tau}\cdot d(n\mathbf{s})/ds = n\boldsymbol{\tau}\cdot d\mathbf{s}/ds = n/D > 0$. Thus, the ray bends toward the direction of increasing refractive index.

For a plane-stratified medium where $n(\mathbf{r}) = n(z)$, the x- and y-component forms of the ray equation (23) yield

$$\frac{d}{ds}\, n\frac{dx}{ds} = 0 = \frac{d}{ds}\, n\frac{dy}{ds}, \tag{25}$$

whence $dy/dx = $ constant along a ray [i.e., the ray is confined to a plane perpendicular to the x, y plane]. It follows from Eq. (25) that $n(z)(d\boldsymbol{\rho}/ds) = $ constant along the ray, where $\boldsymbol{\rho} = \mathbf{x}_0 x + \mathbf{y}_0 y$. On defining $\theta(z) = \sin^{-1}(d\rho/ds)$ as the angle between the ray and the z axis, one then finds that along a ray,†

$$n(z) \sin \theta(z) = a = \text{constant}, \tag{26}$$

which relation expresses Snell's law of refraction and assures that on a ray, each incremental change in wavevector direction is along the direction of maximum rate of change of the refractive index (i.e., along z). The same condition for determining ray paths has been given in Eq. (1.6.29), wherein this ray trajectory interpretation of the saddle-point condition has been used for the graphical construction in Figs. 1.6.7 and 1.6.8. An explicit equation for the ray trajectories follows from $d\rho/dz = \tan\theta$, which, in view of Eq. (26), leads to [see also Eqs. (1.6.28)]

$$\rho(z) = a \int_{z_1}^{z} \frac{d\zeta}{\sqrt{n^2(\zeta) - a^2}}. \tag{27}$$

where $\rho(z_1) = 0$ denotes an arbitrary reference point.

For arbitrary $n(\mathbf{r})$, the second of Eqs. (12b) implies that along the ray trajectory, increments in the wavevector occur in the variable direction of maximum rate of change of the local refractive index. This requirement may be used to adapt the graphical procedure in Fig. 1.6.7 to the case of arbitrary medium variation. The medium along the trajectory is again divided into locally homogeneous segments, with the interfaces between segments now chosen perpendicular to the local refractive index gradient.

---

†The ray parameter a should not be confused with the acoustic speeds a_0 or $a(\mathbf{r})$.

Phase functions

With ray trajectories computed from Eq. (23), one may find the phase function ψ via the integration shown in Eq. (13), or, in view of Eq. (6d),

$$\psi(\mathbf{r}) - \psi(\mathbf{r}_1) = \int_{\mathbf{r}_1}^{\mathbf{r}} n\,ds. \tag{28}$$

The phase integral in Eq. (28) defines the optical path length along the ray. Alternatively, Eq. (28) can be interpreted as the time required for the phase fronts to cover the distance between points $\mathbf{r}_1$ and $\mathbf{r}$ along the ray; on noting that $ds/dt = a_0/n$ is the local propagation speed, one may write

$$\psi(\mathbf{r}) - \psi(\mathbf{r}_1) = a_0 \int_{\mathbf{r}_1}^{\mathbf{r}} dt, \tag{29}$$

with a_0 representing the propagation speed in a medium with $n = 1$. In a regular region where only one ray passes through a given point, it may be shown that for points $\mathbf{r}_1$ and $\mathbf{r}$ along a ray, the optical length in Eq. (28) or the propagation time in Eq. (29) is less along the ray than along any other neighboring curve connecting the two points. This result is known variously as Fermat's principle, the principle of shortest optical path, or the principle of least time.[27]

In a homogeneous medium where $n = $ constant and the rays are straight lines, one has $\psi(\mathbf{r}) - \psi(\mathbf{r}_1) = n|\mathbf{r} - \mathbf{r}_1|$. Thus, the phase accumulation over a distance L along a ray in a homogeneous medium is given by the factor exp $(ik_0 L)$ characteristic of a *plane wave*. For a plane-stratified inhomogeneous medium as in Fig. 1.6.7, use of Eq. (27) and the condition $n \sin \theta = a$ yields

$$\psi(\mathbf{r}) - \psi(\mathbf{r}_1) = \int_{\mathbf{r}_1}^{\mathbf{r}} n(\sin \theta\, d\rho + \cos \theta\, dz)$$

$$= a(\rho - \rho_1) + \int_{z_1}^{z} \sqrt{n^2(z) - a^2}\,dz, \tag{30}$$

an expression for the phase agreeing with that obtained by the saddle-point method [Eq. (1.6.24), with modifications noted after Eq. (1.6.28)]. The above ray solutions in inhomogeneous media are intimately related to the WKB approximation, as shown in Sec. 5.8.

Amplitude variation

To determine the amplitude dependence $u_0(\mathbf{r})$ of the field in Eq. (20), it is convenient to write Eq. (18b) in a somewhat different form. First, the equation is multiplied by the complex-conjugate function $u_0^*(\mathbf{r})$ to yield

$$|u_0|^2 \mathbf{\nabla} \cdot \bar{\mathbf{k}} + 2u_0^* \mathbf{\nabla} u_0 \cdot \bar{\mathbf{k}} = 0. \tag{31a}$$

Recalling that $\bar{\mathbf{k}}$ is real, and adding the complex conjugate of Eq. (31a), one then finds

$$|u_0|^2 \mathbf{\nabla} \cdot \bar{\mathbf{k}} + \mathbf{\nabla}|u_0|^2 \cdot \bar{\mathbf{k}} = 0 = \mathbf{\nabla} \cdot (|u_0|^2 \bar{\mathbf{k}}). \tag{31b}$$

Since this relation yields $\mathbf{\nabla} \cdot \bar{\mathbf{S}} = 0$ [Eq. (22) with $\zeta k_0 = $ constant; for the acoustic-wave-propagation problem, see Eq. (22a)], one obtains by the divergence theorem of vector analysis the conservation-of-energy statement equivalent to the original Eq. (18b):

$$\oint_A \bar{\mathbf{S}} \cdot \mathbf{v}\, dA = 0, \tag{32}$$

where A is a closed surface and $\mathbf{v}$ a unit vector normal to A. If A is chosen as the surface area of a narrow ray tube as shown in Fig. 1.7.1 with $\bar{\mathbf{S}} \cdot \mathbf{v} = 0$ along the side walls of the tube, and $\bar{\mathbf{S}} \cdot \mathbf{v} = \pm \bar{S}_{1,2}$ over the infinitesimal end sections $dA_{1,2}$ then, from Eq. (32), $\bar{S}_1 dA_1 = \bar{S}_2 dA_2$, or

$$\bar{S}_2 = \bar{S}_1 \frac{dA_1}{dA_2}. \tag{33}$$

Thus, the magnitude of the flux density vector (i.e., the intensity) at $\mathbf{r}_2$ along a ray is related to that at $\mathbf{r}_1$ by the ratio of the infinitesimal areas cut out of the wavefronts by a narrow ray bundle, and the total power flow within such a ray tube remains constant. This energy-conservation statement generalizes those in Secs. 1.6a and 1.6b to media with arbitrary (though slow) refractive index variations. From Eqs. (6d), (22), and (23), the field magnitude $|u_0(\mathbf{r})|$, proportional to $[\bar{S}(\mathbf{r})/n(\mathbf{r})]^{1/2}$, at an arbitrary point $\mathbf{r}$ along a ray is related to $|u_0(\mathbf{r}_1)|$ at a reference point $\mathbf{r}_1$ on the same ray as follows:

$$|u_0(\mathbf{r})| = |u_0(\mathbf{r}_1)|\sqrt{\frac{n(\mathbf{r}_1)}{n(\mathbf{r})}\frac{dA(\mathbf{r}_1)}{dA(\mathbf{r})}}. \tag{34}$$

For example, for a spherically spreading disturbance in a homogeneous medium, the rays diverge radially from the origin $r = 0$ and the cross section of a conical ray tube is proportional to r^2. Thus, $dA(\mathbf{r}_1)/dA(\mathbf{r}) = (r_1/r)^2$, whence $|u_0(r)| = |u_0(r_1)|(r_1/r)$.

By an alternative calculation, Eq. (31b) may be written as

$$\frac{1}{|u_0|^2}\frac{d}{ds}|u_0|^2 = \frac{d}{ds}\ln|u_0|^2 = -\frac{1}{n}\boldsymbol{\nabla}\cdot\bar{\mathbf{k}} = -\frac{1}{n}\boldsymbol{\nabla}\cdot(n\mathbf{s}), \tag{35a}$$

so that by integration between the points $\mathbf{r}_1$ and $\mathbf{r}$ along a ray,

$$|u_0(\mathbf{r})| = |u_0(\mathbf{r}_1)|\exp\left[-\frac{1}{2}\int_{\mathbf{r}_1}^{\mathbf{r}}\frac{1}{n}\boldsymbol{\nabla}\cdot(n\mathbf{s})\,ds\right]. \tag{35b}$$

For the preceding example, $\mathbf{s}$ is equal to the radial unit vector $\mathbf{r}_0$, whence $\boldsymbol{\nabla}\cdot\mathbf{s} = 2/r$ and the previous result follows.

It is evident from Eq. (34) that the formula fails when $dA(\mathbf{r}) \to 0$ (i.e., when the rays forming the ray tube converge on a line or point). This condition defines the location of caustics or foci, near which a more detailed analysis of the ray-optical field must be carried out (see Sec. 5.8d for the calculation required in a plane-stratified medium). Thus, the condition

$$dA(\mathbf{r}) \not\approx 0, \tag{36}$$

at each point $\mathbf{r}$ along the ray, must be added to the one in Eq. (21) to assure the validity of the ray-optical formulas.

By collecting results from Eqs. (28) and (34), one may write for the leading term in the asymptotic expansion of a particular constituent of the high-frequency field [Eq. (20)]:

$$u(\mathbf{r}) \sim u(\mathbf{r}_1) \left[\frac{n(\mathbf{r}_1)}{n(\mathbf{r})} \frac{dA(\mathbf{r}_1)}{dA(\mathbf{r})} \right]^{1/2} e^{ik_0[\psi(\mathbf{r}) - \psi(\mathbf{r}_1)]}. \tag{37}$$

In this formula, $\mathbf{r}$ denotes the observation point along a given ray, $\mathbf{r}_1$ is a reference point on the same ray, $dA(\mathbf{r})$ and $dA(\mathbf{r}_1)$ are the ray-tube cross sections at $\mathbf{r}$ and $\mathbf{r}_1$, respectively, $n(\mathbf{r})$ and $n(\mathbf{r}_1)$ are the corresponding refractive indexes, while $\psi(\mathbf{r})$ and $\psi(\mathbf{r}_1)$ are the phase functions. The field at $\mathbf{r}$ is thus specified in terms of a known field at $\mathbf{r}_1$. For applications of these formulas to plane-stratified media, see Sec. 5.8.

1.7c Vector Time-Harmonic Fields

Isotropic media

In an isotropic medium with variable relative permittivity $\epsilon(\mathbf{r}) = n^2(\mathbf{r})$, where n is the refractive index, the steady-state electromagnetic field exterior to source regions is defined by the Maxwell field equations which take the form shown in (6a) with [see Eqs. (1.3.26) and (1.3.28)]

$$iM(\nabla) \to \begin{pmatrix} 0 & -\nabla \times \mathbf{1} \\ \nabla \times \mathbf{1} & 0 \end{pmatrix}, \quad W(\mathbf{r}) \to \begin{pmatrix} \epsilon_0 n^2(\mathbf{r})\mathbf{1} & 0 \\ 0 & \mu_0 \mathbf{1} \end{pmatrix},$$

$$\Psi(\mathbf{r}) \to \begin{pmatrix} \mathbf{E}(\mathbf{r}) \\ \mathbf{H}(\mathbf{r}) \end{pmatrix}, \tag{38}$$

where ϵ_0 and μ_0 are the permittivity and permeability of vacuum and $c = (\mu_0 \epsilon_0)^{-1/2}$. A time factor $\exp(-i\omega t)$ is suppressed. Use of the lowest-order approximation for the wavevector [Eq. (5)] then leads to Eq. (6b), written explicitly as

$$\bar{\mathbf{k}} \times \mathbf{H}_0 = -\frac{n^2}{\zeta} \mathbf{E}_0, \quad \bar{\mathbf{k}} \times \mathbf{E}_0 = \zeta \mathbf{H}_0, \quad \bar{\mathbf{k}} \equiv \nabla\psi, \quad \zeta = \sqrt{\frac{\mu_0}{\epsilon_0}}, \tag{38a}$$

whence the eiconal equation (6c) becomes

$$\bar{\mathbf{k}}^2 = n^2(\mathbf{r}), \qquad \text{or} \qquad \bar{\mathbf{k}} = \mathbf{s}\, n(\mathbf{r}), \tag{38b}$$

as noted in Eq. (6d). The lowest-order eigensolutions $\mathbf{E}(\mathbf{r}) \sim \mathbf{E}_0(\mathbf{r}) \exp[ik_0\psi(\mathbf{r})]$ and $\mathbf{H}(\mathbf{r}) \sim \mathbf{H}_0(\mathbf{r}) \exp[ik_0\psi(\mathbf{r})]$ constitute the geometric-optical field whose polarization properties are evident from Eq. (38a),

$$\mathbf{H}_0 \cdot \bar{\mathbf{k}} \sim 0 \sim \mathbf{E}_0 \cdot \bar{\mathbf{k}}, \qquad \mathbf{E}_0 \cdot \mathbf{H}_0 \sim 0 \tag{38c}$$

i.e., the electric and magnetic field vectors are mutually perpendicular, lie in the equiphase surface $\psi = $ constant, and describe a local plane-wave field in an isotropic medium. After determination of the ray trajectories from Eq. (12b), the phase function ψ can be calculated from Eq. (13). The variation of the amplitudes $E_0(\mathbf{r}) \equiv |\mathbf{E}_0(\mathbf{r})|$ and $H_0(\mathbf{r})$ of the geometric-optical field may be determined from the energy-conservation law

$$\nabla \cdot \bar{\mathbf{S}} = 0, \qquad \bar{\mathbf{S}} = \mathrm{Re}\,(\mathbf{E} \times \mathbf{H}^*) \sim \frac{1}{\zeta} E_0^2 \bar{\mathbf{k}} = \bar{W}v\mathbf{s}, \tag{39}$$

where $\bar{\mathbf{S}}$ is the real time-averaged Poynting vector (power density), $\bar{W} = n^2 \epsilon_0 E_0^2 = (\zeta/c)H_0^2$ is the total average stored energy density, and $v = c/n$ is the local wave propagation speed (rms values for the field quantities are implied). Equation (39) has the same form as Eq. (31b); when applied to a ray tube as shown in Fig. 1.7.1, it yields the amplitude variation specified in Eq. (34). On combining the results for the amplitude and phase variation, one obtains

$$E(\mathbf{r}) \sim E(\mathbf{r}_1) \left[\frac{n(\mathbf{r}_1)\,dA(\mathbf{r}_1)}{n(\mathbf{r})\,dA(\mathbf{r})}\right]^{1/2} \exp\left(ik_0 \int_{\mathbf{r}_1}^{\mathbf{r}} n\,ds\right), \tag{40}$$

where $\mathbf{r}_1$ and $\mathbf{r}$ are two points along a ray defined in Eq. (12b). The direction of the field is determined from a transport equation for the polarization vector [Eq. (49)]. From the so-determined vector electric field at $\mathbf{r}$, one may calculate the vector magnetic field directly from the second equation in (38a).

As for the acoustic field in Sec. 1.7b, one may construct an asymptotic expansion for the high-frequency electromagnetic field in which the geometric-optical field of Eq. (40) et seq. constitutes the leading term.[27] Because of the relative simplicity of the time-harmonic electromagnetic field in an isotropic medium, instead of using the first-order form of the Maxwell field equations in Sec. 1.7a, it is more convenient to perform the calculation from the second-order form. This procedure is instructive even for the lowest-order contribution, since it yields the results in Eqs. (38b) and (39) somewhat more directly and simplifies the formulation of the transport equation for the polarization vectors. We deal explicitly with the electric field $\mathbf{E}$ which satisfies the vector wave equation

$$\nabla^2 \mathbf{E} + k_0^2 n^2 \mathbf{E} + 2\nabla(\mathbf{E} \cdot \nabla \ln n) = 0, \tag{41}$$

with the magnetic field determined from $\mathbf{H} = (ik_0\zeta)^{-1}\nabla \times \mathbf{E}$. As in Eq. (16), we assume that for large k_0,

$$\mathbf{E}(\mathbf{r}) \sim e^{ik_0\psi(\mathbf{r})} \sum_{m=0}^{\infty} \frac{\mathbf{E}_m(\mathbf{r})}{(ik_0)^m}, \tag{42}$$

where the amplitude coefficients $\mathbf{E}_m$ and the phase function ψ (and hence the refractive index n) are regarded as independent of k_0. As in Sec. 1.7d, subsequent considerations will still apply when Eq. (42) is multiplied by a factor containing a fractional power of k_0. Assuming the differentiability of the asymptotic representation, introducing the wavevector $\bar{\mathbf{k}} \equiv \nabla\psi$ normalized with respect to k_0, and substituting Eq. (42) into Eq. (41), one finds

$$e^{ik_0\psi} \sum_{m=0}^{\infty} \frac{1}{(ik_0)^m} \left\{ k_0^2 \mathbf{E}_m[n^2 - \bar{k}^2] + ik_0\left[\mathbf{E}_m\nabla\cdot\bar{\mathbf{k}} + 2\bar{k}\mathbf{E}_m\cdot\nabla\ln n + 2\bar{\mathbf{k}}\cdot\nabla\mathbf{E}_m\right] \right.$$
$$\left. + \nabla^2\mathbf{E}_m + 2\nabla(\mathbf{E}_m\cdot\nabla\ln n) \right\} = 0. \tag{43}$$

On equating to zero the coefficients of each power of k_0, one obtains from the coefficient of k_0^2 the eiconal equation (38b), while from the coefficient of k_0 and Eq. (38b),

$$\mathbf{E}_0 \nabla \cdot \bar{\mathbf{k}} + 2(\mathbf{E}_0 \cdot \nabla \ln n)\bar{\mathbf{k}} + 2(\bar{\mathbf{k}} \cdot \nabla)\mathbf{E}_0 = 0. \tag{44a}$$

As in Eq. (18c), the amplitude functions $\mathbf{E}_m$, $m \geq 1$, can be determined from $\mathbf{E}_{m-1}$ by the recursive system obtained from the coefficient of $k_0^{-\nu}$, $\nu \geq 0$, in Eq. (43):

$$[\nabla \cdot \bar{\mathbf{k}}\mathbf{1} + 2\bar{\mathbf{k}}(\nabla \ln n) + 2\bar{\mathbf{k}} \cdot \nabla \mathbf{1}] \cdot \mathbf{E}_m = -\nabla^2 \mathbf{E}_{m-1} - 2\nabla[(\nabla \ln n) \cdot \mathbf{E}_{m-1}],$$
$$\tag{44b}$$

where $\mathbf{1}$ is the unit dyadic. Equations (44a) and (44b) constitute the transport equations for the electric-field amplitudes. As in Sec. 1.7a, we shall not consider the higher-order terms $m \geq 1$. The $m = 0$ contribution, the geometric-optical field, is specified by Eqs. (38a) and (38c). It represents a valid approximation when $E_1 \ll k_0 E_0$, which condition requires that the medium is slowly varying in the sense defined in Eq. (21).[28] To the order of accuracy of the geometric-optical approximation, the refractive index n may depend weakly on k_0, thereby including dispersive media within the scope of the lowest-order theory.†

In contrast to the abstract procedure, based on the first-order form of the field equations and leading to Eq. (6b), the present approach yields directly separate equations for the phase and amplitude variations in the geometric-optical field. To convert Eq. (44a) into the form (39), it is convenient to introduce the polarization vector $\boldsymbol{\beta}$ parallel to the electric-field direction,

$$\boldsymbol{\beta} = \frac{\mathbf{E}_0}{E_0} = \frac{\mathbf{E}_0}{(\mathbf{E}_0 \cdot \mathbf{E}_0^*)^{1/2}}. \tag{45}$$

On dot-product multiplication of Eq. (44a) by $\mathbf{E}_0^*$ and use of Eqs. (38b) and (38c), one finds

$$\frac{1}{2}(\mathbf{E}_0^* \cdot \mathbf{E}_0)\nabla \cdot \bar{\mathbf{k}} + n\mathbf{E}_0^* \cdot \frac{d}{ds}\mathbf{E}_0 = 0, \tag{46}$$

and, on adding to this equation its complex conjugate,

$$E_0^2 \nabla \cdot \bar{\mathbf{k}} + n\frac{d}{ds}E_0^2 = 0 = \nabla \cdot (E_0^2 \bar{\mathbf{k}}), \tag{47}$$

which expresses $\nabla \cdot \bar{\mathbf{S}} = 0$ as in Eq. (39).

In a similar manner, one may convert Eq. (44a) into a transport equation for the polarization vector $\boldsymbol{\beta}$ along a ray. Since

$$\frac{d\boldsymbol{\beta}}{ds} = \frac{d}{ds}\frac{\mathbf{E}_0}{E_0} = \frac{1}{E_0}\frac{d}{ds}\mathbf{E}_0 - \frac{\boldsymbol{\beta}}{E_0}\frac{d}{ds}E_0, \tag{48}$$

one finds, on use of Eqs. (47) and (44a), recalling that $\bar{\mathbf{k}} \cdot \nabla = n(d/ds)$,

---

†For a cold isotropic plasma, $n^2(\mathbf{r}) = 1 - [\omega_p^2(\mathbf{r})/\omega^2]$, where ω_p is the plasma frequency. Weak dependence on k_0 is taken to imply that ω_p/ω remains finite.

$$\frac{d\boldsymbol{\beta}}{ds} = -\mathbf{s}(\boldsymbol{\beta}\cdot\nabla \ln n). \tag{49}$$

Evidently, $\boldsymbol{\beta}$ is constant in homogeneous regions where $\nabla n = 0$ and also for fields polarized with the electric vector perpendicular to the direction of stratification in the medium. Inclusion of the polarization vector $\boldsymbol{\beta}(\mathbf{r})$ in Eq. (40) completes specification of the geometric-optical field in a medium with slow spatial variation.

Anisotropic media

In an inhomogeneous anisotropic dielectric (e.g., a crystal or a cold magnetoplasma) where the local wave-propagation properties depend on direction as well as location, the time-harmonic electromagnetic field satisfies the Maxwell field equations (6a), with

$$iM(\nabla) \rightarrow \begin{pmatrix} 0 & -\nabla \times \mathbf{1} \\ \nabla \times \mathbf{1} & 0 \end{pmatrix}, \quad W(\mathbf{r}) \rightarrow \begin{pmatrix} \epsilon_0\boldsymbol{\epsilon}(\mathbf{r}) & 0 \\ 0 & \mu_0\mathbf{1} \end{pmatrix}, \quad \Psi(\mathbf{r}) \rightarrow \begin{pmatrix} \mathbf{E}(\mathbf{r}) \\ \mathbf{H}(\mathbf{r}) \end{pmatrix},$$
$$\tag{50}$$

where ϵ_0 and μ_0 are the permittivity and permeability of vacuum and $\boldsymbol{\epsilon}(\mathbf{r})$ is the permittivity dyadic (normalized to ϵ_0) which may depend weakly on $k_0 = \omega/c$, $c = (\mu_0\epsilon_0)^{-1/2}$. A time factor $\exp{(-i\omega t)}$ has been suppressed. The lowest-order field approximation in Eq. (5) satisfies Eq. (6b) written explicitly as

$$\bar{\mathbf{k}} \times \mathbf{H}_0 = -\frac{1}{\zeta}\boldsymbol{\epsilon}\cdot\mathbf{E}_0, \quad \bar{\mathbf{k}} \times \mathbf{E}_0 = \zeta\mathbf{H}_0, \quad \bar{\mathbf{k}} \equiv \nabla\psi, \quad \zeta = \sqrt{\frac{\mu_0}{\epsilon_0}}, \tag{51}$$

from which it follows that

$$\bar{\mathbf{k}}\cdot\mathbf{H}_0 = \bar{\mathbf{k}}\cdot\boldsymbol{\epsilon}\cdot\mathbf{E}_0 = \mathbf{H}_0\cdot\mathbf{E}_0 = \mathbf{H}_0\cdot\boldsymbol{\epsilon}\cdot\mathbf{E}_0 = 0. \tag{52}$$

Thus, $\mathbf{H}$ is perpendicular to $\mathbf{E}$ and $\bar{\mathbf{k}}$, but $\mathbf{E}$ generally is not perpendicular to $\bar{\mathbf{k}}$, so the real time-averaged power flow density vector $\bar{\mathbf{S}} = \mathrm{Re}\,(\mathbf{E} \times \mathbf{H}^*)$ is generally not parallel to $\bar{\mathbf{k}}$. In the absence of spatial dispersion (i.e., when $\boldsymbol{\epsilon}$ is independent of $\mathbf{k}$), the vector $\bar{\mathbf{S}}$ is parallel to the group-velocity (ray) vector $\mathbf{v}_g$ [see Eq. (1.5.24a)]. Since

$$\bar{\mathbf{S}} \sim \frac{1}{\zeta}\,\mathrm{Re}\,[\mathbf{E}_0 \times (\mathbf{k} \times \mathbf{E}_0^*)] \tag{53}$$

and therefore

$$\mathbf{k}\cdot\bar{\mathbf{S}} = \frac{1}{\zeta}\,|\mathbf{k} \times \mathbf{E}_0|^2 \geq 0, \tag{53a}$$

the angle between the ray and wavevector does not exceed $90°$. This observation is important for locating saddle points by the graphical procedures of Secs. 1.6a and 1.6b. From Eqs. (6c) and (50), one finds the eiconal equation

$$\det{[\bar{k}^2\mathbf{1} - \bar{\mathbf{k}}\bar{\mathbf{k}} - \boldsymbol{\epsilon}(\mathbf{r})]} = 0, \tag{54}$$

which permits one to calculate the ray trajectories via Eq. (12a) and thence the phase function ψ via Eq. (13).

Graphical methods for finding the ray direction in a homogeneous or plane-stratified medium have been discussed in connection with Figs. 1.6.5 and 1.6.7, and other observations concerning rays and wavefronts in an anisotropic environment have been made in connection with Figs. 1.6.2 and 1.6.6. Conservation of energy in a tube of rays (Fig. 1.7.1) [inferred from the transport equation mentioned in connection with Eqs. (6)] can be used as in the isotropic case to

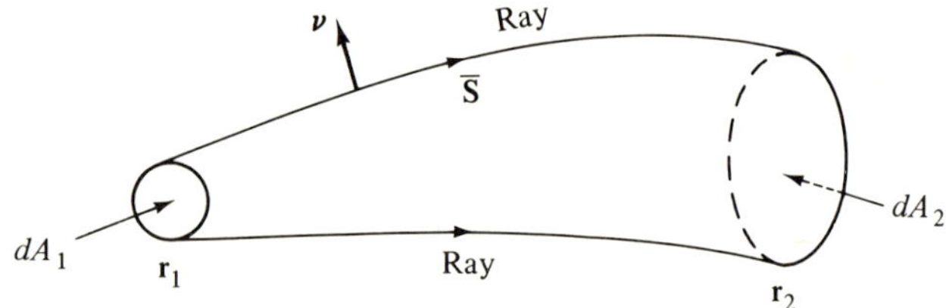

FIG. 1.7.1   Ray tube.

determine the amplitude variation $|\mathbf{E}_0(\mathbf{r})|$, but polarization effects must be found from Eq. (6b), $\bar{L}\Psi_0 = 0$. In the isotropic case $\epsilon(\mathbf{r}) = \mathbf{1}n^2(\mathbf{r})$, and Eq. (54) reduces to Eq. (6d), since in view of Eq. (52), $\mathbf{k}\cdot\mathbf{F}_0 \to 0$ in that limit.

The explicit construction of the field in an inhomogeneous anisotropic medium involves substantial difficulty even for the lowest-order (geometric-optical) approximation (for plane-stratified media, see References 15 and 28); further considerations are restricted to the homogeneous case $\epsilon = $ constant. In this instance, the rays are straight lines with

$$\psi(\mathbf{r}) = \mathbf{k}\cdot\mathbf{r} = Nr = nr\cos\alpha, \tag{55}$$

where $\mathbf{r}$ is the vector coordinate along the ray, $n = |\bar{\mathbf{k}}|$ is the ordinary refractive index descriptive of the speed of the advancing phase front along $\bar{\mathbf{k}}$ (it is recalled that $\bar{\mathbf{k}}$ is normalized to k_0), and $N = n\cos\alpha$ is the "ray refractive index" pertaining to phase propagation along the *ray*. Since $\mathbf{k}$ and the ray are not codirectional but are displaced in general by an angle α (see Fig. 1.6.6), one has instead of the conventional $N = n$ in isotropic media the relation given in Eq. (55). It is to be emphasized that n and N, and hence α, are functions of the polar angles (θ, ϕ) which select a particular ray orientation. In terms of the ray velocity v_r defined in connection with Fig. 1.6.2, one has $v_r = c/N$, whereas for the phase velocity along $\mathbf{k}$, $v_p = c/n$. The phase function in Eq. (55) corresponds exactly to that given in Eq. (1.6.25).

When the field is generated by a point dipole (Sec. 7.3a), the rays diverge radially from the source point. As in the discussion relating to Fig. 1.7.1, energy is conserved in a ray tube, a fact that is exploited for the construction of the amplitude functions $|\mathbf{E}_0|$ and $|\mathbf{H}_0|$ in Eq. (51). One notes from Eq. (53a) that the ray intensity varies like $|\mathbf{E}_0|^2$, so the variation of $|\mathbf{E}_0|$ may be deduced from that of the intensity $\bar{S}$. From conservation of energy in a ray tube,

$$\bar{S}(\mathbf{r})\, dA(\mathbf{r}) = \text{constant}, \tag{55a}$$

where $dA(\mathbf{r})$ is the ray-tube cross section at $\mathbf{r}$, whence for a radially diverging ray system,

$$\bar{S}(\mathbf{r}) = \frac{\text{constant}}{r^2}. \tag{55b}$$

The same observation has been made in Sec. 1.6b, and it has been sharpened by relating the energy density to properties of the wavenumber surface (see Fig. 1.6.6): $\bar{S}(\mathbf{r}) \propto \bar{R}_1 \bar{R}_2 / r^2$, where $\bar{R}_1$ and $\bar{R}_2$ are the principal radii of curvature of the wavenumber surface at the point corresponding to the ray.

From these considerations, one constructs the following ray-optical approximation of the electric field generated by a point source in a homogeneous, unbounded, anisotropic dielectric:

$$\mathbf{E}(\mathbf{r}) = \sum_i \mathbf{E}_i(\mathbf{r}), \quad \mathbf{E}_i(\mathbf{r}) \sim \frac{e^{i\mathbf{k}_i \cdot \mathbf{r}} \sqrt{\bar{R}_{1i} \bar{R}_{2i}}}{r} \mathbf{A}_i, \qquad \mathbf{k}_i \cdot \mathbf{r} = k_0 r N_i, \tag{56}$$

where $\mathbf{r}$ is the vector distance from the source to the observation point, $\mathbf{k}_i$ locates a point on the wavenumber surface which has a normal along the ray direction $\mathbf{r}$ (with $\mathbf{k}_i \cdot \mathbf{r} \geq 0$; see Fig. 1.6.6), and the sum over i includes all $\mathbf{k}_i$ that satisfy this condition. $\bar{R}_{1i}$ and $\bar{R}_{2i}$ are the principal radii of curvature of the wavenumber surface at $\mathbf{k}_.$. The vector $\mathbf{A}_i$ contains the excitation amplitude and polarization of the field along the ray; the determination of the former requires the solution of the source problem (see Secs. 7.3, 8.3c, and 1.3d), whereas the polarization is found from the eigenvalue problem discussed in Sec. 1.3a (see also Reference 24). Multiple ray contributions may arise when the wavenumber surface has inflection points as in Fig. 1.6.6, or has several branches (see Fig. 1.6.10 for analogous phenomena in connection with space–time rays descriptive of transient fields). Since each ray has its own ray refractive index N_i, interference phenomena may be observed when two or more rays have comparable amplitudes. The ray-optical approximation fails when $\bar{R}_{1i}$ and (or) $\bar{R}_{2i} \to \infty$ (i.e., when a ray corresponds to a point of vanishing Gaussian curvature on the wavenumber surface). Calculations for this transition region are presented in Sec. 8.3c.

The form of Eq. (56) agrees with that in Eq. (1.6.25) derived by the modal procedure.

1.7d The Geometrical Theory of Diffraction

The simple ray solutions developed in Secs. 1.7b and 1.7c can be utilized for the study of more complicated high-frequency scattering problems via the geometrical theory of diffraction. The theory, developed initially by Keller for homogeneous isotropic media,[29] has subsequently been extended also to inhomogeneous media,[30] to anisotropic media,[31] and to media supporting several wave species (e.g., a warm plasma[32]). According to the theory, the scattered field at a given observation point consists of contributions from geometrically reflected or refracted, and from diffracted, rays whose individual properties depend on the local nature of the scattering object and the surrounding medium. The incident field is represented by a system of rays which, on striking an obstacle, are scattered; the nature of the scattered rays near the obstacle surface depends on the surface and medium properties at (more properly, in the vicinity of) the point of impact. The formulas developed in Secs. 1.7b and 1.7c can then be employed for tracking the scattered rays away from the obstacle boundary. A ray striking a flat surface element is reflected or refracted according to the relevant (plane-wave) reflection or refraction law, and is propagated with an amplitude specified by the plane-wave reflection coefficient. A ray striking a curved surface element (with principal radii of curvature much greater than the wavelength) is reflected as from a plane tangent to the surface at the point of impact, but the intensity along its path is governed by a "divergence coefficient," due to the surface curvature, which accounts for the spreading of the energy contained in a narrow ray bundle. The reflected and refracted rays are those encountered in the geometrical optics of isotropic and anisotropic media, and their properties are discussed in detail in this section.

In addition, the theory utilizes a class of diffracted rays whose characteristics in an isotropic medium are summarized briefly. A ray incident normally on an edge gives rise to "edge-diffracted" rays that emerge in all directions in the plane perpendicular to the edge at the point of impact; the intensity along an edge-diffracted ray decreases like $\rho^{-1/2}$, where ρ is the distance from the edge (see Sec. 6.4 for a detailed discussion and also the example at the end of this section). The edge-diffracted rays due to an obliquely incident ray lie on a cone having a semiangle at the apex equal to the angle between the incident ray and the edge; the incident and diffracted rays lie on opposite sides of the perpendic-

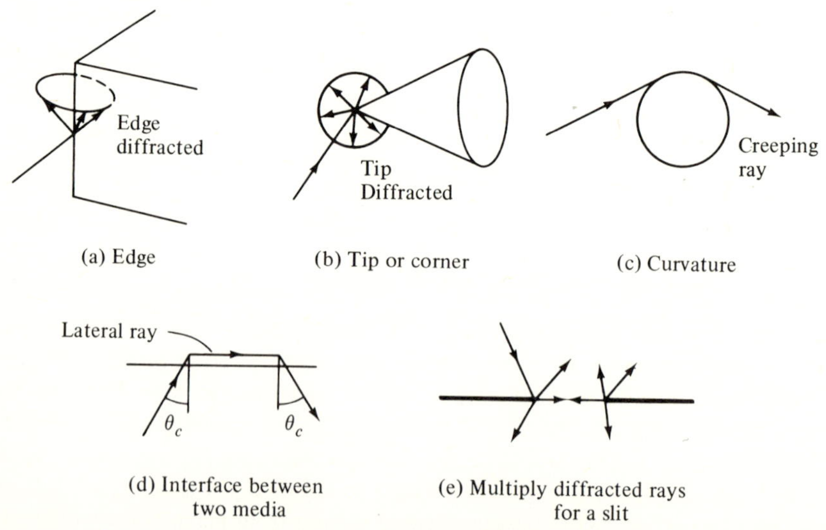

FIG. 1.7.2   Diffracted rays.

ular plane through the point of impact [see Fig. 1.7.2(a)]. A ray incident on a conical tip [Fig. 1.7.2(b)] or on a corner is scattered in all directions, and the intensity along a "tip-diffracted" ray decreases like r^{-1}, where r is the distance from the tip (see Sec. 6.8c). A ray incident tangentially on a curved object [Fig. 1.7.2(c)] creates a "creeping" ray which travels along the obstacle surface into the shadow region and sheds energy continually as it progresses; as a result, its amplitude along the surface is found to decay exponentially with distance (see Sec. 6.7). In the presence of an interface between two media, a ray incident at the critical angle θ_c from the optically denser medium launches a lateral ray that travels parallel to the interface in the optically thinner medium and sheds energy back to the denser medium by refraction [Fig. 1.7.2(d)]; the decay of the lateral ray amplitude due to energy leakage is found to be algebraic (see Sec. 5.5). The trajectories of the reflected and diffracted rays are in accord with Fermat's principle of least propagation time, and the manner of variation of the field, both in amplitude and phase, along a reflected or diffracted ray† can often be predicted from simple geometrical considerations, as illustrated in the example at the end of this section. The starting amplitude and phase along a given diffracted ray cannot in general be ascertained from the geometrical theory but must be taken from the rigorous asymptotic solution for a simple "canonical" scatterer whose contour at the point of diffraction is identical to that of the object under consideration (see the examples discussed elsewhere in this volume for a ray-optical interpretation of various rigorous asymptotic solutions).

Constructed in this manner, the scattered field for a canonical object, as predicted from the geometrical theory of diffraction, manifestly agrees with the asymptotic form of the rigorous solution in the range $k_0 \to \infty$. However, the geometrical theory, exploiting the local character of high-frequency diffraction effects, can also be applied to more complicated shapes for which exact solutions are not readily available. For example, by approximating the vicinity of a point on a variable-curvature cylindrical surface by a circular cylinder whose radius is identical with the local radius of curvature, one can predict the behavior of the diffracted (creeping) rays on such a surface; the validity of this prescription can be confirmed by comparing the result with the rigorous asymptotic solution for parabolic and elliptic cylinders.[33] Verification has also been obtained for the problem of diffraction by a circular aperture and by a slit, wherein one has to take into account multiply diffracted rays that travel from one edge to the other [Fig. 1.7.2(e)].[34,35] Corresponding results have been developed for scattering in anisotropic media, in media capable of supporting multiple wave species, and in wave guide regions.[36,37] An illustrative application of the theory to a non-elementary scattering problem is given later in this section.

---

†The category of diffracted rays following real trajectories in **r** space as shown in Fig. 1.7.2 can be enlarged by inclusion of rays defined on complex trajectories [e.g., rays descriptive of surface waves (see Sec. 5.6) and leaky waves, or of attenuated and evanescent fields].

An appealing feature of the geometrical theory of diffraction is its conceptual simplicity and its inherent broad range of applicability in homogeneous and inhomogeneous, isotropic or anisotropic regions. The theory is, however, based on a series of postulates whose validity has not as yet been proved in general, so it has been necessary to provide verification by recourse to the asymptotic expansions of rigorously solvable problems. In view of the substantial number of such successful comparisons, it appears likely that the geometrical theory of diffraction is capable of predicting at least the dominant effects of high-frequency scattering under relatively arbitrary conditions. The theory has also been extended to transient problems in dispersive media through use of space–time rays instead of time-harmonic rays.[19,24]

Ray reflection and refraction laws

When a ray strikes a surface across which the medium properties change discontinuously, it generates reflected and (in a penetrable medium) refracted rays which generally leave the interface along directions different from that of the incident ray. The new ray directions can be inferred from phase continuity of all relevant wave constituents generated at the boundary, a necessary requirement in order to satisfy field continuity and thence to maintain the local plane-wave character of the complex of incident, reflected and refracted waves. When applied to two ordinary dielectrics, the phase matching yields the conventional Snell's law, but a more general form results when the media are anisotropic and (or) when multiple wave species can be supported.[38] Thus, if m wave constituents are involved, each with a phase dependence $\exp[ik_0\psi_j(\mathbf{r})]$, $j = 1, \ldots, m$ [see Eq. (5)], then on the interface B,

$$\psi_1(\mathbf{r}) = \psi_2(\mathbf{r}) = \cdots = \psi_m(\mathbf{r}), \qquad \mathbf{r} \text{ on } B. \tag{57}$$

It follows from Eq. (57) that the derivatives of $\psi_j(\mathbf{r})$, $j = 1, \ldots, m$, in a direction ξ tangential to the interface are also continuous. Since

$$\frac{\partial \psi_j}{\partial \xi} = \xi_0 \cdot \nabla \psi_j = \bar{\mathbf{k}}_{tj}, \tag{58}$$

where ξ_0 is a unit vector and $\bar{\mathbf{k}}_{tj}$ is the projection of the jth wavevector onto the interface, the phase continuity condition can be written in terms of the normalized wavevectors $\bar{\mathbf{k}}_j = \nabla \psi_j$ as

$$\bar{\mathbf{k}}_{t1} = \bar{\mathbf{k}}_{t2} = \cdots = \bar{\mathbf{k}}_{tm} \qquad \text{on } B. \tag{59}$$

This relation will be used to derive reflected and refracted ray directions in isotropic, anisotropic, and compressible plasma media. The initial amplitude and polarization of the fields along reflected and refracted rays can be deduced either from additional continuity requirements applied to the asymptotic field expression or, equivalently, from the exact solution of plane-wave reflection and refraction at a plane boundary between two homogeneous media having constitutive parameters identical with the local medium parameters at the point of impact of the incident ray.

Isotropic media

At a plane interface between two ordinary dielectrics described by refractive indexes $n_1(\mathbf{r})$ and $n_2(\mathbf{r})$, a ray ($j = 1$) incident from medium 1 generates locally a single ray ($j = 3$) reflected into medium 1 and a single ray ($j = 2$) refracted into medium 2. Since $\bar{\mathbf{k}}_{1,3} = n_1 \mathbf{s}_{1,3}$, $\bar{\mathbf{k}}_2 = n_2 \mathbf{s}_2$ [see Eq. (6d)], and $s_{tj} = \xi_0 \sin \theta_j$, where θ_j is the angle between the ray direction and the surface normal, Eq. (59) yields

$$\sin \theta_1 = \sin \theta_3, \qquad n_1 \sin \theta_1 = n_2 \sin \theta_2. \tag{60}$$

In solving for the angles, the choice of θ_j or $(\pi - \theta_j)$ is made subject to the radiation condition that both the reflected and refracted rays are directed away from the boundary. For the pair of incident and reflected rays $\mathbf{s}_1$ and $\mathbf{s}_3$ in Fig. 1.7.3(a), proceeding into the same medium, the first equation in Eq. (60) yields

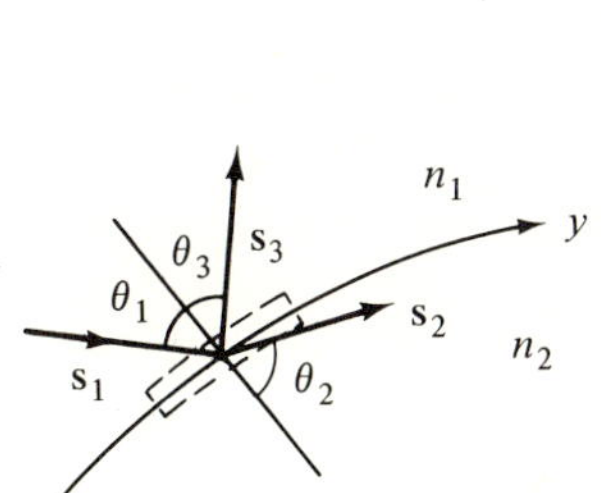

(a) Physical configuration

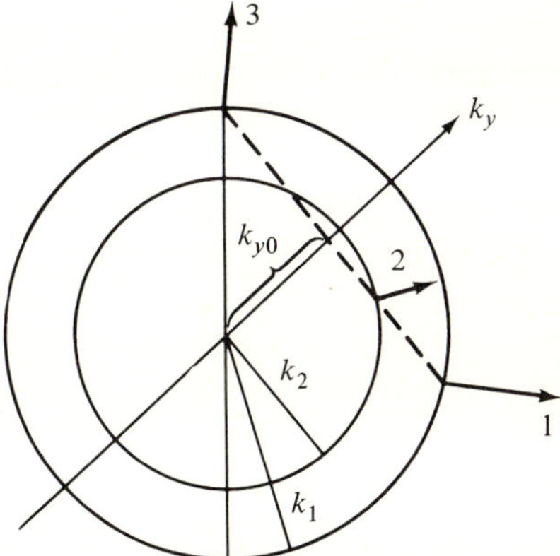

(b) Wavenumber plot (section through the plane containing the rays)

FIG. 1.7.3 Ray reflection and refraction law: $\theta_1 = \theta_3$, $n_1 \sin \theta_1 = n_2 \sin \theta_2$.

the specular reflection law, whereas for the incident and refracted rays $\mathbf{s}_1$ and $\mathbf{s}_2$, one has Snell's refraction law. In this instance Eq. (59) implies that the triplet of incident, reflected, and refracted rays lies in a plane perpendicular to the interface. For a curved interface as in Fig. 1.7.3(a), it must be kept in mind that ray-optical concepts apply only when the radius of curvature is large compared with the local wavelengths $\lambda_{1,2} = 2\pi/k_0 n_{1,2}$.

The refraction and reflection laws contained in Eqs. (59) and (60) may be visualized conveniently with the aid of refractive index or wavenumber surfaces. As noted in Sec. 1.6, for a given medium, such surfaces are obtained by plotting the refractive index n or the wavenumber $k = k_0 n = k_0 \bar{k}$ as a function of direction. The vector from the origin to a point on the surface specifies the direction of phase propagation in the corresponding plane-wave $\exp(i\mathbf{k}\cdot\mathbf{r})$; as shown in Sec. 1.6a, the normal to the surface at this point yields the direction of energy propagation (ray direction) in the wave. In an isotropic medium, n is independent of direction, so the plots of n and k are spherical, whence the phase and ray propagation vectors are codirectional. At an interface between two media, the condition $\bar{\mathbf{k}}_{t1} = \bar{\mathbf{k}}_{t2} = \bar{\mathbf{k}}_{t3}$ leads to the graphical construction in Fig. 1.7.3(b), where the k_y axis is drawn parallel to the local direction of the interface coordinate y in Fig. 1.7.3(a) and k_{y0} specifies the tangential wavenumber of the incident ray 1; the reflected ray 3 and the refracted ray 2 are then determined from those points on the wavenumber surfaces corresponding to the same value of k_{y0}. This graphical procedure has also been employed in Fig. 1.6.7 to chart the progress of a ray in a continuously varying plane-stratified medium.

Anisotropic media

When a ray strikes an interface between two different anisotropic media ϵ_a and ϵ_b, the initial directions of the reflected and refracted rays may be ascertained by the phase-matching procedure given in Eq. (59): all wave constituents must have the same wavevector component $\mathbf{k}_t$ tangential to the interface. An analytical form of the ray reflection and refraction laws is generally quite involved [see Eq. (7.5.9b) for the special case of a uniaxially anisotropic medium], but the ray trajectories may be visualized with the aid of the wavenumber surfaces for the two media. Medium ϵ_a is assumed to have a wavenumber surface with two branches a_1 and a_2; medium ϵ_b has a single branch b, only relevant portions of which are shown in Fig. 1.7.4. Consider an incident ray 1 corresponding to point $\mathbf{k}_1$ on the a_2 branch of the lower medium. When the xy

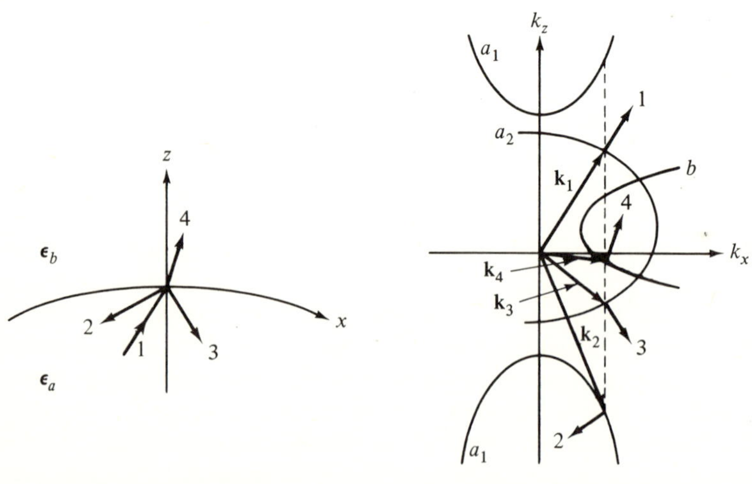

(a) Physical configuration   (b) Wavenumber plot

FIG. 1.7.4 Ray reflection and refraction for anisotropic media (projection on the xz plane).

coordinate system in the tangent plane is oriented appropriately, the incident wavevector tangential to the interface is $\mathbf{k}_t \equiv \mathbf{k}_x$, and this value of $\mathbf{k}_x$ determines other points $\mathbf{k}_i$ on the composite wavenumber diagram. The corresponding rays $\bar{\mathbf{S}}_i$ perpendicular to the surfaces are drawn so as to ensure $\mathbf{k}_i \cdot \bar{\mathbf{S}}_i \geq 0$ [see Eq. (53a)], thereby providing the trajectories in Fig. 1.7.4. Only the projections on the xz plane are shown; although all wavevectors $\mathbf{k}_i$ lie in this plane, the rays are generally not coplanar and possess a y component, unless the wavenumber surfaces are symmetrical about the $k_y = 0$ plane. It is to be noted that the incident ray excites two reflected rays, one of which (ray 2) is turned back toward the direction of incidence. Such phenomena of "backward" reflection or refraction of *rays* may occur in an anisotropic medium because of the general character of the wavenumber surfaces, but no backward characteristics are exhibited by the wavevectors, and hence the directions of advance of the phase fronts, in view of the phase matching condition $\mathbf{k}_t = $ constant. One observes furthermore that the refracted ray 4 carries energy away from the interface, whereas the phase fronts in this ray propagate toward the interface (see direction of $\mathbf{k}_4$), which feature identifies a "backward wave" with respect to the direction z normal to the interface. Of all points $\mathbf{k}_i$ having the same value $\mathbf{k}_t$ on the composite diagram, only those are permissible which satisfy the radiation condition; the latter stipulates that each reflected and refracted ray must carry energy away from the boundary. These constraints eliminate rays corresponding to the upper branch of a_1 and to the upper intersection on branch b.

Warm isotropic plasma

In a homogeneous warm plasma medium, electromagnetic (transverse) and electroacoustic (longitudinal) waves propagate independently with the phase speeds $v = c/n_p$ and $v = a/n_p$, respectively, where c is the speed of light in vacuum, a is the electron-acoustic speed in the plasma, and $n_p = [1 - (\omega_p^2/\omega^2)]^{1/2}$ is the plasma refractive index (see Sec. 1.2c). To a first approximation, these waves also propagate independently in a slowly varying medium, but they are coupled at boundaries across which medium properties change abruptly. Imposition of the phase-matching procedure is illustrated for an electronacoustic ray incident on an interface separating the plasma from an exterior dielectric medium with refractive index n (Fig. 1.7.5). Since the wave processes are isotropic, they are described by spherical wavenumber surfaces having radii $k_0 n_p$, $k_a n_p$, and $k_0 n$ for the electromagnetic and electronacoustic waves in the plasma, and the electromagnetic wave in the dielectric, respectively, with $k_0 = \omega/c$ and $k_a = \omega/a$. The incident electronacoustic ray 1 has a wavevector component $k_{t1} = k_a \sin \theta_1$ parallel to the interface and via Eq. (59) and Fig. 1.7.5(b) establishes the directions of the reflected acoustic ray 2, the reflected electromagnetic ray 3, and the transmitted electromagnetic ray 4, respectively. Since the ray and wavevector directions coincide in the present example, a simple analytical statement of the ray reflection and refraction laws follows at once from Eq. (59):

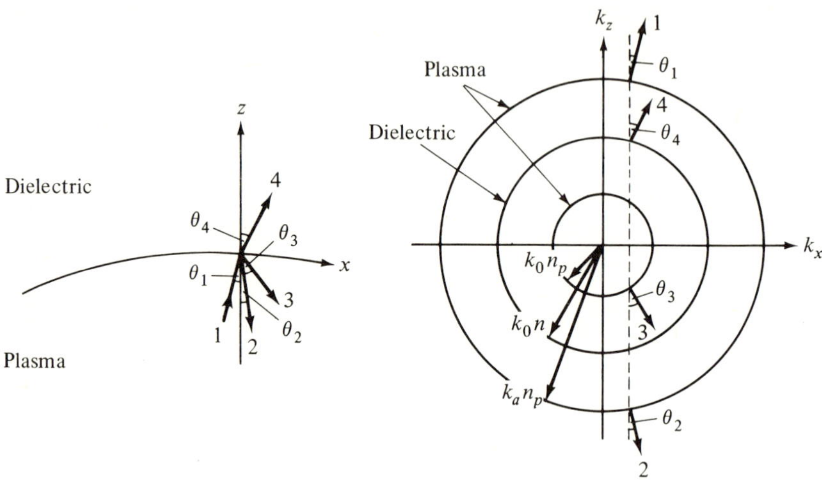

(a) Physical configuration   (b) Wavenumber plot

FIG. 1.7.5   Ray reflection and refraction for warm plasma.

$$\sin \theta_1 = \sin \theta_2 = \frac{a}{c}\sin \theta_3 = \frac{a}{c}\frac{n}{n_p}\sin \theta_4. \tag{61}$$

Analogous considerations apply when an electromagnetic ray is incident either from the plasma or from the dielectric side.

Diffracted rays

Although it is necessary, in general, to solve "canonical" scattering problems for specification of initial field values on diffracted rays, certain general observations can be made concerning the multiplicity and orientation of diffracted rays launched by a prescribed incident ray. For example, to a lowest order of approximation, a ray incident on a slowly curved edge in a slowly varying medium excites the same scattered field as a plane wave in a homogeneous medium incident on a straight edge extended indefinitely along the tangent at the point of impact of the ray. In this canonical problem, the plane-wave and the medium parameters are chosen so as to represent local conditions near the point of impact; because of translational invariance along the direction x parallel to the edge in a homogeneous medium, an incident plane wave with wavenumber k_{xi} along x excites scattered fields characterized by the same value $k_x = k_{xi}$. This condition can be utilized to chart the initial direction of edge-diffracted rays in a manner analogous to that used for reflected and refracted rays. For example, consider an anisotropic medium characterized by a wavenumber surface having three branches, a, b_1, and b_2, symmetrical about an axis w, as in a cold plasma subjected to an external magnetic field along the direction w [Fig. 1.7.6(b)]. An edge is embedded in this medium as shown and is excited by an incident ray $1a$ with $k_x = k_{xi}$ [Fig. 1.7.6(a) and (b)]. Diffracted rays correspond

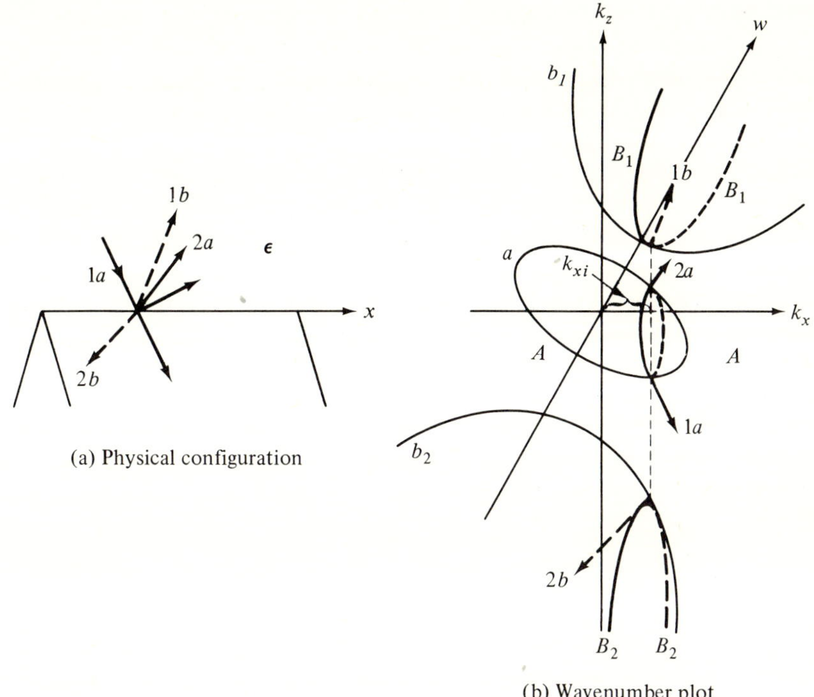

FIG. 1.7.6 Trajectories of edge-diffracted rays.

to **k** values lying on the intersection curves A, B_1, and B_2 of the plane $k_x = k_{xi}$ and the multibranched wavenumber surface; the rays proceed in the direction perpendicular to the wavenumber surface [see Eq. (53a) for constraints on their orientation] to points exterior to the wedge region in Fig. 1.7.6(a). As noted in Fig. 1.7.6, the diffracted rays corresponding to various branches of the wavenumber surface lie along distinct cones (or sections thereof when the relevant branch is open) which are generally non-symmetric about the edge. In the special case of an isotropic medium, the wavenumber surface or surfaces are spherical (see Figs. 1.7.3 and 1.7.5), with consequent x symmetry of the resulting diffracted ray cones. When the medium is an ordinary isotropic dielectric, existence of only a single spherical wavenumber surface implies that the angle of the diffracted ray cone is the same as the angle between the incident ray and the edge, as noted in connection with Fig. 1.7.2(a). It may be noted that the determination of the edge-diffracted ray trajectories is the same as for a line source with phase progression exp $(ik_{xi}x)$ located on the x axis. This aspect is explored in Secs. 5.4d, 7.3c and 7.3d.

The trajectories of lateral rays excited on the interface between two media may be discussed in a similar manner.[39] Lateral rays arise whenever an incident ray of a given species can give rise to a ray of another species directed parallel to the interface; the latter ray may proceed either in the original medium (if several wave types can be supported therein) or in the exterior medium. For an interface between two ordinary dielectrics, one observes from Fig. 1.7.3(b) that $\theta_2 = 90°$ when $k_{y0} = k_2$ [i.e., $\theta_1 = \theta_c = \sin^{-1}(n_2/n_1)$, where θ_c is the critical angle (for $\theta_1 > \theta_c$, the incident rays are totally reflected and no transmission into medium 2 takes place)]. The critically refracted ray 2 travels along the interface in medium 2 and refracts energy continuously back into region 1 along ray 3, as shown in Fig. 1.7.2(d). The triplet of rays 1, 2, 3 is subject to the phase-matching condition in Eq. (59). When anisotropic media are involved as in Fig. 1.7.4, a multiplicity of lateral rays can be excited. For example, when the incident ray 1 on branch a_2 in Fig. 1.7.4 has a direction such that the refracted ray 4 is parallel to the interface in region b, one obtains the lateral-ray configuration schematized in the first sketch in Fig. 1.7.7(a); refraction into medium a occurs in ray species 2 and 3, corresponding to different branches of the wavenumber surface for region a. Alternatively, when the incident ray 1' in region a originates on branch a_1 as shown, it excites a lateral ray 2' on branch a_2 which travels along the interface in medium a and sheds energy continuously into media a and b along rays 3' and 4', respectively. These aspects are treated further in Secs. 5.5a, 5.5b, and 7.5d. One observes from Fig. 1.7.7(b) that lateral ray 4 can also be excited by a critically incident ray 1'' on branch a_1 and that

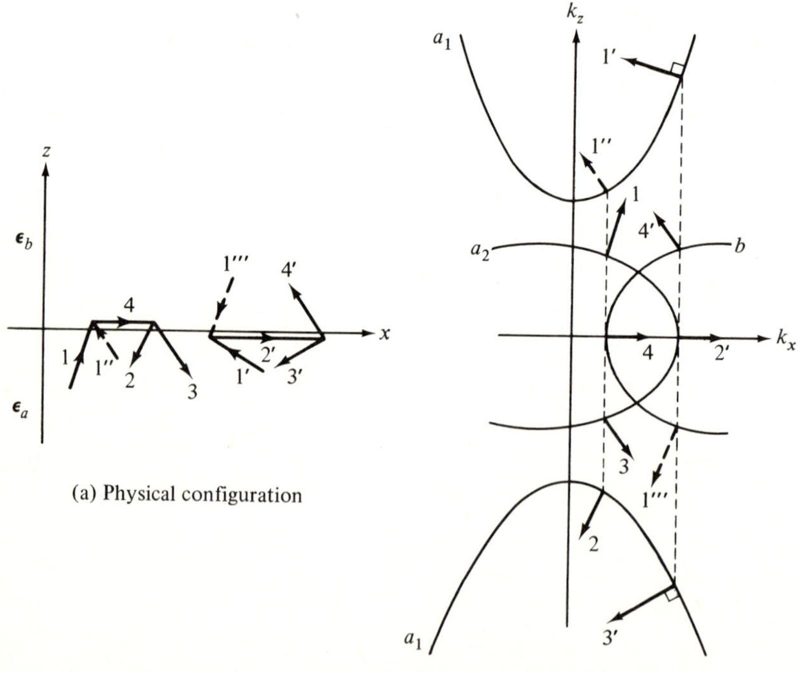

(a) Physical configuration

(b) Wavenumber plot

FIG. 1.7.7 Lateral ray trajectories on interface between aniso-tropic media (projection on xz plane).

lateral ray 2′ can also be excited by a critically incident ray 1‴ from region *b*. Lateral rays occur also on the dielectric–warm plasma interface in Fig. 1.7.5. For example, when the incidence angle for the electron-acoustic ray 1 is $\theta_1 = \sin^{-1}(a/c)$, ray 3 becomes an electromagnetic lateral ray which travels parallel to the interface in the plasma and refracts as the electronacoustic ray 2 in the plasma and the electromagnetic ray 4 in the dielectric. When the incidence angle of ray 1 is $\theta_1 = \sin^{-1}(an/cn_p)$, the electromagnetic ray 4 in the dielectric becomes a lateral ray that refracts as the electronacoustic ray 2 in the plasma. It may be noted that when the dispersion relation is written $k_z = k_z(k_x, k_y)$, appropriate for analysis of propagation in regions with planar stratification along z (see Sec. 1.6b), the lateral ray corresponds to a double root on the dispersion surface and is thus characterized by a branch-point singularity in the k_t integral representation [Eq. (1.6.21)].

Example: Diffraction by a conducting half-plane on the interface between two isotropic dielectrics

The asymptotic representation of the high-frequency field for radiation and diffraction problems comprises in general several constituents, each of which has a dominant behavior of the form exhibited in Eqs. (37), (40), or their equivalents. When reflection and (or) refraction at a boundary is relevant, the associated phenomena are accounted for by the geometric-optical field u_g, whereas shadowing effects, structural singularities, etc., give rise to an additional diffracted field u_d. Thus, the total field at an observation point **r** is given in general by

$$u(\mathbf{r}) \sim u_g(\mathbf{r}) + u_d(\mathbf{r}), \tag{62}$$

where $u_g(\mathbf{r}) = \sum_i u_g^{(i)}(\mathbf{R}_i)$, with $u_g^{(i)}(\mathbf{R}_i)$ representing the direct, reflected, or refracted field that reaches **r** along a ray $\mathbf{R}_i$. Similarly, the diffracted field may also have several distinct species, $u_d(\mathbf{r}) = \sum_j u_d^{(j)}(\mathbf{R}_j)$.

The method of constructing the ray-optical field in this manner is illustrated by the example in Fig. 1.7.8, which shows a plane interface separating two different homogeneous isotropic media having refractive indexes n_1 and $n_2 < n_1$. A semiinfinite, perfectly conducting plane sheet is placed on the interface, and the entire configuration is illuminated by a line source oriented parallel to the edge (this renders the problem two-dimensional). The source emits rays in all directions, and some of the rays reach a given observation point either directly, by reflection, by refraction, or by diffraction. An observation point $\mathbf{r}_1$ in the upper medium is reached by the direct ray 1 and the reflected ray 3 which constitute the geometric-optical field. Contributions arise also from diffraction effects that are associated with the onset of total reflection and with the presence of an edge singularity, respectively. The former gives rise to the lateral ray 5 which travels along the interface in the low-refractive-index region and sheds energy back into the high-refractive-index region along ray 6 (for details, see Sec. 5.5). When the lateral ray strikes the edge, a radially spreading diffracted

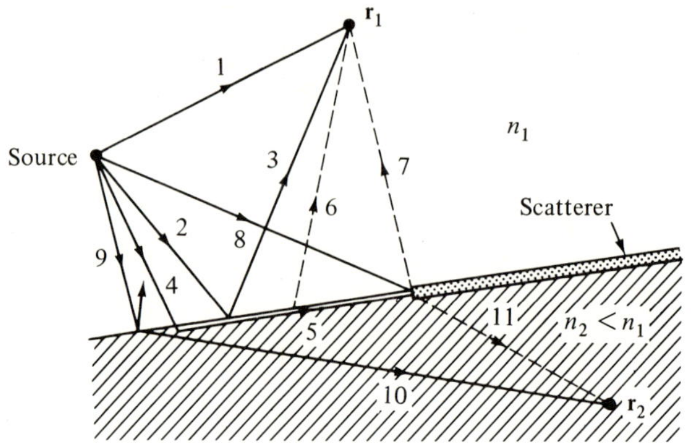

FIG. 1.7.8 Ray-optical construction of the field when a source illuminates a plane interface on which is placed a semiinfinite scatterer (dashed lines denote diffracted rays).

field is generated which reaches point $\mathbf{r}_1$ along ray 7. The edge is also excited directly by the incident ray 8 which sends its own edge-diffracted field along ray 7 (for details of scattering by a half-plane, see Sec. 6.5). Thus

$$u(\mathbf{r}_1) \sim u_g^{(1)} + u_{g2}^{(3)} + u_{d5}^{(6)} + u_{d5}^{(7)} + u_{d8}^{(7)}, \tag{63}$$

where the superscripts identify the rays leading to point $\mathbf{r}_1$, the letter subscripts distinguish geometric-optical and diffracted constituents, and the number subscripts specify the appropriate exciting ray.

The form of the individual field contributions in Eq. (63) can be determined from Eq. (37) or (40). The line-source field is assumed to be so normalized as to generate in an unbounded medium a unit amplitude field at unit radial distance, with the phase taken as zero at the source. Since the rays diverge radially from the source in the plane of the figure, but are parallel at successive cross sections along the line source (see Sec. 5.4d), the ray tubes are wedge shaped and $dA(\mathbf{R}) \sim 1/R$, where R is the distance from the source. Thus, the amplitude along an incident ray varies like $R^{-1/2}$, representative of a cylindrically spreading wave. From Eq. (37), with the above-mentioned normalization, one obtains for the direct field contribution,

$$u_g^{(1)}(\mathbf{R}_1) = \frac{e^{ik_0 n_1 R_1}}{\sqrt{R_1}}, \tag{64a}$$

where it has been recognized that $\psi(R) = k_0 n_1 R$ for the homogeneous medium in the present example. In this equation and the ones to follow, R_i is the length along the ith ray in Fig. 1.7.8.

The reflected field may also be calculated from Eq. (37), with the reference point on ray 3 chosen at the interface. Since the interface is planar and the media are isotropic, the tube of reflected rays appears to originate at the image point of the source [Fig. 1.7.9(a)], whence the ratio of the area at the interface

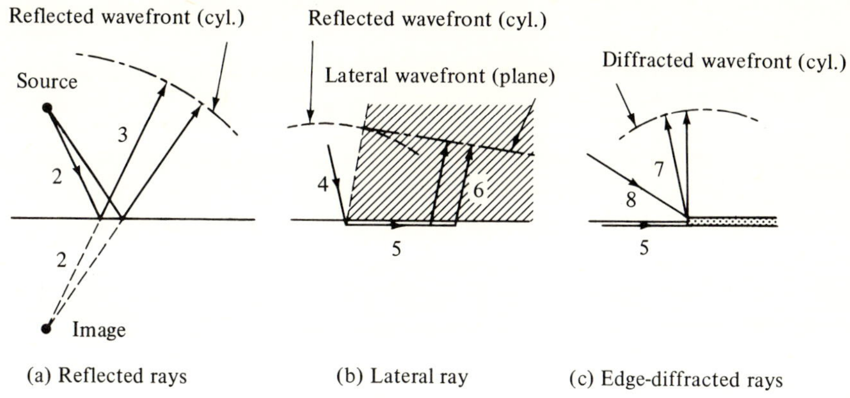

FIG. 1.7.9　Various ray-tube constructions.

to that at $\mathbf{r}_1$ is given by $R_2/(R_2 + R_3)$ (for a general treatment of reflection from a curved surface, see Reference 40). The reflected field at the interface, equal to the incident field multiplied by a reflection coefficient Γ, is given from Eq. (64a) by $\Gamma R_2^{-1/2} \exp(ik_0 n_1 R_2)$. The phase accumulation along ray 3 equals $\exp(ik_0 n_1 R_3)$, whence, from Eq. (37),

$$u_{g2}^{(3)} = \Gamma \frac{e^{ik_0 n_1 (R_2 + R_3)}}{\sqrt{R_2 + R_3}}. \tag{64b}$$

The Fresnel reflection coefficient Γ is calculated by field matching between the incident, reflected, and refracted rays at the interface and, in view of the previously noted plane-wave character of the local field along a ray, is the same as for a *plane-wave* field incident along the direction of ray 2.

The diffracted field along ray 6 is launched by continuous refraction of the field carried along the lateral ray 5, which, in turn, is excited by ray 4 incident at the angle of total reflection $\theta_c = \sin^{-1}(n_2/n_1)$ measured from the normal to the interface. Since the refraction angle equals θ_c and is therefore constant, the emerging tube is bounded by parallel rays [Fig. 1.7.9(b)], thereby furnishing a constant amplitude along ray 6. The phase accumulation along ray 6 is given by $\exp(ik_0 n_1 R_6)$, so $u_{d5}^{(6)} = u_{d5} \exp(ik_0 n_1 R_6)$, where u_{d5}, the reference value of the field at the interface, must again be determined by solving the boundary-value problem (alternatively, see Reference 41). It is found (Sec. 5.5) that the phase of u_{d5} is the one predicted from the trajectory along rays 4 and 5 (note that ray 5 travels in the medium with refractive index n_2) and that the amplitude is proportional to $1/k_0 R_5^{3/2}$. Thus,

$$u_{d5}^{(6)} \propto \frac{e^{ik_0(n_1 R_4 + n_2 R_5 + n_1 R_6)}}{k_0 R_5^{3/2}} U(R_5), \tag{64c}$$

where the Heaviside unit function $U(R_5)$ signifies that this contribution exists only when the ray path has a finite lateral segment R_5 [i.e., for observation points in the shaded region of Fig. 1.7.9(b), which contains the totally reflected geometric-optical rays]. Evidently, the formula fails near the total reflection boundary $R_5 = 0$ and must be replaced there by a transition function that cannot be deduced by ray-optical means (see Sec. 5.5).

When an incident ray strikes an edge, it generates a cylindrically spreading diffraction field that can be described in terms of a set of radial rays originating at the edge [Fig. 1.7.9(c)]. Thus, the field along an edge-diffracted ray 7 decays like $R_7^{-1/2}$ and its phase varies like $\exp(ik_0 n_1 R_7)$. The reference field at unit distance from the edge must now be determined from the solution of the relevant diffraction problem. For the diffracted field excited by ray 8 in Fig. 1.7.8 it suffices, because of the local plane-wave character of the incident field, to treat the simpler plane-wave scattering problem. Its solution furnishes a diffraction coefficient D_{87} which specifies the diffracted reference field generated by a plane wave of unit strength incident along the direction of ray 8. Since the actual incident field has a strength given by Eq. (64a) with R_1 replaced by R_8, one finds that

$$u_{d8}^{(7)} = D_{87} \frac{e^{ik_0 n_1 (R_8 + R_7)}}{\sqrt{R_8 R_7}}. \tag{64d}$$

The diffraction coefficient D_{87} depends on the angle of the incident and diffracted rays with respect to the half-plane; for scattering in a homogeneous medium, it has been calculated in Sec. 6.5.

An analogous calculation must be performed for the incident lateral wave to yield a diffraction coefficient D_{57}. Then, by considerations similar to the above, one may write

$$u_{d5}^{(7)} \propto D_{57} \frac{e^{ik_0 (n_1 R_4 + n_2 R_5 + n_1 R_7)}}{k_0 R_5^{3/2} R_7^{1/2}} U(R_5), \tag{64e}$$

with R_5 representing the entire lateral distance between the point of impact of ray 4 and the edge. The edge also excites a lateral wave traveling to the left along the interface; its contribution has not been shown but should be included in general. Moreover, different wave constituents emerge when the observation point is far to the right of the edge, in which instance the reflected field is modified and the lateral wave field in Eq. (64c) is absent.

The same procedure may be employed for the construction of the ray-optical field in the lower medium of Fig. 1.7.8. Instead of the geometric-optical contributions in Eqs. (64a) and (64b), one now has the refracted field carried along ray 10. The diffracted fields are similar to the above, except for the omission of the lateral-wave constituent in Eq. (64c), which is absent in the optically thinner medium (see Fig. 1.7.3). When the edge of the obstacle is located to the left of the point of impact of the critically refracted ray 4, there exists a geometric-optical shadow region from which the refracted rays 10 are excluded and which is penetrated only by the edge-diffracted rays 11.

Although the dependence on k_0 has not been given explicitly in the reflection and diffraction coefficients in Eqs. (64), the distance dependence has been shown. One observes that for large distances R, the diffraction fields are considerably weaker than the geometric optical fields, thereby complicating the detection of the former in the time-harmonic regime (an exception occurs in shadow regions where the diffraction fields are dominant). This disadvantage is not present under transient conditions since in view of the discussion in Sec. 1.6c, the distinctive phases of the various field constituents in Eq. (63) imply different times of arrival of the transient response at an observation point (see Fig. 1.7.9 for a sketch of the wavefronts, and also Figs. 5.5.8 and 6.4.2). Thus, the time resolution under pulse conditions provides an important means of identifying field contributions, which may be obscured in the time-harmonic steady state.

Finally, it should be emphasized that retention of only the dominant term in Eq. (37) for the asymptotic representation of each of the constituent fields does not assure that the composite solution in Eq. (63) is consistent to a certain order in the small parameter $1/k_0$. The k_0 dependence of a diffraction field amplitude (e.g., that of the lateral wave) may correspond to that of a second term in the asymptotic expansion of the geometric-optical field [see Eq. (16)], so it may be necessary to include such terms if a consistent result to a given order in $1/k_0$ is desired.

1.7e   Transient Fields

Non-time-harmonic fields in a spatially and temporally inhomogeneous medium satisfy the system of equations given abstractly in Eq. (2). To a lowest order of approximation as in Eq. (1), in a parameter regime validating a wave-packet description, the general field equations reduce to the simple geometric-optic form in Eq. (3), which are integrable along space–time ray trajectories defined in Eq. (10). The phase of the assumed field (1) follows from solution of the dispersion equation (4a) as in Eq. (11), while the amplitude and polarization of the field vector $\boldsymbol{\Psi}_0$ requires evaluation of the eigenvectors $\boldsymbol{\Psi}_{0a}$ from Eq. (3). Because of the complicating effects of dispersion, performance of these calculations is generally quite involved. To illustrate salient features, it is instructive to treat the simple example of an isotropic cold plasma for which the lowest-order, as well as higher-order, approximations can be developed without undue difficulty.[19]

As in the simple time-harmonic field problems in Sec. 1.7b and 1.7c, it is convenient to employ the second-order form of the field equations to derive the dispersion and energy-conservation equations for the lowest-order space–time ray amplitudes. Thus, we consider the scalar wave equation

$$\left[\nabla^2 - \frac{1}{c^2} \frac{\partial^2}{\partial t^2} - \frac{\omega_p^2(\mathbf{r})}{c^2} \right] u(\mathbf{r}, t) = 0, \tag{65}$$

satisfied approximately by the scalar electromagnetic potential function u in a spatially inhomogeneous cold plasma characterized by the variable plasma frequency $\omega_p(\mathbf{r})$, with c denoting the speed of light in vacuum. To introduce the parameter v with respect to which the asymptotic development is to be carried out, one scales the space and time coordinates,

$$\mathbf{R} = \frac{\mathbf{r}}{v}, \qquad \tau = \frac{t}{v}, \tag{66}$$

and obtains

$$\left[\nabla_R^2 - \frac{1}{c^2}\frac{\partial^2}{\partial \tau^2} - v^2 \frac{\bar\omega_p^2(\mathbf{R})}{c^2} \right] u(\mathbf{R}, \tau) = 0, \tag{67}$$

where ∇_R is the gradient operator in the $\mathbf{R}$ coordinate system, $\bar\omega_p(\mathbf{R}) = \omega_p(\mathbf{r}) = \omega_p(\mathbf{R}v)$ and $u(\mathbf{R}, \tau)$ is $u(\mathbf{r}, t)$ in the scaled coordinates. v is now taken to be large, so non-vanishing values of $\mathbf{R}$ and τ imply large values of $\mathbf{r}$ and t. As noted previously, this is the range where wavepackets are fully developed; also, $\nabla\omega_p(\mathbf{r}) = (1/v)\nabla_R\bar\omega_p(\mathbf{R}) = O(1/v)$, so $\omega_p(\mathbf{r})$ is slowly varying. A solution for u is assumed in the form of an asymptotic series,

$$u(\mathbf{R}, \tau) = e^{iv\psi(\mathbf{R},\tau)} \sum_{m=0}^{\infty} \frac{u_m(\mathbf{R}, \tau)}{(iv)^m}, \tag{68}$$

with ψ and u_m independent of v; on substitution into Eq. (67), assuming that the orders of summation and differentiation can be exchanged, one finds

$$(iv)^2\left[(\nabla_R\psi)^2 - \frac{1}{c^2}\left(\frac{\partial\psi}{\partial\tau}\right)^2 + \frac{\bar\omega_p^2}{c^2} \right]F + (iv)\left[\nabla_R^2\psi - \frac{1}{c^2}\frac{\partial^2\psi}{\partial\tau^2} + 2\nabla_R\psi\cdot\nabla_R \right.$$
$$\left. - \frac{1}{c^2}2\frac{\partial\psi}{\partial\tau}\frac{\partial}{\partial\tau} \right]F + \left(\nabla_R^2 - \frac{1}{c^2}\frac{\partial^2}{\partial\tau^2} \right)F = 0, \tag{69}$$

where F stands for the sum in Eq. (68). Since this equation must hold for arbitrary (though large) values of v, the coefficient of each power of v must vanish independently. From the v^2 term, one finds

$$(\nabla_R\psi)^2 - \frac{1}{c^2}\left(\frac{\partial\psi}{\partial\tau}\right)^2 + \frac{\bar\omega_p^2(\mathbf{R})}{c^2} = 0; \tag{70}$$

from the v^1 term,

$$\left(\nabla_R^2\psi - \frac{1}{c^2}\frac{\partial^2\psi}{\partial\tau^2} + 2\nabla_R\psi\cdot\nabla_R - \frac{2}{c^2}\frac{\partial\psi}{\partial\tau}\frac{\partial}{\partial\tau} \right)u_0 = 0; \tag{71}$$

and from the v^{-m} term, $m \geq 0$,

$$\left(\nabla_R^2\psi - \frac{1}{c^2}\frac{\partial^2\psi}{\partial\tau^2} + 2\nabla_R\psi\cdot\nabla_R - \frac{2}{c^2}\frac{\partial\psi}{\partial\tau}\frac{\partial}{\partial\tau} \right)u_{m+1} + \left(\nabla_R^2 - \frac{1}{c^2}\frac{\partial^2}{\partial\tau^2} \right)u_m = 0.$$
$$\tag{72}$$

Equation (70) is the dispersion equation for the medium which, on introduction of a temporal frequency $\bar\omega$ and a wavevector $\bar{\mathbf{k}}$ as in Eq. (4b),

$$\bar\omega = -\frac{\partial\psi}{\partial\tau}, \qquad \bar{\mathbf{k}} = \nabla_R\psi, \tag{73}$$

takes on the more familiar form

$$\bar{k}^2 - \frac{\bar{\omega}^2}{c^2} + \frac{\bar{\omega}_p^2(\mathbf{R})}{c^2} = 0, \qquad \text{i.e., } \bar{\omega} = \bar{\omega}(\bar{\mathbf{k}}, \mathbf{R}) = \pm\,[\bar{k}^2 c^2 + \bar{\omega}_p^2(\mathbf{R})]^{1/2}.$$

$$(74)$$

Equation (71) is a transport equation for the leading amplitude term u_0, and Eq. (72) is a transport equation for the higher-order amplitudes. The higher-order amplitudes can be evaluated recursively in terms of the lower-order ones and ψ, but this calculation will not be performed here.

Solution of the dispersion equation

By integrating along space–time ray trajectories defined in Eq. (10), one may solve for the phase function ψ as in Eq. (11). Since the dispersion relation (74) does not contain the time variable explicitly, $d\bar{\omega}/dt = 0$ in Eq. (10), so $\bar{\omega} =$ constant along a ray. Thus, Eq. (11) reduces to

$$\psi(\mathbf{R}, \tau) - \psi(\mathbf{R}_1, \tau_1) = \int_{\mathbf{R}_1}^{\mathbf{R}} \bar{\mathbf{k}} \cdot d\mathbf{R} - \bar{\omega}(\tau - \tau_1), \qquad (75)$$

where $\bar{\mathbf{k}}$ is obtained from Eq. (74). For the special case of planar stratification, this phase function is the same as the one obtained in Eq. (1.6.30) by saddle-point evaluation of the integral representation of the field $u(\mathbf{R}, \tau)$. A graphical procedure for tracking the rays has been discussed in connection with Fig. 1.6.9. For a homogeneous medium, $\bar{\omega}$ and $\bar{\mathbf{k}}$ are constant along a ray and one obtains the phase function in Eq. (1.6.5), with $v\psi$ defined in Eq. (1.6.1).

Solution of the transport equation

In view of Eq. (73), Eq. (71) can be written as

$$\left(\nabla_R \cdot \bar{\mathbf{k}} + \frac{1}{c^2}\frac{\partial \bar{\omega}}{\partial \tau} + 2\bar{\mathbf{k}} \cdot \nabla_R + \frac{2}{c^2}\bar{\omega}\frac{\partial}{\partial \tau}\right)u_0 = 0. \qquad (76)$$

By changing variables from τ to $c\tau$, introducing a four-dimensional group velocity vector $\bar{\mathbf{V}}_g = \bar{\mathbf{v}}_g + \boldsymbol{\tau}_0 c$ as in Fig. 1.6.1(b), and noting from Eq. (74) that $\bar{\mathbf{v}}_g = \nabla_k \bar{\omega} = \bar{\mathbf{k}} c^2/\bar{\omega}$, one may verify that Eq. (76) can be written in the alternative form:

$$\square \cdot \left(\bar{u}_0^2 \frac{\bar{\omega}}{c^2}\bar{\mathbf{V}}_g\right) = 0, \qquad \square \equiv \nabla_R + \boldsymbol{\tau}_0 \frac{\partial}{\partial(c\tau)}, \qquad (77)$$

where $\boldsymbol{\tau}_0$ is a unit vector along the $c\tau$ axis. Equation (77) implies conservation of the energy flux, $\bar{u}_0^2 \bar{V}_g \bar{\omega}/c^2$, in $(\mathbf{R}, c\tau)$ space. $\bar{\mathbf{V}}_g$ is tangent to the space–time rays defined in Eq. (10), which fact has been utilized in Sec. 1.6b for the special case of plane-stratified media. If the divergence theorem in four-dimensional $(\mathbf{R}, c\tau)$ space is applied to a volume generated by a small tube of rays and terminated by hyperplanes $\tau = $ constant (Fig. 1.7.10), the only contribution to the hypersurface integral arises from the cross-section "planes" $\Delta\mathbf{R}_1$ and $\Delta\mathbf{R}$, since on the walls of the tube, $\mathbf{b} \cdot \bar{\mathbf{V}}_g = 0$, $\mathbf{b}$ being a vector normal to the hypersurface. Since the outward normals at $\Delta\mathbf{R}_1$ and $\Delta\mathbf{R}$ have opposite directions along the τ axis and $\boldsymbol{\tau}_0 \cdot \mathbf{V}_g = c$, one finds

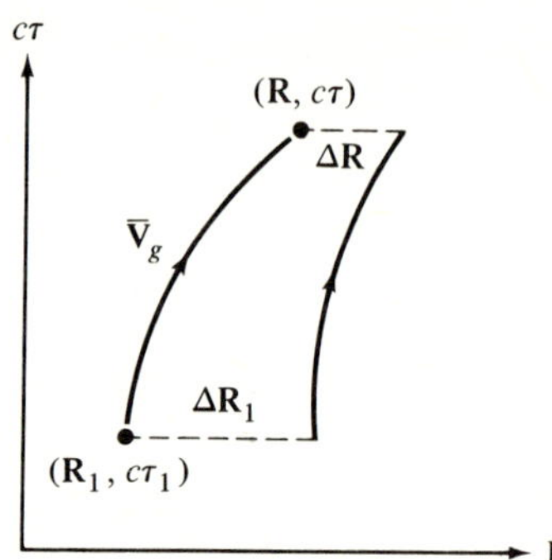

FIG. 1.7.10   Energy conservation in a space–time ray tube.

$$\frac{\bar{\omega}_1}{c^2}\,u_0^2(\mathbf{R}_1, \tau_1)\Delta\mathbf{R}_1 = \frac{\bar{\omega}}{c^2}\,u_0^2(\mathbf{R}, \tau)\Delta\mathbf{R}, \tag{78a}$$

whence, in view of the constancy of $\bar{\omega}$ along a ray, one has for the amplitude u_0 at the point $(\mathbf{R}, \tau)$ along a ray in terms of its value at a reference point $(\mathbf{R}_1, \tau_1)$:

$$u_0(\mathbf{R}, \tau) = u_0(\mathbf{R}_1, \tau_1)\sqrt{\frac{\Delta\mathbf{R}_1}{\Delta\mathbf{R}}}. \tag{78b}$$

Since the hypersurface areas $\Delta\mathbf{R}_1$ and $\Delta\mathbf{R}$ in planes $\tau = $ constant in the four-dimensional $(\mathbf{R}, c\tau)$ space actually represent volumes in the three-dimensional $\mathbf{R}$-space, one recognizes that the energy-conservation statement in Eq. (78a) is equivalent to the statement $\Delta W = |I|^2\Delta\mathbf{R} = $ constant used in connection with Eq. (1.6.9) [see also Eq. (1.6.15)]. Via the present analysis, the validity of this earlier energy-conservation theorem for wavepackets has been extended to propagation in inhomogeneous media. Note that in the inhomogeneous medium, only the frequency spread $\Delta\bar{\omega}$ of waves in the wavepacket is preserved, whereas in a homogeneous medium, both $\Delta\bar{\omega}$ and $\Delta\bar{\mathbf{k}}$ remain invariant. One may also use the considerations of Eqs. (1.6.8)–(1.6.10) for the inhomogeneous case provided that the mapping is carried out locally in such a manner as to conserve energy in the wavepacket; the initial amplitude $u_0(\mathbf{R}_1, \tau_1)$ and phase $\nu\psi(\mathbf{R}_1, \tau_1)$ near the source region can be calculated from Eq. (1.6.5) by assuming the medium to be locally homogeneous, and Eqs. (75) and (78b) are used thereafter, due allowance being made for the variability of $\bar{\mathbf{k}}$ and the constancy of $\bar{\omega}$ along a ray.

Reflection and refraction of space–time rays

As in the time-harmonic case, a space–time ray generates reflected and refracted rays on striking a surface B across which the medium properties vary discontinuously. By considerations analogous to those leading to Eq. (57), one requires continuity of the phase function $\psi(\mathbf{R}, \tau)$ on the surface B for all relevant wave constituents.[38] It then follows that partial derivatives of ψ in the direction tangential to B, and $\partial\psi/\partial\tau$, are also continuous whence from Eq. (73), if m wave constituents are involved,

$$\bar{\mathbf{k}}_{t1} = \bar{\mathbf{k}}_{t2} = \cdots = \bar{\mathbf{k}}_{tm}, \qquad \bar{\omega}_1 = \bar{\omega}_2 = \cdots = \bar{\omega}_m. \qquad (79)$$

Because of constancy of the frequency $\bar{\omega}$ in the complex of incident, reflected, and refracted rays, the matching condition (79) on the tangential wave-vector components $\bar{\mathbf{k}}_t$ is imposed as in the time-harmonic problem, and thus the graphical procedures depicted in Figs. 1.7.3–1.7.5 remain relevant. The initial directions of the time-harmonic rays in $\mathbf{r}$ space determined in this manner represent the projections of the corresponding space–time rays onto hyperplanes perpendicular to the τ axis. The ray directions in $(\mathbf{R}, c\tau)$ space are obtained by constructing the normal vectors to relevant branches of the $(c\bar{\mathbf{k}}, \bar{\omega})$ dispersion surfaces at $(c\bar{\mathbf{k}}_t, \bar{\omega})$ points descriptive of the ray complex (see Sec. 1.6a for details on the graphical construction). For two cold isotropic plasma media characterized by Eq. (74), with local plasma frequencies $\bar{\omega}_{p1}$ and $\bar{\omega}_{p2}$ on opposite sides of a boundary separating them, the matching technique is depicted in Fig. 1.7.11.

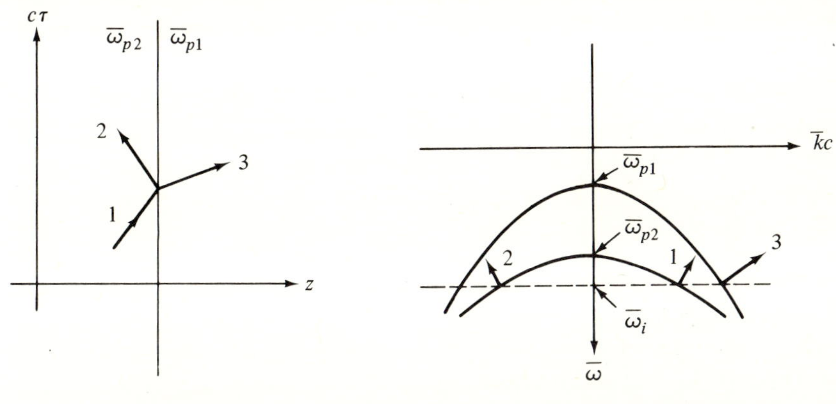

(a) Configuration space　　　　　　　　(b) Dispersion surface

FIG 1.7.11　Ray reflection and refraction at interface between two media.

As in the analogous situation in Fig. 1.6.9, to simplify the graphical construction, only one-dimensional propagation (with $\bar{\mathbf{k}}_t = 0$) is considered. The matching parameter $\bar{\omega}_i$ is determined from the incident wavepacket described by ray 1 and is used as in Fig. 1.7.11 to yield the initial directions of the reflected and refracted rays.

Fields near the wavefront

The asymptotic expansion (68) for the transient field $u(\mathbf{R}, \tau) = u(\mathbf{r}, t)$, valid for the large $(\mathbf{r}, t)$ (wavepacket) regime, is inadequate to describe phenomena at times $t \approx r/c$ corresponding to arrival of the wavefront. As noted in Sec.

1.6c, the wavefront fields are synthesized by the high-frequency time-harmonic wave constituents. The asymptotic expansions of the time-harmonic fields in Secs. 1.7b and 1.7c can be employed to develop corresponding expansions of the transient fields for $t \approx r/c$, thereby improving on the leading term expressions in Eqs. (1.6.45) and (1.6.47).

By the considerations of Sec. 1.6c, it follows that if the time-harmonic field is given by a generalized version of Eq. (16),

$$u(\mathbf{r}) \sim \frac{1}{2\pi} e^{ik_0\psi(\mathbf{r})}\left[A'(\mathbf{r}) + \frac{1}{(-i\omega)^{\beta+1}} \sum_{m=0}^{\infty} \frac{A_m(\mathbf{r})}{(-i\omega)^m}\Gamma(\beta + m + 1)\right], \qquad k_0 = \frac{\omega}{c},$$

$$(80)$$

then, from Eq. (1.6.20), the transient field near $t = \psi(\mathbf{r})/c$ is

$$\hat{u}(\mathbf{r}, t) \sim A'(\mathbf{r})\delta(t') + t'^{\beta} \sum_{m=0}^{\infty} A_m(\mathbf{r})t'^m, \qquad \text{where } t' = t - \frac{\psi(\mathbf{r})}{c} \geq 0.$$

$$(81)$$

The appearance of the phase $\psi(\mathbf{r})$ given in Eq. (28), instead of r, generalizes applicability of corresponding results in Sec. 1.6c to media possessing spatial inhomogeneities at the highest propagation speed c. Equation (81) implies that the wavefronts advance along directions of the wavenormal, which, in the present case of an isotropic medium, are parallel to the rays defined by Eq. (23). (For an asymptotic expansion of the fields covering the range between the wavefront and wavepacket regimes, see reference 22.)

Problems

1. Derive second-order wave equations for the acoustic pressure p and velocity $\mathbf{v}$ as in Eqs. (1.1.3):
   (a) with inclusion of the source terms given in the first-order equations [Eqs. (1.1.1)];
   (b) allowing for either spatial or temporal variability, or both, of the background density n_0 and pressure p_0.
   (c) Find the second-order equations satisfied by the Green's functions G_{11} and $\mathcal{G}_{22}$ in Eqs. (1.1.5) when the background density and pressure are spatially and (or) temporally variable.

2. In a volume V bounded by a surface S whereon the "impedance" boundary conditions in Eq. (1.1.5a) apply, use Green's functions satisfying Eqs. (1.1.5) and simpler boundary conditions $G_{11} = 0 = G_{12}$ on S to derive within V representations for $p(\mathbf{r}, t)$ and $\mathbf{v}(\mathbf{r}, t)$ excited by prescribed sources $s(\mathbf{r}, t)$ and $\mathbf{f}(\mathbf{r}, t)$ in V. *Hint:* In addition to integration over the source region as in Eq. (1.1.4), find and integrate over equivalent source distributions on S to assure satisfaction of Eq. (1.1.5a).

3. While the adjoint problem preserves the original boundary surface S, the boundary conditions on S, as well as the medium filling the volume V enclosed by S, may differ in general. By treating an as yet unspecified adjoint medium

with parameters n_0^+, p_0^+, T_0^+ and boundary parameter α^+, *show* that for derivation of the reciprocity relation in Eq. (1.1.9), the adjoint medium and boundary parameters are identical with the original ones.

4. Show that the reciprocity relations in Eqs. (1.1.11) and (1.1.12) remain valid when the background density and pressure are spatially inhomogeneous.

5. In a volume bounded by a surface S, use electromagnetic dyadic Green's functions satisfying perfect conductor boundary conditions in Eq. (1.1.20c) and derive field representations corresponding to Eqs. (1.1.19), satisfying the impedance boundary conditions in Eq. (1.1.16d). Show that an additional surface integral contribution is required and explain in terms of equivalent electric and magnetic induced currents $\mathbf{H} \times \mathbf{n}$ and $\mathbf{n} \times \mathbf{E}$ on the boundary surface S, where $\mathbf{n}$ is the normal to the boundary. Repeat with use of the free-space dyadic Green's function.

6. (a) Derive the second-order form of the Maxwell field equations (1.1.18) with inclusion of the $\mathbf{J}$ and $\mathbf{M}$ source terms in Eq. (1.1.16). Interpret the source terms in the second-order equations as *equivalent* electric or magnetic currents.
   (b) Use the dyadic Green's functions defined by the second-order equations (1.1.21) to represent the electric and magnetic fields excited by the source distributions in (a). Compare the result with Eqs. (1.1.19), which are based on the first-order form of the field equations and the *actual* source distributions. Explain the difference in the two representations and convert one into the other (*Hint:* Employ the formula $\mathbf{A} \cdot \nabla \times \mathbf{B} = \mathbf{B} \cdot \nabla \times \mathbf{A} - \nabla \cdot (\mathbf{A} \times \mathbf{B})$ and use the divergence theorem). Which formulation is more direct?
   (c) Repeat for an inhomogeneous medium with permittivity $\epsilon(\mathbf{r})$ and permeability $\mu(\mathbf{r})$.

7. Use the time-dependent free space Green's functions in Eqs. (1.1.34) to calculate via Eqs. (1.1.19) the electromagnetic fields radiated by a pulsed circular ring electric current: $\mathbf{J}(\mathbf{r}, t) = \boldsymbol{\phi}_0 A \delta(\rho - a)\delta(z)\delta(t)$. The ring has a radius $\rho = a$ and lies in the $z = 0$ plane; the current direction is along the ϕ coordinate in a (ρ, ϕ, z) circular cylindrical coordinate system. When the ring radius a tends to zero and the source strength A is increased so that (aA) remains finite, show that the fields radiated by the electric ring current are equivalent to those radiated by a magnetic current dipole oriented perpendicularly to the plane of the ring. Relate to $\mathscr{G}_{i2}(i = 1, 2)$ of Eq. (1.1.19).

8. A pulsed, arbitrarily oriented electric current dipole with $\mathbf{J}(\mathbf{r}, t) = \mathbf{p}\delta(\mathbf{r})\delta(t)$ is located in the presence of a perfectly conducting infinite plane. Calculate the "half-space" Green's functions for this region and therefrom the radiated electric and magnetic fields. Interpret the result in terms of the free space field and the field reflected by the plane. Show that the reflected field can be interpreted as arising from a properly oriented image dipole. (*Hint:* Decompose the dipole into components parallel and normal to the boundary and treat each separately.) Repeat for a pulsed magnetic dipole current $\mathbf{M}(\mathbf{r}, t) = \mathbf{m}\delta(\mathbf{r})\delta(t)$.

9. (a) Find the time-dependent fields radiated by an arbitrarily oriented electric current dipole located at the point (x', y', z') in the quarter space region $x > 0$, $y > 0$ formed by two perfectly conducting planes intersecting at right angles. Show that the solution can be obtained by replacing the boundary by three ap-

propriately oriented image dipole sources located at $(-x', y', z')$, $(x', -y', z')$ and $(-x', -y', z')$. Interpret the solution in terms of free space fields and reflected contributions from the bounding surface. Draw relevant wave fronts. Repeat for excitation by a magnetic dipole.

(b) Replace the quarter space region by a trough (sector) of angle $\alpha = \pi/n$, where n is a positive integer. (Problem 8 and part (a) are special cases of this configuration, with $n = 1$ and $n = 2$.) Show that the image concept can be applied to replace the trough by $(n - 1)$ appropriately placed and oriented image sources.

10. By integrating over z' the time-dependent three-dimensional scalar Green's function in Eq. (1.1.31), show that the causal two-dimensional scalar Green's function satisfying the equation

$$\left(\nabla_t^2 - \frac{1}{c^2}\frac{\partial^2}{\partial t^2}\right)g(\boldsymbol{\rho}, \boldsymbol{\rho}'; t, t') = -\delta(\boldsymbol{\rho} - \boldsymbol{\rho}')\delta(t - t'), \qquad \nabla_t^2 = \nabla^2 - \frac{\partial^2}{\partial z^2}, \qquad \text{(1a)}$$

is given by:

$$g(\boldsymbol{\rho}, \boldsymbol{\rho}'; t, t') = \frac{1}{2\pi}\frac{1}{\sqrt{(t - t')^2 - (|\boldsymbol{\rho} - \boldsymbol{\rho}'|^2/c^2)}}\, U\!\left(t - t' - \frac{|\boldsymbol{\rho} - \boldsymbol{\rho}'|}{c}\right), \qquad \text{(1b)}$$

where $\boldsymbol{\rho} = (x, y)$ and the Heaviside unit function $U(\alpha)$ equals unity for $\alpha > 0$ and vanishes for $\alpha < 0$. (*Hint:* Note that $\delta[f(z')] = \sum_i \delta(z' - z_i)/|f'(z_i)|$, where z_i are the zeros of $f(z')$ and $f' = df/dz$.)

11. A flat, perfectly conducting obstacle of area S lying in the $z = 0$ plane is illuminated by a prescribed electromagnetic field $\mathbf{E}_i(\mathbf{r}, t)$, $\mathbf{H}_i(\mathbf{r}, t)$. The electric currents $\mathbf{J}(\mathbf{r}, t)$ induced on the obstacle by the incident field give rise to the scattered field $\mathbf{E}_s(\mathbf{r}, t)$, $\mathbf{H}_s(\mathbf{r}, t)$.

(a) Use Eqs. (1.1.19) to represent the scattered field in terms of the induced currents. (*Note:* The currents flow on both sides of the obstacle.) Invoke the boundary condition $\mathbf{n} \times \mathbf{E} = 0$ on S, where $\mathbf{E} = \mathbf{E}_i + \mathbf{E}_s$ is the total electric field and $\mathbf{n}$ is the outward unit normal on S, to derive an integral equation for the induced currents [include the "edge condition"; see Eq. (1.5.37)].

(b) Assume that the induced currents on S are the same as if S were contained in an infinite perfectly conducting plane; i.e., $\mathbf{J} = \mathbf{n} \times \mathbf{H}_i$ on S^+, $\mathbf{J} = 0$ on S^-, where S^+ and S^- denote the "illuminated" and "dark" sides of S facing toward and away from the incident field, respectively, and $\mathbf{n}$ is the outward unit normal on S. Write explicit expressions for $\mathbf{E}_s$ and $\mathbf{H}_s$. This method of approximation is known as the (Kirchhoff) physical optics approximation and is valid when the obstacle is large compared to the wavelengths contained in the temporal spectral decomposition of the incident field.

12. Consider an aperture of area S in a perfectly conducting screen lying in the plane $z = 0$. A known electromagnetic field $\mathbf{E}_i(\mathbf{r}, t)$, $\mathbf{H}_i(\mathbf{r}, t)$ is incident from the half space $z < 0$.

(a) Show that the required continuity of the total tangential electric and magnetic fields in the aperture can be secured on replacing the aperture by a perfect conductor and placing thereon (on each side) an equivalent magnetic current $\mathbf{M} = \mathbf{E} \times \mathbf{n}$, where $\mathbf{n}$ is the outward unit normal. By this device, the problem of determining the overall fields in the complicated region containing the *perforated* plane is reduced to the problem of first finding the induced magnetic currents $\mathbf{M}$ flowing on an *infinite* perfectly conducting plane. These induced currents give rise to the scattered field $\mathbf{E}_s(\mathbf{r}, t)$, $\mathbf{H}_s(\mathbf{r}, t)$.

(b) Use the half-space dyadic Green's functions of Problem 8 to represent the fields $\mathbf{E}_s$, $\mathbf{H}_s$ radiated by the induced currents $\mathbf{M}$ into the regions $z < 0$ and $z > 0$. Then invoke the required continuity of the total tangential magnetic field $\mathbf{n} \times (\mathbf{H}_i + \mathbf{H}_s)$ in the aperture to derive an integral equation for $\mathbf{M}$ (note the "edge condition," Eq. (1.5.37); see also Problem 6 of Chapter 2). $\mathbf{n} \times \mathbf{H} = \mathbf{n} \times \mathbf{H}_i$ in the aperture, i.e., the tangential component of the incident magnetic field is not modified by the presence of the aperture.

(c) Assume that the tangential electric field in the aperture is the same as the incident electric field, i.e., $\mathbf{n} \times \mathbf{E} = \mathbf{n} \times \mathbf{E}_i$ in S, and give explicit expressions for $\mathbf{E}_s$ and $\mathbf{H}_s$. The resulting "physical optics" approximation is valid under conditions analogous to those stated in Problem 11b.

13. Using the free-space dyadic Green's function representations in Eqs. (1.1.38a) and (1.1.41), derive the response to an electric current excitation $\mathbf{J}(\mathbf{r}, t) = \mathbf{J}\delta(\mathbf{r})e^{-i\omega_0 t} U(t)$, where $U(\alpha) = 1$ for $\alpha > 0$ and $U(\alpha) = 0$ for $\alpha < 0$. Separate the transient contributions from those descriptive of the time-harmonic steady-state $(t \to \infty)$. Identify the resulting steady-state dyadic Green's functions $\mathscr{G}_{21}(\mathbf{r}, \mathbf{r}')$ and compare with those listed in Sec. 5.2. Using Eqs. (1.1.38b) and (1.1.39a,b), find the equations satisfied by the steady-state form of the potential function $\mathscr{S}(\mathbf{r}, \mathbf{r}')$, as well as the steady-state form of the expressions for $\nabla_t^2 \mathscr{S}(\mathbf{r}, \mathbf{r}')$ and $(\partial/\partial\rho)\mathscr{S}(\mathbf{r}, \mathbf{r}')$, and compare with results in Sec. 2.4b. Discuss the relation between causality and the steady-state "radiation condition."

14. Calculate the adjoint (time-reversed, inward-radiating) fields $\mathbf{E}^+(\mathbf{r}, t)$ and $\mathbf{H}^+(\mathbf{r}, t)$ radiated by the analogue of the electric dipole current in Problem 13: $\mathbf{J}^+(\mathbf{r}, t) = \mathbf{J}^+ \delta(\mathbf{r})e^{i\omega_0 t} U(-t)$. Separate the steady-state and transient constituents and interpret the behavior of each. What is the steady-state analogue of the "time reversed radiation condition"? Using complex conjugate notation and referring to the results of Problem 13, find a corresponding phrasing of the time-dependent reciprocity conditions (1.1.28) for the steady-state time-harmonic field.

15. Using a field representation in terms of the potential functions Π' and Π'' analogous to that in Eqs. (1.1.42), derive the E- and H-mode constituents of the fields generated by a pulsed magnetic dipole $\mathbf{M}(\mathbf{r}, t) = \mathbf{y}_0 \delta(\mathbf{r})\delta(t)$, oriented perpendicularly to the symmetry direction $\mathbf{a} \equiv \mathbf{z}_0$. Deduce the result alternatively by duality considerations $(\mathbf{E} \to \mathbf{H}, \mathbf{H} \to -\mathbf{E}, \mathbf{J} \to \mathbf{M}, \mu \longleftrightarrow \epsilon)$ applied to Eqs. (1.1.42)-(1.1.47).

16. Recognizing that an electric current element directed along the symmetry axis $\mathbf{a} \equiv \mathbf{z}_0$ does not generate H-mode fields $[\Pi'' \equiv 0$ in Eq. (1.1.42b)], compute the (E-mode) fields radiated by a so-directed element with an impulsive time dependence. Referring to Eq. (1.1.38a), show that the result does not exhibit the singularity difficulty [as in Eq. (1.1.46a)] across the transverse plane containing the source. If the source currents are spatially distributed in a volume V, does Eq. (1.1.42a) yield a valid field description within V?

17. In a medium which is inhomogeneous but non-dispersive (all wave frequencies propagate at the same speed), ϵ_0 and μ_0 in Eqs. (1.1.16) et seq. are replaced by $\epsilon(\mathbf{r})$ and $\mu(\mathbf{r})$, respectively. Show that reciprocity as in Eq. (1.1.25) remains valid

and derive relations corresponding to Eqs. (1.1.28) and (1.1.29). Repeat for the case of an anisotropic, inhomogeneous, non-dispersive medium wherein $\epsilon_0 \partial E/\partial t$ and $\mu_0 \partial H/\partial t$ are replaced by $\boldsymbol{\epsilon}(\mathbf{r}) \cdot \partial E/\partial t$ and $\boldsymbol{\mu}(\mathbf{r}) \cdot \partial H/\partial t$, respectively, with $\boldsymbol{\epsilon}(\mathbf{r})$ and $\boldsymbol{\mu}(\mathbf{r})$ representing permittivity and permeability dyadics.

18. When defining an adjoint electromagnetic problem, the spatial boundaries S of the region V are preserved but the medium properties in V and the boundary conditions on S must be suitably determined. *Show* that the constitutive parameters ϵ_0^+, μ_0^+, and $\mathscr{L}^+$ of the adjoint problem required to achieve reciprocity as in Eq. (1.1.24) are $\epsilon_0^+ = \epsilon_0$, $\mu_0^+ = \mu_0$, $\mathscr{L}^+ = -\tilde{\mathscr{L}}$, where $\tilde{\mathscr{A}}$ denotes the transpose of the dyadic $\mathscr{A}$. Repeat for an anisotropic medium (see Problem 17) and show that this requires an adjoint medium with $\boldsymbol{\epsilon}^+ = \tilde{\boldsymbol{\epsilon}}$, $\boldsymbol{\mu}^+ = \tilde{\boldsymbol{\mu}}$, $\mathscr{L}^+ = -\tilde{\mathscr{L}}$.

19. An impulsive source acting at time $t = t'$ is distributed uniformly over the plane $z = z'$ in an infinite, homogeneous, non-dispersive medium characterized by the propagation speed c_0. At time $t_0 > t'$, the medium propagation speed changes abruptly (and uniformly throughout all space) from c_0 to c_1. The scalar Green's function $g(z, z'; t, t')$ descriptive of the space–time dependent field is defined as follows:

$$\left[\frac{\partial^2}{\partial z^2} - \frac{1}{c^2(t)} \frac{\partial^2}{\partial t^2} \right] g(z, z'; t, t') = -\delta(t - t')\delta(z - z'), \tag{2}$$

with $c(t) = c_0$ for $t < t_0$, $c(t) = c_1$ for $t > t_0$, and subject to the causality condition $g \equiv 0$ for $t < t'$. To ensure continuity of the electric and magnetic flux densities at $t = t_0$,[†] the boundary condition

$$g, \frac{\partial g}{\partial t} \text{ continuous at } t = t_0 \tag{2a}$$

is imposed.

Show that the solution is given for $t < t_0$ by

$$g = \frac{c_0}{2} U(c_0 T - |Z|), \qquad T = t - t', \qquad Z = z - z', \tag{2b}$$

and for $t > t_0$ by

$$g = \frac{c_0}{4} \left(1 + \frac{c_0}{c_1} \right) U(c_1 \tau + c_0 T_0 - |Z|)$$

$$+ \frac{c_0}{4} \left(1 - \frac{c_0}{c_1} \right) [U(c_0 T_0 - c_1 \tau - |Z|) - U(c_1 \tau - c_0 T_0 - |Z|)], \tag{2c}$$

with $T_0 = t_0 - t'$, $\tau = t - t_0$. The Heaviside unit function $U(x)$ equals unity for $x > 0$ and vanishes for $x < 0$. Explain the various terms in Eqs. (2b) and (2c) as traveling wavefronts and follow their progress in various time intervals. Note that a sudden change in the temporal properties of a medium introduces a reflected as well as a transmitted wave.

20. Repeat the calculation of Problem 19 for an impulsive point source located at $r = 0$, whence the Green's function is defined by:

$$\left[\nabla^2 - \frac{1}{c^2(t)} \frac{\partial^2}{\partial t^2} \right] g(\mathbf{r}; t, t') = \delta(t - t')\delta(\mathbf{r}), \tag{3}$$

---

†M. Kline and I. Kay, *Electromagnetic Theory and Geometrical Optics*, New York: Interscience (1965), Sec. 1.3.

subject to $g \equiv 0$ for $t < t'$, and to the continuity condition in Eq. (2a). Show that the result for $t < t_0$ is

$$g = \frac{c_0 \delta(c_0 T - r)}{4\pi r}, \tag{4a}$$

and for $t > t_0$,

$$g = \frac{c_0}{8\pi r} \left\{ \left(1 + \frac{c_0}{c_1}\right)\delta(c_0 T_0 + c_1 \tau - r) + \left(1 - \frac{c_0}{c_1}\right)[\delta(c_0 T_0 - c_1 \tau - r) \right.$$

$$\left. - \delta(c_1 \tau - c_0 T_0 - r)] \right\}, \tag{4b}$$

where the notation is the same as in **Problem 19**. Interpret the result in terms of spherical wavefronts and follow their progress through various time intervals. Show that the reflected field constituent in Eq. (4b) remains finite at $r = 0$. Examine the field at $r = 0$ in detail by using the Green's functions in Eqs. (4a) and (4b) to synthesize response to a source function $f(t)$ that vanishes for $t < 0$ and acts throughout a time interval $0 < t < \bar{t}$. Distinguish between the cases $\bar{t} < t_0$ and $\bar{t} > t_0$.

21. An electromagnetic current source is distributed over a plane $z = z'$ in a homogeneous, non-dispersive medium, and has the impulsive temporal behavior $\delta'(t - t')$. At time $t = t_0 > t'$, a cold homogeneous plasma is created suddenly (for example, by subjecting a neutral gas to ionizing radiation). The relevant Green's function problem is as follows:

$$\left[\frac{\partial^2}{\partial z^2} - \frac{1}{c^2}\frac{\partial^2}{\partial t^2} - \frac{\omega_p^2(t)}{c^2}\right]\hat{V}(z, z'; t, t') = -\delta'(t - t')\delta(z - z'), \tag{5}$$

where $\delta'(t) = d\delta(t)/dt$ and

$$\omega_p(t) = bU(t - t_0), \qquad b = \text{constant}. \tag{5a}$$

A causality condition $\hat{V} \equiv 0$ for $t < t'$ and a continuity condition as in Eq. (2a) are imposed. Note that $\hat{V}$ represents the time derivative of the Green's function g for $t \neq t'$.
(a) Show that for $t_0 = t'$, i.e., for radiation into a homogeneous, stationary plasma,

$$\hat{V} = \frac{c^2}{2}\delta(cT - |Z|) - \frac{cbT}{2}\frac{J_1(b\sqrt{T^2 - (Z/c)^2})}{\sqrt{T^2 - (Z/c)^2}}U(cT - |Z|). \tag{6}$$

This is the one-dimensional version of the results in Eqs. (1.1.61).
(b) Show that for $t < t_0$, with $t_0 > t'$,

$$\hat{V} = \frac{c^2}{2}\delta(cT - |Z|), \qquad T = t - t', \qquad Z = z - z', \tag{7a}$$

and for $t > t_0$,

$$\hat{V} = \frac{c^2}{2}\delta(cT - |Z|) - \frac{cb}{4}\left\{ \frac{J_1(b\zeta_+^{1/2})}{\zeta_+^{1/2}}\left(\tau - T_0 - \frac{Z}{c}\right)U(\zeta_+) \right.$$

$$\left. + \frac{J_1(b\zeta_-^{1/2})}{\zeta_-^{1/2}}\left(\tau - T_0 + \frac{Z}{c}\right)U(\zeta_-) \right\}, \qquad T_0 = t_0 - t', \tau = t - t_0, \tag{7b}$$

where $\zeta_\pm = \tau^2 - [(Z \pm cT_0)/c]^2$. Explain the results in physical terms, and compare with those for the non-dispersive media in Problem 19 (see also Problem 38).

22. Repeat Problem 21 for excitation by a point source with impulsive behavior $\delta(t - t')$. For $t_0 = t'$, the solution for all $t > t'$ is given in Eq. (1.1.61b). When $t_0 > t'$, the solution for $t' < t < t_0$ is that in Eq. (4a), with $c_0 = c$. Show that for $t > t_0$, one obtains

$$g = \frac{c\delta(cT - r)}{4\pi r} + \frac{b}{8\pi r} \left\{ \frac{J_1(b\zeta_+^{1/2})}{\zeta_+^{1/2}} \left(\tau - T_0 - \frac{r}{c} \right) U(\zeta_+) \right.$$
$$\left. - \frac{J_1(b\zeta_-^{1/2})}{\zeta_-^{1/2}} \left(\tau - T_0 + \frac{r}{c} \right) U(\zeta_-) \right\}, \qquad (8)$$

with $\zeta_\pm$ defined as in Problem 21 provided that Z is replaced by r. Show that for $\tau > T_0$, the Green's function remains finite at $r = 0$ and has the value

$$g = \frac{b^2}{4\pi c} \left[-\frac{J_1(b\sqrt{\tau^2 - T_0^2})}{b\sqrt{\tau^2 - T_0^2}} + \frac{T_0}{T} J_2(b\sqrt{\tau^2 - T_0^2}) \right], \qquad r = 0. \qquad (9)$$

Interpret the results in physical terms and compare with those for the non-dispersive media in Problem 20.

23. Show that in a linear medium the constitutive relations can be expressed in derivative operator form as

$$\mathbf{D}(\mathbf{r}, t) = \boldsymbol{\epsilon}\left(\nabla, \frac{\partial}{\partial t}; \mathbf{r}, t\right) \cdot \mathbf{E}(\mathbf{r}, t), \qquad \mathbf{B}(\mathbf{r}, t) = \boldsymbol{\mu}\left(\nabla, \frac{\partial}{\partial t}; \mathbf{r}, t\right) \cdot \mathbf{H}(\mathbf{r}, t) \qquad (10a)$$

or in integral operator form as

$$\mathbf{D}(\mathbf{r}, t) = \int \boldsymbol{\epsilon}(\mathbf{r}, t; \mathbf{r}', t') \cdot \mathbf{E}(\mathbf{r}', t')\, d\mathbf{r}'\, dt',$$

$$\mathbf{B}(\mathbf{r}, t) = \int \boldsymbol{\mu}(\mathbf{r}, t; \mathbf{r}', t') \cdot \mathbf{H}(\mathbf{r}', t')\, d\mathbf{r}'\, dt' \qquad (10b)$$

if the polarization currents $\mathbf{J}$ and $\mathbf{M}$ in the medium are determined by linear differential equations of the form:

$$\mathscr{L}_e\left(\nabla, \frac{\partial}{\partial t}; \mathbf{r}, t\right) \cdot \mathbf{J}(\mathbf{r}, t) = \mathbf{E}(\mathbf{r}, t), \qquad \mathscr{L}_m\left(\nabla, \frac{\partial}{\partial t}; \mathbf{r}, t\right) \cdot \mathbf{M}(\mathbf{r}, t) = \mathbf{H}(\mathbf{r}, t). \qquad (11)$$

For homogeneous and stationary media, show that the dyadic operators have the form

$$\boldsymbol{\epsilon}(\mathbf{r}, t; \mathbf{r}', t') = \boldsymbol{\epsilon}(\mathbf{r} - \mathbf{r}'; t - t'), \qquad \boldsymbol{\mu}(\mathbf{r}, t; \mathbf{r}', t') = \boldsymbol{\mu}(\mathbf{r} - \mathbf{r}'; t - t'). \qquad (12)$$

24. By rotating coordinates in the four-dimensional $(\mathbf{r}, t)$ space so that the time axis passes through the observation point, show that the asymptotic approximation (for large r) to the integral in Eq. (1.6.1) can be given in the invariant form [alternative to that in Eq. (1.6.5)]:

$$I(\mathbf{r}, t) \sim A(\mathbf{k}_s) e^{i\psi(\mathbf{r}, t;\, \mathbf{k}_s)} (2\pi)^{3/2} e^{-i(\pi/4)\sigma} (R_1 R_2 R_3)^{1/2} (r^2 + t^2)^{-3/4}, \qquad (13)$$

where $\sigma = \sum_{j=1}^{3} \operatorname{sgn} R_j$ and R_j, $j = 1, 2, 3$ are the principal radii of curvature of the four-dimensional $(\mathbf{k}, \omega)$ dispersion surface at the saddle point $\mathbf{k}_s$. [*Hint:* See Eqs. (1.6.24) and (1.6.25).]

25. The time-dependent, two-dimensional scalar Green's function $g(\boldsymbol{\rho}, \boldsymbol{\rho}'; t, t')$ for an isotropic, homogeneous, cold plasma is defined by:

$$\left(\nabla^2 - \frac{1}{c^2}\frac{\partial^2}{\partial t^2} - \frac{\omega_p^2}{c^2}\right) g(\boldsymbol{\rho}, \boldsymbol{\rho}'; t, t') = -\delta(\boldsymbol{\rho} - \boldsymbol{\rho}')\delta(t - t'), \qquad \boldsymbol{\rho} = (y, z), \quad (14)$$

subject to the causality condition $g \equiv 0$ for $t < t'$.

(a) Show that for $\boldsymbol{\rho}' = t' = 0$, g can be represented in the following alternative forms [cf. Eqs. (1.5.51)–(1.5.53)]:

$$g = \frac{1}{(2\pi)^3} \iiint\limits_{-\infty}^{\infty} \frac{e^{i(\eta y + \zeta z - \omega t)}}{\eta^2 + \zeta^2 + (\omega_p^2/c^2) - (\omega^2/c^2)} \, d\eta \, d\zeta \, d\omega \tag{15}$$

$$g = \left(\frac{c}{2\pi}\right)^2 \int_{-\infty}^{\infty} d\omega \int_0^\infty dk \, \frac{k J_0(kR) e^{-i\omega t}}{c^2 k^2 + \omega_p^2 - \omega^2} \tag{16a}$$

$$= \frac{1}{2} \left(\frac{c}{2\pi}\right)^2 \int_{-\infty}^{\infty} d\omega \int_{-\infty}^{\infty} dk \, \frac{k H_0^{(1)}(kR) e^{-i\omega t}}{c^2 k^2 + \omega_p^2 - \omega^2}, \qquad R = \sqrt{y^2 + z^2} \tag{16b}$$

$$g = \frac{c^2}{4\pi} \int_{-\infty}^{\infty} \frac{k H_0^{(1)}(kR) \sin \omega_1 t}{\omega_1} \, dk, \qquad \omega_1 = \sqrt{\omega_p^2 + k^2 c^2} \tag{17}$$

$$g = \frac{i}{8\pi} \int_{-\infty}^{\infty} H_0^{(1)}(k_1 R) e^{-i\omega t} \, d\omega, \qquad k_1 = \frac{\sqrt{\omega^2 - \omega_p^2}}{c}. \tag{18}$$

The transition from Eq. (15) to Eq. (16a) follows on introduction of polar coordinates in (y, z) and (η, ζ) space, and use of the integral representation $J_0(kR) = (1/2\pi)\int_0^{2\pi} e^{ikR \cos \psi} \, d\psi$. Discuss the disposition of the various integration paths with respect to singularities of the integrands. From a table of Laplace transforms, one may show that g is given in closed form by:

$$g = \frac{\cos\left[\omega_p\sqrt{t^2 - (R/c)^2}\,\right]}{2\pi\sqrt{t^2 - (R/c)^2}} U\!\left(t - \frac{R}{c}\right), \tag{19}$$

where $U(x) = 0$ or 1 for $x < 0$, or $x > 0$, respectively.

(b) Substitute into Eqs. (17) and (18) the asymptotic form

$$H_0^{(1)}(w) \sim \sqrt{\frac{2}{\pi w}} \, e^{i[w - (\pi/4)]}, \qquad |w| \gg 1, \tag{20}$$

valid for large R and non-vanishing k or k_1. Decomposing the integrand in Eq. (17) into two parts comprising exponential functions, show that each part gives rise to a single relevant saddle point k_{s1} or k_{s2} as follows:

$$k_{s1,2} = \frac{\pm R\omega_p}{c^2\sqrt{t^2 - (R/c)^2}}, \qquad t > \frac{R}{c}. \tag{21}$$

Show that the sum of the asymptotic approximations of each integral, as obtained from the one-dimensional (single k integral) version of Eq. (1.6.5), yields Eq. (19).

Similarly, proceeding from Eq. (18) with Eq. (20), show that the integrand has two relevant saddle points located at

$$\omega_{s1,2} = \pm\frac{\omega_p}{\sqrt{1 - (R/ct)^2}}, \qquad t > \frac{R}{c}. \tag{22}$$

Use Eq. (1.6.43) for each saddle point to show that the asymptotic approximation of Eq. (18) also yields the result in Eq. (19). Discuss the restrictions on the asymptotic results although they happen to agree with the exact solution.

(c) Use the procedure leading to Eq. (1.6.49), in conjunction with Eqs. (18) and (20), to find the transition function linking the asymptotic approximations with the wavefront regime. Compare with the exact solution in Eq. (19).

26. A line source along the z axis is embedded in a cold, isotropic, homogeneous plasma as in Problem 25. The time dependence of the source is that of a suddenly switched-on harmonic signal

$$f(t) = e^{-i\omega_0 t} U(t), \tag{23a}$$

whence its Fourier transform is

$$F(\omega) = \int_{-\infty}^{\infty} e^{i\omega t} f(t)\, dt = \frac{i}{\omega - \omega_0}. \tag{23b}$$

(a) Show that the response $\bar{g}$ to this excitation is given by Eq. (18) provided that the integrand is multiplied by $F(\omega)$ (discuss the disposition of the integration path). Perform an asymptotic evaluation to show that $\bar{g}$ behaves for $t > R/c$ as follows:

$$\bar{g} \sim \frac{i}{4\pi\tau} \left[\frac{e^{-i\omega_p \tau}}{(\omega_p t/\tau) - \omega_0} - \frac{e^{i\omega_p \tau}}{(\omega_p t/\tau) + \omega_0} \right]$$

$$+ \frac{i}{4} H_0^{(1)}\left(\frac{R}{c} \sqrt{\omega_0^2 - \omega_p^2} \right) e^{-i\omega_0 t} U(t - t_0), \tag{24}$$

where

$$\tau = \sqrt{t^2 - \left(\frac{R}{c}\right)^2}, \qquad t_0 = \frac{R}{c}\, \omega_0 (\omega_0^2 - \omega_p^2)^{-1/2} = \frac{R}{v_{g0}},$$

and with $v_{g0} = [dk_1(\omega_0)/d\omega_0]^{-1}$ denoting the group speed corresponding to the signal frequency ω_0. Interpret this result by referring to Fig. 1.6.12. Note that as $t \to \infty$ there remains only the second term in Eq. (24), the *exact* time-harmonic response [see Eq. (5.4.25)]. Discuss the conditions under which Eq. (24) is valid and note the different behavior of the solution for $\omega_0 > \omega_p$ and $\omega_0 < \omega_p$.

(b) From $(\text{Re } \bar{g})$ and $(\text{Im } \bar{g})$, obtain the response to a suddenly switched-on signal $f_1 = U(t) \cos \omega_0 t$ and $f_2 = U(t) \sin \omega_0 t$, respectively. Referring to Eq. (1.6.49), find the transition functions for $(\text{Re } \bar{g})$ and $(\text{Im } \bar{g})$ near the wavefront $t = R/c$. Discuss the difference in behavior of the two solutions and relate it to the initial time dependence of the source functions f_1 and f_2. Compare with the transition function for the $\delta(t)$ excitation in Problem 25.

27. Consider the one-dimensional form of the wave equation (1.7.15), with $\nabla^2 \to \partial^2/\partial z^2$ and $n(\mathbf{r}) = n(z)$. Assume an asymptotic solution of the form (1.7.16), with $\mathbf{r} \to z$. Show that the second term in the asymptotic expansion is given by [see also Problem 40(a)]:

$$u_1(z) = \frac{-1}{2\sqrt{n(z)}} \int^z \frac{d^2 u_0(\zeta)/d\zeta^2}{\sqrt{n(\zeta)}}\, d\zeta, \tag{25}$$

with $u_0(z) = A/\sqrt{n(z)}$, $A = $ constant. To justify use of only the first term in the asymptotic expansion (1.7.16), it is necessary that $|u_1| \ll |k_0 u_0|$. Show from Eq.

(25) that this requirement is equivalent to condition (1.7.21), viz., $|dn/dz| \ll k_0 n^2$.

28. A source distributed over the entire plane $z = z'$ in vacuum is comprised of electric currents flowing parallel to the y direction, uniformly along y but with a linearly progressing phase $\exp(ik_0\alpha x)$ along x, where $k_0 = \omega/c$, c is the speed of light in vacuum, and α is real. A time dependence $\exp(-i\omega t)$ is implied.

Show that the electromagnetic field has non-vanishing components E_y, H_x, H_z; that H_x and H_z are derivable from E_y; and that E_y is proportional to the Green's function $g(x, z; z')$ defined by:

$$\left(\frac{\partial^2}{\partial x^2} + \frac{\partial^2}{\partial z^2} + k_0^2\right) g(x, z; z') = -\delta(z - z')e^{ik_0\alpha x}, \tag{26}$$

subject to an outward radiation condition at $|z| \to \infty$. Show that the solution is

$$g(x, z; z') = \frac{1}{-2ik_0\sqrt{1 - \alpha^2}} \exp\left[ik_0(\alpha x + \sqrt{1 - \alpha^2}\,|z - z'|)\right]. \tag{27}$$

Discuss the properties of the radiated field for $|\alpha| < 1$ and $|\alpha| > 1$.

Assuming knowledge of the field at $z = z'$, i.e.,

$$g(x, z'; z') \equiv u(x, z') = \frac{e^{ik_0\alpha x}}{-2ik_0\sqrt{1 - \alpha^2}}, \tag{28}$$

derive the solution in Eq. (27) by the ray-optical procedure that leads to Eq. (1.7.40). Why is the ray-optical result exact in this case? Show that the ray configuration for $0 < \alpha < 1$ is as shown in Fig. P1.1.

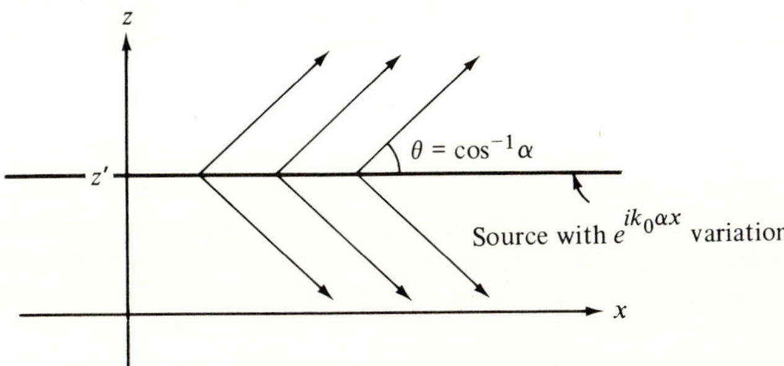

FIG. P1.1   Ray configuration for phased plane sheet source.

29. In vacuum a y-directed electric current source is distributed over a cylindrical surface $\bar{\rho} = a$, uniformly along y and with an azimuthal phase progression $\exp(im\phi)$, where $\bar{\rho}$ and ϕ are polar coordinates in the xz plane and m is a positive integer.

(a) Write $\exp(im\phi) = \exp(ik_0\alpha s)$, where $s = a\phi$ measures linear distance along the source distribution and the wavenumber $k_0\alpha = (m/a)$ describes the source phase. For $k_0 a \gg 1$, a source element at A may be assumed to radiate as though it were contained in an infinite planar source distribution tangent to the cylinder at A. Show that for $\alpha = (m/k_0 a) < 1$, the geometric-optical ray family appears as in Fig. P1.2, with a caustic formed at $\bar{\rho} = b = a \cos \theta = \alpha a$ (see also Fig.

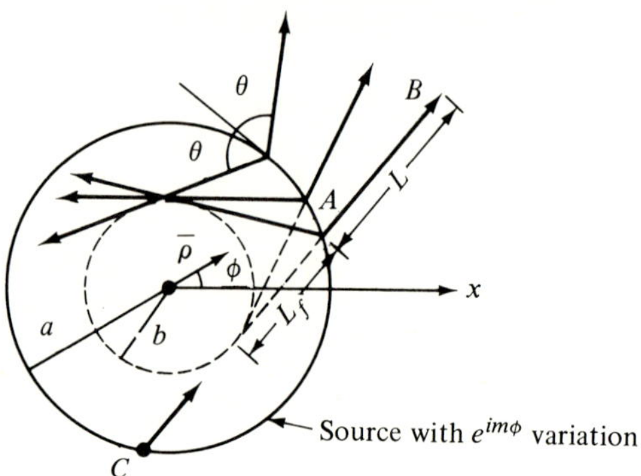

FIG. P1.2   Ray configuration for phased cylindrical sheet source.

5.4.14). Show also that the ray tube cross section ratio $\Delta s_B/\Delta s_A$ at points B and A along a ray is given for $\bar\rho > a$ by:

$$\frac{\Delta s_B}{\Delta s_A} = \frac{L_f + L}{L_f}, \qquad L = \sqrt{\bar\rho^2 - b^2} - \sqrt{a^2 - b^2}, \qquad L_f = \sqrt{a^2 - b^2}. \qquad (29)$$

(b) Use in Eq. (1.7.40) the initial field value from Eq. (28) and the cross section ratio in Eq. (29) to construct the geometric optical field at any point $(\bar\rho, \phi)$, $\bar\rho > a$, as follows:

$$u_1(\bar\rho, \phi) \sim M \exp\left[im(\phi - \gamma + \theta) + ik_0(\bar\rho \sin \gamma - a \sin \theta)\right] \left(\frac{a^2 - b^2}{\bar\rho^2 - b^2}\right)^{1/4}, \qquad (30)$$

where $\gamma = \cos^{-1}(b/\bar\rho)$ and M is a constant. The contribution in Eq. (30) is due to a ray reaching an observation point B from a source point A. The ray originating at C and passing through the interior of the source also contributes at B. Calculate the field due to the ray from C, noting that a phase change $\exp(-i\pi/2)$ must be included due to the contact of this ray with the caustic [see Eq. (5.8.55) for the more general problem of a caustic in an inhomogeneous medium].

(c) In the region $b < \bar\rho < a$, each observation point is also reached by two rays. Using the ray optical formula (1.7.40), show that the resulting field is given by:

$$u(\bar\rho, \phi) \sim 2M \exp\left[im(\phi - \theta) + ik_0 a \sin \theta - \frac{i\pi}{4}\right]$$

$$\times \left(\frac{a^2 - b^2}{\bar\rho^2 - b^2}\right)^{1/4} \cos\left(k_0 \bar\rho \sin \gamma - m\gamma - \frac{\pi}{4}\right). \qquad (31)$$

30. A line source at Q is parallel to a curved interface separating two different homogeneous isotropic media; a time dependence $\exp(-i\omega t)$ is implied.

(a) Referring to Fig. P1.3 for construction of the reflected ray-tube configuration, show that since $\theta_r = \theta_i$ and $\alpha = \beta$,

$$f = r \cos \theta_r \frac{\Delta\varphi}{\Delta\psi}, \qquad \Delta\psi = 2\Delta\varphi + \Delta\theta_0, \qquad \frac{\Delta\theta_0}{\Delta\varphi} = \frac{r \cos \theta_i}{l_1}, \qquad (32a)$$

$$\frac{f}{f+l} = \frac{r l_1 \cos \theta_i}{2 l l_1 + (l + l_1) r \cos \theta_i}, \qquad (32b)$$

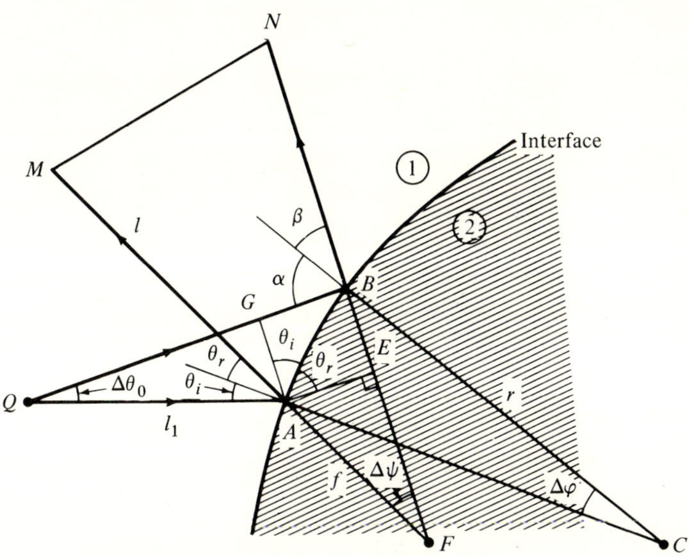

FIG. P1.3   Reflection from a curved boundary.

where r is the radius of curvature of the surface at the point of impact A of the incident ray, M is the observation point, and F is the virtual focus of the reflected ray tube. If the incident field at A is taken as $l_1^{-1/2}\exp(ik_1 l_1)$, show that the reflected field at M is given by:

$$u_{\text{refl}} \sim \Gamma \frac{e^{ik_1(l_1+l)}}{\sqrt{l_1+l}}\, D, \qquad D = \left[\frac{(l_1+l)r\cos\theta_i}{2l_1 l + (l_1+l)r\cos\theta_i}\right]^{1/2}, \tag{33}$$

where Γ is the reflection coefficient appropriate to the incidence angle θ_i and k_1 is the wavenumber in medium 1. The geometrical divergence coefficient D incorporates the effect of interface curvature; for a plane interface, one has $r = \infty$ and $D = 1$, whence Eq. (33) reduces to Eq. (1.7.64b). For a circular cylinder, $r = $ constant. For calculation of fields in the far zone, one lets $l \to \infty$. By letting $l_1 \to \infty$ and renormalizing the source strength so that the incident field is the plane-wave $\exp(ik_0 l_1)$, one obtains the reflected field

$$\bar{u}_{\text{refl}} \sim \Gamma e^{ik_1(l_1+l)}\sqrt{\frac{r\cos\theta_i}{2l + r\cos\theta_i}}. \tag{34}$$

(b) Referring to Fig. P1.4 for construction of the refracted ray-tube configuration, show that

$$f' = \frac{r\cos\theta_R}{\Delta\tau/\Delta\varphi} = \frac{rl_1\cos\theta_R}{(m-1)l_1 + mr\cos\theta_i}, \tag{35}$$

where $m = d\theta_R/d\theta_i = (n_1\cos\theta_i)(n_2\cos\theta_R)^{-1}$, and Snell's law $n_2\sin\theta_R = n_1\sin\theta_i$ has been used. With $k_{1,2} = k_0 n_{1,2}$, show that for an incident field as in (a), the transmitted field at M' is given by [cf. Eqs. (5.5.9)]:

$$E_{\text{trans}} \sim T \frac{e^{i(k_1 l_1 + k_2 l_2)}}{[(ml_2\cos\theta_i + l_1\cos\theta_R)/\cos\theta_R]^{1/2}}\, D', \tag{36}$$

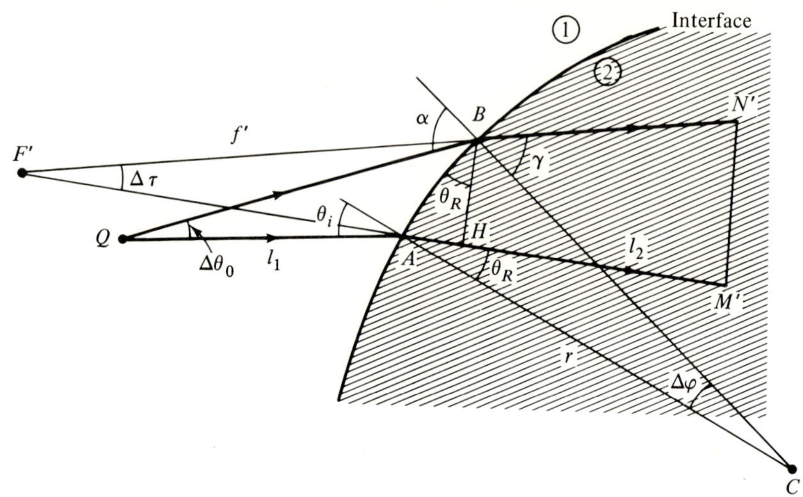

FIG. P1.4   Refraction at a curved boundary.

where T is the plane-wave transmission coefficient. The geometrical divergence coefficient D' incorporates the effect of interface curvature,

$$D' = \left[\frac{(ml_2\cos\theta_i + l_1\cos\theta_R)r}{(m-1)l_1l_2 + (ml_2\cos\theta_i + l_1\cos\theta_R)r}\right]^{1/2}, \tag{36a}$$

whence $D' = 1$ for a plane interface. For an incident plane-wave $\exp(ik_0l_1)$,

$$\bar{u}_{\text{trans}} \sim Te^{i(k_1l_1 + k_2l_2)}\sqrt{\frac{r\cos\theta_R}{(m-1)l_2 + r\cos\theta_R}}. \tag{37}$$

31. A slowly curved interface B separates two isotropic homogeneous or weakly inhomogeneous media 1 and 2 that support, respectively, M and N wave species. Sufficiently close to a point P on the interface, the media may be regarded as locally homogeneous, and the various wave species u_j satisfy there the time-harmonic scalar wave equation $(\nabla^2 + k_j^2)u_j = 0$, $k_j = kn_j$, with k representing a reference wavenumber and n_j the refractive index. Assume all field and geometrical quantities to be independent of the rectilinear coordinate z so that the problem is two-dimensional in the variable $\boldsymbol{\rho} = (x, y)$. Assume that wave type 1 in region 1 is incident on B at P, and that this incident wave u_i has the high-frequency asymptotic approximation

$$u_i(\boldsymbol{\rho}) \sim A_i(\boldsymbol{\rho})e^{ik_i\psi_i'(\boldsymbol{\rho})}\left[1 + O\left(\frac{1}{k}\right)\right], \tag{38a}$$

with an implied $\exp(-i\omega t)$ dependence. The interface couples u_i to M reflected waves in region 1,

$$u_r(\boldsymbol{\rho}) \sim \sum_{m=1}^{M} B_{rm}(\boldsymbol{\rho})e^{ik_m\psi'_{rm}(\boldsymbol{\rho})}\left[1 + O\left(\frac{1}{k}\right)\right], \tag{38b}$$

and to N transmitted waves in region 2,

$$u_t(\boldsymbol{\rho}) \sim \sum_{n=1}^{N} D_{tn}(\boldsymbol{\rho})e^{ik_n\psi'_{tn}(\boldsymbol{\rho})}\left[1 + O\left(\frac{1}{k}\right)\right]. \tag{38c}$$

Each of the phases satisfies near P the eiconal equation (1.7.18a), $(\nabla\psi'_j)^2 = 1$, and the amplitudes satisfy the transport equation (1.7.18b). Note that $n_j\psi'_j$ corresponds to ψ_j in Eq. (1.7.57).

(a) Use the interface matching condition for the tangential wavevector components in Eq. (1.7.59) and the eiconal equation to show that the normal components of the various wavevectors satisfy the relation:

$$k_i\frac{\partial\psi'_i}{\partial v} = -k_m\left[\left(\frac{\partial\psi'_{rm}}{\partial v}\right)^2 + \frac{k_i^2}{k_m^2} - 1\right]^{1/2} = k_n\left[\left(\frac{\partial\psi'_{tn}}{\partial v}\right)^2 + \frac{k_i^2}{k_n^2} - 1\right]^{1/2}, \qquad (39)$$

where v is the normal to the boundary and $m = 1 \cdots M, n = 1 \cdots N$.

(b) Assume that region 1 is a warm plasma that can support two wave types (electromagnetic and electron-acoustic), and that B is a perfect conductor, thereby eliminating region 2 (see the plasma portion of Fig. 1.7.5). The relevant scalar wave equations in region 1 are

$$(\nabla^2 + k_1^2)H(\boldsymbol{\rho}) = 0, \qquad k_1 = kn_p, \qquad k = k_0 \qquad (40a)$$

$$(\nabla^2 + k_2^2)p(\boldsymbol{\rho}) = 0, \qquad k_2 = k_a n_p \qquad (40b)$$

where H is the y component of magnetic field and p the acoustic pressure deviation from the mean. The following boundary conditions are assumed to be satisfied on B [expressive of the vanishing of the tangential component of electric field, and of an "impedance" condition for the normal component of the velocity; cf. Eq. (1.1.1b)]:

$$0 = q\frac{\partial p}{\partial s} - \frac{1}{i\omega\epsilon_0 n_p^2}\frac{\partial H_1}{\partial v},$$

$$(41)$$

$$p\beta = \frac{1}{i\omega m N_0 n_p^2}\frac{\partial p}{\partial v} - q\frac{\partial H_1}{\partial s}, \qquad n_p^2 = 1 - \left(\frac{N_0 e^2}{\epsilon_0 m\omega^2}\right),$$

where q, m, N_0 and β are real parameters and s is the coordinate tangential to B. Derive asymptotic forms of the boundary conditions in Eq. (41) by assuming that H and p can be represented as in Eqs. (38a) and (38b), with $M = 2$, and retaining only dominant terms. Show that if an acoustic ray is incident ($m = 2$), the reflected acoustic ray has an initial amplitude $\Gamma_{22}(\boldsymbol{\rho})$ given by

$$\Gamma_{22}(\boldsymbol{\rho}) = \frac{B_{r2}(\boldsymbol{\rho})}{A_{i2}(\boldsymbol{\rho})}$$

$$= \frac{q[k_1(\partial\psi'_1/\partial v)k_2(\partial\psi'_2/\partial v) - [k_2(\partial\psi'_2/\partial s)]^2(1 - n_p^2)] + \beta k_1(\partial\psi'_1/\partial v)(N_0 e/\omega\epsilon_0)}{q[k_1(\partial\psi'_1/\partial v)k_2(\partial\psi'_2/\partial v) + [k_2(\partial\psi'_2/\partial s)]^2(1 - n_p^2)] - \beta k_1(\partial\psi'_1/\partial v)(N_0 e/\omega\epsilon_0)}.$$

$$(42)$$

Derive analogous expressions for the coupling coefficient $\Gamma_{21} = B_{r1}/A_{i2}$ to the reflected magnetic field. Repeat for an incident electromagnetic field and derive the corresponding coefficients Γ_{11} and Γ_{12}.

(c) Assuming that the boundary is plane and the plasma homogeneous, solve Eqs. (40) and (41) for the exact reflection coefficients when a plane acoustic or electromagnetic wave is incident. Compare the result with Eq. (42), etc.

32. Assume that excitation for the warm plasma in (b) of Problem 31 is provided by a line source and that the medium is homogeneous throughout. Tracking an incident acoustic ray, one finds that the relation between angles of incidence

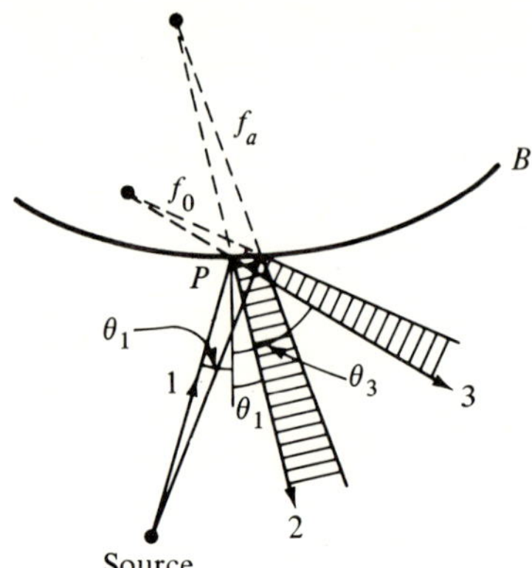

FIG. P1.5 Incident acoustic ray (1) and reflected acoustic (2) and electromagnetic (3) rays.

and reflection is given by Eq. (1.7.61) (see also Fig. 1.7.5). Use these reflection laws and the known radius of curvature r_c of the boundary at P to determine the focal lengths f_a and f_0 for the reflected acoustic and electromagnetic ray tubes, respectively (see Fig. P1.5). If L_2 and L_3 are the distances from P to a point on the reflected acoustic and electromagnetic rays, respectively, show that the ratio of the ray-tube cross sections at P and at the observation point is given for the acoustic ray by (see Problem 30)

$$\frac{f_a}{f_a + L_2} = \frac{r_c L_1 \cos \theta_1}{2L_1 L_2 + (L_1 + L_2)r_c \cos \theta_1},$$
(43a)

and for the electromagnetic ray by

$$\frac{f_0}{f_0 + L_3} = \frac{r_c L_1 \cos^2 \theta_3}{L_1 L_3(\cos \theta_3 + n_a \cos \theta_1) + r_c(L_1 \cos^2 \theta_3 + n_a L_3 \cos^2 \theta_1)},$$
(43b)

where L_1 is the distance from the source to P, and $n_a = k_a/k_0$. Repeat the calculation for an incident electromagnetic ray.

Use the results in (b) of Problem 31 together with the above to construct asymptotic approximations of the fields radiated by a line source in a homogeneous warm plasma bounded by a smoothly curved conducting surface. Assume that the line source generates known initial amplitudes of acoustic and electromagnetic fields.

33. In a gyrotropic medium, with a steady external magnetic field applied along the z axis, the surface described by the wavevector $\mathbf{k} = \mathbf{k}(\bar{\theta}, \omega)$ is rotationally symmetric and depends on the frequency ω and the polar angle $\bar{\theta}$ between $\mathbf{k}$ and the z axis. Instead of this k surface, it is often convenient to use the refractive index surface, defined in terms of k as follows:

$$n(\bar{\theta}, \omega) = \frac{c}{\omega} k(\bar{\theta}, \omega).$$
(44)

By implicit differentiation of the function $f = kc - \omega n = 0$, show that the components of the group velocity vector $\mathbf{v}_g = \nabla_k \omega$ are given by (rotate coordinates so that $k_y = 0$):

$$v_{gx} = c\,\frac{\sin\bar{\theta} - (\cos\bar{\theta}/n)(\partial n/\partial\bar{\theta})}{\partial(n\omega)/\partial\omega} \tag{45a}$$

$$v_{gz} = c\,\frac{\cos\bar{\theta} + (\sin\bar{\theta}/n)(\partial n/\partial\bar{\theta})}{\partial(n\omega)/\partial\omega} \tag{45b}$$

[Note that $\partial k/\partial k_x = \sin\bar{\theta}$, $\partial n/\partial k_x|_{\omega=\text{const.}} = k^{-1}\cos\bar{\theta}(\partial n/\partial\bar{\theta})$, etc.] Show also that

$$v_g = \sqrt{v_{gx}^2 + v_{gz}^2} = \frac{c}{\cos\alpha\,\partial(n\omega)/\partial\omega}, \tag{46}$$

where

$$\alpha = \tan^{-1}\left(\frac{1}{n}\frac{\partial n}{\partial\bar{\theta}}\right) \tag{46a}$$

and that the components of $\mathbf{v}_g$ along and perpendicular to $\mathbf{k}$ are:

$$v_{gk} = \frac{c}{\partial(n\omega)/\partial\omega} = v_g\cos\alpha, \qquad v_{g\bar{\theta}} = -\frac{c\tan\alpha}{\partial(n\omega)/\partial\omega}. \tag{47}$$

Thus, α in Eq. (46a) is the angle between $\mathbf{k}$ and $\mathbf{v}_g$. Show by a graphical construction utilizing Eq. (46a) and the refractive index surface that the vector $\mathbf{v}_g$ is normal to the surface.

34. A source distributed uniformly over the plane $z = 0$ and having the impulsive behavior $\delta(t)$ is embedded in a cold, isotropic, homogeneous plasma described by the constant plasma frequency ω_p. The fields are derivable from a scalar Green's function g that satisfies the wave equation

$$\left(\frac{\partial^2}{\partial z^2} - \frac{1}{c^2}\frac{\partial^2}{\partial t^2} - \frac{\omega_p^2}{c^2}\right)g(z, t) = -\delta(z)\delta(t) \tag{48}$$

subject to the causality requirement $g \equiv 0$ for $t < 0$. Show that the solution for g is given by [cf. the time-integrated form of Eq. (6)]:

$$g = \frac{c}{2}\,J_0\left(\omega_p\sqrt{t^2 - \left(\frac{z}{c}\right)^2}\right)U\left(t - \frac{|z|}{c}\right) \tag{49}$$

(a) The source in Eq. (48) is localizde at the space-time point $(z, t) = (0, 0)$ so that the space-time rays descriptive of the radiation process converge at the origin. Use the space-time ray method of Secs. 1.7a and 1.7e to show that the solution of the ray equation is [cf. notation in Eq. (1.7.67)],

$$Z = v_g(\bar{\omega})\tau, \qquad v_g = \frac{\bar{k}c^2}{\bar{\omega}}, \qquad \bar{\omega} = \pm\sqrt{\bar{k}c^2 + \bar{\omega}_p^2}, \tag{50a}$$

and that the field amplitude along a ray varies like $\tau^{-1/2}$ [cf. Eq. (1.7.78b)]. Show also from Eq. (50a) that the ray parameter $\bar{\omega}$ can be expressed explicitly in the form

$$\bar{\omega} = \frac{\pm\bar{\omega}_p}{\sqrt{1 - (Z/c\tau)^2}}, \qquad \text{whence}\quad \bar{k}c = \frac{\bar{\omega}_p}{\sqrt{(c\tau/Z)^2 - 1}}, \tag{50b}$$

and that the phase at (Z, τ) on a ray correiponding to $\bar{\omega} \gtrless 0$ is given via Eqs. (1.7.75) and (50b) by:

$$\psi_\pm(Z, \tau) - \psi_\pm(0, 0) = \mp\omega_p\tau\sqrt{1 - (Z/c\tau)^2}, \qquad \bar{\omega} \gtrless 0. \tag{50c}$$

Obtain the field by superposition of the ray contributions corresponding to $\bar{\omega} \gtrless 0$ as [$m = 0$ term in Eq. (1.7.68)]:

$$u(Z, \tau) \sim \frac{|A(\bar{\omega})|}{\sqrt{\tau}} \cos \left[v\bar{\omega}_p \sqrt{\tau^2 - \left(\frac{Z}{c}\right)^2} - v\psi_+(0, 0) + \alpha \right], \qquad (50d)$$

where $A(\bar{\omega}) = |A(\bar{\omega})| \exp(i\alpha)$ is a quantity descriptive of the initial field on a ray, and it has been recognized that $\psi_-(0, 0) = -\psi_+(0, 0)$ in order to render $u(Z, \tau)$ real.

(b) To determine the initial values A and $\psi_+(0, 0)$ along a ray, it is necessary to solve the space-time source problem in the vicinity of the source region. Since this cannot be done by the ray method (note that $u \to \infty$ as $\tau \to 0$ along a ray), the initial values must be found from an exact solution valid near the source point, as provided by Eq. (49). For sufficiently large $(ct - z)$ values, the Bessel function in Eq. (49) may be replaced by its large argument asymptotic form, and the quantities A and $\psi_+(0, 0)$ in Eq. (50d) are then determined by the requirement that $u(Z, \tau)$ in Eq. (50d) agree with the asymptotic form of Eq. (49) when Z and τ approach the source region. Perform this evaluation.

Comment: It may be noted that since Eq. (49) provides the exact solution at all space-time points, the ray method offers no advantage for the problem of radiation into a homogeneous plasma. However, if the medium is inhomogeneous with $\omega_p = \omega_p(z)$, closed-form solutions can generally not be obtained. The present results then provide the initial field values in the vicinity of $z = 0$, and the ray method can be employed to yield asymptotic solutions at other space-time points (see Problem 35). The ray solution becomes invalid near the wavefronts $|z| = ct$ where a transition function as in Eq. (1.6.49) must be utilized. For the present problem, Eq. (49) yields the transition function if $\sqrt{t^2 - (z/c)^2}$ is replaced by $\sqrt{2\xi(t - \xi)}$, $\xi = |z|/c$. In view of the dispersion equation $ck = (\omega^2 - \omega_p^2)^{1/2}$, one identifies ω_α^2 in Eq. (1.6.48) as $\omega_\alpha^2 = \omega_p^2/2$, and $v = -1$ in Eq. (1.6.49).

35. Assume that the source configuration in Problem 34 is embedded in an inhomogeneous plasma described by a plasma frequency $\omega_p = \omega_p(z)$. In the following, we employ the notation of Sec. 1.7e.

(a) Show that the space-time rays (up to the turning point, if any) are described by the curves

$$c\tau = \int_0^Z \frac{\bar{\omega}}{\sqrt{\bar{\omega}^2 - \bar{\omega}_p^2(\zeta)}} \, d\zeta, \qquad \bar{\omega} = \text{constant on a ray}. \qquad (51a)$$

(b) Show that for a fixed increment $d\bar{\omega}$ defining two neighboring rays, the ray-tube cross section dZ in planes $\tau = \text{constant}$ is given by:

$$\frac{dZ}{d\bar{\omega}} = \frac{\sqrt{\bar{\omega}^2 - \bar{\omega}_p^2(Z)}}{\bar{\omega}} \int_0^Z \frac{\bar{\omega}_p^2(\zeta) \, d\zeta}{[\bar{\omega}^2 - \bar{\omega}_p^2(\zeta)]^{3/2}}. \qquad (51b)$$

[Note that on writing Eq. (51a) as $F(\tau, Z, \bar{\omega}) = 0$, one has $dZ/d\bar{\omega} = -(\partial F/\partial \bar{\omega})(\partial F/\partial Z)^{-1}$.]

(c) Construct the asymptotic solution given by the $m = 0$ term in Eq. (1.7.68) on use of Eq. (51b) and Eqs. (1.7.75) and (1.7.78b). Obtain the initial field value on a ray by letting the observation point approach the source region and requiring that the asymptotic solution agrees with that for a homogeneous medium having a plasma frequency $\omega_p(0)$; the latter result has been found in Problem 34 (note that in this limiting process, Eq. (51a) becomes $c\tau \to Z\bar{\omega}[\bar{\omega}^2 - \bar{\omega}_p^2(0)]^{-1/2}$; other integrals are reduced in a similar manner). Show that for (Z, τ) values on ray segments between the source and the turning point (if any), the asymptotic form of the Green's function in Eq. (48) with $\omega_p = \omega_p(z)$ is then given by:[†]

$$g(Z, \tau) \sim$$

$$\frac{c^{3/2}}{\sqrt{2\pi v}\,[\bar{\omega}^2 - \bar{\omega}_p^2(Z)]^{1/4}[\bar{\omega}^2 - \bar{\omega}_p^2(0)]^{1/4}\{\int_0^Z (\bar{\omega}_p^2(\zeta)\,d\zeta/[\bar{\omega}^2 - \bar{\omega}_p^2(\zeta)]^{3/2})\}^{1/2}} \cos\left[v/c \int_0^Z \sqrt{\bar{\omega}^2 - \bar{\omega}_p^2(\zeta)}\,d\zeta - v\bar{\omega}\tau + (\pi/4)\right]$$

$$\tag{51c}$$

The turning point is defined by $d\tau/dZ = \infty$ in Eq. (51a). The ray parameter $\bar{\omega}$ depends on (Z, τ) implicitly through Eq. (51a). When Eq. (51a) can be solved for $\bar{\omega} = \bar{\omega}(Z, \tau)$, one may eliminate $\bar{\omega}$ from Eq. (51c).

(d) If a ray has a turning point at (Z_t, τ_t) (see Fig. 1.6.9), show that beyond the turning point, the ray is defined by the equation:

$$c\tau = \int_0^{Z_t} \frac{\bar{\omega}\,d\zeta}{\sqrt{\bar{\omega}^2 - \bar{\omega}_p^2(\zeta)}} + \int_Z^{Z_t} \frac{\bar{\omega}\,d\zeta}{\sqrt{\bar{\omega}^2 - \bar{\omega}_p^2(\zeta)}}. \tag{52}$$

Calculate the field [note the analogous calculation leading to Eq. (5.8.55)].

36. Consider the scalar wave equation

$$\left[\bar{\nabla}^2 - \frac{1}{c^2}\frac{\partial^2}{\partial \tau^2} - v^2 \frac{\bar{\omega}_p^2(\tau)}{c^2}\right] u(\mathbf{R}, \tau) = 0, \tag{53}$$

descriptive of electromagnetic wave processes in a cold, isotropic plasma whose properties are constant in space but vary (sufficiently slowly) with time.

(a) Assuming a solution of the form given in Eq. (1.7.68), show that the disper-

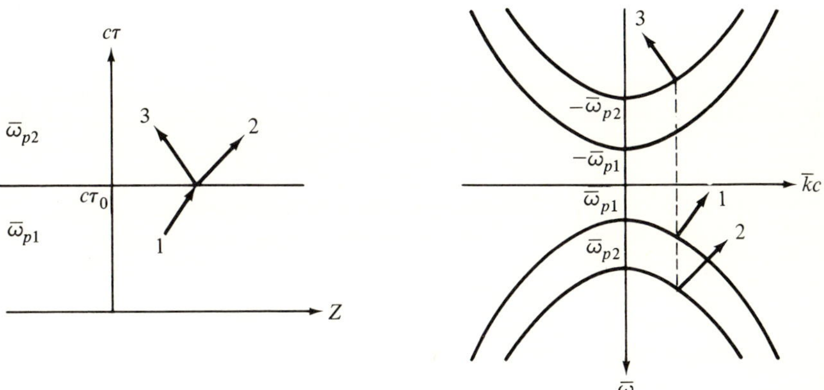

FIG. P1.6 Space–time ray refraction at a time discontinuity in the medium properties.

[†]N. Bleistein and R. M. Lewis, "Space-Time Diffraction for Dispersive Hyperbolic Equations," *SIAM J. of Appl. Math.*, Vol. 14, No. 6, November 1966, pp. 1454–1470.

sion equation and the transport equation for $u_0(\mathbf{R}, \tau)$ are given, respectively, by Eqs. (1.7.70) and (1.7.71), with $\bar{\omega}_p(\mathbf{R})$ replaced by $\bar{\omega}_p(\tau)$. Show also that Eq. (1.7.77) remains valid, but that Eq. (1.7.78b) must be modified since $\bar{\omega} \neq$ constant (whereas $\bar{\mathbf{k}} = $ constant) on a ray [see Eqs. (1.7.10) et seq.]. Discuss the applicability (or non-applicability) of energy conservation in a ray tube for time-varying media.

(b) If the medium properties change abruptly (and homogeneously throughout all space) at time $\tau = \tau_0$, show that instead of Eq. (1.7.79), the phase-matching condition requires

$$\bar{\mathbf{k}}_1 = \bar{\mathbf{k}}_2 = \cdots = \bar{\mathbf{k}}_m \qquad \text{at } \tau = \tau_0 \tag{54}$$

where m denotes the number of wave constituents. This matching condition can be schematized as in Fig. P1.6, with $\bar{\omega}_{p1}$ and $\bar{\omega}_{p2}$ denoting the values of plasma frequency before and after τ_0. Compare analogies and differences between the configuration in Fig. P1.6 and the spatial discontinuity in Fig. 1.7.11. If the medium changes continuously with time, find a graphical construction for the ray trajectories analogous to that in Fig. 1.6.9.

37. Plane-wave propagation in media with temporal inhomogeneity may be characterized by the dispersion equation $\omega = \omega(\mathbf{k}, t)$.

(a) Show that since the eigenfunctions $\exp(i\mathbf{k} \cdot \mathbf{r})$ account properly for the spatial dependence of the fields, the oscillatory representation in Eqs. (1.3.2b) provides the proper basis for synthesis of source-excited fields in such media. Show also that when such oscillatory representations are substituted into the field equations, the reduced equation for the time-dependent modal amplitudes takes the form of Eq. (1.3.13), with the operator W being time-dependent. This equation cannot generally be solved in closed form. However, show that for weak inhomogeneities, one may employ WKB procedures (see Sec. 3.5c) to effect an asymptotic solution of the time-dependent modal amplitudes, and that this leads to consideration of integrals of the form:

$$I(\mathbf{r}, t) \sim \int A(\mathbf{k}, t) e^{i\psi(\mathbf{r}, t; \mathbf{k})} \, d\mathbf{k}, \tag{55}$$

where

$$\psi(\mathbf{r}, t; \mathbf{k}) = \mathbf{k} \cdot \mathbf{r} \pm \int_{t'}^{t} \omega(\mathbf{k}, \eta) \, d\eta \tag{55a}$$

with $A(\mathbf{k}, t)$ and $\omega(\mathbf{k}, t)$ representing weakly time-dependent functions.

(b) Use the asymptotic procedure described in Sec. 1.6a to show that

$$I(\mathbf{r}, t) \sim A(\mathbf{k}_s, t) e^{i\psi(\mathbf{r}, t; \mathbf{k}_s)} \frac{(2\pi)^{3/2} e^{\pm i(\pi/4)\sigma}}{|\det Q(\mathbf{k}_s, t|^{1/2}}, \tag{56}$$

where the saddle points $\mathbf{k}_s$ are defined by

$$\frac{d\mathbf{r}}{dt} \pm \nabla_k \omega(\mathbf{k}, t) = 0 \qquad \text{at } \mathbf{k}_s(\mathbf{r}, t), \tag{56a}$$

and where Q is a matrix whose elements are $(\partial^2 \Omega / \partial k_i \partial k_j)$, with $i, j = x, y, z$; $\Omega(\mathbf{k}, t)$ denotes the integral on the right-hand side of Eq. (55a). As in Eq. (1.6.5), $\sigma = \sum_{j=1}^{3} \text{sgn} \, \bar{R}_i$ where $\bar{R}_i$ are the eigenvalues of the matrix Q.

(c) Relate the saddle-point condition (56a) to the space-time ray equation for a time-varying medium and use the graphical procedure of Problem 36b to locate the saddle points $\mathbf{k}_s$. Show that the magnitude of $I(\mathbf{r}, t)$ can be expressed in the form

$$|I(\mathbf{r}, t)| \sim |I(\mathbf{r}, t_1)|\left|\frac{A^2(\mathbf{k}_s, t)\Delta\mathbf{r}_1}{A^2(\mathbf{k}_s, t_1)\Delta\mathbf{r}}\right|, \tag{57}$$

where $\Delta\mathbf{r} = |\det Q|\Delta\mathbf{k}$ [cf. Eq. (1.6.10)]. Interpret this result in terms of space-time rays and compare with Eq. (1.6.15).

38. To illustrate the use of oscillatory representations for a time-varying medium,† consider the configuration of Problem 21, with $\omega_p(t)$ defined by

$$\omega_p(t) = a, \qquad t < t_0$$
$$= b, \qquad t > t_0 \tag{58}$$

thereby allowing for a homogeneous plasma medium before and after t_0.
(*a*) Writing

$$\hat{V} = \frac{1}{2\pi}\int_{-\infty}^{\infty} e^{ikz}\bar{g}(k, t, t')\,dk, \qquad z' = 0, \tag{59}$$

substituting into Eq. (5) and simplifying, show that

$$\bar{g} = c^2 \cos\omega_1 T, \qquad t < t_0 \tag{60a}$$
$$= c^2[\cos\omega_1 T_0 \cos\omega_2\tau - (\omega_1/\omega_2)\sin\omega_1 T_0 \sin\omega_2\tau], \qquad t > t_0, \tag{60b}$$

where $T = t - t'$, $T_0 = t_0 - t'$, $\tau = t - t_0$, with $\omega_1 = \sqrt{k^2c^2 + a^2}$ and $\omega_2 = \sqrt{k^2c^2 + b^2}$. For $a = 0$, show that Eqs. (59) and (60) yield the result in Eqs. (7) [cf. Eqs. (1.1.61)].
(b) Decompose the trigonometric functions in Eqs. (60) into exponentials and apply to each of the resulting integrals the asymptotic procedure of Problem 37 when modified for a single k integration. Derive the relevant saddle-point equations and interpret them ray-optically. Show that for $a = 0$, the saddle-point (or ray) equation can be solved explicitly for $k_s = k_s(z, t)$; derive the asymptotic solution for $\hat{V}$ and compare it with the asymptotic form of the exact solution in Eq. (7). For $a \neq 0$, show from the space-time ray configuration that the family of reflected rays corresponding to ray 3 in Fig. P1.6 may form a caustic, thereby producing wavepacket contraction (focusing).

39. While dispersion generally acts to spread out an originally narrow pulse, frequency modulated (FM) pulses can be designed so as to interact constructively with a dispersive medium. In particular, a source-pulse frequency spectrum can besought such that pulse compression takes place at a specified space-time point.
(a) To illustrate these aspects, consider FM pulse propagation in a homogeneous, cold, isotropic plasma, with excitation occurring in the plane $z = 0$ for a time interval $0 < t < t_0$; an enhanced and compressed signal is desired at $z = Z$ at time $t = T$, where $T > t_0 + (Z/c)$. For sufficiently large (Z, T), employ asymptotic considerations and schematize the process by the focusing at (Z, T)

---

†L. B. Felsen and G.M. Whitman, "Wave Propagation in Time-Varying Media," *IEEE Transactions on Antennas and Propagation AP-18* (1970), pp. 242–253.

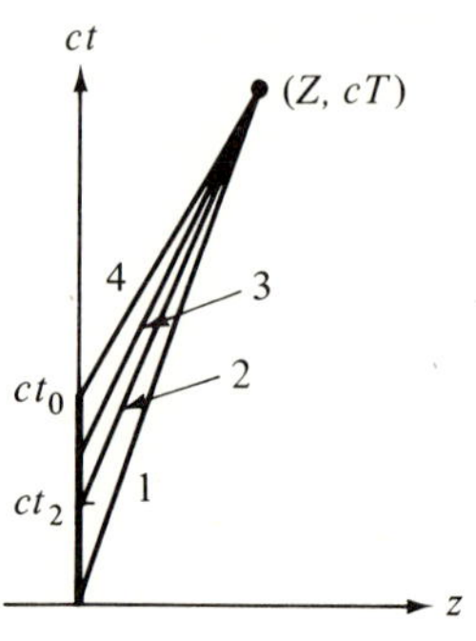

(a) Ray configuration for focusing at (Z, T)

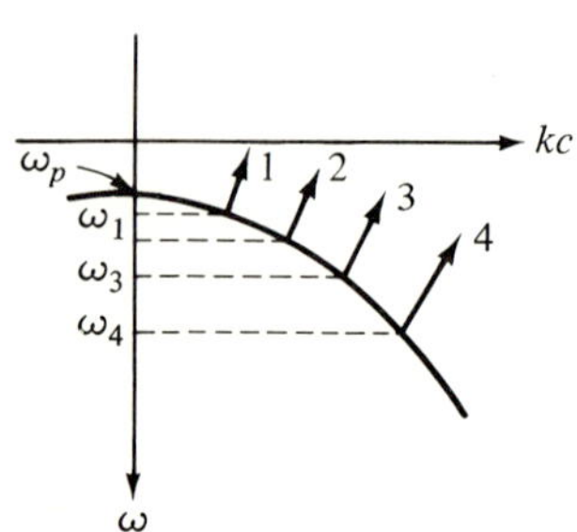

(b) Dispersion curve and frequencies corresponding to different rays

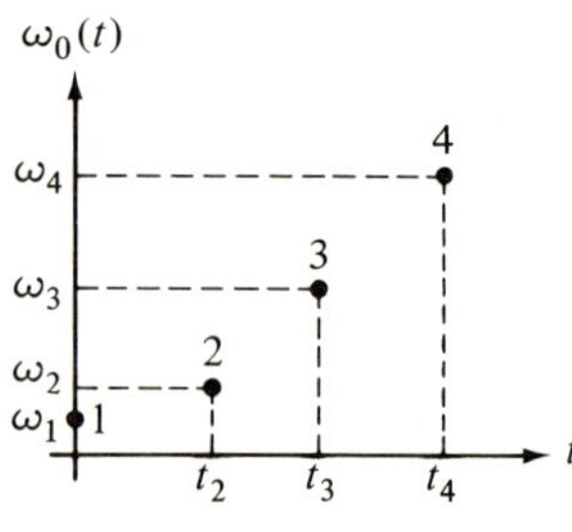

(c) Source frequency profile

FIG. P1.7 Synthesis of source-frequency profile for FM pulse compression.

of the space–time rays emanating from the source region [Fig. P1.7(a)]. Determine the required source frequency spectrum $\omega_0(t)$ by locating each ray α in Fig. P1.7(a) on the dispersion curve in Fig. P1.7(b) and noting the corresponding ray frequency ω_α; the source-frequency profile $\omega_0 = \omega_0(t)$ in Fig. P1.7(c) then follows from the intersections of the various rays with the t axis.

(b) To determine the source profile $\omega_0(t')$ analytically, use the space–time ray equation

$$v_g(\omega)(T - t') = Z \tag{61}$$

where t' denotes the departure time of the ray from the source plane $z = 0$, and $v_g(\omega)$ is the group velocity. Show for the cold isotropic plasma, whose dispersion relation is $\omega^2 - k^2 c^2 - \omega_p^2 = 0$, that

$$\omega_0(t') = \frac{(T - t')\omega_p}{\sqrt{(T - t')^2 - (Z/c)^2}}. \tag{62}$$

When the plasma is inhomogeneous, show that Eq. (61) is replaced by

$$t' = T - \int_0^z \frac{d\zeta}{v_g(\omega, \zeta)}, \qquad (63)$$

where $v_g(\omega, z)$ is the variable group velocity along a ray. Develop a graphical procedure for determination of $\omega_0(t')$, analogous to that of Fig. P1.7. (Note that $\omega = $ constant on a ray in a spatially inhomogeneous medium.)

40. Show that the first-order differential equation

$$\frac{dw(s)}{ds} + \alpha(s)w(s) = \beta(s) \qquad (64)$$

has the solution

$$w(s) = w(s_1) \exp\left(-\int_{s_1}^s \alpha(\eta)\,d\eta\right) + \int_{s_1}^s \beta(\xi) \exp\left(-\int_\xi^s \alpha(\eta)d\eta\right) d\xi. \qquad (65)$$

(a) Referring to the transport equations (1.7.18c) satisfied by the amplitude coefficients in the high-frequency asymptotic expansion (1.7.16) of the time-harmonic field in an inhomogeneous isotropic medium, and using Eq. (65), show that the solution for the mth order coefficient $u_m(\mathbf{r})$ is given by:

$$u_m(\mathbf{r}) = u_m(\mathbf{r}_1)\left[\frac{n(\mathbf{r}_1)\,dA(\mathbf{r}_1)}{n(\mathbf{r})\,dA(\mathbf{r})}\right]^{1/2} - \int_{\mathbf{r}_1}^{\mathbf{r}} \frac{1}{2n(s)}\left[\frac{n(s)\,dA(s)}{n(\mathbf{r})\,dA(\mathbf{r})}\right]^{1/2} \nabla^2 u_{m-1}(s)\,ds, \qquad (66)$$

where $m = 0, 1, 2\cdots$ (with $u_{-1} \equiv 0$); the integration variable s runs along a ray between points $\mathbf{r}_1$ and $\mathbf{r}$, and dA denotes the ray tube cross section. In obtaining the result in Eq. (66), note the equivalent forms (1.7.34) and (1.7.35b) for the lowest-order solution $u_0(\mathbf{r})$. Show that for one-dimensional propagation in a plane-stratified medium, the result for u_1 reduces to that in Problem 27. Show also that for validity of the lowest-order (geometric optical) approximation $u(\mathbf{r}) \sim u_0(\mathbf{r}) \exp[ik_0\psi(\mathbf{r})]$ in Eq. (1.7.16), it is sufficient to require

$$[k_0 u_0(\mathbf{r})n(\mathbf{r})]^{-1} \nabla^2 u_0(\mathbf{r}) \ll 1.$$

(b) Transforming into the form of Eq. (65) the transport equations (1.7.72) satisfied by the amplitude coefficients in the asymptotic expansion (1.7.68) of the time-dependent field in an inhomogeneous, cold, isotropic plasma medium, show that the solution for the mth order coefficient $u_m(\mathbf{R}, \tau)$ is given by:

$$u_m(\mathbf{R}, \tau) = u_m(\mathbf{R}_1, \tau_1)\left[\frac{\Delta\mathbf{R}_1}{\Delta\mathbf{R}}\right]^{1/2} - \frac{c^2}{2\bar{\omega}}\int_{(\mathbf{R}_1,\tau_1)}^{(\mathbf{R},\tau)} \frac{1}{V_g(s)}\left[\frac{\Delta\mathbf{R}(s)}{\Delta\mathbf{R}}\right]^{1/2} \overline{\nabla}^2 u_{m-1}(s)\,ds ,$$

$$(67)$$

where $m = 0, 1, 2\ldots$ (with $u_{-1} \equiv 0$), the integration variable s runs along a space–time ray between points $(\mathbf{R}_1, \tau_1)$ and $(\mathbf{R}, \tau)$, and $\Delta\mathbf{R}$ denotes (in hyperplanes $\tau = $ constant) the ray-tube cross section in the $(\mathbf{R}, \tau)$ four space (i.e., $\Delta\mathbf{R}$ is a volume element in three space). The operator $\overline{\nabla}^2$ is defined as $\overline{\nabla}^2 = \nabla_R^2 - \partial^2/c^2\partial\tau^2$, and $\bar{\omega} = -\partial\psi/\partial\tau$, while V_g is the four-dimensional group speed. In deriving Eq. (67), observe that the lowest-order solution $u_0(\mathbf{R}, \tau)$ may be expressed either as in the first term on the right-hand side of Eq. (67) [see Eq. (1.7.78b)] or in a manner analogous to that of Eq. (1.7.35b). Noting that $ds/V_g = d\tau/c$ (see Fig. 1.7.10), simplify the form of the integral in Eq. (67) by changing the integration variable to τ.

REFERENCES

1. MORSE, P., *Vibration and Sound.* New York: McGraw-Hill, 1948.

2(a). STRATTON, J. A., *Electromagnetic Theory*, Sec. 8.14. New York: McGraw-Hill, 1941.

2(b). *Ibid.*, Secs. 1.3, 9.1, 9.2.

3. CLEMMOW, P. C. and J. P. DOUGHERTY, *Electromagnetics of Particles and Plasmas*, Chapter 5. Reading, Mass. : Addison-Wesley, 1969.

4. CAMPBELL, G. A. and R. M. FOSTER, "Fourier integrals for practical applications," Bell System Technical Publications (1931), B584, Formula 865.1.

5. FRIEDMAN, B., *Principles and Techniques of Applied Mathematics*," Chapter 1. New York: John Wiley Sons, 1956.

6. TITCHMARSH, E. C., *Introduction to the Theory of Fourier Integrals.* London: Oxford University Press, 1937.

7. LIGHTHILL, M. J., *Introduction to Fourier Analysis and Generalized Functions*, Chapter 2. Cambridge, England: Cambridge University Press, 1958.

8. WAIT, J. R., *Electromagnetics and Plasmas*, Chapter 5. New York: Holt, Rinehart and Winston, 1968.

9(a). LANDAU, L. and E. M. LIFSHITZ, *Electrodynamics of Continuous Media*, Sec. 61. New York: Pergamon Press, 1960.

9(b). KURSS, H., "Dispersion relations, stored energy and group velocity for anisotropic electromagnetic media," *Quart. Appl. Math.* **26**, pp. 373–387 (1968).

10. MÜLLER, C., *Grundprobleme der Mathematischen Theorie Elektromagnetischer Schwingungen.* Berlin: Springer Verlag, 1957.

11. MAGNUS. W. and F. OBERHETTINGER, *Formulas and Theorems for the Special Functions of Mathematical Physics*, p. 118. New York: Chelsea Publishing Co., 1954.

12. WHITHAM, G. B., "Group velocity and energy propagation for three-dimensional waves," *Comm. on Pure and Applied Math.* **16** (1961), pp. 675–691.

13. LIGHTHILL, M. J., "Studies on magneto-hydrodynamic waves and other wave motions," *Phil. Trans. Roy. Soc. (London)*, ser. A252 (1960), pp. 397–430.

14. LIGHTHILL, M. J., "Group velocity," *J. Inst. Maths. Applics.* 1 (1965), pp. 1–28.

15. BUDDEN, K. G., *Radio Waves in the Ionosphere*, Chapters 13 and 18. Cambridge, England: Cambridge University Press, 1961.

16. DESCHAMPS, G. A. and O. B. KESLER, "Radiation of an antenna in a compressible magnetoplasma," *Radio Science* 2 (New Series) (1967), pp. 757–767.

17. COPSON, E. T., *Theory of Functions of a Complex Variable*, Sec. 8.4. London: Oxford University Press, 1935.

18. BREKHOVSKIKH, L. M., *Waves in Layered Media*, Sec. 31. New York: Academic Press, 1960. The analogous acoustic wave problem with a group velocity minimum is treated here.

19. LEWIS, R. M., "Asymptotic theory of transients," in *Electromagnetic Wave Theory*, Part 2 (ed. J. Brown), pp. 845-869. New York: Pergamon Press, 1967.

20. KLINE, M. and I. KAY, *Electromagnetic Theory and Geometrical Optics*, Chapter 8. New York: Interscience, 1965.

21. MAGNUS, W. and F. OBERHETTINGER, *Formulas and Theorems for the Special Functions of Mathematical Physics*, New York: Chelsea Publishing Co., 1954. p. 131.

22. HANDELSMAN, R. A. and N. BLEISTEIN, "Uniform asymptotic expansion of integrals that arise in the analysis of precursors," *Arch. for Rat. Mech. and Analysis* 35 (1969), pp. 267-283.

23. BUCHAL, R. N. and J. B. KELLER, "Boundary layer problems in diffraction theory," *Comm. Pure Appl. Math.* 13 (1960), p. 85.

24. LEWIS, R. M. and B. GRANOFF, "Asymptotic theory of electromagnetic wave propagation in an inhomogeneous anisotropic plasma," *Alta Frequenza* (special issue) **38,** May 1969, pp. 51-59.

25. COURANT, R. and D. HILBERT, *Methods of Mathematical Physics*, Vol. II. New York: Interscience, 1962.

26. GARABEDIAN, P. R., *Partial Differential Equations*, Chapter II. New York: John Wiley & Sons, 1964.

27. BORN, M. and E. WOLF, *Principles of Optics*, Chapter 3. New York: Pergamon Press, 1964.

28. GINZBURG, V. L., *The Propagation of Electromagnetic Waves in Plasmas*, Chapters 4 and 5. New York: Pergamon Press, 1964.

29. KELLER, J. B., "A geometrical theory of diffraction," in *Calculus of Variations and its Applications*, pp. 27-52. New York: McGraw-Hill, 1958.

30. SECKLER, B. D. and J. B. KELLER, "Geometrical theory of diffraction in inhomogeneous media, *J. Acoust. Soc. Am.* 31 (1959), pp. 192-205.

31. FELSEN, L. B. and S. ROSENBAUM, "Ray optics for radiation problems in anisotropic regions with boundaries. 1. Line source excitation," *Radio Science* 2 (New Series) (1967), pp. 767-791. H. BERTONI and A. HESSEL, "Ray optics for radiation problems in anisotropic regions with boundaries. 2. Point source excitation," *ibid.*, pp. 793-812.

32. FELSEN, L. B. and F. LABIANCA, "Radiation and scattering problems in compressible plasmas. Part 1. Solution by ray optics," *Radio Science* 2 (New Series) (1967), pp. 29-51.

33. LEVY, B. R. and J. B. KELLER, "Diffraction by a smooth object," *Comm. Pure and Appl. Math.* 12 (1959), pp. 159-209.

34. KELLER, J. B., "Geometrical theory of diffraction," *J. Opt. Soc. Am.* 52 (1962), pp. 116-130.

35. KELLER, J. B. and E. B. HANSEN, "Survey of the theory of diffraction of short waves by edges, " *Acta Phys. Polon.* 27 (1965), pp. 217-234.

36. YEE, H. Y., L. B. FELSEN, and J. B. KELLER, "Ray theory of reflection from the open end of a waveguide," *SIAM J. Appl. Math.* **16** (1968), pp. 268-300. See also J. J. BOWMAN, "Comparison of ray theory with exact theory for scattering by open waveguides," *SIAM J. Appl. Math.* **18** (1970), pp. 818-829.

37. YEE, H. Y. and L. B. FELSEN, "Ray-optical analysis of electromagnetic scattering in waveguides," *IEEE Trans. Microwave Theory and Techniques*, **MTT-17**, No. 9 (September 1969), pp. 671-683.

38. POEVERLEIN, H., "Sommerfeld-Runge law in three and four dimensions," *Phys. Rev.* **128** (1962), pp. 956-964.

39. FELSEN, L. B., "On the use of refractive index diagrams for source excited anisotropic regions," *Radio Science* **69D** (1965), pp. 155-169.

40. FOCK, V. A. *Electromagnetic Diffraction and Propagation Problems*, Chapter 8. New York: Pergamon Press, 1965.

41. KELLER, J. B., R. M. LEWIS, and B. D. Seckler, "Asymptotic solution of some diffraction problems" *Comm. Pure and Appl. Math.* **9** (1956), pp. 207-265.

2. Network Formalism for Time-Harmonic Electromagnetic Fields in Uniform and Spherical Waveguide Regions

2.1 INTRODUCTION

The general theory of time-dependent and time-harmonic linear fields excited by prescribed sources in an infinite homogeneous medium has been given in Secs. 1.1–1.4 primarily for the simple case of unbounded media. The present chapter examines in greater detail the application of these general concepts to the time-harmonic electromagnetic field in waveguides with perfectly conducting walls bounding planes transverse to the rectilinear coordinate z or spherical surfaces transverse to the radial coordinate r. The boundaries for the former uniform waveguide configuration are defined by $f(\boldsymbol{\rho}) = 0$, where $\boldsymbol{\rho}$ is the vector coordinate transverse to z, and for the latter non-uniform waveguide configuration by $F(\theta, \phi) = 0$, where θ and ϕ are angular spherical coordinates; uniform and non-uniform waveguides are distinguished by whether or not the cross sections along the axial direction are constant. The media filling these regions are assumed to be isotropic but may vary abruptly or continuously with z and r, respectively, as schematized in Fig. 2.1.1. As noted in Sec. 1.4 for uniform regions, the waveguide concept is useful even for unbounded cross sections, especially when the medium parameters vary along the axial direction.

The basic procedure for obtaining field solutions is essentially the same as in Sec. 1.4 and utilizes representations of field variables and their sources in terms of a complete set of vector eigenfunctions. Since the orthogonality properties of the eigenfunctions involve only field components transverse to the transmission coordinate [see Eq. (1.4.30) for uniform regions], it is possible to seek modal expansions of the independent transverse fields $\mathbf{E}_t$ and $\mathbf{H}_t$, and to derive therefrom the dependent longitudinal components. A complete represen-

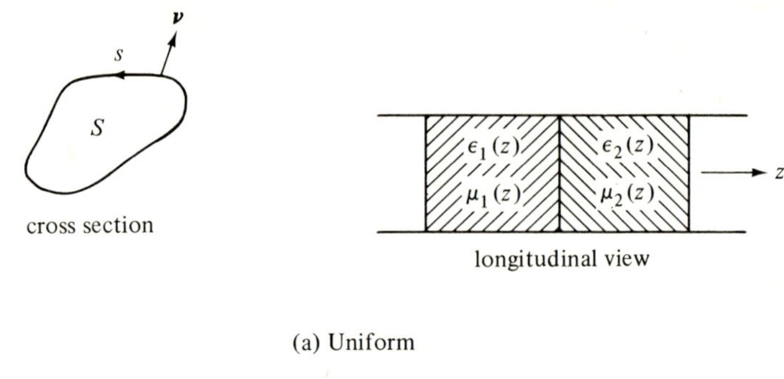

(a) Uniform

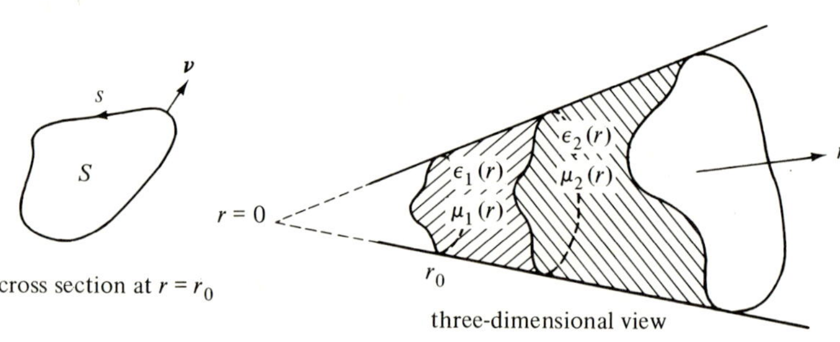

(b) Non-uniform (spherical)

FIG. 2.1.1   Waveguide regions.

tation in terms of E(TM) and H(TE) modes† is utilized in Secs. 2.2b and 2.5b for uniform and spherical waveguides, respectively, to represent the transverse fields and also their equivalent sources. General properties of vector-mode functions in isotropic regions are discussed; their explicit evaluation for various cross-sectional configurations is deferred to Chapter 3, and generalizations to anisotropic regions are considered in Chapters 7 and 8. It is shown that on use of modal representations and orthogonality properties, equations satisfied by the transverse vector field may be reduced to z- or r-dependent scalar transmission-line equations for the modal amplitudes, analogous to those in Eq. (1.4.15) or (1.4.36b).

As noted in Sec. 1.1b, calculation of vector fields is often facilitated through use of scalar potentials. For transversely unbounded uniform regions, this scalarization has been derived, for example, in Eqs. (1.1.38) [see also Eqs. (1.1.49)]. In Sec. 2.3, scalarization of vector fields in transversely bounded uniform regions is achieved by introduction of complete orthogonal sets of

---

†The abbreviations TM and TE stand for transverse magnetic and transverse electric, respectively; in uniform regions, $H_z \equiv 0$ for the former and $E_z \equiv 0$ for the latter, whereas in spherical regions, correspondingly, $H_r \equiv 0$ and $E_r \equiv 0$. See Sec. 8.2 for a general derivation and further discussion of E- and H-mode decompositions.

scalar-mode functions; the latter are then used to yield modal expansions for the scalar potentials. Although this procedure of formulating vector fields in terms of scalar potentials is more circuitous than that in Sec. 1.1b, it has the advantage of furnishing the actual potential solutions. As in Chapter 1, it is recognized that calculations of fields excited by arbitrary source distributions are simplified via dyadic Green's functions. On introduction of one-dimensional transmission-line Green's functions, with reciprocity properties similar to those of three-dimensional dyadic Green's functions, one may develop modal solutions for the latter, both for piecewise homogeneous media and for media exhibiting continuous z variation. These derivations form the substance of Sec. 2.3.

The transmission-line equations formulated in Sec. 2.2 and utilized in Sec. 2.3 are solved in Sec. 2.4 for the special case of media with piecewise homogeneous z dependence (continuous z variation is treated in Secs. 3.3 and 3.5). Emphasis is placed on network schematization of excitation, transmission and reflection processes, and on network methods for calculating transmission line voltages and currents; the latter represent the amplitude coefficients in modal expansions of the transverse electric and magnetic fields, respectively. Section 2.4 concludes with a brief summary of the relation between modes and resonant transmission lines, a subject to be explored more fully in Sec. 3.3.

Fields in spherical waveguide regions are analyzed in a manner similar to the above, with the presentations in Secs. 2.5, 2.6, and 2.7 paralleling those in Secs. 2.2, 2.3, and 2.4, respectively.

2.2 DERIVATION OF TRANSMISSION-LINE EQUATIONS IN UNIFORM REGIONS

2.2a   The Transverse Field Equations

In this section, the transformation of the inhomogeneous, steady-state Maxwell vector field equations into the scalar transmission-line equations for a typical mode is summarized for the case of a uniform waveguide depicted in Fig. 2.1.1a. The procedure is similar to that given in Sec. 1.4 for isotropic linear fields in homogeneous unbounded regions, but it is generalized here to admit regions possessing transverse boundaries and longitudinal stratification. The steady-state electromagnetic vector fields excited by a specified electric current distribution $\mathbf{J}(\mathbf{r})$ and magnetic current distribution $\mathbf{M}(\mathbf{r})$ are defined by the field equations

$$\nabla \times \mathbf{E}(\mathbf{r}) = -j\omega\mu\mathbf{H}(\mathbf{r}) - \mathbf{M}(\mathbf{r}),$$
$$\nabla \times \mathbf{H}(\mathbf{r}) = j\omega\epsilon\mathbf{E}(\mathbf{r}) + \mathbf{J}(\mathbf{r}). \tag{1}$$

On the perfectly conducting boundary of the uniform waveguide, the tangential component of the electric field must vanish, i.e.,

$$\mathbf{v} \times \mathbf{E} = 0 \qquad \text{on } s, \tag{1a}$$

where s denotes the curve bounding the transverse cross section S and $\mathbf{v}$ is a unit vector normal to s and lying in the plane of the cross section. The vanishing of the tangential component of $\mathbf{E}$ on s also implies the vanishing on s of the normal component of $\mathbf{H}$. For a region with infinite cross section, condition (1a) is replaced by a "radiation condition" which requires that, for any source distribution contained in a finite region, the field solution at infinity comprises only "outgoing" waves. The boundary conditions on the longitudinal (z) termini of the region are left open for the moment and will be taken into account in the subsequent solution of the transmission-line equations. A harmonic time-dependence $\exp(+j\omega t)$ is assumed,† whence to obtain the solution for fields with arbitrary time variation, a temporal synthesis as described in Sec. 1.4 must be carried out as well. For the present discussion, the scalar permittivity ϵ and the permeability μ of the medium may both be z dependent.

As noted in Sec. 1.4, orthogonality conditions for modes guided along z involve only field components transverse to z. It is therefore desirable to eliminate the dependent components E_z and H_z from Eqs. (1) and derive field equations for the independent transverse components. To this end, one takes vector and scalar products of Eqs. (1) with the longitudinal unit vector $\mathbf{z}_0$. Thus,

$$j\omega\mu\mathbf{H} \times \mathbf{z}_0 + \mathbf{M} \times \mathbf{z}_0 = \mathbf{z}_0 \times (\nabla \times \mathbf{E}) = -\frac{\partial}{\partial z}\mathbf{E} + \nabla E_z$$

$$= -\frac{\partial}{\partial z}\mathbf{E}_t + \nabla_t E_z \qquad (2a)$$

and

$$-j\omega\mu H_z - M_z = \mathbf{z}_0 \cdot (\nabla \times \mathbf{E}) = -\nabla_t \cdot (\mathbf{z}_0 \times \mathbf{E}), \qquad (2b)$$

where the transverse gradient operator $\nabla_t = \nabla - \mathbf{z}_0\,\partial/\partial z$. Similarly, for the second of Eqs. (1), one has, by duality,

$$j\omega\epsilon\mathbf{z}_0 \times \mathbf{E} + \mathbf{z}_0 \times \mathbf{J} = \nabla_t H_z - \frac{\partial}{\partial z}\mathbf{H}_t, \qquad (3a)$$

$$j\omega\epsilon E_z + J_z = \nabla_t \cdot (\mathbf{H} \times \mathbf{z}_0). \qquad (3b)$$

Upon substituting for E_z in Eq. (2a) from Eq. (3b), one obtains

$$-\frac{\partial}{\partial z}\mathbf{E}_t = j\omega\mu\mathbf{H}_t \times \mathbf{z}_0 - \frac{1}{j\omega\epsilon}(\nabla_t\nabla_t \cdot \mathbf{H}_t \times \mathbf{z}_0 - \nabla_t J_z) + \mathbf{M}_t \times \mathbf{z}_0$$

$$= j\omega\mu\left(1 + \frac{\nabla_t\nabla_t}{k^2}\right) \cdot (\mathbf{H}_t \times \mathbf{z}_0) + \mathbf{M}_{te} \times \mathbf{z}_0, \qquad (4a)$$

and, by duality,

$$-\frac{\partial}{\partial z}\mathbf{H}_t = j\omega\epsilon\left(1 + \frac{\nabla_t\nabla_t}{k^2}\right) \cdot (\mathbf{z}_0 \times \mathbf{E}_t) + \mathbf{z}_0 \times \mathbf{J}_{te}, \qquad (4b)$$

where the equivalent transverse electric and magnetic current distributions are given, respectively, by

---

†An $\exp(j\omega t)$ dependence is customary in applications involving electrical networks and in this chapter is chosen for this reason.

$$\mathbf{J}_{te} = \mathbf{J}_t - \mathbf{z}_0 \times \frac{\nabla_t M_z}{j\omega\mu} = \mathbf{J}_t + \frac{\nabla_t \times \mathbf{M}_z}{j\omega\mu} \tag{5a}$$

$$\mathbf{M}_{te} = \mathbf{M}_t + \mathbf{z}_0 \times \frac{\nabla_t J_z}{j\omega\epsilon} = \mathbf{M}_t - \frac{\nabla_t \times \mathbf{J}_z}{j\omega\epsilon}. \tag{5b}$$

The quantity $k = \omega(\mu\epsilon)^{1/2}$ is the wavenumber in the region and $\mathbf{1}$ is a unit dyadic such that $\mathbf{1} \cdot \mathbf{A} = \mathbf{A} \cdot \mathbf{1} = \mathbf{A}$.

The transverse field equations (4) and (5), which admit a z-dependent ϵ and μ, provide the basis for the treatment of field problems in uniform waveguides.† They are completely descriptive of the total field equations (1), since from Eqs. (2b) and (3b), the longitudinal components are derivable from the transverse components as

$$j\omega\epsilon E_z = \nabla_t \cdot (\mathbf{H}_t \times \mathbf{z}_0) - J_z \tag{6a}$$

$$j\omega\mu H_z = \nabla_t \cdot (\mathbf{z}_0 \times \mathbf{E}_t) - M_z. \tag{6b}$$

The boundary condition (1a), requiring the vanishing of the *total* tangential electric field on the perfectly conducting guide walls, can be restated in terms of the *transverse* field components as

$$\left.\begin{array}{r} \mathbf{v} \times \mathbf{E}_t = 0 \\ \nabla_t \cdot (\mathbf{H}_t \times \mathbf{z}_0) = 0 \end{array}\right\} \quad \text{on } s, \tag{7}$$

where the second relation follows from Eq. (6a) upon assuming that $J_z = 0$ on s. This restriction, which requires the vanishing *on the boundary* of the z component of the *applied* electric current source, is of no practical consequence since an applied tangential electric current source on a perfectly conducting surface is "short-circuited" and cannot radiate a finite field.

2.2b Modal Representation of the Fields and Their Sources

As noted in Sec. 1.4c for the special case of free space, the vector electromagnetic field equations can be transformed into ordinary scalar differential equations on representation of the fields in terms of a complete orthonormal set of "guided" eigenfunctions. Although in Sec. 1.4c, this representation was performed in terms of Ψ eigenvectors indicative of both the electric- and magnetic-field vectors, it could equally well have been effected by individual vector representations of the electric field $\mathbf{E}$ in terms of eigenvectors $\mathbf{e}(\mathbf{\rho})$, of the magnetic field $\mathbf{H}$ in terms of eigenvectors $\mathbf{h}(\mathbf{\rho})$, etc. This latter procedure will be followed for field representations in the bounded regions considered in this chapter. For a perfectly conducting waveguide filled with a homogeneous, isotropic medium, a possible complete eigenvector set comprises both E(TM) mode functions $\mathbf{e}'(\mathbf{\rho})$, $\mathbf{h}'(\mathbf{\rho})$ and H(TE) mode functions $\mathbf{e}''(\mathbf{\rho})$, $\mathbf{h}''(\mathbf{\rho})$; these mode functions are defined below in Eqs. (10) and in Sec. 2.3a; justification of the E- and H-mode decompositions is treated more generally in Sec. 8.2. In terms of

----

†For an analogous treatment of waveguides filled with anisotropic media or with media possessing transverse variation, see Sec. 8.2.

the indicated mode functions, a representation of the independent transverse fields is given as [1,2]

$$\mathbf{E}_t(\mathbf{r}) = \sum_i V_i'(z)\mathbf{e}_i'(\boldsymbol{\rho}) + \sum_i V_i''(z)\mathbf{e}_i''(\boldsymbol{\rho}), \tag{8a}$$

$$\mathbf{H}_t(\mathbf{r}) = \sum_i I_i'(z)\mathbf{h}_i'(\boldsymbol{\rho}) + \sum_i I_i''(z)\mathbf{h}_i''(\boldsymbol{\rho}), \tag{8b}$$

$$\mathbf{J}_{te}(\mathbf{r}) = \sum_i i_i'(z)\mathbf{e}_i'(\boldsymbol{\rho}) + \sum_i i_i''(z)\mathbf{e}_i''(\boldsymbol{\rho}), \tag{8c}$$

$$\mathbf{M}_{te}(\mathbf{r}) = \sum_i v_i'(z)\mathbf{h}_i'(\boldsymbol{\rho}) + \sum_i v_i''(z)\mathbf{h}_i''(\boldsymbol{\rho}), \tag{8d}$$

where i is in general a double index, and

$$\mathbf{h}_i = \mathbf{z}_0 \times \mathbf{e}_i. \tag{8e}$$

In view of Eqs. (6) and (8) one obtains the longitudinal-field representations [by Eqs. (10), only E modes contribute to the representation of E_z, while only H modes contribute to the representation of H_z]

$$j\omega\epsilon E_z(\mathbf{r}) + J_z(\mathbf{r}) = \sum_i I_i'(z)\nabla_t \cdot \mathbf{e}_i'(\boldsymbol{\rho}), \tag{9a}$$

$$j\omega\mu H_z(\mathbf{r}) + M_z(\mathbf{r}) = \sum_i V_i''(z)\nabla_t \cdot \mathbf{h}_i''(\boldsymbol{\rho}). \tag{9b}$$

The specific form of the transverse vector eigenfunctions $\mathbf{e}_i$ and $\mathbf{h}_i$ is dependent on the shape of the guide cross-section and is, in general, defined by the following z-independent equations [see Eqs. (8.2.25)]:

$$\begin{aligned}
\nabla_t\nabla_t \cdot \mathbf{e}_i' = -k_{ti}'^2\mathbf{e}_i' \qquad &\nabla_t\nabla_t \cdot \mathbf{h}_i'' = -k_{ti}''^2\mathbf{h}_i'', \\
\nabla_t\nabla_t \cdot \mathbf{h}_i' = 0, \qquad &\nabla_t\nabla_t \cdot \mathbf{e}_i'' = 0,
\end{aligned} \tag{10}$$

subject, in accord with Eqs. (7), to the boundary conditions on the curve s with normal $\mathbf{v}$ bounding the transverse cross section:

$$\mathbf{v} \times \mathbf{e}_i' = 0 = \nabla_t \cdot (\mathbf{h}_i' \times \mathbf{z}_0), \quad \mathbf{v} \times \mathbf{e}_i'' = 0 = \nabla_t \cdot (\mathbf{h}_i'' \times \mathbf{z}_0) \qquad \text{on } s. \tag{10a}$$

One notes from Eqs. (10a) that the vector mode functions introduced in the representations (8a) and (8b) satisfy individually the appropriate boundary conditions (7) on the transverse electromagnetic fields. Moreover, since *applied* electric and magnetic currents have no tangential or normal components, respectively, at a perfectly conducting surface, the representations for the source currents in Eqs. (8c) and (8d) are likewise meaningful for realizable source current distributions on the boundary.

Upon applying the following transverse form of Green's theorem,†

$$\iint\limits_S dS[\mathbf{A} \cdot \nabla_t\nabla_t \cdot \mathbf{B} - \mathbf{B} \cdot \nabla_t\nabla_t \cdot \mathbf{A}]$$
$$= \oint\limits_s ds[(\mathbf{A} \cdot \mathbf{v})(\nabla_t \cdot \mathbf{B}) - (\mathbf{B} \cdot \mathbf{v})(\nabla_t \cdot \mathbf{A})], \tag{11a}$$

---

†Equation (11a) is obtained by applying the divergence theorem in the transverse cross section to the expression

$$\nabla_t \cdot [\mathbf{A}\nabla_t \cdot \mathbf{B} - \mathbf{B}\nabla_t \cdot \mathbf{A}] = \mathbf{A} \cdot \nabla_t\nabla_t \cdot \mathbf{B} - \mathbf{B} \cdot \nabla_t\nabla_t \cdot \mathbf{A}$$

where **A** and **B** are suitably continuous transverse vector functions, to the vector mode functions defined in Eqs. (10), one deduces the orthogonality conditions over the cross-sectional domain S (normalization to unity is assumed):

$$\iint_S \mathbf{e}'_i \cdot \mathbf{e}'^*_j \, dS = \delta_{ij} = \iint_S \mathbf{e}''_i \cdot \mathbf{e}''^*_j \, dS; \qquad \iint_S \mathbf{e}'_i \cdot \mathbf{e}''^*_i \, dS = 0, \qquad (11b)$$

and similarly for the $\mathbf{h}_i$ functions. The asterisk denotes the complex conjugate,† and the Kronecker delta is defined as follows: $\delta_{ij} = 0, i \neq j; \delta_{ii} = 1$. In view of these orthonormality properties, the mode amplitudes in Eqs. (8) are determined as follows:

$$V_i(z) = \iint_S \mathbf{E}_t(\mathbf{r}) \cdot \mathbf{e}^*_i(\boldsymbol{\rho}) \, dS, \qquad I_i(z) = \iint_S \mathbf{H}_t(\mathbf{r}) \cdot \mathbf{h}^*_i(\boldsymbol{\rho}) \, dS, \qquad (12a)$$

$$v_i(z) = \iint_S \mathbf{M}_{te}(\mathbf{r}) \cdot \mathbf{h}^*_i(\boldsymbol{\rho}) \, dS, \qquad i_i(z) = \iint_S \mathbf{J}_{te}(\mathbf{r}) \cdot \mathbf{e}^*_i(\boldsymbol{\rho}) \, dS, \qquad (12b)$$

where the distinguishing ′ and ″ have been omitted since the equations apply to both mode types. Utilizing the equivalent current definitions in Eq. (5) and employing the vector integration-by-parts formula (divergence theorem in two dimensions)

$$\iint_S dS \nabla_t f \cdot \mathbf{A} = -\iint_S dS \, f \nabla_t \cdot \mathbf{A} + \oint_s ds \, f(\mathbf{A} \cdot \mathbf{v}), \qquad (13)$$

with f and **A** suitably continuous scalar and vector functions, one may reexpress the integrals of Eqs. (12b). The contribution to the gradient integrals from the bounding contour s vanishes in view of the boundary condition $\mathbf{h}_i \cdot \mathbf{v} = 0$ [Eq. (10a)] and the specification $J_z = 0$ on s, so Eqs. (12b) become

$$v_i(z) = \iint_S \mathbf{M}(\mathbf{r}) \cdot \mathbf{h}^*_i(\boldsymbol{\rho}) \, dS + Z^*_i \iint_S \mathbf{J}(\mathbf{r}) \cdot \mathbf{e}^*_{zi}(\boldsymbol{\rho}) \, dS, \qquad (14a)$$

$$i_i(z) = \iint_S \mathbf{J}(\mathbf{r}) \cdot \mathbf{e}^*_i(\boldsymbol{\rho}) \, dS + Y^*_i \iint_S \mathbf{M}(\mathbf{r}) \cdot \mathbf{h}^*_{zi}(\boldsymbol{\rho}) \, dS, \qquad (14b)$$

where

$$Y''_i \mathbf{h}''_{zi}(\boldsymbol{\rho}) \equiv \mathbf{z}_0 \frac{\nabla_t \cdot \mathbf{h}''_i(\boldsymbol{\rho})}{j\omega\mu}, \qquad \mathbf{h}'_{zi} \equiv 0, \qquad (14c)$$

$$Z'_i \mathbf{e}'_{zi}(\boldsymbol{\rho}) \equiv \mathbf{z}_0 \frac{\nabla_t \cdot \mathbf{e}'_i(\boldsymbol{\rho})}{j\omega\epsilon}, \qquad \mathbf{e}''_{zi} \equiv 0. \qquad (14d)$$

The vanishing of $\mathbf{h}'_{zi}$ (for E modes) and of $\mathbf{e}''_{zi}$ (for H modes) follows directly from Eqs. (10). The introduction of the characteristic impedance and admittance Z'_i and Y''_i [defined explicitly in Eqs. (15)] serves to highlight in a physical sense the contributions of the various integrals as either voltages or currents. It is to be noted that the formulations in Eqs. (14) do not require differentiability of J_z and M_z in the cross section S as implied in Eqs. (5) and (12b).

---

†Since the mode functions may be complex, although k'^2_{ti} and k''^2_{ti} are real, the orthogonality condition involves the complex conjugate function.

Upon inserting the modal representations (8) into the transverse field equations (4), interchanging the order of summation and differentiation, making use of Eqs. (10), and equating like coefficients of the mode functions $\mathbf{e}_i$ and $\mathbf{h}_i$, one obtains the desired transmission-line equations for the E- and H-mode amplitudes as [see Eq. (1.4.35b)]

$$-\frac{dV_i}{dz} = j\kappa_i Z_i I_i + v_i, \tag{15a}$$

$$-\frac{dI_i}{dz} = j\kappa_i Y_i V_i + i_i, \tag{15b}$$

where the modal characteristic impedance Z_i (admittance Y_i) and the modal propagation constant κ_i are defined as follows:

E modes:

$$Z_i' = \frac{1}{Y_i'} = \frac{\kappa_i'}{\omega\epsilon}, \qquad \kappa_i' = \sqrt{k^2 - k_{ti}'^2} = -j\sqrt{k_{ti}'^2 - k^2}, \tag{15c}$$

H modes:

$$Z_i'' = \frac{1}{Y_i''} = \frac{\omega\mu}{\kappa_i''}, \qquad \kappa_i'' = \sqrt{k^2 - k_{ti}''^2} = -j\sqrt{k_{ti}''^2 - k^2}. \tag{15d}$$

$k^2 = \omega^2\mu\epsilon$, and both μ and ϵ may be functions of z. The form of Eqs. (15a) and (15b) permits identification of V_i and I_i as transmission-line voltage and current,[3] respectively. The choice of sign on the square roots in Eqs. (15) assures the damping of non-propagating modes (κ_i imaginary) away from the source region for the assumed time dependence exp $(+j\omega t)$. The evaluation of the source voltage v_i and current i_i amplitudes follows directly from the *specified* electric and magnetic source currents $\mathbf{J}$ and $\mathbf{M}$ via Eqs. (14a) and (14b). Solutions of Eqs. (15a) and (15b) for various stratifications and terminations in the z domain are discussed in Secs. 2.4 and 3.3b.

2.3 SCALARIZATION AND MODAL REPRESENTATION OF DYADIC GREEN'S FUNCTIONS IN UNIFORM REGIONS

Solutions for the vector electromagnetic field excited by prescribed sources in a uniform waveguide region bounded by perfectly conducting walls (if any) and filled with a transversely homogeneous material follow from the representations in Eqs. (2.2.8) and (2.2.9); the vector-mode functions are evaluated from Eqs. (2.2.10) and the modal amplitudes from Eqs. (2.2.15) subject to appropriate boundary conditions in the z domain. Solution of the vector eigenvalue problems in Eqs. (2.2.10) is facilitated by introduction of scalar-mode functions. The scalarization achieved in this manner may be utilized to define E- and H-mode (Hertz) potentials from which the electromagnetic fields themselves can be derived. For point-source excitation, these potentials are equivalent to the scalar

Green's functions that have been introduced in Eqs. (1.1.38) and (1.1.49) without intervention of an eigenfunction expansion. The procedure discussed below yields explicit modal representations for these functions, or equivalently for the customarily defined Hertz potentials related to them, and thereby *solves* the scalar potential problems. We first express vector-mode functions in terms of scalar-mode functions, and then scalarize the overall field representation.

2.3a Mode Functions

In representing the transverse electric vector field $\mathbf{E}_t$ in Eq. (2.2.8a) in terms of two independent vector mode sets $\{\mathbf{e}_i'\}$ and $\{\mathbf{e}_i''\}$, use has been made of a theorem which states that any *transverse* vector can be decomposed into two parts, one of which is solenoidal and the other of which is irrotational.[4] The vector set $\{\mathbf{e}_i'\}$ is irrotational (i.e., $\mathbf{\nabla}_t \times \mathbf{e}_i' = 0$ in S), while the vector set $\{\mathbf{e}_i''\}$ is solenoidal (i.e., $\mathbf{\nabla}_t \cdot \mathbf{e}_i'' = 0$ in S) [see also Eqs. (2.2.10)]. In view of these properties, the vector-mode functions $\mathbf{e}_i'$ and $\mathbf{e}_i''$ can be represented as gradients and curls of scalar functions Φ_i and ψ_i as follows:

$$\mathbf{e}_i'(\mathbf{\rho}) = -\frac{\mathbf{\nabla}_t \Phi_i(\mathbf{\rho})}{k_{ti}'}, \tag{1a}$$

$$\mathbf{e}_i''(\mathbf{\rho}) = -\frac{\mathbf{\nabla}_t \psi_i(\mathbf{\rho})}{k_{ti}''} \times \mathbf{z}_0, \tag{1b}$$

and, consequently,

$$\mathbf{h}_i'(\mathbf{\rho}) = -\mathbf{z}_0 \times \frac{\mathbf{\nabla}_t \Phi_i(\mathbf{\rho})}{k_{ti}'}, \tag{1c}$$

$$\mathbf{h}_i''(\mathbf{\rho}) = -\frac{\mathbf{\nabla}_t \psi_i(\mathbf{\rho})}{k_{ti}''}. \tag{1d}$$

By Eqs. (1) and (2.2.10) the mode functions Φ_i and ψ_i are defined by the two scalar eigenvalue problems†

$$\nabla_t^2 \Phi_i + k_{ti}'^2 \Phi_i = 0 \qquad \text{in } S, \tag{2a}$$

$$\Phi_i = 0 \qquad \text{on } s \text{ if } k_{ti}' \neq 0,$$

$$\frac{\partial \Phi_i}{\partial s} = 0 \qquad \text{on } s \text{ if } k_{ti}' = 0 \qquad \text{(TEM mode)}, \tag{2b}$$

and

$$\nabla_t^2 \psi_i + k_{ti}''^2 \psi_i = 0 \qquad \text{in } S, \tag{2c}$$

$$\frac{\partial \psi_i}{\partial v} = 0 \qquad \text{on } s. \tag{2d}$$

---

†The vector-mode functions for the TEM (transverse electromagnetic) case are determined via $\mathbf{e}_0'(\mathbf{\rho}) = \mathbf{h}_0'(\mathbf{\rho}) \times \mathbf{z}_0 = -\mathbf{\nabla}_t \Phi_0(\mathbf{\rho})$, where $\Phi_0(\mathbf{\rho})$ is the solution of Eq. (2a) with $k_{ti}' = 0$, with the normalization $\iint_S e_0'^2(\mathbf{\rho})\, dS = 1$.

Specific solutions of Eqs. (2) for various guide cross sections are tabulated in Chapter 3.[2]†

2.3b   Fields in Source-Free, Homogeneous Regions

Using Eqs. (1) and assuming interchangeability of summation and differentiation operations, one may write Eqs. (2.2.8a) and (2.2.8b) as

$$\mathbf{E}_t(\mathbf{r}) = -\nabla_t V'(\mathbf{r}) - \nabla_t V''(\mathbf{r}) \times \mathbf{z}_0, \tag{3a}$$

$$\mathbf{H}_t(\mathbf{r}) \times \mathbf{z}_0 = -\nabla_t I'(\mathbf{r}) - \nabla_t I''(\mathbf{r}) \times \mathbf{z}_0, \tag{3b}$$

where the potential functions $V'(\mathbf{r})$, $I'(\mathbf{r})$ and $V''(\mathbf{r})$, $I''(\mathbf{r})$ are defined as follows:

$$V'(\mathbf{r}) = \sum_i V'_i(z)\frac{\Phi_i(\boldsymbol{\rho})}{k'_{ti}}, \qquad V''(\mathbf{r}) = \sum_i V''_i(z)\frac{\psi_i(\boldsymbol{\rho})}{k''_{ti}}, \tag{4a}$$

$$I'(\mathbf{r}) = \sum_i I'_i(z)\frac{\Phi_i(\boldsymbol{\rho})}{k'_{ti}}, \qquad I''(\mathbf{r}) = \sum_i I''_i(z)\frac{\psi_i(\boldsymbol{\rho})}{k''_{ti}}. \tag{4b}$$

The potential functions in Eqs. (4) are related to the Hertz potentials Π' and Π'' [Eqs. (1.1.42) and (1.1.53)] by

$$\Pi'(\mathbf{r}) = \frac{I'(\mathbf{r})}{j\omega\epsilon}, \qquad \Pi''(\mathbf{r}) = \frac{V''(\mathbf{r})}{j\omega\mu}. \tag{5}$$

From Eqs. (3) and (2.2.6), the electromagnetic fields can be expressed at any source-free point where ϵ and μ are *non-variable* as[5]

$$\mathbf{E}(\mathbf{r}) = \nabla \times \nabla \times [\mathbf{z}_0\Pi'(\mathbf{r})] - j\omega\mu\nabla \times [\mathbf{z}_0\Pi''(\mathbf{r})], \tag{6a}$$

$$\mathbf{H}(\mathbf{r}) = j\omega\epsilon\nabla \times [\mathbf{z}_0\Pi'(\mathbf{r})] + \nabla \times \nabla \times [\mathbf{z}_0\Pi''(\mathbf{r})]. \tag{6b}$$

The two independent functions $I'(\mathbf{r})$ and $V''(\mathbf{r})$ in Eq. (5) suffice to determine the total fields via Eqs. (6). In a source-free region, $V'(\mathbf{r})$ and $I''(\mathbf{r})$ are obtainable from $I'(\mathbf{r})$ and $V''(\mathbf{r})$, respectively, by differentiation with respect to z, as is evident from the transmission-line equations (2.2.15). Thus,

$$V'(\mathbf{r}) = \sum_i \frac{1}{-j\kappa'_i Y'_i}\frac{dI'_i(z)}{dz}\frac{\Phi_i(\boldsymbol{\rho})}{k'_{ti}} = \frac{1}{-j\omega\varepsilon}\frac{\partial}{\partial z}I'(\mathbf{r}), \tag{7a}$$

and, similarly,

$$I''(\mathbf{r}) = \frac{1}{-j\omega\mu}\frac{\partial}{\partial z}V''(\mathbf{r}). \tag{7b}$$

Equations (2) and (2.2.15) may be used to verify that in a source-free, homogeneous region, the Hertz potentials Π' and Π'', given by Eqs. (4) and (5), satisfy the wave equations

$$(\nabla^2 + k^2)\frac{\Pi'}{\Pi''} = 0. \tag{8}$$

---

†It should be pointed out that the scalar eigenfunctions Φ_i and ψ_i, like the vector eigenfunctions $\mathbf{e}'_i$ and $\mathbf{e}''_i$, each form an orthonormal set (see Sec. 3.2). Normalization of these scalar eigenfunctions differs from that used in Reference 2. The relation between the eigenfunctions here and those in Reference 2 is the following:

$$k'_{ti}[\Phi_i]_{\text{ref.2}} = \Phi_i, \qquad k''_{ti}[\psi_i]_{\text{ref.2}} = \psi_i.$$

2.3c   Green's Functions for the Transmission-Line Equations

To obtain explicit solutions for the Hertz potentials in source regions, it is necessary to relate the modal coefficients in Eqs. (4) to their excitations. For this purpose, it is convenient to introduce modal Green's functions that are one-dimensional scalar analogues of the dyadic Green's functions defined in Sec. 1.1b. In view of the linearity of the transmission-line equations (2.2.15), one can obtain the voltage and current solutions at any point z by superposing separate contributions from suitable point voltage and current generators distributed along points z'. In analogy with Eqs. (1.1.19), one thereby finds[3]

$$V(z) = -\int dz' \, T^V(z, z')v(z') - \int dz' \, Z(z, z')i(z'), \tag{9a}$$

$$I(z) = -\int dz' \, Y(z, z')v(z') - \int dz' \, T^I(z, z')i(z'), \tag{9b}$$

where the mode subscript i has been omitted. Equations (9) reduce the problem to that of determining $T^V(z, z')$, $Y(z, z')$ and $Z(z, z')$, $T^I(z, z')$, whose significance as modal Green's functions is evident: $-T^V(z, z')$ and $-Z(z, z')$ are the component voltage responses at z due, respectively, to a unit voltage and current source (generator) at the point z', while $-Y(z, z')$ and $-T^I(z, z')$ are the corresponding current responses to the same excitations. Thus, if in Eqs. (2.2.15), one sets $v(z) = -\delta(z - z')$ and $i(z) = 0$, there results

$$-\frac{d}{dz} T^V(z, z') = j\kappa Z Y(z, z') - \delta(z - z'), \tag{10a}$$

$$-\frac{d}{dz} Y(z, z') = j\kappa Y T^V(z, z'), \tag{10b}$$

and, if $v = 0, i = -\delta(z - z')$,

$$-\frac{d}{dz} Z(z, z') = j\kappa Z T^I(z, z'), \tag{10c}$$

$$-\frac{d}{dz} T^I(z, z') = j\kappa Y Z(z, z') - \delta(z - z'), \tag{10d}$$

subject to as-yet-unspecified boundary conditions at the z terminations.

The modal Green's functions defined in Eqs. (10) satisfy reciprocity properties when κ and Z are either constant or z dependent. We consider a given terminated transmission line to be excited by two separate source distributions: the first, $v(z), i(z)$, giving rise to $V(z), I(z)$; and the second, $\hat{v}(z), \hat{i}(z)$, giving rise to $\hat{V}(z), \hat{I}(z)$. Both sets satisfy transmission-line equations:

$$-\frac{dV}{dz} = j\kappa Z I + v, \tag{11a}$$

$$-\frac{dI}{dz} = j\kappa Y V + i, \tag{11b}$$

and

$$-\frac{d\hat{V}}{dz} = j\kappa Z\hat{I} + \hat{v}, \tag{11c}$$

$$-\frac{d\hat{I}}{dz} = j\kappa Y\hat{V} + \hat{\imath}. \tag{11d}$$

Upon multiplying Eqs. (11a)–(11d) by $\hat{I}, \hat{V}, I, V$, respectively, subtracting the sum of the resulting Eqs. (11a) and (11d) from the sum of Eqs. (11b) and (11c), and integrating over z between the limits z_1 and z_2, one obtains

$$(\hat{V}I - \hat{I}V)_{z_1}^{z_2} = \int_{z_1}^{z_2} dz(v\hat{I} + \hat{\imath}V - i\hat{V} - \hat{v}I). \tag{12}$$

Both sets of voltages and currents satisfy the same terminal conditions at z_1 and z_2:

$$V(z_{1,2}) = \mp Z(z_{1,2})I(z_{1,2}), \qquad \hat{V}(z_{1,2}) = \mp Z(z_{1,2})\hat{I}(z_{1,2}), \tag{13}$$

where $Z(z_{1,2})$ are terminal impedances† (see Sec. 2.4 for details on the network interpretation of the transmission-line equations). Thus, the left-hand side of Eq. (12), expressing the difference between the values at z_2 and z_1 of the bracketed quantity, vanishes and one obtains the reciprocity relation

$$\int_{z_1}^{z_2} dz(v\hat{I} + \hat{\imath}V - i\hat{V} - \hat{v}I) = 0. \tag{14}$$

To apply the reciprocity condition (14) to the modal Green's functions defined in Eqs. (10), one selects the following special source distributions:

(a) $\qquad v = \hat{v} = 0, \quad i = -\delta(z - z'), \quad \hat{\imath} = -\delta(z - z'');$

$\qquad\qquad\qquad V \to Z(z, z'), \quad \hat{V} \to Z(z, z''),$

(b) $\qquad i = \hat{\imath} = 0, \quad v = -\delta(z - z'), \quad \hat{v} = -\delta(z - z'');$

$\qquad\qquad\qquad I \to Y(z, z'), \quad \hat{I} \to Y(z, z''),$

(c) $\qquad v = \hat{\imath} = 0, \quad i = -\delta(z - z'), \quad \hat{v} = -\delta(z - z'');$

$\qquad\qquad\qquad I \to T^I(z, z'), \quad \hat{V} \to T^V(z, z''),$

whence one obtains the following reciprocity theorems:

(a) $\qquad\qquad\qquad Z(z'', z') = Z(z', z''),$

(b) $\qquad\qquad\qquad Y(z'', z') = Y(z', z''),$

(c) $\qquad\qquad\qquad T^I(z'', z') = -T^V(z', z''),$ $\qquad$ (15)

which bear evident similarities to those for the dyadic Green's functions listed in Eqs. (1.1.29).

In view of the reciprocity relation (15c) between T^I and T^V one deduces from Eqs. (10) the important fact that the general solution for the voltage and current in a source-free region can be expressed *solely* in terms of *either* $Y(z, z')$ or $Z(z, z')$. Suppose we have found $Y(z, z')$; then T^V is obtained from Eq. (10b).

---

†For clarity, it is recalled that in this section, Z, $Z(z_\alpha)$, and $Z(z, z')$ denote, respectively, the characteristic impedance, the terminating impedance at z_α, and the voltage Green's function for the ith mode.

Because of the reciprocity theorem, a knowledge of T^V implies the knowledge of T^I, which in turn determines $Z(z, z')$ via Eq. (10d), provided that $z \neq z'$ (i.e., away from the source). Thus, all the required information is contained in $Y(z, z')$; an alternative statement applies for $Z(z, z')$. Actually, one can also determine the voltage and current in a source-free region from either T^I or T^V; however, as is shown in Sec. 2.4d, the *basic* Green's functions are the transfer impedance $Z(z, z')$ and the transfer admittance $Y(z, z')$, from which T^I and T^V are derivable. Because of the fundamental role played by the current (i.e., the E_z field component) in the case of E modes, it is usually convenient to determine E-mode solutions from $Y(z, z')$; by duality, the Green's function $Z(z, z')$ is usually more convenient for H-mode quantities.

2.3d   Modal Representations of the Dyadic Green's Functions in a Piecewise Homogeneous Medium

As shown in Sec. 1.1, the electromagnetic fields radiated by point current excitations are conveniently expressed in terms of dyadic Green's functions. In this section we derive modal solutions for the dyadic Green's functions in regions whose properties are constant along the z direction and show how the dyadic Green's functions can be related to scalar Green's functions.

From Eqs. (6) one notes that the electromagnetic fields $\mathbf{E}(\mathbf{r})$ and $\mathbf{H}(\mathbf{r})$ exterior to source regions can be expressed in terms of the scalar potential functions $I'(\mathbf{r})$ and $V''(\mathbf{r})$ defined in Eqs. (4). If the assumed sources are electric and magnetic current elements situated at the point $\mathbf{r}'$,

$$\mathbf{J}(\mathbf{r}) = \mathbf{J}^0 \delta(\mathbf{r} - \mathbf{r}'), \qquad \mathbf{M}(\mathbf{r}) = \mathbf{M}^0 \delta(\mathbf{r} - \mathbf{r}'), \tag{16}$$

where $\mathbf{J}^0$ and $\mathbf{M}^0$ are arbitrarily oriented constant vectors, then the modal representations for I' and V'' in Eqs. (4) can be simplified. Consider first the E-mode current $I_i'(z)$ occurring in the representation for the E-mode current potential $I'(\mathbf{r})$ in Eq. (4b). Upon recalling the definitions for the transmission-line Green's functions $Y_i(z, z')$ and $T_i^I(z, z')$ in Eq. (9b), one notes that for a point source

$$I_i'(z, z') = -Y_i'(z, z')v_i'(z') - T_i^{\prime I}(z, z')i_i'(z'), \tag{17}$$

where the dependence of $I_i'(z)$ on z' has been indicated explicitly and the subscripts have been inserted to highlight the modal character of the various quantities (for a network interpretation, see Fig. 2.4.4). It will be desirable to have $T_i^{\prime I}(z, z')$ expressed in terms of $Y_i'(z, z')$. From Eqs. (15c), (10b), and (15b), one finds that

$$T_i^I(z, z') = -T_i^V(z', z) = \frac{1}{j\kappa_i Y_i}\frac{d}{dz'}Y_i(z', z) = \frac{1}{j\kappa_i Y_i}\frac{d}{dz'}Y_i(z, z'). \tag{18}$$

Since $\kappa_i' Y_i' = \omega\epsilon$ for E modes [see Eq. (2.2.15)], one obtains, instead of Eq. (17),

$$I_i'(z, z') = -\left[v_i'(z') + \frac{1}{j\omega\epsilon}i_i'(z')\frac{d}{dz'}\right]Y_i'(z, z'). \tag{19}$$

In a similar manner, one can show that the H-mode voltages $V_i''(z)$, occurring in the representation of the voltage potential function $V''(\mathbf{r})$ in Eq. (4a), can be expressed in a manner dual to that in Eq. (19):

$$V_i''(z, z') = -\left[i_i''(z') + \frac{1}{j\omega\mu}\, v_i''(z')\frac{d}{dz'}\right]Z_i''(z, z').\tag{20}$$

Since $\delta(\mathbf{r} - \mathbf{r}') = \delta(\boldsymbol{\rho} - \boldsymbol{\rho}')\delta(z - z')$ in Eq. (16), the source terms v_i and i_i defined in terms of $\mathbf{J}$ and $\mathbf{M}$ by Eqs. (2.2.14), take on the following simple form:

$$v_i(z) = v_i(z')\delta(z - z'), \qquad i_i(z) = i_i(z')\delta(z - z'),\tag{21}$$

$$v_i(z') = \mathbf{h}_i^*(\boldsymbol{\rho}') \cdot \mathbf{M}^0 + Z_i^* \mathbf{e}_{zi}^*(\boldsymbol{\rho}') \cdot \mathbf{J}^0,\tag{21a}$$

$$i_i(z') = \mathbf{e}_i^*(\boldsymbol{\rho}') \cdot \mathbf{J}^0 + Y_i^* \mathbf{h}_{zi}^*(\boldsymbol{\rho}') \cdot \mathbf{M}^0.\tag{21b}$$

Upon substituting the scalar mode functions via Eqs. (1), one finds that for E modes,

$$-\left[v_i'(z') + \frac{1}{j\omega\epsilon}\, i_i'(z')\frac{d}{dz'}\right]$$
$$= \left[(\mathbf{z}_0 \times \boldsymbol{\nabla}_t')\frac{\Phi_i^*(\boldsymbol{\rho}')}{k_{ti}'}\right] \cdot \mathbf{M}^0 - \frac{1}{j\omega\epsilon}\left[\left(\mathbf{z}_0\boldsymbol{\nabla}_t'^2 - \boldsymbol{\nabla}_t'\frac{\partial}{\partial z'}\right)\frac{\Phi_i^*(\boldsymbol{\rho}')}{k_{ti}'}\right] \cdot \mathbf{J}^0,$$
$$\tag{22}$$

where $\boldsymbol{\nabla}_t'$ denotes differentiation with respect to the primed coordinates $\boldsymbol{\rho}'$. In view of the vector identities

$$\mathbf{z}_0 \times \boldsymbol{\nabla}_t'\varphi = -\boldsymbol{\nabla}' \times (\mathbf{z}_0\varphi) \to -(\boldsymbol{\nabla}' \times \mathbf{z}_0)\varphi\tag{23a}$$

and

$$\left(\boldsymbol{\nabla}_t'\frac{\partial}{\partial z'} - \mathbf{z}_0\boldsymbol{\nabla}_t'^2\right)\varphi = \left(\boldsymbol{\nabla}'\frac{\partial}{\partial z'} - \mathbf{z}_0\boldsymbol{\nabla}'^2\right)\varphi$$
$$= \boldsymbol{\nabla}'(\boldsymbol{\nabla}' \cdot \mathbf{z}_0\varphi) - \boldsymbol{\nabla}'^2(\mathbf{z}_0\varphi) \to (\boldsymbol{\nabla}' \times \boldsymbol{\nabla}' \times \mathbf{z}_0)\varphi,\tag{23b}$$

where φ is a scalar function of $\boldsymbol{\rho}'$, one obtains the following concise expression for $I'(\mathbf{r})$ after substituting Eqs. (19)–(23) into Eq. (4b):

$$I'(\mathbf{r}) = (\boldsymbol{\nabla}' \times \boldsymbol{\nabla}' \times \mathbf{z}_0)\mathcal{S}'(\mathbf{r}, \mathbf{r}') \cdot \mathbf{J}^0 - j\omega\epsilon(\boldsymbol{\nabla}' \times \mathbf{z}_0)\mathcal{S}'(\mathbf{r}, \mathbf{r}') \cdot \mathbf{M}^0,\tag{24}$$

where

$$j\omega\epsilon\mathcal{S}'(\mathbf{r}, \mathbf{r}') = \sum_i \frac{\Phi_i(\boldsymbol{\rho})\Phi_i^*(\boldsymbol{\rho}')}{k_{ti}'^2}\, Y_i'(z, z').\tag{24a}$$

The meaning of the operations $(\boldsymbol{\nabla}' \times \mathbf{z}_0)$ and $(\boldsymbol{\nabla}' \times \boldsymbol{\nabla}' \times \mathbf{z}_0)$ is defined in Eqs. (23a) and (23b), respectively. Equations (24) evidently are valid only when $k_{ti}' \neq 0$ (i.e., any possible TEM modes are excluded).† If the waveguide structure

---

†The interchange of operations of summation and differentiation, assumed valid in deriving Eqs. (24) from Eqs. (4), may not be permissible in certain problems involving continuous spectra of eigenfunctions (see Sec. 5.2b). [Similar remarks apply to Eqs. (25).] In these instances, the above expressions are to be considered as formal and must be properly interpreted [see Eq. (1.1.38) for related comments pertaining to the operator $1/\nabla_t^2$].

can support one or more TEM modes, the contribution to the radiated fields from these modes must be taken into account separately [see footnote to Eq. (2b)].

For the H-mode potential function $V''(\mathbf{r})$ in Eq. (4a) one obtains by analogous considerations the dual representation

$$V''(\mathbf{r}) = j\omega\mu(\nabla' \times \mathbf{z}_0)\mathscr{S}''(\mathbf{r}, \mathbf{r}') \cdot \mathbf{J}^0 + (\nabla' \times \nabla' \times \mathbf{z}_0)\mathscr{S}''(\mathbf{r}, \mathbf{r}') \cdot \mathbf{M}^0, \quad (25)$$

where

$$j\omega\mu\mathscr{S}''(\mathbf{r}, \mathbf{r}') = \sum_i \frac{\psi_i(\boldsymbol{\rho})\psi_i^*(\boldsymbol{\rho}')}{k_{ti}''^2} Z_i''(z, z'). \quad (25a)$$

ψ_i are the scalar H-mode functions defined in Eqs. (2).

Upon substituting the representations for $I'(\mathbf{r})$ and $V''(\mathbf{r})$ from Eqs. (24) and (25) into Eqs. (6), one obtains the desired formulation for the electromagnetic fields observed at $\mathbf{r}$ due to vector point-source excitations of electric and magnetic currents at $\mathbf{r}'$ as in Eq. (16) [see Eq. (3a)]:

$$\mathbf{E}(\mathbf{r}, \mathbf{r}') = -\mathscr{Z}(\mathbf{r}, \mathbf{r}') \cdot \mathbf{J}^0 - \mathscr{T}_e(\mathbf{r}, \mathbf{r}') \cdot \mathbf{M}^0, \quad (26a)$$

$$\mathbf{H}(\mathbf{r}, \mathbf{r}') = -\mathscr{T}_m(\mathbf{r}, \mathbf{r}') \cdot \mathbf{J}^0 - \mathscr{Y}(\mathbf{r}, \mathbf{r}') \cdot \mathbf{M}^0 \quad (26b)$$

where $\mathscr{Z}$, $\mathscr{Y}$ and $\mathscr{T}_e$, $\mathscr{T}_m$ are the dyadic impedance, admittance, and electric and magnetic transfer functions, respectively [see Eqs. (1.1.49) with $\mathbf{r} \neq \mathbf{r}'$][6]:

$$-j\omega\epsilon\mathscr{Z}(\mathbf{r}, \mathbf{r}') = (\nabla \times \nabla \times \mathbf{z}_0)(\nabla' \times \nabla' \times \mathbf{z}_0)\mathscr{S}'(\mathbf{r}, \mathbf{r}')$$
$$+ k^2(\nabla \times \mathbf{z}_0)(\nabla' \times \mathbf{z}_0)\mathscr{S}''(\mathbf{r}, \mathbf{r}') \quad (27a)$$

$$-j\omega\mu\mathscr{Y}(\mathbf{r}, \mathbf{r}') = (\nabla \times \nabla \times \mathbf{z}_0)(\nabla' \times \nabla' \times \mathbf{z}_0)\mathscr{S}''(\mathbf{r}, \mathbf{r}')$$
$$+ k^2(\nabla \times \mathbf{z}_0)(\nabla' \times \mathbf{z}_0)\mathscr{S}'(\mathbf{r}, \mathbf{r}'), \quad (27b)$$

$$\mathscr{T}_e(\mathbf{r}, \mathbf{r}') = (\nabla \times \nabla \times \mathbf{z}_0)(\nabla' \times \mathbf{z}_0)\mathscr{S}'(\mathbf{r}, \mathbf{r}')$$
$$+ (\nabla \times \mathbf{z}_0)(\nabla' \times \nabla' \times \mathbf{z}_0)\mathscr{S}''(\mathbf{r}, \mathbf{r}'), \quad (27c)$$

$$-\mathscr{T}_m(\mathbf{r}, \mathbf{r}') = (\nabla \times \nabla \times \mathbf{z}_0)(\nabla' \times \mathbf{z}_0)\mathscr{S}''(\mathbf{r}, \mathbf{r}')$$
$$+ (\nabla \times \mathbf{z}_0)(\nabla' \times \nabla' \times \mathbf{z}_0)\mathscr{S}'(\mathbf{r}, \mathbf{r}'), \quad (27d)$$

where $k^2 = \omega^2\mu\epsilon = $ constant. Via Eqs. (27), the dyadic Green's functions are expressed in terms of scalar functions $\mathscr{S}'$ and $\mathscr{S}''$ in what appears to be a fundamental form. The symmetry inherent in the expressions is to be noted. In Eqs. (32b) and (33b) the functions $-\nabla_t'^2\mathscr{S}'$ and $-\nabla_t'^2\mathscr{S}''$ are shown to be scalar Green's functions that satisfy Eqs. (36) and (37). Since from Eqs. (15), $Y_i'(z, z') = Y_i'(z', z)$ and $Z_i''(z, z') = Z_i''(z', z)$, it follows from the modal representations for $\mathscr{S}'$ and $\mathscr{S}''$ in Eqs. (24a) and (25a), respectively, that for *real* Φ_i and ψ_i,†

---

†Although not always convenient, the mode functions Φ_i and ψ_i in regions bounded either by perfectly conducting walls, or else unbounded, can always be chosen real. Only such regions, wherein k_{ti}^2 is real, are considered above.

$$\mathscr{S}'(\mathbf{r}, \mathbf{r}') = \mathscr{S}'(\mathbf{r}', \mathbf{r}), \qquad \mathscr{S}''(\mathbf{r}, \mathbf{r}') = \mathscr{S}''(\mathbf{r}', \mathbf{r}), \tag{28}$$

whence, from Eqs. (27),

$$\mathscr{L}(\mathbf{r}, \mathbf{r}') = \tilde{\mathscr{L}}(\mathbf{r}', \mathbf{r}), \quad \mathscr{Y}(\mathbf{r}, \mathbf{r}') = \tilde{\mathscr{Y}}(\mathbf{r}', \mathbf{r}), \quad \mathscr{T}_e(\mathbf{r}, \mathbf{r}') = -\tilde{\mathscr{T}}_m(\mathbf{r}', \mathbf{r}), \tag{29}$$

where the tilde (˜) denotes the transposed dyadics. We therefore verify the time-harmonic form of the reciprocity conditions derived previously in Eqs. (1.1.29), where for ease of identification we have changed the notation to

$$\mathscr{G}_{11} = \mathscr{L}, \quad \mathscr{G}_{12} = \mathscr{T}_e, \quad \mathscr{G}_{21} = \mathscr{T}_m, \quad \mathscr{G}_{22} = \mathscr{Y}. \tag{30}$$

To include also the point $\mathbf{r} = \mathbf{r}'$, Eqs. (27) must be modified as in Eq. (1.1.38) or (1.1.49).

Equations (24) and (25) simplify considerably for the case of longitudinal sources,

$$\mathbf{J}^0 = \mathbf{z}_0 J^0, \qquad \mathbf{M}^0 = \mathbf{z}_0 M^0. \tag{31}$$

From Eq. (23a) one notes that $(\nabla' \times \mathbf{z}_0)\varphi \cdot \mathbf{z}_0 = 0$, while from Eq. (23b), $(\nabla' \times \nabla' \times \mathbf{z}_0)\varphi \cdot \mathbf{z}_0 = -\nabla_t'^2 \varphi$. One may write

$$I'(\mathbf{r}) = J^0 G'(\mathbf{r}, \mathbf{r}'), \tag{32a}$$

where, in view of $\nabla_t'^2 \Phi_i^*(\boldsymbol{\rho}') = -k_{ti}'^2 \Phi_i^*(\boldsymbol{\rho}')$ or $\nabla_t^2 \Phi_i(\boldsymbol{\rho}) = -k_{ti}'^2 \Phi_i(\boldsymbol{\rho})$,

$$G'(\mathbf{r}, \mathbf{r}') \equiv -\nabla_t'^2 \mathscr{S}'(\mathbf{r}, \mathbf{r}') = -\nabla_t^2 \mathscr{S}'(\mathbf{r}, \mathbf{r}')$$

$$= \frac{1}{j\omega\epsilon} \sum_i \Phi_i(\boldsymbol{\rho})\Phi_i^*(\boldsymbol{\rho}')Y_i'(z, z'). \tag{32b}$$

Similarly, one writes

$$V''(\mathbf{r}) = M^0 G''(\mathbf{r}, \mathbf{r}'), \tag{33a}$$

with

$$G''(\mathbf{r}, \mathbf{r}') \equiv -\nabla_t'^2 \mathscr{S}''(\mathbf{r}, \mathbf{r}') = -\nabla_t^2 \mathscr{S}''(\mathbf{r}, \mathbf{r}')$$

$$= \frac{1}{j\omega\mu} \sum_i \psi_i(\boldsymbol{\rho})\psi_i^*(\boldsymbol{\rho}')Z_i''(z, z'). \tag{33b}$$

One notes from Eqs. (32) and (33) that a longitudinal electric current source excites only E modes along z while a longitudinal magnetic current source excites only H modes. The fields are now determined by the following simplified form of Eqs. (26):

$$\mathbf{E}(\mathbf{r}, \mathbf{r}') = \frac{J^0}{j\omega\epsilon}(\nabla \times \nabla \times \mathbf{z}_0)G'(\mathbf{r}, \mathbf{r}') - M^0(\nabla \times \mathbf{z}_0)G''(\mathbf{r}, \mathbf{r}'), \tag{34a}$$

$$\mathbf{H}(\mathbf{r}, \mathbf{r}') = J^0(\nabla \times \mathbf{z}_0)G'(\mathbf{r}, \mathbf{r}') + \frac{M^0}{j\omega\mu}(\nabla \times \nabla \times \mathbf{z}_0)G''(\mathbf{r}, \mathbf{r}'). \tag{34b}$$

We show now that G' and G'' are scalar Green's functions satisfying, subject to appropriate boundary conditions, the scalar wave equation with an inhomogeneous term $-\delta(\mathbf{r} - \mathbf{r}')$. Let the operator $(\nabla^2 + k^2)$ act on G' as represented in Eq. (32b) and assume that the operations of summation and differentiation can be interchanged. Then, since $\nabla_t^2 \Phi_i = -k_{ti}'^2 \Phi_i$ and $\kappa_i'^2 = k^2 - k_{ti}'^2$,

$$\left(\nabla_t^2 + \frac{\partial^2}{\partial z^2} + k^2\right)G'(\mathbf{r}, \mathbf{r}') = \frac{1}{j\omega\epsilon}\sum_i \Phi_i(\boldsymbol{\rho})\Phi_i^*(\boldsymbol{\rho}')\left(\frac{d^2}{dz^2} + \kappa_i'^2\right)Y_i'(z, z') \tag{35a}$$

$$= -\delta(z - z')\sum_i \Phi_i(\boldsymbol{\rho})\Phi_i^*(\boldsymbol{\rho}')$$

$$= -\delta(z - z')\delta(\boldsymbol{\rho} - \boldsymbol{\rho}') = -\delta(\mathbf{r} - \mathbf{r}'). \tag{35b}$$

The transition from Eq. (35a) to Eq. (35b) follows via the differential equation for $Y_i'(z, z')$ obtained on elimination of $T_i^V(z, z')$ from Eqs. (10a) and (10b), while the identification of the mode function series as $\delta(\boldsymbol{\rho} - \boldsymbol{\rho}')$ is discussed in Chapter 3 [see, for example, Eq. (3.2.17a)]. Thus, the E-mode function G' is a scalar three-dimensional Green's function which satisfies the inhomogeneous wave equation

$$(\nabla^2 + k^2)G'(\mathbf{r}, \mathbf{r}') = -\delta(\mathbf{r} - \mathbf{r}') \tag{36}$$

subject on the perfectly conducting waveguide boundary s, to the same boundary condition as $\Phi_i(\boldsymbol{\rho})$ [see Eq. (2b)],

$$G'(\mathbf{r}, \mathbf{r}') = 0, \qquad \mathbf{r} \text{ on } s. \tag{36a}$$

The boundary conditions on G' in the z domain will depend on stratification along the z coordinate. For example, across a dielectric interface at $z = z_1$, the transverse electric and magnetic fields are continuous, so the voltage and current in each mode are continuous [see Eqs. (2.2.8a) and (2.2.8b)]. Since $Y_i'(z, z')$ represents a current, continuity of $Y_i'(z, z')$ across z_1 implies from Eq. (32b) that $G'(\mathbf{r}, \mathbf{r}')$ is likewise continuous across z_1. From the transmission-line equations, the mode voltage is proportional to $(1/\kappa_i' Y_i')(d/dz)Y_i'(z, z')$, and since $\kappa_i' Y_i' = \omega\epsilon$, continuity of voltage implies via Eq. (32b) that $(1/\epsilon)(\partial/\partial z)G'(\mathbf{r}, \mathbf{r}')$ must likewise be continuous at z_1.† Thus, we find that G' and $(1/\epsilon)(\partial G'/\partial z)$ are required to be continuous across a dielectric interface. Similarly, if the region is terminated at z_1 in a perfectly conducting plane on which the transverse electric field vanishes, each modal voltage vanishes and requires that $\partial G'/\partial z = 0$ at z_1, while for an unterminated z domain, a "radiation condition" requiring an outward flow of energy is appropriate. The modal representation for G' in Eq. (32b) thus constitutes the solution of the Green's function problem posed in Eq. (36) subject to the above-discussed boundary conditions.

By analogous considerations, one shows that the H-mode Green's function G'' in Eq. (33b) satisfies the inhomogeneous wave equation

$$(\nabla^2 + k^2)G''(\mathbf{r}, \mathbf{r}') = -\delta(\mathbf{r} - \mathbf{r}'), \tag{37}$$

subject on the perfectly conducting waveguide boundary s to the same condition as $\psi_i(\boldsymbol{\rho})$ [see Eq. (2d)],

---

†ϵ and μ in Eqs. (24a), (25a), (32b), and (33b) have constant values appropriate to the medium containing the source point z'; in Eqs. (24), (25), (27), and (34), ϵ and μ have constant values appropriate to the medium containing the observation point [see also Eqs. (38), (40), and (42)]. These remarks are relevant for analysis of media with piecewise constant ϵ and μ.

$$\frac{\partial G''}{\partial \nu} = 0 \qquad \text{on } s. \tag{37a}$$

The boundary conditions satisfied by G'' in the z domain are dual to those on G'. At an interface plane $z = z_1$, G'' and $(1/\mu)(\partial G''/\partial z)$ must be continuous, while at a perfectly conducting plane, $G'' = 0$.† It follows from Eqs. (32b), (33b), (36), and (37) that the scalar functions $\mathscr{S}'(\mathbf{r}, \mathbf{r}')$ and $\mathscr{S}''(\mathbf{r}, \mathbf{r}')$ satisfy the time-harmonic form of the differential equation (1.1.38b), subject to boundary conditions identical with those on G' and G'', respectively. The recovery of $\mathscr{S}'$ and $\mathscr{S}''$ from G' and G'', respectively, requires the inversion of Eqs. (32b) and (33b). For $k_{ti}^2 \neq 0$, this inversion is accomplished readily in a basis wherein $-\nabla_t^2 \rightarrow k_{ti}'^2$ or $k_{ti}''^2$, and leads directly to the representations in Eqs. (24a) and (25a).

2.3e Modal Representations of the Dyadic Green's Functions in an Inhomogeneous Medium

The formulas derived in Sec. 2.3d apply to homogeneous media and must be modified if ϵ and μ are functions of z. In this instance, the results of Secs. 2.2, 2.3a, 2.3b, and 2.3c remain valid with the exception of Eqs. (6), which should be written at a source-free point as

$$\mathbf{E}(\mathbf{r}) = \frac{1}{j\omega\epsilon(z)}\,(\nabla \times \nabla \times \mathbf{z}_0)I'(\mathbf{r}) - (\nabla \times \mathbf{z}_0)V''(\mathbf{r}), \tag{38a}$$

$$\mathbf{H}(\mathbf{r}) = \frac{1}{j\omega\mu(z)}\,(\nabla \times \nabla \times \mathbf{z}_0)V''(\mathbf{r}) + (\nabla \times \mathbf{z}_0)I'(\mathbf{r}), \tag{38b}$$

with $I'(\mathbf{r})$ and $V''(\mathbf{r})$ defined in Eqs. (4). As regards the results in Sec. 2.3d, one notes from the method of derivation that Eqs. (19)–(23) still apply provided that ϵ and μ are replaced by $\epsilon(z')$ and $\mu((z')$, respectively. It then follows that Eq. (24) should be written as

$$I'(\mathbf{r}) = -L_1'\mathscr{S}_d' \cdot \mathbf{M}^0 + \frac{1}{j\omega\epsilon(z')}\,L_2'\mathscr{S}_d' \cdot \mathbf{J}^0, \tag{39}$$

where the vector operators L_1' and L_2' are defined as

$$L_1' \equiv \nabla' \times \mathbf{z}_0, \qquad L_2' \equiv \nabla' \times \nabla' \times \mathbf{z}_0, \tag{39a}$$

and

$$\mathscr{S}_d' = \sum_i \frac{\Phi_i(\boldsymbol{\rho})\Phi_i^*(\boldsymbol{\rho}')}{k_{ti}'^2}\,Y_i'(z, z'). \tag{39b}$$

Dual considerations apply to Eq. (25).

With the above modifications, the dyadic Green's functions in Eqs. (27) are now written in the following form:

$$\mathscr{L}(\mathbf{r}, \mathbf{r}') = \frac{1}{\omega^2\epsilon(z)\epsilon(z')}\,L_2 L_2'\mathscr{S}_d' + L_1 L_1'\mathscr{S}_d'', \tag{40a}$$

---

†See the preceding footnote.

$$\mathscr{Y}(\mathbf{r}, \mathbf{r}') = \frac{1}{\omega^2 \mu(z)\mu(z')} L_2 L_2' \mathscr{S}_d'' + L_1 L_1' \mathscr{S}_d', \tag{40b}$$

$$\mathscr{T}_e(\mathbf{r}, \mathbf{r}') = \frac{1}{j\omega\epsilon(z)} L_2 L_1' \mathscr{S}_d' + \frac{1}{j\omega\mu(z')} L_1 L_2' \mathscr{S}_d'', \tag{40c}$$

$$-\mathscr{T}_m(\mathbf{r}, \mathbf{r}') = \frac{1}{j\omega\mu(z)} L_2 L_1' \mathscr{S}_d'' + \frac{1}{j\omega\epsilon(z')} L_1 L_2' \mathscr{S}_d', \tag{40d}$$

where

$$L_1 \equiv \nabla \times \mathbf{z}_0, \quad L_2 \equiv \nabla \times \nabla \times \mathbf{z}_0, \quad \mathscr{S}_d'' = \sum_i \frac{\psi_i(\boldsymbol{\rho})\psi_i^*(\boldsymbol{\rho}')}{k_{ti}''^2} Z_i''(z, z').$$

$$\tag{40e}$$

It is readily verified that these more general expressions satisfy, as they must, the reciprocity relations (29).

The modal Green's functions $Y_i'(z, z')$ and $Z_i''(z, z')$ are defined in Eqs. (10). Because $\kappa(z) = [\omega^2 \mu(z)\epsilon(z) - k_{ti}^2]^{1/2}$ is now variable, the characteristic impedances $Z_i(z)$ and admittances $Y_i(z)$ are also functions of z, so the associated transmission lines are non-uniform.† On elimination of T_i^V and T_i^I from Eqs. (10a), (10b) and (10c), (10d), respectively, one finds that the modal Green's functions satisfy the following second-order differential equations [note from Eqs. (2.2.15) that $\kappa_i'(z)Y_i'(z) = \omega\epsilon(z)$, $\kappa_i''(z)Z_i''(z) = \omega\mu(z)$]:

$$[D_\epsilon^2(z) + \kappa_i'^2(z)]Y_i'(z, z') = -j\omega\epsilon(z')\delta(z - z'), \tag{41a}$$

$$[D_\mu^2(z) + \kappa_i''^2(z)]Z_i''(z, z') = -j\omega\mu(z')\delta(z - z'), \tag{41b}$$

where

$$D_\alpha^2(z) = \alpha(z) \frac{d}{dz} \frac{1}{\alpha(z)} \frac{d}{dz}, \qquad \alpha = \epsilon \text{ or } \mu. \tag{41c}$$

The boundary conditions at the endpoints of the transmission line are phrased as in Eq. (13). Note that the E-mode terminal impedance is given via Eqs. (10a) and (10b) by $[(d/dz)Y_i'(z, z')/-j\kappa_i' Y_i' Y_i'(z, z')]_{z_1, z_2}$; the spatially varying characteristic impedance here should not be confused with the terminal impedance in Sec. 2.3c. At a junction between two transmission lines with parameters $\kappa_{i1}(z)$, $Z_{i1}(z)$ and $\kappa_{i2}(z)$, $Z_{i2}(z)$, respectively, the voltage and current are continuous. Thus, from Eqs. (10), $Y_i'(z, z')$, $[1/\epsilon(z)](d/dz)Y_i'(z, z')$, and $Z_i''(z, z')$, $[1/\mu(z)]$ $(d/dz)Z_i''(z, z')$ are continuous across the junction point. (See also Secs. 3.3a and 3.3b for a solution of the non-uniform transmission-line equations.)

If the sources are longitudinal, Eqs. (40) simplify and lead to expressions analogous to those in Sec. 2.3d. In fact, one obtains expressions similar to Eqs. (34):

---

†Although the waveguide region is geometrically uniform in that successive geometrical cross sections transverse to z are identical, an electrical non-uniformity is introduced by the longitudinal variability of the medium constants. Consequently, the network representation discussed in Sec. 2.4 involves non-uniform transmission lines representative of the z behavior of a typical mode (see also Secs. 3.2d and 3.3b).

$$\mathbf{E}(\mathbf{r}, \mathbf{r}') = \frac{J^0}{j\omega\epsilon(z)} L_2 G'(\mathbf{r}, \mathbf{r}') - M^0 L_1 G''(\mathbf{r}, \mathbf{r}'),\tag{42a}$$

$$\mathbf{H}(\mathbf{r}, \mathbf{r}') = J^0 L_1 G'(\mathbf{r}, \mathbf{r}') + \frac{M^0}{j\omega\mu(z)} L_2 G''(\mathbf{r}, \mathbf{r}'),\tag{42b}$$

where

$$G'(\mathbf{r}, \mathbf{r}') = \frac{1}{j\omega\epsilon(z')} \sum_i Y_i'(z, z')\Phi_i(\boldsymbol{\rho})\Phi_i^*(\boldsymbol{\rho}') = -\frac{1}{j\omega\epsilon(z')} \nabla_t'^2 \mathscr{S}_d',\tag{43a}$$

$$G''(\mathbf{r}, \mathbf{r}') = \frac{1}{j\omega\mu(z')} \sum_i Z_i''(z, z')\psi_i(\boldsymbol{\rho})\psi_i^*(\boldsymbol{\rho}') = -\frac{1}{j\omega\mu(z')} \nabla_t'^2 \mathscr{S}_d''.\tag{43b}$$

The differential equations for the scalar Green's functions G' and G'' are now in view of Eqs. (41):

$$[\mathscr{D}_\epsilon^2(z) + \nabla_t^2 + k^2(z)]G'(\mathbf{r}, \mathbf{r}') = -\delta(\mathbf{r} - \mathbf{r}'), \quad k^2(z) = \omega^2\mu(z)\epsilon(z),\tag{44a}$$

$$[\mathscr{D}_\mu^2(z) + \nabla_t^2 + k^2(z)]G''(\mathbf{r}, \mathbf{r}') = -\delta(\mathbf{r} - \mathbf{r}'),\tag{44b}$$

where

$$\mathscr{D}_\alpha^2(z) = \alpha(z)\frac{\partial}{\partial z}\frac{1}{\alpha(z)}\frac{\partial}{\partial z}.\tag{44c}$$

It may also be verified that the Green's function $G'(\mathbf{r}, \mathbf{r}')/\sqrt{\epsilon(z)}$ satisfies the wave equation with the modified wavenumber $\bar{k}(z)$:

$$[\nabla^2 + \bar{k}^2(z)]\frac{G'(\mathbf{r}, \mathbf{r}')}{\sqrt{\epsilon(z)}} = -\frac{\delta(\mathbf{r} - \mathbf{r}')}{\sqrt{\epsilon(z')}}, \qquad \bar{k}^2(z) = k^2(z) - \sqrt{\epsilon(z)}\frac{d^2}{dz^2}\frac{1}{\sqrt{\epsilon(z)}},$$

$$\tag{45}$$

with a dual relation applicable to $G''(\mathbf{r}, \mathbf{r}')/\sqrt{\mu(z)}$. Corresponding equations for $\mathscr{S}_d'$ and $\mathscr{S}_d''$ follow on use of Eqs. (43). The conditions satisfied by G' and G'' on the transverse and longitudinal boundaries of the region are the same as those deduced in connection with Eqs. (36) and (37). These boundary conditions, in conjunction with Eqs. (44), render the specification of G' and G'' unique. The modal representations in Eqs. (43) constitute solutions for G' and G'' and are directly deducible from a z-transmission analysis. Alternative representations of the solution for G' and G'' can be constructed by the procedure discussed in Sec. 3.3c.

All the above relations reduce to those in Sec. 2.3d when ϵ and μ are constant.

2.4 SOLUTION OF UNIFORM TRANSMISSION-LINE EQUATIONS (NETWORK ANALYSIS)

2.4a Source-Free Case

As shown in Secs. 2.2 and 2.3, the solution of an electromagnetic field problem in a uniform waveguide region requires knowledge of the (vector or scalar) eigenfunctions in the transverse domain, and of the modal voltage and current amplitudes along the transmission coordinate z. Mode functions for various transverse cross-sectional configurations are given in Chapter 3. This section is concerned with solutions of the transmission-line equations (2.2.15) or (2.3.10) when the medium filling the waveguide region is piecewise constant along z.[2,3] Continuous z variation leads to Eqs. (2.3.41) which are studied in detail in Sec. 3.3.

As a preliminary, we shall consider a region that contains no sources. Thus, $v = i = 0$ in Eq. (2.2.15) and omitting the modal subscript i, one obtains the *homogeneous* (*source-free*) transmission-line equations

$$-\frac{d}{dz} V(z) = j\kappa Z I(z),$$

$$-\frac{d}{dz} I(z) = j\kappa Y V(z). \tag{1}$$

Since κ and Z are assumed to be constant, the two first-order differential equations (1) are reduced to the following second-order equation for either V or I:

$$\left(\frac{d^2}{dz^2} + \kappa^2\right)\frac{V(z)}{I(z)} = 0, \qquad \kappa, Z \text{ constant.} \tag{2}$$

The solution of Eqs. (2) may be expressed either in traveling-wave or standing-wave form. For a *traveling-wave* description one writes

$$V(z) = V_{\text{inc}}(z')e^{-j\kappa(z-z')} + V_{\text{refl}}(z')e^{+j\kappa(z-z')}, \tag{3a}$$

$$ZI(z) = V_{\text{inc}}(z')e^{-j\kappa(z-z')} - V_{\text{refl}}(z')e^{+j\kappa(z-z')}, \tag{3b}$$

$V_{\text{inc}}(z')$ and $V_{\text{refl}}(z')$ are the (generally complex) amplitudes of the incident and reflected voltage waves at a point z' on the transmission line. The first term on the right-hand side of Eq. (3a) represents a wave traveling in the $+z$ direction [for the assumed time dependence $\exp(+j\omega t)$]; the second term represents a wave traveling in the $-z$ direction. Equation (3b) follows from Eq. (3a) in view of Eqs. (1). If one defines the voltage reflection coefficient $\Gamma(z)$ as

$$\Gamma(z) = \frac{V_{\text{refl}}(z)}{V_{\text{inc}}(z)}, \tag{4}$$

one may rewrite Eqs. (3) as

$$V(z) = V_{\text{inc}}(z')[e^{-j\kappa(z-z')} + \Gamma(z')e^{+j\kappa(z-z')}], \tag{5a}$$

$$I(z) = YV_{\text{inc}}(z')[e^{-j\kappa(z-z')} - \Gamma(z')e^{+j\kappa(z-z')}]. \tag{5b}$$

It is evident from Eqs. (3a) and (4) that the reflection coefficients at two points z and z' on the transmission line are related as follows:

$$\Gamma(z) = \Gamma(z')e^{+j2\kappa(z-z')}. \tag{6}$$

In the *standing-wave* description one expresses the voltage and current solutions as

$$V(z) = V(z')\cos\kappa(z - z') - jZI(z')\sin\kappa(z - z'), \tag{7a}$$

$$I(z) = I(z')\cos\kappa(z - z') - jYV(z')\sin\kappa(z - z'). \tag{7b}$$

The duality between the solutions in Eqs. (7a) and (7b) is to be noted. By defining the absolute impedance $Z(z)$ [admittance $Y(z)$] and the relative (normalized) impedance $Z'(z)$ [admittance $Y'(z)$] at the point z as†

$$Z(z) = \frac{V(z)}{I(z)} = \frac{1}{Y(z)}, \qquad Z'(z) = \frac{V(z)}{ZI(z)} = \frac{1}{Y'(z)}, \tag{8}$$

one may write Eqs. (7) as

$$V(z) = V(z') \left[\cos \kappa(z - z') - jY'(z') \sin \kappa(z - z')\right], \tag{9a}$$

$$I(z) = I(z') \left[\cos \kappa(z - z') - jZ'(z') \sin \kappa(z - z')\right]. \tag{9b}$$

Upon taking the ratio of Eqs. (9a) and (9b), one finds for the relation between impedances at two points on a transmission line,

$$Z'(z) = \frac{1 + jZ'(z') \cot \kappa(z - z')}{Z'(z') + j \cot \kappa(z - z')}. \tag{10}$$

From Eqs. (3) the relation between the standing-wave and traveling-wave formulations is evidently

$$V(z) = V_{\text{inc}}(z) + V_{\text{refl}}(z),$$
$$ZI(z) = V_{\text{inc}}(z) - V_{\text{refl}}(z), \tag{11a}$$

and

$$V_{\text{inc}}(z) = \tfrac{1}{2} \left[V(z) + ZI(z)\right],$$
$$V_{\text{refl}}(z) = \tfrac{1}{2} \left[V(z) - ZI(z)\right]. \tag{11b}$$

The relation between the impedance and reflection coefficient at z then follows readily as

$$Z'(z) = \frac{1 + \Gamma(z)}{1 - \Gamma(z)} = \frac{1}{Y'(z)}, \qquad \Gamma(z) = \frac{Z'(z) - 1}{Z'(z) + 1}. \tag{12}$$

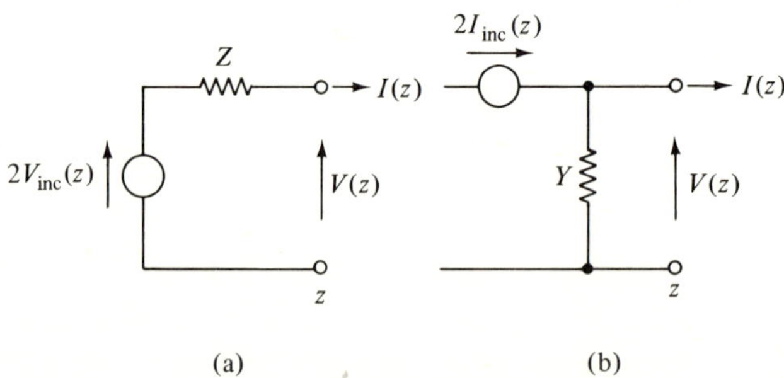

FIG. 2.4.1   Network representation of an incident wave.

---

†The prime in this section denotes normalized quantities and is not to be confused with the prime denoting E-mode quantities in Secs. 2.2 and 2.3. Also $Z(z)$ and $Y(z)$ denote herein the input impedance and admittance at the point z, while Z and Y represent the constant characteristic impedance and admittance, respectively.

The first of Eqs. (11b) permits the network representation of an incident wave at the point z in terms of either a (zero internal impedance) voltage generator of strength $2V_{\text{inc}}(z)$ or a (infinite internal impedance) current generator of strength $2I_{\text{inc}}(z) = 2YV_{\text{inc}}(z)$ shown in Figs. 2.4.1(a) and 2.4.1(b), respectively. Positive directions for voltage and current are assumed as indicated.

2.4b   Point Source on an Infinite Transmission Line

In Sec. 2.4a we have obtained the solution of the homogeneous transmission-line equations with as-yet-unspecified boundary conditions. To obtain a solution of the inhomogeneous equations (2.2.15) requires the ability to describe source regions on the transmission line. As pointed out in Sec. 2.3c, the description of source regions is simplified by first considering point-source excitations and then obtaining the total response by use of the superposition theorem. The representation of the distributed voltage and current sources $v(z)$ and $i(z)$ in terms of Point sources is accomplished as follows:

$$v(z) = \int v(z')\delta(z - z')\,dz', \qquad i(z) = \int i(z')\delta(z - z')\,dz', \tag{13}$$

where the integration over z' extends over the source region. If the voltage and the current at z due to point sources $v(z')\delta(z - z')$ and $i(z')\delta(z - z')$ at z' are denoted by $V(z, z')$ and $I(z, z')$, respectively, one obtains the total voltage and current by superposition as[†]

$$V(z) = \int V(z, z')\,dz', \qquad I(z) = \int I(z, z')\,dz'. \tag{14}$$

In view of Eqs. (14), the basic problem is that of finding the point-source responses $V(z, z')$ and $I(z, z')$. The modal voltage and current sources $v(z)$ and $i(z)$ are known in terms of the *specified* electromagnetic source currents [see Eqs. (2.2.14)]. The problem can be reduced further by introduction of modal Green's functions as in Eqs. (2.3.9), but this aspect is deferred for the present.

The transmission-line equations appropriate to the indicated point-source excitations are

$$-\frac{d}{dz}V(z, z') = j\kappa Z I(z, z') + v\delta(z - z'),$$

$$-\frac{d}{dz}I(z, z') = j\kappa Y V(z, z') + i\delta(z - z'), \tag{15}$$

where the notation $v(z') \equiv v, i(z') \equiv i$ has been adopted. By integrating Eqs. (15) over z between the limits $z' - \alpha$ and $z' + \alpha$, where $\alpha \to 0$, one obtains the following discontinuity relations for $V(z, z')$ and $I(z, z')$ at $z = z'$,

$$[V(z, z')]_{z'-\alpha}^{z'+\alpha} = -v, \qquad [I(z, z')]_{z'-\alpha}^{z'+\alpha} = -i. \tag{16}$$

[†]Note that the functions $V(z, z')$ and $I(z, z')$ are related to the modal Green's functions defined in Eqs. (2.3.9); $V(z, z') = -T^V(z, z')v(z') - Z(z, z')i(z')$ and $I(z, z') = -Y(z, z')v(z') - T^I(z, z')i(z')$ [see also Eq. (2.3.17)].

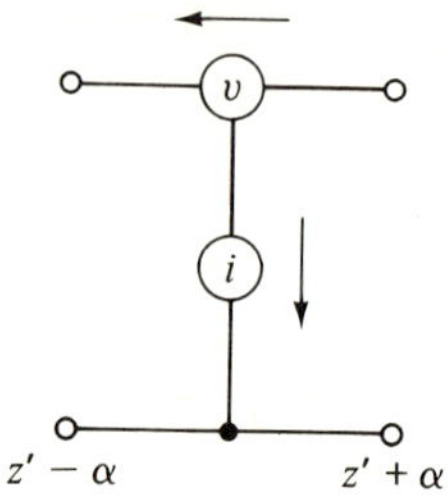

FIG. 2.4.2. Network representation for point- source excitation

It is assumed in Eq. (16), subject to subsequent verification, that $V(z, z')$ and $I(z, z')$, although discontinuous, are bounded at z'. The discontinuity relations (16) admit the simple network representation shown in Fig. 2.4.2, where the point-source voltage and current generators have zero and infinite internal impedance, respectively. In a strict network sense the voltage generator in Fig. 2.4.2 is to be interpreted as two generators of strength $v/2$, located on each side of the current generator.

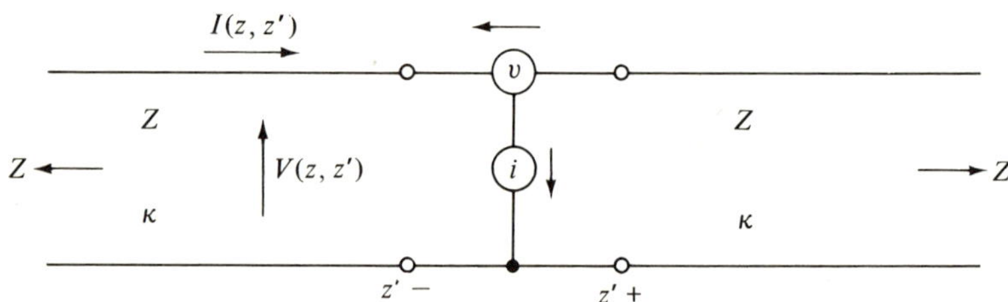

FIG. 2.4.3 Point source in an infinite transmission line.

To illustrate the use of the network in Fig. 2.4.2, we consider a simple example of a point source situated in an infinite transmission line (Fig. 2.4.3). (The general problem of arbitrary terminations is treated in Sec. 2.4c.) Since the line is infinitely long with no reflections, waves travel *away* from the source and boundary conditions at $z = \pm\infty$ are provided thereby. By inspection, the solution is written as

$$V(z, z') = \begin{cases} V(z'+, z')e^{-j\kappa(z-z')}, & z > z', \qquad \text{(17a)} \\ V(z'-, z')e^{+j\kappa(z-z')}, & z < z', \qquad \text{(17b)} \end{cases}$$

and, similarly for $I(z, z')$,

$$I(z, z') = \begin{cases} I(z'+, z')e^{-j\kappa(z-z')}, & z > z', \qquad \text{(17c)} \\ I(z'-, z')e^{+j\kappa(z-z')}, & z < z'. \qquad \text{(17d)} \end{cases}$$

To determine the amplitude of the voltage and current at $z'-$ and $z'+$, one notes that the impedance looking away from the source in either direction is equal to Z. Thus, the voltage generator v operates into a series impedance $2Z$, while the current generator i operates into a shunt admittance $2Y$. By superposing the responses from v and i one has

$$V(z'\pm, z') = -\tfrac{1}{2}(\pm v + Zi), \qquad I(z'\pm, z') = -\tfrac{1}{2}(\pm i + Yv), \qquad (18)$$

so that anywhere on the transmission line

$$V(z, z') = -\tfrac{1}{2}[u(z, z')v + Zi]e^{-j\kappa|z-z'|}, \qquad (19a)$$

$$I(z, z') = -\tfrac{1}{2}[u(z, z')i + Yv]e^{-j\kappa|z-z'|}, \qquad (19b)$$

where

$$u(z, z') = \begin{cases} 1, & z > z', \\ -1, & z < z'. \end{cases} \qquad (19c)$$

2.4c Excitation of General Transmission-Line Network by a Point Source

We consider now the general problem of a point source in a transmission-line system terminated by arbitrary, but prescribed, impedances. The terminating impedances represent the boundary conditions on the z termini of the transmission line and are determined directly from the corresponding boundary conditions on the electromagnetic fields. As a typical example we consider the network problem shown in Fig. 2.4.4, where $\overset{\leftarrow}{Z}_T$ and $\vec{Z}_T$ are the terminating

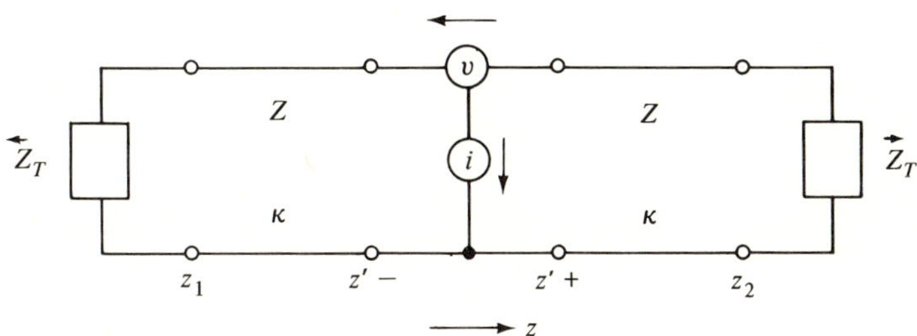

FIG. 2.4.4 Network representation for general point-source-excited transmission line.

impedances at the end points z_1 and z_2, respectively, and Z and κ are the (constant) characteristic impedance and propagation constant of the line. To the left and right, respectively, of the terminating points z_1 and z_2, the transmission-line system may be relatively arbitrary and is described by the impedances $\overset{\leftarrow}{Z}_T$ and $\vec{Z}_T$, which suffice to characterize the effect of the terminating structure on the transmission-line behavior in the interval $z_1 < z < z_2$. In passing from the transmission line with constants Z, κ to a line with constants Z_1, κ_1 connected at, say, z_2, one makes use of the relations in Sec. 2.4a, together with the continuity of the voltage and current across a direct junction between transmission lines. The latter property follows directly from the continuity of the transverse electric and magnetic fields across a (source-free) dielectric interface [see Eqs. (2.2.8a) and (2.2.8b) and recall the orthogonality of the vector-mode functions and the fact that the mode functions are independent of the parameters $\epsilon_{1,2}$ and $\mu_{1,2}$ descriptive of media 1 and 2 separated by an interface $z = $ constant].

The solution of the above network problem is accomplished by first finding the voltages and currents at the generator terminals, and determining therefrom the voltages and currents anywhere in the interval $z_1 < z < z_2$ by means of the homogeneous transmission-line solutions Eqs. (5) or (9). The generator circuit shown in Fig. 2.4.5 is representative of a general problem wherein $\overrightarrow{Z}(z')$ and

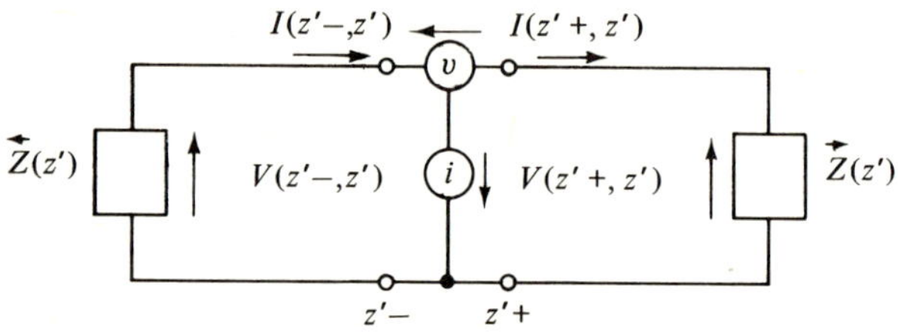

FIG. 2.4.5   Generator circuit for general transmission-line problem.

$\overleftarrow{Z}(z')$ are the impedances seen looking to the right and to the left, respectively, from the generator terminals. One computes $\overrightarrow{Z}(z')$ and $\overleftarrow{Z}(z')$ in terms of the given terminations $\overleftrightarrow{Z}_T$ of Fig. 2.4.4 by means of Eq. (10).† By conventional network analysis employing the superposition of the responses from v and i, the solution of the network problem in Fig. 2.4.5 is seen to be the following:

$$V(z'+,z') = \frac{-v\overrightarrow{Z}(z')}{\overleftrightarrow{Z}(z')} - \frac{i}{\overleftrightarrow{Y}(z')}, \tag{20a}$$

$$V(z'-,z') = \frac{v\overleftarrow{Z}(z')}{\overleftrightarrow{Z}(z')} - \frac{i}{\overleftrightarrow{Y}(z')}, \tag{20b}$$

$$I(z'+,z') = \frac{-v}{\overleftrightarrow{Z}(z')} - \frac{i\overrightarrow{Y}(z')}{\overleftrightarrow{Y}(z')}, \tag{20c}$$

$$I(z'-,z') = \frac{-v}{\overleftrightarrow{Z}(z')} + \frac{i\overleftarrow{Y}(z')}{\overleftrightarrow{Y}(z')}, \tag{20d}$$

where

$$\overleftrightarrow{Z}(z') = \overrightarrow{Z}(z') + \overleftarrow{Z}(z'), \qquad \overleftrightarrow{Y}(z') = \overrightarrow{Y}(z') + \overleftarrow{Y}(z'). \tag{20e}$$

The determination of V and I at any point z on a transmission line with constants κ and Z follows from corresponding values at z' via either Eqs. (5) or (9). Thus, from Eqs. (5), (11b), and (20), one obtains in the traveling-wave

---

†In view of the choice of positive directions for voltage and current noted in Fig. 2.4.1, $Z' \equiv \overrightarrow{Z}'$ in Eq. (10) represents a normalized impedance looking to the right. To apply Eq. (10) to an impedance $\overleftarrow{Z}'$ looking to the left, a minus sign must be placed before every Z' (i.e., $Z' \rightarrow -\overleftarrow{Z}'$).

description for $z > z'$,

$$V(z, z') = -\frac{1}{2}\left[v\,\frac{\vec{Z}(z') + Z}{\overleftrightarrow{Z}(z')} + Zi\,\frac{\vec{Y}(z') + Y}{\overleftrightarrow{Y}(z')}\right][e^{-j\kappa(z-z')} + \vec{\Gamma}(z')e^{+j\kappa(z-z')}],$$

$$(21\text{a})$$

$$I(z, z') = -\frac{1}{2}\left[i\,\frac{\vec{Y}(z') + Y}{\overleftrightarrow{Y}(z')} + Yv\,\frac{\vec{Z}(z') + Z}{\overleftrightarrow{Z}(z')}\right][e^{-j\kappa(z-z')} - \vec{\Gamma}(z')e^{+j\kappa(z-z')}],$$

$$(21\text{b})$$

while for $z < z'$,†

$$V(z, z') = -\frac{1}{2}\left[-v\,\frac{\overleftarrow{Z}(z') + Z}{\overleftrightarrow{Z}(z')} + Zi\,\frac{\overleftarrow{Y}(z') + Y}{\overleftrightarrow{Y}(z')}\right][e^{+j\kappa(z-z')} + \overleftarrow{\Gamma}(z')e^{-j\kappa(z-z')}],$$

$$(21\text{c})$$

$$I(z, z') = -\frac{1}{2}\left[-i\,\frac{\overleftarrow{Y}(z') + Y}{\overleftrightarrow{Y}(z')} + Yv\,\frac{\overleftarrow{Z}(z') + Z}{\overleftrightarrow{Z}(z')}\right][e^{+j\kappa(z-z')} - \overleftarrow{\Gamma}(z')e^{-j\kappa(z-z')}].$$

$$(21\text{d})$$

The reflection coefficients $\vec{\Gamma}$ and $\overleftarrow{\Gamma}$ looking to the right and left, respectively, are obtained in terms of the corresponding impedances $\vec{Z}'$ and $\overleftarrow{Z}'$ as in Eq. (12).

In the standing-wave description, one has, from Eqs. (9) and (20) for $z > z'$,

$$V(z, z') = -\left[\frac{v\vec{Z}(z')}{\overleftrightarrow{Z}(z')} + \frac{i}{\overleftrightarrow{Y}(z')}\right][\cos \kappa(z - z') - j\vec{Y}'(z') \sin \kappa(z - z')],$$

$$(22\text{a})$$

$$I(z, z') = -\left[\frac{i\vec{Y}(z')}{\overleftrightarrow{Y}(z')} + \frac{v}{\overleftrightarrow{Z}(z')}\right][\cos \kappa(z - z') - j\vec{Z}'(z') \sin \kappa(z - z')],$$

$$(22\text{b})$$

and, for $z < z'$,

$$V(z, z') = -\left[-\frac{v\overleftarrow{Z}(z')}{\overleftrightarrow{Z}(z')} + \frac{i}{\overleftrightarrow{Y}(z')}\right][\cos \kappa(z - z') + j\overleftarrow{Y}'(z') \sin \kappa(z - z')],$$

$$(22\text{c})$$

---

†The traveling-wave description for $z < z'$ has the form (incident wave in the $-z$ direction)

$$V(z) = \overleftarrow{V}_{\text{inc}}(z')[e^{+j\kappa(z-z')} + \overleftarrow{\Gamma}(z')e^{-j\kappa(z-z')}],$$
$$-I(z) = Y\overleftarrow{V}_{\text{inc}}(z')[e^{+j\kappa(z-z')} - \overleftarrow{\Gamma}(z')e^{-j\kappa(z-z')}],$$

so

$$V_{\text{inc}}(z') = \tfrac{1}{2}[V(z') - ZI(z')].$$

$$I(z, z') = -\left[-\frac{i\overleftarrow{Y}(z')}{\overleftrightarrow{Y}(z')} + \frac{v}{\overleftrightarrow{Z}(z')} \right] [\cos \kappa(z - z') + j\overleftarrow{Z}'(z') \sin \kappa(z - z')].$$

$$(22\text{d})$$

Equations (21) and (22) are general solutions of the point source-excited transmission-line problem. For special cases, appreciable simplification is achieved. The case of an infinite transmission line, for which $\overrightarrow{Z}(z') = \overleftarrow{Z}(z') = Z$ [i.e., $\overrightarrow{\Gamma}(z') = \overleftarrow{\Gamma}(z') = 0$; see Eq. (12)] has already been discussed [Eqs. (19)]. If the transmission line is terminated on the right but extends undisturbed to infinity on the left, then $\overleftarrow{Z}(z') = Z$, and $\overleftarrow{\Gamma}(z') = 0$. One finds for the traveling-wave description, particularly appropriate in this case:

$$V(z, z') = -\tfrac{1}{2}(v + Zi)[e^{-j\kappa(z-z')} + \overrightarrow{\Gamma}(z')e^{+j\kappa(z-z')}] \Big|_{} \qquad (23\text{a})$$

$$I(z, z') = -\tfrac{1}{2}(i + Yv)[e^{-j\kappa(z-z')} - \overrightarrow{\Gamma}(z')e^{+j\kappa(z-z')}] \Big|_{z > z';} \qquad (23\text{b})$$

$$V(z, z') = -\left[-v\frac{Z}{\overrightarrow{Z}(z') + Z} + i\frac{1}{\overrightarrow{Y}(z') + Y} \right]e^{+j\kappa(z-z')} \qquad (23\text{c})$$

$$I(z, z') = -\left[-i\frac{Y}{\overrightarrow{Y}(z') + Y} + v\frac{1}{\overrightarrow{Z}(z') + Z} \right]e^{+j\kappa(z-z')} \Big|_{z < z'.} \qquad (23\text{d})$$

Additional special terminations for this case are

1. Short circuit at z_2: $\Gamma(z_2) = -1$, $Z(z_2) = 0$,
whence, from Eq. (10),

$$\overrightarrow{Z}'(z') = j \tan \kappa(z_2 - z'), \qquad (24\text{a})$$

and

2. Open circuit at z_2: $\Gamma(z_2) = +1$, $Z(z_2) = \infty$,
whence

$$\overrightarrow{Z}'(z') = -j \cot \kappa(z_2 - z'). \qquad (24\text{b})$$

The total voltage $V(z)$ and current $I(z)$ excited by a distributed source $v(z)$ and $i(z)$ are obtained from $V(z, z')$ and $I(z, z')$ by integration as in Eq. (14).

2.4d Green's Functions for Transmission-Line Equations

The point-source excitation discussed above can be simplified on introduction of unit amplitude generators of voltage and current to represent the excitations v and i employed in Figs. 2.4.2–2.4.5. The transmission-line voltages and currents excited by these point sources constitute the transmission-line (modal) Green's functions $Z(z, z')$, $Y(z, z')$, $T^I(z, z')$, and $T^V(z, z')$ introduced in Sec. 2.3c and defined in Eqs. (2.3.10). We shall evaluate the modal Green's functions by two alternative procedures, the first of which is based on the network representation of the transmission-line problem, while the second is inferred directly from the differential equations (2.3.10).

The network procedure is illustrated in the following calculation of $Y(z, z')$. Since $Y(z, z')$ is defined as the current at a point z on a transmission line excited at z' by a series voltage source of amplitude $v = -1$ (see Sec. 2.3c), the pertinent network problem is shown in Fig. 2.4.6. The general solution for the voltage and current on a point-source-excited and arbitrarily terminated

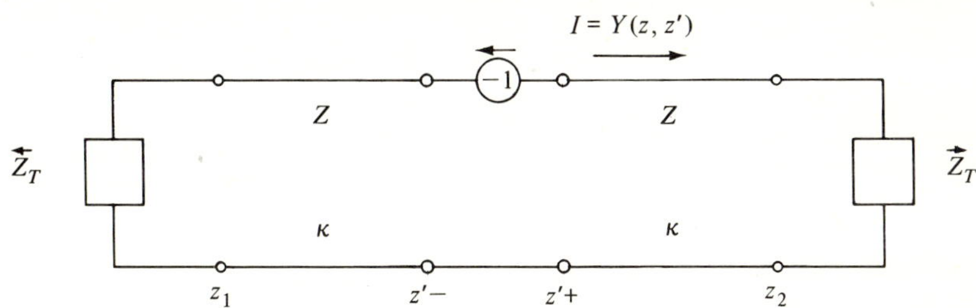

FIG. 2.4.6 Network problem for the determination of $Y(z, z')$.

transmission line of propagation constant κ and characteristic impedance Z has been given in Eqs. (21) and (22), and is readily specialized to the determination of $Y(z, z')$. In particular, the standing-wave representation yields

$$Y(z, z') = \begin{cases} Y(z', z')\,[\cos \kappa(z - z') - j\overrightarrow{Z'}(z') \sin \kappa(z - z')], & z > z', \quad (25a) \\[2mm] Y(z', z')\,[\cos \kappa(z - z') + j\overleftarrow{Z'}(z') \sin \kappa(z - z')], & z < z', \quad (25b) \end{cases}$$

where

$$Y(z', z') = \frac{1}{\overleftrightarrow{Z}(z')} = \frac{1}{\overrightarrow{Z}(z') + \overleftarrow{Z}(z')}. \tag{25c}$$

It is to be noted that the evaluation of $Y(z', z')$ is source independent and depends only on the impedances seen to the right and left of the point z'.

It is desirable on occasion to evaluate $\overleftrightarrow{Z}$ not at the source point z' but at some other point z_0 at which its determination is simpler. The relation between $\overleftrightarrow{Z}(z')$ and $\overleftrightarrow{Z}(z_0)$ may be inferred from Eq. (10) (see also the footnote to Fig. 2.4.5), which yields

$$\frac{1}{\overleftrightarrow{Z}(z')} = \frac{[\cos \kappa(z' - z_0) + j\overleftarrow{Z'}(z_0) \sin \kappa(z' - z_0)]}{\overleftrightarrow{Z}(z_0)}$$

$$\times [\cos \kappa(z' - z_0) - j\overrightarrow{Z'}(z_0) \sin \kappa(z' - z_0)]. \tag{26}$$

For $z > z'$, the term inside the brackets in Eq. (25a) can be written as $I(z)/I(z')$ [see Eq. (9b)], whence the second term inside the brackets in Eq. (26) is written as $I(z')/I(z_0)$. Upon substituting Eq. (26) for $Y(z', z')$ into Eq. (25a) and interpreting the resulting ratio $I(z)/I(z_0)$, one obtains the desired reformulation for $z > z'$:

$$Y(z, z') = \frac{[\cos \kappa(z' - z_0) + j\overleftarrow{Z}'(z_0) \sin \kappa(z' - z_0)]}{\overleftrightarrow{Z}(z_0)}$$

$$\times [\cos \kappa(z - z_0) - j\overrightarrow{Z}'(z_0) \sin \kappa(z - z_0)]. \qquad (27a)$$

In view of the reciprocity relation $Y(z, z') = Y(z', z)$ [Eq. (2.3.15)], the corresponding expression valid for $z < z'$ follows as

$$Y(z, z') = \frac{[\cos \kappa(z - z_0) + j\overleftarrow{Z}'(z_0) \sin \kappa(z - z_0)]}{\overleftrightarrow{Z}(z_0)}$$

$$\times [\cos \kappa(z' - z_0) - j\overrightarrow{Z}'(z_0) \sin \kappa(z' - z_0)]. \qquad (27b)$$

Equations (27a) and (27b), can be subsumed into a single expression valid for all z as

$$Y(z, z') = \frac{[\cos \kappa(z_< - z_0) + j\overleftarrow{Z}'(z_0) \sin \kappa(z_< - z_0)]}{\overleftrightarrow{Z}(z_0)}$$

$$\times [\cos \kappa(z_> - z_0) - j\overrightarrow{Z}'(z_0) \sin \kappa(z_> - z_0)], \qquad (28a)$$

where $z_<$ stands for z when $z < z'$, and for z' when $z > z'$; the converse holds for $z_>$.

By duality, the Green's function $Z(z, z')$ is obtained as

$$Z(z, z') = \frac{[\cos \kappa(z_< - z_0) + j\overleftarrow{Y}'(z_0) \sin \kappa(z_< - z_0)]}{\overleftrightarrow{Y}(z_0)}$$

$$\times [\cos \kappa(z_> - z_0) - j\overrightarrow{Y}'(z_0) \sin \kappa(z_> - z_0)], \qquad (28b)$$

and by virtue of its definition as the voltage excited on a transmission line by a point current source of amplitude $i = -1$, evidently represents the voltage solution for the network problem shown in Fig. 2.4.7. Since

$$\cos \kappa(z - z_0) \mp j\overrightarrow{\overleftarrow{Z}}'(z_0) \sin \kappa(z - z_0)$$

$$= \tfrac{1}{2}[1 + \overrightarrow{\overleftarrow{Z}}'(z_0)]\{e^{\mp j\kappa(z-z_0)} - \overrightarrow{\overleftarrow{\Gamma}}(z_0)\, e^{\pm j\kappa(z-z_0)}\}, \qquad (29a)$$

where

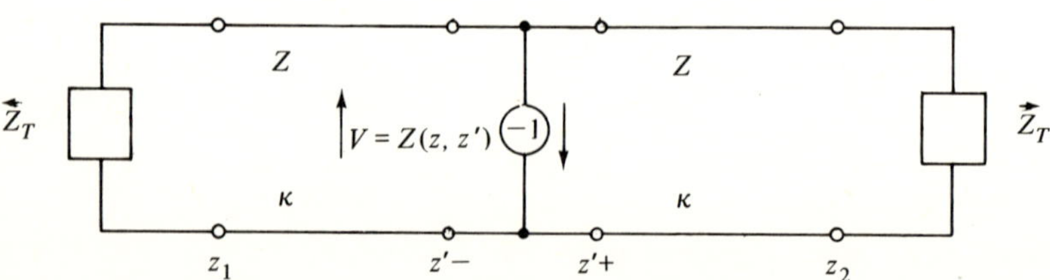

FIG. 2.4.7   Network problem for the determination of $Z(z, z')$.

$$\overset{\rightleftarrows}{\Gamma}(z_0) = \frac{\overset{\rightleftarrows}{Z'}(z_0) - 1}{\overset{\rightleftarrows}{Z'}(z_0) + 1}, \tag{29b}$$

one may write Eq. (28a) in the alternative traveling-wave form,

$$Y(z, z') = \frac{1}{\overset{\rightleftarrows}{Z}(z_0)} \frac{e^{+j\kappa(z_< - z_0)} - \overset{\leftarrow}{\Gamma}(z_0)\, e^{-j\kappa(z_< - z_0)}}{1 - \overset{\leftarrow}{\Gamma}(z_0)}$$

$$\times \frac{e^{-j\kappa(z_> - z_0)} - \overset{\rightarrow}{\Gamma}(z_0)\, e^{+j\kappa(z_> - z_0)}}{1 - \overset{\rightarrow}{\Gamma}(z_0)}. \tag{29c}$$

Similarly, one has, for Eq. (28b),

$$Z(z, z') = \frac{1}{\overset{\rightleftarrows}{Y}(z_0)} \frac{e^{+j\kappa(z_< - z_0)} + \overset{\leftarrow}{\Gamma}(z_0)\, e^{-j\kappa(z_< - z_0)}}{1 + \overset{\leftarrow}{\Gamma}(z_0)}$$

$$\times \frac{e^{-j\kappa(z_> - z_0)} + \overset{\rightarrow}{\Gamma}(z_0)\, e^{+j\kappa(z_> - z_0)}}{1 + \overset{\rightarrow}{\Gamma}(z_0)}. \tag{29d}$$

Alternative to the network derivation of the admittance and impedance Green's functions is an evaluation based on their defining differential equations (2.3.10) and on the source-free wave solutions thereof, each of the latter satisfying the boundary condition to the left or right of the source point. By eliminating T^V from Eqs. (2.3.10a) and (2.3.10b), one obtains a second-order differential equation for $Y(z, z')$ in the form

$$\left(\frac{d^2}{dz^2} + \kappa^2\right) Y(z, z') = -j\kappa Y \delta(z - z'), \tag{30}$$

valid for constant κ and Z. The boundary conditions at the endpoints z_1 and z_2 of the z domain distinguish a unique solution of Eq. (30), and are stated in terms of the terminating "logarithmic derivatives" or normalized impedances (i.e., ratios of voltage to current):

$$\left[\frac{(d/dz)\, Y(z, z')}{(-j\kappa Y)\, Y(z, z')}\right]_{z_1, z_2}, \tag{30a}$$

as seen from Eq. (2.3.10).

The construction of the Green's function of a second-order differential equation can be carried out by a well-established mathematical formalism (see also Sec. 3.3) which is conveniently and significantly viewed from a network standpoint. For all points $z \neq z'$, $Y(z, z')$ must satisfy the homogeneous equation (30) (i.e., it is a source-free wave solution of the transmission-line equations). Since $Y(z, z')$ is actually a (normalized) current on a transmission line, then for $z > z'$, a homogeneous solution that satisfies the boundary conditions at z_2 can be written from Eq. (9b) as

$$\vec{I}(z) = \cos \kappa(z - z_0) - j\overset{\rightarrow}{Z'}(z_0) \sin \kappa(z - z_0), \tag{31a}$$

where an arbitrary point z_0 has been introduced as a reference, and a normalization $\vec{I}(z_0) = 1$ has been adopted. The homogeneous network problem for which $\vec{I}(z)$ represents a solution is shown in Fig. 2.4.8; it should be stressed that the above wave solution makes no reference to the presence of a source.

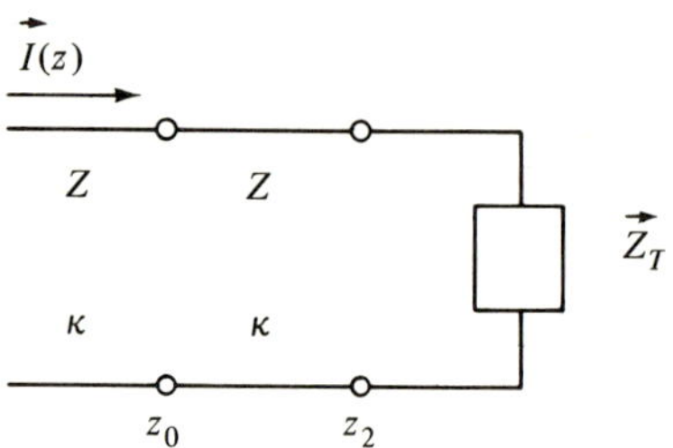

FIG. 2.4.8 Homogeneous network problem for $\vec{I}(z)$.

In an entirely analogous manner, one obtains for $z < z'$ a homogeneous solution which satisfies the boundary conditions at z_1 as

$$\overleftarrow{I}(z) = \cos \kappa(z - z_0) + j\overleftarrow{Z}'(z_0) \sin \kappa(z - z_0). \tag{31b}$$

Again, $\overleftarrow{I}(z_0) = 1$. In the notation of Figs. 2.4.6 and 2.4.8, the homogeneous network problem is that shown in Fig. 2.4.9. The normalized impedances (logarithmic derivatives) at z_0 are related bilinearly to those specified at the endpoints z_1 and z_2 [see Eq. (10), with footnote on p. 208].

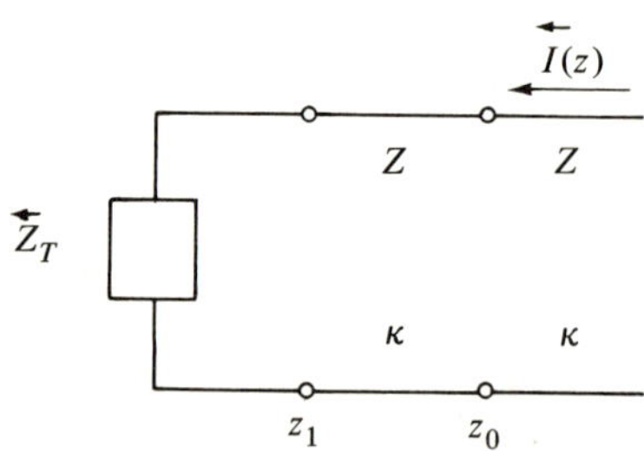

FIG. 2.4.9 Homogeneous network problem for $\overleftarrow{I}(z)$.

To construct the Green's function, one imposes the following requirements on $Y(z, z')$:

1. It must have a z dependence of the form $\vec{I}(z)$ for $z > z'$, and $\overleftarrow{I}(z)$ for $z < z'$.

2. From the symmetry condition $Y(z, z') = Y(z', z)$ [Eq. (2.3.15)], $Y(z, z')$ depends symmetrically on $\vec{I}$ and $\overleftarrow{I}$.

3. From the defining differential equation (30),† or the network picture

---

†From Eq. (30) one infers by integration over an infinitesimal interval centered at z' that $(d/dz)Y(z, z')$ has a jump discontinuity, with $Y(z, z')$ bounded and continuous, at $z = z'$.

(Fig. 2.4.6), it is evident that $Y(z, z')$ is continuous at $z = z'$ and has a jump in its derivative (i.e., the voltage) at $z = z'$.

From the above it follows that $Y(z, z')$ must be of the form

$$Y(z, z') = \begin{cases} A \overrightarrow{I}(z)\overleftarrow{I}(z'), & z > z', & \text{(32a)} \\ A \overrightarrow{I}(z')\overleftarrow{I}(z), & z < z', & \text{(32b)} \end{cases}$$

or, in the notation of Eq. (28),

$$Y(z, z') = A \overleftarrow{I}(z_<)\overrightarrow{I}(z_>). \tag{33}$$

A is a constant *independent* of z (or z') and must have the correct magnitude to satisfy the discontinuity requirement on $dY(z, z')/dz$ at z'. Since A is not a function of z (or z'), we may choose the source point at $z' = z_0$ so as to permit a particularly simple evaluation of A. Since $\overleftarrow{I}(z_0) = \overrightarrow{I}(z_0) = 1$, a simple network evaluation of the current at $z = z_0$ yields (see Fig. 2.4.6, with $z' = z_0$)

$$Y(z_0, z_0) = \frac{1}{\overleftrightarrow{Z}(z_0)} = A, \tag{34}$$

and the Green's function $Y(z, z')$ is given by

$$Y(z, z') = \frac{\overleftarrow{I}(z_<, z_0)\overrightarrow{I}(z_>, z_0)}{\overleftrightarrow{Z}(z_0)}, \tag{35}$$

where the dependence on z_0 has been explicitly exhibited in $\overleftarrow{I}$ and $\overrightarrow{I}$. Equation (35) is identical with Eq. (28). It is to be noted that the evaluation of A in Eq. (34) by a simple network scheme is equivalent to calculating the reciprocal of the Wronskian of the two solutions $\overleftarrow{I}$ and $\overrightarrow{I}$. (See Sec. 3.3b.)

2.4e Resonance Properties of Terminated Transmission Lines

In the preceding sections we have evaluated the response of a terminated transmission line to prescribed excitation. This involves the determination of a modal Green's function $Y(z, z')$ or $Z(z, z')$ from which the desired response can be inferred. The form of these Green's functions makes evident the interesting possibility of obtaining a response even in the absence of excitation. Such a circumstance characterizes a resonance and is achieved whenever $Y(z, z')$ or $Z(z, z')$ becomes infinite, either for some mode wavenumber k_t with fixed k, or for some k with k_t fixed.

Since a terminated transmission line is representative of a cavity or of a transversely viewed waveguide, resonant situations provide information about the complete sets of modes in cavities or in waveguides. In the former case we consider a waveguide of finite cross section so terminated as to constitute a (closed) cavity resonator (see Fig. 2.4.10). In the latter, the original waveguide must be of infinite extent along at least one of its cross-sectional directions and terminated in the longitudinal direction; it may then be viewed alternatively as an infinitely long waveguide with axis along the infinitely extended cross-sec-

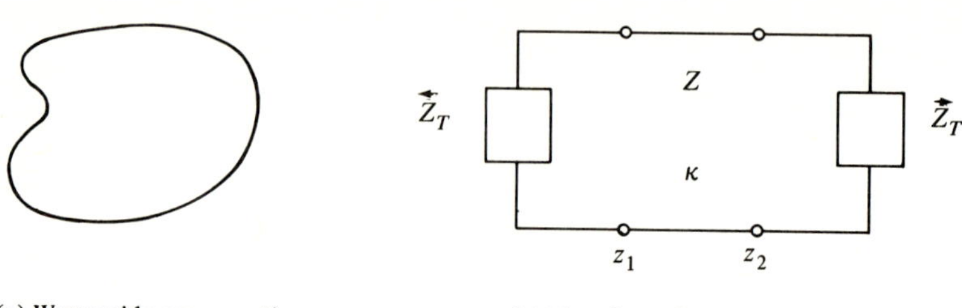

(a) Waveguide cross section
(k_t prescribed)

(b) Terminated transmission line

FIG. 2.4.10   Cavity problem.

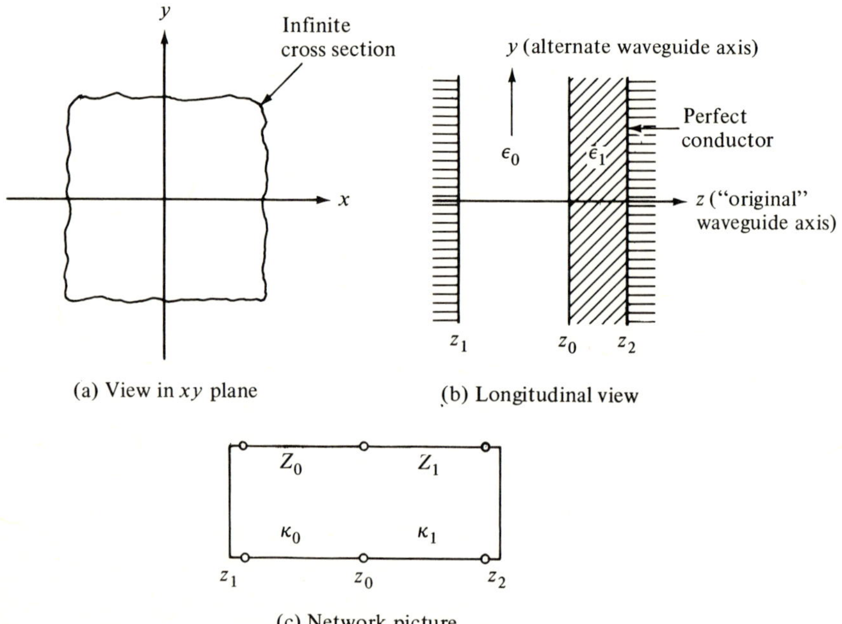

(a) View in xy plane

(b) Longitudinal view

(c) Network picture

FIG. 2.4.11   "Transverse resonance" problem.

tional direction. The special case of a parallel-plate waveguide containing a stratified dielectric is depicted in Fig. 2.4.11. In each event the resonances characterize the possible modes either of oscillation or of propagation that are capable of being excited in such a structure. More generally and mathematically, it can be shown that the singularities (poles or branch points and associated branch cuts) of modal Green's functions distinguish the complete spectrum of possible modes in cavities or waveguides, and the residues at the poles and/or suitable branch-cut contributions, characterize the corresponding mode functions. (See Sec. 3.3a.)

By inspection of Eqs. (27)–(29), it is apparent that the modal Green's functions become infinite at the zeros of the total impedance $\overleftrightarrow{Z}(z_0)$, or of the total admittance $\overleftrightarrow{Y}(z_0)$. One observes that the impedances $\overleftarrow{Z}(z_0)$ and $\overrightarrow{Z}(z_0)$ seen looking to the left and right, respectively, of the point z_0 are functions of the propagation constant $\kappa = (k^2 - k_t^2)^{1/2}$ and hence depend implicitly on the free-space wavenumber $k = \omega/c$ ($c =$ velocity of light in the waveguide medium) and mode wavenumber k_t. The values $\kappa = \kappa_r$ for which

$$\overleftrightarrow{Z}(z_0) = 0 \qquad \text{or} \qquad \overleftarrow{Z}(z_0, \kappa) = -\overrightarrow{Z}(z_0, \kappa) \tag{36}$$

are independent of the choice of z_0 and distinguish the resonances of the transmission-line system (i.e., the wavenumbers κ_r for which the line current or voltage may be finite even with vanishing excitation). One distinguishes between real and complex values of κ_r; the former obtain in reactive (nondissipative) systems, the latter in lossy systems.

The resonant frequencies of modes in a cavity, formed by a uniform guide terminated at both ends, may be determined from the equivalent modal network. For a specified cross-sectional mode wavenumber k_t, the values of κ_r at which $\overleftrightarrow{Z}(z_0)$ vanishes determine the resonant angular frequencies ω_r of the given mode k_t in the cavity. For a non-dissipative cavity, ω_r is real and hence the resonance is undamped (in time). For a dissipative or loaded cavity, ω_r is complex and the resonance is damped, the ratio of the real and imaginary parts of κ_r being indicative of the Q (quality factor) of the particular resonance.

The modes in an infinite uniform waveguide may be inferred from the resonances of an appropriate "transverse network" if the guide cross section has the appropriate symmetry. In this application, one prescribes the frequency ω and determines mode wavenumbers k_t corresponding to the resonant values κ_r of the terminated network representative of the transversely viewed waveguide under consideration; this is the so-called "transverse resonance" procedure for mode evaluations. The k_t so determined are the propagation constants of the modes capable of propagating along the axis of the waveguide in question. The relation between the problem of resonances on a terminated transmission-line network and the eigenvalue problem for modes in the waveguide cross section, of which the network is representative, are treated in more detail in the Sec. 3.3a.

A resonant situation provides information not only about the resonant frequencies of a network but also about the corresponding source-free voltage and current distribution that can exist at these frequencies. Although the distributions $\overrightarrow{I}(z)$ and $\overleftarrow{I}(z)$ in Eqs. (31a) and (31b) arise from sources to the left or right, respectively, of the network terminations, it is evident from Eq. (36) that at resonance, the wave solutions $\overrightarrow{I}(z)$ and $\overleftarrow{I}(z)$ are identical and satisfy the boundary conditions at both terminations of the network. These source-free resonant solutions characterize the eigensolutions or mode functions of the eigenvalue problems mentioned above (see Sec. 3.3b).

2.5 DERIVATION OF TRANSMISSION-LINE EQUATIONS IN SPHERICAL REGIONS

To describe fields in the spherical waveguide region of Fig. 2.1.1(b), the vector field equations (2.2.1) are conveniently expressed in spherical polar coordinates (r, θ, ϕ). The electric and magnetic field components $\mathbf{E}_t$ and $\mathbf{H}_t$ transverse to the radial direction r are independent field quantities from which the longitudinal components E_r and H_r may be determined. The field equations may therefore be reduced to equations for just the transverse fields. On introduction of a complete set of transverse vector eigenfunctions, one may derive modal representations for $\mathbf{E}_t$ and $\mathbf{H}_t$, with r-dependent mode amplitudes that satisfy spherical transmission-line equations. The procedure is directly analogous to that in Sec. 2.2 for uniform waveguide regions with the direction z replaced by r.

2.5a The Transverse Field Equations

In the spherical coordinate system of Fig. 2.5.1, the location of a point P is specified by the radial distance r from the origin; the angle θ measures the inclination of the radius vector $\mathbf{r}$ with respect to the z axis, and the azimuthal

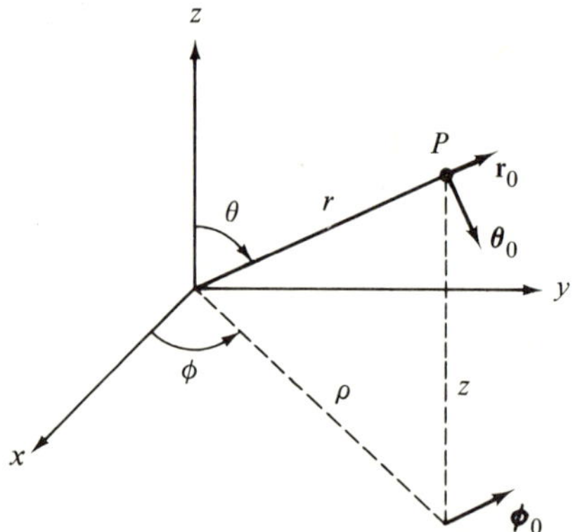

FIG. 2.5.1 Spherical coordinates.

angle ϕ locates the xy plane projection ρ of r with respect to the x axis. $\mathbf{r}_0$, $\boldsymbol{\theta}_0$, and $\boldsymbol{\phi}_0$ are defined as unit vectors along the directions of increasing r, θ, and ϕ, respectively, whence $\mathbf{r}_0 \times \boldsymbol{\theta}_0 = \boldsymbol{\phi}_0$. It will be convenient to employ two alternative representations of the operator ∇ in a spherical coordinate system[7]:

$$\nabla = \mathbf{r}_0 \frac{\partial}{\partial r} + {}_t\nabla = \frac{1}{r^2} \frac{\partial}{\partial r} r^2 \mathbf{r}_0 + \nabla_t, \tag{1}$$

where

$${}_t\nabla = \boldsymbol{\theta}_0 \frac{1}{r} \frac{\partial}{\partial \theta} + \boldsymbol{\phi}_0 \frac{1}{r \sin \theta} \frac{\partial}{\partial \phi},$$

$$\nabla_t = \frac{1}{r \sin \theta} \frac{\partial}{\partial \theta} \sin \theta \, \boldsymbol{\theta}_0 + \frac{1}{r \sin \theta} \frac{\partial}{\partial \phi} \boldsymbol{\phi}_0. \tag{1a}$$

The first and second formulations are convenient for vector operations to the left and right, respectively, with their difference arising from the non-vanishing angular derivatives of the unit vectors in a curvilinear system.

To reduce the Maxwell field equations (2.2.1) to their transverse to $\mathbf{r}$ form, we take scalar and vector products with the radial unit vector $\mathbf{r}_0$ and find, after manipulations analogous to those leading to Eqs. (2.2.4),[8]

$$-\frac{\partial}{\partial r}(r\mathbf{E}_t) = j\omega\mu(r)\left[1 + \frac{1}{k^2(r)}{}_t\nabla\nabla_t\right]\cdot(\mathbf{H}_t \times \mathbf{r}) + \mathbf{M}_{te} \times \mathbf{r}, \qquad (2a)$$

$$-\frac{\partial}{\partial r}(r\mathbf{H}_t) = j\omega\epsilon(r)\left[1 + \frac{1}{k^2(r)}{}_t\nabla\nabla_t\right]\cdot(\mathbf{r} \times \mathbf{E}_t) + \mathbf{r} \times \mathbf{J}_{te}, \qquad (2b)$$

where the permittivity $\epsilon(r)$ and permeability $\mu(r)$ are permitted to be functions of r but not of θ or ϕ, $k^2 = \omega^2\mu\epsilon$, $\mathbf{r} = \mathbf{r}_0 r$, and $\mathbf{J}_{te}$, $\mathbf{M}_{te}$ are the equivalent transverse source distributions:

$$\mathbf{J}_{te} = \mathbf{J}_t - \frac{r}{j\omega\mu(r)}\mathbf{r} \times {}_t\nabla M_r, \qquad \mathbf{M}_{te} = \mathbf{M}_t + \frac{r}{j\omega\epsilon(r)}\mathbf{r} \times {}_t\nabla J_r. \qquad (2c)$$

A subscript t distinguishes a vector transverse to r. The longitudinal (radial) field components are derivable from the transverse fields and the longitudinal source currents via

$$E_r = \frac{1}{j\omega\epsilon(r)}[\nabla_t \cdot (\mathbf{H}_t \times \mathbf{r}_0) - J_r], \qquad H_r = \frac{1}{j\omega\mu(r)}[\nabla_t \cdot (\mathbf{r}_0 \times \mathbf{E}_t) - M_r].$$

$$(3)$$

If a perfectly conducting transverse boundary s described by the equation $f(\theta, \phi) = 0$ is present, the tangential electric field must vanish so that

$$\mathbf{v} \times \mathbf{E}_t = 0, \quad \nabla_t \cdot (\mathbf{H}_t \times \mathbf{r}_0) = 0 \qquad \text{on } s, \qquad (4)$$

where $\mathbf{v}$ is a unit vector normal to s. Equations (2) and (3) are completely equivalent to the original field equations (2.2.1).

2.5b　Modal Representation of the Fields and Their Sources

As in the uniform waveguide case in Sec. 2.2b, solutions of the source-free transverse field equations, individually satisfying the transverse boundary conditions, constitute a complete set of functions that may be employed to represent an arbitrary field. In view of the assumed variability of ϵ and μ with r only, a typical source-free solution may be written in the separable form

$$r\mathbf{E}_{ti} = V_i(r)\mathbf{e}_i(\boldsymbol{\rho}), \qquad r\mathbf{H}_{ti} = I_i(r)\mathbf{h}_i(\boldsymbol{\rho}), \qquad (5)$$

where $\boldsymbol{\rho} = (\theta, \phi)$. Substitution into Eqs. (2) (with $\mathbf{J} = \mathbf{M} = 0$) yields separate equations for the radially and transversely dependent factors. The transverse vector dependence may be eliminated from these equations in a simple manner if: (1) operation on a vector by ${}_t\nabla\nabla_t \cdot$ produces a result proportional to the vector, and (2) $\mathbf{h}_i = \mathbf{r}_0 \times \mathbf{e}_i$. As in Eqs. (2.2.10), these two requirements may be made consistent by separately imposing the auxiliary conditions $\nabla_t \cdot (\mathbf{r}_0 \times \mathbf{e}_i) = 0$ or $\nabla_t \cdot (\mathbf{h}_i \times \mathbf{r}_0) = 0$, thereby splitting the mode set into two parts. It is then implied from Eq. (3) that $H_{ri} = 0$ and $E_{ri} = 0$, respectively, thus identifying the two sets as E and H modes with respect to r. This argument is again related to the theorem that permits decomposition of any vector into a curlless and a divergenceless part.[4] The curlless part may subsequently be expressed as the gradient of a scalar function, whereas the divergenceless part may be written as the curl of a longitudinal vector [see Eqs. (2.6.1) and Sec. 2.3a].

The mode functions $\mathbf{e}_i$ and $\mathbf{h}_i = \mathbf{r}_0 \times \mathbf{e}_i$ are chosen to satisfy the following eigenvalue problems and subsidiary conditions in the transverse cross section S:[9]

$$r^2{}_t\nabla\nabla_t \cdot \mathbf{e}_i' = -k_{ti}'^2\mathbf{e}_i', \qquad \nabla_t \cdot (\mathbf{r}_0 \times \mathbf{e}_i') = 0, \tag{6a}$$

$$r^2{}_t\nabla\nabla_t \cdot \mathbf{h}_i'' = -k_{ti}''^2\mathbf{h}_i'', \qquad \nabla_t \cdot (\mathbf{h}_i'' \times \mathbf{r}_0) = 0, \tag{6b}$$

where r^2 in $(r^2{}_t\nabla\nabla_t)$ has been included to make the operator dependent on the transverse coordinates only, and the single and double primes denote E and H modes, respectively. The boundary conditions on both mode sets are, from Eq. (4),

$$\mathbf{v} \times \mathbf{e}_i = 0 = \nabla_t \cdot (\mathbf{h}_i \times \mathbf{r}_0) \qquad \text{on } s. \tag{6c}$$

On use of Eq. (2.2.11a), one may verify that the vector eigenfunctions satisfy orthogonality relations analogous to those in Eq. (2.2.11b):

$$\iint_S \mathbf{e}_i' \cdot \mathbf{e}_j'^* \, d\Omega = \delta_{ij} = \iint_S \mathbf{e}_i'' \cdot \mathbf{e}_j''^* \, d\Omega; \qquad \iint_S \mathbf{e}_i' \cdot \mathbf{e}_i''^* \, d\Omega = 0, \tag{7}$$

and similarly for the $\mathbf{h}_i$, with $d\Omega = \sin\theta \, d\theta \, d\phi = dS/r^2$ denoting the angular surface element. The eigenfunctions are normalized by setting the integral equal to unity when $i = j$.

The complete set of vector eigenfunctions defined in Eqs. (6) may now be utilized to represent the electromagnetic fields and the excitation functions in Eqs. (2a) and (2b):

$$r\mathbf{E}_t(\mathbf{r}) = \sum_i V_i'(r)\mathbf{e}_i'(\boldsymbol{\rho}) + \sum_i V_i''(r)\mathbf{e}_i''(\boldsymbol{\rho}), \tag{8a}$$

$$r\mathbf{H}_t(\mathbf{r}) = \sum_i I_i'(r)\mathbf{h}_i'(\boldsymbol{\rho}) + \sum_i I_i''(r)\mathbf{h}_i''(\boldsymbol{\rho}), \qquad \mathbf{h}_i = \mathbf{r}_0 \times \mathbf{e}_i, \tag{8b}$$

$$r\mathbf{J}_{te}(\mathbf{r}) = \sum_i i_i'(r)\mathbf{e}_i'(\boldsymbol{\rho}) + \sum_i i_i''(r)\mathbf{e}_i''(\boldsymbol{\rho}), \tag{8c}$$

$$r\mathbf{M}_{te}(\mathbf{r}) = \sum_i v_i'(r)\mathbf{h}_i'(\boldsymbol{\rho}) + \sum_i v_i''(r)\mathbf{h}_i''(\boldsymbol{\rho}). \tag{8d}$$

Correspondingly, from Eq. (3),

$$rE_r + \frac{rJ_r}{j\omega\epsilon} = \sum_i Z_i'I_i'e_{ri}', \qquad rH_r + \frac{rM_r}{j\omega\mu} = \sum_i Y_i''V_i''h_{ri}'', \tag{8e}$$

with $Z_i e_{ri}$ and $Y_i h_{ri}$ defined below in Eqs. (10c) and (10d). When these expansions are substituted into Eqs. (2a) and (2b), and use is made of Eqs. (6a) and (6b) together with the orthogonality relations (7), one obtains spherical transmission-line equations for the modal voltage and current amplitudes V_i and I_i:

$$-\frac{dV_i}{dr} = j\kappa_i Z_i I_i + v_i, \qquad -\frac{dI_i}{dr} = j\kappa_i Y_i V_i + i_i, \tag{9}$$

where κ_i, Z_i, and Y_i are the modal propagation constant, characteristic impedance, and characteristic admittance, respectively. For the E modes,

$$\kappa_i'(r) = \sqrt{k^2(r) - \frac{k_{ti}'^2}{r^2}}, \qquad Z_i'(r) = \frac{1}{Y_i'(r)} = \frac{\kappa_i'(r)}{\omega\epsilon(r)}; \tag{9a}$$

for the H modes,

$$\kappa_i''(r) = \sqrt{k^2(r) - \frac{k_{ti}''^2}{r^2}}, \qquad Z_i''(r) = \frac{1}{Y_i''(r)} = \frac{\omega\mu(r)}{\kappa_i''(r)}, \tag{9b}$$

with $k^2(r) = \omega^2 \mu(r)\epsilon(r)$. One observes that even when the medium is homogeneous (ϵ, μ constant), the propagation constant and characteristic impedance are r dependent (i.e., the transmission lines are non-uniform). This is a consequence of the non-uniformity of the transverse sections, which are not identical but only similar, at different radial distances. The source terms v_i and i_i in Eqs. (9) are calculated from the known currents $\mathbf{J}$ and $\mathbf{M}$ on inversion of Eqs. (8c) and (8d):

$$\frac{1}{r}\, v_i(r) = \iint_S \mathbf{M(r)} \cdot \mathbf{h}_i^*(\boldsymbol{\rho})\, d\Omega + Z_i^* \iint_S \mathbf{J(r)} \cdot \mathbf{e}_{ri}^*\, d\Omega, \tag{10a}$$

$$\frac{1}{r}\, i_i(r) = \iint_S \mathbf{J(r)} \cdot \mathbf{e}_i^*(\boldsymbol{\rho})\, d\Omega + Y_i^* \iint_S \mathbf{M(r)} \cdot \mathbf{h}_{ri}^*\, d\Omega, \tag{10b}$$

where

$$Z_i'\mathbf{e}_{ri}' = \mathbf{r}_0 \frac{\nabla_t \cdot \mathbf{e}_i'}{j\omega\epsilon} = \mathbf{r}_0 \frac{k_{ti}'\Phi_i}{j\omega\epsilon r}, \qquad \mathbf{e}_{ri}'' \equiv 0, \tag{10c}$$

$$Y_i''\mathbf{h}_{ri}'' = \mathbf{r}_0 \frac{\nabla_t \cdot \mathbf{h}_i''}{j\omega\mu} = \mathbf{r}_0 \frac{k_{ti}''\Psi_i}{j\omega\mu r}, \qquad \mathbf{h}_{ri}' \equiv 0. \tag{10d}$$

The scalar mode functions Φ_i and ψ_i are defined in Sec. 2.6a. Solutions of the transmission line equations for various radial domains are given in Sec. 2.7.

The average power $\bar{S}$ transferred across a spherical surface with radius r is given by

$$\bar{S} = \mathrm{Re}\int_S \mathbf{E} \times \mathbf{H}^* \cdot \mathbf{r}_0\, dS = \mathrm{Re}\left[\sum_i V_i'(r)I_i'^*(r) + \sum_i V_i''(r)I_i''^*(r)\right], \tag{11a}$$

where the last expression follows from Eqs. (8a) and (8b) and the orthogonality conditions (7). The total power is therefore equal to the sum of the contributions carried by the individual modes. When the observation point is moved to infinity ($r \to \infty$), and all sources (real or induced) are assumed to be confined within the region $r < a$, the contributing modes have $kr \gg k_{ti}$, whence $\kappa_i \to k$ $Z_i(r) \to \zeta = \sqrt{\mu/\epsilon}$, and $V_i(r) \to \zeta I_i(r)$ [see Eqs. (9) and Sec. 2.7]. Since modes with $k_{ti} \gg ka$ have strongly damped amplitudes, the mode series contains only a finite number of terms in the effective range. The total radiated power may then be expressed in terms of the "far-field" voltages as follows:

$$\bar{S} = \frac{1}{\zeta}\sum_i |V_i'(r)|^2 + \frac{1}{\zeta}\sum_i |V_i''(r)|^2, \qquad r \to \infty, \tag{11b}$$

which formulation is useful for the calculation of such quantities as the scattering cross section of an obstacle, or the transmission cross section of an aperture.

2.6 SCALARIZATION AND MODAL REPRESENTATION OF DYADIC GREEN'S FUNCTIONS IN SPHERICAL REGIONS

As for uniform waveguide regions, the vector-field and eigenvalue problems in Sec. 2.5 may be reduced to scalar problems through use of E- and H-mode (Hertz) potentials. The scalarization procedure and the definition of scalar Green's functions $G'(\mathbf{r}, \mathbf{r}')$, $G''(\mathbf{r}, \mathbf{r}')$, $\mathscr{S}'(\mathbf{r}, \mathbf{r}')$, $\mathscr{S}''(\mathbf{r}, \mathbf{r}')$ equivalent to these potentials are summarized below, in analogy with the discussion in Sec. 2.3.

2.6a Mode Functions

Since from the second of Eqs. (2.5.6a) and (2.5.6b), the vector set $\{\mathbf{e}'_i\}$ is irrotational ($\nabla_t \times \mathbf{e}'_i = 0$) while the vector set $\{\mathbf{e}''_i\}$ is solenoidal ($\nabla_t \cdot \mathbf{e}''_i = 0$), the mode functions $\mathbf{e}'_i$ and $\mathbf{e}''_i$ can be represented as gradients and curls of scalar functions Φ_i and ψ_i as follows:

$$\mathbf{e}'_i = -\frac{r_t \nabla \Phi_i}{k'_{ti}} = \mathbf{h}'_i \times \mathbf{r}_0, \qquad \mathbf{h}''_i = -\frac{r_t \nabla \psi_i}{k''_{ti}} = \mathbf{r}_0 \times \mathbf{e}''_i. \tag{1}$$

Then from the first of Eqs. (2.5.6a) and (2.5.6b) there result the following scalar eigenvalue problems subject to boundary conditions deduced from Eq. (2.5.6c):

$$r^2 \nabla_t \cdot {}_t\nabla \Phi_i + k'^2_{ti} \Phi_i = 0 \qquad \text{in } S,$$

$$r^2 \nabla_t \cdot {}_t\nabla = \frac{1}{\sin\theta} \frac{\partial}{\partial\theta} \sin\theta \frac{\partial}{\partial\theta} + \frac{1}{\sin^2\theta} \frac{\partial^2}{\partial\phi^2}, \tag{2}$$

$$\left.\begin{aligned}
\Phi_i &= 0, & k'_{ti} &\neq 0 \\
\frac{\partial\Phi_i}{\partial s} &= 0, & k'_{ti} &= 0\dagger
\end{aligned}\right\} \quad \text{on } s \tag{2a}$$

and

$$r^2 \nabla_t \cdot {}_t\nabla \psi_i + k''^2_{ti} \psi_i = 0 \qquad \text{in } S, \tag{3}$$

$$\frac{\partial\psi_i}{\partial v} = 0 \qquad \text{on } s. \tag{3a}$$

2.6b Fields in Source-Free, Homogeneous Regions

As for uniform regions (Sec. 2.3b), the definition and modal representation of the scalar radial Hertzian (or Debye) potentials $\Pi'(\mathbf{r})$ and $\Pi''(\mathbf{r})$ follows directly on substitution of Eq. (1) into Eqs. (2.5.8) with the result that at any source-free point $\mathbf{r}$,[10-12]

$$\mathbf{E}(\mathbf{r}) = \mathbf{E}'(\mathbf{r}) + \mathbf{E}''(\mathbf{r}), \qquad \mathbf{H}(\mathbf{r}) = \mathbf{H}'(\mathbf{r}) + \mathbf{H}''(\mathbf{r}). \tag{4}$$

---

†This case corresponds to a TEM mode ($E_r = H_r = 0$), which is treated separately.

The E-mode constituents are given by

$$\mathbf{E}'(\mathbf{r}) = \nabla \times \nabla \times [\mathbf{r}\Pi'(\mathbf{r})], \qquad \mathbf{H}'(\mathbf{r}) = j\omega\epsilon\nabla \times [\mathbf{r}\Pi'(\mathbf{r})], \tag{4a}$$

and the H-mode constituents by

$$\mathbf{E}''(\mathbf{r}) = -j\omega\mu\nabla \times [\mathbf{r}\Pi''(\mathbf{r})], \qquad \mathbf{H}''(\mathbf{r}) = \nabla \times \nabla \times [\mathbf{r}\Pi''(\mathbf{r})]. \tag{4b}$$

ϵ and μ have been assumed constant, for simplicity. In reducing these expressions, the following relations are useful:

$$\nabla \times \mathbf{r}_0 A = -\mathbf{r}_0 \times {}_t\nabla A, \qquad \nabla \times \nabla \times \mathbf{r}_0 A = {}_t\nabla \frac{\partial A}{\partial r} - \mathbf{r}_0 \nabla_t \cdot {}_t\nabla A. \tag{5}$$

The scalar potential functions have the modal representation

$$\Pi'(\mathbf{r}) = \frac{1}{j\omega\epsilon r} \sum_i \frac{I_i'(r)\Phi_i(\boldsymbol{\rho})}{k_{ti}'}, \tag{6a}$$

$$\Pi''(\mathbf{r}) = \frac{1}{j\omega\mu r} \sum_i \frac{V_i''(r)\psi_i(\boldsymbol{\rho})}{k_{ti}''}, \tag{6b}$$

and satisfy at any source-free point the scalar wave equation

$$(\nabla^2 + k^2)\frac{\Pi'(\mathbf{r})}{\Pi''(\mathbf{r})} = 0,$$

$$\nabla^2 A = \left(\frac{1}{r^2}\frac{\partial}{\partial r}r^2\frac{\partial}{\partial r} + \nabla_t \cdot {}_t\nabla\right)A = \left(\frac{1}{r}\frac{\partial^2}{\partial r^2} + \frac{1}{r}\nabla_t \cdot {}_t\nabla\right)(rA), \tag{7}$$

where $\nabla_t \cdot {}_t\nabla$ is defined in Eq. (2). As for uniform waveguides, one may alternatively define Π' and Π'' directly by Eq. (7) subject to appropriate boundary conditions, and then derive the electromagnetic fields from Eqs. (4); this sequence, reversed herein, yields in the process the modal solutions in Eqs. (6a) and (6b).

The boundary conditions satisfied by the potentials Π' and Π'' are easily inferred from the modal representations (6a) and (6b). On a transverse boundary, the E mode potential behaves as Φ_i, and across a radial boundary, I_i' and V_i' must be continuous [see Eqs. (2.5.8a) and (2.5.8b)]. Thus, if the region is bounded transversely by a perfectly conducting surface s described by the equation $f(\theta, \phi) = 0$, and comprises homogeneous spherical layers along the radial direction, then from Eqs. (2a) and (2.5.9),

$$\Pi'(\mathbf{r}) = 0 \qquad \text{on } s,$$

$$\epsilon\Pi' \text{ and } \frac{\partial}{\partial r}(r\Pi') \qquad \text{continuous at } r = r_a, \tag{8a}$$

where r_a locates an interface between two different homogeneous regions. Similarly, for the H-mode potential,

$$\frac{\partial\Pi''(\mathbf{r})}{\partial v} = 0 \qquad \text{on } s,$$

$$\mu\Pi'' \text{ and } \frac{\partial}{\partial r}(r\Pi'') \qquad \text{continuous at } r = r_a, \tag{8b}$$

where v denotes the normal to s. For a perfectly conducting radial boundary at $r = r_a$, the vanishing of V_i'' and dI_i'/dr implies that these conditions reduce to $\partial(r\Pi')/\partial r = 0$ and $\Pi'' = 0$, respectively.

2.6c Modal Representations of the Dyadic Green's Functions

The potentials in Eqs. (6a) and (6b) may be represented in terms of modal Green's functions $Z_i(r, r')$, $Y_i(r, r')$, $T_i^I(r, r')$, and $T_i^V(r, r')$, defined in Eqs. (2.3.10) as voltage or current responses to either a voltage or current point generator. The considerations in Sec. 2.3c hold for variable κ_i and Z_i so that the results apply directly to the present discussion, with z, z' replaced by r, r', respectively. By proceeding as in Sec. 2.3d, one may show that the potential functions Π' and Π'', corresponding to excitation by point current elements $\mathbf{M}(\mathbf{r}) = \mathbf{M}^0\delta(\mathbf{r} - \mathbf{r}')$ and $\mathbf{J}(\mathbf{r}) = \mathbf{J}^0\delta(\mathbf{r} - \mathbf{r}')$, may be expressed in terms of auxiliary functions $\mathscr{S}'(\mathbf{r}, \mathbf{r}')$ and $\mathscr{S}''(\mathbf{r}, \mathbf{r}')$ for $r \neq r'$ as

$$j\omega\epsilon r\Pi'(\mathbf{r}, \mathbf{r}') = (\nabla' \times \nabla' \times \mathbf{r}_0')\mathscr{S}'(\mathbf{r}, \mathbf{r}') \cdot \mathbf{J}^0 - j\omega\epsilon(\nabla' \times \mathbf{r}_0')\mathscr{S}'(\mathbf{r}, \mathbf{r}') \cdot \mathbf{M}^0,$$

$$(9a)$$

$$j\omega\mu r\Pi''(\mathbf{r}, \mathbf{r}') = (\nabla' \times \nabla' \times \mathbf{r}_0')\mathscr{S}''(\mathbf{r}, \mathbf{r}') \cdot \mathbf{M}^0 + j\omega\mu(\nabla' \times \mathbf{r}_0')\mathscr{S}''(\mathbf{r}, \mathbf{r}') \cdot \mathbf{J}^0,$$

$$(9b)$$

with†

$$j\omega\epsilon\mathscr{S}'(\mathbf{r}, \mathbf{r}') = \sum_i \frac{\Phi_i(\boldsymbol{\rho})\Phi_i^*(\boldsymbol{\rho}')}{k_{ti}'^2} Y_i'(r, r'),$$

$$j\omega\mu\mathscr{S}''(\mathbf{r}, \mathbf{r}') = \sum_i \frac{\psi_i(\boldsymbol{\rho})\psi_i^*(\boldsymbol{\rho}')}{k_{ti}''^2} Z_i''(r, r').$$

$$(9c)$$

In these equations, ∇' denotes differentiation with respect to the source point coordinates (r', θ', ϕ') and $\mathbf{r}_0'$ is the radial unit vector in the primed coordinate system which describes the orientation of the source vectors $\mathbf{J}^0$ and $\mathbf{M}^0$ (i.e., $\mathbf{J}^0 = \mathbf{r}_0'J_r^0 + \boldsymbol{\theta}_0'J_\theta^0 + \boldsymbol{\phi}_0'J_\phi^0$, etc.). It is to be noted that the vectors $\mathbf{J}^0$ and $\mathbf{M}^0$ are treated as *constant* and that the curl operations are to be interpreted as in Eq. (5), with the dot product carried out subsequently. The form of Eqs. (9a) and (9b) permits the dyadic Green's functions at a source-free point to be written symmetrically as follows:

$$-j\omega\epsilon\mathscr{Z}(\mathbf{r}, \mathbf{r}') = (\nabla \times \nabla \times \mathbf{r}_0)(\nabla' \times \nabla' \times \mathbf{r}_0')\mathscr{S}'(\mathbf{r}, \mathbf{r}')$$

$$+ k^2(\nabla \times \mathbf{r}_0)(\nabla' \times \mathbf{r}_0')\mathscr{S}''(\mathbf{r}, \mathbf{r}'), \qquad (10a)$$

$$-j\omega\mu\mathscr{Y}(\mathbf{r}, \mathbf{r}') = (\nabla \times \nabla \times \mathbf{r}_0)(\nabla' \times \nabla' \times \mathbf{r}_0')\mathscr{S}''(\mathbf{r}, \mathbf{r}')$$

$$+ k^2(\nabla \times \mathbf{r}_0)(\nabla' \times \mathbf{r}_0')\mathscr{S}'(\mathbf{r}, \mathbf{r}'), \qquad (10b)$$

with similar formulas resulting for $\mathscr{T}_e$ and $\mathscr{T}_m$. These results are directly analogous to those in Eqs. (2.3.27).

---

†It is implied that $k_{ti} \neq 0$. The case $k_{ti}' = 0$ arises in connection with the TEM mode, whose contribution must be included separately.

As in uniform regions, the preceding formulas simplify substantially when the sources are longitudinal (radial). One observes from Eqs. (9) that a radial electric current element excites only E modes and a radial magnetic current element only H modes, thereby making the fields excited separately by these sources derivable from a single scalar function. The potential functions Π' and Π'' are now expressed as follows:

$$\Pi'(\mathbf{r}) = J_r^0 \frac{G'(\mathbf{r}, \mathbf{r}')}{j\omega\epsilon r'}, \qquad \Pi''(\mathbf{r}) = M_r^0 \frac{G''(\mathbf{r}, \mathbf{r}')}{j\omega\mu r'}, \tag{11}$$

where, in view of Eqs. (2), (3), and (9c),

$$rr'G'(\mathbf{r}, \mathbf{r}') \equiv -r'^2 \nabla_t' \cdot {}_t\nabla' \mathscr{S}'(\mathbf{r}, \mathbf{r}') = \sum_i \Phi_i(\mathbf{\rho})\Phi_i^*(\mathbf{\rho}') \frac{Y_i'(r, r')}{j\omega\epsilon}, \tag{11a}$$

$$rr'G''(\mathbf{r}, \mathbf{r}') \equiv -r'^2 \nabla_t' \cdot {}_t\nabla' \mathscr{S}''(\mathbf{r}, \mathbf{r}') = \sum_i \psi_i(\mathbf{\rho})\psi_i^*(\mathbf{\rho}') \frac{Z_i''(r, r')}{j\omega\mu}. \tag{11b}$$

It may be verified that G' and G'' are scalar Green's functions which satisfy the inhomogeneous wave equations

$$(\nabla^2 + k^2)\frac{G'(\mathbf{r}, \mathbf{r}')}{G''(\mathbf{r}, \mathbf{r}')} = -\delta(\mathbf{r} - \mathbf{r}'), \qquad \delta(\mathbf{r} - \mathbf{r}') = \frac{\delta(r - r')\delta(\theta - \theta')\delta(\phi - \phi')}{r'^2 \sin\theta'},$$

$$\tag{11c}$$

subject to boundary conditions identical with those stated for Π' and Π'', respectively, in Eqs. (8a) and (8b). This follows on performing the operation $(\nabla^2 + k^2)$ on the modal expansions for G' and G'', recalling Eqs. (2), (3), and (2.5.9), and recognizing that $\sum_i \Phi_i(\mathbf{\rho})\Phi_i^*(\mathbf{\rho}') = \delta(\theta - \theta')\delta(\phi - \phi')/\sin\theta'$ (see Eq. 3.3.32a).

2.7 SOLUTION OF SPHERICAL TRANSMISSION-LINE EQUATIONS (NETWORK ANALYSIS)

2.7a   Source-Free and Source-Excited Transmission Lines

As noted in Sec. 2.6c, the solution of the transmission-line equations (2.5.9) is facilitated by introduction of the modal Green's functions $Z_i(r, r')$, $Y_i(r, r')$, $T_i^I(r, r')$, and $T_i^V(r, r')$ defined in Eqs. (2.3.10). Since details on non-uniform transmission lines are given in Secs. 3.3a and 3.3b, only a summary for spherical lines is presented here. We observe that the source-free equations for the voltage and the current are different, and that it is usually preferable [in view of the relations $\kappa_i' Y_i' = \omega\epsilon$, $\kappa_i'' Z_i'' = \omega\mu$ from Eqs. (2.5.9a) and (2.5.9b)] to solve these equations for the E-mode current and for the H-mode voltage, with the E-mode voltage and H-mode current derived from Eqs. (2.5.9). The E-mode current Green's function $Y_i'(r, r')$ and the H-mode voltage Green's function $Z_i''(r, r')$ satisfy the second-order differential equations [see Eqs. (2.3.10)]†

---

†One should not confuse the Green's function $Y_i'(r, r')$ with the characteristic admittance Y_i', and similarly for $Z_i''(r, r')$ and Z_i''.

$$\left[\kappa_i' Y_i' \frac{d}{dr} \frac{1}{\kappa_i' Y_i'} \frac{d}{dr} + \kappa_i'^2\right] Y_i'(r, r') = -j\kappa_i' Y_i' \delta(r - r'), \tag{1a}$$

$$\left[\kappa_i'' Z_i'' \frac{d}{dr} \frac{1}{\kappa_i'' Z_i''} \frac{d}{dr} + \kappa_i''^2\right] Z_i''(r, r') = -j\kappa_i'' Z_i'' \delta(r - r'), \tag{1b}$$

which simplify substantially in a homogeneous medium wherein $(\kappa_i' Y_i')$ and $(\kappa_i'' Z_i'')$ are constant. This reduction does not occur for $(\kappa_i'' Y_i'')$ or $(\kappa_i' Z_i')$, which enter into the equations for the H-mode current and E-mode voltage, respectively.

In a source-free region where μ and ϵ are constant, both I_i' and V_i'' are solutions of

$$\left[\frac{d^2}{dr^2} + \kappa_i^2(r)\right]\begin{matrix} I_i'(r) \\ V_i''(r) \end{matrix} = 0, \qquad \kappa_i(r) = k\sqrt{1 - \frac{k_{ti}^2}{(kr)^2}}. \tag{2}$$

A distinction arises in the use of $k_{ti} = k_{ti}'$ and k_{ti}'' for the E and H modes, respectively. These equations are solved in terms of the spherical Bessel functions [see also Eq. (5.9.3)]

$$j_p(kr), \quad n_p(kr), \quad h_p^{(1)}(kr), \quad h_p^{(2)}(kr), \quad k_{ti}^2 = p(p + 1), \tag{3}$$

any two of which are linearly independent. The spherical Bessel function $z_p(kr)$ is related to the cylindrical Bessel function $Z_{p+1/2}(kr)$ as follows:

$$z_p(kr) = \sqrt{\frac{\pi kr}{2}}\, Z_{p+1/2}(kr). \tag{3a}$$

It is noted from Eqs. (2.5.9) that the spatial dependence of the E-mode voltage V_i' and the H-mode current I_i'' is given by the r derivatives of the functions in Eq. (3). The spherical Bessel functions have the following asymptotic behavior for reasonably large p and $kr \gg p$:

$$j_p(kr) \sim \sin\left(kr - \frac{p\pi}{2}\right), \qquad n_p(kr) \sim -\cos\left(kr - \frac{p\pi}{2}\right),$$

$$h_p^{(1,2)}(kr) \sim \mp j e^{\pm j(kr - p\pi/2)}, \qquad kr \gg p;^\dagger \tag{4a}$$

for $kr \ll p$, one has, to within constant factors,

$$j_p(kr) \sim \left(\frac{kr}{p}\right)^{p+1}, \quad n_p(kr) \sim \mp j h_p^{(1,2)}(kr) \sim -\left(\frac{p}{kr}\right)^{p}, \qquad kr \ll p. \tag{4b}$$

The expression for $j_p(kr)$ when p is arbitrary and $r \to 0$ is

$$j_p(kr) \sim \frac{\sqrt{\pi}}{\Gamma(p + \frac{3}{2})}\left(\frac{kr}{2}\right)^{p+1}, \tag{4c}$$

from which the first of Eqs. (4b) follows on employing Stirling's formula for the gamma function [Eq. (3.6.52b)].

Evidently, for observation points with $kr \gg p$, a mode characterized by the

---

†Here and in the following, the first and second of $(1, 2)$ superscripts or subscripts refer to the upper and lower symbols, respectively.

index p propagates locally like a uniform plane wave in the radial direction, whereas no propagation takes place in the region $kr \ll p$. This behavior conforms with the transmission-line viewpoint since the propagation constant κ_i and the characteristic impedance Z_i reduce to their free-space values when $kr \gg k_{ti}$, but are imaginary when $kr \ll k_{ti}$ (below cutoff mode), with the change-over occurring in the vicinity of the turning point $kr = k_{ti}$ (see also Sec. 3.5c). Thus, depending on the location of the observation point, a given spherical mode may exhibit either propagating or evanescent characteristics. To amplify on these observations, consider a point source in free space located at $r = r'$ in a coordinate system not centered at the source; the equivalent network includes a point generator at r' on the non-uniform transmission line. As will be seen below in Eq. (11), the voltage and current in a mode with index p behave according to $j_p(kr)$ (or its derivative) when $r < r'$, and $h_p^{(2)}(kr)$ (or its derivative) when $r > r'$. Two regimes may now be distinguished: $p \ll kr'$ and $p \gg kr'$. When $p \ll kr'$, then $kr \gg p$ in the region $r > r'$ so that the mode field propagates undamped in the outward direction. Propagation obtains as well in the region $r' > r > p/k$ but changes to decay when $r < p/k$, with the latter inequalities to be understood in an approximate sense only. Alternatively, when $p \gg kr'$, the mode fields decay on both sides of the source but propagation takes place eventually when $r > p/k$. If various modes have the same amplitude at the source, the important ones in the radiation field are those with $p < kr'$. Modes with $p > kr'$ contribute primarily to energy storage in the neighborhood of the source region.

It may also be mentioned that in the radiation zone $kr \gg k_{ti}$, where the ith mode propagates locally like a radial plane wave, the associated electromagnetic field is transverse. This follows from Eqs. (4a) and (2.5.8) from which it is noted that $\mathbf{E}_{ti}$ and $\mathbf{H}_{ti}$ are $O(1/r)$, whereas E_{ri} and H_{ri} are $O(1/r^2)$ (the scalar and vector mode functions have no radial dependence). Thus, the well-known far-field behavior of a localized source distribution emerges naturally from the spherical transmission-line analysis.

Methods for solving the E- or H-mode equations (1) in the presence of sources are detailed in Secs. 3.3a and 3.3b. From the spherical Bessel functions in (3) and their Wronskian $[j_p(x)n_p'(x) - n_p(x)j_p'(x)] = 1$, one finds that source-free "standing-wave" solutions of Eqs. (1) are expressible in terms of the functions c and s of Eq. (3.3.18); for a radially homogeneous medium with constant ϵ and μ, these have the form

$$c(kr, kr_0) = j_p(kr)n_p'(kr_0) - n_p(kr)j_p'(kr_0), \tag{5a}$$

$$s(kr, kr_0) = \frac{1}{k}\,[n_p(kr)j_p(kr_0) - j_p(kr)n_p(kr_0)], \tag{5b}$$

where the prime denotes the derivative with respect to the argument. The H-mode input admittance of a line terminated at $r = r_{2,1}$, as seen from r_0, may then be determined from Eqs. (3.3.20) and (3.3.21) as

$$\mp \frac{j\omega\mu}{k}\,\overleftrightarrow{Y}_i''(r_0) =$$

$$\frac{[j_p'(x_0)n_p'(x_{2,1}) - n_p'(x_0)j_p'(x_{2,1})] \mp (j\omega\mu/k)\overrightarrow{Y}_{Ti}''[n_p'(x_0)j_p(x_{2,1}) - j_p'(x_0)n_p(x_{2,1})]}{[j_p(x_0)n_p'(x_{2,1}) - n_p(x_0)j_p'(x_{2,1})] \mp (j\omega\mu/k)\overrightarrow{Y}_{Ti}''[n_p(x_0)j_p(x_{2,1}) - j_p(x_0)n_p(x_{2,1})]},$$

$$(6a)$$

with $x_\nu = kr_\nu$, $\nu = 1, 2$. $\overrightarrow{Y}_{Ti}''$ and $\overleftarrow{Y}_{Ti}''$ denote terminal admittances at the end-points $r_2 > r_0$ and $r_1 < r_0$, respectively. The solution for the H-mode voltage Green's function $Z_i''(r, r')$ then follows from Eq. (3.3.22) as

$$Z_i''(r, r') = \frac{[c(x_<, x_0) + j\omega\mu\overleftarrow{Y}_i''(r_0)s(x_<, x_0)][c(x_>, x_0) - j\omega\mu\overrightarrow{Y}_i''(r_0)s(x_>, x_0)]}{\overleftarrow{Y}_i''(r_0) + \overrightarrow{Y}_i''(r_0)}.$$

$$(6b)$$

From Eq. (3.3.26b), the expression for the E-mode input impedances $\overleftrightarrow{Z}_i'(r_0)$ has the same form as Eq. (6a) except for the duality replacements $\mu \leftrightarrow \epsilon$, $Y \to Z$; the E-mode current Green's function $Y_i'(r, r') = j\omega\epsilon g'(r, r')$ then follows from Eq. (3.3.26).

An interesting feature of Eq. (6a) may be noted. When the region includes the origin (i.e., $r_1 = 0$), the input admittance or impedance reduces to [see Eq. (4b)]

$$\zeta\overleftarrow{Y}_i''(r_0) = \frac{\overleftarrow{Z}_i'(r_0)}{\zeta} = -j\frac{j_p'(x_0)}{j_p(x_0)}, \qquad \zeta = \sqrt{\frac{\mu}{\epsilon}}, \tag{7}$$

which expression is independent of the termination $\overleftarrow{Y}_{Ti}''$ or $\overleftarrow{Z}_{Ti}'$. Thus, no boundary condition need be specified at the singular point $r = 0$, the only requirement being the finiteness of the solutions for the voltage and current. As noted in Sec. 3.3b, this point is termed a "limit point" in the theory of differential equations.

An alternative formulation may be carried out in terms of traveling-wave functions as in Eqs. (3.3.28)–(3.3.30). For constant μ, ϵ, wave functions traveling in the $(+r)$ direction are given, for a time dependence $\exp(j\omega t)$, by

$$\left.\begin{array}{c} I_{i+}' \\ V_{i+}'' \end{array}\right\} \propto h_p^{(2)}(kr), \qquad \left.\begin{array}{c} I_{i+}'' \\ V_{i+}' \end{array}\right\} \propto h_p^{\prime(2)}(kr); \tag{8a}$$

in the $-r$ direction,

$$\left.\begin{array}{c} I_{i-}' \\ V_{i-}'' \end{array}\right\} \propto h_p^{(1)}(kr), \qquad \left.\begin{array}{c} I_{i-}'' \\ V_{i-}' \end{array}\right\} \propto h_p^{\prime(1)}(kr). \tag{8b}$$

Thus, for matched (reflectionless) terminations,† the input impedances for the E and H modes are

$$\overrightarrow{\zeta}_i'(r) = j\zeta\,\frac{h_p'^{(2)}(kr)}{h_p^{(2)}(kr)}, \qquad \overleftarrow{\zeta}_i'(r) = -j\zeta\,\frac{h_p'^{(1)}(kr)}{h_p^{(1)}(kr)}, \qquad \zeta = \sqrt{\frac{\mu}{\epsilon}}, \qquad (9a)$$

$$\overrightarrow{\zeta}_i''(r) = -j\zeta\,\frac{h_p^{(2)}(kr)}{h_p'^{(2)}(kr)}, \qquad \overleftarrow{\zeta}_i''(r) = j\zeta\,\frac{h_p^{(1)}(kr)}{h_p'^{(1)}(kr)}, \qquad (9b)$$

and these quantities, together with the input impedances at $r_0 = r$ obtained from Eq. (6) and its dual, may be employed in the calculation of the reflection coefficients in Eqs. (3.3.28), etc.

2.7b Special Terminations

Bilaterally matched region

Bilateral matching occurs in a suitably idealized spherical region wherein the field solution is comprised entirely of waves traveling away from the source (bilateral radiation condition); such a region constitutes the spherical analogue of a biinfinite uniform region. While the matched condition is automatically satisfied at $r \to \infty$ in an unbounded region, it does not obtain at the lower endpoint $r = 0$, which introduces reflection. It is therefore necessary to shield the origin by a reflectionless termination, for example, a "perfectly absorbing" sphere having a radius a [Fig. 2.7.1(a)].‡ As in the analogous problem of propagation

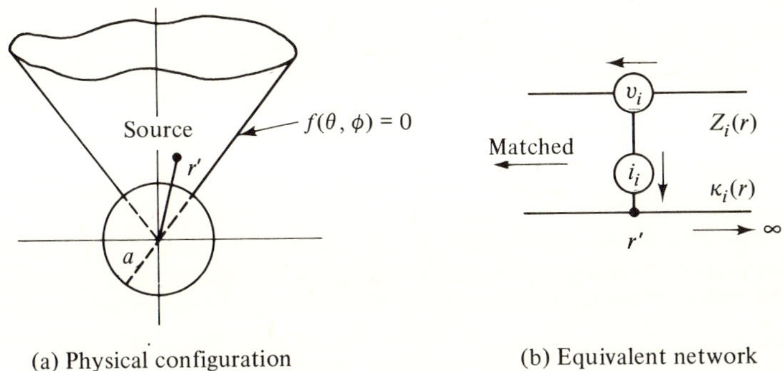

(a) Physical configuration   (b) Equivalent network

FIG. 2.7.1 Bilaterally matched radial region.

on an angular transmission line (Sec. 3.4b), such a boundary, which must absorb completely *all* radially propagating modes, is generally non-physical but forms a useful prototype for subsequent considerations; it may be synthesized approxi-

---

†When observed for $kr \gg p$, a matched termination is taken to imply negligible reflected field; no separate identification of waves traveling along the $+r$ and $-r$ directions is possible when $p \geq kr$.

‡The present discussion concerns the radial domain only and remains valid in the presence of radially independent, perfectly conducting boundaries as shown in Fig. 2.7.1(a).

mately in sufficiently lossy regions. The modal network problem is sketched in Fig. 2.7.1(b), with the voltage and (or) current generators representing the excitation. The input impedances $\overset{\rightleftarrows}{Z}_i(r')$ are now given by their matched values $\overset{\rightleftarrows}{\zeta}_i(r')$, and the reflection coefficients $\overset{\rightarrow}{\Gamma}_i(r')$ equal zero. Equations (6) and their dual then yield, for the modal Green's functions,

$$\frac{Z_i''(r,r')}{\zeta} = Y_i'(r,r')\zeta = \frac{1}{2}\, h_p^{(1)}(kr_<)h_p^{(2)}(kr_>). \tag{10}$$

Homogeneous region, $0 < r < \infty$

When the source is located in an infinite homogeneous region, the boundary at $r = a$ is absent and the modal network problem is the one shown in Fig. 2.7.2. The input impedance seen looking toward $r = \infty$ is still equal to $\overset{\rightarrow}{\zeta}_i(r')$,

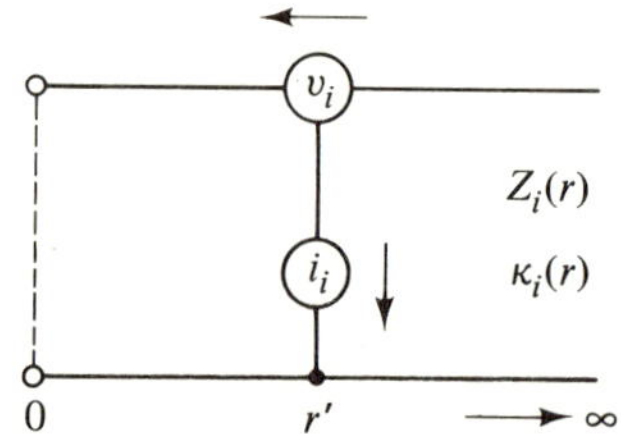

FIG. 2.7.2   Radial region $0 < r < \infty$.

but its value in the other direction must now be taken from Eq. (7). Then from Eq. (6b) and its dual, the modal Green's functions follow as

$$\frac{Z_i''(r,r')}{\zeta} = Y_i'(r,r')\zeta = j_p(kr_<)h_p^{(2)}(kr_>). \tag{11}$$

Semiinfinite homogeneous region, $0 < a \leq r < \infty$

When the obstacle in Fig. 2.7.1(a) is perfectly conducting, the expression in Eq. (11) must be modified to assure the vanishing of the voltage when $r = a$. This may be accomplished either from Eqs. (6a) and (6b), with $\overset{\leftarrow}{Y}_{Ti}'' = \infty$ (see Fig. 2.7.3, where a short circuit appears at $r = a$), or directly by inspection. The following result is obtained:

$$\frac{Z_i''(r,r')}{\zeta} = \left[j_p(kr_<) - \frac{j_p(ka)}{h_p^{(2)}(ka)}\, h_p^{(2)}(kr_<) \right] h_p^{(2)}(kr_>). \tag{12a}$$

For the E modes, the requirement $V_i'(a) = 0$ implies via Eq. (2.5.9) the vanishing of $(dI_i'/dr)_{r=a}$, whence

$$Y_i'(r,r')\zeta = \left[j_p(kr_<) - \frac{j_p'(ka)}{h_p'^{(2)}(ka)}\, h_p^{(2)}(kr_<) \right] h_p^{(2)}(kr_>). \tag{12b}$$

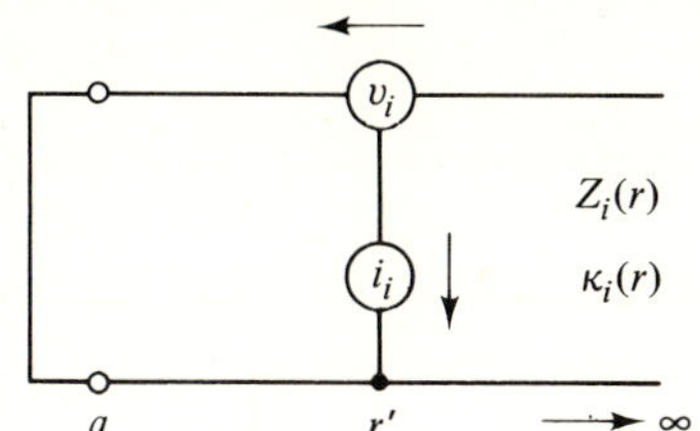

FIG. 2.7.3 Perfectly conducting sphere.

The expressions for $Z_i''(r, r')$ and $Y_i'(r, r')$ are no longer identical as in Eqs. (10) and (11) because the perfectly conducting boundary is not its own dual.

Composite region, $0 < r < \infty$

If the region $r > a$ is filled with a homogeneous medium having constitutive parameters ϵ, μ, while the region $0 < r < a$ is characterized by ϵ_1, μ_1, one has the network problem shown in Fig. 2.7.4. The continuity of the tangential components of **E** and **H** at $r = a$ is assured from the continuity of V_i and I_i

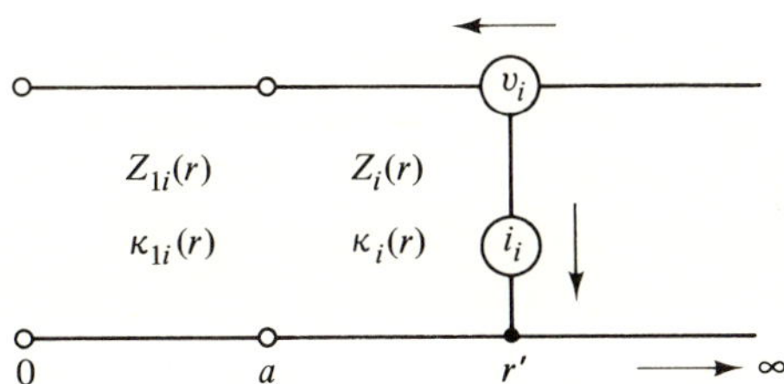

FIG. 2.7.4 Dielectric sphere.

[see Eqs. (2.5.8a) and (2.5.8b)], so the equivalent network involves only a simple junction. When the source is located outside the sphere, the transmission line is matched for $r > r'$ and reflection occurs only at $r = a$. The solution may then be constructed by adding to the bilaterally matched Green's functions in Eq. (10) the reflected wave contribution; there is no need to analyze the problem from the beginning. The situation is more involved when the source is located inside the sphere, in which instance reflections occur from both ends of the transmission line.

The E-mode current reflection coefficient $\overleftarrow{\Gamma}'_{Ii}(a+)$ seen slightly to the left of the point $r = a$ in Fig. 2.7.4 is given from Eqs. (3.3.29a) and (3.3.29c) by

$$\overleftarrow{\Gamma}'_{Ii}(a+) = \frac{[\overleftarrow{\zeta}'_i(a)/\overleftarrow{Z}'_i(a)] - 1}{[\overrightarrow{\zeta}'_i(a)/\overleftarrow{Z}'_i(a)] + 1}, \tag{13}$$

where the $\overrightarrow{\zeta}'_i(a)$ represent the matched impedances seen from $r = a$ in the transmission-line descriptive of the region exterior to the sphere,

$$\overleftarrow{\zeta'_i}(a) = -j\zeta \frac{h_p'^{(1)}(ka)}{h_p^{(1)}(ka)}, \quad \overrightarrow{\zeta'_i}(a) = j\zeta \frac{h_p'^{(2)}(ka)}{h_p^{(2)}(ka)}, \quad k = \omega\sqrt{\mu\epsilon}, \quad \zeta = \sqrt{\frac{\mu}{\epsilon}},$$

$$(13a)$$

while $\overleftarrow{Z'_i}(a)$ denotes the input impedance seen to the left from $r = a$ [see Eq. (7)],

$$\overleftarrow{Z'_i}(a) = -j\zeta_1 \frac{j'_p(k_1 a)}{j_p(k_1 a)}, \quad k_1 = \omega\sqrt{\mu_1\epsilon_1}, \quad \zeta_1 = \sqrt{\frac{\mu_1}{\epsilon_1}}. \qquad (13b)$$

If the incident current is taken as $h_p^{(1)}(kr)$, the reflected current at $r = a$ has an amplitude $\overleftarrow{\Gamma'_{Ii}} h_p^{(1)}(ka)$, so its value at any point must be $[h_p^{(1)}(ka)/h_p^{(2)}(ka)]\overleftarrow{\Gamma'_{Ii}} h_p^{(2)}(kr)$. Upon combining this result with Eq. (10), one finds

$$Y'_i(r, r')\zeta = \frac{1}{2}\left[h_p^{(1)}(kr_<) + \frac{h_p^{(1)}(ka)}{h_p^{(2)}(ka)}\overleftarrow{\Gamma'_{Ii}}(a)h_p^{(2)}(kr_<) \right]h_p^{(2)}(kr_>), \qquad a \leq r < \infty,$$

$$(14)$$

and an analogous expression for $Z''_i(r, r')$. The behavior of the modal Green's function inside the sphere may in this simple case be obtained by inspection from the requirements that the current remains finite at $r = 0$ and continuous at $r = a$:

$$Y'_i(r, r') = Y'_i(a, r')\frac{j_p(k_1 r)}{j_p(k_1 a)}, \qquad 0 \leq r \leq a. \qquad (15)$$

The previously derived results may be recovered as special cases from Eq. (14). For the bilaterally matched configuration, $\overleftarrow{\Gamma'_{Ii}} = 0$. For the perfectly conducting sphere, obtained in the limit as $|\epsilon_1| \to \infty$, $\arg \epsilon_1 \to -\pi/2$, one has $\overleftarrow{Z'_i}(a) \to 0$, and, therefrom, Eq. (12b). Equation (11) follows on letting $a \to 0$.

PROBLEMS

1. A homogeneously filled uniform waveguide has a cross section S bounded by the curve $s = s_1 + s_2$. The following boundary conditions are assumed on the two portions s_1 and s_2 which comprise the waveguide boundary: s_1 is a perfect electric conductor ($\mathbf{E}_{\text{tan}} = 0$) while s_2 is a perfect magnetic conductor ($\mathbf{H}_{\text{tan}} = 0$).

    (a) Show that the fields in this waveguide may be decomposed into E and H modes, and formulate the eigenvalue problem for the scalar mode functions from which the vector modes may be derived.

    (b) Apply these results to the determination of the mode functions in a rectangular waveguide, one side wall of which is a perfect magnetic conductor while the remaining walls are perfect electric conductors.

2. The dyadic Green's functions $\mathscr{Z}(\mathbf{r}, \mathbf{r}')$ and $\mathscr{Y}(\mathbf{r}, \mathbf{r}')$ defined in Eqs. (2.3.27) are given in an unbounded homogeneous medium with constant ϵ, μ by

$$\sqrt{\frac{\epsilon}{\mu}}\,\mathscr{Z}(\mathbf{r}, \mathbf{r}') = \sqrt{\frac{\mu}{\epsilon}}\,\mathscr{Y}(\mathbf{r}, \mathbf{r}') = jk\left[\mathbf{1} + \frac{\nabla\nabla}{k^2}\right]G_f(\mathbf{r}, \mathbf{r}'), \qquad (1a)$$

where $G_f = (4\pi|\mathbf{r} - \mathbf{r}'|)^{-1}\exp[-jk|\mathbf{r} - \mathbf{r}'|]$ [assumed time dependence $\exp(j\omega t)$]. **1** is the unit dyadic $(\mathbf{1} \cdot \mathbf{A} = \mathbf{A} \cdot \mathbf{1} = \mathbf{A})$. When $r \to \infty$, show that Eq. (1a) reduces to the far field relation

$$\sqrt{\frac{\epsilon}{\mu}}\,\mathscr{L} = \sqrt{\frac{\mu}{\epsilon}}\,\mathscr{Y} \sim jk\,\frac{e^{-jkr+jk(\mathbf{r}\cdot\mathbf{r}')/r}}{r}(\mathbf{1} - \mathbf{r}_0\mathbf{r}_0), \qquad r \gg r', \tag{1b}$$

where $\mathbf{r}_0$ is a unit vector in the r direction.

3.   (a) Show that for an $\exp(j\omega t)$ dependence, the free-space dyadic Green's functions in Eqs. (1) have for $r \neq r'$ the spherical mode representation

$$\zeta\mathscr{Y}(\mathbf{r}, \mathbf{r}') = \sum_i \mathscr{H}'^{(1)}_i(\mathbf{r}_<)\mathscr{H}'^{(+)}_i(\mathbf{r}_>) + \sum_i \mathscr{H}''^{(1)}_i(\mathbf{r}_<)\mathscr{H}''^{(+)}_i(\mathbf{r}_>), \tag{2}$$

$$\frac{1}{\zeta}\mathscr{L}(\mathbf{r}, \mathbf{r}') = \sum_i \mathscr{E}'^{(1)}_i(\mathbf{r}_<)\mathscr{E}'^{(+)}_i(\mathbf{r}_>) + \sum_i \mathscr{E}''^{(1)}_i(\mathbf{r}_<)\mathscr{E}''^{(+)}_i(\mathbf{r}_>), \qquad \zeta = \sqrt{\frac{\mu}{\epsilon}}, \tag{3}$$

where $\mathbf{r}_< = \mathbf{r}$ when $r < r'$, $\mathbf{r}_< = \mathbf{r}'$ when $r > r'$, with the converse applying to $\mathbf{r}_>$. Also, if $h_p(kr)$ denotes the spherical Hankel function,

$$r\mathscr{H}'^{(+,-)}_i(\mathbf{r}) = h^{(2,1)}_p(kr)\mathbf{h}'_i(\theta, \phi), \qquad k'^2_{ti} = p(p+1),$$

$$r\mathscr{H}''^{(+,-)}_i(\mathbf{r}) = h'^{(2,1)}_p(kr)\mathbf{h}''_i(\theta, \phi) - jh^{(2,1)}_p(kr)\zeta Y''_i\mathbf{h}''_{ri}(\theta, \phi), \tag{4a}$$

$$\mathscr{H}^{(1)}_i(\mathbf{r}) = \frac{1}{2}[\mathscr{H}^{(-)}_i(\mathbf{r}) + \mathscr{H}^{(+)}_i(\mathbf{r})], \quad \mathscr{H}^{(2)}_i(\mathbf{r}) = \frac{1}{2j}[\mathscr{H}^{(-)}_i(\mathbf{r}) - \mathscr{H}^{(+)}_i(\mathbf{r})],$$

and

$$r\mathscr{E}'^{(+,-)}_i(\mathbf{r}) = h'^{(2,1)}_p(kr)\mathbf{e}'_i(\theta, \phi) - \frac{j}{\zeta}h^{(2,1)}_p(kr)Z'_i\mathbf{e}'_{ri}(\theta, \phi), \qquad k''^2_{ti} = p(p+1),$$

$$r\mathscr{E}''^{(+,-)}_i(\mathbf{r}) = h^{(2,1)}_p(kr)\mathbf{e}''_i(\theta, \phi), \tag{4b}$$

$$\mathscr{E}^{(1)}_i(\mathbf{r}) = \frac{1}{2}[\mathscr{E}^{(-)}_i(\mathbf{r}) + \mathscr{E}^{(+)}_i(\mathbf{r})], \qquad \mathscr{E}^{(2)}_i(\mathbf{r}) = \frac{1}{2j}[\mathscr{E}^{(-)}_i(\mathbf{r}) - \mathscr{E}^{(+)}_i(\mathbf{r})].$$

     The transverse and longitudinal mode functions are defined in Eqs. (2.6.1) and (2.5.10c, 2.5.10d), and their explicit form is given by Eqs. (3.4.63), (3.2.51b) and (3.4.79a).

   (b) Apply Eqs. (1)–(3) to calculate the field of an electric current element with strength J located at r' on the z axis and oriented parallel to the x axis. By letting $r' \to \infty$, show that the identification

$$-\frac{j\omega\mu e^{-jkr'}J}{4\pi r'} \to 1 \tag{5a}$$

yields an incident plane wave of unit amplitude.

     Note: Show first from Eq. (6.8.2a), with $r' \to \infty$ and $\theta = 0$, that

$$kre^{-jkr} = \sum_{n=0}^{\infty}(2n+1)(-j)^n j_n(kr). \tag{5b}$$

   (c) What modification must be included in Eqs. (2)–(4) when $r = r'$ [see Eq. (2.5.8e)]?

   (d) Show that the preceding formulas remain valid in the presence of conical boundaries provided that one employs the appropriate eigenfunctions in Sec. 3.4b.

4. Show that the normalized vector mode functions for the TEM mode in a biconical region bounded by the surface $\theta = \theta_{1,2}$ are as follows:

$$\mathbf{e}'_{0,0}(\theta, \phi) = \mathbf{\theta}_0 \frac{1}{\sin \theta} \left[2\pi \ln \left(\frac{\tan (\theta_2/2)}{\tan (\theta_1/2)} \right) \right]^{-1/2}, \qquad 0 < \theta_1 \le \theta \le \theta_2 < \pi, \qquad (6)$$

$$\mathbf{h}'_{0,0} = \mathbf{r}_0 \times \mathbf{e}'_{0,0}, \qquad e'_{ri} = h'_{ri} = 0. \qquad (6a)$$

5. Derive the modal Green's functions (radial transmission-line solutions) when the source in Fig. 2.7.4 is situated in the region $0 < r' < a$.

6. The electric field in an aperture of arbitrary shape perforating a perfectly conducting infinite screen of negligible thickness may be represented equivalently by a magnetic current distribution $\mathbf{M}(\mathbf{r}') = \mathbf{E}(\mathbf{r}') \times \mathbf{n}$ placed on the unperforated screen (Sec. 1.5b). $\mathbf{E}(\mathbf{r}')$ denotes the aperture electric field and $\mathbf{n}$ the unit normal pointing into the half-space region. The power $\bar{S}$ radiated into the half-space is given by Eq. (2.5.11b).

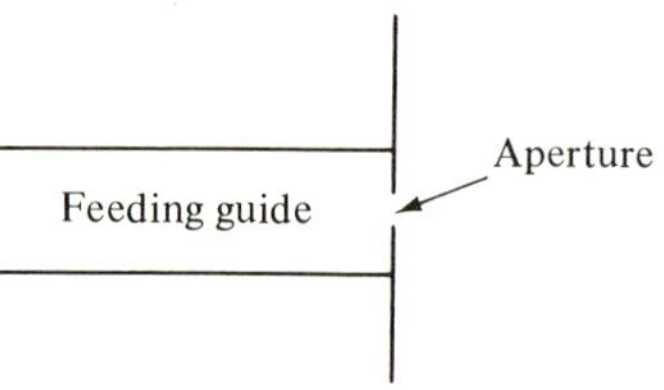

FIG. P2.1   Radiating aperture.

(a) Show that the voltages required in Eq. (2.5.11b) are given by:

$$|V_i(r)| = \left| \iint_A \mathbf{M}(\mathbf{r}') \cdot \mathscr{H}_i^{(1)}(\mathbf{r}') \, dS' \right|, \qquad (7)$$

where $\mathscr{H}_i^{(1)}$ is defined in Eq. (4a) and A is the area of the aperture.

(b) Assume that the aperture is fed on one side by a waveguide propagating a single mode (Fig. P2.1); at a reference plane removed "many" guide wavelengths from the aperture, the voltage in this mode is given by

$$\hat{V} = \iint_A \mathbf{M}(\mathbf{r}') \cdot \hat{\mathbf{h}}(\mathbf{r}') \, dS', \qquad (8)$$

where $\hat{\mathbf{h}}$ is the real transverse vector mode function for the dominant mode in the guide (see Sec. 2.3a). Assume that $\mathbf{M}$, and therefore $\hat{V}$, is real. Show that if $\hat{I}$ is the dominant mode current in the waveguide referred to the aperture plane, the radiation conductance G of the aperture as seen from the waveguide may be defined as

$$G = \frac{\operatorname{Re} \hat{I}}{\hat{V}} = \frac{1}{\zeta} \sum_i \frac{|V_i'(r)|^2}{\hat{V}^2} + \frac{1}{\zeta} \sum_i \frac{|V_i''(r)|^2}{\hat{V}^2}, \qquad (9)$$

which expression is obtained from power flow considerations at $r \to \infty$ in the half-space, or by

$$G = \frac{\text{Re}\,\hat{I}}{\hat{V}} = \frac{\text{Re} \displaystyle\iint_A \mathbf{E} \times \mathbf{H}^* \cdot \mathbf{n}\, dS}{\hat{V}^2}$$

$$= \frac{\displaystyle\iint_A dS \iint_A dS'\,\mathbf{M}(\mathbf{r}) \cdot [\text{Re}\,\mathscr{Y}_h^-(\mathbf{r}, \mathbf{r}')] \cdot \mathbf{M}(\mathbf{r}')}{\hat{V}^2}, \tag{10}$$

which results from the total power flow calculated at the aperture. The half-space dyadic admittance $\mathscr{Y}_h(\mathbf{r}, \mathbf{r}')$ for $\mathbf{r}'$ on A is equal to twice the free space admittance $\mathscr{Y}(\mathbf{r}, \mathbf{r}')$ in Eq. (1a). Show that the aperture field satisfies the integral equation

$$(\text{Re}\,\hat{I})\hat{\mathbf{h}}(\mathbf{r}) = \iint_A [\text{Re}\,\mathscr{Y}_h(\mathbf{r}, \mathbf{r}')] \cdot \mathbf{M}(\mathbf{r}')\, dS' \tag{11a}$$

$$= \frac{1}{\zeta} \sum_i |V_i'(r)|\mathscr{H}_i'^{(1)}(\mathbf{r}) + \frac{1}{\zeta} \sum_i |V_i''(r)|\mathscr{H}_i''^{(1)}(\mathbf{r}), \quad \mathbf{r} \text{ in } A. \tag{11b}$$

Show also that the expressions for G are stationary (variational) with respect to small variations of $\mathbf{M}$ about the correct value specified in Eqs. (11), and that the result obtained for G by substituting an approximate "trial function" for $\mathbf{M}$ is larger than the true conductance. (See R. E. Collin, *Field Theory of Guided Waves*, McGraw-Hill Book Co., New York (1960), Chapter 8.)

7. Various elementary distributions of magnetic current as shown in Fig. P2.2 are placed on a perfectly conducting infinite plane. Consider two possible coordinate systems, both having the common origin shown in Fig. P2.2. In system I, the z axis is perpendicular to the plane whereas in system II, the z axis is taken along the horizontal axis in Fig. P2.2. Show that the following spherical E modes (E_{mn}) and H modes (H_{mn}) are excited by these arrangements [refer to Problem 6 of Chapter 3 for forms of the scalar mode functions]:

System I

(a) H_{11};

(b) H_{mn}, m odd, n odd (lowest: H_{11});

(c) H_{mn}, m even, n even (lowest: H_{02}, H_{22});

(d) E_{mn}, m odd, n even and H_{mn}, m, n odd (lowest: H_{11});

(e) E_{mn}, m even, n odd and H_{mn}, m, n even (lowest: E_{01}, H_{22});

(f) E_{mn}, m even, n odd and H_{mn}, m, n even (lowest: E_{01}, H_{22});

(g) E_{0n}, n odd (lowest: E_{01});

(h) H_{0n}, n even (lowest: H_{02}).

System II

(a) H_{01};

(b) H_{0n}, n odd (lowest: H_{01});

(c) H_{0n}, n even (lowest: H_{02});

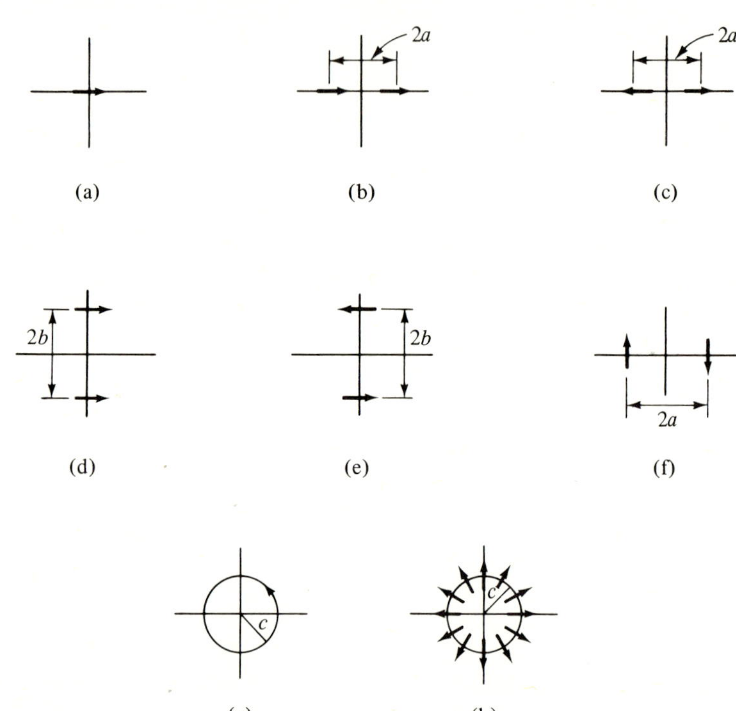

FIG. P2.2 Elementary source distributions: (a) single element; (b)-(f) various symmetrical arrangements of two elements; (g) ring current; (h) ring of radial elemets.

(d) E_{mn}, m, n even and H_{mn}, m even, n odd (lowest: H_{01});

(e) E_{mn}, m, n odd and H_{mn}, m odd, n even (lowest: E_{11}, H_{12});

(f) E_{1n}, n odd and H_{1n}, n even (lowest: E_{11}, H_{12});

(g) E_{mn}, m, n odd (lowest: E_{11});

(h) H_{mn}, m, n even (lowest: H_{02}, H_{22}).

The modes designated as "lowest" remain dominant when the dimensions a, b, c are small compared to the wavelength.

8.   When the aperture in Problem 6 is small compared to the wavelength, the spherical mode series in Eq. (9) is rapidly convergent. If the field distribution in the aperture is representable in terms of one of the elementary source configurations in Fig. P2.2, the dominant contribution to the conductance G arises from the corresponding lowest mode(s).

(a) Rectangular slot terminating a rectangular waveguide:
A symmetrically placed rectangular slot having dimensions a', b' terminates a rectangular waveguide with dimensions a, b. When the waveguide is ex-

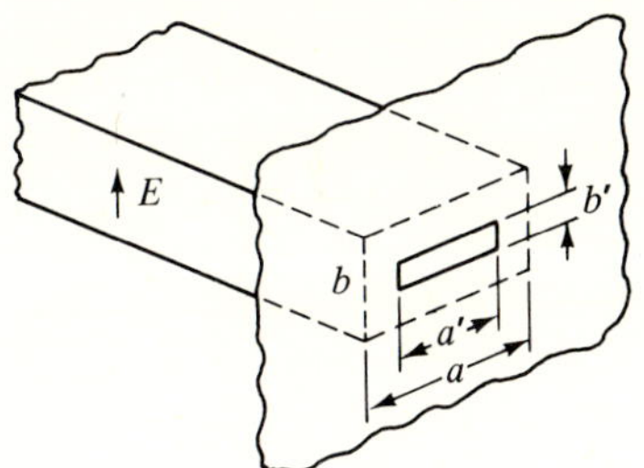

FIG. P2.3   Rectangular configuration.

cited in the dominant mode, a reasonable "guess value" for the induced aperture field is

$$\mathbf{E}(\mathbf{r}') \times \mathbf{n} = \mathbf{M}(\mathbf{r}') = \mathbf{z}_0 \cos \frac{\pi z'}{a'}, \tag{12}$$

where the z axis has been chosen parallel to the a dimension of the guide. This source distribution excites predominantly the H_{01} spherical mode (see Problem 7). When the slot dimensions are small, show that

$$\frac{G}{\hat{Y}_0} \cong \frac{1}{\zeta \hat{Y}_0} \frac{|V_{01}''(r)|^2}{\hat{V}^2} \approx \frac{2\pi}{3} ab \frac{\lambda_g}{\lambda^3} \frac{[1-(a'/a)^2]^2}{[\cos(\pi a'/2a)]^2} \left\{1 + O\left[\left(\frac{a'}{\lambda}\right)^2\right]\right\}, \tag{13}$$

where $\hat{Y}_0$ and λ_g are the admittance and the guide wavelength of the dominant (H_{10}) mode in the rectangular waveguide, and λ is the free-space wavelength [for $\hat{Y}_0$ and $\lambda_g = 2\pi/\kappa$, see Eqs. (2.2.15d)].

(b) Annular slot terminating a coaxial waveguide:

A coaxial waveguide with dimensions a, b and excited in the TEM mode is terminated in a concentric annular slot having dimensions a', b' (Fig. P2.4).

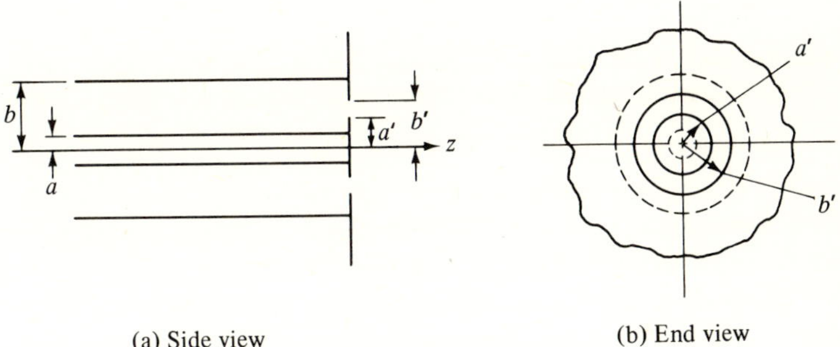

(a) Side view   (b) End view

FIG. P2.4   Coaxial configuration.

A simple trial function for the electric field in the slot is $\mathbf{E}(\mathbf{r}') = \rho_0(1/\rho')$, the variation in the incident TEM mode, so that

$$\mathbf{M}(\mathbf{r}') = -\boldsymbol{\phi}_0 \frac{1}{\rho'}, \tag{14}$$

where ρ is the radial coordinate transverse to the guide axis which has been chosen coincident with z (in view of the rotational symmetry of the aperture field, this coordinate choice is natural). Show that for small values of (b'/λ) and (a'/λ),

$$\frac{G}{\hat{Y}_0} \approx \frac{|V'_{10}(r)|^2}{\hat{V}^2} \approx \frac{2\ln(b/a)}{3[\ln(b'/a')]^2}\left[\left(\frac{\pi b'}{\lambda}\right)^2 - \left(\frac{\pi a'}{\lambda}\right)^2\right]^2. \tag{15}$$

Show that the complete spherical mode expansion for the conductance is [refer to Problem 6 of Chapter 3 for the scalar mode functions]:

$$\frac{G}{\hat{Y}_0} = \frac{\ln(b/a)}{\left[\int_{a'}^{b'} M(r')\,dr'\right]^2} \sum_{n=0}^{\infty} \frac{(4n+3)}{(2n+1)(2n+2)}\left[\frac{d}{d\theta}P_{2n+1}(\cos\theta)\right]^2_{\theta=\pi/2}$$

$$\times \left[\int_{a'}^{b'} M(r')j_n(kr')\,dr'\right]^2. \tag{16}$$

REFERENCES

1. MARCUVITZ, N. and J. SCHWINGER, "On the representation of the electric and magnetic fields produced by currents and discontinuities in waveguides. I," *J. Appl. Phys.* **22**, No. 6 (1951), pp. 806–819.

2. MARCUVITZ, N., *Waveguide Handbook*, Secs. 1.2 and 1.3. New York: McGraw-Hill, 1951.

3. SCHELKUNOFF, S. A., *Electromagnetic Waves*, Chapter 7. Princeton: D. Van Nostrand Co., 1943.

4. PHILIPS, H. B., *Vector Analysis*, Sec. 8.3. New York: John Wiley & Sons, 1933.

5. STRATTON, J. A., *Electromagnetic Theory*, Chapter 6. New York: McGraw-Hill, 1941.

6. BORGNIS, F. E., and C. H. PAPAS, "*Electromagnetic Waveguides*," Chapter in Vol. 16, in *Handbuch der Physik*. Berlin: Springer Verlag, 1958.

7. STRATTON, J. A., *Electromagnetic Theory*, Sec. 1.18. New York: McGraw-Hill, 1941.

8. MARCUVITZ, N., "Field representations in spherically stratified regions," *Comm. Pure and Appl. Math.* **4** (1951), pp. 263–315.

9. MARCUVITZ, N., *Waveguide Handbook*, Sec. 1.8. New York: McGraw-Hill, 1951.

10. STRATTON, J. A., *Electromagnetic Theory*, Chapter 7. New York: McGraw-Hill, 1941.

11. BREMMER, H., *Terrestrial Radio Waves*, Chapter 2. New York: Elsevier Publishing Co., 1949.

12. SCHELKUNOFF, S. A., *Electromagnetic Waves*, Secs. 10.10 and 10.11. Princeton: D. Van Nostrand Co., 1943.

3. Mode Functions in Closed and Open Waveguides

3.1 INTRODUCTION

By the modal analysis and synthesis procedure of Chapter 2, electromagnetic fields in isotropically filled uniform waveguide regions are representable as superpositions of modal fields. The latter have z-dependent amplitudes described by transmission-line voltages and currents, and transverse vector characteristics determined by the shape of the guide cross section. Solutions of the transmission-line equations in media piecewise homogeneous along the waveguide axis z have been given in Sec. 2.4, and their network interpretation has been emphasized. The present chapter deals with evaluation of the transverse vector eigenfunction characteristics. As noted in Sec. 2.3a, the solution of the vector eigenvalue problem is facilitated by introduction of scalar eigenfunctions satisfying appropriate scalar eigenvalue problems in the cross-sectional domain. This scalarization, equivalent to a field decomposition into E and H modes, is possible for homogeneously filled cross sections but fails when the medium properties depend on the cross-sectional vector coordinate, $\boldsymbol{\rho}$. Nevertheless, for $\boldsymbol{\rho}$-dependent medium parameters, the vector-field problem may still be scalarized if the waveguide region admits a transmission-line analysis along one of the transverse coordinates. This procedure is illustrated in Sec. 3.2d. Section 3.2d also contains a generalized derivation of the transverse field equations and modal representations of the electromagnetic field in transversely *inhomogeneous* regions.

Scalarization permits evaluation of vector-mode functions by solution of scalar eigenvalue problems in the transverse domain.† Little analytical progress

---

†It should be emphasized that scalarization is automatic in an acoustic field whose properties are derivable from a scalar pressure p. While solutions of scalar eigenvalue problems in this chapter are phrased in electromagnetic terms, they are applicable to the acoustic field on simple reinterpretation of relevant variables.

can be made unless the geometry is separable, in which event a two-dimensional eigenvalue problem can be reduced to two one-dimensional problems. Treatment of the one-dimensional case is thus of fundamental importance and is of primary concern in this of chapter. The relevant eigenvalue equation is of the Sturm–Liouville type and exhibits certain general features that are reviewed in Sec. 3.2a. Eigenfunctions for closed and open regions describable in rectangular and circular cylindrical coordinates are derived in Secs. 3.2b and 3.2c by classical techniques, and orthonormality and completeness are expressed succinctly by spectral representations of the (unit operator) delta function $\delta(\boldsymbol{\rho} - \boldsymbol{\rho}')$. Anticipating subsequent function-theoretic applications to the construction of alternative field representations, some attention is given in Secs 3.2b and 3.2c to analytic continuation of modal representations into the complex plane of the spectral variable.

The classical procedure for solving eigenvalue problems leads to difficulties in open regions (one or both domain endpoints at infinity) since the mode spectra may then be continuous and the eigenfunctions improper. To normalize the mode set under these circumstances, one may pass to the infinite limit from an originally bounded domain, as illustrated in Secs. 3.2b and 3.2c for rectangular and circular geometries, respectively. A more powerful and direct technique is provided by the characteristic Green's function (resolvent) procedure of Sec. 3.3, based on the intimate relation between resonant solutions (eigenfunctions) and the response to point-source excitation already noted in Sec. 2.4e. This method, formulated in Secs. 3.3.1 and 3.3.2 for the Sturm–Liouville differential operator describing propagation on a general non-uniform transmission line, is relevant not only for one-dimensional eigenvalue problems but also for source problems in inhomogeneous media. Questions of mode completeness and normalization for both discrete and continuous eigenfunctions are answered systematically by exploring the singularities of the characteristic Green's function in the complex plane. The above-mentioned analytic continuation of spectral representations forms an integral part of this procedure, thereby facilitating the construction of alternative field representations as is shown in Sec. 3.3c. Detailed applications of the method to eigenvalue problems in various geometrical configurations are given in Sec. 3.4.

Although the theory of the Sturm–Liouville differential equation or, equivalently, propagation on a non-uniform transmission line, can be discussed in some generality, explicit solutions in terms of known functions are possible only for special inhomogeneity profiles. A variety of such special solutions is presented in Secs. 3.2 and 3.4. Under more general conditions, explicit construction of the field behavior requires approximation procedures whose success relies on the ability to represent the solution to the given problem as a weak perturbation of a known solution to a related problem. If the unperturbed problem is described by a differential equation that exhibits in certain critical regions (near turning points, singularities, etc.) the same analytical behavior as the desired problem, solution of the latter can be constructed by systematic techniques detailed in Sec. 3.5. One aspect of this procedure involves reformulation of the differential equation problem as an integral equation whose kernel constitutes an unperturbed ("comparison") Green's function, and subsequent solution by iteration. Various techniques are illustrated by discussion of specific examples.

3.2 CLASSICAL EVALUATION OF MODE FUNCTIONS

3.2a General One-Dimensional Eigenvalue Problem

Before proceeding with the calculation of scalar two-dimensional eigenfunctions for various separable geometries encompassed by Eqs. (2.3.2), we consider briefly some characteristics of one-dimensional eigenfunctions and the associated eigenvalues. The determination of the eigenfunctions f_m and the eigenvalues λ_m in the domain $x_1 \leq x \leq x_2$ poses a problem of the Sturm–Liouville type:[1]

$$\left[\frac{d}{dx}p(x)\frac{d}{dx} - q(x) + \lambda_m w(x)\right]f_m(x) = 0, \qquad x_1 \leq x \leq x_2, \qquad (1)$$

subject to the homogeneous boundary conditions

$$p\frac{df_m}{dx} + \alpha_{1,2}f_m = 0, \qquad x = x_{1,2}, \qquad (1a)$$

where p, q, and the weight function w are assumed to be piecewise continuous functions of x in $x_1 \leq x \leq x_2$. The boundary condition in Eq. (1a) is of the "impedance" type, as may be noted from the discussion relating to Eqs. (2.3.41) [see also Eqs. (3.3.5)].

We show first that the eigenvalues λ_m are real for real p, q, w, and $\alpha_{1,2}$, the so-called Hermitian case corresponding to a non-dissipative medium. Upon multiplying Eq. (1) by f_m^*, where * denotes the complex conjugate, integrating over x between the limits of x_1 and x_2, and using integration by parts and the boundary conditions in Eq. (1a) on the first integral, one finds

$$\lambda_m = \frac{\displaystyle\int_{x_1}^{x_2} dx\, p|(df_m/dx)|^2 + \int_{x_1}^{x_2} dx\, q|f_m|^2 - \alpha_1|f_m(x_1)|^2 + \alpha_2|f_m(x_2)|^2}{\displaystyle\int_{x_1}^{x_2} dx\, w|f_m|^2}. \qquad (2)$$

Since the right-hand side of Eq. (2) is real for real values of p, q, w, $\alpha_{1,2}$, it follows that λ_m is real in this case.

To deduce the orthogonality property of the eigenfunctions for the Hermitian case, one multiplies Eq. (1) by the eigenfunction f_n^* belonging to the eigenvalue $\lambda_n^* = \lambda_n$ and integrates over the x domain to obtain

$$\int_{x_1}^{x_2} dx\, f_n^* \frac{d}{dx}\left(p\frac{df_m}{dx}\right) - \int_{x_1}^{x_2} dx\, qf_n^* f_m + \lambda_m \int_{x_1}^{x_2} dx\, wf_n^* f_m = 0. \qquad (3)$$

Similarly, starting from the defining equation for f_n^*, multiplying by f_m and integrating, one obtains Eq. (3) except for the interchange of m and n. Upon subtracting the second equation from the first, rearranging the terms and carrying out a simple integration by parts, one finds

$$(\lambda_m - \lambda_n) \int_{x_1}^{x_2} dx\, w f_n^* f_m = \left[p \left(f_m \frac{df_n^*}{dx} - f_n^* \frac{df_m}{dx} \right) \right]_{x_1}^{x_2}. \tag{4}$$

In view of the boundary conditions (1a), the right-hand side of Eq. (4) vanishes, leading to the orthogonality property of the eigenfunctions f_m and f_n^* relative to the weight factor w:

$$\int_{x_1}^{x_2} dx\, w f_m f_n^* = 0, \qquad m \neq n. \tag{5a}$$

The eigenfunctions are normalized to unity by the requirement that

$$\int_{x_1}^{x_2} dx\, w |f_m|^2 = 1. \tag{5b}$$

The set of eigenfunctions f_m comprising all possible solutions of Eq. (1) constitutes a complete set that can be employed to represent a permissible function $F(x)$ in the interval $x_1 < x < x_2$ [a permissible function is one for which the representations below exist; i.e., the sums or integrals in Eqs. (6) converge]:

$$F(x) = \sum_m F_m f_m(x), \tag{6a}$$

where the sum extends over all eigenfunctions f_m. From the orthonormality property of the f_m functions in Eqs. (5) one evaluates the transform F_m as

$$F_m = \int_{x_1}^{x_2} d\xi\, w(\xi) F(\xi) f_m^*(\xi). \tag{6b}$$

The completeness and orthonormality of the set f_m can be expressed concisely in a symbolic manner by choosing for $F(x)$ the delta function $\delta(x - x')$. Then

$$F_m = \int_{x_1}^{x_2} d\xi\, w(\xi) \delta(\xi - x') f_m^*(\xi) = w(x') f_m^*(x'), \tag{7a}$$

so that from Eq. (6a) one infers the completeness relation

$$\frac{\delta(x - x')}{w(x')} = \sum_m f_m(x) f_m^*(x'), \qquad x_1 < \frac{x}{x'} < x_2. \tag{7b}$$

The representation of a permissible function $F(x)$ as in Eq. (6a) follows from Eq. (7b) upon multiplication by $F(x')w(x')$ and integration over x' between the limits x_1 and x_2.

The considerations above are based on the assumption that the eigenvalue spectrum is simple and discrete. Continuous spectra, when applicable, can be derived therefrom by the limiting procedure of Secs. 3.2b (semiinfinite region) and 3.2c (open angular sector). A more direct treatment in terms of characteristic Green's functions is given in Sec. 3.3.

3.2b Homogeneously Filled Rectangular Cross Sections

Finite rectangular region

The cross-sectional geometry is bounded by perfectly conducting walls and illustrated in Fig. 3.2.1. The transverse operator ∇_t^2 is represented in this case by

$$\nabla_t^2 = \frac{\partial^2}{\partial x^2} + \frac{\partial^2}{\partial y^2}. \tag{8}$$

Upon assuming a solution of the form

$$\Phi_i(\boldsymbol{\rho}) = \Phi_p(x)\Phi_q(y), \qquad \Phi_i(\boldsymbol{\rho}) = 0 \quad \text{on s}, \tag{9}$$

where Φ_p and Φ_q are functions of x and y, respectively, one reduces Eqs. (2.3.2a) and (2.3.2b) to the two one-dimensional equations

$$\left(\frac{d^2}{dx^2} + p^2\right)\Phi_p(x) = 0, \qquad \Phi_p(0) = \Phi_p(a) = 0, \tag{10a}$$

$$\left(\frac{d^2}{dy^2} + q^2\right)\Phi_q(y) = 0, \qquad \Phi_q(0) = \Phi_q(b) = 0, \tag{10b}$$

where p^2 and q^2 are "separation" constants in terms of which the transverse wave-number k'_{ti} of Eq. (2.3.2a) is given by

$$k_{ti}^{\prime 2} = p^2 + q^2. \tag{10c}$$

Similarly, one assumes for the *H*-mode functions defined in Eqs. (2.3.2c) and (2.3.2d),

$$\psi_i(\boldsymbol{\rho}) = \psi_p(x)\psi_q(y), \qquad \frac{\partial \psi_i}{\partial v} = 0 \quad \text{on } s, \tag{11}$$

from which it follows that ψ_p and ψ_q satisfy one-dimensional equations as in Eqs. (10) with the boundary conditions

$$\frac{\partial \psi_p}{\partial x} = 0 \quad \text{at } x = 0, a; \qquad \frac{\partial \psi_q}{\partial y} = 0 \quad \text{at } y = 0, b. \tag{12}$$

The solution of the eigenvalue problems posed in Eqs. (10) is readily found to be

$$\Phi_p(x) = \sqrt{\frac{2}{a}}\sin px, \quad p = \frac{m\pi}{a}, \qquad m = 1, 2, 3, \cdots, \tag{13a}$$

$$\Phi_q(y) = \sqrt{\frac{2}{b}}\sin qy, \quad q = \frac{n\pi}{b}, \qquad n = 1, 2, 3, \cdots, \tag{13b}$$

where the multiplicative constants have been so chosen as to normalize the mode sets to unity, i.e.,

$$\int_0^a \Phi_p^2(x)dx = 1 = \int_0^b \Phi_q^2(y)dy. \tag{14a}$$

One readily verifies the orthogonality of the eigenfunctions from the relation

$$\int_0^a \sin\frac{m\pi x}{a}\sin\frac{m'\pi x}{a}\,dx = 0, \qquad m \neq m'; \quad m, m' = 1, 2, 3, \cdots, \tag{14b}$$

and similarly for the $\Phi_q(y)$, so Eqs. (14a) and (14b) can be subsumed into the single orthonormality relation

$$\int_0^a \Phi_p(x)\Phi_{p'}(x)\,dx = \delta_{pp'} = \begin{cases} 1, & p = p' \\ 0, & p \neq p'. \end{cases} \tag{14c}$$

Since Eqs. (10) and (12) are evident specializations of Eqs. (1), the orthonormality properties of the one-dimensional eigenfunctions could have been anticipated from Eqs. (5) without the explicit calculation in Eqs. (14). As in Eq. (7b), completeness and orthonormality can be expressed concisely as

$$\delta(x - x') = \sum_{m=1}^{\infty} \Phi_p(x)\Phi_p(x') = \frac{2}{a}\sum_{m=1}^{\infty}\sin\frac{m\pi x}{a}\sin\frac{m\pi x'}{a}, \qquad 0 < \frac{x}{x'} < a \tag{15}$$

whence the representation of a permissible function $F(x)$ follows from Eq. (15) upon multiplication by $F(x')$ and integration over x' between the limits 0 and a, while the orthonormality relation (14) is deduced upon choosing for $F(x)$ the eigenfunction $\Phi_{p'}(x)$. In a directly analogous manner, one has for the y domain,

$$\delta(y - y') = \sum_{n=1}^{\infty} \Phi_q(y)\Phi_q(y') = \frac{2}{b}\sum_{n=1}^{\infty}\sin\frac{n\pi y}{b}\sin\frac{n\pi y'}{b}, \qquad 0 < \frac{y}{y'} < b. \tag{16}$$

The representation of the two-dimensional delta function $\delta(\boldsymbol{\rho} - \boldsymbol{\rho}') = \delta(x - x')\delta(y - y')$ for the rectangular domain in Fig. 1 follows directly from the knowledge of the two one-dimensional representations above as

$$\delta(\boldsymbol{\rho} - \boldsymbol{\rho}') = \begin{cases} \displaystyle\sum_i \Phi_i(\boldsymbol{\rho})\Phi_i(\boldsymbol{\rho}'), & 0 < \dfrac{x}{x'} < a,\ 0 < \dfrac{y}{y'} < b, & (17a) \\[2ex] \displaystyle\frac{4}{ab}\sum_{m=1}^{\infty}\sum_{n=1}^{\infty}\sin\frac{m\pi x}{a}\sin\frac{n\pi y}{b}\sin\frac{m\pi x'}{a}\sin\frac{n\pi y'}{b}, & (17b) \end{cases}$$

from which the desired two-dimensional mode functions in Eqs. (2.3.2) are

$$\Phi_i(\boldsymbol{\rho}) = \frac{2}{\sqrt{ab}}\sin\frac{m\pi x}{a}\sin\frac{n\pi y}{b}, \qquad m, n = 1, 2, 3, \cdots, \tag{18a}$$

and the transverse wavenumbers

$$k_{ti}'^2 = \left(\frac{m\pi}{a}\right)^2 + \left(\frac{n\pi}{b}\right)^2. \tag{18b}$$

$\sum_i$ in this instance represents the double sum $\sum_{m=1}^{\infty}\sum_{n=1}^{\infty}$. The product representation of the delta function in Eq. (17a) is obtained from the requirement

$$1 = \iint_S \delta(\boldsymbol{\rho} - \boldsymbol{\rho}')\,dS = \int_0^a dx \int_0^b dy\, \delta(x - x')\delta(y - y'), \qquad \boldsymbol{\rho}' \text{ in } S. \tag{19}$$

The orthonormality of the two-dimensional set $\Phi_i(\rho)$ is assured from that of $\Phi_p(x)$ and $\Phi_q(y)$ since for $i = (p, q), j = (p', q')$,

$$\iint_S \Phi_i(\rho)\Phi_j(\rho)\, dS = \int_0^a dx \int_0^b dy\, \Phi_p(x)\Phi_q(y)\Phi_{p'}(x)\Phi_{q'}(y) = \delta_{pp'}\delta_{qq'} = \delta_{ij}.$$

$$(20)$$

A permissible function $F(\rho)$ in the rectangular domain can now be represented via Eq. (17a) as

$$F(\rho) = \iint_S F(\rho')\delta(\rho - \rho')\, dS' = \sum_i F_i\Phi_i(\rho), \tag{21a}$$

$$F_i = \iint_S F(\rho')\Phi_i(\rho')\, dS'. \tag{21b}$$

Similarly, one deduces the mode functions ψ_i appropriate to the rectangular cross 'section in Fig. 3.2.1 and to the boundary conditions in Eq. (11). The orthonormal functions $\psi_p(x)$ and $\psi_q(y)$ are given by

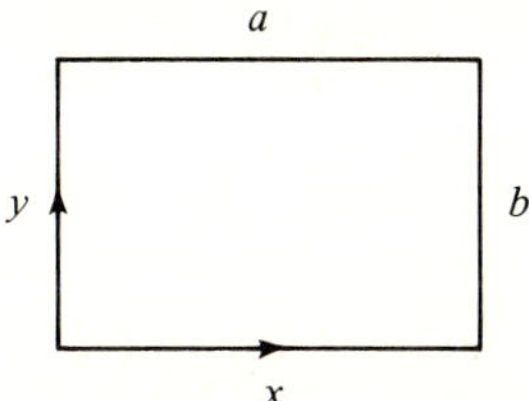

FIG. 3.2.1   Finite rectangular region.

$$\psi_p(x) = \sqrt{\frac{2}{a}}\cos px, \quad p = \frac{m\pi}{a}, \qquad m = 1, 2, \cdots; \qquad \psi_0(x) = \frac{1}{\sqrt{a}}; \tag{22a}$$

$$\psi_q(y) = \sqrt{\frac{2}{b}}\cos qy, \quad q = \frac{n\pi}{b}, \qquad n = 1, 2, \cdots; \qquad \psi_0(y) = \frac{1}{\sqrt{b}}. \tag{22b}$$

As in Eq. (15), the completeness and orthonormality of the mode set $\psi_p(x)$ is conveniently expressed via the delta-function representation

$$\delta(x - x') = \sum_{m=0}^{\infty} \psi_p(x)\psi_p(x') = \frac{1}{a}\sum_{m=0}^{\infty}\epsilon_m \cos px \cos px', \qquad 0 < \frac{x}{x'} < a,$$

$$(23)$$

with $\epsilon_0 = 1, \epsilon_m = 2, m \geq 1$, while for the two-dimensional mode set $\psi_i(\rho)$,

$$\delta(\rho - \rho') = \sum_i \psi_i(\rho)\psi_i(\rho'), \qquad 0 < \frac{x}{x'} < a, \quad 0 < \frac{y}{y'} < b,$$

$$= \frac{1}{ab}\sum_{m=0}^{\infty}\sum_{n=0}^{\infty}\epsilon_m\epsilon_n \cos\frac{m\pi x}{a}\cos\frac{n\pi y}{b}\cos\frac{m\pi x'}{a}\cos\frac{n\pi y'}{b}. \tag{24}$$

Thus,

$$\psi_i(\mathbf{\rho}) = \sqrt{\frac{\epsilon_m \epsilon_n}{ab}} \cos\frac{m\pi x}{a} \cos\frac{n\pi y}{b}, \qquad m, n = 0, 1, 2, \cdots, \tag{25a}$$

and

$$k_{ti}^{\prime\prime 2} = \left(\frac{m\pi}{a}\right)^2 + \left(\frac{n\pi}{b}\right)^2. \tag{25b}$$

Semiinfinite rectangular region

If the *a* dimension of the waveguide in Fig. 3.2.1 is allowed to become infinite, the rectangular waveguide configuration goes over into the semiinfinite rectangular trough shown in Fig. 3.2.2. The transition to the open region is traced out by defining

$$p \equiv \xi_m = \frac{m\pi}{a}, \qquad \Delta\xi_m = \xi_{m+1} - \xi_m = \frac{\pi}{a}. \tag{26}$$

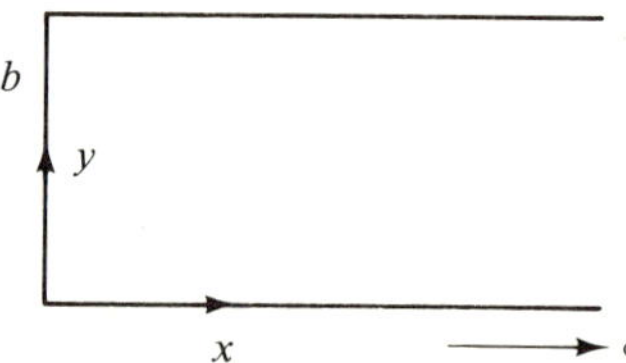

FIG. 3.2.2   Semi-infinite rectangular region.

As *a* becomes very large, one notes that the eigenvalues $p = m\pi/a, m = 1, 2,$ $\cdots$, fall closer and closer together until they coalesce in the limit into a continuous spectrum. Upon substituting Eq. (26) into Eq. (15), and letting $a \to \infty$ (i.e., $\Delta\xi_m \to 0$), one obtains

$$\delta(x - x') = \lim_{\Delta\xi_m \to 0} \frac{2}{\pi} \sum_{\xi_m = \Delta\xi_m}^{\infty} \sin(\xi_m x) \sin(\xi_m x') \Delta\xi_m \tag{27a}$$

$$= \frac{2}{\pi} \int_0^{\infty} \sin \xi x \sin \xi x' \, d\xi, \qquad 0 < \frac{x}{x'} < \infty, \tag{27b}$$

so

$$\Phi_\xi(x) \equiv \Phi(\xi, x) = \sqrt{\frac{2}{\pi}} \sin \xi x, \qquad 0 < \xi < \infty. \tag{27c}$$

The sum $\sum_{m=1}^{\infty}$ in Eq. (15) is replaced here by the integral $\int_0^{\infty} d\xi$, since the eigenvalues are continuous.

The continuous eigenfunctions in Eq. (27c) are improper [i.e., the normalizing integral $(2/\pi)\int_0^{\infty} dx \sin^2 \xi x$ does not exist (is infinite)]. The normalization constant $\delta_{pp'}$ in Eqs. (14) is replaced in this case by the delta function $\delta(\xi - \xi')$, as is verified upon multiplying Eq. (27b) by $\sin \xi' x'$, integrating from $x' = 0$ to $x' = \infty$, and interchanging the order of integration on the right-hand side (the interchange of the order of integration is essential for the deduction of a transform theorem and is assumed valid for the class of permissible functions):

$$\sin \xi' x = \int_0^\infty \sin \xi x \left(\frac{2}{\pi} \int_0^\infty \sin \xi' x' \sin \xi x' \, dx' \right) d\xi$$

$$= \int_0^\infty \sin \xi x \, \delta(\xi - \xi') \, d\xi, \tag{28a}$$

i.e.,

$$\frac{2}{\pi} \int_0^\infty \sin \xi' x' \sin \xi x' \, dx' = \delta(\xi - \xi'), \tag{28b}$$

which relation is evidently of the same form as Eq. (27b). $\Phi_\xi(x)$ in Eq. (27c) is seen to satisfy the required boundary condition $\Phi_\xi(0) = 0$. The lack of a boundary condition at $x = \infty$ is a consequence of the singular (limit point) character of the endpoint at infinity[2] [see the footnote to Eq. (3.3.21)].

The two-dimensional completeness statement for *E modes* in the semiinfinite region of Fig. 3.2.2 can be written as

$$\delta(\boldsymbol{\rho} - \boldsymbol{\rho}') = \begin{cases} \displaystyle\sum_i \Phi_i(\boldsymbol{\rho})\Phi_i(\boldsymbol{\rho}'), & 0 < \dfrac{x}{x'} < \infty, \ \ 0 < \dfrac{y}{y'} < b, & (29a) \\[2em] \displaystyle\frac{4}{\pi b} \int_0^\infty d\xi \sum_{n=1}^\infty \sin \xi x \sin \frac{n\pi y}{b} \sin \xi x' \sin \frac{n\pi y'}{b}, & & (29b) \end{cases}$$

i.e.,

$$\Phi_i(\boldsymbol{\rho}) = \frac{2}{\sqrt{\pi b}} \sin \xi x \sin \frac{n\pi y}{b}, \qquad 0 < \xi < \infty, \quad n = 1, 2, 3, \cdots,$$

$$k_{ti}'^2 = \xi^2 + \left(\frac{n\pi}{b} \right)^2. \tag{29c}$$

It is noted that $\sum_i$ in Eq. (29a) stands in this instance for $\int_0^\infty d\xi \sum_{n=1}^\infty$ since the mode set in the x domain is continuous.

For the *H-mode functions* one obtains, by a similar limiting process,

$$\delta(x - x') = \frac{2}{\pi} \int_0^\infty \cos \xi x \cos \xi x' \, d\xi, \qquad 0 < \frac{x}{x'} < \infty, \tag{30a}$$

so

$$\psi_\xi(x) = \sqrt{\frac{2}{\pi}} \cos \xi x, \qquad 0 < \xi < \infty. \tag{30b}$$

Thus,

$$\delta(\boldsymbol{\rho} - \boldsymbol{\rho}') = \begin{cases} \displaystyle\sum_i \psi_i(\boldsymbol{\rho})\psi_i(\boldsymbol{\rho}'), & 0 < \dfrac{x}{x'} < \infty, \ \ 0 < \dfrac{y}{y'} < b, & (31a) \\[2em] \displaystyle\frac{2}{\pi b} \int_0^\infty d\xi \sum_{n=0}^\infty \epsilon_n \cos \xi x \cos \frac{n\pi y}{b} \cos \xi x' \cos \frac{n\pi y'}{b}, & & (31b) \end{cases}$$

i.e.,

$$\psi_i(\boldsymbol{\rho}) = \sqrt{\frac{2\epsilon_n}{\pi b}} \cos \xi x \cos \frac{n\pi y}{b}, \qquad 0 < \xi < \infty, \quad n = 0, 1, 2, \cdots,$$

$$k_{ti}''^2 = \xi^2 + \left(\frac{n\pi}{b}\right)^2. \tag{31c}$$

Quarter-space region

Upon letting the b dimension in Fig. 3.2.2 increase indefinitely, one obtains the quarter-space region in Fig. 3.2.3 and, from the preceding section, the following completeness statements (the continuously variable index q is denoted by η):

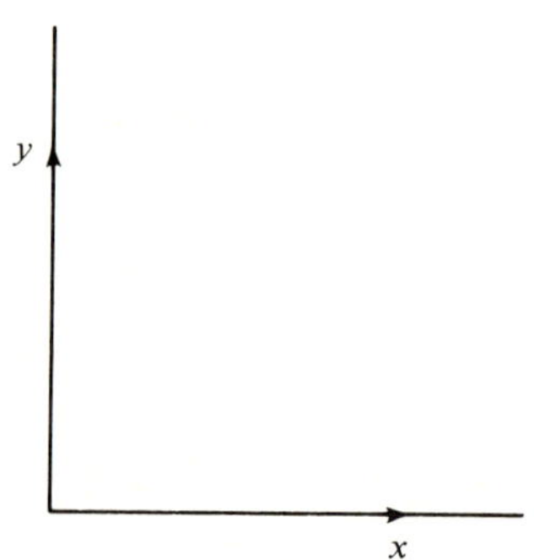

FIG. 3.2.3   Quarter-space region.

E modes:

$$\delta(\boldsymbol{\rho} - \boldsymbol{\rho}') = \begin{cases} \sum_i \Phi_i(\boldsymbol{\rho})\Phi_i(\boldsymbol{\rho}'), & 0 < \dfrac{x}{x'} < \infty, \quad 0 < \dfrac{y}{y'} < \infty, & (32a) \\[2ex] \dfrac{4}{\pi^2} \displaystyle\int_0^\infty d\xi \int_0^\infty d\eta \, \sin \xi x \sin \eta y \sin \xi x' \sin \eta y', & (32b) \end{cases}$$

i.e.,

$$\Phi_i(\boldsymbol{\rho}) = \frac{2}{\pi} \sin \xi x \sin \eta y, \quad 0 < \xi < \infty, \quad 0 < \eta < \infty; \quad k_{ti}'^2 = \xi^2 + \eta^2.$$

$$\tag{32c}$$

The symbol $\sum_i$ denotes in this case the double integral $\displaystyle\int_0^\infty d\xi \int_0^\infty d\eta$.

H modes:

$$\delta(\boldsymbol{\rho} - \boldsymbol{\rho}') = \begin{cases} \sum_i \psi_i(\boldsymbol{\rho})\psi_i(\boldsymbol{\rho}'), & 0 < \dfrac{x}{x'} < \infty, \quad 0 < \dfrac{y}{y'} < \infty, & (33a) \\[2ex] \dfrac{4}{\pi^2} \displaystyle\int_0^\infty d\xi \int_0^\infty d\eta \, \cos \xi x \cos \eta y \cos \xi x' \cos \eta y', & (33b) \end{cases}$$

i.e.,

$$\psi_i(\mathbf{\rho}) = \frac{2}{\pi} \cos \xi x \cos \eta y, \quad 0 < \xi < \infty, \ \ 0 < \eta < \infty; \quad k_{ti}''^2 = \xi^2 + \eta^2.$$

$$(33c)$$

Half-space region

A half-space region $x > 0$ is shown in Fig. 3.2.4. Eigenfunctions for the infinite interval $-\infty < y < \infty$ are deduced most simply by considering not the bounded domain $0 < y < b$ as above, but instead $-b/2 < y < b/2$, and

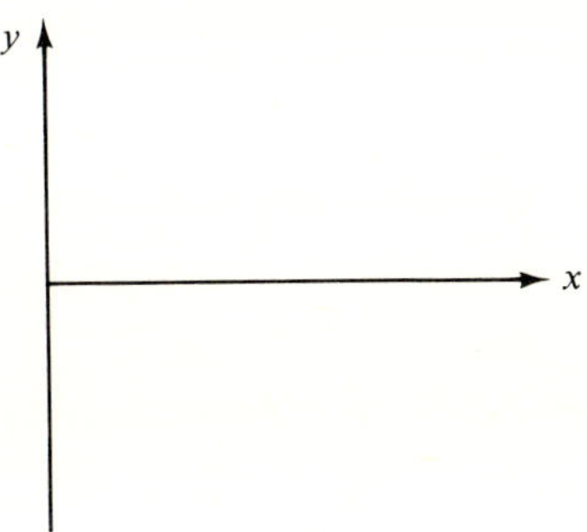

FIG. 3.2.4 Half-space region.

then letting $b \to \infty$. One finds the delta-function representation for the latter range directly from Eq. (16) upon introducing the change of variable $y \to y - b/2$, $y' \to y' - b/2$:

$$\delta(y - y') = \frac{2}{b} \sum_{n=1}^{\infty} \cos \frac{(2n-1)\pi y}{b} \cos \frac{(2n-1)\pi y'}{b}$$

$$+ \frac{2}{b} \sum_{n=1}^{\infty} \sin \frac{2n\pi y}{b} \sin \frac{2n\pi y'}{b}, \qquad -\frac{b}{2} < \frac{y}{y'} < \frac{b}{2} \qquad (34a)$$

i.e., the eigenfunctions Φ_q in this case comprise the two mutually orthogonal sets

$$\Phi_q(y) = \sqrt{\frac{2}{b}} \left\{ \begin{array}{c} \cos \dfrac{(2n-1)\pi y}{b} \\[2mm] \sin \dfrac{2n\pi y}{b} \end{array} \right\}, \qquad n = 1, 2, 3, \cdots, \quad -\frac{b}{2} < y < \frac{b}{2},$$

$$(34b)$$

which are even and odd in y, respectively. Upon defining in the first sum in Eq. (34a),

$$\eta_n = \frac{(2n-1)\pi}{b}, \qquad \Delta\eta_n = \eta_{n+1} - \eta_n = \frac{2\pi}{b}, \qquad (35a)$$

and in the second sum,

$$\eta_n = \frac{2n\pi}{b}, \qquad \Delta\eta_n = \frac{2\pi}{b}, \tag{35b}$$

and going to the limit $b \to \infty$, one obtains

$$\delta(y - y') = \begin{cases} \dfrac{1}{\pi}\displaystyle\int_0^\infty \cos \eta y \cos \eta y' \, d\eta + \dfrac{1}{\pi}\int_0^\infty \sin \eta y \sin \eta y' \, d\eta, & \text{(36a)} \\[2ex] \dfrac{1}{\pi}\displaystyle\int_0^\infty \cos \eta(y - y') \, d\eta, & -\infty < \dfrac{y}{y'} < \infty. \quad\text{(36b)} \end{cases}$$

For the H-mode functions, a similar result is obtained if the transition $b \to \infty$ is carried out for the $\psi_q(y)$ in Eq. (22b) after defining the latter over the interval $-b/2 < y < b/2$ as above. The lack of dependence of the infinite interval eigenfunctions $\Phi_n(y) = \psi_n(y)$ on the boundary conditions for the finite interval is a consequence of the limit-point type of singularlity at $y = \pm\infty$ [see the footnote to Eq. (3.3.21)]. Thus we have, from Eq. (36a),

$$\Phi_n(y) = \psi_n(y) = \sqrt{\frac{1}{\pi}} \begin{Bmatrix} \sin \eta y \\ \cos \eta y \end{Bmatrix}, \qquad -\infty < y < \infty, \quad 0 < \eta < \infty. \tag{36c}$$

An alternative and frequently more useful representation of the delta function is obtained upon expressing the cosine terms in Eq. (36b) as the sum of two exponentials:†

$$\begin{aligned} \delta(y - y') &= \frac{1}{2\pi}\int_0^\infty e^{j\eta(y-y')}\, d\eta + \frac{1}{2\pi}\int_0^\infty e^{-j\eta(y-y')}\, d\eta \\ &= \frac{1}{2\pi}\int_{-\infty}^\infty e^{j\eta(y-y')}\, d\eta = \frac{1}{2\pi}\int_{-\infty}^\infty e^{j\eta(y'-y)}\, d\eta \qquad\text{(37a)} \\ &= \sum_n \Phi_n(y)\Phi_n^*(y') = \sum_n \psi_n(y)\psi_n^*(y'). \end{aligned}$$

In this Fourier integral representation, the eigenfunctions are complex and the delta-function representation involves the complex conjugate [see Eq. (7b)]. The eigenvalue parameter η ranges from $-\infty$ to $+\infty$. Thus,

$$\Phi_n(y) = \psi_n(y) = \frac{1}{\sqrt{2\pi}} e^{-j\eta y}, \qquad -\infty < y < \infty, \quad -\infty < \eta < \infty. \tag{37b}$$

The orthogonality condition analogous to Eq. (28b) is now, with respect to the complex conjugate,

$$\frac{1}{2\pi}\int_{-\infty}^\infty e^{j\eta' y'} e^{-j\eta y'} \, dy' = \delta(\eta - \eta'). \tag{37c}$$

An advantage of the representation in Eq. (37a) is the simple deformability of the contour of integration into the complex η plane. If $y > y'$, the integrand in the first integral in Eq. (37a) decays exponentially in the upper half of the complex η plane (Im $\eta > 0$), so the endpoints of the integration contour can be shifted from the real η axis to $\eta = -\infty + j\epsilon$ and $\eta = +\infty + j\delta$, with ϵ, $\delta > 0$. Similarly, if $y < y'$, the contour of integration in the second integral

---

†Note that the use of $j = \sqrt{-1}$ does not imply a preferred time dependence at this stage.

representation in Eq. (37a) can be deformed into the upper half of the η plane. The utility of these formulations in the complex plane for transformation of a given field representation into an alternative one will be emphasized in Sec. 3.3.

In summary, two-dimensional delta-function representations for the half-space region are as follows:

E modes

$$\delta(\boldsymbol{\rho} - \boldsymbol{\rho}') = \sum_i \Phi_i(\boldsymbol{\rho})\Phi_i^*(\boldsymbol{\rho}'), \qquad 0 > \frac{x}{x'} < \infty, \quad -\infty < \frac{y}{y'} < \infty,$$

$$= \frac{1}{\pi^2} \int_0^\infty d\xi \int_{-\infty}^\infty d\eta \, \sin \xi x \, e^{-j\eta y} \sin \xi x' e^{+j\eta y'}, \tag{38a}$$

i.e.,

$$\Phi_i(\boldsymbol{\rho}) = \frac{1}{\pi} \sin \xi x \, e^{-j\eta y}, \qquad 0 < \xi < \infty, \quad -\infty < \eta < \infty; \quad k_{ti}'^2 = \xi^2 + \eta^2. \tag{38b}$$

H modes

$$\delta(\boldsymbol{\rho} - \boldsymbol{\rho}') = \sum_i \psi_i(\boldsymbol{\rho})\psi_i^*(\boldsymbol{\rho}'), \qquad 0 < \frac{x}{x'} < \infty, \quad -\infty < \frac{y}{y'} < \infty,$$

$$= \frac{1}{\pi^2} \int_0^\infty d\xi \int_{-\infty}^\infty d\eta \, \cos \xi x \, e^{-j\eta y} \cos \xi x' e^{+j\eta y'}, \tag{39a}$$

i.e.,

$$\psi_i(\boldsymbol{\rho}) = \frac{1}{\pi} \cos \xi x \, e^{-j\eta y}, \qquad 0 < \xi < \infty, \quad -\infty < \eta < \infty, \quad k_{ti}''^2 = \xi^2 + \eta^2. \tag{39b}$$

Free-space region

For the free-space region shown in Fig. 3.2.5, the E- and H-mode scalar eigenfunctions $\Phi_i(\boldsymbol{\rho})$ and $\psi_i(\boldsymbol{\rho})$ are seen to be identical. The eigenvalue problem in x leads to the same Fourier integral representation as in Eq. (37a), so the two-dimensional free-space representation constitutes the two-dimensional Fourier integral theorem:

$$\delta(\boldsymbol{\rho} - \boldsymbol{\rho}') = \sum_i \Phi_i(\boldsymbol{\rho})\Phi_i^*(\boldsymbol{\rho}') = \sum_i \psi_i(\boldsymbol{\rho})\psi_i^*(\boldsymbol{\rho}'), \qquad -\infty < \frac{x}{x'} < \infty,$$

$$-\infty < \frac{y}{y'} < \infty,$$

$$= \frac{1}{4\pi^2} \int_{-\infty}^\infty d\xi \int_{-\infty}^\infty d\eta \, e^{-j(\xi x + \eta y)} e^{+j(\xi x' + \eta y')}, \tag{40a}$$

i.e.,

$$\Phi_i(\boldsymbol{\rho}) = \psi_i(\boldsymbol{\rho}) = \frac{1}{2\pi} e^{-j(\xi x + \eta y)}, \qquad -\infty < \xi < \infty, \quad -\infty < \eta < \infty;$$

$$k_{ti}^2 = \xi^2 + \eta^2. \tag{40b}$$

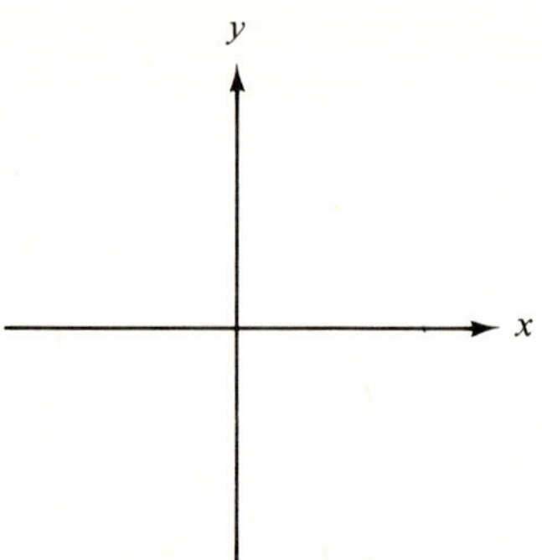

FIG. 3.2.5 Free-space region.

Parallel-plate region

For the parallel-plate region shown in Fig. 3.2.6, the y-dependent eigenfunctions are those in Eq. (37b), while those for the finite x domain are given in Eqs. (13a) and (22a), respectively. Thus, for E modes,

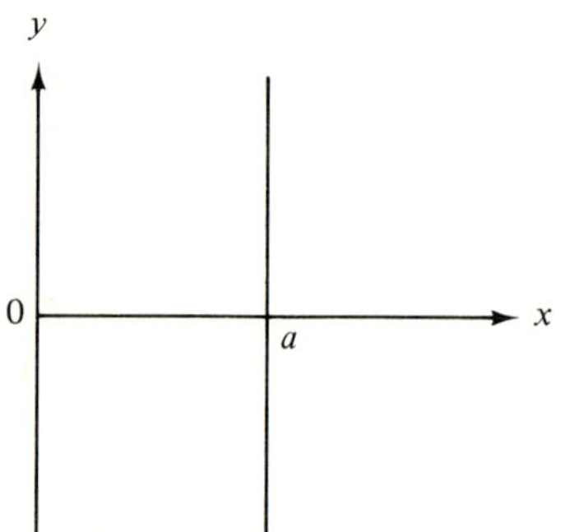

FIG. 3.2.6 Parallel-plate region.

$$\delta(\boldsymbol{\rho} - \boldsymbol{\rho}') = \sum_i \Phi_i(\boldsymbol{\rho})\Phi_i^*(\boldsymbol{\rho}'), \qquad 0 < \frac{x}{x'} < a, \quad -\infty < \frac{y}{y'} < \infty,$$

$$= \frac{1}{\pi a} \sum_{m=1}^{\infty} \int_{-\infty}^{\infty} d\eta \, \sin\frac{m\pi x}{a} e^{-jny} \sin\frac{m\pi x'}{a} e^{jny'}, \tag{41a}$$

i.e.,

$$\Phi_i(\boldsymbol{\rho}) = \frac{1}{\sqrt{\pi a}} \sin\frac{m\pi x}{a} e^{-jny}, \qquad m = 1, 2, 3, \cdots, \quad -\infty < \eta < \infty;$$

$$k_{ti}'^2 = \left(\frac{m\pi}{a}\right)^2 + \eta^2; \tag{41b}$$

for H modes,

$$\delta(\mathbf{\rho} - \mathbf{\rho}') = \sum_i \psi_i(\mathbf{\rho})\psi_i^*(\mathbf{\rho}'), \qquad 0 < \frac{x}{x'} < a, \quad -\infty < \frac{y}{y'} < \infty,$$

$$= \frac{1}{2\pi a} \sum_{m=0}^{\infty} \int_{-\infty}^{\infty} d\eta \, \epsilon_m \cos\frac{m\pi x}{a} \, e^{-jny} \cos\frac{m\pi x'}{a} \, e^{jny'}, \qquad (42a)$$

i.e.,

$$\psi_i(\mathbf{\rho}) = \sqrt{\frac{\epsilon_m}{2\pi a}} \cos\frac{m\pi x}{a} \, e^{-jny}, \qquad m = 0, 1, 2, \cdots, \quad -\infty < \eta < \infty;$$

$$k_{ti}''^2 = \left(\frac{m\pi}{a}\right)^2 + \eta^2, \qquad \epsilon_0 = 1, \qquad \epsilon_m = 2, \quad m \geq 1. \qquad (42b)$$

Transmission-line interpretation of one-dimensional eigenvalue problem

The one-dimensional eigenvalue problems in the preceding sections can be interpreted in transmission-line terms as defining resonant voltage or current solutions on an appropriately terminated source-free transmission line. Consider, for example, the eigenvalue problem in the x domain as stated in Eq. (10a). If x is taken as the transmission coordinate, the equivalent transmission-line configuration is that shown in Fig. 2.4.10 with $z_1 \equiv x_1 = 0, z_2 \equiv x_2 = a, \overleftarrow{Z}_T = 0 = \overrightarrow{Z}_T$. The source-free voltage solutions on such a transmission line are defined by the homogeneous second-order differential equation (2.4.2), which is identical with Eq. (10a), provided that $z \equiv x, \kappa \equiv p$. In view of the requirement that $\Phi_p(0) = \Phi_p(a) = 0$ in Eq. (10a) (i.e., the vanishing of the transmission-line solutions at the short-circuit terminations), one identifies $\Phi_p(x)$ as the resonant voltage $V(x)$. Equation (2.4.7a), with $z' \equiv x' = 0$, leads upon imposition of the boundary conditions $V(0) = V(a) = 0$ to the resonant solutions in Eq. (13a), to within an arbitrary multiplying constant. Similarly, from the source-free transmission-line equations, V is proportional to dI/dx [see Eq. (2.4.1)]. Hence, the eigenfunction $\psi_p(x)$ satisfying the boundary conditions in Eq. (12) can be identified as the current $I(x)$ on a short-circuited transmission line. The eigensolutions in Eq. (22a) then follow directly from Eq. (2.4.7b). Alternatively, one may employ the dual configuration of a transmission line terminated in open circuits at $x = 0, a$ (i.e., $\overleftarrow{Z}_T = \overrightarrow{Z}_T = \infty$). In this instance, the currents vanish at $x = 0, a$, so $\Phi_p(x)$ and $\psi_p(x)$ can be identified as the resonant current and voltage solutions, respectively.

As emphasized in Sec. 2.4e, the source-free solutions on a terminated transmission line are intimately related to the singularities of the modal Green's function $Z(x, x')$ or $Y(x, x')$ defined in Eqs. (2.4.28) and (2.4.29) (with z, z' replaced by x, x'). This connection is highlighted further by an examination of the voltage and current solutions $V(x, x')$ and $I(x, x')$ in Eqs. (2.4.22) excited by voltage and current sources v and i at x'. To obtain a finite response in the absence of excitation (i.e., when $v = i = 0$), one requires via Eqs. (2.4.22) the vanishing of the total impedance $\overleftrightarrow{Z}(x')$ and admittance $\overleftrightarrow{Y}(x')$ [see also Eq. (2.4.36)]:

$$\overset{\leftrightarrow}{Z}(x') = 0 = \overset{\leftarrow}{Z}(x') + \vec{Z}(x'), \tag{43a}$$

or, equivalently,

$$\overset{\leftrightarrow}{Y}(x') = 0 = \overset{\leftarrow}{Y}(x') + \vec{Y}(x'), \tag{43b}$$

where the choice of x' is arbitrary. Equations (43) constitute "transverse resonance" relations that can be satisfied only for resonant values of the propagation constant $k_{xr} = p$. For the case of a simple transmission line of length a short-circuited at both ends, one chooses $x' = 0$ so that $\overset{\leftarrow}{Z}(0) = 0$; then one has, via Eqs. (2.4.24a) and (43a), $\vec{Z}(0) = jZ_0 \tan pa = 0$, i.e.,

$$p = \frac{m\pi}{a}, \qquad m = 1, 2, \cdots, \tag{43c}$$

as in Eq. (13a). The corresponding source-free solutions (eigenfunctions) are then given by the expressions inside the second set of brackets of Eqs. (2.4.22b). For eigenvalue problems of a general type, the above relation between the singularities (resonances) of the modal Green's functions and the source-free solutions on a terminated transmission line will be elaborated in Sec. 3.3a.

3.2c Homogeneously Filled Cylindrical Cross Sections

Coaxial cross-sectional waveguide configurations of interest are the coaxial sector and annulus shown in Fig. 3.2.7. The sector is bounded by perfectly conducting cylindrical segments at $\rho = a, b$ and radial plane segments at $\phi =$

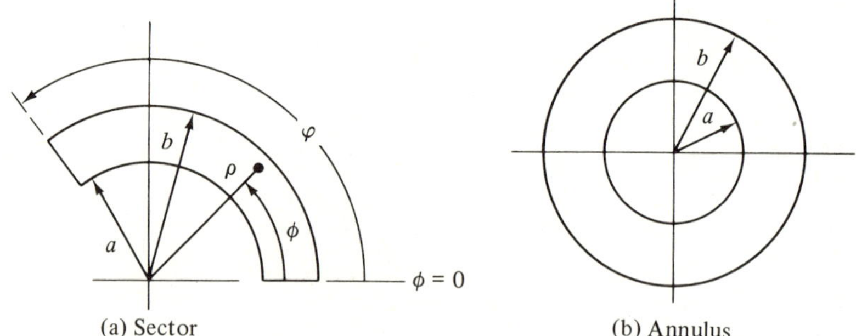

(a) Sector   (b) Annulus

FIG. 3.2.7 Coaxial regions.

$0, \varphi$; the annulus is bounded by perfectly conducting cylinders at $\rho = a, b$. It is to be noted that a sector of angle $\varphi = 2\pi$ still contains a septum along the $\phi = 0$ axis and is not equivalent to the annular region.

The cylindrical ρ, ϕ coordinate representation for the transverse operator ∇_t^2 is

$$\nabla_t^2 = \frac{1}{\rho}\frac{\partial}{\partial \rho}\rho\frac{\partial}{\partial \rho} + \frac{1}{\rho^2}\frac{\partial^2}{\partial \phi^2}. \tag{44}$$

Assuming product solutions for E and H modes,

$$\Phi_i(\boldsymbol{\rho}) = \Phi_p(\rho)\Phi_q(\phi), \qquad \Phi_i(\boldsymbol{\rho}) = 0 \quad \text{on } s, \tag{45a}$$

$$\psi_i(\boldsymbol{\rho}) = \psi_p(\rho)\psi_q(\phi), \qquad \frac{\partial \psi_i}{\partial v} = 0 \quad \text{on } s, \tag{45b}$$

one reduces Eqs. (2.3.2) to two one-dimensional equations:

$$\left(\frac{d^2}{d\phi^2} + q^2\right)\frac{\Phi_q(\phi)}{\psi_q(\phi)} = 0, \qquad \left.\begin{array}{c}\Phi_q \\ \dfrac{\partial \psi_q}{\partial \phi}\end{array}\right\} = 0 \quad \text{at } \phi = 0, \varphi, \tag{46a}$$

$$\left(\frac{d}{d\rho}\rho\frac{d}{d\rho} - \frac{q^2}{\rho} + p^2\rho\right)\frac{\Phi_p(\rho)}{\psi_p(\rho)} = 0, \qquad \left.\begin{array}{c}\Phi_p \\ \dfrac{\partial \psi_p}{\partial \rho}\end{array}\right\} = 0 \quad \text{at } \rho = a, b, \tag{46b}$$

where $p \equiv k_{ti}$; the eigenvalues p^2 and q^2 are, of course, different for E and H modes. For the annular region in Fig. 3.2.7(b), the boundary condition in the ϕ-domain is replaced by a periodicity requirement on Φ_q and $d\Phi_q/d\phi$ (similarly for ψ_q and $d\psi_q/d\phi$).

As for the rectangular-cross-section case, the two-dimensional eigenvalue problems in Eqs. (46) can be interpreted as resonant transmission-line problems for the ρ and ϕ domains. The eigenvalue problems in the ϕ domain in Eq. (46a) are identical in form to those encountered in the rectangular geometry in Eqs. (10)–(12), so the corresponding solutions $\Phi_q(\phi)$ and $\psi_q(\phi)$, are representative of the resonances on a uniform angular transmission line of length $\phi = \varphi$, with the eigenvalue parameter q distinguishing the resonant values of the propagation constant. In the radial eigenvalue problem (46b), distinguished by the Bessel differential operator, p is the eigenvalue parameter with q fixed by Eq. (46a). Upon comparison with Eqs. (2.3.41) (see also Sec. 3.3a), one notes that the resonant transmission line of length $b - a$, representative of the radial domain, is non-uniform, since both the characteristic impedance and propagation constant are functions of the transmission coordinate.

If the ϕ domain is bounded by radial planes at $\phi = 0, \varphi$, the complete set of eigenfunctions Φ_q and ψ_q can be employed as in Eqs. (15) and (23), to represent the delta function $\delta(\phi - \phi')$:

E modes

$$\delta(\phi - \phi') = \sum_q \Phi_q(\phi)\Phi_q(\phi') = \frac{2}{\varphi}\sum_{m=1}^{\infty} \sin\frac{m\pi\phi}{\varphi}\sin\frac{m\pi\phi'}{\varphi}, \qquad 0 < \frac{\phi}{\phi'} < \varphi, \tag{47a}$$

i.e.,

$$\Phi_q(\phi) = \sqrt{\frac{2}{\varphi}}\sin q\phi, \qquad q = \frac{m\pi}{\varphi}, \qquad m = 1, 2, \cdots. \tag{47b}$$

H modes

$$\delta(\phi - \phi') = \sum_q \psi_q(\phi)\psi_q(\phi') = \frac{1}{\varphi} \sum_{m=0}^{\infty} \epsilon_m \cos \frac{m\pi\phi}{\varphi} \cos \frac{m\pi\phi'}{\varphi}, \qquad 0 < \genfrac{}{}{0pt}{}{\phi}{\phi'} < \varphi,$$

$$(48a)$$

i.e.,

$$\psi_q(\phi) = \sqrt{\frac{\epsilon_m}{\varphi}} \cos q\phi, \quad q = \frac{m\pi}{\varphi}, \qquad m = 0, 1, 2, \cdots, \quad \epsilon_m = \begin{cases} 1, & m = 0, \\ 2, & m \geq 1. \end{cases}$$

$$(48b)$$

For the complete annular region, $0 \leq \phi \leq 2\pi$, the eigenfunctions and their derivatives must be periodic with period 2π, i.e.,

$$\Phi_q(\phi)\Big|_\phi^{\phi+2\pi} = 0, \qquad \frac{d}{d\phi}\Phi_q(\phi)\Big|_\phi^{\phi+2\pi} = 0, \tag{49}$$

and similarly for $\psi_q(\phi)$. The *E*- and *H*-mode functions are identical in this case and can be inferred from the delta-function representation

$$\delta(\phi - \phi') = \begin{cases} \dfrac{1}{2\pi} \displaystyle\sum_{m=0}^{\infty} \epsilon_m \cos m\phi \cos m\phi' + \dfrac{1}{\pi} \displaystyle\sum_{m=1}^{\infty} \sin m\phi \sin m\phi', & (50a) \\[12pt] \dfrac{1}{2\pi} \displaystyle\sum_{m=0}^{\infty} \epsilon_m \cos m(\phi - \phi'), & (50b) \\[12pt] \dfrac{1}{2\pi} \displaystyle\sum_{m=-\infty}^{\infty} e^{-jm(\phi - \phi')}, & 0 \leq \phi \leq 2\pi. \quad (50c) \end{cases}$$

The eigenfunctions can therefore be given either in the real form,

$$\Phi_q(\phi) = \psi_q(\phi) = \sqrt{\frac{\epsilon_m}{2\pi}} \begin{Bmatrix} \cos q\phi \\ \sin q\phi \end{Bmatrix}, \qquad q = m = 0, 1, 2, \cdots, \tag{51a}$$

or in the complex form [see Eq. (37b)],

$$\Phi_q(\phi) = \psi_q(\phi) = \frac{1}{\sqrt{2\pi}} e^{-jq\phi}, \qquad q = m = 0, \pm 1, \pm 2, \cdots. \tag{51b}$$

The radial equation (46b) is satisfied by Bessel functions $Z_q(p\rho)$, where q and $(p\rho)$ denote the order and argument, respectively, and Z stands for the Bessel function J, the Neumann function N, or the Hankel functions $H^{(1)}$ and $H^{(2)}$. Two linearly independent solutions must generally be employed to satisfy the required boundary conditions at $\rho = a$ and $\rho = b$. Detailed calculations for various geometrical configurations are given in the examples below.

Finite angular sector

    E modes.   The angular sector configuration shown in Fig. 3.2.8 is obtained from Fig. 3.2.7(a) by letting $a \to 0$. The eigenfunctions for the ϕ domain are those listed in Eq. (47b). Concerning the radial domain $0 \leq \rho \leq b$, the eigenfunctions $\Phi_p(\rho)$ satisfy the differential equation and boundary conditions specified in Eq. (46b). At the singular endpoint $\rho = 0$, the boundary con-

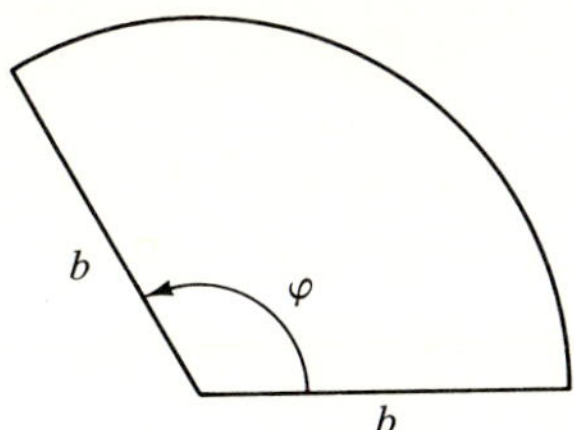

FIG. 3.2.8　Finite angular sector.

dition $\Phi_p(0) = 0$ is replaced by a finiteness condition. This requirement is satisfied by the Bessel functions J_q:

$$\Phi_p(\rho) = A_p J_q(p\rho); \qquad J_q(x_{nq}) = 0, \quad x_{nq} \equiv pb, \tag{52}$$

where A_p is an as-yet-undetermined normalization constant and x_{nq} is the nth positive root of the Bessel function $J_q(x)$, satisfying the boundary condition $\Phi_p(b) = 0$. Finiteness at $\rho = 0$ follows from the small-argument behavior of the Bessel function $J_q(x) \sim x^q$ for $x \to 0$, where $q > 0$ from Eq. (47b). Since Eq. (46b) is a special case of Eq. (1) with $x \to \rho$, $p(x) = w(x) \to \rho$, $q(x) \to q^2/\rho$, $\lambda_m \to p^2$, it follows from Eq. (2) that p is real and that $\Phi_p(\rho)$ satisfies the orthogonality condition (5a). Indeed, proceeding as in Eq. (4),

$$(p^2 - p'^2) \int_{\rho_1}^{\rho_2} d\rho \, \rho J_q(p\rho) J_q(p'\rho) \tag{53}$$

$$= [\rho p' J_q(p\rho) J_q'(p'\rho) - \rho p J_q(p'\rho) J_q'(p\rho)]_{\rho_1}^{\rho_2},$$

where $J_q'(x) \equiv (d/dx) J_q(x)$, and recalling the behavior of $J_q(x)$ near $x \to 0$, one affirms, for two unequal eigenvalues p and p', that

$$\int_0^b \rho J_q(p\rho) J_q(p'\rho) d\rho = 0, \qquad p \neq p'. \tag{54}$$

The normalization constant A_p can be determined from Eq. (53) by letting p be an eigenvalue satisfying Eq. (52) and treating p' as a variable parameter that approaches p. The resulting indeterminate form is evaluated by L'Hopital's rule to yield

$$\int_0^b \rho J_q^2(p\rho) \, d\rho = \lim_{p' \to p} \frac{[\rho p' J_q(p\rho) J_q'(p'\rho) - \rho p J_q(p'\rho) J_q'(p\rho)]_0^b}{p^2 - p'^2}$$

$$= -b \frac{(d/dp')[p J_q(p'b) J_q'(pb)]}{(d/dp')(p^2 - p'^2)} \bigg|_{p'=p}$$

$$= \frac{b^2}{2} J_q'^2(pb) = \frac{b^2}{2} J_{q+1}^2(pb),$$

so the identification

$$A_p = \frac{\sqrt{2}}{b} \frac{1}{J_q'(x_{nq})} = \frac{\sqrt{2}}{b} \frac{-1}{J_{q+1}(x_{nq})} \tag{55a}$$

assures the desired orthonormality

$$\int_0^b \rho \Phi_p(\rho)\Phi_{p'}(\rho)\, d\rho = \delta_{pp'}. \tag{55b}$$

The relation

$$J_q'^2(x_{nq}) = J_{q+1}^2(x_{nq}), \qquad \text{when } J_q(x_{nq}) = 0, \quad n = 1, 2, 3, \ldots, \tag{55c}$$

is a consequence of the recurrence formula[3]

$$J_{q+1}(x) = \frac{q}{x} J_q(x) - J_q'(x). \tag{55d}$$

It then follows from Eq. (7b) that we may represent the delta function $\delta(\rho - \rho')$ with weight factor ρ' as

$$\frac{\delta(\rho - \rho')}{\rho'} = \sum_p \Phi_p(\rho)\Phi_p(\rho') = \frac{2}{b^2} \sum_{n=1}^{\infty} \frac{J_q(p\rho)J_q(p\rho')}{J_{q+1}^2(pb)}, \qquad 0 \le \rho \le b, \tag{56a}$$

where

$$p = \frac{x_{nq}}{b}; \qquad J_q(x_{nq}) = 0, \quad n = 1, 2, 3, \ldots, \quad q \text{ fixed}. \tag{56b}$$

The two-dimensional delta function $\delta(\boldsymbol{\rho} - \boldsymbol{\rho}')$ can be represented in cylindrical coordinates as the product of $\delta(\rho - \rho')/\rho'$ and $\delta(\phi - \phi')$ since

$$1 = \int_S \delta(\boldsymbol{\rho} - \boldsymbol{\rho}')\, dS' = \int_0^{\varphi} d\phi' \int_0^b d\rho' \, \rho' \, \frac{\delta(\rho - \rho')}{\rho'} \delta(\phi - \phi'), \qquad \boldsymbol{\rho} \text{ in } S. \tag{57}$$

Thus, for $0 \le \phi \le \varphi, 0 \le \rho \le b$,

$$\begin{aligned}
\delta(\boldsymbol{\rho} - \boldsymbol{\rho}') &= \sum_i \Phi_i(\boldsymbol{\rho})\Phi_i(\boldsymbol{\rho}') \\
&= \frac{4}{\varphi b^2} \sum_{m=1}^{\infty} \sum_{n=1}^{\infty} \frac{\sin q\phi \, J_q(p\rho) \sin q\phi' \, J_q(p\rho')}{J_{q+1}^2(pb)}, \qquad q = \frac{m\pi}{\varphi}, \qquad p = \frac{x_{nq}}{b}
\end{aligned} \tag{58a}$$

i.e.,

$$\Phi_i(\boldsymbol{\rho}) = \frac{2}{b\sqrt{\varphi}\, J_{q+1}(pb)} \sin q\phi \, J_q(p\rho), \qquad k_{ti}'^2 = \left(\frac{x_{nq}}{b}\right)^2. \tag{58b}$$

H modes. The angular eigenfunctions for the *H*-mode problem are given in Eqs. (48). Upon following the same procedure as in Eqs. (52)–(55), one obtains, for the radial eigenfunctions,

$$\psi_p(\rho) = p\sqrt{\frac{2}{(pb)^2 - q^2}} \frac{J_q(p\rho)}{J_q(pb)}, \tag{59a}$$

where for q fixed,

$$p = \frac{x_{nq}'}{b}; \qquad J_q'(x_{nq}') = 0, \quad n = 1, 2, \ldots, \tag{59b}$$

x_{nq}' being the nth positive zero of $J_q'(x)$. Thus, one may represent the two-dimensional delta function as

$$\delta(\mathbf{\rho} - \mathbf{\rho}') = \sum_i \psi_i(\mathbf{\rho})\psi_i(\mathbf{\rho}'), \qquad 0 \leq \genfrac{}{}{0pt}{}{\rho}{\rho'} \leq b, \quad 0 \leq \genfrac{}{}{0pt}{}{\phi}{\phi'} \leq \varphi,$$

$$= \frac{2}{\varphi} \sum_{m=0}^{\infty} \sum_{n=1}^{\infty} \epsilon_m \frac{p^2}{[(pb)^2 - q^2]J_q^2(pb)} \cos q\phi\, J_q(p\rho) \cos q\phi'\, J_q(p\rho'),$$

$$q = \frac{m\pi}{\varphi}, \qquad p = \frac{x'_{nq}}{b}, \tag{60a}$$

i.e.,

$$\psi_i(\mathbf{\rho}) = \left[\frac{2\epsilon_m}{\varphi[(pb)^2 - q^2]}\right]^{1/2} \frac{p}{J_q(pb)} \cos q\phi J_q(p\rho), \qquad k_{ti}''^{2} = \left(\frac{x'_{nq}}{b}\right)^2, \tag{60b}$$

with $\epsilon_0 = 1, \epsilon_m = 2$ for $m \geq 1$.

Open angular sector

E modes.   If the b dimension in Fig. 3.2.8 is allowed to become infinite, one approaches in the limit the wedge geometry of Fig. 3.2.9. To obtain the radial spectrum from that for the finite sector, we introduce into Eqs. (56) the change of variable

$$p \equiv \xi_n = \frac{x_{nq}}{b}. \tag{61a}$$

FIG. 3.2.9   Open angular sector.

As $b \to \infty$, the contributing terms in Eq. (56a) are those for which x_{nq} is large so that the asymptotic form for the Bessel function of large argument can be employed [see Eq. (4.2.22b)]:[3]

$$J_q(\xi_n b) \sim \sqrt{\frac{2}{\pi \xi_n b}} \cos\left(\xi_n b - \frac{q\pi}{2} - \frac{\pi}{4}\right), \qquad b \longrightarrow \infty, \tag{61b}$$

from which one obtains for the zeros $(\xi_n b)$ in Eq. (56b),

$$\xi_n b \sim \left(n + \frac{3}{4}\right)\pi + \frac{q\pi}{2}, \qquad n = \text{large positive integer}. \tag{61c}$$

One notes that as $b \to \infty$, the discrete eigenvalues ξ_n coalesce into a continuum along the positive real axis. Also,

$$J_{q+1}(\xi_n b) \sim \sqrt{\frac{2}{\pi \xi_n b}} (-1)^n. \tag{61d}$$

If one defines the increment

$$\Delta \xi_n = \xi_{n+1} - \xi_n \sim \frac{\pi}{b}, \qquad b \longrightarrow \infty, \tag{61e}$$

then the radial delta function in Eq. (56a) transforms into

$$\frac{\delta(\rho - \rho')}{\rho'} = \lim_{\Delta \xi_n \to 0} \sum_{\xi_n = 0}^{\infty} \xi_n J_q(\xi_n \rho) J_q(\xi_n \rho') \Delta \xi_n = \int_0^\infty \xi J_q(\xi \rho) J_q(\xi \rho') \, d\xi,$$

$$0 < \frac{\rho}{\rho'} < \infty. \tag{62}$$

The transform theorem associated with Eq. (62) is referred to as the Fourier-Bessel or Hankel transformation.[3] The lack of a definite boundary condition on the eigenfunctions $\Phi_\xi(\rho) = \sqrt{\xi}\, J_q(\xi \rho)$ at $\rho \to \infty$ is a consequence of the limit-point singularity of the Bessel differential equation at this endpoint[2] [see the footnote to Eq. (3.3.21)].

The two-dimensional scalar representation theorem for the wedge-shaped region in Fig. 3.2.9 then becomes

$$\delta(\boldsymbol{\rho} - \boldsymbol{\rho}') = \sum_i \Phi_i(\boldsymbol{\rho}) \Phi_i(\boldsymbol{\rho}'), \qquad 0 < \frac{\rho}{\rho'} < \infty, \quad 0 < \frac{\phi}{\phi'} < \varphi,$$

$$= \frac{2}{\varphi} \sum_{m=1}^{\infty} \int_0^\infty \xi \sin q\phi \, J_q(\xi \rho) \sin q\phi' \, J_q(\xi \rho') \, d\xi, \tag{63a}$$

i.e.,

$$\Phi_i(\boldsymbol{\rho}) = \sqrt{\frac{2\xi}{\varphi}} \sin q\phi \, J_q(\xi \rho), \qquad q = \frac{m\pi}{\varphi}, \quad m = 1, 2, \ldots, \quad 0 < \xi < \infty;$$

$$k_{ti}'^2 = \xi^2. \tag{63b}$$

As in the transition from Eq. (36a) to the Fourier integral representation in Eq. (37a), it is often desirable to cast the Fourier-Bessel transform theorem in Eq. (62) into an alternative form in which the integration over ξ extends from $-\infty$ to $+\infty$. Upon introducing

$$J_q(\xi \rho) = \tfrac{1}{2} [H_q^{(1)}(\xi \rho) + H_q^{(2)}(\xi \rho)], \tag{64}$$

where $H_q^{(1,2)}(x)$ is the Hankel function of the first (second) kind of order q and argument x, one may write Eq. (62) as

$$\frac{\delta(\rho - \rho')}{\rho'} = I_1 + I_2,$$

$$I_1 = \tfrac{1}{2} \int_0^\infty \xi H_q^{(1)}(\xi \rho) J_q(\xi \rho') \, d\xi, \qquad I_2 = \tfrac{1}{2} \int_0^\infty \xi H_q^{(2)}(\xi \rho) J_q(\xi \rho') \, d\xi. \tag{65}$$

If the range of ξ is to be extended to $\xi = -\infty$, account must be taken in the integrands of I_1 and I_2 of the $\xi = 0$ branch-point singularity arising from the presence of the Hankel function [and also of the Bessel function in Eq. (62), since q is not integral]. To assure single-valuedness of the integrands when continued into the complex ξ plane, we introduce a branch cut along the negative real ξ axis (see Fig. 3.2.10). The following circuital relations[3] then provide the means for changing ξ into $-\xi (0 < \xi < \infty)$:

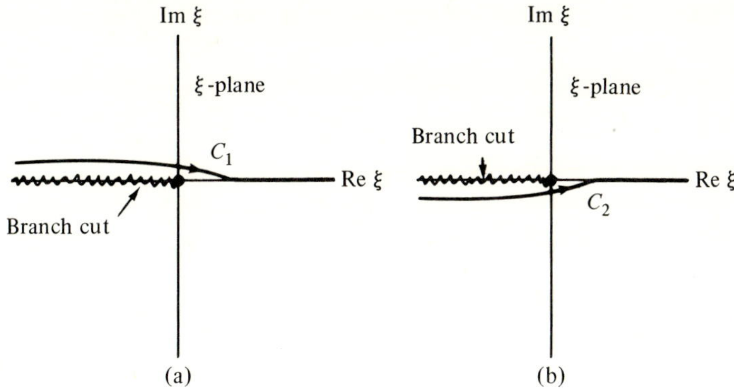

FIG. 3.2.10   Contours of integration in the complex ξ plane.

$$J_q(xe^{\pm j\pi}) = e^{\pm jq\pi} J_q(x), \tag{66a}$$

$$H_q^{(1)}(xe^{j\pi}) = -e^{-jq\pi} H_q^{(2)}(x), \qquad H_q^{(2)}(xe^{-j\pi}) = -e^{jq\pi} H_q^{(1)}(x). \tag{66b}$$

Suppose we introduce the change of variable $\bar{\xi} = \xi \exp(j\pi)$ in the integrand of I_2 in Eq. (65). Then

$$I_2 = \frac{1}{2} \int_0^{\infty e^{j\pi}} \bar{\xi} H_q^{(2)}(\bar{\xi}\rho e^{-j\pi}) J_q(\bar{\xi}\rho' e^{-j\pi})\, d\bar{\xi}, \tag{67a}$$

$$= \frac{1}{2} \int_{\infty e^{j\pi}}^0 \bar{\xi} H_q^{(1)}(\bar{\xi}\rho) J_q(\bar{\xi}\rho')\, d\bar{\xi}, \tag{67b}$$

where Eq. (67b) follows from Eq. (67a) upon use of Eqs. (66). Thus, Eq. (65) may be written as

$$\frac{\delta(\rho - \rho')}{\rho'} = \frac{1}{2} \int_{\infty e^{j\pi}}^{\infty} \xi H_q^{(1)}(\xi\rho) J_q(\xi\rho')\, d\xi = \frac{1}{2} \int_{\infty e^{j\pi}}^{\infty} \xi H_q^{(1)}(\xi\rho') J_q(\xi\rho)\, d\xi. \tag{68}$$

The contour of integration extends along the upper shore of the branch cut and the positive real axis as shown in Fig. (3.2.10a). The alternative expression, with ρ and ρ' interchanged, given by the second integral in Eq. (68), is deduced by representing $J_q(\xi\rho')$ in Eq. (62) in terms of the Hankel function combination in Eq. (64), and proceeding as above. It is noted that the spectral representation for the weighted delta function in Eq. (68) is not given in the Hermitean (complex-conjugate) form as in Eq. (7b), but rather in the "biorthogonal" form

$$\frac{\delta(\rho - \rho')}{\rho'} = \int_{\infty e^{j\pi}}^{\infty} \Phi_\xi(\rho)\bar{\Phi}_\xi(\rho')\, d\xi, \tag{68a}$$

$$\Phi_\xi(\rho) = \sqrt{\frac{\xi}{2}} J_q(\xi\rho), \qquad \bar{\Phi}_\xi(\rho) = \sqrt{\frac{\xi}{2}}\, H_q^{(1)}(\xi\rho), \tag{68b}$$

where $\bar{\Phi}_\xi(\rho)$ represents an "adjoint" function.

In a completely analogous manner, one may proceed by introducing the new variable $\bar{\xi} = \xi \exp(-j\pi)$ into I_1 of Eq. (65) and simplifying in accordance with the circuital formulas in Eqs. (66) to obtain the representation

$$\frac{\delta(\rho - \rho')}{\rho'} = \frac{1}{2} \int_{\infty e^{-j\pi}}^{\infty} \xi H_q^{(2)}(\xi\rho) J_q(\xi\rho')\, d\xi = \frac{1}{2} \int_{\infty e^{-j\pi}}^{\infty} \xi H_q^{(2)}(\xi\rho') J_q(\xi\rho)\, d\xi$$

$$(69)$$

with the path of integration as shown in Fig. (3.2.10b).

The integral representations in Eqs. (68) and (69) permit deformation of the contours of integration into the complex ξ plane, results of which will be useful for subsequent function-theoretic manipulations. To show this, we examine the asymptotic behavior of the Bessel and Hankel functions for large values of their argument:[3]

$$J_q(w) \sim \sqrt{\frac{2}{\pi w}} \cos\left(w - \frac{q\pi}{2} - \frac{\pi}{4}\right), \qquad |w| \longrightarrow \infty, \quad -\pi < \arg w < \pi,$$

$$(70a)$$

$$H_q^{(1,2)}(w) \sim \sqrt{\frac{2}{\pi w}} \exp\left[\pm j\left(w - \frac{q\pi}{2} - \frac{\pi}{4}\right)\right], \quad |w| \longrightarrow \infty, \quad -\pi < \arg w < \pi,$$

$$(70b)$$

where w is a complex variable whose argument lies between $-\pi$ and $+\pi$. If $\operatorname{Im} w > 0$, the magnitude of $H_q^{(1)}(w)$ decays like $\exp(-\operatorname{Im} w)$, while that of $J_q(w)$ increases like $\exp(\operatorname{Im} w)$, so

$$|H_q^{(1)}(\xi\rho) J_q(\xi\rho')| \sim \frac{1}{\pi\sqrt{\rho\rho'|\xi|}}\, e^{-(\operatorname{Im} \xi)(\rho - \rho')}, \qquad |\xi| \longrightarrow \infty, \quad 0 < \arg \xi < \pi.$$

$$(71)$$

For $\rho > \rho'$, the integrand of the first integral in Eq. (68) therefore decays exponentially over an infinite semicircle in the upper half of the complex ξ plane, while for $\rho < \rho'$, the integrand of the second integral exhibits a similar behavior. For a suitable class of functions representable by the transform theorem in Eq. (69), the contour C_1 in Fig. 3.2.10a can thus be deformed away from the real axis at $|\xi| \to \infty$ into a path C_1' in the upper half of the ξ plane; in consequence, the delta function in Eq. (68) can be represented as

$$\frac{\delta(\rho - \rho')}{\rho'} = \frac{1}{2} \int_{C_1'} \xi H_q^{(1)}(\xi\rho_>) J_q(\xi\rho_<)\, d\xi, \qquad 0 < \frac{\rho}{\rho'} < \infty, \qquad (72)$$

where $\rho_<$ and $\rho_>$ denote the lesser and greater, respectively, of the quantities ρ and ρ'.

It should be emphasized that the above deformation of the contour C_1 into the contour C_1' must be examined in detail when a function $F(\rho)$ is being represented. The representation for $F(\rho)$ becomes, via Eq. (68),

$$F(\rho) = \frac{1}{2} \int_{\infty e^{j\pi}}^{\infty} \xi H_q^{(1)}(\xi\rho)\hat\varphi(\xi)\, d\xi, \qquad \hat\varphi(\xi) = \int_0^{\infty} \rho'\, F(\rho') J_q(\xi\rho')\, d\rho'. \quad (72a)$$

Although it is assumed that the representation exists (integrals converge) and that $\hat\varphi(\xi)$ is a regular function of ξ on the contour C_1, $\hat\varphi(\xi)$ may have singularities in the upper half of the ξ plane. The presence of such singularities (whether poles or branch points) must be taken into account in any path deformation. This aspect is given further attention in Sec. 3.3.

In a directly analogous manner, one may show that the delta-function representation in Eq. (69) involving the contour C_2 in Fig. (3.2.10b) can be expressed in terms of an integral over the contour C_2' in the lower half of the ξ plane as

$$\frac{\delta(\rho - \rho')}{\rho'} = \frac{1}{2} \int_{C_2'} \xi H_q^{(2)}(\xi\rho_>) J_q(\xi\rho_<)\, d\xi, \qquad 0 < \frac{\rho}{\rho'} < \infty, \qquad (73)$$

with similar remarks applying to contour deformation when a function $F(\rho)$ is being represented.

H modes.   Upon going to the limit $b \to \infty$ in Eq. (60a) in a manner analogous to that employed in Eqs. (61) and (62), one finds that the radial eigenfunctions $\psi_\xi(\rho)$ become identical with $\Phi_\xi(\rho)$ so that the delta-function representation in Eq. (62) applies here as well. This lack of dependence of the eigenfunctions for the infinite interval $b \to \infty$ on the boundary conditions at the finite endpoint b is a consequence of the limit-point type of singularity at $\rho \to \infty$. Thus, one obtains

$$\delta(\boldsymbol{\rho} - \boldsymbol{\rho}') = \sum_i \psi_i(\boldsymbol{\rho})\psi_i(\boldsymbol{\rho}'), \qquad 0 < \frac{\phi}{\phi'} < \varphi, \;\; 0 < \frac{\rho}{\rho'} < \infty,$$

$$= \frac{1}{\varphi} \sum_{m=0}^{\infty} \int_0^{\infty} \xi\epsilon_m \cos q\phi\, J_q(\xi\rho) \cos q\phi'\, J_q(\xi\rho')\, d\xi, \qquad (74a)$$

i.e.,

$$\psi_i(\boldsymbol{\rho}) = \sqrt{\frac{\epsilon_m \xi}{\varphi}} \cos q\phi\, J_q(\xi\rho), \quad q = \frac{m\pi}{\varphi}, \qquad m = 0, 1, 2, \ldots, \quad 0 < \xi < \infty;$$

$$k_{ti}^{\prime\prime 2} = \xi^2, \quad \epsilon_m = \begin{cases} 1, & m = 0, \\ 2, & m \geq 1. \end{cases} \qquad (74b)$$

Circular waveguide

The eigenfunctions for the circular waveguide region shown in Fig. 3.2.11 differ from those for the finite angular sector in Fig. 3.2.8 only in that the one-dimensional eigenfunctions for the ϕ domain are those given in Eqs. (50) instead of Eqs. (47) and (48). The results can then be written down directly from Eqs. (56) and (59) for E and H modes, respectively.

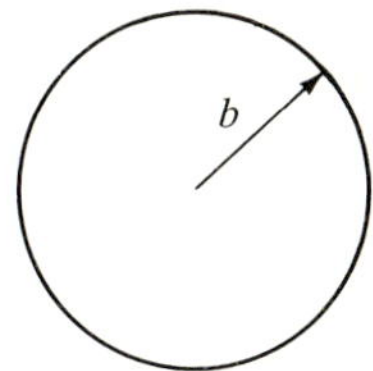

FIG. 3.2.11   Circular waveguide.

E modes

$$\delta(\mathbf{\rho} - \mathbf{\rho'}) = \sum_i \Phi_i(\mathbf{\rho})\Phi_i^*(\mathbf{\rho'}), \qquad 0 \le \frac{\phi}{\phi'} \le 2\pi, \quad 0 \le \frac{\rho}{\rho'} \le \infty,$$

$$= \frac{1}{\pi b^2} \sum_{m=-\infty}^{\infty} \sum_{n=1}^{\infty} \frac{e^{-jm\phi}J_m(p\rho)e^{jm\phi'}J_m(p\rho')}{J_{m+1}^2(pb)}, \tag{75a}$$

where

$$p = \frac{x_{nm}}{b}; \qquad J_m(x_{nm}) = 0, \quad n = 1, 2, \ldots.$$

Thus,

$$\Phi_i(\mathbf{\rho}) = \frac{1}{b\sqrt{\pi}\,J_{m+1}(pb)}\, e^{-jm\phi}J_m(p\rho), \qquad m = 0, \pm 1, \pm 2, \ldots,$$

$$n = 1, 2, \ldots; \qquad k_{ti}'^2 = \left(\frac{x_{nm}}{b}\right)^2. \tag{75b}$$

H modes

$$\delta(\mathbf{\rho} - \mathbf{\rho'}) = \sum_i \psi_i(\mathbf{\rho})\psi_i^*(\mathbf{\rho'})$$

$$= \frac{1}{\pi} \sum_{m=-\infty}^{\infty} \sum_{n=1}^{\infty} \frac{p^2}{[(pb)^2 - m^2]J_m^2(pb)}\, e^{-jm\phi}J_m(p\rho)e^{jm\phi'}J_m(p\rho'), \tag{76a}$$

where

$$p = \frac{x_{nm}'}{b}; \qquad J_m'(x_{nm}') = 0, \quad n = 1, 2, \ldots.$$

Thus,

$$\psi_i(\mathbf{\rho}) = \frac{p}{\sqrt{\pi[(pb)^2 - m^2]}J_m(pb)}\, e^{-jm\phi}J_m(p\rho), \qquad m = 0, \pm 1, \pm 2, \ldots,$$

$$n = 1, 2, \ldots; \qquad k_{ti}'^2 = \left(\frac{x_{nm}'}{b}\right)^2. \tag{76b}$$

Although the radial eigenfunctions $\Phi_q(\rho)$ and $\psi_q(\rho)$ in Eqs. (56) and (59) were obtained on the assumption that $q > 0$, the extension of the above to negative integer values of q follows from the relation

$$J_m(x) = (-1)^m J_{-m}(x), \qquad m = 0, 1, 2, \ldots. \tag{77}$$

Free space

The free-space region in Fig. 3.2.5 can be analyzed in a polar-coordinate representation upon letting $b \to \infty$ in Fig. 3.2.11. The E- and H-mode problems become identical in this case, so

$$\delta(\mathbf{\rho} - \mathbf{\rho'}) = \sum_i \Phi_i(\mathbf{\rho})\Phi_i^*(\mathbf{\rho'}) = \sum_i \psi_i(\mathbf{\rho})\psi_i^*(\mathbf{\rho'}), \qquad 0 < \frac{\phi}{\phi'} \le 2\pi, \quad 0 \le \frac{\rho}{\rho'} < \infty,$$

$$= \frac{1}{2\pi} \sum_{m=-\infty}^{\infty} \int_0^\infty \xi e^{-jm\phi}J_m(\xi\rho)e^{jm\phi'}J_m(\xi\rho')\, d\xi, \tag{78a}$$

i.e.,

$$\Phi_i(\mathbf{\rho}) = \psi_i(\mathbf{\rho}) = \sqrt{\frac{\xi}{2\pi}}\, e^{-jm\phi} J_m(\xi\rho), \qquad m = 0, \pm 1, \pm 2, \ldots,$$

$$0 < \xi < \infty, \qquad k_{ti}^2 = \xi^2. \tag{78b}$$

Alternatively, one may employ the radial delta function representation in Eqs. (68) or (69) instead of the symmetric form (62) implied above.

3.2d Inhomogeneously Filled Cross Sections

Transverse field equations and modal representations

The representation of electromagnetic fields in uniform waveguide regions filled with a homogeneous medium has been discussed in Sec. 2.2 and has led to the eigenvalue problems in Secs. 3.2b and 3.2c. In this section we consider the more general problem wherein the cross-sectional medium is inhomogeneous (i.e., the dielectric constant ϵ and the permeability μ are functions of the cross-sectional variable $\mathbf{\rho}$). To derive the transverse field equations for this case, one modifies the procedure of Sec. 2.2 since $\epsilon = \epsilon(\mathbf{\rho})$, $\mu = \mu(\mathbf{\rho})$. Instead of Eqs. (2.2.4) and (2.2.5), one has for a suppressed time dependence $\exp{(j\omega t)}$:

$$-\frac{\partial \mathbf{E}_t}{\partial z} = j\omega\left(\mu\mathbf{1} + \frac{1}{\omega^2}\mathbf{\nabla}_t\frac{1}{\epsilon}\mathbf{\nabla}_t\right) \cdot \mathbf{H}_t \times \mathbf{z}_0 + \mathbf{M}_{te} \times \mathbf{z}_0, \tag{79a}$$

$$\mathbf{M}_{te} = \mathbf{M}_t - \frac{1}{j\omega}\mathbf{\nabla}_t \times \frac{\mathbf{J}_z}{\epsilon}, \tag{79b}$$

$$-\frac{\partial \mathbf{H}_t}{\partial z} = j\omega\left(\epsilon\mathbf{1} + \frac{1}{\omega^2}\mathbf{\nabla}_t\frac{1}{\mu}\mathbf{\nabla}_t\right] \cdot \mathbf{z}_0 \times \mathbf{E}_t + \mathbf{z}_0 \times \mathbf{J}_{te}, \tag{79c}$$

$$\mathbf{J}_{te} = \mathbf{J}_t + \frac{1}{j\omega}\mathbf{\nabla}_t \times \frac{\mathbf{M}_z}{\mu}. \tag{79d}$$

Although Eqs. (79) are valid even in the general case $\epsilon = \epsilon(\mathbf{\rho}, z)$, $\mu = \mu(\mathbf{\rho}, z)$, the restriction to variation only in the cross section permits the use of simple field representations with z-independent vector-mode functions.

As in Sec. 2.2, it is assumed that the transverse electromagnetic fields defined by Eqs. (79) can be represented in terms of complete sets of transverse vector modes, characterizing the possible guided waves that can be propagated along the z direction:

$$\mathbf{E}_t(\mathbf{r}) = \sum_i V_i(z)\mathbf{e}_i(\mathbf{\rho}), \tag{80a}$$

$$\mathbf{H}_t(\mathbf{r}) = \sum_i I_i(z)\mathbf{h}_i(\mathbf{\rho}). \tag{80b}$$

In contrast to the homogeneous medium, the transverse vector-mode functions $\mathbf{e}_i$ and $\mathbf{h}_i$, in general, are not related via $\mathbf{h}_i = \mathbf{z}_0 \times \mathbf{e}_i$ as in Eqs. (2.2.8e); moreover, no separability into E and H modes relative to the z direction exists (see Sec. 8.2 for a treatment of arbitrary media).

To determine vector modes $\mathbf{e}_i(\mathbf{\rho})$ and $\mathbf{h}_i(\mathbf{\rho})$, which constitute the transverse z-independent parts of the three-dimensional mode fields $\mathbf{E}_i(\mathbf{r})$ and $\mathbf{H}_i(\mathbf{r})$, we observe that the latter represent solutions of the source-free Maxwell field equations

$$\nabla \times \mathbf{E}_i(\mathbf{r}) = -j\omega\mu\mathbf{H}_i(\mathbf{r}), \qquad \nabla \times \mathbf{H}_i(\mathbf{r}) = j\omega\epsilon\mathbf{E}_i(\mathbf{r}). \tag{81}$$

Since ϵ and μ are z independent, a typical mode field propagating in the $+z$ direction is characterized by a z-dependence $\exp(-j\kappa_i z)$, so one may write

$$\mathbf{E}_i(\mathbf{r}) = \mathcal{E}_i(\mathbf{\rho})e^{-j\kappa_i z}, \qquad \mathbf{H}_i(\mathbf{r}) = \mathcal{H}_i(\mathbf{\rho})e^{-j\kappa_i z}. \tag{82a}$$

The transverse mode functions $\mathbf{e}_i$ and $\mathbf{h}_i$ are then identified as

$$\mathbf{E}_{ti}(\mathbf{r}) = \mathbf{e}_i(\mathbf{\rho})e^{-j\kappa_i z}, \qquad \mathbf{H}_{ti}(\mathbf{r}) = Y_i\mathbf{h}_i(\mathbf{\rho})e^{-j\kappa_i z}, \tag{82b}$$

where the characteristic admittance $Y_i = 1/Z_i$ is introduced as a convenient normalization parameter, to obtain the $\mathbf{e}_i$ and $\mathbf{h}_i$ relations previously derived in Sec. 2.2 for ϵ and μ constant. The defining equations for the eigenfunctions $\mathbf{e}_i$ and $\mathbf{h}_i$ and corresponding eigenvalue κ_i are obtained upon substituting Eqs. (82) into the source-free transverse field equations (79):

$$Z_i\kappa_i\mathbf{e}_i = \omega\left(\mu\mathbf{1} + \frac{1}{\omega^2}\nabla_t\frac{1}{\epsilon}\nabla_t\right) \cdot \mathbf{h}_i \times \mathbf{z}_0, \tag{83a}$$

$$Y_i\kappa_i\mathbf{h}_i = \omega\left(\epsilon\mathbf{1} + \frac{1}{\omega^2}\nabla_t\frac{1}{\mu}\nabla_t\right) \cdot \mathbf{z}_0 \times \mathbf{e}_i. \tag{83b}$$

On the perfectly conducting cross-section boundary s, $\mathbf{e}_i$ and $\mathbf{h}_i$ satisfy the boundary conditions

$$\mathbf{v} \times \mathbf{e}_i = 0 = \nabla_t \cdot (\mathbf{h}_i \times \mathbf{z}_0) \qquad \text{on } s. \tag{83c}$$

In the previously considered case of constant ϵ and μ, Eqs. (83) were decoupled by letting $\mathbf{h}_i = \mathbf{z}_0 \times \mathbf{e}_i$; this choice leads to two separate and essentially scalar eigenvalue problems for E and H modes given in Eqs. (2.2.10) and (2.3.2), with Z_i and Y_i defined as in Eqs. (2.2.15). No such simplification obtains in this more general case. However, if the transverse cross-sectional configuration is describable by separable coordinates and the variability of ϵ and μ depends only on a single transverse coordinate, it is sometimes possible to decouple Eqs. (83) into two scalar eigenvalue problems. For the case of rectangular waveguide containing a dielectric stratified along one of the transverse coordinates, this procedure is illustrated below. It may be noted that Eqs. (83a) and (83b) and hence $\mathbf{e}_i$ and $\mathbf{h}_i$ are unchanged if Y_i and κ_i are replaced by $-Y_i$ and $-\kappa_i$, respectively; $-Y_i$ and $-\kappa_i$ describe a field traveling in the $-z$ direction [see Eq. (82b)].

Upon use of the transverse form of Green's theorem [see Eq. (2.2.11a)],

$$\iint_s dS\left[\mathbf{A}_i \cdot \nabla_t\left(\frac{1}{\epsilon}\nabla_t \cdot \mathbf{A}_j^*\right) - \mathbf{A}_j^* \cdot \nabla_t\left(\frac{1}{\epsilon}\nabla_t \cdot \mathbf{A}_i\right)\right]$$

$$= \oint_s ds\frac{1}{\epsilon}[(\mathbf{A}_i \cdot \mathbf{v})(\nabla_t \cdot \mathbf{A}_j^*) - (\mathbf{A}_j^* \cdot \mathbf{v})(\nabla_t \cdot \mathbf{A}_i)],$$

$$\mathbf{A}_{i,j} = \mathbf{h}_{i,j} \times \mathbf{z}_0, \tag{84}$$

and Eqs. (83a) and (83c), one deduces the following relation between the mode functions (ϵ and μ are assumed real so that the waveguide is non-dissipative):

$$Z_j^* \kappa_j^* \iint_S \mathbf{h}_i \times \mathbf{z}_0 \cdot \mathbf{e}_j^* \, dS = Z_i \kappa_i \iint_S \mathbf{h}_j^* \cdot \mathbf{z}_0 \times \mathbf{e}_i \, dS. \tag{85a}$$

By changing ϵ into μ and $\mathbf{h}_{i,j}$ into $-\mathbf{e}_{i,j}$ in Eq. (84), one obtains, via Eqs. (83b) and (83c), the dual relation

$$Y_j^* \kappa_j^* \iint_S \mathbf{z}_0 \times \mathbf{e}_i \cdot \mathbf{h}_j^* \, dS = Y_i \kappa_i \iint_S \mathbf{z}_0 \times \mathbf{e}_j^* \cdot \mathbf{h}_i \, dS, \tag{85b}$$

and, from combining Eqs. (85a) and (85b),

$$(\kappa_j^{*2} - \kappa_i^2) \iint_S \mathbf{h}_i \times \mathbf{z}_0 \cdot \mathbf{e}_j^* \, dS = 0. \tag{85c}$$

If $i = j$ in Eq. (85c), the integral is non-vanishing and represents the complex power carried by a typical mode. To satisfy the equation in this case, it is necessary that $\kappa_j^2 = \kappa_j^{*2}$ so that κ_j^2 is real. The appearance of κ_j^2 in Eq. (85c) implies the existence of propagation constants $\pm \kappa_j$ and is indicative of the z reflection symmetry in this configuration. Upon choosing an appropriate normalization, we may therefore write the biorthogonality relation between $\mathbf{e}_i$ and $\mathbf{h}_i$ for the non-degenerate case ($\kappa_i \neq \kappa_j$ when $i \neq j$) as

$$\iint_S \mathbf{h}_i \times \mathbf{z}_0 \cdot \mathbf{e}_j^* \, dS = \delta_{ij} \qquad \text{(non-dissipative case)}. \tag{86}$$

The orthogonality property in Eq. (86) permits the simple evaluation of the modal amplitudes in Eqs. (80) as

$$V_i(z) = \iint_S \mathbf{E}_t(\mathbf{r}) \cdot \mathbf{h}_i^*(\boldsymbol{\rho}) \times \mathbf{z}_0 \, dS, \tag{87a}$$

$$I_i(z) = \iint_S \mathbf{H}_t(\mathbf{r}) \cdot \mathbf{z}_0 \times \mathbf{e}_i^*(\boldsymbol{\rho}) \, dS. \tag{87b}$$

Multiplication of Eqs. (79a) and (79b) by $\mathbf{h}_i^* \times \mathbf{z}_0$ and $\mathbf{z}_0 \times \mathbf{e}_i^*$, respectively, integration over the cross-sectional domain S, reduction of the resulting integrals involving the gradient operators via Eqs. (83) and (84), and use of Eqs. (87) finally leads to transmission-line equations for the determination of V_i and I_i:

$$-\frac{dV_i}{dz} = j\kappa_i Z_i I_i + v_i, \tag{88a}$$

$$-\frac{dI_i}{dz} = j\kappa_i Y_i V_i + i_i, \tag{88b}$$

where

$$v_i(z) = \iint_S \mathbf{M}_{te}(\mathbf{r}) \cdot \mathbf{h}_i^*(\boldsymbol{\rho}) \, dS = \iint_S \mathbf{M}(\mathbf{r}) \cdot \mathbf{h}_i^*(\boldsymbol{\rho}) \, dS + Z_i^* \iint_S \mathbf{J}(\mathbf{r}) \cdot \mathbf{e}_{zi}^*(\boldsymbol{\rho}) \, dS,$$

$$\tag{89a}$$

$$Z_i \mathbf{e}_{zi}(\boldsymbol{\rho}) = \mathbf{z}_0 \frac{\nabla_t \cdot \mathbf{h}_i \times \mathbf{z}_0}{j\omega\epsilon}, \tag{89b}$$

and

$$i_i(z) = \iint_S \mathbf{J}_{te}(\mathbf{r}) \cdot \mathbf{e}_i^*(\boldsymbol{\rho})\, dS = \iint_S \mathbf{J}(\mathbf{r}) \cdot \mathbf{e}_i^*(\boldsymbol{\rho})\, dS + Y_i^* \iint_S \mathbf{M}(\mathbf{r}) \cdot \mathbf{h}_{zi}^*(\boldsymbol{\rho})\, dS,$$

$$\tag{89c}$$

$$Y_i \mathbf{h}_{zi}(\boldsymbol{\rho}) = \mathbf{z}_0 \frac{\nabla_t \cdot \mathbf{z}_0 \times \mathbf{e}_i}{j\omega\mu}. \tag{89d}$$

The expressions for the source terms in Eqs. (89a) and (89c) are derived upon use of the integration-by-parts formula (2.2.13) and the specification $J_z = 0$ on the perfectly conducting boundary s. As in the analogous equations (2.2.14), no assumption concerning the differentiability of J_z/ϵ and M_z/μ, implied initially in Eqs. (79), is required. It is noted that the mode amplitudes V_i and I_i are determined from the solution of the uniform transmission-line equations (88) as for the case of homogeneously filled waveguide cross sections.

Evaluation of vector-mode functions by transverse transmission analysis

As for the homogeneously filled waveguide regions discussed in Chapter 2, field solutions in waveguides with inhomogeneously filled cross sections may be synthesized via Eqs. (80) from knowledge of the modal amplitudes $V_i(z)$ and $I_i(z)$ and the transverse vector-mode functions $\mathbf{e}_i(\boldsymbol{\rho})$ and $\mathbf{h}_i(\boldsymbol{\rho})$. The former satisfy the transmission-line equations (88), whereas the latter are defined by Eqs. (83). Solutions of vector field problems such as those in Eqs. (83) are facilitated if scalar potentials can be introduced; the vector partial differential field equations are transformed thereby into more easily treated scalar equations. Such a scalarization procedure has been employed in Sec. 2.3a for reduction of the vector eigenfunction equations (2.2.10) in homogeneously filled regions and has led to the decomposition of a general vector field into E and H modes with respect to z. As noted in connection with Eqs. (80), this procedure is inapplicable when the waveguide medium depends on the cross-sectional coordinates; in fact, it does not seem possible to scalarize the vector eigenvalue problem in the general case of inhomogeneously filled cross sections. Scalarization can, however, be achieved for special cross-sectional configurations, included among which are rectangular waveguides filled with a medium whose properties depend only on one of the cross-sectional coordinates, x. In this practical case one may decompose the field into E and H modes with respect to x; although these modes are characterized by $H_x \equiv 0$ and $E_x \equiv 0$, respectively, they are in general "hybrid" with respect to z (i.e., they possess z components of both the electric and magnetic field).

In the solution of the constituent eigenvalue problems, it is convenient to view the cross-sectional region as a resonant transmission line, as was noted in Sec. 2.4e. To illustrate the "transverse transmission" method for determining mode functions, we consider first the homogeneously filled rectangular cross section. Although this simple configuration is analyzed more easily by the conventional E- and H-mode decomposition with respect to z as in Secs. 2.3a and 3.2b, the transverse transmission analysis yields an alternative set of modes, hybrid with respect to z, comprised of a superposition of conventional E and H modes. The hybrid modes satisfy the orthogonality condition in Eq. (86) and are not characterized by the simple conventional relation $\mathbf{h}_i = \mathbf{z}_0 \times \mathbf{e}_i$. Considerations for the homogeneously filled cross section are then generalized to the inhomogeneous case for which scalarization cannot be achieved by the method of Sec. 2.2.

Homogeneous cross section

Referring to Fig. 3.2.12, we seek to evaluate mode fields characteristic of propagation along the z axis with a dependence $\exp(-j\kappa_i z)$, with the wavenumber κ_i to be determined. Viewed in terms of transverse propagation, the configuration in Fig. 3.2.12 can be regarded as a parallel plate waveguide in the

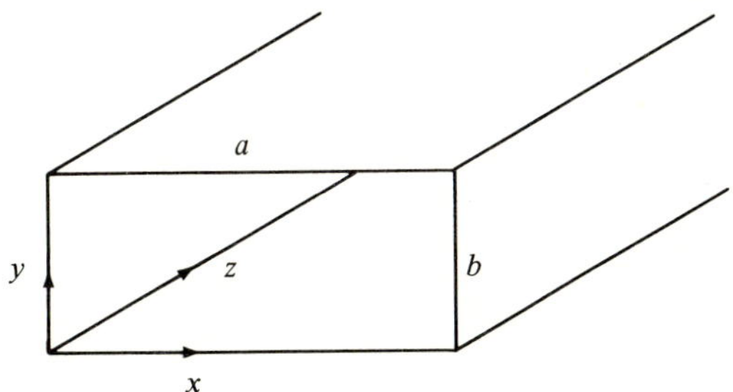

FIG. 3.2.12   Infinite rectangular waveguide.

x direction, terminated in short circuits at $x = 0$ and $x = a$. For a typical parallel-plate-guide mode, the transverse (to x) vector-mode functions (E modes and H modes with respect to x) have the form

$$\bar{\mathbf{e}}_T(y, z) = \bar{\mathbf{e}}_T(y)e^{-j\kappa z} = [\mathbf{y}_0\bar{e}_y(y) + \mathbf{z}_0\bar{e}_z(y)]e^{-j\kappa z}, \tag{90a}$$

$$\bar{\mathbf{h}}_T(y, z) = \bar{\mathbf{h}}_T(y)e^{-j\kappa z} = [\mathbf{y}_0\bar{h}_y(y) + \mathbf{z}_0\bar{h}_z(y)]e^{-j\kappa z}, \tag{90b}$$

$$\bar{\mathbf{h}}_T = \mathbf{x}_0 \times \bar{\mathbf{e}}_T, \tag{90c}$$

where the bar $^-$ denotes quantities relating to the x transmission analysis, and the subscript T distinguishes vectors transverse to x; the mode index i has been suppressed and $\mathbf{x}_0$, $\mathbf{y}_0$, and $\mathbf{z}_0$ are unit vectors along the x, y, and z directions. The mode functions $\bar{\mathbf{e}}_T$ and $\mathbf{h}_T$ are derivable directly from Eqs. (41b) and (42b) (with x, a, y, η replaced by y, b, z, κ, respectively) and Eqs. (2.3.1) (with $\nabla_t, \mathbf{z}_0$ replaced by $\nabla_T, \mathbf{x}_0$). The transverse (to x) electric and magnetic mode fields can thus be represented as

$$\mathbf{E}_T(x, y, z) = \bar{V}(x)\bar{\mathbf{e}}_T(y)e^{-j\kappa z}, \tag{91a}$$

$$\mathbf{H}_T(x, y, z) = \bar{I}(x)\bar{\mathbf{h}}_T(y)e^{-j\kappa z}, \tag{91b}$$

while the x components of the fields are given by [see Eqs. (2.2.6)]

$$E_x(x, y, z) = \frac{1}{j\omega\epsilon}\nabla_T \cdot (\mathbf{H}_T \times \mathbf{x}_0) = \frac{\bar{I}(x)}{j\omega\epsilon}e^{-j\kappa z}\overline{\nabla}_T \cdot \bar{\mathbf{e}}_T(y), \tag{91c}$$

$$H_x(x, y, z) = \frac{1}{j\omega\mu}\nabla_T \cdot (\mathbf{x}_0 \times \mathbf{E}_T) = \frac{\bar{V}(x)}{j\omega\mu}e^{-j\kappa z}\overline{\nabla}_T \cdot \bar{\mathbf{h}}_T(y), \tag{91d}$$

where

$$\nabla_T = \mathbf{y}_0\frac{\partial}{\partial y} + \mathbf{z}_0\frac{\partial}{\partial z}, \qquad \overline{\nabla}_T = \mathbf{y}_0\frac{\partial}{\partial y} - j\kappa\mathbf{z}_0. \tag{91e}$$

$\bar{V}(x)$ and $\bar{I}(x)$ are the resonant transmission-line solutions discussed at the end of Sec. 3.2b.

The desired mode functions $\mathbf{e}(x, y)$ and $\mathbf{h}(x, y)$ corresponding to propagation in the z direction are obtained upon comparing Eqs. (91) with the alternative representation of the fields transverse to z:

$$\mathbf{E}_t(x, y, z) = V(z)\mathbf{e}(x, y) = \frac{1}{N}e^{-j\kappa z}\mathbf{e}(x, y), \tag{92a}$$

$$\mathbf{H}_t(x, y, z) = I(z)\mathbf{h}(x, y) = \frac{1}{NZ_0}e^{-j\kappa z}\mathbf{h}(x, y). \tag{92b}$$

Since we seek a traveling-wave solution in the z direction, the choice

$$V(z) = Z_0 I(z) = \frac{1}{N}e^{-j\kappa z}, \tag{92c}$$

has been made, with Z_0 representing the as yet unspecified characteristic impedance and N a normalization constant indicative of the arbitrary amplitude associated with the resonant solutions $\bar{V}(x)$ and $\bar{I}(x)$. Comparison of Eqs. (91) and (92) yields

$$\mathbf{e}(x, y) = N\left[\mathbf{x}_0\frac{\bar{I}(x)}{j\omega\epsilon}\overline{\nabla}_T \cdot \bar{\mathbf{e}}_T(y) + \mathbf{y}_0\bar{V}(x)\bar{e}_y(y)\right], \tag{93a}$$

$$\mathbf{h}(x, y) = NZ_0\left[\mathbf{x}_0\frac{\bar{V}(x)}{j\omega\mu}\overline{\nabla}_T \cdot \bar{\mathbf{h}}_T(y) + \mathbf{y}_0\bar{I}(x)\bar{h}_y(y)\right]. \tag{93b}$$

Since all mode quantities (except N and Z_0) in Eqs. (93) are known, it is evident that the mode functions are solely dependent on the transmission-line solutions $\bar{V}(x)$ and $\bar{I}(x)$. In the simple example appropriate to the configuration in Fig. 3.2.12, the transverse transmission-line solutions are

$$\bar{V}(x) = \sin px, \qquad p \equiv k_x \equiv \frac{m\pi}{a}, \tag{94a}$$

$$\bar{I}(x) = \frac{1}{-jk_x\bar{Z}_0}\frac{d\bar{V}(x)}{dx} = j\bar{Y}_0\cos px,$$

with

$$\bar{Y}_0 = \begin{cases} \dfrac{\omega\epsilon}{p} & \text{for } E \text{ modes in } x, \\[3mm] \dfrac{p}{\omega\mu} & \text{for } H \text{ modes in } x. \end{cases} \qquad (94\text{b})$$

The propagation constant κ of the mode propagating in the z direction is given by

$$\kappa = \sqrt{k^2 - k_t^2} = \sqrt{k^2 - k_x^2 - k_y^2}, \qquad k_x = p = \frac{m\pi}{a}, \quad k_y = \frac{n\pi}{b}, \quad (94\text{c})$$

where k_y is the originally given eigenvalue for the y domain of the parallel-plate configuration. It is noted that

$$\begin{aligned} \bar{\nabla}_T \cdot \bar{\mathbf{e}}_T(y) = 0 & \qquad \text{for } H \text{ modes along } x, \\ \bar{\nabla}_T \cdot \bar{\mathbf{h}}_T(y) = 0 & \qquad \text{for } E \text{ modes along } x, \end{aligned} \qquad (94\text{d})$$

while, from Eqs. (90) and (91),

$$\begin{aligned} E_z(x, y, z) &= \bar{V}(x)\bar{e}_z(y)e^{-j\kappa z}, \\ H_z(x, y, z) &= \bar{I}(x)\bar{h}_z(y)e^{-j\kappa z}. \end{aligned} \qquad (94\text{e})$$

The total vector-mode fields whose transverse parts are given by $\mathbf{e}(x, y)$ and $\mathbf{h}(x, y)$ in Eqs. (93) separate naturally into E and H modes along x but contain, in general, both E_z and H_z components. Moreover, $\mathbf{h} \neq \mathbf{z}_0 \times \mathbf{e}$, and the orthogonality properties are those given in Eq. (86). While N is chosen to achieve orthonormality of the mode set, the choice of Z_0 is a matter of convenience. If $\partial/\partial y \equiv 0$ in Eqs. (93) and $\bar{e}_z = 0 = \bar{h}_y$, then one notes that $\mathbf{h} = \mathbf{z}_0 \times \mathbf{e}$ provided that $Z_0\kappa/\omega\mu = 1$. The corresponding modes are H modes with respect to both the x and z directions, Z_0 is the conventional H mode characteristic impedance [see Eq. (2.2.15d)], and the $\mathbf{e}$- and $\mathbf{h}$-mode functions can be made identical with those in Eqs. (2.3.1).

Inhomogeneous cross section

To generalize the preceding results, we consider the evaluation of the vector-mode functions when the rectangular waveguide configuration in Fig. 3.2.1 is filled with a medium whose dielectric constant ϵ and permeability μ are x dependent [i.e., $\epsilon = \epsilon(x)$ and $\mu = \mu(x)$]. As mentioned previously, this structure cannot, in general, be described in terms of the conventional E and H modes with respect to the z direction. However, an analysis in terms of E and H modes with respect to the x coordinate can be carried out since the separability of the Maxwell field equations into transverse and longitudinal parts, and the subsequent derivation of the transmission-line equations, is not affected by a variability of the medium along the x transmission coordinate (see Sec. 2.2). Thus, the transverse (to x) and the x components of the electric and magnetic fields for a typical mode can be expressed as in Eqs. (91), leading to the transverse (to z) vector mode functions $\mathbf{e}(x, y)$ and $\mathbf{h}(x, y)$ in Eqs. (93), provided that $\bar{V}(x)$ and

$\bar{I}(x)$ satisfy the source-free transmission-line equations (2.2.15) descriptive of a variable medium:

$$-\frac{dV(x)}{dx} = jk_x(x)Z(x)I(x), \tag{95a}$$

$$-\frac{dI(x)}{dx} = jk_x(x)Y(x)V(x), \qquad Y(x) = \frac{1}{Z(x)}, \tag{95b}$$

where k_x denotes the propagation constant in the x direction. The bar $^-$ has been omitted for convenience.

Second-order differential equations for either $V(x)$ or $I(x)$ are obtained from the two coupled first-order differential equations in Eq. (95) as

$$\frac{d}{dx}\left[\frac{1}{k_x(x)Z(x)}\frac{dV(x)}{dx}\right] + k_x(x)Y(x)V(x) = 0, \tag{96a}$$

$$\frac{d}{dx}\left[\frac{1}{k_x(x)Y(x)}\frac{dI(x)}{dx}\right] + k_x(x)Z(x)I(x) = 0. \tag{96b}$$

Since, from Eqs. (2.2.15),

$$k_x(x)Z(x) = \omega\mu(x) \qquad \text{for } H \text{ modes in } x, \tag{97a}$$

$$k_x(x)Y(x) = \omega\epsilon(x) \qquad \text{for } E \text{ modes in x}, \tag{97b}$$

the differential equations (96a) and (96b) have a simple form when applied, respectively, to H modes and E modes. Thus, the resonant transmission-line solutions are defined by:

H modes (along x)

$$\left[\mu(x)\frac{d}{dx}\frac{1}{\mu(x)}\frac{d}{dx} + k_x''^2(x)\right]V''(x) = 0, \qquad k_x''^2(x) = \omega^2\mu(x)\epsilon(x) - k_T''^2,$$
$$\tag{98a}$$

$$V''(0) = V''(a) = 0. \tag{98b}$$

I'' is determined from a knowledge of V'' via

$$I''(x) = \frac{1}{-j\omega\mu(x)}\frac{dV''(x)}{dx}, \tag{98c}$$

E modes (along x)

$$\left[\epsilon(x)\frac{d}{dx}\frac{1}{\epsilon(x)}\frac{d}{dx} + k_x'^2(x)\right]I'(x) = 0, \qquad k_x'^2(x) = \omega^2\mu(x)\epsilon(x) - k_T'^2, \tag{99a}$$

$$\frac{dI'(x)}{dx} = 0 \qquad \text{at } x = 0, a, \tag{99b}$$

with

$$V'(x) = \frac{1}{-j\omega\epsilon(x)}\frac{dI'(x)}{dx}. \tag{99c}$$

The primes and double primes distinguish E- and H-mode quantities on the

x-directed transmission line. The resonance (eigenvalue) problems above are of the general form discussed in Sec. 3.2a.

We note again that the vector-mode fields obtained on substitution of the above H- and E-mode resonance solutions from Eqs. (98) and (99) into Eqs. (91) are characterized by the vanishing of E_x and H_x, respectively, but they possess, in general, both E_z and H_z components.

3.3 CHARACTERISTIC GREEN'S FUNCTION (RESOLVENT) PROCEDURE AND ALTERNATIVE REPRESENTATIONS

For finite intervals and piecewise constant p, q, and w, the mode spectrum associated with the general Sturm–Liouville eigenvalue problem of Eqs. (3.2.1) is discrete.[1] Hence, the direct solution of these equations as well as the subsequent normalization of the eigenfunctions f_m in Eq. (3.2.5b) are straightforward. Moreover, the superposition of all possible discrete solutions constitutes a complete set of functions as noted in Eq. (3.2.7b). However, for open intervals and singular p, q, or w, which may admit continuous mode spectra, the direct solution of the eigenvalue problem offers difficulties; boundary conditions on f_m at singular endpoints are often ill defined, normalization of the now improper mode functions [see Eq. (3.2.28b)] may become formidable, and completeness of the modal set is not evident. To circumvent these problems, the normalized continuous mode spectra for a number of open configurations have been deduced in Sec. 3.2 as a limiting case of the known discrete spectra for corresponding finite domains in which one of the dimensions becomes infinite. In this section we examine an alternative and direct procedure for the determination of both discrete and continuous complete sets of modal representations.

The present method exploits an intimate relation between source-free (eigenvalue) solutions and the characteristic Green's function. In network terms, the eigenvalue problem defines the resonant voltage or current waves on a terminated non-uniform transmission line while the Green's function displays the similar resonant responses to excitation by a point voltage or current generator; these analogies are useful in subsequent discussion. After a description of the general procedure in Sec. 3.3a, construction of Green's functions for the Sturm–Liouville problem is given in Sec. 3.3b.

The theory of one-dimensional characteristic Green's functions facilitates consideration of alternative representations for two- and three-dimensional electromagnetic (or acoustic) fields. As noted in Sec. 2.3, electromagnetic field solutions for a variety of closed and open regions can be constructed via guided-wave representations along a preferred direction z. Such representations are predicated upon the availability of E- and H-mode eigenfunctions $\Phi_i(\boldsymbol{\rho})$ and $\psi_i(\boldsymbol{\rho})$ for separable cross sections transverse to z (Secs. 3.2 and 3.4) and the ability to determine the corresponding z-dependent modal amplitudes. For field calculations in certain parameter ranges, alternative representations are often desired since their convergence properties may be better than those for the z-

guided formulation. These observations have already been made in Sec. 1.5 and have been illustrated there for simple space- and time-dependent field representations either in space-guided or time-guided form. It has been shown that one representation may be derived from another by function-theoretic techniques involving analytic continuation and contour deformation in the complex plane. These function-theoretic aspects are treated naturally and concisely by use of one-dimensional characteristic Green's functions as shown in Sec. 3.3c. The method is illustrated in detail for three-dimensional scalar Green's functions in rectangular or cylindrical geometries, and more briefly for spherical regions.

3.3a   Relation between Characteristic Green's Function and Eigenvalue Problems

The characteristic Green's function $g(x, x'; \lambda)$ for the Sturm–Liouville problem of Eqs. (3.2.1) is defined by the equation[4,5]

$$\left[\frac{d}{dx}p(x)\frac{d}{dx} - q(x) + \lambda w(x)\right]g(x, x'; \lambda) = -\delta(x - x'), \qquad x_1 < \frac{x}{x'} < x_2,$$

$$(1)$$

subject to the boundary conditions

$$\left(p\frac{d}{dx} + \alpha_{1,2}\right)g(x, x'; \lambda) = 0, \qquad x = x_{1,2}.$$ $$(1a)$$

The parameter λ is arbitrary but so restricted as to assure a unique solution of Eqs. (1); discussion of these restrictions is deferred until later. The distinction between characteristic Green's functions and the modal Green's functions discussed in Sec. 2.4 is that the parameter λ is left unspecified for the former.

A network schematization of the characteristic Green's function problem

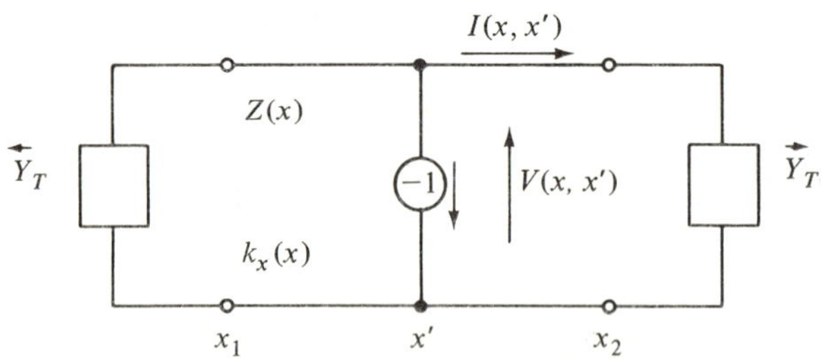

FIG. 3.3.1   Non-uniform transmission line excited with a unit current generator.

is shown in Fig. 3.3.1. The voltage and current on this transmission line satisfy the non-uniform transmission-line equations (3.2.95) with the addition of a current source term $i(x') = -\delta(x - x')$ [see Eqs. (2.3.10) or (2.4.15)]:[†]

[†]When referring to wave solutions, a time-dependence $\exp(+j\omega t)$ will be used in this section.

$$-\frac{dV(x, x')}{dx} = jk_x(x)Z(x)I(x, x'), \tag{2a}$$

$$-\frac{dI(x, x')}{dx} = jk_x(x)Y(x)V(x, x') - \delta(x - x'), \tag{2b}$$

where $Z(x) = 1/Y(x)$ and $k_x(x)$ are the characteristic impedance and propagation constant. For an H-mode transmission line with distinguishing double-prime superscripts, $k_x''Z'' = \omega\mu$, so the corresponding second-order differential equation for $V''(x, x')$ has the form [see Eq. (3.2.98a)]

$$\left\{\frac{d}{dx}\left[\frac{1}{\mu'(x)}\frac{d}{dx}\right] + k_0^2\epsilon'(x) - \frac{k_T''^2}{\mu'(x)}\right\}V''(x, x') = -j\omega\mu_0\delta(x - x'), \tag{3}$$

where $k_0^2 = \omega^2\mu_0\epsilon_0$ and

$$\mu'(x) = \frac{\mu(x)}{\mu_0}, \qquad \epsilon'(x) = \frac{\epsilon(x)}{\epsilon_0}, \tag{3a}$$

μ_0 and ϵ_0 being convenient constant reference values for the permeability and dielectric constant, respectively. Upon comparing Eqs. (1) and (3) one makes the identifications:

$$p(x) = w(x) = \frac{1}{\mu'(x)}, \quad q(x) = -k_0^2\epsilon'(x), \quad k_T''^2 = -\lambda,$$

$$V''(x, x') = j\omega\mu_0 g''(x, x'; \lambda). \tag{4}$$

The boundary conditions in Eq. (1a) are rephrased by Eqs. (2a) and (4) in terms of

$$\frac{I''}{V''} = j\frac{p(dg''/dx)}{\omega\mu_0 g''}, \tag{5a}$$

and replaced by terminating admittances $\overrightarrow{Y}_T''$ and $\overleftarrow{Y}_T''$ at x_1 and x_2:

$$\overleftarrow{Y}_T'' = -\frac{I''(x_1, x')}{V''(x_1, x')} = \frac{j\alpha_1}{\omega\mu_0}, \qquad \overrightarrow{Y}_T'' = \frac{I''(x_2, x')}{V''(x_2, x')} = \frac{-j\alpha_2}{\omega\mu_0}. \tag{5b}$$

The behavior of g and dg/dx in the vicinity of the source at $x = x'$ is readily inferred from the network representation in Fig. 3.3.1. V'' is continuous across the current generator, i.e.,†

$$V''(x, x')\big|_{x'-\Delta}^{x'+\Delta} = 0, \qquad \Delta \longrightarrow 0, \tag{6a}$$

while the discontinuity in the current is given by

$$I''(x, x')\big|_{x'-\Delta}^{x'+\Delta} = 1. \tag{6b}$$

Thus, the corresponding conditions on g are

$$g''(x, x'; \lambda)\big|_{x'-\Delta}^{x'+\Delta} = 0, \qquad p(x)\frac{d}{dx}g''(x, x'; \lambda)\big|_{x'-\Delta}^{x'+\Delta} = -1. \tag{6c}$$

---

†Note that $f(x)\big|_a^b \equiv f(b) - f(a)$.

For the E-mode problem denoted by prime superscripts, $k'_x Y' = \omega\epsilon$, and the current $I'(x, x')$ is the most readily determined quantity [see Eqs. (3.2.99)]. The appropriate network problem, dual to that in Fig. 3.3.1, is shown in Fig. 3.3.2. Network equations and their connection with the characteristic Green's function are obtained upon making the following duality replacements in Eqs. (3)–(6):

$$V'' \to I', \quad I'' \to V', \quad \mu' \leftrightarrow \epsilon', \quad \mu_0 \leftrightarrow \epsilon_0, \quad k''_T \to k'_T, \quad Y''_T \to Z'_T, \quad g'' \to g'.$$

$$(7)$$

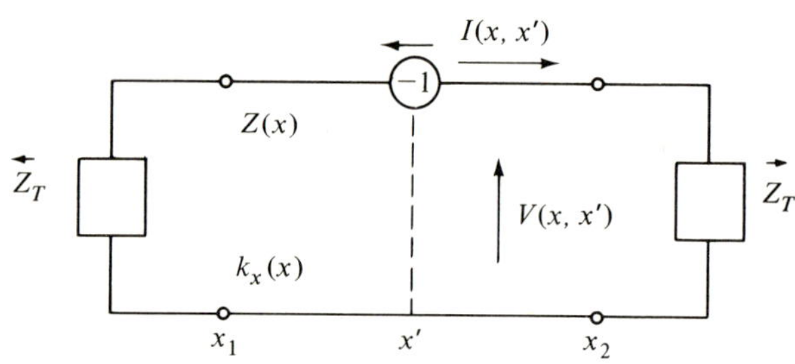

FIG. 3.3.2.   Non-uniform transmission line excited with a unit voltage generator.

Since the configuration in Fig. 3.3.1 (or Fig. 3.3.2) can be viewed as a cavity (lossless if $j\vec{Y}_T$, $j\overleftarrow{Y}_T$, k_x^2, and Z^2 are real), it is physically manifest that the voltage (or current) response g will be finite and well defined unless the choice of parameter λ is such that a resonance can exist. For fixed values of $\alpha_{1,2}$ and $x_{1,2}$, resonances will exist for parameter values λ_m at which the corresponding voltage or current will be infinite. To assure a unique solution of the network problem it is therefore necessary to restrict the parameter λ in Eq. (1) so that $\lambda \neq \lambda_m$. For the lossless situation, denoted mathematically as the Hermitian case wherein the resonant values λ_m are real [see Eq. (3.2.2)], this restriction can be stated more weakly as Im $\lambda \neq 0$. If λ in Eq. (1) is now regarded as a general complex parameter, $g(x, x'; \lambda)$ is a regular function of λ in the complex λ plane except at points $\lambda = \lambda_m$, where it becomes infinite and possesses simple pole singularities. Since the resonant condition $\lambda = \lambda_m$ implies the persistence of a response even when the source is removed, the functional form of the resonant solution satisfies the homogeneous equation (3.2.1). Thus, information about the desired eigensolutions of Eq. (3.2.1) is contained in the singularities of the characteristic Green's function g, and the problem of determining all possible resonances (i.e., a complete set of eigenfunctions) is directly related to the complete investigation of the singularities of $g(x, x'; \lambda)$ in the complex λ plane. In the above discussion it has been assumed that the dimensions x_1 and x_2 are finite so that cavity resonances, which occur for discrete values of λ_m, characterize simple pole singularities of g. If one of the dimensions becomes infinite, or if p, q, or w possesses a singularity, the discrete resonances may coalesce into a continuous spectrum;

in this instance, $g(x, x'; \lambda)$ possesses a branch-point singularity giving rise to the necessity of introducing a branch cut in the complex λ plane to ensure uniqueness of g.

To relate the complete eigenmode set f_m in Eq. (3.2.1) to the characteristic Green's function in Eq. (1), it will be assumed for the moment that the mode set is known so that any suitable function $F(x)$ can be represented as in Eqs. (3.2.6). In particular, $g(x, x'; \lambda)$ can be represented as

$$g(x, x'; \lambda) = \sum_m g_m(x'; \lambda)f_m(x), \qquad x_1 < x < x_2, \tag{8a}$$

where

$$g_m(x', \lambda) = \int_{x_1}^{x_2} w(\xi)f_m^*(\xi)g(\xi, x'; \lambda)\, d\xi. \tag{8b}$$

Upon substituting Eq. (8a) into Eq. (1), utilizing the delta-function representation from Eq. (3.2.7b), interchanging the orders of summation and differentiation, employing Eq. (3.2.1), and equating like coefficients of the (linearly independent) eigenfunctions f_m on both sides of the equation, one obtains

$$g_m(x', \lambda) = -\frac{f_m^*(x')}{\lambda - \lambda_m}, \tag{9}$$

so, from Eq. (8a),

$$g(x, x'; \lambda) = -\sum_m \frac{f_m(x)f_m^*(x')}{\lambda - \lambda_m}. \tag{10}$$

This representation for g highlights the existence of singularities in the complex λ plane at the resonant (eigen) values λ_m.

If Eq. (10) is integrated in the complex λ plane about a contour C enclosing *all* the singularities of g, then an application of Cauchy's theorem yields the following formal relation [see Eq. (3.2.7b)]:

$$-\frac{1}{2\pi j}\oint_C g(x, x'; \lambda)\, d\lambda = -\sum_m f_m(x)f_m^*(x')\left(-\frac{1}{2\pi j}\right)\oint_C \frac{d\lambda}{\lambda - \lambda_m}$$

$$= \sum_m f_m(x)f_m^*(x') = \frac{\delta(x - x')}{w(x')}, \tag{11a}$$

i.e.,†

$$\frac{\delta(x - x')}{w(x')} = -\frac{1}{2\pi j}\oint_C g(x, x'; \lambda)\, d\lambda. \tag{11b}$$

The integration is carried out in the positive (counterclockwise) sense. Although the restriction $\lambda \neq \lambda_m$ has been imposed to yield a unique solution of Eq. (1),

---

†An alternative proof of Eq. (11b) follows from Eq. (1) on multiplication by $1/\lambda$ and integration over a closed contour C such that

$$\oint_C \left[\frac{d}{dx}p\frac{d}{dx} - q\right]\frac{g(x, x', \lambda)}{\lambda}\, d\lambda$$

vanishes.[6]

this solution can be extended by analytic continuation to apply as $\lambda \to \lambda_m$. The procedure sketched above has assumed the presence of simple pole singularities of g (i.e., a discrete spectrum of eigenvalues). Equation (11b) can also be employed when g has branch-point singularities representative of a continuous spectrum; in this case, the integration contour must lie on the "spectral sheet" of the complex λ plane, whereon g decays as $|\lambda| \to \infty$.[4,6] Moreover, the technique may apply to problems with dissipative media or boundaries (non-Hermitian case with p, q, w, or $\alpha_{1,2}$ not real) for which the eigenvalues λ_m may be complex. In this instance, $f_m^*(x')$ in the delta-function representation of Eq. (11) is generally replaced by an adjoint function $\bar{f}_m(x')$.

Thus, the problem of finding a complete orthonormal set of functions is reduced systematically to determining the solution of the corresponding inhomogeneous differential equation (1), completely investigating its singularities, and then inferring the desired representation theorem by carrying out the contour integration in Eq. (11) about all the singularities of the characteristic Green's function in the complex λ plane. After discussing a general procedure for evaluation of g in the next section, the above technique will be illustrated in Sec. 3.4 by various examples.

3.3b Construction of the Characteristic Green's Function

The characteristic H-mode Green's function can be constructed from a knowledge of the two solutions $\overleftarrow{V}(x)$ and $\overrightarrow{V}(x)$ of the homogeneous equation (1) satisfying the required boundary conditions at x_1 and x_2, respectively:

$$\left(\frac{d}{dx}p\frac{d}{dx} - q + \lambda w\right)\overleftarrow{V}(x) = 0, \qquad \left(p\frac{d}{dx} + \alpha_1\right)\overleftarrow{V} = 0 \quad \text{at } x = x_1, \qquad (12a)$$

$$\left(\frac{d}{dx}p\frac{d}{dx} - q + \lambda w\right)\overrightarrow{V}(x) = 0, \qquad \left(p\frac{d}{dx} + \alpha_2\right)\overrightarrow{V} = 0 \quad \text{at } x = x_2. \qquad (12b)$$

The solution for g satisfying Eqs. (1) when $x \neq x'$ and the required continuity condition at $x = x'$ [see Eq. (6c)] is thus

$$g''(x, x'; \lambda) = A\overleftarrow{V}(x_<)\overrightarrow{V}(x_>), \qquad (13a)$$

The symbol V (voltage) has been retained to emphasize the network interpretation of the Green's function, and $x_<$ or $x_>$ denote, respectively, the lesser and greater of the quantities x and x'; the double prime on V, distinctive of an H-mode problem, has been omitted. The constant A must be so determined as to satisfy the jump condition (6c) on $p\,dg/dx$ (i.e., the current) at $x = x'$:

$$Ap(x')\left[\overleftarrow{V}(x')\frac{d}{dx'}\overrightarrow{V}(x') - \overrightarrow{V}(x')\frac{d}{dx'}\overleftarrow{V}(x')\right] = -1. \qquad (13b)$$

Thus,

$$g''(x, x'; \lambda) = \frac{\overleftarrow{V}(x_<)\overrightarrow{V}(x_>)}{-pW(\overleftarrow{V}, \overrightarrow{V})}, \qquad (14a)$$

where W is the Wronskian determinant,

$$W(\overleftarrow{V}, \overrightarrow{V}) = \left(\overleftarrow{V} \frac{d\overrightarrow{V}}{dx} - \overrightarrow{V} \frac{d\overleftarrow{V}}{dx} \right). \tag{14b}$$

If one multiplies Eq. (12a) by $\overrightarrow{V}$ and Eq. (12b) by $\overleftarrow{V}$, and subtracts the resulting equations, one finds

$$0 = \overleftarrow{V} \frac{d}{dx} p \frac{d\overrightarrow{V}}{dx} - \overrightarrow{V} \frac{d}{dx} p \frac{d\overleftarrow{V}}{dx} = \frac{d}{dx} [pW(\overleftarrow{V}, \overrightarrow{V})], \tag{15a}$$

so

$$pW(\overleftarrow{V}, \overrightarrow{V}) = \text{constant for all } x, \tag{15b}$$

and can be evaluated at any convenient point x_0 in the interval $x_1 \leq x \leq x_2$. Upon introducing homogeneous solutions of Eqs. (12), $\overleftarrow{V}(x, x_0)$ and $\overrightarrow{V}(x, x_0)$, normalized to unity at x_0,

$$\overleftarrow{V}(x, x_0) = \frac{\overleftarrow{V}(x)}{\overleftarrow{V}(x_0)}, \qquad \overrightarrow{V}(x, x_0) = \frac{\overrightarrow{V}(x)}{\overrightarrow{V}(x_0)}, \tag{15c}$$

and recalling Eqs. (5), one may write the Green's function solution in Eq. (14a) as

$$g''(x, x'; \lambda) = \frac{\overleftarrow{V}(x_<, x_0)\overrightarrow{V}(x_>, x_0)}{j\omega\mu_0 \overleftrightarrow{Y}(x_0)}, \tag{16}$$

where $\overleftrightarrow{Y}(x_0)$ denotes the sum of the admittances seen looking to the left and right from x_0:

$$\overleftrightarrow{Y}(x_0) = \overrightarrow{Y}(x_0) + \overleftarrow{Y}(x_0) = \frac{\overrightarrow{I}(x_0)}{\overrightarrow{V}(x_0)} + \frac{\overleftarrow{I}(x_0)}{\overleftarrow{V}(x_0)}, \qquad \overleftrightarrow{I}(x_0) = \pm\frac{p}{j\omega\mu_0} \frac{d\overleftrightarrow{V}}{dx}\bigg|_{x_0}. \tag{16a}$$

Although we have employed network notation for the convenience of those familiar with such concepts, it should be noted that the network identifications in the derivation of $g(x, x'; \lambda)$ via Eqs. (12)–(16) et seq. are not essential.

The above considerations for non-uniform transmission lines are generalizations of those carried out in Sec. 2.4 for uniform transmission lines. It is noted that the resonant condition determining the singularities of g in the complex λ plane is given as in Eq. (2.4.36) by

$$\overleftrightarrow{Y}(x_0) \equiv \overleftrightarrow{Y}(x_0, \lambda_m) = 0, \tag{17}$$

where the dependence on λ has been indicated explicitly. The Green's function representation in Eq. (16) can be cast into a form analogous to that in Eq. (2.4.28) if one introduces standing-wave solutions $c(x, x_0)$ and $s(x, x_0)$ (regular in the complex λ plane) satisfying the homogeneous equation

$$\left(\frac{d}{dx}p\frac{d}{dx} - q + \lambda w\right)\frac{c(x, x_0)}{s(x, x_0)} = 0, \tag{18}$$

with the boundary conditions

$$c(x_0, x_0) = 1, \qquad s(x_0, x_0) = 0,$$

$$p(x_0)c'(x_0, x_0) = 0, \qquad p(x_0)s'(x_0, x_0) = 1, \tag{18a}$$

where the prime denotes the derivative with respect to the first argument of c and s [i.e., $c'(x_0, x_0) \equiv (d/dx)c(x, x_0)|_{x=x_0}$]. For the uniform transmission line in Sec. 2.4, $\mu' = \epsilon' = 1$ [i.e., $p = w = 1$, $q = -k_0^2$, $c(x, x_0) = \cos k_{x0}(x - x_0)$, $s(x, x_0) = (1/k_{x0}) \sin k_{x0}(x - x_0)$ with $k_{x0}^2 = k_0^2 - \lambda = k_0^2 - k_T^2$]. In analogy to Eqs. (2.4.22a) and (2.4.22c) we may write

$$\overrightarrow{V}(x, x_0) = c(x, x_0) - j\omega\mu_0 \overrightarrow{Y}(x_0)s(x, x_0), \tag{19a}$$

whence, by Eq. (2a),

$$\omega\mu_0\overrightarrow{I}(x, x_0) = jp(x)c'(x, x_0) + \omega\mu_0 p(x)\overrightarrow{Y}(x_0)s'(x, x_0). \tag{19b}$$

Equations (19) can be employed to deduce the relation between the admittance at x_0 and the terminal admittance $\overrightarrow{Y}_T''$ at x_2 in Eq. (5b):

$$\omega\mu_0 \overrightarrow{Y}(x_0) = \frac{jp(x_0)c'(x_0, x_2) + \omega\mu_0 p(x_0)\overrightarrow{Y}_T''s'(x_0, x_2)}{c(x_0, x_2) - j\omega\mu_0 \overrightarrow{Y}_T''s(x_0, x_2)}. \tag{20}$$

Similarly,†

$$\overleftarrow{V}(x, x_0) = c(x, x_0) + j\omega\mu_0 \overleftarrow{Y}(x_0)s(x, x_0), \tag{21a}$$

and

$$-\omega\mu_0\overleftarrow{Y}(x_0) = \frac{jp(x_0)c'(x_0, x_1) - \omega\mu_0 p(x_0)\overleftarrow{Y}_T''s'(x_0, x_1)}{c(x_0, x_1) + j\omega\mu_0 \overleftarrow{Y}_T''s(x_0, x_1)}. \tag{21b}$$

The expression for the characteristic Green's function in Eq. (16) can now be written in the general form analogous to that in Eq. (2.4.28), which is particularly useful for the analysis of stratified regions:‡

$$g''(x, x'; \lambda)$$

$$= \frac{[c(x_<, x_0) + j\omega\mu_0 \overleftarrow{Y}(x_0)s(x_<, x_0)][c(x_>, x_0) - j\omega\mu_0 \overrightarrow{Y}(x_0)s(x_>, x_0)]}{j\omega\mu_0 \overleftrightarrow{Y}(x_0)}. \tag{22}$$

---

†If x_1 or x_2 are singular points of the differential equations (12), the necessity or not, of specifying a boundary condition may be assessed from the dependence of $\overleftrightarrow{Y}(x_0)$ on $\overleftrightarrow{Y}_T$. If $\overrightarrow{Y}(x_0)$ is independent of the terminating admittance at the singular point ("limit-point" condition), requirement of a finite solution at x_1 or x_2 suffices.[4]

‡In the absence of stratification, it may be more convenient to use Eq. (14a) directly.

Since for the resonant values λ_m of λ one has

$$\overleftrightarrow{Y}(x_0, \lambda_m) = \overleftarrow{Y}(x_0, \lambda_m) + \overrightarrow{Y}(x_0, \lambda_m) = 0, \tag{23a}$$

the distinction between $\overleftarrow{V}$ and $\overrightarrow{V}$ in the expressions in the numerator of Eq. (22) can be suppressed so that each bracketed expression satisfies simultaneously the boundary conditions at x_1 and x_2. Thus, the discrete H-mode eigenfunctions $\psi_m(x)$ (if they occur) have the form

$$\psi_m(x) = A_m \overleftarrow{V}(x, x_0; \lambda_m) = A_m \overrightarrow{V}(x, x_0; \lambda_m)$$

$$= A_m[c(x, x_0; \lambda_m) + j\omega\mu_0 \overleftarrow{Y}(x_0, \lambda_m)s(x, x_0; \lambda_m)], \qquad x_1 \leq x \leq x_2 \tag{23b}$$

where A_m is a normalization constant, and the dependence of all functions on λ_m has been exhibited explicitly.

A virtue of the characteristic Green's function procedure is that the mode set, obtained via the integration in Eq. (11b). is automatically normalized; the classical normalization procedure, as illustrated in Eq. (3.2.53) et seq., can become quite cumbersome if the eigenfunctions have a complicated form. If the only singularities of g in Eq. (16) [or Eq. (22)] in the complex λ plane are simple poles situated at the zeros λ_m of $\overleftrightarrow{Y}(x_0)$, then the behavior of the denominator in the neighborhood of a typical zero at λ_m is given by the Taylor series expansion

$$\overleftrightarrow{Y}(x_0, \lambda) = \overleftrightarrow{Y}(x_0, \lambda_m) + (\lambda - \lambda_m)\frac{\partial}{\partial\lambda_m}\overleftrightarrow{Y}(x_0, \lambda_m) + \cdots, \tag{24a}$$

where

$$\frac{\partial}{\partial\lambda_m}\overleftrightarrow{Y}(x_0, \lambda_m) \equiv \frac{\partial}{\partial\lambda}\overleftrightarrow{Y}(x_0, \lambda)\big|_{\lambda=\lambda_m}. \tag{24b}$$

From Eqs. (11), (4), and (16), one thus has the following delta-function representation for the H-mode problem (in the non-dissipative case):

$$\mu'(x)\delta(x - x') = -\frac{1}{2\pi j}\oint_c g''(x, x'; \lambda)\, d\lambda = \sum_m \hat{\psi}_m(x)\hat{\psi}_m^*(x') \tag{25a}$$

$$= -\frac{1}{2\pi j}\oint_c \frac{\overleftarrow{V}(x_<, x_0; \lambda)\overrightarrow{V}(x_>, x_0; \lambda)}{j\omega\mu_0\overleftrightarrow{Y}(x_0, \lambda)}\, d\lambda \tag{25b}$$

$$= \sum_m \frac{\overleftarrow{V}(x_<, x_0; \lambda_m)\overleftarrow{V}^*(x_>, x_0; \lambda_m)}{-j\omega\mu_0(\partial/\partial\lambda_m)\overleftrightarrow{Y}(x_0, \lambda_m)}\frac{1}{2\pi j}\oint_c\frac{d\lambda}{\lambda - \lambda_m}$$

$$= \sum_m \frac{\overleftarrow{V}(x, x_0; \lambda_m)\overleftarrow{V}^*(x', x_0; \lambda_m)}{\omega\mu_0(\partial/\partial\lambda_m)\overleftrightarrow{B}(x_0, \lambda_m)}, \qquad \overleftrightarrow{Y} = j\overleftrightarrow{B}. \tag{25c}$$

The normalized mode functions $\hat{\psi}_m(x)$ are therefore given by

$$\hat{\psi}_m(x) = \frac{1}{\sqrt{\omega\mu_0(\partial/\partial\lambda_m)\overleftrightarrow{B}(x_0,\lambda_m)}}\,\overleftarrow{V}(x,x_0;\lambda_m). \tag{25d}$$

A typical contour of integration in the complex λ plane is sketched in Fig. 3.3.3. The caret has been introduced to emphasize that the $\hat{\psi}_m$ are the scalar eigenfunctions appropriate to an H mode decomposition with respect to the transverse (x) coordinate. The vector eigenfunctions relative to the z direction are obtained as in Sec. 3.2d. For non-dissipative configurations, the eigenvalues

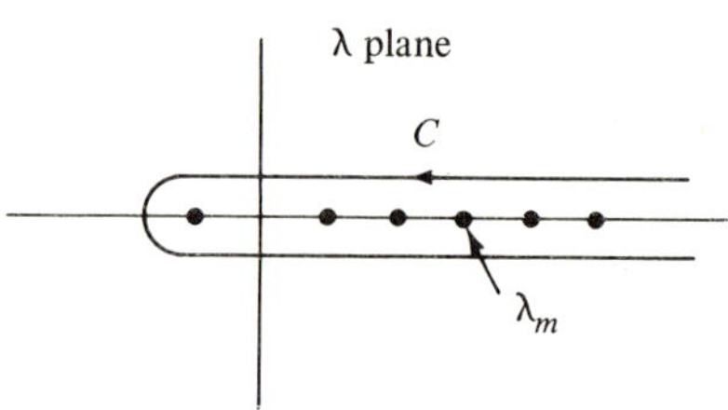

FIG. 3.3.3   Contour of integration.

λ_m and the susceptance function $\overleftrightarrow{B}$ are real, so that, in view of Eq. (21a), the eigenfunctions $\hat{\psi}_m(x)$ are also real. However, an alternative representation employing the traveling-wave solutions in Eqs. (28), instead of the standing-wave functions in Eqs. (18), leads as in Eq. (25a) to a delta-function representation involving complex-conjugate functions $\hat{\psi}_m^*(x')$ [see Eqs. (3.2.76)]. These possibilities are illustrated by the examples in Sec. 3.4.

For the E-mode problem, a corresponding representation theorem is deduced upon use of the duality relations in Eq. (7). The E-mode characteristic Green's function $g'(x, x'; \lambda)$ is given by [see Eq. (16)]

$$g'(x, x'; \lambda) = \frac{\overleftarrow{I}(x_<, x_0)\,\overrightarrow{I}(x_>, x_0)}{j\omega\epsilon_0\overleftrightarrow{Z}(x_0)}, \qquad \overleftrightarrow{Z}(x_0) = \overleftarrow{Z}(x_0) + \overrightarrow{Z}(x_0), \tag{26}$$

where the primes, distinctive of the E-mode problem, have been omitted from $\overleftrightarrow{I}$ and the total impedance function $\overleftrightarrow{Z}(x_0)$ [see the footnote for (Eq. 22)]. By duality with Eqs. (19)–(21),

$$\overleftrightarrow{I}(x, x_0) = c(x, x_0) \mp j\omega\epsilon_0\overleftrightarrow{Z}(x_0)s(x, x_0) \tag{26a}$$

and

$$\pm\omega\epsilon_0\overleftrightarrow{Z}(x_0) = \frac{jp(x_0)c'(x_0, x_{2,1}) \pm \omega\epsilon_0 p(x_0)\overrightarrow{Z}_T s'(x_0, x_{2,1})}{c(x_0, x_{2,1}) \mp j\omega\epsilon_0\overleftrightarrow{Z}_T s(x_0, x_{2,1})}. \tag{26b}$$

The eigenvalues λ_m are specified implicitly by the resonance equation

$$\overleftrightarrow{Z}(x_0, \lambda_m) = 0, \tag{26c}$$

and the delta function can be represented in terms of the E-mode eigenfunctions $\hat{\Phi}_m$ as

$$\epsilon'(x)\delta(x - x') = -\frac{1}{2\pi j}\oint_C g'(x, x'; \lambda)d\lambda = \sum_m \hat{\Phi}_m(x)\hat{\Phi}_m^*(x')$$

$$= \sum_m \frac{\overleftarrow{I}(x, x_0; \lambda_m)\overleftarrow{I}^*(x, x_0; \lambda_m)}{\omega\epsilon_0(\partial/\partial\lambda_m)\overleftrightarrow{X}(x_0', \lambda_m)}, \qquad \overleftrightarrow{Z} = j\overleftrightarrow{X}. \tag{27}$$

Thus, the discrete orthonormal E-mode eigenfunctions $\hat{\Phi}_m$ are given by

$$\hat{\Phi}_m(x) = \frac{1}{\sqrt{\omega\epsilon_0(\partial/\partial\lambda_m)\overleftrightarrow{X}(x_0, \lambda_m)}}\overleftarrow{I}(x, x_0; \lambda_m), \qquad x_1 \leq x \leq x_2, \tag{27a}$$

with $\overleftarrow{I}$ taken from Eq. (26a).

It is sometimes convenient to deal with modal reflection coefficients instead of modal impedances. The appropriate generalizations of the uniform line relations (2.4.12) are obtained on writing

$$V(x) = V_+(x) + V_-(x) = V_+(1 + \overrightarrow{\Gamma}_V) = V_-(1 + \overleftarrow{\Gamma}_V), \tag{28a}$$

$$I(x) = I_+(x) + I_-(x) = I_+(1 + \overrightarrow{\Gamma}_I) = I_-(1 + \overleftarrow{\Gamma}_I), \tag{28b}$$

where the subscripts $_+$ and $_-$ on V or I denote wave components traveling in the $+x$ and $-x$ directions, respectively, and $\overrightarrow{\Gamma}_V(\overrightarrow{\Gamma}_I)$ are the voltage (current) reflection coefficients seen when looking along the $\pm x$ directions:

$$\overrightarrow{\Gamma}_V = \frac{V_\mp}{V_\pm}, \qquad \overrightarrow{\Gamma}_I = \frac{I_\mp}{I_\pm}. \tag{28c}$$

If $\overrightarrow{\zeta} = V_+/I_+$ denotes the input impedance of a matched transmission line looking in the $+x$ direction, and $\overleftarrow{\zeta} = -V_-/I_-$ represents the corresponding quantity for the $-x$ direction, then

$$\overrightarrow{\Gamma}_I = -\frac{\overrightarrow{\zeta}}{\overleftarrow{\zeta}}\overrightarrow{\Gamma}_V, \qquad \overleftarrow{\Gamma}_I = -\frac{\overleftarrow{\zeta}}{\overrightarrow{\zeta}}\overleftarrow{\Gamma}_V, \tag{29a}$$

and

$$\overrightarrow{Z}(x) = \overrightarrow{\zeta}(x)\frac{1 + \overrightarrow{\Gamma}_V(x)}{1 - \frac{\overrightarrow{\zeta}(x)}{\overleftarrow{\zeta}(x)}\overrightarrow{\Gamma}_V(x)}, \qquad \overleftarrow{Z}(x) = \overleftarrow{\zeta}(x)\frac{1 + \overleftarrow{\Gamma}_V(x)}{1 - \frac{\overleftarrow{\zeta}(x)}{\overrightarrow{\zeta}(x)}\overleftarrow{\Gamma}_V(x)}.$$

$$\tag{29b}$$

Conversely,

$$\vec{\Gamma}_V(x) = \frac{\dfrac{\vec{Z}(x)}{\vec{\zeta}(x)} - 1}{\dfrac{\vec{Z}(x)}{\overleftarrow{\zeta}(x)} + 1}, \qquad \overleftarrow{\Gamma}_V(x) = \frac{\dfrac{\overleftarrow{Z}(x)}{\overleftarrow{\zeta}(x)} - 1}{\dfrac{\overleftarrow{Z}(x)}{\vec{\zeta}(x)} + 1}. \tag{29c}$$

The transverse resonance relation

$$\vec{Z}(x) + \overleftarrow{Z}(x) = 0$$

can then be expressed alternatively as

$$\vec{\Gamma}_V(x)\overleftarrow{\Gamma}_V(x) = 1 = \vec{\Gamma}_I(x)\overleftarrow{\Gamma}_I(x). \tag{30}$$

The above traveling-wave formulation leads to a set of eigenfunctions alternative to that in Eq. (27).

3.3c   Alternative Representations

The theory of alternative Green's function representations, alluded to at the beginning of Sec. 3.3, is best discussed in conjunction with the one-dimensional characteristic Green's functions introduced in Sec. 3.3a. The starting point is provided by the completeness theorem in Eq. (11) relating the characteristic Green's function to a complete orthonormal set of eigenfunctions via an integration in the "characteristic" complex plane. We consider first uniform waveguide regions describable in a $(\boldsymbol{\rho}, z)$ coordinate system, and then summarize corresponding results for spherical waveguides. To extend the considerations of Sec. 3.3a to two-dimensional eigenfunctions $\Phi_i(\boldsymbol{\rho})$ of the form

$$\Phi_i(\boldsymbol{\rho}) = \Phi_p(u)\Phi_q(v), \qquad \boldsymbol{\rho} = (u, v), \tag{31}$$

where $\Phi_p(u)$ and $\Phi_q(v)$ are one-dimensional orthonormal functions in separable u and v coordinate spaces transverse to z, we note from Sec. 3.2 that the completeness relation for the $\Phi_i(\boldsymbol{\rho})$ is as follows:

$$\delta(\boldsymbol{\rho} - \boldsymbol{\rho}') = \frac{\delta(u - u')\delta(v - v')}{h_u h_v} = \sum_i \Phi_i(\boldsymbol{\rho})\Phi_i^*(\boldsymbol{\rho}') \tag{32a}$$

$$= \sum_p \Phi_p(u)\Phi_p^*(u') \sum_q \Phi_q(v)\Phi_q^*(v'), \tag{32b}$$

where the curvilinear metric parameters h_u and h_v in Eq. (32a) are defined via the relation $dS = h_u h_v \, du \, dv$, dS being an area element in the cross section. Then from Eq. (11) applied to the u-dependent functions,

$$\frac{\delta(u - u')}{h_u} = \sum_p \Phi_p(u)\Phi_p^*(u') = \frac{1}{2\pi j} \sum_p \oint_{C_u} \frac{\Phi_{\lambda_u}(u)\Phi_{\lambda_u}^*(u')}{\lambda_u - \lambda_p} \, d\lambda_u$$

$$= -\frac{1}{2\pi j} \oint_{C_u} g_u(u, u'; \lambda_u) \, d\lambda_u, \tag{33}$$

where $\Phi_{\lambda_p} \equiv \Phi_p$, g_u is the characteristic Green's function associated with the eigenvalue problem in the u domain, and the contour C_u in the complex λ_u plane encloses in the positive sense all the singularities (poles or branch points, with associated branch cuts) of g_u. The first representation in Eq. (33), involving the discrete or continuous sum over the eigenvalues λ_p, results on evaluation of the contour integral from the singularities of g_u, the contour integral being the more general form of the completeness relation. On employing the analogue of Eq. (33) for the v domain,

$$\frac{\delta(v - v')}{h_v} = \sum_q \Phi_q(v)\Phi_q^*(v') = -\frac{1}{2\pi j} \oint_{C_v} g_v(v, v'; \lambda_v)\, d\lambda_v \tag{34}$$

with C_v defined similarly to C_u, one has

$$\delta(\boldsymbol{\rho} - \boldsymbol{\rho}') = \left[-\frac{1}{2\pi j} \oint_{C_u} g_u(u, u'; \lambda_u)\, d\lambda_u\right]\left[-\frac{1}{2\pi j} \oint_{C_v} g_v(v, v'; \lambda_v)\, d\lambda_v\right] \tag{35a}$$

$$= \frac{1}{(-2\pi j)^2} \oint_{C_u} \oint_{C_v} g_u(u, u'; \lambda_u)g_v(v, v'; \lambda_v)\, d\lambda_u\, d\lambda_v. \tag{35b}$$

When the eigenfunctions in Eqs. (33) or (34) are used to represent a three-dimensional Green's function in (u, v, z) space, one obtains

$$G(\mathbf{r}, \mathbf{r}') = \sum_i \Phi_i(\boldsymbol{\rho})\Phi_i^*(\boldsymbol{\rho}')g_z(z, z'; \lambda_{zi}). \tag{36}$$

The z-dependent modal Green's function g_z satisfies a one-dimensional equation resulting on elimination of the (u, v) dependence from the corresponding three-dimensional equation. Examples of a reduction of this kind have been given in Sec. 1.3, where Eq. (1.3.16) defines the one-dimensional time-dependent Green's function in terms of $\mathbf{r}$-dependent eigenfunctions; in Sec. 1.4, where Eq. (1.4.15) defines essentially the function g_z in Eq. (36)†; and in Sec. 2.3, where $(j\omega\epsilon)^{-1}Y_i'(z, z')$ in Eqs. (2.3.43a) and (2.3.41) can be identified with g_z. On comparing Eqs. (32), (35), and (36), one observes that the three-dimensional scalar Green's function G can be represented in terms of the one-dimensional characteristic Green's functions‡ g_u and g_v and the modal Green's function g_z as follows:

$$G(\mathbf{r}, \mathbf{r}') = \frac{1}{(-2\pi j)^2} \oint_{C_u} \oint_{C_v} g_u(u, u'; \lambda_u)g_v(v, v'; \lambda_v)g_z(z, z'; \lambda_z)\, d\lambda_v\, d\lambda_u. \tag{37}$$

The contour C_u in the complex λ_u plane encloses in the positive sense all singularities of g_u *but no others*, while the contour C_v in the complex λ_v plane encloses in the positive sense all singularities of g_v *but no others*. Additional singularities in the λ_u and (or) λ_v planes arise due to $g_z(z, z'; \lambda_z)$; it is recognized that generally $\lambda_z = \lambda_z(\lambda_u, \lambda_v)$, where the detailed dependence of λ_z on λ_u and λ_v is dictated by the particular coordinate representation in the u, v domain. For example, from Eq. (3.2.10c), with $\lambda_z \equiv \kappa_i'^2 = k^2 - k_{ti}'^2$,

---

†Note use of a time-dependence $\exp(-i\omega t)$ in Sec. 1.4.

‡As pointed out in Sec. 3.3a, the modal and characteristic Green's functions differ only in that the parameter λ is specified for the former ($\lambda = \lambda_i$), but unspecified for the latter.

$$\lambda_z = k^2 - \lambda_u - \lambda_v \qquad \text{for rectangular coordinates } u \equiv x, v \equiv y, \tag{38a}$$

and from Eq. (3.2.46b), with $k_{ti}'^2 \equiv p^2 \to \lambda_u$,

$$\lambda_z = k^2 - \lambda_u \qquad \text{for cylindrical coordinates } u \equiv p, v \equiv \varphi.\dagger \tag{38b}$$

The contour integral representation in Eq. (37), involving the one-dimensional Green's functions g_u, g_v, and g_z, can be considered as the most general separable representation for the three-dimensional Green's function G. Upon evaluating the contour integrals in Eq. (37) in terms of the discrete and (or) continuous spectra arising from the pole or branch-cut singularities, respectively, of g_u and g_v, and noting that g_z has no singularities inside the contours C_u and C_v, one recovers the original z-transmission formulation in Eq. (36). Different representations are also obtainable by contour deformations in the λ_u and λ_v planes. Typical examples wherein g_u, g_v, and g_z have singularities in the λ_u and λ_v planes are shown in Fig. 3.3.4. The functions g_u, g_v, and g_z are so defined as to vanish sufficiently rapidly at infinity in the λ_u and λ_v planes. This is achieved

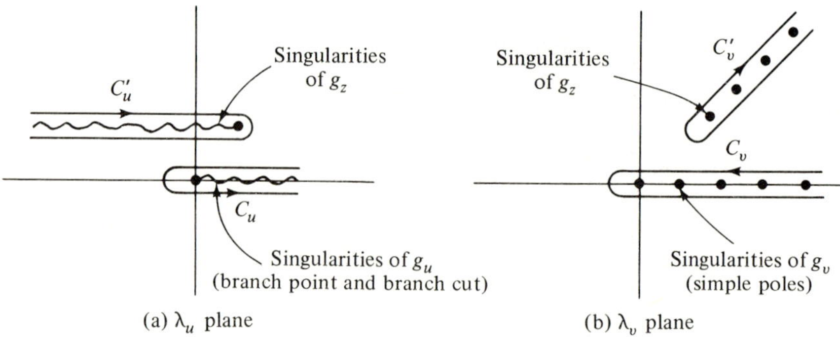

FIG. 3.3.4   Contours and singularities in λ_u and λ_v planes.

by an appropriate choice of branch cuts and Riemann surfaces, associated with any existing branch-point singularities of the g functions, so as to result in no contribution to the integral in Eq. (37) from closed contours as $|\lambda_u| \to \infty$ and $|\lambda_v| \to \infty$. The path C_u in Fig. 3.3.4(a) can therefore be deformed into the path C_u' enclosing the singularities of g_z in the λ_u plane, to yield

$$G(\mathbf{r}, \mathbf{r}') = \frac{1}{(-2\pi j)^2} \oint_{C_u'} \oint_{C_v} g_u(u, u'; \lambda_u) g_v(v, v'; \lambda_v) g_z(z, z'; \lambda_z)\, d\lambda_v\, d\lambda_u \tag{39a}$$

$$= \sum_q \Phi_q(v)\Phi_q^*(v') \sum_r \Phi_r(z)\Phi_r^*(z') g_u(u, u'; \lambda_{urq}), \tag{39b}$$

---

†In this case, $g_u \equiv g_p$ depends also on λ_v [see Eq. (3.2.46b)], so one should write $g_u \to g_u(u, u'; \lambda_u, \lambda_v)$. Thus, g_u has singularities in both the λ_u and λ_v planes, while g_z has singularities in the λ_u plane only. Only the singularities of g_u enclosed by the contour C_u in the complex λ_u plane contribute to the modal representation for G as in Eq. (36).

where the modal representation in Eq. (39b) is obtained upon evaluating the integrals over the contours C_u' and C_v in Eq. (39a). The $\Phi_r(z)$ denote the eigenfunctions in the z-domain arising from the eigenvalue problem associated with g_z, λ_u being the characteristic parameter.† In Eq. (39a), g_v and g_z are now characteristic Green's functions, while g_u is a modal Green's function wherein λ_u takes on the values specified along C_u'. Because of the explicit presence of $g_u(u, u'; \lambda_{urq})$ in Eq. (39b), one identifies this representation as arising from a guided-wave analysis in which the transmission direction is taken along the u coordinate. The above procedure has evident similarities with that used in connection with Eqs. (1.5.55) and (1.5.59); with reference to Figs. 3.3.4(a) and (1.5.3), λ_u may be identified with ω^2.

Alternatively, one may deform the contour C_v into the contour C_v' in the complex λ_v plane as shown in Fig. 3.3.4(b) to obtain

$$G(\mathbf{r}, \mathbf{r}') = \frac{1}{(-2\pi j)^2} \oint_{C_u} \oint_{C_v'} g_u(u, u'; \lambda_u) g_v(v, v'; \lambda_v) g_z(z, z'; \lambda_z)\, d\lambda_v\, d\lambda_u, \quad (40a)$$

$$= \sum_s \Phi_s(z)\Phi_s^*(z') \sum_p \Phi_p(u)\Phi_p^*(u') g_v'(v, v'; \lambda_{vsp}). \quad (40b)$$

The modal representation in Eq. (40b) is derived by considerations analogous to the above and is identified as a v-transmission formulation. The $\Phi_s(z)$ are the eigenfunctions in the z domain arising from the eigenvalue problem associated with g_z as the characteristic Green's function and λ_v as the characteristic parameter. Additional representations are possible wherein, say, only the integral C_u in Eq. (40a) is evaluated in terms of the mode spectrum in u while the integral C_v' remains unchanged. It is to be emphasized that all of the above alternative representations are to be considered as formal in that the deformability of contours must be verified in each case.

Equations (40) are valid for the case wherein g_z is a function of both λ_u and λ_v. For a radial transmission formulation, as in Eq. (38b), g_z is not a function of λ_v; instead, g_u, as noted thereunder, is a function of both λ_u and λ_v. In this instance, the contour C_v' encloses the singularities of g_u in the λ_v plane, with λ_u treated as a fixed parameter. From Eqs. (3.2.46b) and (1) [see also Eq. (3.4.91)], one notes that

$$\left(\frac{d}{d\rho} \rho \frac{d}{d\rho} + \lambda_u \rho - \frac{\lambda_v}{\rho}\right) g_u(\rho, \rho'; \lambda_u, \lambda_v) = -\delta(\rho - \rho'). \quad (41)$$

Then one has, instead of Eq. (40a),

$$G(\mathbf{r}, \mathbf{r}') = \frac{1}{(-2\pi j)^2} \oint_{C_u'} \oint_{C_v'} g_u'(u, u'; \lambda_u; \lambda_v) g_v'(v, v'; \lambda_v) g_z'(z, z'; k^2 - \lambda_u)\, d\lambda_v\, d\lambda_u.$$

$$(42)$$

---

†For non-Hermitian problems with complex eigenvalues, the spectral representation involves the symmetric form wherein $\Phi_r^*(z')$ is replaced by $\Phi_r(z')$, or more generally by an "adjoint" function $\bar{\Phi}_r(z')$.

Equation (40b) still applies formally, except that $\Phi_s(z)$ are the eigenfunctions in the z domain arising now from the eigenvalue problem associated with g_z in the λ_u plane, while $\Phi_p(u)$ are eigenfunctions in the u domain arising from the eigenvalue problem associated with g_u in the λ_v plane (in the latter, λ_u is held fixed at the eigenvalues arising from the eigenvalue problem in the z domain). As for Eq. (39b), the remarks concerning the form of the spectral representation apply here as well.

Alternative representations for Green's functions in spherical regions are constructed in a similar manner. On defining radial and angular characteristic Green's functions g_r, g_ϕ, and g_θ as in Eqs. (3.4.92) and (3.4.64), respectively, one may rewrite the E-mode Green's function in Eq. (2.6.11a) in the following forms [see also Eq. (11)][5]:

$$rr'G'(\mathbf{r}, \mathbf{r}') = \begin{cases} -\dfrac{1}{2\pi j} \sum_q \Phi_q(\phi)\Phi_q^*(\phi') \oint_{C_\theta} g_\theta(\theta, \theta'; q^2; \lambda_\theta)g_r(r, r'; \lambda_\theta)\, d\lambda_\theta, \\[4pt] \hfill (43a) \\[8pt] \left(-\dfrac{1}{2\pi j}\right)^2 \oint_{C_\phi} \oint_{C_\theta} g_\phi(\phi, \phi'; \lambda_\phi)g_\theta(\theta, \theta'; \lambda_\phi; \lambda_\theta)g_r(r, r'; \lambda_\theta)\, d\lambda_\phi\, d\lambda_\theta, \\[4pt] \hfill (43b) \\[8pt] +\dfrac{1}{2\pi j} \sum_q \Phi_q(\phi)\Phi_q^*(\phi') \oint_{C_r} g_\theta(\theta, \theta'; q^2; \lambda_\theta)g_r(r, r'; \lambda_\theta)\, d\lambda_\theta, \\[4pt] \hfill (43c) \\[8pt] \sum_q \Phi_q(\phi)\Phi_q^*(\phi') \sum_s R_s(r)\bar{R}_s(r')g_\theta(\theta, \theta'; q^2; \lambda_s), \qquad \text{etc.} \quad (43d) \end{cases}$$

The dependence of g_θ on the two parameters $\lambda_\phi = q^2$ and $\lambda_\theta = p(p+1)$ has been exhibited explicitly, and C_θ, C_r, and C_ϕ denote contours that enclose in the positive sense all of (and only) the singularities of g_θ, g_r, and g_ϕ in the complex λ_θ and λ_ϕ planes, respectively. Equation (43c) follows from Eq. (43a) by contour deformation about the singularities of g_r, and Eq. (43d) results by evaluating the integral in terms of the radial eigenfunctions $R_s(r)$ and the adjoint functions $\bar{R}_s(r)$ via [see Eqs. (3.4.100) and (3.4.101)]:

$$r'^2\delta(r - r') = \frac{1}{2\pi j} \oint_{C_r} g_r(r, r'; \lambda)\, d\lambda = \sum_s R_s(r)\bar{R}_s(r'). \qquad (44)$$

In addition to Sec. 1.5, detailed applications of the characteristic Green's function method for construction of alternative representations for $G(\mathbf{r}, \mathbf{r}')$ may be found in Secs. 5.6a, 5.7b, 6.7, and 6.8. Directly analogous considerations can be applied to the scalar function $\mathscr{S}'$ defined in Eqs. (2.3.24) or (2.3.39), in which case an additional pole singularity exists in the complex λ_u and (or) λ_v plane because of the presence of the $1/k_{ti}^2$ factor. Although the examples above involve primarily the electromagnetic E-mode problem, construction of the electromagnetic H-mode Green's functions, or of the acoustic Green's function defined in Sec. 1.1a, proceeds similarly.

3.4 ONE-DIMENSIONAL CHARACTERISTIC GREEN'S FUNCTION AND EIGENFUNCTION SOLUTIONS

The characteristic Green's function method of Sec. 3.3a for solving eigenvalue problems in closed and open regions is applied in this section to rectangular, cylindrical, and spherical cross sections. To illustrate the method, we begin with a detailed discussion of closed rectangular geometries and then pass on to open regions, first by a limiting procedure and subsequently by a direct approach. Examples thereafter are treated more concisely, with the presentation comprising essentially the characteristic Green's functions, their analytic properties, and corresponding completeness relations.

3.4a Rectangular Cross Sections

Bounded x domains

We consider the composite cross sections shown in Fig. 3.4.1, which are all characterized by the same one-dimensional eigenvalue problem in the x domain. The various media contain a piecewise constant lossless dielectric with

$$\epsilon(x) = \begin{cases} \epsilon_1, & -d < x < 0 \\ \\ \epsilon_2, & 0 < x < a \end{cases}, \qquad \epsilon_1 > \epsilon_2, \tag{1}$$

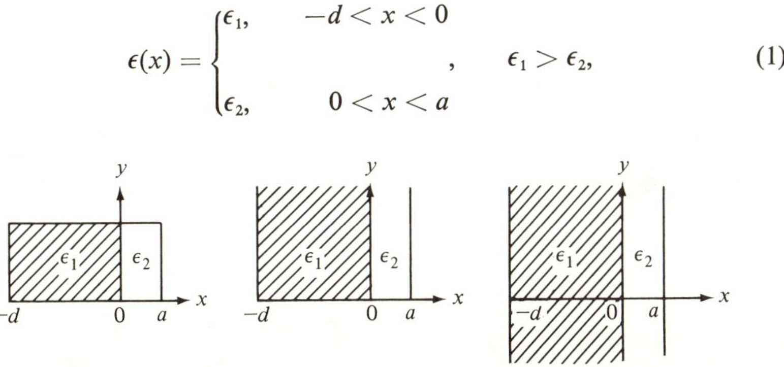

FIG. 3.4.1 Rectangular regions (bounded in x) partially filled with dielectric.

the discontinuous nature of which leads to a discontinuous representation of the eigenfunctions, as will be seen below. The eigenvalue problems in the y domain are those appropriate to a homogeneous medium and have been solved in Sec. 3.2b. A constant, free-space permeability μ_0 is assumed, so $\mu'(x) = 1$ in Eq. (3.3.3a), and the surfaces at $x = a, -d$ are assumed to be perfectly conducting.

H modes (in x)

The network configuration descriptive of the H-mode characteristic Green's function problem is shown in Fig. 3.4.2, where we distinguish between source locations in mediums 1 and 2, respectively. The relevant propagation constants and characteristic admittances are denoted, respectively, by $k_{x1} \equiv \kappa_1, Y_1$, and

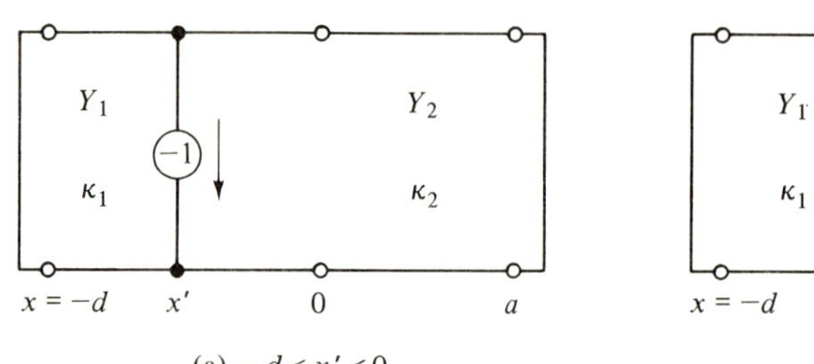
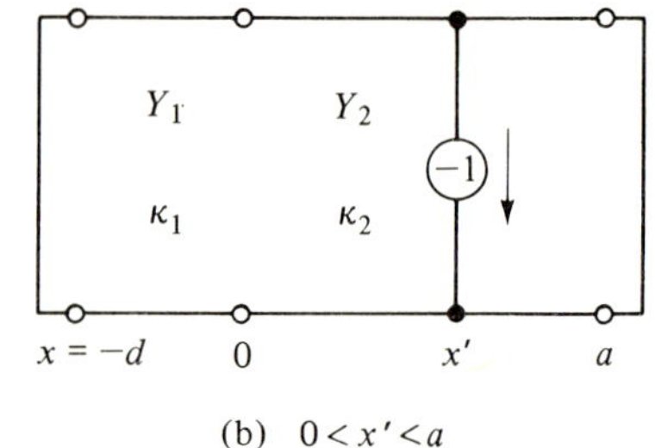

FIG. 3.4.2   Equivalent transmission-line problems (H modes along x).

$k_{x2} \equiv \kappa_2$, Y_2. From Eq. (3.3.4) it is noted that the homogeneous equation (3.3.18) for the standing-wave functions c and s becomes

$$\left[\frac{d^2}{dx^2} + \kappa^2(x, \lambda)\right]\begin{matrix}c(x)\\s(x)\end{matrix} = 0, \tag{2}$$

where

$$\kappa^2(x, \lambda) = \begin{cases} \kappa_1^2(\lambda) = k_1^2 + \lambda, & -d < x < 0 \\[2mm] \kappa_2^2(\lambda) = k_2^2 + \lambda, & 0 < x < a \end{cases}, \quad k_{1,2}^2 = \omega^2 \mu_0 \epsilon_{1,2} > 0.$$

$$\tag{2a}$$

If one chooses $x_0 = 0$ in Eq. (3.3.18a), he finds that

$$c(x) = \cos \kappa_1 x, \quad s(x) = \frac{1}{\kappa_1} \sin \kappa_1 x, \quad -d < x < 0,$$

$$c(x) = \cos \kappa_2 x, \quad s(x) = \frac{1}{\kappa_2} \sin \kappa_2 x, \quad 0 < x < a, \tag{3}$$

where the explicit dependence of the c and s functions on $x_0 = 0$ will not be exhibited henceforth. Since $\overset{\leftarrow}{Y}_T = \infty = \vec{Y}_T$ for the perfectly conducting terminations at $x = -d, a$, it follows from Eqs. (3.3.20) and (3.3.21b) that

$$\omega\mu_0 \vec{Y}(0) = -j\kappa_2 \cot \kappa_2 a, \quad \omega\mu_0 \overset{\leftarrow}{Y}(0) = -j\kappa_1 \cot \kappa_1 d, \tag{4}$$

$\kappa/\omega\mu_0$ being the H-mode characteristic admittance, whence from Eq. (3.3.19a) and (3.3.21a),

$$\vec{V}(x) = \begin{cases} \vec{V}_2(x) = \dfrac{\sin \kappa_2(a - x)}{\sin \kappa_2 a}, & 0 < x < a, \quad \text{(5a)} \\[4mm] \vec{V}_1(x) = \cos \kappa_1 x - \dfrac{\kappa_2}{\kappa_1} \cot \kappa_2 a \sin \kappa_1 x, & -d < x < 0, \quad \text{(5b)} \end{cases}$$

$$\overset{\leftarrow}{V}(x) = \begin{cases} \overset{\leftarrow}{V}_2(x) = \cos \kappa_2 x + \dfrac{\kappa_1}{\kappa_2} \cot \kappa_1 d \sin \kappa_2 x, & 0 < x < a, \quad \text{(5c)} \\[4mm] \overset{\leftarrow}{V}_1(x) = \dfrac{\sin \kappa_1(x + d)}{\sin \kappa_1 d}, & -d < x < 0. \quad \text{(5d)} \end{cases}$$

For subsequent application it will be convenient to employ, instead of Eq. (5c), the traveling-wave formulation:

$$\overleftarrow{V}_2(x) = \frac{1}{1 + \overleftarrow{\Gamma}_2(0)}\,[e^{j\kappa_2 x} + \overleftarrow{\Gamma}_2(0)e^{-j\kappa_2 x}], \qquad 0 < x < a, \tag{6}$$

where the reflection coefficient $\overleftarrow{\Gamma}_2(0)$ looking to the left at $x = +0$ is given by

$$\overleftarrow{\Gamma}_2(0) = \frac{Y_{02} - \overleftarrow{Y}(0)}{Y_{02} + \overleftarrow{Y}(0)} = \frac{\kappa_2 + j\kappa_1 \cot \kappa_1 d}{\kappa_2 - j\kappa_1 \cot \kappa_1 d}, \qquad Y_{02} = \frac{\kappa_2}{\omega\mu_0}. \tag{6a}$$

The H-mode characteristic Green's function $g''(x, x'; \lambda)$ can now be written down directly from Eq. (3.3.16).† In view of the discontinuous representation of $\overrightarrow{\overleftarrow{V}}(x)$ for $x > 0$ and $x < 0$, g'' is represented discontinuously about $x = 0$. For a source location as in Fig. 3.4.2(a),

$$g''(x, x'; \lambda) = \begin{cases} \dfrac{\overleftarrow{V}_1(x_<)\overrightarrow{V}_1(x_>)}{j\omega\mu_0\,\overleftrightarrow{Y}(0)}, & -d < x < 0, \quad -d < x' < 0, \tag{7a} \\[2em] \dfrac{\overleftarrow{V}_1(x')\overrightarrow{V}_2(x)}{j\omega\mu_0\,\overleftrightarrow{Y}(0)}, & 0 < x < a, \quad -d < x' < 0; \tag{7b} \end{cases}$$

for the source location in Fig. 3.4.2(b),

$$g''(x, x'; \lambda) = \begin{cases} \dfrac{\overleftarrow{V}_1(x)\overrightarrow{V}_2(x')}{j\omega\mu_0\,\overleftrightarrow{Y}(0)}, & -d < x < 0, \quad 0 < x' < a, \tag{7c} \\[2em] \dfrac{\overleftarrow{V}_2(x_<)\overrightarrow{V}_2(x_>)}{j\omega\mu_0\,\overleftrightarrow{Y}(0)}, & 0 < x < a, \quad 0 < x' < a. \tag{7d} \end{cases}$$

Equations (7) can be subsumed under the single formula

$$g''(x, x'; \lambda) = \frac{\overleftarrow{V}_\beta(x_<)\overrightarrow{V}_\beta(x_>)}{j\omega\mu_0\,\overleftrightarrow{Y}(0)}, \qquad \overleftrightarrow{Y}(0) = \overleftarrow{Y}(0) + \overrightarrow{Y}(0), \tag{8}$$

where the subscript β stands for 1 or 2 if the corresponding variable x or x' lies in the range $-d$ to 0 or 0 to a, respectively. To assure that the solution for g'' is unique, the restriction Im $\lambda \neq 0$ (i.e., Im $\kappa_1^2 \neq 0$, Im $\kappa_2^2 \neq 0$) is implied.

The singularities of g'' in the complex λ plane consist of real simple poles at the zeros of $\overleftrightarrow{Y}(0)$. Although g'' is a function of $\kappa_{1,2}$, and, from Eq. (2a),

$$\kappa_{1,2} = \sqrt{\lambda + k_{1,2}^2}, \tag{9}$$

---

† $\overrightarrow{V}(x)$ and $\overleftarrow{V}(x)$, as given in Eqs. (5) and (6), are identically the normalized quantities defined in Eq. (3.3.15c). To simplify the present notation the dependence on $x_0 = 0$ has not been exhibited.

no branch-point singularities exist at $\lambda = -k_{1,2}^2$ since $\overset{\rightleftarrows}{V_\beta}$, $\overset{\leftrightarrow}{Y}(0)$ and therefore g'' are even functions of $\kappa_{1,2}$ [see Eqs. (4)–(6)]. Thus, a power-series expansion about $\kappa_1 = 0$ or $\kappa_2 = 0$ comprises only integral powers of κ_1^2 or κ_2^2 and hence integral powers of λ, so the regularity of g'' in the neighborhood of the points $\lambda = -k_{1,2}^2$ is assured. From Eq. (4) the zeros λ_m of $\overset{\leftrightarrow}{Y}(0, \lambda)$ are specified implicitly by the transcendental equation

$$\kappa_2 \cot \kappa_2 a = -\kappa_1 \cot \kappa_1 d, \tag{10}$$

$$\kappa_2^2 = \lambda + k_2^2 = \hat{\lambda}, \qquad \kappa_1^2 = \lambda + k_1^2 = \hat{\lambda} + h, \qquad h = k_1^2 - k_2^2 > 0. \tag{10a}$$

For real values of κ_1 and κ_2 (i.e., $\hat{\lambda} > 0$), Eq. (10) has an infinite number of solutions to be denoted by κ_{1m}, κ_{2m} (only positive roots κ_{1m} and κ_{2m} need be considered since negative values leads to the same λ_m). For imaginary values of κ_1 and κ_2 ($\hat{\lambda} < -h$), Eq. (10) becomes

$$|\kappa_2| \coth [|\kappa_2|a] = -|\kappa_1| \coth [|\kappa_1|d], \qquad \kappa_1, \kappa_2 \text{ imaginary.} \tag{11a}$$

Since the left-hand side of Eq. (11a) is positive while the right-hand side is negative, no solution exists. However, for real κ_1 and imaginary κ_2 ($-h < \hat{\lambda} < 0$), Eq. (10) can have roots κ_{1v}, $|\kappa_{2v}|$:

$$r_v \cot r_v = -t_v \coth \left(\frac{a}{d} t_v\right), \qquad r_v^2 + t_v^2 = hd^2 = \left(1 - \frac{\epsilon_2}{\epsilon_1}\right)(k_1 d)^2, \tag{11b}$$

where

$$\kappa_{1v} d \equiv r_v > 0; \qquad |\kappa_{2v}| d \equiv t_v, \quad \kappa_{2v} \text{ imaginary.} \tag{11c}$$

Equation (11b) can be interpreted graphically as in Fig. 3.4.3. It is noted that N roots exist for $(2N - 1)\pi/2 < \sqrt{h}\,d < (2N + 1)\pi/2$, with no solution possible when $\sqrt{h}\,d < \pi/2$. These modes therefore possess a low-frequency cutoff.

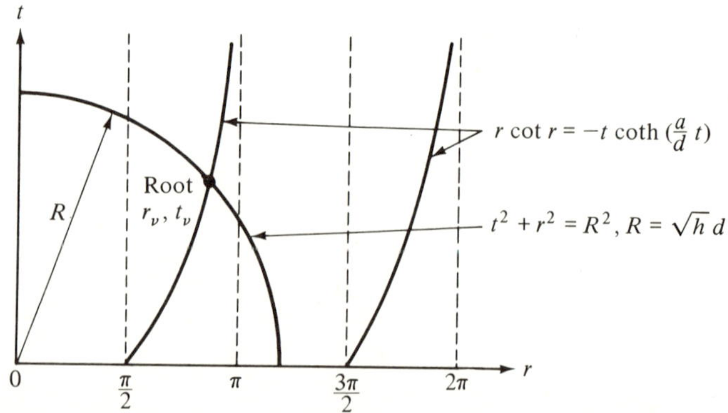

FIG. 3.4.3 Graphical solution of transcendental equation (11b).

The spectral representation of the delta function is now obtained, as in Eqs. (3.3.25), by integrating the characteristic Green's function g'' in Eq. (8) along the contour C shown in Fig. 3.4.4 enclosing all singularities:

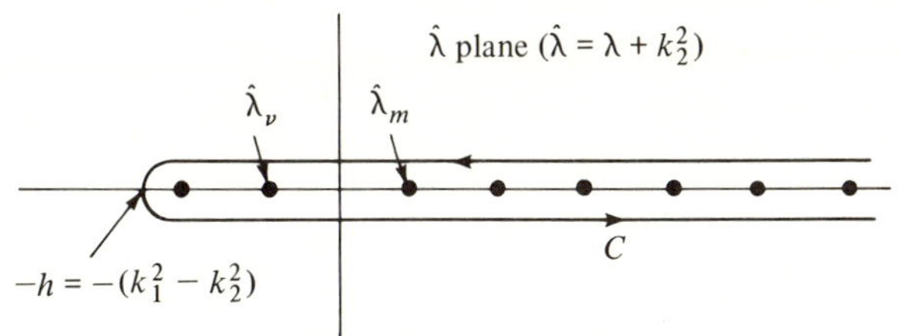

FIG. 3.4.4   Contour of integration and singularities in the λ plane.

$$\delta(x - x') = -\frac{1}{2\pi j} \oint_C g''(x, x'; \lambda)\, d\lambda, \tag{12a}$$

$$= \sum_\nu \hat{\psi}_{\nu\beta}(x)\hat{\psi}^*_{\nu\beta}(x') + \sum_m \hat{\psi}_{m\beta}(x)\hat{\psi}^*_{m\beta}(x'), \qquad -d < \frac{x}{x'} < a, \tag{12b}$$

where the contributions for $-h < \hat{\lambda}_\nu < 0$ and $\hat{\lambda}_m > 0$ have been exhibited separately. From Eq. (3.3.25d) and Eqs. (5) and (6) one obtains, for the ortho-normal eigenfunctions $\hat{\psi}_{\nu\beta}$ and $\hat{\psi}_{m\beta}$,

$$\hat{\psi}_{\nu 1}(x) = \frac{1}{A_\nu} \frac{\sin [r_\nu(x/d + 1)]}{\sin r_\nu}, \qquad 0 < r_\nu < \sqrt{h}\,d, \quad -d < x < 0, \tag{13a}$$

$$\hat{\psi}_{\nu 2}(x) = \frac{1}{A_\nu} \frac{\sinh [t_\nu\alpha(1 - x/a)]}{\sinh (t_\nu\alpha)}, \qquad \alpha = \frac{a}{d}, \quad 0 < x < a, \tag{13b}$$

where

$$A_\nu^2 = \omega\mu_0 \frac{\partial}{\partial\lambda_\nu} \overset{\leftrightarrow}{B}(0, \lambda_\nu) = \frac{d}{2}\left[\frac{\coth (t_\nu\alpha)}{t_\nu r_\nu^2} hd^2 + \csc^2 r_\nu - \alpha \operatorname{csch}^2 (t_\nu\alpha)\right]. \tag{13c}$$

Similarly,

$$\hat{\psi}_{m1}(x) = \frac{1}{A_m} \frac{\sin \kappa_{1m}(x + d)}{\sin \kappa_{1m}d}, \qquad \kappa_{1m} > 0, \quad -d < x < 0, \tag{14a}$$

$$\hat{\psi}_{m2}(x) = \begin{cases} \dfrac{1}{A_m} \dfrac{\sin \kappa_{2m}(a - x)}{\sin \kappa_{2m}a}, & \tag{14b} \\[2ex] \hspace{3em} \kappa_{2m} > 0, \quad 0 < x < a, & \\[2ex] \dfrac{1}{A_m[1 + \overset{\leftarrow}{\Gamma}_m(0)]} [e^{j\kappa_{2m}x} + \overset{\leftarrow}{\Gamma}_m(0)^{-j\kappa_{2m}x}], & \tag{14c} \end{cases}$$

with

$$A_m^2 = \omega\mu_0 \frac{\partial}{\partial\lambda_m} \overset{\leftrightarrow}{B}(0, \lambda_m) = \begin{cases} \frac{a}{2}\left[\left(1 + \frac{1}{\alpha}\right) + \left(\frac{\kappa_{1m}^2}{\kappa_{2m}^2} + \frac{1}{\alpha}\right)\cot^2\kappa_{1m}d \right. \\ \qquad\qquad \left. + \frac{h}{a\kappa_{1m}\kappa_{2m}^2}\cot\kappa_{1m}d\right], \qquad (14d) \\[2ex] \frac{d}{2}\left[(1+\alpha) + \left(\frac{\kappa_{2m}^2}{\kappa_{1m}^2} + \alpha\right)\cot^2\kappa_{2m}a \right. \\ \qquad\qquad \left. - \frac{h}{d\kappa_{2m}\kappa_{1m}^2}\cot\kappa_{2m}a\right]. \qquad (14e) \end{cases}$$

It is verified that Eqs. (13) and (14) reduce to the results obtained previously for a homogeneously filled waveguide [Sec. 3.2b] when (a) $h = 0$ ($\epsilon_1 = \epsilon_2$), (b) $d = 0$, or (c) $a = 0$. Attention should be called to the different behavior of the eigenfunctions $\hat{\psi}_\nu(x)$ in Eqs. (13) and $\hat{\psi}_m(x)$ in Eqs. (14). While $\hat{\psi}_m(x)$ is represented by an oscillating function over the entire region $-d < x < a$, $\hat{\psi}_\nu(x)$ behaves in this manner only in the dielectric ϵ_1 (note: $\epsilon_1 > \epsilon_2$). In the remaining interval $0 < x < a$, $\hat{\psi}_\nu$ decays away from the interface $x = 0$. Viewed in modal terms with respect to propagation along z, the fields corresponding to the $\hat{\psi}_\nu$ are essentially confined within the dielectric slab while the fields derived from the $\hat{\psi}_m$ fill the entire waveguide cross section. The former are termed "trapped" modes and their existence depends entirely on the presence of the dielectric; the latter may be regarded as perturbations about the dielectric-free case.

E modes (*in x*).   The solution for the E-mode characteristic Green's function $g'(x, x'; \lambda)$ and the associated orthonormal eigenfunctions is similar to the above except for duality replacements [see Eqs. (3.3.26)–(3.3.30)]. The results are summarized below.

Characteristic Green's function

$$g'(x, x'; \lambda) = \frac{\overset{\leftarrow}{I}_\beta(x_<)\vec{I}_\beta(x_>)}{j\omega\epsilon_0 \overset{\leftrightarrow}{Z}(0)}, \qquad \overset{\leftrightarrow}{Z}(0) = \overset{\leftarrow}{Z}(0) + \vec{Z}(0). \qquad (15)$$

$$c(x) = \cos\kappa_1 x, \quad s(x) = \frac{\epsilon_1'}{\kappa_1}\sin\kappa_1 x, \qquad -d < x < 0,$$
$$(15a)$$
$$c(x) = \cos\kappa_2 x, \quad s(x) = \frac{\epsilon_2'}{\kappa_2}\sin\kappa_2 x, \qquad 0 < x < a,$$

$$\omega\epsilon_0 \vec{Z}(0) = j\frac{\kappa_2}{\epsilon_2'}\tan\kappa_2 a, \quad \omega\epsilon_0 \overset{\leftarrow}{Z}(0) = j\frac{\kappa_1}{\epsilon_1'}\tan\kappa_1 d, \quad \epsilon_{1,2}' = \frac{\epsilon_{1,2}}{\epsilon_0},$$

$$\kappa_1^2 = k_1^2 + \lambda, \quad \kappa_2^2 = k_2^2 + \lambda. \qquad (15b)$$

$$\left.\begin{aligned} \vec{I}_1(x) &= \cos\kappa_1 x + \frac{\epsilon_1'\kappa_2}{\epsilon_2'\kappa_1}\tan\kappa_2 a \sin\kappa_1 x \\[1ex] \overset{\leftarrow}{I}_1(x) &= \frac{\cos\kappa_1(x + d)}{\cos\kappa_1 d} \end{aligned}\right\} -d < x < 0, \qquad (15c)$$

$$\vec{I_2}(x) = \frac{\cos \kappa_2(a - x)}{\cos \kappa_2 a}$$

$$\overleftarrow{I_2}(x) = \cos \kappa_2 x - \frac{\epsilon_2' \kappa_1}{\epsilon_1' \kappa_2} \tan \kappa_1 d \sin \kappa_2 x \qquad \Bigg\} \quad 0 < x < a. \qquad (15d)$$

Singularities of g': Simple real poles at

$$\frac{\epsilon_1'}{\epsilon_2'} \kappa_{2m} \tan \kappa_{2m} a = -\kappa_{1m} \tan \kappa_{1m} d, \qquad \kappa_{1m}, \kappa_{2m} > 0, \qquad (16a)$$

and at

$$\frac{\epsilon_2'}{\epsilon_1'} r_v \tan r_v = t_v \tanh (t_v \alpha), \quad r_v^2 + t_v^2 = hd^2, \quad \alpha = \frac{a}{d}, \ h = k_1^2 - k_2^2,$$

$$\kappa_{1v} d \equiv r_v > 0; \qquad |\kappa_{2v}|d \equiv t_v, \qquad \kappa_{2v} \text{ imaginary.} \qquad (16b)$$

Equation (16a) has an infinite number of solutions and Eq. (16b) has a finite number. The graphical representation of Eq. (16b) resembles that in Fig. 3.4.3, provided that all curves (except the circle) are shifted to the left through an interval $\pi/2$. The low-frequency cutoff found for the *H*-mode solutions $\hat{\psi}_v$ is therefore absent in the *E*-mode case.

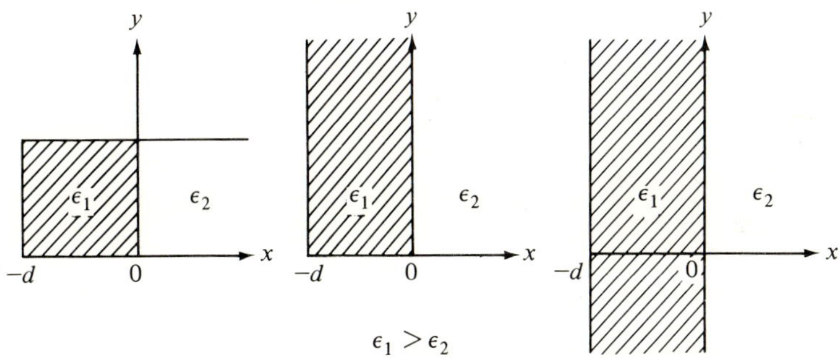

FIG. 3.4.5 Semi-infinite regions (in *x*) partially filled with dielectric.

Delta-function representation $(-d < \frac{x}{x'} < a)$,

$$\epsilon'(x')\delta(x - x') = -\frac{1}{2\pi j} \oint_C g'(x, x'; \lambda) \, d\lambda$$

$$= \sum_v \hat{\Phi}_{v\beta}(x)\hat{\Phi}^*_{v\beta}(x') + \sum_m \hat{\Phi}_{m\beta}(x)\hat{\Phi}^*_{m\beta}(x'), \qquad (17)$$

with the subscript β defined as under Eq. (8), and

$$\hat{\Phi}_{v1}(x) = \frac{\cos [r_v(x/d + 1)]}{A_v \cos r_v}, \qquad 0 < r_v < \sqrt{h} \, d, \quad -d < x < 0, \qquad (17a)$$

$$\hat{\Phi}_{v2}(x) = \frac{\cosh [t_v \alpha(1 - x/a)]}{A_v \cosh t_v \alpha}, \qquad 0 < x < a, \qquad (17b)$$

$$A_v^2 = \omega \epsilon_0 \frac{\partial}{\partial \lambda_v} \overleftrightarrow{X}(0, \lambda_v) = \frac{d}{2} \left[\frac{\tanh (t_v \alpha)}{r_v^2 t_v \epsilon_2'} hd^2 + \frac{\sec^2 r_v}{\epsilon_1'} + \frac{\alpha}{\epsilon_2'} \operatorname{sech}^2 (t_v \alpha) \right], \qquad (17c)$$

while

$$\hat{\Phi}_{m1}(x) = \frac{\cos \kappa_{1m}(x + d)}{A_m \cos \kappa_{1m} d}, \qquad \kappa_{1m} > 0, \quad -d < x < 0, \tag{17d}$$

$$\hat{\Phi}_{m2}(x) = \frac{\cos \kappa_{2m}(a - x)}{A_m \cos \kappa_{2m} a}, \qquad \kappa_{2m} > 0, \qquad 0 < x < a, \tag{17e}$$

$$A_m^2 = \frac{a}{2\epsilon_1'}\left[\frac{\epsilon_1'}{\epsilon_2'} + \frac{1}{\alpha} + \left(\frac{\kappa_{1m}^2 \epsilon_2'}{\kappa_{2m}^2 \epsilon_1'} + \frac{1}{\alpha}\right)\tan^2 \kappa_{1m} d - \frac{h}{a\kappa_{1m}\kappa_{2m}^2}\tan \kappa_{1m} d\right]. \tag{17f}$$

The physical distinction between the mode fields corresponding to the $\hat{\Phi}_\nu$ and $\hat{\Phi}_m$ is the same as discussed in connection with the H modes.

Employing Eq. (17), one may represent a suitable function $F(x)$ in the interval $-d < x < a$ as follows:

$$F(x) = \int_{-d}^{a} \frac{F(x')}{\epsilon'(x')}\epsilon'(x')\delta(x - x')\,dx'$$

$$= \begin{cases} \sum_\nu f_\nu \hat{\Phi}_{\nu1}(x) + \sum_m f_m \hat{\Phi}_{m1}(x), & -d < x \leq 0, \\[2mm] \sum_\nu f_\nu \hat{\Phi}_{\nu2}(x) + \sum_m f_m \hat{\Phi}_{m2}(x), & 0 \leq x < a, \end{cases} \tag{18}$$

where

$$f_\nu = \frac{1}{\epsilon_1'}\int_{-d}^{0} F(x')\hat{\Phi}_{\nu1}^*(x')\,dx' + \frac{1}{\epsilon_2'}\int_{0}^{a} F(x')\hat{\Phi}_{\nu2}^*(x')\,dx', \tag{18a}$$

$$f_m = \frac{1}{\epsilon_1'}\int_{-d}^{0} F(x')\hat{\Phi}_{m1}^*(x')\,dx' + \frac{1}{\epsilon_2'}\int_{0}^{a} F(x')\hat{\Phi}_{m2}^*(x')\,dx', \tag{18b}$$

and the asterisk denotes the complex conjugate.

Semiinfinite x domain

As $a \to \infty$ in Fig. 3.4.1, one obtains the open cross-section configurations in Fig. 3.4.5. The eigenfunctions appropriate to this case can be obtained as a limiting case of those for finite a.

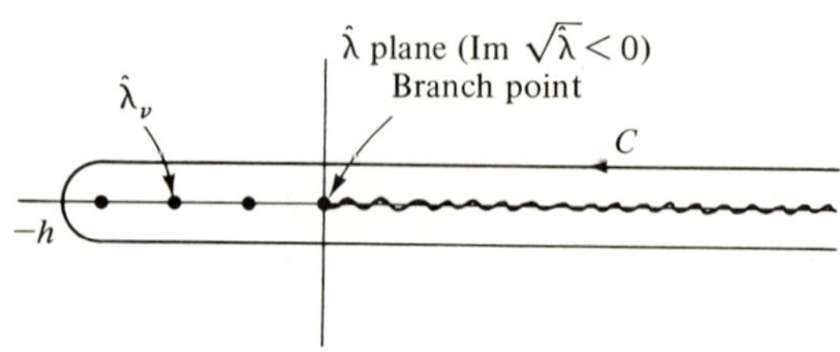

FIG. 3.4.6 Path of integration and singularities in a complex $\hat{\lambda}$ plane.

H modes (in x) (a → ∞). As $a \to \infty$, the resonances κ_{1m} and κ_{2m} in Eq. (10), with $\kappa_{2m} > 0$, coalesce into a continuous spectrum, while those in Eq. (11) remain discrete and satisfy the equation

$$r_\nu \cot r_\nu = -t_\nu, \quad r_\nu^2 + t_\nu^2 = hd^2, \qquad \text{as } a \to \infty. \tag{19a}$$

Moreover, from Eq. (13c),

$$A_\nu^2 \to \frac{d}{2} \frac{hd^2}{r_\nu^2} \left(1 + \frac{1}{t_\nu} \right), \tag{19b}$$

while, from Eq. (14d),

$$A_m^2 \to A_\xi^2 = \frac{a}{2} \left(1 + \frac{\xi_1^2}{\xi^2} \cot^2 \xi_1 d \right) = \frac{2a}{[1 + \overleftarrow{\Gamma}_2(\xi, 0)][1 + \overleftarrow{\Gamma}_2(\xi, 0)^*]}, \tag{19c}$$

where $\overleftarrow{\Gamma}_2(\xi, 0)$ is given in Eq. (6). In the last equation the continuous variables ξ_1 and ξ have been defined as the limiting values of κ_{1m} and κ_{2m} as $a \to \infty$:

$$\kappa_{2m} \to \xi, \quad \kappa_{1m} \to \xi_1 = \sqrt{\xi^2 + h}, \qquad 0 < \xi < \infty, \quad a \to \infty. \tag{19d}$$

Upon noting that the increment $\Delta \xi$ is given as in Eq. (3.2.26) by $\Delta \xi = \pi/a$, $a \to \infty$, one may write Eq. (12b) as

$$\delta(x - x') = \sum_\nu \hat{\psi}_{\nu\beta}(x) \hat{\psi}_{\nu\beta}^*(x') + \int_0^\infty \hat{\psi}_\beta(\xi, x) \hat{\psi}_\beta^*(\xi, x') \, d\xi,$$

$$-d < \frac{x}{x'} < \infty, \quad \beta = 1, 2, \tag{20}$$

where, in view of Eqs. (13), (14), and (19), one has, for the discrete spectrum,

$$\hat{\psi}_{\nu 1}(x) = \frac{1}{A_\nu} \frac{\sin [r_\nu((x/d) + 1)]}{\sin r_\nu}, \qquad 0 < r_\nu < \sqrt{h}\, d, \quad -d < x < 0, \tag{20a}$$

$$\hat{\psi}_{\nu 2}(x) = \frac{1}{A_\nu} e^{-t_\nu x/d}, \qquad\qquad 0 < x < \infty. \tag{20b}$$

As in Eq. (8), $\beta = 1$ for x or x' between $-d$ and 0, while $\beta = 2$ for x or x' between 0 and ∞. Just as in the closed region, the magnitude of $\hat{\psi}_{\nu 1}$ oscillates while that of $\hat{\psi}_{\nu 2}$ decreases exponentially for $x > 0$. Thus, the field of such a mode is confined again to the region $-d < x < 0$ occupied by the dielectric ϵ_1. Modes traveling in the z direction with this transverse field behavior are characterized as "trapped waves," or "surface waves," since the field appears to be trapped inside the dielectric with the larger permittivity and guided by the dielectric surface.

For the continuous spectrum,

$$\hat{\psi}_1(\xi, x) = \frac{\sin \xi_1(x + d)}{\sqrt{2\pi} \sin \xi_1 d} [1 + \overleftarrow{\Gamma}_2(0, \xi)], \qquad 0 < \xi < \infty, \quad -d < x < 0, \tag{20c}$$

$$\hat{\psi}_2(\xi, x) = \frac{1}{\sqrt{2\pi}}\,[e^{j\xi x} + \overset{\leftarrow}{\Gamma}_2(0, \xi)e^{-j\xi x}], \qquad 0 < x < \infty, \tag{20d}$$

where

$$\overset{\leftarrow}{\Gamma}_2(0, \xi) = \frac{\xi + j\xi_1 \cot \xi_1 d}{\xi - j\xi_1 \cot \xi_1 d}, \qquad \xi_1^2 = h + \xi^2 = (k_1^2 - k_2^2) + \xi^2. \tag{20e}$$

The traveling-wave representation for $\hat{\psi}_2$, derived as a limiting case of Eq. (14c), has a significant physical interpretation. For the assumed time dependence $\exp(+j\omega t)$, the contribution from the first term inside the brackets in Eq. (20d) constitutes a properly normalized (incident) free-space plane-wave mode traveling in the $-x$ direction [see Eq. (3.2.37b)], while the second term comprises the wave reflected at $x = 0$ with reflection coefficient $\overset{\leftarrow}{\Gamma}_2(0, \xi)$. Thus, the continuous spectrum for $x > 0$ is obtained by adding to a properly normalized incident wave a reflected wave so adjusted that the boundary conditions at $x = 0$ are satisfied.

The delta-function representation in Eq. (20) could also have been deduced directly from the characteristic Green's function. As $a \to \infty$ and since $\mathrm{Im}\,\kappa_2 \neq 0$, the standing wave in Eq. (5a) goes over into a traveling wave. In this transition, the restriction $\mathrm{Im}\,\kappa_2 < 0$ appropriate to the assumed time dependence $\exp(+j\omega t)$ must be observed [see Eqs. (2.2.15)] and yields the following (bounded) result for $x > 0$:

$$\vec{V}_2(x) \to e^{-j\kappa_2 x}, \quad \kappa_2 = \sqrt{k_2^2 + \lambda} = \sqrt{\hat{\lambda}}, \quad \mathrm{Im}\,\kappa_2 < 0, \tag{21a}$$

and

$$\vec{V}_1(x) \to \cos \kappa_1 x - j\frac{\kappa_2}{\kappa_1} \sin \kappa_1 x, \quad \kappa_1 = \sqrt{\hat{\lambda} + h}, \quad h = k_1^2 - k_2^2. \tag{21b}$$

Moreover, from Eq. (4),

$$\vec{Y}(0) \to \frac{\kappa_2}{\omega\mu_0}, \qquad \text{i.e., } j\omega\mu_0\overset{\leftrightarrow}{Y}(0) = j\kappa_2 + \kappa_1 \cot \kappa_1 d. \tag{21c}$$

The $\overset{\leftarrow}{V}_{1,2}(x)$ are still given by Eqs. (5c) and (5d). $\vec{V}_\beta$ and $\overset{\leftrightarrow}{Y}(0)$ remain even functions of κ_1 but not of κ_2. $\lambda = -k_1^2$ is therefore a regular point in the complex λ-plane. On the other hand, an expansion of $g''(x, x'; \lambda)$ about the point $\lambda = -k_2^2$ contains integral powers of κ_2, so $\lambda + k_2^2 = \hat{\lambda} = 0$ is a branch point of order 1. If we define

$$\hat{\lambda} = |\hat{\lambda}|e^{j\gamma}, \qquad \sqrt{\hat{\lambda}} = |\sqrt{\hat{\lambda}}|e^{j\gamma/2}, \tag{22}$$

the convergence requirement $\mathrm{Im}\,\sqrt{\hat{\lambda}} < 0$ in Eq. (21a) restricts the argument γ to the range $0 > \gamma > -2\pi$. To impose this condition on the entire top sheet, the spectral sheet, of the two-sheeted complex $\hat{\lambda}$ plane, one chooses a branch cut along the positive real axis as shown in Fig. 3.4.6.

The Green's function g'' may also have relevant pole singularities at the zeros of $\overleftrightarrow{Y}(0)$, namely when

$$jκ_2 = -κ_1 \cot κ_1 d. \tag{23}$$

Solutions of Eq. (23) exist only for real values of $κ_1$ and imaginary values of $κ_2 = -j|κ_2|$ (i.e., $0 > \hat{λ} > -h$), leading to the transcendental equation (19a). The location of possible pole singularities is shown in Fig. 3.4.6. Upon performing an integration as in Eq. (12a) about the contour C in Fig. 3.4.6 enclosing all the singularities of g'' in the complex $λ$ plane, one obtains after residue evaluation at the poles $λ_v$ the series in Eq. (20), with $g''(x, x'; λ)$ given by Eq. (8) and subject to the modifications in Eqs. (21). The remaining contour integral about the branch cut can be written as

$$I = -\frac{1}{2\pi j} \int_{\infty e^{-j2\pi}}^{0} g''(x, x'; λ)\, d\hat{λ} - \frac{1}{2\pi j} \int_{0}^{\infty e^{-j0}} g''(x, x; λ)\, d\hat{λ} \tag{24a}$$

$$= -\frac{1}{2\pi j} \int_{0}^{\infty e^{-j0}} [g''(x, x'; \hat{λ} - k_2^2) - g''(x, x'; \hat{λ}e^{-j2\pi} - k_2^2)]\, d\hat{λ}$$

$$= -\frac{1}{\pi} \operatorname{Im} \int_{0}^{\infty e^{-j0}} g''(x, x'; \hat{λ} - k_2^2)\, d\hat{λ}$$

$$= -\frac{2}{\pi} \operatorname{Im} \int_{0}^{\infty} ξ\, g''(x, x'; ξ^2 - k_2^2)\, dξ, \qquad ξ^2 = \hat{λ}. \tag{24b}$$

The transition from Eq. (24a) to (24b) is based on the property

$$g''(x, x'; \hat{λ}e^{-j2\pi}) = g''(x, x'; \hat{λ}^*) = g''^*(x, x'; \hat{λ}), \qquad \hat{λ} = |\hat{λ}|e^{-j0},$$

$$\tag{24c}$$

satisfied by g''. Upon substituting the appropriate representations for g'' into Eq. (24b), one obtains directly the continuous spectrum as in Eq. (20).

E modes (in x) ($a \to \infty$)

The results for the E-mode problem, obtained in direct analogy to those above, are summarized below:

$$ε'(x')δ(x - x') = -\frac{1}{2\pi j} \oint_C g'(x, x'; λ)\, dλ \tag{25a}$$

$$= \sum_v \hat{Φ}_{vβ}(x)\hat{Φ}_{vβ}^*(x') + \int_0^{\infty} \hat{Φ}_β(ξ, x)\hat{Φ}_β^*(ξ, x')\,dξ, \tag{25b}$$

$$-d < \frac{x}{x'} < \infty, \quad β = 1, 2,$$

where, for the discrete spectrum [see Eq. (20) for definition of domains corresponding to $β = 1, 2$],

$$\hat{Φ}_{v1}(x) = \frac{\cos\left[r_v((x/d) + 1)\right]}{A_v \cos r_v}, \qquad 0 < r_v < \sqrt{h}\,d, \quad -d < x < 0,$$

$$\tag{26a}$$

$$\hat{\Phi}_{\nu 2}(x) = \frac{e^{-t_\nu x/d}}{A_\nu}, \qquad 0 < x < \infty, \tag{26b}$$

$$A_\nu^2 = \frac{d}{2}\left\{\left[1 + \left(\frac{t_\nu}{r_\nu}\right)^2\right]\frac{1}{t_\nu \epsilon_2'} + \left[1 + \left(\frac{t_\nu \epsilon_1'}{r_\nu \epsilon_2'}\right)^2\right]\frac{1}{\epsilon_1'}\right\}. \tag{26c}$$

Also, r_ν and t_ν are the solutions of the transcendental equations

$$\frac{\epsilon_2'}{\epsilon_1'} r_\nu \tan r_\nu = t_\nu, \qquad r_\nu^2 + t_\nu^2 = hd^2. \tag{26d}$$

The continuous spectrum is given by

$$\hat{\Phi}_1(\xi, x) = \sqrt{\frac{\epsilon_2'}{2\pi}} \frac{\cos \xi_1(x + d)}{\cos \xi_1 d} [1 - \overleftarrow{\Gamma}_2(0, \xi)], \qquad \xi_1^2 = h + \xi^2,$$

$$0 < \xi < \infty, \quad -d < x < 0, \tag{27a}$$

$$\hat{\Phi}_2(\xi, x) = \sqrt{\frac{\epsilon_2'}{2\pi}} [e^{j\xi x} - \overleftarrow{\Gamma}_2(0, \xi)e^{-j\xi x}], \qquad 0 < x < \infty, \tag{27b}$$

$$\overleftarrow{\Gamma}_2(0, \xi) = \frac{\overleftarrow{Z}(0) - Z_{02}}{\overleftarrow{Z}(0) + Z_{02}} = \frac{j\xi_1 \tan \xi_1 d - \xi(\epsilon_1'/\epsilon_2')}{j\xi_1 \tan \xi_1 d + \xi(\epsilon_1'/\epsilon_2')}. \tag{27c}$$

If $d \to \infty$ in Fig. 3.4.1, one obtains the semiinfinite configurations shown in Fig. 3.4.7, which differ from those in Fig. 3.4.5 in that the medium with the larger dielectric constant (ϵ_1) extends to infinity in the x direction.

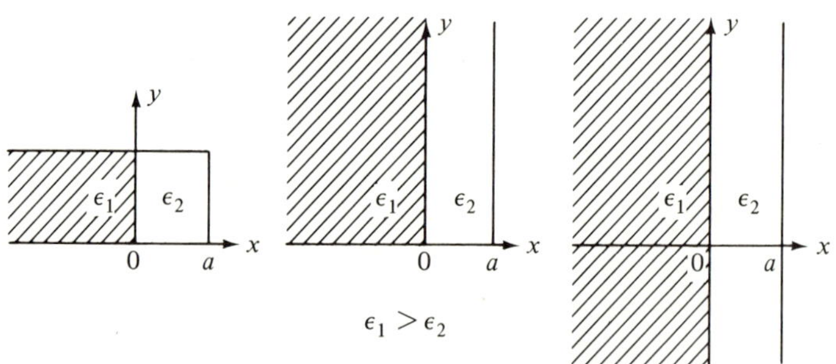

FIG. 3.4.7   Semi-infinite regions (in x) partially filled with dielectric.

H modes (in x) (d → ∞)

As $d \to \infty$ in Eq. (10), the resonances $\kappa_{1m}, \kappa_{2m} > 0$, coalesce into a continuous spectrum and the second series in the delta-function representation (12b) transforms into an integral analogous to that in Eq. (20). However, in distinction to the case $a \to \infty$, the resonance parameters $\kappa_{1\nu}$ and $|\kappa_{2\nu}|$ in Eq. (11b) become continuous as $d \to \infty$. In tracing out the transition $d \to \infty$, one employs instead of Eq. (5b) the traveling-wave formulation similar to that in Eq. (6):

$$\overrightarrow{V}_1(x) = \frac{1}{1 + \overrightarrow{\Gamma}_1(0)} [e^{-j\kappa_1 x} + \overrightarrow{\Gamma}_1(0)e^{j\kappa_1 x}], \qquad -d < x < 0, \tag{28}$$

where the reflection coefficient $\overrightarrow{\Gamma}_1(0)$ seen to the right at $x = 0^-$ is given by

$$\overrightarrow{\Gamma}_1(0) = \frac{\kappa_1 + j\kappa_2 \cot \kappa_2 a}{\kappa_1 - j\kappa_2 \cot \kappa_2 a}. \tag{28a}$$

Since from Eqs. (13c) and (14e),

$$A_m^2 \to A_{\xi_1}^2 = \frac{2d}{[1 + \overrightarrow{\Gamma}_1(\xi_1, 0)][1 + \overrightarrow{\Gamma}_1(\xi_1, 0)^*]}, \qquad d \to \infty, \quad \sqrt{h} < \xi_1 < \infty, \tag{29a}$$

with

$$\overrightarrow{\Gamma}_1(\xi_1, 0) = \frac{\xi_1 + j\xi \cot \xi a}{\xi_1 - j\xi \cot \xi a}, \qquad \xi = \sqrt{\xi_1^2 - h}, \quad h = k_1^2 - k_2^2 > 0, \tag{29b}$$

and

$$A_v^2 \to A_{\xi_1}^2, \qquad d \to \infty, \quad 0 < \xi_1 < \sqrt{h}, \tag{29c}$$

one obtains via Eqs. (12)–(14) and (28) the delta-function representation:

$$\delta(x - x') = \int_0^\infty \hat{\psi}_\beta(\xi_1, x)\hat{\psi}_\beta^*(\xi_1, x') \, d\xi_1, \qquad -\infty < \frac{x}{x'} < a, \quad \beta = 1, 2, \tag{30}$$

where $-\infty < (x \text{ or } x') < 0$ for $\beta = 1$ and $0 < (x \text{ or } x') < a$ for $\beta = 2$, with

$$\hat{\psi}_1(\xi_1, x) = \frac{1}{\sqrt{2\pi}} [e^{-j\xi_1 x} + \overrightarrow{\Gamma}_1(\xi_1, 0)e^{j\xi_1 x}], \qquad 0 < \xi_1 < \infty, \quad -\infty < x < a, \tag{30a}$$

$$\hat{\psi}_2(\xi_1, x) = \frac{\sin \xi(a - x)}{\sqrt{2\pi} \sin \xi a} [1 + \overrightarrow{\Gamma}_1(\xi_1, 0)], \qquad 0 < x < a. \tag{30b}$$

It is noted that ξ is imaginary for $0 < \xi_1 < \sqrt{h}$.

To deduce Eqs. (30) directly from a characteristic Green's function analysis, one notes from Eqs. (5) that as $d \to \infty$, with $\text{Im } \kappa_1 < 0$ appropriate to an $\exp(j\omega t)$ time dependence,

$$\overleftarrow{V}_1(x) \to e^{j\kappa_1 x}, \qquad -\infty < x < 0, \tag{31a}$$

$$\overleftarrow{V}_2(x) \to \cos \kappa_2 x + j\frac{\kappa_1}{\kappa_2} \sin \kappa_2 x, \qquad 0 < x < a, \tag{31b}$$

$$\omega\mu_0 \overleftarrow{Y}(0) \to \kappa_1. \tag{31c}$$

Since $g''(x, x'; \lambda)$, by Eqs. (8) and (31), is an even function of κ_2 but not of κ_1, a branch-point singularity exists at $\kappa_1 = 0$ (i.e., $\lambda = -k_1^2$) in the complex λ

plane. In analogy to Eq. (22), the restriction on the argument of λ on the spectral sheet is

$$\text{Im } \sqrt{\hat{\lambda} + h} < 0, \qquad \text{i.e., } -2\pi < \arg(\hat{\lambda} + h) < 0, \quad \hat{\lambda} = \lambda + k_2^2 = \kappa_2^2, \tag{32}$$

so that the branch cut is drawn from $\hat{\lambda} = -h$ to ∞ along the positive real axis in the $\hat{\lambda}$ plane (see Fig. 3.4.8). To determine possible pole singularities we examine the resonance condition

$$j\omega\mu_0 \overset{\leftrightarrow}{Y}(0) = 0 = j\kappa_1 + \kappa_2 \cot \kappa_2 a. \tag{33}$$

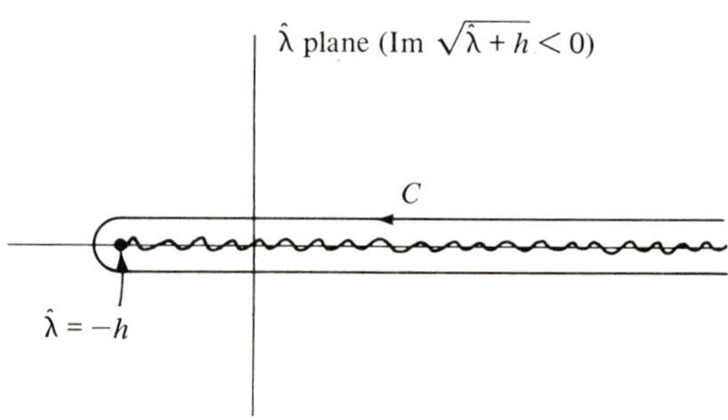

FIG. 3.4.8   Singularities and path of integration in $\hat{\lambda}$ plane.

Since Eq. (33) has no real solution λ_ν on the branch Im $\kappa_1 < 0$†, no pole singularities exist, and the contour of integration is that shown in Fig. 3.4.8. Thus, in analogy with Eqs. (24),

$$\delta(x - x') = -\frac{1}{2\pi j} \oint_C g''(x, x'; \lambda) \, d\hat{\lambda} \tag{34a}$$

$$= -\frac{2}{\pi} \text{Im} \int_0^\infty d\xi_1 \xi_1 g''(x, x'; \xi_1^2 - k_1^2), \qquad -\infty < \frac{x}{x'} < a, \tag{34b}$$

which, upon insertion of g'' from Eqs. (8), (5), and (31), yields Eq. (30).

E modes (*in* x) ($d \rightarrow \infty$)

Spectral representation of delta function:

$$\epsilon'(x')\delta(x - x') = -\frac{1}{2\pi j} \oint_C g'(x, x'; \lambda) \, d\hat{\lambda}, \tag{35a}$$

$$= \int_0^\infty \hat{\Phi}_\beta(\xi_1, x)\hat{\Phi}_\beta^*(\xi_1, x') \, d\xi_1, \qquad \beta = 1, 2, \tag{35b}$$

---

†The corresponding discrete eigenfunctions, if they exist, must be square integrable (i.e., vanish at $x \rightarrow -\infty$), so Sec. 3.2a applies. Since the problem is non-dissipative, any discrete eigenvalues must be real.

where $\beta = 1$ when $-\infty < (x \text{ or } x') < 0$ while $\beta = 2$ when $0 < (x \text{ or } x') < a$. The contour C in the complex $\hat{\lambda}$ plane is as shown in Fig. 3.4.8, and from Eqs. (17) as $d \to \infty$,

$$\hat{\Phi}_1(\xi_1, x) = \sqrt{\frac{\epsilon_1'}{2\pi}} \, [e^{-j\xi_1 x} - \vec{\Gamma}_1(\xi_1, 0)e^{j\xi_1 x}], \qquad 0 < \xi_1 < \infty, \quad -\infty < x < 0,$$

(36a)

$$\hat{\Phi}_2(\xi_1, x) = \sqrt{\frac{\epsilon_1'}{2\pi}} \, [1 - \vec{\Gamma}_1(\xi_1, 0)] \frac{\cos \xi(a - x)}{\cos \xi a}, \qquad 0 < x < a, \tag{36b}$$

$$\vec{\Gamma}_1(\xi_1, 0) = \frac{j\xi \tan \xi a - \xi_1(\epsilon_2'/\epsilon_1')}{j\xi \tan \xi a + \xi_1(\epsilon_2'/\epsilon_1')}, \qquad \xi = \sqrt{\xi_1^2 - h}, \quad h = k_1^2 - k_2^2 > 0.$$

(36c)

Infinite x domain

Configurations comprising two dielectrics, semiinfinite in x, are shown in Fig. 3.4.9.

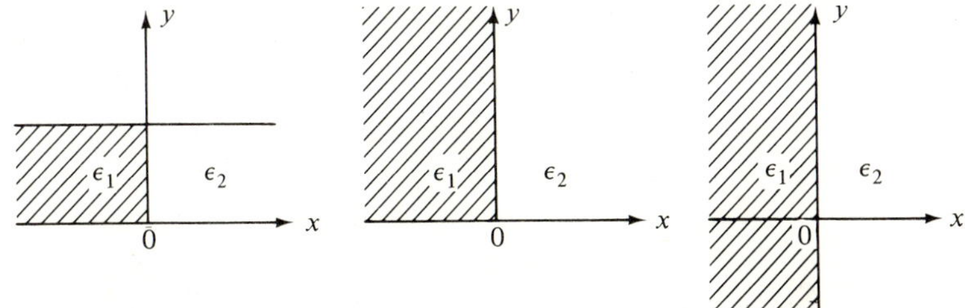

FIG. 3.4.9   Rectangular domains (infinite in x) partially filled with dielectric.

H modes (in x).   The characteristic Green's function for this case is given by

$$g''(x, x'; \lambda) = \frac{\overleftarrow{V}_\beta(x_<)\vec{V}_\beta(x_>)}{j\omega\mu_0 \overleftrightarrow{Y}(0)}, \tag{37}$$

with

$$\overleftarrow{V}_1(x) = e^{j\kappa_1 x}, \quad \kappa_1^2 = k_1^2 + \lambda, \quad \text{Im } \kappa_1 < 0, \tag{37a}$$

$$\vec{V}_2(x) = e^{-j\kappa_2 x}, \quad \kappa_2^2 = k_2^2 + \lambda, \quad \text{Im } \kappa_2 < 0, \tag{37b}$$

$$\vec{V}_1(x) = \frac{1}{1 + \vec{\Gamma}_1(0)} \, [e^{-j\kappa_1 x} + \vec{\Gamma}_1(0)e^{j\kappa_1 x}], \quad \vec{\Gamma}_1(0) = \frac{\kappa_1 - \kappa_2}{\kappa_1 + \kappa_2}, \tag{38a}$$

$$\overleftarrow{V}_2(x) = \frac{1}{1 + \overleftarrow{\Gamma}_2(0)} \, [e^{j\kappa_2 x} + \overleftarrow{\Gamma}_2(0)e^{-j\kappa_2 x}], \quad \overleftarrow{\Gamma}_2(0) = -\vec{\Gamma}_1(0), \tag{38b}$$

$$j\omega\mu_0 \overleftrightarrow{Y} = j(\kappa_1 + \kappa_2). \tag{38c}$$

Since $g''(x, x'; \lambda)$ is not an even function of either κ_1 or κ_2, branch points exist in the complex λ plane at $\lambda = -k_1^2$ and $\lambda = -k_2^2$. The argument of $\hat{\lambda} = \lambda + k_2^2$ is then restricted in accordance with Im $\kappa_1 < 0$, Im $\kappa_2 < 0$, as follows [see Eqs. (22) and (32)]:

$$0 > \arg \hat{\lambda} > -2\pi, \quad 0 > \arg(\hat{\lambda} + h) > -2\pi, \quad h = k_1^2 - k_2^2, \qquad (39)$$

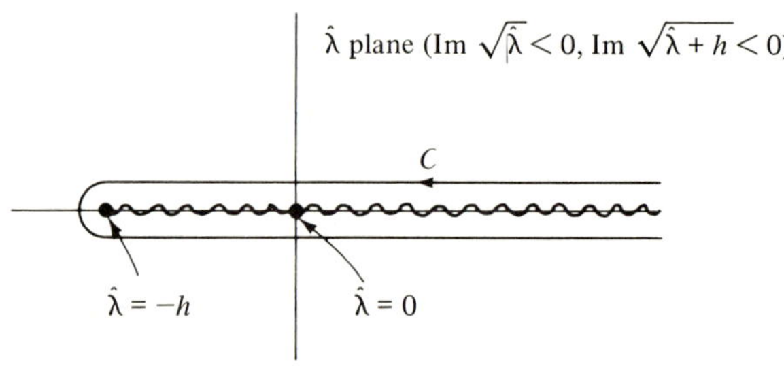

FIG. 3.4.10   Contour of integration and singularities in the complex $\hat{\lambda}$-plane.

with corresponding branch cuts along the real $\hat{\lambda}$ axis shown in Fig. 3.4.10. Since g'' possesses no pole singularities on the branch of the Riemann surface for which Im $\kappa_1 < 0$ and Im $\kappa_2 < 0$, the contour of integration is drawn as in Fig. 3.4.10. Since the replacement of $\hat{\lambda}$ by $\hat{\lambda} e^{-j2\pi}$ in g'' yields g''^* [see Eq. (24c)], we may write

$$\delta(x - x') = -\frac{1}{2\pi j} \oint_C g''(x, x'; \lambda) \, d\hat{\lambda}$$

$$= -\frac{2}{\pi} \operatorname{Im} \int_0^{\sqrt{h}} \xi_1 g''(x, x'; \xi_1^2 - k_1^2) \, d\xi_1 - \frac{2}{\pi} \operatorname{Im} \int_0^\infty \xi \, g''(x, x'; \xi^2 - k_2^2) \, d\xi$$

$$\tag{40a}$$

$$= \int_0^{\sqrt{h}} \hat{\psi}_\beta(\xi_1, x) \hat{\psi}_\beta^*(\xi_1, x') \, d\xi_1 + \int_0^\infty \hat{\psi}_\beta'(\xi, x) \hat{\psi}_\beta'^*(\xi, x) \, d\xi,$$

$$-\infty < \frac{x}{x'} < \infty, \qquad (40b)$$

where $\beta = 1$ and $\beta = 2$ correspond to $-\infty < (x \text{ or } x') < 0$ and $0 < (x \text{ or } x') < \infty$, respectively. For $0 < \xi < \infty$, one has the two mutually orthogonal sets

$$\hat{\psi}_1'(\xi, x) = \sqrt{\frac{1 - \vec{\Gamma}_1(\xi, 0)}{\pi}} \begin{Bmatrix} \cos \xi_1 x \\ \sqrt{\dfrac{\xi}{\xi_1}} \sin \xi_1 x \end{Bmatrix}, \qquad -\infty < x < 0, \qquad (41a)$$

$$\hat{\psi}_2'(\xi, x) = \sqrt{\frac{1 - \vec{\Gamma}_1(\xi, 0)}{\pi}} \begin{Bmatrix} \cos \xi x \\ \sqrt{\frac{\xi_1}{\xi}} \sin \xi x \end{Bmatrix}, \qquad 0 < x < \infty, \qquad (41b)$$

with

$$\xi_1 = \sqrt{\xi^2 + h} > 0, \qquad \vec{\Gamma}_1(\xi, 0) = \frac{\xi_1 - \xi}{\xi_1 + \xi}. \qquad (41c)$$

For $0 < \xi_1 < \sqrt{h}$ (i.e., $\xi = -j|\xi|$), the reflection coefficient $\vec{\Gamma}_1$ is complex and of unit magnitude; one has

$$\hat{\psi}_1(\xi_1, x) = \frac{1}{\sqrt{2\pi}} [e^{-j\xi_1 x} + \vec{\Gamma}_1(-j|\xi|, 0)e^{j\xi_1 x}], \qquad -\infty < x < 0, \quad (42a)$$

$$\hat{\psi}_2(\xi_1, x) = \frac{1}{\sqrt{2\pi}} [1 + \vec{\Gamma}_1(-j|\xi|, 0)]e^{-|\xi| x}, \qquad 0 < x < \infty. \qquad (42b)$$

E modes (in x)

Characteristic Green's function:

$$g'(x, x'; \lambda) = \frac{\overleftarrow{I}_\beta(x_<)\overrightarrow{I}_\beta(x_>)}{j\omega\epsilon_0 \overleftrightarrow{Z}(0)}, \qquad (43)$$

with

$$\overleftarrow{I}_1(x) = e^{j\kappa_1 x}, \qquad \mathrm{Im}\, \kappa_1 < 0, \quad \kappa_1^2 \equiv \xi_1^2 = k_1^2 + \lambda, \qquad (43a)$$

$$\overrightarrow{I}_2(x) = e^{-j\kappa_2 x}, \qquad \mathrm{Im}\, \kappa_2 < 0, \quad \kappa_2^2 \equiv \xi^2 = k_2^2 + \lambda = \hat{\lambda}, \qquad (43b)$$

$$\overrightarrow{I}_1(x) = \cos \kappa_1 x - j\frac{\epsilon_1' \kappa_2}{\epsilon_2' \kappa_1} \sin \kappa_1 x, \qquad (43c)$$

$$\overleftarrow{I}_2(x) = \cos \kappa_2 x + j\frac{\epsilon_2' \kappa_1}{\epsilon_1' \kappa_2} \sin \kappa_2 x, \qquad (43d)$$

$$j\omega\epsilon_0 \overleftrightarrow{Z}(0) = j\left(\frac{\kappa_2}{\epsilon_2'} + \frac{\kappa_1}{\epsilon_1'}\right). \qquad (43e)$$

Spectral representation of delta function for $-\infty < \dfrac{x}{x'} < \infty$:

$$\epsilon'(x')\delta(x - x') = -\frac{1}{2\pi j} \oint_C d\hat{\lambda}\, g'(x, x'; \lambda)$$

$$= \int_0^{\sqrt{h}} d\xi_1\, \hat{\Phi}_\beta(\xi_1, x)\hat{\Phi}_\beta^*(\xi_1, x') + \int_0^\infty d\xi\, \hat{\Phi}_\beta'(\xi, x)\hat{\Phi}_\beta'^*(\xi, x'),$$

$$(44)$$

where the contour C is given as in Fig. 3.4.10, and with $0 < \xi < \infty$,

$$\hat{\Phi}'_1(\xi, x) = \sqrt{\frac{\epsilon'_2}{\pi}[1 + \vec{\Gamma}_1(\xi, 0)]}\begin{cases}\cos \xi_1 x \\[2ex] \sqrt{\dfrac{\xi \epsilon'_1}{\xi_1 \epsilon'_2}}\sin \xi_1 x\end{cases}, \qquad -\infty < x < 0, \qquad (45a)$$

$$\hat{\Phi}'_2(\xi, x) = \sqrt{\frac{\epsilon'_2}{\pi}[1 + \vec{\Gamma}_1(\xi, 0)]}\begin{cases}\cos \xi x \\[2ex] \sqrt{\dfrac{\xi_1 \epsilon'_2}{\xi \epsilon'_1}}\sin \xi x\end{cases}, \qquad 0 < x < \infty, \qquad (45b)$$

with

$$\vec{\Gamma}_1(\xi, 0) = \frac{\xi - \xi_1(\epsilon'_2/\epsilon'_1)}{\xi + \xi_1(\epsilon'_2/\epsilon'_1)}, \qquad \xi_1 = \sqrt{\xi^2 + h}. \qquad (45c)$$

Also, with $0 < \xi_1 < \sqrt{h}$,

$$\hat{\Phi}_1(\xi_1, x) = \frac{1}{\sqrt{2\pi}}[e^{-j\xi_1 x} - \vec{\Gamma}_1(-j|\xi|, 0)e^{j\xi_1 x}], \qquad -\infty < x < 0, \qquad (46a)$$

$$\hat{\Phi}_2(\xi_1, x) = \frac{1}{\sqrt{2\pi}}[1 - \vec{\Gamma}_1(-j|\xi|, 0)]e^{-|\xi|x}, \qquad 0 < x < \infty. \qquad (46b)$$

3.4b Angular Transmission Lines

When waves propagate along an angular coordinate in a curvilinear coordinate system, their characteristics differ from those associated with rectilinear propagation. A distinguishing feature in angular regions is the invisibility of a source point from an observation point displaced far enough along the curved axis; the consequent division of an angular propagation region into illuminated and shadow zones is explored in detail in Chapter 6 for diffraction in cylindrical and spherical geometries. Despite these differences, some angular regions are described by one-dimensional angular transmission lines that are indistinguishable from uniform lines. Angular intervals are finite in extent, owing either to the presence of angular boundaries or to periodicity requirements in the absence of boundaries; corresponding angular transmission lines therefore have finite length and are terminated in impedances representative of boundary conditions at the endpoints of the interval. However, as in rectilinear regions, wave motion is analyzed most directly on an infinite reflectionless (bilaterally "matched") transmission line. Reflective terminations are then accounted for by superposition of multiply reflected waves or, alternatively, by auxiliary (image) sources located outside the physical section of the infinite line. Because of their importance for cylindrical and spherical scattering problems, image representations on angular transmission lines are emphasized in this section; the ray-optical interpretation of image contributions is presented in Chapter 6.

Cylindrical regions

Angular boundary-value problems encountered in the analysis of scattering in cylindrical (ρ, ϕ, z) regions are schematized in Fig. 3.4.11; Fig. 3.4.11(a) shows a wedge-shaped domain with radial planes at $\phi = 0$, φ, while Fig. 3.4.11(b) indicates the angular periodicity required in the absence of ϕ-dependent boundaries. The radial coordinate ρ and the axial coordinate z are perpendicular to

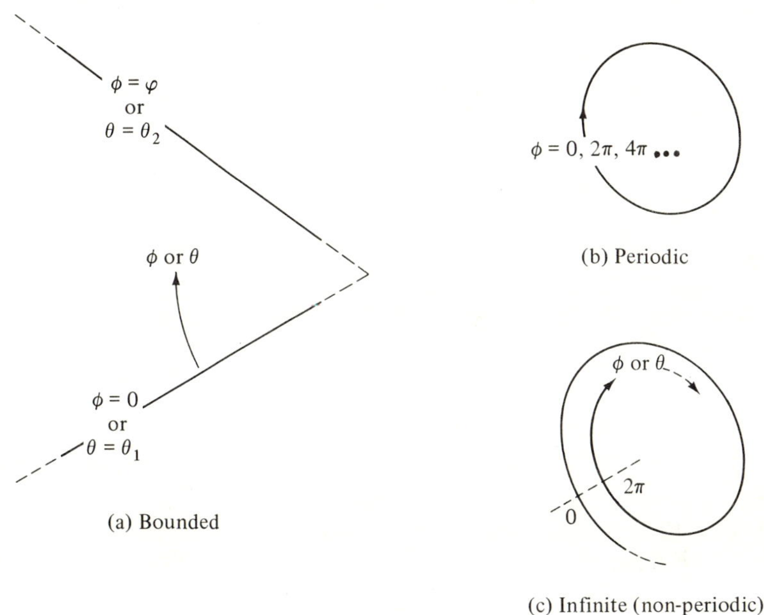

FIG. 3.4.11   Angular domains (ϕ=azimuthal coordinate, θ=latitudinal coordinate).

ϕ in the plane of each figure and perpendicular to that plane, respectively. As noted in Sec. 6.2, information relevant to angular transmission is obtained most simply by considering z-independent fields which decompose into E-mode and H-mode constituents with respect to z. The corresponding angular characteristic Green's functions are denoted, respectively, by $g'_\phi(\phi, \phi'; \lambda)$ and $g''_\phi(\phi, \phi'; \lambda)$, and are defined by the differential equation [see Eqs. (3.2.46a) and (3.3.1)]

$$\left(\frac{d^2}{d\phi^2} + \lambda\right) g_\phi(\phi, \phi'; \lambda) = -\delta(\phi - \phi'). \tag{47}$$

While Eq. (47) is of the same form as Eq. (2.4.30) for a uniform rectilinear region and the transmission-line solutions for the two systems are therefore identical, the physical attributes of angularly and rectilinearly propagating waves differ substantially. In the absence of angular boundaries [see Fig. 3.2.7(b)], g_ϕ must meet the periodicity requirements [Fig. 3.4.11(b)]

$$g_\phi(\phi + \pi, \phi'; \lambda) = g_\phi(\phi - \pi, \phi'; \lambda);$$

$$\frac{d}{d\phi} g_\phi(\phi + \pi, \phi'; \lambda) = \frac{d}{d\phi} g_\phi(\phi - \pi, \phi'; \lambda), \tag{48a}$$

with $-\pi \leq (\phi, \phi') \leq \pi$. The E- and H-mode functions are identical in this case, thereby making the distinguishing primes on g_ϕ unnecessary. If the ϕ interval is taken as $-\pi < (\phi - \phi') < \pi$ (i.e., symmetric about the source location), condition (48a) can be written in the simpler form

$$\frac{d}{d\phi} g_\phi(\phi, \phi'; \lambda) = 0 \qquad \text{at} \quad (\phi - \phi') = \pm\pi. \tag{48b}$$

If perfectly conducting plane boundaries are present at $\phi = 0$ and at $\phi = \varphi$, $0 < \varphi \leq 2\pi$ [see Figs. 3.2.7(a), 3.2.9, and 3.4.11(a)], the range of ϕ and ϕ' is restricted to $0 \leq (\phi, \phi') \leq \varphi$ and the corresponding boundary conditions on g_ϕ are

E modes (along z)

$$g_\phi'(\phi, \phi'; \lambda) = 0 \qquad \text{at} \quad \phi = 0, \varphi. \tag{49}$$

H modes (along z)

$$\frac{d}{d\phi} g_\phi''(\phi, \phi'; \lambda) = 0 \qquad \text{at} \quad \phi = 0, \varphi. \tag{50}$$

Equivalent network configurations representing the various boundary-value problems above are shown in Figs. 3.4.12 and 3.4.13 (see Figs. 3.3.1 and 3.3.2). For two-dimensional diffraction problems involving E modes along z, the source is an electric line current and the only electric field component, E_z, is proportional to the two-dimensional Green's function $\bar{G}'$ (see Sec. 6.2). Hence, it is appropriate to treat g_ϕ' as a measure of E_z (i.e., as a voltage), leading via Fig. 3.3.1 and Eq. (3.3.4) to the equivalent networks in Fig. 3.4.12.† Analogous considerations apply to the H-mode problem and yield the network representations in Fig. 3.4.13. It is noted that the network problems in Figs. 3.4.12(a) and

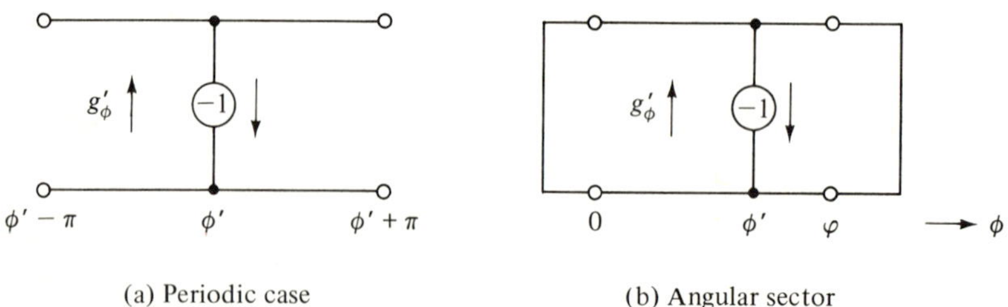

(a) Periodic case   (b) Angular sector

FIG. 3.4.12 E modes along $z(g_\phi' \to$ voltage).

---

†E- and H-mode designations are with respect to the z axis. The E-mode problem along z is actually an H-mode problem along the transmission coordinate ϕ since $E_\phi \equiv 0$, thereby corresponding to the designation in Eq. (3.3.4).

3.4.13(a) are exact duals of one another, thereby yielding the previous result $g'_\phi = g''_\phi$ for the periodic case. Perfectly conducting boundaries at the endpoints of the sectoral region give rise to the short-circuit terminations shown in Figs. 3.4.12(b) and 3.4.13(b). From Eq. (47), the ϕ-independent propagation constant on the angular transmission lines is given by $\sqrt{\lambda}$ and a characteristic impedance (admittance) for the H-mode (E-mode) problems can be chosen proportional to $\sqrt{\lambda}$ [see Eqs. (3.2.97), (3.2.98), (3.3.4) and the footnote on p. 000]. The voltage–current solutions on these transmission lines can therefore be taken directly from Sec. 2.4.

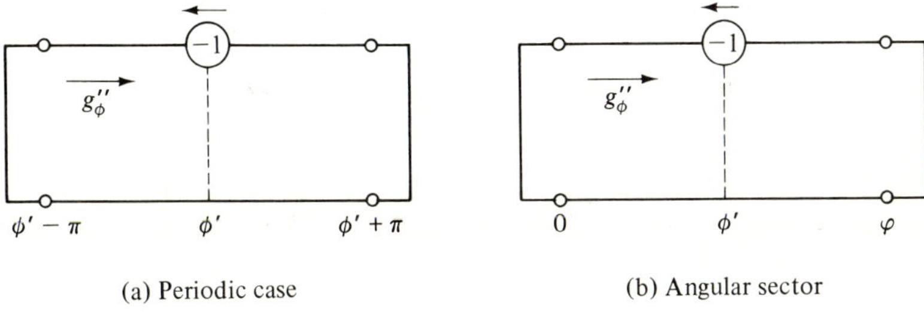

(a) Periodic case   (b) Angular sector

FIG. 3.4.13 H modes along $z(g''_\phi \rightarrow$ current$)$.

The solution for the network problems in Figs. 3.4.12 and 3.4.13 can be expressed in terms of the traveling-wave representation in Eq. (2.4.29d) (with $z_0 = 0$) appropriate to a point-source-excited, terminated transmission line extending from $\phi = 0$ to $\phi = \varphi$:†

$$g_\phi(\phi, \phi'; \lambda) = \frac{(e^{j\mu\phi_<} + \overleftarrow{\Gamma}e^{-j\mu\phi_<})(e^{j\mu(\varphi - \phi_>)} + \overrightarrow{\Gamma}e^{-j\mu(\varphi - \phi_>)})}{2j\mu(e^{j\mu\varphi} - \overleftarrow{\Gamma}\overrightarrow{\Gamma}e^{-j\mu\varphi})} \tag{51}$$

where $\overleftarrow{\Gamma}$ and $\overrightarrow{\Gamma}$ are the reflection coefficients at the left and right endpoints, respectively [see Eq. (2.4.6) for relation between $\overrightarrow{\Gamma}(0)$ and $\overrightarrow{\Gamma}$)], and $\mu = \sqrt{\lambda}$. To have g'_ϕ vanish at $\phi = 0, \varphi$, as in Fig. 3.4.12(b), it is required that $\overrightarrow{\Gamma} = \overleftarrow{\Gamma} = -1$, whence

$$g'_\phi(\phi, \phi'; \lambda) = \frac{\sin \mu\phi_< \sin \mu(\varphi - \phi_>)}{\mu \sin \mu\varphi} \tag{51a}$$

while the vanishing of $dg''_\phi/d\phi$ at the endpoints in Fig. 3.4.13(b) requires $\overrightarrow{\Gamma} = \overleftarrow{\Gamma} = 1$, whence

$$g''_\phi(\phi, \phi'; \lambda) = \frac{\cos \mu\phi_< \cos \mu(\varphi - \phi_>)}{-\mu \sin \mu\varphi}. \tag{51b}$$

---

†In this section μ is a complex parameter and is not to be confused with the same symbol used elsewhere for the permeability.

For the periodic case, one notes from Fig. 3.4.13(a) that the symmetric network can be short-circuit bisected at ϕ'. Assume for the moment that $\phi' = 0$. The resulting problem is then identical with that in Fig. 3.4.13(b), provided that the source (of strength $-\frac{1}{2}$ because of the bisection) is moved to $\phi' = 0$, and that φ is set equal to π. Displacement of the origin to ϕ' merely implies that ϕ is replaced by $\phi - \phi'$. Finally, from symmetry considerations, the same response holds in the region $\phi < \phi'$, whence $\phi - \phi'$ can be replaced by $|\phi - \phi'|$ to yield an expression valid in the entire interval $|\phi - \phi'| < \pi$. Therefore, twice the angular Green's function for the periodic case is deduced from Eq. (51) on writing $\overrightarrow{\Gamma} = \overleftarrow{\Gamma} = 1$, $\varphi = \pi$, $\phi_< = 0$, $\phi_> \to |\phi - \phi'|$, whence

$$g_\phi(\phi, \phi'; \lambda) = -\frac{\cos \mu \, [\pi - |\phi - \phi'|]}{2\mu \sin \mu\pi}. \tag{51c}$$

The Green's function solution in Eq. (51) is more general than that required for the problems discussed above since it applies also when the reflection coefficients $\overrightarrow{\overleftarrow{\Gamma}}$ have a value different from $+1$ or -1. In fact, if one employs instead of the simple boundary conditions in Eqs. (48)–(50) the more general "impedance condition"

$$\frac{d}{d\phi} g_\phi = \pm j c_{1,2} g_\phi \qquad \text{at} \quad \phi = \begin{cases} 0 \\ \varphi, \end{cases} \tag{52}$$

where c_1 and c_2 are constants independent of ϕ and μ, one finds that Eq. (51) also represents the solution in this case provided that the following values for the reflection coefficients are employed:

$$\overleftarrow{\Gamma} = \frac{\mu - c_1}{\mu + c_1}, \qquad \overrightarrow{\Gamma} = \frac{\mu - c_2}{\mu + c_2}. \tag{53}$$

Upon recalling that μ represents the characteristic impedance (admittance) for the H-mode (E-mode) problems, one notes that the voltage (current) reflection coefficients in Eqs. (53) have precisely the form (2.4.12), with $c_{1,2}$ representing the terminal impedances (admittances) (for passive terminations, Re $c_{1,2} \geq 0$). The boundary condition (52) is utilized in Sec. 6.6 for analysis of diffraction by a wedge with a variable surface impedance.

For the above terminations, one verifies readily that g_ϕ is an even function of μ (i.e., a regular function of λ near $\lambda = 0$), and hence no special care need be taken in the definition of $\sqrt{\lambda}$. The only singularities are simple poles located at the zeros of the denominator of Eq. (51) (with the possible exception of $\mu = 0$). At infinity in the complex λ plane,

$$|g_\phi| \sim \frac{e^{-|\text{Im}\,\mu||\phi-\phi'|}}{2|\mu|}, \qquad \text{as} \quad |\text{Im}\,\mu| \to \infty. \tag{54}$$

The representation in Eq. (51) is directly suited to the development of the image formulation described previously. The Green's function $g_\phi^\infty(\phi, \phi'; \mu)$ on an "infinite" (bilaterally matched) angular line is obtained upon letting $\overrightarrow{\overleftarrow{\Gamma}} = 0$.

Since complex μ corresponds to a dissipative line, in which the response is required to decay away from the source, Eq. (51) (for $\Gamma = 0$) is valid only when Im $\mu < 0$. Conversely, since g_ϕ is an even function of μ, upon changing μ into $-\mu$ in Eq. (51), one derives a limiting form valid when Im $\mu > 0$, so

$$g_\phi^\infty(\phi, \phi'; \mu) = \frac{e^{\mp j\mu|\phi - \phi'|}}{\pm 2j\mu}, \qquad \text{Im } \mu \lessgtr 0. \tag{55}$$

Unlike g_ϕ, g_ϕ^∞ is not an even function of μ and is, in fact, discontinuous across the real μ axis. Hence, viewed as a function of $\lambda = \mu^2$, g_ϕ^∞ possesses a first-order branch point singularity at $\lambda = 0$, with branch cut along the positive real x axis [the top sheets in Eq. (55) are defined by Im $\sqrt{\lambda} \lessgtr 0$]. Its behavior as $|\text{Im } \mu| \to \infty$ is identical with that of g_ϕ in Eq. (54). Upon employing the convergent power-series expansion†

$$\frac{1}{e^{j\mu\varphi} - \overleftarrow{\Gamma}\overrightarrow{\Gamma}e^{-j\mu\varphi}} = \sum_{n=1}^{\infty} (\overleftarrow{\Gamma}\overrightarrow{\Gamma})^n e^{-j(2n+1)\mu\varphi}, \qquad \text{Im } \mu < 0, \tag{56}$$

in Eq. (51) and rearranging terms, one may write

$$g_\phi(\phi, \phi'; \lambda) = g_\phi^\infty(\phi, \phi') + \sum_{n=1}^{\infty} \overleftarrow{\Gamma}^{n-1}\overrightarrow{\Gamma}^n g_\phi^\infty(\phi, 2n\varphi - \phi')$$

$$+ \sum_{n=1}^{\infty} (\overleftarrow{\Gamma}\overrightarrow{\Gamma})^n [g_\phi^\infty(\phi, 2n\varphi + \phi') + g_\phi^\infty(\phi, -2n\varphi + \phi')]$$

$$+ \sum_{n=0}^{\infty} \overrightarrow{\Gamma}^n\overleftarrow{\Gamma}^{n+1} g_\phi^\infty(\phi, -2n\varphi - \phi'), \qquad \text{Im } \mu < 0, \tag{57}$$

with g_ϕ^∞ given in Eq. (55), and $\overleftarrow{\Gamma}, \overrightarrow{\Gamma}$ defined as in Eq. (53). On replacing μ by $-\mu$ in Eq. (57), one obtains a series representation valid when Im $\mu > 0$.

Equation (57) comprises contributions that can be interpreted as arising from a set of sources located at the points

$$\phi = 2n\varphi - \phi', \qquad n = 0, \pm 1, \pm 2, \ldots$$
$$\phi = 2n\varphi + \phi', \qquad n = \pm 1, \pm 2, \ldots \tag{58}$$

on an infinitely extended transmission line. Only the actual source at $\phi = \phi'$ is situated in the physical angular domain $0 \leq (\phi, \phi') \leq \varphi$; all the other (image) sources lie outside this range. The amplitude of a given image source is identical with that of the corresponding multiply reflected wave; with each reflection at $\phi = 0$ and $\phi = \varphi$, a reflected wave acquires an amplitude factor $\overleftarrow{\Gamma}$ and $\overrightarrow{\Gamma}$, respectively. Hence, image 1, arising from the first reflection at $\phi = 0$, has an amplitude $\overleftarrow{\Gamma}$; image 1', arising from the first reflection at $\phi = \varphi$, has an amplitude $\overrightarrow{\Gamma}$; image 2, arising from the second reflection at $\phi = 0$, has an amplitude $\overleftarrow{\Gamma}\overrightarrow{\Gamma}$; etc. The angular coordinate ϕ in Fig. 3.4.14 has been plotted along a straight-line scale which makes the meaning of the extended non-periodic angular domain $-\infty < (\phi, \phi') < \infty$ unambiguous. If ϕ is plotted on its con-

---

†It is assumed that $|\overleftarrow{\Gamma}\overrightarrow{\Gamma}|e^{-2|\text{Im } \mu|\varphi} < 1$.

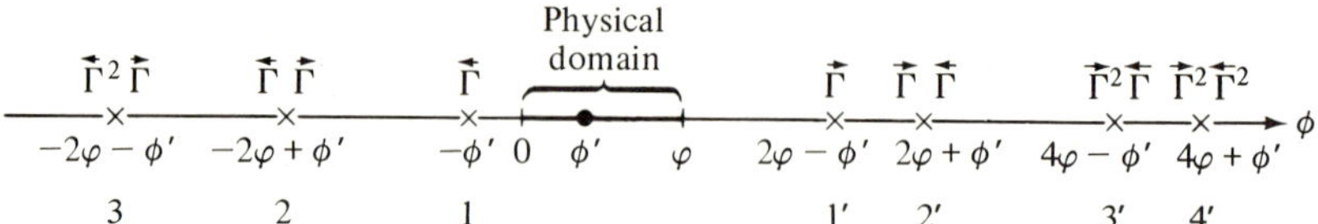

FIG. 3.4.14 Image representation for reflecting terminations at $\phi = 0, \varphi$.

ventional circular cylindrical coordinate scale whereon an increment of 2π corresponds to a complete circuit, it is necessary to view the ϕ domain as a complex Riemann surface[7,8] having an infinite number of sheets [Fig. 3.4.11(c)]; the passage out of the physical domain $0 \le \phi \le \varphi$ into the image region then proceeds via branch cuts introduced along the lines $\phi = 0$ and $\phi = \varphi$. However, the extension of ϕ into a complex Riemann surface is not essential if all subsequent considerations are confined to the physical range $0 \le (\phi, \phi') \le \varphi$, over which Eq. (57) is defined unambiguously. In this instance, one interprets the image contributions as multiply reflected waves between the boundaries $\phi = 0, \varphi$. Although we shall adopt this latter, simpler interpretation to avoid the necessity of defining ϕ in a multisheeted complex plane, the equivalent image formulation schematized in Fig. 3.4.14 may be kept in mind as a simple schematization of the multiple reflection process.

If the terminations at $\phi = 0, \varphi$ comprise short or open circuits (i.e., $\Gamma = \pm 1$), Eqs. (57) simplify and can be written more compactly as

$$g_\phi(\phi, \phi'; \lambda) = \sum_{n=-\infty}^{\infty} g_\phi^\infty(\phi, 2n\varphi + \phi') \pm \sum_{n=-\infty}^{\infty} g_\phi^\infty(\phi, 2n\varphi - \phi'), \qquad (59)$$

where the upper and lower signs correspond to $\overleftarrow{\Gamma} = \overrightarrow{\Gamma} = +1$ and $\overleftarrow{\Gamma} = \overrightarrow{\Gamma} = -1$, respectively. For the periodic case, one obtains in view of the remarks preceding Eq. (51c):

$$g_\phi(\phi, \phi'; \lambda) = \sum_{n=-\infty}^{\infty} g_\phi^\infty(\phi - \phi', 2n\pi) = \sum_{n=-\infty}^{\infty} g_\phi^\infty(\phi, 2n\pi + \phi'), \qquad (60)$$

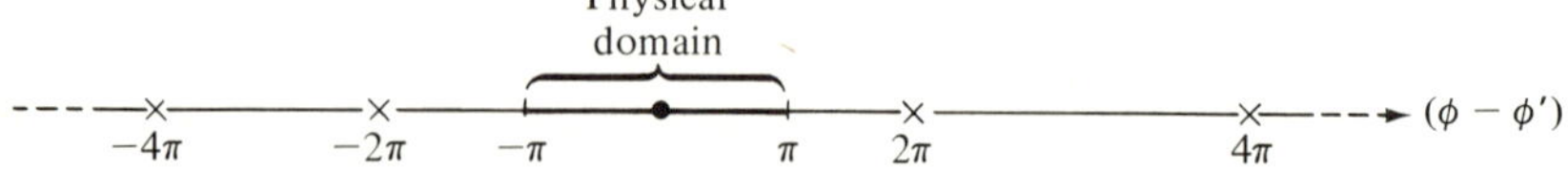

FIG. 3.4.15 Image representation for periodic case.

with an image representation as shown in Fig. 3.4.15. Equations (59) and (60) are used in Secs. 6.5 and 6.7 for an angular transmission analysis of high-frequency diffraction by a perfectly conducting wedge and cylinder, respectively.

Completeness relations for the domain $0 \le \phi \le \varphi$, with the boundary conditions of Eqs. (51a)–(51c), have already been given in Eqs. (3.2.47), (3.2.48), and (3.2.50). For the impedance boundary condition in Eq. (52), with $c_2 = 0$

for convenience, one finds that g_ϕ in Eq. (51) has simple singularities at $\mu = \xi$, where

$$\cot \xi\varphi = -\frac{j\xi}{c_1}. \tag{61}$$

If the surface impedance parameter is reactive so that $c_1 = j\gamma$, with $\gamma > 0$, then Eq. (61) has a discrete infinity of real roots $\xi_1, \xi_2, \ldots$, and a single imaginary root $\xi_0 = j\eta$, η real, satisfying $\coth \eta\varphi = \eta/\gamma$; these roots may be determined from the graphical construction in Fig. 3.4.16. Then from Eq. (3.3.11),

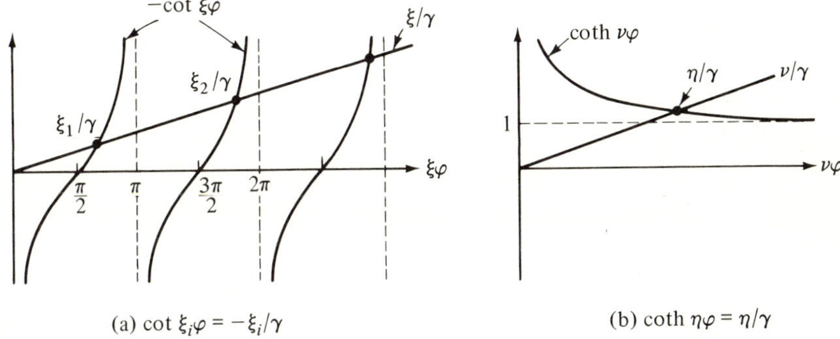

(a) cot $\xi_i\varphi = -\xi_i/\gamma$ (b) coth $\eta\varphi = \eta/\gamma$

FIG. 3.4.16   Graphical solution of Eq. (61), with $c_1 = j\gamma, \gamma > 0$.

$$\delta(\phi - \phi') = \begin{cases} -\dfrac{1}{2\pi j} \oint g_\phi(\phi, \phi'; \lambda)\, d\lambda, & (62a) \\[2mm] \displaystyle\sum_q \psi_q(\phi)\psi_q(\phi') + \psi_0(\phi)\psi_0(\phi')U(\gamma), & (62b) \end{cases}$$

where $U(\gamma) = 1$ or 0 when $\gamma > 0$ or $\gamma < 0$, respectively, and

$$\psi_q(\phi) = \left[\frac{2}{\varphi[1 - (\sin^2 \xi_i\varphi)/\gamma\varphi]}\right]^{1/2} \cos\left[\xi_i(\varphi - \phi)\right], \qquad q \equiv \xi_i > 0, \tag{62c}$$

$$\psi_0(\phi) = \left[\frac{2}{\varphi[1 + (\sinh^2 \eta\varphi)/\gamma\varphi]}\right]^{1/2} \cosh\left[\eta(\varphi - \phi)\right], \qquad \eta > 0. \tag{62d}$$

The eigenfunctions $\psi_q(\phi)$ in Eq. (62c) are oscillatory in the interval $0 \le \phi \le \varphi$, while $\psi_0(\phi)$ in Eq. (62d) decays away from the reactive boundary at $\phi = 0$ and is therefore analogous to the surface-wave modes in Eq. (17b). The occurrence of only a single surface-wave mode is attributable to the assumption at $\phi = 0$ of a surface impedance that idealizes an implied medium in $\phi < 0$; in the analogous problem shown in Fig. 3.4.1(a), the medium in $x < 0$ is exhibited explicitly. It should be emphasized that constancy of the parameters $c_{1,2}$ does not imply constancy of surface impedance on the angular boundaries $\phi = 0, \varphi$ since the cylindrical (ρ, ϕ, z) coordinate system is curvilinear. This aspect is explored in detail in Sec. 6.6.

Spherical regions

Two-dimensional eigenvalue problems in the (θ, ϕ) cross section transverse to the radial coordinate r in a spherical coordinate system are defined by the differential equations (2.6.2) and (2.6.3). For separable boundary conditions [i.e., in regions bounded by plane surfaces $\phi = $ constant and (or) conical surfaces $\theta = $ constant], the two-dimensional eigenvalue problems may be reduced to two one-dimensional eigenvalue problems:

$$\left(\frac{d^2}{d\phi^2} + q^2\right)\frac{\Phi_q(\phi)}{\psi_q(\phi)} = 0, \qquad \left.\begin{array}{r}\Phi_q \\ \dfrac{\partial \psi_q}{\partial \phi}\end{array}\right\} = 0 \quad \text{at } \phi = 0, \varphi \tag{63a}$$

$$\left(\frac{d}{d\theta}\sin\theta\frac{d}{d\theta} - \frac{q^2}{\sin\theta} + p(p+1)\sin\theta\right)\frac{\Phi_p^q(\theta)}{\psi_p^q(\theta)} = 0, \qquad \left.\begin{array}{r}\Phi_p^q \\ \dfrac{\partial \psi_p^q}{\partial \theta}\end{array}\right\} = 0 \quad \text{at } \theta = \theta_{1,2}, \tag{63b}$$

where the product solutions $\Phi_i(\mathbf{\rho}) = \Phi_q(\phi)\Phi_p^q(\theta)$ and $\psi_i(\mathbf{\rho}) = \psi_q(\phi)\psi_p^q(\theta)$ have been introduced, and where $p(p+1) = k_{ti}^2$. Since Eqs. (63a) and (3.2.46a) are identical, the eigenvalue and characteristic Green's function problems for the azimuthal (ϕ) domain are the same as in cylindrical geometry. For the θ domain $\theta_1 \le \theta \le \theta_2$, comparison of Eqs. (63b) and (3.3.1) yields the characteristic Green's function problem:

$$\left(\frac{d}{d\theta}\sin\theta\frac{d}{d\theta} - \frac{q^2}{\sin\theta} + \lambda\sin\theta\right)g_\theta(\theta, \theta'; q^2, \lambda) = -\delta(\theta - \theta'), \tag{64}$$

where q is a fixed parameter which, in spherical boundary-value problems, represents the azimuthal eigenvalue [see Eq. (63a)]. On replacing x by θ, comparison with Eq. (3.3.1) shows that the θ transmission line is non-uniform, since the parameters $p(\theta) = w(\theta) = \sin\theta$, $q(\theta) = q^2/\sin\theta$ are θ dependent. For an angularly unbounded spherical region, the endpoints of the θ interval are $\theta_1 = 0$ and $\theta_2 = \pi$; $\theta_1 = 0, 0 < \theta_2 < \pi$ represents a single cone at $\theta = \theta_2$, while $0 < \theta_1 < \theta_2 < \pi$ defines cones at $\theta = \theta_1, \theta_2$, respectively. The boundary conditions for electromagnetic E- and H-mode problems with respect to the radial direction, distinguished by primes and double primes, respectively, are [see Eqs. (2.6.2) and (2.6.3)]

$$g_\theta'(\theta, \theta'; q^2; \lambda) = 0 \qquad \text{at } \theta = \theta_{1,2} \quad (E \text{ modes}), \tag{64a}$$

$$\frac{d}{d\theta}g_\theta''(\theta, \theta'; q^2; \lambda) = 0 \qquad \text{at } \theta = \theta_{1,2} \quad (H \text{ modes}). \tag{64b}$$

Note $\theta = 0, \pi$ are "limit point" singularities of the differential operator in Eq. (63). If these points terminate the θ interval, a sufficient boundary condition is

$$g_\theta \text{ finite at } \theta_1 = 0 \text{ and (or) } \theta_2 = \pi. \tag{64c}$$

These angular boundary-value problems may be schematized as in Fig. 3.4.11.

The various $g_\theta(\theta, \theta'; q^2; \lambda)$ may be constructed from the linearly independent solutions of the homogeneous equation (64). Since $\lambda = \nu(\nu + 1)$ is a complex parameter and q may also be non-real (see discussion on alternative representations, Sec. 3.3c), the classical Legendre polynomials are inadequate to describe this more general situation. Instead, it is best to express the associated Legendre functions as special cases of hypergeometric functions, the theory of which is well advanced. It may be shown that $P_\nu^{-q}(\cos\theta)$ and $P_\nu^{-q}[\cos(\pi - \theta)] \equiv P_\nu^{-q}(-\cos\theta)$ are linearly independent solutions of the associated Legendre equation and are related to the hypergeometric function $F(\alpha, \beta; \gamma; z)$ as follows[9]:

$$P_\nu^{-q}(\cos\theta) = \frac{1}{\Gamma(1+q)} \tan^q\left(\frac{\theta}{2}\right) F\left(-\nu, \nu + 1; 1 + q; \sin^2\frac{\theta}{2}\right)$$

$$= P_{-\nu-1}^{-q}(\cos\theta), \tag{65a}$$

$$P_\nu^{-q}(-\cos\theta) = \frac{1}{\Gamma(1+q)} \cot^q\left(\frac{\theta}{2}\right) F\left(-\nu, \nu + 1; 1 + q; \cos^2\frac{\theta}{2}\right), \tag{65b}$$

where $\Gamma(x)$ is the gamma function and ν and q are arbitrary. Some properties of the hypergeometric function are given in Sec. 3.6b. Important for the present discussion is the relation

$$F(a, b; c; 0) = 1, \tag{65c}$$

which permits the study of the behavior of $P_\nu^{-q}(\pm\cos\theta)$ near the singular points $\theta = 0, \pi$. It is also useful to recall that

$$F(a, b; c; z) = \frac{\Gamma(c)\Gamma(c - a - b)}{\Gamma(c - a)\Gamma(c - b)} F(a, b; a + b - c + 1; 1 - z)$$

$$+ (1 - z)^{c-a-b} \frac{\Gamma(c)\Gamma(a + b - c)}{\Gamma(a)\Gamma(b)} F(c - a, c - b; c - a - b + 1; 1 - z), \tag{65d}$$

which relation permits study of the behavior of the hypergeometric function near $z = 1$. If $\text{Re } q > 0$, then $P_\nu^{-q}(\cos\theta)$ and $P_\nu^{-q}(-\cos\theta)$ are bounded at $\theta = 0$ and $\theta = \pi$, respectively, but not at the opposite endpoint. We shall also have occasion to employ the Wronskian

$$P_\nu^{-q}(\cos\theta) \frac{d}{d\theta} P_\nu^{-q}(-\cos\theta) - P_\nu^{-q}(-\cos\theta) \frac{d}{d\theta} P_\nu^{-q}(\cos\theta)$$

$$= \frac{2}{\pi} \frac{\sin(\nu - q)\pi}{\sin\theta} \frac{\Gamma(\nu - q + 1)}{\Gamma(\nu + q + 1)}$$

$$= -\frac{2}{\sin\theta\, \Gamma(q - \nu)\Gamma(\nu + q + 1)}; \tag{66a}$$

the asymptotic formula

$$P_\nu^{-q}(\cos\theta) \sim \sqrt{\frac{2}{\pi\sin\theta}}\,\frac{\Gamma(\nu-q+1)}{\Gamma(\nu+\frac{3}{2})}\cos\left[\left(\nu+\frac{1}{2}\right)\theta - \frac{\pi}{4} - \frac{q\pi}{2}\right]$$

$$\times\left[1 + O\left(\frac{1}{\nu}\right)\right], \tag{66b}$$

valid when $|\nu| \gg |q|$, $|\arg\nu| < \pi$, $|\nu|\sin\theta \gg 1$; the Stirling approximation

$$\Gamma(\nu+\alpha) \sim \sqrt{\frac{2\pi}{\nu}}\left(\frac{\nu}{e}\right)^\nu \nu^\alpha, \qquad |\nu|\to\infty,\ |\arg\nu| < \pi,\ \alpha > 0; \tag{66c}$$

and the relation

$$F(a,b;c;z) \sim 1 + O\left(\frac{1}{c}\right), \qquad c \to \infty. \tag{66d}$$

Like the azimuthal Green's function g_ϕ defined in Eq. (47), g_θ may be synthesized in terms of images on an infinitely extended θ transmission line. As noted earlier, such a representation is useful for analysis of high-frequency diffraction by conical and spherical obstacles. We illustrate the procedure for the domain bounded by $\theta_1 = 0$ and $\theta_2 \equiv \theta_0 < \pi$ since closed-form expressions for g_θ' or g_θ'', analogous to Eq. (51), then have a relatively simple form. By Eq. (3.3.14) and Eqs. (64)–(66), the desired solution can be written in terms of the Green's function $g_\theta^0(\theta,\theta';q^2;\lambda)$ for the interval $0 \le \theta \le \pi$, plus a correction term. Thus, g_θ^0 is given by

$$g_\theta^0(\theta,\theta';q^2;\lambda) = -\frac{\pi}{2}\frac{\Gamma(\nu+q+1)}{\Gamma(\nu-q+1)\sin(\nu-q)\pi}P_\nu^{-q}(\cos\theta_<)P_\nu^{-q}(-\cos\theta_>),$$

$$0 \le \frac{\theta}{\theta'} \le \pi, \tag{67}$$

with $\lambda = \nu(\nu+1)$, $\mathrm{Im}\,\lambda \ne 0$, and $\mathrm{Re}\,q > 0$. Then, for E modes,

$$g_\theta'(\theta,\theta';q^2;\lambda) = g_\theta^0(\theta,\theta';q^2;\lambda)$$

$$+ \frac{\pi}{2}P_\nu^{-q}(\cos\theta)P_\nu^{-q}(\cos\theta')\frac{\Gamma(\nu+q+1)P_\nu^{-q}(-\cos\theta_0)}{\Gamma(\nu-q+1)P_\nu^{-q}(\cos\theta_0)\sin(\nu-q)\pi}, \tag{68}$$

and, for H modes,

$$g_\theta''(\theta,\theta';q^2;\lambda) = g_\theta^0(\theta,\theta';q^2;\lambda)$$

$$+ \frac{\pi}{2}P_\nu^{-q}(\cos\theta)P_\nu^{-q}(\cos\theta')\frac{\Gamma(\nu+q+1)(d/d\theta_0)P_\nu^{-q}(-\cos\theta_0)}{\Gamma(\nu-q+1)(d/d\theta_0)P_\nu^{-q}(\cos\theta_0)\sin(\nu-q)\pi}. \tag{69}$$

For positive real q, one verifies that g_θ^0, g_θ', and g_θ'' behave asymptotically as

$$|g_\theta(\theta,\theta';q^2;\lambda)| \to \frac{1}{|\nu|\sqrt{\sin\theta\sin\theta'}}\,e^{-|\mathrm{Im}\,\nu||\theta-\theta'|}, \qquad |\nu|\to\infty, \tag{70}$$

whence the characteristic Green's functions decay at infinity in the complex λ plane. Since $\lambda = \nu(\nu+1) = (\nu+\frac{1}{2})^2 - \frac{1}{4}$, and $(\sin\pi z)\Gamma(z)\Gamma(1-z) = \pi$, one observes that g_θ^0, g_θ', and g_θ'' are even functions of $(\nu+\frac{1}{2})$.

To obtain traveling-wave solutions on a bilaterally matched θ transmission line, it is necessary to introduce traveling-wave functions instead of the standing-wave functions $P_\nu^{-q}(\pm\cos\theta)$ [see Eq. (66b)]. The traveling-wave functions $E_q^{(1,2)}(\xi,\theta)$ for arbitrary ν and q are defined as follows†:

$$E_q^{(1,2)}(\xi,\theta) = \pm\frac{j}{\sin(\nu-q)\pi}[P_\nu^{-q}(-\cos\theta) - e^{\pm j(\nu-q)\pi}P_\nu^{-q}(\cos\theta)], \qquad (71)$$

$$= E_q^{(1,2)}(\xi,\pi-\theta)e^{\pm j(\nu-q)\pi}, \qquad \xi = \nu + \tfrac{1}{2}, \qquad (71a)$$

whence

$$P_\nu^{-q}(\cos\theta) = \tfrac{1}{2}[E_q^{(1)}(\xi,\theta) + E_q^{(2)}(\xi,\theta)]. \qquad (71b)$$

The relation in Eq. (71a) is evident from the definition (71). On use of hypergeometric function representations for $P_\nu^{-q}(\pm\cos\theta)$, one may show that[10]

$$E_q^{(1,2)}(\xi,\theta) = \sqrt{\frac{2}{\pi\sin\theta}}\,\frac{\Gamma(\xi+\tfrac{1}{2}-q)}{\Gamma(\xi+1)}\,e^{\pm j(\xi\theta-q\pi/2-\pi/4)}R_q^{(1,2)}(\xi,\theta), \qquad (72)$$

where

$$R_q^{(1,2)}(\xi,\theta) = \begin{cases} F\!\left(\tfrac{1}{2}+q,\tfrac{1}{2}-q;\xi+1;\dfrac{e^{\pm j\theta}}{\pm 2j\sin\theta}\right), & (72a) \\[2ex] (1-e^{\pm j2\theta})^{(1/2)-q}F(\tfrac{1}{2}-q,\xi+\tfrac{1}{2}-q;\xi+1;e^{\pm j2\theta}). & (72b) \end{cases}$$

Equation (72b) follows from Eq. (72a) via the transformation

$$F\!\left(a,b;c;\frac{z}{1-z}\right) = (1-z)^b F(b,c-a;c;z). \qquad (72c)$$

The traveling-wave character (for large $\xi\sin\theta$) of the functions $E_q^{(1,2)}(\xi,\theta)$ becomes evident from Eq. (72) on use of the asymptotic formula (66d), which yields

$$R_q^{(1,2)}(\xi,\theta) \sim 1, \qquad |\xi| \to \infty, \qquad \sin\theta \neq 0. \qquad (72d)$$

The characteristic Green's functions in Eqs. (68) and (69) can now be represented in terms of multiply reflected traveling waves if one utilizes Eq. (71b) and the power-series expansion [note analogy with Eq. (56)]

$$\frac{1}{1+b_q(\xi,\theta)} = \sum_{n=0}^{\infty}[-b_q(\xi,\theta)]^n, \qquad (73)$$

where, for $\mathrm{Im}\,\xi > 0$,

$$b_q(\xi,\theta_0) = \frac{LE_q^{(1)}(\xi,\theta_0)}{LE_q^{(2)}(\xi,\theta_0)} \equiv -e^{j(2\xi\theta_0-q\pi-\pi/2)}f_{\xi q}(\theta_0), \qquad (73a)$$

with $L \equiv 1$ for the E-mode case in Eq. (68), and $L \equiv d/d\theta_0$ for the H-mode case in Eq. (69). Thus, one finds

---

†In the present discussion, j is the imaginary unit and does not refer to a particular harmonic time dependence.

$$g_\theta(\theta, \theta'; q^2; \xi^2 - \tfrac{1}{4}) = \left[\frac{\Gamma(\xi + q + \tfrac{1}{2})\Gamma(\xi - q + \tfrac{1}{2})}{\xi\Gamma^2(\xi)}\right]\frac{1}{\sqrt{\sin\theta \sin\theta'}}\left\{\ \right\},$$

$$(74)$$

where

$$\left\{\ \right\} = g_\theta^\infty(\theta, \theta')R_q^{(2)}(\theta_<)R_q^{(1)}(\theta_>)$$

$$+ \sum_{n=1}^\infty \overleftarrow{\Gamma}^{n-1}\overrightarrow{\Gamma}^n g_\theta^\infty(\theta, 2n\theta_0 - \theta')f_{\xi q}^n R_q^{(2)}(\xi, \theta)R_q^{(2)}(\xi, \theta')$$

$$+ \sum_{n=1}^\infty (\overleftarrow{\Gamma}\overrightarrow{\Gamma})^n f_{\xi q}^n [g_\theta^\infty(\theta, -2n\theta_0 + \theta')R_q^{(1)}(\xi, \theta)R_q^{(2)}(\xi, \theta')$$

$$+ g_\theta^\infty(\theta, 2n\theta_0 + \theta')R_q^{(2)}(\xi, \theta)R_q^{(1)}(\xi, \theta')]$$

$$+ \sum_{n=0}^\infty \overleftarrow{\Gamma}^{n+1}\overrightarrow{\Gamma}^n f_{\xi q}^n g_\theta^\infty(\theta, -2n\theta_0 - \theta')R_q^{(1)}(\xi, \theta)R_q^{(1)}(\xi, \theta'), \qquad (74a)$$

$$g_\theta^\infty(\theta, \theta') = \frac{e^{j\xi|\theta-\theta'|}}{-2j\xi}, \qquad \overleftarrow{\Gamma} = e^{-j(q+1/2)\pi}, \qquad \overrightarrow{\Gamma} = \begin{cases} -1 & \text{for } g'_\theta, \\ +1 & \text{for } g''_\theta. \end{cases} \qquad (74b)$$

Equation (74) is valid when $\mathrm{Im}\,\xi > 0$; since g_θ is an even function of ξ [see remarks following Eq. (70)], a series representation for $\mathrm{Im}\,\xi < 0$ is obtained on replacing ξ by $-\xi$.

The series in Eq. (74a) has been written in a manner that permits direct identification with corresponding terms in the azimuthally traveling-wave expansion (57). Recalling Eq. (72d) and recognizing from Eq. (73a) that for $\sin\theta$, $\sin\theta'$, $\sin\theta_0 \not\approx 0$,

$$f_{\xi q} \sim 1 + O\left(\frac{1}{\xi}\right), \qquad |\xi| \to \infty, \qquad (75)$$

one finds, that the large ξ asymptotic behavior of Eq. (74a) coincides with Eq. (57), whence the image representation in Fig. 3.4.14 applies directly. $\overleftarrow{\Gamma}$ in Eq. (74b) expresses the effective reflection coefficient at the singular endpoint $\theta = 0$, while $\overrightarrow{\Gamma}$ accounts for reflection at $\theta = \theta_0$. Since the large-ξ value of the term in brackets in Eq. (74) equals unity, the asymptotic Green's function in θ space differs from that in ϕ space only through the factor $[\sin\theta \sin\theta']^{-1/2}$. g_θ^∞ represents the asymptotic form of the Green's function on a bilaterally matched θ transmission line (i.e., when $\overleftarrow{\Gamma} = \overrightarrow{\Gamma} = 0$). For arbitrary ξ values, the image representation retains its validity, but it is no longer possible to represent the propagation process in terms of waves traveling solely along $+\theta$ and $-\theta$. Thus, the first term in Eq. (74a) [with Eq. (74)] yields the exact Green's function in an angularly matched (i.e., infinitely extended) θ space, but the presence of the functions $R_q^{(2)}(\theta_<)$ and $R_q^{(1)}(\theta_>)$ distorts the purely traveling wave character of $g_\theta^\infty(\theta, \theta')$.

Completeness relations involving eigenfunctions that satisfy the differential equation (63b) are obtained directly on use of Eq. (3.3.11) and the appropriate characteristic Green's function. The results are summarized below, with $\lambda_p = p(p+1)$ denoting the eigenvalues while $\Phi_p^q(\theta)$ and $\psi_p^q(\theta)$ represent the E- and H-mode eigenfunctions, respectively.

$$0 \leq \theta \leq \pi$$

The characteristic Green's function g_θ^0 for the complete θ interval between 0 and π is given in Eq. (67). As noted after Eq. (70), g_θ^0 is an even function of $\nu + \frac{1}{2}$ and hence contains no branch-point singularities in the complex $\lambda = \nu(\nu + 1)$ plane. $\Gamma(w)$ has no zeros and has simple poles with residues $(-1)^n/n!$ at $w = -n, n = 0, 1, 2, \ldots$, while the Legendre functions $P_\nu^{-q}(\pm \cos \theta)$ are regular in any finite part of the complex ν plane. Thus, the singularities of g_θ^0 are simple poles located at

$$\lambda_p = p(p+1), \quad p + \tfrac{1}{2} = \pm(q + n + \tfrac{1}{2}), \qquad n = 0, 1, 2, \ldots. \tag{76}$$

From Eq. (3.3.11), on substitution of Eq. (67) and a residue evaluation,

$$\frac{\delta(\theta - \theta')}{\sin \theta'} = \tfrac{1}{2} \sum_{n=0}^{\infty} [2(n+q) + 1] \frac{\Gamma(n + 2q + 1)}{n!} P_{n+q}^{-q}(\cos \theta) P_{n+q}^{-q}(\cos \theta'), \tag{77}$$

so the normalized scalar eigenfunctions $\Phi_p^q(\theta)$ and $\psi_p^q(\theta)$ for the E- and H-mode problems, respectively, are identical and are given by

$$\Phi_p^q(\theta) = \psi_p^q(\theta) = \left\{ \frac{[2(n+q) + 1]\Gamma(n + 2q + 1)}{2n!} \right\}^{1/2} P_{n+q}^{-q}(\cos \theta),$$
$$n = 0, 1, 2, \ldots. \tag{77a}$$

When $q = m$ is an integer, one may employ the relation[9]

$$P_\nu^{-m}(x) = (-1)^m \frac{\Gamma(\nu - m + 1)}{\Gamma(\nu + m + 1)} P_\nu^m(x) \tag{78}$$

to reduce Eq. (77) to

$$\frac{\delta(\theta - \theta')}{\sin \theta'} = \tfrac{1}{2} \sum_{n=m}^{\infty} (2n + 1) \frac{(n - m)!}{(n + m)!} P_n^m(\cos \theta) P_n^m(\cos \theta'), \tag{79}$$

where $P_n^m(\cos \theta)$ is the conventional Legendre polynomial which vanishes identically when $n < m$. The eigenfunctions are now

$$\Phi_p^q(\theta) = \psi_p^q(\theta) = \left[\frac{(n + \tfrac{1}{2})(n - m)!}{(n + m)!} \right]^{1/2} P_n^m(\cos \theta), \qquad n - m = 0, 1, 2, \ldots, \tag{79a}$$

where

$$P_n(\cos \theta) = \frac{1}{2^n n!} \frac{d^n}{dz^n} (z^2 - 1)^n \bigg|_{z = \cos \theta},$$

$$P_n^m(\cos \theta) = (-1)^m \sin^m \theta \frac{d^m}{dz^m} P_n(z) \bigg|_{z = \cos \theta}, \tag{79b}$$

or, for special choices of n and m,

$$P_0(\cos \theta) = 1, \quad P_1(\cos \theta) = \cos \theta,$$

$$P_2(\cos \theta) = \tfrac{1}{2}(3 \cos^2 \theta - 1) = \tfrac{1}{4}(3 \cos 2\theta + 1), \text{ etc.,} \tag{79c}$$

$$P_1^1(\cos \theta) = -\sin \theta, \quad P_2^1(\cos \theta) = -\tfrac{3}{2} \sin 2\theta,$$

$$P_2^2(\cos \theta) = \tfrac{3}{2}(1 - \cos 2\theta), \quad \text{etc.}$$

Also, the following relations are useful when n and m are positive integers:

$$P_n^m(0) = 0 = \int_0^\pi P_n^m(\cos \theta)\, d\theta, \qquad n + m \text{ odd,}$$

$$\frac{d}{d\theta} P_n^m(\cos \theta)\Big|_{\theta = \pi/2} = 0 = \int_0^\pi P_n^m(\cos \theta) \cos \theta\, d\theta, \qquad n + m \text{ even}$$

$$\frac{d}{d\theta} P_n^m(\cos \theta)\Big|_{\theta = 0, \pi} = \frac{P_n^m(\cos \theta)}{\sin \theta}\Big|_{\theta = 0, \pi} = 0, \qquad\qquad m > 1, \tag{79d}$$

$$\frac{d}{d\theta} P_n^1(\cos \theta)\Big|_{\theta = 0} = \frac{P_n^1(\cos \theta)}{\sin \theta}\Big|_{\theta = 0} = -\frac{n(n + 1)}{2}.$$

$$0 \leq \theta_1 \leq \theta \leq \theta_2$$

E modes

g_θ' in Eq. (68) is an even function of $v + \tfrac{1}{2}$ and thus exhibits pole singularities only. The poles are simple and are located at

$$\lambda_p = p(p + 1), \qquad \text{where } P_p^{-q}(\cos \theta_0) = 0. \tag{80}$$

No poles arise when $v - q$ equals an integer since under these conditions, $P_v^{-q}(-x) = (-1)^{v-q} P_v^{-q}(x)$; a similar relation obtains when $v + q + 1$ is an integer (note: $P_v^{-q} \equiv P_{-v-1}^{-q}$). The weighted delta function then has the spectral representation

$$\frac{\delta(\theta - \theta')}{\sin \theta'} = -\frac{\pi}{2} \sum_{p>0} (2p + 1) \frac{\Gamma(p + q + 1)}{\Gamma(p - q + 1)}$$

$$\times \frac{P_p^{-q}(-\cos \theta_0)}{[\sin (p - q)\pi][(\partial/\partial p)P_p^{-q}(\cos \theta_0)]} P_p^{-q}(\cos \theta) P_p^{-q}(\cos \theta'), \tag{81}$$

and the orthonormal eigenfunctions are

$$\Phi_p^q(\theta) = \left[-\frac{\pi(2p + 1)\Gamma(p + q + 1)P_p^{-q}(-\cos \theta_0)}{2\Gamma(p - q + 1) \sin (p - q)\pi[(\partial/\partial p)\, P_p^{-q}(\cos \theta_0)]} \right]^{1/2} P_p^{-q}(\cos \theta), \tag{81a}$$

where p is any positive zero of $P_p^{-q}(\cos \theta_0)$.

H modes

g_θ'' in Eq. (69) is an even function of $v + \tfrac{1}{2}$ and exhibits pole singularities only. The poles are simple and are located at

$$\lambda_p = p(p + 1), \qquad \text{where} \qquad \frac{d}{d\theta_0} P_p^{-q}(\cos \theta_0) = 0. \tag{82}$$

The weighted delta function has the spectral representation

$$\frac{\delta(\theta - \theta')}{\sin \theta'} = \tfrac{1}{2} \csc^2 \left(\frac{\theta_0}{2}\right) \delta_{q0}$$

$$- \frac{\pi}{2} \sum_{p>0} (2p + 1) \frac{\Gamma(p + q + 1)}{\Gamma(p - q + 1)}$$

$$\times \frac{(d/d\theta_0)P_p^{-q}(-\cos \theta_0)}{\sin (p - q)\pi[(\partial^2/\partial p\, \partial\theta_0)P_p^{-q}(\cos \theta_0)]} P_p^{-q}(\cos \theta)P_p^{-q}(\cos \theta')$$

$$(83)$$

where $\delta_{\alpha\beta} = 0$, $\alpha \neq \beta$, and $\delta_{\alpha\alpha} = 1$; the orthonormal eigenfunctions are

$$\psi_p^q(\theta) = \left[\frac{-\pi(2p + 1)\Gamma(p + q + 1)[(d/d\theta_0)P_p^{-q}(-\cos \theta_0)]}{2\Gamma(p - q + 1)[\sin (p - q)\pi][(\partial^2/\partial p\, \partial\theta_0)P_p^{-q}(\cos \theta_0)]}\right]^{1/2} P_p^{-q}(\cos \theta),$$

$$p > 0, \quad q \neq 0. \qquad (83a)$$

When $q = 0$, $\lambda_p = 0$ is an eigenvalue, and the constant term $2^{-1/2} \csc (\theta_0/2)$ must be included. In this instance, the functions $P_p^{-q}(x)$ reduce to the ordinary Legendre functions $P_p^0(x) \equiv P_p(x)$.

$$0 \leq \theta_1 \leq \theta \leq \theta_2 < \pi$$

E modes

The E-mode characteristic Green's function $g'_\theta(\theta, \theta'; q^2; \lambda)$ which vanishes at $\theta = \theta_{1,2}$, may be expressed in the form

$$g'_\theta(\theta, \theta'; q^2; \lambda) = \frac{\pi}{2} \frac{C(\nu, q; \theta_<, \theta_1)C(\nu, q; \theta_>, \theta_2)\Gamma(\nu + q + 1)}{[\sin (\nu - q)\pi]\Gamma(\nu - q + 1)C(\nu, q; \theta_2, \theta_1)}, \qquad (84)$$

where

$$C(\nu, q; \alpha, \beta) = P_\nu^{-q}(\cos \alpha)P_\nu^{-q}(-\cos \beta) - P_\nu^{-q}(-\cos \alpha)P_\nu^{-q}(\cos \beta). \qquad (84a)$$

The behavior at $|\lambda| \to \infty$ is still specified by Eq. (70), and the singularities of g'_θ are simple poles located at

$$\lambda_p = p(p + 1), \qquad \text{where} \qquad C(p, q; \theta_2, \theta_1) = 0. \qquad (85)$$

Then

$$\frac{\delta(\theta - \theta')}{\sin \theta'} = \frac{\pi}{2} \sum_p \frac{(2p + 1)\Gamma(p + q + 1)}{[\sin (p - q)\pi]\Gamma(p - q + 1)} \frac{P_p^{-q}(\cos \theta_1)}{P_p^{-q}(\cos \theta_2)}$$

$$\times \frac{C(p, q; \theta, \theta_2)C(p, q; \theta', \theta_2)}{(\partial/\partial p)C(p, q; \theta_1, \theta_2)}, \qquad (86)$$

and

$$\Phi_p^q(\theta) = \left[\frac{\pi(2p + 1)\Gamma(p + q + 1)P_p^{-q}(\cos \theta_1)}{2[\sin (p - q)\pi]\Gamma(p - q + 1)P_p^{-q}(\cos \theta_2)[(\partial/\partial p)C(p, q; \theta_1, \theta_2)]}\right]^{1/2}$$

$$\times C(p, q; \theta, \theta_2). \qquad (86a)$$

H mode

$$g_\theta''(\theta, \theta'; q^2; \lambda) = \frac{\pi}{2} \frac{B(\nu, q; \theta_<, \theta_1)B(\nu, q; \theta_>, \theta_2)\Gamma(\nu + q + 1)}{[\sin (\nu - q)\pi]\Gamma(\nu - q + 1)[(\partial/\partial\theta_2)B(\nu, q; \theta_2, \theta_1)]},$$

(87)

where

$$B(\nu, q; \alpha, \beta) = P_\nu^{-q}(\cos \alpha)\frac{d}{d\beta} P_\nu^{-q}(-\cos \beta) - P_\nu^{-q}(-\cos \alpha) \frac{d}{d\beta} P_\nu^{-q}(\cos \beta).$$

(87a)

Simple poles are located at

$$\lambda_p = p(p + 1), \qquad \text{where} \qquad \frac{\partial}{\partial\theta_2} B(p, q; \theta_2, \theta_1) = 0.$$

(88)

The delta-function representation is

$$\begin{aligned}
\frac{\delta(\theta - \theta')}{\sin \theta'} = &\frac{1}{2} \csc \left(\frac{\theta_2 - \theta_1}{2}\right) \csc \left(\frac{\theta_2 + \theta_1}{2}\right)\delta_{q0} \\
&- \frac{\pi}{2} \sum_p (2p + 1) \frac{\Gamma(p + q + 1)[(d/d\theta_1)P_p^{-q}(\cos \theta_1)]}{[\sin (p - q)\pi]\Gamma(p - q + 1)[(d/d\theta_2)P_p^{-q}(\cos \theta_2)]} \\
&\times \frac{B(p, q; \theta, \theta_2)B(p, q; \theta', \theta_2)}{(\partial^2/\partial p \, \partial\theta_2)B(p, q; \theta_2, \theta_1)}
\end{aligned}$$

(89)

whence for $q \neq 0$ and $p > 0$,

$$\begin{aligned}
\psi_p^q(\theta) = &\left[-\frac{\pi(2p + 1)\Gamma(p + q + 1)[(d/d\theta_1)P_p^{-q}(\cos \theta_1)]}{2[\sin (p - q)\pi]\Gamma(p - q + 1)(d/d\theta_2)P_p^{-q}(\cos \theta_2)}\right]^{1/2} \\
&\times \frac{B(p, q; \theta, \theta_2)}{[(\partial^2/\partial p \, \partial\theta_2)B(p, q; \theta_2, \theta_1)]^{1/2}}
\end{aligned}$$

(89a)

When $q = 0$, the constant term

$$2^{-1/2}\left[\csc \left(\frac{\theta_2 - \theta_1}{2}\right) \csc \left(\frac{\theta_2 + \theta_1}{2}\right)\right]^{1/2}$$

must be included.

Equations (86) or (89) lend themselves to the study of various special cases. First, one may derive the results in Eqs. (81) and (83) by letting $\theta_1 \to 0$. Next, consider the symmetrical case $\theta_2 = \pi - \theta_1$. In this instance, the transcendental equation (85) may be separated into

$$\begin{aligned}
0 = &C(p, q; \pi - \theta_1, \theta_1) \\
= &[P_p^{-q}(\cos \theta_1) + P_p^{-q}(-\cos \theta_1)][P_p^{-q}(\cos \theta_1) - P_p^{-q}(-\cos \theta_1)],
\end{aligned}$$

(90)

so that the eigenvalues occur in two sets p' and p'' corresponding to the vanishing of the first and second factors, respectively. It may be verified that the modes described by p' and p'' possess even or odd symmetry about the bisecting plane $\theta = \pi/2$ so that a source problem in a symmetrical biconical region may be decomposed into two separate simpler problems arising from even and odd excitation about the symmetry plane. Analogous considerations apply to the *H*-mode case.

While the mode functions described above are complete for the representation of scalar functions, azimuthally symmetric vector fields may contain in addition a TEM mode that must be derived separately.

3.4c   Radial Transmission Lines

Because of the non-uniformity of successive cross sections transverse to the radial direction, radial transmission problems in cylindrical and spherical geometries involve non-uniform transmission lines, as noted in Sec. 2. 7 for the spherical case. Although vector separability into E and H modes with respect to the radial direction is possible for spherical geometries, it is generally not possible for electromagnetic fields in cylindrical geometries; nevertheless, cylindrical regions may be viewed as waveguides in the direction parallel to the cylinder axis. Radial transmission is then relevant for alternative representations of the z-separated solution (see Secs. 3.3a and 6.2). Since the radial problem in spherical coordinates, as specified by Eq. (2.7.2), is related to that in cylindrical coordinates in Eq. (3.2.46b) by the transformation in Eqs. (3.5.1) and (3.5.2) [see also Eq. (2.7.3a)], results for the cylindrical case are readily obtained from the spherical transmission-line analysis in Sec. 2.7.

The radial characteristic Green's-function problem in cylindrical regions is defined as follows [see Eqs. (3.2.46b) and (3.3.1)] :

$$\left(\frac{d}{d\rho}\,\rho\,\frac{d}{d\rho} + \tau\rho - \frac{\lambda}{\rho}\right)g_\rho(\rho,\rho';\tau,\lambda) = -\delta(\rho - \rho'), \tag{91}$$

subject to the boundary conditions

$$g'_\rho(\rho,\rho';\tau,\lambda) = 0 \qquad \text{at } \rho = \rho_{1,2}, \tag{91a}$$

$$\frac{dg''_\rho(\rho,\rho';\tau,\lambda)}{d\rho} = 0 \qquad \text{at } \rho = \rho_{1,2}, \tag{91b}$$

for E modes (single primes) and H modes (double primes), respectively. When $\rho_1 = 0$, the boundary condition at the origin is replaced by a finiteness requirement [see also Eq. (2.7.7)], while an unbounded domain $\rho_2 \to \infty$ requires imposition of a radiation condition. The eigenvalue problems associated with the characteristic Green's functions g_ρ in Eqs. (91) differ according to whether λ or τ is the characteristic parameter. In a z-transmission representation of three-dimensional Green's functions involving the eigenfunctions $\Phi_q(\phi)$ or $\Psi_q(\phi)$ in Eq. (3.2.46a) [see also Sec. 3.3c], $\lambda = q^2$ is prescribed from the angular eigenvalue problem and τ is the characteristic parameter for the radial domain. On the other hand, in a ϕ-transmission representation of two- or three-dimensional Green's functions, τ is prescribed and λ is the characteristic parameter [see Secs. 3.3c and 6.2b; for two-dimensional problems, $\tau = k^2$, while for the three-dimensional case, $\tau = (k^2 - \gamma^2)$, where γ is the eigenvalue for the z-dependent eigenfunctions]. Mode spectra and completeness relations for various radial domains, required for the z-transmission representation, have already been given in Sec. 3.2c. It follows from Eqs. (2.7.1-2) (with $\mu, \epsilon = \text{constant}$, $k^2 \rightarrow \tau$, $k_{ti}^2 \rightarrow \lambda$) and from Eqs. (91) that the characteristic Green's functions $g_r(r, r'; \tau, \lambda)$ and $g_\rho(\rho, \rho'; \tau, \lambda)$ for the spherical and cylindrical geometries, respectively, are related as follows:

$$g_\rho(\rho, \rho'; \tau, \lambda) = \frac{1}{\sqrt{rr'}} g_r(r, r'; \tau, \lambda + \tfrac{1}{4}) \Big|_{\substack{r=\rho \\ r'=\rho'}} \tag{92}$$

where for the E-mode case,

$$g_r' = \frac{i\zeta}{k} Y_i'(r, r'); \qquad k^2 \rightarrow \tau, \quad k_{ti}'^2 \rightarrow \lambda, \tag{92a}$$

and, for H-mode case,

$$g_r'' = \frac{i}{k\zeta} Z_i''(r, r'), \qquad k^2 \rightarrow \tau, \quad k_{ti}''^2 \rightarrow \lambda, \tag{92b}$$

with $Y_i'(r, r')$ and $Z_i''(r, r')$ denoting the modal Green's functions in Sec. 2.7, and $\zeta = \sqrt{\mu/\epsilon}$.

For the unbounded radial domain $0 < \rho < \infty$ wherein the E-and H-mode solutions coincide, one has from Eqs. (92), (2.7.11), and (2.7.3a) (with $j \rightarrow -i$),[†]

$$g_\rho(\rho, \rho'; \tau, \lambda) = \frac{\pi i}{2} J_\nu(\sqrt{\tau}\,\rho_<) H_\nu^{(1)}(\sqrt{\tau}\,\rho_>), \qquad \nu = \sqrt{\lambda}, \quad 0 < \frac{\rho}{\rho'} < \infty. \tag{93}$$

If $\nu = q > 0$ is prescribed, then τ is the characteristic variable and g_ρ possesses a branch-point singularity at $\tau = 0$ in the complex τ plane. It then follows from Eq. (3.2.70b) that the condition $\text{Im} \sqrt{\tau} > 0$ must be imposed in order for g_ρ to decay as $\rho \rightarrow \infty$. By choosing a branch cut along the positive real τ axis and defining $0 < \arg \tau < 2\pi$, one may enforce $\text{Im} \sqrt{\tau} > 0$ on the entire top sheet, the "spectral" sheet, of the complex τ plane. From Eq. (3.3.11) and the change of variable $\xi = \sqrt{\tau}$, one obtains the spectral representation of $\delta(\rho - \rho')/\rho'$ in Eq. (3.2.72), which can be written in the alternative forms in Eqs. (3.2.68) or (3.2.62).

When τ is prescribed and λ is the characteristic variable, a different spectral representation is derived; as noted earlier, $\sqrt{\tau} = k$ for angular transmission formulations of z-independent two-dimensional Green's functions (see Sec. 6.2). g_ρ now has a branch-point singularity at $\lambda = 0$ in the complex λ plane. The Hankel function of the first kind (with $\text{Im} \, k > 0$ for small dissipation) satisfies the radiation condition at $\rho \rightarrow \infty$ for fields with time dependence $\exp(-i\omega t)$. To enforce finiteness at $\rho = 0$, it is necessary to impose $\text{Re} \, \nu > 0$ since $J_\nu(k\rho) \sim (k\rho)^\nu$ as $\rho \rightarrow 0$; by drawing the branch cut along the negative real axis as in Fig. 17(a) and defining $-\pi < \arg \lambda < \pi$ with $\nu = \sqrt{\lambda}$, one has $\text{Re} \, \nu > 0$

[†]For the discussion of characteristic Green's functions, j or i merely denotes the imaginary unit; however, subsequent results apply directly to radiation problems with time dependence $\exp(-i\omega t)$.

on the spectral sheet of the complex λ plane. Since g_ρ behaves like $(\rho_</\rho_>)^{\sqrt{\lambda}}$ as $\lambda \to \infty$ [see Eqs. (2.7.4b)], the characteristic Green's function decays at infinity on the spectral sheet, thereby providing the basis for application of Eq. (3.3.11). Comparison of Eqs. (3.3.1) and (91) shows that the weight function w equals $-1/\rho$, whence one obtains as the desired completeness relation in the domain $0 < (\rho, \rho') < \infty$ [note that the imaginary unit j in Eq. (3.3.11) is replaced by i in the present discussion] :

$$\rho'\delta(\rho - \rho') = \frac{1}{2\pi i}\oint_C g_\rho(\rho, \rho'; k^2, \lambda)\, d\lambda \tag{94a}$$

$$= \frac{1}{2}\int_{-i\infty}^{i\infty} \nu J_\nu(k\rho_<)H_\nu^{(1)}(k\rho_>)\, d\nu = \frac{1}{2}\int_{-i\infty}^{i\infty} \nu J_\nu(k\rho)H_\nu^{(1)}(k\rho')\, d\nu \tag{94b}$$

$$= \frac{1}{4}\int_{-i\infty}^{i\infty} \nu H_\nu^{(1)}(k\rho)H_\nu^{(1)}(k\rho')\, d\nu \tag{94c}$$

$$= \frac{1}{4}\int_0^{i\infty} \nu(1 - e^{i2\nu\pi})H_\nu^{(1)}(k\rho)H_\nu^{(1)}(k\rho')\, d\nu \tag{94d}$$

$$= \sum_p \Phi_p(k\rho)\bar{\Phi}_p(k\rho'), \qquad 0 < \frac{\rho}{\rho'} < \infty, \tag{94e}$$

where C is the integration contour in Fig. 3.4.17(a). Equation (94c) follows from Eq. (94b) on use of the relations

$$J_\nu(x) = \tfrac{1}{2}[H_\nu^{(1)}(x) + H_\nu^{(2)}(x)], \qquad H_{-\nu}^{(1,2)}(x) = e^{\pm i\nu\pi}H_\nu^{(1,2)}(x), \tag{94f}$$

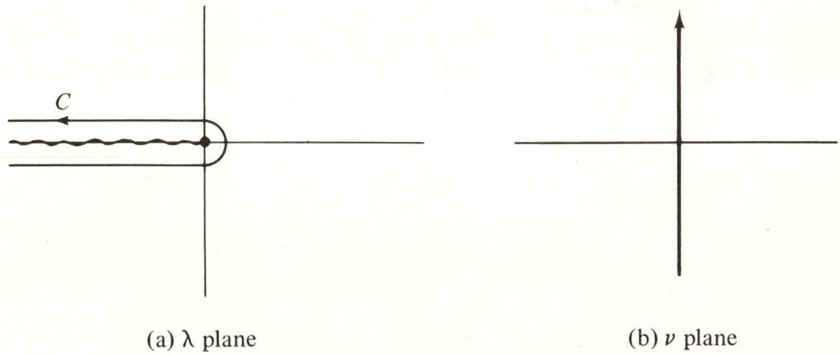

(a) λ plane   (b) ν plane

FIG. 3.4.17   Integration paths and singularities.

whence $\nu H_\nu^{(2)}(x)H_\nu^{(1)}(y)$ is an odd function of ν that does not contribute to the integral along the imaginary ν axis. The symmetrical form of Eq. (94c) justifies suppression of $\rho_<$ and $\rho_>$ in the second of Eqs. (94b). The representation theorem contained in Eqs. (94) gives rise to the Kontorovitch–Lebedev transform pair[11]

$$f(\rho) = \tfrac{1}{4}\int_{-i\infty}^{i\infty} \nu H_\nu^{(1)}(k\rho)F(\nu)\, d\nu, \qquad F(\nu) = \int_0^\infty \frac{1}{\rho} f(\rho)H_\nu^{(1)}(k\rho)\, d\rho, \tag{95}$$

the convergence properties of which are discussed in connection with Eqs. (6.

3.1). From the generic representation in Eq. (94e), one may identify on comparison with Eqs. (94b)–(94d) the orthonormal radial eigenfunctions $\Phi_p(k\rho)$, the adjoint functions $\bar{\Phi}_p(k\rho)$ [see Eq. (94b)], and the meaning of $\sum_p$.

When the radial domain extends from $\rho = a$ to $\rho = \infty$, boundary conditions at $\rho = a$ must be included. For generality, we assume that

$$\frac{dg_\rho}{d\rho} = -ik\bar{C}g_\rho \quad \text{at } \rho = a, \qquad \bar{C} = \text{constant}, \tag{96}$$

from which follow the special cases in Eqs. (91a) ($\bar{C} = \infty$) and (91b) ($\bar{C} = 0$). As noted in Sec. 6.7a, $\bar{C}$ may be related to the surface impedance on a cylindrical scatterer with radius a. The solution for g_ρ satisfying Eq. (96) and the radiation condition at infinity (for Im $k > 0$) is now given by [see also Eq. (2.7.14)]

$$g_\rho(\rho, \rho'; k^2; \lambda) = \frac{\pi i}{2}\left[J_\nu(k\rho_<) - \frac{b(\nu)}{d(\nu)}H_\nu^{(1)}(k\rho_<)\right]H_\nu^{(1)}(k\rho_>), \qquad \nu = \sqrt{\lambda}, \tag{97}$$

where $a \leq (\rho, \rho') < \infty$ and

$$b(\nu) = J_\nu'(ka) + i\bar{C}J_\nu(ka), \qquad d(\nu) = H_\nu'^{(1)}(ka) + i\bar{C}H_\nu^{(1)}(ka), \tag{97a}$$

with the prime denoting the derivative with respect to the argument. One may verify on use of Eq. (94f) that g_ρ in Eq. (97) is an even function of ν, so $\lambda = 0$ is a regular point in the complex λ plane. g_ρ decays as $\lambda \to \infty$ [see Eqs. (6.7.14) et seq.] and has complex pole singularities ν_p defined by the transcendental equation

$$d(\nu_p) = 0, \tag{97b}$$

located in the first and third quadrants of the complex ν plane (see Sec. 6.A5 for the special cases $\bar{C} = 0$ or $\bar{C} = \infty$); the pole locations $\lambda_p = \nu_p^2$ in the complex λ plane are shown in Fig. 3.4.18. Equation (3.3.11) then yields the completeness relation[12]:

$$\rho'\delta(\rho - \rho') = \frac{1}{2\pi i}\oint_C g_\rho(\rho, \rho'; k^2, \lambda)\, d\lambda, \tag{98a}$$

$$= -\pi i \sum_p \frac{\nu_p b(\nu_p)}{\left[\dfrac{\partial}{\partial \nu} d(\nu)\right]_{\nu_p}} H_{\nu_p}^{(1)}(k\rho)H_{\nu_p}^{(1)}(k\rho'), \tag{98b}$$

$$= \sum_p \Phi_p(k\rho)\Phi_p(k\rho'), \qquad a \leq \frac{\rho}{\rho'} < \infty. \tag{98c}$$

Equation (97) may also be used to derive a spectral theorem for the case where $\nu = q > 0$ is prescribed and $k^2 \to \tau$ is the characteristic parameter. One may verify that g_ρ possesses a branch point at $\tau = 0$ in the complex τ plane and that the condition Im $\sqrt{\tau} > 0$ must be imposed to satisfy the radiation condition at infinity [see discussion following Eq. (93)]. For the special cases $\bar{C} = 0$ and $\bar{C} = \infty$, no pole singularities [i.e., zeros of $H_q^{(1)}(\sqrt{\tau}\,a)$ or $H_q'^{(1)}(\sqrt{\tau}\,a)$] exist on the spectral sheet $0 < \arg \tau < 2\pi$ [see discussion preceding Eq. (5.9.12)].

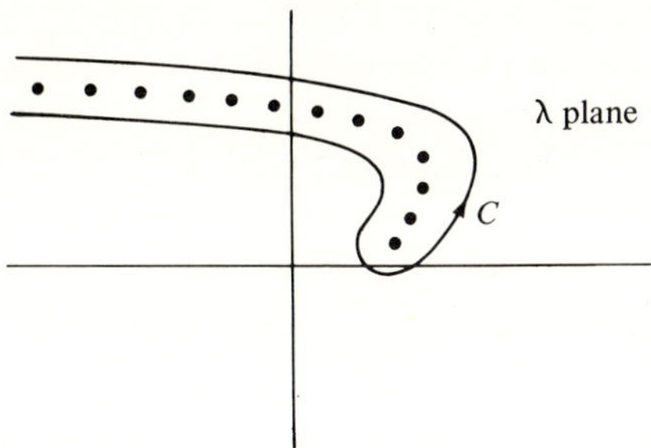

FIG. 3.4.18 Singularities of g_ρ in complex λ plane.

By considerations analogous to those mentioned earlier, one then obtains the following completeness relations in the domain $a \leq (\rho, \rho') < \infty$ [see also Eqs. (3.4.24)]:

$$\frac{\delta(\rho - \rho')}{\rho'} = -\frac{1}{2\pi i} \oint_C g_\rho(\rho, \rho'; \tau, q^2) \, d\tau \tag{99a}$$

$$= \int_0^\infty \Phi_\xi(\rho)\Phi_\xi^*(\rho') \, d\xi, \qquad \bar{C} = \infty, \tag{99b}$$

$$= \int_0^\infty \Psi_\xi(\rho)\Psi_\xi^*(\rho') \, d\xi, \qquad \bar{C} = 0, \tag{99c}$$

where

$$\Phi_\xi(\rho) = \sqrt{\xi}\left[J_q(\xi\rho) - \frac{J_q(\xi a)}{H_q^{(1)}(\xi a)} H_q^{(1)}(\xi\rho) \right], \tag{99d}$$

$$\Psi_\xi(\rho) = \sqrt{\xi}\left[J_q(\xi\rho) - \frac{J_q'(\xi a)}{H_q'^{(1)}(\xi a)} H_q^{(1)}(\xi\rho) \right]. \tag{99e}$$

C is a contour enclosing the branch cut along the positive real λ axis, and $\xi = \sqrt{\tau}$. It may be noted that the normalized eigenfunctions in Eqs. (99d) and (99e) are obtainable from those for the interval $0 < \rho < \infty$ in Eq. (3.2.62) on adding to the latter a solution of the homogeneous equation (91) so as to satisfy the boundary conditions $\Phi_\xi(a) = 0$ and $d\Psi_\xi(\rho)/d\rho|_{\rho=a} = 0$, respectively.

The corresponding characteristic Green's functions and completeness relations for the radial domain in spherical regions follow from the preceding results for cylindrical regions and from Eq. (92). In particular, for the unbounded domain $0 < (r, r') < \infty$:

$$r'^2\delta(r - r') = \frac{1}{2\pi i} \oint_C g_r(r, r'; \lambda) \, d\lambda \tag{100a}$$

$$= \frac{1}{2\pi k} \int_{-(1/2)-i\infty}^{-(1/2)+i\infty} j_v(kr_<)h_v^{(1)}(kr_>)(2v + 1) \, dv \tag{100b}$$

$$= \frac{1}{4\pi k} \int_{-(1/2)-i\infty}^{-(1/2)+i\infty} h_v^{(1)}(kr)h_v^{(1)}(kr')(2v + 1) \, dv, \tag{100c}$$

which formulas are a restatement of the Kontorovitch–Lebedev transform theorem in Eqs. (94) in terms of spherical Bessel functions as defined in Eq. (2.7. 3a). Similarly, for the domain $a < r < \infty$ with a perfectly conducting surface located at $r = a$, the H-mode Green's function in Eq. (2.7.12a) yields [see Eq. (98b) with $\bar{C} = \infty$]

$$r'^2\delta(r - r') = \sum_s (2s + 1)\frac{h_s^{(1)}(kr)h_s^{(1)}(kr')j_s(ka)}{ik[(\partial/\partial s)h_s^{(1)}(ka)]}, \qquad h_s^{(1)}(ka) = 0, \qquad (101a)$$

while the spectrum corresponding to the E-mode Green's function in Eq. (2.7. 12b) leads to

$$r'^2\delta(r - r') = \sum_\sigma (2\sigma + 1)\frac{h_\sigma^{(1)}(kr)h_\sigma^{(1)}(kr')j_\sigma'(ka)}{ik[(\partial/\partial\sigma)h_\sigma'^{(1)}(ka)]}, \qquad h_\sigma'^{(1)}(ka) = 0. \qquad (101b)$$

In these relations Im $k \geq 0$, thereby making the results applicable to problems with exp $(-i\omega t)$ dependence. For exp $(+j\omega t)$ dependence, one replaces $h_a^{(1)}$ by $h_a^{(2)}$ throughout and lets $i \to -j$ in Eqs. (101).

3.5 APPROXIMATE METHODS FOR SOLVING THE NON-UNIFORM TRANSMISSION-LINE EQUATIONS

As presented in Chapter 2, the solution of field problems by modal analysis and synthesis requires knowledge of the eigenfunctions transverse to a chosen transmission coordinate, and of the modal amplitudes along that coordinate. By the characteristic Green's function method of Sec. 3.3a, eigenvalue and transmission-line problems are shown to be closely related, whence solution of separable multidimensional field problems rests essentially on the ability to solve one-dimensional transmission-line equations. These equations are of the Sturm–Liouville type and admit solutions in terms of known functions only for special forms of the x-dependent parameters p, q, and w in Eqs. (3.3.1) or (3.3.18). It is therefore relevant to discuss approximation or perturbation procedures that can be applied to more general parameter variations. If the parameters vary "slowly" over a distance interval equal to the local wavelength, the WKB approximation furnishes an unperturbed solution with a broad range of validity. For rapid variations, one must resort to special functions that approximate the desired one as closely as possible in the interval in question. While application of these methods, as described below, is successful in many cases of practical interest, the detailed calculation of the Green's functions or eigenfunctions for specific medium variations must frequently be accomplished by numerical means.

The Sturm–Liouville equation (3.3.18) may be reduced to a standard form by renormalization of the c, s solutions (3.3.18a) into modified $\hat{c}, \hat{s}$ solutions defined by

$$\left.\begin{matrix} c(x, x_0) \\ s(x, x_0) \end{matrix}\right\} = \frac{1}{\sqrt{p}} \left\{\begin{matrix} \hat{c}(x, x_0) \\ \hat{s}(x, x_0) \end{matrix}\right. , \tag{1}$$

whence

$$\left[\frac{d^2}{dx^2} + \gamma(x)\right]\begin{matrix} \hat{c}(x, x_0) \\ \hat{s}(x, x_0) \end{matrix} = 0, \tag{2}$$

where

$$\gamma(x) = -\frac{q(x)}{p(x)} + \frac{\lambda w(x)}{p(x)} - \frac{1}{\sqrt{p(x)}}\frac{d^2}{dx^2}\sqrt{p(x)}. \tag{2a}$$

For convenience, one chooses as initial conditions:

$$\begin{aligned} \hat{c}(x_0, x_0) &= 1, & \hat{s}(x_0, x_0) &= 0, \\ \hat{c}'(x_0, x_0) &= 0, & \hat{s}'(x_0, x_0) &= 1, \end{aligned} \tag{2b}$$

whence the functions c and s derived therefrom differ in general from those defined by Eqs. (3.3.18). When applied to propagation in an inhomogeneous medium, as in Eqs. (3.2.98) and (3.2.99), one has $p(x) = 1/\mu(x)$ for H modes along x (i.e., fields with $E_x \equiv 0$), and $p(x) = 1/\epsilon(x)$ for E modes along x.† If the medium is a variable dielectric $\epsilon(x)$ with constant permeability μ, the H-mode functions c, s are identical with $\hat{c}$ and $\hat{s}$; similarly, Eq. (3.3.18) reduces to the standard form for E modes in a medium with constant ϵ. Moreover, for the planar stratification considered in these examples, $w = p$, so the form of the functions c and s for arbitrary λ can be inferred from that for $\lambda = 0$ upon replacing $\gamma(x)|_{\lambda=0}$ by $(\gamma + \lambda)$. Since $\hat{c}$ and $\hat{s}$ are linearly independent solutions of Eq. (2), as verified from the non-vanishing of the Wronskian $\hat{c}\hat{s}' - \hat{s}\hat{c}'$ in Eq. (2b), a linear superposition of $\hat{c}$ and $\hat{s}$ may be employed to satisfy initial conditions of a type more general than those in Eq. (2b).

In the first approximation method to be described, Eq. (2) is rephrased as a Fredholm integral equation, with the kernel chosen to be as "small" as possible in order to yield a rapidly convergent iterative solution.

3.5a Integral Equation Formulation

Suppose that for given $\gamma(x)$, Eq. (2) has no known solution but that a solution is known for some other function $\zeta(x)$ which is chosen so that it resembles $\gamma(x)$ over the interval $x_1 \leq x \leq x_2$ [see Sec. 3.5b for a method of choosing

---

†In an inhomogeneous dielectric where $\epsilon = \epsilon(x)$ and $\mu = $ constant, $\sqrt{\gamma(x)} = \omega\sqrt{\mu\epsilon(x)}$ represents the actual wavenumber descriptive of an H-mode field. For an E mode, $\sqrt{\gamma(x)} = \left[\omega^2\mu\epsilon(x) - \sqrt{\epsilon(x)}\frac{d^2}{dx^2}\frac{1}{\sqrt{\epsilon(x)}}\right]^{1/2}$ represents an *effective* wavenumber [see also Eq. (2.3.45)]. Substantial differences in the propagation characteristics of the two wave types may arise when $\epsilon(x)$ varies rapidly or vanishes at some point in the interval. Dual considerations apply when $\mu = \mu(x)$ and ϵ is constant.

$\zeta(x)$]. We construct the Green's function $G(x, x')$ that satisfies the differential equation

$$\left[\frac{d^2}{dx^2} + \zeta(x)\right] G(x, x') = -\delta(x - x'), \qquad x_1 \leq (x, x') \leq x_2, \tag{3}$$

with as yet unspecified boundary conditions. Then from the one-dimensional form of Green's theorem of the second kind,[†] applied to the functions $G(x, x')$ and $u(x) = \hat{c}$ or $\hat{s}$, one derives the integral equation

$$u(x) = \left[G(x', x)\frac{du(x')}{dx'} - u(x')\frac{d}{dx'}G(x', x)\right]_{x'=x_1}^{x'=x_2}$$

$$+ \int_{x_1}^{x_2} [\gamma(x') - \zeta(x')]G(x', x)u(x')\,dx'. \tag{4}$$

If $\gamma(x) \simeq \zeta(x)$ over the interval in question, the integral on the right-hand side of Eq. (4) is expected to be small and can be regarded as a correction to the unperturbed contribution in the first term. This highlights the importance of a proper choice of $\zeta(x)$, to be described in Sec. 3.5b. If $\gamma \equiv \zeta$ over one or more subintervals $x_3 \leq x \leq x_4$, where $x_3 > x_1$ and $x_4 < x_2$, then these ranges are excluded from the integration in the correction integral.

Initial conditions on u are specified at some point x_1. The perturbation method of solution will be simplified if the Green's function satisfies the boundary condition

$$G(x', x) = \frac{d}{dx'}G(x', x) = 0, \qquad x' > x, \tag{5}$$

whence Eq. (4) reduces to

$$u(x) = \left\{u(x_1)\frac{d}{dx_1}G(x_1, x) - G(x_1, x)\frac{du(x_1)}{dx_1}\right\}$$

$$+ \int_{x_1}^{x} [\gamma(x') - \zeta(x')]\,G(x', x)u(x')\,dx', \tag{6}$$

a Fredholm integral equation of the second kind, with a variable upper limit (Volterra integral equation).[13] If $g_1(x)$ and $g_2(x)$ denote any two linearly independent solutions of the homogeneous equation (3),

$$\left[\frac{d^2}{dx^2} + \zeta(x)\right] g_{1,2}(x) = 0, \tag{7}$$

then it is verified that[‡]

[†] $\displaystyle \int_{x_1}^{x_2}\left[u(x')\frac{d^2}{dx'^2}G(x', x) - G(x', x)\frac{d^2}{dx'^2}u(x')\right]dx'$

$\displaystyle \qquad\qquad = \left[u(x')\frac{d}{dx'}G(x', x) - G(x', x)\frac{d}{dx'}u(x')\right]_{x_1}^{x_2}.$

[‡] See Eq. (3.3.14a), from which one subtracts the homogeneous solution $\vec{V}(x)\overleftarrow{V}(x')/(-pW)$ in order to satisfy the boundary condition (5). We have here utilized the fact that two Green's functions satisfying the same differential equation, but different boundary conditions, differ only by a solution of the homogeneous equation.

$$G(x', x) = \begin{cases} \dfrac{g_2(x')g_1(x) - g_1(x')g_2(x)}{W(g_1, g_2)}, & x_1 \leq x' \leq x, \\ 0, & x' > x. \end{cases} \qquad (8)$$

$W(g_1, g_2)$ represents the Wronskian $(g_1 g_2' - g_2 g_1')$, where the prime on $g_{1,2}$ denotes the derivative with respect to the argument. Thus, one finds for $\hat{c}(x, x_1)$

$$\hat{c}(x, x_1) = \frac{d}{dx_1} G(x_1, x) + \int_{x_1}^{x} [\gamma(x') - \zeta(x')]G(x', x)\hat{c}(x', x_1)\, dx', \qquad (9a)$$

while

$$\hat{s}(x, x_1) = - G(x_1, x) + \int_{x_1}^{x} [\gamma(x') - \zeta(x')]G(x', x)\hat{s}(x', x_1)\, dx'. \qquad (9b)$$

For suitable kernels, these integral equations can be solved by the method of successive approximations. Let $u_0(x)$ denote the expression inside the braces on the right-hand side of Eq. (6) ; it constitutes the zeroth-order approximation to $u(x)$. Substitution of $u_0(x')$ for $u(x')$ in the integral yields the first approximation $u_1(x)$; substitution of $u_1(x')$ for $u(x')$ yields the second approximation $u_2(x)$; etc. Thus, the solution for $u(x)$ can be expressed in the form of a series,

$$u(x) = \sum_{n=0}^{\infty} u_n(x), \qquad (10a)$$

$$u_n(x) = \int_{x_1}^{x} [\gamma(x') - \zeta(x')]G(x', x)u_{n-1}(x')\, dx', \qquad n \geq 1. \qquad (10b)$$

If γ, ζ, and G are (real) continuous functions of x in the interval $x_1 \leq x \leq x_2$ [if γ and ζ are piecewise continuous, the determination of $u(x)$ above is carried out separately in each interval segment wherein γ and ζ are continuous], then $u(x)$ is a real continuous function in the interval. Let the absolute value of $u_0(x)$ in the interval be smaller than some positive number N, and assume also that $|(\gamma - \zeta)G| \leq M$ in the interval, where M is a positive number. Then, if $K(x', x)$ denotes the kernel $(\gamma - \zeta)G$,

$$|u_n(x)| = \left| \int_{x_1}^{x} dx' K(x', x) \int_{x_1}^{x'} dx_1' K(x_1', x') \int_{x_1}^{x_1'} \cdots dx_\alpha' \int_{x_1}^{x_\alpha'} dx_\beta'\, K(x_\beta', x_\alpha')u_0(x_\beta') \right|,$$

$$(11)$$

where $\alpha = n - 2$, $x_0' \equiv x'$, $x_{-1}' \equiv x$. In view of the above-mentioned inequalities, the absolute value of the last integral is bounded by $NM(x_\alpha' - x_1)$, that of the last two integrals by $(NM^2/2!)(x_{\alpha-1}' - x_1)^2$, and, finally,

$$|u_n(x)| \leq \frac{NM^n}{n!}(x - x_1)^n, \qquad (12)$$

for all x in the interval $x_1 \leq x \leq x_2$. Hence, if $|K|$ is bounded in the interval, the series (10a) converges uniformly. The rapidity of convergence depends on the smallness of M (i.e., on $|\gamma - \zeta|$). If the function $\gamma(x)$ in Eq. (2) is well approximated by $\zeta(x)$ in Eq. (3), the first correction $u_1(x)$ may be sufficiently accurate in furnishing an acceptable solution for $u(x)$. The inequality $\sum_{n=0}^{\infty} |u_n|$ $\leq N \exp[M(x - x_1)]$ derived from (12) illustrates that convergence of the method depends not only on M but also on the boundedness of the interval $x - x_1$.

In many scattering problems, especially those involving unbounded geometrical cross sections, the inhomogeneity is confined to a finite region in space; in the exterior, the medium is homogeneous. Under these circumstances, the perturbed and unperturbed wave functions should be chosen to satisfy identical boundary conditions at the endpoints of the region, and it is convenient to employ an alternative formulation. Suppose that we wish to determine the Green's function $\hat{g}(x, x')$ associated with Eq. (2),

$$\left[\frac{d^2}{dx^2} + \gamma(x)\right] \hat{g}(x, x') = -\delta(x - x'), \tag{13}$$

subject to the boundary conditions

$$\frac{d\hat{g}}{dx} + \hat{\alpha}_{1,2}\hat{g} = 0 \qquad \text{at } x_{1,2}, \tag{13a}$$

where $\hat{\alpha}_{1,2}$ are constants that may be complex [thereby admitting a radiation condition if $x_1 \to -\infty$ and (or) $x_2 \to \infty$]. Upon writing Eq. (13) as

$$\left[\frac{d^2}{dx^2} + \zeta(x)\right] \hat{g}(x, x') = \left[\zeta(x) - \gamma(x)\right] \hat{g}(x, x') - \delta(x - x'), \tag{14}$$

and treating the right-hand side of (14) as a "known" inhomogeneous term $Q(x, x')$, one may construct the solution for $\hat{g}$ in terms of the unperturbed Green's function G by multiplying Eq. (3) by $-Q(x', x'')$ and integrating over x':

$$\hat{g}(x, x'') = G(x, x'') - \int_{x_1}^{x_2} [\zeta(x') - \gamma(x')]G(x, x')\hat{g}(x', x'') \, dx'. \tag{15}$$

The absence of endpoint contributions in this (Fredholm) integral equation of the second kind for $\hat{g}$ implies that G must satisfy the boundary condition (13a). The solution of the integral equation can again be found by an iteration procedure. However, since the integration limits are fixed, the resulting series converges under more restrictive conditions. A convergent series for any continuous (and therefore bounded) K can be obtained by the more general, but more cumbersome, method of Fredholm determinants.[13]

The Green's function G in the present case differs from that in Eq. (8) and is given via Eq. (3.3.14) by

$$G(x, x') = \frac{g_1(x_<)g_2(x_>)}{-W}, \qquad W = g_1 g_2' - g_2 g_1', \tag{16}$$

where g_2 and g_1 now denote solutions of the homogeneous equation (7) that satisfy the boundary conditions (13a) at x_2 and x_1, respectively. $\hat{g}$ can similarly be written as

$$\hat{g}(x, x') = \frac{\hat{g}_1(x_<)\hat{g}_2(x_>)}{-\hat{W}}, \qquad \hat{W} = \hat{g}_1\hat{g}_2' - \hat{g}_2\hat{g}_1', \tag{17}$$

where $\hat{g}_2$ and $\hat{g}_1$ are solutions of the homogeneous equation (13) [or of Eq. (2)] and satisfy the boundary conditions (13a) at x_2 and x_1, respectively. Upon substituting these expressions into Eq. (15), one obtains, for $x < x''$,

$$\hat{g}(x, x'') = \frac{g_1(x)g_2(x'')}{-W} - \frac{g_2(x)\hat{g}_2(x'')}{W\hat{W}} \int_{x_1}^{x} [\zeta(x') - \gamma(x')]g_1(x')\hat{g}_1(x')\,dx'$$

$$- \frac{g_1(x)\hat{g}_2(x'')}{W\hat{W}} \int_{x}^{x''} [\zeta(x') - \gamma(x')]g_2(x')\hat{g}_1(x')\,dx'$$

$$- \frac{g_1(x)\hat{g}_1(x'')}{W\hat{W}} \int_{x''}^{x_2} [\zeta(x') - \gamma(x')]g_2(x')\hat{g}_2(x')\,dx', \tag{18a}$$

while, for $x'' < x$,

$$\hat{g}(x, x'') = \frac{g_2(x)g_1(x'')}{-W} - \frac{g_2(x)\hat{g}_2(x'')}{W\hat{W}} \int_{x_1}^{x''} [\zeta(x') - \gamma(x')]g_1(x')\hat{g}_1(x')\,dx'$$

$$- \frac{g_2(x)\hat{g}_1(x'')}{W\hat{W}} \int_{x''}^{x} [\zeta(x') - \gamma(x')]g_1(x')\hat{g}_2(x')\,dx'$$

$$- \frac{g_1(x)\hat{g}_1(x'')}{W\hat{W}} \int_{x}^{x_2} [\zeta(x') - \gamma(x')]g_2(x')\hat{g}_2(x')\,dx'. \tag{18b}$$

Suppose now that $x'' = x_1$. Then Eqs. (17) and (18b) yield

$$\hat{g}_2(x) = C_2 g_2(x) + \frac{g_2(x)}{W} \int_{x_1}^{x} (\zeta - \gamma)g_1\hat{g}_2\,dx' + \frac{g_1(x)}{W} \int_{x}^{x_2} (\zeta - \gamma)g_2\hat{g}_2\,dx', \tag{19}$$

where the constant C_2 is given by

$$C_2 = \frac{g_1(x_1)}{W}\frac{\hat{W}}{\hat{g}_1(x_1)} = \frac{\hat{g}_2\hat{g}_1'/\hat{g}_1 - \hat{g}_2'}{g_2g_1'/g_1 - g_2'}\bigg|_{x_1} = \frac{\alpha_1\hat{g}_2(x_1) + \hat{g}_2'(x_1)}{\alpha_1 g_2(x_1) + g_2'(x_1)}. \tag{19a}$$

The last expression in Eq. (19a) results upon imposition of the boundary condition $g_1'(x_1) + \alpha_1 g_1(x_1) = 0$, and similarly for $\hat{g}_1$. Equation (19) represents an integral equation for the wave function $\hat{g}_2$ that satisfies the homogeneous equation (13) and the required boundary conditions at x_2. Since we are interested only in the functional form of $\hat{g}_2$, and the integral equation is linear, we may put $C_2 = 1$. If $\zeta \approx \gamma$ for $x_1 < x < x_3 < x_2$, the contribution from the first integral is negligible in this range and one finds that

$$\hat{g}_2(x) = g_2(x) + \frac{g_1(x)}{W} \int_{x_1}^{x_2} (\zeta - \gamma)g_2\hat{g}_2\,dx', \qquad x_1 \le x \le x_3. \tag{20}$$

This formulation is useful for scattering problems (with $x_1 \to -\infty$, $x_2 \to \infty$) wherein one seeks to assess the influence on the scattered field of a deviation of the medium parameter $\gamma(x)$ from its unperturbed value $\zeta(x)$ for which the solution g_2 is known. To a first approximation, one finds, from Eq. (20),

$$\hat{g}_2(x) \cong g_2(x) + g_1(x) \left\{ \frac{1}{W} \int_{x_1}^{x_2} [\zeta(x') - \gamma(x')] g_2^2(x') \, dx' \right\} \tag{21}$$

for observation points $x < x_3$. Similarly, if $\zeta \approx \gamma$ for $x_4 < x \leq x_2$, one has, from Eq. (19),

$$\hat{g}_2(x) = g_2(x) \left\{ 1 + \frac{1}{W} \int_{x_1}^{x_2} [\zeta(x') - \gamma(x')] g_1(x') \hat{g}_2(x') \, dx' \right\}, \qquad x_4 < x. \tag{22}$$

To a first approximation, $\hat{g}_2(x')$ in the integral is replaced by $g_2(x')$.

Equation (21) yields the correction to the wave reflected from the inhomogeneity described by $\zeta - \gamma$, while Eq. (22) shows the perturbation of the transmitted wave. As an illustration, let us consider an infinite medium ($x_1 \to -\infty$, $x_2 \to \infty$) with a variable permittivity $\epsilon(x)$ that approaches the constant value ϵ_0 as $x \to \pm\infty$, or in terms of (13), a medium for which

$$\gamma(x) = k_0^2, \qquad x \to \pm\infty, \tag{23a}$$

where k_0 is the propagation constant corresponding to ϵ_0. We choose as the unperturbed problem

$$\zeta(x) = k_0^2, \qquad -\infty < x < \infty. \tag{23b}$$

The unperturbed solution $g_2(x)$ which satisfies the radiation condition at $x_2 = \infty$ is given by†

$$g_2(x) = e^{ik_0 x} \qquad (e^{-i\omega t} \text{ dependence}), \tag{24a}$$

while the corresponding solution for $x_1 = -\infty$ is

$$g_1(x) = e^{-ik_0 x}. \tag{24b}$$

Thus, $W = 2ik_0$, and from Eqs. (20) and (22),‡

$$\hat{g}_2(x) = e^{ik_0 x} + \bar{\Gamma} e^{-ik_0 x}, \qquad \bar{\Gamma} = \frac{1}{2ik_0} \int_{-\infty}^{\infty} [k_0^2 - \gamma(x')] e^{ik_0 x'} \hat{g}_2(x') \, dx',$$

$$x \to -\infty, \tag{25a}$$

$$\hat{g}_2(x) = T e^{ik_0 x}, \qquad T = 1 + \frac{1}{2ik_0} \int_{-\infty}^{\infty} [k_0^2 - \gamma(x')] e^{-ik_0 x'} \hat{g}_2(x') \, dx',$$

$$x \to +\infty. \tag{25b}$$

$\bar{\Gamma}$ in Eq. (25a) represents the plane-wave reflection coefficient of the inhomogeneous medium, while T in Eq. (25b) is the transmission coefficient. To a first approximation, the Born approximation,[14] one has for a slightly inhomogeneous medium,

---

† In applications to wave propagation throughout this section, the time dependence is $\exp(-i\omega t)$.

‡ To avoid confusion with the gamma function $\Gamma(x)$, the reflection coefficient in this section is denoted by $\bar{\Gamma}$.

$$\bar{\Gamma} \simeq \frac{1}{2ik_0} \int_{-\infty}^{\infty} [k_0^2 - \gamma(x')]e^{i2k_0 x'}\, dx', \tag{26a}$$

$$T \cong 1 + \frac{1}{2ik_0} \int_{-\infty}^{\infty} [k_0^2 - \gamma(x')]\, dx', \tag{26b}$$

where it is implied that $(k_0^2 - \gamma)$ is integrable over the infinite interval.

Since $\hat{g}_2(x)$ plays the role of a voltage V (or current I) on a non-uniform transmission line (see Figs. 3.3.1 and 3.3.2), the constancy of real power flow on a non-dissipative line (k_0, γ real) implies that

$$\mathrm{Re}\,(VI^*) = \mathrm{Re}\left[\frac{-1}{ik_x^*(x)Z^*(x)}\, V(x)\frac{dV^*(x)}{dx}\right] = \text{constant}, \tag{26c}$$

where k_x and Z are the propagation constant and characteristic impedance, respectively [see Eqs. (3.3.2)]. Application of Eq. (26c) to Eqs. (25) yields the conservation condition

$$1 - |\bar{\Gamma}|^2 = |T|^2, \tag{26d}$$

satisfied by the magnitudes of the reflection and transmission coefficients. Because k_0 and γ are real, T in Eq. (26b) has a magnitude greater than unity, so that this approximation does not satisfy the conservation-of-energy requirement. This is not surprising since, to a first order of approximation, $1 + i\alpha = (1 - i\alpha)^{-1}$, where α is a small quantity, and the two results are equivalent to $O(\alpha)$. If one writes $T = (1 - i\alpha)^{-1}$, where $i\alpha$ represents the second term on the right-hand side of Eq. (26b), the resulting expression satisfies $|T| < 1$. These considerations highlight the fact that approximate results obtained by a perturbation method do not necessarily obey all the conditions satisfied by the exact solution.

For an alternative derivation of T and $\bar{\Gamma}$, consider Eq. (4) with the Green's function $G(x', x)$ now chosen so that Eq. (5) is satisfied when $x > x'$. One obtains instead of Eq. (6),

$$u(x) = \left\{G(x_2, x)\frac{du(x_2)}{dx_2} - u(x_2)\frac{d}{dx_2} G(x_2, x)\right\} + \int_x^{x_2} [\gamma - \zeta]G(x', x)u(x')\, dx', \tag{27a}$$

with $G(x', x)$ given by the negative of the first expression in Eq. (8). If $x_2 \to \infty$ and we choose for $u(x_2)$ the solution that behaves like $T \exp(ik_0 x_2)$, where T is a constant, then via Eqs. (23), (24), and (27a),

$$u(x) = Te^{ik_0 x} + \frac{e^{ik_0 x}}{2ik_0}\int_x^{\infty} (\gamma - \zeta)e^{-ik_0 x'}u(x')\, dx' - \frac{e^{-ik_0 x}}{2ik_0}\int_x^{\infty} (\gamma - \zeta)e^{ik_0 x'}u(x')\, dx', \tag{27b}$$

whence, for the first-order approximation as $x \to -\infty$,

$$u(x) \cong e^{ik_0 x}\left[1 + \frac{1}{2ik_0}\int_{-\infty}^{\infty} (\gamma - \zeta)\, dx'\right]T - e^{-ik_0 x}\left[\frac{T}{2ik_0}\int_{-\infty}^{\infty} (\gamma - \zeta)e^{i2k_0 x'}\, dx'\right]. \tag{28}$$

For a unit amplitude incident wave, the coefficient of $\exp(ik_0 x)$ is set equal to unity and the resulting expression for T, with $\zeta = k_0^2$, is equal to the complex conjugate of the reciprocal of Eq. (26b); as observed above, they agree to first order. The coefficient of the $\exp(-ik_0 x)$ term is equal to the reflection coefficient $\bar{\Gamma}$ and agrees to first order with that in Eq. (26a).

3.5b   The Comparison Equation

The success of the method described in the preceding section rests upon the ability to choose a comparison function $\zeta(x)$ so that the difference function $[\gamma(x) - \zeta(x)]$ is small. The idea of "smallness" is to be defined more precisely in this section, and it will be convenient to exhibit explicitly a large positive parameter Ω with respect to which an asymptotic solution of Eq. (2) may be found. For example, if Ω is the free-space wavenumber k_0, the largeness of Ω describes short wavelength propagation phenomena for which the medium parameters appear to be slowly varying. Thus, we assume that

$$\gamma(x) = \Omega^2[\alpha_0(x) + \alpha_1(x, \Omega)] \equiv \Omega^2\alpha(x, \Omega),$$

$$\zeta(x) = \Omega^2[\beta_0(x) + \beta_1(x, \Omega)] \equiv \Omega^2\beta(x, \Omega), \tag{29}$$

where $\alpha_1(x, \Omega)$ and $\beta_1(x, \Omega)$ are functions of x and Ω that tend to zero as $\Omega \to \infty$.

To facilitate the determination of ζ or β, it is convenient to transform the differential equation (7) so that the resulting $\zeta(x)$ involves an arbitrary transformation function $\varphi(x)$ chosen so as to make $\zeta(x)$ a good comparison function.[15-17] To this purpose, assume that

$$\left[\frac{d^2}{d\xi^2} + \Omega^2\bar{\beta}(\xi, \Omega)\right]\bar{g}(\xi) = 0 \tag{30}$$

defines known solutions $\bar{g}(\xi)$ and introduce *a* variable x via

$$\xi = \varphi(x), \qquad \text{i.e., } d\xi = \varphi'(x)\,dx, \tag{31}$$

where the prime denotes differentiation with respect to x. With

$$\bar{g}(\xi) = \sqrt{\varphi'}\, g(x), \tag{32}$$

substitution of Eqs. (31) and (32) into Eq. (30) yields

$$\left[\frac{d^2}{dx^2} + \Omega^2\beta(x, \Omega)\right]g(x) = 0, \tag{33}$$

where

$$\beta(x, \Omega) = \bar{\beta}(\varphi, \Omega)\varphi'^2 + \frac{1}{2\Omega^2}\{\varphi, x\}, \tag{33a}$$

$\{\varphi, x\}$ being the Schwarzian derivative of φ such that

$$\{\varphi, x\} = \frac{\varphi'''}{\varphi'} - \frac{3}{2}\left(\frac{\varphi''}{\varphi'}\right)^2 = -2\sqrt{\varphi'}\,\frac{d^2}{dx^2}\frac{1}{\sqrt{\varphi'}}. \tag{33b}$$

Since the solutions of Eq. (30) are assumed to be known, the solutions of Eq. (33) are likewise known for any function $\varphi(x)$ that is thrice differentiable and has $\varphi' \neq 0$ in the interval $x_2 - x_1$.

It is now recognized that for large Ω, Eq. (33) resembles the desired equation

$$\left[\frac{d^2}{dx^2} + \Omega^2 \alpha(x, \Omega)\right] \hat{g}(x) = 0, \tag{34}$$

if $\varphi(x)$ is chosen so that

$$\varphi'^2 \bar{\beta}_0(\varphi) = \alpha_0(x), \qquad \text{i.e., } \int_{\varphi_1}^{\varphi} \sqrt{\bar{\beta}_0(\varphi)} \, d\varphi = \int_{x_1}^{x} \sqrt{\alpha_0(x)} \, dx, \tag{35}$$

where $\bar{\beta}_0(\varphi) = \lim_{\Omega \to \infty} \bar{\beta}(\varphi, \Omega)$. This choice assures that the dominant terms in Eqs. (33) and (34) are identical and that the difference function $\zeta - \gamma$ is small compared to γ [i.e., $(\zeta - \gamma)/\gamma \to 0$ as $\Omega \to \infty$]. Thus, an approximate solution for $\hat{g}$ is given by

$$\hat{g}(x) \approx g(x) = \frac{\bar{g}[\varphi(x)]}{\sqrt{d\varphi(x)/dx}}. \tag{36}$$

where

$$\varphi(x) = \int_{x_0}^{x} \sqrt{\frac{\alpha_0(\eta)}{\bar{\beta}_0[\varphi(\eta)]}} \, d\eta. \tag{36a}$$

Since $\varphi'(x)$ and $1/\varphi'(x)$ are to remain bounded in the interval, the comparison function $\bar{\beta}_0[\varphi(x)]$ must be chosen so that its zeros and poles occur at the same locations as those of the given function $\alpha_0(x)$.[†] Whether $g(x)$ in Eq. (36) is, in fact, a good approximation to $\hat{g}(x)$ can be assessed by examining the order of magnitude of the correction terms via the integral equation procedure in Sec. 3.4a. Although it is difficult to make a general statement,[18] one observes that $\{\varphi, x\}$ in Eq. (33b) is small when $\varphi'(x)$ is a slowly varying function over the x interval under consideration; Eq. (36) may then be expected to apply. The previously mentioned matching of the singularities and zeros of $\alpha(x, \Omega)$ plays an important role in this connection.

3.5c Various Comparison Functions

$\alpha_0(x)$ *has no zeros or poles* (WKB solution)

As an illustration, let us suppose that $\alpha(x, \Omega) \equiv \alpha_0(x)$ has no zeros and is analytic in the interval under consideration (see Fig. 3.5.1). $\bar{\beta}$ may then be chosen as a constant,[‡] $\bar{\beta} = 1$, whence $\bar{g}(\xi) = \exp(\pm i\Omega\xi)$. Equation (36) yields

---

[†] Analogous considerations enter into the asymptotic evaluation of integrals where a given unknown integral is compared to a known and simpler one having a similar integrand in the vicinity of the stationary points (Sec. 4.1).

[‡] A better approximation may be obtained by choosing a comparison function which provides a "better fit" than $\beta = $ constant (e.g., a piecewise linear or a parabolic profile; see also Sec 3.6b).

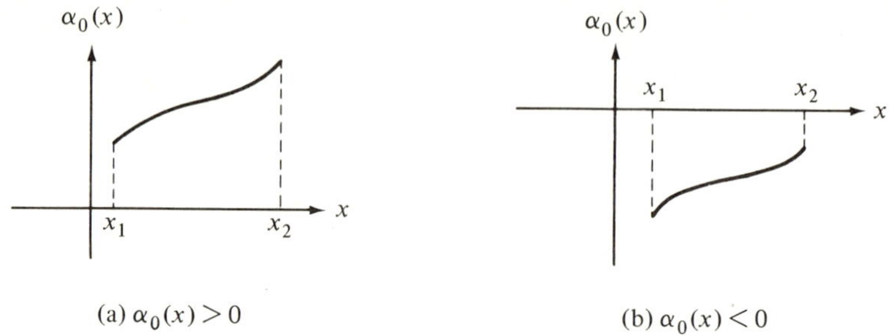

(a) $\alpha_0(x) > 0$ (b) $\alpha_0(x) < 0$

FIG. 3.5.1   Profiles without zeros or singularities.

$$\hat{g}(x) \cong g(x) = \frac{\exp\left[\pm i\Omega \int_{x_0}^{x} \sqrt{\alpha_0(\eta)}\, d\eta\right]}{\sqrt[4]{\alpha_0(x)}}, \qquad \Omega \to \infty, \qquad (37)$$

where x_0 is arbitrary. We choose arg $\alpha_0(x) = 0$ if $\alpha_0(x) > 0$, arg $\alpha_0(x) = \pi$ if $\alpha_0(x) < 0$, and take the principal value of all roots. This form of $\hat{g}(x)$ represents the *WKB approximation*, and satisfies *exactly* the differential equation (33), with $\bar{\beta} = 1$, $\varphi' = \sqrt{\alpha_0}$; moreover,

$$\gamma(x) - \zeta(x) = -\frac{1}{2\Omega^2}\{\varphi, x\}. \qquad (38)$$

The Green's function $G(x', x)$ of Eq. (8) becomes

$$G(x', x) = \frac{\sin \int_{x}^{x'} \sqrt{\gamma(\eta)}\, d\eta}{\sqrt[4]{\alpha_0(x)\alpha_0(x')}\,\Omega}, \qquad \gamma = \Omega^2\alpha_0. \qquad (39a)$$

The Green's function in Eq. (16) can be constructed in a similar manner. For an unbounded region, α_0 is positive, and for a time dependence $\exp(-i\omega t)$, the positive and negative signs in Eq. (37) distinguish outgoing wave solutions at $x = +\infty$ and $x = -\infty$, respectively (Ω represents the wavenumber k_0). Thus, from Eq. (16),

$$G(x, x') = \frac{\exp\left[i\Omega \int_{x_<}^{x_>} \sqrt{\alpha_0(\eta)}\, d\eta\right]}{-2i\Omega[\alpha_0(x)\alpha_0(x')]^{1/4}}. \qquad (39b)$$

On the other hand, if the region is bounded at $x = d$, and if $g = 0$ at $x = d$, the solution g_1 (which satisfies the boundary condition at $x = d$) is of the form

$$g_1(x) = \frac{\sin\left[\Omega \int_{d}^{x} \sqrt{\alpha_0(\eta)}\, d\eta\right]}{[\alpha_0(x)]^{1/4}}, \qquad (39c)$$

while g_2 remains as in Eq. (37) (with $x_0 = d$, for convenience). Thus,

$$G(x, x') = \frac{\exp\left[i\Omega \int_{x_<}^{x_>} \sqrt{\alpha_0(\eta)}\, d\eta\right]}{-2i\Omega[\alpha_0(x)\alpha_0(x')]^{1/4}} - \frac{\exp\left[i\Omega \left\{\int_{d}^{x} + \int_{d}^{x'}\right\} \sqrt{\alpha_0(\eta)}\, d\eta\right]}{-2i\Omega[\alpha_0(x)\alpha_0(x')]^{1/4}}, \qquad (39d)$$

which formulation exhibits directly the breakup into incident and reflected waves; the factor (-1) multiplying the second term represents the reflection coefficient at the boundary. If $\alpha_0(x_0) = 0$ and $\alpha_0(x) \gtrless 0$ when $x \gtrless x_0$, the solution g_1 that remains bounded for $x < x_0$ is given below in Eq. (47a); it has the same form as in Eq. (39c) except that $d = x_0$ and a phase shift of $\pi/4$ is added to the sine argument. Thus, for $x, x' > x_0$,

$$G(x, x') = \frac{\exp\left[i\Omega \int_{x_<}^{x_>} \sqrt{\alpha_0(\eta)}\, d\eta\right]}{-2i\Omega[\alpha_0(x)\alpha_0(x')]^{1/4}} + e^{-i\pi/2}\, \frac{\exp\left[i\Omega\left\{\int_{x_0}^{x} + \int_{x_0}^{x'}\right\} \sqrt{\alpha_0(\eta)}\, d\eta\right]}{-2i\Omega[\alpha_0(x)\alpha_0(x')]^{1/4}}.$$

$$(39e)$$

In this instance, the effective reflection coefficient at the turning point† x_0 is $-i$. The formulas in Eqs. (37)–(39) evidently fail when α_0 has a zero in the interval. The continuation of a given form of the solution in Eq. (37) from positive to negative values of $\alpha_0(x)$ requires the knowledge of appropriate connection formulas, which are also discussed below [see Eqs. (47)].

$\alpha_0(x)$ *has a simple zero*

If $\alpha(x, \Omega) \equiv \alpha_0(x)$ is analytic in the interval and has a simple zero, conveniently chosen at $x = 0$ (see Fig. 3.5.2), the simplest comparison function is $\bar{\beta}(\xi, \Omega) \equiv \bar{\beta}_0(\xi) = \xi = \varphi(x)$, with $\varphi(0) = 0$. From Eq. (35),

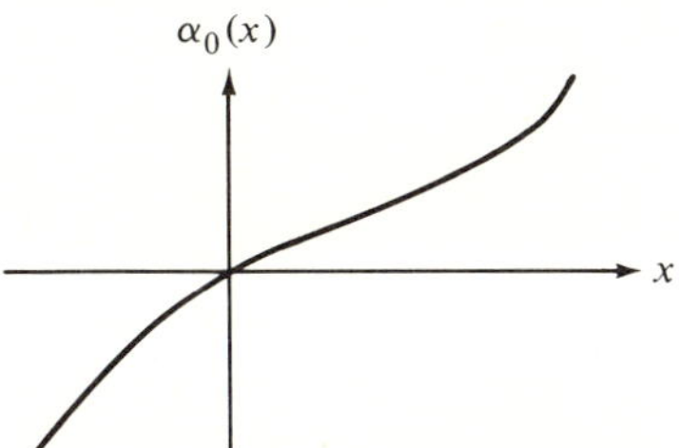

FIG. 3.5.2   Profile with simple zero.

$$\alpha_0^{1/2} = \varphi'\varphi^{1/2} = \frac{2}{3}\frac{d}{dx}\varphi^{3/2}, \tag{40}$$

so

$$\varphi(x) = \left[\frac{3}{2}\int_0^x \sqrt{\alpha_0(\eta)}\, d\eta\right]^{2/3}, \tag{41}$$

where the condition $\varphi(0) = 0$ has been utilized in determining the lower integration limit. To resolve the ambiguity introduced by $\sqrt{\alpha_0(x)}$ we draw a branch cut from zero to $-\infty$ along the negative real axis in the complex x plane and define arg $\alpha_0(x) = 0$ when x is positive real. When arg x increases by π, x is located on the upper shore of the cut and

---

†If $\alpha_0(x)$ has a zero at x_0, this point is called a "turning point" of the differential equation (34).

$$\int_0^{|x|e^{i\pi}} \sqrt{\alpha_0(\eta)}\, d\eta = e^{i\pi} \int_0^{|x|} \sqrt{\alpha_0(\zeta e^{i\pi})}\, d\zeta = e^{i3\pi/2} \int_0^{|x|} |\sqrt{\alpha_0(\zeta)}|\, d\zeta, \qquad x < 0,$$

$$(42a)$$

so

$$\varphi(x) \gtrless 0 \qquad \text{when } x \gtrless 0. \tag{42b}$$

The solution of the differential equation

$$\left(\frac{d^2}{d\xi^2} + \Omega^2 \xi\right) \bar{g}(\xi) = 0 \tag{43}$$

is given by

$$\bar{g}(\xi) = \frac{\text{Ai}}{\text{Bi}}(-\Omega^{2/3}\xi), \tag{44}$$

where $\text{Ai}(\sigma)$ and $\text{Bi}(\sigma)$ are the Airy functions, whose properties are discussed in detail in Secs. 4.2e and 6.A5. Thus, from Eq. (36),

$$\hat{g}(x) \approx g(x) = \left[\frac{3}{2}\int_0^x \sqrt{\alpha_0(\eta)}\, d\eta\right]^{1/6} [\alpha_0(x)]^{-1/4} \frac{\text{Ai}}{\text{Bi}}\left\{-\left[\frac{3}{2}\Omega\int_0^x \sqrt{\alpha_0(\eta)}\, d\eta\right]^{2/3}\right\},$$

$$(45)$$

with the principal value taken in all roots. Since $\text{Ai}(\sigma)$ and $\text{Bi}(\sigma)$ are single-valued functions of σ, arg σ may be taken as $\pm\pi$ when $\sigma < 0$.

Since $\alpha_0(x) \sim x$ near $x = 0$, it follows that $\varphi(x) \sim x$ near $x = 0$ so that the expression for $\hat{g}(0)$ exists. For $|x|$ sufficiently large so that $\Omega^{2/3}|\varphi| \gg 1$, one may employ the asymptotic formulas (see Sec. 4.2e),

$$\text{Ai}(-\sigma) \sim \begin{cases} \dfrac{1}{\sqrt{\pi}\,\sigma^{1/4}} \sin\left(\dfrac{2}{3}\sigma^{3/2} + \dfrac{\pi}{4}\right), & \sigma \to +\infty, \\[2ex] \dfrac{1}{2\sqrt{\pi}\,(-\sigma)^{1/4}} e^{-(2/3)(-\sigma)^{3/2}}, & -\sigma \to +\infty \end{cases} \tag{46a}$$

$$\text{Bi}(-\sigma) \sim \begin{cases} \dfrac{1}{\sqrt{\pi}\,\sigma^{1/4}} \cos\left(\dfrac{2}{3}\sigma^{3/2} + \dfrac{\pi}{4}\right), & \sigma \to +\infty, \\[2ex] \dfrac{1}{\sqrt{\pi}\,(-\sigma)^{1/4}} e^{(2/3)(-\sigma)^{3/2}}, & -\sigma \to +\infty, \end{cases} \tag{46b}$$

to reduce the Ai part of the expression for $\hat{g}(x)$ in Eq. (45) for $x > 0$ to

$$\hat{g}(x) \approx \frac{1}{\sqrt{\pi}} \frac{1}{[\alpha_0(x)]^{1/4}\Omega^{1/6}} \sin\left[\Omega\int_0^x \sqrt{\alpha_0(\eta)}\, d\eta + \frac{\pi}{4}\right], \qquad \Omega^{2/3}\varphi \gg 1, \tag{47}$$

and similarly for the Bi part. Thus, if the right-hand side of Eq. (45) is multiplied by $\Omega^{1/6}$, the resulting expression can be made to go smoothly from $x = 0$ to $|x|$ large, coinciding in the latter range with the WKB formula in Eq. (37). We then obtain the previously mentioned connection relations for the WKB formulas when the differential equation has a simple turning point. Application of Eqs. (46) to the general solution $\hat{g}(x)$ in Eq. (45) comprising both the Ai and Bi functions in the combination $[C_1\text{Ai} + C_2\text{Bi}]$ yields, to within irrelevant constants,

$$\hat{g}(x) \cong \begin{cases} \dfrac{1}{\sqrt[4]{\alpha_0(x)}} \left\{ C_1 \sin\left[\Omega\psi(x) + \dfrac{\pi}{4}\right] + C_2 \cos\left[\Omega\psi(x) + \dfrac{\pi}{4}\right] \right\}, & \alpha_0(x) > 0, \\[4mm] \dfrac{1}{\sqrt[4]{|\alpha_0(x)|}} \left\{ \dfrac{C_1}{2} e^{-\Omega|\psi(x)|} + C_2 e^{\Omega|\psi(x)|} \right\}, & \alpha_0(x) < 0, \end{cases}$$

$$\tag{47a}$$

where

$$\psi(x) = \int_0^x \sqrt{\alpha_0(\eta)}\, d\eta, \tag{47b}$$

and $C_{1,2}$ are arbitrary constants that are determined from the boundary conditions. For example, if the region extends to $\pm\infty$, the requirement of a bounded solution leads to $C_2 = 0$; the corresponding physical phenomenon involves propagating waves in the region $x > 0$ which are totally reflected in the vicinity of $x = 0$. Since the argument of the Airy functions in Eq. (45) is positive when $x < 0$, Ai yields an exponentially decaying, and Bi an exponentially increasing, solution for $x < 0$. Upon multiplication by $\Omega^{1/6}$, these results can be made to coincide with the corresponding WKB solutions in Eq. (37).

To construct the Green's function in Eq. (16) we select solutions $g_1(x)$ and $g_2(x)$ that satisfy boundary conditions at the endpoints $x = x_1$ and $x = x_2 > x_1$, respectively. Let us suppose that the region is infinitely extended (i.e., $x_1 = -\infty$ and $x_2 = +\infty$) and that $\alpha_0(x)$ approaches the value of unity as $x \to +\infty$. Since a source placed at the point x' radiates fields which are bounded at $x \to \pm\infty$, the Ai function must be chosen for $g_1(x)$. To satisfy a radiation condition at $x = +\infty$, the wave function $g_2(x)$ must vary like $\exp(+i\Omega x)$ and therefore involves via Eqs. (46) both the Ai and Bi functions. Thus,

$$g_1(x) = \frac{1}{[\alpha_0(x)]^{1/4}} [Y(x)]^{1/6} \mathrm{Ai}[-(\Omega Y)^{2/3}], \qquad Y(x) = \frac{3}{2}\int_0^x \sqrt{\alpha_0(\eta)}\, d\eta, \tag{48a}$$

$$g_2(x) = \frac{1}{[\alpha_0(x)]^{1/4}} Y^{1/6}\{\mathrm{Ai}[-(\Omega Y)^{2/3}] - i\,\mathrm{Bi}[-(\Omega Y)^{2/3}]\}, \tag{48b}$$

where the values of $Y(x)$ and $[\alpha_0(x)]^{1/4}$ for $x \gtrless 0$ have been defined in Eqs. (42). From the asymptotic form of $\mathrm{Ai}(z)$ and $\mathrm{Bi}(z)$ in Eqs. (46) as $z \to \infty$, one finds for the Wronskian

$$\mathrm{Ai}(z)\frac{d}{dz}\mathrm{Bi}(z) - \mathrm{Bi}(z)\frac{d}{dz}\mathrm{Ai}(z) = \frac{1}{\pi}, \tag{49}$$

whence for the functions in Eqs. (48),

$$W(g_1, g_2) = \frac{i}{\pi}\left(\frac{2}{3}\right)^{1/3}. \tag{50}$$

$\alpha_0(x)$ *has two neighboring simple zeros*

If $\alpha(x, \Omega) = \alpha_0(x)$ has isolated simple zeros (see Fig. 3.5.3), the asymptotic approximation of $\hat{g}(x)$ as $\Omega \to \infty$ can be constructed by a combination of the WKB- and Airy-function formulations. The WKB approximation is valid in those regions where $\alpha_0(x) \neq 0$, and the Airy-function approximation is used

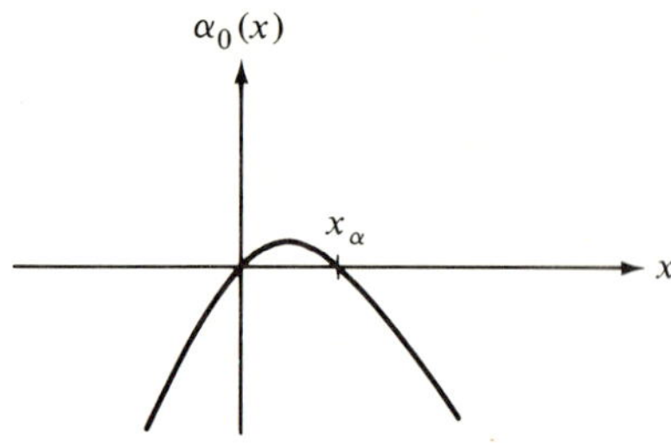

FIG. 3.5.3   Profile with two simple zeros.

to match smoothly through each turning point region $\alpha_0(x) \approx 0$. However, if two simple zeros of $\alpha_0(x)$ located at $x = 0$ and $x = x_\alpha$, respectively, tend to coalesce, the expression in Eq. (45) loses its validity and diverges as $x_\alpha \to 0$. The simplest comparison function when $\alpha_0(x)$ has two adjacent zeros is

$$\bar{\beta}(\xi, \Omega) \equiv \bar{\beta}_0(\xi) = (a^2 - \xi^2), \tag{51}$$

with no loss in generality incurred by the symmetrical location of the zeros at $\xi = \pm a$. To ensure that these zeros in the ξ plane correspond to those at $x = 0$, x_α in the x plane, we return to Eq. (35):

$$\int_0^x \sqrt{\alpha_0(x)}\, dx = \int_{-a}^\varphi \sqrt{\bar{\beta}_0(\varphi)}\, d\varphi = \frac{1}{2}\left\{ \varphi\sqrt{a^2 - \varphi^2} + a^2 \sin^{-1}\frac{\varphi}{a} + \frac{a^2\pi}{2} \right\}. \tag{52a}$$

The choice of the lower integration limits implies the correspondence of $\varphi = -a$ and $x = 0$. To enforce that $\varphi = a$ when $x = x_\alpha$, set

$$\int_0^{x_\alpha} \sqrt{\alpha_0(x)}\, dx = \int_{-a}^a \sqrt{\bar{\beta}_0(\varphi)}\, d\varphi = \frac{\pi a^2}{2}, \tag{52b}$$

which equation serves to determine the value of a.[19,20]

The solution of the differential equation

$$\left[\frac{d^2}{d\xi^2} + \Omega^2(a^2 - \xi^2) \right] \bar{g}(\xi) = 0 \tag{53}$$

is given by Weber's parabolic cylinder functions

$$\bar{g}(\xi) = D_v(\pm\sqrt{2\Omega}\,\xi), \qquad v = \tfrac{1}{2}(\Omega a^2 - 1). \tag{54}$$

When $\xi \not\approx \pm a$, these functions have the asymptotic approximation

$$D_v(\sqrt{2\Omega}\xi) \sim \frac{e^{-\Omega a^2(\tau + 1/4)}}{[2\Omega(\xi^2 - a^2)]^{1/4}(\Omega a^2/2)^{-\Omega a^2/4}}\left[1 + O\left(\frac{1}{\Omega a^2}\right) \right],$$

$$|\arg(\xi\sqrt{\Omega})| \le \frac{\pi}{2}, \tag{55a}$$

$$a^2\tau = a^2 \int_1^{\xi/a} \sqrt{t^2 - 1}\, dt = \frac{\xi}{2}\sqrt{\xi^2 - a^2} - \frac{a^2}{2}\ln\left[\frac{\xi}{a} + \sqrt{\left(\frac{\xi}{a}\right)^2 - 1} \right], \tag{55b}$$

where the latter expression constitutes the analytic continuation of the right-hand side of Eq. (52a) to the range $(\xi = \varphi) > a$. This result is nothing but the WKB approximation in Eq. (37), normalized so that the function $D_\nu(z)$ reduces to the large argument form $z^\nu \exp(-z^2/4)$ as $|z/\nu| \to \infty$.[21] Alternatively, when $\pi/2 \leq \arg(\xi\sqrt{\Omega}) \leq \pi$, one may utilize the large z behavior

$$D_\nu(z) \sim e^{-z^2/4} z^\nu \left[1 + O\left(\frac{1}{z}\right)\right] - \frac{\sqrt{2\pi}}{\Gamma(-\nu)} e^{\nu\pi i} \frac{e^{z^2/4}}{z^{\nu+1}} \left[1 + O\left(\frac{1}{z}\right)\right], \qquad (56)$$

to select the proper WKB approximation in this range. In the vicinity of the turning points $\xi = a$ or $\xi = -a$, one may employ the Airy-function approximation in the preceding section, and only when $a \to 0$ is it necessary to retain the parabolic cylinder function intact. The various asymptotic representations for $D_\nu(z)$ for large (and possibly complex) values of ν and z have been discussed in detail by Olver.[22]

$\alpha_0(x)$ *has a simple pole*

If $\alpha(x, \Omega) = \alpha_0(x)$ has a simple pole at $x = 0$ (see Fig. 3.5.4), the simplest comparison function is†

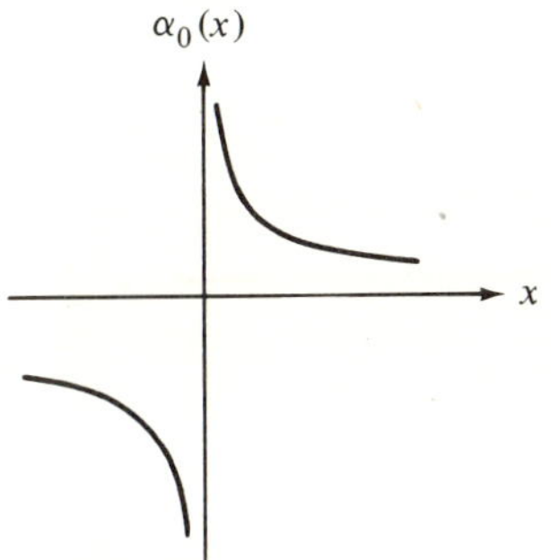

FIG. 3.5.4   Profile with simple pole.

$$\bar{\beta}(\xi, \Omega) = \frac{1}{\xi}, \qquad (57)$$

which yields the following solutions of Eq. (30):

$$\bar{g}(\xi) = \sqrt{\xi}\, H_1^{(1,2)}(2\Omega\sqrt{\xi}). \qquad (58)$$

Despite the singularity of $\bar{\beta}$, the solution $\bar{g}(\xi)$ is bounded at $\xi = 0$. Since $\bar{\beta}_0(\varphi) = 1/\varphi$, one obtains, from Eq. (35),

$$\frac{d\varphi}{\sqrt{\varphi}} = \sqrt{\alpha_0(x)}\, dx, \qquad (59)$$

or, upon integration,

---

†A related comparison function is $\bar{\beta}(\xi, \Omega) = (1/\xi) - [(1 - \nu^2)/4\xi^2\Omega^2]$, which leads to the solutions $\bar{g}(\xi) = \sqrt{\xi}\, H_\nu^{(1,2)}(2\Omega\sqrt{\xi})$. This comparison function has a double pole at $\xi = 0$ and a simple zero at $(1 - \nu^2)/4\Omega^2$.

$$\varphi(x) = \tfrac{1}{4}\left[\int_0^x \sqrt{\alpha_0(\eta)}\, d\eta\right]^2. \tag{60}$$

Thus, from Eq. (36),

$$\hat{g}(x) \approx g(x) = \left[\frac{\int_0^x \sqrt{\alpha_0(\eta)}\, d\eta}{\sqrt{\alpha_0(x)}}\right]^{1/2} H_1^{(1,2)}\left[\Omega \int_0^x \sqrt{\alpha_0(\eta)}\, d\eta\right]. \tag{61}$$

For $x \not\approx 0$, one may employ the asymptotic formula for the Hankel function,

$$H_1^{(1,2)}(z) \sim \sqrt{\frac{2}{\pi z}}\, e^{\pm i(z - 3\pi/4)}, \qquad |z| \gg 1, \quad -\pi < \arg z < \pi. \tag{61a}$$

If Eq. (61) is multiplied by $\Omega^{1/2}$, the resulting expression transforms smoothly into the WKB approximation (37), apart from x and Ω independent factors.

To determine the appropriate continuation of $\xi = \varphi(x)$ around the singularity at $\xi = 0$, the permissible range of $\arg \xi$ must be specified. If $\bar{\beta}(\xi, \Omega)$ represents the dielectric constant in the medium and slight loss is assumed, then $\operatorname{Im} \bar{\beta}(\xi, \Omega) \geq 0$ [for a time dependence $\exp(-i\omega t)$]. The function $\bar{\beta} = 1/\xi$ may be replaced by $\bar{\beta} = 1/(\xi - i\delta)$, where δ is a small positive constant, so that the singularity is located above the real ξ axis. As $\delta \to 0$, the path along which ξ varies must be indented below the pole, whence $\arg \xi$ changes from zero to $-\pi$ as ξ passes through zero. Thus, $0 \geq \arg \sqrt{\xi} \geq -\pi/2$; hence $\arg \sqrt{\xi} = -\pi/2$ when $\xi < 0$. To obtain a solution that decays when $\xi < 0$, it is necessary [see Eq. (61a)] to select the $H_1^{(2)}$ function in Eq. (61). Since $\arg \sqrt{\xi} = 0$ when $\xi > 0$, this function represents a wave traveling in the $-\xi$ direction [for the assumed $\exp(-i\omega t)$ dependence], i.e., a wave incident from $\xi = +\infty$. The absence of an $H_1^{(1)}$ contribution implies that there is no reflected wave, and the incident wave is "completely absorbed" by the singularity, although the medium itself is assumed to be lossless. For a possible physical realization of this result, see Reference 23.

$\alpha_0(x)$ *has neighboring simple pole and simple zero*

The confluent hypergeometric function of Whittaker, $W_{l,m}(z)$, which satisfies the differential equation

$$\frac{d^2 W_{l,m}}{dz^2} + \left(-\frac{1}{4} + \frac{l}{z} - \frac{m^2 - \frac{1}{4}}{z^2}\right) W_{l,m} = 0, \tag{62}$$

may serve as the prototype for a general class of comparison functions. For example, the previously mentioned parabolic cylinder function $D_\nu(z)$ may be derived from $W_{l,m}(z)$ via the relation

$$D_\nu(z) = 2^{(2\nu+1)/4} z^{-1/2} W_{l,m}\left(\frac{z^2}{2}\right), \qquad l = \frac{2\nu + 1}{4}, \quad m = -\frac{1}{4}. \tag{63}$$

Of particular interest is the case $m = \pm\frac{1}{4}$, in which instance the coefficient $[(l/z) - \frac{1}{4}]$ has a simple pole at $z = 0$ and a simple zero at $z = 4l$. In the notation of the present chapter, the following comparison function is appropriate when $\alpha(x, \Omega) = \alpha_0(x)$ has a simple pole and an adjacent simple zero:

$$\left[\frac{d^2}{d\xi^2} + \Omega^2\left(1 + \frac{a}{\xi}\right)\right]\bar{g}(\xi) = 0, \tag{64}$$

$$\bar{g}(\xi) = W_{-ia\Omega/2,\,1/2}(i2\Omega\xi). \tag{64a}$$

The resulting profile is similar to that shown in Fig. 3.5.4, except that the branches are asymptotic to the line $\alpha_0(x) = 1$ at $|x| \to \infty$; the left branch crosses the x axis at $x = -a$. For a wave incident from the right, the singularity shields the turning point while the converse is true when the wave impinges from the left. The solution $W_{-l,m}(-z)$ is linearly independent of $W_{l,m}(z)$, and for large z, $W_{l,m}(z)$ has the asymptotic approximation

$$W_{l,m}(z) \sim e^{-z/2}z^l\left[1 + O\left(\frac{1}{z}\right)\right], \qquad |\arg z| < \pi. \tag{65}$$

Detailed asymptotic formulas for various ranges of l, m, and z may be found in References 24 and 25.

An effective permittivity $\alpha(x)$ of the type described here arises in a plasma medium with linearly varying electron density, $(\omega_p/\omega)^2 = 1 + b\xi$, when an infinite static magnetic field is impressed along ξ [see Eq. (7.3.1) with $\epsilon - 1 - (\omega_p^2/\omega^2)$]. E modes in such a medium propagate according to Eqs. (7.2.7), which may be written in the form of Eq. (64), with $\bar{g}$ representing the current.[26]

3.5d Error Bounds on the Approximate Solutions

It can be shown that the expressions in Eqs. (37), (45), (54), and (61) represent the asymptotic approximations of the corresponding functions $\hat{g}(x)$ as $\Omega \to \infty$, i.e.,†

$$\hat{g}(x) \sim g(x)[1 + \epsilon(\Omega)], \qquad \frac{d\hat{g}}{dx} \sim \frac{dg}{dx}[1 + \bar{\epsilon}(\Omega)], \tag{66}$$

where ϵ and $\bar{\epsilon}$ approach zero as $\Omega \to \infty$. While asymptotic estimates are very useful and usually sufficiently accurate to infer the basic physical behavior of the wave solution as $\Omega \to \infty$, it is desirable to have exact error bounds when detailed numerical calculations are involved. Estimates for various approximating functions have been supplied by Olver,[27] whose results for the WKB approximation are given below.

Let $\alpha(x, \Omega)$ be negative real, let $d^2\alpha/dx^2$ be a continuous function of x in the interval $a \leq x \leq b$ (which may be infinite), and let Ω be a positive parameter; then the differential equation (34) has solutions $\hat{g}_{1,2}$ such that

$$\hat{g}_{1,2}(x) = g_{1,2}(x)[1 + \Delta_{1,2}], \tag{67a}$$

$$g_{1,2}(x) = \frac{1}{\sqrt[4]{|\alpha(x, \Omega)|}} \exp\left[\pm\Omega\int_{x_0}^{x}\sqrt{|\alpha(\eta, \Omega)|}\,d\eta\right], \tag{67b}$$

---

†The derivation of formal asymptotic series involving higher-order terms in the WKB approximation is illustrated in Sec. 3.5e.

$$\frac{d}{dx}\hat{g}_{1,2}(x) =$$

$$\pm\Omega\sqrt[4]{|\alpha(x,\Omega)|}\exp\left[\pm\Omega\int_{x_0}^{x}\sqrt{|\alpha(\eta,\Omega)|}\,d\eta\right]\left[1 + 2\chi_{1,2} \pm \frac{d\alpha/dx}{4\Omega|\alpha|^{3/2}}(1 + \Delta_{1,2})\right],$$

$$(68)$$

where the magnitudes of $\Delta_{1,2}$ and $\chi_{1,2}$ are bounded by the numbers

$$|\Delta_{1,2}| \text{ or } |\chi_{1,2}| \leq \exp\left[\frac{1}{2\Omega}F_{1,2}(x,\Omega)\right] - 1,$$

$$(69a)$$

with

$$F_1(x,\Omega) = \int_a^x |\alpha(\eta,\Omega)|^{-1/4}\left|\frac{d^2}{d\eta^2}|\alpha(\eta,\Omega)|^{-1/4}\right|\,d\eta.$$

$$(69b)$$

F_2 is given by the same integral except that the interval of integration runs from x to b. $-a$ and (or) b may be infinite provided that the integrals $F_{1,2}$ converge. A somewhat weaker but more easily applied bound replaces Eq. (69a) by

$$|\Delta_{1,2}| \text{ or } |\chi_{1,2}| \leq \frac{F_{1,2}(x,\Omega)}{2\Omega - \frac{1}{2}F_{1,2}(x,\Omega)},$$

$$(69c)$$

provided that $4\Omega > F_{1,2}$. This formulation makes evident the validity of Eq. (66).

If $\alpha(x,\Omega)$ is positive real, one has instead of the above,

$$\hat{g}_{1,2}(x) = g_{1,2}(x)(1 + \Delta_{1,2}),$$

$$(70a)$$

$$g_{1,2}(x) = \frac{1}{\sqrt[4]{\alpha(x,\Omega)}}\exp\left[\pm i\Omega\int_{x_0}^{x}\sqrt{\alpha(\eta,\Omega)}\,d\eta\right],$$

$$(70b)$$

$$\frac{d}{dx}\hat{g}_{1,2}(x) =$$

$$\pm i\Omega\sqrt[4]{\alpha(x,\Omega)}\exp\left[\pm i\Omega\int_{x_0}^{x}\sqrt{\alpha(\eta,\Omega)}\,d\eta\right]\left[1 + \chi_{1,2} \pm \frac{i\,d\alpha/dx}{4\Omega\alpha^{3/2}}(1 + \Delta_{1,2})\right],$$

$$(71)$$

where $\Delta_1 = \Delta_2^*$, $\chi_1 = \chi_2^*$, and

$$|\Delta_{1,2}| \text{ or } |\chi_{1,2}| \leq \exp\left[\frac{1}{\Omega}F(x,\Omega)\right] - 1 \leq \frac{F(x,\Omega)}{\Omega - \frac{1}{2}F(x,\Omega)},$$

$$(72a)$$

$$F(x,\Omega) = \left|\int_c^x [\alpha(\eta,\Omega)]^{-1/4}\left|\frac{d^2}{d\eta^2}[\alpha(\eta,\Omega)]^{-1/4}\right|\,d\eta\right|.$$

$$(72b)$$

c is an arbitrary point such that $a \leq c \leq b$. The interval (a, b) and the value of c may be infinite provided that the integral F converges. The second inequality in Eq. (72a) represents the weaker bound as in Eq. (69c). Uniform bounds valid in the entire interval $a \leq x \leq b$ are obtained upon replacing the lower and upper limits of integration in (69b) and (72b) by a and b, respectively. From Eqs. (69), $\Delta_1 = 0$ when $x = a$, so the function $\hat{g}_1(x)$ in Eq. (67a) satisfies the boundary condition $\hat{g}_1(a) = g_1(a)$. The boundary conditions on the other functions or their derivatives are ascertained in a similar manner.

3.5e Corrections to the WKB Approximation

In Sec. 3.5c, appropriate comparison functions $g(x)$ are selected as a first asymptotic approximation to the desired function $\hat{g}(x)$ in Eq. (34). Corrections to the first approximation can then be obtained via the iteration procedure in Eq. (10a) based on the integral equations (9). Corrections can alternatively be derived directly from the differential equation. The latter procedure is illustrated here for the WKB approximation when $\alpha(x, \Omega) \equiv \alpha(x)$ is positive (i.e., it has no zeros in the interval under consideration).[28] It is convenient to start from the first-order, source-free transmission line equations (3.3.2) for H modes [for an $\exp(-i\omega t)$ dependence; i.e., $j \rightarrow -i$]:

$$\frac{d\hat{V}(x)}{dx} = i\omega\mu\hat{I}(x), \qquad \frac{d\hat{I}(x)}{dx} = i\omega\epsilon(x)\hat{V}(x), \tag{73}$$

with constant permeability μ and x-dependent permittivity ϵ. Evidently,

$$\left[\frac{d^2}{dx^2} + \Omega^2\alpha(x)\right]\hat{V}(x) = 0, \qquad \Omega^2 = k_0^2 = \omega^2\mu\epsilon_0, \quad \alpha(x) = \frac{\epsilon(x)}{\epsilon_0}, \tag{74}$$

so the voltage function $\hat{V}(x)$ can be identified with the wave function $\hat{g}$, while k_0 and $\epsilon(x)/\epsilon_0$ correspond to Ω and α, respectively (ϵ_0 = constant).

The first approximation to $\hat{V}(x) \equiv \hat{g}(x)$ is given in Eq. (37). To derive corrections, assume that $\hat{V}(x)$ is represented by

$$\hat{V}(x) = [A(x)e^{i\Omega\psi(x)} - B(x)e^{-i\Omega\psi(x)}]\frac{1}{\sqrt[4]{\alpha(x)}}, \tag{75}$$

where $\psi(x)$ is the phase integral

$$\psi(x) = \int_{x_0}^{x} \sqrt{\alpha(\eta)}\, d\eta. \tag{75a}$$

$A(x)$ and $B(x)$ are slowly varying amplitude functions which are to be determined; if they are constant, Eq. (75) reduces to the WKB approximation. Similarly, let

$$\hat{I}(x) = \frac{\Omega}{\omega\mu}\sqrt[4]{\alpha(x)}[A(x)e^{i\Omega\psi(x)} + B(x)e^{-i\Omega\psi(x)}]. \tag{76}$$

Note that the expression for $\hat{I}(x)$ in Eq. (76) is the first-order asymptotic approximation to $(1/i\omega\mu)d\hat{V}/dx$ as $\Omega \rightarrow \infty$, with $\hat{V}$ taken from Eq. (75). For a self-consistent determination of $A(x)$ and $B(x)$, substitute Eqs. (75) and (76) into Eqs. (73) to obtain

$$Ae^{i\Omega\psi} - Be^{-i\Omega\psi} = 4\frac{\alpha}{\alpha'}[A'e^{i\Omega\psi} - B'e^{-i\Omega\psi}], \tag{77a}$$

$$Ae^{i\Omega\psi} + Be^{-i\Omega\psi} = -4\frac{\alpha}{\alpha'}[A'e^{i\Omega\psi} + B'e^{-i\Omega\psi}], \tag{77b}$$

where a prime denotes the x derivative. Addition of these equations yields

$$B'(x) = -\frac{\alpha'(x)}{4\alpha(x)} e^{i2\Omega\psi(x)} A(x), \tag{78a}$$

while subtraction of the second from the first leads to

$$A'(x) = -\frac{\alpha'(x)}{4\alpha(x)} e^{-i2\Omega\psi(x)} B(x). \tag{78b}$$

Since $\psi(x)$ increases with x, the first term in Eq. (75) represents a wave traveling in the $+x$ direction while the second represents a wave traveling in the $-x$ direction. $A(x)$ and $B(x)$ are the corresponding wave amplitudes, and Eqs. (78) show that the variability of the medium parameter $\alpha(x)$ produces coupling between these two wave types. When α'/α is very small, the right-hand sides of Eqs. (78) can be set equal to zero and the resulting $A, B = $ constant yields the lowest-order approximation to $\hat{V}(x)$ and $\hat{I}(x) = (1/i\omega\mu)d\hat{V}/dx$; in this approximation, the waves traveling in the $+x$ and $-x$ directions are uncoupled.

Equations (78) can be solved by the method of successive approximations. They are first converted to coupled integral equations by integrating over x:

$$B(x) = B_0 + \int_{x_2}^{x} \delta_2(x)A(x)\,dx, \qquad A(x) = A_0 + \int_{x_1}^{x} \delta_1(x)B(x)\,dx, \tag{79}$$

with

$$\delta_{1,2}(x) = -\frac{\alpha'}{4\alpha} e^{\mp i2\Omega\psi}, \tag{79a}$$

and $B(x_2) = B_0$, $A(x_1) = A_0$. B_0, A_0 are arbitrary but specified, and $x_{1,2}$ are arbitrary points in the interval. Iteration of the two integrals then leads to the series

$$A(x) = \sum_{n=0}^{\infty} A_n(x), \qquad B(x) = \sum_{n=0}^{\infty} B_n(x), \tag{80}$$

where $A_0(x) \equiv A_0$, $B_0(x) \equiv B_0$, and, for $n \geq 1$,

$$A_n(x) = \int_{x_1}^{x} \delta_1(\xi)B_{n-1}(\xi)\,d\xi, \qquad B_n(x) = \int_{x_2}^{x} \delta_2(\xi)A_{n-1}(\xi)\,d\xi. \tag{81}$$

For conditions analogous to those discussed in connection with Eq. (10), these series are convergent.

Equations (81) have an important physical interpretation: they show that a zeroth-order wave traveling in the $+x$ direction gives rise to a first-order wave traveling in the $-x$ direction, which in turn generates a second-order wave in the $+x$ direction, etc. Thus, each term in the series (80) can be viewed as arising from continuous internal reflections caused by those portions in the inhomogeneous medium which lie between the observation point x and the points x_1 and x_2 where the amplitudes of the waves traveling in the positive and negative x directions, respectively, are assumed to be specified. This process is schematized in Fig. 3.5.5.

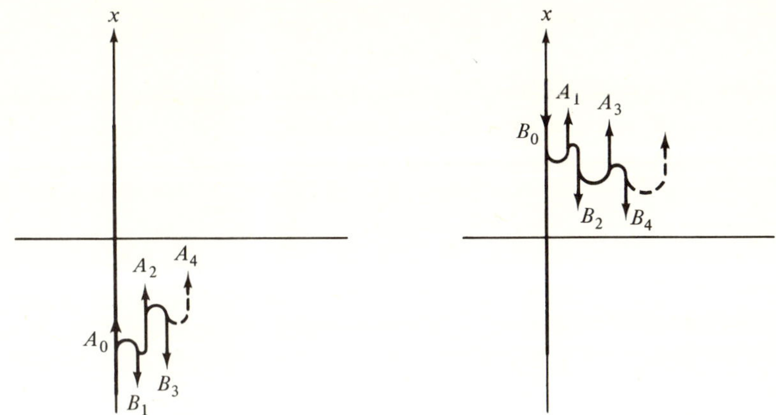

(a) Incident wave in $+x$ direction   (b) Incident wave in $-x$ direction

FIG. 3.5.5 Wave coupling by continuous reflection.

The lowest-order (WKB) approximation is adequate if $|A_1| \ll B_0$ or $|B_1| \ll A_0$. From Eq. (81),

$$|A_1| = \frac{|B_0|}{4}\left|\int_{x_1}^{x} \frac{\alpha'(\xi)}{\alpha(\xi)} e^{-i2\Omega\psi(\xi)} \, d\xi\right| = \frac{|B_0|}{8\Omega}\left|\int_{x_1}^{x} \frac{\alpha'(\xi)}{\alpha(\xi)\psi'(\xi)} \left[\frac{d}{d\xi} e^{-i2\Omega\psi(\xi)}\right] d\xi\right|, \quad (82)$$

where it has been assumed that $\psi'(x) = \sqrt{\alpha(x)}$ [see Eq. (75a)] does not vanish in the interval. Integration by parts yields

$$|A_1| = \frac{|B_0|}{8\Omega}\left|\frac{\alpha'(\xi)}{[\alpha(\xi)]^{3/2}}\right|_{x_1}^{x} + O\left(\frac{1}{\Omega^2}\right), \quad (83)$$

with the validity of the order estimate verified upon repeated integration by parts. Thus, A_1 is $O(1/\Omega)$ as $\Omega \to \infty$, thereby confirming the asymptotic validity of the WKB approximation (see Sec. 4.2b for a discussion of asymptotic expansions). That the maximum value of the first term in Eq. (83) provides a suitable bound for $|A_1|$ when Ω is large can be observed from the fact that the exponential in Eq. (82) is a very rapidly oscillating function while the remaining terms are slowly varying over a period of this oscillation; hence, if $\alpha(\xi)$ varies monotonically, the integral between the limits x_1 and x can be bounded by an integral over a half period, with the function $(\alpha'/\alpha\psi')_{\max}$ removed from within the integration sign. The resulting estimate

$$\left|\frac{\alpha'(x)}{\Omega[\alpha(x)]^{3/2}}\right|_{\max} \ll 1 \quad (84)$$

can thus be taken as a criterion for the validity of the WKB approximation $A(x) \approx A_0$ or $B(x) \approx B_0$ in a medium with monotonically varying properties.

The above formulation allows the voltage reflection coefficient $\bar{\Gamma}$ to be expressed as follows:

$$\bar{\Gamma} = -\frac{B(x)}{A(x)} e^{-i2\Omega\psi(x)} = -\frac{B_1(x) + B_3(x) + B_5(x) + \cdots}{A_0 + A_2(x) + A_4(x) + \cdots} e^{-i2\Omega\psi(x)}, \qquad (85)$$

where it has been assumed that the incident wave travels along the $+x$ direction so that $B_0 \equiv 0$. From this condition it follows that $A(x)$ contains only even-order terms while $B(x)$ contains only odd-order terms. For large Ω, the reflection coefficient magnitude arising from internal reflections is given to a lowest order by $|B_1(x)/A_0|$, which in turn behaves like $|\alpha'/\Omega\alpha^{3/2}|$.

3.6 APPLICATION TO VARIOUS INHOMOGENEITY PROFILES

3.6a Reflection from a Continuous Transition

For illustration of the results in Sec. 3.5e, consider the profile in Fig. 3.6.1, where $\alpha(x) = \alpha_1$ when $x < x_\alpha$, $\alpha(x) = \alpha_2$ when $x > x_\beta$, and α_1 and α_2 are positive constants. $\alpha(x)$ is assumed to be analytic and monotonic in the region

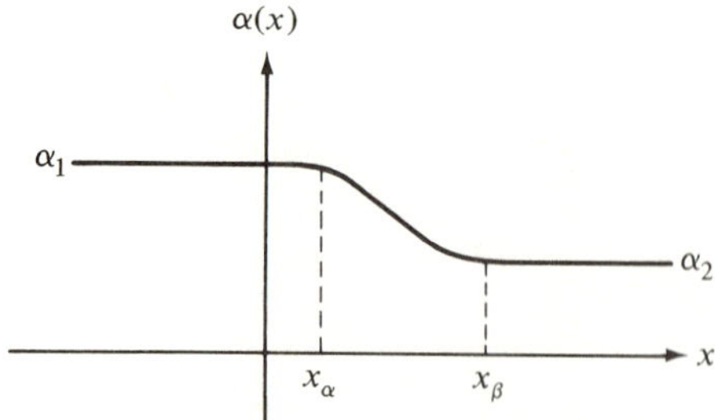

FIG. 3.6.1 Smooth transition.

$x_\alpha < x < x_\beta$. When a wave is incident from the left, the reflection coefficient $\bar{\Gamma}$ observed in the region $x < x_\alpha$ is given in the first approximation by [see Eqs. (3.5.85) and (3.5.81)]

$$\bar{\Gamma}(x) = -\frac{B_1(x)}{A_0} = \left[\frac{1}{4} \int_{x_2}^{x_\alpha} \frac{\alpha'(\xi)}{\alpha(\xi)} e^{i2\Omega\psi(\xi)} \, d\xi \right] e^{-i2\Omega\psi(x)}, \qquad (1)$$

with the upper limit replaced by x_α since $\alpha' = 0$ when $x < x_\alpha$. If the phase reference is chosen so that $\psi(x_1) = 0$, then

$$\bar{\Gamma}(x_1) = \frac{1}{4} \int_{x_\beta}^{x_\alpha} \frac{\alpha'(\xi)}{\alpha(\xi)} e^{i2\Omega\psi(\xi)} \, d\xi. \qquad (2)$$

The choice of the lower limit results from the requirement $B_1 = 0$ at $x = x_2$ in Eq. (3.5.81), whence the reference point x_2 is located in the region where the internal reflections tend to zero (i.e., $x_2 \geq x_\beta$); since $\alpha' = 0$ when $x > x_\beta$, the lower limit may be replaced by x_β. Physically, Eq. (2) implies that the first-order internal reflections are taken into account from the entire region of inhomogeneity lying to the right of the observation point.

The order of magnitude of the reflection coefficient depends on the smoothness of the junction of the inhomogeneous and homogeneous regions at $x = x_{\alpha,\beta}$, as can be verified by repeated integration by parts as in Eq. (3.5.82). If $\alpha(x)$ is continuous at $x_{\alpha,\beta}$ but $\alpha'(x)$ is not, then $\bar{\Gamma}$ is $O(1/\Omega)$; if $\alpha(x)$ and $\alpha'(x)$ are continuous at $x_{\alpha,\beta}$ but $\alpha''(x)$ is not, then $\bar{\Gamma}$ is $O(1/\Omega^2)$. In general, continuity of $\alpha(x)$ and its first N derivatives at $x_{\alpha,\beta}$ implies that $\bar{\Gamma}$ is $O(1/\Omega^{N+1})$. If $\alpha(x)$ is constant for $x < x_\alpha$ or $x > x_\beta$ (i.e., all its derivatives vanish identically in these regions), it is non-analytic when regarded in the entire interval $-\infty < x < \infty$, and thus the reflection coefficient is proportional to some finite inverse power of Ω determined by the order of the first discontinuous derivative. If $\alpha(x)$ is analytic in the entire interval (this requires that $x_\alpha \to -\infty$, $x_\beta \to \infty$), the reflection coefficient vanishes faster than any finite inverse power of Ω, as $\Omega \to \infty$, and is, in fact, exponentially small.

As an example of a smooth transition characterized by an analytic function of x, consider[29]

$$\alpha(x) = \alpha_0 - a \tanh \frac{x}{b}, \tag{3}$$

where α_0, a, and b are positive numbers, with $a \ll \alpha_0$. Since $|\tanh x| \leq 1$ in the interval $-\infty < x < \infty$, we have, to $O(a)$,

$$\sqrt{\alpha(x)} \cong \left(1 - \frac{a}{2\alpha_0} \tanh \frac{x}{b}\right)\sqrt{\alpha_0}, \tag{4a}$$

$$\frac{\alpha'}{\alpha} \cong -\frac{a}{b\alpha_0} \operatorname{sech}^2 \frac{x}{b}. \tag{4b}$$

Thus, from Eq. (1), with x_2 and x_α replaced by $+\infty$ and x, respectively,

$$\bar{\Gamma}(x) \cong \frac{-a}{4b\alpha_0} \int_\infty^x \operatorname{sech}^2 \frac{\xi}{b} \exp\left[i2\Omega \int_x^\xi \sqrt{\alpha(\eta)}\, d\eta\right] d\xi, \tag{5}$$

which expression is to be evaluated for $x \to -\infty$. With Eq. (4a), the integral over η is elementary and yields

$$\frac{1}{\sqrt{\alpha_0}} \int_x^\xi \sqrt{\alpha(\eta)}\, d\eta = (\xi - x) - \frac{ab}{2\alpha_0}\left(\ln \cosh \frac{\xi}{b} - \ln \cosh \frac{x}{b}\right). \tag{6}$$

Since $|x|$ is very large, we may approximate $\cosh(x/b)$ by $\frac{1}{2}\exp(|x|/b)$, whence $\ln \cosh(x/b) \approx |x|/b - \ln 2$. Upon imposing the restriction

$$\Omega a b \ll 1, \tag{7}$$

and recognizing that the major contribution to the integral in Eq. (5) arises from the vicinity of $\xi = 0$, one may write

$$\bar{\Gamma}(x) \cong \frac{a}{4b\alpha_0} \exp\left[-i2\bar{\Omega}\left(1 + \frac{a}{2\alpha_0}\right)x\right] I \qquad \text{as } x \to -\infty, \tag{8a}$$

where

$$I = \int_{-\infty}^{\infty} \operatorname{sech}^2 \frac{\xi}{b} e^{i2\bar{\Omega}\xi}\, d\xi, \qquad \bar{\Omega} = \Omega \sqrt{\alpha_0}. \tag{8b}$$

To evaluate I, introduce the change of variable

$$t = e^{2\xi/b} \tag{9}$$

to transform Eq. (8b) into

$$I = 2b \int_0^\infty \frac{t^{ib\bar\Omega}}{(1+t)^2}\, dt = 2b \frac{\Gamma(1 + ib\bar\Omega)\Gamma(1 - ib\bar\Omega)}{\Gamma(2)}. \tag{10}$$

The expression in terms of the gamma functions results from the formula[30]

$$\int_0^\infty \frac{t^z\, dt}{(1+t)^{w+1}} = \frac{\Gamma(1+z)\Gamma(w-z)}{\Gamma(w+1)}, \qquad \mathrm{Re}\, w > \mathrm{Re}\, z > -1. \tag{11}$$

Since $\Gamma(2) = 1$, $\Gamma(z)\Gamma(1-z) = \pi/\sin \pi z$, and $\Gamma(z+1) = z\Gamma(z)$, one obtains

$$\Gamma(1 + ib\bar\Omega)\Gamma(1 - ib\bar\Omega) = \frac{\pi}{i \sinh(\pi b\bar\Omega)} \frac{\Gamma(1 + ib\bar\Omega)}{\Gamma(ib\bar\Omega)} = \frac{\pi b\bar\Omega}{\sinh(\pi q\bar\Omega)}, \tag{12}$$

and the reflection coefficient is given by

$$\bar\Gamma(x) \simeq \frac{a}{2\alpha_0} \frac{\pi b\bar\Omega}{\sinh(\pi b\bar\Omega)} \exp\left[-i2\bar\Omega\left(1 + \frac{a}{2\alpha_0}\right)x\right], \qquad x \to -\infty. \tag{13}$$

This result, valid under the restriction in Eq. (7), demonstrates the previous assertion that the reflection coefficient vanishes exponentially when $\alpha(x)$ is analytic and Ω is large.

The x dependence of the reflection coefficient is the same as for a medium with constant $\alpha(x) = [\sqrt{\alpha_0} + (a/2\sqrt{\alpha_0})]^2 \approx (\alpha_0 + a)$ for $a \ll \alpha_0$. To within the approximations made,† this is the expected phase dependence since $\alpha(-\infty) = \alpha_0 + a$. If b is very small, the transition region, wherein $\alpha(x)$ changes from $(\alpha_0 + a)$ to $(\alpha_0 - a)$, is localized near $x = 0$, and the limiting case $b = 0$ corresponds to an abrupt transition. In this case, Eq. (13) yields

$$|\bar\Gamma(x)| \simeq \frac{a}{2\alpha_0}, \qquad b = 0, \tag{14}$$

which, upon evaluating to $O(a)$ the rigorous expression $[\sqrt{\alpha(-\infty)} - \sqrt{\alpha(\infty)}]$ $\times [\sqrt{\alpha(-\infty)} + \sqrt{\alpha(\infty)}]^{-1}$, is easily seen to be the correct result.

The exact value of $|\bar\Gamma|$ for arbitrary a and b is given by the formula [see Eq. (33), with $\nu = 2a$, $\tau = 2/b$, $\alpha_0 = 1 + a$]

$$|\bar\Gamma| = \frac{\sinh[\pi b\bar\Omega(\sqrt{1+\bar a} - \sqrt{1-\bar a})/2]}{\sinh[\pi b\bar\Omega(\sqrt{1+\bar a} + \sqrt{1-\bar a})/2]}, \qquad \bar a = \frac{a}{\alpha_0}, \quad \bar\Omega = \Omega\sqrt{\alpha_0}, \tag{15}$$

which reduces to Eq. (13) when $\bar a \ll 1$, $\Omega ab \ll 1$.

If the transition between two media with slightly different, constant values of α is confined to a small region wherein the phase factor in Eq. (2) is essentially constant, then

$$|\bar\Gamma(x_1)| \simeq \left|\tfrac{1}{4} e^{i2\Omega\psi(x_\alpha)} \int_{x_\beta}^{x_\alpha} \left[\frac{d}{d\xi} \ln \alpha(\xi)\right] d\xi\right| = \left|\tfrac{1}{4} \ln \frac{\alpha_2}{\alpha_1}\right| \simeq \frac{|\alpha_2 - \alpha_1|}{4\alpha_1}, \tag{16}$$

---

†The phase dependence is given more correctly by $e^{-i2\Omega\sqrt{\alpha_0 + ax}}$.

where $|\alpha_2 - \alpha_1| \ll \alpha_1$. When applied to the special case in Eq. (3), one obtains the limiting value in Eq. (14).

3.6b The Epstein Solution for a Continuous Transition

For the specific $\alpha(x)$ in Eq. (3), $\hat{V}(x)$ in Eq. (3.5.74) can be expressed in closed form in terms of hypergeometric functions. This solution, given below, can also be employed as a comparison function in the determination of the scattering properties of other smooth transition profiles. It can be shown that Eq. (3.5.74), with

$$\alpha(x) \equiv \frac{\epsilon(x)}{\epsilon_0} = (1+a) - a\tanh\frac{x}{b} = 1 + \frac{v}{1+e^{\tau x}}, \qquad v = 2a, \quad \tau = \frac{2}{b},$$

$$\tag{17}$$

where v and τ are constant parameters, can be solved by[31,32]†

$$\hat{V}_i(x) = \zeta^{(\gamma-1)/2}(1-\zeta)^{(\alpha+\beta+1-\gamma)/2}u_i(\alpha,\beta;\gamma;\zeta), \qquad \zeta = -e^{\tau x}. \tag{18}$$

While v and τ may in general be complex, they will be assumed positive real in this section. The functions u_i are solutions of the hypergeometric equation:

$$\left\{\zeta(1-\zeta)\frac{d^2}{d\zeta^2} + [\gamma - (\alpha+\beta+1)\zeta]\frac{d}{d\zeta} - \alpha\beta\right\} u_i(\alpha,\beta;\gamma;\zeta) = 0, \tag{19}$$

and

$$\alpha = 1 + i\frac{\Omega}{\tau}(\sqrt{1+v}-1), \quad \beta = 1 + i\frac{\Omega}{\tau}(\sqrt{1+v}+1),$$

$$\gamma = 1 + \frac{2i\Omega}{\tau}\sqrt{1+v}. \tag{20}$$

From these definitions,

$$\alpha + 1 - \beta = 1 - i\frac{2\Omega}{\tau}, \quad \alpha + 1 - \gamma = 1 - i\frac{\Omega}{\tau}\left[\sqrt{1+v}+1\right],$$

$$\gamma - \beta = \frac{i\Omega}{\tau}\left[\sqrt{1+v}-1\right], \tag{20a}$$

The differential equation (19) has singular points at $\zeta = 0, 1, \infty$, near each of which its two independent solutions can be expressed in terms of convergent hypergeometric series. Near $\zeta = 0$, the appropriate solutions are

$$u_1 = F(\alpha,\beta;\gamma;\zeta), \tag{21a}$$

$$u_5 = \zeta^{1-\gamma}F(\alpha-\gamma+1,\beta-\gamma+1;2-\gamma;\zeta), \tag{21b}$$

where the hypergeometric series

$$F(\alpha,\beta;\gamma;\zeta) = 1 + \frac{\alpha\beta}{1!\gamma}\zeta + \frac{\alpha(\alpha+1)\beta(\beta+1)}{2!\gamma(\gamma+1)}\zeta^2 + \cdots \tag{21c}$$

---

†The notation is the same as in H. Bateman et al., *Higher Transcendental Functions*, Vol. 1, McGraw-Hill Book Company, 1953. Note that the index α in u_i, etc., should not be confused with $\alpha(x)\equiv\epsilon(x)/\epsilon_0$.

converges inside the circle $|\zeta| < 1$. Near $\zeta = 1$, the pertinent functions are

$$u_2 = F(\alpha, \beta; \alpha + \beta - \gamma + 1; 1 - \zeta), \tag{22a}$$

$$u_6 = (1 - \zeta)^{\gamma - \alpha - \beta} F(\gamma - \alpha, \gamma - \beta; \gamma - \alpha - \beta + 1; 1 - \zeta), \tag{22b}$$

and, near $\zeta = \infty$,

$$u_3 = \frac{1}{(-\zeta)^{\alpha}} F\left(\alpha, \alpha - \gamma + 1; \alpha - \beta + 1; \frac{1}{\zeta}\right), \qquad -\zeta = \zeta e^{i\pi}, \tag{23a}$$

$$u_4 = \frac{1}{(-\zeta)^{\beta}} F\left(\beta, \beta - \gamma + 1; \beta - \alpha + 1; \frac{1}{\zeta}\right). \tag{23b}$$

While the various hypergeometric *series* converge only in the ranges $|\zeta| < 1$, $|1 - \zeta| < 1$, and $|\zeta| > 1$, respectively, the corresponding hypergeometric *functions* denoted by the same symbols can be continued analytically beyond the ranges of convergence of the series representations. Since the resulting two solutions in each set are linearly independent, they can be superposed (with appropriate constants) to represent any of the remaining functions. The ensuing formulas are called the "connection" or "circuit" relations for these functions, and the ones pertinent for the discussion herein are as follows:

$$u_1 = \frac{\Gamma(\gamma)\Gamma(\beta - \alpha)}{\Gamma(\gamma - \alpha)\Gamma(\beta)} u_3 + \frac{\Gamma(\gamma)\Gamma(\alpha - \beta)}{\Gamma(\gamma - \beta)\Gamma(\alpha)} u_4, \tag{24a}$$

$$u_5 = \frac{\Gamma(2 - \gamma)\Gamma(\beta - \alpha)e^{i\pi(1-\gamma)}}{\Gamma(1 - \alpha)\Gamma(\beta + 1 - \gamma)} u_3 + \frac{\Gamma(2 - \gamma)\Gamma(\alpha - \beta)e^{i\pi(1-\gamma)}}{\Gamma(1 - \beta)\Gamma(\alpha + 1 - \gamma)} u_4, \tag{24b}$$

$$u_3 = \frac{\Gamma(1 - \gamma)\Gamma(\alpha + 1 - \beta)}{\Gamma(1 - \beta)\Gamma(\alpha + 1 - \gamma)} u_1 - \frac{\Gamma(\gamma)\Gamma(1 - \gamma)\Gamma(\alpha + 1 - \beta)}{\Gamma(2 - \gamma)\Gamma(\gamma - \beta)\Gamma(\alpha)} e^{i\pi(\gamma - 1)} u_5,$$

$$\tag{25a}$$

$$u_4 = \frac{\Gamma(1 - \gamma)\Gamma(\beta + 1 - \alpha)}{\Gamma(1 - \alpha)\Gamma(\beta + 1 - \gamma)} u_1 - \frac{\Gamma(\gamma)\Gamma(1 - \gamma)\Gamma(\beta + 1 - \alpha)}{\Gamma(2 - \alpha)\Gamma(\gamma - \alpha)\Gamma(\beta)} e^{i\pi(\gamma - 1)} u_5.$$

$$\tag{25b}$$

The asymptotic behavior of the solutions $u_{3,4}$ at $x = \infty$ (i.e., $\zeta = -e^{\tau x} = -\infty$) is obtained directly from Eqs. (23) and the hypergeometric-series representation in Eq. (21c):

$$u_3 \sim (-\zeta)^{-\alpha}, \quad u_4 \sim (-\zeta)^{-\beta}, \qquad x \to \infty, \tag{26a}$$

whence, from Eqs. (18) and (20),

$$\hat{V}_3(x) \sim e^{i\pi(\gamma - 1)/2} e^{\tau x(\beta - \alpha)/2} = e^{i\pi(\gamma - 1)/2} e^{i\Omega x}, \qquad x \to \infty, \tag{26b}$$

$$\hat{V}_4(x) \sim e^{i\pi(\gamma - 1)/2} e^{\tau x(\alpha - \beta)/2} = e^{i\pi(\gamma - 1)/2} e^{-i\Omega x}, \qquad x \to \infty. \tag{26c}$$

Thus, $\hat{V}_3$ and $\hat{V}_4$ represent, respectively, outgoing and incoming waves as $x \to \infty$. At $x \to -\infty$, the behavior of $u_{1,5}$ follows from Eqs. (21) as

$$u_1 \to 1, \quad u_5 \to \zeta^{1-\gamma}, \qquad x \to -\infty, \tag{27a}$$

so

$$\hat{V}_1(x) \sim e^{i\pi(\gamma - 1)/2} (e^{\tau x})^{(\gamma - 1)/2} = e^{i\pi(\gamma - 1)/2} e^{i\Omega \sqrt{1+\nu} \, x}, \qquad x \to -\infty, \tag{27b}$$

$$\hat{V}_5(x) \sim e^{-i\pi(\gamma-1)/2} e^{-i\Omega\sqrt{1+\nu}\,x} \qquad\qquad x \to -\infty. \qquad (27c)$$

At $x = -\infty$, $\hat{V}_1$ and $\hat{V}_5$ represent incoming and outgoing waves, respectively. It is noted from the above that the asymptotic wave character of the $\hat{V}_i$ is contained entirely in the simple factors multiplying the hypergeometric functions, whence the latter serve as correction terms accounting for the medium inhomogeneity in the range of finite x. One observes also that the wave functions at $x = +\infty$ are those for a medium with $\epsilon(x)/\epsilon_0 = 1$, while those at $x = -\infty$ are appropriate to a medium with $\epsilon(x)/\epsilon_0 = 1 + \nu$, in accord with expectations from Eq. (17).

Let us suppose that a plane wave of unit amplitude is incident from $x = -\infty$ in the medium of Eq. (17), and that we seek the amplitude and phase of the reflected and transmitted waves. Since the transmitted wave must be outgoing at $x = +\infty$, it is necessary to choose the solution $\hat{V}_3(x)$ whose properties as $x \to \infty$ [see Eq. (26b)] can be calculated directly from its hypergeometric-series representation. This hypergeometric series cannot be employed, however, to calculate the behavior of $\hat{V}_3(x)$ at $x = -\infty$, and it is necessary to utilize the connection relation (25a) expressing u_3 in terms of u_1 and u_5, which latter functions can be evaluted near $x = -\infty$ from Eqs. (21). If $\hat{V}_1(x)$ in Eq. (27b) is to represent an incident wave of unit amplitude, the expression in Eq. (18) must be multiplied by the factor $\exp\left[i\pi(1 - \gamma)/2\right]$. Let

$$\bar{V}_i(x) = \hat{V}_i(x)e^{i\pi(1-\gamma)/2}; \qquad (28)$$

then, from Eq. (25a),

$$\bar{V}_3(x)\frac{\Gamma(1 - \beta)\Gamma(\alpha + 1 - \gamma)}{\Gamma(1 - \gamma)\Gamma(\alpha + 1 - \beta)}$$

$$= \bar{V}_1(x) - \bar{V}_5(x)\frac{\Gamma(\gamma)\Gamma(1 - \beta)\Gamma(\alpha + 1 - \gamma)}{\Gamma(2 - \gamma)\Gamma(\gamma - \beta)\Gamma(\alpha)}\, e^{i\pi(\gamma-1)}. \qquad (29)$$

The right-hand side of this equation can be used to evaluate the wave function near $x = -\infty$, while the left-hand side is appropriate for $x = +\infty$. Since $\bar{V}_1(-\infty) \sim \exp\left[i\Omega\sqrt{1 + \nu}\,x\right]$, the second term on the right-hand side represents directly the reflected-wave contribution. If we write

$$\bar{V}(-\infty) = e^{i\Omega\sqrt{1+\nu}\,x} + \bar{\Gamma}(-\infty)e^{-i\Omega\sqrt{1+\nu}\,x}, \qquad (30a)$$

then the voltage reflection coefficient $\bar{\Gamma}(-\infty)$ is given by

$$\bar{\Gamma}(-\infty)$$

$$= -\frac{\Gamma(\gamma)\Gamma(1 - \beta)\Gamma(\alpha + 1 - \gamma)}{\Gamma(2 - \gamma)\Gamma(\gamma - \beta)\Gamma(\alpha)} = \frac{\Gamma(\gamma - 1)}{\Gamma(1 - \gamma)}\frac{\Gamma(1 - \beta)\Gamma(\alpha + 1 - \gamma)}{\Gamma(\gamma - \beta)\Gamma(\alpha)} \qquad (30b)$$

$$= \frac{\Gamma\left[\dfrac{i2\Omega}{\tau}\sqrt{1 + \nu}\right]\Gamma\left[-\dfrac{i\Omega}{\tau}(\sqrt{1 + \nu} + 1)\right]\Gamma\left[1 - i\dfrac{\Omega}{\tau}(\sqrt{1 + \nu} + 1)\right]}{\Gamma\left[\dfrac{-i2\Omega}{\tau}\sqrt{1 + \nu}\right]\Gamma\left[i\dfrac{\Omega}{\tau}(\sqrt{1 + \nu} - 1)\right]\Gamma\left[1 + i\dfrac{\Omega}{\tau}(\sqrt{1 + \nu} - 1)\right]}.$$

$$(30c)$$

Similarly, if

$$\bar{V}(\infty) = T(\infty)e^{i\Omega x}, \tag{31a}$$

then the transmission coefficient T is obtained from the left-hand side of Eq. (29) as

$$T(\infty) = \frac{\Gamma(1 - \beta)\Gamma(\alpha + 1 - \gamma)}{\Gamma(1 - \gamma)\Gamma(\alpha + 1 - \beta)}$$

$$= \frac{\Gamma\left[-\dfrac{i\Omega}{\tau}(\sqrt{1 + \nu} + 1)\right]\Gamma\left[1 - i\dfrac{\Omega}{\tau}(\sqrt{1 + \nu} + 1)\right]}{\Gamma\left[-\dfrac{i2\Omega}{\tau}\sqrt{1 + \nu}\right]\Gamma\left[1 - i\dfrac{2\Omega}{\tau}\right]}. \tag{31b}$$

The magnitude of the reflection coefficient can be expressed in a simple manner. Since $\Gamma(z)$ is real for real z, it follows from the Schwarz reflection principle[33] that $\Gamma(z^*) = \Gamma^*(z)$. Thus, the ratio of the first two gamma functions in Eq. (30c) has unit magnitude since Ω, ν, and τ are real. Also, from

$$\Gamma(z)\Gamma(1 - z) = \frac{\pi}{\sin \pi z}, \tag{32a}$$

it follows that for imaginary $z = ih$, h real,

$$|\Gamma(ih)\Gamma(1 + ih)| = |\Gamma(ih)\Gamma(1 - ih)| = \frac{\pi}{|\sinh \pi h|}. \tag{32b}$$

Thus, from Eq. (30c), the reflection coefficient is

$$|\bar{\Gamma}(-\infty)| = \frac{\sinh\left[\pi\dfrac{\Omega}{\tau}(\sqrt{1 + \nu} - 1)\right]}{\sinh\left[\pi\dfrac{\Omega}{\tau}(\sqrt{1 + \nu} + 1)\right]}. \tag{33}$$

The limiting cases of an abrupt transition ($\tau = \infty$) and of a very gradual transition (τ small) have been discussed in the preceding section.

The preceding considerations are appropriate for a plane wave propagating along the x direction, with no variation along y or z, and must be modified for oblique incidence. The simplest case arises when the wave in question is an H mode (i.e., $E_x \equiv 0$), in which case $p(x)$ in Eq. (3.5.2a) is equal to the (constant) permeability. For a plane-stratified medium, $w = p$, and the obliquity manifests itself through the appearance of the constant parameter λ in Eq. (3.5.2a). If Eq. (3.5.2) is written as

$$\left[\frac{d^2}{dx^2} + \Omega^2\frac{\epsilon(x)}{\epsilon_0} - \xi^2\right]\bar{V}_\xi(x) = 0, \tag{34}$$

with $\xi^2 = -\lambda$, then, from Eq. (17),

$$\Omega^2\frac{\epsilon(x)}{\epsilon_0} - \xi^2 = (\Omega^2 - \xi^2)\left[1 + \frac{\nu\Omega^2/(\Omega^2 - \xi^2)}{1 + e^{\tau x}}\right]. \tag{35}$$

This equation is equivalent to Eq. (3.5.74), with Eq. (17), provided that one makes the following replacements:

$$\Omega \to \sqrt{\Omega^2 - \xi^2}, \qquad v \to v\,\frac{\Omega^2}{\Omega^2 - \xi^2}, \tag{36}$$

from which the solutions $\bar{V}_\xi(x)$ for $\xi \neq 0$ are obtainable at once from those for $\bar{V}(x)$ above. In particular, the parameters α, β, and γ in Eqs. (20) are, for $\xi \neq 0$,

$$\alpha = 1 + \frac{i}{\tau}\,[\sqrt{\Omega^2(1 + v) - \xi^2} - \sqrt{\Omega^2 - \xi^2}], \tag{37a}$$

$$\beta = 1 + \frac{i}{\tau}\,[\sqrt{\Omega^2(1 + v) - \xi^2} + \sqrt{\Omega^2 - \xi^2}], \tag{37b}$$

$$\gamma = 1 + \frac{2i}{\tau}\,\sqrt{\Omega^2(1 + v) - \xi^2}. \tag{37c}$$

To satisfy the radiation condition, all square roots are defined to have positive imaginary parts [see Eqs. (26) and (27)], and reduce to positive numbers when $\xi = 0$. These results, valid for arbitrary ξ, are required for the representation of the field of a concentrated source in Sec. 5.9d. For a plane wave whose propagation vector makes an angle θ with the positive x axis at $x \to \infty$, $\xi = \Omega \sin \theta$ and $-\pi/2 < \theta < \pi/2$.

The reflection coefficient magnitude in Eq. (33) is given for real $\xi \leq \Omega$ by

$$|\bar{\Gamma}(-\infty)| = \frac{\sinh \left\{\dfrac{\pi}{\tau}\,[\sqrt{\Omega^2(1 + v) - \xi^2} - \sqrt{\Omega^2 - \xi^2}]\right\}}{\sinh \left\{\dfrac{\pi}{\tau}\,[\sqrt{\Omega^2(1 + v) - \xi^2} + \sqrt{\Omega^2 - \xi^2}]\right\}}. \tag{38}$$

For a plane wave incident at an angle φ at $x = -\infty$, one has $\xi = \Omega\sqrt{1 + v} \times \sin \varphi$, and Eq. (38) becomes

$$|\bar{\Gamma}(-\infty)| = \frac{\sinh \left\{\dfrac{\pi\Omega}{\tau}\,[\sqrt{1 + v}\cos \varphi - \sqrt{1 - (1 + v)\sin^2 \varphi}]\right\}}{\sinh \left\{\dfrac{\pi\Omega}{\tau}\,[\sqrt{1 + v}\cos \varphi + \sqrt{1 - (1 + v)\sin^2 \varphi}]\right\}}, \tag{39}$$

with the angle φ restricted so that $\varphi < \varphi_c$, where φ_c is the critical angle $\varphi_c = \sin^{-1}[(1 + v)^{-1/2}]$. The reflection coefficient magnitude increases from its minimum value at $\varphi = 0$ to unity at $\varphi = \varphi_c$. Since the angle of refraction θ at $x = \infty$ is related to the angle of incidence at $x = -\infty$ via

$$\xi = \Omega\sqrt{1 + v}\sin \varphi = \Omega \sin \theta, \qquad \text{i.e.,}\quad \sin \theta = \sqrt{1 + v}\sin \varphi, \tag{40}$$

one has $\theta = \pi/2$ when $\varphi = \varphi_c$. When $\varphi > \varphi_c$, the wave is totally reflected and it is to be expected that $|\bar{\Gamma}(-\infty)| = 1$. This can be verified from Eq. (30c) when modified in accord with Eq. (35) since for $(\pi/2) \geq \varphi > \varphi_c$, the gamma functions in the numerator are the complex conjugates of those in the denominator.

If the wave is incident from $x = +\infty$, it has to be outgoing at $x = -\infty$, and one must choose the solution $\hat{V}_s$ [see Eq. (27c)] which is continued to $x =$

$+\infty$ via the connection relation in Eq. (24b). The incident wave is of unit amplitude if $\bar{V}_4$ is the corresponding wave function; the resulting reflection and transmission coefficients are then given by

$$\bar{\Gamma}(\infty) = \frac{\Gamma(1 - \beta)\Gamma(\alpha + 1 - \gamma)\Gamma(\beta - \alpha)}{\Gamma(1 - \alpha)\Gamma(\beta + 1 - \gamma)\Gamma(\alpha - \beta)},$$

$$T(-\infty) = \frac{\Gamma(1 - \beta)\Gamma(\alpha + 1 - \gamma)}{\Gamma(2 - \gamma)\Gamma(\alpha - \beta)}, \tag{41}$$

with α, β, and γ taken from Eqs. (37). In this case, with $\xi = \Omega \sin \theta$, $-\pi/2 \leq \theta \leq \pi/2$,

$$|\bar{\Gamma}(\infty)| = \frac{\sinh\left\{\dfrac{\pi\Omega}{\tau}[\sqrt{\cos^2 \theta + \nu} - \cos \theta]\right\}}{\sinh\left\{\dfrac{\pi\Omega}{\tau}[\sqrt{\cos^2 \theta + \nu} + \cos \theta]\right\}}. \tag{42}$$

The function $\alpha(x) = \epsilon(x)/\epsilon_0$ in Eq. (17) may be used as a comparison function for some other smooth transition profile for which no known solution of the differential equation (3.5.34) exists. If the problem of determining the unknown wave function is phrased as in Eq. (3.5.15), it is necessary to evaluate the Green's function in Eq. (3.5.16). To satisfy the boundary conditions (radiation condition) at $x = \pm\infty$, we choose $g_2(x) = \hat{V}_3(x)$ and $g_1(x) = \hat{V}_5(x)$. The Wronskian of these two solutions is a constant independent of x and can be evaluated at any convenient point, for example at $x = -\infty$. $\hat{V}_5(-\infty)$ is given in Eq. (27c). To calculate $\hat{V}_3(-\infty)$, it is necessary the employ the continuation formula (25a), which yields

$$\hat{V}_3(-\infty) = e^{i\pi(\gamma-1)/2}[Ae^{i\Omega\sqrt{1+\nu}\,x} - Be^{-i\Omega\sqrt{1+\nu}\,x}], \tag{43}$$

where A and B denote the gamma-function factors multiplying u_1 and $-u_5 \exp[i\pi(\gamma - 1)]$, respectively, in Eq. (25a). Thus,

$$W(g_2, g_1) = \hat{V}_3 \frac{d\hat{V}_5}{dx} - \hat{V}_5 \frac{d\hat{V}_3}{dx} = -2i\Omega\sqrt{1+\nu}\frac{\Gamma(1 - \gamma)\Gamma(\alpha + 1 - \beta)}{\Gamma(1 - \beta)\Gamma(\alpha + 1 - \gamma)}, \tag{44}$$

which expression has been derived for $\xi = 0$ but is easily transformed to apply as well when $\xi \neq 0$ [see Eqs. (35) and (36)].

Hypergeometric functions can also be employed to solve the more general profile associated with an asymmetrical layer wherein $\alpha(x)$ in Fig. 3.6.1 has a minimum at x_β and approaches the value $\alpha_2 = \alpha(\infty) > \alpha(x_\beta)$, $\alpha_1 \neq \alpha_2$. The symmetrical layer or duct is included herein as the special case $\alpha_1 = \alpha_2$.

3.6c Dielectric Constant Profile with Simple Zero

To illustrate the use of the WKB and Airy function formulations for a profile having a simple zero, consider the example

$$\alpha(x, \Omega) = 1 - \frac{\mu^2 - \frac{1}{4}}{\Omega^2 x^2}, \tag{45}$$

where μ is a constant but may be complex.† If Ω denotes the wavenumber $k_0 = \omega/c$ in vacuum and μ^2 is positive and greater than $\frac{1}{4}$, the function $\alpha(x, \Omega)$ is of the form $[1 - (\omega_p^2/\omega^2)]$, which represents the permittivity of an ionized plasma model having an electron density distribution $N(x) \propto \omega_p^2(x)$[see Eq. (1.5.20) with $\omega_c = 0$]. In most of the subsequent considerations, μ will be assumed positive and large enough so that $\mu^2 - \frac{1}{4} \approx \mu^2$. The simple zero of $\Omega^2\alpha(x)$ then occurs at $x \cong \mu/\Omega$; the singularity at $x = 0$ is of no interest for the present discussion and x is restricted to a range of positive values sufficiently far from the origin. Eq. (3.5.34), with Eq. (47), is solved exactly by the functions

$$\hat{g}(x) = \sqrt{\Omega x}\, C_\mu(\Omega x), \tag{46}$$

where C_μ is any linear combination of solutions of the Bessel equation.

If μ is small and Ω is very large, one may put $\alpha(x, \Omega) \approx 1$, so the resulting asymptotic approximation in Eq. (3.5.37) yields $\hat{g}(x) \sim \exp(\pm i\Omega x)$. If μ is large, this result still applies to observation points for which $\Omega x \gg \mu$. A more accurate description is obtained, however, upon retaining the complete expression in Eq. (45). By an elementary integration,

$$\int_{x_0}^{x} \sqrt{a(\eta)}\, d\eta \cong \int_{x_0}^{x} \cos\phi(\eta)\, d\eta = x\left[\cos\phi(x) + \left(\phi(x) - \frac{\pi}{2}\right)\sin\phi(x)\right],$$
$$\tag{47}$$

where the approximation arises from $\mu^2 - \frac{1}{4} \approx \mu^2$, and

$$\sin\phi(x) = \frac{\mu}{\Omega x}, \qquad \phi(x_0) \equiv \frac{\pi}{2}, \qquad \mu < \Omega x. \tag{47a}$$

Similarly,

$$\int_{x}^{x_0} |\sqrt{a(\eta)}|\, d\eta \cong \int_{x}^{x_0} \sinh\bar{\phi}(\eta)\, d\eta = x[\bar{\phi}(x)\cosh\bar{\phi}(x) - \sinh\bar{\phi}(x)]. \tag{48}$$

with

$$\cosh\bar{\phi}(x) = \frac{\mu}{\Omega x}, \qquad \bar{\phi}(x_0) = 0, \qquad \mu > \Omega x. \tag{48a}$$

Equation (48) can be obtained from Eq. (47) by analytic continuation of $\phi(x)$ to a range of complex values $\phi(x) = \pi/2 - i\bar{\phi}(x)$, $\bar{\phi}(x) > 0$.

To identify the expression for $\hat{g}(x)$ in Eq. (3.5.37) with any one of the solutions of the Bessel equation, appropriate normalization factors must be included. For example, if the solution is known for $\Omega x \to \infty$, the required constants can be deduced from this limiting result. Since for $\Omega x \gg \mu$, the Hankel functions may be approximated as [see Eq. (5.3.13)]

$$\sqrt{\Omega x}\, H_\mu^{(1,2)}(\Omega x) \sim A^{(1,2)} e^{\pm i\Omega x}, \qquad A^{(1,2)} = \sqrt{\frac{2}{\pi}}\, e^{\mp i(\mu\pi/2 + \pi/4)}, \tag{49}$$

one finds from Eqs. (3.5.37) and (47),

---

†Here μ is not to be confused with the same symbol used elsewhere for permeability.

$$\sqrt{\Omega x}\, H_\mu^{(1,2)}(\Omega x) \sim \sqrt{\frac{2}{\pi}}\, e^{\mp i\pi/4} \frac{\exp\{\pm i\Omega x[\cos\phi + (\phi - (\pi/2))\sin\phi]\}}{\sqrt{\cos\phi}},$$

$$\mu < \Omega x. \qquad (50a)$$

The constants in the range $\mu > \Omega x$ are now chosen in accord with Eq. (3.5.47a) and yield

$$\sqrt{\Omega x}\, H_\mu^{(1,2)}(\Omega x) \sim \mp i\sqrt{\frac{2}{\pi}}\, \frac{e^{\Omega x[\bar\phi \cosh \bar\phi - \sinh \bar\phi]}}{\sqrt{\sinh \bar\phi}}, \qquad \mu > \Omega x. \qquad (50b)$$

For the Bessel function, application of the formula $2J_\mu(y) = [H_\mu^{(1)}(y) + H_\mu^{(2)}(y)]$ yields

$$\sqrt{\Omega x}\, J_\mu(\Omega x) \sim \sqrt{\frac{2}{\pi \cos\phi}} \cos\left\{\Omega x\left[\cos\phi + \left(\phi - \frac{\pi}{2}\right)\sin\phi\right] - \frac{\pi}{4}\right\},$$

$$\mu < \Omega x, \qquad (51a)$$

while, from Eq. (3.5.47b),

$$\sqrt{\Omega x}\, J_\mu(\Omega x) \sim \frac{1}{\sqrt{2\pi \sinh \bar\phi}}\, e^{-\Omega x[\bar\phi \cosh \bar\phi - \sinh \bar\phi]}, \qquad \mu > \Omega x. \qquad (51b)$$

Alternatively, the normalization factors could have been deduced from the known behavior of the functions when $\mu \gg \Omega x$. In this instance,

$$J_\mu(\Omega x) \sim \left(\frac{\Omega x}{2}\right)^\mu \frac{1}{\Gamma(\mu + 1)} \sim \frac{1}{\sqrt{2\pi\mu}}\left(\frac{e\Omega x}{2\mu}\right)^\mu, \qquad (52a)$$

where we have employed the asymptotic formula [see Eq. (3.4.66c)],

$$\Gamma(\mu + 1) = \mu\Gamma(\mu) \sim \sqrt{2\pi\mu}\left(\frac{\mu}{e}\right)^\mu, \qquad \mu \gg 1. \qquad (52b)$$

Since for $\mu \gg \Omega x$,

$$\frac{\mu}{\Omega x} = \cosh \bar\phi \sim \sinh \bar\phi \sim \frac{e^{\bar\phi}}{2}, \qquad \bar\phi \sim \ln \frac{2\mu}{\Omega x}, \qquad (53)$$

one has

$$\frac{e^{\pm \Omega x(\bar\phi \cosh \bar\phi - \sinh \bar\phi)}}{\sqrt{\sinh \bar\phi}} \sim \sqrt{\frac{\Omega x}{\mu}}\left(\frac{2\mu}{\Omega x e}\right)^{\pm\mu}. \qquad (54)$$

Upon comparing Eqs. (54) and (52a), one finds that the decaying exponential must be chosen and that a factor $(2\pi)^{-1/2}$ must be included, whence the proper representation for the Bessel function is the one given in Eq. (51b). The appropriate form for the range $\mu < \Omega x$ is then obtained via Eq. (3.5.47a) and leads to Eq. (51a).

In the transition region $\mu \approx \Omega x$, it is necessary to employ the more elaborate formula (3.5.45). From a comparison of Eqs. (3.5.45), (3.5.47), and (51), one finds that the proper representation for the Bessel function near the turning point is

$$\sqrt{\Omega x}\, J_\mu(\Omega x) \sim \sqrt{\frac{2}{P}}\,[\tfrac{3}{2}\,\Omega x Q]^{1/6}\,\mathrm{Ai}[-(\tfrac{3}{2}\,\Omega x Q)^{2/3}], \tag{55}$$

where

$$P = \cos\phi, \qquad Q = \cos\phi + \left(\phi - \frac{\pi}{2}\right)\sin\phi, \qquad \mu \le \Omega x, \tag{55a}$$

$$P = i\sinh\bar\phi, \quad Q = e^{i3\pi/2}[\bar\phi\cosh\bar\phi - \sinh\bar\phi], \qquad \mu \ge \Omega x. \tag{55b}$$

Expressions for the Hankel functions are deduced in a similar manner. The results obtained agree with those derived in Sec. 4.5a [see Eq. (4.5.33)] from the Sommerfeld integral representation of the cylinder function (see also Sec. 6.A).

A criterion delimiting the range of validity of the WKB approximation has been given in Eq. (3.5.84). For the present problem, with $\mu < \Omega x$, $\alpha \approx \cos^2\phi$, and

$$\alpha'(x) \cong \frac{2\mu^2}{\Omega^2 x^3} = \frac{2}{x}\sin^2\phi, \tag{56}$$

whence the required restriction in Eq. (3.5.84) is

$$\frac{2}{\mu}\tan^3\phi \ll 1. \tag{57}$$

Since μ is assumed to be large, the condition is satisfied except when $\phi \to \pi/2$. Upon noting that

$$\mu \approx \Omega x, \quad \tan\phi \approx \sec\phi \approx \sqrt{\frac{\Omega x}{2(\Omega x - \mu)}}, \qquad \text{as } \phi \to \frac{\pi}{2}, \tag{58}$$

and omitting irrelevant factors, one may rewrite the inequality as

$$(\Omega x - \mu) \gg (\Omega x)^{1/3}, \tag{59}$$

which agrees with the condition $\Omega^{2/3}\varphi \gg 1$ in Eq. (3.5.47). For $\mu > \Omega x$, the left-hand side of the inequality is replaced by its absolute value. Thus, Eqs. (50) and (51) are valid in this range but must be replaced by Eq. (55) or its equivalent when $|\Omega x - \mu| = O[(\Omega x)^{1/3}]$.

PROBLEMS

1. Use the free-space spectral representation in Eq. (3.2.40),

$$\delta(\boldsymbol{\rho} - \boldsymbol{\rho}') = \frac{1}{(2\pi)^2}\int\!\!\int_{-\infty}^{\infty} e^{-j\xi(x-x')-j\eta(y-y')}\,d\xi\,d\eta, \tag{1}$$

where $-\infty < (x, x', y, y') < \infty$, to construct via the image principle E- and H-mode spectral representations for the half-space region $0 < (x, x') < \infty$, $-\infty < (y, y') < \infty$ [see Eqs. (3.2.38) and (3.2.39) and Fig. 3.2.4]. Repeat for the quarter-space region $0 < (x, x', y, y') < \infty$ [see Eqs. (3.2.32) and (3.2.33) and Fig. 3.2.3].

2. Utilizing the traveling wave formulation in Eqs. (3.3.28) and (3.3.29), show that the modal Green's function $Y(x, x')$ may be represented as [note that $Y(x, x')$ is the current in the network of Fig. 3.3.2; see also Eqs. (3.3.3)-(3.3.7)]:

$$Y(x, x') = \frac{\overleftarrow{I}(x_<, x_1)\overrightarrow{I}(x_>, x_2)}{\overleftrightarrow{\zeta}(x')\left[\dfrac{I^+(x')}{I^+(x_2)}\dfrac{I^-(x')}{I^-(x_1)} - \overrightarrow{\Gamma}_l(x_2)\overleftarrow{\Gamma}_l(x_1)\dfrac{I^-(x')}{I^-(x_2)}\dfrac{I^+(x')}{I^+(x_1)}\right]}, \tag{2}$$

where, with $\overleftrightarrow{\zeta} = \overrightarrow{\zeta} + \overleftarrow{\zeta}$,

$$\overleftarrow{I}(x, x_1) = \frac{I^-(x)}{I^-(x_1)} + \overleftarrow{\Gamma}_l(x_1)\frac{I^+(x)}{I^+(x_1)}.$$

$$\overrightarrow{I}(x, x_2) = \frac{I^+(x)}{I^+(x_2)} + \overrightarrow{\Gamma}_l(x_2)\frac{I^-(x)}{I^-(x_2)}.$$

x_1 and x_2 are two conveniently chosen points in the x interval which, for example, coincide with the left and right termini of the region, respectively. Show that the formula reduces to Eq. (2.4.29c), with $x \to z$, when $x_1 = x_2 = x_0$ and when the transmission line is uniform [i.e., $I^\pm = \exp(\mp j\kappa x)$]. Show that an analogous formula for $Z(x, x')$ is dual to the above with $\overrightarrow{I} \to \overrightarrow{V}$, etc.

3. By examining the input admittance in Eq. (3.3.21b) seen to the left from a finite point ρ in a region extending to the origin $\rho = 0$ in a cylindrical coordinate system, show that the boundary condition for the differential equation (3.2.46b) at $\rho = 0$ is of the limit point type; i.e., the boundary condition is independent of the termination at $\rho = 0$.

4. Calculate by the classical procedure the orthonormal H-mode eigenfunctions $\hat{\psi}_{v\beta}(x), \hat{\psi}_{m\beta}(x), \beta = 1, 2$, for the slab-loaded waveguide defined in Eqs. (3.4.13) and (3.4.14). Start from the general relation in Eq. (3.2.4) satisfied by any two eigenfunctions $f_m(x)$ and $f_n(x)$ with eigenvalues λ_m and λ_n,

$$(\lambda_m - \lambda_n)\int_{x_1}^{x_2} wf_n^* f_m\, dx = \left[p\left(f_m\frac{df_n^*}{dx} - f_n^*\frac{df_m}{dx}\right)\right]_{x_1}^{x_2}. \tag{3}$$

When $n \neq m$, vanishing of the right-hand side in Eq. (3) via the boundary condition $p\, df_i/dx + \alpha_{1,2}f_i = 0$ at $x = x_{1,2}$, with $i = n, m$, implies the orthogonality of the mode set. To find the normalization constant when $n = m$, retain f_m as an eigenfunction with eigenvalue λ_m but treat λ_n as a variable parameter that approaches λ_m (since f_n still satisfies the differential equation (3.2.1), though not the boundary conditions, Eq. (3) remains valid). Evaluate the integral in Eq. (3) when $\lambda_n = \lambda_m$ by applying L'Hôpital's rule to the resulting indeterminate form [see Eqs. (3.2.55)].

5. To achieve a ϕ transmission formulation of a scattering problem in a spherical coordinate system (r, θ, ϕ), the contour C_θ in Eq. (3.3.43b) is deformed about the singularities of $g_r(r, r'; \lambda_\theta)$ in the complex λ_θ plane. This specifies the λ_θ in terms of the radial eigenvalues, with λ_θ subsequently replaced by λ_r [for the radial domain $0 < r < \infty$, one has $\lambda_r = v_r(v_r + 1)$, $v_r = -\frac{1}{2} + iv_i$, v_i real; see Eq. (3.4.100b)]. The contour C_ϕ is then deformed about the singularities of $g_\theta(\theta, \theta'; \lambda_\phi; \lambda_r)$ in the complex λ_ϕ-plane. For the corresponding eigenvalue in the θ-domain, λ_ϕ replaced by $\hat{\lambda}_\theta$ plays the role of the eigenvalue, with λ_r remain-

ing a fixed parameter. These eigenfunctions may therefore be determined from a characteristic Green's function

$$\left(\frac{d}{d\theta}\sin\theta\frac{d}{d\theta} + v_r(v_r + 1)\sin\theta - \frac{\hat{\lambda}}{\sin\theta}\right)g_\theta(\theta,\theta';\hat{\lambda};\lambda_r) = -\delta(\theta - \theta'), \quad (4)$$

wherein v_r is assumed to be specified and $\hat{\lambda}$ is the characteristic parameter.

(a) Show that if λ_r is negative real ($v_r = -\frac{1}{2} + iv_i$, v_i real), then the eigenvalues $\hat{\lambda} = \hat{\lambda}_\mu$ are negative real. Assume this condition in the following.

(b) Show that the mode spectrum for the domain $0 \leq \theta \leq \pi$ is continuous and gives rise to the completeness relation

$$\sin\theta'\delta(\theta - \theta') =$$

$$-\frac{1}{2i}\int_{-i\infty}^{i\infty} \mu \frac{P_{v_r}^{-\mu}(\cos\theta)P_{v_r}^{-\mu}(-\cos\theta')}{[\Gamma(v_r - \mu + 1)/\Gamma(v_r + \mu + 1)]\sin(v_r - \mu)\pi}\,d\mu. \quad (5)$$

(c) Show that the mode spectrum for the domain $0 < \theta_1 \leq \theta \leq \theta_2 < \pi$ is discrete and gives rise to the completeness relation

$$\sin\theta'\delta(\theta - \theta') = \pi\sum_\mu \frac{\mu}{\sin\mu\pi}\frac{(d/d\theta_2)P_v^\mu(\cos\theta_2)}{(d/d\theta_1)P_{v_r}^\mu(\cos\theta_1)}\frac{A(v_r,\mu;\theta,\theta_1)A(v_r,\mu;\theta',\theta_1)}{(\partial^2/\partial\mu\,\partial\theta_2)A(v_r,\mu;\theta_2,\theta_1)},$$

$$(6a)$$

where μ is imaginary and is determined from a solution of the equation $(\partial/\partial\theta_2)A(v_r,\mu;\theta_2,\theta_1) = 0$, with

$$A(v,\mu;\varphi,\psi) \equiv P_v^\mu(\cos\varphi)\frac{d}{d\psi}P_v^{-\mu}(\cos\psi) - P_v^{-\mu}(\cos\varphi)\frac{d}{d\psi}P_v^\mu(\cos\psi). \quad (6b)$$

These eigenfunctions have vanishing derivatives at $\theta = \theta_{1,2}$.

(d) Repeat for eigenfunctions which vanish at $\theta_{1,2}$.

6. A half-space is bounded by an infinite perfectly conducting plane. In a coordinate system (I) whose z axis is perpendicular to the plane, the equation of the boundary may be described by $\theta = \pi/2$, whereas in a coordinate system (II) whose z-axis lies in the plane, the description is $\phi = 0, \pi$.

(a) Show that the normalized scalar mode functions in System I are:

$$\Phi_i(\theta,\phi) = \frac{1}{N_i}P_n^m(\cos\theta)\begin{Bmatrix}\cos m\phi\\\sin m\phi\end{Bmatrix}, \qquad (n+m)\ \text{odd}, \qquad (7a)$$

$$\psi_i(\theta,\phi) = \frac{1}{N_i}P_n^m(\cos\theta)\begin{Bmatrix}\cos m\phi\\\sin m\phi\end{Bmatrix}, \qquad (n+m)\ \text{even}, \qquad (7b)$$

where $0 \leq \theta \leq \pi/2$, $0 \leq \phi \leq 2\pi$, n and m are positive integers (or zero), and

$$N_i^2 = \frac{2\pi}{\epsilon_m}\frac{1}{(2n+1)}\frac{(n+m)!}{(n-m)!}, \qquad (7c)$$

with $\epsilon_m = 1$, $m = 0$, and $\epsilon_m = 2$, with $m \geq 1$.

(b) Show that the normalized scalar mode functions in System II are:

$$\Phi_i(\theta,\phi) = \frac{1}{N_i}P_n^m(\cos\theta)\sin m\phi, \qquad 0 \leq \theta \leq \pi, 0 \leq \phi \leq \pi, \qquad (8a)$$

$$\psi_i(\theta,\phi) = \frac{1}{N_i}P_n^m(\cos\theta)\cos m\phi. \qquad (8b)$$

7. (a) Use Eqs. (3.5.45) and the formula $D_\nu(-z) \sim (-z)^\nu \exp(-z^2/4)$ as $z \to \infty$ to derive the following uniform asymptotic representation for the parabolic cylinder function $D_\nu(-\sqrt{2\Omega}\, x)$, $2\nu = (\Omega a^2 - 1)$, valid for large Ω, and for all x in the interval $-\infty < x < a$:

$$D_\nu(-\sqrt{2\Omega}\, x) \sim \frac{2^{3/4}\pi^{1/2}}{\Omega^{1/12}} \left(\frac{\Omega a^2}{2e}\right)^{\Omega a^2/4} \frac{[\tfrac{3}{2}\,\psi(x)]^{1/6}}{[\alpha(x)]^{1/4}} \, \mathrm{Ai}\{-[\tfrac{3}{2}\,\Omega\psi(x)]^{2/3}\}, \qquad (9)$$

where $\psi(x) = \int_{-a}^{x} \sqrt{a^2 - \eta^2}\, d\eta$ and $\alpha(x) = a^2 - x^2$. Verify that this formula reduces to Eq. (3.5.55a) when $x \ll -a$.

(b) Use Eq. (9) with $a \to \infty$ to derive an asymptotic representation valid for large ν. By letting Ω take on complex values, show that the resulting formula may be written as

$$D_\nu(z) \sim \frac{1}{\sqrt{2}} \left(\frac{-\nu}{e}\right)^{\nu/2} e^{-z\sqrt{-\nu}}, \qquad |\arg(-\nu)| < \pi,\ |\nu| \to \infty. \qquad (10)$$

8. A uniform plane wave with electric vector parallel to y is incident on an inhomogeneous slab with refractive index $n(x)$. The electric field satisfies the wave equation [see Eq. (1.7.41)]:

$$(\nabla^2 + k_0^2 n^2)E_y = 0. \qquad (11)$$

(a) If the profile is smooth and the inhomogeneity is confined to the interval $0 < x_1 < x < x_2$, show that the first-order reflection coefficient $\bar{\Gamma}(x)$ is given by [see Eq. (3.6.2)]:

$$\bar{\Gamma}(0) = \frac{1}{2} \int_{x_1}^{x_2} \frac{n(x)(dn/dx)}{n^2(x) - \sin^2\theta} \left(\exp\left[i2k_0 \int_0^x \sqrt{n^2(\eta) - \sin^2\theta}\, d\eta\right]\right) dx, \qquad (12)$$

where θ is the angle of incidence with respect to the x axis ($\partial/\partial y = 0$).

(b) If $n = 1 + \Delta$ in the interval, where Δ is very small, show that $\bar{\Gamma}(0)$ is given approximately by:

$$\bar{\Gamma}(0) \approx \frac{1}{2\cos^2\theta} \int_{x_1}^{x_2} \frac{d\Delta}{dx}\, e^{i2k_0 x \cos\theta}\, dx. \qquad (13)$$

State the conditions for the validity of Eq. (13).

9. If the wave in Problem 8 is incident on an inhomogeneous dielectric slab bounded by plane interfaces at $x = 0$ and $x = d$, and if the field solutions inside the slab are assumed to have the form $E_y = \exp[(ik_0 \sin\theta)y]f_\pm(x)$, where $f_\pm(x)$ are two linearly independent solutions of the x-dependent part of the wave equation, derive exact expressions for the reflection coefficient $\bar{\Gamma}$ and transmission coefficient T.

10. Let $\hat{V}(x)$ satisfy the non-uniform wave equation,

$$\left[\frac{d^2}{dx^2} + \Omega^2\alpha(x)\right]\hat{V}(x) = 0, \qquad (14)$$

where Ω is a large positive parameter.

(a) Assume that

$$\hat{V}(x) = e^{i\Omega\hat{\psi}(x,\Omega)}, \qquad (15)$$

and derive the differential equation satisfied by $\hat{\psi}$. Solve this equation by the method of successive approximations to derive an asymptotic expansion for $\hat{V}(x)$ in inverse powers of Ω.

(b) By an alternative procedure, assume that

$$\hat{V}(x) = \sum_{n=0}^{\infty} \frac{A_n(x)}{(i\Omega)^n}\, e^{i\Omega\psi(x)}, \tag{16}$$

where A_n and ψ are Ω-independent. Determine ψ and A_n and compare with the procedure in (a) [see also Eq. (1.7.16) and Problem 27 of Chapter 1]. Which procedure is more convenient for determining the asymptotic expansion of $\hat{V}(x)$?

11. The discussion of linear differential equations of higher order is systematized by expressing such equations as a system of first-order differential equations wherein the desired function and its derivatives appear as dependent variables. For example, show that the second-order differential equation (3.5.74) can be written equivalently as in Eq. (3.5.73) (see also Sec. 1.1d),

$$\frac{d}{dx}\,\Psi(x) = \mathscr{A}(x)\Psi(x), \tag{17}$$

where the wave vector Ψ and the matrix $\mathscr{A}$ have the following representation,

$$\Psi(x) \to \begin{pmatrix} \hat{V}(x) \\ i\hat{I}(x) \end{pmatrix}, \qquad \mathscr{A}(x) \to \begin{pmatrix} 0 & \omega\mu(x) \\ -\omega\epsilon(x) & 0 \end{pmatrix}. \tag{18}$$

For generality, assume both μ and ϵ to be x-dependent. The solution of Eq. (17) is to be effected subject to the initial condition

$$\Psi(x) = \Psi_0 \text{ at } x = x_0, \tag{19}$$

with the assumption that the $N \times N$ elements of the square matrix $\mathscr{A}(x)$ are real, single-valued, bounded, and integrable in an interval surrounding x_0.

Show that if a matrix propagator (Green's function) $G(x, x_0)$ satisfies the equations

$$\frac{d}{dx}\,G(x, x_0) = \mathscr{A}(x)G(x, x_0) \text{ for } x \geq x_0, \qquad G(x_0, x_0) = \mathbf{1}, \tag{20}$$

where $\mathbf{1}$ is the unit matrix, then the solution for $\Psi(x)$ is given by

$$\Psi(x) = G(x, x_0)\Psi_0, \qquad \Psi_0 \equiv \Psi(x_0). \tag{21}$$

Show that Eq. (20) is solved by the ordered exponential having the series expansion

$$G(x, x_0) = \mathbf{1} + \int_{x_0}^{x} \mathscr{A}(\xi)\, d\xi + \int_{x_0}^{x} d\xi\, \mathscr{A}(\xi) \int_{x_0}^{\xi} \mathscr{A}(\xi_1)\, d\xi_1 + \cdots \tag{22}$$

Construct the solution for $\Psi(x)$, and discuss the rate of convergence of the solution for various ranges of a parameter ω which may be contained in $\mathscr{A}$ [see Eq. (18)].

12. To construct an alternative solution to the problem posed in Eqs. (17) and (19),† introduce via

---

†See H. B. Keller and J. B. Keller, "Exponential-Like Solutions of Systems of Linear Ordinary Differential Equations," *J. Soc. Indust. Appl. Math.*, **10** (1962), p. 246–259. Also I. Kay, "Some Remarks Concerning the Bremmer Series," *J. Math. Anal. Appl.*, **3** (1961), p. 40.

$$\Psi(x) = \mathscr{P}(x)\mathbf{W}(x) \tag{23}$$

a non-singular transformation matrix $\mathscr{P}(x)$ chosen so that

$$\mathscr{L} = \mathscr{P}^{-1}\mathscr{A}\mathscr{P} \tag{24}$$

is a diagonal matrix.

(a) If a prime denotes the derivative with respect to x, show that

$$\mathbf{W}' = (\mathscr{L} - \mathscr{P}^{-1}\mathscr{P}')\mathbf{W}. \tag{25}$$

Show also that if a certain $\mathscr{P} = \mathscr{P}_1$ diagonalizes $\mathscr{A}$ as in Eq. (24), then so does the matrix $\mathscr{P}_1\mathscr{D}$, where $\mathscr{D}$ is a non-singular diagonal matrix. ($\mathscr{D}$ may therefore be chosen to "normalize" the matrix $\mathscr{P}_1$, and $\mathscr{P}$ may be written in the form $\mathscr{P} = \mathscr{P}_1\mathscr{D}$.) If $\mathscr{D}$ is to be chosen so that $\mathscr{P}^{-1}\mathscr{P}'$ in Eq. (25) has no diagonal terms, show that

$$\mathscr{D}(x) = \mathscr{D}_0 \exp\left[-\int_{x_0}^{x} \text{diag.}\,\mathscr{P}_1^{-1}(\xi)\mathscr{P}_1'(\xi)\,d\xi\right], \tag{26}$$

where $\mathscr{D}_0$ is a constant diagonal matrix and diag. $\mathscr{T}(x)$ denotes the diagonal part of $\mathscr{T}(x)$.

(b) If $\mathscr{A}(x)$ is slowly varying with x, show that $\mathbf{W}$ is given approximately by the traveling wave (exponential) solution

$$\mathbf{W}(x) \cong \begin{pmatrix} e^{\int_{x_0}^{x}L_{11}(\xi)\,d\xi} & & 0 \\ & \ddots & \\ 0 & & e^{\int_{x_0}^{x}L_{nn}(\xi)\,d\xi} \end{pmatrix}\mathbf{W}(x_0), \tag{27}$$

where $L_{ii}(\xi)$ are the elements of the matrix $\mathscr{L}(\xi)$.

(c) Returning again to the general case, introduce into Eq. (25) the "traveling wave" transformation

$$\mathbf{W}(x) = e^{\int_{x_0}^{x}\mathscr{L}(\xi)\,d\xi}\,\mathbf{Z}(x). \tag{28}$$

and show that the wave vector $\mathbf{Z}(x)$ is then given by the convergent series [see Eq. (22)]

$$\mathbf{Z}(x) = G_M(x, x_0)\mathbf{W}(x_0) = \left[1 + \int_{x_0}^{x}\mathscr{M}(\xi_1)\,d\xi_1 + \cdots\right]\mathbf{W}(x_0), \tag{29}$$

where

$$\mathscr{M}(\xi_1) = -e^{-\int_{x_0}^{\xi_1}\mathscr{L}(\xi)\,d\xi}\,\mathscr{P}^{-1}(\xi_1)\mathscr{P}'(\xi_1)e^{\int_{x_0}^{\xi_1}\mathscr{L}(\xi)\,d\xi} \tag{29a}$$

is small when the medium is slowly varying. Show that the resulting solution for $\Psi(x)$ is

$$\Psi(x) = \mathscr{P}(x)e^{\int_{x_0}^{x}\mathscr{L}(\xi)\,d\xi}\,G_M(x, x_0)\,\mathscr{P}^{-1}(x_0)\Psi_0, \tag{30}$$

where it is recalled that $\mathscr{P}$ has been chosen so that diag. $\mathscr{P}^{-1}\mathscr{P}' = 0$. Compare this solution with the one in Eq. (23) and discuss its utility.

13. Apply the analysis in Problem 12 to the matrix $\mathscr{A}$ in Eq. (18).

(a) Show that the eigenvalues of the matrix $\mathscr{A}$ are $\pm ik$, $k = \omega\sqrt{\mu\epsilon}$; that a set of eigenvectors is $\begin{pmatrix} \pm i\zeta \\ 1 \end{pmatrix}$, $\zeta = \sqrt{\mu/\epsilon}$; and that a matrix $\mathscr{P}_1$ which diagonalizes $\mathscr{A}$ is given by

$$\mathscr{P}_1 \rightarrow \begin{pmatrix} i\zeta(x) & -i\zeta(x) \\ 1 & 1 \end{pmatrix}. \tag{31}$$

Show also that to within a constant diagonal matrix, one has $\mathscr{D}(x) = \mathbf{1}\zeta^{-1/2}$, and

$$\mathscr{P}(x) \rightarrow \begin{pmatrix} i\zeta^{1/2} & -i\zeta^{1/2} \\ \zeta^{-1/2} & \zeta^{-1/2} \end{pmatrix}. \tag{32}$$

Then show that the solution for $\Psi(x)$ obtained by setting $\mathscr{M}(\xi) \approx 0$ in Eq. (30) is given by:

$$\Psi(x) = \frac{1}{2} \begin{pmatrix} i\zeta^{1/2} & -i\zeta^{1/2} \\ \zeta^{-1/2} & \zeta^{-1/2} \end{pmatrix} \begin{pmatrix} e^{-i\int_{x_0}^x k(\xi)d\xi} & 0 \\ 0 & e^{i\int_{x_0}^x k(\xi)d\xi} \end{pmatrix} \begin{pmatrix} -i\zeta^{-1/2} & \zeta^{1/2} \\ i\zeta^{-1/2} & \zeta^{1/2} \end{pmatrix}_{x_0} \begin{pmatrix} \hat{V}_0 \\ i\hat{I}_0 \end{pmatrix}. \tag{33}$$

Compare this result with the WKB solution in Eqs. (3.5.75) and (3.5.76) (with $A, B = $ constant).

(b) Consider the "matched" initial condition, $\hat{V}_0 = \zeta(x_0)\hat{I}_0$, and show that this implies $\hat{V}(x) = \zeta(x)\hat{I}(x)$ to a lowest order of approximation (WKB). By retaining the second term in the series representation of $G_M(x, x_0)$ in Eq. (30), derive a correction to the WKB result and show that it agrees with that obtained from Eq. (3.5.80).

REFERENCES

1. MORSE, P.M. and H. FESHBACH, *Methods of Theoretical Physics*, Sec. 6.3. New York: McGraw-Hill, 1953.

2. FRIEDMAN, B., *Principles and Techniques of Applied Mathematics*, p. 231. New York: John Wiley & Sons, 1956.

3. MAGNUS, W. and F. OBERHETTINGER, *Formulas and Theorems for the Special Functions of Mathematical Physics*, pp. 16-18, 136. New York: Chelsea Publishing Co., 1954.

4. FRIEDMAN, B., *Principles and Techniques of Applied Mathematics*, pp. 213-250. New York: John Wiley & Sons, 1956.

5. MARCUVITZ, N., "Field representations in spherically stratified regions," *Comm. Pure and Appl. Math.* 4 (1951), pp. 263-315 (Secs. 3a and 5).

6. LORCH, E. R., *Spectral Theory*, Chapter 4. London: Oxford University Press, 1962. Also I. Stakgold, *Boundary Value Problems of Mathematical Physics*, Vol. I, Chapter 4. New York: Macmillan, 1967.

7. FRIEDLANDER, F. G., "Diffraction of pulses by a circular cylinder," *Comm. Pure Appl. Math.* 7 (1954), pp. 705-732.

8. CLEMMOW, P.C., "Infinite integral transforms in diffraction theory," *IRE Trans. on Antennas and Propagation*, **AP-7** (1959), pp. S7-11.

9. MAGNUS, W. and F. OBERHETTINGER, *Formulas and Theorems for the Special Functions of Mathematical Physics.*, pp. 1, 7, 9, 106, 112; Chapter 4. New York: Chelsea Publishing Co., 1954.

10. ERDELYI, A., W. MAGNUS, F. OBERHETTINGER, and F. TRICOMI, *Higher Transcendental Functions*, Vol. I, pp. 142-144. New York: McGraw-Hill, 1953.

11. KONTOROVITCH, M. J. and N. H. LEBEDEV, "On a method for solution of certain diffraction problems," *J. of Physics (Moscow)* **1** (1939), pp. 229-241.

12. SOMMERFELD, A., *Partial Differential Equations in Physics*, Sec. 20. New York: Academic Press, 1949.

13. WHITTAKER, E. T. and G. N. WATSON, *A Course of Modern Analysis*, Chapter XI. Cambridge, England: Cambridge University Press, 1952.

14. MORSE, P. M. and H. FESHBACH, *Methods of Theoretical Physics*, p. 1073. New York: McGraw-Hill, 1953.

15. LANGER, R. E., "The asymptotic solutions of ordinary linear differential equations of the second order, with special reference to turning points," *Trans. Am. Math. Soc.* **67** (1949), p. 461.

16. ERDELYI, A., "Asymptotic solutions of differential equations with transition points or singularities," *J. of Math. Phys.* **1** (1960), p. 16.

17. OLVER, F. W. J., "Uniform asymptotic expansions of solutions of linear second-order differential equations for large values of a parameter," *Phil. Trans. Roy. Soc. London*, **250A** (1958), p. 479.

18. PIKE, E.R., "On the related-equation method of asymptotic approximation (W. K. B. or A-A Method). 1. A proposed new existence theorem, "*Quart. J. Mech. and Appl. Math.* **17** (1964), p. 105.

19. LANGER, R. E., "The asymptotic solution of a linear differential equation of the second order with two turning points," *Trans. Amer. Math. Soc.* **90** (1959), pp. 113-142.

20. KAZARINOFF, N. D., "Asymptotic theory of second order differential equations with two simple turning points," *Arch. for Rat. Mech. and Anal.* **2** (1958-1959), pp. 129-150.

21. MAGNUS, W. and. F. OBERHETTINGER, *Formulas and Theorems for the Special Functions of Mathematical Physics*, p. 92. New York: Chelsea Publishing Co., 1954.

22. OLVER, F. W. J., "Uniform asymptotic expansions for Weber parabolic cylinder functions of large orders," *J. Res. NBS*, **63B** (1959), p. 131.

23. BUDDEN, K. G., *Radio Waves in the Ionosphere*, Sec. 21.13. Cambridge, England: Cambridge University Press, 1961.

24. ERDELYI, A. and C. A. SWANSON, "Asymptotic forms of Whittaker's Confluent Hypergeometric functions," *Memoirs of the Am. Math. Soc.*, No. 25, 1957.

25. SLATER, L.J., *Confluent Hypergeometric Functions*, Cambridge, England: Cambridge University Press, 1960.

26. BUDDEN, K. G., *Radio Waves in the Ionosphere.*, Sec. 21.15. Cambridge, England: Cambridge University Press, 1961.

27. OLVER, F. W. J., "Error bounds for the Liouville-Green (or WKB) approximation," *Proc. Camb. Phil. Soc.* **57** (1961), pp. 790-810. F. W. J. OLVER, "Error

bounds for first approximations in turning point problems," *J. Soc. Ind. Appl. Math.* **11** (1963), pp. 748–772. F. W. J. OLVER and F. STENGLER, "Error bounds for asymptotic solutions of second-order differential equations having an irregular singularity of arbitrary rank," *J. SIAM Numer. Anal.* **B2** (1965), pp. 244–249.

28. BREMMER, H., "The WKB approximation as the first term of a geometric-optical series," *Comm. Pure and Appl. Math.* **4** (1951), p. 105.

29. RYDBECK, O. E. H., "On the propagation of waves in an inhomogeneous medium," *Trans. of Chalmers Univ. of Technology*, Gothenburg, Sweden, No. 74 (1948).

30. MAGNUS, W. and F. OBERHETTINGER, *Formulas and Theorems for the Special Functions of Mathematical Physics*, p. 4. New York: Chelsea Publishing Co., 1954.

31. EPSTEIN, P. S., "Reflection of waves in an inhomogeneous medium," *Proc. Natl. Acad. Sci. (USA)*, **16** (1930), p. 627.

32. BREKHOVSKIKH, L. M., *Waves in Layered Media*, Secs. 14 and 16. New York: Academic Press (1960).

33. COPSON, E. T., *Theory of Functions of a Complex Variable*, Sec. 8.4. London: Oxford University Press, 1935.

4. Asymptotic Evaluation of Integrals

4.1 GENERAL CONSIDERATIONS

4.1a Transformation to a Canonical Form

Infinite integrals

Radiation and diffraction fields in open regions (with infinite cross section) are usually expressed by integral representations that cannot be evaluated in closed form. In many applications, however, the integrands contain a large parameter, Ω, in terms of which the integrals may be approximated. While such an evaluation can be treated for rather general functional dependences of the integrand on Ω, it will suffice within the present context to consider integrals of the following type:

$$I(\Omega) = \int_{\bar{P}} f(z) e^{\Omega q(z)} \, dz, \tag{1}$$

where f and q are analytic functions of the complex variable z along a path of integration $\bar{P}$ with endpoints at infinity, and where the large parameter Ω is assumed to be positive.†

Suppose that at the point z_s on $\bar{P}$, $\operatorname{Re} q(z)$ has a maximum value so that $\operatorname{Re} q(z) < \operatorname{Re} q(z_s)$ on the remainder of the path. Since Ω is very large, it follows that $A = |\exp[\Omega q(z)]|$ likewise has a maximum at z_s and decreases very rapidly away from z_s. It is then suggestive to approximate $I(\Omega)$ only by the path contribution from the vicinity of z_s, since the contribution from the remainder of the path will be exponentially small in comparison. If $f(z)$ is regular and slowly varying in the vicinity of z_s, it may be approximated there by $f(z_s)$

---

†If $\Omega = |\Omega| \exp(i \arg \Omega)$ is complex, the phase term is included in the definition of $q(z)$. Alternatively, it may be more convenient to obtain an asymptotic evaluation of $I(\Omega)$ for real values of Ω and then continue Ω analytically into a range of permitted complex values.

370

and taken outside the integrand in Eq. (1), thereby leaving in the integrand only the exponential. Approximate integration of the latter can be effected by expanding $q(z)$ in a power series about z_s and retaining only the first few terms; in a more accurate procedure described below, the integral is compared with a "canonical" one having similar properties and a simpler structure. This, in rough outline, constitutes the basis for an asymptotic approximation of $I(\Omega)$ for large values of Ω. In general, A will not have the above-described behavior along the given contour $\bar{P}$, but along some other path $\bar{P}_z$. One then attempts to deform $\bar{P}$ into $\bar{P}_z$, proper account being taken of any interfering singularities of $f(z)$ (such as poles or branch points) in the complex z plane.

It is shown in Sec. 4.1b that the point(s) z_s, where there occur maximum contributions to the integral, are "saddle" or "stationary points" of the function $q(z)$, defined by the vanishing of one or more of the derivatives of $q(z)$. The desired path $\bar{P}_z$ is one on which Im $q(z)$ remains constant. Since only the vicinity of the various saddle points traversed by the path $\bar{P}_z$ is relevant, it is unnecessary to deal completely with the generally complicated function $q(z)$; instead, as mentioned above, a finite number of terms in the power-series expansion of $q(z)$ about z_s characterizes the asymptotic evaluation for large Ω. This feature may be exploited in a rigorous manner by transforming the given integral into a "canonical" form, wherein the function $q(z)$ is replaced by another function, a polynomial that describes in the simplest fashion the relevant saddle-point arrangement at z_s. The transformation will be phrased in terms of a new variable s and the polynomial $\tau(s)$:

$$\tau(s) = q(z), \tag{2}$$

the point z_s in the complex z plane being chosen to correspond to $s = 0$ in the complex s plane. Thus, Eq. (1) leads to the transformed integral

$$I(\Omega) = \int_{P'} G(s)e^{\Omega\tau(s)} \, ds \tag{3}$$

where

$$G(s) = f(z)\frac{dz}{ds} \quad \text{and} \quad \frac{dz}{ds} = \frac{\tau'(s)}{q'(z)}, \tag{3a}$$

with the prime denoting the derivative with respect to the argument. P' represents the mapping onto the s plane of the path $\bar{P}$ in the z plane.

By considerations analogous to those mentioned above, $I(\Omega)$ in the s plane is approximated most simply on a contour P, which passes in the vicinity of the origin and along which the magnitude of $\exp[\Omega\tau(s)]$ decreases rapidly away from $s = 0$. The desired contour P in the s plane, along which Im $\tau(s) = $ constant and Re $\tau(s) < $ Re $\tau(0)$ away from a region about $s = 0$, constitutes a mapping into the s plane of the path $\bar{P}_z$ in the z plane. Generally, P' will not be identical with P, and in any ensuing deformation of P' into P, the presence of singularities of the integrand in Eq. (3) may have to be taken into account. A typical case is shown in Fig. 4.1.1, where the desired path P and the given path P' are deformable one into the other at $|s| = \infty$. In view of the pole

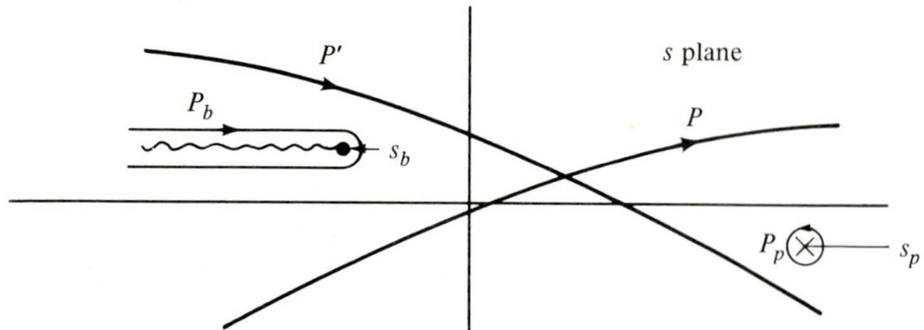

FIG. 4.1.1 Deformation of P' into P in the s plane.

singularity at s_p and the branch-point singularity at s_b, the integrals over the contours P' and P are related by Cauchy's theorem as follows:

$$\int_{P'} = \int_{P} + \int_{P_p} + \int_{P_b}, \tag{4}$$

with P_p and P_b denoting contours surrounding the pole and branch-cut singularities, respectively.

Assume now that the integral in Eq. (3) is to be evaluated over the desired contour P and that $G(s)$ is regular and slowly varying in the vicinity of $s = 0$. For large values of Ω the dominant contribution to the integral in Eq. (3) arises from the vicinity of the origin since the exponential term decreases rapidly away from the region $s \approx 0$. One may therefore write†

$$I(\Omega) = \int_{P} G(s)e^{\Omega\tau(s)}\,ds \sim G(0)\int_{P} e^{\Omega\tau(s)}\,ds, \qquad \Omega \to \infty, \tag{5}$$

where the symbol $\sim$ means "asymptotically represented by" and $G(0)$ is the value of the regular function $G(s)$ at $s = 0$. The last integral in Eq. (5) represents a "canonical" integral that provides a first-order approximation to the unknown integral $I(\Omega)$ of Eq. (1). Asymptotic approximations are discussed further in Sec. 4.2b.

The above remarks serve to highlight the motivation for a proper choice of the transformation (2). First, Re $\tau(s)$ along P should decrease most rapidly away from $s = 0$ so that the major contribution to the integral arises from the vicinity of the origin in the s plane, and Re $\tau(0)$ should be as small as possible; the latter condition facilitates the asymptotic evaluation. As shown in Sec. 4.1b, these requirements imply that the stationary or saddle points of $\tau(s)$ [the zeros of $\tau'(s)$] are situated near $s = 0$.‡ Viewed in the z plane, the path $\bar{P}_z$ (corresponding to P in the s plane) passes near one or more of the pertinent saddle points z_s of $q(z)$ [zeros of $q'(z)$], and Re $q(z)$ decreases along $\bar{P}_z$ away from the neighborhood of z_s. Thus, the transformation (2) should be chosen so that the

---

†This simple asymptotic formula is valid provided that Ω is the only relevant parameter. If $I(\Omega)$ depends on other parameters, uniform asymptotic approximations with respect to these are generally somewhat more involved (see Secs. 4.4–4.6).

‡If $\tau'(s)$ has an Mth-order zero at s_s, $\tau(s)$ is said to have a saddle point of order M at s_s.

neighborhood of the pertinent saddle points z_s in the z plane is mapped into the neighborhood of $s = 0$ in the s plane. Second, the mapping derivative dz/ds in Eq. (3a) must be finite near $s = 0$ in order to assure the assumed regularity of $G(s)$ near $s = 0$. Therefore, $\tau(s)$ must be selected so that at the points s_s the derivative $\tau'(s)$ possesses zeros of the same order as those of $q'(z)$ at z_s, where the points s_s in the s plane correspond to z_s in the z plane. The simplest $\tau(s)$ meeting these requirements will yield the simplest comparison integral on the right-hand side of Eq. (5). If $f(z)$ has singularities near z_s, $G(s)$ has singularities near $s = 0$; these must be isolated in their simplest form and require consideration of a new class of comparison integrals. A number of important canonical (comparison) integrals constructed in this manner may be expressed in terms of known functions. Included among these, and employed in subsequent discussion, are the gamma function, the error function or Fresnel integral, the Airy function, and the parabolic cylinder function.

We shall investigate, in detail, infinite integrals with the following configurations of saddle points and singularities:

1. $q'(z)$ has a simple (or multiple) zero at z_s and no other zero near z_s; $f(z)$ is regular near z_s (ordinary saddle-point procedure) (Secs. 4.2 and 4.3).

2. $q'(z)$ has simple zeros at $z_s = z_1$ and z_2, where z_1 lies arbitrarily near z_2; $f(z)$ is regular near z_1 and z_2 (double-saddle-point procedure) (Sec. 4.5a).

3. $q'(z)$ has equally spaced, collinear simple zeros at $z_s = z_1, z_2, z_3$ (i.e., $z_1 - z_2 = z_2 - z_3$) which may be arbitrarily near one another; $f(z)$ is regular near $z_{1,2,3}$ (triple-saddle-point procedure) (Sec. 4.5b).

4. $q'(z)$ has a simple zero at z_s and no other zero near z_s; $f(z)$ has a simple (or multiple) pole near z_s (saddle-point integration in the vicinity of a pole) (Secs. 4.4a and 4.4b).

5. $q'(z)$ has a simple zero at z_s and no other zeros near z_s; $f(z)$ has an algebraic branch-point singularity at or near z_s (saddle-point integration in the vicinity of a branch point) (Sec. 4.4c).

6. $q'(z)$ has a simple (or multiple) zero at z_s and no other zeros near z_s; $f(z)$ has an algebraic branch-point singularity at z_s. [By following the procedure in Sec. 4.3, this problem is reduced to the evaluation of integrals as in Eq. (4.3.5), except that n is not an integer. The asymptotic result therefore involves the gamma function and will not be discussed in detail.]

Emphasis is placed on uniform representations that reduce correctly to simpler ones under appropriate limiting conditions. For example, when $z_1 \not\approx z_2$, case 2 must reduce to the simple case 1; similarly, cases 4, 5, and 6 must reduce to case 1 when the singularity moves away from the saddle point. Corresponding integrals with finite endpoints are discussed separately below.

To illustrate the choice of the transformation in Eq. (2), and the principles set out above, we consider further the various cases.

Case 1. When $q'(z)$ has an Mth-order zero at z_s and no other zeros nearby, one may select the polynomial $\tau(s)$ as

$$q(z) = \tau(s) = q(z_s) - s^{M+1}, \tag{6}$$

whence $s = 0$ corresponds to $z = z_s$, and $dz/ds = \tau'(s)/q'(z)$ is finite at $s = 0$. The corresponding integral in Eq. (5) is expressible in terms of the gamma function (details are given in Sec. 4.2). The desired path P is one along which $s^{M+1} > 0$, or less stringently, Re $s^{M+1} > 0$.

Case 2. when $q'(z)$ has two simple neighboring zeros $z_{1,2}$, one may select

$$q(z) = \tau(s) = a_0 + \sigma s - \frac{s^3}{3}, \tag{7}$$

where a_0 and σ are constants. Since $\tau'(s)$ has two simple zeros at $s_{1,2} = \pm\sqrt{\sigma}$, we let $s_1 = \sqrt{\sigma}$ correspond to z_1 and $s_2 = -\sqrt{\sigma}$ correspond to z_2 to assure the regularity of dz/ds at $s_{1,2}$. From the ensuing relations

$$q(z_1) = \tau(\sqrt{\sigma}) = a_0 + \tfrac{2}{3}\sigma^{3/2}, \tag{7a}$$

$$q(z_2) = \tau(-\sqrt{\sigma}) = a_0 - \tfrac{2}{3}\sigma^{3/2}, \tag{7b}$$

one obtains the following expressions for a_0 and σ in terms of $q(z_{1,2})$:

$$a_0 = \tau(0) = \tfrac{1}{2}[q(z_1) + q(z_2)], \tag{8a}$$

$$\tfrac{2}{3}\sigma^{3/2} = \tfrac{1}{2}[q(z_1) - q(z_2)]. \tag{8b}$$

The integral in Eq. (5) with $\tau(s)$ given in Eq. (7) is expressible in terms of the Airy function (details are given in Sec. 4.5a). It should be emphasized that the case of two neighboring simple zeros of $q'(z)$ must be treated separately from case 1 only when these zeros are arbitrarily close (i.e., when $\sigma \to 0$). If σ remains finite, the formal asymptotic evaluation of the integral in Eq. (1) as $\Omega \to \infty$ can be carried out in terms of the separate contributions from each relevant zero treated in isolation. Most of the problems encountered in practice belong to this class and can be treated by the techniques in category 1. Just how close the zeros of $\tau'(s)$ can approach each other before they must be considered jointly will be illustrated in the detailed discussion carried out in Sec. 4.5a. Analogous remarks apply to the remaining cases wherein saddle points are not isolated.

Case 3. It suffices in the case of three collinear, equally spaced zeros to select

$$q(z) = \tau(s) = a_0 - (a + s^2)^2, \tag{9}$$

since $\tau'(s)$ then has zeros at $s = 0, \pm i\sqrt{a}$ which correspond to z_2 and to $z_{1,3}$, respectively. Evidently,

$$a_0 = q(z_1) = q(z_3), \qquad a^2 = q(z_1) - q(z_2), \tag{9a}$$

and the integral resulting upon substitution of Eq. (9) into Eq. (5) may be expressed in terms of the parabolic cylinder function. Details of this evaluation are given in Sec. 4.5b.

Case 4. If $f(z)$ in Eq. (1) has a pole singularity near a simple zero z_s of $q'(z)$ [i.e., $G(s)$ in Eq. (3) has a pole near $s = 0$], procedure 1 can no longer be applied directly, since $G(s)$ is now rapidly varying near $s = 0$. A modified approach is sketched here for the case where $G(s)$ has a simple pole at $s = b$, with $b \to 0$. Suppose that $G(s)(s - b) \to a$ as $s \to b$, where a is a constant. Then one represents $G(s)$ as

$$G(s) = \frac{a}{s - b} + T(s), \qquad (10)$$

so $T(s)$ is regular at both $s = b$ and $s = 0$, and is slowly varying near $s = 0$. The asymptotic approximation of $I(\Omega)$ in Eq. (3) as $\Omega \to \infty$ is now given by [see Eq. (5)]

$$I(\Omega) \sim T(0) \int_P e^{\Omega \tau(s)} \, ds + a \int_P \frac{e^{\Omega \tau(s)}}{s - b} \, ds, \qquad (11)$$

with $\tau(s)$ given in Eq. (6) with $M = 1$. The first integral in Eq. (11) is identical with that encountered in case 1; the second integral containing the effect of the pole singularity can be evaluated in terms of the error function or Fresnel integral (see Sec. 4.4a).

Case 5. If $f(z)$ has a first-order branch-point singularity near a first-order saddle point z_s, the resulting canonical integral [see Eq. (4.4.28)] may be transformed into an integral as for case 3, so the parabolic cylinder function describes this case as well.

In the preceding discussion we have been concerned only with the first-order asymptotic evaluation [see Eq. (5)] for the integral $I(\Omega)$ in Eq. (1) as $\Omega \to \infty$. It is possible, however, to obtain complete asymptotic expansions for $I(\Omega)$ involving successively decreasing terms as $\Omega \to \infty$ so that higher-order approximations can also be calculated. Complete expansions are presented for several of the cases described above (for a general treatment, see Reference 1).

Integrals with finite endpoints

If the interval of integration in Eq. (1) ranges between the finite limits z_α and z_β, the asymptotic evaluation of the integral can be carried out analogously to that over the infinite path $\bar{P}$. The given finite path $\bar{P}_{\alpha\beta}$ with endpoints at z_α and z_β can be deformed to pass in the vicinity of one (or more) pertinent saddle point(s) z_s of $q(z)$. If $\mathrm{Re}\, q(z) < \mathrm{Re}\, q(z_s)$ along the portions of the path away from the saddle point(s), the dominant contribution to the integral as $\Omega \to \infty$ again arises from the vicinity of the point(s) z_s, and its asymptotic evaluation is identical with that for the infinite integral.

The finite integral

$$I_{\alpha\beta}(\Omega) = \int_{z_\alpha}^{z_\beta} f(z) e^{\Omega q(z)} \, dz \qquad (12)$$

transforms into the s plane via Eq. (2) as

$$I_{\alpha\beta}(\Omega) = \int_{P_{\alpha\beta}} G(s) e^{\Omega \tau(s)} \, ds, \qquad \int_{P_{\alpha\beta}} = \int_{s_\alpha}^{s_\beta}, \qquad (13)$$

where s_α and s_β correspond to z_α and z_β, respectively. The integral in Eq. (13) can be written as a sum of the following three integrals:

$$I_{\alpha\beta}(\Omega) = \left(\int_P + \int_{P_\alpha} - \int_{P_\beta}\right) G(s)e^{\Omega\tau(s)}\,ds = I(\Omega) + I_\alpha(\Omega) + I_\beta(\Omega), \qquad (14)$$

where the paths P, P_α, and P_β are those shown in Fig. 4.1.2. P is the previously encountered infinite path along which $\tau(s)$ decreases most rapidly away from the vicinity of the pertinent saddle point(s) near the origin, and P_α and P_β are paths connecting the endpoints of P at $\pm\infty$ with the given endpoints s_α and

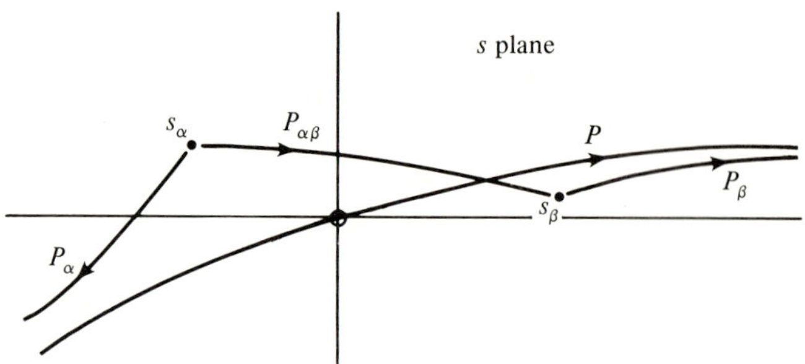

FIG. 4.1.2   Contours for evaluation of a finite integral.

s_β, respectively. It is assumed herein that $\operatorname{Re}\tau(s) < \operatorname{Re}\tau(0)$ along all path segments leading away from the origin $s = 0$, and that $\operatorname{Re}\tau(s) < \operatorname{Re}\tau(s_{\alpha,\beta})$ along the contours $P_{\alpha,\beta}$, respectively. It should also be emphasized that the presence of any singularities of $G(s)$ near $s = 0$, such as poles or branch points, must be taken into account in the deformation of $P_{\alpha\beta}$ into the three paths in Eq. (14) [see Eq. (4) and Fig. 4.1.1].

The asymptotic evaluation of the integral $I(\Omega)$ in Eq. (14) along the infinite path P proceeds as before. To estimate the contribution from the remaining integrals as $\Omega \to \infty$, we make use of the fact that $\tau'(s) \neq 0$ on P_α and P_β (i.e., there are no saddle points on the contours P_α and P_β). By an integration by parts one obtains

$$I_\alpha(\Omega) = \int_{P_\alpha} G(s)e^{\Omega\tau(s)}\,ds = \frac{1}{\Omega}\int_{P_\alpha} \frac{G(s)}{\tau'(s)}\frac{d}{ds}e^{\Omega\tau(s)}\,ds, \qquad (15)$$

$$= -\frac{1}{\Omega}\frac{G(s_\alpha)e^{\Omega\tau(s_\alpha)}}{\tau'(s_\alpha)} - \frac{1}{\Omega}\int_{P_\alpha} \frac{d}{ds}\left[\frac{G(s)}{\tau'(s)}\right]e^{\Omega\tau(s)}\,ds,$$

$$(16)$$

where it has been recognized that the exponential term vanishes at infinity. The same procedure can be repeated for the integral in Eq. (16). In fact, the integration-by-parts technique is frequently an effective means for asymptotic evaluation of integrals whose integrands do not contain a saddle point on or near the contour of integration. One notes from Eq. (16) that the magnitude of $I_\alpha(\Omega)$ as $\Omega \to \infty$ is determined by the factor $(1/\Omega)e^{\Omega\tau(s_\alpha)}$. Similarly, the magnitude of the integral $I_\beta(\Omega)$ over the path P_β is determined by the factor

$(1/\Omega)e^{\Omega\tau(s_\beta)}$. On the other hand, the contribution to $I(\Omega)$ over the path P traversing a saddle point has an exponential dependence $e^{\Omega\tau(0)}$ [see Eqs. (5), (6), and (7)]. If $\mathrm{Re}\,\tau(s_{\alpha,\beta}) < \mathrm{Re}\,\tau(0)$, one notes that $|I_{\alpha,\beta}|/|I|$ is proportional to $e^{\Omega\mathrm{Re}[\tau(s_{\alpha,\beta})-\tau(0)]}$ (i.e., the ratio is exponentially small as $\Omega \to \infty$). Thus, just as for the infinite integral, the predominant contribution to the finite integral in Eq. (13) arises from the neighborhood of the saddle point(s) of $\tau(s)$, and we may write

$$I_{\alpha\beta}(\Omega) \sim I(\Omega) \qquad \text{as } \Omega \to \infty. \tag{17}$$

If the exponential term in the integrand of Eq. (13) has the same magnitude at the endpoint s_α as at the saddle point $s = 0$ [i.e., $\mathrm{Re}\,\tau(s_\alpha) = \mathrm{Re}\,\tau(0)$], the endpoint contribution in Eq. (14) can still be estimated as in Eq. (16) and its magnitude is proportional to $(1/\Omega)e^{\Omega\mathrm{Re}\,\tau(0)}$. In this instance one requires for the above comparison a better estimate of the asymptotic behavior of $I(\Omega)$. It is shown in Eq. (4.3.7) that for an Mth-order saddle point of $q(z)$ at $z = z_s$ [see Eq. (6)], $I(\Omega)$ behaves like $\Omega^{-1/(M+1)}e^{\Omega\tau(0)}$. Thus, the major contribution to $I_{\alpha\beta}(\Omega)$ as $\Omega \to \infty$ still arises from the vicinity of the saddle point(s), although the endpoint contribution is now smaller in magnitude only by an algebraic factor of the form $\Omega^{-\gamma}$, where γ is a positive number less than 1 whose exact value depends on the particular form of $\tau(s)$.

The applicability of the integration-by-parts technique fails when the endpoint approaches the saddle point, since $\tau'(s_\alpha) \to 0$ in Eq. (16). Under these circumstances, the saddle-point and endpoint contributions can no longer be treated separately and the asymptotic evaluation involves new canonical integrals. Two specific cases are considered: a single first-order saddle point (Sec. 4.6a), and two adjacent first-order saddle points (Sec. 4.6b) near an endpoint of the integration interval. The former leads to a canonical integral

$$\int_{s_\alpha}^{``\infty"} e^{-\Omega s^2}\,ds,$$

which is expressible in terms of the Fresnel integral, whereas the latter requires knowledge of the "incomplete" Airy function

$$\int_{s_\alpha}^{``\infty"} e^{\tau s - s^3/3}\,ds.$$

4.1b Saddle Points and Paths of Constant Level and Constant Phase[2-4]

Saddle points

The various classes of integrals of the type shown in Eq. (1) can be distinguished by the form of the argument function $q(z)$. A pictorial characterization is effected by a "level map" of the behavior of $q(z)$ over the complex z plane. The resulting picture of the $q(z)$ terrain with its mountains, valleys, and interconnecting passes provides an appropriate physical basis for distinguishing and selecting the integration paths $\bar{P}$ of Sec. 4.1a. Since the stationary points z_s of

$q(z)$ distinguish the passes, it is appropriate to investigate first the behavior of the analytic function $q(z)$ in the neighborhood of such a stationary point. Let us write

$$q(z) = u(x, y) + iv(x, y), \qquad z = x + iy, \tag{18}$$

where u, v, x, and y are real. A stationary point z_s of $q(z)$ is determined by

$$\frac{dq}{dz} = 0 \qquad \text{at } z = z_s; \tag{19}$$

in view of the analytic behavior of q about z_s, it is thereby implied that the functions u and v are also stationary at (x_s, y_s),

$$\frac{\partial u}{\partial x} = \frac{\partial u}{\partial y} = \frac{\partial v}{\partial x} = \frac{\partial v}{\partial y} = 0 \qquad \text{at } x = x_s, \quad y = y_s. \tag{20}$$

Although Eqs. (20) are satisfied, the surfaces $u(x, y) = $ constant and $v(x, y) = $ constant do not have absolute maximum or minimum values at (x_s, y_s). This is seen readily from the Cauchy–Riemann equations, satisfied by u and v,

$$\frac{\partial u}{\partial x} = \frac{\partial v}{\partial y}, \qquad \frac{\partial u}{\partial y} = -\frac{\partial v}{\partial x}, \tag{21}$$

from which it follows that for a first-order saddle point $(d^2q/dz_s^2 \neq 0)$,

$$\frac{\partial^2 u}{\partial x^2} = -\frac{\partial^2 u}{\partial y^2}, \qquad \frac{\partial^2 v}{\partial x^2} = -\frac{\partial^2 v}{\partial y^2}. \tag{22}$$

Thus, if at (x_s, y_s) the curvature of the surface $u(x, y) = $ constant, or $v(x, y) = $ constant, is positive along the x direction, it is negative along the (perpendicular) y direction. The stationary points z_s are therefore "saddle points," as noted in Fig. 4.1.3, and locate pass regions in the $q(z)$ terrain. Depending on the choice of path through z_s, the magnitude of $u(x, y)$ and therefore of $\exp|\Omega q(z)|$ may increase, decrease, or remain constant along the path.

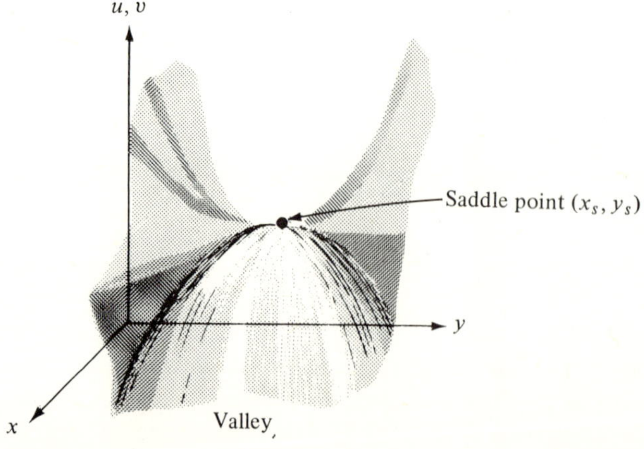

FIG. 4.1.3 Relief map of the functions u or v in the vicinity of a first-order saddle point.

Paths of constant level and constant phase

We now seek a criterion that determines the selection of the "steepest paths" through a saddle point (i.e., those paths along which the magnitude of $\exp[\Omega q(z)]$ changes most rapidly). In view of Eq. (18) these will be the paths along which $u(x, y)$ changes most rapidly. If, as shown in Fig. 4.1.4, ds denotes

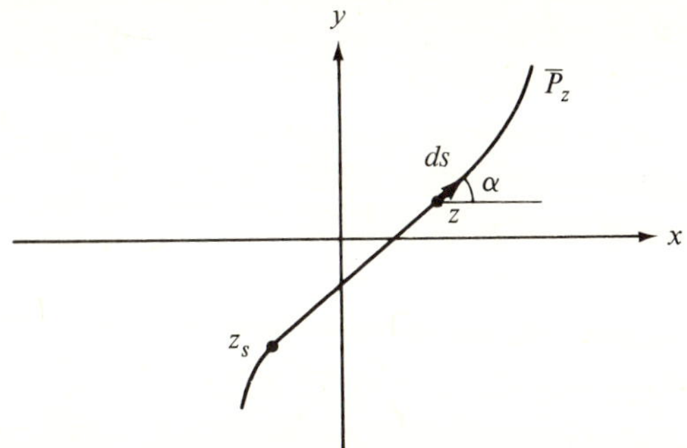

FIG. 4.1.4 Steepest-descent path in the z plane.

an element of length along a path of integration $\bar{P}_z$ through the saddle point z_s, the rate of change of u along $\bar{P}_z$ is given by

$$\frac{du}{ds} = \frac{\partial u}{\partial x}\frac{dx}{ds} + \frac{\partial u}{\partial y}\frac{dy}{ds} = \frac{\partial u}{\partial x}\cos\alpha + \frac{\partial u}{\partial y}\sin\alpha, \tag{23}$$

where α is the angle between the element ds and the x axis. The values of α for which du/ds is a maximum are defined by

$$\frac{\partial^2 u}{\partial\alpha\,\partial s} = 0 = -\frac{\partial u}{\partial x}\sin\alpha + \frac{\partial u}{\partial y}\cos\alpha, \tag{24a}$$

whence via the Cauchy–Riemann equations (21),

$$0 = -\frac{\partial v}{\partial y}\frac{dy}{ds} - \frac{\partial v}{\partial x}\frac{dx}{ds} = -\frac{dv}{ds}. \tag{24b}$$

Thus, $v = $ constant along the path on which u changes most rapidly, so the steepest path is a constant-phase path. Similarly, one can show from an examination of the maximum values of dv/ds that the magnitude of the exponential (i.e., the "level") is constant along the paths of most rapid phase variation. Although the path in Fig. 4.1.4 is shown passing through a saddle point, the above conclusions apply as well to constant-level and constant-phase paths through any given point z.

To examine the disposition of the constant-phase and constant-level paths in the neighborhood of a saddle point z_s, we expand $q(z)$ in a power series about the point z_s:

$$q(z) = q(z_s) + \frac{q''(z_s)}{2!}(z - z_s)^2 + \cdots, \tag{25}$$

where the double prime denotes the second derivative with respect to the argument. Then, for the case of a first-order saddle point,

$$e^{\Omega q(z)} \approx e^{\Omega q(z_s)} e^{(1/2)\Omega q''(z_s)(z-z_s)^2}, \tag{26a}$$

or

$$e^{\Omega q(z)} \approx e^{\Omega q(z_s)} e^{|(1/2)\Omega q''(z_s)(z-z_s)^2|(\cos 2\psi + i\sin 2\psi)}, \tag{26b}$$

where

$$\psi = \arg(z - z_s) + \tfrac{1}{2}\arg q''(z_s). \tag{26c}$$

Since $\arg q''(z_s)$ is constant, ψ changes only with $\arg(z - z_s)$. One notes the following behavior of $\exp[\Omega q(z)]$ along the various paths $\psi = $ constant originating at z_s:

$\psi = \pm\pi/4, \pm 3\pi/4$ (*paths of constant level*): Phase of $e^{\Omega q(z)}$ varies most rapidly, $|e^{\Omega q(z)}| = $ constant.

$\psi = 0, \pi$ (*paths of steepest ascent*): $|e^{\Omega q(z)}|$ increases most rapidly, phase of $e^{\Omega q(z)} = $ constant.

$\psi = \pm\pi/2$ (*paths of steepest descent*): $|e^{\Omega q(z)}|$ decreases most rapidly, phase of $e^{\Omega q(z)} = $ constant.

The above relations are illustrated in Fig. 4.1.5(a); the regions wherein the magnitude of $\exp[\Omega q(z)]$ increases or decreases characterize the mountain and valley regions, respectively.

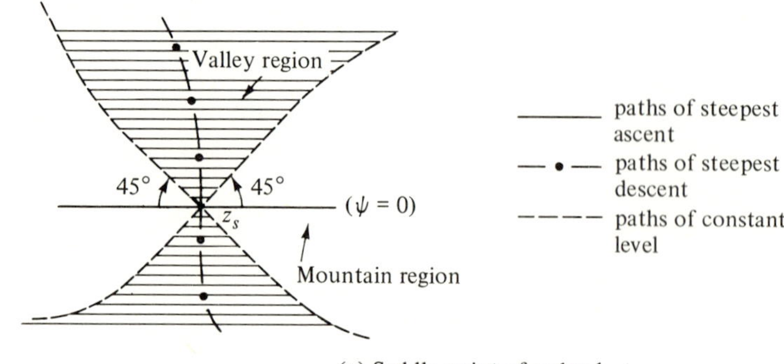

(a) Saddle point of order 1

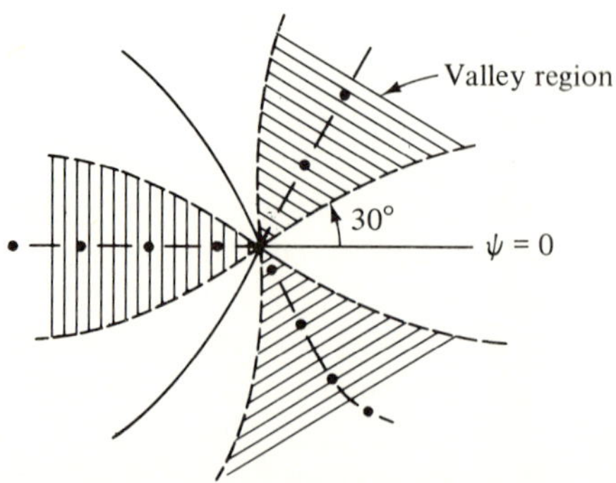

(b) Saddle point of order 2

FIG. 4.1.5   Behavior in the vicinity of a saddle point.

For a second-order saddle point, both $q'(z_s) = 0$ and $q''(z_s) = 0$. In this instance the expansion of $q(z)$ near the saddle point has the form

$$q(z) = q(z_s) + \frac{q^{(3)}(z_s)}{3!}(z - z_s)^3 + \cdots, \tag{27}$$

so

$$e^{\Omega q(z)} \approx e^{\Omega q(z_s)} \exp\left[\left|\frac{1}{3!}\, q^{(3)}(z_s)(z - z_s)^3\right|(\cos 3\psi + i\sin 3\psi)\right], \tag{27a}$$

where

$$\psi = \arg(z - z_s) + \tfrac{1}{3}\arg q^{(3)}(z_s). \tag{27b}$$

There exist now three mountain regions and three valley regions in the vicinity of the saddle point, as shown in Fig. 4.1.5(b). The angle between any two adjacent lines at the saddle point is 30°. A contour plot in the vicinity of an *m*th-order saddle point is constructed analogously and exhibits $(m + 1)$ mountain and valley regions.

If the given path of integration can be deformed into paths of steepest descent through one or more saddle points, then since the magnitude of the exponential term decreases most rapidly away from the saddle points, an approximate evaluation of the integral in Eq. (1) for large values of Ω can be phrased in terms of contributions only from the vicinity of the saddle points. On the other hand, if a constant-level path is chosen, an approximate evaluation for large Ω can also be carried out in terms of contributions only from the vicinity of the saddle points since the phase is stationary there and the exponential term oscillates so rapidly over the remainder of the path that the corresponding integral contributes negligibly to the overall result ("stationary-phase" procedure; see Sec. 4.2c). Since different paths of integration are involved in the two procedures, the asymptotic evaluation of the integral in Eq. (1) along a steepest-descent path need not yield the same result as that along a path of constant level. The two results will be identical only if the constant-level path can be continuously deformed into the steepest-descent path; this is possible if the two paths have the same termination (for example, at "infinity") and if $f(z)$ has no singularities in the region between the paths.

For rather general forms of the function $q(z)$, the determination of the complete steepest-descent paths [i.e., those paths along which Im $q(z)$ = constant] may be quite complicated. In this instance, one may proceed less stringently by utilizing steepest-descent paths only in the immediate vicinity of the saddle points, where their progress is easily ascertained (see Fig. 4.1.5). The requirement on the remaining path segments L is merely that they proceed along a level lower than that at the saddle points. If a pertinent saddle point is at the level $q(z_s)$ and if Re $q(z) \leq$ Re $q(z_1)$ along L, where Re $q(z_1) <$ Re $q(z_s)$, then the relative error incurred in the asymptotic approximation by neglecting the contribution from L is $O\{\exp[\Omega q(z_1) - \Omega q(z_s)]\}$ (see Sec. 4.1a).† The difference

---

†The notation $g(\Omega) = O[h(\Omega)]$ as $\Omega \to \infty$ implies that $[g(\Omega)/h(\Omega)]$ remains bounded as $\Omega \to \infty$. The notation $g(\Omega) = o[h(\Omega)]$ as $\Omega \to \infty$ implies that $[g(\Omega)/h(\Omega)] \to 0$ as $\Omega \to \infty$.

between the asymptotic evaluation of the integral in the present case and the result obtained with the complete steepest-descent path SDP lies in the exponential error estimate, based here on the level at z_1, whereas for the complete SDP it involves the distance to the nearest singularity (see Sec. 4.2b). The results are therefore asymptotically equivalent if exponentially small terms are neglected. It must be emphasized that the retention of exponentially small contributions (e.g., from another saddle point, or from a singularity, situated at a lower level than the dominant saddle point) may be justified only if its magnitude exceeds that of the error inherent in the SDP approximation.

The simplified procedure described above has been employed in various problems discussed in this volume (see Figs. 7.5.9 and 8.3.4).

4.2 ISOLATED FIRST-ORDER SADDLE POINTS

4.2a First-order Approximation

If in the integrand of

$$I(\Omega) = \int_{\mathrm{SDP}} f(z)e^{\Omega q(z)}\, dz, \tag{1}$$

the function $f(z)$ has no singularities near an isolated first-order saddle point z_s of $q(z)$, where $q'(z_s) = 0$, $q''(z_s) \neq 0$, the asymptotic approximation of $I(\Omega)$ is given by [see Eq. (17) below for a complete asymptotic expansion]

$$I(\Omega) \sim \sqrt{\frac{-2\pi}{\Omega q''(z_s)}}\, f(z_s)e^{\Omega q(z_s)}, \qquad \Omega \to \infty. \tag{1a}$$

One must choose $\arg(\sqrt{\ \ }) = \varphi \equiv \arg(dz)_{z_s}$, with dz denoting an element at z_s along the steepest-descent path SDP [see Fig. 4.2.1 and Eq. (4.1.26)]. When

FIG. 4.2.1 Integration paths in the z plane.

$q(z) = i\hat{q}(z)$, with $\hat{q}$ denoting a real function of z, and z_s is real, Eq. (1a) may be written as [see Eq. (20a) below]

$$I(\Omega) = \int_{\mathrm{SDP}} f(z)e^{i\Omega\hat{q}(z)}\, dz \sim \sqrt{\frac{2\pi}{\Omega|\hat{q}''(z_s)|}}\, f(z_s)e^{i\Omega\hat{q}(z_s)\pm i\pi/4}, \qquad \hat{q}''(z_s) \gtrless 0, \tag{1b}$$

provided that $\mathrm{Re}\,(dz)$ increases along the SDP near z_s (see Fig. 4.2.1).

Analytical details

Since the isolated saddle point at z_s is of the first order, the pertinent change of variable from z to s is that given in Eq. (4.1.6), with $M = 1$:

$$q(z) = \tau(s) = q(z_s) - s^2. \tag{2}$$

The steepest-descent path P in the s plane, along which $\operatorname{Im}\tau(s) = $ constant, is clearly the real s axis. Thus, the integral in Eq. (4.1.5) becomes

$$I(\Omega) = e^{\Omega q(z_s)} \int_{-\infty}^{\infty} G(s)e^{-\Omega s^2}\, ds, \tag{3}$$

with $G(s)$ given via Eq. (4.1.3a) as

$$G(s) = f(z)\frac{dz}{ds}, \qquad \frac{dz}{ds} = \frac{-2s}{q'(z)}. \tag{3a}$$

Since $G(s)$ is assumed regular near $s = 0$, it can be expanded into a power series

$$G(s) = G(0) + G'(0)s + G''(0)\frac{s^2}{2!} + \cdots + G^{(n)}(0)\frac{s^n}{n!} + \cdots, \tag{4}$$

which converges uniformly inside a circle with finite radius r centered at $s = 0$, r being the distance to the nearest singularity of $G(s)$. Upon applying L'Hôpital's rule to the indeterminate form for dz/ds in Eq. (3a) when $s = 0$ (i.e., $z = z_s$), one evaluates the first coefficient of the expansion as

$$G(0) = f(z_s)\left(\frac{dz}{ds}\right)_{s=0}, \qquad \left(\frac{dz}{ds}\right)_{s=0} = \sqrt{\frac{-2}{q''(z_s)}}, \tag{5}$$

where $q''(z_s) \neq 0$ at the first-order saddle point. Since ds is positive along the path of integration and, in particular, at $s = 0$, $\arg(dz/ds)$ at $s = 0$ must be chosen equal to $\arg(dz)$ at z_s along the steepest descent path (see Fig. 4.2.1). This requirement specifies the square root function in Eq. (5). If $G(s)$ is approximated by $G(0)$ only (see Sec. 4.2b for a discussion of implications thereof), one obtains as the first-order asymptotic approximation for $I(\Omega)$ in Eq. (3):

$$I(\Omega) \sim G(0)e^{\Omega q(z_s)} \int_{-\infty}^{+\infty} e^{-\Omega s^2}\, ds \tag{6}$$

or, by Eq. (5), noting that the integral in Eq. (6) equals $\sqrt{\pi/\Omega}$:

$$I(\Omega) \sim \sqrt{\frac{-2\pi}{\Omega q''(z_s)}}\, f(z_s)e^{\Omega q(z_s)}, \qquad \Omega \to \infty. \tag{7}$$

These results continue to be applicable to integrals with finite endpoints if the conditions noted in Sec. 4.1a are satisfied.

Examples. To illustrate the use of Eq. (7), we treat two simple examples. First, consider the gamma function $\Gamma(\Omega + 1)$ defined by the integral

$$\Gamma(\Omega + 1) = \int_0^\infty e^{-x}x^\Omega\, dx = \Omega^{\Omega+1}\int_0^\infty e^{-\Omega z}z^\Omega\, dz = \Omega^{\Omega+1}\int_0^\infty e^{\Omega(\ln z - z)}\, dz, \tag{8}$$

whence $f(z) = 1$, $q(z) = \ln z - z$, $z_s = 1$, $q''(z_s) = -1$, and $\operatorname{Re} q(z) < -1$ for $z \neq 1$. Since the steepest-descent path in the z plane extends along the real z axis and the element of integration dz is positive along the path, one notes that $(dz/ds)_{s=0}$ is likewise positive. Thus, from Eq. (7),

$$\Gamma(\Omega + 1) \sim \Omega^{\Omega+1} \sqrt{\frac{2\pi}{\Omega}}\, e^{-\Omega}, \qquad \Omega \to \infty. \tag{9}$$

Since $q(0) < q(z_s)$, the endpoint contribution is negligible (see Sec. 4.1a).

Next, consider the modified Hankel function defined by the integral

$$K_0(\Omega) = \int_0^\infty e^{-\Omega \cosh z}\, dz, \qquad \Omega > 0. \tag{10}$$

Here, $f(z) = 1$, $q(z) = -\cosh z$, $z_s = 0$, $q''(z_s) = -1$. Again, Eq. (7) can be used directly except that we must include a factor $\frac{1}{2}$ since the stationary point is located at the endpoint $z = 0$ and hence the integral in Eq. (3) now runs only over a semiinfinite interval. Thus,

$$K_0(\Omega) \sim \sqrt{\frac{\pi}{2\Omega}}\, e^{-\Omega}, \qquad \Omega \to \infty. \tag{11}$$

4.2b Complete Asymptotic Expansion

A complete asymptotic expansion† for $I(\Omega)$ in Eq. (3) is obtained upon substituting for $G(s)$ the power-series expansion in Eq. (4) and integrating term by term:

$$I(\Omega) \sim e^{\Omega q(z_s)} \sum_{n=0}^{\infty} \frac{G^{(n)}(0)}{n!}\, I_n(\Omega). \tag{12}$$

The integral $I_n(\Omega)$ is evaluated in terms of the gamma function $\Gamma(z)$, defined in Eq. (8), as

$$I_n(\Omega) = \begin{cases} \displaystyle\int_{-\infty}^{\infty} s^n e^{-\Omega s^2}\, ds = \dfrac{\Gamma[(1+n)/2]}{\Omega^{(1+n)/2}}, & n \text{ even,} \tag{13a} \\[2ex] 0, & n \text{ odd,} \tag{13b} \end{cases}$$

where Eq. (13b) is a consequence of the symmetrical integration interval and the fact that the integrand is an odd function of s. The values of $\Gamma(n + \frac{1}{2})$, $n = 0, 1, 2, \cdots$ are readily inferred from the recursion formula

$$\Gamma(z + 1) = z\Gamma(z), \qquad \Gamma(\tfrac{1}{2}) = \sqrt{\pi}. \tag{14}$$

Alternatively, one may express $I_n(\Omega)$ as

---

†A function $I(\Omega)$ is said to have an asymptotic expansion

$$I(\Omega) \sim \sum_{n=0}^{\infty} a_n f_n(\Omega) \qquad \text{as } \Omega \to \infty$$

if, for any N and for $\arg \Omega$ in a given interval,

$$\frac{I(\Omega) - I_N(\Omega)}{f_N(\Omega)} \to 0 \qquad \text{as } \Omega \to \infty,$$

where $I_N(\Omega) = \sum_{n=0}^{N} a_n f_n(\Omega)$ [i.e., $f_{n+1}(\Omega)/f_n(\Omega) \to 0$ as $\Omega \to \infty$].[2-5]

$$I_n(\Omega) = (-1)^{n/2} \frac{d^{n/2}}{d\Omega^{n/2}} \int_{-\infty}^{\infty} e^{-\Omega s^2}\, ds \tag{15a}$$

$$= \left(-\frac{d}{d\Omega}\right)^{n/2} \sqrt{\frac{\pi}{\Omega}}, \qquad n \text{ even.} \tag{15b}$$

From Eqs. (15b) and (13a), one infers the recursion relation between I_n and I_{n+2}:

$$I_{n+2}(\Omega) = \left(-\frac{d}{d\Omega}\right) I_n(\Omega), \qquad n = 0, 2, 4, \ldots . \tag{16}$$

Thus, the complete asymptotic expansion for $I(\Omega)$ as $\Omega \to \infty$ is given by

$$I(\Omega) \sim \frac{e^{\Omega q(z_s)}}{\sqrt{\Omega}} \sum_{n=0}^{\infty} \frac{G^{(2n)}(0)}{(2n)!} \frac{\Gamma(n + \frac{1}{2})}{\Omega^n}, \tag{17}$$

or, alternatively,

$$I(\Omega) \sim e^{\Omega q(z_s)} \sum_{n=0}^{\infty} \frac{G^{(2n)}(0)}{(2n)!} \left(\sqrt{-\frac{d}{d\Omega}}\right)^{2n} \sqrt{\frac{\pi}{\Omega}}. \tag{18a}$$

Equation (18a) can be written in a convenient operator notation as

$$I(\Omega) \sim e^{\Omega q(z_s)} G_e\left(\sqrt{-\frac{d}{d\Omega}}\right) \sqrt{\frac{\pi}{\Omega}}, \tag{18b}$$

where the even function $G_e(x)$ is expressed in terms of a power series about $x = 0$ as follows:

$$G_e(x) = \sum_{n=0}^{\infty} \frac{G_e^{(2n)}(0)}{(2n)!} x^{2n}. \tag{18c}$$

That Eqs. (17) or (18a) indeed constitute the asymptotic expansion of $I(\Omega)$ as $\Omega \to \infty$ follows from the recognition that the ratio between successive terms of the series [i.e., between the $(N + 1)$th and Nth terms], approaches zero as $\Omega \to \infty$, for any N (see footnote on p. 384)].

The term-by-term integration in Eq. (12) is not rigorously justifiable since the radius of convergence r of the power-series expansion is generally finite, so the series representation for $G(s)$ cannot be employed over the infinite range in s. The error incurred by this procedure may be estimated on dividing the integration interval into three parts. In the first, with $|s| = r_0 < r$, Eq. (4) is applicable while in the other two, with $-\infty < s < -r_0$ and $r_0 < s < \infty$, the function $G(s)$ is retained intact. The saddle-point contribution from the first interval is given by Eq. (12) or (17); the endpoint contribution is $O[\exp(-\Omega r_0^2)]$ [see Eq. (4.1.16)]. The remaining integrals from $|s| = r_0$ to ∞ are of the same exponential order of magnitude since the integrand decays along the entire path [$G(s)$ in Eq. (3) is dominated by $\exp(-\Omega s^2)$]. This exponentially small error, less significant than any of the algebraically small terms in the expansion of Eq. (17), assures the "asymptotic" validity of Eqs. (12) or (17) as $\Omega \to \infty$, provided that r_0 is finite [i.e., the saddle point is separated from singularities of $G(s)$].

One notes that the lowest-order approximation to $I(\Omega)$, which arises from the $n = 0$ term in Eqs. (17) or (18a), has been given in Eq. (7). For the evaluation of the higher-order terms, one requires a knowledge of the higher-order derivatives of $G(s) = f(z)dz/ds$ evaluated at $s = 0$. Formally, the derivatives $d^n z/ds^n$ can be obtained by successive differentiation of Eq. (3a) and evaluation of the resulting indeterminate forms at $s = 0$. An alternative procedure is to expand $q(z)$ in Eq. (2) in a power series about $z = z_s$ and thereby find s as a function of $z - z_s$. To obtain the required behavior of $z - z_s$ as a function of s, this power series must be inverted. One of the several possible techniques for the inversion of power series is presented in Appendix A. An explicit expression for $G^{(2)}(0)$ is given in Eq. (A6).

An important consideration in the use of asymptotic series is the error made on stopping the series after N terms. It may be shown[6] that if $\varphi(\zeta)$ is analytic on the real axis and, nearest the origin, possesses a singularity at $\zeta_1 = r \exp(i\alpha)$, $\alpha \neq 0$, where r is the modulus of ζ_1 and α its phase, then the integral

$$\hat{I} = \int_0^\infty e^{-\Omega\zeta} \zeta^\mu \varphi(\zeta)\, d\zeta \tag{19a}$$

has the exact representation

$$\hat{I} = \sum_{n=0}^{N-1} u_n + R_N, \qquad u_n = \frac{\varphi^{(n)}(0)\Gamma(n + \mu + 1)}{\Omega^{n+\mu+1} n!}, \tag{19b}$$

where the u_n result from termwise integration of the power-series expansion (to N terms) of $\varphi(\zeta)$. The remainder R_N may be shown to be given approximately by $u_N(1 - e^{-i\alpha})^{-1}$, so the best approximation to $\hat{I}$ is obtained by stopping the series at the smallest term. The integrals considered in Eq. (3), or in Eq. (4.3.5a) for the more general case of a saddle point of order M, are readily transformed into $\hat{I}$ in Eq. (19a). In addition, there exist exponentially small error terms, as noted above.

Although it has been assumed throughout that Ω is real, it is evident that $I_n(\Omega)$ in Eq. (13a) can be continued analytically into the range $|\arg \Omega| < \pi/2$ since the integral converges in the extended range of Ω for which $\mathrm{Re}\,(\Omega s^2) > 0$. If Ω is complex, this process of analytic continuation is frequently more convenient than the inclusion of the factor $\exp(i \arg \Omega)$ in $q(z)$. Thus, in the range $|\arg \Omega| < \pi/2$, the asymptotic expansions in Eqs. (17) or (18a) are also valid for complex Ω, uniformly in $\arg \Omega$.

4.2c   First-order, "Stationary-Phase" Evaluation of Finite Integrals

It was pointed out in Sec. 4.1a that the first-order asymptotic formula in Eq. (7), as well as the complete asymptotic expansion in Eq. (17), is valid for finite integrals, provided that $\mathrm{Re}\,q(z) < \mathrm{Re}\,q(z_s)$ on the portions of the path away from the saddle point z_s and that the integrand has no singularities near z_s. If the endpoint of a finite integral coincides with a saddle point z_s, then in the s plane one obtains integrals as in Eq. (13a) except that one of the limits of integration in zero. Thus, the contribution from such a saddle point is one half that given in Eq. (17). If $\operatorname{Re} q(z) = \operatorname{Re} q(z_s)$ along the contour, we are dealing with a constant-level path along which the phase of the exponential term $\exp[\Omega q(z)]$ varies most rapidly. As mentioned in Sec. 4.1b, a first-order asymptotic evaluation can then be carried out by a stationary-phase argument and yields the same result as the first-order steepest-descent evaluation, provided that $f(z)$ has no relevant singularities.

To highlight these remarks, let us introduce into the integral in Eq. (13a) (with $n = 0$) the change of variable $\bar{s} = s \exp(-i\pi/4)$. The resulting contour of integration in the $\bar{s}$ plane proceeds along a $-45°$ line which can be deformed into the real $\bar{s}$ axis. Thus, one obtains an integral as in Eq. (13a) except that Ω is replaced by $i\Omega$ so that the real axis in the $\bar{s}$ plane is a constant-level path. The desired result follows from Eq. (7) upon continuing the real variable Ω to imaginary values. In particular, if $q(x)$ is a real function of the real variable x, one obtains the "stationary-phase" formula [omitting the caret used in Eq. (1b)]

$$I(\Omega) = \int_{x_a}^{x_b} f(x)e^{i\Omega q(x)}\,dx \sim I_s(\Omega)U[(x_s - x_a)(x_b - x_s)]$$

$$+ I_e(\Omega) + O\left(\frac{1}{\Omega^{3/2}}\right), \qquad \Omega \to \infty, \qquad (20)$$

where $\Omega > 0$, and $U(\alpha) = 1$ or 0 for $\alpha > 0$ and $\alpha < 0$, respectively. I_s is the lowest-order contribution from the stationary-phase point, namely,

$$I_s(\Omega) = \sqrt{\frac{2\pi}{\Omega|q''(x_s)|}}\, f(x_s)e^{i\Omega q(x_s)\pm i\pi/4}, \qquad q''(x_s) \gtrless 0, \qquad (20a)$$

and $I_e(\Omega)$ is the lowest-order contribution from the endpoints [see Eq. (4.1.16)],

$$I_e(\Omega) = \frac{1}{i\Omega}\left[\frac{f(x_b)}{q'(x_b)}\,e^{i\Omega q(x_b)} - \frac{f(x_a)}{q'(x_a)}\,e^{i\Omega q(x_a)}\right]. \qquad (20b)$$

The stationary point x_s [at which $q'(x_s) = 0$] is assumed to lie on the interval $x_a < x_s < x_b$; if the interval contains several stationary points, $I_s(\Omega)$ is a sum comprising terms representative of each x_s. When the saddle point coincides with either of the endpoints x_a or x_b, the corresponding endpoint contribution in Eq. (20b) is omitted and one takes $\frac{1}{2}$ times the stationary-point contribution in Eq. (20a). When the interval $x_a < x < x_b$ does not contain a stationary point, the Heaviside function vanishes and the integral $I(\Omega)$ is approximated by $I_e(\Omega)$ only. When an endpoint moves to infinity, the corresponding contribution is omitted. Equation (20) fails when the saddle point x_s approaches one of the endpoints; for example, when x_s moves continuously toward and across x_b, the $I_s(\Omega)$ term is discontinuous since the saddle point disappears from the integration interval, and the $I_e(\Omega)$ term diverges, since $q'(x_b) \to 0$ as $x_b \to x_s$. In this instance, one requires a more careful asymptotic evaluation that involves the error function or the Fresnel integral (see Sec. 4.6a).

Example

Use of Eq. (20) is illustrated by the asymptotic evaluation of the following integral representation for the Bessel function $J_n(\Omega)$:

$$J_n(\Omega) = \frac{e^{-in\pi/2}}{\pi} \int_0^\pi e^{i\Omega \cos x} \cos nx \, dx, \qquad n = 0, 1, 2, \cdots. \tag{21}$$

In this case, $f(x) = \cos nx$, $q(x) = \cos x$, $x_s = m\pi$, $m = 0, \pm 1, \pm 2, \cdots$. The interval of integration contains the two saddle points at $x_1 = 0$, $x_2 = \pi$, with $q''(x_1) = -1$, $q''(x_2) = 1$. Since the saddle points are situated at the endpoints of the interval of integration, a factor $\frac{1}{2}$ must be included in Eq. (20). Thus, as $\Omega \to \infty$,

$$J_n(\Omega) \sim \frac{e^{-in\pi/2}}{\pi} \frac{1}{2} \sqrt{\frac{2\pi}{\Omega}} [e^{i\Omega - i\pi/4} + e^{in\pi} e^{-i\Omega + i\pi/4}], \tag{22a}$$

which simplifies to

$$J_n(\Omega) \sim \sqrt{\frac{2}{\pi\Omega}} \cos\left[\Omega - \frac{n\pi}{2} - \frac{\pi}{4}\right], \qquad \Omega \to \infty. \tag{22b}$$

4.2d Steepest-Descent Evaluation of a Typical Diffraction Integral

Consider the integral

$$I_1(\Omega, \alpha, \beta) = \int_{\bar{P}} \frac{e^{i\Omega \cos (z-\alpha)}}{z - \beta} \, dz, \qquad 0 \le \alpha < \frac{\pi}{2}, \tag{23}$$

taken over the path $\bar{P}$ shown in Fig. 4.2.2. This integral arises in several of the diffraction problems to be studied in the following chapters. Since

$$\mathrm{Im} \cos (z - \alpha) = -\sin (x - \alpha) \sinh y, \tag{24}$$

and Ω is assumed positive, the exponential term decays in the regions $y > 0$, $-\pi < x - \alpha < 0$, and $y < 0$, $0 < x - \alpha < \pi$, shown shaded in Fig. 4.2.2. The integrand contains a simple pole at $z = \beta$, where β is arbitrary. Upon

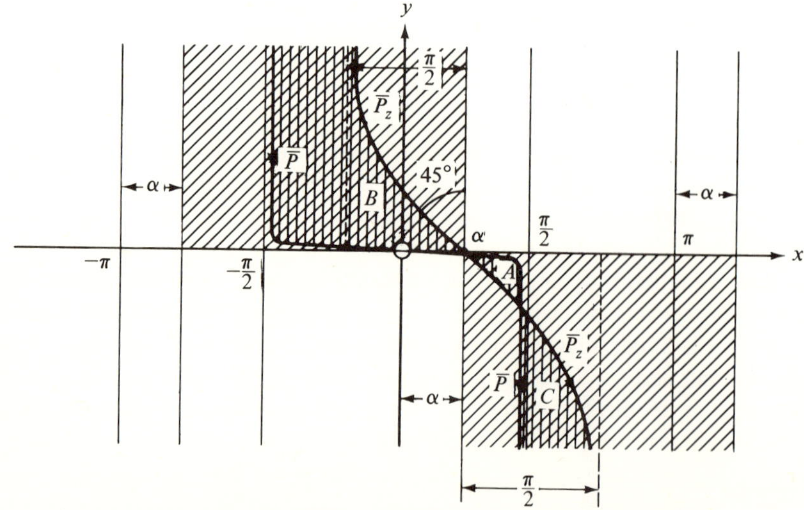

FIG. 4.2.2 Contours of integration in the z plane $(z = x + iy)$.

comparison with Eq. (4.1.1) one notes that

$$q(z) = i \cos (z - \alpha), \qquad f(z) = \frac{1}{z - \beta}. \tag{25}$$

The pertinent saddle point z_s in the z plane is located at

$$\frac{d}{dz} q(z) = -i \sin (z - \alpha) = 0 \qquad \text{or at } z_s = \alpha. \tag{26a}$$

The steepest-descent path $\bar{P}_z$ through the saddle point is defined by

$$\text{Im } q(z) = \text{Im } q(z_s) = i. \tag{26b}$$

The change of variable to the s plane is then effected via Eq. (2) by

$$i \cos (z - \alpha) = i - s^2, \tag{27a}$$

or

$$s = \pm \sqrt{2}\, e^{i\pi/4} \sin \frac{z - \alpha}{2}. \tag{27b}$$

Although we could proceed immediately to the s plane via the transformation in Eq. (27b), we study first the nature of the steepest-descent path $\bar{P}_z$ in the z plane. Along $\bar{P}_z$, s is real, and the slope of $\bar{P}_z$ at $z = z_s = \alpha$ is inferred from Eq. (5) or Eq. (27b) to be

$$\left. \frac{dz}{ds} \right|_{s=0} = \pm \sqrt{2}\, e^{-i\pi/4}. \tag{28}$$

Thus, $\bar{P}_z$ makes an angle of $(-45°)$ with the x axis at $z = \alpha$. The direction of integration along $\bar{P}$ in Fig. 4.2.2 suggests a choice of the indicated direction along $\bar{P}_z$, so that arg $(z - z_s) = -\pi/4$ along $\bar{P}_z$ near $z = z_s$, and the positive sign is chosen in Eqs. (28) and (27b). The complete steepest-descent path can be plotted readily from Eq. (26b) and requires that Im $[i \cos (z - \alpha)] = i$ along $\bar{P}_z$, or

$$x - \alpha = \cos^{-1} (\text{sech } y) \qquad \text{along } \bar{P}_z. \tag{29}$$

Equation (29) yields the x coordinate of any point on $\bar{P}_z$ for an assumed value of y. The resulting path is shown in Fig. 4.2.2 and is asymptotic to the lines $x = \alpha \pm \pi/2$.

The transformation of the path $\bar{P}$ from the z plane to the s plane is accomplished via Eq. (27b), where it is recalled that the positive sign is chosen. The resulting contour P' is shown in Fig. 4.2.3 and has as its asymptotes in the lower half of the s plane the lines arg $s = -\alpha/2$ and arg $s = \pi + \alpha/2$. Corresponding regions in the z and s planes where the exponential term decays in magnitude are shown shaded with slanted lines. Since $ds/dz = 0$ at $z - \alpha = \pm\pi$, the transformation in Eq. (27b) gives rise to first-order branch-point singularities at $s = \pm s_b = \pm \sqrt{2} \exp (i\pi/4)$ in the s plane. These branch points and the associated choice of branch cuts are shown in Fig. 4.2.3. The steepest-descent path P in the s plane, corresponding to the path $\bar{P}_z$ in the z plane, extends along the real s axis.

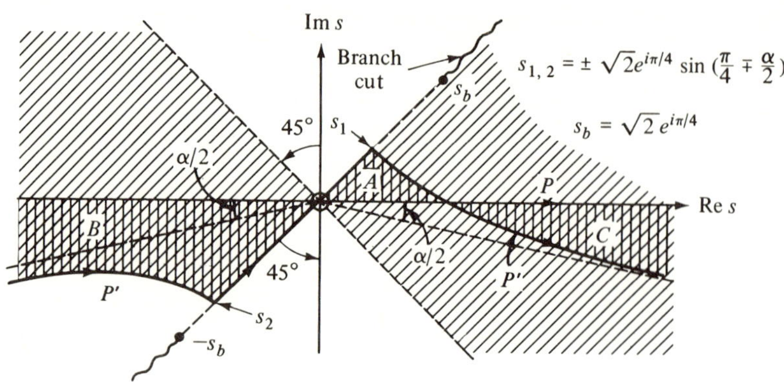

FIG. 4.2.3 Contours of integration in the *s* plane.

Since all paths considered begin and terminate in a shaded region of the *z* or *s* plane, it is evident that the contours $\bar{P}$ and P' can be deformed at infinity into the contours $\bar{P}_z$ and P, respectively. For deformation of the paths in the remainder of the *z* or *s* planes, attention must be given to the location of the pole singularity at $z = \beta$ in Eq. (23). If $z = \beta$ [or $s = s_\beta = \sqrt{2}\,e^{i\pi/4} \sin(\frac{1}{2}\beta - \frac{1}{2}\alpha)$] is situated in any of the vertically shaded regions *A, B, C* in Fig. 4.2.2 (or Fig. 4.2.3), the residue at the pole must be taken into account in the contour deformation. Thus,

$$I_1(\Omega, \alpha, \beta) = 2\pi i e^{i\Omega \cos(\beta - \alpha)} \epsilon(\beta) + e^{i\Omega} \int_{-\infty}^{\infty} G(s)e^{-\Omega s^2}\, ds, \qquad (30)$$

where

$$\epsilon(\beta) = \begin{cases} +1, & \text{if } \beta \text{ (or } s_\beta) \text{ lies in regions } B \text{ or } C, \\ -1, & \text{if } \beta \text{ (or } s_\beta) \text{ lies in region } A, \\ 0, & \text{if } \beta \text{ (or } s_\beta) \text{ lies outside regions } A, B, C. \end{cases} \qquad (30a)$$

In view of the change of variable in Eqs. (27), $G(s)$ is given by

$$G(s) = \frac{1}{z - \beta}\frac{dz}{ds}, \qquad \frac{dz}{ds} = \frac{-2is}{\sin(z - \alpha)} = \frac{-2is}{\sqrt{1 - \cos^2(z - \alpha)}}. \qquad (30b)$$

The simple form of dz/ds permits a direct determination of the complete series expansion by the binomial theorem. One has

$$\frac{dz}{ds} = \sqrt{2}\,e^{-i\pi/4}\left(1 - \frac{is^2}{2}\right)^{-1/2}$$

$$= \sqrt{2}\,e^{-i\pi/4}\left(1 + \frac{is^2}{4} - \frac{3}{32}s^4 + \cdots\right), \qquad (30c)$$

which converges in the interior of a circle of radius $|s| = \sqrt{2}$ passing through the branch-point singularities. Thus, the range of convergence of the power-series expansion for $G(s)$ is $|s| < \sqrt{2}$ if $|s_\beta| > \sqrt{2}$ and is $|s| < |s_\beta|$ if $|s_\beta| < \sqrt{2}$ (i.e., within a circle of finite radius provided $|s_\beta| > 0$). The first two terms in the asymptotic expansion of $I_1(\Omega, \alpha, \beta)$ in Eq. (23) as $\Omega \to \infty$ are therefore given via Eqs. (30), (17), and (A2) by

$$I_1(\Omega, \alpha, \beta) \sim 2\pi i e^{i\Omega \cos(\beta - \alpha)} \epsilon(\beta)$$

$$+ \frac{e^{i(\Omega - \pi/4)}}{\alpha - \beta} \sqrt{\frac{2\pi}{\Omega}} \left\{ 1 - \frac{i}{4\Omega} \left[\frac{4}{(\alpha - \beta)^2} + \frac{1}{2} \right] + \cdots \right\}, \qquad \Omega \to \infty.$$

$$(31)$$

Concerning the residue contribution from the first term in Eq. (31), one notes that the magnitude of the exponential term behaves like

$$\exp\left[-\Omega \left|\sin(\beta_r - \alpha) \sinh \beta_i\right|\right],$$

where β_r and β_i are the real and imaginary parts of β, respectively. If $\beta_i \neq 0$ and $\beta_r \neq \alpha$, the pole contribution is exponentially small and can be neglected in comparison with the remaining terms. On the other hand, if $\beta_i \to 0$ or if $\beta_r \to \alpha$, the residue contribution may be the dominant one since its magnitude then remains constant as $\Omega \to \infty$. Care should be exercised in the explicit retention of an exponentially small residue contribution when β is complex, since the asymptotic expansion in Eq. (31) has itself an exponentially small error (see p. 385). If this error term decays more slowly than $\exp[i\Omega \cos(\beta - \alpha)]$, the residue term is not significant.

Although the asymptotic representation in Eq. (31) remains valid as $\Omega \to \infty$ for any $\beta \neq \alpha$, the proximity of the pole and the saddle point is seen to influence the accuracy of the approximation for large fixed values of Ω. If the pole approaches the saddle point $[(\alpha - \beta) \to 0]$, the radius of convergence of the series representation for $G(s)$ shrinks to zero. The representation in Eq. (31) then becomes inapplicable since examination of the second term in the expansion indicates that the error incurred by use of even the first term becomes large [see Eq. (19b)]. One notes in this connection that the quantity $\sqrt{\Omega} |\alpha - \beta|$ plays a special role in assessing whether the pole is near enough to the saddle point to invalidate Eq. (31). If for $\Omega \gg 1$, one also has $\sqrt{\Omega} |\alpha - \beta| \gg 1$, the pole can be considered to be far from the saddle point and Eq. (31) applies; on the other hand, if $\sqrt{\Omega} |\alpha - \beta| \leq 1$, with $\Omega \gg 1$, the terms in the asymptotic series are no longer small and the validity of the expansion is in question. The dependence of the asymptotic expansion of $I_1(\Omega, \alpha, \beta)$ on the magnitude of $\sqrt{\Omega} |\alpha - \beta|$ is studied in detail in Sec. 4.4a, where we explore the evaluation of integrals whose integrands contain a pole near a saddle point.

4.2e Integrands with Two Relevant Isolated Saddle Points: Asymptotic Expansion of the Airy Integral

Consider the Airy function $\mathrm{Ai}(\sigma)$ defined by the integral

$$\mathrm{Ai}(\sigma) = \frac{1}{\pi} \int_0^\infty \cos\left(\frac{1}{3} z^3 + \sigma z\right) dz, \tag{32}$$

which can also be written as

$$\mathrm{Ai}(\sigma) = \frac{1}{2\pi} \int_{-\infty}^{\infty} e^{i(z^3/3 + \sigma z)} \, dz, \tag{32a}$$

or, upon transforming z into $-iz$, as

$$\mathrm{Ai}(\sigma) = \frac{1}{2\pi i} \int_{-i\infty}^{i\infty} e^{\sigma z - z^3/3} \, dz. \tag{32b}$$

The contour of integration in Eq. (32b) can be deformed away from the imaginary axis upon recognizing that the behavior of the integrand as $|z| \to \infty$ is determined by $\exp(-z^3/3)$. $\mathrm{Re}\,(z^3) > 0$ in the following sectors of the complex z plane: (1) $|\arg z| < \pi/6$; (2) $\pi/2 < \arg z < 5\pi/6$; (3) $-\pi/2 > \arg z > -5\pi/6$. Thus, by Cauchy's theorem, the integral in Eq. (32b) can be evaluated over any contour L_{32} in the complex z plane which, as shown in Fig. 4.2.4,

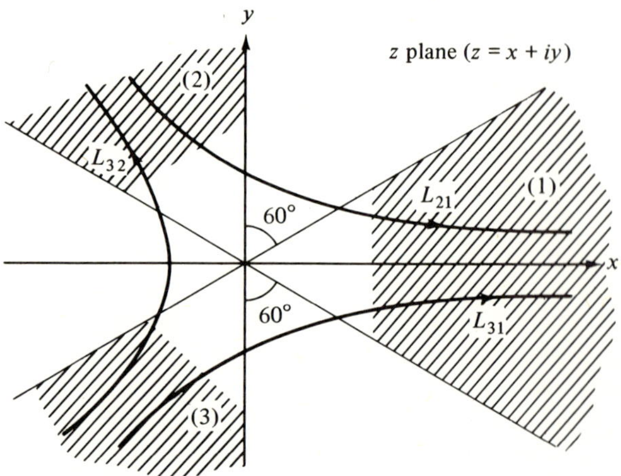

FIG. 4.2.4   Airy integral paths (Re $z^3 > 0$ in shaded regions).

begins at $|z| \to \infty$ in sector 3 and ends at $|z| \to \infty$ in sector 2. From Eq. (32b) one notes that $\mathrm{Ai}(\sigma)$ satisfies the differential equation

$$\left(\frac{d^2}{d\sigma^2} - \sigma\right)\mathrm{Ai}(\sigma) = -\frac{1}{2\pi i}\int_{L_{32}} e^{\sigma z - z^3/3} \, d\!\left(\sigma z - \frac{z^3}{3}\right) = 0, \tag{33}$$

where it has been recognized that the integrand is a perfect differential and vanishes at the endpoints of the path L_{32}.

A second independent solution of the differential equation (33) is defined as

$$\mathrm{Bi}(\sigma) = \frac{1}{2\pi}\int_{L_{21}+L_{31}} e^{\sigma z - z^3/3} \, dz \tag{34}$$

where the paths L_{21} and L_{31} are shown in Fig. 4.2.4.

The exponent in Eq. (32b) or (34) possesses stationary points of order 1 at

$$\sigma - z_s^2 = 0, \qquad \text{i.e., } z_s = \pm\sqrt{\sigma}. \tag{35}$$

If σ is large, the saddle points are widely separated and as $\sigma \to \infty$, one may obtain an asymptotic expansion of $\mathrm{Ai}(\sigma)$ and $\mathrm{Bi}(\sigma)$ by treating separately the contribution from each relevant saddle point.[4] To transform the exponent into the form $\Omega q(z)$, with $q(z)$ independent of Ω, we introduce the parameter

$$\Omega = \sigma^{3/2} \tag{36}$$

and change the variable in Eqs. (32b) and (34) from z to $\bar{z} = \Omega^{-1/3} z$ to obtain (assuming Ω and σ positive for the present)

$$\mathrm{Ai}(\sigma) = \frac{\Omega^{1/3}}{2\pi i} \int_{L_{32}} e^{\Omega(z - z^3/3)} \, dz, \tag{37a}$$

$$\mathrm{Bi}(\sigma) = \frac{\Omega^{1/3}}{2\pi} \int_{L_{21}+L_{31}} e^{\Omega(z - z^3/3)} \, dz. \tag{37b}$$

In Eqs. (37), the notation z instead of $\bar{z}$ has been retained for convenience. The saddle points of the exponent in the integrands are now located at

$$z_s = \pm 1. \tag{38}$$

The steepest-descent and steepest-ascent paths through the saddle points are determined by the constant-phase requirement: $\mathrm{Im}\, q(z) = \mathrm{Im}\, q(z_s)$. Since $q(z) = z - z^3/3$, where $z = x + iy$, one obtains for the equation of the steepest paths,

$$y(y^2 - 3x^2 + 3) = 0. \tag{39}$$

Thus, the steepest paths extend along the real z axis and along the hyperbola $3x^2 - y^2 = 3$. The steepest-descent paths through the saddle points at $z_s = \pm 1$ are shown in Fig. 4.2.5. It is evident from examination of $q(z_s)$ that the saddle point at $z_s = +1$ lies on a higher level than that at $z_s = -1$. Since the hyperbola in Fig. 4.2.5 is asymptotic to the lines $y = \pm \sqrt{3}\, x$ [i.e., $\arg z = \mp 2\pi/3$], one verifies from a comparison with Fig. 4.2.4 that the steepest-descent paths end in the shaded regions. Therefore, any of the contours in Fig. 4.2.4 can be deformed into appropriate steepest-descent paths through a saddle point.

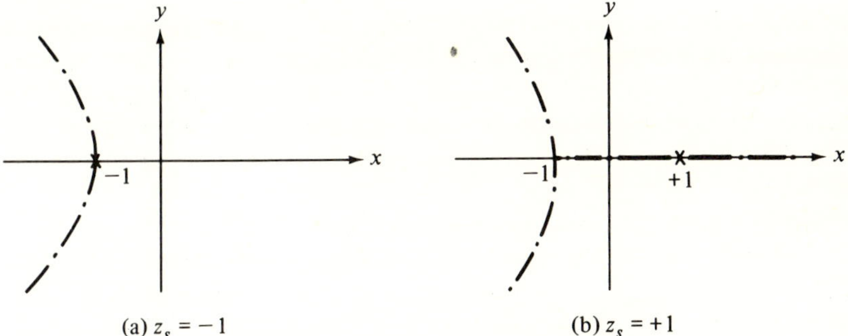

FIG. 4.2.5   Steepest-descent paths through saddle points $z_s = \pm 1$.

The formal asymptotic expansion of the integrals in Eqs. (37) can now be written down directly from Eq. (17). For $\mathrm{Ai}(\sigma)$, the pertinent saddle point is located at $z_s = -1$ and the steepest-descent paths are shown in Fig. 4.2.5(a). The transformation to the complex s plane is accomplished via

$$q(z) = z - \frac{z^3}{3} = -\frac{2}{3} - s^2, \tag{40}$$

where s is real and increases from $-\infty$ to $+\infty$ along the steepest-descent path. Thus,

$$\frac{dz}{ds} = \frac{-2s}{1 - z^2}, \tag{41a}$$

and, from Eq. (5),

$$\left.\frac{dz}{ds}\right|_{s=0} = \pm i. \tag{41b}$$

If it is assumed that $s = -\infty$ and $s = +\infty$ correspond, respectively, to the lower and upper endpoints of the path L_{32} in Fig. 4.2.4, then from the direction of integration along the path, one notes that $\arg(dz/ds) = +\pi/2$ at $z_s = -1$, so the plus sign is chosen in Eq. (41b). Thus, from Eqs. (18b), (36), and (37a), one obtains the formal asymptotic expansion

$$\mathrm{Ai}(\sigma) \sim \frac{\sqrt{\sigma}\, e^{-(2/3)\sigma^{3/2}}}{2\pi i} \left[G_e\left(\sqrt{-\frac{d}{d\Omega}}\right)\sqrt{\frac{\pi}{\Omega}} \right]_{\Omega = \sigma^{3/2}}, \qquad \sigma \to \infty, \tag{42}$$

whose first term is given in view of Eq. (41b) by

$$\mathrm{Ai}(\sigma) \sim \frac{1}{2\sqrt{\pi}\,\sigma^{1/4}} e^{-(2/3)\sigma^{3/2}}. \tag{42a}$$

It is relatively simple in the present case to obtain the complete expansion of $G(s) = dz/ds$ as a function of s and therefore the value of the general term $G^{(2n)}(0)$. The complete asymptotic expansion of $\mathrm{Ai}(\sigma)$ as in Eq. (42) is then found to be given explicitly by[4,5]

$$\mathrm{Ai}(\sigma) \sim \frac{1}{2\sqrt{\pi}\,\sigma^{1/4}} e^{-(2/3)\sigma^{3/2}} \sum_{n=0}^{\infty} \frac{\Gamma(3n + \frac{1}{2})}{\sqrt{\pi}\,(2n)!(-9\sigma^{3/2})^n}. \tag{43}$$

Although it has been assumed above that σ, and therefore Ω, are positive real, the results apply as well for complex Ω in the range $|\arg \Omega| < \pi/2$, as pointed out in Sec. 4.2b. In view of Eq. (36), the validity of Eqs. (42) and (43) can therefore be extended to complex values of σ lying in the sector $|\arg \sigma| < \pi/3$ [i.e., the expansion in Eq. (43) holds uniformly for $\arg \sigma$ in this range as $|\sigma| \to \infty$].

For the asymptotic evaluation of $\mathrm{Bi}(\sigma)$, defined in Eq. (37b), one deforms the contours L_{21} and L_{31} in Fig. 4.2.4 into the appropriate steepest paths through the saddle point at $z_s = +1$ in Fig. 4.2.5(b). For L_{21}, the path consists of the segment of the real z axis indicated in Fig. 4.2.5(b) and the upper branch of the hyperbola; for L_{31}, one employs the lower branch of the hyperbola in conjunction with the segment along the real axis. The transformation to the s plane for both L_{21} and L_{31} is now

$$q(z) = z - \frac{z^3}{3} = \frac{2}{3} - s^2, \qquad -\infty < s < \infty, \tag{44}$$

from which dz/ds is still given by Eq. (41a). However, from Eq. (5), $dz/ds = \pm 1$ at $s = 0$; from the direction of integration along L_{21} and L_{31} one notes that the positive sign must be chosen. Thus, as $\sigma \to \infty$, one obtains, to a first order,

$$\text{Bi}(\sigma) \sim \frac{1}{\sqrt{\pi}\,\sigma^{1/4}}\,e^{(2/3)\sigma^{3/2}}, \tag{45}$$

or, as in Eq. (43), the complete asymptotic expansion[4,5]

$$\text{Bi}(\sigma) \sim \frac{1}{\sqrt{\pi}\,\sigma^{1/4}}\,e^{(2/3)\sigma^{3/2}} \sum_{n=0}^{\infty} \frac{\Gamma(3n + \tfrac{1}{2})}{\sqrt{\pi}\,(2n)!}\,\frac{1}{(9\sigma^{3/2})^n}, \qquad \sigma \to \infty. \tag{45a}$$

As for $\text{Ai}(\sigma)$, Eqs. (45) apply uniformly in $\arg \sigma$ for complex values of σ lying in the sector $|\arg \sigma| < \pi/3$.

To obtain the asymptotic behavior of $\text{Ai}(\sigma)$ and $\text{Bi}(\sigma)$ for large negative real values of σ, it is convenient to consider the functions $\text{Ai}(-\sigma)$ and $\text{Bi}(-\sigma)$, where $\sigma > 0$. These functions are defined by Eq. (32b) (with L_{32} as the contour of integration) and Eq. (34), and can be expressed analogously to Eqs. (37) as

$$\text{Ai}(-\sigma) = \frac{\Omega^{1/3}}{2\pi i} \int_{L_{32}} e^{-\Omega(z + z^3/3)}\,dz, \qquad \Omega = \sigma^{3/2} > 0, \tag{46a}$$

$$\text{Bi}(-\sigma) = \frac{\Omega^{1/3}}{2\pi} \int_{L_{21}+L_{31}} e^{-\Omega(z + z^3/3)}\,dz. \tag{46b}$$

The saddle points are now located at $z_s = \pm i$, so $|\exp \Omega q(z_s)| = 1$ at both saddle points, which thus have the same level. The steepest paths through the saddle points satisfy the equations $\text{Im}\, q(z) = \text{Im}\, q(z_s) = \pm 2/3$, or

$$y^3 - 3x^2 y - 3y = \pm 2. \tag{47}$$

For large values of x and y, the cubic curves defined by Eq. (47) are asymptotic to the lines $y = 0$ and $y = \pm\sqrt{3}\,x$. Since the steepest-descent contours must end in the shaded regions in Fig. 4.2.4, one selects as the steepest-descent paths the curves shown in Fig. 4.2.6. Upon comparing Figs. 4.2.4 and 4.2.6, one notes that contours L_{21} and L_{31} can be deformed directly into the steepest-descent paths P_{21} and P_{31}, respectively, while the contour L_{32} must be deformed first along P_{31} and then along P_{21} so that it traverses both saddle points.

The transformation from the z to the s plane is given by

$$q(z) = -z - \frac{z^3}{3} = \mp \frac{2}{3}i - s^2, \qquad -\infty < s < \infty, \tag{48}$$

where the upper and lower signs apply for $z_s = \pm i$. Thus,

$$\frac{dz}{ds} = \frac{2s}{1 + z^2}, \tag{49a}$$

and, from Eq. (5),

$$\left.\frac{dz}{ds}\right|_{s=0} = \begin{cases} e^{-i\pi/4} & \text{on } P_{21}, \\ e^{i\pi/4} & \text{on } P_{31}. \end{cases} \tag{49b}$$

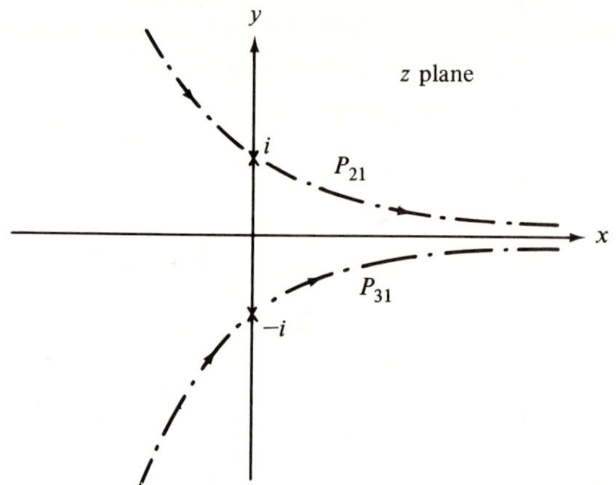

FIG. 4.2.6   Steepest-descent paths through the saddle points $z_s = \pm i$.

The asymptotic evaluation of the integrals in Eqs. (46) as $\Omega \to \infty$ can now be carried out directly from Eq. (17). To a first order one obtains

$$I_{21} = \int_{P_{21}} e^{-\Omega(z+z^3/3)} \, dz \sim e^{-i(2/3)\Omega - i\pi/4} \sqrt{\frac{\pi}{\Omega}}, \tag{50a}$$

$$I_{31} = \int_{P_{31}} e^{-\Omega(z+z^3/3)} \, dz \sim e^{i(2/3)\Omega + i\pi/4} \sqrt{\frac{\pi}{\Omega}}. \tag{50b}$$

Accordingly, the integral in Eq. (46a) is equal to $I_{31} - I_{21}$ while that in Eq. (46b) is equal to $I_{31} + I_{21}$; the positive sense on P_{21} and P_{31} in Fig. 4.2.6 is to be noted. Thus, one obtains the following first-order asymptotic representations as $\sigma \to \infty$ [see Eq. (36)]:

$$\mathrm{Ai}(-\sigma) \sim \frac{1}{\sqrt{\pi}\,\sigma^{1/4}} \sin\left(\frac{2}{3}\sigma^{3/2} + \frac{\pi}{4}\right), \tag{51a}$$

$$\mathrm{Bi}(-\sigma) \sim \frac{1}{\sqrt{\pi}\,\sigma^{1/4}} \cos\left(\frac{2}{3}\sigma^{3/2} + \frac{\pi}{4}\right). \tag{51b}$$

The complete asymptotic expansion of $\mathrm{Ai}(-\sigma)$ and $\mathrm{Bi}(-\sigma)$ as $\sigma \to \infty$ can be shown to be as follows[4,5]:

$$\mathrm{Ai}(-\sigma) = \frac{1}{\sqrt{\pi}\,\sigma^{1/4}} \left\{ P(\sigma) \sin\left(\frac{2}{3}\sigma^{3/2} + \frac{\pi}{4}\right) - Q(\sigma) \cos\left(\frac{2}{3}\sigma^{3/2} + \frac{\pi}{4}\right) \right\}, \tag{52a}$$

$$\mathrm{Bi}(-\sigma) = \frac{1}{\sqrt{\pi}\,\sigma^{1/4}} \left\{ P(\sigma) \cos\left(\frac{2}{3}\sigma^{3/2} + \frac{\pi}{4}\right) + Q(\sigma) \sin\left(\frac{2}{3}\sigma^{3/2} + \frac{\pi}{4}\right) \right\}, \tag{52b}$$

where

$$P(\sigma) \sim \frac{1}{\sqrt{\pi}} \sum_{n=0}^{\infty} \frac{\Gamma(6n + \tfrac{1}{2})}{(4n)!} \frac{(-1)^n}{(9\sigma^{3/2})^{2n}}, \tag{52c}$$

$$Q(\sigma) \sim \frac{1}{\sqrt{\pi}} \sum_{n=0}^{\infty} \frac{\Gamma(6n + \tfrac{7}{2})}{(4n + 2)!} \frac{(-1)^n}{(9\sigma^{3/2})^{2n+1}}. \tag{52d}$$

As previously, Eqs. (51) and (52) also apply if $|\sigma| \to \infty$ in the sector $|\arg \sigma| < \pi/3$.

When the argument of the Airy function is not negative real, the trigonometric functions may be approximated by the dominant exponential. It is then found that for $\pi/3 < \arg \sigma < \pi$, the dominant exponential in Eq. (51a) yields the same formula for $\mathrm{Ai}(\sigma)$ as Eq. (42a). Thus, it is suggestive to employ Eq. (42a) over the entire range $0 < \arg \sigma < \pi$ and then switch to Eq. (51a) when $\arg \sigma \geq \pi$. The validity of this argument may be established by a consideration of the Stokes phenomenon, a detailed discussion of which is contained in references 7 and 8. It may then be inferred that Eq. (42a) remains valid when $-2\pi/3 \leq \arg \sigma \leq 2\pi/3$, whereas Eq. (51a) is employed when $2\pi/3 \leq \arg \sigma \leq 4\pi/3$. The matching up of the two formulas has been performed at $\arg \sigma = 2\pi/3$, where the dominant exponential has its maximum value.

4.3   ISOLATED SADDLE POINTS OF HIGHER ORDER

If in the integrand of

$$I(\Omega) = \int_{z_s}^{``\infty"} f(z)e^{\Omega q(z)}\, dz, \tag{1}$$

the function $f(z)$ has no singularities near the isolated Mth-order saddle point z_s of $q(z)$, with $q^{(n)}(z_s) = 0, n = 1 \cdots M, q^{(M+1)}(z_s) \neq 0$, the asymptotic approximation of $I(\Omega)$ is given by

$$I(\Omega) \sim \left[\frac{-(M+1)!}{q^{(M+1)}(z_s)}\right]^{1/(M+1)} f(z_s)e^{\Omega q(z_s)} \frac{\Gamma[1/(M+1)]}{(M+1)\Omega^{1/(M+1)}}, \qquad \Omega \to \infty, \tag{2}$$

where the $(M+1)$th root in the first factor is chosen so that $\arg\{[\ ]^{1/(M+1)}\} = \varphi \equiv \arg (dz)_{z_s}$. dz denotes an element along the steepest-descent path, SDP, which begins at z_s and ends in an appropriate valley at "∞" (not necessarily along the real z axis) (Fig. 4.3.1). Since $M + 1$ valleys are accessible from an

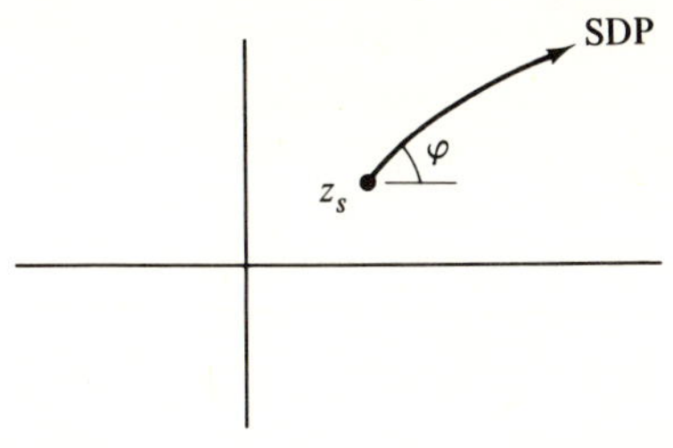

FIG. 4.3.1   Integration path in the z plane.

Mth-order saddle point, it is possible (for $M > 1$) to have non-collinear segments of the SDP leading to and from the saddle point, respectively. For this reason, each segment is treated separately and provides a contribution as in Eq. (1).

Analytical details

If $q(z)$ in Eq. (1) has one pertinent Mth-order saddle point at z_s [i.e., $q^{(n)}(z_s) = 0, n = 1, \ldots, M, q^{(M+1)}(z_s) \neq 0$], the appropriate transformation to the s plane is given by [see Eq. (4.1.6)]

$$q(z) = \tau(s) = q(z_s) - s^{M+1}. \tag{3}$$

The requirement Im $q(z) = $ constant (i.e., s^{M+1} positive along the steepest-descent paths) leads to the following possible contours in the s plane:

$$\arg s = 2\mu\pi/(M+1), \qquad \mu = 0, 1, 2, \ldots, M. \tag{4}$$

Thus, the $M + 1$ steepest-descent paths originating at the saddle point z_s in the z plane [see Fig 4.1.5(b) for $M = 2$] map in the s plane into the $M + 1$ straight lines defined in Eq. (4). The latter originate at $s = 0$ and extend to $|s| = \infty$. Since s^{M+1} is positive along any of the paths defined in Eq. (4) we may consider without loss of generality the case $\mu = 0$, wherein the steepest-descent path maps into the positive real s axis. Then the integrals $I_n(\Omega)$ in Eq. (4.2.13) are given in view of Eq. (3) by

$$I_n(\Omega) = \int_0^\infty s^n e^{-\Omega s^{M+1}} \, ds \tag{5a}$$

$$= \Gamma\left(\frac{1+n}{M+1}\right) \frac{\Omega^{-(1+n)/(M+1)}}{M+1}, \tag{5b}$$

where the integral in Eq. (5a) has been evaluated in terms of the gamma function defined in Eq. (4.2.8). The asymptotic nature of the resulting expansion in Eq. (4.2.17) as $\Omega \to \infty$ follows by the same considerations as those mentioned in Sec. 4.2b.

Upon expanding $q(z)$ in Eq. (3) in a power series about the point z_s, one obtains directly

$$\left.\frac{dz}{ds}\right|_{s=0} = \zeta\left[\frac{-(M+1)!}{q^{(M+1)}(z_s)}\right]^{1/(M+1)}, \qquad \zeta = e^{i2\pi\mu/(M+1)}, \tag{6}$$

where ζ is the appropriate $(M+1)$th root of unity. The choice of root is determined by the given steepest-descent path [see Eq. (4)], and the principal value of the $(M+1)$th root is taken in the remaining term. From Eqs. (4.2.12) and (5), one obtains the following first-order approximation to the integral $\bar{I}(\Omega)$ (for $\mu = 0$):

$$\bar{I}(\Omega) \equiv \int_0^\infty G(s)e^{\Omega\tau(s)} \, ds \sim \left[\frac{-(M+1)!}{q^{(M+1)}(z_s)}\right]^{1/(M+1)} f(z_s)e^{\Omega q(z_s)} \frac{\Gamma[1/(M+1)]}{(M+1)\Omega^{1/(M+1)}} \tag{7}$$

where it is recalled that $G(s) = f(z)(dz/ds)$. When $M = 1$, one recovers from Eq. (7) the previously derived result in Eq. (4.2.7), save for a factor of 2 which arises since the interval of integration in Eq. (4.2.6) extends from $s = -\infty$ to

$s = \infty$. If $G(s)$ has a branch-point singularity at $s = 0$, the integrand in Eq. (5a) involves instead of s^n the factor $s^{n+\beta}$, where $\beta > -1$ is non-integral. The resulting formula for $I_n(\Omega)$ is still given in terms of the gamma function.

4.4 FIRST-ORDER SADDLE POINT AND NEARBY SINGULARITIES

4.4a Simple Pole Singularity

If in the integrand of

$$I(\Omega) = \int_{SDP} f(z)e^{\Omega q(z)}\, dz, \tag{1}$$

the function $f(z)$ has a simple pole singularity at $z = z_0$ near the isolated first-order saddle point z_s of $q(z)$ [i.e., $q'(z_s) = 0$, $q''(z_s) \neq 0$], the asymptotic approximation of $I(\Omega)$, valid uniformly as $z_0 \to z_s$, is given by[9] [for a complete asymptotic expansion, see Eq. (16)]

$$I(\Omega) \sim e^{\Omega q(z_s)}\left\{\pm i 2a\sqrt{\pi}\, e^{-\Omega b^2} Q(\mp ib\sqrt{\Omega}) + \sqrt{\frac{\pi}{\Omega}} T(0)\right\},$$

$$\operatorname{Im} b \gtrless 0, \quad \Omega \to \infty, \tag{2}$$

where

$$a = \lim_{z \to z_0} [(z - z_0)f(z)], \qquad b = \sqrt{q(z_s) - q(z_0)}, \tag{2a}$$

$$T(0) = hf(z_s) + \frac{a}{b}, \qquad h = \sqrt{\frac{-2}{q''(z_s)}}, \tag{2b}$$

$$Q(y) = \int_y^\infty e^{-x^2}\, dx, \qquad Q(y) + Q(-y) = \sqrt{\pi}, \tag{2c}$$

The square root in Eq. (2b) is defined so that $\arg h = \varphi \equiv (\arg dz)_{z_s}$, where dz is an element along the steepest-descent path SDP, while $\arg b$ is defined so that $b \to (z_0 - z_s)/h$ as $z_0 \to z_s$. When $q(z) = i\hat{q}(z)$, with $\hat{q}$ denoting a real function of z, and z_s is real, $\arg h = \pm \pi/4$ when $\hat{q}''(z_s) \gtrless 0$, provided that $\operatorname{Re}(dz)$ increases along the SDP near z_s (see Fig. 4.2.1). The discontinuity in $I(\Omega)$ as $\operatorname{Im} b$ changes from positive to negative values is exactly equal to the residue of the integrand at the pole $z = z_0$ (see Fig. 4.4.1). In view of the last relation in Eq. (2c), the expression in Eq. (2) can be written either in terms of $Q(-ib\sqrt{\Omega})$ or $Q(+ib\sqrt{\Omega})$.

When $\operatorname{Im} b = 0$, Eq. (2) is replaced by

$$I(\Omega) \sim e^{\Omega q(z_s)}\left\{i 2a\sqrt{\pi}\, e^{-\Omega b^2} Q(-ib\sqrt{\Omega}) - \pi i a e^{-\Omega b^2} + \sqrt{\frac{\pi}{\Omega}} T(0)\right\}. \tag{3}$$

In this instance, the pole z_0 lies *on* the SDP since in order to render b real, $\operatorname{Im} q(z_0) = \operatorname{Im} q(z_s)$, $\operatorname{Re} q(z_0) < \operatorname{Re} q(z_s)$. Equation 3 represents an asymptotic approximation of the principal value of the integral $I(\Omega)$ in Eq. (1). When the pole and the saddle point coincide, $b = 0$ and $Q(0) = \frac{1}{2}\sqrt{\pi}$, so

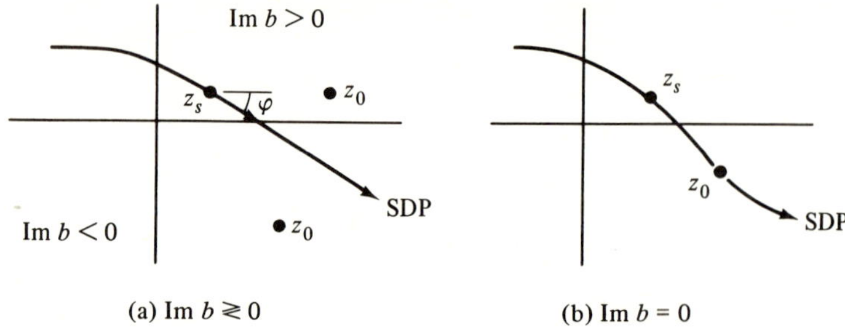

FIG. 4.4.1   Integration paths and pole locations in the z plane.

$$I(\Omega) \sim e^{\Omega q(z_s)} \sqrt{\frac{\pi}{\Omega}}\, T(0), \qquad z_0 = z_s, \tag{4a}$$

where $T(0)$ remains bounded as $z_0 \to z_s$ and is, in fact, given by

$$T(0) = hf(z_s) + \frac{ah}{(z_0 - z_s)\{1 + [q^{(3)}(z_s)/3q''(z_s)](z_0 - z_s) + \cdots\}^{1/2}}$$

$$= h\left[g(z_s) - a\frac{q^{(3)}(z_s)}{6q''(z_s)} \right], \qquad z_0 = z_s, \tag{4b}$$

with $g(z) = f(z) - a(z - z_0)^{-1}$.

Analytical details

If $f(z)$ in the integrand of Eq. (1) has a simple pole singularity at $z = z_0$ near a first-order saddle point z_s, $G(s)$ in Eq. (4.1.3) will possess correspondingly a simple pole singularity at $s = b$ in the vicinity of $s = 0$. Suppose that $G(s)(s - b) \to a$ as $s \to b$; then $G(s)$ can be represented in the vicinity of $s = 0$ by

$$G(s) = \frac{a}{s - b} + T(s). \tag{5}$$

It will be convenient to employ the identity

$$\frac{a}{s - b} = \frac{as}{s^2 - b^2} + \frac{ab}{s^2 - b^2}, \tag{5a}$$

and to expand

$$T(s) = T(0) + T'(0)s + T''(0)\frac{s^2}{2!} + \cdots, \tag{5b}$$

which is regular at $s = b$ and has a radius of convergence uninfluenced by the presence of the pole.

Since the saddle point is of order 1, the transformation in Eq. (4.2.2) still applies and the integral in Eq. (1) can be written as (note that $s = [q(z_s) - q(z)]^{1/2}$, $h = (dz/ds)_{s=0}$, and $a = \lim_{z \to z_0} [(z - z_0)f(z)] = \lim_{s \to b} [(s - b)G(s)]$)

$$I(\Omega, b) = e^{\Omega q(z_s)} \int_{-\infty}^{\infty} G(s)e^{-\Omega s^2}\, ds, \tag{6}$$

whence the formal result in Eq. (4.2.18b) yields directly

$$I(\Omega, b) \sim e^{\Omega q(z_s)} G_e\left(\sqrt{-\frac{d}{d\Omega}}\right)\sqrt{\frac{\pi}{\Omega}}, \tag{7}$$

with G_e obtained from Eqs. (5) as

$$G_e\left(\sqrt{-\frac{d}{d\Omega}}\right) = \frac{-ab}{(d/d\Omega) + b^2} + T_e\left(\sqrt{-\frac{d}{d\Omega}}\right). \tag{8}$$

The operation on $\sqrt{\pi/\Omega}$ implied by the last term in Eq. (8) is the same as in Eq. (4.2.18b) and leads to a formal asymptotic expansion as in Eq. (4.2.17).

It is to be noted that the integral $I(\Omega, b)$ in Eq. (6), with $G(s)$ given in Eq. (5), is defined only for $\text{Im } b \neq 0$ and does not exist when b is real or zero. Viewed as a function of b, a study of the analytic properties of $I(\Omega, b)$ as $\text{Im } b \rightarrow 0$ reveals that the integral is discontinuous across the real b axis. Suppose that b approaches the real b axis from above [i.e., $b \rightarrow b_r + i\bar{\delta}$, with $b_r, \bar{\delta}$ real and $\bar{\delta} > 0$]. Then the path of integration is indented at $s = b_r + i\bar{\delta}$ as shown in Fig. 4.4.2(a). Similarly, when $b \rightarrow b_r - i\bar{\delta}$, the appropriate path is

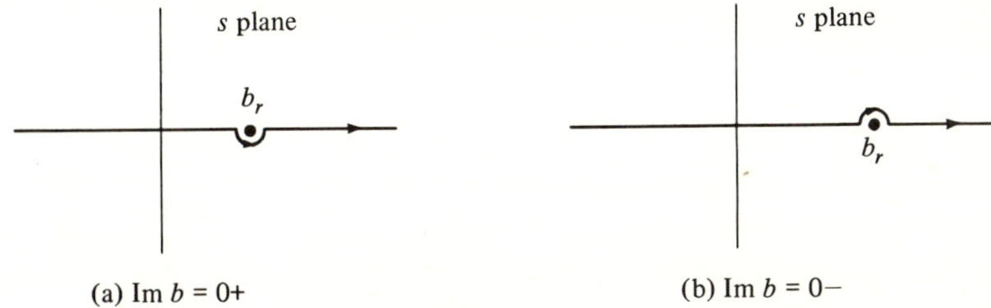

(a) Im b = 0+   (b) Im b = 0−

FIG. 4.4.2   Contours of integration.

that in Fig. 4.4.2(b). To exhibit the discontinuity in $I(\Omega, b)$ across the real b axis, one constructs the difference $I(\Omega, b_r + i\bar{\delta}) - I(\Omega, b_r - i\bar{\delta})$ and notes that the contributions to the integrals from the straight portions of the paths in Fig. 4.4.2 cancel; there then remains only a small circular contour enclosing the pole at $s = b_r$ in the positive sense. Since $T(s)$ in Eq. (5) is regular inside this circle, its contribution vanishes and one obtains from the residue at $s = b$,[†]

$$I(\Omega, b_r + i\bar{\delta}) - I(\Omega, b_r - i\bar{\delta}) = 2\pi i a e^{-\Omega b_r^2} e^{\Omega q(z_s)}. \qquad \bar{\delta} \rightarrow 0. \tag{9}$$

The operation

$$A(\Omega, b) = \frac{-ab}{(d/d\Omega) + b^2}\sqrt{\frac{\pi}{\Omega}} \tag{10}$$

implied by the first term in Eq. (8) can be interpreted in terms of the ordinary first-order differential equation

---

†To obtain Eq. (9), one also may use the generalized function[see Eq. (1.2.6c)]

$$\lim_{\bar{\delta}\to 0} \frac{1}{s - b \mp i\bar{\delta}} = P\frac{1}{s - b} \pm \pi i \delta(s - b),$$

where $\delta(x)$ is the delta function and P denotes the principal value.

$$\left(\frac{d}{d\Omega} + b^2\right) A(\Omega, b) = -ab\sqrt{\frac{\pi}{\Omega}}. \tag{10a}$$

To find a particular integral of Eq. (10), substitute

$$A(\Omega, b) = e^{-\Omega b^2} B(\Omega, b) \tag{11}$$

into Eq.(10a), whence

$$\frac{dB}{d\Omega} = -abe^{\Omega b^2}\sqrt{\frac{\pi}{\Omega}}. \tag{12}$$

Upon integrating Eq. (12) over Ω between the limits $\bar{\Omega}$ and ∞, one obtains

$$B(\bar{\Omega}, b) = ab\sqrt{\pi} \int_{\bar{\Omega}}^{\infty} e^{\Omega b^2}\Omega^{-1/2}\, d\Omega, \tag{13}$$

where it has been assumed for the moment that $b^2 < 0$,[†] so in view of Eq. (11) and for $A(\infty, b)$ finite, $B(\infty, b) = 0$.[‡] A change of variable in Eq. (13) from Ω to $-x^2/b^2$, or $x = \mp ib\sqrt{\Omega}$, yields

$$B(\Omega, b) = 2a\sqrt{\pi} \,\frac{b}{\mp ib}\, Q[\mp ib\sqrt{\Omega}\,], \tag{14}$$

where $Q(y)$ is the "error-function complement"

$$Q(y) = \int_{y}^{\infty} e^{-x^2}dx. \tag{14a}$$

The ambiguity in sign introduced into Eq. (14) by the change of variable is resolved by the previously imposed requirement $B(\infty, b) = 0$ for $b^2 < 0$. Since the error-function integral in Eq. (14a) will vanish if the lower limit approaches infinity along the positive real axis, we require for $b^2 < 0$ a choice of sign such that

$$\mp ib > 0 \tag{14b}$$

(i.e., the minus sign when $b = i|b|$ and the plus sign when $b = -i|b|$).

The validity of Eq. (14) can be extended by analytic continuation to values other than $b^2 < 0$. By direct substitution of Eqs. (5) into the integral in Eq. (6) and comparison with Eqs. (8), (10), and (11), one notes that $B(\Omega, b)$ is also given in terms of the definite integral[10]

$$B(\Omega, b) = abe^{\Omega b^2} \int_{-\infty}^{\infty} \frac{1}{s^2 - b^2}\, e^{-\Omega s^2}\, ds = ae^{\Omega b^2} \int_{-\infty}^{+\infty} \frac{e^{-\Omega s^2}}{s - b}\, ds. \tag{15}$$

One verifies readily that $B(\Omega, b)$ as defined in Eq. (15) satisfies the differential equation (12). Since the expressions for B in Eqs. (14) and (15) represent

---

[†]More generally Re $b^2 < 0$.

[‡]By an alternative procedure, if $b^2 < 0$,

$$\frac{1}{ab} B(\Omega, b) \equiv \int_{-\infty}^{\infty} \frac{e^{-\Omega\,(s^2-b^2)}}{s^2 - b^2}\, ds = \int_{-\infty}^{\infty} ds \int_{\Omega}^{\infty} e^{-\xi(s^2-b^2)}\, d\xi = \int_{\Omega}^{\infty} d\xi\, e^{\xi b^2} \int_{-\infty}^{\infty} e^{-\xi s^2}\, ds$$

$$= \sqrt{\pi} \int_{\Omega}^{\infty} \frac{e^{\xi b^2}}{\sqrt{\xi}}\, d\xi.$$

identical functions of b^2 when $b^2 < 0$, and since Eq. (15) clearly remains valid for all except positive and zero values of b^2, it follows that the expression in Eq. (14) can likewise be continued analytically for all values of b^2 except $b^2 \geq 0$. The analytic continuation from pure imaginary values of b ($b^2 < 0$) to complex values must be consistent with condition (14b). In view of the analytic properties of the error-function complement, this implies that the sign in Eq. (14) must be chosen according to the more general condition $\mathrm{Re}\,(\mp ib) > 0$ (i.e., the minus sign applies when $\mathrm{Im}\,b > 0$ and the plus sign when $\mathrm{Im}\,b < 0$).

From Eqs. (7), (10), and (14), we can now write down the asymptotic expansion of the integral in Eq. (6) for $\Omega \gg 1$ and for arbitrary values of b:

$$I(\Omega, b) \sim e^{\Omega q(z_s)}\left\{\pm\, i2a\sqrt{\pi}\; e^{-\Omega b^2} Q(\mp ib\sqrt{\Omega}) + T_e\left(\sqrt{-\frac{d}{d\Omega}}\right)\sqrt{\frac{\pi}{\Omega}}\right\},$$

$$\mathrm{Im}\,b \gtrless 0, \qquad (16)$$

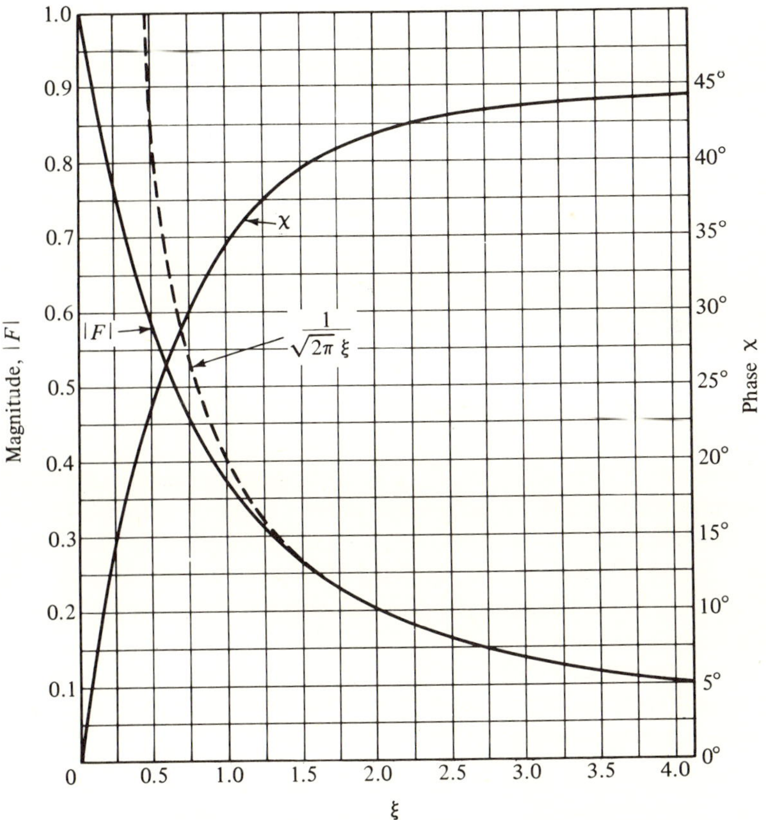

FIG. 4.4.3(a)   Plot of $F = |F|e^{i\chi} = \dfrac{2}{\sqrt{\pi}}\, e^{-i2\xi^2} \displaystyle\int_{(1-i)\xi}^{\infty} e^{-y^2}\, dy$

$$\left[\text{for large } \xi : F \sim \frac{e^{i\pi/4}}{\sqrt{2\pi}\,\xi} + O\left(\frac{1}{\xi^3}\right)\right].$$

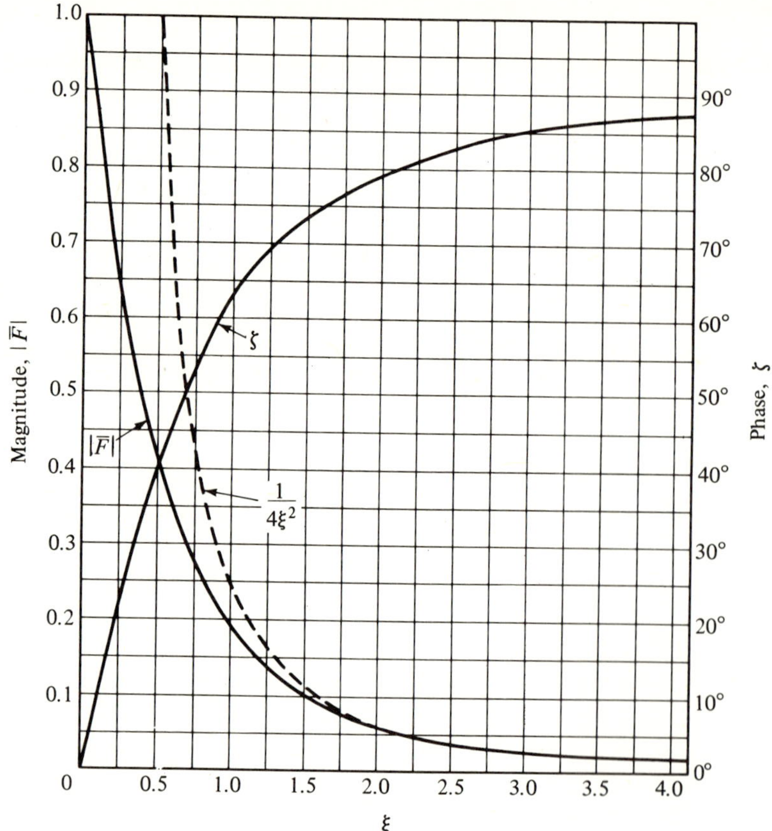

FIG. 4.4.3(b)   Plot of $\bar{F} = |\bar{F}|e^{i\zeta} = 1 - 2\sqrt{2}\,\xi e^{-i(2\xi^2+\pi/4)}\int_{(1-i)\xi}^{\infty} e^{-y^2}\,dy$

$\left[\text{for large } \xi : \bar{F} \sim \dfrac{i}{4\xi^2} + O\left(\dfrac{1}{\xi^4}\right)\right].$

the lowest-order term of which $[T_e \to T(0)]$ furnishes the result in Eq. (2). Thus, the asymptotic expansion of an integral whose integrand contains a simple pole near a saddle point has the same form as that for an integrand without a pole except for an additional term involving the error function Q. The function $e^{-\Omega b^2}Q(\mp ib\sqrt{\Omega})$ is tabulated for real and complex values of $b\sqrt{\Omega}$ (see Fig. 4.4.3)

It is of interest to verify from Eq. (16) the previously noted expression for the discontinuity in the value of $I(\Omega, b)$[Eq. (9)] when Im b changes from positive to negative values. As before, we define

$$b_{1,2} = b_r \pm i\bar{\delta}, \qquad b_r, \bar{\delta} \text{ real}, \quad \bar{\delta} \to +0. \tag{17}$$

Since $T_e(\sqrt{-d/d\Omega})\sqrt{\pi/\Omega}$ is continuous for all b, the jump $[I(\Omega, b_1) - I(\Omega, b_2)]$ in the value of $I(\Omega, b)$ is given by

$$I(\Omega, b)\Big|_{b=b_2}^{b=b_1} = 2ia\sqrt{\pi}\,e^{-\Omega b_r^2}[Q(-ib_r\sqrt{\Omega}) + Q(ib_r\sqrt{\Omega})]e^{\Omega q(z_s)}. \tag{18}$$

To treat the sum $[Q(i\alpha) + Q(-i\alpha)]$, we choose a path of integration for $Q(\pm i\alpha)$ in Eq. (14a) from $\pm i\alpha$ to 0 and then from 0 to ∞ along the real axis. Thus,

$$Q(i\alpha) + Q(-i\alpha) = \int_{-i\alpha}^{0} e^{-x^2}\, dx + \int_{i\alpha}^{0} e^{-x^2}\, dx + 2\int_{0}^{\infty} e^{-x^2}\, dx = \sqrt{\pi}, \qquad (18a)$$

since the first two terms in the first equality of Eq. (18a) cancel. Thus, Eq. (18) reduces to the previous result in Eq. (9).

It may be noted that the function

$$\hat{B}(\Omega, b) = i2a\sqrt{\pi}\, Q(-ib\sqrt{\Omega}) \qquad (19a)$$

represents the integral in Eq. (15) for *all* complex values of b provided that the integral is taken as the definition of an *analytic function* $\hat{B}$ of b. This assertion may be verified by assuming initially that $\operatorname{Im} b > 0$, for which case the equivalence of Eqs. (15) and (19a) has been demonstrated. To extend the range of validity, we first deform the integration path in Eq. (15) into the lower half of the complex s plane. This deformation does not alter the value of the function $\hat{B}$, since no singularities are located between the two paths [it is convenient for the present argument to replace $ab/(s^2 - b^2)$ in the integrand by $a/(s - b)$]. The form of the function $\hat{B}$ defined along the new path remains unchanged even when the pole at $s = b$ moves across the real s axis as long as it does not intercept the deformed integration contour (evidently, the contour may be distorted sufficiently to accommodate in this manner an arbitrarily situated pole). Thus, the integral is represented by the error-function complement in Eq. (19a) for *arbitrary* b provided that the integration contour always passes *beneath* the pole. To obtain a representation involving an integration path along the real axis, one performs the path deformation in reverse and must now exhibit the residue from the pole at $s = b$. Thus, for $\operatorname{Im} b < 0$,

$$\hat{B}(\Omega, b) = i2a\sqrt{\pi}\, Q(-ib\sqrt{\Omega}) = i2\pi a + a\int_{-\infty}^{\infty} \frac{e^{-\Omega s^2}}{s - b}\, ds$$

$$= i2\pi a - i2a\sqrt{\pi}\, Q(ib\sqrt{\Omega}), \qquad (19b)$$

the last relation following from Eq. (14). Since $Q(x) + Q(-x) = \sqrt{\pi}$ [see Eq. (18a)], the two alternative forms in Eq. (19b) are equivalent and each may be used to represent the analytic function $\hat{B}(\Omega, b)$ for arbitrary values of b. These considerations may be utilized to simplify the formulas for the asymptotic representation of analytic functions defined by integrals containing pole singularities in the integrand [for example, Eq. (34) may be written in a simpler form].

When b is large enough so that $b\sqrt{\Omega}$ is likewise large, one may employ an asymptotic expansion for $Q(\mp ib\sqrt{\Omega})$ in Eq. (16). This expansion is obtained directly from the representation in Eq. (13) by repeated integrations by parts or by expansion of $(s^2 - b^2)^{-1}$ in Eq. (15) in powers of s^2:

$$B(\Omega, b) = \pm 2ia\sqrt{\pi}\, Q(\mp ib\sqrt{\Omega}) \sim -\frac{a}{b} e^{\Omega b^2} \sqrt{\frac{\pi}{\Omega}} \left[1 + \frac{1}{2b^2\Omega} + O\left(\frac{1}{b^4\Omega^2}\right)\right].$$

$$(20)$$

The first-order asymptotic representation for $I(\Omega, b)$ in Eq. (16) is then given
by

$$I(\Omega, b) \sim e^{\Omega q(z_s)} \sqrt{\frac{\pi}{\Omega}}\, G(0), \qquad |b|\sqrt{\Omega} \gg 1, \tag{21}$$

where

$$G(0) = -\frac{a}{b} + T(0). \tag{21a}$$

In this instance, the pole is situated "far" from the origin in the s plane and the
expression in Eq. (21) is identical with that obtained in Eq. (4.2.7). Just how
large $|b|\sqrt{\Omega}$ has to be, before Eq. (20) can be employed to retain a given
accuracy, can be assessed by comparing Eq. (20) with the exact expression (14)
whose values for a given b are found from numerical tables. A detailed
comparison is made in Eqs. (38) et seq. for the special case $\arg(\pm b) = \pi/4$.
   When $|b| \to 0$, $Q(\mp ib\sqrt{\Omega}) \to \frac{1}{2}\sqrt{\pi}$.

4.4b Multiple Pole Singularity

   If $G(s)$ has a pole of order N at $s = b$, $G(s)$ is represented by the expression
[see Eq. (5)]

$$G(s) = \frac{a_{-N}}{(s-b)^N} + \frac{a_{-N+1}}{(s-b)^{N-1}} + \cdots + \frac{a_{-1}}{s-b} + \bar{T}(s), \tag{22}$$

where $\bar{T}(s)$ is regular at $s = b$. To infer the asymptotic expansion of $I(\Omega, b)$ in
this case, one must investigate integrals of the form

$$A_{-N}(\Omega, b) = \int_{-\infty}^{\infty} \frac{1}{(s-b)^N}\, e^{-\Omega s^2}\, ds. \tag{23}$$

For $N = 1$ (with $a_{-1} = 1$), the result has been obtained in Eq. (14)[with Eq.
(11)] as

$$A_{-1}(\Omega, b) = \pm 2i\sqrt{\pi}\, e^{-\Omega b^2} Q(\mp ib\sqrt{\Omega}), \qquad \mathrm{Im}\, b \gtrless 0. \tag{24}$$

Since one notes from Eq. (23) that

$$A_{-N}(\Omega, b) = \frac{1}{N-1}\frac{d}{db} A_{-N+1}(\Omega, b), \qquad N = 2, 3, \ldots, \tag{25}$$

all A_{-N}, $N \geq 2$, can be inferred from A_{-1} in Eq. (24) by successive differentiation with respect to b. The integral in Eq. (23) is uniformly convergent for
$\mathrm{Im}\, b \neq 0$, so the differentiation under the integral sign implied in Eq. (25) is
permitted. Evaluation of the derivatives of Q is readily accomplished via the
formula [see Eq. (14a)]

$$\frac{d}{dy} Q(y) = -e^{-y^2}, \tag{26}$$

so

$$A_{-2}(\Omega, b) = -2\sqrt{\Omega\pi} - 2b\Omega A_{-1}(\Omega, b), \tag{27a}$$

etc.

One notes that as $b \to 0$, the dependence on Ω is $O(\sqrt{\Omega})$ for A_{-2} and $O(1)$ for A_{-1}. Thus, a higher-order pole in the vicinity of the saddle point yields a larger value for the integral than a first-order pole. If b is large enough so that $|b|\sqrt{\Omega} \gg 1$, one can employ in Eq. (27a) the asymptotic expansion in Eq. (20). The result is found to be

$$A_{-2}(\Omega, b) \sim \sqrt{\frac{\pi}{\Omega}} \frac{1}{b^2}, \qquad \sqrt{\Omega}\, |b| \gg 1, \tag{27b}$$

which agrees with that obtained from Eq. (23) by a direct asymptotic evaluation.

4.4c Branch-Point Singularity

If $f(z)$ in Eq. (1) has an algebraic branch-point singularity z_b near a first-order saddle point z_s, then in the transformed integral of Eq. (4.2.3), G(s) will have a corresponding singularity at $s = b$ in the complex s plane. It is then required to evaluate integrals of the form

$$I(\Omega, b) = \int_{-\infty}^{\infty} s^n (s - b)^\alpha e^{-\Omega s^2}\, ds, \tag{28}$$

where $n = 0, 1, 2 \dots$ and $0 < \alpha < 1$. For $\alpha > 1$, factors of $(s - b)^m$, $m=$ integer, may be split off to keep α in the range $0 < \alpha < 1$; a dependence on $(s - b)^{-\beta}$, $\beta > 0$, may be accommodated by differentiating I with respect to b. By the change of variable

$$u = (s - b)^\alpha, \tag{29}$$

the integral in Eq. (28) is transformed into

$$I(\Omega, b) = \frac{1}{\alpha} \int (u^{1/\alpha} + b)^n u^{1/\alpha} e^{-\Omega(u^{1/\alpha} + b)^2}\, du, \tag{30}$$

with the endpoints of the integration path placed appropriately at infinity. Saddle points in the complex u plane are now located at $u = 0, (-b)^\alpha$. If $\alpha = 1/m$, $m = 2, 3, \dots$, the $(m - 1)$th-order branch-point singularity in the s plane is seen to give rise to a cluster of m saddle points surrounding $u = 0$. For the special case $m = 2$, $I(\Omega, b)$ is expressible in terms of the parabolic cylinder function in Eq. (4.5.36). For a more general discussion, see Reference 11.

4.4d Uniform Asymptotic Evaluation of a Typical Diffraction Integral

We return now to the evaluation of the integral in Eq. (4.2.23) for the case when the pole at $z = \beta$ is situated near the saddle point $z_s = \alpha$. The representation of $I_1(\Omega, \alpha, \beta)$ as in Eq. (4.2.30) still applies. However, G(s) in Eq. (4.2.30b) should now be represented as in Eq. (5), with the pole contribution exhibited separately. To determine the behavior of $(z - \beta)^{-1}$ as a functions of s, we first expand s in Eq. (4.2.27b) (with the plus sign) in a power series about $z = \beta$:

$$s = \sqrt{2}\, e^{i\pi/4}\left[\left(\sin\frac{\beta - \alpha}{2}\right) + \tfrac{1}{2}\left(\cos\frac{\beta - \alpha}{2}\right)(z - \beta)\right.$$

$$\left. - \tfrac{1}{8}\left(\sin\frac{\beta - \alpha}{2}\right)(z - \beta)^2 + \cdots\right], \tag{31}$$

and invert the series (see Appendix A) to obtain

$$z - \beta = \frac{\sqrt{2}}{e^{i\pi/4}\cos\left[(\beta - \alpha)/2\right]}(s - b) + \frac{\tan\left[(\beta - \alpha)/2\right]}{2i\cos^2\left[(\beta - \alpha)/2\right]}(s - b)^2 + \cdots,$$

$$\tag{31a}$$

where

$$b = \sqrt{2}\, e^{i\pi/4}\sin\frac{\beta - \alpha}{2}. \tag{31b}$$

Thus,

$$\frac{1}{z - \beta} = \frac{e^{i\pi/4}\cos\left[(\beta - \alpha)/2\right]}{\sqrt{2}}\frac{1}{s - b} + \text{(terms finite at } s = b) \tag{32}$$

and a in Eq. (5) is given by [see also Eq. (2a)]

$$a = \lim_{s\to b}\left[G(s)(s - b)\right] = \frac{e^{i\pi/4}\cos\left[(\beta - \alpha)/2\right]}{\sqrt{2}}\left(\frac{dz}{ds}\right)_{s=b} = 1, \tag{33}$$

with the value of dz/ds at $s = b$ obtained from Eq. (31a). By Eq. (16) the asymptotic expansion for the integral in Eqs. (4.2.23) or (4.2.30) as $\Omega \to \infty$ can be written down directly as

$$I_1(\Omega, \alpha, \beta) \sim 2\pi i e^{i\Omega\,\cos(\beta - \alpha)}\,\epsilon(\beta)$$

$$+ e^{i\Omega}\left[\pm i2\sqrt{\pi}\, e^{-\Omega b^2}Q(\mp ib\sqrt{\Omega}) + T_e\left(\sqrt{-\frac{d}{d\Omega}}\right)\sqrt{\frac{\pi}{\Omega}}\right], \qquad \text{Im } b \gtrless 0.$$

$$\tag{34}$$

where $\epsilon(\beta)$ is defined in Eq. (4.2.30), b is given in Eq. (31b), and

$$T(s) = \frac{1}{z - \beta}\frac{dz}{ds} - \frac{1}{s - b}, \tag{34a}$$

$$T(0) = \frac{1}{\alpha - \beta}\sqrt{2}\, e^{-i\pi/4} + \frac{e^{-i\pi/4}}{\sqrt{2}\,\sin\left[(\beta - \alpha)/2\right]} \tag{34b}$$

It is of interest to note that Eq. (34) is a continuous function of b, although various terms therein are discontinuously represented. This is verified upon inspection of Fig. 4.2.3, Eq. (4.2.30), and Eq. (9). If the pole at $s = b$ crosses the real axis in Fig. 4.2.3, the term inside the brackets in Eq. (34) experiences a jump as in Eq. (9). However, the first term on the right-hand side of Eq. (34) also changes discontinuously under these conditions and compensates exactly for the first-mentioned discontinuity. For values of b such that $|b|\sqrt{\Omega} \gg 1$, Eq. (34) reduces via Eq. (20) to Eq. (4.2.31). For a more compact way of writing Eq. (34), see the remarks following Eqs. (19).

The special case where β is real is of particular importance in several of the diffraction problems to be treated subsequently. In this instance, $\sin\left[(\beta-\alpha)/2\right]$ in Eq. (31b) is real and, for $|\beta-\alpha|<\pi$,

$$\mp ib\sqrt{\Omega} = (1-i)\xi, \qquad \xi = \sqrt{\Omega}\left|\sin\frac{\beta-\alpha}{2}\right|, \qquad \operatorname{Im} b \gtrless 0. \tag{35}$$

Moreover, $\operatorname{Im} b > 0$ if $\beta-\alpha>0$, and $\operatorname{Im} b < 0$ if $\beta-\alpha<0$. Thus, the following term in Eq. (34) can be rewritten for $\operatorname{Im} b \gtrless 0$ as

$$\pm i2\sqrt{\pi}\, e^{-\Omega b^2}Q(\mp ib\sqrt{\Omega}) = 2i\sqrt{\pi}\,\operatorname{sgn}(\beta-\alpha)e^{-2i\xi^2}Q[(1-i)\xi], \tag{36}$$

where

$$\operatorname{sgn}(\beta-\alpha) = \pm 1, \qquad \beta-\alpha \gtrless 0. \tag{36a}$$

The function

$$e^{i\pi/4}Q[(1-i)\xi] = e^{i\pi/4}\int_{(1-i)\xi}^{\infty} e^{-x^2}\,dx = e^{i\pi/4}\left[\frac{1}{2}\sqrt{\pi} - \int_0^{(1-i)\xi} e^{-x^2}\,dx\right] \tag{37a}$$

can be expressed via the change of variable

$$x = \frac{\sqrt{\pi}}{2}(1-i)t \tag{37b}$$

in terms of the well-tabulated Fresnel integrals $C(x)$ and $S(x)$ as

$$e^{i\pi/4}Q[(1-i)\xi] = \frac{1}{2}\sqrt{\pi}\, e^{i\pi/4} - \sqrt{\frac{\pi}{2}}\left[C\left(\frac{2\xi}{\sqrt{\pi}}\right) + iS\left(\frac{2\xi}{\sqrt{\pi}}\right)\right], \tag{37c}$$

where

$$C(x) = \int_0^x \cos\left(\frac{\pi}{2}t^2\right)dt, \qquad S(x) = \int_0^x \sin\left(\frac{\pi}{2}t^2\right)dt. \tag{37d}$$

To provide an estimate of how large ξ has to be before the asymptotic representation for $Q[(1-i)\xi]$ in Eq. (20) can be employed, we have plotted the function[12]

$$F(\xi) = \frac{2}{\sqrt{\pi}}e^{-i2\xi^2}Q[(1-i)\xi], \qquad \xi \geq 0. \tag{38}$$

For $\xi \to 0$, $F(\xi) \to 1$, while for $\xi \gg 1$ one has, from Eq. (20),

$$F(\xi) \sim \frac{e^{i\pi/4}}{\xi\sqrt{2\pi}}, \qquad \xi \gg 1 \tag{38a}$$

Upon comparing the plots of Eqs. (38) and (38a) shown in Fig. 4.4.3(a), one notes that the lowest-order asymptotic formula in Eq. (38a) holds with very good accuracy when $\xi \geq 3$. In terms of this estimate one finds from Eq. (35) that the "transition region," inside which the simple asymptotic representation in Eq. (4.2.31) fails, is given approximately by $|\beta-\alpha| \leq 6/\sqrt{\Omega}$. An analogous estimate can be found for the case of a double pole singularity, in which event one requires values for the derivative of $F(\xi)$[see Eq. (25)]. The function[13]

$$\bar{F}(\xi) = 1 - 2\sqrt{2}\,\xi e^{-i(2\xi^2+\pi/4)}Q[(1-i)\xi], \tag{39}$$

which occurs in this connection, as well as its asymptotic approximation

$$\bar{F}(\xi) = \frac{i}{4\xi^2}, \qquad \xi \gg 1, \tag{39a}$$

are plotted in Fig. 4.4.3(b).

4.5 NEARBY FIRST-ORDER SADDLE POINTS

4.5a Two Saddle Points

If in the integrand of

$$I(\Omega) = \int_{\text{SDP}} f(z)e^{\Omega q(z)}\, dz, \tag{1}$$

the function $f(z)$ is regular near the two adjacent first-order saddle points $z_{1,2}$ of $q(z)$ [i.e., $q'(z_{1,2}) = 0$ and $q''(z_{1,2}) \neq 0$ unless $z_1 = z_2$], the asymptotic approximation of $I(\Omega)$, uniformly valid as $z_1 \to z_2$, is given by [see Eq. (23) for a complete asymptotic expansion]:

$$I(\Omega) \sim \frac{1}{2}\,[f(z_1)h_1 + f(z_2)h_2]\,\frac{e^{\Omega a_0}}{\Omega^{1/3}}\,C(\sigma\Omega^{2/3})$$

$$+ \frac{1}{2\sigma^{1/2}}\,[f(z_1)h_1 - f(z_2)h_2]\,\frac{e^{\Omega a_0}}{\Omega^{2/3}}\,C'(\sigma\Omega^{2/3}), \qquad \Omega \to \infty, \tag{2}$$

where

$$a_0 = \tfrac{1}{2}[q(z_1) + q(z_2)], \qquad \sigma^{1/2} = \{\tfrac{3}{4}[q(z_1) - q(z_2)]\}^{1/3}, \tag{2a}$$

$$h_{1,2} = \sqrt{\frac{\mp 2\sigma^{1/2}}{q''(z_{1,2})}}, \qquad C(\zeta) = \int_{\text{SDP}} e^{\zeta t - t^3/3}\, dt, \qquad C'(\zeta) = \frac{dC(\zeta)}{d\zeta}. \tag{2b}$$

When $\sigma \to 0$ (i.e., $z_1 \to z_2$), one finds that $q''(z_{1,2}) \to 0$ at the resulting second-order saddle point. In this instance, $h_{1,2}$ takes on the limiting form

$$h_1 = h_2 = \left[\frac{-2}{q^{(3)}(z_s)}\right]^{1/3}, \qquad z_1 = z_2 = z_s. \tag{2c}$$

The multivalued expressions for $h_{1,2}$ and $\sigma^{1/2}$ may be defined by requiring that when $z_1 = z_2 = z_s$, $\arg h_{1,2} = \varphi \equiv \arg (dz)_{z_s}$, where dz is the integration element along the steepest-descent path SDP leading away from the second-order saddle point z_s (see Fig. 4.5.1). Arg $\sigma^{1/2}$ is then chosen to satisfy Eq. (2a) and to make the expression in Eq. (2b) compatible with Eq. (2c) when $z_1 \to z_2$. Alternatively, one may choose $\arg \sigma^{1/2}$ in a convenient manner and then deduce $\arg h_{1,2}$ by referring to the integration path. The integral in Eq. (2b) is identifiable in terms of the Airy integrals in Eqs. (4.2.32) and (4.2.34) for any specified allowable path P in Fig. 4.2.4 [see Eq. (14)]. When convenient, the integration variable t is defined to be real along the SDP leading away from z_s in Fig. 4.5.1(b); this implies that paths terminating in regions 2 or 3 of Fig. 4.2.4 are rotated into the real t axis.

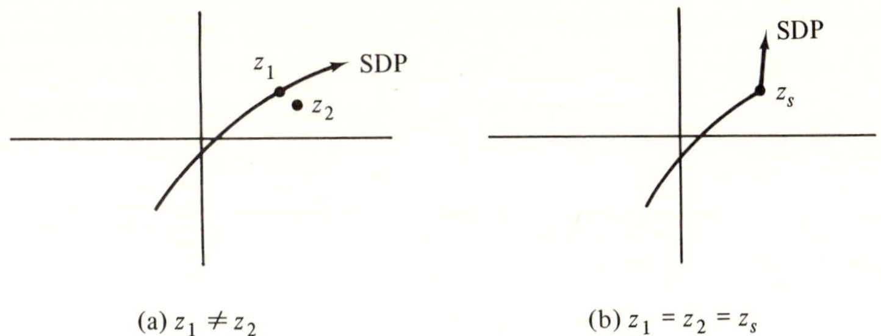

FIG. 4.5.1   Integration paths and saddle points.

When $z_1 = z_2$ $(\sigma = 0)$, one employs Eq. (2c) and substitutes for $C(0)$ the following values into Eq. (2) [see Eq. (4.2.8)]:

$$C(0) = \begin{cases} i3^{-1/6}\Gamma(\tfrac{1}{3}), & \text{if } P = L_{32}, \\ e^{-i\pi/6}3^{-1/6}\Gamma(\tfrac{1}{3}), & \text{if } P = L_{21}, \\ e^{i\pi/6}3^{-1/6}\Gamma(\tfrac{1}{3}), & \text{if } P = L_{31}. \end{cases} \tag{3}$$

One verifies, on use of Eq. (3), that the leading term in Eq. (2) for $z_1 = z_2$ agrees with the second-order saddle-point result in Eq. (4.3.7) (with $M = 2$) provided that the path P for $C(0)$ is taken from $s = 0$ to $s = \infty$. In this case, the value for $C(0)$ as obtained from Eq. (4.2.8) is $3^{-2/3}\Gamma(\tfrac{1}{3})$. The expressions in Eq. (3) are given in terms of this result as follows:

$$\int_{L_{32}} e^{-s^3/3}\,ds = \int_{\infty e^{-i2\pi/3}}^{0} + \int_{0}^{\infty e^{i2\pi/3}} = (e^{i2\pi/3} - e^{-i2\pi/3})\int_{0}^{\infty} = i3^{-1/6}\Gamma(\tfrac{1}{3}), \tag{4}$$

etc.

When $|\sigma|\Omega^{2/3} \gg 1$, one may employ the large-argument asymptotic approximation for the Airy-type function $C(\sigma\Omega^{2/3})$. To be specific, let it be assumed that the path SDP in Eq. (2b) is the same as L_{32} in Fig. 4.2.4. Then, from Eq. (14) and Eq. (4.2.42),

$$C(\sigma\Omega^{2/3}) = 2\pi i\, \mathrm{Ai}(\sigma\Omega^{2/3}) \sim \frac{i\sqrt{\pi}}{\sigma^{1/4}\Omega^{1/6}} e^{-(2/3)\,\Omega\sigma^{3/2}}, \tag{5a}$$

$$C'(\sigma\Omega^{2/3}) \sim -i\sqrt{\pi}\,\sigma^{1/4}\Omega^{1/6} e^{-(2/3)\,\Omega\sigma^{3/2}}, \tag{5b}$$

thereby reducing Eq. (2) to

$$I(\Omega) \sim \sqrt{\frac{-2\pi}{\Omega q''(z_2)}}\, f(z_2) e^{\Omega q(z_2)}, \tag{6}$$

the correct formula appropriate to an *isolated* first-order saddle point at z_2. Similar results are obtained when any of the other integration contours in Fig. 4.2.4 are involved.

Since $\Omega \to \infty$, the condition $|\sigma|\Omega^{2/3} \to \infty$, which defines widely separated saddle points, is satisfied when $|\sigma| \sim \Omega^{-(2/3)+\alpha}$, where α is positive but may be small. Thus, the uniform approximation in Eq. (2) need be employed only when

$\sigma = O(\Omega^{-2/3})$; for larger values of σ, each of the two saddle points may be treated independently of the other. In the range of small σ, the first term in Eq. (2) dominates and one may employ the simpler formula resulting from direct application of Eq. (4.1.5):

$$I(\Omega) \sim f(z_i)h_i \frac{e^{\Omega a_0}}{\Omega^{1/3}} C(\sigma\Omega^{2/3}), \qquad \sigma \text{ small}, \tag{7}$$

where the subscript i stands for either 1 or 2 since $f(z_i)h_i$ is slowly varying in the vicinity of $z_1 \approx z_2$. It must, however, be kept in mind that Eq. (7) is valid only when σ is small (although $\sigma\Omega^{2/3}$ may be large), whereas no such restriction is imposed on the more general formula in Eq. (2). The transition from Eq. (7) to Eq. (6) follows on use of Eqs. (5a) and (2b), subject to σ remaining small. Equation (6) may then be used for larger values of σ. Together, Eqs. (7) and (6), or some other relevant form for different integration paths, constitute non-uniform (though overlapping) approximations of $I(\Omega)$ with respect to the parameter σ, in contrast to the more complicated uniform approximation in Eq. (2) which applies for arbitrary σ.

An important special case of Eq. (2) pertains to $q(z) = i\hat{q}(z)$, where $\hat{q}(z)$ is a real function of z with real saddle points $z_{1,2}$. Let us assume that $\hat{q}''(z_1) < 0$, $\hat{q}''(z_2) > 0$, and that the integration path traverses the saddle point z_2 as shown in Fig. 4.5.2(a). It then follows from Fig. 4.5.2(b) that when $z_1 = z_2 = z_s$, $\varphi = \arg(dz)_{z_s} = \pi/6$, and, from Eq. (16c), $\hat{q}^{(3)}(z_s) > 0$. Thus, from Eq. (2c),

$$h_1 = h_2 = \left[\frac{2}{\hat{q}^{(3)}(z_s)}\right]^{1/3} e^{i\pi/6}, \text{ when } z_1 = z_2 = z_s, \tag{8}$$

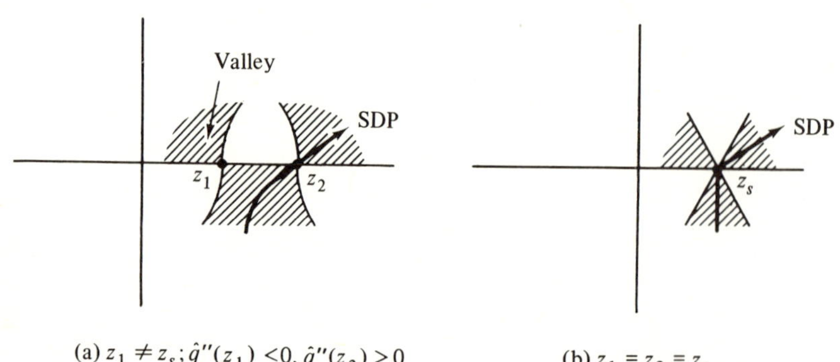

(a) $z_1 \neq z_s$; $\hat{q}''(z_1) < 0$, $\hat{q}''(z_2) > 0$ (b) $z_1 = z_2 = z_s$

FIG. 4.5.2 Integration paths when $q(z) = i\hat{q}(z)$.

where the positive value of the cube root is taken. From Eq. (16b), $\arg \sigma^{3/2} = \pi/2$, whence $\arg \sigma^{1/2} = (\pi/6) + (2n\pi/3)$, $n = 0, 1, 2$; comparison with Eqs. (2b) and (8) requires the choice $n = 1$. Taking the integration variable t in the integral formula for $C(\zeta)$ in Eq. (2b) as positive along the path segment leading away from z_s in Fig. 4.5.2(b), one finds from Eq. (14) that $C(\zeta) = \pi[\text{Bi}(\zeta) + i\,\text{Ai}(\zeta)]$, where in view of the definition of $\arg \sigma^{1/2}$, $\zeta = \sigma\Omega^{2/3} = |\sigma|\Omega^{2/3} \times \exp(-i\pi/3) = -|\sigma|\Omega^{2/3} \exp(+i2\pi/3)$. Use of Eq. (B20c) then yields a formula wherein the arguments of Ai and Bi are negative real, so Eq. (2) becomes

$$I(\Omega) \sim \frac{\pi}{2} [f(z_1)h_1 + f(z_2)h_2] \frac{e^{\Omega a_0}}{\Omega^{1/3}} e^{-i\pi/6} [\mathrm{Ai}(-|\sigma|\Omega^{2/3}) + i\,\mathrm{Bi}(-|\sigma|\Omega^{2/3})]$$

$$+ \text{[analogous term with } C'], \tag{9}$$

where $q = i\hat{q}$ is to be employed in the definitions of a_0 and σ, and it is recalled that arg $h_{1,2} = \pi/6$. For an extension of these results to the case where the saddle points move into the complex plane after coalescing at z_s, see Eqs. (5.8.59)–(5.8.61).

Analytical details

To transform the integral in Eq. (1) into the canonical form in Eq. (4.1.5), we use Eq. (4.1.7), as is appropriate when $q(z)$ has two adjacent saddle points at z_1 and z_2,

$$q(z) = \tau(s) = a_0 + \sigma s - \frac{s^3}{3}, \tag{10}$$

and obtain, as in Eq. (4.1.8),

$$a_0 = \tau(0) = \tfrac{1}{2} [q(z_1) + q(z_2)], \tag{11}$$

$$\tfrac{2}{3}\sigma^{3/2} = \tfrac{1}{2} [q(z_1) - q(z_2)]. \tag{12}$$

Special attention must be given to the choice of the proper branch of $(\sigma^{3/2})^{2/3}$ required for the evaluation of σ from Eq. (12). Upon substituting Eq. (10) into Eq. (4.1.5), one obtains the first-order asymptotic approximation for $I(\Omega)$ in Eq. (1) as $\Omega \to \infty$, valid for small values of σ:

$$I(\Omega) \sim \Omega^{-1/3} G(0) e^{a_0 \Omega} C(\sigma \Omega^{2/3}), \tag{13}$$

with a_0 and σ defined as in Eqs. (11) and (12) and

$$G(0) = \left[f(z) \frac{dz}{ds} \right]_{s=0}, \tag{13a}$$

$$C(\zeta) = \int_P e^{\zeta t - t^3/3}\, dt. \tag{13b}$$

The integral in Eq. (13b) is readily identified in terms of the Airy integrals in Eqs. (4.2.37) for any specified allowable path P in Fig. 4.2.4. Since $\int_{L_{21}} - \int_{L_{31}} + \int_{L_{32}} = 0$, one notes that

$$C(\zeta) = \begin{cases} 2\pi i\,\mathrm{Ai}(\zeta), & \text{if } P = L_{32}, \\ \pi[\mathrm{Bi}(\zeta) - i\,\mathrm{Ai}(\zeta)], & \text{if } P = L_{21}, \\ \pi[\mathrm{Bi}(\zeta) + i\,\mathrm{Ai}(\zeta)], & \text{if } P = L_{31}. \end{cases} \tag{14}$$

Equation (13) is valid for small values of σ (i.e., for $z_1 \approx z_2$). Since $f(z)$ and dz/ds are assumed to be regular and slowly varying functions in the vinicity of $s = 0$, we may then write approximately

$$f(z)|_{s=0} \approx f(z_1) \approx f(z_2), \tag{15a}$$

and

$$\left(\frac{dz}{ds}\right)_{s=0} \approx \left(\frac{dz}{ds}\right)_{s=\sqrt{\sigma}} \approx \left(\frac{dz}{ds}\right)_{s=-\sqrt{\sigma}}. \tag{15b}$$

From Eq. (10) one finds that

$$h_1 \equiv \left(\frac{dz}{ds}\right)_{s=\sqrt{\sigma}} = \sigma^{1/4}\sqrt{\frac{-2}{q''(z_1)}}, \qquad h_2 \equiv \left(\frac{dz}{ds}\right)_{s=-\sqrt{\sigma}} = \sigma^{1/4}\sqrt{\frac{2}{q''(z_2)}},$$

$$\tag{15c}$$

where the sign ambiguity introduced by the square roots is resolved on determining arg (dz/ds), as mentioned after Eq. (2c). This completes the derivation of the result in Eq. (7).

If the two first-order saddle points $z_{1,2}$ coincide, the point $z_1 = z_2 = z_s$ is a second-order saddle point, since both $q'(z_s)$ and $q''(z_s)$ vanish. The derivative $(dz/ds)_{s=0}$ must have a unique value, implying that $q''(z_2) \approx -q''(z_1)$ for $z_1 \approx z_2$ (i.e., $\sigma \approx 0$); it then follows that $q''(z_2) = q''(z_1) = 0$ if $z_1 = z_2$. Thus, in Eq. (10) for $\sigma = 0$, $\tau'(0) = \tau''(0) = 0$ and hence the origin in the s plane is likewise a second-order saddle point. To evaluate the resulting indeterminate form for $(dz/ds)_{s=0}$ in Eq. (15c), we observe first that, as $z_1 \to z_2$,

$$q^{(3)}(z_1) \approx q^{(3)}(z_2) \approx q^{(3)}(z_s), \qquad z_s = \frac{z_1 + z_2}{2}, \tag{16a}$$

where z_s is the second-order saddle point for which $q''(z_s) = 0$. This is verified from a power-series expansion of $q''(z_{1,2})$ about z_s after recalling that $q''(z_1) \approx -q''(z_2)$. Also, from Eq. (12),

$$\frac{4}{3}\sigma^{3/2} = q(z_1) - q(z_2) = q'(z_2)(z_1 - z_2) + \frac{q''(z_2)}{2}(z_1 - z_2)^2$$

$$+ \frac{q^{(3)}(z_2)}{6}(z_1 - z_2)^3 + \cdots = -\frac{1}{12}q^{(3)}(z_s)(z_1 - z_2)^3 \tag{16b}$$

where we have utilized Eq. (16a), $q'(z_2) = 0$, and

$$q''(z_2) = q''(z_s) + q^{(3)}(z_s)(z_2 - z_s) + \cdots = -\tfrac{1}{2}q^{(3)}(z_s)(z_1 - z_2) + \cdots.$$

$$\tag{16c}$$

Thus,

$$\sigma^{1/2} = \tfrac{1}{2}\left[-\tfrac{1}{2}q^{(3)}(z_s)\right]^{1/3}(z_1 - z_2) + \cdots, \tag{17}$$

and substitution of Eqs. (16c) and (17) into (15c) yields the result in Eq. (2c):

$$\left(\frac{dz}{ds}\right)_{s=0} = \left[\frac{-2}{q^{(3)}(z_s)}\right]^{1/3}, \qquad \text{when } z_1 = z_2 = z_s. \tag{18}$$

When $\sigma\Omega^{2/3} \gg 1$, one may employ Eq. (5a) to reduce Eq. (13), the integration path being taken along L_{32} in Fig. 4.2.4. Subject to Eqs. (15), the resulting form is that given in Eq. (6), so the large-argument asymptotic approximation of Eq. (13) connects onto the isolated saddle-point result. In all the considerations based on Eqs. (13) or (7), it has been necessary to keep σ small.

For larger σ, descriptive of isolated saddle points, one must deal not with Eq. (13) but with Eq. (6) and its counterpart for the saddle point at z_1. Thus, as noted earlier, the expression in Eq. (13) is non-uniform in the sense that it cannot accommodate all ranges of σ. A uniform expression valid for arbitrary values of σ is given in Eq. (2) and represents the first terms of a complete asymptotic expansion of the integral in Eq. (1).

To obtain such an expansion, we employ the canonical form in Eq. (4.1.5) with $\tau(s)$ given by Eq. (10). To achieve uniformity with respect to σ, one expands $G(s)$ not in a power series about $s = 0$, as in Eq. (4.2.4), but rather in series involving powers of the polynomial $\xi = \tau'(s) = \sigma - s^2$ and the coefficients $b_{n,m}$:[14]

$$G(s) = \sum_{n=0}^{\infty} b_{n,0} \xi^n + \sum_{n=0}^{\infty} b_{n,1} \xi^n s. \tag{19a}$$

This type of expansion permits an asymptotic-series representation for $I(\Omega)$,

$$I(\Omega) \sim \sum_{m=0}^{1} \sum_{n=0}^{\infty} b_{n,m} I_{n,m}(\Omega), \tag{19b}$$

that involves the integrals

$$I_{n,m}(\Omega) = e^{\Omega \tau(0)} \bar{I}_{n,m}(\Omega) \tag{19c}$$

$$\bar{I}_{n,m}(\Omega) = \int_P s^m \xi^n e^{\Omega(\sigma s - s^3/3)} \, ds, \qquad m = 0, 1. \tag{19d}$$

The lowest-order term ($m = n = 0$), $\bar{I}_{0,0}$, is identically the function $\Omega^{-1/3} C(\zeta)$, with $C(\zeta)$ defined in Eq. (13b). Upon introducing the change of variable $t = \Omega^{1/3} s$, one obtains

$$\bar{I}_{n,m}(\Omega) = \Omega^{-(2n+m+1)/3} \int_P t^m (\zeta - t^2)^n e^{\zeta t - t^3/3} \, dt, \qquad \zeta = \sigma \Omega^{2/3}. \tag{20}$$

From Eq. (13b) one finds

$$C^{(n)}(\zeta) \equiv \frac{d^n}{d\zeta^n} C(\zeta) = \int_P t^n e^{\zeta t - t^3/3} \, dt, \tag{21a}$$

and, from Eq. (4.2.33),

$$C''(\zeta) = \zeta C(\zeta). \tag{21b}$$

Expressions for the first few $C^{(n)}$ follow as

$$C^{(3)} = C + \zeta C'; \quad C^{(4)} = \zeta^2 C + 2C'; \quad C^{(5)} = 4\zeta C + \zeta^2 C';$$
$$C^{(6)} = (4 + \zeta^3)C + 6\zeta C'; \quad C^{(7)} = 9\zeta^2 C + (10 + \zeta^3)C'; \tag{21c}$$
$$C^{(8)} = (28\zeta + \zeta^4)C + 12\zeta^2 C'.$$

It is not difficult to prove the recursion relations

$$\bar{I}_{n,0} = -\frac{2n-1}{\Omega} \bar{I}_{n-2,1}, \qquad \bar{I}_{n,1} = -\frac{1}{\Omega} \bar{I}_{n-1,0} - \frac{2(n-1)}{\Omega} \bar{I}_{n-2,2},$$

$$\bar{I}_{n,2} = \sigma \bar{I}_{n,0} - \bar{I}_{n+1,0}, \qquad \bar{I}_{n,1} = -\frac{1 - 2(n-1)}{\Omega} \bar{I}_{n-1,0} - \frac{2\sigma(n-1)}{\Omega} \bar{I}_{n-2,0}.$$

$$\tag{21d}$$

Upon expanding the factor $(\zeta - t^2)^n$ in Eq. (20) by the binomial theorem and employing Eqs. (21), one finds that

$$\bar{I}_{0,0} = \frac{1}{\Omega^{1/3}} C(\zeta), \quad \bar{I}_{1,0} = 0, \quad \bar{I}_{2,0} = \frac{2}{\Omega^{5/3}} C'(\zeta),$$

$$\bar{I}_{3,0} = \frac{-4}{\Omega^{7/3}} C(\zeta), \quad \bar{I}_{4,0} = \frac{12\zeta}{\Omega^3} C(\zeta) = \frac{12\sigma}{\Omega^{7/3}} C(\zeta), \tag{22a}$$

and

$$\bar{I}_{0,1} = \frac{1}{\Omega^{2/3}} C'(\zeta), \quad \bar{I}_{1,1} = \frac{-1}{\Omega^{4/3}} C(\zeta),$$

$$\bar{I}_{2,1} = \frac{2\zeta}{\Omega^2} C(\zeta) = \frac{2\sigma}{\Omega^{4/3}} C(\zeta), \quad \bar{I}_{3,1} = -\frac{10}{\Omega^{8/3}} C'(\zeta), \tag{22b}$$

where ζ is defined in Eq. (20). The largest contribution arises from the $\bar{I}_{0,0}$ term, which involves the factor $\Omega^{-1/3}$, while all other terms contain at least a factor $\Omega^{-2/3}$. For any given path P which begins and ends in a shaded region in Fig. 4.2.4, the integrals $C(\zeta)$ and $C'(\zeta)$ are readily identified in terms of the Airy functions in Eq. (14) and their derivatives, respectively.

In view of Eqs. (22) and the recursion relations in Eqs. (21), one notes that the asymptotic expansion in Eq. (19b) for the integral in Eq. (4.1.5) has the form

$$I(\Omega) \sim e^{\Omega\tau(0)} \left\{ \frac{b_{0,0}}{\Omega^{1/3}} C(\zeta) + \frac{b_{0,1}}{\Omega^{2/3}} C'(\zeta) + C(\zeta) \sum_{n=1}^{\infty} \left[\frac{c_n}{\Omega^{(6n+1)/3}} + \frac{d_n}{\Omega^{(6n-2)/3}} \right] \right.$$

$$\left. + C'(\zeta) \sum_{n=1}^{\infty} \left[\frac{e_n}{\Omega^{(6n+2)/3}} + \frac{f_n}{\Omega^{(6n-1)/3}} \right] \right\}, \qquad \Omega \to \infty, \tag{23}$$

where the coefficients c_n, d_n, e_n, and f_n are polynomials in σ. The asymptotic character of the expansion is verified on referring to the comments following Eq. (4.2.18c). Since from Eq. (19a), $G(s_i) = b_{0,0} + b_{0,1}s_i$, $i = 1, 2$, one finds that $b_{0,0} = \frac{1}{2}[G(\sqrt{\sigma}) + G(-\sqrt{\sigma})]$ represents the average of the values of $G(s)$ at the saddle points $s_{1,2} = \pm\sqrt{\sigma}$, whereas $b_{0,1} = 1/(2\sqrt{\sigma})[G(\sqrt{\sigma}) - G(-\sqrt{\sigma})]$. Since $\sigma = O(\Omega^{-2/3})$ in the range where the double-saddle-point procedure must be employed (i.e., where σ is very small), $b_{0,0}$ can be approximated by $G(0)$. The first term in the expansion in Eq. (23) thus yields the first-order result in Eqs. (7) or (13). By retaining the $b_{0,0}$ and $b_{0,1}$ terms, one obtains the uniform approximation given in Eq. (2).

Example: Asymptotic evaluation of Hankel function

Consider the Hankel function $H_\nu^{(1)}(\Omega)$ defined by the integral

$$H_\nu^{(1)}(\Omega) = \frac{1}{\pi} \int_{\bar{P}} e^{\Omega q(z)}\, dz, \qquad (\nu, \Omega \text{ positive}), \tag{24}$$

$$q(z) = i\left[\cos z + \left(z - \frac{\pi}{2}\right)\sin\alpha\right], \qquad \sin\alpha = \frac{\nu}{\Omega}, \tag{24a}$$

where the path $\bar{P}$ is the same as in Fig. 4.2.2. The saddle points of $q(z)$ are located at

$$q'(z) = -i[\sin z - \sin \alpha] = 0, \tag{25}$$

whence the pertinent solutions are

$$z_1 = \alpha, \qquad z_2 = \pi - \alpha. \tag{26}$$

We assume that $v/\Omega \leq 1$, so $0 < \alpha \leq \pi/2$, and seek a first-order asymptotic evaluation of the integral in Eq. (24) as $\Omega \to \infty$ and $\alpha \to \pi/2$ (i.e., when the order and argument of the Hankel function are both large and almost equal).

From Eqs. (24),

$$q(z_{1,2}) = \pm i\left[\cos \alpha - \left(\frac{\pi}{2} - \alpha\right)\sin \alpha\right]. \tag{27}$$

Since $q(z_{1,2})$ is imaginary, $\sigma^{3/2}$ as defined in Eq. (12) is imaginary and can be satisfied by $\sigma < 0$. In the following, we choose

$$\sigma = e^{-i\pi}|\sigma| = e^{-i\pi}\eta, \qquad \eta > 0, \tag{28a}$$

and subsequently determine arg $h_{1,2}$ in Eq. (2) by referring to the integration path. From Eq. (28a),

$$s_{1,2} = \pm\sqrt{\sigma} = \mp i\sqrt{\eta}, \qquad \sqrt{\eta} > 0. \tag{28b}$$

Then, from Eq. (12),

$$\frac{2}{3}\eta^{3/2} = \cos \alpha - \left(\frac{\pi}{2} - \alpha\right)\sin \alpha \geq 0, \tag{29a}$$

$$\tau(0) = 0, \tag{29b}$$

while, from Eq. (15c),

$$\left(\frac{dz}{ds}\right)_{s_1} = \left(\frac{dz}{ds}\right)_{s_2} = \pm i\eta^{1/4}\left(\frac{2}{\cos \alpha}\right)^{1/2}, \tag{29c}$$

with the sign ambiguity noted explicitly. Since $f(z) = 1$ in Eq. (24) and in view of Eq. (29c), Eqs. (15a) and (15b) are satisfied exactly in this case. It is implied thereby that the expressions for $I(\Omega)$ in Eqs. (2) and (7) are equivalent, with a consequent removal of the restriction on σ in Eq. (7). Near $\alpha = \pi/2$, one obtains from Eq. (29a) and the requirement that $\sqrt{\eta} \geq 0$:

$$\frac{2}{3}\eta^{3/2} \approx \frac{1}{3}\left(\frac{\pi}{2} - \alpha\right)^3, \qquad \text{i.e., } \eta^{1/2} \approx 2^{-1/3}\left(\frac{\pi}{2} - \alpha\right), \tag{30a}$$

whence, from Eq. (29c),

$$\left(\frac{dz}{ds}\right)_{s_1} = \left(\frac{dz}{ds}\right)_{s_2} = \pm i 2^{1/3} \qquad \text{when } \eta = 0. \tag{30b}$$

To determine the contour of integration in the s plane, it suffices to consider the transformation in Eq. (10) as $\sigma \to 0$ (i.e., $\eta \to 0$) in order to establish the location of the endpoints of the transformed path. Thus, we examine

$$q(z) = i\left[\cos z - \left(\frac{\pi}{2} - z\right)\right] = -\frac{s^3}{3}, \qquad (\eta = 0), \tag{31}$$

which can be written near $z = \pi/2$ as

$$\frac{i}{6}\left(\frac{\pi}{2} - z\right)^3 \approx \frac{s^3}{3}. \tag{31a}$$

Upon taking the cube root of Eq. (31a), one obtains

$$s \approx 2^{-1/3}\left(\frac{\pi}{2} - z\right)e^{i\pi/6}e^{i2n\pi/3}, \qquad n = 0, 1, 2, \tag{32}$$

where the last factor expresses the three possible values of the cube root of unity. The branch to be chosen is that for which ds/dz evaluated at $z = \pi/2$ assumes one of the values in Eq. (30b). The proper choice is $n = 2$, so that

$$s \approx i2^{-1/3}\left(z - \frac{\pi}{2}\right), \qquad z \approx \frac{\pi}{2}. \tag{32a}$$

It is thereby implied that selection of the minus sign in Eqs. (30b) and (29c) resolves the sign ambiguity. One notes from Eq. (32a) that arg $s = 0$ when $z = (\pi/2) + iz_i$, $z_i < 0$, and that arg $s = -\pi/2$ when z is real and $z < \pi/2$. It then follows from Eq. (31) and a continuity argument that the entire line Re $z = \pi/2$, Im $z < 0$ maps into the positive real s axis, while the remainder of the path $\bar{P}$ terminates in the shaded region in the third quadrant of Fig. 4.2.4. Thus, the contour $\bar{P}$ is transformed in the s plane into a path L_{31}, as shown in Fig. 4.2.4. From the direction of integration along $\bar{P}$ in the neighborhood of $z = \pi/2$, one confirms the choice of sign noted above.

The first-order asymptotic representation as $\Omega \to \infty$ for $H_\nu^{(1)}(\Omega)$ in Eq. (24) can now be written down directly from Eqs. (2) or (7), (14), (27), and (29):

$$H_\nu^{(1)}(\Omega) \sim \left(\frac{2}{\cos\alpha}\right)^{1/2}\frac{\eta^{1/4}}{\Omega^{1/3}}[\mathrm{Ai}(-\eta\Omega^{2/3}) - i\,\mathrm{Bi}(-\eta\Omega^{2/3})], \tag{33}$$

where α and η are defined in Eqs. (24a) and (29a), respectively. When $\alpha = \pi/2$, use of Eqs. (3) and (30b) yields

$$H_\Omega^{(1)}(\Omega) \sim \left(\frac{2}{\Omega}\right)^{1/3}[\mathrm{Ai}(0) - i\,\mathrm{Bi}(0)] = \left(\frac{2}{\Omega}\right)^{1/3}\frac{3^{-1/6}}{\pi}\Gamma\left(\frac{1}{3}\right)e^{-i\pi/3}. \tag{34a}$$

When α is sufficiently different from $\pi/2$ to yield $\eta\Omega^{2/3}$ large, we may employ the asymptotic expressions for the Airy functions in Eqs. (4.2.51) and obtain via Eq. (29a) the Debye formula[15]

$$H_\nu^{(1)}(\Omega) \sim \left(\frac{2}{\pi\Omega\cos\alpha}\right)^{1/2}\exp\left\{i\Omega\left[\cos\alpha - \left(\frac{\pi}{2} - \alpha\right)\sin\alpha\right] - i\pi/4\right\}. \tag{34b}$$

Since from Eq. (30a),

$$\eta \approx 2^{-2/3}\cos^2\alpha = 2^{-2/3}\left[1 - \left(\frac{\nu}{\Omega}\right)^2\right] \approx 2^{1/3}\frac{\Omega - \nu}{\Omega}, \tag{35a}$$

the condition $\eta\Omega^{2/3} \gg 1$, required for the validity of Eq. (34b), can be phrased as

$$\Omega - v \gg \Omega^{1/3}. \tag{35b}$$

If $\Omega - v = O(\Omega^{1/3})$, one must employ Eq. (33).

4.5b Three Saddle Points

When $q(z)$ in Eq. (1) has three neighboring collinear, equidistant, saddle points $z_{1,2,3}$, a suitable comparison function $\tau(s)$ is given in Eq. (4.1.9). Substitution into Eq. (4.1.5) and comparison with the integral representation for the parabolic cylinder function[15]

$$D_{-v}(t) = \frac{2e^{t^2/4}}{\Gamma(v)} \int_0^\infty p^{2v-1} e^{-(1/2)(t+p^2)^2} \, dp, \qquad \mathrm{Re}\, v > 0, \tag{36}$$

then yields the lowest-order (non-uniform) asymptotic approximation as $\Omega \to \infty$ (see reference 1 for a uniform approximation)

$$\bar{I}(\Omega) \sim \frac{G(0)\sqrt{\pi}}{2(2\Omega)^{1/4}} \, e^{\Omega(a_0 - a^2/2)} D_{-1/2}(\sqrt{2\Omega}\, a), \qquad a \text{ small}, \tag{37}$$

where $G(0) = f(z_2)(dz/ds)|_{s=0}$ while a and a_0 are obtained from Eq. (4.1.9a). This result contains only the contribution from the portion of the integration path that begins at $s = 0$ and ends in a valley region at $s = \infty$ wherein the integral converges. For an evaluation of the mapping derivative

$$\frac{dz}{ds} = \frac{-4s(a + s^2)}{q'(z)}, \qquad a = \sqrt{q(z_1) - q(z_2)}, \tag{38a}$$

at $s = 0$, L'Hôpital's rule may be applied and furnishes alternative expressions that remain valid when $z_1 \to z_2$ (i.e., $a \to 0$) and verify the slowly varying character of dz/ds near $s = 0$,

$$\left.\frac{dz}{ds}\right|_{s=0} = \sqrt{\frac{-4a}{q''(z_2)}} \approx \sqrt{\frac{8a}{q''(z_1)}} \approx i\left[\frac{24}{-q^{(4)}(z_2)}\right]^{1/4}; \tag{38b}$$

the proper definition of the radicals [i.e., $\arg (dz/ds)$] remains to be carried out from a study of the transformation of the original integration path. The approximate equality of the various formulas in Eq. (38b) follows from the series expansions of $q(z)$ and its derivatives in the vicinity of the interior saddle-point location $z = z_2$; since the saddle points are assumed to be collinear and equidistant, $q(z - z_2) \cong q(z_2 - z)$ when $z \approx z_2$:

$$q(z) = q(z_2) + q''(z_2)\frac{(z - z_2)^2}{2} + q^{(4)}(z_2)\frac{(z - z_2)^4}{4!} + \cdots \tag{39a}$$

$$q'(z) = q''(z_2)(z - z_2) + q^{(4)}(z_2)\frac{(z - z_2)^3}{6} + \cdots \tag{39b}$$

$$q''(z) = q''(z_2) + q^{(4)}(z_2)\frac{(z - z_2)^2}{2} + \cdots \tag{39c}$$

Using $q'(z_1) = 0$, one finds from Eq. (39b) by letting $z = z_1$,

$$q''(z_2) \approx -\frac{q^{(4)}(z_2)(z_1 - z_2)^2}{6}, \tag{40a}$$

and from this result combined with Eq. (39c),

$$q''(z_1) \approx \tfrac{1}{3} q^{(4)}(z_2)(z_1 - z_2)^2 \approx -2q''(z_2). \tag{40b}$$

When $|\sqrt{2\Omega}\, a| \gg 1$, the parabolic cylinder function in Eq. (37) may be replaced by its large-argument approximation in Eqs. (7.5.66), and the resulting formula may be reduced to the conventional one for isolated saddle points.[†] Since $\Omega \gg 1$, $a = [q(z_1) - q(z_2)]^{1/2}$ may be small and still yield $|\sqrt{2\Omega}\, a| \gg 1$, thereby requiring the above analysis to be carried out only for $a = O(\Omega^{-1/2})$. The transition to the conventional result occurs when z_1, z_2, and z_3 are still approximately equal, under which circumstances the slowly varying function $f(z)$ may be approximated by $f(z_1) \approx f(z_2) \approx f(z_3)$.

In the limit $z_1 = z_2 = z_3$, the formula

$$D_{-\nu}(0) = \frac{\sqrt{\pi}}{2^{\nu/2}\Gamma[(1 + \nu)/2]}, \qquad \Gamma(\alpha)\Gamma(1 - \alpha) = \frac{\pi}{\sin \pi\alpha}, \tag{41}$$

may be employed to reduce Eq. (37) to the correct asymptotic expression for $\bar{I}(\Omega)$ when the integrand has a third-order saddle point at z_2 [see Eq. (4.3.7)]:

$$\bar{I}(\Omega) \sim i\left[\frac{24}{-q^{(4)}(z_2)}\right]^{1/4} \frac{\Gamma(\tfrac{1}{4})}{4\Omega^{1/4}} f(z_2)e^{\Omega q(z_2)}. \tag{42}$$

Again, the proper choice of the radical remains to be clarified in any given situation.

Implicit in the preceding analysis is the assumption that $G(0) \neq 0$. If this condition is not satisfied and the first non-vanishing term in the power-series expansion of $G(s)$ about $s = 0$ is $G^{(n)}(0)s^n/n!$, where n is a positive integer, then one has, instead of Eq. (4.1.5),

$$\bar{I}(\Omega) \sim \frac{G^{(n)}(0)}{n!}\int_0^\infty s^n e^{\Omega\tau(s)}\, ds, \tag{43}$$

with $\tau(s)$ given in Eq. (4.1.9). From a comparison with Eq. (36), one derives the result

$$\bar{I}(\Omega) \sim \frac{G^{(n)}(0)\Gamma[(n + 1)/2]}{2(2\Omega)^{(n+1)/4}n!}\, e^{\Omega(a_0 - a^2/2)}\, D_{-(n+1)/2}(\sqrt{2\Omega}\, a), \tag{44}$$

which contains Eq. (37) as the special case $n = 0$. Applications of these formulas may be found in Secs. 7.5d and 7.5e.

If the three saddle points of $q(z)$ are not equidistant and collinear, the canonical integral involves the more general exponent

$$q(z) = \tau(s) = \hat{a}_0 + a_2 s^2 + a_3 s^3 - s^4, \tag{45a}$$

whence $\tau'(s)$ has zeros at $s = 0$ and at the arbitrary points $s_{1,2}$. Alternatively, by not requiring that one of the zeros of $\tau'(s)$ is located at the origin, one may employ

---

†When a is positive, the integration path passes only through the saddle point at $s = 0$, while for negative real a another saddle point at $s = \pm i\sqrt{a}$ also lies on the path segment extending from $s = 0$ to $s = \infty$.

$$q(z) = \tau(s) = \hat{a}_0 + a_1 s + a_2 s^2 - s^4 \tag{45b}$$

The resulting canonical integral has been studied in Reference 16, and a summary of results and certain numerical data have been presented in Reference 17.

4.6 SADDLE POINTS NEAR AN ENDPOINT

4.6a Single Saddle Point

If in the integrand of

$$I(\Omega) = \int_{z_a}^{``\infty``} f(z)e^{\Omega q(z)}\, dz \tag{1}$$

the function $q(z)$ has a first-order saddle point at z_s [i.e. $q'(z_s) = 0$, $q''(z_s) \neq 0$] and the function $f(z)$ is regular in the vicinity of z_a and z_s, the asymptotic approximation of $I(\Omega)$, valid uniformly as $z_s \to z_a$, is given by

$$I(\Omega) \sim e^{\Omega q(z_s)}\left\{\frac{f(z_s)h_s}{\sqrt{\Omega}}Q(\sqrt{\Omega}\,s_a) + \frac{e^{-\Omega s_a^2}}{2\Omega}\frac{[f(z_a)h_a - f(z_s)h_s]}{s_a}\right\}, \qquad \Omega \to \infty, \tag{2}$$

where

$$s_a = \sqrt{q(z_s) - q(z_a)}, \quad Q(y) = \int_y^\infty e^{-x^2}\,dx, \quad h_a = \frac{-2s_a}{q'(z_a)},$$

$$h_s = \sqrt{\frac{-2}{q''(z_s)}} = h_a\Big|_{z_a=z_s}. \tag{2a}$$

$\arg h_s \equiv \varphi = \arg(dz)_{z_s}$, where dz is an element along the steepest-descent path SDP through z_s, and $\arg s_a$ is defined so that $h_a \to h_s$ as $s_a \to 0$.

When $z_a = z_s$, one has $Q(0) = \frac{1}{2}\sqrt{\pi}$, whence Eq. (2) reduces to the result for coincident saddle point and endpoint [see Eq. (4.2.7) with inclusion of the factor $\frac{1}{2}$ to account for the semiinfinite integration interval]:

$$I(\Omega) \sim \frac{1}{2}\sqrt{\frac{-2\pi}{\Omega q''(z_s)}}\, f(z_s)e^{\Omega q(z_s)}\left[1 + O\left(\frac{1}{\sqrt{\Omega}}\right)\right], \qquad z_s = z_a; \tag{3}$$

for $\sqrt{\Omega}\,|s_a| \gg 1$, use of the asymptotic formula [Eq. (4.4.20)],

$$Q(\sqrt{\Omega}\,s_a) \sim \frac{e^{-\Omega s_a^2}}{2s_a\sqrt{\Omega}} + \sqrt{\pi}\,U(-\mathrm{Re}\,s_a) \tag{4}$$

yields the result for isolated saddle point and endpoint,

$$I(\Omega) \sim \sqrt{\frac{-2\pi}{\Omega q''(z_s)}}\, f(z_s)e^{\Omega q(z_s)}U(-\mathrm{Re}\,s_a) - \frac{f(z_a)e^{\Omega q(z_a)}}{\Omega q'(z_a)}, \tag{5}$$

where $U(\alpha)$ equals unity or zero for $\alpha > 0$ and $\alpha < 0$ respectively.

An important special case arises when $q(z) = i\hat{q}(z)$, where $\hat{q}$ is a real function of z so that $\hat{q}(z_{s,a})$ is real when $z_{s,a}$ is real. In this instance, for real $z_{s,a}$,

$$I(\Omega) = \int_{z_a}^{\infty} f(z) e^{i\Omega \hat{q}(z)} \, dz$$

$$\sim e^{i\Omega \hat{q}(z_s)} \left\{ \frac{f(z_s) h_s}{\sqrt{\Omega}} Q(\hat{s}_a \sqrt{\Omega} \, e^{\mp i\pi/4}) + \frac{e^{\pm i(\Omega \hat{s}_a^2 + \pi/4)}}{2\Omega \hat{s}_a} [f(z_a) h_a - f(z_s) h_s] \right\},$$

$$\hat{q}''(z_s) \gtrless 0 \qquad (6)$$

where $\hat{s}_a = \pm |\hat{q}(z_s) - \hat{q}(z_a)|^{1/2}$ for $(z_a - z_s) \gtrless 0$, and

$$h_s = \sqrt{\frac{2}{|\hat{q}''(z_s)|}} \, e^{\pm i\pi/4}, \quad h_a = \frac{2|\hat{s}_a| e^{\pm i\pi/4}}{|\hat{q}'(z_a)|}, \qquad \text{for } \hat{q}''(z_s) \gtrless 0. \qquad (6a)$$

The reduction for $|\hat{s}_a| \sqrt{\Omega} \gg 1$ follows from Eq. (4) or (5),

$$I(\Omega) \sim \sqrt{\frac{2\pi}{\Omega |\hat{q}''(z_s)|}} f(z_s) e^{i\Omega \hat{q}(z_s) \pm i\pi/4} U(z_s - z_a) - \frac{f(z_a) e^{i\Omega \hat{q}(z_a)}}{i\Omega \hat{q}'(z_a)}, \qquad \hat{q}''(z_s) \gtrless 0,$$

$$(6b)$$

and agrees with the result in Eq. (4.2.20).

Analytical details

Since $q(z)$ is assumed to have a first-order saddle point, the relevant transformation to the canonical form is $q(z) = \tau(s) = q(z_s) - s^2$, $s^2 > 0$ [see Eq. (4.1.6)] Thus, if s_a corresponds to z_a,

$$I(\Omega) = e^{\Omega q(z_s)} \int_{s_a}^{\infty} G(s) e^{-\Omega s^2} \, ds, \qquad G(s) = f(z) \frac{dz}{ds} = f(z) \frac{-2s}{q'(z)}, \qquad (7)$$

with s_a defined in Eq. (2a). Upon writing the integral in Eq. (7) as

$$\bar{I}(\Omega) = G(0) \int_{s_a}^{\infty} e^{-\Omega s^2} \, ds - \frac{1}{2\Omega} \int_{s_a}^{\infty} \frac{[G(s) - G(0)]}{s} \frac{d}{ds} e^{-\Omega s^2} \, ds \qquad (7a)$$

and, integrating the second term by parts, one finds

$$\bar{I}(\Omega) = \frac{G(0)}{\sqrt{\Omega}} Q(\sqrt{\Omega} \, s_a) + \frac{e^{-\Omega s_a^2}}{2\Omega s_a} [G(s_a) - G(0)]$$

$$+ \frac{1}{2\Omega} \int_{s_a}^{\infty} e^{-\Omega s^2} \frac{d}{ds} \left[\frac{G(s) - G(0)}{s} \right] ds, \qquad (7b)$$

which furnishes Eq. (2) via the definitions $h_s = (dz/ds)_{s=0}$ and $h_a = (dz/ds)_{s_a}$. The last term in Eq. (7b), omitted in Eq. (2), provides higher-order corrections; an asymptotic expansion of $I(\Omega)$ may be constructed by repeated decomposition of the remainder integrals as in Eq. (7a), with subsequent integration by parts.[18] When $s_a \to 0$, $z_a \to z_s$, one has $q'(z_a) \approx q''(z_s)(z_a - z_s)$, $s_a \approx \sqrt{-q''(z_s)/2} \times (z_a - z_s)$, so $h_a \to h_s$.

For the special case $q(z) = i\hat{q}(z)$, with $\hat{q}$ a real function of z, one has, as in Eq. (4.2.20),

$$h_s = \sqrt{\frac{2}{|\hat{q}''(z_s)|}} \, e^{\pm i\pi/4}, \qquad \text{for } \hat{q}''(z_s) \gtrless 0. \qquad (8a)$$

Since $\hat{q}'(z_s) = 0$, it follows that $\operatorname{sgn} \hat{q}'(z_a) = \pm \operatorname{sgn}(z_a - z_s)$ for $\hat{q}''(z_s) \gtrless 0$. Thus, from Eq. (2a) with $s_a = \hat{s}_a \exp(\mp i\pi/4)$ when $\hat{q}''(z_s) \gtrless 0$,

$$h_a = \frac{\pm i2s_a \operatorname{sgn}(z_a - z_s)}{|\hat{q}'(z_a)|} = \frac{2\hat{s}_a \operatorname{sgn}(z_a - z_s)e^{\pm i\pi/4}}{|\hat{q}'(z_a)|}, \qquad \text{for } \hat{q}''(z_s) \gtrless 0, \qquad (8b)$$

where the definition of $\hat{s}_a$ in Eq. (6) ensures that $\arg h_a = \arg h_s$. Use of these relations in Eq. (2) furnishes the formula in Eq. (6).

4.6b Two First-Order Saddle Points

In the integrand of

$$I(\Omega) = \int_{z_a}^{``\infty"} f(z)e^{\Omega q(z)}\, dz = \int_{s_a}^{``\infty"} G(s)e^{\Omega \tau(s)}\, ds, \qquad (9)$$

the function $q(z)$ has two first-order saddle points at $z_{1,2}$ [i.e., $q'(z_{1,2}) = 0$ and $q''(z_{1,2}) \neq 0$ unless $z_1 = z_2$]. If the function $f(z)$ is regular in the vicinity of z_1, z_2, and z_a, the asymptotic approximation of $I(\Omega)$, valid uniformly as $z_1 \to z_2 \to z_a$, is given by[19]

$$I(\Omega) \sim [G(s_1) + G(s_2)]\frac{e^{\Omega a_0}}{2\Omega^{1/3}} C_a(s_1^2 \Omega^{2/3}) + [G(s_1) - G(s_2)]\frac{e^{\Omega a_0}}{2\Omega^{2/3}s_1} C_a'(s_1^2 \Omega^{2/3})$$

$$- \left[G(s_a) - \left(1 + \frac{s_a}{s_1}\right)\frac{G(s_1)}{2} - \left(1 - \frac{s_a}{s_1}\right)\frac{G(s_2)}{2} \right]\frac{e^{\Omega \tau(s_a)}}{\Omega(s_1^2 - s_a^2)}, \qquad \Omega \to \infty, \qquad (10)$$

where the variable s is related to z by the transformation

$$q(z) = \tau(s) = a_0 + \sigma s - \frac{s^3}{3}, \quad q(z_i) = \tau(s_i), \qquad i = 1, 2, a, \qquad (10a)$$

$$a_0 = \tfrac{1}{2}[q(z_1) + q(z_2)], \qquad \sigma^{1/2} = \{ \tfrac{3}{4}[q(z_1) - q(z_2)]\}^{1/3} \equiv s_1 = -s_2,$$

and

$$G(s) = f(z)\frac{dz}{ds}, \quad C_a(\zeta) = \int_{\Omega^{1/3}s_a}^{``\infty"} e^{\zeta t - t^3/3}\, dt, \quad C_a'(\zeta) = \frac{dC_a(\zeta)}{d\zeta},$$

$$\left(\frac{dz}{ds}\right)_{s_{1,2}} = \sqrt{\frac{\mp 2s_1}{q''(z_{1,2})}}, \qquad \left(\frac{dz}{ds}\right)_{s_a} = \frac{s_1^2 - s_a^2}{q'(z_a)}. \qquad (10b)$$

In order not to complicate the notation, the results in this section are given in terms of the variable s rather than the original variable z. The multivaluedness in the definition of σ and $(dz/ds)_{s_i}$ is resolved as in Sec. 4.5. The function $C_a(\zeta)$ may be termed an "incomplete" Airy function, in comparison with the Airy function $C(\zeta)$ in Eq. (4.5.14) for which both endpoints of the integration path lie at infinity; the location of the upper limit "∞" in the complex t plane is determined by comparison with the original path in the complex z plane as in Sec. 4.5 [for functions related to $C_0(\zeta)$ see Reference 20].

A number of special cases may be deduced from Eq. (10), which applies for arbitrary values of $z_{1,2,a}$ (or $s_{1,2,a}$). When $z_1 = z_2 = z_s$ (corresponding to $s_{1,2} =$

0), the two first-order saddle points coalesce into a single second-order saddle point. In this instance, one derives the following uniform asymptotic approximation when a second-order saddle point is situated near an endpoint of the integration interval:

$$I(\Omega) \sim G(0)\frac{e^{\Omega q(z_s)}}{\Omega^{1/3}} C_a(0) + G'(0)\frac{e^{\Omega q(z_s)}}{\Omega^{2/3}} C_a'(0)$$

$$+ \frac{e^{\Omega[q(z_s)-s_a^3/3]}}{\Omega s_a^2} [G(s_a) - G(0) - s_a G'(0)], \tag{11}$$

with $(dz/ds)_0 = [-2/q^{(3)}(z_s)]^{1/3}$ as in Eq. (4.5.18). When $z_1 = z_2 = z_a$ (corresponding to $s_a = s_1 = s_2 = 0$), the second-order saddle point coincides with the endpoint and Eq. (11) furnishes the expression

$$I(\Omega) \sim \frac{G(0)e^{\Omega q(z_s)}}{\Omega^{1/3}} \int_0^{``\infty"} e^{-t^3/3}\, dt + O\left(\frac{1}{\Omega^{2/3}}\right) \tag{12}$$

which agrees with the result in Eq. (4.3.7) since the integral can be expressed in terms of the gamma function of order $\frac{1}{3}$ [see Eq. (4.3.5)].

When neither z_1 nor z_2 is near z_a (i.e., $s_{1,2} \not\approx s_a$), the endpoint contribution to the incomplete Airy function may be evaluated separately from the saddle-point contribution, so by an integration by parts,

$$C_a(s_1^2 \Omega^{2/3}) \sim \frac{\exp\left[\Omega\tau(s_a) - \Omega a_0\right]}{\Omega^{2/3}(s_a^2 - s_1^2)} + UC(\sigma\Omega^{2/3}), \tag{13}$$

where $C(\zeta)$ is the complete Airy function defined in Eq. (4.5.13b), descriptive of two adjacent saddle points situated far from other critical points in the integrand. U is a discontinuous factor which equals unity when the original integration path traverses the saddle-point region but equals zero when the original integration path proceeds from z_a to infinity in a valley without passing through the saddle-point region. Examples are shown in Fig. 4.6.1, where the integration path is assumed to end in the valley designated by A. When z_a is situated as in Fig. 4.6.1(a), it is necessary to pass through the saddle point at

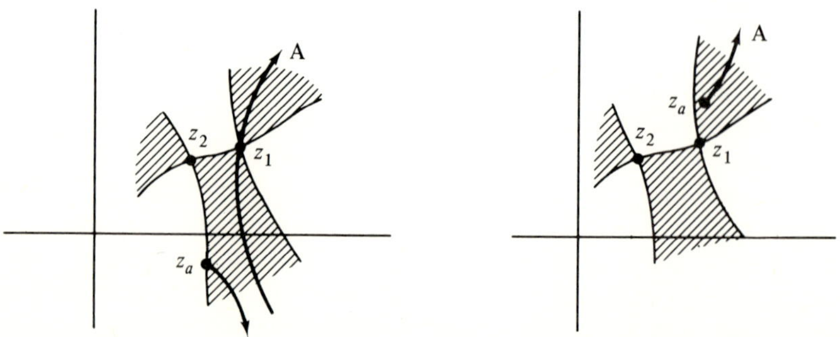

(a) Endpoint plus saddle point contribution   (b) Endpoint contribution only

FIG. 4.6.1   Integration paths in the complex z plane.

z_1, whence $U = 1$ in Eq. (13), whereas $U = 0$ for the case depicted in Fig. 4.6.1(b). The analogous formula for C_a' differs from Eq. (13) in that C is replaced by C' and the endpoint contribution is multiplied by $\Omega^{1/3}s_a$. Thus, Eq. (10) reduces for $s_{1,2} \not\approx s_a$ to

$$I(\Omega) \sim -G(s_a) \frac{e^{\Omega\tau(s_a)}}{\Omega(s_1^2 - s_a^2)} + \left[\frac{G(s_1) + G(s_2)}{2\Omega^{1/3}} e^{\Omega a_0} C(s_1^2\Omega^{2/3})\right.$$

$$\left. + \frac{G(s_1) - G(s_2)}{2\Omega^{2/3} s_1} e^{\Omega a_0} C'(s_1^2\Omega^{2/3})\right]U. \tag{14}$$

This formula accounts correctly for an isolated endpoint and for the contribution from two arbitrarily closely spaced saddle points (see Eqs. (5), (10b), and (4.5.2); it is noted that the first term in Eq. (14) can be written as $-f(z_a)[\Omega q'(z_a)]^{-1}$ $\exp[\Omega\tau(s_a)])$. When s_1 and s_2 are not adjacent, the last two terms in Eq. (14) may be simplified as in Eqs. (4.5.5) to furnish the isolated first-order saddle-point contributions. When s_1 and s_2 coalesce into a second-order saddle point at $s = 0$, the last two terms in Eq. (14) may be expressed as in Eqs. (4.5.2), with (4.5.2c) and (4.5.3).

A different reduction occurs when the two saddle points are not near one another ($z_1 \not\approx z_2$, or $s_{1,2} \not\approx 0$), but one of the saddle points is near the endpoint ($z_1 \approx z_a$, or $s_1 \approx s_a$). In this instance, it is found that [see Eqs. (26) and (27)]

$$C_a(s_1^2\Omega^{2/3}) - U_2\Omega^{1/3}\sqrt{\frac{\pi}{\Omega s_2}} e^{\Omega[\tau(s_2) - a_0]} \sim \frac{1}{\Omega^{1/3} s_1} C_a'(s_1^2\Omega^{2/3}) - \frac{e^{\Omega[\tau(s_a) - a_0]}}{\Omega^{2/3} s_1(s_1 + s_a)}$$

$$\sim \Omega^{1/3} e^{\Omega[\tau(s_1) - a_0]}\left[\frac{Q(\alpha\sqrt{\Omega})}{\sqrt{\Omega s_1}} - \frac{e^{-\Omega\alpha^2}}{2\Omega\alpha}\left(\frac{2\alpha}{s_1^2 - s_a^2} + \frac{1}{\sqrt{s_1}}\right)\right], \qquad s_1 \not\approx 0, \tag{15}$$

where

$$\alpha = \sqrt{\tau(s_1) - \tau(s_a)}, \qquad Q(y) = \int_y^\infty e^{-x^2}\, dx, \tag{15a}$$

and U_2 equals 1 or 0, depending on whether or not the integration path traverses the isolated saddle point at s_2. Substitution of these formulas into Eq. (10) yields the reduced expression

$$I(\Omega) \sim \frac{G(s_1)e^{\Omega\tau(s_1)}Q(\alpha\sqrt{\Omega})}{\sqrt{\Omega s_1}} - \frac{e^{\Omega\tau(s_a)}}{\Omega}\left[\frac{G(s_1)}{2\alpha\sqrt{s_1}} + \frac{G(s_a)}{s_1^2 - s_a^2}\right] + U_2 G(s_2)e^{\Omega\tau(s_2)}\sqrt{\frac{\pi}{\Omega s_2}}, \tag{16}$$

which is the correct result for a first-order saddle point near an endpoint of the integration interval [see Eq. (2)], the last term providing the contribution from the isolated saddle point at s_2 when required. When $\alpha \to 0$, the saddle point at s_1 coincides with the endpoint at s_a, and Eq. (16) leads correctly to the expression [Note: $Q(0) = \sqrt{\pi}/2$]

$$I(\Omega) \sim \tfrac{1}{2} G(s_1)e^{\Omega\tau(s_1)}\sqrt{\frac{\pi}{\Omega s_1}}\left[1 + O\left(\frac{1}{\Omega}\right)\right] + U_2 G(s_2)e^{\Omega\tau(s_2)}\sqrt{\frac{\pi}{\Omega s_2}}. \tag{17}$$

When $\alpha\sqrt{\Omega} \gg 1$, the saddle point at s_1 and the endpoint at s_a are "widely separated" and permit use of the asymptotic formula [see Eqs. (4.4.18a) and (4.4.20)]

$$Q(\alpha\sqrt{\Omega}) \sim \frac{e^{-\Omega\alpha^2}}{2\alpha\sqrt{\Omega}} + \sqrt{\pi}\,U(-\operatorname{Re}\alpha); \tag{18}$$

hence Eq. (16) reduces to the conventional result where the two saddle points are separated from one another and from the endpoint:

$$I(\Omega) \sim U_1 G(s_1)e^{\Omega\tau(s_1)}\sqrt{\frac{\pi}{\Omega s_1}} - G(s_a)\frac{e^{\Omega\tau(s_a)}}{\Omega(s_1^2 - s_a^2)} + U_2 G(s_2)^{\Omega\tau(s_2)}\sqrt{\frac{\pi}{\Omega s_2}}. \tag{19}$$

$U_{1,2}$ equals unity when the integration path traverses the saddle points s_1 and s_2, respectively, but vanishes otherwise. In terms of functions defined in the original z plane of Eq. (9), Eq. (19) may be written equivalently as

$$I(\Omega) \sim U_1 f(z_1)e^{\Omega q(z_1)}\sqrt{\frac{-2\pi}{\Omega q''(z_1)}} - f(z_a)\frac{e^{\Omega q(z_a)}}{\Omega q'(z_a)} + U_2 f(z_2)e^{\Omega q(z_2)}\sqrt{\frac{-2\pi}{\Omega q''(z_2)}}. \tag{20}$$

Analytical details

Since $q(z)$ in Eq. (9) is assumed to have two neighboring first-order saddle points, the relevant transformation to the canonical form in the complex s plane is given as in Eq. (4.5.10), thereby leading to Eq. (10a) and to the second of Eqs. (9). The transformed integral in the variable s may now be written as

$$I(\Omega) = b_{0,0}\int_{s_a}^{``\infty"} e^{\Omega\tau(s)}\,ds + b_{0,1}\int_{s_a}^{``\infty"} se^{\Omega\tau(s)}\,ds$$
$$+ \frac{1}{\Omega}\int_{s_a}^{``\infty"} \frac{G(s) - b_{0,0} - sb_{0,1}}{s_1^2 - s^2}\frac{d}{ds}\,e^{\Omega\tau(s)}\,ds, \tag{21}$$

where $b_{0,0}$ and $b_{0,1}$ are the first two coefficients in the uniform asymptotic expansion of the infinite double-saddle-point integral [see Eq. (4.5.23)]:

$$b_{0,0} = \tfrac{1}{2}[G(s_1) + G(s_2)], \qquad b_{0,1} = \frac{1}{2s_1}[G(s_1) - G(s_2)], \qquad s_2 = -s_1. \tag{21a}$$

Since $[G(s) - b_{0,0} - sb_{0,1}](s_1^2 - s^2)^{-1}$ remains bounded at $s = \pm s_1$, integration by parts on the third integral I_3 is allowed, with the result

$$I_3 = -\frac{1}{\Omega}\frac{G(s_a) - b_{0,0} - s_a b_{0,1}}{s_1^2 - s_a^2}e^{\Omega\tau(s_a)}$$
$$- \frac{1}{\Omega}\int_{s_a}^{``\infty"} \frac{d}{ds}\left[\frac{G(s) - b_{0,0} - sb_{0,1}}{s_1^2 - s^2}\right]e^{\Omega\tau(s)}\,ds. \tag{22}$$

By regarding the function $(d/ds)\,[\;]$ in the integrand of Eq. (22) as a new function $G(s)$, one may apply the decomposition in Eq. (21) successively to generate terms with increasing inverse powers of Ω. To a lowest order in Ω, the integral in Eq. (22) is omitted and Eq. (21) thus furnishes the result in Eq. (10), after changes of variable from s to t and the introduction of the parameter ζ

as in Eq. (4.5.2). From $dz/ds = \tau'(s)/q'(z) = (s_1^2 - s^2)/q'(z)$, one obtains the various forms of $(dz/ds)_{s_i}$ specified in Eq. (10b). As mentioned earlier, due regard must be given to the definition of the radicals appearing in these expressions.

The formulas for $I(\Omega)$ accommodating various special cases are derived from the general result in Eq. (10) by substitution of appropriate reductions of the canonical integrals $C_a(\zeta)$ and $C_a'(\zeta)$. The derivation of Eqs. (11) and (12) pertaining to $s_{1,2} = 0$ and $s_{1,2,a} = 0$, respectively, is evident. When $s_{1,2} \not\approx s_a$, the endpoint and saddle points are isolated and may be treated separately. As in Eq. (4.1.15), the endpoint contribution follows from integration by parts,

$$\frac{e^{\Omega a_0} C_a(s_1^2 \Omega^{2/3})}{\Omega^{1/3}} = \int_{s_a}^{``\infty"} e^{\Omega \tau(s)}\, ds = \frac{1}{\Omega} \int_{s_a}^{``\infty"} \frac{ds}{s_1^2 - s^2} \frac{d}{ds} e^{\Omega \tau(s)} \sim -\frac{1}{\Omega} \frac{e^{\Omega \tau(s_a)}}{s_1^2 - s_a^2},$$

$$(23a)$$

which leads to the first term in Eq. (13). As noted in the discussion following Eq. (13), it may be necessary to add to this contribution the saddle-point result if the integration path traverses the vicinity of $s_{1,2}$. Similarly,

$$\frac{e^{\Omega a_0} C_a'(s_1^2 \Omega^{2/3})}{\Omega^{2/3}} = \int_{s_a}^{``\infty"} s e^{\Omega \tau(s)}\, ds \sim -\frac{1}{\Omega} \frac{s_a e^{\Omega \tau(s_a)}}{s_1^2 - s_a^2}, \qquad (23b)$$

plus the saddle-point contribution, when required.

When the two saddle points are widely spaced ($s_{1,2} \not\approx 0$), but one saddle point is near the endpoint ($s \approx s_a$), the incomplete Airy integral may be reduced to an incomplete gamma function (Fresnel integral). Since $s_1 \not\approx s_2$, only first-order saddle points occur and one may transform the function $\tau(s)$ into the canonical function $\hat{\tau}(\mu)$ to describe the contribution from the first-order saddle point at s_1:

$$\tau(s) = a_0 + s_1^2 s - \frac{s^3}{3} = \tau(s_1) - \mu^2, \qquad \tau(s_1) = a_0 + \tfrac{2}{3} s_1^3, \qquad (24)$$

whence $\mu = 0$ corresponds to $s = s_1$. If $\mu = \alpha$ is taken to correspond to $s = s_a$, then

$$\hat{I} = \int_{s_a}^{``\infty"} e^{\Omega \tau(s)}\, ds = e^{\Omega \tau(s_1)} \int_{\alpha}^{``\infty"} e^{-\Omega \mu^2} \left(\frac{ds}{d\mu}\right) d\mu, \qquad \frac{ds}{d\mu} = -\frac{2\mu}{s_1^2 - s^2}. \qquad (25)$$

The uniform asymptotic approximation of the μ integral, valid as the endpoint $\mu = \alpha$ approaches the saddle point $\mu = 0$, is given in Eq. (2) and yields

$$\hat{I} \sim e^{\Omega \tau(s_1)} \left\{ \left(\frac{ds}{d\mu}\right)_0 \frac{Q(\sqrt{\Omega}\, \alpha)}{\sqrt{\Omega}} + \frac{e^{-\Omega \alpha^2}}{2\Omega \alpha} \left[\left(\frac{ds}{d\mu}\right)_\alpha - \left(\frac{ds}{d\mu}\right)_0 \right] \right\}, \qquad (26)$$

where $\alpha = [\tau(s_1) - \tau(s_a)]^{1/2}$ from Eq. (24), and $Q(y)$ is the Fresnel integral defined in Eq. (15a). The mapping derivative $(ds/d\mu)_\alpha$ is calculated by direct substitution into Eq. (25), whereas from an evaluation of the resulting indeterminate form, $(ds/d\mu)_0 = s_1^{-1/2}$. Recalling the definitions of C_a and $\hat{I}$ in Eqs. (23a) and (25) and adding the conventional contribution from the isolated saddle point at s_2 when relevant, one is led from Eq. (26) to Eq. (15). In the analogous treatment via Eq. (25) of

$$\hat{I}' = \int_{s_a}^{``\infty"} s e^{\Omega \tau(s)} \, ds = \frac{d\hat{I}}{\Omega d(s_1^2)} \sim s_1 \hat{I} - \frac{e^{\Omega[\tau(s_1) - \alpha^2]}}{\Omega} \left(\frac{ds}{d\mu}\right)_\alpha \frac{d\alpha}{d(s_1^2)}, \qquad (27)$$

one obtains via the definition of C'_a in Eq. (23b) the result in Eq. (15) since in view of $\alpha^2 = \tau(s_1) - \tau(s_a)$,

$$\frac{d\alpha}{d(s_1^2)} = \frac{1}{2\alpha} \frac{d}{d(s_1^2)} [\tau(s_1) - \tau(s_a)] = \frac{s_1 - s_a}{2\alpha}. \qquad (27a)$$

In the last portion of Eq. (27), a term involving the derivative $d^2 s/d(s_1^2) d\mu$ in the integrand has been omitted since it may be shown to represent a higher-order contribution.

4.7 MULTIPLE INTEGRALS

If in the integrand of the real multiple integral

$$I_n(\Omega) = \int_{-\infty}^{\infty} \cdots \int f(\mathbf{x}) e^{-\Omega q(\mathbf{x})} \, dx_1 \cdots dx_n, \qquad \Omega > 0, \qquad (1)$$

where $\mathbf{x} = (x_1, \ldots, x_n)$, the function q has a simple stationary point $\mathbf{x}_s = (x_{1s}, \ldots, x_{ns})$ defined by

$$\frac{\partial q}{\partial x_i} = 0 \text{ at } x_{is}, \qquad i = 1, \cdots, n, \qquad (2)$$

and the function f is regular near $\mathbf{x}_s$, then the asymptotic approximation of $I_n(\Omega)$ for large Ω is given by

$$I_n(\Omega) \sim f(\mathbf{x}_s) e^{-\Omega q(\mathbf{x}_s)} \left(\frac{2\pi}{\Omega}\right)^{n/2} \frac{1}{|\det (\partial^2 q/\partial x_{is} \partial x_{js})|^{1/2}}. \qquad (3)$$

One encounters frequently a modified form $\hat{I}_n(\Omega)$ of Eq. (1). If in the integrand of the multiple integral

$$\hat{I}_n(\Omega) = \int_{-\infty}^{\infty} \cdots \int f(\mathbf{x}) e^{i\Omega q(\mathbf{x})} \, dx_1 \cdots dx_n, \qquad \Omega > 0, \qquad (4)$$

the real function q has a simple, real stationary point $\mathbf{x}_s$ defined as in Eq. (2), and if f is regular near $\mathbf{x}_s$, then the asymptotic approximation of $\hat{I}_n(\Omega)$ for large Ω is given by

$$\hat{I}_n(\Omega) \sim f(\mathbf{x}_s) e^{i\Omega q(\mathbf{x}_s)} \left(\frac{2\pi}{\Omega}\right)^{n/2} \frac{e^{i(\pi/4)\sigma}}{|\det (\partial^2 q/\partial x_{is} \partial x_{js})|^{1/2}}, \qquad (5)$$

where

$$\sigma = \sum_{i=1}^{n} \operatorname{sgn} d_i, \qquad (5a)$$

and d_i are the eigenvalues of the matrix comprising the elements $\partial^2 q/\partial x_{is} \partial x_{js}$, $i, j = 1, \ldots, n$. Equation (5) is a generalization of the stationary-phase formula in Eq. (4.2.20a) for single integrals. If several stationary points exist, each contributes as in Eqs. (3) or (5).

Analytical details

The stationary points $\mathbf{x}_s$ of a function q of n variables $(x_1, \ldots, x_n)$ are defined as in Eq. (2) and, for large Ω, the principal contribution to the integrals arises from the vicinity of $\mathbf{x}_s$ [possible contributions from relevant singularities of $f(\mathbf{x})$ are not included in the present discussion]. Near $\mathbf{x}_s$ one may write $f(\mathbf{x}) \sim f(\mathbf{x}_s)$ and expand

$$q(\mathbf{x}) \sim q(\mathbf{x}_s) + \tfrac{1}{2} \sum_{i=1}^{n} \sum_{j=1}^{n} a_{ij}(x_i - x_{is})(x_j - x_{js}), \qquad a_{ij} \equiv \frac{\partial^2 q}{\partial x_{is} \partial x_{js}}, \tag{6}$$

with linear terms absent in view of Eq. (2). On substituting into Eqs. (1) or (4) and translating the origin to $\mathbf{x}_s$, one obtains

$$\bar{J}_n(\Omega) \sim f(\mathbf{x}_s) e^{\alpha \Omega q(\mathbf{x}_s)} \int_{-\infty}^{\infty} \cdots \int \exp\left(\frac{\alpha}{2}\Omega \sum_{i,j=1}^{n} a_{ij} x_i x_j\right) dx_1 \cdots dx_n, \tag{7}$$

where $\bar{J}_n \equiv I_n$ when $\alpha = -1$ and $\bar{J}_n \equiv \hat{I}_n$ when $\alpha = i$. To evaluate the multiple integral in Eq. (7), we introduce a transformation such that

$$\sum_{i,j=1}^{n} a_{ij} x_i x_j = \sum_{i=1}^{n} d_i y_i^2, \qquad x_i = \sum_{j=1}^{n} r_{ij} y_j, \tag{8a}$$

or, in matrix notation, with A denoting the $n \times n$ matrix comprising the elements a_{ij},

$$\tilde{\mathbf{x}} A \mathbf{x} = \tilde{\mathbf{y}} D \mathbf{y}, \qquad \mathbf{x} = R\mathbf{y}, \tag{8b}$$

where $\sim$ indicates the transposed quantity and D is a diagonal matrix. Since the diagonalization of A involves a coordinate rotation, R is an orthogonal matrix with $\det R = 1$. It follows from Eq. (8b) that

$$D = \tilde{R} A R, \qquad \det D = (\det R)^2 \det A = \det A, \tag{9a}$$

and that the Jacobian of the transformation from $\mathbf{x}$ to $\mathbf{y}$ is

$$\frac{\partial(x_1, \ldots, x_n)}{\partial(y_1, \ldots, y_n)} = \det R = 1. \tag{9b}$$

Thus, on changing variables to $\mathbf{y}$, one has, for the integral of Eq. (7),

$$\int_{-\infty}^{\infty} \cdots \int \exp\left(\frac{\alpha}{2}\Omega \sum_{i=1}^{n} d_i y_i^2\right) dy_1 \cdots dy_n = \prod_{i=1}^{n} \int_{-\infty}^{\infty} \exp\left(\frac{\alpha}{2}\Omega d_i y_i^2\right) dy_i. \tag{10}$$

When $\alpha = -1$ and $d_i > 0$, the single integral in Eq. (10) has the value $(2\pi)^{1/2}(\Omega d_i)^{-1/2}$; when $\alpha = i$, its value is $(2\pi)^{1/2}(\Omega|d_i|)^{-1/2} \exp[i(\pi/4)\,\text{sgn}\,d_i]$ [see Eq. (4.2.20)]. Since $\prod_{i=1}^{n} d_i = \det D = \det A$, the results in Eqs. (3) and (5) are established. For higher-order terms in the asymptotic expansion, see Reference 21.

4.8 INTEGRATION AROUND A BRANCH POINT

Integrands of diffraction integrals often exhibit branch-point singularities that contribute to the asymptotic solution if they are crossed during the deformation of the original integration path into the steepest-descent path [see

Fig. 4.1.1 and Eq. (4.1.4)]. It is then necessary to calculate the contribution from the branch-cut integral, a generic form of which is given by

$$I_b = \int_{\bar{P}_b} f(z) e^{\Omega q(z)}\, dz, \tag{1}$$

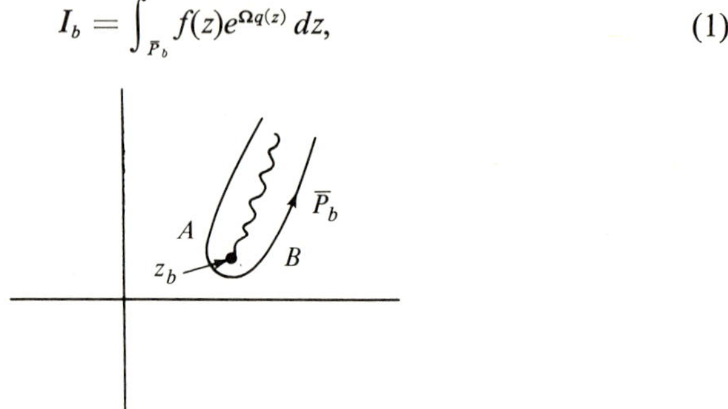

FIG. 4.8.1 First-order branch point singularity in the complex z plane.

where $\bar{P}_b$ is a contour encircling the branch cut in the positive sense (Fig. 4.8.1). In typical problems, $f(z)$ behaves like

$$f(z) \cong a + b\sqrt{z - z_b} \quad \text{near } z_b, \qquad \text{i.e., } b = 2[\sqrt{z - z_b}\, f'(z)]_{z_b}, \tag{2}$$

where a and b are constants; $q(z)$ is regular in the vicinity of z_b, and $\exp[\Omega q(z)]$ decays along $\bar{P}_b$. For large Ω, I_b has the following asymptotic approximation:

$$I_b \sim \frac{2\sqrt{\pi}}{[\Omega|q'(z_b)|]^{3/2}} [\sqrt{z - z_b}\, f'(z)]_{z_b} \exp\{\Omega q(z_b) - i\,\tfrac{3}{2}\arg[-q'(z_b)]\}. \tag{3}$$

Analytical details

The asymptotic evaluation of I_b in Eq. (1) may be performed in a manner analogous to that employed for a saddle-point integration if the following change of variable to s is introduced:

$$s^2 = q(z_b) - q(z), \tag{4}$$

with the integration path near z_b adjusted so that $s^2 > 0$ along $\bar{P}_b$. (Since $\exp[\Omega q(z)]$ is assumed to decay along $\bar{P}_b$, one has $\operatorname{Re} s^2 > 0$; the condition $s^2 > 0$ requires a legitimate path distortion which facilitates the analysis and defines essentially the path of steepest descent away from the branch point). Thus, for $z \approx z_b$ or $s \approx 0$,

$$s = \sqrt{-q'(z_b)}\,\sqrt{z - z_b}\,[1 + O(z - z_b)], \tag{5a}$$

or, upon inversion,

$$\sqrt{z - z_b} = \frac{s}{\sqrt{-q'(z_b)}}[1 + O(s^2)], \tag{5b}$$

and

$$\frac{dz}{ds} = \frac{2s}{-q'(z_b)} + O(s^3), \tag{6a}$$

$$f(z) = a + \frac{bs}{\sqrt{-q'(z_b)}} + O(s^2). \tag{6b}$$

For definiteness, we choose $s < 0$ on portion A and $s > 0$ on portion B of $\bar{P}_b$ in Fig. 4.8.1, implying that

$$\arg(z - z_b) = -\arg[-q'(z_b)] \qquad \text{on } B,$$
$$\arg(z - z_b) = -\arg[-q'(z_b)] - 2\pi \qquad \text{on } A. \tag{7}$$

Upon substitution of Eq. (4) into Eq. (1) one obtains

$$I_b = e^{\Omega q(z_b)} \int_{-``\infty"}^{``\infty"} \left[f(z) \frac{dz}{ds} \right] e^{-\Omega s^2} \, ds, \tag{8}$$

which may be evaluated asymptotically by power series expansion and termwise integration of the bracketed quantity. The first term in the expansion involves the odd power s^1, which integrates out to zero. The first non-vanishing contribution derives from the s^2 term and yields

$$I_b \sim \frac{b\sqrt{\pi}}{[-\Omega q'(z_b)]^{3/2}} e^{\Omega q(z_b)}, \tag{9}$$

from which Eq. (3) follows. Higher-order terms in the asymptotic series may be generated by carrying further the power-series expansion of $[f(z)dz/ds]$.

APPENDIX 4A.   HIGHER-ORDER DERIVATIVES OF $G(s) = f(z)dz/ds$

Use of the asymptotic expansion in Eq. (4.2.12) requires knowledge of the derivatives of

$$G(s) = f(z)\varphi(s), \qquad \varphi(s) = \frac{dz}{ds}, \tag{A1}$$

evaluated at $s = 0$ ($z = z_s$). By direct differentiation of Eq. (A1) one finds

$$G(0) = f(z_s)\varphi(0), \tag{A2a}$$
$$G'(0) = f'(z_s)[\varphi(0)]^2 + f(z_s)\varphi'(0), \tag{A2b}$$
$$G^{(2)}(0) = f^{(2)}(z_s)[\varphi(0)]^3 + 3f'(z_s)\varphi(0)\varphi'(0) + f(z_s)\varphi^{(2)}(0), \tag{A2c}$$

etc.

To obtain the value of the derivatives of $\varphi(s)$ at $s = 0$, we expand the function $q(z) = \tau(s)$ about the saddle point z_s for which $q'(z_s) = 0$ and invert the resulting series. The procedure is illustrated for the case of a first-order saddle point at z_s, in which case the relation $q(z) = q(z_s) - s^2$ applies. Thus, if $\zeta = z - z_s$,

$$-s^2 = q^{(2)}(z_s)\frac{\zeta^2}{2!} + q^{(3)}(z_s)\frac{\zeta^3}{3!} + q^{(4)}(z_s)\frac{\zeta^4}{4!} + \cdots. \tag{A3}$$

Equation (A3) has an inverse solution which is regular and vanishes at $s = 0$,†

---

†For an alternative procedure utilizing Cauchy's theorem, see Reference 22.

so in view of the definition of $\varphi(s)$ in Eq. (A1), we may employ the power-series expansion

$$\zeta(s) = \varphi(0)s + \varphi'(0)\frac{s^2}{2!} + \varphi^{(2)}(0)\frac{s^3}{3!} + \varphi^{(3)}(0)\frac{s^4}{4!} + \cdots, \qquad (A4)$$

whence

$$\zeta^2 = [\varphi(0)]^2 s^2 + \varphi(0)\varphi'(0)s^3 + \left\{\left[\frac{\varphi'(0)}{2!}\right]^2 + \frac{\varphi^{(2)}(0)}{3}\varphi(0)\right\}s^4 + \cdots, \qquad (A5a)$$

$$\zeta^3 = [\varphi(0)]^3 s^3 + \{\tfrac{3}{2}[\varphi(0)]^2\varphi'(0)\}\, s^4 + \cdots, \qquad (A5b)$$

$$\zeta^4 = [\varphi(0)]^4 s^4 + \cdots, \qquad (A5c)$$

etc. Upon substituting Eqs. (A5) into Eq. (A3) and equating coefficients of like powers of s on both sides of the resulting equation, one obtains

Coefficient of s^2

$$-1 = \tfrac{1}{2}[\varphi(0)]^2 q^{(2)}(z_s) \qquad \text{or} \qquad \varphi(0) = \pm\sqrt{\frac{-2}{q^{(2)}(z_s)}}. \qquad (A6a)$$

Coefficient of s^3

$$0 = \frac{q^{(2)}(z_s)}{2!}\varphi(0)\varphi'(0) + \frac{q^{(3)}(z_s)}{3!}[\varphi(0)]^3,$$

or

$$\varphi'(0) = -\frac{1}{3}\frac{q^{(3)}(z_s)}{q^{(2)}(z_s)}[\varphi(0)]^2 = \frac{2}{3}\frac{q^{(3)}(z_s)}{[q^{(2)}(z_s)]^2}. \qquad (A6b)$$

Coefficient of s^4

$$0 = \frac{q^{(2)}(z_s)}{2!}\left\{\left[\frac{\varphi'(0)}{2}\right]^2 + \frac{\varphi^{(2)}(0)\varphi(0)}{3}\right\} + \frac{q^{(3)}(z_s)}{3!}\frac{3}{2}[\varphi(0)]^2\varphi'(0) + \frac{q^{(4)}(z_s)}{4!}[\varphi(0)]^4,$$

or

$$\varphi^{(2)}(0) = \frac{-3}{\varphi(0)}\left\{\left[\frac{\varphi'(0)}{2}\right]^2 + \frac{2}{q^{(2)}(z_s)}\left[\frac{q^{(4)}(z_s)}{4!}[\varphi(0)]^4 + \frac{q^{(3)}(z_s)}{4}[\varphi(0)]^2\varphi'(0)\right]\right\},$$

$$(A6c)$$

etc.

APPENDIX 4B. PROPERTIES OF THE AIRY FUNCTIONS

The Airy functions $\mathrm{Ai}(\sigma)$ and $\mathrm{Bi}(\sigma)$ may be defined either by the differential equation (4.2.33) (with appropriate boundary conditions), or by the integral representations in Eqs. (4.2.32) and (4.2.34). Section 4.2e contains a discussion of the properties of these functions when the argument σ is large. Other properties are discussed in this appendix. For tabulated results, see Reference 23.

To obtain a convergent series expansion useful for evaluation of $\mathrm{Ai}(\sigma)$ for small σ, one may substitute the power series for the exponential function $\exp(\sigma z)$ in the integrand of Eq. (4.2.32b) to find

$$\mathrm{Ai}(\sigma) = \frac{1}{2\pi i} \sum_{n=0}^{\infty} \frac{\sigma^n}{n!} a_n, \qquad a_n = \int_{L_{32}} z^n e^{-z^3/3}\, dz, \tag{B1}$$

after a permissible interchange of summation and integration. The coefficient a_n is defined by the convergent integral over the contour L_{32} in Fig. 4.2.4, which can be chosen along the straight lines $\arg z = 4\pi/3$ and $\arg z = 2\pi/3$. Since

$$\int_{\infty e^{i4\pi/3}}^{0} z^n e^{-z^3/3}\, dz = e^{i2(n+1)\pi/3} \int_{\infty e^{i2\pi/3}}^{0} z^n e^{-z^3/3}\, dz, \tag{B2a}$$

one may write

$$a_n = [1 - e^{i2(n+1)\pi/3}] \int_{0}^{\infty e^{i2\pi/3}} z^n e^{-z^3/3}\, dz, \tag{B2b}$$

or, upon changing variables to $\eta = z\exp(-i2\pi/3)$ and employing Eq. (4.2.8),

$$a_n = 2i(-1)^n 3^{(n-2)/3}\, \Gamma\!\left(\frac{n+1}{3}\right) \sin\left[\frac{(n+1)\pi}{3}\right]. \tag{B3}$$

Thus,

$$\mathrm{Ai}(\sigma) = \frac{1}{3^{2/3}\pi} \sum_{n=0}^{\infty} \frac{\sin[(n+1)\pi/3]\,\Gamma[(n+1)/3]3^{n/3}}{n!}(-\sigma)^n, \tag{B4}$$

and, upon use of the formula

$$\sin\left[(n+1)\frac{\pi}{3}\right]\Gamma\!\left(\frac{n+1}{3}\right) = \frac{\pi}{\Gamma[1-(n+1)/3]} = \frac{\pi}{\Gamma[(2-n)/3]}, \tag{B5}$$

one has, alternatively,

$$\mathrm{Ai}(\sigma) = \frac{1}{3^{2/3}} \sum_{n=0}^{\infty} \frac{3^{n/3}(-\sigma)^n}{n!\,\Gamma[(2-n)/3]}. \tag{B6}$$

Since $n! = \Gamma(n+1)$ and [see Eq. (4.2.9)]

$$\Gamma(v+\alpha) \sim \sqrt{\frac{2\pi}{v}}\left(\frac{v}{e}\right)^v v^\alpha, \qquad |v| \to \infty, \quad |\arg v| < \pi, \tag{B7}$$

the power-series expansions (B4) or (B6) are convergent for all σ. By a similar procedure, one derives

$$\mathrm{Bi}(\sigma) = \frac{1}{3^{1/6}} \sum_{n=0}^{\infty} \frac{3^{n/3}\sigma^n}{n!\,|\Gamma[(2-n)/3]|}. \tag{B8}$$

Because $\Gamma(-n) = \infty,\, n = 0, 1, 2, \ldots,$ the series in Eqs. (B6) and (B8) separate into two parts—one with powers σ^{3n} and the other with σ^{3n+1}. Repeated use of $\Gamma(y+1) = y\Gamma(y)$ leads to the alternative formulations

$$\mathrm{Ai}(\sigma) = \frac{1}{3^{2/3}\Gamma(\tfrac{2}{3})} W_1(\sigma) - \frac{1}{3^{1/3}\Gamma(\tfrac{1}{3})} W_2(\sigma), \tag{B9a}$$

$$\mathrm{Bi}(\sigma) = \frac{1}{3^{1/6}\Gamma(\tfrac{2}{3})} W_1(\sigma) + \frac{3^{1/6}}{\Gamma(\tfrac{1}{3})} W_2(\sigma), \tag{B9b}$$

where

$$W_1(\sigma) = \sum_{n=0}^{\infty} \frac{1 \cdot 4 \cdot 7 \cdots (3n+1)}{(3n+1)!} \sigma^{3n},$$

$$W_2(\sigma) = \sum_{n=0}^{\infty} \frac{2 \cdot 5 \cdot 8 \cdots (3n+2)}{(3n+2)!} \sigma^{3n+1}.$$

(B9c)

The functions $W_1(-\sigma)$ and $W_2(-\sigma)$ can be expressed in terms of Bessel functions of order $\frac{1}{3}$. We begin with the series

$$J_{1/3}(y) = \left(\frac{y}{2}\right)^{1/3} \sum_{n=0}^{\infty} \left(\frac{iy}{2}\right)^{2n} \frac{1}{n!\Gamma(\frac{1}{3}+n+1)}.$$

(B10)

and use for $n!\Gamma(\frac{1}{3}+n+1)$ the product formula for the gamma function,

$$\Gamma(3x) = \frac{1}{2\pi\sqrt{3}} 3^{3x} \Gamma(x)\Gamma\left(x+\tfrac{1}{3}\right)\Gamma\left(x+\tfrac{2}{3}\right), \qquad x = n+1.$$

(B11)

In view of Eq. (B5) with $n = 0$, one obtains

$$J_{1/3}(\tfrac{2}{3}\sigma^{3/2}) = \frac{\sqrt{\sigma}}{2\pi 3^{1/3}\sqrt{3}} \sum_{n=0}^{\infty} \frac{(-1)^n \Gamma(n+1+\tfrac{2}{3})3^{n+3}}{\Gamma(3n+3)} \sigma^{3n}$$

$$= -\frac{3^{2/3}}{\Gamma(\frac{1}{3})} \frac{1}{\sqrt{\sigma}} W_2(-\sigma),$$

(B12a)

and, in a similar manner,

$$J_{-1/3}(\tfrac{2}{3}\sigma^{3/2}) = \frac{3^{1/3}}{\Gamma(\frac{2}{3})} \frac{1}{\sqrt{\sigma}} W_1(-\sigma).$$

(B12b)

Thus, Eqs. (B9) may be written as

$$\mathrm{Ai}(-\sigma) = \frac{\sqrt{\sigma}}{3} [J_{-1/3}(\tfrac{2}{3}\sigma^{3/2}) + J_{1/3}(\tfrac{2}{3}\sigma^{3/2})],$$

(B13a)

$$\mathrm{Bi}(-\sigma) = \sqrt{\frac{\sigma}{3}} [J_{-1/3}(\tfrac{2}{3}\sigma^{3/2}) - J_{1/3}(\tfrac{2}{3}\sigma^{3/2})].$$

(B13b)

The combination $\mathrm{Ai}(-\sigma) \pm i\,\mathrm{Bi}(-\sigma)$ can likewise be expressed simply in terms of Hankel functions of order $\frac{1}{3}$, or more conveniently in terms of an Airy function with shifted argument. Via the formula

$$H_\nu^{(1,2)}(y) = \frac{J_{-\nu}(y) - e^{\mp i\nu\pi} J_\nu(y)}{\pm i \sin \nu\pi},$$

(B14)

one obtains from Eqs. (B13), after a simple calculation [see also Eq. (6. A32)],

$$A_{2,1}(-\sigma) \equiv \mathrm{Ai}(-\sigma) \pm i\,\mathrm{Bi}(-\sigma) = \sqrt{\frac{\sigma}{3}} e^{\mp i\pi/6} H_{1/3}^{(2,1)}(\tfrac{2}{3}\sigma^{3/2})$$

(B15a)

or, from Eqs. (B18)–(B20),

$$A_{2,1}(-\sigma) = -2e^{\pm i2\pi/3}\,\mathrm{Ai}(-\sigma e^{\pm i2\pi/3}).$$

(B15b)

The Wronskian of the Airy functions $\mathrm{Ai}(\sigma)$ and $\mathrm{Bi}(\sigma)$ can be calculated from the asymptotic formulas in Eqs. (4.2.42) and (4.2.45) as $\sigma \to \infty$, or from the power-series expansions (B4) and (B8) when $\sigma \to 0$. Either procedure is valid since the Wronskian is a constant, and the calculation by one method serves as a check on the result derived by the other. From the expressions for $\sigma \to \infty$, one finds directly,

$$\text{Ai}(\sigma)\text{Bi}'(\sigma) - \text{Bi}(\sigma)\text{Ai}'(\sigma) = \frac{1}{\pi}, \tag{B16}$$

while the calculation at $\sigma \to 0$ likewise yields $2\,[3^{1/2}\Gamma(\tfrac{1}{3})\Gamma(\tfrac{2}{3})] = 1/\pi$, via Eq. (B5).

The two linearly independent solutions $\text{Ai}(\sigma)$ and $\text{Bi}(\sigma)$ are convenient because both are real when σ is real. However, it is evident from Fig. 4.2.4 that the contour integral taken only over path L_{21} or L_{31} appears simpler than the one for $\text{Bi}(\sigma)$ in Eq. (4.2.34), and that the resulting functions are likewise linearly independent with respect to $\text{Ai}(\sigma)$. If we define

$$\int_{L_{21}} e^{\sigma z - z^3/3}\, dz \equiv -\pi i A_2(\sigma), \tag{B17a}$$

$$\int_{L_{31}} e^{\sigma z - z^3/3}\, dz \equiv \pi i A_1(\sigma), \tag{B17b}$$

then, from Cauchy's theorem,

$$2\text{Ai}(\sigma) = A_1(\sigma) + A_2(\sigma), \tag{B18a}$$

and, from Eq. (4.2.34),

$$2\text{Bi}(\sigma) = i[A_1(\sigma) - A_2(\sigma)]. \tag{B18b}$$

Conversely, $A_{1,2}$ is expressed in terms of Ai and Bi as

$$A_{1,2}(\sigma) = \text{Ai}(\sigma) \mp i\,\text{Bi}(\sigma). \tag{B18c}$$

Figure 4.2.4 also shows that A_1, A_2, and Ai can be related to one another by changing the phase of the integration variable by $\pm\,(2\pi/3)$. Thus, if $\xi = z\exp(-i2\pi/3)$,

$$\int_{L_{32}} e^{\sigma z - z^3/3}\, dz = e^{i2\pi/3}\int_{L_{21}} e^{\sigma e^{i2\pi/3}\xi - \xi^3/3}\, d\xi, \tag{B19}$$

whence

$$\text{Ai}(\sigma) = -\frac{e^{i2\pi/3}}{2}\, A_2(\sigma e^{i2\pi/3}). \tag{B20a}$$

Similarly,

$$\text{Ai}(\sigma) = -\frac{e^{-i2\pi/3}}{2}\, A_1(\sigma e^{-i2\pi/3}), \tag{B20b}$$

and

$$A_1(\sigma) = e^{-i2\pi/3} A_2(\sigma e^{-i2\pi/3}). \tag{B20c}$$

Also, from Eqs. (B16) and (B18c),

$$A_{1,2}(\sigma)\text{Ai}'(\sigma) - \text{Ai}(\sigma)A_{1,2}'(\sigma) = \pm\frac{i}{\pi}, \tag{B21a}$$

whence

$$\mathrm{Ai}(\sigma)\frac{d}{d\sigma}\mathrm{Ai}(\sigma e^{\pm i2\pi/3}) - \mathrm{Ai}'(\sigma)\mathrm{Ai}(\sigma e^{\pm i2\pi/3}) = \frac{e^{\mp i\pi/6}}{2\pi}. \tag{B21b}$$

PROBLEMS

1. For large Ω, perform an asymptotic evaluation of the integral

$$I(\Omega) = \int_{\bar{P}_z} f(z)e^{\Omega q(z)}\,dz, \tag{1}$$

where $\bar{P}_z$ is the steepest-descent path through the first-order saddle point z_s of $q(z)$ (i.e., $q'(z_s) = 0$), by writing

$$e^{\Omega q(z)} = \exp\left[\Omega q(z_s) + \frac{\Omega q''(z_s)(z-z_s)^2}{2!}\right]M(\Omega, z) \tag{2}$$

with

$$M(\Omega, z) = \exp\left[\Omega \sum_{n=3}^{\infty} \frac{q^{(n)}(z_s)(z-z_s)^n}{n!}\right]. \tag{3}$$

Expand the regular functions $f(z)$ and $M(\Omega, z)$ in power series about $z = z_s$ and perform termwise integration to obtain the asymptotic expansion for $I(\Omega)$. Compare the results with those obtained by the procedure of Secs. 4.2a and 4.2b.

2. Through use of continued integration by parts, obtain the expansion for the exponential integral,

$$E_1(\Omega) = \int_{\Omega}^{\infty} \frac{e^{-y}}{y}\,dy \sim \frac{e^{-\Omega}}{\Omega} \sum_{n=0}^{\infty} \frac{n!}{\Omega^n}, \tag{4}$$

in inverse powers of Ω. Show that this series diverges in the ordinary sense but is a valid asymptotic expansion as $\Omega \to \infty$.

3. The following contour integral representation† for the parabolic cylinder function is valid for arbitrary values of v:

$$D_{-v}(z) = -\frac{\Gamma(1-v)}{2\pi i}e^{-z^2/4}\int_L e^{-z\xi-\xi^2/2}(-\xi)^{v-1}\,d\xi, \qquad |\arg(-\xi)| \le \pi, \tag{5}$$

where the contour L encircles in the counterclockwise sense a branch cut extending from $\xi = 0$ to $\xi = \infty$ along the positive real ξ axis.

(a) Show that this expression may be reduced to the one in Eq. (4.5.36) when $\mathrm{Re}\, v > 0$. Introduce into Eq. (5) the new variable $\zeta = \sqrt{\xi}$ to show that for $(2v - 1)$ equal to a positive even integer or zero,

$$D_{-v}(z) = -\frac{1}{\Gamma(v)}e^{-z^2/4}I_{13}, \tag{6}$$

where

$$I_{13} = \int_{L_{13}} \zeta^{2v-1}e^{-z\zeta^2-\zeta^4/2}\,d\zeta = 2I_{10} = 2I_{03}, \tag{7}$$

---

†E. T. Whittaker and G. N. Watson, *A Course of Modern Analysis*, Cambridge Univ. Press, 1952, pp. 348–349.

and L_{ij} is a contour that begins at infinity in sector i, and ends at infinity in sector j, of the complex ζ plane as shown in Fig. P4.1. When i or j is equal to zero, the corresponding endpoint of the path is at the origin $\zeta = 0$.

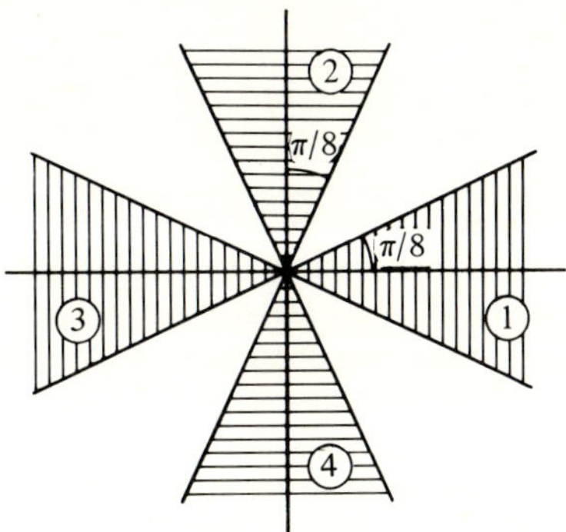

FIG. P4-1   Regions in the complex ζ plane.

Show from the integral representations that the parabolic cylinder functions satisfy the differential equation

$$\left[\frac{d^2}{dz^2} + \left(\frac{1}{2} - v - \frac{z^2}{4}\right)\right]D_{-v}(z) = 0. \tag{8}$$

(b) Assume that $(2v - 1) = $ positive even integer, i.e., $v = m + \frac{1}{2}$, where m is a positive integer or zero. Observe that since $I_{13} \exp(z^2/4)$ in Eq. (6) satisfies the differential equation (8), it follows that $I_{24} \exp(z^2/4)$ also satisfies the differential equation and furnishes a function linearly independent of $D_{-v}(z)$ (since the contour L_{24} cannot be deformed into the contour L_{13}, different solutions are involved). Show that

$$I_{02} = I_{40} = \frac{e^{iv\pi}}{2}\Gamma(v)e^{z^2/4}D_{-v}(-z). \tag{9}$$

Note that if $D_{-v}(z)$ satisfies Eq. (8), then so does $D_{v-1}(\pm iz)$, and that since $D_{-v}(\pm z)$ are two linearly independent solutions of the differential equation, it must be possible to employ them to represent $D_{v-1}(\pm iz)$. The relation may be shown to be

$$D_{-v}(z) - e^{\pm v\pi i}D_{-v}(-z) = \frac{\sqrt{2\pi}}{\Gamma(v)}e^{\mp i(1-v)\pi/2}D_{v-1}(\pm iz). \tag{10}$$

Show that

$$I_{12} = i\sqrt{\frac{\pi}{2}}\,e^{iv\pi/2}e^{z^2/4}D_{v-1}(iz), \tag{11a}$$

$$I_{14} = i\sqrt{\frac{\pi}{2}}\,e^{iv\pi/2}e^{z^2/4}D_{v-1}(iz) - \Gamma(v)e^{iv\pi}e^{z^2/4}D_{-v}(-z). \tag{11b}$$

4.   (a) Explain why, for an asymptotic evaluation of I_{01} (and generally of I_{ij}) in Problem 3 for large values of z, it is convenient to transform Eq. (7) into

$$I_{01} = \Omega^{v/2}e^{iv\alpha}\int_0^{\infty e^{-i\alpha/2}}\mu^{2v-1}e^{-\Omega e^{i2\alpha}[\mu^2 + (\mu^4/2)]}\,d\mu, \tag{12}$$

where $\Omega = |z|^2$, $\alpha = \arg z$, $\mu = \zeta z^{-1/2}$. The integrand has saddle points at $\mu_s = 0, \pm i$; show that the direction of the steepest-descent paths at the saddle points is as follows:

$$\arg (d\mu)|_{\mu_s=0} = \begin{cases} -\alpha \\ -\alpha \pm \pi \end{cases}, \quad \arg (d\mu)|_{\mu_s=i} = -\alpha \pm \frac{\pi}{2},$$

$$\arg(d\mu)|_{\mu_s=-i} = -\alpha \pm \frac{\pi}{2}. \tag{13}$$

(b) Show that for $0 < \alpha < \pi/2$, the integration path can be deformed into the steepest-descent path without passing the saddle point at $\mu_s = -i$ ($\mu_s = +i$ is irrelevant for the range of α considered); that for $\pi/2 \le \alpha < 3\pi/4$, both saddle points are traversed but that the dominant contribution arises from $\mu_s = 0$; that the saddle points $\mu_s = 0$ and $\mu_s = -i$ contribute equally for $\alpha = 3\pi/4$; and that the contribution from $\mu_s = -i$ is dominant when $3\pi/4 < \alpha < 5\pi/4$. Carry out the analogous considerations for $\alpha < 0$.

(c) From the information in (b) and the consequent asymptotic evaluation of the integral I_{01} for large $|z|$, show that

$$D_{-\nu}(z) \sim e^{-z^2/4} z^{-\nu} \qquad -\pi/2 < \arg z < \pi/2, \tag{14a}$$

$$D_{-\nu}(z) \sim e^{-z^2/4} z^{-\nu} - \frac{\sqrt{2\pi}\, e^{-i\nu\pi}}{\Gamma(\nu)} e^{z^2/4} z^{\nu-1}, \qquad \pi/2 \le \arg z < 5\pi/4. \tag{14b}$$

(Note: Since $\nu = m + \frac{1}{2}$, $m = 0, 1, 2, \ldots$, the asymptotic result for $D_{-\nu}(z) \exp(-z^2/4)$ may be inferred from that for $m = 0$ by repeated differentiation with respect to z.) Show that formula (14b) applies also when $-\pi/2 \ge \arg z > -5\pi/4$ provided that $\exp(-i\nu\pi)$ is replaced by $\exp(i\nu\pi)$. (Equations (14a) and (14b) may actually be shown to hold for arbitrary ν satisfying the inequality $|\nu| \ll |z|$.)

(d) Deduce asymptotic expressions for the ranges $\pi/2 \le \arg z < 5\pi/4$ and $-\pi/2 \ge \arg z > -5\pi/4$ from the one in Eq. (14a) through use of Eq. (10).

5.   The associated Legendre function $P_\nu^{-\mu} [\cos (t/\nu)]$ may be defined by the integral†

$$P_\nu^{-\mu}\left(\cos \frac{t}{\nu}\right) = \frac{1}{\Gamma(\nu + \mu + 1)} \int_0^\infty e^{-z\, \cos(t/\nu)} z^\nu J_\mu\left(z \sin \frac{t}{\nu}\right) dz, \tag{15}$$

valid for $0 < (t/\nu) < \pi/2$, $\mathrm{Re}(\mu + \nu + 1) > 0$. Evaluate the integral asymptotically as $\nu \to \infty$ to show that

$$\lim_{\nu \to \infty} \nu^\mu P_\nu^{-\mu}\left(\cos \frac{t}{\nu}\right) = J_\mu(t). \tag{16}$$

$$\left(\text{Note: } \int_0^\infty e^{-z} z^\nu dz = \Gamma(\nu + 1), \qquad \mathrm{Re}(\nu + 1) > 0.\right)$$

6.   Using the integral representation for the Legendre function†

$$P_n(\cos \theta) = \frac{1}{2\pi} \int_0^{2\pi} (\cos \theta + i \sin \theta \cos w)^n\, dw, \tag{17}$$

derive the asymptotic approximation for large n:

---

†W. Magnus and F. Oberhettinger, *Formulas and Theorems for the Functions of Mathematical Physics*, Chelsea Publishing Co., New York, 1954, pp. 67–68.

$$P_n(\cos\theta) \sim \sqrt{\frac{2}{\pi n \sin\theta}}\cos\left[\left(n + \frac{1}{2}\right)\theta - \frac{\pi}{4}\right], \qquad \sin\theta \not\approx 0. \tag{18}$$

7. The free-space Green's function has the following integral representation [cf. Eq. (5.4.7a)]:

$$G_f = \frac{e^{ikr}}{4\pi r} = \frac{i}{8\pi}\int_{\infty e^{i\pi}}^{\infty} \xi H_0^{(1)}(\xi\rho)\frac{e^{i\sqrt{k^2-\xi^2}|z|}}{\sqrt{k^2-\xi^2}}\,d\xi, \tag{19}$$

where $r = (\rho^2 + z^2)^{1/2}$, and the integration path passes above the branch points at $\xi = -k, 0$ and below the branch point at $\xi = k$. Assuming ρ large, use the asymptotic expansion for $H_0^{(1)}(\xi\rho)$ given in Eq. (6.4.8a), and then perform the asymptotic expansion of the integral in Eq. (19) by the methods of Secs. 4.2a and 4.2b. Show that the higher-order terms in the expansion vanish so that the leading term furnishes the exact result.

8. Using the integral representation [cf. Eq. (5.4.36c)],

$$H_0^{(1)}(k\rho) = \frac{1}{\pi}\int_{\bar{P}} e^{ik\rho\,\cos w}\,dw, \tag{20}$$

where $\bar{P}$ is the integration contour in Fig. 5.3.6(b), derive the asymptotic expansion in Eq. (6.4.8a) by the method of Sec. 4.2b.

9. Perform the operations implied in Eq. (4.2.42) to obtain the complete asymptotic expansion in Eq. (4.2.43).

10. Utilize the double integral representation in Eq. (5.4.12b),

$$G_f = \frac{i}{8\pi^2}\int_{-\infty}^{\infty} d\xi \int_{-\infty}^{\infty} d\eta \frac{\exp\left[i(\xi x + \eta y) + i\sqrt{k^2 - \xi^2 - \eta^2}|z|\right]}{\sqrt{k^2 - \xi^2 - \eta^2}} \tag{21}$$

to derive the asymptotic approximation for large $r = \sqrt{x^2 + y^2 + z^2}$.

(a) Perform the asymptotic evaluation by using the results of Sec. 4.7.
(b) Perform first the asymptotic evaluation of the η integral via the procedure of Sec. 4.2a, and then perform the evaluation of the ξ integral. Compare results of the two procedures.

REFERENCES

1(a). BLEISTEIN, N., "Uniform asymptotic expansions of integrals with many nearby stationary points and algebraic singularities," *J. Math. Mech.* **17** (1967), pp. 533–559.

1(b). RICE, S. O., "Uniform asymptotic expansions for saddle point integrals," *Bell System Tech. Jour.* **47**, (1968), pp. 1971–2013.

2. JEFFREYS, H., *Asymptotic Approximations*, Chapter 2. London: Oxford University Press, 1962.

3. DEBRUIJN, N. G., *Asymptotic Methods in Analysis*, Chapters 4–6. New York: Interscience Publishing Co., 1958.

4. COPSON, E. T., *Asymptotic Expansions*, Chapter 7. Cambridge, England: Cambridge University Press, 1965.

5. ERDELYI, A., *Asymptotic Expansions*, Chapter 2. New York: Dover Publishing Co., 1956.

6. JEFFREYS, H., *Asymptotic Approximations.*, Chapter 7. London: Oxford University Press, 1962.

7. BUDDEN, K. G., *Radio Waves in the Ionosphere*, Chapter 15. Cambridge, England: Cambridge University Press, 1961.

8. HEADING, J., *An Introduction to Phase-Integral Methods*, Chapter 3. New York: John Wiley & Sons, 1962.

9. VAN DER WAERDEN, B. L., "On the method of saddle points," *Applied Sci. Rsch.* **B2** (1951), pp. 33-45. [For a related method of evaluating steepest-descent integrals in the vicinity of a pole, see: W. PAULI, *Phys. Rev.* **54** (1938): H. OTT, *Annalen d. Physik (Leipzig)*, **43** (1943) p. 393; P. C. CLEMMOW, *Quart. J. Mech. Appl. Math.* 3 (1950) p. 241, and *Proc. Roy. Soc. London, Sec. A* **203** (1951).]

10. FRIED, B. D. and S. D. CONTE, *The Plasma Dispersion Function*. New York: Academic Press, 1961.

11. BLEISTEIN, N., "Uniform asymptotic expansion of integrals with stationary point near algebraic singularity," *Comm. Pure and Appl. Math.* **19** (1966), pp. 353-370.

12. CLEMMOW, P. C. and C. M. MUNFORD, "A Table of $\frac{1}{2}\sqrt{\pi}\, e^{i(1/2)\pi\rho^2}\int_{\rho}^{\infty} e^{-(1/2)i\pi\lambda^2}\, d\lambda$ for complex values of ρ," *Phil. Trans. Roy. Soc. London, Vol.* **A** (1952), pp. 189-211.

13. HORNER, F., "A table of a function used in radio-propagation theory," *Proc. IEE (London)* **102**, Part C, No. 1, March 1955, pp. 134-137.

14. CHESTER, C., B. FRIEDMAN, and F. URSELL, "An extension of the method of steepest descents," *Proc. of Cambridge Phil. Soc.* **53** (1957), pp. 599-611.

15. MAGNUS, W. and F. OBERHETTINGER, *Formulas and Theorems for the Special Functions of Mathematical Physics*, p. 93. New York: Chelsea Publishing Co., 1954.

16. PEARCEY, T., "The structure of an electromagnetic field in the neighborhood of a cusp or caustic," *Phil. Mag.* **37** (1946), p. 311.

17. BREKHOVSKIKH, L. M., *Waves in Layered Media*, pp. 492-496. New York: Academic Press, 1960.

18. LEWIS, R. M., "Asymptotic theory of transients," in *Electromagnetic Wave Theory*, Part 2 (ed. J. Brown), p. 864. New York: Pergamon Press, 1967.

19. LEVEY, L. and L. B. FELSEN, "On incomplete Airy functions and their application to diffraction problems," *Radio Science* **4**, (1969), pp. 959-969.

20(a). ROTHMAN, M., "Tables of the integrals and differential coefficients of $Gi(x)$ and $Hi(-x)$," *Quart. J. Mech. Appl. Math.* **7** (1954), pp. 379-384.

20(b). NOSOVA, L. N. and S. A. TUMARKIN, "Tables of generalized Airy Functions for the asymptotic solution of differential equations $e(py')' + (q + \epsilon r)y = f$," (in Russian). Moscow: Computing Center, Acad. of Sciences (1961).

21. JONES, D. S., and M. KLINE, "Asymptotic expansion of multiple integrals and the method of stationary phase," *J. Math. Phys.* 37 (1958), pp. 1–28.

22. COPSON, E. T., *Theory of Functions of a Complex Variable*, Sec. 6.23. London: Oxford University Press, 1936.

23. MILLER, J. C. P., *The Airy Integral* (British Association Mathematical Tables). Cambridge, England: Cambridge University Press, 1946.

5. Fields in Plane-Stratified Regions

5.1 INTRODUCTION

Representations of Green's functions for regions with planar stratification along z, derived in Sec. 2.3, require knowledge of eigenfunctions in the domain transverse to z, and of their z-dependent modal amplitudes. Eigenfunction solutions for various cross-sectional domains have been given in Chapter 3, while Sec. 2.4 contains information on modal voltage and current amplitudes. For transversely unbounded regions, the eigenfunctions form a continuous spectrum (see Sec. 3.2). The resulting Green's function representations then involve single or double integrals that must be evaluated for extraction of explicit information on field behavior. Of special interest in radiation and diffraction theory are far fields, for which the integrals may be reduced by the asymptotic methods discussed in Chapter 4. Contributions from saddle points and singularities in the integrands result in wave constituents that can be interpreted as geometric-optical and diffracted ray fields. This chapter explores these aspects in detail for time-harmonic and impulsive excitation of a number of plane-stratified configurations.

The discussion begins in Sec. 5.2 with a summary of steady-state and time-dependent electromagnetic field representations and their scalarization. These formal results for arbitrary longitudinal stratification are specialized to unbounded cross sections viewed either as rectangular or cylindrical waveguides excited by time-harmonic and pulsed point or line sources, and also by moving sources. Analytic properties of typical representation integrals and asymptotic forms for the far-zone (or high-frequency) solution are summarized in Sec. 5.3, with emphasis on a physical interpretation of wave species resulting from saddle points, poles, and branch points.

Attention is given to specific geometrical configurations, the simplest being the unbounded homogeneous dielectric medium in Sec. 5.4. While the Green's

442

functions for various source configurations in such a medium can be derived directly in closed form, we treat modal representations and their asymptotic reduction to illustrate relevant concepts and techniques for simple examples. The discussion proceeds to sources in the presence of a semiinfinite dielectric medium (Sec. 5.5) for which the far-zone field contains not only the direct, reflected, and refracted contributions of geometrical optics but also diffracted constituents in the form of surface waves and (or) lateral waves (the latter are associated with phenomena of total reflection). Consideration is given to transition regions wherein the field cannot be described in terms of these distinct wave types; analytically, transition effects arise from a confluence of saddle points and pole or branch-point singularities in the integral representation. The nature of geometric-optical and diffraction fields becomes evident from the study of transient propagation (Sec. 5.5d), which permits tracking of various wavefronts and thus clarifies corresponding time-harmonic phenomena.

Section 5.6 is concerned with fields excited by sources in the presence of a dielectric slab. This configuration exhibits effects of multiple reflection between the slab boundaries and, when the dielectric constant in the slab exceeds that in the exterior medium, of wave trapping. The energy in the trapped or surface waves is confined to the slab region and transported in the direction parallel to the boundaries, whence a field representation in terms of modes guided along a transverse coordinate ρ is appropriate for emphasizing such wave phenomena. The ρ-transmission representation can be constructed directly on use of the slab eigenfunctions developed in Sec. 3.3c, or by contour deformation from the z-transmission representation; the latter procedure involves the characteristic Green's functions of Sec. 3.3a and illustrates the general theory of alternative representations presented in Sec. 3.3c. Analogous considerations apply to the constant-impedance boundary treated in Sec. 5.7, which may, under suitable conditions, also guide a surface wave. Section 5.7d contains an example of excitation by an aperture, thereby demonstrating how Green's function solutions are used for synthesizing distributed source configurations.

The preceding examples comprise plane-stratified regions with piecewise constant properties along z for which the modal Green's function solutions can be obtained in terms of trigonometric or exponential functions as in Sec. 2.4. For continuous stratification, treated in Secs. 5.8 and 5.9, the non-uniform transmission-line theory of Sec. 3.3b is applicable and yields formal results for arbitrary inhomogeneity profiles, as summarized in Sec. 5.8b. Explicit solutions can be obtained either for "slowly varying" inhomogeneities or for special profiles. For the former, the geometrical-optics method introduced in Sec. 1.7 is applicable and is illustrated in detail in Sec. 5.8c. Alternatively, one may proceed as in Sec. 5.8d from modal representation integrals, simplified on use of the WKB approximations (Sec. 3.5c) for modal Green's functions. When applicable, asymptotic evaluation of the integrals yields the geometric-optical field but unlike the direct geometric-optical method in Sec. 5.8c, also provides field solutions in transition regions near caustics; to accommodate these regions, the asymptotic evaluation must account for the confluence of two saddle points in the integrand (see Sec. 4.5a).

Special inhomogeneous stratifications are exemplified in Sec. 5.9 by the inverse-square and continuous-transition (Epstein) profiles. Particular attention is given to the inverse-square medium, which possesses a number of interesting properties. Its transmission-line solutions involve the well-explored Bessel functions, so various analytical and asymptotic aspects of the general procedure in Sec. 5.8 can be verified in detail. The simplicity of the solution carries over to the study of ducted propagation in Sec. 5.9b. Two-dimensional radiation and diffraction problems in the inverse-square medium are closely related to a class of rotationally symmetric scattering problems in three dimensions; utilization of this analogy in Sec. 5.9c provides insight into a number of two-dimensional and three-dimensional radiation and scattering processes.

5.2 FIELD REPRESENTATIONS IN REGIONS WITH PIECEWISE CONSTANT PROPERTIES

5.2a Derivation of the Time-Harmonic Field From Scalar Potentials

The electromagnetic fields excited by time-harmonic electric point currents $\hat{\mathbf{J}}(\mathbf{r}, t) = \mathbf{J}^o\delta(\mathbf{r} - \mathbf{r}')e^{j\omega t}$ and magnetic point currents $\hat{\mathbf{M}}(\mathbf{r}, t) = \mathbf{M}^o\delta(\mathbf{r} - \mathbf{r}')e^{j\omega t}$ may be represented at $\mathbf{r} \neq \mathbf{r}'$ as [see Eq (2.3.6)]†

$$\mathbf{E}(\mathbf{r}, \mathbf{r}') = \nabla \times \nabla \times \mathbf{z}_0\Pi'(\mathbf{r}, \mathbf{r}') - j\omega\mu\nabla \times \mathbf{z}_0\Pi''(\mathbf{r}, \mathbf{r}'), \tag{1a}$$

$$\mathbf{H}(\mathbf{r}, \mathbf{r}') = j\omega\epsilon\nabla \times \mathbf{z}_0\Pi'(\mathbf{r}, \mathbf{r}') + \nabla \times \nabla \times \mathbf{z}_0\Pi''(\mathbf{r}, \mathbf{r}'), \tag{1b}$$

where the *E*- and *H*-mode Hertz potentials Π' and Π'', respectively, are related via Eqs. (2.3.24) and (2.3.25) to the scalar functions $\mathscr{S}'$ and $\mathscr{S}''$,

$$\Pi'(\mathbf{r}, \mathbf{r}') = \frac{1}{j\omega\epsilon}\mathbf{J}^o\cdot\nabla' \times \nabla' \times \mathbf{z}_0\mathscr{S}'(\mathbf{r}, \mathbf{r}') - \mathbf{M}^o\cdot\nabla' \times \mathbf{z}_0\mathscr{S}'(\mathbf{r}, \mathbf{r}'), \tag{1c}$$

$$\Pi''(\mathbf{r}, \mathbf{r}') = \mathbf{J}^o\cdot\nabla' \times \mathbf{z}_0\mathscr{S}''(\mathbf{r}, \mathbf{r}') + \frac{1}{j\omega\mu}\mathbf{M}^o\cdot\nabla' \times \nabla' \times \mathbf{z}_0\mathscr{S}''(\mathbf{r}, \mathbf{r}'). \tag{1d}$$

It may be recalled that the operations in Eqs. (1c) and (1d) imply that

$$\mathbf{A}\cdot\nabla' \times \mathbf{z}_0 = -\mathbf{A}_t \times \mathbf{z}_0\cdot\nabla'_t, \tag{1e}$$

$$\mathbf{A}\cdot\nabla' \times \nabla' \times \mathbf{z}_0 = \mathbf{A}_t\cdot\nabla'_t\frac{\partial}{\partial z'} - A_z\nabla'^2_t, \tag{1f}$$

where $\mathbf{A}$ is any vector and ∇'_t operates on the source-point (primed) coordinates. Equations (1e) and (1f) illuminate the role played by the transverse and longitudinal vector components of the source configuration. In particular, a longitudinal electric current element $\mathbf{J}^o = \mathbf{z}_0 J^o_z$, $\mathbf{M}^o = 0$, contributes only to

---

†Time-dependent quantities in this chapter will be distinguished by a superscript circumflex.

$\Pi'(\mathbf{r}, \mathbf{r}')$, thereby exciting E modes only, and a longitudinal magnetic current element $\mathbf{M}^o = \mathbf{z}_0 M_z^o$, $\mathbf{J}^o = 0$, generates only H modes, whereas both mode types are excited by a transversely directed source of either electric or magnetic current.

It has been shown [Eqs. (2.3.32) and (2.3.33)] that if $\mathbf{E}(\mathbf{r}, \mathbf{r}')$ and $\mathbf{H}(\mathbf{r}, \mathbf{r}')$ satisfy the Maxwell field equations for point current excitation, then $\mathscr{S}'(\mathbf{r}, \mathbf{r}')$ and $\mathscr{S}''(\mathbf{r}, \mathbf{r}')$ satisfy the differential equation

$$(\nabla^2 + k^2)\nabla_t^2 \genfrac{}{}{0pt}{}{\mathscr{S}'(\mathbf{r}, \mathbf{r}')}{\mathscr{S}''(\mathbf{r}, \mathbf{r}')} = \delta(\mathbf{r} - \mathbf{r}'), \qquad k^2 = \omega^2 \mu \epsilon, \tag{2}$$

subject to appropriate boundary conditions. By defining†

$$-\nabla_t^2 \mathscr{S}'(\mathbf{r}, \mathbf{r}') = G'(\mathbf{r}, \mathbf{r}'), \qquad -\nabla_t^2 \mathscr{S}''(\mathbf{r}, \mathbf{r}') = G''(\mathbf{r}, \mathbf{r}'), \tag{3a}$$

one may relate $\mathscr{S}'$ and $\mathscr{S}''$ to the corresponding Green's functions G' and G'' of the scalar wave equation since

$$(\nabla^2 + k^2)\genfrac{}{}{0pt}{}{G'(\mathbf{r}, \mathbf{r}')}{G''(\mathbf{r}, \mathbf{r}')} = -\delta(\mathbf{r} - \mathbf{r}'). \tag{3b}$$

It may be noted from Eqs. (1c)–(1f) that the derivation of the Hertz potentials and thence of the fields for arbitrary source orientation does not require a knowledge of $\mathscr{S}'$ and $\mathscr{S}''$ but rather of $\nabla_t' \mathscr{S}'$ and $\nabla_t' \mathscr{S}''$. This aspect is of importance since the latter quantities are sometimes determined more easily than $\mathscr{S}'$ and $\mathscr{S}''$ and they also do not exhibit convergence difficulties associated with certain spectral representations of $\mathscr{S}'$ and $\mathscr{S}''$. Furthermore, for purely longitudinal sources, the significant quantities are $\nabla_t'^2 \mathscr{S}'$ and (or) $\nabla_t'^2 \mathscr{S}''$, whence in this instance the fields may be derived from the scalar Green's functions in Eqs. (3). Since an arbitrarily oriented source may be decomposed into a transverse and longitudinal part, it is useful to list the corresponding reduction of Eqs. (1c) and (1d) for these separate cases. When the source is transverse ($J_z^o = M_z^o = 0$),

$$\Pi'(\mathbf{r}, \mathbf{r}') = \left(\frac{1}{j\omega\epsilon}\mathbf{J}_t^o\frac{\partial}{\partial z'} + \mathbf{M}_t^o \times \mathbf{z}_0\right) \cdot \nabla_t' \mathscr{S}'(\mathbf{r}, \mathbf{r}'), \tag{4a}$$

$$\Pi''(\mathbf{r}, \mathbf{r}') = \left(\mathbf{z}_0 \times \mathbf{J}_t^o + \frac{1}{j\omega\mu}\mathbf{M}_t^o\frac{\partial}{\partial z'}\right) \cdot \nabla_t' \mathscr{S}''(\mathbf{r}, \mathbf{r}'), \tag{4b}$$

whereas for a longitudinal source ($\mathbf{J}_t^o = \mathbf{M}_t^o = 0$),

$$\Pi'(\mathbf{r}, \mathbf{r}') = \frac{J_z^o}{j\omega\epsilon}G'(\mathbf{r}, \mathbf{r}'), \qquad \Pi''(\mathbf{r}, \mathbf{r}') = \frac{M_z^o}{j\omega\mu}G''(\mathbf{r}, \mathbf{r}'). \tag{4c}$$

For certain simple problems, Eqs. (2) and (3) may be solved directly, but generally it is necessary to resort to appropriate representations. From a transmission-line analysis along the z axis, with eigenfunctions evaluated in the cross section transverse to z, one has the following solutions in terms of a modal expansion [Eqs. (2.3.24a) and (2.3.25a)]:

---

†The operator ∇_t^2 could have been replaced by $\nabla_t'^2$ [see Eqs. (2.3.32b) and (2.3.33b)].

$$\mathscr{S}'(\mathbf{r}, \mathbf{r}') = \frac{1}{j\omega\epsilon(z')} \sum_i \frac{\Phi_i(\boldsymbol{\rho})\Phi_i^*(\boldsymbol{\rho}')}{k_{ti}'^2} Y_i'(z, z'), \qquad k_{ti}' \neq 0, \qquad (5a)$$

$$\mathscr{S}''(\mathbf{r}, \mathbf{r}') = \frac{1}{j\omega\mu(z')} \sum_i \frac{\psi_i(\boldsymbol{\rho})\psi_i^*(\boldsymbol{\rho}')}{k_{ti}''^2} Z_i''(z, z'), \qquad k_{ti}'' \neq 0, \qquad (5b)$$

and also, from Eqs. (2.3.32b) and (2.3.33b),

$$G'(\mathbf{r}, \mathbf{r}') = \frac{1}{j\omega\epsilon(z')} \sum_i \Phi_i(\boldsymbol{\rho})\Phi_i^*(\boldsymbol{\rho}') Y_i'(z, z'), \qquad (5c)$$

$$G''(\mathbf{r}, \mathbf{r}') = \frac{1}{j\omega\mu(z')} \sum_i \psi_i(\boldsymbol{\rho})\psi_i^*(\boldsymbol{\rho}') Z_i''(z, z'). \qquad (5d)$$

The notation $\epsilon(z')$ and $\mu(z')$ for a multilayered region implies that these quantities are to be evaluated in the medium containing the source point z'. When no argument is indicated, ϵ and μ refer to the medium properties at the observation point. It is recalled that $\Phi_i(\boldsymbol{\rho})$ and $\psi_i(\boldsymbol{\rho})$ are the scalar eigenfunctions that have been listed in Chapter 3 for a variety of cross sections. $Y_i'(z, z')$ and $Z_i''(z, z')$ are, respectively, the E-mode current excited by a unit voltage generator and the H-mode voltage excited by a unit current generator (see Figs. 2.6 and 2.7), and they are related as follows to the one-dimensional E- and H-mode Green's functions:

$$Y_i'(z, z') = j\omega\epsilon(z')g_{zi}'(z, z'), \qquad Z_i''(z, z') = j\omega\mu(z')g_{zi}''(z, z'), \qquad (6a)$$

where $g_{zi}'(z, z')$ and $g_{zi}''(z, z')$ satisfy the equation

$$\left(\frac{d^2}{dz^2} + \kappa_i^2\right) g_{zi}(z, z') = -\delta(z - z'), \qquad \kappa_i^2 = k^2 - k_{ti}^2, \qquad (6b)$$

subject to appropriate boundary conditions at the endpoints of the z domain. It is to be anticipated from Eqs. (5a) and (5b) that difficulties in the representation arise when the eigenvalue $k_{ti} = 0$ is admitted, as may be the case for continuous spectral distributions [see the footnote to Eq. (2.3.24a)].

The preceding considerations make evident that the basic problem involves the determination of the scalar functions $\mathscr{S}'$, $\mathscr{S}''$ or G', G'' specified by the differential equations (2) and (3b), respectively; some simple boundary conditions permit the direct integration of these equations in closed form, whereas more general cases require the series or integral solutions resulting from an eigenfunction expansion.

5.2b   Modal Representations for Unbounded Cross Sections

Since the configurations to be analyzed possess transversely unbounded cross sections, the transverse eigenvalue problem is highly degenerate and many alternative choices of coordinate systems are possible. The most useful for point-source and transverse-line-source excitation are the circular cylindrical and rectangular, respectively, since they account in the most direct manner for the symmetry properties of the associated fields. In view of the absence of transverse boundaries, the eigenvalue problems for the E and H modes are identical so that $\Phi_i = \psi_i$, and the distinction between the E- and H-mode Green's functions resides solely in their longitudinal dependence. For the rectangular waveguide description, one obtains via Eqs. (5c), (5d), (6a), (6b), and (3.2.40) the double Fourier integral representation,

$$G(\mathbf{r}, \mathbf{r}') = \frac{1}{4\pi^2} \int_{-\infty}^{\infty} \int_{-\infty}^{\infty} e^{-j\xi(x-x')-j\eta(y-y')} g_{zi}(z, z') \, d\xi \, d\eta. \tag{7}$$

While the circular waveguide description can be obtained in a similar manner on use of Eqs. (3.2.78) instead of Eqs. (3.2.40), it is instructive to develop the resulting representation directly from Eq. (7).[1] We consider an integral of the form

$$I = \int_{-\infty}^{\infty} \int_{-\infty}^{\infty} f(k_t) e^{-j\mathbf{k}_t \cdot (\mathbf{\rho}-\mathbf{\rho}')} d\mathbf{k}_t, \qquad \mathbf{k}_t = \mathbf{x}_0 \xi + \mathbf{y}_0 \eta, \tag{7a}$$

with $k_t = |\mathbf{k}_t|$ and $d\mathbf{k}_t = d\xi \, d\eta$. On introducing polar coordinates (ρ, ϕ) and (k_t, α) in the $\mathbf{\rho}$ and $\mathbf{k}_t$ spaces via $\mathbf{\rho} = \rho(\mathbf{x}_0 \cos \phi + \mathbf{y}_0 \sin \phi)$, $\mathbf{k}_t = k_t(\mathbf{x}_0 \cos \alpha + \mathbf{y}_0 \sin \alpha)$, and noting that $d\mathbf{k}_t = k_t \, dk_t \, d\alpha$, one rewrites Eq. (7a) as

$$I = \int_0^{\infty} dk_t \, k_t \, f(k_t) \int_0^{2\pi} d\alpha \, e^{-jk_t\rho \cos(\alpha-\phi)} e^{jk_t\rho' \cos(\alpha-\phi')}. \tag{7b}$$

Replacement of each of the exponentials in Eq. (7b) by the series expansion

$$e^{j\beta \cos \zeta} = \sum_{m=-\infty}^{\infty} j^m J_m(\beta) e^{jm\zeta}, \tag{7c}$$

interchange of the orders of summation and integration, and reduction of the α integral, yields the Fourier–Bessel representation for I in Eq. (7a):

$$I = 2\pi \sum_{m=-\infty}^{\infty} e^{-jm(\phi-\phi')} \int_0^{\infty} dk_t \, k_t \, f(k_t) J_m(k_t\rho) J_m(k_t\rho'). \tag{7d}$$

Thus, the circular waveguide description of $G(\mathbf{r}, \mathbf{r}')$ is in the notation of Eq. (3.2.78) (with ξ denoting k_t),

$$G(\mathbf{r}, \mathbf{r}') = \frac{1}{2\pi} \sum_{m=-\infty}^{\infty} e^{-jm(\phi-\phi')} \int_0^{\infty} \xi J_m(\xi\rho) J_m(\xi\rho') g_{zi}(z, z') \, d\xi. \tag{8a}$$

For subsequent application, it will be useful to employ instead of Eq. (8a) a representation involving a range of integration in ξ from $-\infty$ to $+\infty$. Provided that $g_{zi}(z, z')$ is an even function of ξ, a requirement satisfied by the longitudinal Green's functions to be encountered subsequently, the desired representation may be derived as in Eqs. (3.2.64)–(3.2.73) and yields

$$G(\mathbf{r}, \mathbf{r}') = \frac{1}{4\pi} \sum_{m=-\infty}^{\infty} e^{-jm(\phi-\phi')} \int_{\infty e^{-j\pi}}^{\infty} \xi J_m(\xi\rho_<) H_m^{(2)}(\xi\rho_>) g_{zi}(z, z') \, d\xi. \tag{8b}$$

The lower integration limit indicates that the branch point at $\xi = 0$ introduced by the transformation is avoided as in Fig. 3.2.10(b). The choice of transformation involving $H_m^{(2)}$ instead of $H_m^{(1)}$ is motivated by the fact that, for a time dependence $\exp(j\omega t)$, the former satisfies the radiation condition at $\rho \rightarrow \infty$ in a natural manner and facilitates subsequent asymptotic evaluation of the integral (see Sec. 5.3d).

Since $k_{ti}^2 = \xi^2 + \eta^2$ and $k_{ti}^2 = \xi^2$ for the rectangular and circular waveguide representations, respectively, it follows from Eqs. (5a), (5b), and (6a) that in rectangular coordinates,

$$\mathscr{S}(\mathbf{r}, \mathbf{r}') = \frac{1}{4\pi^2} \int_{-\infty}^{\infty} \int_{-\infty}^{\infty} \frac{e^{-j\xi(x-x') - j\eta(y-y')}}{\xi^2 + \eta^2} \, g_{zi}(z, z') \, d\xi \, d\eta, \qquad (9)$$

whereas in circular cylindrical coordinates,

$$\mathscr{S}(\mathbf{r}, \mathbf{r}') = \begin{cases} \dfrac{1}{2\pi} \displaystyle\sum_{m=-\infty}^{\infty} e^{-jm(\phi-\phi')} \int_0^{\infty} \dfrac{1}{\xi} J_m(\xi\rho) J_m(\xi\rho') g_{zi}(z, z') \, d\xi, & (10a) \\[3mm] \dfrac{1}{4\pi} \displaystyle\sum_{m=-\infty}^{\infty} e^{-jm(\phi-\phi')} \int_{\infty e^{-j\pi}}^{\infty} \dfrac{1}{\xi} J_m(\xi\rho_<) H_m^{(2)}(\xi\rho_>) g_{zi}(z, z') \, d\xi, & (10b) \end{cases}$$

where $g_{zi}(z, z')$ depends on k_{ti} via the longitudinal propagation constant $\kappa_i = (k^2 - k_{ti}^2)^{1/2}$. The occurrence of $k_{ti} = 0$ in the continuous spectrum of eigenvalues leads to the previously mentioned convergence difficulties in Eqs. (9) and (10a), so these results must be regarded as purely formal. However, the functions $\nabla_t' \mathscr{S}$, from which the Hertz potentials are calculated via Eqs. (4a) and (4b), have valid representations, as will be demonstrated for Eq. (10a). In cylindrical coordinates, $\nabla_t' = \boldsymbol{\rho}_0'(\partial/\partial\rho') + \boldsymbol{\phi}_0'(\partial/\rho' \, \partial\phi')$. The convergence problem arises from the $m = 0$ term only, since for $m \neq 0$, the behavior $J_m(\xi\rho) \sim (\xi\rho)^m$ as $\xi \to 0$ assures regularity of the integrand at $\xi = 0$. The differentiation of the $m = 0$ terms with respect to ρ' results in the occurrence of $\xi J_0'(\xi\rho') \sim \xi^2$ in the numerator of the integrand, so the integral representation for $(\partial/\partial\rho')\mathscr{S}$ is regular at $\xi = 0$. The derivative of the $m = 0$ term with respect to ϕ' vanishes identically since this term is independent of ϕ'. Thus, the modal representation for $\nabla_t' \mathscr{S}$ is regular at $\xi = 0$, with the implication that the differentiations are performed on the summands (or integrands) of Eqs. (5a) and (5b).†

Point-source excitation

The suitability of the circular waveguide representation for point-source-excitation problems may be made manifest if the coordinate system is chosen so that $\rho' = 0$ (i. e., the source is located on the z axis). For a longitudinal dipole, the electromagnetic fields may be derived directly from $G(\mathbf{r}, \mathbf{r}')$ [see Eq. (4c)], which reduces in view of $J_0(0) = 1$, $J_m(0) = 0$, $m \neq 0$, to

$$G(\mathbf{r}, \mathbf{r}') = \frac{1}{4\pi} \int_{\infty e^{-j\pi}}^{\infty} \xi H_0^{(2)}(\xi\rho) g_{zi}(z, z') \, d\xi, \qquad \rho' = 0. \qquad (11)$$

For transverse source vector distributions, the requisite scalar functions are the potentials Π' and Π'' in Eqs. (4a) and (4b). When $\rho' \to 0$ in the previously discussed representation for $\nabla_t' \mathscr{S}$ derived from Eqs. (10a) and (10b), one may

---

†In the derivation of Eqs. (2.3.24) and (2.3.25), it was assumed that derivative operations can be commuted with the summation in Eqs. (2.3.24a) and (2.3.25a). This is not the case for integral representations as in Eq. (10a), whence one must consider $\nabla_t' \mathscr{S}'$ and $\nabla_t' \mathscr{S}''$, as noted above.

verify that only the $m = 1$ terms contribute, whence

$$\nabla'_t \mathscr{S}(\mathbf{r}, \mathbf{r}') = \begin{cases} \boldsymbol{\rho}_0 \dfrac{1}{2\pi} \displaystyle\int_0^\infty J_1(\xi\rho)g_{zi}(z, z')\,d\xi, & (12a) \\[4mm] & \quad \rho' = 0, \\[2mm] \boldsymbol{\rho}_0 \dfrac{1}{4\pi} \displaystyle\int_{\infty e^{-j\pi}}^\infty H_1^{(2)}(\xi\rho)g_{zi}(z, z')\,d\xi, & (12b) \end{cases}$$

where $\boldsymbol{\rho}_0$ is the radial unit vector in the plane transverse to z: $\boldsymbol{\rho}_0 = \mathbf{x}_0 \cos\phi + \mathbf{y}_0 \sin\phi$.

Line-source excitation

When the excitation is in the form of a line distribution of sources confined to the plane $z = z'$, it is convenient to choose the rectangular waveguide representation and to orient one of the transverse coordinate axes, say x, along the source direction., The resulting simplification of Eqs. (7) and (9), a consequence of integration over x' to synthesize the line distribution, is not impaired when the source elements are phased progressively according to exp $(-j\alpha x')$, where α is a real constant. Moreover, the orientation of the elements with respect to the line axis is arbitrary (Fig. 5.2.1). It follows that the relevant forms of

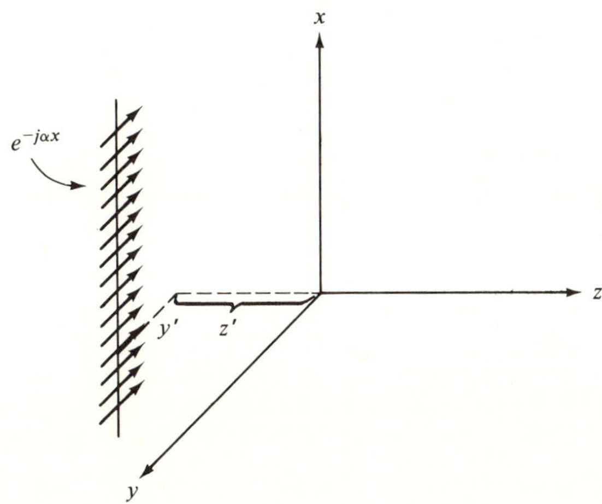

FIG. 5.2.1 Line source of arbitrarily oriented, progressively phased current elements.

$G(\mathbf{r}, \mathbf{r}')$ and $\mathscr{S}(\mathbf{r}, \mathbf{r}')$ are obtained by multiplying Eqs. (7) and (9) by exp $(-j\alpha x')$ and integrating between $x' = -\infty$ and $x' = +\infty$; a subsequent interchange of the orders of integration and the recognition that $\int_{-\infty}^\infty \exp\left[-j(\alpha - \xi)x'\right] dx' = 2\pi\delta(\alpha - \xi)$ leads to the two-dimensional forms

$$\bar{G}(\mathbf{r}, \hat{\boldsymbol{\rho}}') = \int_{-\infty}^\infty e^{-j\alpha x'} G(\mathbf{r}, \mathbf{r}')\,dx' = e^{-j\alpha x}\bar{G}(\hat{\boldsymbol{\rho}}, \hat{\boldsymbol{\rho}}'), \qquad \hat{\boldsymbol{\rho}} = (y, z), \qquad (13)$$

where $\bar{G}(\hat{\boldsymbol{\rho}}, \hat{\boldsymbol{\rho}}')$ is the two-dimensional Green's function

$$\bar{G}(\hat{\boldsymbol{\rho}}, \hat{\boldsymbol{\rho}}') = \frac{1}{2\pi} \int_{-\infty}^{\infty} e^{-j\eta(y-y')} g_{zi}(z, z') \, d\eta. \tag{13a}$$

Similarly, from Eq. (9),

$$\bar{\mathscr{S}}(\mathbf{r}, \hat{\boldsymbol{\rho}}') = e^{-j\alpha x} \bar{\mathscr{S}}(\hat{\boldsymbol{\rho}}, \hat{\boldsymbol{\rho}}'), \tag{14}$$

where

$$\bar{\mathscr{S}}(\hat{\boldsymbol{\rho}}, \hat{\boldsymbol{\rho}}') = \frac{1}{2\pi} \int_{-\infty}^{\infty} \frac{e^{-j\eta(y-y')}}{\eta^2 + \alpha^2} g_{zi}(z, z') \, d\eta. \tag{14a}$$

In Eqs. (13a) and (14a), the longitudinal propagation constant occurring in $g_{zi}(z, z')$ is to be written as $\kappa_i = (k^2 - \alpha^2 - \eta^2)^{1/2}$.

Unless $\alpha = 0$, no difficulty at $\eta = 0$ is encountered in the integral representation in Eq. (14a), and one may derive the potential functions $\overline{\Pi}'(\mathbf{r}, \hat{\boldsymbol{\rho}}')$ and $\overline{\Pi}''(\mathbf{r}, \hat{\boldsymbol{\rho}}')$ from Eqs. (4a) and (4b) by substituting Eq. (14) and replacing $\mathbf{V}'_t$ by $-j\alpha \mathbf{x}_0 + \mathbf{y}_0(\partial/\partial y')$. When $\alpha = 0$, the integral representation for $\mathbf{V}'_t \bar{\mathscr{S}}$ possesses a simple pole at $\eta = 0$. This pole is of no consequence since it does not contribute to fields obtained from $\mathscr{S}$ by differentiation operations involving single or double derivatives with respect to y [see Eqs. (1) and Eqs. (5.4.31)].

5.2c Fields Excited by Impulsive Sources

When the sources are stationary in space but have an impulsive time dependence

$$\hat{\mathbf{J}}(\mathbf{r}, t) = \mathbf{J}(\mathbf{r})\delta(t - t'), \qquad \hat{\mathbf{M}}(\mathbf{r}, t) = \mathbf{M}(\mathbf{r})\delta(t - t'), \tag{15}$$

then since $\delta(t - t') = (2\pi)^{-1} \int_{-\infty}^{\infty} \exp(j\omega t - j\omega t') \, d\omega$, the resulting transient electromagnetic fields may be inferred from the time-harmonic solutions in Sec. 5.2a upon multiplication by $(2\pi)^{-1} \exp(j\omega t - j\omega t')$ and integration between $-\infty$ and $+\infty$ over the frequency variable $\omega = k\bar{c} = k/\sqrt{\mu\epsilon}$, where $\bar{c}$ is the speed of light in a medium with *non-dispersive* dielectric constant ϵ and permeability μ.† This implies via Eqs. (1) that the vector fields $\hat{\mathbf{E}}(\mathbf{r}, \mathbf{r}'; t, t')$ and $\hat{\mathbf{H}}(\mathbf{r}, \mathbf{r}'; t, t')$ for an impulsive point source are derivable from the time-dependent scalar Hertz potentials $\hat{\Pi}(\mathbf{r}, \mathbf{r}'; t, t')$ or from their time derivatives (a multiplicative factor $j\omega$ is replaced by $\partial/\partial t$), and the Hertz potentials may in turn be obtained from $\hat{\mathscr{S}}(\mathbf{r}, \mathbf{r}'; t, t')$ or $\hat{G}(\mathbf{r}, \mathbf{r}'; t, t')$. It is sometimes more convenient to deal not with the source current densities $\hat{\mathbf{J}}$ and $\hat{\mathbf{M}}$ but rather with the dipole moments $\hat{\mathbf{p}}$ and $\hat{\mathbf{m}}$ defined as

$$\hat{\mathbf{J}}(\mathbf{r}, t) = \frac{\partial}{\partial t} \hat{\mathbf{p}}(\mathbf{r}, t), \qquad \hat{\mathbf{M}}(\mathbf{r}, t) = \frac{\partial}{\partial t} \hat{\mathbf{m}}(\mathbf{r}, t), \tag{16}$$

since one avoids thereby the necessity of time integration of the space- and

---

†While all physical media exhibit dispersion, the results derived on the basis of dispersionless materials are meaningful when the transient source distribution has a *confined* frequency spectrum over which ϵ and μ are essentially frequency independent. The impulse excitations discussed here may be used to synthesize physically realizable source functions with these spectral characteristics.

time-dependent functions $\hat{\mathscr{G}}$ or $\hat{G}$ [factor of $(j\omega)^{-1}$ in Eqs. (1c), (1d), and (4c)]. Thus, the fields generated by a longitudinally directed electric or magnetic dipole having a moment density

$$\hat{\mathbf{p}}(\mathbf{r}, t) = \mathbf{z}_0 \hat{p}\delta(\mathbf{r} - \mathbf{r}')\delta(t - t'), \qquad \hat{\mathbf{m}}(\mathbf{r}, t) = \mathbf{z}_0 \hat{m}\delta(\mathbf{r} - \mathbf{r}')\delta(t - t') \quad (17)$$

are obtained from Eqs. (1a), (1b), and (4c) as [see also Eqs. (1.1.53)]

$$\hat{\mathbf{E}}(\mathbf{r}, \mathbf{r}'; t, t') = \frac{\hat{p}}{\epsilon} \boldsymbol{\nabla} \times \boldsymbol{\nabla} \times \mathbf{z}_0 \hat{G}'(\mathbf{r}, \mathbf{r}'; t, t') - \hat{m}\boldsymbol{\nabla} \times \mathbf{z}_0 \frac{\partial}{\partial t} \hat{G}''(\mathbf{r}, \mathbf{r}'; t, t'),$$

$$(18a)$$

$$\hat{\mathbf{H}}(\mathbf{r}, \mathbf{r}'; t, t') = \hat{p}\boldsymbol{\nabla} \times \mathbf{z}_0 \frac{\partial}{\partial t} \hat{G}'(\mathbf{r}, \mathbf{r}'; t, t') + \frac{\hat{m}}{\mu} \boldsymbol{\nabla} \times \boldsymbol{\nabla} \times \mathbf{z}_0 \hat{G}''(\mathbf{r}, \mathbf{r}'; t, t'),$$

$$(18b)$$

where the dielectric constant ϵ and the permeability μ are assumed to be frequency independent (non-dispersive medium). $\hat{G}$ is the space- and time-dependent Green's function

$$\hat{G}(\mathbf{r}, \mathbf{r}'; t, t') = \frac{1}{2\pi}\int_{-\infty}^{\infty} G_\omega(\mathbf{r}, \mathbf{r}')e^{j\omega\,(t-t')}\,d\omega, \qquad (18c)$$

which satisfies the differential equation

$$\left(\boldsymbol{\nabla}^2 - \frac{1}{\bar{c}^2}\frac{\partial^2}{\partial t^2}\right)\hat{G}(\mathbf{r}, \mathbf{r}'; t, t') = -\delta(\mathbf{r} - \mathbf{r}')\delta(t - t'),$$

$$(18d)$$

$$\bar{c} = \frac{1}{\sqrt{\mu\epsilon}} = c\sqrt{\frac{\mu_0\epsilon_0}{\mu\epsilon}},$$

subject to the same spatial boundary conditions as $G_\omega(\mathbf{r}, \mathbf{r}')$, and to the temporal (causality) condition that all fields vanish prior to $t = t'$. μ_0 and ϵ_0 are the constitutive parameters in vacuum. The subscript ω has been added to the time-harmonic solution $G(\mathbf{r}, \mathbf{r}')$ to emphasize its dependence on frequency. Analogous expressions may be obtained for transversely directed dipole sources. The temporal profiles corresponding to the excitation functions in Eq. (17) are shown in Fig. 5.2.2.

It has been shown in Sec. 1.6b that the recovery of the transient solution from the time-harmonic result is accomplished directly if the integral represen-

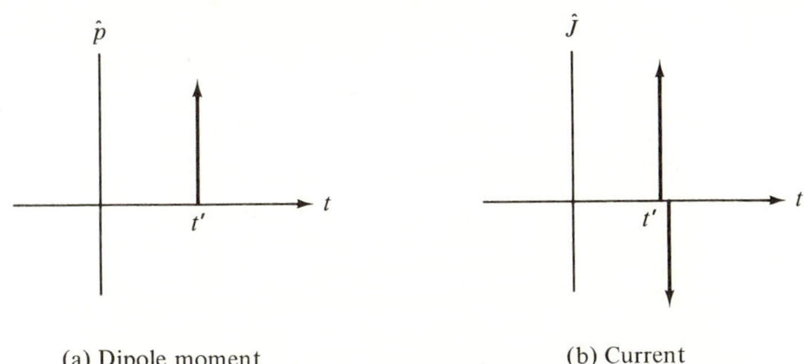

(a) Dipole moment   (b) Current

FIG. 5.2.2 Corresponding dipole moment and current profiles.

tation for the latter can be cast into a special form. In particular, if it is possible to write G_ω as a Laplace integral,

$$G_\omega(\mathbf{r}, \mathbf{r}') = \int_0^\infty e^{-s\tau} A(\mathbf{r}, \mathbf{r}'; \tau)\, d\tau, \qquad s = j\omega, \tag{19a}$$

where A is independent of s, then from the definition of the Laplace transform,

$$\hat{G}(\mathbf{r}, \mathbf{r}'; t, t') = A(\mathbf{r}, \mathbf{r}'; t - t'). \tag{19b}$$

Included in the category of integrals expressible as in Eq. (19a) is the generic radiation integral in Eq. (5.3.14),

$$I_\omega(L, \bar{\alpha}) = \int_{\bar{P}} e^{-jkL \cos(w - \bar{\alpha})} f(w)\, dw, \tag{20}$$

where $\bar{P}$ is the contour shown in Fig. 5.3.5b. The parameters L and $\bar{\alpha}$ are assumed to be positive, $\bar{\alpha}$ being restricted to the range $0 < \bar{\alpha} < \pi/2$, and the function $f(w)$ is independent of $k = \omega/\bar{c}$, where $\bar{c} = (\mu\epsilon)^{-1/2}$ is the speed of light in the medium. As observed in this chapter and in Chapter 6, a number of time-harmonic radiation and diffraction problems can be expressed in this form. Since the analysis in Eqs. (1.6.34)–(1.6.41) is based on an assumed time dependence $\exp(-i\omega t)$, we repeat the principal steps for the presently used $\exp(j\omega t)$. Introducing $\omega \to -js$ in Eq. (20), one may write

$$I_\omega(L, \bar{\alpha}) = \int_{-j\infty}^{j\infty} e^{-s(L/\bar{c}) \cos w} f(w + \bar{\alpha})\, dw, \tag{21}$$

where it has been assumed that the function $f(w)$ has no singularities in the strip $0 < |\mathrm{Re}\, w| < \pi/2$. The contour deformation employed in achieving Eq. (21) is justified as in Eq. (1.6.38).

On successive changes of variable $\beta = -jw$ and $\tau = (L/\bar{c}) \cosh \beta$, one obtains†

$$I_\omega(L, \bar{\alpha}) = j \int_{L/\bar{c}}^\infty e^{-s\tau} \frac{b(\tau)}{\sqrt{\tau^2 - (L/\bar{c})^2}}\, d\tau, \tag{22a}$$

$$b(\tau) = f\left[\bar{\alpha} + j \cosh^{-1}\left(\frac{\bar{c}\tau}{L}\right)\right] + f\left[\bar{\alpha} - j \cosh^{-1}\left(\frac{\bar{c}\tau}{L}\right)\right], \tag{22b}$$

whence on comparison with Eq. (19a),

$$A(\tau) = \begin{cases} 0, & \tau < \dfrac{L}{\bar{c}}, \\[2ex] \dfrac{jb(\tau)}{\sqrt{\tau^2 - L^2/\bar{c}^2}}, & \tau > \dfrac{L}{\bar{c}}. \end{cases} \tag{23}$$

If $v(w) = jf(w)$ is real for real values of w, then $v(w^*) = v^*(w)$ and

$$jb(\tau) = 2\mathrm{Re}\left\{ jf\left[\bar{\alpha} + j \cosh^{-1}\left(\frac{\bar{c}\tau}{L}\right)\right] \right\}. \tag{23a}$$

Several applications of this result are given in the following sections.

---

†If the exponential in Eq. (21) has some other form $\exp[-s(L/\bar{c})h(w)]$, the variable τ is defined as $\tau = (L/\bar{c})h(w)$.

5.2d   Fields Excited by Charges in Uniform Rectilinear Motion

Transient fields may also be excited when the location of the source changes with time, even though the source strength itself is non-varying. The simplest class of such problems involves uniform motion of an electric charge along a straight-line path. The results are of interest for studies of the interaction of high-speed charged particles with material media of various types (dielectrics, plasmas, etc.), and bear on such physical applications as the absorption of protons in "swimming-pool" atomic reactors, or the excitation of low-frequency noise in the earth's exosphere by streams of charges emanating from the sun.

A point charge of strength q is assumed to move with constant speed v parallel to the x axis of a rectangular coordinate system. The current density $\hat{\mathbf{J}}(\mathbf{r}, t)$ associated with this moving charge is

$$\hat{\mathbf{J}}(\mathbf{r}, t) = \mathbf{x}_0 qv\,\delta(x - vt)\,\delta(y - y')\,\delta(z - z'); \tag{24}$$

its Fourier spectrum function $\mathbf{J}(\mathbf{r}, \omega)$ is obtained as

$$\mathbf{J}(\mathbf{r}, \omega) = \int_{-\infty}^{\infty} \hat{\mathbf{J}}(\mathbf{r}, t)e^{-j\omega t}\,dt = \mathbf{x}_0 q e^{-j(k_0/\beta)x}\delta(\hat{\boldsymbol{\rho}} - \hat{\boldsymbol{\rho}}'),$$

$$\hat{\boldsymbol{\rho}} = (y, z), \qquad \beta = \frac{v}{c} < 1,\ k_0 = \frac{\omega}{c} = \omega\sqrt{\mu_0 \epsilon_0}, \tag{25}$$

where c is the speed of light in vacuum. Thus, the associated time-harmonic source distribution $\mathbf{J}(\mathbf{r}, \omega)$ is a line current with a linearly varying phase [see Eqs. (13) and (14)], and the radiation from the point charge moving in various environments can be obtained from the time-harmonic line-source solutions via the inverse Fourier transform. Applications to special geometries, and discussion of concomitant Cerenkov-type effects, may be found in Secs. 5.4e and 5.5j.

While application of the inverse Fourier transform is essential for the determination of the real, time-dependent electromagnetic fields $\hat{\mathbf{E}}(\mathbf{r}, t)$ and $\hat{\mathbf{H}}(\mathbf{r}, t)$ excited by moving charges, it can be avoided for the evaluation of the total radiated energy, which is often of more interest than the fields themselves. The Poynting vector

$$\hat{\mathbf{P}}(\mathbf{r}, t) = \hat{\mathbf{E}}(\mathbf{r}, t) \times \hat{\mathbf{H}}(\mathbf{r}, t) \tag{26}$$

represents the flow of electromagnetic field energy per unit area per unit time. To calculate the radiated energy, one defines a total energy flow vector $\mathbf{W}(\mathbf{r})$ per unit area as

$$\mathbf{W}(\mathbf{r}) = \int_{-\infty}^{\infty} \hat{\mathbf{P}}(\mathbf{r}, t)\,dt. \tag{27}$$

We now substitute for $\hat{\mathbf{E}}$ and $\hat{\mathbf{H}}$ their Fourier integral representations; these may be written conveniently in terms of $f(\mathbf{r}, \omega)$ and a complex-conjugate function $f^*(\mathbf{r}, \omega)$. Since $\hat{f}(\mathbf{r}, t)$ is real, one notes from

$$f(\mathbf{r}, \omega) = \int_{-\infty}^{\infty} \hat{f}(\mathbf{r}, t) \exp(-j\omega t)\,dt$$

that $f(\mathbf{r}, -\omega) = f^*(\mathbf{r}, \omega)$:

$$\hat{f}(\mathbf{r}, t) = \frac{1}{2\pi} \int_0^\infty f(\mathbf{r}, \omega)e^{j\omega t}\, d\omega + \frac{1}{2\pi} \int_0^\infty f^*(\mathbf{r}, \omega)e^{-j\omega t}\, d\omega. \tag{28}$$

Thus, upon assuming the interchangeability of the t and ω integrations,

$$4\pi^2 \mathbf{W}(\mathbf{r}) = \int_0^\infty d\omega \int_0^\infty d\omega' \left[\mathbf{E}(\mathbf{r}, \omega) \times \mathbf{H}(\mathbf{r}, \omega') \int_{-\infty}^\infty e^{j(\omega+\omega')t}\, dt \right.$$

$$+ \mathbf{E}^*(\mathbf{r}, \omega) \times \mathbf{H}(\mathbf{r}, \omega') \int_{-\infty}^\infty e^{-j(\omega-\omega')t}\, dt$$

$$+ \mathbf{E}(\mathbf{r}, \omega) \times \mathbf{H}^*(\mathbf{r}, \omega') \int_{-\infty}^\infty e^{j(\omega-\omega')t}\, dt$$

$$\left. + \mathbf{E}^*(\mathbf{r}, \omega) \times \mathbf{H}^*(\mathbf{r}, \omega') \int_{-\infty}^\infty e^{-j(\omega-\omega')t}\, dt \right], \tag{29}$$

and, since $\int_{-\infty}^\infty e^{-j\alpha t}\, dt = 2\pi\delta(\alpha)$,

$$\mathbf{W}(\mathbf{r}) = \frac{1}{\pi} \operatorname{Re} \int_0^\infty \mathbf{P}(\mathbf{r}, \omega)\, d\omega, \qquad \mathbf{P}(\mathbf{r}, \omega) = \mathbf{E}(\mathbf{r}, \omega) \times \mathbf{H}^*(\mathbf{r}, \omega).\dagger \tag{30}$$

Hence,

$$\mathbf{W}_\omega(\mathbf{r}) = \frac{1}{\pi} \operatorname{Re} \mathbf{P}(\mathbf{r}, \omega), \tag{31}$$

represents the total energy flow per unit area in the frequency interval between ω and $\omega + d\omega$.

The total energy flowing through a plane $z = $ constant, per unit angular frequency at ω, is given by

$$W_\omega(z) = \int_S \mathbf{W}_\omega(\mathbf{r}) \cdot \mathbf{z}_0 dS, \tag{32}$$

where S is the cross-sectional area transverse to z. Upon substituting the modal representations [see Eqs. (2.2.8)]

$$\mathbf{E}_t(\mathbf{r}, \omega) = \sum_i V_i'(z, \omega)\mathbf{e}_i'(\boldsymbol{\rho}) + \sum_i V_i''(z, \omega)\mathbf{e}_i''(\boldsymbol{\rho}), \tag{33a}$$

$$\mathbf{H}_t(\mathbf{r}, \omega) \times \mathbf{z}_0 = \sum_i I_i'(z, \omega)\mathbf{e}_i'(\boldsymbol{\rho}) + \sum_i I_i''(z, \omega)\mathbf{e}_i''(\boldsymbol{\rho}), \tag{33b}$$

into Eq. (32), inverting the orders of summation and integration, and recalling the orthonormality properties (2.2.11b) of the vector-mode functions, one finds

$$W_\omega(z) = \frac{1}{\pi} \operatorname{Re}\left[\sum_i V_i'(z, \omega)I_i'^*(z, \omega) + \sum_i V_i''(z, \omega)I_i''^*(z, \omega) \right]. \tag{34}$$

Thus, W_ω is given as a superposition of individual mode powers and involves only the modal amplitudes V_i and I_i. Finally, the total energy W flowing through a plane $z = $ constant is given by

$$W(z) = \int_S \mathbf{W}(\mathbf{r}) \cdot \mathbf{z}_0 dS = \int W_\omega(z)\, d\omega. \tag{35}$$

---

†The equality of Eqs. (27) and (30) is known as Parseval's formula [2] [see also Eq. (1.2. 27)].

Although the analysis in Eqs. (26)–(35) has been performed in the context of radiation from moving charges, it should be noted that the above energy calculation, and in particular the modal result in Eq. (34), is more generally applicable.

5.3　INTEGRATION TECHNIQUES

5.3a　Analytical Properties of the Representation Integrals

Calculation of the multidimensional Green's functions from the integral representations in Eqs. (5.2.7)–(5.2.14) requires knowledge of the one-dimensional modal Green's functions $g_{zi}(z, z')$; determination of the latter from Eq. (5.2.6b) implies specification of the detailed properties of the region along the z coordinate. While g_{zi} thus depends on the nature of the z stratification, general asymptotic properties of the field can be inferred from the analytic behavior of the integrands without restriction to specific geometries. The discussion in this section is concerned with the contribution to the asymptotic field solution from stationary (saddle) points, poles, and branch points in the integral representations, and with the physical interpretation of the corresponding wave types. These general results are applied to specific geometrical configurations in the remaining sections of this chapter.

The longitudinal Green's functions $g_{zi}(z, z')$, left unspecified so far, depend on the integration variables ξ or η via the longitudinal propagation constant $\kappa_i = (k^2 - k_{ti}^2)^{1/2}$. In a multilayered region with piecewise constant ϵ_β, μ_β, $\beta = 1, 2, \ldots, N$, g_{zi} is a function of the various propagation constants $\kappa_{i\beta} = (k_\beta^2 - k_{ti}^2)^{1/2}$, where $k_\beta = \omega(\mu_\beta \epsilon_\beta)^{1/2}$. The functional dependence of g_{zi} will be shown to be even with resqect to $\kappa_{i\beta}$ in a βth layer of finite width along z, whereas this is not the case in a semiinfinite region. Since an even function of $\kappa_{i\beta}$ possesses a series expansion in powers of $(\kappa_{i\beta})^{2n}$, $n = 0, 1, 2, \ldots$, g_{zi} is regular at the point $\kappa_{i\beta} = 0$ in the complex k_{ti} plane. On the other hand, for a semiinfinite section characterized by ϵ_1, μ_1 (extending to $z = -\infty$) or ϵ_N, μ_N (extending to $z = +\infty$), the lack of evenness implies the existence of first-order branch points at $k_{ti} = \pm k_1$ or $k_{ti} = \pm k_N$, respectively. Thus, the longitudinal Green's function g_{zi} for a region containing an arbitrary number of layers exhibits branch points at $k_{ti} = \pm k_1$, $\pm k_N$ when both $z = -\infty$ and $z = +\infty$ are accessible, branch points at either $\pm k_1$ or $\pm k_N$ when an impenetrable boundary prevents accessibility to $z = +\infty$ or $z = -\infty$, respectively (see Sec. 5.7), and no branch points when two impenetrable boundaries confine the region to a finite z interval (see Fig. 5.3.1). The presence of boundaries also gives rise, in general, to simple pole singularities in g_{zi} (see Secs. 5.5 and 5.6).

The precise form of $g_{zi}(z, z')$ depends on the details of the stratification along the z direction, and methods for its construction are discussed in Secs. 2.4 and 3.3b. In view of the differential equation (5.2.6b), the solution is synthesized in terms of trigonometric or exponential functions. While for a mul-

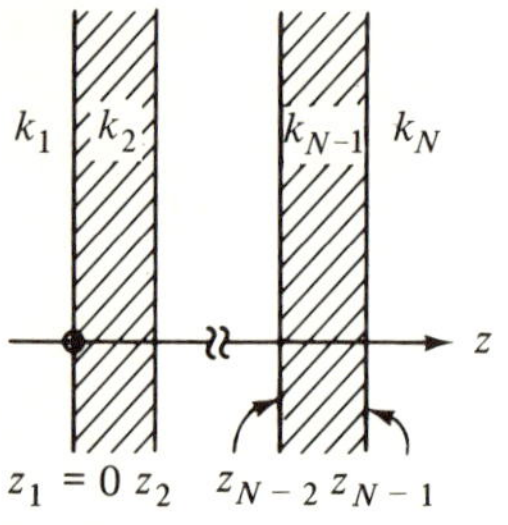
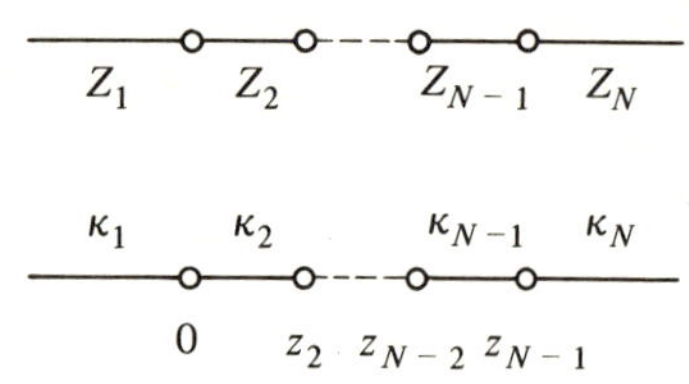

(a) Branch points at $k_t = \pm k_1$, $\pm k_N$

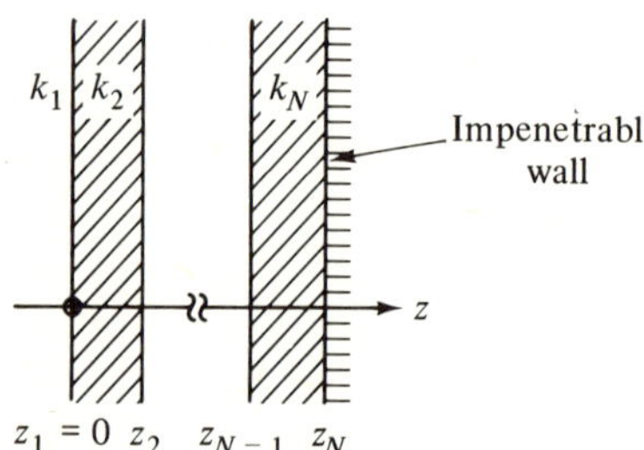
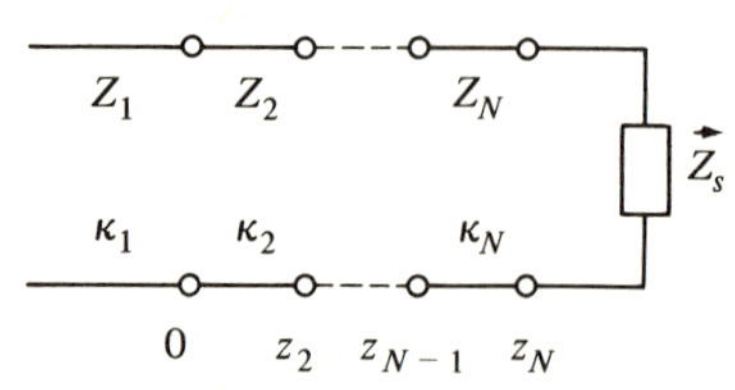

(b) Branch points at $k_t = \pm k_1$

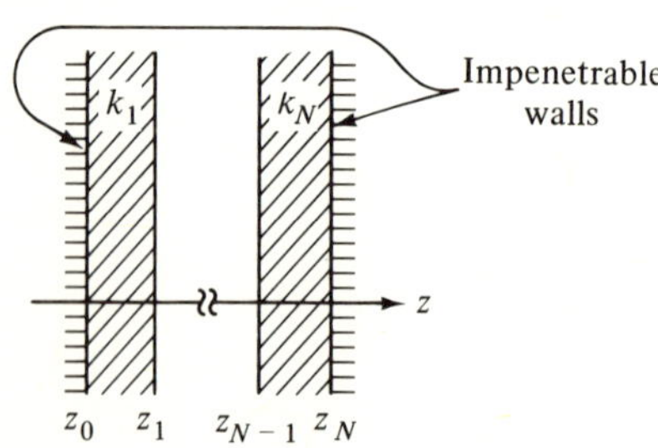
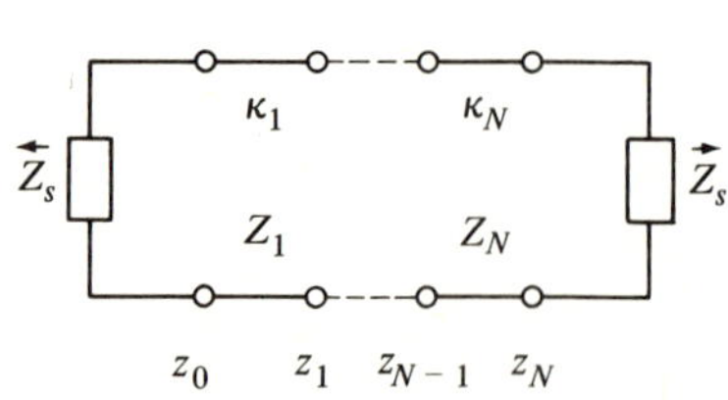

(c) No branch points

FIG. 5.3.1 Various physical regions, modal network representations, and the corresponding branch-point singularities of the modal Green's function in the complex k_t plane. The modal subscript i has been omitted and $\kappa_\beta = \sqrt{k_\beta^2 - k_t^2}$.

tilayered configuration $g_{zi}(z, z')$ may be quite complicated [see Eqs. (2.4.28)], it has a simple generic form when both the source point and the observation point are located in a seminfinite region [e. g., in the region $z < 0$ in Figs.

5.3.1(a) or 5. 3. 1(b)]. In this instance, the Green's function is composed of an "incident" wave appropriate to an infinite region with wavenumber k_1 and a "reflected" wave whose amplitude is determined by the details of the configuration in the half-space $z > 0$. In particular [see Eqs. (2.4.29c) and (2.4.29d) with $\overleftarrow{\Gamma}_i(z_0) = 0$],

$$g_{zi}(z, z') = \frac{1}{2j\kappa_{i_1}} [e^{-j\kappa_{i_1}|z-z'|} \mp \overrightarrow{\Gamma}_i(0)e^{j\kappa_{i_1}(z+z')}], \qquad z, z' < 0, \tag{1}$$

where $\overrightarrow{\Gamma}_i(0)$, the modal (voltage) reflection coefficient looking into the region $z > 0$ from $z = 0$ in region 1, is expressed in terms of the modal input impedance $\overrightarrow{Z}_i(0)$, as follows:

$$\overrightarrow{\Gamma}_i(0) = \frac{\overrightarrow{Z}_i(0) - Z_{i_1}}{\overrightarrow{Z}_i(0) + Z_{i_1}}. \tag{1a}$$

The $-$ and $+$ signs in Eq. (1) are associated with the E-mode and H-mode problems, respectively [see Eqs. (5.2.6a), (2.4.29c) and (2.4.29d)]. $\overrightarrow{Z}_i(0)$ may be determined by a repeated application of Eq. (2.4.10), which relates the input impedance $\overrightarrow{Z}_i(z_{\beta-1})$ at $z_{\beta-1}$ to the input impedance $\overrightarrow{Z}_i(z_\beta)$ at z_β:

$$\frac{\overrightarrow{Z}_i(z_{\beta-1})}{Z_{i\beta}} = \frac{1 + j[\overrightarrow{Z}_i(z_\beta)/Z_{i\beta}] \cot [\kappa_{i\beta}(z_{\beta-1} - z_\beta)]}{[\overrightarrow{Z}_i(z_\beta)/Z_{i\beta}] + j \cot [\kappa_{i\beta}(z_{\beta-1} - z_\beta)]}, \tag{1b}$$

with $Z_{i\beta}$ denoting the characteristic impedance in the region between $z_{\beta-1}$ and z_β [Eq. (2.2.15)];

$$Z_{i\beta} = \frac{\kappa_{i\beta}}{\omega\epsilon_\beta} \text{ for } E \text{ modes}; \qquad Z_{i\beta} = \frac{\omega\mu_\beta}{\kappa_{i\beta}} \text{ for } H \text{ modes}. \tag{1c}$$

The primes and double primes that distinguish E- and H-mode quantities, respectively, have been omitted for convenience.

For a determination of the singularities of $g_{zi}(z, z')$ in the complex k_{ti} plane, it suffices to examine the singularities of the reflection coefficient $\overrightarrow{\Gamma}_i(0)$ or equivalently, the branch points and zeros of the total impedance function $\overleftrightarrow{Z}_i(0)$ $= [\overrightarrow{Z}_i(0) + Z_{i1}]$.† The zeros of $\overleftrightarrow{Z}_i(0)$ locate pole singularities which are descriptive of the waves guided in the direction transverse to z [see Secs. 5.5 and 5.6; these waves may be either of the surface-wave (modal) or leaky-wave (usually non-modal) type]. Concerning branch-point singularities, one observes from Eq. (1b) that $\overrightarrow{Z}_i(z_{\beta-1})$ is an even function of $\kappa_{i\beta}$ and therefore regular at $k_{ti} = \pm k_\beta$ since $Z_{i\beta}$ is an odd function of $\kappa_{i\beta}$ and $\overrightarrow{Z}_i(z_\beta)$ is a function only of $\kappa_{in}, n \geq \beta + 1$. Since the expression for $\overrightarrow{Z}_i(0)$ is deduced by repeated application of Eq. (1b), it follows that $\overrightarrow{Z}_i(0)$ is an even function of all $\kappa_{i\beta}$ associated

---

†For configurations containing stratification or boundaries in the region $z < 0$ as well, the total impedance function is replaced dy $\overleftrightarrow{Z}_i(0) = \overleftarrow{Z}_i(0) + \overrightarrow{Z}_i(0)$ [see Eqs. (2.4.28)].

with layers of finite width. If the last (Nth) layer extends to infinity, its input impedance is given by Z_{iN}, which is an odd function of κ_{iN} and hence possesses branch-point singularities at $k_t = \pm k_N$. On the other hand, for an impenetrable wall at z_N, $\overrightarrow{Z}_i(z_N) = \overrightarrow{Z}_s$, and no branch points are introduced since the surface impedance $\overrightarrow{Z}_s$ is a constant independent of k_{ti}. Analogous considerations apply to $\overleftarrow{Z}_i(0)$. The results stated in Fig. 5.3.1 are therefore verified.

Since the branch points lie on the integration paths when k_β is real (lossless medium), the disposition of the contours in Eqs. (5.2.7)–(5.2.14) near the singularities must be clarified. For example, in Eq. (5.2.8b), where $k_{ti} = \xi$, branch points due to κ_{i1} occur at $\xi = \pm k_1$. While the determination of the proper integration paths can be carried out for real k_1 (see Sec. 5.3b), it is simpler for the present to assume slight dissipation, so that for an exp $(j\omega t)$ time dependence $k = \omega[\mu(\epsilon_r - j\epsilon_i)]^{1/2}$ has a small negative imaginary part, where ϵ_r and ϵ_i are the real and imaginary parts, respectively, of the dielectric constant ϵ. Thus the branch points are displaced from the real ξ and η axes into the fourth and second quadrants of the complex ξ and η planes. If one now lets $\epsilon_i \to 0$, it follows that the path of integration should be indented around the singularities into the first or third quadrants, respectively, for branch-point locations on the positive or negative axes (Fig. 5.3.2). Concerning the convergence of the integral (for a lossless medium) with real k^2, we recall the restriction that for non-propagating modes (imaginary κ_i), $\kappa_i = -j|\kappa_i|$, so finiteness of the integral is assured [see Eqs. (2.2.15)]. Phrased more generally, for complex values of

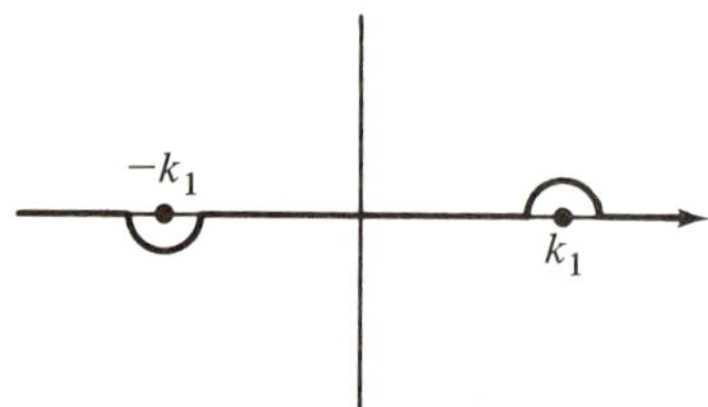

FIG. 5.3.2 Avoidance of branch-point singularities in the complex ξ plane [exp $(j\omega t)$ dependence].

ξ relevant to subsequent contour deformation, we impose on the square root the condition $\mathrm{Im}\sqrt{k^2 - \xi^2} < 0$ along the path, which then guarantees convergence of the exponential terms in Eq. (1). The regions in the complex ξ plane for which $\mathrm{Im}\sqrt{k^2 - \xi^2}$ is negative are examined in Sec. 5.3b for various choices of branch cuts that make the integrand single valued on a multisheeted Riemann surface. Analogous considerations may be applied to the double integral in Eq. (5.2.7), for example, where $k_{ti}^2 = \xi^2 + \eta^2$. Branch points in the ξ and η integrals now occur at $\pm\sqrt{k_1^2 - \eta^2}$ and $\pm\sqrt{k_1^2 - \xi^2}$, respectively.

Pole singularities of $g_{zi}(z, z')$ may also lie on the integration path, and their avoidance may be clarified by considerations analogous to the above, either by displacement from the real axis through the assumption of small loss, or by imposition of the radiation condition on possible residue contributions to the field at infinity.

5.3b Definition of $\kappa(\xi) = \sqrt{k^2 - \xi^2}$ in the Complex ξ Plane

As observed previously, the Green's function integrands in Sec. 5.2b may contain branch-point singularities arising, for example, from the mode wavenumber $\kappa_i = \sqrt{k^2 - \xi^2}$. To assure a unique specification of integrands in the complex ξ plane it is necessary to discuss in detail the analytic properties of the square-root function $\kappa_i \equiv \kappa(\xi)$. When ξ is real and $|\xi| < k$, k being assumed real for the moment, the guided wave along z is propagating and hence the propagation constant κ is real and positive, consistent with a positive modal characteristic impedance [see Eq. (2.2.15)]. Thus, we require a definition of $\kappa(\xi)$ such that

$$\sqrt{k^2 - \xi^2} > 0, \qquad -k < \xi < k. \tag{2}$$

To ensure that integrands remain bounded as $|\kappa(\xi)|\,|z - z'| \to \infty$, it is necessary to impose restrictions on the imaginary part of κ. For the time dependence $\exp(j\omega t)$, the required restriction for real ξ is $\kappa = -j|\kappa|$ (i.e., Im $\kappa < 0$ when $|\xi| > k$).† If ξ is allowed to be complex, the condition Im $\kappa < 0$ will be imposed for all permitted complex values of κ. The analytic continuation of ξ from real to complex values is required for subsequent deformation of the integration contours.

To make the definition of the double-valued function $\kappa(\xi)$ unique, a two-sheeted complex ξ plane is necessary, with branch cuts providing the means of passing from one Riemann sheet to the other.[3] The selection of branch cuts is arbitrary but determines the disposition of those regions of the complex ξ plane in which Im $\kappa < 0$, or Im $\kappa > 0$. Three particularly useful choices are shown in Fig. 5.3.3. Let us define

$$(k - \xi) = |k - \xi|e^{i\alpha}, \quad (k + \xi) = |k + \xi|e^{i\beta}, \qquad \alpha, \beta \text{ real}, \tag{3a}$$

with the angles α and β so selected as to make $\alpha = 0$ and $\beta = 0$ when ξ is real and $|\xi| < k$ on the top sheet. Hence,

$$\sqrt{k^2 - \xi^2} = |\sqrt{k^2 - \xi^2}|e^{i\,(\alpha + \beta)/2}, \tag{3b}$$

where we have chosen the positive sign of the square root. i is the imaginary unit and does not refer in this context to a harmonic time dependence of the field. To satisfy condition (2) it is required that

$$\alpha + \beta = 0, \qquad \text{when } -k < \xi < k. \tag{4a}$$

Consider now the choice of branch cuts in the first part of Fig. 5.3.3 (a). With the angles α and β defined as shown, it is evident that condition (4a) is

---

†This requirement, and also (2), follows from the radiation condition, which demands that the energy radiated by the source to distant observation points is bounded and outgoing (see Sec. 1.5.b).

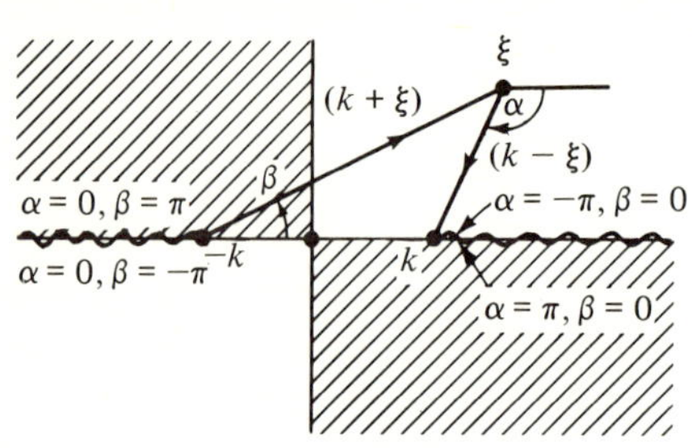

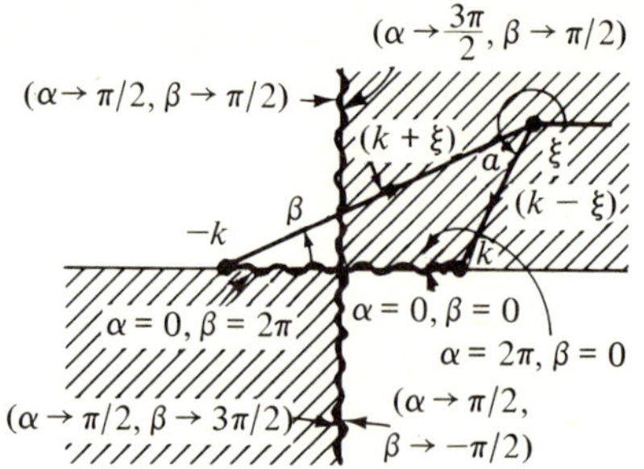

(1)   Im $\kappa > 0$ in shaded region
      Im $\kappa < 0$ in unshaded region
      Re $\kappa > 0$ on entire top sheet

(2)   Im $\kappa > 0$ on entire top sheet
      Re $\kappa > 0$ in unshaded region
      Re $\kappa < 0$ in shaded region

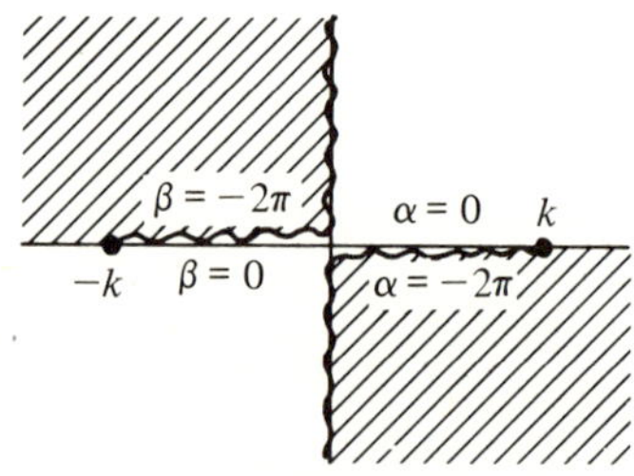

(3)   Im $\kappa < 0$ on entire top sheet
      Re $\kappa > 0$ in unshaded region
      Re $\kappa < 0$ in shaded region

(a) k real

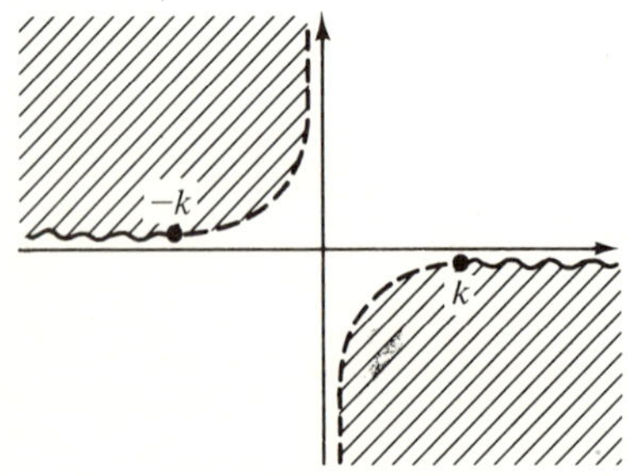

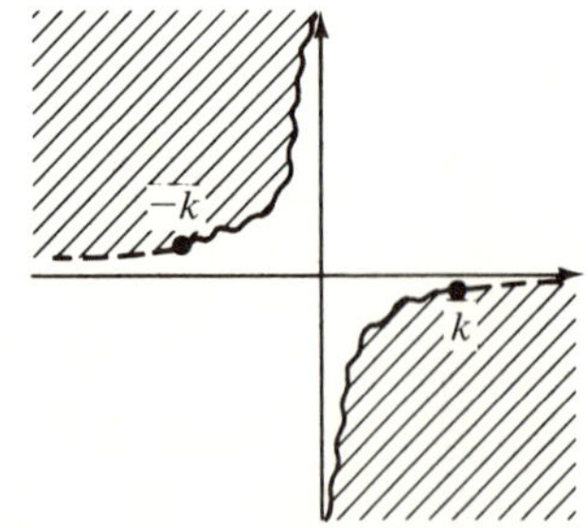

(1)   Im $\kappa > 0$ in shaded region
      Im $\kappa < 0$ in unshaded region
      Re $\kappa > 0$ on entire top sheet

(2)   Im $\kappa < 0$ on entire top sheet
      Re $\kappa > 0$ in unshaded region
      Re $\kappa < 0$ in shaded region

(b) Im $k < 0$

FIG. 5.3.3 Analytic properties of $\sqrt{k^2 - \xi^2}$ in various complex ξ planes.

met since $\alpha = \beta = 0$ when $-k < \xi < k$. This position of the branch cuts restricts the ranges of α and β on the top sheet to[†]

$$-\pi < \alpha < \pi, \qquad -\pi < \beta < \pi. \tag{4b}$$

In accord with the usual convention, an angle is counted positive or negative if measured in a counterclockwise or clockwise sense, respectively, from the horizontal axis. The arguments α and β of $(k - \xi)$ and $(k + \xi)$ on the upper and lower sides of the branch cuts have been shown explicitly. Since $0 > (\alpha + \beta) > -\pi$ in the first and third quadrants of the complex ξ plane, one notes from Eq. (3b) that $\sin [(\alpha + \beta)/2] < 0$ (i.e., $\operatorname{Im} \kappa < 0$ in these regions). Conversely, $0 < \alpha + \beta < \pi$ in the second and fourth quadrants, implying that $\operatorname{Im} \kappa > 0$. It also follows that $\operatorname{Re} \kappa > 0$ on the entire top sheet. All these conditions are reversed in the corresponding quadrants of the second Riemann sheet. The branch cuts in the first part of Fig. 5.3.3(a) evidently follow the contours $\operatorname{Re} \kappa = 0$.

In an alternative choice of branch cuts in the second part of Fig. 5.3.3(a), drawn along contours on which $\operatorname{Im} \kappa = 0$, one notes again that α and β are so defined as to satisfy Eq. (4a). In this instance, $0 < \alpha + \beta < 2\pi$ in all quadrants, so $\operatorname{Im} \kappa > 0$ on the entire top sheet. [This choice is pertinent for field problems involving an assumed $\exp (-i\omega t)$ time dependence.] The behavior of $\operatorname{Re} \kappa$ in the different quadrants is also shown. If the branch cuts are drawn along the contours $\operatorname{Im} \kappa = 0$ in the manner shown in the third part of Fig. 5.3.3(a) then $\operatorname{Im} \kappa < 0$ on the entire top sheet and the algebraic sign of $\operatorname{Re} \kappa$ is as indicated. Again, the algebraic signs of $\operatorname{Re} \kappa$ and $\operatorname{Im} \kappa$ in the various quadrants on the second sheet are the reverse of those on the first sheet.

In the above discussion, k was assumed to be real. Since all physical media have some loss, it is also pertinent to treat the case of a lossy dielectric wherein, for an assumed time dependence $\exp (j\omega t)$, $\epsilon = \epsilon_r - j\sigma/\omega$, $\sigma =$ conductivity of medium. Correspondingly, $k = \omega\sqrt{\mu\epsilon}$ has a negative imaginary part. As for real k, it is convenient to choose the branch cuts along the contours on which $\operatorname{Re} \kappa$ or $\operatorname{Im} \kappa$ vanishes. These contours are determined by enforcing $\alpha + \beta = 0, \pm 2\pi, \ldots$ for $\operatorname{Im} \kappa = 0$ and $\alpha + \beta = \pm\pi, \pm 3\pi, \ldots$ for $\operatorname{Re} \kappa = 0$. Since

$$\tan \alpha = \frac{k_i - \xi_i}{k_r - \xi_r}, \qquad \tan \beta = \frac{k_i + \xi_i}{k_r + \xi_r}, \tag{5a}$$

one finds that both $\operatorname{Re} \kappa = 0$ and $\operatorname{Im} \kappa = 0$ along the hyperbolas

$$\xi_r \xi_i = k_r k_i.$$

ξ_r, k_r and ξ_i, k_i denote the real and imaginary parts, respectively, of ξ and k. One verifies readily that $\operatorname{Re} \kappa = 0$ on those portions of the curves for which $|\xi_r| > k_r$, while $\operatorname{Im} \kappa = 0$ for $|\xi_r| < k_r$. The corresponding behavior of κ in the complex ξ plane is shown in Figs. 5.3.3(b) for $k_i < 0$. One notes that as

---

[†]Conditions (4a) and (4b) constitute the definition of $\sqrt{k^2 - \xi^2}$ on the first branch of the two-sheeted Riemann surface. A definition on the second branch would involve $\pi < \alpha < 3\pi$ or $\pi < \beta < 3\pi$, for example.

$k_i \to 0$ in Fig. 5.3.3(b), one obtains the configurations in the first and third parts of Fig. 5.3.3(a). If $k_i > 0$ [for fields having an assumed time dependence $\exp(-i\omega t)$], one has a subdivision of the ξ plane which is obtainable from that in Fig. 5.3.3(b) by a reflection of all curves about either the real or the imaginary ξ axis [see Fig. 5.5.4(b)]. The various regions must then be so assigned as to reduce to those in Figs. 5.3.3(a) in the limit $k_i \to 0$.

A simple rule may be stated for the assignment of the algebraic signs of Re κ and Im κ in various portions of the complex ξ plane. It follows from the recognition that sign changes in Re κ or Im κ can occur only when ξ crosses the contours Re $\kappa = 0$ or Im $\kappa = 0$, respectively. Thus, if branch cuts are chosen along the contours Im $\kappa = 0$, for example, the sign of Im κ is constant on either the top sheet or the bottom sheet, since the crossing of the Im $\kappa = 0$ curve leads from one sheet to the other. It then suffices to specify Re κ and Im κ *consistently* at a single point on the top sheet, for example at $\xi = 0$, since one may then deduce the sign alternations in Re κ from the crossings of the Re $\kappa = 0$ contours. In this manner, one may arrive directly at the designation in the second part of Fig. 5.3.3(b) if the top sheet is chosen so that $\kappa_i = +k$ at $\xi = 0$.

5.3c The Transformation $\xi = k \sin w$

To facilitate subsequent function-theoretic manipulations involving integrals of the type occurring in Eqs. (5.2.11)–(5.2.14), it is desirable to introduce a new complex variable w via the transformation

$$\xi = k \sin w, \tag{6}$$

which identifies w as a complex angle variable and makes $\kappa_i = (k^2 - \xi^2)^{1/2} = 0$ a regular point in the w plane. The transcendental function $\sin w$ is single-valued. From its periodicity property $\sin(w + 2n\pi) = \sin w$, $n = \pm 1, \pm 2, \ldots$, it is evident that a multiplicity of w values corresponds to the same value of ξ. Thus, the entire ξ plane can be mapped into various adjacent sections of "width" 2π in the w plane. The inverse function $\sin^{-1}(\xi/k)$ in the ξ plane is multiple valued, implying the existence of branch points in that plane. The branch points occur at those points in the ξ plane for which the mapping derivative $d\xi/dw$ vanishes in the w plane and are of the order of the zero of $d\xi/dw$. Since $d\xi/dw = k \cos w$, the branch points are situated at $\xi = \pm k$ and are of the first order.

To investigate in detail the properties of the mapping from the ξ to the w plane, we separate Eq. (6) into its real and imaginary parts (k is assumed to be real):

$$\xi_r = k \sin w_r \cosh w_i, \qquad \xi_i = k \cos w_r \sinh w_i, \tag{7}$$

where

$$\xi = \xi_r + i\xi_i, \qquad w = w_r + iw_i, \tag{7a}$$

with ξ_r, ξ_i, w_r, and w_i real. As shown in Fig. 5.3.4(a), the four quadrants in the ξ plane map in the w plane into corresponding regions, identified via Eq. (7),

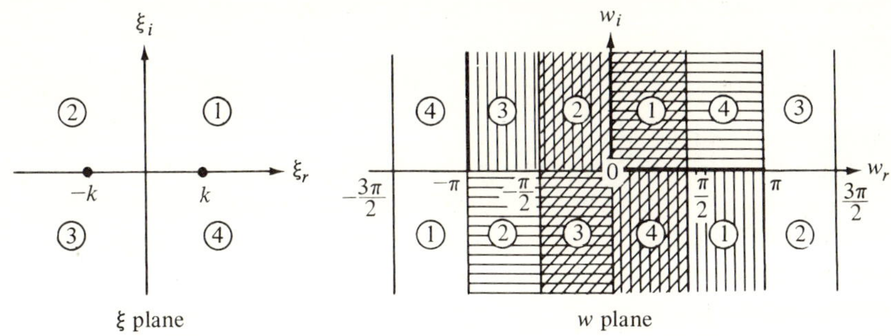

(a) k real

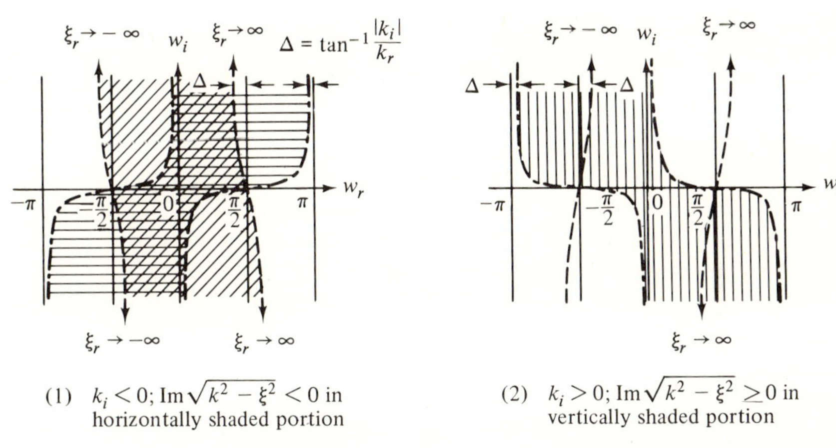

(1) $k_i < 0$; $\mathrm{Im}\sqrt{k^2 - \xi^2} < 0$ in horizontally shaded portion

(2) $k_i > 0$; $\mathrm{Im}\sqrt{k^2 - \xi^2} \geq 0$ in vertically shaded portion

(b) k complex ($--\kappa_r = 0$; $---\kappa_i = 0$)

FIG. 5.3.4 Mapping from the complex ξ plane to the complex w plane.

which bear the same numbers as those in the ξ plane. One notes the repetitious character of the regions in the w plane as w_r changes by multiples of 2π. Thus, the two-sheeted ξ plane can be mapped, for example, into the strip $-\pi < w_r < \pi$ in the w plane. The top sheet of the Riemann surface in the ξ plane as shown in Fig. 5.3.4(a) maps, for example, into the strip $-\pi/2 < w_r < \pi/2$ in the w plane; each neighboring strip of width π in the w plane can then represent the other sheet of the Riemann surface in the complex ξ plane. Thus, in the mapping from the ξ plane to the w plane, the Riemann sheets in the former appear as adjacent regions in the latter.

The detailed correspondence between regions in the complex w plane and those in the two-sheeted ξ plane depends on the choice of branch cuts in the ξ

plane. To assess this correspondence in a simple manner we utilize the behavior of

$$\kappa = \sqrt{k^2 - \xi^2} = + k \cos w = k(\cos w_r \cosh w_i - i \sin w_r \sinh w_i). \quad (8)$$

With $\kappa(0) = k$, the plus sign in Eq. (8) is chosen to make the point $\xi = 0$ correspond to $w = 0$. For a choice of branch cuts as in the first part of Fig. 5.3.3(a), Im κ is positive in quadrants 2 and 4 and negative in quadrants 1 and 3. From Eq. (8), Im $\kappa = -k \sin w_r \sinh w_i$, so one can identify the regions in the w plane of Fig. 5.3.4(a), shaded with slanted lines, as corresponding to those on the entire top sheet of the Riemann surface in the first part of Fig. 5.3.3(a). Similarly, the vertically shaded regions in the w plane correspond to the choice of branch cuts in the second part of Fig. 5.3.3(a) with Im $\kappa > 0$ on the entire top sheet; the horizontally shaded regions correspond to the third part of Fig. 5.3.3(a) with Im $\kappa < 0$ on the entire top sheet.

When k is complex,

$$\kappa_r = k_r \cos w_r \cosh w_i + k_i \sin w_r \sinh w_i,$$

$$\kappa_i = k_i \cos w_r \cosh w_i - k_r \sin w_r \sinh w_i, \quad (8a)$$

so the boundary curves $\kappa_r = 0$, $\kappa_i = 0$ in the complex w plane appear as in Fig. 5.3.4(b). As $|w_i| \to \infty$, these curves are asymptotic to the lines $w_r = \pi/2 \pm \Delta$, etc., as shown. The various shaded regions in the w plane of Fig. 5.3.4(a) are therefore distorted. The slanted and horizontally shaded regions in the first part of Fig. 5.3.4(b) correspond to the entire top sheets of the Riemann surfaces in the ξ planes in the first and second parts of Fig. 5.3.3(b), respectively. In the horizontally shaded domain of Fig. 5.3.4(b), Im $\kappa < 0$. Analogous considerations apply to the second part of Fig. 5.3.4(b), wherein Im $\kappa > 0$ in the vertically shaded domain.

5.3d Asymptotic Evaluation of a Typical Radiation Integral for the Incident and Reflected Fields

The preliminaries of Secs. 5.3a–5.3c facilitate the asymptotic evaluation of integrals of the form given in Eqs. (5.2.11)–(5.2.14), with g_{zi} specified as in Eq. (1). For an assumed exp $(j\omega t)$ time dependence, generic forms of these integrals are, for the point-source problem,

$$I_1 = \int_{\infty e^{-j\pi}}^{\infty} H_n^{(2)}(\xi\rho)e^{-j\kappa_{11}\bar{z}}f_1(\xi)\,d\xi, \qquad n = 0, 1, \quad (9a)$$

and, for the line-source problem,

$$I_2 = \int_{-\infty}^{\infty} e^{-j\eta(y-y')}e^{-j\kappa_{11}\bar{z}}f_2(\eta)\,d\eta, \quad (9b)$$

where the positive distance parameter $\bar{z}$ may represent either $|z - z'|$ or $|z + z'|$, and the functions $f_{1,2}$ are independent of the space coordinates but possess branch-point and (or) pole singularities in the complex ξ or η plane. A typical integration path is shown in Fig. 5.3.5(a), wherein the branch-point singularities

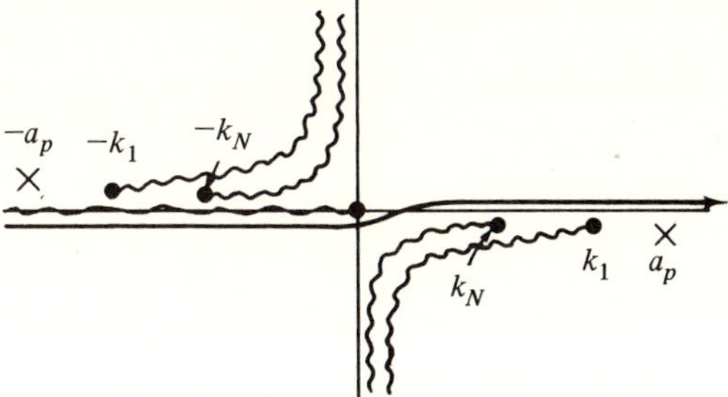

(a) Complex ξ or η planes (in η plane, omit branch point at $\eta = 0$)

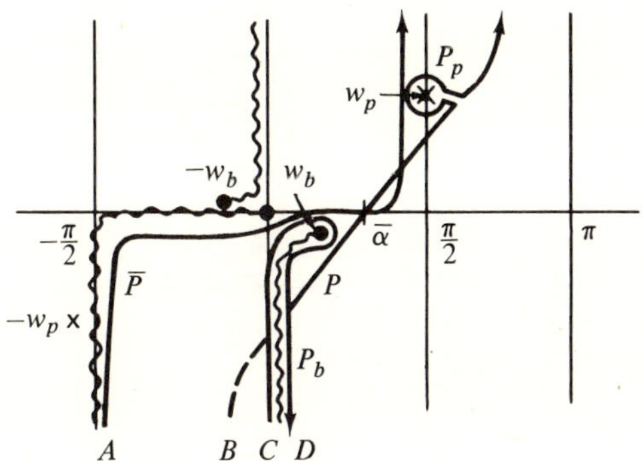

(b) w plane (w_b corresponds to k_N, w_p corresponds to a_p, branch point at $w = 0$ corresponds to branch point at $\xi = 0$)

FIG. 5.3.5   Integration paths ($e^{j\omega t}$ dependence).

at $\pm k_1$, and $\pm k_N$ and the pole singularities at $\pm a_p$ have been slightly displaced from the real axis to signify the presence of small loss: In the lossless limit, the required indentations of the integration path around the singularities are thereby evident. The branch point at the origin arises from the Hankel function in Eq. (9a) and is not present in the corresponding η-plane description of Eq. (9b). Branch cuts have been drawn so that Im κ_1 and Im κ_N are negative on the entire top sheet of the multisheeted Riemann surface (see Fig. 5.3.3).

Since one of the aims of the presentation in this book is to provide familiarity with field solutions corresponding to either an exp ($j\omega t$) or an exp ($-i\omega t$) time dependence, it is appropriate to give as well the exp ($-i\omega t$) forms of Eqs. (9a) and (9b). To pass from one formulation to the other, one replaces j by $-i$, thereby performing essentially a complex conjugation. The resulting modification affects not only the appearance of the integrands but also the paths of integration on which the radiation condition is satisfied. Thus, it is now necessary to have Im $\kappa > 0$ to provide for the decay of non-propagating wave solutions $\exp(i\kappa_i \bar{z})$. The generic radiation integrals for the $\exp(-i\omega t)$ variation are given by

$$I_1 = \int_{\infty e^{i\pi}}^{\infty} H_n^{(1)}(\xi\rho) e^{i\kappa_{i1}\bar{z}} f_1(\xi)\, d\xi, \tag{10a}$$

and

$$I_2 = \int_{-\infty}^{\infty} e^{i\eta(y-y')} e^{i\kappa_{i1}\bar{z}} f_2(\eta)\, d\eta, \tag{10b}$$

and the corresponding integration paths and singularities are shown in Fig. 5.3.6(a) [see Fig. 3.2.10(a) in connection with Eq. (10a)].

It is convenient to transform from the complex wavenumber variables ξ or η to the complex angle variable w via $\xi = k_1 \sin w$ or $\eta = k_1 \sin w$, since the branch-point pair at $\pm k_1$ in Figs. 5.3.5(a) or 5.3.6(a) may thereby be removed

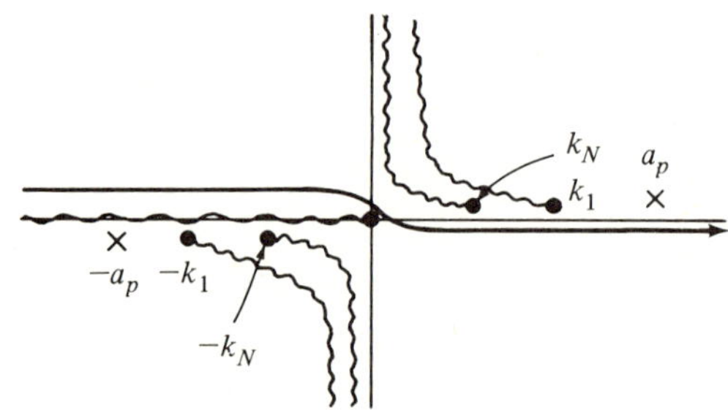

(a) Complex ξ or η planes
(in η plane, omit branch point at $\eta = 0$)

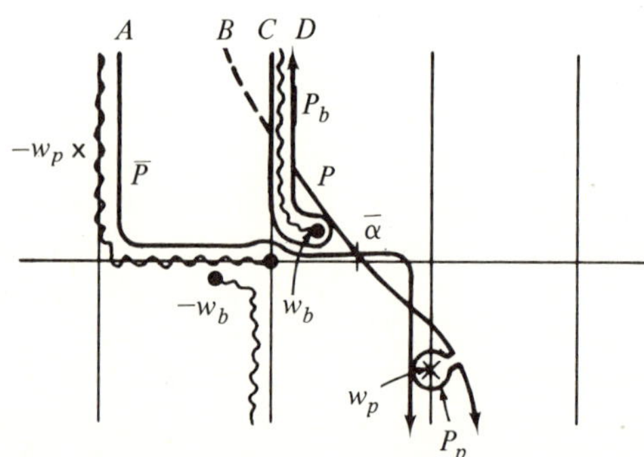

(b) w plane (w_b corresponds to k_N, w_p corresponds to a_p, branch point at $w = 0$ corresponds to branch point at $\xi = 0$)

FIG. 5.3.6   Integration paths ($e^{-i\omega t}$ dependence).

and the subsequent calculation facilitated. In view of the mapping properties described in Sec. 5.3c, the configuration in the complex w plane is as shown in Figs. 5.3.5(b) and 5.3.6(b), with $\bar{P}$ representing the transformed integration contour. Upon introducing polar coordinates L and $\bar{\alpha}$ via

$$\left.\begin{array}{c} \rho \\ y - y' \end{array}\right\} = L \sin \bar{\alpha}, \qquad \bar{z} = L \cos \bar{\alpha}, \tag{11}$$

one obtains the following expressions for the point-source problem with an exp $(j\omega t)$ dependence,

$$I_1 = k_1 \int_{\bar{P}} H_n^{(2)}(k_1 L \sin \bar{\alpha} \sin w) e^{-jk_1 L \cos \bar{\alpha} \cos w} f_1(k_1 \sin w) \cos w \, dw, \tag{12a}$$

and, for the line-source problem,

$$I_2 = k_1 \int_{\bar{P}} e^{-jk_1 L \cos(w - \bar{\alpha})} f_2(k_1 \sin w) \cos w \, dw. \tag{12b}$$

Analogous results are found for the exp $(-i\omega t)$ dependence. In Eq. (12a), L is the spherical polar distance from $\rho = 0$, $\bar{z} = 0$, and $\bar{\alpha} \leq \pi/2$ is the angle between L and the $\bar{z}$ axis, whereas in Eq. (12b), L is the cylindrical polar distance from $y = y'$, $\bar{z} = 0$, and the polar angle $\bar{\alpha}$ may range from $-\pi/2$ to $+\pi/2$.

Equation (12b) is already in the form suitable for an asymptotic evaluation by the method of steepest descent (Sec. 4.1), since for observation points in the far zone, the large parameter $k_1 L$ appears in the exponential term, and the function $f_2 \cos w$ is slowy varying by comparison. To achieve a similar formulation for Eq. (12a) and its counterpart for the exp $(-i\omega t)$ case, one may employ the large-argument approximation for the Hankel functions [see also Eqs. (5.4.37)][4] :

$$H_n^{(2)}(u) \sim \sqrt{\frac{2}{\pi u}} e^{-j(u - n\pi/2 - \pi/4)} \left[1 + O\left(\frac{1}{u}\right)\right],$$
$$|u| \gg n, \qquad -2\pi < \arg u < \pi, \tag{13a}$$

$$H_n^{(1)}(u) \sim \sqrt{\frac{2}{\pi u}} e^{i(u - n\pi/2 - \pi/4)} \left[1 + O\left(\frac{1}{u}\right)\right],$$
$$|u| \gg n, \qquad -\pi < \arg u < 2\pi, \tag{13b}$$

provided that $k_1 L \sin \bar{\alpha} \gg 1$ and that the integration paths are deformed away from the origin sufficiently to assure $|k_1 L \sin \bar{\alpha} \sin w| \gg 1$ along $\bar{P}$. The restrictions on arg $(\sin w)$ are met on $\bar{P}$ and various deformed contours employed subsequently. Equations (12a) and (12b) are then both represented by an integral of the form

$$I = \int_{\bar{P}} f(w) \exp\left[\begin{array}{c} -j \\ +i \end{array}\right. k_1 L \cos(w - \bar{\alpha})\right] dw \qquad \left(\begin{array}{c} e^{j\omega t} \\ e^{-i\omega t} \end{array} \text{dependence}\right), \tag{14}$$

where f takes on values appropriate to the point-source or line-source excitation.

The asymptotic evaluation of I for large values of $k_1 L$ is performed directly by the methods decribed in Chapter 4. The integration path $\bar{P}$ is first deformed into the steepest-descent path P through the saddle point at $w = \bar{\alpha}$ [see Fig.

4.2.2, where $P \equiv \bar{P}_z$ is shown for the exp $(-i\omega t)$ case]; this path distortion is permissible since both $\bar{P}$ and P terminate at $|w| \to \infty$ in regions wherein the exponential in Eq. (14) decays, thereby eliminating contributions from contour segments connecting the endpoints. For certain saddle-point locations, it may happen that singularities in the finite portion of the w plane must be crossed during the path deformation. These pole or branch-point singularities must then be surrounded by appropriate contours P_p and P_b, respectively, as shown in Figs. 5.3.5(b) and 5.3.6(b). When the branch point at w_b is intercepted, the steepest-descent path passes through the branch cut onto the second Riemann sheet; connection between the endpoints A and B of $\bar{P}$ and P is brought about by the path segments AC, CD (along P_b) and DB, with no contribution arising from AC and DB.† Thus, for the exp $(-i\omega t)$ dependence, Eq. (14) may be rewritten in terms of a "steepest-descent representation" as

$$I = \int_P \cdots dw + U(\bar{\alpha} - \bar{\alpha}_b) \int_{P_b} \cdots dw$$

$$+ U(\bar{\alpha} - \bar{\alpha}_p) 2\pi i [(w - w_p) f(w)]_{w_p} \exp [ik_1 L \cos (w_p - \bar{\alpha})], \quad (15)$$

where $U(x)$ is the Heaviside unit function, which equals unity when x is positive and vanishes when x is negative, while $\bar{\alpha}_b$ and $\bar{\alpha}_p$ are those values of $\bar{\alpha}$ for which the steepest-descent path crosses the branch-point and pole singularities, respectively. From Eq. (4.2.29) it is concluded that

$$\bar{\alpha}_{p,b} = \operatorname{Re} w_{p,b} - \cos^{-1} \operatorname{sech} (\operatorname{Im} w_{p,b}). \quad (15a)$$

The residue contribution arising from a (simple) pole at w_p has been exhibited explicitly in the last term of Eq. (15).

The integrals over the paths P and P_b cannot generally be reduced further, but they may be simplified by asymptotic techniques when $(k_1 L)$ is large. From Eqs. (4.2.1a), one has, for the saddle-point contribution,

$$\int_P f(w) e^{ik_1 L \cos (w - \bar{\alpha})} \, dw \sim \sqrt{\frac{2\pi}{k_1 L}} f(\bar{\alpha}) e^{i(k_1 L - \pi/4)} \left[1 + O\left(\frac{1}{k_1 L}\right) \right]. \quad (16a)$$

Similarly, when P_b encircles the branch point in the positive sense, Eq. (4.8.3) yields for the branch-cut integral contribution,

$$\int_{P_b} f(w) \exp [ik_1 L \cos (w - \bar{\alpha})] \, dw$$

$$\sim \frac{2\sqrt{\pi}}{|k_1 L \sin (\bar{\alpha} - w_b)|^{3/2}} \left[\sqrt{w - w_b} \frac{df(w)}{dw} \right]_{w_b}$$

$$\exp [ik_1 L \cos (w_b - \bar{\alpha})] \exp \left(-i \frac{3}{2} \left\{ \frac{\pi}{2} + \arg [k_1 \sin (w_b - \bar{\alpha})] \right\} \right), \quad (16b)$$

where it has been assumed that $\sqrt{w - w_b} \, df/dw$ is finite at w_b, a behavior ex-

---

†The Riemann surface is associated with the branch points of $f(w)$ at $\pm w_b$ that arise from the corresponding singularities at $\pm k_N$ in the ξ or η planes. Since these singularities do not occur in the exponential term in Eq. (14), this function has identical decay characteristics on either Riemann sheet.

hibited by functions encountered in an actual calculation. For the $\exp(+j\omega t)$ dependence, i in Eqs. (15) and (16) is replaced by $-j$. One observes that for real values of w_p and w_b, the residue contribution in Eq. (15) is dominant at large distances since it behaves like $O(1)$ in $k_1 L$, whereas the saddle-point and branch-point contributions decay like $(k_1 L)^{-1/2}$ and $(k_1 L)^{-3/2}$, respectively. On the other hand, when w_p is complex, the residue contribution has an exponential decay factor which may render the saddle-point result dominant. It may also be remarked that the field due to the saddle point is observed everywhere, while the pole or branch-point fields are generally confined to special regions delimited by the Heaviside unit functions in Eq. (15). For the class of problems to be considered in this chapter, the poles account for surface waves and leaky waves, the branch points yield lateral waves, and the saddle points furnish the direct, reflected, and refracted fields of geometrical optics; the result in Eq. (16a) may be interpreted as a local plane-wave field propagating along L with varying amplitude.

An interesting schematization of the relation between propagating wave solutions in the wavenumber ζ or η and angle w planes, and their relevance to the saddle-point condition, may be realized with the aid of the wavenumber surface. This surface, traced out by the endpoints of the wave vector $\mathbf{k}_1$ for plane waves propagating in all possible directions, is a sphere since the medium is isotropic. The outward normal to the surface, evidently parallel to $\mathbf{k}_1$ and defining the direction of the power-flow vector $\bar{\mathbf{S}}$, is also called the *ray direction* (see the discussion in Sec. 1.6b). Propagating plane-wave solutions are characterized by real values of the wavenumbers ζ (or η) and κ, and their relation to the propagation angle w is as shown in Fig. 5.3.7(a). The saddle point in the radiation integral is situated at $w = \bar{\alpha}$ (or ζ, $\eta = k_1 \sin \bar{\alpha}$), where $\bar{\alpha}$ is the angle between the $\bar{z}$ axis and the radius vector from the origin at $L = 0$ to the observation point [Fig. 5.3.7(b)]. Thus, one may find the saddle point graphically from the wavenumber plot by constructing a normal to the surface in the direction parallel to L and by reading off the corresponding values of w, ζ, or η.

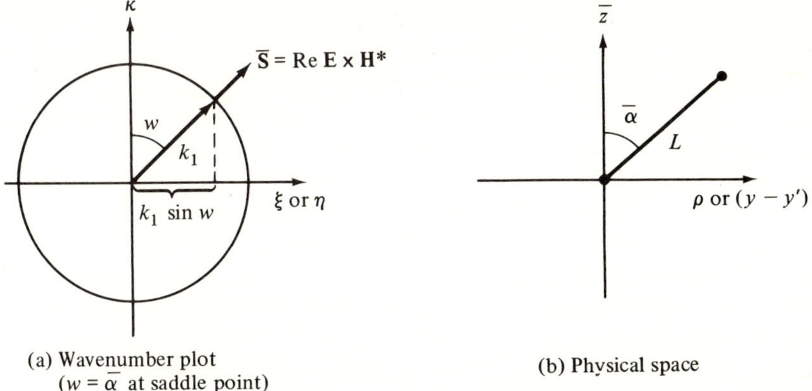

(a) Wavenumber plot
($w = \bar{\alpha}$ at saddle point)

(b) Physical space

FIG. 5.3.7 Wavenumber surface and saddle-point condition.

Although this graphical visualization is not too important in the present case of an isotropic medium since the propagation characteristics are so simple, it aids greatly in an anisotropic environment (see Secs. 1.6b and 8.3). The saddle point locates the region of major contribution to the steepest-descent integral, and the saddle-point value, $w = \bar{\alpha}$ or ξ, $\eta = k_1 \sin \bar{\alpha}$ selects that particular plane wave which establishes the local field and carries energy along the radial direction L to the corresponding observation point; more precisely, the field is estabished by a wavepacket having a small spread of angles or wavenumbers about the saddle-point value. This physical interpretation gives insight into the mechanism of energy transfer to the radiation field and is obtained directly from the wave-number diagram.

The simple formulas in Eqs. (16a) and (16b) fail when $\bar{\alpha} \to w_p$ and w_b, respectively; in this instance, the saddle point is situated near a singularity, and the asymptotic evaluation must be performed in a more refined manner. As noted in Secs. 4.4a and 4.4c, the modified formula, valid for arbitrary proximity of a pole and a saddle point, involves Fresnel integrals, whereas the modified result for neighboring saddle point and branch point is given in terms of a Weber (parabolic cylinder) function.

5.3e General Properties of Pole and Branch-Point Wave Contributions

The asymptotic results in Eqs. (15) and (16) contain wave contributions that arise from poles or branch points of the reflection coefficient $\vec{\Gamma}_i(0)$ in Eq. (1); these singularities in the complex w plane are located at $w_\mu = w_{\mu r} + i w_{\mu_i} = w_p$ or w_b, where $w_{\mu r}$ and w_{μ_i} denote the real and imaginary parts of w_μ, respectively. Apart from a possible algebraic dependence on the space coordinates, a typical wave constituent is characterized by the exponential behavior

$$\exp\left[ik_1 L \cos(w_\mu - \bar{\alpha})\right] =$$

$$\exp\left[ik_1 L \cosh w_{\mu_i} \cos(w_{\mu r} - \bar{\alpha})\right] \exp\left[-k_1 L \sinh w_{\mu_i} \sin(\bar{\alpha} - w_{\mu r})\right], \quad (17)$$

representing for real k_1 a non-uniform plane wave whose phase fronts advance along the direction $\bar{\alpha} = w_{\mu r}$ (direction of constant amplitude) and whose amplitudes change most rapidly in the perpendicular direction $\bar{\alpha} = w_{\mu r} \pm \pi/2$ (direction of constant phase). The phase velocity $v = \omega/(k_1 \cosh w_{\mu_i})$ is seen to be smaller than the ω/k_1 value for a uniform plane wave. In particular, when $w_{\mu r} = \pi/2$, $w_{\mu_i} < 0$, the phase fronts propagate transverse to $\bar{z}$ and the amplitude decays along $\bar{z}$ only [see the definitions of L and $\bar{\alpha}$ in Eq. (11)], thereby characterizing the field as a *surface wave* carrying energy in the direction parallel to an interface at $\bar{z} = 0$ (Secs. 5.6 and 5.7). On the other hand, when $0 < w_{\mu r} < \pi/2$, $w_{\mu_i} > 0$, energy is transferred out of the interface, and yields a *leaky wave* (Sec. 5.6). Important in a physical interpretation of these plane-wave solutions is their domain of existence in the asymptotic field, described essentially by that range of observation angles $\bar{\alpha}$ for which the singularity w_μ is encountered during the deformation of the original integration contour $\bar{P}$ into the steepest-descent path P in Fig. 5.3.6(b); the explicit separation of the contributions from a singularity and from the saddle point applies only when these points are "widely separated" (see Sec. 4.1). If $\bar{\alpha}_\mu$ denotes the value of $\bar{\alpha}$ for which P just passes through w_μ [see Eq. (15a)], then when w_μ is situated above and (or) to the right of the original path $\bar{P}$ in Fig. 5.3.8(a), the singularity contributes in the range $\bar{\alpha}_\mu < \bar{\alpha} < \pi/2$, with $(\bar{\alpha}_\mu - w_{\mu r}) \gtrless 0$ when $w_{\mu_i} \gtrless 0$. For the remaining locations

Im w
Re w
B
A
C
E
D
F
π/2
π
−π/2
0
P̄

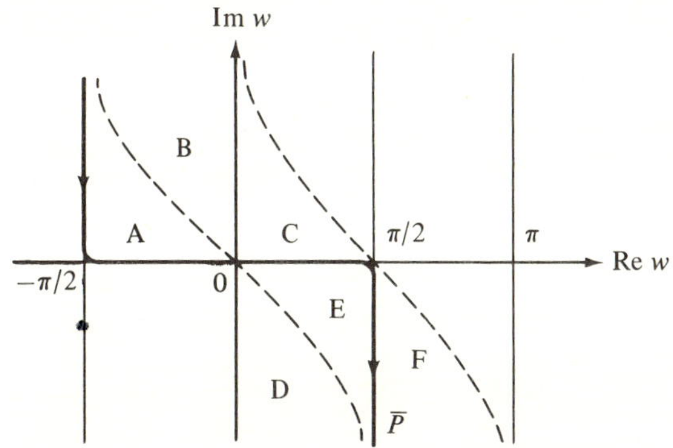

(a) Location of singularities [exp $(-i\omega t)$ dependence].

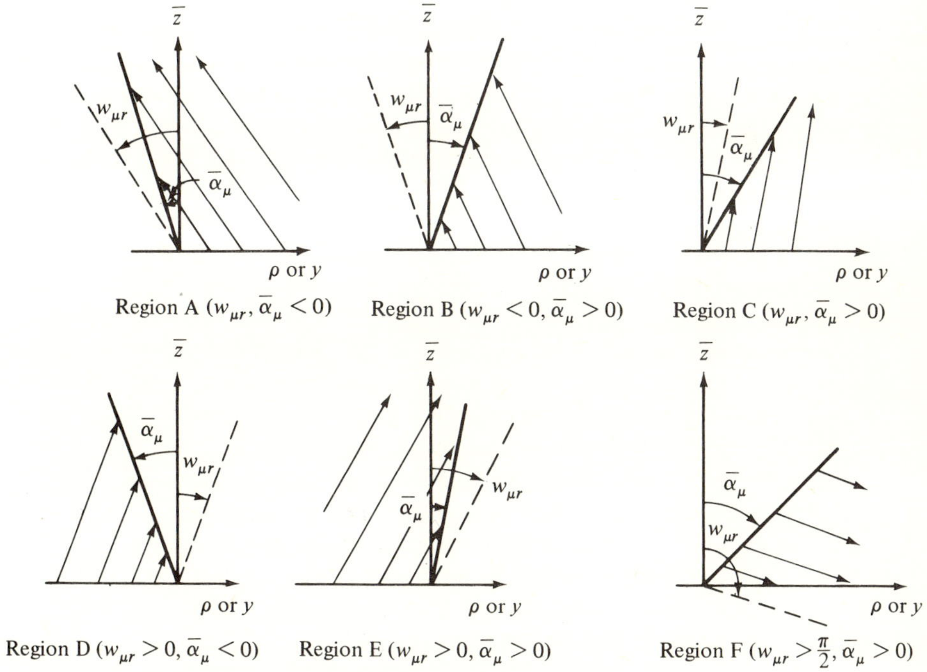

z̄
$w_{\mu r}$
$\bar{\alpha}_\mu$
ρ or y
Region A ($w_{\mu r}, \bar{\alpha}_\mu < 0$)
Region B ($w_{\mu r} < 0, \bar{\alpha}_\mu > 0$)
Region C ($w_{\mu r}, \bar{\alpha}_\mu > 0$)
Region D ($w_{\mu r} > 0, \bar{\alpha}_\mu < 0$)
Region E ($w_{\mu r} > 0, \bar{\alpha}_\mu > 0$)
Region F ($w_{\mu r} > \frac{\pi}{2}, \bar{\alpha}_\mu > 0$)

(b) Wave types

FIG. 5.3.8 Physical characteristics of waves arising from singularities of the modal reflection coefficient $\overrightarrow{\Gamma}_i(0)$ [Eq. (1)] in the complex w plane.

of w_μ in the semistrip $0 < w_r < \pi/2$, $w_i < 0$, the contribution appears in the complementary region $0 < \bar{\alpha} < \bar{\alpha}_\mu$ and $(\bar{\alpha}_\mu - w_{\mu r}) < 0$ with $w_{\mu i} < 0$. One observes from Eq. (17) that the corresponding field decays exponentially in its domain of existence along any radial direction leading from $L = 0$ to a distant observation point; inclusion and far-zone interpretation of the wave in Eq. (17) is therefore significant, although, taken by itself without reference to the mechanism of excitation, this field may be an improper solution of the Maxwell equations. In this connection one should note that improper, or non-modal (non-spectral), singularities lie in the region $\text{Im}\,\kappa_\mu = \text{Im}\,\sqrt{k^2 - k_{t\mu}^2} < 0$ corresponding, for the assumed $\exp(-i\omega t)$ dependence, to the improper sheet of the complex transverse wavenumber k_t plane [see, for example, region C in Fig. 5.3.8(a)].

These remarks are illustrated in Fig. 5.3.8, which exhibits the wave types arising from singularity contributions in various parts of the complex w plane.[5] The dashed lines in the first figure represent the steepest-descent paths through saddle points at $w = 0$ and $w = \pi/2$, respectively, and provide the partitioning of the w plane. The parallel lines (rays) in the remaining plots represent the direction of phase propagation (and energy flow) in the inhomogeneous plane wave whose domain of existence is bounded by the observation angle $\bar{\alpha}_r$; the rays are parallel to the line $\bar{\alpha} = w_{\mu r}$ and perpendicular to the equiphase planes (not shown); the wave amplitude is constant along a ray but changes elsewhere. Increased spacing between the rays indicates that the amplitude decreases in this direction. While the physical domain for radiation from a point source is confined to $\rho > 0$ or $0 \leq \bar{\alpha} \leq \pi/2$, some of the plots extend also into the region $\bar{\alpha} < 0$ relevant for line-source excitation [see Eq. (11)]. One observes that with the exception of waves arising from region C, all wave types have decreasing intensities as the observation point moves either along the $\bar{z}$ or ρ directions away from the source region $\rho = \bar{z} = 0$. It is also noted that the waves corresponding to regions A through E carry energy away from the interface (i.e., they "leak" into the exterior region $\bar{z} > 0$), while the wave from region F supplies energy to overcome losses in the medium occupying the half-space $\bar{z} < 0$ (see Secs. 5.5–5.7). Furthermore, the waves arising from regions A, B, D, and E are backward, since their energy-flow characteristics are toward the source plane $\rho = 0$; these features are of importance in connection with the imposition of a radiation condition. While these aspects are valid for lossy or lossless stratification in $\bar{z} < 0$, it is worth mentioning explicitly the non-dissipative case. In this instance, w_μ in region F moves onto the line $w_r = \pi/2$ (i.e., $w_{\mu r} = \pi/2$) and one obtains a forward surface wave, as mentioned earlier. If the structure exhibits a singularity that moves onto the line $w_r = \pi/2$ from region E, the corresponding surface wave is backward.

Wave types arising from regions A through E in the complex w plane of Fig. 5.3.8 carry power in the direction $w_{\mu r}$ toward the vicinity of the surface $\bar{\alpha} = \bar{\alpha}_\mu$, which bounds their domain of existence. It is natural to ask what happens to this energy. We recall at this point that the singularity contribution constitutes only a portion of the asymptotic field solution and that there is present in all observation regions a "space wave", arising form the saddle-point evaluation in Eq. (16a), which transports the radiated energy into the far zone $k_1 L \to \infty$. Because of their exponential decay with L, the waves arising from w_μ are not observable at very great distances from the source and it appears, therefore, that the energy carried by them is converted into the radially propagating space wave in the vicinity of the angle $\bar{\alpha} = \bar{\alpha}_\mu$. The radial decay along $\bar{\alpha}_\mu$ is smallest when $\bar{\alpha}_\mu \approx w_{\mu r}$ (i.e., when the complex singularity w_μ lies near the real w axis); one may then expect its presence to be strongly noticeable in the radiation field since appreciable coupling occurs between the two wave types when they propagate essentially along the direction $\bar{\alpha}_\mu$. For a pole singularity, it is simple to verify the validity of this statement by observing that, as a function of the observation angle $\bar{\alpha}$, the space-wave amplitude in Eq. (16a) may become quite large near $\bar{\alpha}_\mu$ when $f(w)$ has a pole at $w_\mu = w_{\mu r} + i w_{\mu_i}$, with $|w_{\mu_i}| \ll 1$. Under these conditions, the strongly excited complex pole dominates the field near the source and gives rise to a peak in the radiation pattern near $\bar{\alpha} = \bar{\alpha}_\mu$.[4] For a branch-point contribution, the additional algebraic decay factor $[k_1 L \sin(\bar{\alpha} - w_b)]^{-3/2}$ in Eq. (16b) implies that strong interaction near $\bar{\alpha}_\mu \approx w_{\mu r}$ takes place with the $O(1/k_1 L)^{-3/2}$ term in the saddle-point result of Eq. (16a).

5.3f   Asymptotic Evaluation of a Typical Radiation Integral for the Transmitted Fields

The calculation of fields transmitted across an interface is more involved than that of direct or reflected fields since the modal Green's function $g_{zi}(z, z')$ has a more complicated form. This is true even for the case of a single plane interface at $z = 0$ separating two semiinfinite dielectrics (Fig. 5.5.1), for which $\vec{\Gamma}_i(0)$ in Eq. (1) is given in Eqs. (5.5.12). The form of $g_{zi}(z, z')$ for $z > 0$ may be inferred from that for $z < 0$ by invoking the required continuity at $z = 0$ and recognizing that to satisfy the radiation condition at $z = +\infty$ for a time dependence $\exp(+j\omega t)$, the solution must have the dependence $\exp(-j\kappa_{i2}z)$. Thus, for $z' < 0$, $z > 0$,

$$g_{zi}(z, z') = g_{zi}(0, z')e^{-j\kappa_{i2}z} = \frac{1}{2j\kappa_{i1}}[1 \mp \vec{\Gamma}_i(0)]e^{-j\kappa_{i1}|z'| - j\kappa_{i2}z}, \qquad (18)$$

where κ_{i1} and κ_{i2} are the modal propragation constants in regions 1 and 2, respectively, and the $+$ and $-$ signs refer, respectively, to the E and H modes. Since the generic forms of the resulting radiation integrals in Eqs. (9) contain integrands wherein both $\kappa_{i1} = \sqrt{k_1^2 - k_t^2}$ and $\kappa_{i2} = \sqrt{k_2^2 - k_t^2}$ appear in the exponent, k_t being either ξ or η, it follows that the transformations $k_t = k_1 \sin w$ or $k_t = k_2 \sin w$ do not bring about simplifications analogous to those in Eq. (14). The difficulty can be circumvented by assuming that $k_2 z \gg k_1 |z'|$ (observation point located very far from the interface), in which instance the $\exp(-j\kappa_{i1}|z'|)$ term may be considered part of the slowly varying amplitude function f_1 or f_2 in Eqs. (9). The change of variable $k_t = k_2 \sin w$ then leads to a formulation directly analogous to the one in Eq. (14), thereby making the subsequent discussion applicable. In a similar manner, one may treat the case $k_2 z \ll k_1 |z'|$, with $\exp(-j\kappa_{i2}z)$ now playing the role of a slowly varying factor.

An asymptotic evaluation based on the preceding assumptions, although adequate for far-field calculations, does not yield information about the important case where $k_{1,2}$ are large but z and z' are unrestricted. In this regime of geometrical optics, neither of the exponential terms in Eq. (18) may be regarded as slowly varying, and both must be retained in the saddle-point calculation. There is no simple transformation from the ξ or η plane to the w plane furnishing an exponential as in Eq. (14), so an alternative approach is required. The procedure to be described is carried out directly in the plane of the transverse wavenumbers. As before, we consider the typical integral which, for the $\exp(j\omega t)$ dependence, is

$$I = \int_{-\infty}^{\infty} e^{-jq(\eta)} \bar{f}(\eta) \, d\eta, \qquad q(\eta) = \eta y + \kappa_{i1}|z'| + \kappa_{i2}z, \tag{19a}$$

or, alternatively, for the $\exp(-i\omega t)$ dependence,

$$I = \int_{-\infty}^{\infty} e^{iq(\eta)} \bar{f}(\eta) \, d\eta. \tag{19b}$$

The integration paths are specified in Figs. 5.3.5(a) and 5.3.6(a). The saddle points η_s in the integrand, located at $dq/d\eta = 0$, are determined by the implicit relation

$$y + \frac{d\kappa_{i1}}{d\eta}|z'| + \frac{d\kappa_{i2}}{d\eta} z = 0 \qquad \text{at} \quad \eta = \eta_s. \tag{19c}$$

Explicit specification of the *complete* steepest-descent path SDP through a saddle point at η_s, defined by the requirement $\mathrm{Re}\, q(\eta) = \mathrm{Re}\, q(\eta_s)$ (see Sec. 4.1b), is rather difficult. However, its progress *near* η_s is determined simply: Along a $\pm 45°$ line with respect to the real η axis if the saddle point is of the first order and real [see Fig. 4.2.1; the present discussion treats only non-evanescent waves arising from real saddle points with $\kappa_{i1,2}(\eta_s)$ real, i.e., $|\eta_s| < k_{1,2}$]. One may now consider the alternative path in Fig. 5.3.9, which proceeds along SDP near η_s; its remaining portions C, drawn parallel to the real η axis for simplicity, lie in

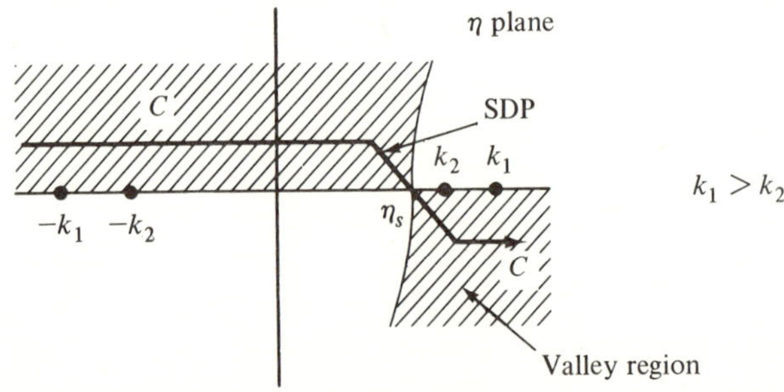

FIG. 5.3.9 Modified integration path when $d^2q/d\eta^2|_{\eta_s} < 0$ ($e^{-i\omega t}$ dependence).

"valley" regions of the complex η plane, wherein exp $[iq(\eta)]$ decays (drawing of the branch cuts has been omitted). Since $|\eta_s| < k_{1,2}$, no branch points are intercepted during the path deformation. Possible residue contributions due to intercepted pole singularities are treated as before.

Since Im $q(\eta) > 0$ in the shaded region of Fig. 5.3.9, the principal contribution to the integral arises from the path segment SDP in the vicinity of η_s. Although the SDP does not go to infinity, the error due to truncation at some finite point in a valley region is exponentially small and may be neglected [see Eq. (4.1.17)]. Equation (4.2.1) or (4.2.20a) then yields the saddle-point approximation of Eq. (19b),

$$I_{SDP} \sim \sqrt{\frac{2\pi}{|d^2q/d\eta^2|_{\eta_s}}}\, \bar{f}(\eta_s)\, \exp\left[iq(\eta_s) + i(\pi/4)\,\mathrm{sgn}\,(d^2q/d\eta^2)_{\eta_s}\right]. \qquad (20)$$

When the $\kappa_{i2}z$ term in Eq. (19c) is omitted and the substitution $\eta_s = k_1 \sin \bar{\alpha}$ is made, this result reduces to the one in Eq. (16a). Since the procedure leading to Eq. (20) does not involve the *complete* steepest-descent path, the exponentially small error bound on Eq. (20) is weaker than that for Eq. (16a) [see remarks following Eq. (4.2.18)]. However, when exponentially small contributions are neglected, Eq. (20) is sufficiently accurate.

A significant graphical interpretation of the saddle-point condition may be given. Upon writing

$$\tan \bar{\alpha}_1 = -\left.\frac{\partial \kappa_{i1}}{\partial \eta}\right|_{\eta_s}, \qquad \tan \bar{\alpha}_2 = -\left.\frac{\partial \kappa_{i2}}{\partial \eta}\right|_{\eta_s}, \qquad \kappa_{i1,2} = \sqrt{k_{1,2}^2 - \eta^2} \qquad (21)$$

one obtains, from Eq. (19c),

$$y - z \tan \bar{\alpha}_2 - |z'| \tan \bar{\alpha}_1 = 0 \qquad \text{at} \quad \eta_s. \qquad (22)$$

This equation for a straight line may be interpreted graphically: fields reach an observation point (y, z) in the region $z > 0$ via the ray trajectories 1 and 2 in Fig. 5.3.10(b) in such a manner that the real angles $\bar{\alpha}_1$ and $\bar{\alpha}_2$ satisfy the required plane-wave refraction law (Snell's law) at the interface. This follows from Eq. (21), whence $\bar{\alpha}_1$ and $\bar{\alpha}_2$ define, respectively, the inclination of the normals to the κ_{i1} versus η and κ_{i2} versus η wavenumber plots in Fig. 5.3.10(a); since $\eta_s = k_1 \sin \bar{\alpha}_1 = k_2 \sin \bar{\alpha}_2$, Snell's law is satisfied by these angles. The saddle-point condition, in conjunction with the wavenumber surface, therefore permits a ray-optical interpretation consistent with the constructs of geometrical optics. The saddle-point condition for the reflected field may be interpreted in a similar manner from the integral in Eq. (10b) and the resulting ray trajectory is shown dashed in Fig. 5.3.10.

It may easily be verified, in accord with expectations from ray optics, that the term exp $[iq(\eta_s)]$ in Eq. (20) expresses precisely the phase change for a wave propagating along the trajectories 1 and 2 in Fig. 5.3.10. As noted above, $\eta_s = k_1 \sin \bar{\alpha}_1 = k_2 \sin \bar{\alpha}_2$, and similarly, $\kappa_{i1}(\eta_s) = k_1 \cos \bar{\alpha}_1$, $\kappa_{i2}(\eta_s) = k_2 \cos \bar{\alpha}_2$. When these relations as well as Eqs. (19c) and (21) are employed, one may

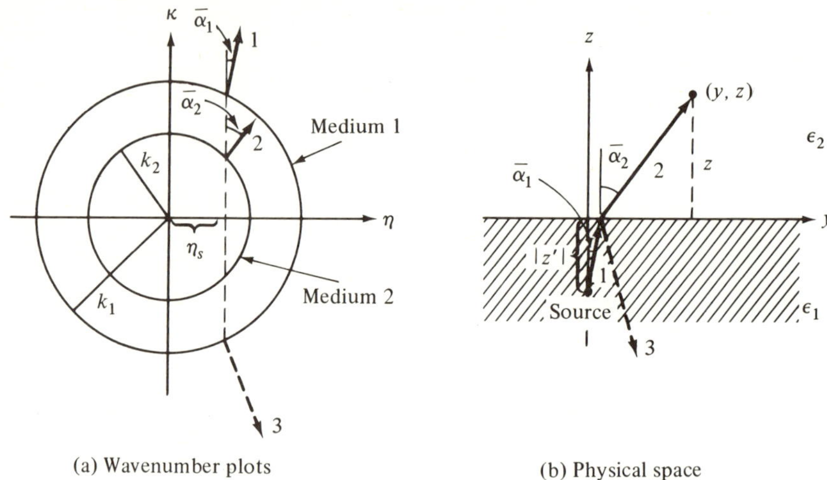

(a) Wavenumber plots　　　　　　　　(b) Physical space

FIG. 5.3.10 Interpretation of the saddle-point condition for $k_1 > k_2$.

write

$$q(\eta_s) = k_1 L_1 + k_2 L_2, \tag{23}$$

where L_1 and L_2 are the lengths along rays 1 and 2, respectively. The amplitude term in Eq. (20) also conforms with ray optics as shown in Eqs. (5.5.9).

5.4 SOURCES IN AN UNBOUNDED DIELECTRIC

The results of the general discussion in Secs. 5.2 and 5.3 are now applied to problems involving various source configurations in an unbounded homogeneous medium. As noted in Sec. 5.2, the electromagnetic fields can be derived from the scalar functions G', G'' or $\mathscr{S}'$, $\mathscr{S}''$ so primary emphasis is placed on the determination of these functions. Although some of the solutions in this simple environment may be obtained by direct integration of the differential equations, the method of modal representations is employed to illustrate implications of this more general procedure in elementary terms. In the format to be adopted, the statement of the problem and its solution are listed in the beginning, followed by a discussion of the physical interpretation of the result, and finally by a section summarizing relevant analytical steps. In this manner, the essential properties of the field may be surveyed succinctly, without the danger of hiding them in analytical details. To conform with general practice in the theory of radiation and diffraction, the final results are presented for the $\exp(-i\omega t)$ dependence, although some of the preliminary work involving modal analysis and network concepts is performed for the $\exp(j\omega t)$ dependence conventional in the latter context.

5.4a Dipoles Oriented Along z

Time-harmonic electric source current density

$$\hat{\mathbf{J}}(\mathbf{r},\,t) = Il\delta(\mathbf{r})e^{-i\omega t}\mathbf{z}_0. \tag{1}$$

The fields of a source oriented as shown in Fig. 5.4.1 can be derived from

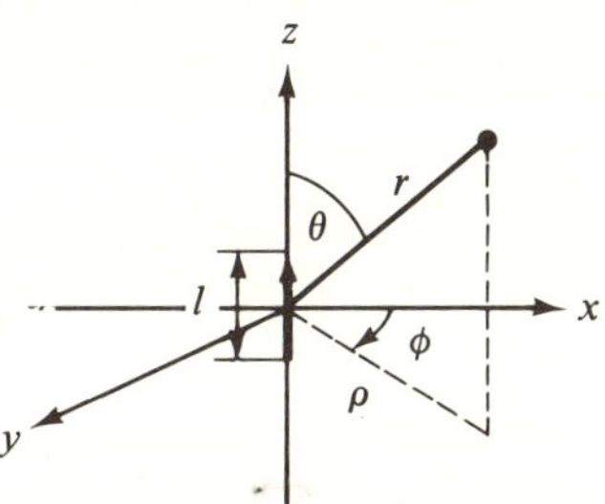

FIG. 5.4.1 Coordinate system.

the three-dimensional E-mode Green's function $G' = G_f$ defined by [see Eq. (5.2.3b)]

$$(\nabla^2 + k^2)G_f(\mathbf{r},\,\mathbf{r}') = -\delta(\mathbf{r} - \mathbf{r}'), \tag{2a}$$

subject to a radiation condition at infinity. The solution for $\mathbf{r}' = 0$, obtained by direct integration of Eq. (2a), is given by [see Eq. (1.1.31), on multiplication by $\exp(-i\omega t')$ and integration over t']

$$G_f(\mathbf{r},\,\mathbf{r}') = \frac{e^{ikr}}{4\pi r}, \tag{2b}$$

where $k = \omega\sqrt{\mu\epsilon}$ while ϵ and μ are the dielectric constant and permeability of the medium, respectively. The root-mean-square field components, obtained via Eqs. (5.2.1a), (5.2.1b), and (5.2.4c) (with $G' = G_f$) are, in spherical coordinates,

$$E_r = \sqrt{\frac{\mu}{\epsilon}}\,\frac{Ikl\cos\theta\,e^{ikr}}{2\pi kr^2}\left(1 + \frac{i}{kr}\right), \tag{3a}$$

$$E_\theta = -\sqrt{\frac{\mu}{\epsilon}}\,\frac{iIkl\sin\theta\,e^{ikr}}{4\pi r}\left(1 + \frac{i}{kr} - \frac{1}{(kr)^2}\right), \tag{3b}$$

$$H_\phi = -\frac{iIkl\sin\theta\,e^{ikr}}{4\pi r}\left(1 + \frac{i}{kr}\right), \tag{3c}$$

$$H_r = H_\theta = E_\phi = 0, \tag{3d}$$

where I is the current in the element and l is its length. For arbitrary source location at $\mathbf{r}' \neq 0$, one replaces r by $|\mathbf{r} - \mathbf{r}'|$ in Eq. (2b) and rederives the equivalent of Eqs. (3). The results remain valid in the dissipative case $\mathrm{Im}\,k > 0$. The average *radiated power density* $\bar{\mathbf{S}}$ is given by

$$\bar{S}_r = \text{Re}(E_\theta H_\phi^*) = \sqrt{\frac{\mu}{\epsilon}} \frac{|Ikl|^2 \sin^2 \theta}{(4\pi)^2 r^2}, \tag{4a}$$

$$\bar{S}_\theta = \bar{S}_\phi = 0, \tag{4b}$$

and the total radiated power P is [see Eq. (1.2.29)]

$$P = \int_0^\pi \bar{S}_r \cdot 2\pi r^2 \sin\theta \, d\theta = \sqrt{\frac{\mu}{\epsilon}} \frac{|Ikl|^2}{6\pi}. \tag{5}$$

The asymptotic field solution in the far zone $kr \gg 1$ contains to $O(1/kr)$ the components E_θ and H_ϕ only; their values are inferred from Eqs. (3b) and (3c) upon retaining only the first term inside the parentheses.

Discussion

Only E modes (with respect to z) are excited. The far field is transverse to the radius vector from the source to the observation point, and the transverse components account for the radiation of energy [see Eqs. (4)]. The mechanism of energy transport into the far zone may be schematized in ray-optical terms as in Fig. 5.4.2, where it is recalled (see Sec. 1.6) that the direction of a ray at

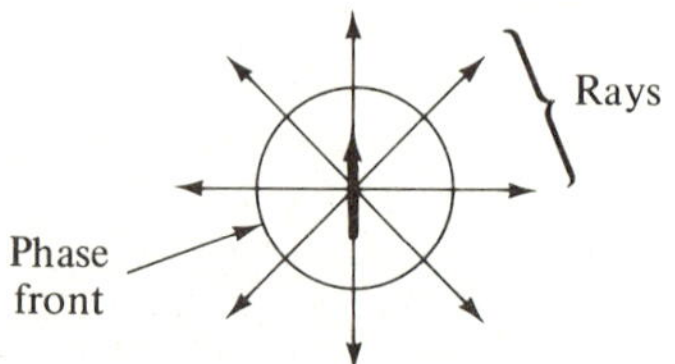

FIG. 5.4.2 Ray diagram.

a given point coincides with that of the power density vector $\bar{\mathbf{S}}$ (see also Fig. 5.3.7). The equiphase surfaces (phase fronts) perpendicular to the rays are spherical, and the field strength along a ray decays according to $1/r$.

Normalization for plane wave incidence

If the source point $\mathbf{r}'$ moves to infinity the field observed at finite $\mathbf{r}$ behaves like a plane wave since the curvature of the resulting spherical wavefront is negligible over any finite region. To make these remarks quantitative, we examine Eq. (2), with $r \to |\mathbf{r} - \mathbf{r}'|$, in the limit $\mathbf{r}' \to \infty$ and also introduce the source strength $J^o = Il$. Since

$$|\mathbf{r} - \mathbf{r}'| = [r^2 + r'^2 - 2rr' \cos\gamma]^{1/2}, \tag{6}$$

with γ representing the angle between the vectors $\mathbf{r}$ and $\mathbf{r}'$, and

$$\cos\gamma = \frac{\mathbf{r} \cdot \mathbf{r}'}{rr'} = \cos\theta \cos\theta' + \sin\theta \sin\theta' \cos(\phi - \phi'), \tag{6a}$$

one obtains, upon expansion by the binomial theorem,†

---

†From Eq. (6a), one obtains $|\mathbf{r} - \mathbf{r}'| \sim r' - r\cos\gamma + O(1/r')$ as $r' \to \infty$. For the determination of the amplitude, it is sufficient to approximate $|\mathbf{r} - \mathbf{r}'|$ by r' as r' becomes very large. However, for the determination of the phase we can neglect only those terms which tend to zero as $r' \to \infty$ [i.e., terms of $O(1/r')$].

$$J^o G_f(\mathbf{r}, \mathbf{r}') \sim A \exp\{-ikr[\cos\theta\cos\theta' + \sin\theta\sin\theta'\cos(\phi - \phi')]\},$$

$$r' \to \infty, \quad (6b)$$

where

$$A = \frac{J^o e^{ikr'}}{4\pi r'}. \quad (6c)$$

The right-hand side of Eq. (6b) represents a plane wave with amplitude A incident along a vector defined by the angular spherical coordinates (θ', ϕ'). To keep A finite as $r' \to \infty$, it is necessary to increase the source strength J^o proportionally with r'. If the incident plane wave is to have unit amplitude, one sets $A = 1$. From the above one can formulate a simple rule for passing from a Green's function corresponding to a *unit* strength point-source excitation at $\mathbf{r}'$ to excitation by a *unit* amplitude plane wave: One first lets $r' \to \infty$ along the desired direction (θ', ϕ') and then sets $(1/4\pi r') \exp(ikr') = 1$.

Modal procedure

Although Eq. (2a) may be integrated directly to yield the closed-form expression for the free-space Green's function in Eq. (2b), such a procedure is generally not successful for other geometrical configurations containing interfaces perpendicular to z. It is therefore instructive to study the solution of Eq. (2a) by the modal analysis and synthesis procedure employed in these more general cases. The relevant representation is the one in Eq. (5.2.11). Since the z domain is bilaterally unbounded, the modal Green's function $g'_{zi}(z, z')$ is given by the first term in Eq. (5.3.1) [note $\exp(+j\omega t)$ dependence],

$$g_{zi}(z, z') = \frac{e^{-j\sqrt{k^2-\xi^2}|z|}}{2j\sqrt{k^2 - \xi^2}}, \qquad z' = 0. \quad (7)$$

Thus,

$$G_f = -\frac{j}{8\pi}\int_{\infty\,\exp(-j\pi)}^{\infty} \xi H_0^{(2)}(\xi\rho)\frac{\exp(-j\sqrt{k^2 - \xi^2}|z|)}{\sqrt{k^2 - \xi^2}}\, d\xi, \quad (7a)$$

with the integration path given in Fig. 5.3.5a (the singularities shown in the figure at $\pm k_N$, $\pm a_p$ are omitted). In the w plane, with $\xi = k\sin w$,

$$G_f = -\frac{jk}{8\pi}\int_{\bar{P}} \sin w\, H_0^{(2)}(kr\sin\theta\sin w)e^{-jkr\cos\theta\cos w}\, dw, \quad (7b)$$

where the contour $\bar{P}$ is shown in Fig. 5.3.5b and (r, θ) are the spherical polar coordinates. Comparison of Eqs. (7b), (5.3.13a), and (5.3.14) leads to the identification

$$f(w) = -\frac{jk}{8\pi}\sqrt{\frac{2\sin w}{\pi kr\sin\theta}}\, e^{j\pi/4}, \quad (8)$$

so that, from Eq. (5.3.16a), for the $\exp(+j\omega t)$ dependence,

$$G_f \sim \frac{e^{-jkr}}{4\pi r}\left[1 + O\!\left(\frac{1}{kr}\right)\right], \qquad kr\sin\theta \gg 1. \quad (9)$$

The corresponding result for the $\exp(-i\omega t)$ dependence follows on letting $j \to -i$. It is of interest to note that the expression in Eq. (9) is independent of

$\sin\theta$, although in the derivation it had to be assumed that $\sin\theta \neq 0$. This lack of dependence on $\sin\theta$ indicates that the approximate result in Eq. (9) should remain valid even as $\sin\theta \to 0$. [By orienting the polar axis along a new direction $\hat{\theta} = 0$, one derives Eq. (9) subject to $kr\sin\hat{\theta} \gg 1$, whence the original direction $\theta = 0$ is now included in the permitted range.]

Upon comparing Eq. (9) (with $j \to -i$) and Eq. (2b), one notes that the approximate and exact solutions agree perfectly. It is implied thereby that in the asymptotic representation of the integral in Eq. (7b), all higher-order terms involving powers of $1/(kr)^n$, $n = 2, 3, \ldots$, vanish. This may be verified upon evaluation of the first few higher-order terms via the procedure described in Sec. 4.2b. It should be emphasized that in this more accurate evaluation, the asymptotic representation of the Hankel function must include higher-order terms.

Alternative representations

Since the integral in Eq. (5.2.8b), and the one in Eq. (7a) derived therefrom for the special case of a source at the coordinate origin in an unbounded medium, has the generic form given in Eq. (3.3.37), it is expected to be useful for the derivation of an alternative result by contour deformation. To illustrate these manipulations for the $\exp(-i\omega t)$ dependence, we rewrite Eq. (7a) as

$$G_f = \frac{i}{8\pi} \int_{\infty\,\exp(i\pi)}^{\infty} \xi H_0^{(1)}(\xi\rho) \frac{e^{i\sqrt{k^2-\xi^2}\,|z|}}{\sqrt{k^2-\xi^2}} \, d\xi, \tag{10}$$

with the path defined in Fig. 5.4.3 [see also Fig. 5.3.6(a)]. Since the branch cuts have been chosen so that $\operatorname{Im}\sqrt{k^2-\xi^2} > 0$ on the entire top sheet of the Riemann surface [see Fig. 5.3.3(a)], the exponential term in Eq. (10) decays

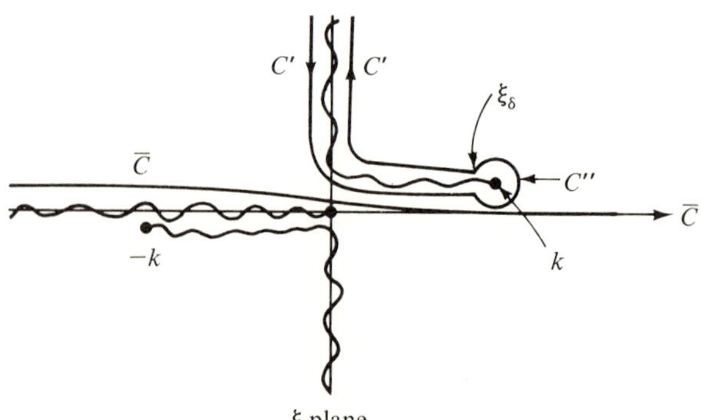

FIG. 5.4.3 Contours employed for alternative representations.

everywhere at $|\xi| \to \infty$. Since $H_0^{(1)}(\xi\rho)$ decays exponentially when $|\xi| \to \infty$, $\operatorname{Im}\xi > 0$ [see Eq. (5.3.13b)], the contour $\bar{C}$ can be deformed for any ρ and $|z|$ away from the origin into the contour $C' + C''$ enclosing the upper branch cut. C'' is a circular contour centered at $\xi = k$ with a small, but finite, radius $\delta = |\xi_\delta - k|$. Although the integrand in Eq. (10) grows like $1/\sqrt{\delta}$ on C'', the integral itself is $O(\sqrt{\delta})$, as can be verified upon expanding the regular portion of the integrand in a power series about $\xi = k$ and carrying out the integration over C''. Hence, the contribution from the integration over C'' vanishes as $\delta \to 0$.

Upon introducing the change of variable

$$\zeta = \sqrt{k^2 - \xi^2}, \qquad \text{i.e., } \xi \, d\xi = -\zeta \, d\zeta,$$

in the remaining integral over C' (with $\delta = 0$), one notes that ζ is real and varies from $+\infty$ to $-\infty$ as ξ moves along C' in the direction shown in Fig. 5.4.3 [see Fig. 5.3.3(a) for determination of the sign of Re ζ]. Thus, Eq. (10) transforms into

$$G_f = \frac{i}{8\pi} \int_{-\infty}^{+\infty} H_0^{(1)}[\sqrt{k^2 - \zeta^2}\rho]e^{i\zeta|z|}d\zeta, \tag{11}$$

where the restriction Im $\sqrt{k^2 - \zeta^2} > 0$ must be imposed to assure the decay of the Hankel function as $|\zeta| \to \infty$, and $\sqrt{k^2 - \zeta^2}$ is chosen positive for $-k < \zeta < k$. The path of integration relative to the branch points at $\zeta = \pm k$ proceeds in the same manner as shown in Fig. 5.4.3 for the ξ plane. Upon changing ζ into $(-\zeta)$ in Eq. (11), one obtains the same integral except that $\exp(i\zeta|z|)$ is replaced by $\exp(-i\zeta|z|)$. Thus, the algebraic sign of the exponent is of no consequence and we may replace $|z|$ by z.

The various representations for G_f in the parameter range $0 < \rho < \infty$, $-\infty < z < \infty$, obtained in Eqs. (2b), (10), (11), and (5.2.7) are summarized below for the case where the source point is situated at $\rho' = z' = 0$ and for an $\exp(-i\omega t)$ time dependence:

$$G_f(\mathbf{r}, 0) = \begin{cases} \dfrac{e^{ik\sqrt{\rho^2+z^2}}}{4\pi\sqrt{\rho^2+z^2}}, & (12a) \\[2em] \dfrac{i}{8\pi^2} \displaystyle\int_{-\infty}^{\infty}\int_{-\infty}^{\infty} \dfrac{e^{i(\xi x+\eta y)+i\sqrt{k^2-\xi^2-\eta^2}|z|}}{\sqrt{k^2-\xi^2-\eta^2}}\, d\xi\, d\eta, & (12b) \\[2em] \dfrac{i}{8\pi} \displaystyle\int_{\infty\exp(i\pi)}^{\infty} \xi H_0^{(1)}(\xi\rho)\dfrac{e^{i\sqrt{k^2-\xi^2}|z|}}{\sqrt{k^2-\xi^2}}\, d\xi, & (12c) \\[2em] \dfrac{i}{8\pi} \displaystyle\int_{-\infty}^{\infty} H_0^{(1)}[\sqrt{k^2-\zeta^2}\rho]e^{i\zeta z}\, d\zeta. & (12d) \end{cases}$$

Equation (12a) is the closed-form result, Eqs. (12b) and (12c) are z-transmission modal representations involving plane-wave and cylindrical-wave mode functions, respectively, in the cross-sectional domain, while Eq. (12d) is a radial transmission formulation involving a plane-wave modal representation in the z domain. Equation (12b) is the most cumbersome since it involves a double integral. The more convenient integral representations in Eqs. (12c) and (12d) have different convergence properties. In the z-transmission representation in Eq. (12c), the integrand decays rapidly for $|\xi| > k$ when $|z|$ is large, while in the radial transmission formulation in Eq. (12d), the integrand decays rapidly for $|\zeta| > k$ when ρ is large. Although these properties are inconsequential in the present instance where an exact solution is known in closed form, analogous considerations are important for more general problems that cannot be solved explicitly. Moreover, the radial transmission formulation is frequently more useful for the derivation of results appropriate to a source distributed along z. In this case, one first replaces z by $z - z'$ in Eqs. (12) to obtain the expression for a point source at z'. If the source distribution ranges between the limiting values z_1 and z_2 along the z direction, then $|z - z'|$ must be represented discontinuously for observation points z lying in the interval $z_1 < z < z_2$; that is $|z - z'| = \pm (z - z')$ for $z \gtrless z'$. The integration over z' required for the response to the distributed source is then performed more conveniently in Eq. (12d) than in Eq. (12c), since the former has a dependence on $z - z'$ which does not require a breakup of the integration interval into regions $z' < z$ and $z' > z$.

Pulsed electric source current density

$$\hat{\mathbf{J}}(\mathbf{r}, t) = \hat{p}\delta(\mathbf{r})\frac{d}{dt}\delta(t)\mathbf{z}_0. \tag{13}$$

The fields excited by this source distribution with impulsive *dipole moment* $\hat{\mathbf{p}}(\mathbf{r}, t) = \hat{p}\delta(\mathbf{r})\delta(t)\mathbf{z}_0$ can be derived via Eqs. (5.2.18) (with $\hat{G}' = \hat{G}_f$) from the space- and time-dependent Green's function defined by

$$\left(\nabla^2 - \frac{1}{\bar{c}^2}\frac{\partial^2}{\partial t^2}\right)\hat{G}_f(\mathbf{r}, \mathbf{r}'; t, t') = -\delta(\mathbf{r} - \mathbf{r}')\delta(t - t'), \qquad \bar{c} = \frac{1}{\sqrt{\mu\epsilon}}, \tag{14a}$$

subject to quiescence when $t < t'$. The solution for $\mathbf{r}' = 0$, $t' = 0$, is given by

$$G_f(\mathbf{r}, \mathbf{r}'; t, t') = \frac{\delta(t - r/\bar{c})}{4\pi r}, \tag{14b}$$

and follows at once from Eqs. (2) and (5.2.19) [with $A = (1/4\pi r)\delta(\tau - r/\bar{c})$] [see also Eqs. (1.1.15)]. The (scalar) disturbance is spherically symmetric about the source (see Fig. 5.4.2), propagates outward with velocity $\bar{c}$, and at r has a functional dependence on time identical with that at the source (Fig. 5.4.4).

If the source is moved to infinity along the x axis so that the incident field is in the form of a plane wave propagating along the x direction, the steady-state behavior for the exp $(j\omega t)$ variation is expressed in terms of the potential

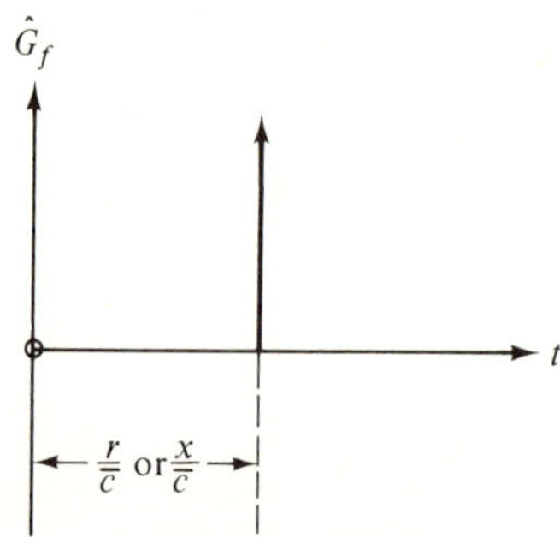

FIG. 5.4.4   Impulsive scalar point source or plane-wave response observed at a distance r and x, respectively.

function $G = \exp(-jk|x|)$. Since the k dependence in this expression is the same as in Eq. (2b), it is concluded that the field of an impulsive plane wave is characterized by the potential function (see Fig. 5.4. 4.),†

$$\hat{G}_f(x, t) = \delta\left(t - \frac{|x|}{\bar{c}}\right). \tag{15}$$

Magnetic dipole source

The fields for the time-harmonic magnetic current density

$$\hat{\mathbf{M}}(\mathbf{r}, t) = Vl\delta(\mathbf{r})e^{-i\omega t}\mathbf{z}_0, \tag{16}$$

where V is the voltage across the element and l is its length, are derived from the H-mode Green's function $G'' = G_f$ in Eqs. (2) since in an unbounded medium, the E- and H-mode Green's functions satisfy the same boundary conditions (radiation condition at infinity) and are therefore identical. The results are dual to those obtained for the electric dipole ($E \to H$, $H \to -E$, $I \to V$, $\mu \leftrightarrow \epsilon$). A similar statement applies to the pulsed excitation.

5.4b Dipoles Oriented Transverse to *z*

Time-harmonic electric source current density

$$\hat{\mathbf{J}}(\mathbf{r}, t) = Il\delta(\mathbf{r})e^{-i\omega t}\mathbf{y}_0. \tag{17}$$

Evaluation of the fields is carried out in terms of the steady-state scalar functions $\mathscr{S}' = \mathscr{S}'' = \mathscr{S}_f$ defined by [see Eq. (5.2.2)]

$$(\nabla^2 + k^2)\nabla_t^2\mathscr{S}_f(\mathbf{r}, \mathbf{r}') = \delta(\mathbf{r} - \mathbf{r}'), \qquad k^2 = \omega^2\mu\epsilon, \tag{18}$$

subject to a radiation condition at infinity. The following solutions obtain for $\mathbf{r}' = 0$ [see Eq. (2b)]:

$$\nabla_t^2\mathscr{S}_f(\mathbf{r}, \mathbf{r}') = -\frac{e^{ik\sqrt{\rho^2+z^2}}}{4\pi\sqrt{\rho^2 + z^2}}, \quad r = \sqrt{\rho^2 + z^2}, \quad \rho = \sqrt{x^2 + y^2}, \tag{19a}$$

and, by direct integration [see Eq. (21)],

$$\frac{\partial\mathscr{S}_f}{\partial\rho} = \frac{i}{4\pi k\rho}\left[e^{ik\sqrt{\rho^2+z^2}} - e^{ik|z|}\right]. \tag{19b}$$

The electromagnetic fields may be calculated from $\nabla_t'\mathscr{S}_f(\mathbf{r}, \mathbf{r}') = -\nabla_t\mathscr{S}_f(\mathbf{r}, \mathbf{r}') = -\boldsymbol{\rho}_0(\partial/\partial\rho)\mathscr{S}_f(\mathbf{r}, \mathbf{r}')$ via Eqs. (5.2.4a), (5.2.4b) and (5.2.1a), (5.2.1b). (Note: $\partial/\partial z' \to -\partial/\partial z$, and $\mathbf{x}_0\cdot\boldsymbol{\rho}_0 = x/\rho$, $\mathbf{y}_0\cdot\boldsymbol{\rho}_0 = y/\rho$). One finds for the x component of the magnetic field at the point $\mathbf{r} = (x, y, z)$,

$$H_x = H_x' + H_x'', \tag{20}$$

where H_x' is the E-mode contribution,

$$H_x' = Il\left[\frac{x^2 - y^2}{4\pi\rho^4}\left(e^{ikz} - \frac{z}{r}e^{ikr}\right) + \frac{y^2ze^{ikr}}{4\pi\rho^2r^2}\left(\frac{1}{r} - ik\right)\right], \tag{20a}$$

---

†Note from Eqs. (7) and (5.2.6b) that $G = \exp(-jk|x|)$ satisfies the equation $[d^2/dx^2 + k^2]G = -2jk\delta(x)$, whence in view of $k = \omega/\bar{c}$, $j\omega \to (\partial/\partial t)$, the source giving rise to $\hat{G}_f$ in Eq. (15) is $(2/\bar{c})\delta(x)\,d\delta(t)/dt$.

and H_x'' is the H-mode contribution,

$$H_x'' = Il\left[\frac{y^2 - x^2}{4\pi\rho^4}\left(e^{ikz} - \frac{z}{r}e^{ikr}\right) + \frac{x^2 z e^{ikr}}{4\pi\rho^2 r^2}\left(\frac{1}{r} - ik\right)\right], \qquad (20b)$$

or, upon combining,

$$H_x = \frac{z e^{ikr} Ikl}{4\pi r^2}\left(\frac{1}{kr} - i\right). \qquad (20c)$$

Analogous expressions are obtained for the other field components. The radiated power density is the same as for the z-directed current element in Eqs. (4) and (5) except that θ is now measured from the positive x axis. The physical interpretation of the result is the same as in Fig. 5.4.2.

Discussion

While the total fields in the present case are identical with those in Eqs. (3), provided that the y axis is chosen as the polar axis of the spherical coordinate system, the calculation shows that both E modes and H modes with respect to a preferred direction are excited by a transverse current element.[6] The results become significant when the medium is stratified along the z direction since the E- and H-mode constituents of the fields are then affected unequally (i.e., $\mathscr{S}' \neq \mathscr{S}''$). Moreover, fields radiated by a current element transverse to the optic axis in a uniaxially anisotropic region may be inferred by applying to Eqs. (20a) and (20b) *different* coordinate scaling transformations, thereby necessitating the availability of these explicit formulas (see Sec. 7.2c).

Equation (19b) is obtained from Eq. (19a) by recognizing that, in view of the rotational symmetry of $\mathscr{S}_f$ with respect to the z axis, $\nabla_t^2 = (1/\rho)(\partial/\partial\rho)\cdot(\rho\partial/\partial\rho)$. Thus, by integration, noting that $\partial\mathscr{S}_f/\partial\rho$ is bounded at $\rho = 0$,

$$\rho\frac{\partial\mathscr{S}_f}{\partial\rho} = -\int_0^\rho \rho\frac{e^{ik\sqrt{\rho^2+z^2}}}{4\pi\sqrt{\rho^2+z^2}}d\rho = \frac{i}{4\pi k}\int_0^\rho \frac{d}{d\rho}[e^{ik\sqrt{\rho^2+z^2}}]d\rho, \qquad (21)$$

thereby confirming the result in Eq. (19b). It may also be noted that with g_{zi} given by Eq. (7), the resulting integrals in Eqs. (5.2.12) may be evaluated in closed form via Eq. (19b). (Note: $j \to -i$.)

Time-harmonic magnetic source current density

$$\hat{\mathbf{M}}(\mathbf{r}, t) = Vl\delta(\mathbf{r})e^{-i\omega t}\mathbf{y}_0. \qquad (22)$$

The results in this case are obtainable from those for the electric current element by the duality replacements $E \to H$, $H \to -E$, $I \to V$, and $\mu \leftrightarrow \epsilon$.

Pulsed electric or magnetic source currents

These problems are treated in detail in Sec. 1.1b [see Eqs. (1.1.43)–(1.1.46)].

5.4c   Line Currents Oriented Transverse to z

Time-harmonic electric source current density

$$\hat{\mathbf{J}}(\mathbf{r}, t) = I\delta(\hat{\boldsymbol{\rho}} - \hat{\boldsymbol{\rho}}')e^{-i\omega t}\mathbf{x}_0. \qquad (23)$$

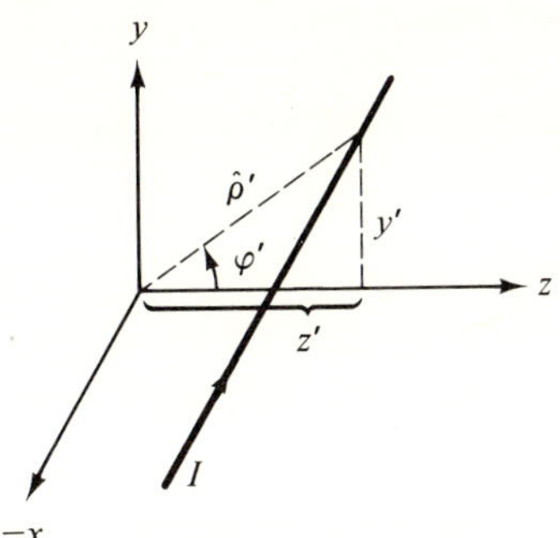

FIG. 5.4.5 Line source.

The electromagnetic fields of a source oriented as in Fig. 5.4.5 are derived from the two-dimensional Green's function $\bar{G}_f$, which satisfies the differential equation [two-dimensional analogue of Eq. (5.2.3b)]

$$(\nabla^2 + k^2)\bar{G}_f(\hat{\boldsymbol{\rho}}, \hat{\boldsymbol{\rho}}') = -\delta(\hat{\boldsymbol{\rho}} - \hat{\boldsymbol{\rho}}'),$$

$$\hat{\boldsymbol{\rho}} = (y, z) = (\hat{\rho}, \varphi), \qquad \frac{\partial}{\partial x} \equiv 0, \tag{24}$$

subject to a radiation condition at infinity. The solution is given by

$$\bar{G}_f(\hat{\boldsymbol{\rho}}, \hat{\boldsymbol{\rho}}') = \frac{i}{4} H_0^{(1)}(k|\hat{\boldsymbol{\rho}} - \hat{\boldsymbol{\rho}}'|), \tag{25}$$

where $k = \omega\sqrt{\mu\epsilon}$. For $\hat{\boldsymbol{\rho}}' = 0$, the fields are

$$E_x = i\omega\mu I\bar{G}_f = -\frac{\omega\mu}{4} IH_0^{(1)}(k\hat{\rho}), \qquad E_{\hat{\rho}} = E_{\varphi} = 0, \tag{26a}$$

$$H_{\varphi} = -\frac{i}{\omega\mu}\frac{\partial E_x}{\partial\hat{\rho}} = -\frac{ik}{4} IH_1^{(1)}(k\hat{\rho}), \qquad H_{\hat{\rho}} = H_x = 0, \tag{26b}$$

where I is the source current and $\hat{\rho}$ and φ represent polar coordinates in the yz plane. (Note the cyclic order: $\mathbf{x}_0 \times \boldsymbol{\varphi}_0 = \hat{\boldsymbol{\rho}}_0$.) The radiated power density $\bar{\mathbf{S}}$ per unit length along x is

$$\bar{S}_{\hat{\rho}} = \text{Re}\ (E_x H_{\varphi}^*) = \sqrt{\frac{\mu}{\epsilon}}\frac{k}{8\pi\hat{\rho}}|I|^2, \tag{27a}$$

$$\bar{S}_{\varphi} = \bar{S}_x \equiv 0, \tag{27b}$$

and the total radiated power P per unit length along x,

$$P = \sqrt{\frac{\mu}{\epsilon}}\frac{k}{4}|I|^2. \tag{28}$$

The asymptotic field solution at large distance from the source is obtained upon replacing the Hankel functions by their large-argument approximation in Eq. (5.3.13b):

$$E_x \sim -\frac{\omega\mu I}{2\sqrt{2\pi k\hat{\rho}}} e^{i(k\hat{\rho}-\pi/4)}\left[1 + O\left(\frac{1}{k\hat{\rho}}\right)\right], \tag{29a}$$

$$H_{\varphi} \sim -\frac{1}{2}\sqrt{\frac{k}{2\pi\hat{\rho}}} Ie^{i(k\hat{\rho}-\pi/4)}\left[1 + O\left(\frac{1}{k\hat{\rho}}\right)\right]. \tag{29b}$$

The inclusion of a phase progression in the transverse current distribution in Eq. (23) leads to a more complicated description [see Eqs. (5.2.1) and (5.2.14)] which is avoided by the alternative coordinate orientation in Eq. (45) (see also Sec. 5.4e).

Normalization for plane-wave incidence

As in Eqs. (6), we can readily formulate the rule for passing from the line-source excitation in Eq. (23) to a plane-wave excitation by letting the source point $\hat{\rho}' \to \infty$. Since

$$|\hat{\mathbf{\rho}} - \hat{\mathbf{\rho}}'| = [\hat{\rho}^2 + \hat{\rho}'^2 - 2\hat{\rho}\hat{\rho}' \cos(\varphi - \varphi')]^{1/2}, \tag{30a}$$

one obtains, upon use of Eq. (5.3.13b) in Eq. (25), by considerations analogous to those leading from Eq. (6a) to Eq. (6b),

$$\bar{G}_f(\hat{\mathbf{\rho}}, \hat{\mathbf{\rho}}') \sim A \exp[-ik\hat{\rho}\cos(\varphi - \varphi')],$$

$$A = \frac{1}{4}\sqrt{\frac{2}{\pi k \hat{\rho}'}} \exp\left[i\left(k\hat{\rho}' + \frac{\pi}{4}\right)\right], \qquad \hat{\rho}' \to \infty. \tag{30b}$$

The term $\exp[-ik\hat{\rho}\cos(\varphi - \varphi')]$ in Eq. (30b) represents a plane wave of unit amplitude incident along the direction φ'. Thus, to pass from a result derived for a *unit* strength line source located at $\hat{\mathbf{\rho}}'$ to the result for a *unit* amplitude plane wave incident along the direction φ', one first lets $\hat{\rho}' \to \infty$ and then sets $A = 1$.

Discussion

The configuration in Fig. 5.4.5 follows from that in Fig. 5.2.1 when the current elements are oriented along the line axis. The electric field generated by the line source has a single axial component, whereas the magnetic field is circular about the source axis. Energy flows out radially from the source and the energy transport mechanism may be schematized as in Fig. 5.4.2, with the dipole replaced by the line current. The far fields decay according to $\hat{\rho}^{-1/2}$, characteristic for the two-dimensional case.

While transverse currents generally excite both E and H modes with respect to z, the lack of a phase progression along the source implies that $\partial/\partial x = \partial/\partial x' = 0$ in Eqs. (5.2.1) or (5.2.4a), (5.2.4b), with $\mathscr{S}'$ and $\mathscr{S}''$ replaced by their two-dimensional analogues $\bar{\mathscr{S}}' = \bar{\mathscr{S}}'' = \bar{\mathscr{S}}_f$ in Eqs. (5.2.14) (with $\alpha = 0$). It then follows that the E-mode Hertz potential $\bar{\Pi}'(\mathbf{r}, \hat{\mathbf{\rho}}) = 0$, and the H-mode potential

$$\bar{\Pi}''(\mathbf{r}, \hat{\mathbf{\rho}}') = I\frac{\partial}{\partial y'}\bar{\mathscr{S}}_f(\mathbf{r}, \hat{\mathbf{\rho}}'). \tag{31}$$

Thus, from Eqs. (5.2.1),

$$\mathbf{E} = -\mathbf{x}_0 j\omega\mu I\frac{\partial^2}{\partial y\,\partial y'}\bar{\mathscr{S}}_f = \mathbf{x}_0 j\omega\mu I\frac{\partial^2\bar{\mathscr{S}}_f}{\partial y^2} = -\mathbf{x}_0 j\omega\mu I\bar{G}_f, \tag{31a}$$

thereby confirming Eq. (26a) (with $j \to -i$). Similarly, Eq. (26b) follows from

$$\mathbf{H} = -I\left(\mathbf{y}_0 \frac{\partial}{\partial z} - \mathbf{z}_0 \frac{\partial}{\partial y}\right)\frac{\partial^2 \overline{\mathscr{S}}_f}{\partial y^2} = -I\mathbf{x}_0 \times \hat{\mathbf{V}}_t \bar{G}_f. \tag{31b}$$

The solution in Eq. (25) may be obtained by direct integration of Eq. (24). With $\hat{\boldsymbol{\rho}}' = 0$, $\bar{G}_f$ is recognized to be symmetrical about $\hat{\rho} = 0$, so $\nabla^2 \rightarrow (\partial/\hat{\rho}\partial\hat{\rho})(\hat{\rho}\partial/\partial\hat{\rho})$. For $\hat{\rho} \neq 0$, Eq. (24) is solved by

$$\bar{G}_f(\hat{\rho}) = AH_0^{(1)}(k\hat{\rho}) + BH_0^{(2)}(k\hat{\rho}), \tag{32a}$$

where A and B are constants. To satisfy the radiation condition at infinity for an assumed time dependence $\exp(-i\omega t)$, we must retain only the function $H_0^{(1)}(k\hat{\rho})$ [see Eqs. (5.3.13)], so $B = 0$. For evaluation of A, Eq. (24) is integrated over a circular region S with radius $\hat{\rho}$, where $\hat{\rho} \rightarrow 0$. Since

$$H_0^{(1)}(k\hat{\rho}) \rightarrow i\frac{2}{\pi}\ln\hat{\rho}, \qquad \hat{\rho} \rightarrow 0, \tag{32b}$$

one notes that, as $\hat{\rho} \rightarrow 0$,

$$-1 = \int_S (\nabla^2 + k^2)\bar{G}_f dS = \oint_s \frac{\partial\bar{G}_f}{\partial\hat{\rho}}\, ds = i4A, \tag{32c}$$

where s is the circumference of the circular region S. Thus, $A = i/4$, and $\bar{G}_f(\hat{\rho}) = (i/4)H_0^{(1)}(k\hat{\rho})$. Equation (25) follows upon shifting the source point to an arbitrary location $\hat{\boldsymbol{\rho}}'$. Correspondingly, for the $\exp(+j\omega t)$ dependence, $\bar{G}_f(\hat{\rho}) = -(j/4)H_0^{(2)}(k\hat{\rho})$.

Modal procedure

Since the source distribution is comprised of transverse electric currents, the modal network problem may be schematized as in Fig. 5.4.6. The resulting modal Green's function is the same as in Eq. (7), so from Eq. (5.2.13a),

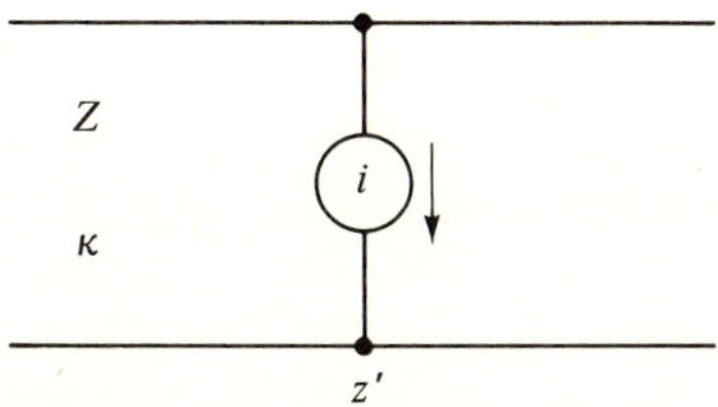

FIG. 5.4.6   Modal network problem.

$$\bar{G}_f(\hat{\boldsymbol{\rho}}, \hat{\boldsymbol{\rho}}') = \frac{-j}{4\pi}\int_{-\infty}^{\infty}\frac{\exp[-j\eta(y-y') - j\sqrt{k^2 - \eta^2}|z - z'|]}{\sqrt{k^2 - \eta^2}}\, d\eta. \tag{33}$$

The path of integration proceeds relative to the branch points at $\eta = \pm k$ as in Fig. 5.3.5 (a); the branch point at $\eta = 0$ in the figure is absent in this case. Upon transforming variables $\eta = k\sin w$, and employing the polar coordinates

$$y - y' = R\sin\varphi, \quad |z - z'| = R\cos\varphi, \qquad 0 < \varphi < \frac{\pi}{2}, \tag{34}$$

one obtains the contour-integral representation

$$\bar{G}_f(\hat{\boldsymbol{\rho}}, \hat{\boldsymbol{\rho}}') = \begin{cases} -\dfrac{j}{4}\, H_0^{(2)}(k|\hat{\boldsymbol{\rho}} - \hat{\boldsymbol{\rho}}'|) \equiv -\dfrac{j}{4}\, H_0^{(2)}(kR), & \text{(35a)} \\[2ex] -\dfrac{j}{4\pi} \int_{\bar{P}} e^{-jkR\,\cos(w-\varphi)}\, dw, & \text{(35b)} \\[2ex] -\dfrac{j}{4\pi} \int_{\bar{P}} e^{-jkR\,\cos w}\, dw. & \text{(35c)} \end{cases}$$

The path $\bar{P}$ in the complex w plane is shown in Fig. 5.3.5(b). If one introduces $w - \varphi$ as a new variable in Eq. (35b), one obtains the representation in Eq. (35c). Since $0 < \varphi < \pi/2$ and the integrand contains no singularities [i.e., the branch cut in Fig. 5.3.5(b) is absent], the path $\bar{P}$ is equivalent to any contour which begins in the w plane in the section $0 > \operatorname{Re} w > -\pi$, $\operatorname{Im} w < 0$, and ends in the section $0 < \operatorname{Re} w < \pi$, $\operatorname{Im} w > 0$. The right-hand side of Eq. (35c) is independent of φ, as it must be in view of the closed-form result in Eq. (35a). The integral representation (35c) and that in Eq. (36c) below for the Hankel function was first derived by Sommerfeld (for a detailed discussion see reference 7).

Upon starting from the $\exp(-i\omega t)$ formulation, one deduces, in a similar manner,

$$\bar{G}_f(\hat{\boldsymbol{\rho}}, \hat{\boldsymbol{\rho}}') = \begin{cases} -\dfrac{i}{4}\, H_0^{(1)}(kR), & \text{(36a)} \\[2ex] \dfrac{i}{4\pi} \int_{-\infty}^{\infty} \dfrac{\exp\left[i\eta(y - y') + i\sqrt{k^2 - \eta^2}\,|z - z'|\right]}{\sqrt{k^2 - \eta^2}}\, d\eta, & \text{(36b)} \\[2ex] \dfrac{i}{4\pi} \int_{\bar{P}} e^{ikR\,\cos w}\, dw, & \text{(36c)} \end{cases}$$

where the contours of integration in Eqs. (36b) and (36c) are the same as those in Figs.5.3.6(a) and 5.3.6(b), respectively. Via the integral expressions in Eqs. (35) and (36), the outgoing cylindrical waves as given by the Hankel functions are represented as a superposition of plane waves with complex angles of incidence.

From an asymptotic evaluation of the Sommerfeld integral representations in Eqs. (35c) and (36c), one readily derives by considerations analogous to those following Eq. (5.3.14) the first-order asymptotic formulas in Eqs. (5.3.13). For a complete asymptotic expansion of $H_0^{(1)}(kR)$ for large values of kR, one employs Eqs. (4.2.18) instead of the first-order asymptotic approximation in Eq. (4.2.1a). Upon introducing into Eq. (36c) the change of variable $\cos w = 1 + is^2$, $-\infty < s < \infty$, one obtains, via Eq. (4.2.18a),

$$H_0^{(1)}(kR) = \frac{1}{\pi} \int_{\bar{P}} e^{ikR\,\cos w}\, dw = \frac{1}{\pi} e^{ikR} \int_{-\infty}^{\infty} G(s) e^{-kRs^2}\, ds$$

$$\sim \frac{e^{ikR}}{\pi} G_e\left[\sqrt{-\frac{d}{d(kR)}}\right]\sqrt{\frac{\pi}{kR}}. \tag{37a}$$

From Eqs. (4.2.27a) and (4.2.30c), $G(s) = dw/ds = \sqrt{2}\, e^{-i\pi/4}[1 - (is^2/2)]^{-1/2}$, whence, from Eq. (4.2.18c),

$$G_e\left[\sqrt{-\frac{d}{d(kR)}}\right] = \sqrt{2}\, e^{-i\pi/4}\left[1 + \frac{i}{4}\frac{d}{d(kR)} - \frac{3}{32}\frac{d^2}{d(kR)^2} + \cdots\right]. \quad (37b)$$

Time-harmonic electric source current density

$$\hat{\mathbf{J}}(\mathbf{r}, t) = I\delta(\hat{\boldsymbol{\rho}} - \hat{\boldsymbol{\rho}}')e^{-i\omega t}\mathbf{z}_0. \quad (38)$$

The electromagnetic fields for a $\mathbf{z}_0$-directed line-source configuration at $\hat{\boldsymbol{\rho}}'$ are derived from the Green's function $\bar{G}_f(\hat{\boldsymbol{\rho}}, \hat{\boldsymbol{\rho}}')$ in Eq. (25). For $\hat{\boldsymbol{\rho}}' = 0$, the fields are as follows:

$$H_x = I\frac{\partial}{\partial y}\bar{G}_f(\hat{\boldsymbol{\rho}}, \hat{\boldsymbol{\rho}}') = -\frac{iIk}{4}\sin\varphi\, H_1^{(1)}(k\hat{\rho}), \qquad H_y = H_z = 0, \quad (39a)$$

$$E_{\hat{\rho}} = -\frac{i}{\omega\epsilon\hat{\rho}}\frac{\partial H_x}{\partial\varphi}, \qquad E_\varphi = \frac{i}{\omega\epsilon}\frac{\partial H_x}{\partial\hat{\rho}}, \qquad E_x = 0. \quad (39b)$$

$\hat{\rho} = \sqrt{y^2 + z^2}$ is the distance from the source to the observation point, and φ is the angle between $\hat{\rho}$ and the positive z axis. Asymptotic solutions for the field are obtained by substitution of the formula in Eq. (5.3.13b).

Discussion

The configuration in Fig. 5.4.7 follows from that in Fig. 5.2.1 when the current elements are oriented along z. The fields are not symmetrical about the line-source axis but exhibit a pattern similar to that for the point dipole in Eqs. (3). Energy into the far field is transported along straight-line trajectories emanating radially from the source (see Fig. 5.4.2). Since the electric currents in Eq. (38) are longitudinal, only E modes with respect to z are excited and the Hertz potential function $\bar{\Pi}'(\mathbf{r}, \hat{\boldsymbol{\rho}}')$ is proportional to $\bar{G}_f(\hat{\boldsymbol{\rho}}, \hat{\boldsymbol{\rho}}')$ [see Eq. (5.2.4c)]. Equations (39) then follow directly from Eqs. (5.2.1a) and (5.2.1b).

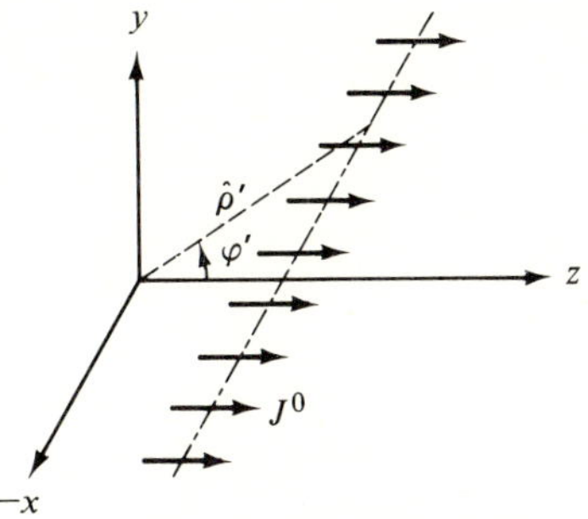

FIG. 5.4.7　Dipole line source.

In this instance a modal-analysis procedure involves the network problem schematized in Fig. 5.4.8 and leads to an integral representation as in Eq. (33).

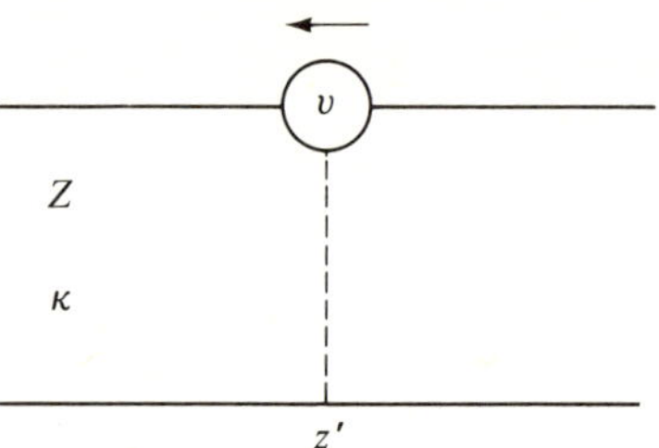

FIG. 5.4.8　Modal network problem.

Time-harmonic magnetic current density

Problems involving the magnetic current densities

$$\hat{\mathbf{M}}(\mathbf{r}, t) = V\delta(\hat{\boldsymbol{\rho}} - \hat{\boldsymbol{\rho}}')e^{-i\omega t}\mathbf{x}_0, \tag{40a}$$

or

$$\hat{\mathbf{M}}(\mathbf{r}, t) = V\delta(\hat{\boldsymbol{\rho}} - \hat{\boldsymbol{\rho}}')e^{-i\omega t}\mathbf{z}_0, \tag{40b}$$

are dual to the corresponding cases in Eqs. (23) and (38), respectively. Results are obtained via the duality replacements $E \to H$, $H \to -E$, and $I \to V$. Also, the modal network problems in Figs. 5.4.6 and 5.4.8 are interchanged.

Pulsed source currents

Since the time-harmonic fields excited by the current distributions in Eqs. (23), (38), (40a) and (40b) are derivable from the Green's function $\bar{G}_f(\hat{\boldsymbol{\rho}}, \hat{\boldsymbol{\rho}}')$ in Eq. (25), the transient fields may be obtained from a knowledge of the corresponding time-dependent Green's function $\hat{\bar{G}}_f(\hat{\boldsymbol{\rho}}, \hat{\boldsymbol{\rho}}'; t, t')$, which satisfies the differential equation

$$\left(\frac{\partial^2}{\partial y^2} + \frac{\partial^2}{\partial z^2} - \frac{1}{\bar{c}^2}\frac{\partial^2}{\partial t^2}\right)\hat{\bar{G}}_f(\hat{\boldsymbol{\rho}}, \hat{\boldsymbol{\rho}}'; t, t') = -\delta(\hat{\boldsymbol{\rho}} - \hat{\boldsymbol{\rho}}')\delta(t - t'),$$

$$\bar{c} = \frac{1}{\sqrt{\mu\epsilon}} \tag{41}$$

subject to quiescence when $t < t'$. The solution for $t' = 0$, $\hat{\boldsymbol{\rho}}' = 0$, is

$$\hat{\bar{G}}_f = \begin{cases} \dfrac{1}{2\pi}\dfrac{1}{\sqrt{t^2 - (\hat{\rho}/\bar{c})^2}}, & t > \dfrac{\hat{\rho}}{\bar{c}}, \tag{42a} \\[4mm] 0, & t < \dfrac{\hat{\rho}}{\bar{c}}, \tag{42b} \end{cases}$$

where $\hat{\rho} = \sqrt{y^2 + z^2}$. The solution for arbitrary $\boldsymbol{\rho}'$, t' is obtained by the replacements $\hat{\rho} \to |\hat{\boldsymbol{\rho}} - \hat{\boldsymbol{\rho}}'|$, $t \to t - t'$. For example, when the source configuration in Fig. 5.4.7 has the temporal dependence

$$\hat{\mathbf{J}}(\mathbf{r}, t) = \delta(\hat{\boldsymbol{\rho}})I(t)\mathbf{z}_0, \qquad I(t) = \frac{d\hat{p}(t)}{dt} = \hat{p}\frac{d}{dt}\delta(t), \tag{43}$$

where $\hat{p}$ is the dipole moment strength, it follows from Eqs. (39) (see also Sec. 5.2c) that the transient fields are given by

$$\hat{H}_x = \hat{p}\,\frac{\partial^2}{\partial t\,\partial y}\,\hat{\bar{G}}_f, \quad \hat{E}_\rho = -\hat{p}\,\frac{1}{\epsilon\hat{\rho}}\,\frac{\partial^2}{\partial\varphi\,\partial y}\,\hat{\bar{G}}_f, \quad \hat{E}_\varphi = \hat{p}\,\frac{1}{\epsilon}\,\frac{\partial^2}{\partial\hat{\rho}\,\partial y}\,\hat{\bar{G}}_f. \tag{44}$$

[Note that $(\partial^2/\partial\varphi\,\partial y) \neq (\partial^2/\partial y\,\partial\varphi)$ since $y = y(\hat{\rho}, \varphi)$.] The fields due to the other source arrangements may be obtained in a similar manner.

$\hat{\bar{G}}_f$ in Eqs. (42) represents a cylindrically symmetrical disturbance spreading outward from the source with speed $\bar{c}$ and reaching an observation point $\hat{\rho}$ at a time $t = \hat{\rho}/\bar{c}$. Although the action of the source is confined to the instant $t = 0$, a response of decreasing intensity persists at $\hat{\rho}$ after the passing of the initial wave front (Fig.5.4.9).

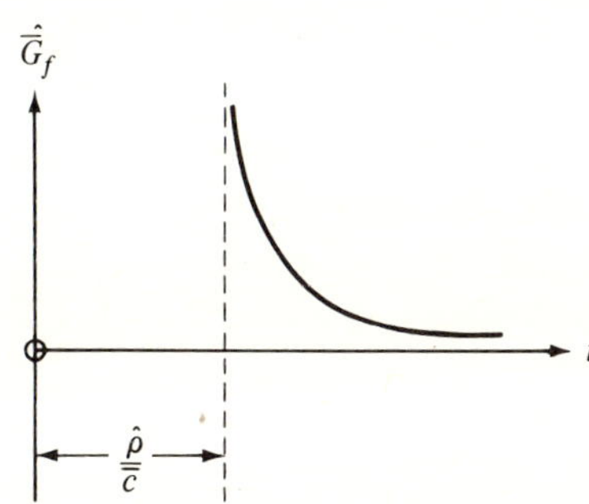

FIG. 5.4.9   Pulsed scalar line-source response observed at a distance $\hat{\rho}$.

The solution in Eq. (42) may be deduced directly by the method described in Sec. 5.2c. From the integral representation for the time-harmonic Green's function in Eq. (35c) with $\hat{\boldsymbol{\rho}}' = 0$, one identifies $\bar{\alpha} = 0$, $L = \hat{\rho}$, $f(w) = -j/4\pi$ in Eq. (5.2.20). The transient result then follows from Eqs. (5.2.19) and (5.2.23) (see also See. 1.6b).

5.4d   Line Currents Oriented Along z

Time-harmonic electric current density

$$\hat{\mathbf{J}}(\mathbf{r}, t) = I\delta(\boldsymbol{\rho} - \boldsymbol{\rho}')e^{i\alpha z}e^{-i\omega t}\mathbf{z}_0. \tag{45}$$

Since the current distribution is oriented along z, the fields may be derived from the scalar Green's function

$$\bar{G}'(\mathbf{r}, \boldsymbol{\rho}') = e^{i\alpha z}\bar{G}_f(\boldsymbol{\rho}, \boldsymbol{\rho}'), \qquad \boldsymbol{\rho} = (x, y) = (\rho, \phi), \tag{46a}$$

where $\bar{G}_f$ satisfies the differential equation

$$(\nabla_t^2 + \hat{\kappa}^2)\bar{G}_f(\boldsymbol{\rho}, \boldsymbol{\rho}') = -\delta(\boldsymbol{\rho} - \boldsymbol{\rho}'), \quad \hat{\kappa}^2 = k^2 - \alpha^2,$$

$$\nabla_t^2 = \frac{1}{\rho}\frac{\partial}{\partial\rho}\,\rho\,\frac{\partial}{\partial\rho} + \frac{1}{\rho^2}\frac{\partial^2}{\partial\phi^2}, \tag{46b}$$

subject to a radiation condition at infinity. The solution for $\boldsymbol{\rho}' = 0$ is given via Eq. (25) as

$$\bar{G}_f(\boldsymbol{\rho}, \boldsymbol{\rho}') = \frac{i}{4} H_0^{(1)}(\hat{\kappa}\rho), \qquad \text{Im } \hat{\kappa} \geq 0, \tag{47}$$

and the fields are then obtained from Eqs. (5.2.1a), (5.2.1b) and (5.2.4c):

$$E_\rho = \frac{I}{-i\omega\epsilon} \frac{\partial^2 \bar{G}_f}{\partial\rho\,\partial z} = \frac{I\alpha\hat{\kappa}}{-i4\omega\epsilon} e^{i\alpha z} H_1^{(1)}(\hat{\kappa}\rho), \tag{48a}$$

$$E_z = \frac{I}{i\omega\epsilon}\left[\nabla_t^2 \bar{G}_f + e^{i\alpha z}\delta(\boldsymbol{\rho} - \boldsymbol{\rho}')\right] = \frac{-\hat{\kappa}^2 I}{4\omega\epsilon} e^{i\alpha z} H_0^{(1)}(\hat{\kappa}\rho), \tag{48b}$$

$$H_\phi = \frac{i\hat{\kappa}}{4} I e^{i\alpha z} H_1^{(1)}(\hat{\kappa}\rho), \tag{48c}$$

$$E_\phi = H_\rho = H_z = 0. \tag{48d}$$

When $\alpha = 0$ (i.e., $\hat{\kappa} = k$) one has $E_\rho = 0$, and the expressions for E_z and H_ϕ agree with those for E_x and H_φ, respectively, in Eqs. (26), allowance being made for different coordinate designations.

For large values of $|\hat{\kappa}\rho|$, one can employ the asymptotic approximation (5.3.13) for the Hankel functions and obtain for the fields observed at large distances from the source:

$$E_\rho \sim \frac{I\alpha\hat{\kappa}e^{i\pi/4}}{2\omega\epsilon\sqrt{2\pi\hat{\kappa}\rho}} e^{i(\hat{\kappa}\rho + \alpha z)} \sim -\frac{\alpha}{\hat{\kappa}} E_z \sim \frac{\alpha}{\omega\epsilon} H_\phi. \tag{49}$$

Upon defining for $\alpha < k$ an angle ψ and unit vector $\mathbf{R}_0$ (see Fig. 5.4.10) as

$$\alpha = k\cos\psi, \quad \hat{\kappa} = k\sin\psi, \quad \mathbf{R}_0 = \boldsymbol{\rho}_0\sin\psi + \mathbf{z}_0\cos\psi, \tag{50}$$

one notes that

$$\mathbf{E} \sim (\mathbf{R}_0 \times \boldsymbol{\phi}_0)\frac{k}{\hat{\kappa}} E_z; \quad \mathbf{H} \sim -\boldsymbol{\phi}_0\frac{\omega\epsilon}{\hat{\kappa}} E_z \sim \sqrt{\frac{\epsilon}{\mu}}\,\mathbf{R}_0 \times \mathbf{E}. \tag{51}$$

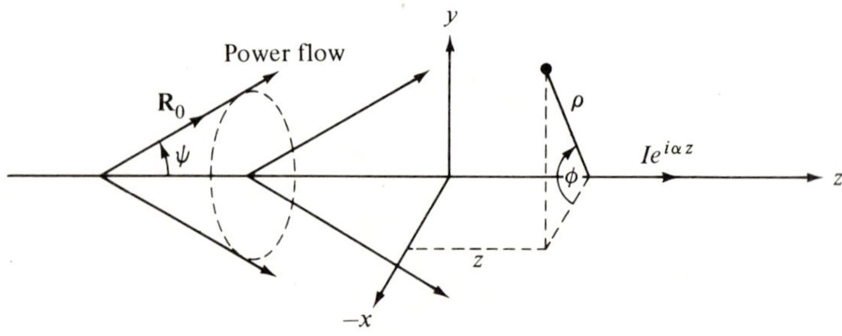

FIG. 5.4.10   Progressively phased line current.

Discussion

The line-source Green's function in Eq. (46a) is related to the point-source Green's function $G_f(\mathbf{r}, \mathbf{r}')$ in Eq. (2) by a continuous superposition of source elements along the z axis:

$$\bar{G}'(\mathbf{r}, \boldsymbol{\rho}') = \int_{-\infty}^{\infty} e^{i\alpha z'} G_f(\mathbf{r}, \mathbf{r}')\, dz' = \int_{-\infty}^{\infty} e^{i\alpha z'}\, \frac{e^{ik\sqrt{\rho^2 + (z-z')^2}}}{4\pi\sqrt{\rho^2 + (z - z')^2}}\, dz', \quad (52)$$

and this integral is expressible in closed form via Eq. (47) [see Eqs. (12a) and (12d), which are Fourier transforms of this result]. The transition from Eq. (5.2.3b) to Eqs. (46) is evident.

The far fields behave locally like plane waves propagating in the $\mathbf{R}_0$ direction at an angle ψ with the line-source axis. This observation can be highlighted by writing Eq. (49) as

$$E_z \sim A\, \frac{e^{i\mathbf{k}\cdot\mathbf{R}}}{\sqrt{kR}}, \qquad A = -\frac{I\omega\mu \sin \psi e^{-i\pi/4}}{2\sqrt{2\pi}}, \qquad (53a)$$

where the propagation vector $\mathbf{k}$ and the distance vector $\mathbf{R}$ are defined as

$$\mathbf{k} = k\mathbf{R}_0, \qquad \mathbf{R} = \boldsymbol{\rho}_0\rho + \mathbf{z}_0 z. \qquad (53b)$$

The factor A is a constant for a given line-source distribution, and the factor $1/\sqrt{kR}$ accounts for the cylindrical spreading of the wave as it progresses outward from the source. The wave observed at (ρ, z) appears to originate from a point on the line source that is connected to the observation point by a vector of length R making an angle ψ with the source axis (see Fig. 5.4.10). The angle ψ decreases to zero as $\alpha \to k$ (i.e., as the phase velocity along the line source approaches that of a wave in free space).

When $\alpha > k$, $\hat{\kappa}$ in Eq. (46b) is imaginary and one must choose $\hat{\kappa} = i\sqrt{\alpha^2 - k^2}$. In this instance, the fields at large distances from the source decay exponentially with ρ. It is to be noted that when $\alpha < k$ the phase velocity along the z direction is given by $v_p = \omega/\alpha > c$, where $c = \omega/k$ is the velocity of light in the medium, while $\alpha > k$ yields $v_p < c$. These ranges of α characterize "fast" and "slow" waves, respectively, as compared with the velocity of light. From the above results it is recognized that fast waves radiate energy into the space surrounding the current distribution, while the energy in the slow waves is confined to the immediate vicinity of the source currents and travels along the direction of current flow [see Eqs. (48), with $\hat{\kappa}$ imaginary, whence E_ρ and H_ϕ are in time phase, while E_z and H_ϕ are 90° out of phase]. These considerations apply not only to the present problem, wherein the source distribution is specified, but also to diffraction problems, wherein one encounters induced currents with fast or slow phase variations.

The above-described radiation characteristics of the phased line source may also be inferred directly from the wavenumber diagram in Fig. 5.3.7. In the present instance, the longitudinal wavenumber κ has the prescribed value α, so only the intersection of the plane $\kappa = \alpha$ with the surface is relevant. When $\alpha < k$, the $\kappa = \alpha$ plane intersects the wavenumber surface [Fig. 5.4.11(a)], and the plane-wave solutions corresponding to this intersection (i.e., the normals to the sphere) give rise to the conical ray diagram in Fig. 5.4.10. When $\alpha > k$, no intersection occurs [Fig. 5.4.11(b)], thereby eliminating propagating wave solutions with real values of the transverse wavenumber $\hat{\kappa}$.

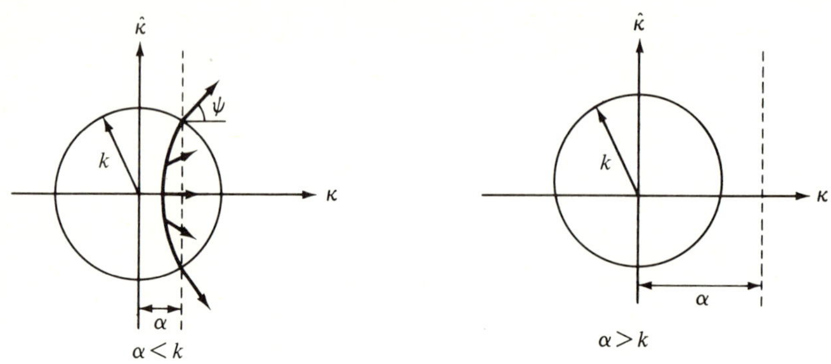

FIG. 5.4.11   Use of wavenumber plot.

5.4e   Point Charge in Uniform Straight Motion

$$\hat{\mathbf{J}}(\mathbf{r}, t) = qv\delta(x - vt)\delta(\hat{\boldsymbol{\rho}} - \hat{\boldsymbol{\rho}}')\mathbf{x}_0. \tag{54}$$

The fields produced by a point charge q moving, as shown Fig. 5.4.12, with constant speed v along a straight-line path parallel to the x axis in a medium

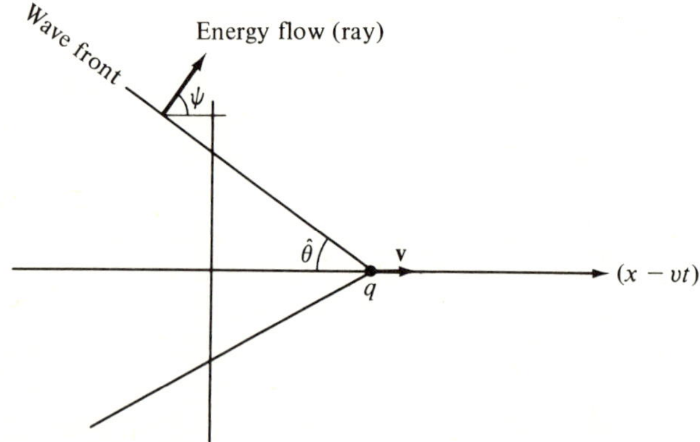

FIG. 5.4.12   Point charge moving in an infinite medium with $\epsilon > c^2/v^2$.

with frequency independent dielectric constant $\epsilon = \bar{\epsilon}\epsilon_0$ and vacuum permeability $\mu = \mu_0$ are derivable from a function $\hat{G}(\mathbf{r}, t)$ which satisfies the differential equation

$$\left(\nabla^2 - \frac{\bar{\epsilon}}{c^2}\frac{\partial^2}{\partial t^2}\right)\hat{G}(\mathbf{r}, t) = -\delta(\hat{\boldsymbol{\rho}} - \hat{\boldsymbol{\rho}}')\delta\left(t - \frac{x}{v}\right), \tag{55}$$

$$\hat{\boldsymbol{\rho}} = (y, z) = (\hat{\rho}, \varphi)$$

subject to initial conditions discussed subsequently. The solution for $\hat{\boldsymbol{\rho}}' = 0$ is given by

$$\hat{G}(\mathbf{r}, t) = \begin{cases} 0, & t - \dfrac{x}{v} < \dfrac{\hat{\rho}\gamma}{c}, & \text{(55a)} \\[3mm] \dfrac{1}{2\pi\sqrt{(t - x/v)^2 - (\hat{\rho}\gamma/c)^2}}, & t - \dfrac{x}{v} > \dfrac{\hat{\rho}\gamma}{c}, & \text{(55b)} \end{cases}$$

when $\gamma = \sqrt{\bar{\epsilon} - (1/\beta)^2}$ is positive. $\bar{\epsilon} = \epsilon/\epsilon_0$ is the dielectric constant normalized to its value ϵ_0 in vacuum and $\beta = v/c$ the ratio of the particle speed to the speed of light in vacuum (i.e., $\beta < 1$). It is thereby implied that Eqs. (55) are valid only when the particle speed is high enough to assure $\bar{\epsilon} > (c/v)^2$. At low speeds, where $\bar{\epsilon} < (c/v)^2$, one has, instead,

$$\hat{G}(\mathbf{r}, t) = \frac{1}{4\pi\sqrt{(t - x/v)^2 + (\hat{\rho}|\gamma|/c)^2}}, \qquad |\gamma| = \sqrt{\frac{1}{\beta^2} - \bar{\epsilon}}. \qquad \text{(56)}$$

For $\hat{\rho} \neq 0$, one replaces $\hat{\rho}$ by $|\hat{\rho} - \hat{\rho}'|$.

To determine the electromagnetic fields, one performs the differentiations

$$\hat{E}_{\hat{\rho}} = \frac{-q}{v\epsilon_0\bar{\epsilon}} \frac{\partial}{\partial\hat{\rho}} \hat{G}, \qquad \text{(57a)}$$

$$\hat{E}_x = \frac{-q\gamma^2}{c^2\epsilon_0\bar{\epsilon}} \frac{\partial}{\partial t} \hat{G}, \qquad \text{(57b)}$$

$$\hat{H}_\varphi = -q\frac{\partial}{\partial\hat{\rho}} \hat{G}, \qquad \text{(57c)}$$

where $\hat{\rho}$ and φ are cylindrical coordinates in the $y - z$ plane. In particular, for the low-speed case $\bar{\epsilon} < 1/\beta^2$,

$$\hat{E}_{\hat{\rho}} = \frac{q}{4\pi\epsilon_0\bar{\epsilon}} \frac{\hat{\rho}}{\sqrt{1 - \beta^2\bar{\epsilon}}\{[(x - vt)^2/(1 - \beta^2\bar{\epsilon})] + \hat{\rho}^2\}^{3/2}}, \qquad \text{(58a)}$$

$$\hat{E}_x = \frac{q}{4\pi\epsilon_0\bar{\epsilon}} \frac{x - vt}{\sqrt{1 - \beta^2\bar{\epsilon}}\{[(x - vt)^2/(1 - \beta^2\bar{\epsilon})] + \hat{\rho}^2\}^{3/2}}, \qquad \text{(58b)}$$

$$\hat{H}_\varphi = \frac{qv}{4\pi} \frac{\hat{\rho}}{\sqrt{1 - \beta^2\bar{\epsilon}}\{[(x - vt)^2/(1 - \beta^2\bar{\epsilon})] + \hat{\rho}^2\}^{3/2}}. \qquad \text{(58c)}$$

The radial energy flow through a unit area in a cylindrical surface with radius $\hat{\rho}$ surrounding the particle trajectory is, in the frequency interval from ω to $\omega + d\omega$,

$$W_\omega(\mathbf{r}) = \begin{cases} 0, & \bar{\epsilon} < \left(\dfrac{c}{v}\right)^2, & \text{(59a)} \\[4mm] \dfrac{q^2\omega(1 - 1/\bar{\epsilon}\beta^2)}{8\pi^2c^2\epsilon_0\hat{\rho}}, & \bar{\epsilon} > \left(\dfrac{c}{v}\right)^2. & \text{(59b)} \end{cases}$$

The *total* energy radiated per unit length of particle trajectory is

$$W = 2\pi\hat{\rho} \int_0^{\hat{\omega}} W_\omega(\mathbf{r}) \, d\omega = \frac{q^2}{4\pi\epsilon_0c^2} \int_0^{\hat{\omega}} \omega \left(1 - \frac{1}{\bar{\epsilon}\beta^2}\right) d\omega, \qquad \text{(60)}$$

where $\hat{\omega}$ is the limiting frequency for which $\bar{\epsilon}(\hat{\omega})\beta^2 = 1$ (i.e., $\bar{\epsilon}\beta^2 < 1$ for $\omega > \hat{\omega}$). If $\bar{\epsilon}$ is assumed frequency independent, W becomes infinite; the dielectric constant of all *physical* media is frequency dependent and approaches that of vacuum as the frequency increases without limit; it therefore exhibits a cutoff frequency $\hat{\omega}$.

Discussion

As shown in Fig. 5.4.12, Eqs. (55) reveal that the electromagnetic fields trail behind the moving charge inside a cone making an angle $\hat{\theta}$ with the x axis; since the speed of the particle in this regime is greater than the speed $c/\sqrt{\bar{\epsilon}}$ of a wave propagating in the medium, the electromagnetic disturbance cannot run ahead of the particle. The cone is defined by the equation $vt - x = \hat{\rho}\gamma\beta$, so

$$\cot\hat{\theta} = \gamma\beta = \sqrt{\bar{\epsilon}\beta^2 - 1}, \qquad \beta = \frac{v}{c}. \tag{61}$$

In a coordinate system fixed to the moving charge, the electromagnetic fields are zero when $\theta > \hat{\theta}$ and are derivable from Eq. (55b) when $\theta < \hat{\theta}$. The normal direction to the conical wave front is given by the previously defined angle $\psi = (\pi/2) - \hat{\theta} = \cos^{-1}(1/\beta\sqrt{\bar{\epsilon}})$. The presence of the wave front indicates that radiation takes place, the associated phenomena being generally referred to as Cerenkov radiation. These results were first derived by Frank and Tamm[8] to explain radiation observed from fast charged particles in media with large refractive index.

No radiation occurs in the slow-speed case $\bar{\epsilon} < 1/\beta^2$; this parameter range includes the vacuum $\bar{\epsilon} = 1$. The fields are similar to those in the electrostatic problem $v = 0$ and may, in fact, be derived therefrom by an application of the Lorentz transformation.[9] While for $v = 0$, $\hat{E}_x/\hat{E}_\rho = x/\hat{\rho}$, the effect of the motion is to shrink this ratio to $(x - vt)/\hat{\rho}$, so the field intensity seen by a stationary observer is no longer symmetrical about the charge.

To construct the solutions in Eqs. (55) and (56), it is useful to observe that the fields produced by the source distribution in Eq. (54) are related intimately to the fields of a progressively phased line current extending over the entire trajectory. This follows from the recognition that the Fourier spectrum of the current distribution $\hat{\mathbf{J}}$ in Eq. (54) is given for $\hat{\rho}' = 0$ by [see Eq. (5.2.25)]

$$\mathbf{J}(\mathbf{r}, \omega) = \int_{-\infty}^{\infty} \hat{\mathbf{J}}(\mathbf{r}, t)e^{-j\omega t}\,dt = q\delta(\hat{\rho})e^{-j(\omega/v)x}\mathbf{x}_0, \tag{62}$$

representative of a line distribution with a phase progression given by $\alpha = \omega/v$ [see Eq. (45)]. The results in Sec. 5.4d are therefore directly applicable, allowance being made for the different time dependence ($i \to -j$) and also for the differently oriented coordinate system; the charge travels along the x axis in order to facilitate subsequent anlaysis when the medium exhibits stratification along z. From Eqs. (46) and (47), one has for the relevant time-harmonic Green's function,

$$G(\mathbf{r}, \omega) = \frac{-j}{4} e^{-j(k_0/\beta)x} H_0^{(2)}(k_0 \hat{\rho} \gamma), \qquad k_0 = \frac{\omega}{c}, \qquad \beta = \frac{v}{c}, \qquad (63)$$

with the ω dependence exhibited explicitly. The corresponding time-harmonic fields are then calculated as in Eqs. (48) and lead (with $\partial/\partial t \longleftrightarrow j\omega$) to Eqs. (57).

$G(\mathbf{r}, \omega)$ satisfies the differential equation [see Eq. (46b)]

$$\left(\hat{\nabla}_t^2 + \frac{\omega^2}{c^2}\bar{\epsilon} - \frac{\omega^2}{v^2}\right) G(\mathbf{r}, \omega) = -\delta(\hat{\boldsymbol{\rho}} - \hat{\boldsymbol{\rho}}')e^{-j(\omega/v)x}, \qquad \hat{\nabla}_t^2 = \nabla^2 - \frac{\partial^2}{\partial x^2}, \quad (63a)$$

which can be written in the following alternative forms upon inclusion of the time factor $\exp(j\omega t)$:

$$\left.\begin{array}{c} \left(\nabla^2 - \dfrac{\bar{\epsilon}}{c^2}\dfrac{\partial^2}{\partial t^2}\right) \\[2ex] \left(\hat{\nabla}_t^2 - \dfrac{\gamma^2}{c^2}\dfrac{\partial^2}{\partial t^2}\right) \\[2ex] \left(\hat{\nabla}_t^2 + \beta^2|\gamma|^2\dfrac{\partial^2}{\partial x^2}\right) \end{array}\right\} G(\mathbf{r}, \omega)e^{j\omega t} = -\delta(\hat{\boldsymbol{\rho}} - \hat{\boldsymbol{\rho}}')e^{j\omega(t-x/v)}, \qquad (63b)$$

where $\gamma^2 = \bar{\epsilon} - (1/\beta)^2$. The time-dependent Green's function

$$\hat{G}(\mathbf{r}, t) = \frac{1}{2\pi}\int_{-\infty}^{\infty} G(\mathbf{r}, \omega)e^{j\omega t}\, d\omega \qquad (64)$$

is recovered from Eqs. (63b) upon multiplication by $1/2\pi$ and integration over ω, and the first of Eqs. (63b) is then transformed into Eq. (55).

The second of Eqs. (63b),

$$\left(\hat{\nabla}_t^2 - \frac{\gamma^2}{c^2}\frac{\partial^2}{\partial t^2}\right)\hat{G}(\mathbf{r}, t) = -\delta(\hat{\boldsymbol{\rho}} - \hat{\boldsymbol{\rho}}')\delta\left(t - \frac{x}{v}\right), \qquad (64a)$$

is of the same form as Eq. (41) provided that the positive constant c^2 is replaced by the positive constant c^2/γ^2, and $t' = x/v$. x can be regarded as a parameter since it does not occur in the differential operator on the left-hand side of Eq. (64a). The solution in Eqs. (42) then furnishes directly the result in Eqs. (55a) and (55b).

When γ^2 is negative, the third of Eqs. (63b),

$$\left(\hat{\nabla}_t^2 + \beta^2|\gamma|^2\frac{\partial^2}{\partial x^2}\right)\hat{G}(\mathbf{r}, t) = -\delta(\hat{\boldsymbol{\rho}} - \hat{\boldsymbol{\rho}}')\delta\left(t - \frac{x}{v}\right), \qquad (64b)$$

is useful. Upon introducing a new variable $\bar{x} = x/\beta|\gamma|$, the differential operator on the left-hand side becomes the Laplacian in the $\bar{x}, y, z$ coordinate space. Furthermore, $\delta(t - x/v) \to \delta(t - \bar{x}|\gamma|/c) = (c/|\gamma|)\delta(\bar{x} - \bar{x}')$, where $\bar{x}' = ct/|\gamma|$ plays the role of a parameter. The transformed problem therefore requires the evaluation of the static Green's function G_S in the $\bar{x}, y, z$ space, which, upon inclusion of the multiplicative factor $c/|\gamma|$, yields, for $\hat{\boldsymbol{\rho}}' = 0$,

$$G_S = \frac{c/|\gamma|}{4\pi\sqrt{\hat{\rho}^2 + [\bar{x} - (ct/|\gamma|)]^2}}. \qquad (65)$$

This expression, when transformed back into the x, y, z space, leads to Eq. (56).

As in Sec. 5.4d, the parameter $\alpha = k_0/\beta = \omega/v$ in the time-harmonic problem determines whether radiation does, or does not, take place. When $\alpha < k_0\sqrt{\bar{\epsilon}}$, radiation occurs, whereas no energy escapes from the vicinity of the source when $\alpha > k_0\sqrt{\bar{\epsilon}}$. These conditions are equivalent to the previously stated $\beta^2\bar{\epsilon} > 1$ and $\beta^2\bar{\epsilon} < 1$, respectively. The direction of propagation of the waves in the radiating case is given via Fig. 5.4.11 by the angle ψ. This figure demonstrates furthermore that the radiation characteristics of the moving particle may be inferred *directly* from the wavenumber surface for the medium: One first constructs the plane $\kappa = k_0/\beta$, where in this instance, κ is taken as the wavenumber along x. When this plane intersects the sphere $k = k_0\sqrt{\bar{\epsilon}}$, radiation leaves the particle axis along the conical trajectories shown. When no intersection occurs, there is no radiation. If $\bar{\epsilon}$ is assumed to be frequency independent, all plane waves in the radiating case leave at the same angle ψ, thereby establishing the wave-front in Fig. 5.4.12. From a physical viewpoint, the particle excites those plane waves whose phase speeds $v_p > c/\sqrt{\bar{\epsilon}}$ along its trajectory are equal to v; evidently, this condition, the "Cerenkov coherence condition," cannot be met when $v < c/\sqrt{\bar{\epsilon}}$.

The energy flow in the frequency interval from ω to $\omega + d\omega$ is evaluated from the time-harmonic fields via Eqs. (48) (modified as noted above) and (5.2.31) and leads directly to Eqs. (59) upon use of the Wronskian relation: $J_0(w)(d/dw)N_0(w) - N_0(w)(d/dw)J_0(w) = 2/\pi w$. Formula (59a) confirms that no radiation takes place when $\bar{\epsilon}\beta^2 < 1$.

Modal representation

While the most direct method of determining the radiation from a moving charge utilizes a cylindrical coordinate system centered on the charge trajectory along x, it is possible by an alternative approach to view the configuration as a waveguide along z in which the moving charge sets up a transverse electric current. This procedure is unnecessarily complicated in an unbounded homogeneous medium but becomes essential when the region is stratified along z. Since a problem in the latter category is to be examined subsequently, it is useful to consider the required formulation even for the simple case of infinite space. Again, the Fourier transform is taken with respect to time, thereby rendering the phased line source in Eq. (62) as the relevant excitation. Since in many applications the radiated energy rather than the fields is the quantity of primary interest, we shall formulate the problem in terms of the modal voltages and currents in the relevant formula [Eq. (5.2.34)].

The steady-state modal network problem is sketched in Fig. 5.4.6, whence for $z > z'$,

$$V_i(z, \omega) = -\frac{Z_i(\omega)i_i(\omega)}{2} e^{-j\kappa_i(\omega)\,(z-z')} = Z_i(\omega)I_i(z, \omega), \qquad (66)$$

where i_i, Z_i, and κ_i are the current generator strength, modal characteristic impedance, and propagation constant, respectively, and the dependence on ω

has been indicated explicitly. The vector-mode functions are [Eqs. (2.3.1)]

$$\mathbf{e}_i'(\boldsymbol{\rho}) = -\frac{\nabla_t \Phi_i(\boldsymbol{\rho})}{k_{ti}'}, \qquad \mathbf{e}_i''(\boldsymbol{\rho}) = \mathbf{z}_0 \times \frac{\nabla_t \psi_i(\boldsymbol{\rho})}{k_{ti}''}. \tag{67}$$

Since all field quantities excited by the current distribution in Eq. (62) have an x dependence given by $\exp(-jk_0 x/\beta)$, one may define the scalar mode functions Φ_i and ψ_i as

$$\Phi_i(\boldsymbol{\rho}) = \psi_i(\boldsymbol{\rho}) = \frac{1}{\sqrt{2\pi}} e^{-jn y} e^{-j(k_0/\beta)x}, \qquad -\infty < \eta < \infty, \tag{67a}$$

whence

$$k_{ti}' = k_{ti}'' = \sqrt{\left(\frac{k_0}{\beta}\right)^2 + \eta^2}, \qquad \sum_i \to \int_{-\infty}^{\infty} d\eta. \tag{67b}$$

Upon substituting Eqs. (62) and (67) into Eq. (2.2.14b), one finds, with $y' = 0$,

$$i_i' = \frac{-jq(k_0/\beta)}{\sqrt{2\pi}\, k_{ti}'}, \qquad i_i'' = \frac{-jq\eta}{\sqrt{2\pi}\, k_{ti}''} \tag{68a}$$

while, from Eqs. (2.2.15),

$$Z_i' = \frac{\kappa_i'}{\omega\epsilon_0\bar\epsilon}, \quad Z_i'' = \frac{\omega\mu_0}{\kappa_i''}, \quad \kappa_i' = \sqrt{k_0^2\bar\epsilon - k_{ti}'^2} = \kappa_i''. \tag{68b}$$

To calculate $W_\omega(z)$ from Eq. (5.2.34), we note first that

$$V_i'I_i'^* + V_i''I_i''^* = \frac{Z_i'|i_i'|^2}{4} + \frac{Z_i''|i_i''|^2}{4} = \frac{q^2 k_0^2 \gamma^2}{8\pi\omega\epsilon_0\bar\epsilon\kappa_i''}, \tag{69}$$

where γ is defined in connection with Eq. (55b). This expression is real only when κ_i' is real, so the integration over η in Eq. (5.2.34) extends only over the interval $|\eta| \le k_0\gamma$:

$$W_\omega(z) = \frac{q^2 k_0^2 \gamma^2}{8\pi^2\omega\epsilon_0\bar\epsilon} \int_{-k_0\gamma}^{k_0\gamma} \frac{d\eta}{\sqrt{k_0^2\gamma^2 - \eta^2}} = \frac{q^2 k_0^2 \gamma^2}{8\pi\omega\epsilon_0\bar\epsilon}. \tag{70}$$

This result represents the energy radiated through a plane $z > z'$. By symmetry, an equal amount is radiated through a plane $z < z'$, so the total radiated energy per unit length in x, in the frequency interval between ω and $\omega + d\omega$, is given by $2W_\omega(z)$, which is identical with the result $2\pi\bar\rho W_\omega(\mathbf{r})$, with $W_\omega(\mathbf{r})$ given in Eq. (59b).

5.4f　Ring Currents

Circular ring currents represent another important simple form of field excitation arising in various physical applications. We consider two types of ring currents (Fig. 5.4.13) distinguished by currents flowing parallel to the ring axis or along the ring periphery, respectively. The former is termed a "dipole ring source"; the latter is merely called a "ring source." Unlike the field due to a scalar point or line source of constant strength, which depends only on the

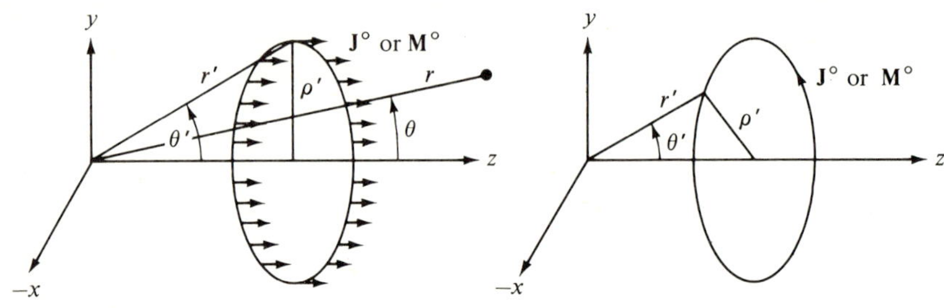

FIG. 5.4.13 Ring currents.

distance measured from the source, the ring-source field is also a function of the angle of observation measured from the ring axis.

Time-harmonic longitudinal electric source current density

$$\hat{\mathbf{J}}(\mathbf{r}, t) = J^\circ \delta(\rho - \rho')\delta(z - z')e^{in\phi}e^{-i\omega t}\mathbf{z}_0. \tag{71}$$

Since the currents are longitudinal, the fields may be derived via Eqs. (5.2.1) and (5.2.4c) from a scalar Green's function $\mathring{G}_f(\mathbf{r}; r', \theta')$, obtained by integration of the free-space Green's function $G_f(\mathbf{r}, \mathbf{r}')$ in Eq. (2b) (with $r \to |\mathbf{r} - \mathbf{r}'|$) around the contour of the ring:

$$\mathring{G}_f(\mathbf{r}, r', \theta') = \rho'\int_{-\pi}^{\pi} e^{in\phi'}G_f(\mathbf{r}, \mathbf{r}')d\phi', \tag{72a}$$

$$= \frac{\rho'}{4\pi}\int_{-\pi}^{\pi} e^{in\phi'}\frac{e^{ik|\mathbf{r}-\mathbf{r}'|}}{|\mathbf{r} - \mathbf{r}'|}d\phi', \tag{72b}$$

where $|\mathbf{r} - \mathbf{r}'|$ in a spherical coordinate representation is defined in Eq. (6). It is observed from Eq. (71) that the source distribution may possess a progressive phase shift given by $\exp(in\phi)$, where n is an integer or zero. It does not appear to be possible to evaluate the integral in Eq. (72b) in closed form. However, for a far-field evaluation (see also Sec. 5.9c) we may employ the asymptotic approximation $|\mathbf{r} - \mathbf{r}'| \sim r - r'\cos\gamma$, whence [see Eqs. (6)]

$$\frac{e^{ik|\mathbf{r}-\mathbf{r}'|}}{|\mathbf{r} - \mathbf{r}'|} \sim \frac{e^{ik(r-r'\cos\gamma)}}{r}, \qquad r \gg r'. \tag{73a}$$

Since

$$\int_{-\pi}^{\pi} e^{-ix\cos\psi-in\psi}\,d\psi = 2\pi e^{-in\pi/2}J_n(x), \tag{73b}$$

and the upper and lower limits of integration in Eq. (72b) can be replaced by $\pi + \phi$ and $-\pi + \phi$, respectively, one obtains, via Eqs. (72b) and (73a), the far-field form of $\mathring{G}_f$,

$$\mathring{G}_f(\mathbf{r}, r', \theta') \sim \frac{e^{ik(r-r'\cos\theta\cos\theta')}}{2r}\rho'e^{in(\phi-\pi/2)}J_n(kr'\sin\theta\sin\theta'), \qquad r \gg r'. \tag{74}$$

The radial variation $(1/r) \exp(ikr)$ indicates that the far field progresses as a spherical wave.

Discussion

The physical interpretation of Eq. (74) differs for various values of source radius $\rho' = r' \sin \theta'$. If $x \ll 1$, then $J_n(x) \sim x^n$, so $\mathring{G}_f \propto L(k\rho')^n$ for very small values of $k\rho'$, where $L = 2\pi\rho'$ is the length of the ring source. Thus, the far fields radiated by a ring source with small $k\rho'$ are very small except when $n = 0$, as for a constant current distribution; in this case $J_0(x) \sim 1$, $x \to 0$, and $\mathring{G}_f \to LG_f$. If $k\rho'$ is large but much smaller than kr, one distinguishes the separate ranges $k\rho' \sin \theta > n$ and $k\rho' \sin \theta < n$. Let us assume that n is likewise large; then, for $x \ll n$, $J_n(x) \sim (x/n)^n$, while for $x \gg n$, one employs $2J_n(x) = H_n^{(1)}(x) + H_n^{(2)}(x)$, with the asymptotic formulas for the Hankel functions given in Eqs. (37) or (5.3.13). Thus, for observation angles θ small enough so that $k\rho' \sin \theta \ll n$, the fields radiated by a ring source become very small.

On the other hand, when $k\rho' \sin \theta \gg n$, use of Eqs. (5.3.13) yields

$$\mathring{G}_f \sim \frac{e^{i(n\phi + \pi/4)} \sqrt{\rho'}}{2r\sqrt{2\pi k \sin \theta}} \{e^{ik[r - r'\cos(\theta - \theta')]} + e^{-i(n+1/2)\pi} e^{ik[r - r'\cos(\theta + \theta')]}\} \quad (75)$$

which may be written as

$$\mathring{G}_f \sim e^{in\phi} \sqrt{\frac{r'\sin\theta'}{r\sin\theta}} \{\bar{G}_f[r - r'\cos(\theta - \theta')] + \Gamma\bar{G}_f[r - r'\cos(\theta + \theta')]\}, \quad (76)$$

where

$$\Gamma = (-i)(-1)^n \quad (76a)$$

and $\bar{G}_f(r)$ is the line-source Green's function in Eq. (25),

$$\bar{G}_f(r) = \frac{i}{4} H_0^{(1)}(kr) \sim \frac{e^{i(kr + \pi/4)}}{2\sqrt{2\pi kr}}, \qquad kr \gg 1. \quad (76b)$$

Equation (76) admits a direct ray-optical interpretation. Since $k\rho'$ is large, the ring source may be regarded as a superposition of linear elements. Introducing $s = \rho'\phi$ as the distance variable along the ring, one may write $\exp(in\phi) = \exp(i\alpha s)$, $\alpha = n/\rho'$ where α is the wavenumber descriptive of the phased excitation in Eq. (71). The radiation characteristics of linearly phased line currents are discussed in Sec. 5.4d, and one notes from Fig. 5.4.10 that each element emits geometric optical rays at an angle $\psi = \cos^{-1}(\alpha/k)$ with respect to the element axis. For the case considered, $k\rho' \gg n$, whence $\psi \approx \pi/2$ and the rays leave each element almost perpendicularly. Thus, a distant observation point P is reached by two rays originating, respectively, at the nearest and farthest points P_1 and P_2 on the ring [Fig. 5.4.14(a)]; the corresponding distances measured along these rays are $[r - r'\cos(\theta - \theta')]$ and $[r - r'\cos(\theta + \theta')]$. The two $\bar{G}_f$ functions in Eq. (76) therefore represent the scalar fields radiated by unit strength linear current elements located at P_1 and P_2. The factor Γ multiplying the contribution from P_2 is separated into two parts as in Eq. (76a). The $(-1)^n$ term accounts for the phase difference $\exp(in\pi)$ between the source elements at P_1 and P_2. The $-i$ term represents the phase change experienced by a ray on crossing a caustic; such a caustic (ray envelope) exists on the ring axis since all rays emanating from the source converge thereon. When

(a) $kr' \sin\theta \sin\theta' >> n$

(b) $kr' > n, \quad \theta = \theta' = \pi/2$ – general ray picture

(c) $kr' > n, \quad \theta = \theta' = \pi/2$ – rays reaching distant observation point

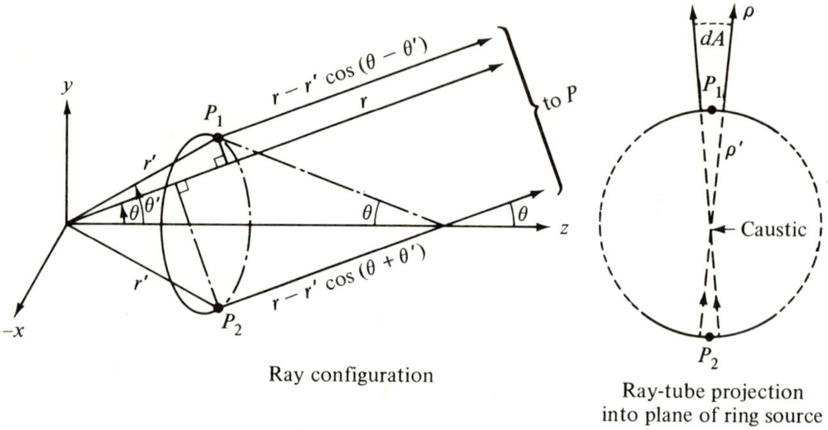

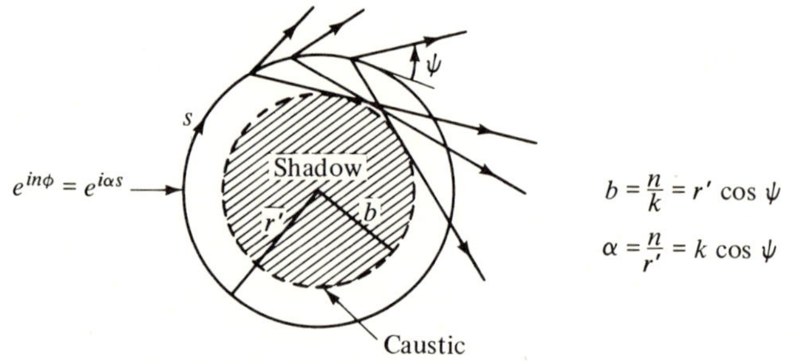

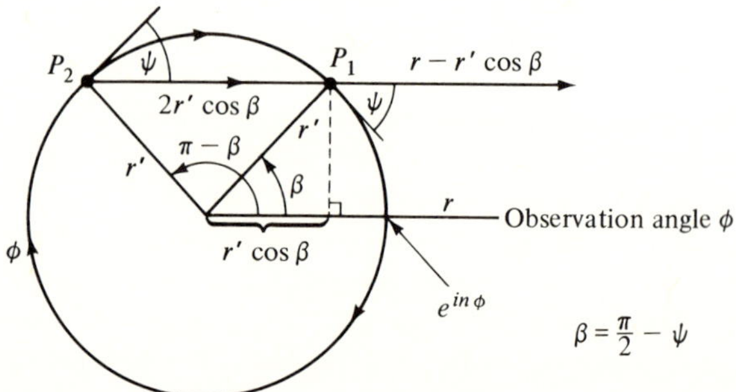

FIG. 5.4.14 Geometrical interpretation of radiation from a large ring source.

the observation point is located in a plane described by the azimuthal coordinate ϕ, the source element at A in Fig. 5.4.14(a) has the phase $\exp(in\phi)$, thereby explaining the occurrence of this factor in Eq. (76). Finally, the $\sqrt{\rho'/\rho}$ term in Eq. (76) accounts for the fact that source elements in the vicinity of P_1 and P_2 are not collinear but lie on the ring; when projected into the plane of the ring, ray-tube cross sections dA exhibit a divergence ρ/ρ' in addition to that descriptive of a straight-line current. Since the field amplitude along a ray varies inversely with the square root of the ray-tube cross section, the result in Eq. (76) follows [see Eq. (1.7.37) wherein ψ denotes a phase function]. Evidently, the divergence factor transforms the cylindrical wave near a ring element into a spherical wave far from the element.

Equations (75) or (76) become invalid as $\sin\theta \to 0$, as evidenced by the resulting divergence of the formulas. In this transition range, where observation points lie near or on the caustic, one must employ Eq. (74).

In Eq. (75) it was assumed that $kr'\sin\theta'\sin\theta \gg n$, so the simple asymptotic approximation in Eq. (37) for the Hankel functions could be used. If one seeks an approximate representation of Eq. (74) which is valid for both n and $k\rho'\sin\theta$ large, one must employ the more accurate asymptotic representations below [see Eqs. (4.5.33) and (4.5.34)]. For $1 - n/x \gg x^{-2/3}$,

$$H_n^{(1,2)}(x) \sim \sqrt{\frac{2}{\pi x \cos\beta}}\, e^{\pm ix[\cos\beta + (\beta-\pi/2)\sin\beta]\mp i\pi/4}, \qquad \sin\beta = \frac{n}{x}. \quad (77a)$$

For $(n/x) - 1 \gg x^{-2/3}$,

$$H_n^{(1,2)}(x) \sim \mp\, i\sqrt{\frac{2}{\pi x \sinh\psi}}\, e^{x(\psi\cosh\psi - \sinh\psi)}, \qquad \cosh\psi = \frac{n}{x},$$

$$J_n(x) \sim \frac{1}{\sqrt{2\pi x \sinh\psi}}\, e^{-x(\psi\cosh\psi - \sinh\psi)}. \quad (77b)$$

For $|1 - n/x| = O(x^{-2/3})$,

$$H_n^{(1,2)}(x) \sim \frac{2^{1/3}}{n^{1/3}}[\mathrm{Ai}(-2^{1/3}\tau) \mp i\,\mathrm{Bi}(-2^{1/3}\tau)], \qquad \tau = n^{-1/3}(x - n), \quad (77c)$$

where $\mathrm{Ai}(\alpha)$ and $\mathrm{Bi}(\alpha)$ are the Airy functions defined in Sec. 4.2e. The above representations, though written for real positive n and x, hold also for selected complex ranges. Formulas for $J_n(x)$ are obtained from Eqs. (77a) or (77c) via $2J_n = H_n^{(1)} + H_n^{(2)}$. For the range $n/x > 1$, a special formula is required since $H_n^{(1)} + H_n^{(2)} \sim 0$ from Eq. (77b). Equations (77) highlight the fact that $J_n(x)$ is a propagating wave function when $n < x$, an attenuating wave function when $n > x$, and goes through a transition range when $n \approx x$.

If one examines Eq. (74) in the range where n is large and $k\rho' > n$, but where it is not required that $k\rho' \gg n$ then use of Eq. (77a) reveals a more complicated asymptotic behavior than that exhibited in Eq. (75). Owing to the presence of additional phase terms, the two ray contributions at P do not appear to originate from the nearest and farthest points on the ring but from shifted positions. The radiation mechanism in this case is illustrated in Fig. 5.4.14(b), where we restrict ourselves for simplicity to observation points lying in the plane of the ring source ($\theta = \theta' = \pi/2$). As before, rays leave each ring-source element with an inclination measured by the angle $\psi = \cos^{-1}(\alpha/k)$. For wavenumbers α characterizing "slow" waves ($\alpha = n/r' > k$), the angle ψ is complex, no real rays exist, and the fields are confined to the vicinity of the ring source and propagate along the direction of current flow. In the "fast" wave case $\alpha < k$, radiation occurs in accord with the ray picture shown in Fig. 5.4.14(b). $\alpha \approx k$ corresponds to the transition range $n \approx kr'$ noted in connection with Eqs. (77). When $kr' \gg n$, [i.e., the spatial period $(2\pi r')/n$ comprises many wavelengths], one obtains the situation depicted in Fig. 5.4.14(a), wherein the angle $\psi \approx \pi/2$. The validity of this interpretation is verified upon examining in Eq. (74) the exponential terms $\exp(i\chi_{1,2})$ resulting after substitution of Eq. (77a) for $J_n(kr')$. One finds that

$$\chi_1 = k(r - r' \cos \beta) + n(\phi - \beta) + \frac{\pi}{4},$$

$$\chi_2 = k(r + r' \cos \beta) + n(\phi - \pi + \beta) - \frac{\pi}{4}.$$

The physical significance of these phase functions is appreciated from Fig. 5.4.14(c), wherein the field radiated to a distant point (r, ϕ) is examined, r being the distance from the center of the ring. According to the ray picture in Fig. 5.4.14(b), the field comprises contributions from two rays originating at P_1 and P_2 in Fig. 5.4.14(c). The ray emanating from P_1 travels a distance $(r - r' \cos \beta)$ with an associated phase $k(r - r' \cos \beta)$. The phase of the source distribution at P_1 is $n(\phi - \beta)$, yielding a total phase at (r, ϕ) identical with that in χ_1, save for the $\pi/4$ term, which arises in conjunction with the radiation from a line source as noted in Eq. 76(b). The same interpretation applies to χ_2 relative to the ray from P_2, except for an additional phase shift of $-\pi/2$ which is contributed when this ray crosses the caustic at the shadow boundary [see Fig. 5.4.14(b)].

Figure 5.4.14(b) schematizes the radiation mechanism not only for the case $kr \gg kr' > n$, but also for $kr' \gg kr > n$, wherein the field is observed in the interior of the ring. Unless $\psi \approx \pi/2$, there exists a "shadow" region which is not penetrated by real rays. The envelope of the ray family (caustic), a circle with radius $b = r' \cos \psi$, bounds the shadow region in the plane of the ring source. The analytical formulation for the field in the interior of the ring is obtained from Eq. (74) by interchanging r and r' and letting $\theta = \theta' = \pi/2$. The field behavior is then governed primarily by the factor $J_n(kr)$, which gives rise to propagating waves (real rays) when $n < kr$, attenuating waves when $n > kr$ (shadow region), with the transition occurring at $r \approx b = n/k$. An alternative, and more extensive, ray-optical treatment proceeding directly from Eqs. (72) is given in Sec. 5.9c.

Modal representation (circular waveguide)

A circular-waveguide representation for the scalar ring-source Green's function $\overset{\circ}{G}_f$ defined in Eq. (72a) is obtained from Eqs. (5.2.8a) or (5.2.8b), with Eq. (7), upon carrying out the integration over ϕ' [exp($j\omega t$) dependence]:

$$\mathring{G}_f(\mathbf{r}; \rho', z') = \frac{-j\rho'}{2} e^{-jn\phi} \int_0^\infty \xi J_n(\xi\rho) J_n(\xi\rho') \frac{e^{-j\sqrt{k^2-\xi^2}\,|z-z'|}}{\sqrt{k^2-\xi^2}}\, d\xi, \tag{78a}$$

$$= \frac{-j\rho'}{4} e^{-jn\phi} \int_{\infty e^{-j\pi}}^\infty \xi H_n^{(2)}(\xi\rho_>) J_n(\xi\rho_<) \frac{e^{-j\sqrt{k^2-\xi^2}\,|z-z'|}}{\sqrt{k^2-\xi^2}}\, d\xi. \tag{78b}$$

The singularities and path of integration appropriate to the integral representation in Eq. (78b) are shown in Fig. 5.3.5(a). Upon letting $z' = 0$ for convenience, introducing the change of variable $\xi = k \sin w$ and the polar coordinates (r, θ), one obtains for $\rho > \rho'$ the following representation in the w plane;

$$\mathring{G}_f = \frac{-jk\rho'}{4} e^{-jn\phi} \int_{\bar{P}} \sin w\, H_n^{(2)}(kr \sin \theta \sin w) J_n(k\rho' \sin w) e^{-jkr \cos \theta \cos w}\, dw, \tag{79}$$

with the path $\bar{P}$ shown in Fig. 5.3.5(b). For $\rho < \rho'$, ρ and ρ' in Eq. (79) are interchanged.

To effect an approximate evaluation of the integral in Eq. (79) for the case where the observation point is situated very far from the ring source (i.e., $kr \to \infty$) one may represent the Hankel function by its asymptotic form in Eq. (5.3.13) subject to the condition that $\sin \theta \sin w \neq 0$ along the path of integration. Moreover, since the source radius ρ' is finite, we may treat $J_n(k\rho' \sin w)$ as a slowly varying function, compared with $H_n^{(2)}(kr \sin \theta \sin w)$, as $kr \to \infty$, or, less stringently, when $r \sin \theta \gg \rho'$, $kr \gg 1$. Thus, the pertinent formulation becomes

$$\mathring{G}_f \sim \frac{-j\rho' e^{-jn(\phi-\pi/2)+j\pi/4}}{4} \sqrt{\frac{2k}{\pi r \sin \theta}} \int_{\bar{P}} \sqrt{\sin w}\, J_n(k\rho' \sin w) e^{-jkr \cos(w-\theta)}\, dw. \tag{80}$$

The asymptotic evaluation of the integral in Eq. (80) proceeds in direct analogy with that in Eq. (5.3.14), with $f(w) = \sqrt{\sin w}\, J_n(k\rho' \sin w)$, etc. Use of $j \to -i$ and Eq. (5.3.16a) [for $\exp(-i\omega t)$ dependence] then yields the first-order asymptotic approximation in Eq. (74). It is to be noted that the asymptotic evaluation is carried out under the restriction $r \sin \theta \gg \rho'$, while the direct derivation leading to Eq. (74) shows that this result is valid for all θ in the range $0 \leq \theta \leq \pi$.

Time-harmonic azimuthal electric source current density

$$\hat{\mathbf{J}}(\mathbf{r}, t) = I\delta(\rho - \rho')\delta(z - z')e^{in\phi}e^{-i\omega t}\boldsymbol{\phi}_0. \tag{81}$$

For a calculation of the radiation from the ring source in Fig. 5.4.13(b), wherein source currents with a progressive phase variation $\exp(in\phi)$ flow in the azimuthal direction, one may utilize the previously derived results for a transverse (to z) current element in Sec. 5.4b. For an azimuthal electric current element, for example, we let $\mathbf{J}^o = \boldsymbol{\phi}_0 J^o$, $\mathbf{M}^o \equiv 0$, and obtain for the potential functions in Eqs. (5.2.1) [$\exp(j\omega t)$ dependence],

$$j\omega\epsilon\Pi'(\mathbf{r}, \mathbf{r}') = J^o \frac{1}{\rho'} \frac{\partial^2}{\partial\phi'\,\partial z'} \mathscr{S}_f, \qquad \Pi''(\mathbf{r}, \mathbf{r}') = -J^o \frac{\partial \mathscr{S}_f}{\partial \rho'}, \tag{82}$$

with $\mathscr{S}_f$ given in Eq. (5.2.10b), with Eq. (7); it is recalled that the representation for $\mathscr{S}_f$ in Eq. (5.2.10b) is to be considered as purely formal, but that the functions $\partial \mathscr{S}_f/\partial \phi'$ and $\partial \mathscr{S}_f/\partial \rho'$ are meaningful after the partial differentiations have been carried out on the integrand of $\mathscr{S}_f$.

The potential functions for the ring source are now deduced from an integration as in Eq. (72a), with $i \to -j$. One notes that, for $n \neq 0$, both E and H modes are excited (i.e., the radiation from the ring source gives rise to longitudinal electric and magnetic fields). For the special case of a constant ring-source distribution $n = 0$, the potential $\overset{\circ}{\Pi}'$ vanishes. The associated electric and magnetic fields are then obtained via Eqs. (5.2.1) and (82). The electric field has only a ϕ component, which is given by

$$\overset{\circ}{\mathbf{E}}(\mathbf{r}) = \boldsymbol{\phi}_0 \frac{j\omega\mu \, \partial \overset{\circ}{\Pi}''}{\partial \rho} = -\boldsymbol{\phi}_0 \, j\omega\mu J^\circ \frac{\partial^2 \overset{\circ}{\mathscr{S}}_f}{\partial \rho \, \partial \rho'}, \tag{83a}$$

where from an integration of Eq. (5.2.10b), with Eq. (7),

$$\frac{\partial^2 \overset{\circ}{\mathscr{S}}_f}{\partial \rho \, \partial \rho'} = \frac{-j\rho'}{2} \int_0^\infty \frac{\xi J_1(\xi\rho) J_1(\xi\rho') \exp(-j\sqrt{k^2 - \xi^2}\,|z - z'|)}{\sqrt{k^2 - \xi^2}} \, d\xi. \tag{83b}$$

Upon comparing Eqs. (78a) and (83b) one notes that

$$\left.\frac{\partial^2 \overset{\circ}{\mathscr{S}}_f}{\partial \rho \, \partial \rho'}\right|_{n=0} = (e^{jn\phi} \overset{\circ}{G}_f)_{n=1}. \tag{83c}$$

Thus, the previously obtained asymptotic results for $\overset{\circ}{G}_f$ can be employed directly for the determination of the far fields in the present problem. Results for the $\exp(-i\omega t)$ dependence are obtained on letting $j \to -i$.

Time-harmonic magnetic current distributions

The results in this instance are obtained from those above by the duality replacements $E \to H$, $H \to -E$, $J^\circ \to M^\circ$, $I \to V$, and $\mu \leftrightarrow \epsilon$.

5.5 SOURCES IN THE PRESENCE OF A SEMIINFINITE DIELECTRIC MEDIUM

5.5a Time-harmonic Longitudinal Electric Current Element

$$\hat{\mathbf{J}}(\mathbf{r}, t) = Il\delta(\boldsymbol{\rho})\delta(z - z')e^{-i\omega t}\mathbf{z}_0. \tag{1}$$

We consider the physical configuration in Fig. 5.5.1, wherein a longitudinal electric current element of strength $J^\circ = Il$, with I the current in the element and l its infinitesimal length, is situated at the point $z' < 0$ on the z axis in the presence of a dielectric interface at $z = 0$. The medium in the half-space $z < 0$ is characterized by a dielectric constant ϵ_1 and a permeability μ, while the analogous constants for the half-space $z > 0$ are ϵ_2 and μ. Unlike the problem of the dipole in free space, this problem cannot be solved in terms of a single simple function of the spatial coordinates so that one must resort to a modal technique. With z as the transmission coordinate, the electromagnetic fields are derived from a scalar E-mode Green's function $G(\mathbf{r}, \mathbf{r}')$ [see Eqs. (5.2.1) and (5.2.4c)], which satisfies the separate differential equations

$$(\nabla^2 + k_1^2)G_1'(\mathbf{r}, \mathbf{r}') = -\delta(\mathbf{r} - \mathbf{r}'), \qquad z < 0, \tag{2a}$$

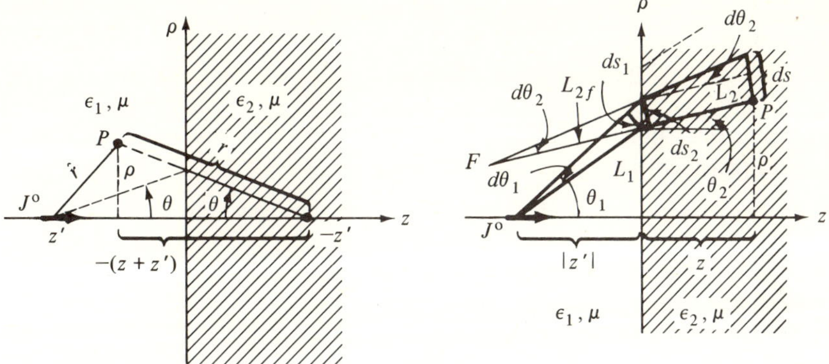

(a) Ray-optical coordinates for observation points in region 1

(b) Ray-optical coordinates for observation points in region 2 ($\sqrt{\epsilon_1}\,\sin\theta_1 = \sqrt{\epsilon_2}\,\sin\theta_2$)

FIG. 5.5.1 Semiinfinite dielectric medium with a longitudinal current element.

$$(\nabla^2 + k_2^2)G_2'(\mathbf{r}, \mathbf{r}') = 0, \qquad z > 0, \qquad (2b)$$

subject to a radiation condition at infinity in both regions, and to the following continuity requirements at $z = 0$ [see Eqs. (2.3.36) et seq.]:

$$G_1' = G_2', \qquad \frac{1}{\epsilon_1}\frac{\partial G_1'}{\partial z} = \frac{1}{\epsilon_2}\frac{\partial G_2'}{\partial z} \quad \text{at} \quad z = 0, \qquad (2c)$$

where $k_{1,2}^2 = \omega^2\mu\epsilon_{1,2}$. In cylindrical coordinates with $\mathbf{r} = (\rho, z)$ and $\mathbf{r}' = (0, z')$, one obtains, for $z < 0$,

$$G_1'(\mathbf{r}, \mathbf{r}') = G_{f1}(\mathbf{r}, \mathbf{r}') + G_s'(\mathbf{r}, \mathbf{r}'), \qquad (3)$$

where $G_{f1} = (1/4\pi\hat{r})\exp(ik_1\hat{r})$, with $\hat{r} = |\mathbf{r} - \mathbf{r}'|$, is the free-space Green's function in Eq. (5.4.2) (Fig. 5.5.1). G_s', expressive of the interface effect, is represented by an integral extending either over the transverse wavenumber ξ or over the complex angle variable w:

$$G_s' = \frac{-i}{8\pi}\int_{\infty e^{i\pi}}^{\infty}\xi H_0^{(1)}(\xi\rho)\frac{\exp[-i\sqrt{k_1^2 - \xi^2}(z + z')]}{\sqrt{k_1^2 - \xi^2}}\Gamma(\xi)\,d\xi, \qquad (3a)$$

$$= \frac{-ik_1}{8\pi}\int_{\bar{P}}\sin w\, H_0^{(1)}(k_1 r\sin\theta\sin w)e^{ik_1 r\cos\theta\cos w}\Gamma(k_1\sin w)\,dw, \qquad (3b)$$

with the integration paths given in Fig. 5.3.6. The two representations are related via the transformation $\xi = k_1\sin w$; spherical coordinates with respect to the image point have been introduced through the definitions

$$\rho = r\sin\theta, \quad -(z + z') = r\cos\theta > 0, \qquad 0 \leq \theta \leq \frac{\pi}{2}, \qquad (3c)$$

and Γ is the E-mode reflection coefficient

$$\Gamma(\xi) = \frac{\sqrt{k_1^2\epsilon - \xi^2} - \epsilon\sqrt{k_1^2 - \xi^2}}{\sqrt{k_1^2\epsilon - \xi^2} + \epsilon\sqrt{k_1^2 - \xi^2}}, \qquad \epsilon = \frac{\epsilon_2}{\epsilon_1}, \qquad (3d)$$

whence

$$\Gamma(k_1 \sin w) = \frac{\sqrt{\epsilon - \sin^2 w} - \epsilon \cos w}{\sqrt{\epsilon - \sin^2 w} + \epsilon \cos w}. \tag{3e}$$

The Green's function $G_2'(\mathbf{r}, \mathbf{r}')$ in the region $z > 0$ can be represented as in Eq. (3a) provided that

$$-\exp\left[-i\sqrt{k_1^2 - \xi^2}(z + z')\right]\Gamma(\xi) \text{ is replaced by}$$

$$[1 - \Gamma(\xi)] \exp\left(i\sqrt{k_2^2 - \xi^2}\,z - i\sqrt{k_1^2 - \xi^2}\,z'\right). \tag{4}$$

For large values of $k_1 r$ but for arbitrary values of $k_1 \hat{r}$, where r and $\hat{r}$ are the distances from the image and source points, respectively [Fig. 5.5.1(a)], the integrals for G_s' in Eqs. (3a) and (3b) may be evaluated asymptotically. Saddle-point integration yields the following result, which in the region $z < 0$ generally dominates other contributions mentioned subsequently:

$$G_1' \sim \frac{e^{ik_1\hat{r}}}{4\pi\hat{r}} - \Gamma(k_1 \sin \theta)\frac{e^{ik_1 r}}{4\pi r}\left[1 + O\!\left(\frac{1}{k_1 r}\right)\right], \qquad \theta \not\approx 0, \tag{5}$$

where

$$\Gamma(k_1 \sin \theta) = \frac{\sqrt{\epsilon - \sin^2 \theta} - \epsilon \cos \theta}{\sqrt{\epsilon - \sin^2 \theta} + \epsilon \cos \theta}. \tag{5a}$$

Equation (5) is valid provided that the observation point is not too near the interface ($\theta \approx \pi/2$) or the dipole axis ($\theta \not\approx 0$). This result, with the exclusion of the generically indicated higher-order terms, can be deduced *directly* by arguments of *geometrical optics* (see Discussion).

In addition to the geometric-optical field, which arises from the saddle point at $w = \theta$ in the integrand, there are other possible contributions to G_s' from branch-point and pole singularities representative of distinct *diffracted* field types. The branch point singularity at $w_b = \sin^{-1}\sqrt{\epsilon}$ furnishes the following *additional* term G_{sb}' to the asymptotic expression for G_s':

$$G_{sb}' \sim$$

$$-\frac{e^{i\pi/4}}{8\pi}\sqrt{\frac{2k_1}{\pi r \sin \theta}}\frac{2\exp[ik_1(\sqrt{1 - \epsilon}|z + z'| + \sqrt{\epsilon}\,\rho) - i\pi/4]}{[k_1(\sqrt{\epsilon}\,|z + z'| - \sqrt{1 - \epsilon}\,\rho)]^{3/2}(1 - \epsilon)^{1/4}}\sqrt{\frac{2\pi}{\epsilon}}\,U(\theta - \hat{\theta}).$$

$$\tag{6}$$

The Heaviside unit function U vanishes when $\theta < \hat{\theta}$ and equals unity when $\theta > \hat{\theta}$, while $\hat{\theta}$ is that value of the observation angle for which the steepest-descent path through the saddle point $w = \theta$ crosses the branch point at w_b [see Eq. (5.3.15a)]:

$$\hat{\theta} = \text{Re }w_b - \cos^{-1}\text{sech}\,(\text{Im }w_b), \qquad w_b = \sin^{-1}\sqrt{\epsilon}, \qquad \epsilon = \frac{\epsilon_2}{\epsilon_1}. \tag{6a}$$

This "lateral wave" field (see Fig. 5.5.2) decays according to $(k_1 \cdot \text{distance})^{-2}$ [see Eq. (6)], more rapidly than the geometric-optical field in Eq. (5). Moreover, when the medium in $z > 0$ is denser than in $z < 0$ ($\epsilon > 1$), or when losses are present (Im $\epsilon > 0$), there are also exponential decays, so G'_{sb} is usually negligible under these conditions. An important exception occurs when medium 1 is lossy and medium 2 is lossless, so Im $k_1 > 0$, Im $k_2 = 0$. The geometric-optical field in Eq. (5) is then strongly damped, whereas the damping in Eq. (6) arises only from the factor multiplying $|z + z'|$, but not from the factor $k_1\sqrt{\epsilon} = k_2$ multiplying p. Thus, G'_{sb} *dominates* when the lateral distance p is large and medium 1 has dissipation (see the discussion relating to Fig. 5.5.2). It may also be noted that $\Gamma(k_1 \sin \theta) \to 1$ and $r \to \hat{r}$ when $\theta \to \pi/2$, so the geometric-optical field vanishes near the interface; the $O(1/k_1^2 r^2)$ term, next in the asymptotic expansion in Eq. (5), then becomes relevant together with G'_{sb} when the region is lossless.

For highly dissipative dielectrics filling the half-space $z > 0$ (i.e., $|\epsilon| \gg 1$, arg $\epsilon \approx \pi/2$), G'_{sb} is negligible but a further modification must be considered for observation points near the interface. The pole singularity† in Eq. (3e) is then found to be located near the saddle point and the composite contribution to the field becomes

$$G'_1 \sim \frac{e^{ik_1\hat{r}}}{4\pi\hat{r}} + \frac{e^{ik_1 r}}{4\pi r} - \frac{\theta - w_p}{8\pi}\hat{\Gamma}(k_1 \sin \theta)\sqrt{\frac{2k_1}{r}}\, e^{i(k_1 r + \pi/4)}\{\ \}, \tag{7}$$

$$\{\ \} = i2 e^{-k_1 r b^2} Q(-ib\sqrt{k_1 r}) + T(0)\frac{1}{\sqrt{k_1 r}}, \tag{7a}$$

where $\hat{\Gamma} = 1 + \Gamma$, and

$$b = \sqrt{2}\, e^{i\pi/4} \sin\frac{w_p - \theta}{2}, \qquad w_p \approx \frac{\pi}{2} + \frac{e^{-i\pi/4}}{\sqrt{|\epsilon|}},$$

$$T(0) = \frac{\sqrt{2}\, e^{-i\pi/4}}{\theta - w_p} + \frac{e^{-i\pi/4}}{\sqrt{2}\sin[(w_p - \theta)/2]}, \tag{7b}$$

$$Q(y) = \int_y^\infty e^{-x^2}\, dx.$$

This expression is uniformly valid for *all* observation angles $0 < \theta \leq \pi/2$ in the half-space $z < 0$ and for *arbitrary* ϵ_2; it may be shown to reduce to Eq. (5) when $\theta \not\approx \pi/2$. For the special case of large dissipation in medium 2 (arg $\epsilon \approx \pi/2$) and no loss in medium 1, and for $\theta = \pi/2$ (source and observation point *on* the interface), the contribution from the second term in Eq. (7a) is small compared with the first term and can therefore be neglected. Via Eq. (7b), Eq. (7) can then be written in the following form to yield for G'_1 along the interface,

$$G'_1 \sim \frac{e^{ik_1 r}}{2\pi r}[1 + i2\sqrt{\zeta}\, e^{-\zeta} Q(-i\sqrt{\zeta})], \quad k_1 r \gg 1, \quad |\epsilon| \gg 1, \quad \text{arg}\, \epsilon \approx \frac{\pi}{2}, \tag{8}$$

where the parameter ζ is Sommerfeld's "numerical distance"[10]

$$\zeta = \frac{k_1 r}{2|\epsilon|}. \tag{8a}$$

---

†When medium 2 is a plasma for which it is possible to have $\epsilon_2 < -\epsilon_1$, with $\epsilon_1 > 0$, then the pole may contribute in a manner different from that described here; see Sec. 5.5k.

Alternatively, upon recognizing that $Q(-i\sqrt{\zeta}) = Q(0) - \int_0^{-i\sqrt{\zeta}} e^{-x^2} dx$, with $Q(0) = \sqrt{\pi}/2$, one can rewrite Eq. (8) as

$$G_1' \sim \frac{e^{ik_1 r}}{2\pi r}\left[1 + i\sqrt{\pi\zeta}\, e^{-\zeta} - i2\sqrt{\zeta}\, e^{-\zeta} \int_0^{-i\sqrt{\zeta}} e^{-x^2}\, dx\right]. \qquad (8b)$$

The first term on the right-hand side of Eq. (8b) represents the solution for waves propagating along a perfectly conducting surface ($|\epsilon| = \infty$); the second term stems from the fact that $|\epsilon|$ is large but finite. While for small ζ the first term is predominant, the second and third terms grow in importance with increasing ζ and actually approach -1 when $\zeta \gg 1$ [see Eq. (4.4.20)]. Thus, the wave character depends quite strongly on the numerical distance. The second term in Eq. (8b) represents a "surface wave," as is evident upon identifying its contribution to G_1' as proportional $(1/\sqrt{\xi_p r})\exp(i\xi_p r)$, where $\xi_p r = k_1 r \sin w_p \approx k_1 r + i\zeta$ (see the discussion to follow). However, at very large distances (ζ large) the effect of the surface-wave contribution is nullified by the third term in Eq. (8b), whence the discussion of its predominant existence in the dipole field has only limited validity.

In the region $z > 0$ a saddle-point evaluation of the integral in Eq. (3a) for large real values of $k_{1,2}$ yields by Eq. (4) the first-order result

$$G_2'(\mathbf{r},\, \mathbf{r}') \sim \frac{e^{i(k_1 L_1 + k_2 L_2)}}{4\pi\sqrt{k_1 \rho}} \frac{\sqrt{\sin\theta_1}}{\cos\theta_1} \frac{1}{\sqrt{\dfrac{L_1}{k_1 \cos^2\theta_1} + \dfrac{L_2}{k_2 \cos^2\theta_2}}} [1 - \Gamma(k_1 \sin\theta_1)], \qquad (9)$$

where the distances $L_{1,2}$ and the angles $\theta_{1,2}$ are defined in Fig. 5.5.1(b). Higher-order contributions arising from the saddle-point calculation have not been included in Eq. (9). This expression for G_2' may be deduced completely by considerations of *geometrical optics* (see Discussion). When $\theta_1 \approx \pi/2$ and L_2 is small the geometric-optical formula must be corrected in a manner similar to that described for the region $z < 0$.[11]

Discussion

The approximate evaluation of the integral representations for G_s' in Eqs. (3) has received a great deal of attention in the literature because of its relevance to the study of the propagation of electromagnetic waves along the earth's surface (to a first approximation, the earth's sphericity and the effect of the ionosphere are neglected). Difficulties in an asymptotic evaluation of the appropriate ξ plane integrals arise because in addition to branch points at $\xi = 0$ and $\xi = \pm k_1$, the integrand in Eq. (3a) possesses first-order branch points at $\xi_b = \pm k_1 \sqrt{\epsilon}$ and simple pole singularities at the zeros ξ_p of the denominator on the right-hand side of Eq. (3d). Sommerfeld[10] obtained the first formal solution in terms of the circular waveguide (cylindrical coordinate) representation in Eq. (3a) and carried out an asymptotic evaluation for large values of $k_1 r$. In this evaluation he chose to deform the contour of integration and thereby to cross a pole of the integrand. The resulting residue contribution, as noted from Eq. (3a) for $z' = 0$ (source point located on the interface), has a coordinate dependence given by

$$H_0^{(1)}(\xi_p \rho) \exp[+i\sqrt{k_1^2 - \xi_p^2}\,|z|] \sim (1/\sqrt{\xi_p \rho}) \exp[i\sqrt{k_1^2 - \xi_p^2}\,|z| + i\xi_p \rho],$$

where, if ϵ_2 is complex, $\mathrm{Im}\sqrt{k_{1,2}^2 - \xi_p^2} > 0$ for the $\exp(-i\omega t)$ time convention. The associated fields decay exponentially away from the interface for both $z < 0$ and $z > 0$, as noted from Eq. (4); the residue contribution thus has the character of a "surface wave," first discussed by Zenneck, "guided" by the interface. Because of its inverse $\sqrt{\rho}$ radial decay at large distances, slower than the $1/\hat{r}$ variation of the primary field G_{f1}, a great deal of discussion has ensued about the independent existence of this wave and its utilization for the transmission of radio waves over long distances. In view of the voluminous literature on this subject, we abstain from discussing the historical development of this problem; for an appreciation thereof, the reader is referred to Reference 12. As seen from the asymptotic evaluation for source and observation points located near the interface [Eq. (8b)], the far field radiated by a dipole antenna can be represented so as to exhibit a surface-wave term. However, the simultaneous presence of other terms nullifies the effect of the surface wave except over a restricted range of values of Sommerfeld's numerical distance parameter $\zeta = k_1 r / 2\,|\epsilon|$. Hence despite the fact that the surface wave represents a field type that can be independently sustained on a lossy interface, its isolation from the other contributions to the field radiated by a dipole source has only limited validity.

The aforementioned complications are absent when the source point and (or) the observation point are located far from the interface, in which instance the asymptotic formulas in Eqs. (5) and (9) describe the field behavior. These partial results may be derived completely via considerations of geometrical optics. With reference to an observation point P in Fig. 5.5.1(a), the first and second terms of the field response, as expressed in Eq. (5), evidently comprise a direct-wave and a reflected-wave contribution, respectively. The contribution from the direct wave is identical with that observed in an infinite medium with wavenumber k_1. The reflected-wave contribution can be interpreted as a geometric-optical term arising from a ray which reaches the observation point after being reflected from the interface at the angle of incidence (dashed lines in Fig. 5.5.1). The distance from the source to P along this trajectory is r [see also Fig. 1.7.9(a) and Eq. (1.7.64b) for the analogous line-source problem]. Alternatively, the reflected-wave contribution can be viewed as arising from a weighted image source situated as in Fig. 5.5.1 in an infinite medium with wavenumber k_1. The amplitude function of the reflected ray appearing to originate at the image point contains the factor $-\Gamma(k_1 \sin\theta)$, the reflection coefficient for a plane wave, polarized with magnetic vector parallel to the interface (this is the relevant polarization for the dipole in Fig. 5.5.1) and incident on the interface at an angle θ with respect to the normal direction.[13] The geometric-optical field is modified by terms of $O(1/k_1^2 r^2)$, which represent higher-order terms in the asymptotic evaluation (these terms are not evaluated explicitly here).

The ray-optical interpretation of Eq. (9) is somewhat more involved since the ray paths now proceed in both regions. With reference to Fig. 5.5.1(b), the evaluation of the field according to geometrical optics requires the following steps, motivated by the fact that the field along a ray propagates locally like that of a plane wave having the appropriate polarization (see Sec. 1.7b):

1. Determination of a ray path connecting the source point and observation point in a manner such that the ray reflection and refraction laws (same as for plane waves) are satisfied at the interface.

2. Evaluation of the initial field G'_{20} on the refracted ray by calculating the incident field on the interface and multiplying by the plane-wave transmission coefficient.

3. Evaluation of the phase at P by adding to the phase of G'_{20} the appropriate phase increment along L_2.

4. Evaluation of the amplitude at P by multiplying $|G'_{20}|$ by the square root of the refracted ray-tube cross-section ratio at the interface and at P.

Item 1 leads to the ray path in Fig. 5.5.1(b), with the angles θ_1 and θ_2 related by the plane-wave refraction law (Snell's law); both rays are contained in a plane normal to the interface, a feature not necessarily satisfied when anisotropic media are involved (see Fig. 1.7.4). To construct the field G'_{20} in item 2, we recognize that the incident field on the interface is given by $(4\pi L_1)^{-1} \cdot \exp{(ik_1 L_1)}$ and the plane-wave transmission coefficient by $[1 - \Gamma(k_1 \sin \theta)]$ [see also Eq. (5), with $\hat{r} = r = L_1$, and the continuity requirement $G'_1 = G'_2$ at $z = 0$]. The phase increment along the refracted ray path (item 3) is given by $k_2 L_2$. For determination of the amplitude in item 4, reference is made to the incident and refracted tubes formed by closely neighboring rays [Fig. 5.5.1(b)]. The intensity at P is related to the intensity at the interface by the area ratio

$$\frac{dA_2}{dA} = \frac{2\pi|z'| \tan \theta_1 \, ds_2}{2\pi p \, ds}, \tag{9a}$$

where ds_2 and ds are the cross-sectional length elements in a pz section through the conically spreading ray tube; the rotational symmetry of the ray structure with respect to the z axis has been used in deriving Eq. (9a). Thus, the geometrical optics solution at an observation point (p, z) in region 2 is [see Eq. (1.7.37)].[11]

$$G'_2\big|_{\text{opt}} = \left\{ \frac{e^{ik_1 L_1}}{4\pi L_1} [1 - \Gamma(k_1 \sin \theta_1)] \right\} e^{ik_2 L_2} \sqrt{\frac{dA_2}{dA}}. \tag{9b}$$

The quantity in the braces represents the field G'_{20} at the interface, and the remaining factors account for the phase and amplitude change along the refracted ray away from the interface.

To establish the equivalence of Eqs. (9) and (9b), one may employ the following geometrical relations deducible from Fig. 5.5.1(b):

$$ds = ds_2 + L_2\, d\theta_2 = (L_2 + L_{2f})d\theta_2,$$

$$ds_2 = ds_1 \frac{\cos\theta_2}{\cos\theta_1} = L_1 \frac{\cos\theta_2}{\cos\theta_1} d\theta_1 = L_{2f}\, d\theta_2, \tag{9c}$$

so that in view of $d\theta_1/d\theta_2 = (k_2/k_1)(\cos\theta_2/\cos\theta_1)$, derived from Snell's law $k_1 \sin\theta_1 = k_2 \sin\theta_2$,

$$\frac{ds_2}{ds} = \frac{L_{2f}}{L_2 + L_{2f}}, \qquad L_{2f} = L_1 \frac{k_2}{k_1}\frac{\cos^2\theta_2}{\cos^2\theta_1}. \tag{9d}$$

L_{2f} represents the length of the refracted ray tube from the virtual focus F in Fig. 5.5.1(b) to the interface. When Eqs. (9d) and (9a) are substituted into Eq. (9b), one obtains Eq. (9). It has therefore been shown that the first-order saddle-point contributions to the reflected or refracted fields are identical with results predicted from geometrical optics.

An interesting ray-optical interpretation may also be given to the diffraction field expressed by the formula for G'_{sb} in Eq. (6) when both dielectrics are lossless and the medium containing the source is *denser* than the exterior region (i.e., $\epsilon_1 > \epsilon_2$, or $\epsilon < 1$ and real).[11] The phase of the exponential term, $k_1\sqrt{1-\epsilon}\,|z + z'| + k_1\sqrt{\epsilon}\,\rho$, can be reexpressed as $k_1(L_1 + L_3) + k_2 L_2$, where L_1, L_2, and L_3 are the distances defined in Fig. 5.5.2(b). This is verified readily upon noting that $k_2 = k_1\sqrt{\epsilon}$, $\rho = (L_1 + L_3)\sin\hat{\theta} + L_2$, $|z + z'| = (L_1 + L_3)$ $\cdot \cos\hat{\theta}$, where $\hat{\theta} = \sin^{-1}\sqrt{\epsilon}$ is the angle of total reflection, or of critical refrac-

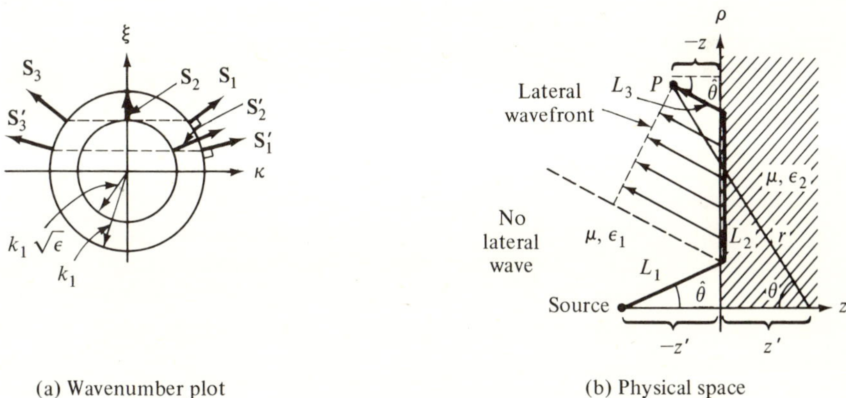

(a) Wavenumber plot   (b) Physical space

FIG. 5.5.2   The lateral wave.

tion, when $\epsilon < 1$ and real [see Eq. (6a)]. Thus, the phase of the branch-cut wave can be interpreted as arising from a ray that propagates from the source to the interface at the angle $\hat{\theta}$, is refracted parallel to the interface in the *second* medium (since the wavenumber associated with L_2 is k_2), and leaves the interface by refraction at the angle $\hat{\theta}$ to reach the observation point P. The name "lateral" arises from the "sideways" nature of the propagation of this wave along the interface. Furthermore, $L_2 = \rho - |z + z'|\tan\hat{\theta}$, with $\sin\hat{\theta} = \sqrt{\epsilon}$,

$\cos \hat{\theta} = \sqrt{1 - \epsilon}$, so the spatially dependent portion of Eq. (6) may be written as

$$G'_{sb} \propto \frac{e^{i(k_1 L_1 + k_2 L_2 + k_1 L_3)}}{\sqrt{\rho/k_1}(k_1 L_2)^{3/2}} U(\theta - \hat{\theta}). \tag{10}$$

One observes that the domain of existence of G'_{sb} coincides precisely with that predicted from Fig. 5.5.2(b) if the need for a finite lateral segment L_2 along the interface is recognized. The solution evidently fails when $L_2 \to 0$ (dashed line), corresponding to observation points along the angle of total reflection. In this instance, the saddle point at $w = \theta$ in the integrand of Eq. (3b) moves near the branch point $w = \hat{\theta}$, thereby necessitating a more elaborate calculation in terms of parabolic cylinder functions (see Secs 4.4c and 7.5d, where lateral waves on an anisotropic interface are treated in detail). The equiphase surface $(k_2 L_2 + k_1 L_3) = $ constant is conical (planar in the ρz plane), in contrast to the spherical equiphase surfaces descriptive of the direct and reflected fields [see Fig. 1.7.9(b)]. Owing to the continuous leakage of energy along the lateral path, the amplitude of the wave decays more rapidly with lateral distance than that of the direct or reflected fields. Equation (10) emphasizes succinctly why G'_{sb} may dominate the geometric-optical field in Eq. (5) when ϵ_1 is lossy and ϵ_2 is lossless. The geometric-optical field trajectory is confined entirely to the dissipative region, whereas the lateral portion of the lateral-wave field trajectory proceeds in the lossless region.

The disposition of the incident–reflected–refracted ($\bar{S}'_{1,2,3}$) and incident–lateral–refracted ($\bar{S}_{1,2,3}$) ray groupings is conveniently inferred from the wave-number plot in Fig. 5.5.2(a). For further discussion of these plots, see Sec. 1.7d with Figs. 1.7.3 and 1.7.7.

Analytical details

The three-dimensional E-mode Green's function $G'(\mathbf{r}, \mathbf{r}')$ defined in Eqs. (2), from which the electromagnetic fields can be derived via Eqs. (5.2.1) and (5.2.4c), has the generic integral representation given in Eq. (5.2.11).[14,15] The one-dimensional modal Green's function $g_{zi}(z, z')$ satisfies a differential equation of the type (5.2.6b) in each region (the subscript i is suppressed):†

$$\left(\frac{d^2}{dz^2} + \kappa_1^2\right) g_{z1}(z, z') = -\delta(z - z'), \qquad -\infty < \frac{z}{z'} < 0, \tag{11a}$$

$$\left(\frac{d^2}{dz^2} + \kappa_2^2\right) g_{z2}(z, z') = 0, \qquad 0 < z < \infty, \tag{11b}$$

subject to a radiation condition at $|z| \to \infty$ and to the following continuity conditions at $z = 0$:

$$g_{z1} = g_{z2}, \qquad \frac{1}{\epsilon_1}\frac{dg_{z1}}{dz} = \frac{1}{\epsilon_2}\frac{dg_{z2}}{dz}. \tag{11c}$$

---

†All modal quantities in this section refer to E modes. To simplify the notation, the distinguishing primes will be omitted.

The propagation constants are defined as $\kappa_{1,2} = \sqrt{k_{1,2}^2 - \xi^2}$, with Im $\kappa_{1,2} < 0$ for the exp $(j\omega t)$ dependence and Im $\kappa_{1,2} > 0$ for the exp $(-i\omega t)$ dependence. The network analogue of Eqs. (11) is shown in Fig. 5.5.3 (g_z is proportional

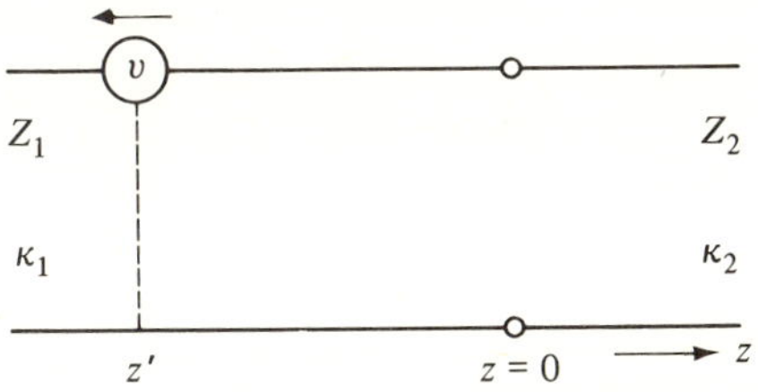

FIG. 5.5.3   Equivalent modal network (E modes).

to the current), and either directly from the differential equation (see Sec. 3.4) or from a network calculation,[†] one may derive the solution for $z < 0$ as in Eq. (5.3.1). Since region 2 is semiinfinite, the input impedance seen to the right from $z = 0$ is equal to the characteristic impedance $Z_2 = \kappa_2/\omega\epsilon_2$, so

$$\overrightarrow{\Gamma}_i(0) \equiv \Gamma(\xi) = \frac{Z_2 - Z_1}{Z_2 + Z_1} = \frac{(\kappa_2/\epsilon_2) - (\kappa_1/\epsilon_1)}{(\kappa_2/\epsilon_2) + (\kappa_1/\epsilon_1)}, \tag{12}$$

thus confirming Eq. (3d). Equation (3a) is thereby established (with $j \to -i$) and Eq. (3b) follows by transforming into the w plane; one recalls that Eq. (5.4.7a) permits the extraction of the infinite space contribution in closed form. Equation (4) follows from the recognition that, from Eqs. (11b) and (11c) and the radiation condition at $z = \infty$,

$$g_{z2}(z, z') = g_{z2}(0, z')e^{-j\kappa_2 z} = g_{z1}(0, z')e^{-j\kappa_2 z}. \tag{13}$$

The integral in Eq. (3b) is in the generic form of the exp$(-i\omega t)$ equivalent of Eq. (5.3.12a), with $n = 0$. Upon employing the large-argument approximation for the Hankel function in Eq. (5.3.13a), one may reduce the integral for G'_s to the one given in Eq. (5.3.14), with the following identification:

$$f(w) = -\frac{e^{i\pi/4}}{8\pi} \sqrt{\frac{2k_1}{\pi r \sin\theta}} \sqrt{\sin w}\, \Gamma(k_1 \sin w), \qquad \bar{\alpha} = \theta, \; L = r. \tag{14}$$

An asymptotic evaluation by the steepest-descent procedure then furnishes for real k_1 the contributions listed in Eqs. (5.3.15) and (5.3.16); in particular, the asymptotic formula for the steepest-descent-path integral in Eq. (5.3.16a) yields the result listed in Eq. (5).

To account for the possible contributions from singularities, their location must be clarified. For convenience, the original integration path in the w plane has been redrawn in Fig. 5.5.4(a). In addition to the branch point at $w = 0$ arising from the Hankel function, there are branch point and pole singularities due to the reflection coefficient $\Gamma(k_1 \sin w)$. Let us assume that μ and ϵ_1 are

---

[†]See Eqs. (2.4.29); Eqs. (11c) specify continuity of voltage and current at $z = 0$.

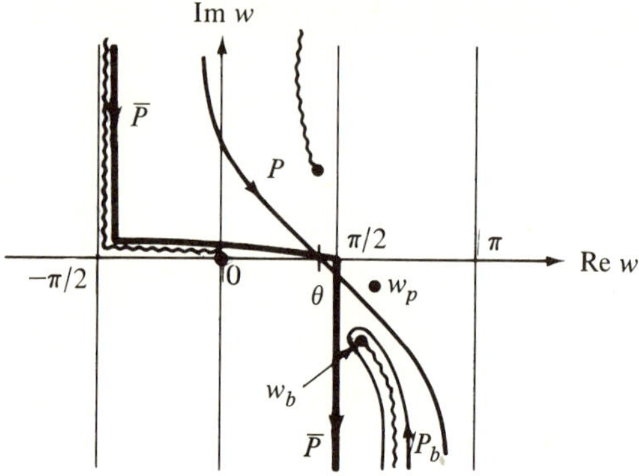

(a) Integration paths in w plane

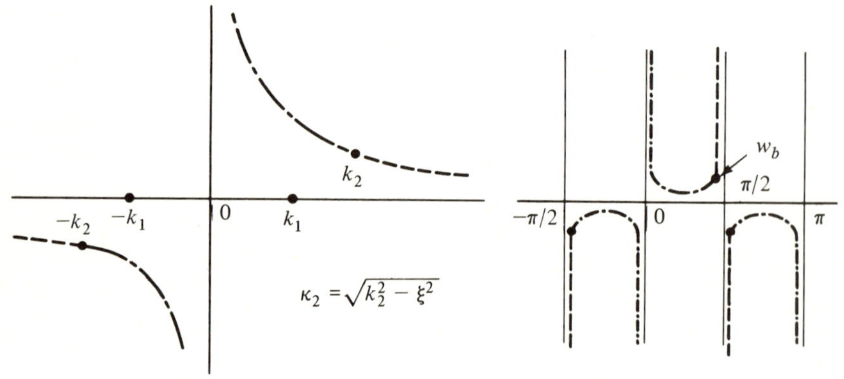

(b) Singularities in ξ plane

$$\kappa_2 = \sqrt{k_2^2 - \xi^2}$$

(c) Singularities in w plane

$$(\xi = k_1 \sin w)$$

$$-\!-\!- \ \operatorname{Re} \kappa_2 = 0$$
$$-\!\cdot\!- \ \operatorname{Im} \kappa_2 = 0$$

FIG. 5.5.4 Contours of integration and singularities.

real and that ϵ_2 may be dissipative; that is, for an $\exp(-i\omega t)$ time dependence, $\epsilon_2 = \epsilon_2' + i\epsilon_2''$, where ϵ_2' and ϵ_2'' are real and $\epsilon_2' > 0$, $\epsilon_2'' \geq 0$. Thus,

$$\epsilon = \frac{\epsilon_2}{\epsilon_1} = |\epsilon|\, e^{i2\psi}, \qquad 0 \leq \psi < \frac{\pi}{4}, \tag{15a}$$

and

$$\sqrt{\epsilon} = \sqrt{|\epsilon|}\, e^{i\psi}. \tag{15b}$$

The branch points are located at $w_b = \pm\sin^{-1}\sqrt{\epsilon}$, and with $w_b = w_{br} + iw_{bi}$, w_{br}, w_{bi} real, the relation

$$\sin w_b = \sin w_{br} \cosh w_{bi} + i \cos w_{br} \sinh w_{bi} = \sqrt{|\epsilon|}\cos\psi + i\sqrt{|\epsilon|}\sin\psi,$$

$$\tag{15c}$$

implies their location as in Fig. 5.5.4(a). If $\psi \to 0$, with $|\epsilon| > 1$, then $w_b \to \pi/2 + iw_{bi}$, while if $\psi \to 0$, with $|\epsilon| < 1$, $w_{bi} \to 0$. It is recalled that on the top sheet of the four-sheeted Riemann surface shown in Fig. 5.3.6(a) with $k_N = k_2 = k_1\sqrt{\epsilon}$, $\sqrt{k_1^2 - \xi^2} \equiv k_1$ when $\xi = 0$, and $\sqrt{k_1^2\epsilon - \xi^2} \equiv k_1\sqrt{\epsilon}$ when $\xi = 0$. Consequently, on the top sheet of the two-sheeted w plane of Fig. 5.5.4(a), $\sqrt{\epsilon - \sin^2 w} \equiv \sqrt{\epsilon}$ when $w = 0$ (i. e., the square root is positive when the radicand is positive). The behavior of Re κ_2 and Im κ_2 is clarified in Fig. 5.5.4(b) and (c), which exhibit curves along which Re $\kappa_2 = 0$ and Im $\kappa_2 = 0$. Since the algebraic sign of Re κ_2 (Im κ_2) can change only upon crossing a curve on which Re $\kappa_2 = 0$ (Im $\kappa_2 = 0$), a choice of branch cut along the curve(s) Re $\kappa_2 = 0$ (Im $\kappa_2 = 0$) assures that the algebraic sign of Re κ_2 (Im κ_2) is constant on the entire top sheet of the two-sheeted Riemann surface. The sign of Im κ_2 (Re κ_2), however, changes whenever a curve Im $\kappa_2 = 0$ (Re $\kappa_2 = 0$) is crossed.

The pole singularities w_p of $\Gamma(k_1 \sin w)$ in Eq. (3e) are located at

$$\sqrt{\epsilon - \sin^2 w_p} = -\epsilon \cos w_p. \tag{16}$$

Since the real and imaginary parts of $\epsilon(1 + \epsilon)^{-1/2}$ are positive, this relation locates the poles in those regions of the w plane in which the real and imaginary parts of $\kappa_2 = k_1\sqrt{\epsilon - \sin^2 w}$ are also positive. For the choice of branch cuts in Fig. 5.5.4(a), Re $\kappa_2 > 0$ everywhere on the top sheet. Upon squaring both sides of Eq. (16) and simplifying, and noting from above that the square root is defined to be positive when the radicand is positive, one finds for the pole singularity pertinent to the subsequent asymptotic calculation,

$$\cos w_p = -\frac{1}{\sqrt{1 + \epsilon}}, \qquad \sin w_p = \sqrt{\frac{\epsilon}{1 + \epsilon}}. \tag{16a}$$

Since $0 < \arg\sqrt{1 + \epsilon} < \pi/4$, the pole is situated in the strip $\pi/2 < \text{Re } w < \pi$, Im $w < 0$, in Fig. 5.5.4(a). One may verify that the pole is not intercepted by the steepest-descent path P but that it may move close to the point $w = \pi/2$ when $|\epsilon| \gg 1$. It is therefore effective only when the dielectric in region 2 is very lossy and when the observation point is near the interface $\theta \approx \pi/2$.

A branch point may, however, be intercepted provided that the observation angle $\theta > \hat{\theta}$, where $\hat{\theta}$ is given in Eq. (6a) [see also Eq. (5.3.15a)]. Although its contribution may be written down directly from the general formula (5.3.16b), with Eq. (14), it is instructive to carry out the derivation in detail for the special case at hand. We consider the integral

$$I_b = \int_{P_b} \sqrt{\sin w} \, e^{ik_1 r \cos(w - \theta)} \Gamma(k_1 \sin w) \, dw, \tag{17}$$

where P_b is the path in Fig. 5.5.4(a). It is convenient to introduce the transformation,

$$\cos(w - \theta) = \cos(w_b - \theta) + is^2, \qquad 0 < s^2 < \infty, \tag{18}$$

which maps the branch point w_b into the origin in the s plane. s is a real variable running from $-\infty$ to $+\infty$. Upon taking the real part of Eq. (18) and employing Eq. (6a), one finds that

$$\cos (w_r - \theta) \cosh w_i = \frac{\cos (w_{br} - \theta)}{\cos (w_{br} - \hat{\theta})}, \qquad \theta \geq \hat{\theta}. \tag{18a}$$

Along P_b, which follows the steepest-descent contour through $w = \hat{\theta}$, one has $\cos (w_r - \hat{\theta}) \cosh w_i = 1$. To satisfy Eq. (18a) for $\theta > \hat{\theta}$, the path in Fig. 5.5.4(a) must be distorted toward the right. Since there are no singularities near P_b, such a distortion can be carried out readily and assures the validity of the mapping in Eq. (18). Upon substituting Eq. (18) into Eq. (17), one obtains

$$I_b = e^{ik_1 r \cos (w_b - \theta)} \int_{-\infty}^{\infty} G(s) e^{-k_1 r s^2} \, ds, \tag{19}$$

where

$$G(s) = \sqrt{\sin w} \, \Gamma(k_1 \sin w) \frac{dw}{ds}. \tag{19a}$$

An asymptotic evaluation of the integral in Eq. (19) can now be carried out directly via Eq. (4.2.17) provided that $G(s)$ in Eq. (19a) is regular in the vicinity of $s = 0$. To determine the properties of the mapping in Eq. (18), it will be convenient to assume that ϵ is real. If $\epsilon > 1$, the pertinent branch point is located at $w_b = \pi/2 + iw_{bi}$, $w_{bi} < 0$, with $\cosh w_{bi} = \sqrt{\epsilon}$. Then, from Eq. (18),

$$\frac{dw}{ds} = \frac{-i2s}{\pm\sqrt{1 - [\cos(w_b - \theta) + is^2]^2}} \simeq \frac{-i2s}{\sin(w_b - \theta)} + O(s^3), \text{ near } s = 0. \tag{20a}$$

The choice of sign in the last term of Eq. (20a) assures that s increases from $-\infty$ to $+\infty$ as w moves along the deformed path (to be called P_b') in the direction shown in Fig. 5.5.4(a). This is verified upon noting that

$$\sin (w_b - \theta) = \cosh w_{bi}[\cos \theta + i \sin \theta \tanh w_{bi}],$$

whence $0 > \arg \sin (w_b - \theta) > -\theta$. Since $ds > 0$ along P_b', Eq. (20a) yields, near $w = w_b$,

$$\arg dw = -\arg \sin (w_b - \theta) - \frac{\pi}{2} + \arg s, \tag{20b}$$

which behavior is consistent with Fig. 5.5.4(a) when P_b is distorted as noted above; note that $\arg s = \pi$ when $s < 0$, and $\arg s = 2\pi$ when $s > 0$. Upon expanding $\cos (w - \theta)$ in Eq. (18) in a power series about $w = w_b$, one finds that

$$s \cong \sqrt{i \sin (w_b - \theta)} \sqrt{w - w_b} \, e^{i\pi}. \tag{20c}$$

The choice of sign in Eq. (20c) is consistent with Eq. (20b), as is verified upon noting that, on the left side of the distorted branch cut in Fig. 5.5.4(a), $s > 0$ and $\arg(w - w_b) = \arg dw$ near w_b. Hence, the first-order branch point at $w = w_b$ maps into the regular point $s = 0$. Since

$$\sqrt{\epsilon - \sin^2 w} = \sqrt{w - w_b}\,\sqrt{-\sin 2w_b}\,[1 + O(w - w_b)], \tag{21a}$$

one has

$$\Gamma(k_1 \sin w) = -1 - \frac{2s}{\epsilon \cos w_b}\left[\frac{-\sin 2w_b}{i \sin (w_b - \theta)}\right]^{1/2} + O(s^2). \tag{21b}$$

It is recalled that $\sqrt{\epsilon - \sin^2 w} > 0$ when $\sin w < \sqrt{\epsilon}$ (i. e., for $w = \pi/2 - i|w_i|$, $|w_i| < |w_{bi}|$). This condition is satisfied by the right-hand side of Eq. (21a) since $-\sin 2w_b = -i|\sinh 2w_{bi}|$, while $w - w_b = i[|w_{bi}| - |w_i|]$. Finally,

$$\sqrt{\sin w} = \sqrt{\sin w_b} + O(s^2). \tag{21c}$$

These results apply also to complex ϵ, provided that the square roots are so defined as to reduce to the above when arg $\epsilon \to 0$.

Upon substituting Eqs. (21) into Eq. (19a), one notes that $G(0) = 0$. Since only the even-powered terms in the power-series expansion of $G(s)$ about $s = 0$ contribute to the integral in Eq. (19), one finds from Eq. (4.2.17) that the asymptotic approximation of I_b is given for large values of $k_1 r$ by

$$I_b \sim \frac{e^{ik_1 r \cos(w_b - \theta)}}{(k_1 r)^{3/2}} \, \frac{2\sqrt{2\pi}\, e^{-i\pi/4} \sin w_b}{\epsilon \sqrt{\cos w_b}\, [\sin(w_b - \theta)]^{3/2}}, \tag{22}$$

which leads to the form in Eq. (6) since $\sin w_b = \sqrt{\epsilon}$, $\cos w_b = \sqrt{1 - \epsilon}$. The above considerations apply to $|\epsilon| > 1$. If $|\epsilon| < 1$, the branch point w_{b1} in Fig. 5.5.4(a) moves toward the real axis in the interval $\pi/2 < w_r < \pi$, while w_{b2} moves toward the real axis in the interval $0 < w_r < \pi/2$. Hence, w_{b2} is the important branch-point singularity in this range and its contribution can be evaluated using the same procedure as above.

It is to be noted that Eq. (22) is inapplicable when $\theta \to w_b$ (i.e., when $\theta \to \hat{\theta}$ for the case $\epsilon < 1$). This is caused by the proximity of the branch-point singularity and the saddle point $w = \theta$, thereby invalidating Eqs. (20) and (21). The behavior of I_b at $\theta = w_b$ is ascertained readily. Upon making the transformation to s via Eq. (18), one finds that dw/ds is finite near $s = 0$, $s \propto w - \theta$, and that I_b becomes proportional to $e^{ik_1 r}\int_0^\infty e^{-k_1 r s^2}\sqrt{s}\, ds$, whence, upon changing variables to $t = -k_1 r s^2$, I_b is seen to vary like $e^{ik_1 r}(k_1 r)^{-3/4}$. The complete transition from the radial dependence $(k_1 r)^{-3/2}$ when $\theta \neq w_b$ to the dependence $(k_1 r)^{-3/4}$ when $\theta = w_b$ can be expressed in terms of parabolic cylinder functions (for further details, see Secs. 4.4c and 7.5d). It should be emphasized that the steepest-descent evaluation of the reflected-wave integral is likewise influenced when $\theta \to w_b$, although the geometric-optical result in Eq. (5) remains valid; the effect occurs in the $O(1/k_1^2 r^2)$ term.

While the pole at w_p in Eq. (16a) is not intercepted by the steepest descent path, its proximity to the saddle point when $\theta \approx \pi/2$ and $|\epsilon| \gg 1$ influences the saddle-point calculation; as noted earlier, the branch-point contribution is negligible under these conditions. To appreciate some of the difficulties, we consider an approximation to the reflection coefficient in the saddle-point result (5) valid when $\theta \not\approx \pi/2$:

$$\Gamma(k_1 \sin \theta) \cong \frac{1 - \sqrt{\epsilon}\,\cos\theta}{1 + \sqrt{\epsilon}\,\cos\theta} \cong -1 + \frac{2}{\sqrt{\epsilon}\,\cos\theta}, \qquad |\epsilon| \gg 1, \tag{23}$$

where the last relation applies only when $\sqrt{|\epsilon|}\,\cos\theta \gg 1$. As $\theta \to \pi/2$, i.e., for source and observation point locations near the interface, the reflection coefficient in Eq. (23) changes very rapidly from $\Gamma \approx -1$ to $\Gamma \approx +1$, so the function $\Gamma(k_1 \sin w)$ in the integrand of Eq. (3b) can no longer be considered as slowly varying in the vicinity of the saddle point $w = \theta \approx \pi/2$. In fact, both the numerator and denominator of $\Gamma(k_1 \sin w)$ may become very small. To circumvent this difficulty in the numerator it is convenient to write

$$\Gamma(k_1 \sin w) = -1 + \hat{\Gamma}(k_1 \sin w), \quad \hat{\Gamma}(k_1 \sin w) = \frac{2\sqrt{\epsilon - \sin^2 w}}{\sqrt{\epsilon - \sin^2 w} + \epsilon\cos w}. \tag{24}$$

The contribution to the integral in Eq. (3b) from the (-1) term on the right-hand side of Eq. (24) is just the free-space Green's function relative to the image point. The $\hat{\Gamma}$ term in Eq. (24) possesses a pole singularity w_p as defined in Eq. (16a), which, for $|\epsilon| \gg 1$, $\arg \epsilon \approx \pi/2$, is located at

$$w_p \cong \frac{\pi}{2} + \frac{e^{-i\pi/4}}{\sqrt{|\epsilon|}} \tag{25}$$

(i.e., near the saddle point). Consequently, the saddle-point evaluation of Eq. (3a) must be modified to account explicitly for the presence of the pole.

The appropriate procedure has been discussed in Sec. 4.4a. One notes first that the function $(w - w_p)\hat{\Gamma}(k_1 \sin w)$ is regular near w_p and hence near the saddle point $w = \theta \approx \pi/2$. Therefore, for a first-order asymptotic evaluation, the contribution to the steepest-descent integral arising from $\hat{\Gamma}$ can be represented as

$$\hat{I}_s = \int_P \sqrt{\sin w}\, e^{ik_1 r \cos(w-\theta)} \hat{\Gamma}(k_1 \sin w)\, dw$$

$$\sim (\theta - w_p)\hat{\Gamma}(k_1 \sin \theta)\sqrt{\sin \theta} \int_P \frac{e^{ik_1 r \cos(w-\theta)}}{w - w_p}\, dw. \tag{26}$$

The integral in Eq. (26) is identical in form with that in Eq. (4.2.23), the result for which is listed in Eq. (4.4.34). Thus, for large $k_1 r$ and any θ in the range $0 < \theta \leq \pi/2$, one finds [10,16]

$$\hat{I}_s \sim \sqrt{\sin \theta}\,(\theta - w_p)\hat{\Gamma}(k_1 \sin \theta)e^{ik_1 r}\left[i2\sqrt{\pi}\, e^{-k_1 r b^2} Q(-ib\sqrt{k_1 r}) + T(0)\sqrt{\frac{\pi}{k_1 r}} \right] \tag{27}$$

where the relevant quantities are defined in Eq. (7b). The result in Eq. (7) follows. For large $\sqrt{k_1 r}\,|b|$ values (i.e., $w_p - \theta \not\approx 0$), $Q(-ib\sqrt{k_1 r})$ can be approximated by its asymptotic representation (4.4.20) and then yields for G_1' the ordinary saddle-point result.

For a derivation of Eqs. (8) via the approximate characterization of the medium by a normalized surface impedance $Z_s = 1/\sqrt{\epsilon}$, see Sec. 5.7.

The asymptotic formula for the transmitted field G_2' in Eq. (9) results from a saddle-point evaluation of the integral in Eq. (3a), modified according to Eq. (4). Use of the large-argument approximation for the Hankel function renders the form of the integral identical with that in Eq. (5.3.19b), subject to the identifications

$$\eta \to \xi, \quad y \to \rho, \quad \bar{f} \to \frac{e^{i\pi/4}}{4\pi\sqrt{2\pi\rho}}\sqrt{\frac{\xi}{k_1^2 - \xi^2}}[1 - \Gamma(\xi)], \quad \bar{\alpha}_{1,2} \to \theta_{1,2}.$$

The expression in Eq. (9) follows by substitution into Eq. (5.3.20) [see also Eq. (5.3.23)], with the recognition that the saddle point is given by $\xi_s = k_1 \sin\theta_1 = k_2 \sin\theta_2$. Attention may be called to the use of the wavenumber diagram in Fig. 5.3.10(a) in facilitating the interpretation of the saddle-point condition and of the ray-optical character of the fields.

When the source and observation points are near the interface, complications in the behavior of G_2' may arise due to the previously mentioned pole and branch-point singularities. These aspects are not pursued further.

5.5b Time-harmonic Transverse Electric Current Element

$$\hat{\mathbf{J}}(\mathbf{r}, t) = Il\delta(\boldsymbol{\rho})\delta(z - z')e^{-i\omega t}\mathbf{x}_0. \tag{28}$$

A transverse current element of strength $J^o = Il$, where I is the current in the element and l is its infinitesimal length, in the presence of a semiinfinite dielectric medium is shown in Fig. 5.5.5. In this instance the source excites

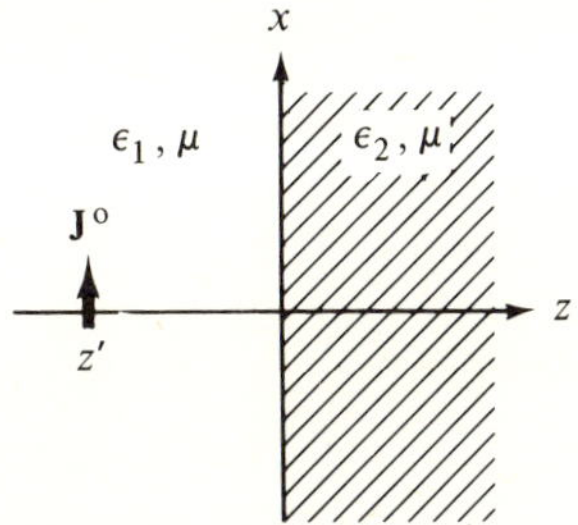

FIG. 5.5.5 Dielectric interface with transverse current element.

both E and H modes along the z direction. The electromagnetic fields are inferred from the functions $\nabla_t'\mathscr{S}'$ and $\nabla_t'\mathscr{S}''$ as in Eqs. (5.2.1) and (5.2.4), where $\mathscr{S}'$ and $\mathscr{S}''$ satisfy the separate differential equations [see Eqs. (5.2.2)]

$$(\nabla^2 + k_1^2)\nabla_t^2 \begin{cases} \mathscr{S}_1'(\mathbf{r}, \mathbf{r}') \\ \mathscr{S}_1''(\mathbf{r}, \mathbf{r}') \end{cases} = \delta(\mathbf{r} - \mathbf{r}'), \qquad z < 0, \tag{29a}$$

$$(\nabla^2 + k_2^2)\nabla_t^2 \begin{cases} \mathscr{S}_2'(\mathbf{r}, \mathbf{r}') \\ \mathscr{S}_2''(\mathbf{r}, \mathbf{r}') \end{cases} = 0, \qquad z > 0, \tag{29b}$$

subject to a radiation condition at infinity in both regions, and to the following continuity requirement at $z = 0$ [see Eqs. (2.3.36) et seq.]:

$$\mathscr{S}'_1 = \mathscr{S}'_2, \qquad \frac{1}{\epsilon_1}\frac{\partial \mathscr{S}'_1}{\partial z} = \frac{1}{\epsilon_2}\frac{\partial \mathscr{S}'_2}{\partial z} \qquad \text{for } E \text{ modes}, \tag{30a}$$

$$\mathscr{S}''_1 = \mathscr{S}''_2, \qquad \frac{\partial \mathscr{S}''_1}{\partial z} = \frac{\partial \mathscr{S}''_2}{\partial z} \qquad \text{for } H \text{ modes}. \tag{30b}$$

It has been recognized that G and $\mathscr{S}$ have the same continuity conditions across an interface perpendicular to the z axis since they are related by the *transverse* operator $-\nabla_t^2 \mathscr{S} = G$. In a cylindrical-coordinate description of the cross section transverse to z, $\nabla_t' \mathscr{S}$ is given for $z < 0$ and for $\rho' = 0$ by the integral [Eq. (5.2.12)]

$$\nabla_t' \mathscr{S}_1(\mathbf{r},\,\mathbf{r}') = \boldsymbol{\rho}_0 \frac{i}{8\pi} \int_{\infty e^{i\pi}}^{\infty} \frac{H_1^{(1)}(\xi\rho)}{\sqrt{k_1^2 - \xi^2}} \left[\exp(i\sqrt{k_1^2 - \xi^2}\,|z - z'|) \right.$$

$$\left. - f(\xi)\exp[-i\sqrt{k_1^2 - \xi^2}(z + z')] \right] d\xi, \tag{31}$$

where for the E–mode case $\mathscr{S} = \mathscr{S}'$,

$$f(\xi) = \Gamma'(\xi) = \frac{\sqrt{k_1^2\epsilon - \xi^2} - \epsilon\sqrt{k_1^2 - \xi^2}}{\sqrt{k_1^2\epsilon - \xi^2} + \epsilon\sqrt{k_1^2 - \xi^2}}, \qquad \epsilon = \frac{\epsilon_2}{\epsilon_1}, \tag{31a}$$

and for the H–mode case $\mathscr{S} = \mathscr{S}''$,

$$f(\xi) = -\Gamma''(\xi) = \frac{\sqrt{k_1^2\epsilon - \xi^2} - \sqrt{k_1^2 - \xi^2}}{\sqrt{k_1^2\epsilon - \xi^2} + \sqrt{k_1^2 - \xi^2}}, \tag{31b}$$

with $k_{1,2}^2 = \omega^2\mu\epsilon_{1,2}$.

The integration path proceeds as in Fig. 5.3.6(a) on the top sheet of the Riemann surface on which all square roots are defined to have positive imaginary parts. The representation for $\nabla_t' \mathscr{S}_2(\mathbf{r},\,\mathbf{r}')$ is as in Eq. (31) provided that one replaces the quantity inside the brackets by

$$[\quad] \rightarrow [1 - f(\xi)]\exp(i\sqrt{k_2^2 - \xi^2}\,z - i\sqrt{k_1^2 - \xi^2}\,z'). \tag{32}$$

The solution in an unbounded region with wavenumber k_1, represented by the contribution $\nabla_t' \mathscr{S}_{1f}$ from the first term inside the brackets in Eq. (31), can be expressed in closed form [Eq. (5.4.19b)]; as in Eq. (3), this leaves an integral representation for the interface effect $\nabla_t' \mathscr{S}_s$ observed in region 1. It may also be remarked that the unbounded-space (primary) portion of the field solution is calculated more directly from the formulas

$$\mathbf{H}_{1f} = J^o \nabla \times \mathbf{x}_0 G_{f1}, \qquad \mathbf{E}_{1f} = \frac{J^o}{-i\omega\epsilon_1}\nabla \times \nabla \times \mathbf{x}_0 G_{f1}, \tag{33}$$

which follow from Eqs. (5.2.1) and (5.2.4c), with the recognition that the current element in the unbounded region excites only E modes with respect to its axial direction. G_{f1} is defined in Eq. (3).

An asymptotic evaluation of the integral representation for $\nabla_t' \mathscr{S}'_s$ and $\nabla_t' \mathscr{S}''_s$ may be carried out by the techniques described in Sec. 5.5a and yields wave constituents analogous to those encountered in the longitudinal dipole case.

The derivation of Eq. (31) beyond the general form given in Eq. (5.2.12) requires substitution of the modal Green's functions $g'_{zi}(z, z')$ for the E modes and $g''_{zi}(z, z')$ for the H modes. The modal network problem is schematized in Fig. 5.5.6 and the result for the modal Green's function g'_{zi} is the same as in

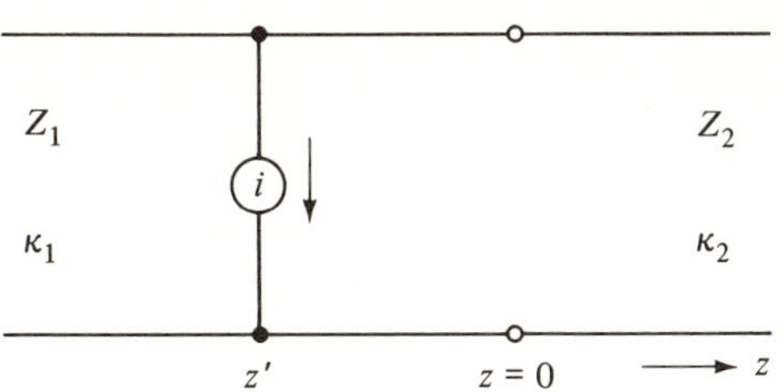

FIG. 5.5.6 Equivalent modal network (E or H modes).

Eqs. (11)–(13). For the H-mode case, g''_{zi} satisfies Eqs. (11) with $g_z \to g''_z$, provided that the second equation in (11c) is replaced by $dg''_{z1}/dz = dg''_{z2}/dz$ at $z = 0$; this is a consequence of the fact that both regions in Fig. 5.5.5 have the same permeability μ. The H-mode reflection coefficient $\Gamma''(\xi)$ is given by

$$\Gamma''(\xi) = \frac{Z''_2 - Z''_1}{Z''_2 + Z''_1} \tag{34}$$

and yields Eq. (31b) since $Z''_{1,2} = \omega\mu/\kappa''_{1,2}$, $\kappa''_{1,2} = \sqrt{k^2_{1,2} - \xi^2}$.

An alternative evaluation of the horizontal dipole field has been described by Sommerfeld,[17,18] wherein he utilizes a Hertzian potential function $\Pi = x_0\Pi_x + z_0\Pi_z$, instead of the z-directed E-and H-mode potentials employed above.

5.5c   Time-Harmonic Magnetic Current Element

The electromagnetic fields excited by a magnetic source current density

$$\hat{\mathbf{M}}(\mathbf{r}, t) = Vl\delta(\boldsymbol{\rho})\delta(z - z')e^{-i\omega t}\mathbf{u}_0, \tag{35}$$

where V is the source element voltage, l is its infinitesimal length, and $\mathbf{u}_0 = z_0$ or x_0 for the longitudinal and transverse orientations, respectively, can be derived via Eqs. (5.2.1) from the scalar functions $G''_{1,2}$ or $\mathscr{S}'_{1,2}$, $\mathscr{S}''_{1,2}$. The relevant functions $\nabla'_t\mathscr{S}$ have already been determined in Eqs. (31). The H-mode Green's function G'' is defined as in Eqs. (2) with $G' \to G''$, except that the second equation in (2c) is replaced by $\partial G''_1/\partial z = \partial G''_2/\partial z$ at $z = 0$. In the resulting representations, as in Eqs. (3) and (4), the E-mode reflection coefficient $\Gamma(\xi) \equiv \Gamma'(\xi)$ in Eq. (3d) is replaced by $[-\Gamma''(\xi)]$ in Eq. (31b). Additional details are left as an exercise for the reader.

5.5d   Pulsed Longitudinal Electric Current Element

$$\hat{\mathbf{J}}(\mathbf{r}, t) = \hat{p}\delta(\boldsymbol{\rho})\delta(z - z') \frac{d}{dt} \delta(t)z_0. \tag{36}$$

When the current source in Fig. 5.5.1 has the pulsed behavior prescribed by the impulsive dipole moment of strength $\hat{p}$ in Eq. (36) (see Fig. 5.2.2), and when the dielectrics ϵ_1 and ϵ_2 are assumed to be frequency independent (non-dispersive), the electromagnetic fields may be derived via Eqs. (5.2.18) from the space and time dependent scalar E-mode Green's function $\hat{G}'(\mathbf{r}, \mathbf{r}'; t, t')$ defined by

$$\left(\nabla^2 - \frac{1}{\bar{c}_1^2}\frac{\partial^2}{\partial t^2}\right)\hat{G}_1'(\mathbf{r}, \mathbf{r}'; t, t') = -\delta(\mathbf{r} - \mathbf{r}')\delta(t - t'), \qquad z < 0, \qquad (37a)$$

$$\left(\nabla^2 - \frac{1}{\bar{c}_2^2}\frac{\partial^2}{\partial t^2}\right)\hat{G}_2'(\mathbf{r}, \mathbf{r}'; t, t') = 0 \qquad\qquad z > 0, \qquad (37b)$$

subject to quiescence for $t < t'$, and the following continuity requirements at $z = 0$ [see Eqs. (2)]:

$$\hat{G}_1' = \hat{G}_2', \quad \frac{1}{\epsilon_1}\frac{\partial \hat{G}_1'}{\partial z} = \frac{1}{\epsilon_2}\frac{\partial \hat{G}_2'}{\partial z} \qquad \text{at} \quad z = 0, \qquad (37c)$$

with $\bar{c}_{1,2}^2 = (\mu\epsilon_{1,2})^{-1}$. The solution in the region $z < 0$ is given by

$$\hat{G}_1'(\mathbf{r}, \mathbf{r}'; t, t') = \hat{G}_{f1}(\mathbf{r}, \mathbf{r}'; t, t') + \hat{G}_s'(\mathbf{r}, \mathbf{r}'; t, t'), \qquad (38)$$

where for $\rho' = 0, t' = 0$, $\hat{G}_{f1}$ is the free-space Green's function in Eq. (5.4.14),

$$\hat{G}_{f1}(\mathbf{r}, \mathbf{r}'; t, t') = \frac{\delta(t - \hat{r}/\bar{c}_1)}{4\pi\hat{r}}. \qquad (38a)$$

$\hat{G}_s'$ accounts for the reflection from the interface and is given for $\epsilon_2 > \epsilon_1$ by

$$\hat{G}_s'(\mathbf{r}, \mathbf{r}'; t, t') = \begin{cases} -\dfrac{1}{4\pi^2}\dfrac{d}{dt}\, A(t), & t > \dfrac{r}{\bar{c}_1}, \\[2ex] 0, & t < \dfrac{r}{\bar{c}_1}, \end{cases} \qquad (38b)$$

where

$$A(t) = \frac{2}{r}\int_0^{\pi/2} \text{Re}\, \Gamma\left[\theta - i\cosh^{-1}\frac{t}{[t^2 + (r/\bar{c}_1)^2\cos^2 v]^{1/2}}\right]dv, \qquad (39a)$$

with

$$\Gamma(w) = \frac{\epsilon_2\cos w - \epsilon_1\sqrt{(\Omega_2/\Omega_1)^2 - \sin^2 w}}{\epsilon_2\cos w + \epsilon_1\sqrt{(\Omega_2/\Omega_1)^2 - \sin^2 w}}, \qquad (39b)$$

$$\Omega_{1,2} = \sqrt{\frac{1}{\bar{c}_{1,2}^2} + \left[\left(\frac{t}{r}\right)^2 - \frac{1}{\bar{c}_1^2}\right]\sin^2 v}. \qquad (39c)$$

As in Fig. 5.5.1, $\hat{r}$ and r denote the distance to the observation point (ρ, z) from the source point and the image point, respectively, and θ is the angle between r and the negative z axis. These rather complicated exact results, valid for arbitrary observation times, reduce substantially when $t \approx r/\bar{c}_1$, in which instance one may employ the more direct formulas derivable from the time-harmomic solution [see Eqs. (1.7.80) and (1.7.81)]. The reflected response $\hat{G}_s'$ appears to originate at the image point $r = 0$ and first reaches the observation point after a time interval $t = r/\bar{c}_1$ required for propagation over the dashed-line path in Fig. 5.5.1. When $\epsilon_2 < \epsilon_1$, an additional contribution, the transient counterpart of the lateral wave in Fig. 5.5.2, must be taken into account (see also Fig. 5.5.8).

Analytical details

Because of the occurrence of the Hankel function in the integral representation (3b), the (cylindrical waveguide) time-harmonic solution is not in the form of Eq. (5.2.20), which admits a direct inversion for recovery of the transient response. Another approach must therefore be taken, and it seems best to proceed from a rectangular waveguide representation as in Eq. (5.2.7).[19] For an exp ($j\omega t$) dependence, the rectangular waveguide analogue of Eq. (3b) is

$$G_s'(\mathbf{r}, \mathbf{r}', \omega) = \frac{j}{8\pi^2} \int_{-\infty}^{\infty} d\xi \int_{-\infty}^{\infty} d\eta \, \frac{e^{-j\xi x - j\eta y - j\kappa_1 Z}}{\kappa_1} \, \Gamma(\kappa_1, \kappa_2), \qquad (40)$$

where $Z = |z + z'|$ and

$$\Gamma(\kappa_1, \kappa_2) = -\frac{\epsilon_2 \kappa_1 - \epsilon_1 \kappa_2}{\epsilon_2 \kappa_1 + \epsilon_1 \kappa_2}, \qquad \kappa_{1,2} = \sqrt{\frac{\omega^2}{\bar{c}_{1,2}^2} - \xi^2 - \eta^2}, \qquad \mathrm{Im}\,\kappa_{1,2} \leq 0. \tag{40a}$$

Next, we let $\omega \to -js$, $s > 0$, and introduce the change of scale $\xi = \alpha's$, $\eta = \beta's$, which renders Γ independent of s and also allows the factor s to appear explicitly in the exponential:

$$G_s' = -\frac{s}{8\pi^2} \int_{-\infty}^{\infty} \int_{-\infty}^{\infty} \frac{e^{s(-j\alpha'x - j\beta'y - \gamma_1'Z)}}{\gamma_1'} \, \Gamma(\gamma_1', \gamma_2') \, d\alpha' \, d\beta', \qquad (41)$$

where

$$\gamma_{1,2}' = \sqrt{\frac{1}{\bar{c}_{1,2}^2} + \alpha'^2 + \beta'^2}, \qquad \mathrm{Re}\,\gamma_{1,2}' \geq 0. \tag{41a}$$

Cylindrical coordinates (ρ, ϕ) in the xy plane are now introduced via a change of variable that makes the direction of the vector $\mathbf{x}_0\alpha' + \mathbf{y}_0\beta'$ identical to that of the radius vector $\boldsymbol{\rho} = \rho(\mathbf{x}_0 \cos\phi + \mathbf{y}_0 \sin\phi) \equiv \mathbf{x}_0 x + \mathbf{y}_0 y$.

$$\alpha' = \alpha \cos\phi - \beta \sin\phi, \qquad \beta' = \alpha \sin\phi + \beta \cos\phi. \tag{42}$$

Thus,

$$\alpha'x + \beta'y = \alpha\rho, \qquad \alpha'^2 + \beta'^2 = \alpha^2 + \beta^2, \qquad d\alpha' \, d\beta' = d\alpha \, d\beta, \tag{42a}$$

so Eq. (41) can be written as

$$G_s' = -\frac{s}{8\pi^2} \int_{-\infty}^{\infty} d\beta \int_{-\infty}^{\infty} \frac{e^{s(-j\alpha\rho - \gamma_1 Z)}}{\gamma_1} \, \Gamma(\gamma_1, \gamma_2) \, d\alpha, \qquad (43)$$

where

$$\gamma_{1,2} = \sqrt{\alpha^2 + \Omega_{1,2}^2}, \qquad \Omega_{1,2} = \sqrt{\frac{1}{\bar{c}_{1,2}^2} + \beta^2}, \qquad \mathrm{Re}\,\gamma_{1,2} \geq 0, \quad \Omega_{1,2} > 0. \tag{43a}$$

The integral over α can be cast into the form given in Eq. (5.2.20). First, apply the transformation

$$\alpha = \Omega_1 \sinh\chi, \qquad (44)$$

so $\gamma_1 = \Omega_1 \cosh \chi$. Upon also utilizing spherical coordinates (r, θ) via $\rho = r \sin \theta$, $Z = r \cos \theta$, $0 < \theta < \pi/2$, one obtains

$$I = \int_{-\infty}^{\infty} \frac{e^{-s(j\alpha\rho+\gamma_1 Z)}}{\gamma_1} \Gamma(\gamma_1, \gamma_2)\, d\alpha = \int_{-\infty}^{\infty} e^{-sr\Omega_1 \cosh(\chi+j\theta)} \Gamma(\gamma_1, \gamma_2)\, d\chi. \qquad (45)$$

With $\chi = -jw$,

$$I = -j \int_{-j\infty}^{j\infty} e^{-sr\Omega_1 \cos(w-\theta)} \Gamma(w)\, dw,$$

$$\Gamma(w) = \frac{\epsilon_2 \cos w - \epsilon_1 \sqrt{\Omega_2^2/\Omega_1^2 - \sin^2 w}}{\epsilon_2 \cos w + \epsilon_1 \sqrt{\Omega_2^2/\Omega_1^2 - \sin^2 w}}, \qquad (46a)$$

or equivalently,

$$I = -j \int_{-j\infty}^{j\infty} e^{-sr\Omega_1 \cos w} \Gamma(w + \theta)\, dw. \qquad (46b)$$

The transition from Eq. (46a) to Eq. (46b) is permitted if Γ has no singularities in the strip $0 < |\mathrm{Re}\, w| < \pi/2$. If $\Omega_1 < \Omega_2$ (i. e., $\epsilon_2 > \epsilon_1$), the branch-point singularities w_b lie on the lines $\mathrm{Re}\, w = \pm\pi/2$. The pole singularities w_p of Γ are located at

$$\cos w_p = -\sqrt{\frac{(\Omega_2/\Omega_1)^2 - 1}{\epsilon - 1}}, \qquad \epsilon = \frac{\epsilon_2}{\epsilon_1} = \frac{\bar{c}_1^2}{\bar{c}_2^2}. \qquad (47)$$

$\cos w_p$ is real when $\epsilon > 1$, and must be chosen negative since $\sqrt{(\Omega_2/\Omega_1)^2 - \sin^2 w} > 0$ when the radicand is positive. Hence, the poles also lie outside the range $|\mathrm{Re}\, w| < \pi/2$, and Eq. (46b) is valid for $0 < \theta < \pi/2$ when $\epsilon > 1$. Thus, from Eq. (5.2.22),

$$I = 2 \int_{r\Omega_1}^{\infty} \frac{e^{-s\tau}}{\sqrt{\tau^2 - (r\Omega_1)^2}} \mathrm{Re}\, \Gamma\left[\theta + j \cosh^{-1}\left(\frac{\tau}{r\Omega_1}\right)\right] d\tau, \qquad \epsilon > 1, \qquad (48)$$

with Ω_1 defined in Eq.(43a).

The desired formulation in Eq. (5.2.19a) is achieved after substituting Eq. (48) into Eq. (43) and interchanging the orders of integration.[19] The τ integration extends from the curve $\tau = r\sqrt{(1/\bar{c}_1)^2 + \beta^2}$ to $\tau = \infty$, while $-\infty < \beta < \infty$. If the β integration is performed first, one has $[(\tau/r)^2 - 1/\bar{c}_1^2] < |\beta|$, while $r/\bar{c}_1 < \tau < \infty$. Thus,

$$G_s' = -\frac{s}{4\pi^2} \int_0^{\infty} e^{-s\tau} A(\tau)\, d\tau, \qquad (49)$$

where

$$A(\tau) = \begin{cases} 0 & \tau < \dfrac{r}{\bar{c}_1}, & (50a) \\[2ex] \displaystyle\int_{-b}^{b} \frac{\mathrm{Re}\, \Gamma[\theta + j \cosh^{-1}(\tau/r\Omega_1)]}{\sqrt{\tau^2 - (r\Omega_1)^2}}\, d\beta, & \tau > \dfrac{r}{\bar{c}_1}, & (50b) \end{cases}$$

and $b = [(\tau/r)^2 - 1/\bar{c}_1^2]^{1/2}$, $\Omega_1^2 = (1/\bar{c}_1^2) + \beta^2$. A final change of variable

$$\beta = \left[\left(\frac{\tau}{r}\right)^2 - \frac{1}{\bar{c}_1^2}\right]^{1/2} \sin \nu, \qquad (51)$$

transforms the expression for $A(\tau)$ into the one in Eq. (39), and Eq. (38) follows from Eq. (5.2.19b) and from the equivalence $s \rightarrow d/dt$.

5.5e   Time-Harmonic Transverse Electric Line Current

$$\hat{\mathbf{J}}(\mathbf{r}, t) = I\delta(\hat{\boldsymbol{\rho}} - \hat{\boldsymbol{\rho}}')e^{-i\omega t}\mathbf{x}_0 \tag{52}$$

The fields excited by an electric line current with constant strength I located at $\hat{\boldsymbol{\rho}}' = (y', z')$, $z' < 0$, in the presence of a dielectric interface at $z = 0$ can be derived from the scalar two-dimensional (H-mode) Green's function $\bar{G}(\hat{\boldsymbol{\rho}}, \hat{\boldsymbol{\rho}}')$, which satisfies the differential equations†

$$\left(\frac{\partial^2}{\partial y^2} + \frac{\partial^2}{\partial z^2} + k_1^2\right)\bar{G}_1(\hat{\boldsymbol{\rho}}, \hat{\boldsymbol{\rho}}'), = -\delta(\hat{\boldsymbol{\rho}} - \hat{\boldsymbol{\rho}}') \qquad z < 0, \tag{53a}$$

$$\left(\frac{\partial^2}{\partial y^2} + \frac{\partial^2}{\partial z^2} + k_2^2\right)\bar{G}_2(\hat{\boldsymbol{\rho}}, \hat{\boldsymbol{\rho}}') = 0 \qquad z > 0, \tag{53b}$$

subject to a radiation condition at infinity in both regions, and to the continuity requirements

$$\bar{G}_1 = \bar{G}_2, \qquad \frac{\partial \bar{G}_1}{\partial z} = \frac{\partial \bar{G}_2}{\partial z} \quad \text{at} \quad z = 0, \tag{53c}$$

with $k_{1,2}^2 = \omega^2\mu\epsilon_{1,2}$. The solution is given by

$$\bar{G}_1(\hat{\boldsymbol{\rho}}, \hat{\boldsymbol{\rho}}') = \bar{G}_{f1}(\hat{\boldsymbol{\rho}}, \hat{\boldsymbol{\rho}}') + \bar{G}_s(\hat{\boldsymbol{\rho}}, \hat{\boldsymbol{\rho}}'), \qquad z < 0, \tag{54}$$

where $\bar{G}_{f1}$ is the free-space Green's function for medium 1 [Eq. (5.4.24) with $k = k_1$] and $\bar{G}_s$ contains the interface effect,

$$\bar{G}_s(\hat{\boldsymbol{\rho}}, \hat{\boldsymbol{\rho}}') = \begin{cases} \dfrac{i}{4\pi}\displaystyle\int_{-\infty}^{\infty} \dfrac{e^{i\eta(y-y')}}{\sqrt{k_1^2 - \eta^2}}\Gamma''(\eta) \exp\left[-i\sqrt{k_1^2 - \eta^2}(z + z')\right] d\eta, & (54a) \\[2em] \dfrac{i}{4\pi}\displaystyle\int_{\bar{P}} e^{ik_1 R \cos(w-\varphi)}\Gamma''(k_1 \sin w) \, dw, & (54b) \end{cases}$$

where

$$\Gamma''(\eta) = \frac{\sqrt{k_1^2 - \eta^2} - \sqrt{k_1^2\epsilon - \eta^2}}{\sqrt{k_1^2 - \eta^2} + \sqrt{k_1^2\epsilon - \eta^2}}, \qquad \epsilon = \frac{\epsilon_2}{\epsilon_1}, \tag{54c}$$

and the transformation $\eta = k_1 \sin w$ has been employed. The integration path $\bar{P}$ is shown in Fig. 5.3.6(b). For the region $z > 0$,

$$\bar{G}_2(\hat{\boldsymbol{\rho}}, \hat{\boldsymbol{\rho}}') =$$

$$\frac{i}{4\pi}\int_{-\infty}^{\infty} \frac{\exp[i\eta(y - y') + i\sqrt{k_1^2\epsilon - \eta^2}z]}{\sqrt{k_1^2 - \eta^2}} \exp\left(-i\sqrt{k_1^2 - \eta^2}\, z'\right)[1 + \Gamma''(\eta)] \, d\eta.$$

$$\tag{55}$$

The electromagnetic fields are calculated from [see Eq. (5.4.31)]

$$E_{x1,2} = i\omega\mu I\bar{G}_{1,2}, \qquad E_y = E_z = 0, \tag{56a}$$

---

† For simplicity, the distinguishing superscript $''$ for H modes is omitted on $\bar{G}$.

$$H_{y1,2} = I\frac{\partial}{\partial z}\bar{G}_{1,2}, \qquad H_{z1,2} = -I\frac{\partial}{\partial y}\bar{G}_{1,2}, \qquad H_x = 0. \qquad (56b)$$

An asymptotic evaluation for large values of $k_1 R$, where R is the distance from the image point in Fig. 5.5.7, leads to the following approximation for $\bar{G}_s$:

$$\bar{G}_s = I_s + U(\varphi - \hat{\varphi})I_b, \qquad (57)$$

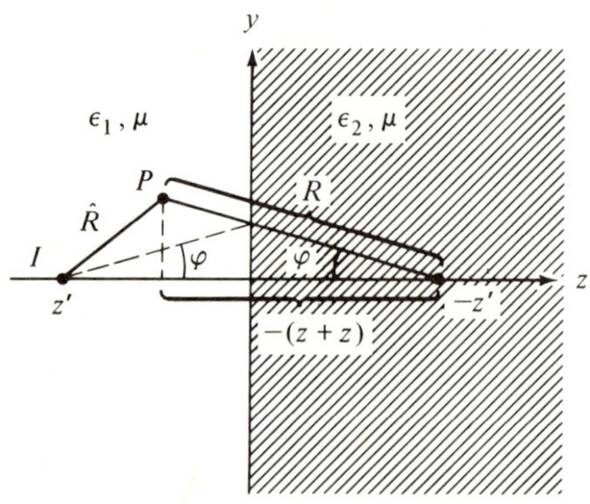

FIG. 5.5.7   Dielectric interface with parallel line current.

U being the unit step function, with the steepest-descent path contribution I_s given by

$$I_s \sim \frac{i}{4\pi}\sqrt{\frac{2\pi}{k_1 R}}\, e^{i\,(k_1 R - \pi/4)}\,\Gamma''(k_1 \sin\varphi)\left[1 + O\left(\frac{1}{k_1 R}\right)\right], \qquad (57a)$$

$$\Gamma''(k_1 \sin\varphi) = \frac{\cos\varphi - \sqrt{\epsilon - \sin^2\varphi}}{\cos\varphi + \sqrt{\epsilon - \sin^2\varphi}}. \qquad (57b)$$

The branch-point contribution I_b is

$$I_b \sim \frac{1}{\sqrt{2\pi}}\frac{\exp[ik_1(\sqrt{1-\epsilon}\,|z+z'| + \sqrt{\epsilon}\,y) + i\pi/4]}{[k_1(\sqrt{\epsilon}\,|z+z'| - \sqrt{1-\epsilon}\,y)]^{3/2}}\left(\frac{\epsilon}{1-\epsilon}\right)^{1/4}. \qquad (57c)$$

$\hat{\varphi}$ is the angle given on the right-hand side of Eq. (6a), while φ is the polar angle shown in Fig. 5.5.7 [see also Eq. (10) and Fig. 5.5.2].

When $k_1\hat{R}$ is also large, one may employ the asymptotic form of the free-space Green's function [see Eq. (5.3.13b)], and the result for $\bar{G}_1$ may then be written as

$$\bar{G}_1 \sim \frac{e^{i\pi/4}}{2\sqrt{2\pi}}\left[\frac{e^{ik_1\hat{R}}}{\sqrt{k_1\hat{R}}} + \Gamma''(k_1\sin\varphi)\frac{e^{ik_1 R}}{\sqrt{k_1 R}}\right]\left[1 + O\left(\frac{1}{k_1 R}\right) + O\left(\frac{1}{k_1\hat{R}}\right)\right]$$

$$+ U(\varphi - \hat{\varphi})I_b, \qquad k_1\hat{R} \gg 1, \quad k_1 R \gg 1. \qquad (58)$$

The first group of terms on the right-hand side of Eq. (58) represents the two-dimensional (cylindrical wave) geometric-optical field while the contribution from I_b may be interpreted as a lateral wave [see also Fig. 1.7.9 and Eqs. (1.7.64b) and (1.7.64c)]. The relative importance of these constituents is the same as for the longitudinal dipole problem in Sec. 5.5a and the discussion will therefore not be repeated. The branch-cut integral contribution is $O[(k_1 R)^{-3/2}]$ and therefore smaller than the geometric-optical field except when ϵ is real and $\varphi \to \pi/2$ (glancing incidence) in which case $\hat{R} \to R$ and $\Gamma''(k_1) = -1$. In this instance, the entire field is $O[(k_1 R)^{-3/2}]$, and for a calculation of $\bar{G}$, the asymptotic evaluation of $\bar{G}_{f1}$ and $\bar{G}_s$ must be carried out to the next higher order. The higher-order terms in the asymptotic representation for $\bar{G}_{f1}$ can be obtained from Eq. (5.4.37). For a higher-order evaluation of the steepest-descent integral in Eq. (54b) over the path $\bar{P}$, one employs Eqs. (4.2.17), (4.A2c), and (4.A6c), after first introducing the change of variable $\cos(w - \varphi) = 1 + is^2$, $-\infty < s < \infty$.

It may be emphasized that Γ'' in Eq. (57b) is well behaved for $0 \leq \varphi \leq \pi/2$ even when $|\epsilon| \gg 1$, so the field excited by the electric line source does not exhibit a behavior analogous to that described in Eqs. (7) and (8).

Analytical details

The discussion in Eqs. (5.4.31) applies here as well, therefore justifying Eqs. (56). Since the electric field has only an x component, only H modes with respect to z are relevant, so $\bar{G} \equiv \bar{G}''$ is the two-dimensional H-mode Green's function. The differential equations (53a) and (53b) follow from Eqs. (5.2.3b) with $\partial/\partial x \equiv 0$, and the boundary conditions at $z = 0$ may be inferred from the discussion following Eq. (2.3.36).

The integral representations for $\bar{G}_s$ follow from Eqs. (5.2.13a) and (5.3.1) (with $j \to -i$), the calculation of the H-mode reflection coefficient being described in connection with Eq. (34). It may be noted that the modal network problem is the same as in Fig. 5.5.6 since the present electric source current distribution is also transverse; only the strength of the current generator i differs in the two cases. The integral in Eq. (54b) is exactly in the form given in Eq. (5.3.12b), allowance being made for the different time dependence $\exp(-i\omega t)$, so the results in Eqs. (5.3.15) and (5.3.16) apply directly [see also the analogous treatment of I_b in Eq. (17)]. It may be noted that while $\Gamma''(k_1 \sin w)$ has branch-point singularities at $w_b = \pm \sin^{-1}\sqrt{\epsilon}$, no pole singularities analogous to those for the E-mode case in Eq. (16) are present. Difficulties from an adjacent pole and saddle point therefore do not arise here.

In the construction of $\bar{G}_s$ in Eq. (54a) by the modal network technique, simplification results on direct recognition of the nondependence on x of the total field, and consequent use of x-independent transverse eigenfunctions. Alternatively, and less directly, one may proceed by using the two-dimensional transverse eigenfunctions in Eq. (3.2.40b): $\psi_i(\boldsymbol{\rho}) = (1/2\pi) \exp(-j\xi x - j\eta y)$, $-\infty < (\xi, \eta) < \infty$. On evaluating the source current generator strengths from Eqs. (2.2.14), with $\mathbf{J} = \mathbf{x}_0 I \delta(y - y')\delta(z - z')$, one finds that for the E modes, $i'_i \propto \xi\delta(\xi) = 0$, while for the H modes, $i''_i = -jI\delta(\xi) \exp(j\eta y')$. Recalling from Eq. (2.3.9a) that $V''_i(z, z') = -i''_i Z''_i(z, z')$, evaluating $\Pi''(\mathbf{r})$ from Eq.

(2.3.5) (with $\sum_i \to \iint_{-\infty}^{\infty} d\xi\, d\eta$) and noting from Eq. (2.3.6a) that $\mathbf{E} = \mathbf{x}_0 E_x = -\mathbf{x}_0 j\omega\mu\partial\Pi''(\mathbf{r})/\partial y$, one arrives at Eqs. (54).

5.5f   Time-harmonic Transverse Line Distribution of Longitudinally Directed Electric Current Elements

$$\hat{\mathbf{J}}(\mathbf{r}, t) = J^o\delta(\hat{\boldsymbol{\rho}} - \hat{\boldsymbol{\rho}}')e^{-i\omega t}\mathbf{z}_0. \tag{59}$$

The source configuration at $\hat{\boldsymbol{\rho}}' = (0, z')$ in Fig. 5.5.7 has the form sketched in Fig. 5.4.7, and the electromagnetic fields may be derived as in Eqs. (5.4.39) [see also Eqs. (5.2.1) and (5.2.4c)] from a two-dimensional E-mode Green's function $\bar{G}'(\hat{\boldsymbol{\rho}}, \hat{\boldsymbol{\rho}}')$ that satisfies Eqs. (53a) and (53b) subject to a radiation condition at infinity and to boundary conditions as in Eq. (2c) at $z = 0$. The solution for $\bar{G}'$ is the same as in Eqs. (54) except that $-\Gamma''(\eta)$ is replaced by the E-mode reflection coefficient $\Gamma(\eta)$ in Eq. (3d) [see Eqs. (77) and (78)]. Asymptotic results valid in the far zone may be derived as in preceding examples.

5.5g   Time-harmonic Progressively Phased Transverse Electric Line Currents

Transversely directed current elements

$$\hat{\mathbf{J}}(\mathbf{r}, t) = I\delta(\hat{\boldsymbol{\rho}} - \hat{\boldsymbol{\rho}}')e^{i\alpha x}e^{-i\omega t}\mathbf{x}_0. \tag{60}$$

If the line source in Fig. 5.5.7 has a linear phase variation $\exp(i\alpha x)$, then the electric field everywhere can no longer be represented by the single component E_x as in Eq. (56a), but both E and H modes, relative to the z direction, are excited [see Eqs. (5.4.68a)]. The electromagnetic fields in this case are inferred from the functions $\nabla'_t\bar{\mathcal{S}}'$ and $\nabla'_t\bar{\mathcal{S}}''$ via Eqs. (5.2.1) and (5.2.4), where

$$\bar{\mathcal{S}}(\mathbf{r}; \hat{\boldsymbol{\rho}}') = \int_{-\infty}^{\infty} \mathcal{S}(\mathbf{r}, \mathbf{r}')e^{i\alpha x'}\, dx', \tag{61}$$

and $\mathcal{S}'(\mathbf{r}, \mathbf{r}')$, $\mathcal{S}''(\mathbf{r}, \mathbf{r}')$ are the three-dimensional functions defined in Eqs. (29). The functions $\mathcal{S}$ have the integral representation given in Eq. (5.2.14), with $j \to -i$, g_{zi} taken from Eqs. (5.3.1) and (13) for $z < 0$ and $z > 0$, respectively, and $\Gamma(\eta)$ taken from Eqs. (31a) and (31b) for the E-mode case ($\mathcal{S}'$) and H-mode case ($\mathcal{S}''$), respectively.

Asymptotic approximations for $\bar{\mathcal{S}}$ may be determined as in Sec. 5.5a.

Longitudinally directed current elements

$$\hat{\mathbf{J}}(\mathbf{r}, t) = J^o\delta(\hat{\boldsymbol{\rho}} - \hat{\boldsymbol{\rho}}')e^{i\alpha x}e^{-i\omega t}\mathbf{z}_0. \tag{62}$$

In this case, the fields are derivable from the E-mode Green's function $\bar{G}'(\mathbf{r}; \hat{\boldsymbol{\rho}}')$ defined in terms of $G'(\mathbf{r}, \mathbf{r}')$ [Eqs. (2)] as in Eq. (61). An integral representation analogous to Eq. (61) is obtained by substitution into Eq. (5.2.13).

5.5h   Time-harmonic Ring Currents

When ring currents are situated at a dielectric interface perpendicular to the z axis (see Fig. 5.4.13), representations for the fields may be obtained as in Sec. 5.4f, provided that the modal Green's function g_{zi} for the unbounded medium [Eq. (5.4.7)] is replaced by that for the semiinfinite medium described in connection with Eq. (61).

5.5i Pulsed Transverse Electric Line Currents

When the line current in Fig. 5.5.7 has the temporal dependence

$$\hat{\mathbf{J}}(\mathbf{r}, t) = \delta(\hat{\boldsymbol{\rho}} - \hat{\boldsymbol{\rho}}')\hat{I}(t)\mathbf{x}_0, \qquad \hat{I}(t) = I\delta(t) \tag{63}$$

the electromagnetic fields may be inferred from Eqs. (56) (with $i\omega \to -\partial/\partial t$),

$$\hat{E}_{x1,2} = -\mu I \frac{\partial}{\partial t} \hat{\bar{G}}_{1,2}, \qquad \hat{E}_y = \hat{E}_z = 0, \tag{64a}$$

$$\hat{H}_{y1,2} = I \frac{\partial}{\partial z} \hat{\bar{G}}_{1,2}, \qquad \hat{H}_{z1,2} = -I \frac{\partial}{\partial y} \hat{\bar{G}}_{1,2}, \qquad \hat{H}_x = 0, \tag{64b}$$

where the time-dependent two-dimensional H-mode Green's function $\hat{\bar{G}}'' \equiv \hat{\bar{G}}$ satisfies the differential equations

$$\left(\frac{\partial^2}{\partial y^2} + \frac{\partial^2}{\partial z^2} - \frac{1}{\bar{c}_1^2}\frac{\partial^2}{\partial t^2}\right)\hat{\bar{G}}_1(\hat{\boldsymbol{\rho}}, \hat{\boldsymbol{\rho}}'; t, t') = -\delta(\hat{\boldsymbol{\rho}} - \hat{\boldsymbol{\rho}}')\delta(t - t'), \qquad z < 0, \tag{65a}$$

$$\left(\frac{\partial^2}{\partial y^2} + \frac{\partial^2}{\partial z^2} - \frac{1}{\bar{c}_2^2}\frac{\partial^2}{\partial t^2}\right)\hat{\bar{G}}_2(\hat{\boldsymbol{\rho}}, \hat{\boldsymbol{\rho}}'; t, t') = 0, \qquad z > 0, \tag{65b}$$

subject to quiescence for $t < t'$, and to the continuity requirements

$$\hat{\bar{G}}_1 = \hat{\bar{G}}_2, \qquad \frac{\partial}{\partial z}\hat{\bar{G}}_1 = \frac{\partial}{\partial z}\hat{\bar{G}}_2 \quad \text{at} \quad z = 0. \tag{65c}$$

The electromagnetic propagation speeds in the two regions are $\bar{c}_{1,2} = (\mu\epsilon_{1,2})^{-1/2}$; also $\hat{\boldsymbol{\rho}} = (y, z)$.

The solution in the region $z < 0$ may be written for $t' = 0$ as

$$\hat{\bar{G}}_1(\hat{\boldsymbol{\rho}}, \hat{\boldsymbol{\rho}}'; t, t') = \hat{\bar{G}}_{f1}(\hat{\boldsymbol{\rho}}, \hat{\boldsymbol{\rho}}'; t, t') + \hat{\bar{G}}_s(\hat{\boldsymbol{\rho}}, \hat{\boldsymbol{\rho}}'; t, t'), \tag{66}$$

where $\hat{\bar{G}}_{f1}$, the Green's function for the unbounded medium 1, is given in Eqs. (5.4.42) with $\bar{c} \to \bar{c}_1$, $\hat{\rho} \to \hat{R}$ (see Fig. 5.5.7), while

$$\hat{\bar{G}}_s = \begin{cases} \dfrac{1}{2\pi} \dfrac{\operatorname{Re} \Gamma[\varphi + j\cosh^{-1}(\bar{c}_1 t/R)]}{\sqrt{t^2 - R^2/\bar{c}_1^2}}, & t > \dfrac{R}{\bar{c}_1}, & (67a) \\[4ex] 0, & t < \dfrac{R}{\bar{c}_1}, & (67b) \end{cases}$$

with

$$\Gamma(w) = \frac{\cos w - \sqrt{\epsilon - \sin^2 w}}{\cos w + \sqrt{\epsilon - \sin^2 w}}, \qquad \epsilon = \frac{\epsilon_2}{\epsilon_1} = \frac{\bar{c}_1^2}{\bar{c}_2^2}. \tag{67c}$$

For $t' \neq 0$, one replaces t by $t - t'$. R and φ are polar coordinates with respect to the image point (Fig. 5.5.7), and the dielectric constants $\epsilon_{1,2}$ are assumed to be frequency independent, with $\epsilon_2 > \epsilon_1$. Thus, when $\epsilon_2 > \epsilon_1$, the field due to an impulsive source located at $(0, z')$, $z' < 0$, observed at the observation point (y, z), $z < 0$, comprises the direct wave $\hat{\bar{G}}_{f1}$ plus a reflected contribution $\hat{\bar{G}}_s$ which has an amplitude given by Re Γ and which appears to emanate from the image point $(0, -z')$ located in a medium with wave velocity $\bar{c}_1$. When $\epsilon_1 > \epsilon_2$, an additional contribution may arise which represents the transient counterpart of the lateral wave sketched in Fig. 5.5.2 (see also Fig. 5.5.8).

Analytical details

The time-harmonic solution for $\bar{G}_s$ in Eq. (54b) is in the form given in Eq. (5.2.20) (with $j \to -i$) so that the recovery of the transient result in Eq. (67) follows at once from Eqs. (5.2.19) and (5.2.23). The restriction $\epsilon_2 > \epsilon_1$ (i.e., $\epsilon > 1$) assures that the branch points $w_b = \pm\sin^{-1}\sqrt{\epsilon}$ of $\Gamma(k_1 \sin w)$ are not crossed during the path deformation leading to Eq. (5.2.21); for convenience, $\Gamma(k_1 \sin w)$ has been abbreviated by $\Gamma(w)$ in Eq. (67c).

When $\epsilon < 1$, the branch points lie on the real w axis in the interval $|\text{Re } w| < \pi/2$, and they may contribute a lateral wave (Figs 5.5.2 and 5.5.8) in certain spatial regions. The occurrence of an additional wave solution may be appre-

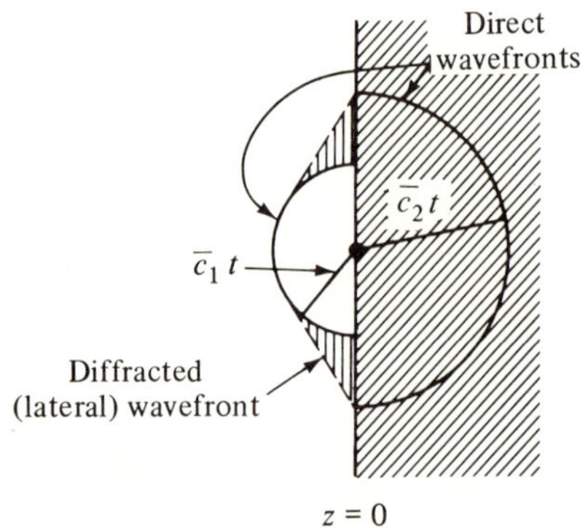

FIG. 5.5.8   Wavefronts when source is on interface. Points in vertically shaded region are reached first by diffracted wave traveling a ray path corresponding to that of lateral wave in Fig. 5.5.2.

ciated on simple physical grounds when the source lies in the interface. The disturbance propagates radially outward in each region with propagation speeds $\bar{c}_1$ and $\bar{c}_2$, where $\bar{c}_2 > \bar{c}_1$. In view of the required continuity across the interface [Eq. (65c)], the field spills over from region 2 into region 1, giving rise to the diffracted wave front. It is not difficult to verify that the diffracted wave front exists precisely in the same spatial domain wherein the lateral wave in Fig. 5.5.2 (with $L_1 = 0$) is present, and that the ray trajectories L_2 and L_3 remain valid.[20]

5.5j   Point Charge in Uniform Straight Motion Parallel to Interface

A point charge q moving with constant speed v parallel to the x axis as in Fig. 5.5.9 may be characterized by the source current density

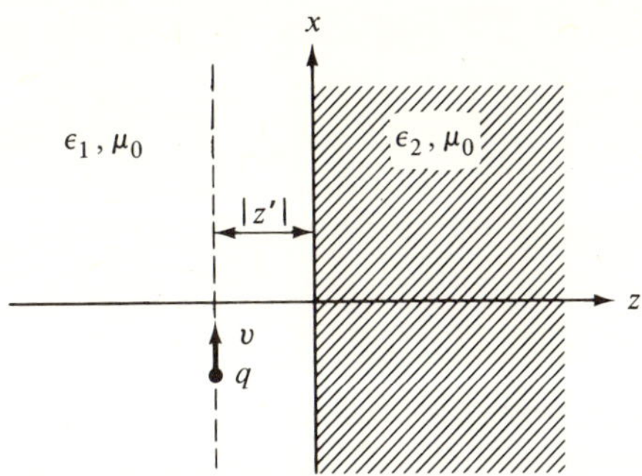

FIG. 5.5.9 Particle moving parallel to a plane interface between two media.

$$\hat{\mathbf{J}}(\mathbf{r}, t) = qv\delta(x - vt)\delta(\hat{\boldsymbol{\rho}} - \hat{\boldsymbol{\rho}}')\mathbf{x}_0, \qquad \hat{\boldsymbol{\rho}} = (y, z). \qquad (68)$$

As in the latter portion of Sec. 5.4e, we shall emphasize here the radiated energy rather than the actual fields, and consequently deal directly with Eq. (5.2.34), which requires a knowledge of the modal voltage and current solutions $V_i(z, \omega)$ and $I_i(z, \omega)$. The dielectric constants ϵ_1 and ϵ_2 are kept arbitrary although in many practical applications, medium 1 is taken as vacuum ($\epsilon_1 = \epsilon_0$). In the latter instance, radiation may be produced in the denser medium 2 if $(\epsilon_2/\epsilon_0)\beta^2 > 1$, $\beta = v/c$, in which instance the evanescent waves incident from medium 1 (see Sec. 5.4e) are converted into propagating waves in medium 2.

The associated steady-state problem, wherein the line current distribution in Eq. (5.4.62) flows along the particle trajectory in Fig. 5.5.9, has been solved in Sec. 5.4e for an unbounded medium. In the present case, the equivalent modal network problem is shown in Fig. 5.5.6, and its solution is obtained from Eqs. (2.4.23) and (2.4.6) as

$$V_i(z) = -\frac{Z_i i_i}{2}\left[e^{-j\kappa_{t1}|z-z'|} + \Gamma_i(0)e^{+j\kappa_{t1}(z+z')}\right], \qquad z \leq 0, \quad (69a)$$

$$I_i(z) = -\frac{i_i}{2}\left[\operatorname{sgn}(z - z')e^{-j\kappa_{t1}|z-z'|} - \Gamma_i(0)e^{+j\kappa_{t1}(z+z')}\right], \quad z \leq 0, \quad (69b)$$

$$V_i(z) = V_i(0)e^{-j\kappa_{t2}z}, \qquad I_i(z) = I_i(0)e^{-j\kappa_{t2}z}, \qquad z \geq 0, \quad (69c)$$

when $\operatorname{sgn} x = \pm 1$, $x \gtrless 0$. Upon defining the relative dielectric constants

$$\bar{\epsilon}_1 = \frac{\epsilon_1}{\epsilon_0}, \qquad \bar{\epsilon}_2 = \frac{\epsilon_2}{\epsilon_0}, \qquad (70)$$

one may utilize the definitions for Z_i and κ_i given in Eq. (5.4.68b), with $\bar{\epsilon}$ replaced by $\bar{\epsilon}_1$ and $\bar{\epsilon}_2$ in the regions $z < 0$ and $z > 0$, respectively. The modal current generator strengths are the same as in Eq. (5.4.68a), while the reflection coefficients are, from Eq. (2.4.12),

$$\Gamma_i(0) = \frac{Z_{i2} - Z_{i1}}{Z_{i2} + Z_{i1}}. \qquad (71)$$

This completes the solution of the modal network problem.

To calculate the energy flow W_ω into region 2, it is convenient to employ Eq. (69c). Since

$$V_i(0) = -\frac{Z_{i1}Z_{i2}i_i}{Z_{i1} + Z_{i2}} e^{j\kappa_{i1}z'}, \qquad I_i(0) = \frac{V_i(0)}{Z_{i2}}, \tag{72}$$

one finds that for imaginary κ_{i2},

$$\vec{P}_i = V_i'I_i'^* + V_i''I_i''^* = 0, \tag{73}$$

while for (positive) real κ_{i2},

$$\vec{P}_i = \begin{cases} \dfrac{q^2\omega\mu_0\kappa_{i2}}{2\pi k_{ti}^2}\left[\dfrac{\kappa_{i1}^2\bar{\epsilon}_2}{\beta^2(\bar{\epsilon}_2\kappa_{i1} + \bar{\epsilon}_1\kappa_{i2})^2} + \dfrac{\eta^2}{(\kappa_{i1} + \kappa_{i2})^2}\right], & \kappa_{i1} \text{ real,} \tag{73a} \\[2em] \dfrac{q^2\omega\mu_0\kappa_{i2}e^{-2|\kappa_{i1}z'|}}{2\pi k_{ti}^2}\left[\dfrac{|\kappa_{i1}|^2\bar{\epsilon}_2}{\beta^2(\bar{\epsilon}_2^2|\kappa_{i1}|^2 + \bar{\epsilon}_1^2\kappa_{i2}^2)} + \dfrac{\eta^2}{|\kappa_{i1}|^2 + \kappa_{i2}^2}\right], & \kappa_{i1} \text{ imaginary.} \tag{73b} \end{cases}$$

Equation (73b) applies when $\bar{\epsilon}_2 > 1/\beta^2 > \bar{\epsilon}_1$, and only this case is considered further. Upon substituting for $\kappa_{i1,2}$ from Eqs. (5.4.68) and (5.4.67b), one finds after some manipulation that the expression inside the brackets in Eq. (73b) may be written in the form $ak_{ti}^2(\eta^2 + f)[b(\eta^2 + g)]^{-1}$, where a, b, f, and g are quantities independent of η, so one obtains the following result for the real power carried in a combined E- and H-mode field characterized by the index η:

$$\vec{P}_i \equiv \vec{P}_\eta =$$

$$\frac{q^2\omega\mu_0}{2\pi k_0^2(\bar{\epsilon}_2 - \bar{\epsilon}_1)}\sqrt{k_0^2\left(\bar{\epsilon}_2 - \frac{1}{\beta^2}\right) - \eta^2}\,\frac{\eta^2 + f}{\eta^2 + g}\exp\left[-2\sqrt{\eta^2 + k_0^2\left(\frac{1}{\beta^2} - \bar{\epsilon}_1\right)}|z'|\right] \tag{74}$$

where $k_0 = \omega/c$ and

$$f = \frac{k_0^2\bar{\epsilon}_2(1 - \bar{\epsilon}_1\beta^2)}{\beta^2(\bar{\epsilon}_1 + \bar{\epsilon}_2)}, \quad g = \frac{k_0^2[\bar{\epsilon}_2(1 - \bar{\epsilon}_1\beta^2) + \bar{\epsilon}_1]}{\beta^2(\bar{\epsilon}_1 + \bar{\epsilon}_2)}, \quad \bar{\epsilon}_2 > \frac{1}{\beta^2} > \bar{\epsilon}_1. \tag{74a}$$

Equation (5.2.34) then yields the total energy flowing into the region $z > 0$ in a small frequency interval $d\omega$ centered about ω; it is recalled that the η integration extends only over those values which render $\kappa_{1,2}$ real[21]:

$$W_\omega = \frac{1}{\pi}\int_{-\eta_0}^{\eta_0}\vec{P}_\eta\,d\eta = \frac{q^2\omega\mu_0(\bar{\epsilon}_2 - 1/\beta^2)}{2\pi^2(\bar{\epsilon}_2 - \bar{\epsilon}_1)}$$

$$\times\int_{-1}^{1}\sqrt{1 - \xi^2}\,\frac{\xi^2 + \hat{f}}{\xi^2 + \hat{g}}\exp\left[-2k_0|z'|\sqrt{\left(\bar{\epsilon}_2 - \frac{1}{\beta^2}\right)\xi^2 + \left(\frac{1}{\beta^2} - \bar{\epsilon}_1\right)}\right]d\xi, \tag{75}$$

where the change of variable $\eta = \eta_0\xi$, $\eta_0 = k_0[\bar{\epsilon}_2 - (1/\beta^2)]^{1/2}$, has been introduced, with

$$\hat{f} = \frac{f}{k_0^2(\bar{\epsilon}_2 - 1/\beta^2)}, \qquad \hat{g} = \frac{g}{k_0^2(\bar{\epsilon}_2 - 1/\beta^2)}. \tag{75a}$$

The integral in Eq. (75) can be evaluated when the exponential term may be replaced by unity.[21] This happens when the charge trajectory lies in the interface $z' = 0$, or less stringently, when the parameters in question are such as to make $2k_0|z'|\sqrt{\bar{\epsilon}_2 - \bar{\epsilon}_1} \ll 1$, $k_0 = \omega/c$. The change of variable $\zeta = \xi(\xi^2 - 1)^{-1/2}$ leads to

$$W_\omega = \frac{q^2 \omega \mu_0 (\bar{\epsilon}_2 - 1/\beta^2)}{2\pi^2 (\bar{\epsilon}_2 - \bar{\epsilon}_1)} \int_{-\infty}^{\infty} \frac{d\zeta}{(1 + \zeta^2)^2} \frac{\zeta^2(1 + \hat{f}) + \hat{f}}{\zeta^2(1 + \hat{g}) + \hat{g}}, \tag{76}$$

which may be evaluated in terms of the residues at the poles in the upper or lower halves of the complex ζ plane. The details are left as an exercise for the reader.

5.5k Phenomena in Bounded Regions with Negative Real Dielectric Constant (Time-Harmonic Regime)

The time-harmonic radiation problems considered in the preceding sections have been concerned with conventional dielectric media characterized by a complex dielectric constant with a positive real part. However, the effect of a macroscopically neutral ionized plasma medium on an electromagnetic field can under certain conditions be represented by a complex dielectric constant whose real part may be negative [see Eq. (1.1.64), with $\partial^2/\partial t^2 \to -\omega^2$, $a \equiv 0$]. To assess the influence of such a medium on the radiation field of an electromagnetic source, and to deal in particular with interface effects associated with a bounded region, we examine the field of a line source of longitudinally directed electric current elements [Eq. (59)][†] situated at the point $\hat{\rho}' = (0, z')$, $z' < 0$, as in Fig. 5.5.7. The fields may be derived from a two-dimensional E-mode Green's function $\bar{G}'(\hat{\rho}, \hat{\rho}') \equiv \bar{G}(\hat{\rho}, \hat{\rho}')$, represented as follows [see discussion following Eq. (59)]:

$$\bar{G}_1 = \bar{G}_{f1} + \bar{G}_s, \qquad z < 0, \tag{77}$$

where $\bar{G}_{f1}$ is the free-space Green's function for medium 1 given in Eq. (5.4.25), while $\bar{G}_s$ represents the reflected contribution

$$\bar{G}_s = -\frac{i}{4\pi} \int_{-\infty}^{\infty} \frac{e^{i\eta y - i\kappa_1(z+z')}}{\kappa_1} \Gamma(\eta)\, d\eta, \qquad \kappa_1 = \sqrt{k_1^2 - \eta^2}, \tag{77a}$$

and $\Gamma(\eta) = (\epsilon_1 \kappa_2 - \epsilon_2 \kappa_1)/(\epsilon_1 \kappa_2 + \epsilon_2 \kappa_1)$ is the reflection coefficient given in Eq. (3d). In region 2 ($z > 0$),

$$\bar{G}_2 = \frac{i}{4\pi} \int_{-\infty}^{\infty} e^{i\eta y + i\kappa_2 z - i\kappa_1 z'} \left\{ \frac{[1 - \Gamma(\eta)]}{\kappa_1} \right\} d\eta, \qquad \kappa_2 = \sqrt{k_2^2 - \eta^2}, \quad \epsilon = \frac{\epsilon_2}{\epsilon_1}. \tag{78}$$

ϵ_1 is taken as real, but ϵ_2 may be arbitrary.

Upon transforming to the w plane via $\eta = k_1 \sin w$, k_1 positive real, one observes that the contour of integration maps into the path $\bar{P}$ in Fig. 5.3.6(b);

[†]Surface-wave phenomena encountered subsequently are associated with E modes. A line current as in Eq. (52) excites only H modes.

the branch cut leading to the branch point at $w = 0$ is absent for the line-source problem. The singularities of the integrands in Eqs. (77a) and (78) have been discussed in Eqs. (15) and (16). Since Re ϵ may be negative, it is convenient to draw the branch cuts in the w plane along the contours Im $\kappa_2 = 0$ (see Fig. 5.5.4), whence Im κ_2 can be chosen positive on the entire top Riemann sheet. Since Im $[\epsilon(1 + \epsilon)^{-1/2}] > 0$, $0 \leq \arg \epsilon \leq \pi$, Eq. (16a) still locates the pole singularities on the top sheet, with Im $\cos w_p > 0$, Re $\cos w_p \leq 0$, whence the pertinent pole is confined to the strip $0 < $ Re $w < \pi$, Im $w < 0$. If ϵ is real, the pole is located on the borders of the strip $\pi/2 < $ Re $w < \pi$, Im $w < 0$, in accord with the following:

$$\frac{\pi}{2} < \text{Re } w_p \leq \pi, \qquad \text{Im } w_p = 0, \qquad \text{for } 0 \leq \epsilon < \infty, \tag{79a}$$

$$\text{Re } w_p = \pi, \qquad -\infty < \text{Im } w_p \leq 0, \qquad \text{for } -1 < \epsilon \leq 0, \tag{79b}$$

$$\text{Re } w_p = \frac{\pi}{2}, \qquad -\infty < \text{Im } w_p \leq 0, \qquad \text{for } -\infty < \epsilon < -1. \tag{79c}$$

If Re $\epsilon < 0$, Im $\epsilon = 0$, the locations of the pole and branch-point singularities are shown in Fig. 5.5.10. For an asymptotic field evaluation in region 1, it is convenient to introduce cylindrical coordinates as in Fig. 5.5.7, whence the integrand takes on the form shown in Eq. (5.3.14). The steepest-descent path P through the saddle point $w = \varphi$ is also shown in Fig. 5.5.10. For sufficiently

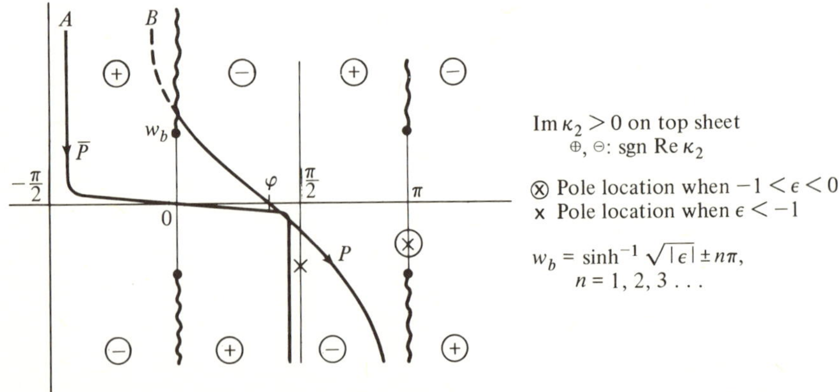

FIG. 5.5.10   Paths of integration and singularities in the w plane when $\epsilon < 0$.

large φ, P passes through the branch cut along the positive imaginary w axis and continues on to the lower Riemann sheet; although Im $\kappa_2 < 0$ on the lower sheet, the exponential convergence of the integral in Eq. (77a) is not affected, since the exponent contains κ_1 only. The deformation of $\bar{P}$ into P in the lower half of the w plane can be carried out directly provided that one accounts for the possible pole contribution when $\epsilon < -1$. At Im $w \to +\infty$, one follows the procedure in Fig. 5.3.6(b). Thus, for large enough φ, whose exact value can be assessed with reference to Eq. (5.3.15a), the original integral can be written as

$$\int_{\bar{P}} = \int_{P} + \int_{P_b'} + (\text{residue at pole}). \tag{80}$$

The steepest-descent integral yields the geometrically reflected wave contribution

$$\bar{G}_{\text{ref}} = -A_2(k_1 R)\Gamma(k_1 \sin \varphi), \qquad k_1 R \gg 1, \tag{81}$$

where $A_2(k_1 R)$ is the factor multiplying Γ'' in Eq. (57a). For real $\epsilon < 0$, $|\Gamma(k_1 \sin \varphi)| = 1$, and the half-space region is totally reflecting; this observation follows at once from the imaginary wavenumber $k_2 = k_1\sqrt{\epsilon}$, which does not permit wave propagation in medium 2. The branch-cut integral contribution has a form similar to that in Eq. (57c) and is negligible because of the presence of $\exp(-|k_2|y)$; this decay factor arises since the lateral wave (see Fig. 5.5.2) travels in the attenuating medium 2. The pole contribution, possible only when $\epsilon < -1$, represents a surface wave that propagates parallel to the interface. This wave is truly guided by the interface since it decays away from the $z = 0$ plane in both media, as will be demonstrated. In region 1, its spatial dependence is given by

$$\bar{G}_{\text{pole}} \sim \exp\left[ik_1\left(\frac{\sqrt{|\epsilon|}}{\sqrt{|\epsilon| - 1}}\right)y \right] \exp\left[\frac{k_1(z + z')}{\sqrt{|\epsilon| - 1}}\right], \qquad z, z' < 0, \quad |\epsilon| > 1 \tag{82}$$

(i.e., the velocity of propagation in the y direction is smaller than that of a uniform plane wave).

Since when $\epsilon < 0$, κ_2 in Eq. (78) is imaginary for all η, the fields in region 2 decay exponentially with increasing z. We therefore restrict our attention to observation points lying near the interface and treat $\exp(i\kappa_2 z)$ as an amplitude factor that does not exhibit rapid fluctuation. While an asymptotic evaluation of the integral in Eq. (78) is readily performed for large values of y and z' [see Eq. (5.3.19)], we assume for convenience that the source point lies near the interface and treat $\exp(-i\kappa_1 z')$ also as a slowly varying factor when compared with $\exp(i\eta y)$. Transforming to the w plane via $\eta = k_1 \sin w$, one then obtains a steepest-descent path as in Fig. 5.5.10, with the saddle point located at $w = \pi/2$. The deformation of $\bar{P}$ into P is carried out as above. The steepest-descent integral yields a space-wave contribution which varies like

$$\bar{G}_{\text{SDP}} \propto (k_1 y)^{-3/2} \exp(ik_1 y - k_1\sqrt{1 + |\epsilon|}\, z), \qquad k_1 y \gg 1, \quad z > 0, \tag{83}$$

owing to the fact that $\Gamma(k_1 \sin \varphi) \to 1$ as $\varphi \to \pi/2$, so the $O[(k_1 y)^{-1/2}]$ term derived from the saddle-point evaluation vanishes. The branch-cut integral contribution is certainly negligible since it decays exponentially both with y and z. If $\epsilon < -1$, the pole contribution is the most significant and represents a surface wave with a spatial dependence

$$G_{\text{pole}} \sim \exp\left[ik_1\left(\frac{\sqrt{|\epsilon|}}{\sqrt{|\epsilon| - 1}}\right)y \right] \exp\left[-k_1\left(\frac{|\epsilon|}{\sqrt{|\epsilon| - 1}}\right)z + \frac{k_1 z'}{\sqrt{|\epsilon| - 1}} \right],$$

$$z > 0, \quad z' < 0, \quad |\epsilon| > 1. \tag{84}$$

As noted above, the surface wave decays away from the interface in both regions.

The existence of a surface wave on a plane interface between two media, with $\epsilon < -1$, can also be deduced from a transverse resonance argument (see Sec. 2.4e). The resonance relation $Z_1 + Z_2 = 0$ becomes in the present case [see Eq. (12)],

$$\kappa_2 = -\epsilon\kappa_1, \quad \kappa_{1,2} = \sqrt{k_{1,2}^2 - \eta^2}, \quad k_2 = \sqrt{\epsilon}\, k_1. \tag{85}$$

Since the region extends to $z = \pm\infty$, a proper modal solution satisfying the radiation condition at infinity is possible only if $\kappa_{1,2} = i|\kappa_{1,2}|$. Consequently, Eq. (85) can be satisfied only for negative ϵ and for $\eta^2 > k_1^2$. Solving for the resonant values η_p^2, one finds

$$\eta_p^2 = \frac{|\epsilon|k_1^2}{|\epsilon| - 1}, \tag{86}$$

which, since $\eta_p^2 > 0$, imposes the restriction $\epsilon < -1$.

5.6 TIME-HARMONIC SOURCES IN THE PRESENCE OF A DIELECTRIC SLAB

5.6a Longitudinal Electric Current Element

$$\hat{\mathbf{J}}(\mathbf{r}, t) = Il\delta(\boldsymbol{\rho})\delta(z - z')e^{-i\omega t}\mathbf{z}_0. \tag{1}$$

The physical configuration of a longitudinal electric current element of strength $J^o = Il$, where I is the current in the element and l is its infinitesimal length, situated at the point $z' < 0$ on the z axis in the presence of a grounded dielectric slab is shown in Fig. 5.6.1. The slab is characterized by a dielectric

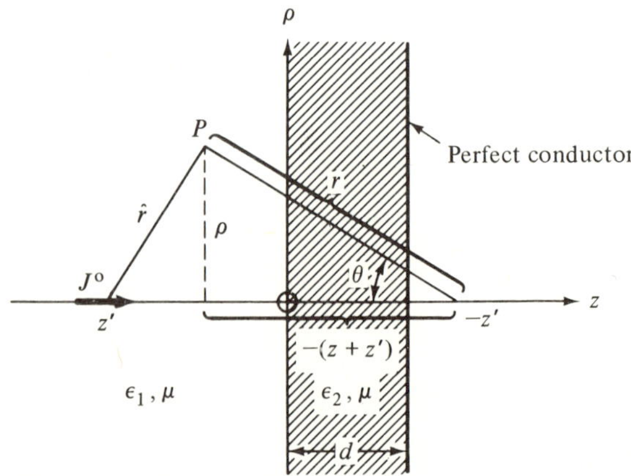

FIG. 5.6.1 Longitudinal current element and grounded dielectric slab.

constant ϵ_2 and permeability μ and occupies the region between the planes $z = 0$ and $z = d$, with a perfect conductor located at $z = d$. The medium properties in the exterior region $z < 0$ are characterized by ϵ_1 and μ. The electromagnetic fields may be derived via Eqs. (5.2.1) and (5.2.4c) from a scalar

E-mode Green's function $G'(\mathbf{r}, \mathbf{r}')$ which satisfies in the two regions the separate differential equations

$$(\nabla^2 + k_1^2)G'_1(\mathbf{r}, \mathbf{r}') = -\delta(\mathbf{r} - \mathbf{r}'), \qquad z < 0, \tag{2a}$$

$$(\nabla^2 + k_2^2)G'_2(\mathbf{r}, \mathbf{r}') = 0, \qquad\qquad 0 < z < d, \tag{2b}$$

subject to a radiation condition at infinity, and to the following boundary conditions at $z = 0$ and $z = d$ [see Eqs. (2.3.36) et seq.]:

$$G'_1 = G'_2; \quad \frac{1}{\epsilon_1}\frac{\partial G'_1}{\partial z} = \frac{1}{\epsilon_2}\frac{\partial G'_2}{\partial z} \qquad \text{at} \quad z = 0; \tag{2c}$$

$$\frac{\partial G'_2}{\partial z} = 0 \qquad\qquad \text{at} \quad z = d. \tag{2d}$$

$k_{1,2} = \omega\sqrt{\mu\epsilon_{1,2}}$ are the wavenumbers in the two regions. In a cylindrical coordinate description of the cross section transverse to z, $G'(\mathbf{r}, \mathbf{r}')$ in the region $z < 0$ may be represented as follows:

$$G'_1(\mathbf{r}, \mathbf{r}') = G_{f_1}(\mathbf{r}, \mathbf{r}') + G'_s(\mathbf{r}, \mathbf{r}'), \tag{3}$$

where $G_{f_1} = (1/4\pi\hat{r})\exp(ik_1\hat{r})$ is the free-space Green's function for region 1 (see Sec. 5.4a). The scattered component G'_s has the integral representation

$$G'_s = \frac{-ik_1}{8\pi}\int_{\bar{P}} \sin w \, H_0^{(1)}(k_1 r \sin\theta \sin w)e^{ik_1 r\cos\theta\cos w}\,\Gamma(k_1\sin w)\,dw \tag{4}$$

with the *E*-mode reflection coefficient given by

$$\Gamma(k_1\sin w) = \frac{\sqrt{\epsilon - \sin^2 w}\,\sin\psi - i\epsilon\cos w\cos\psi}{\sqrt{\epsilon - \sin^2 w}\,\sin\psi + i\epsilon\cos w\cos\psi}, \tag{4a}$$

$$\psi = k_1 d\sqrt{\epsilon - \sin^2 w}, \qquad \epsilon = \frac{\epsilon_2}{\epsilon_1}.$$

The contour of integration $\bar{P}$ is that in Fig. 5.3.6(b) (see also Fig. 5.6.3), and the polar coordinates r, θ relative to the image point $-z'$ are depicted in Fig. 5.6.1 with $\rho = r\sin\theta$, $|z + z'| = r\cos\theta$. In the complex wavenumber plane $\xi = k_1\sin w$, the integral representation takes the form given in Eq. (5.5.3a), with $\Gamma(\xi)$ taken from Eq. (4a).

For observation points in the slab region $0 < z \le d$, $G'_2(\mathbf{r}, \mathbf{r}')$ has an integral representation as in Eq. (5.5.3a) provided that one replaces

$$-\exp\left[-i\sqrt{k_1^2 - \xi^2}(z + z')\right]\Gamma(\xi)$$

$$\text{by} \quad [1 - \Gamma(\xi)]\exp\left(-i\sqrt{k_1^2 - \xi^2}z'\right)\frac{\cos[\sqrt{k_1^2\epsilon - \xi^2}(d - z)]}{\cos(\sqrt{k_1^2\epsilon - \xi^2}d)}, \tag{5}$$

with $\Gamma(\xi)$ given in Eq. (4a).

For large values of $k_1 r \sin\theta$, the secondary contribution G'_s to the Green's function in the region $z < 0$ may be evaluated asymptotically by the saddle-point method. The resulting $G'_1(\mathbf{r}, \mathbf{r}')$ for arbitrary distances $\hat{r}$ from the source, but for large distances r from the image point, is given by

$$G_1' \sim \frac{e^{ik_1\hat{r}}}{4\pi\hat{r}} - \Gamma(k_1 \sin \theta)\frac{e^{ik_1r}}{4\pi r}\left[1 + O\!\left(\frac{1}{k_1r}\right)\right]$$

$$+ \frac{e^{i\pi/4}}{2\sqrt{2\pi k_1 r \sin \theta}}\left[\sum_{\nu} \frac{e^{ik_1r\cos(w_\nu - \theta)}}{\epsilon_1 A_\nu^2 \sqrt{\sin w_\nu}}\, U(\theta - \theta_\nu)\right.$$

$$\left.+ \sum_{\mu} \frac{e^{ik_1r\cos(w_\mu - \theta)}}{\epsilon_1 A_\mu^2 \sqrt{\sin w_\mu}}\, U(\theta - \theta_\mu)\right]\left[1 + O\!\left(\frac{1}{k_1r}\right)\right], \qquad (6)$$

where $U(\gamma)$ is the Heaviside unit function, which equals unity for $\gamma > 0$ and vanishes for $\gamma < 0$, while

$$\Gamma(k_1 \sin \theta) = \frac{\cos \theta' \tan(k_1 d\sqrt{\epsilon}\cos \theta') - i\sqrt{\epsilon}\cos \theta}{\cos \theta' \tan(k_1 d\sqrt{\epsilon}\cos \theta') + i\sqrt{\epsilon}\cos \theta}, \qquad \sin \theta' = \frac{\sin \theta}{\sqrt{\epsilon}}, \quad (6a)$$

and

$$\epsilon_1 A_\alpha^2 = \frac{d}{2}\left\{\left[1 - \left(\frac{q}{p}\right)^2\right]\frac{i}{q} + \frac{1}{\epsilon}\left[1 - \epsilon^2\left(\frac{q}{p}\right)^2\right]\right\}, \qquad \epsilon = \frac{\epsilon_2}{\epsilon_1}, \quad \alpha = \nu, \mu. \;(6b)$$

p and q satisfy the transcendental equations

$$\epsilon q = ip \tan p,$$
$$p^2 - q^2 = (k_1 d)^2(\epsilon - 1) \equiv l^2, \qquad\qquad (7)$$

and are related as follows to $w_\alpha = w_\nu$ or $w_\alpha = w_\mu$:

$$p = k_1 d\sqrt{\epsilon - \sin^2 w_\alpha}, \qquad q = k_1 d\cos w_\alpha. \qquad (7a)$$

While the terms in the series have also been approximated for the far zone $k_1\rho \gg 1$, they may easily be given exactly for arbitrary values of $k_1\rho$ [see the discussion preceding Eqs. (15)]. The pole singularities w_α of the reflection coefficient in Eq. (4a), defined by Eqs. (7), may be grouped into two categories: (1) surface-wave poles $w_\alpha \equiv w_\nu$ and (2) leaky-wave poles $w_\alpha \equiv w_\mu$. The sums in Eq. (6) denote corresponding residue contributions to the asymptotic field solution, and they arise at observation angles $\theta > \theta_\alpha$, where

$$\theta_\alpha = \mathrm{Re}\, w_\alpha - \cos^{-1}\mathrm{sech}\,(\mathrm{Im}\, w_\alpha), \qquad \alpha = \nu, \mu. \qquad (8)$$

The results are valid exterior to the transition regions $\theta \approx \theta_\alpha$. For modifications in transition regions, see the discussion following Eq. (16).

Discussion

The first two terms on the right-hand side of Eq. (6) (without the quantity in brackets) represent the geometric optical field; they may be interpreted exactly as for the semiinfinite medium in Sec. 5.5a, except that $\Gamma(k_1 \sin \theta)$ now accounts for the reflection from the slab configuration. It may in fact be shown that for lossless media, the slab reflection coefficient in Eq. (6a) may be synthesized by summing up the multiply reflected- and refracted-ray contributions sketched in Fig. 5.6.2.

The residue terms constitute diffraction fields which contribute for observation angles θ sufficiently near the interface. When the region is almost lossless,

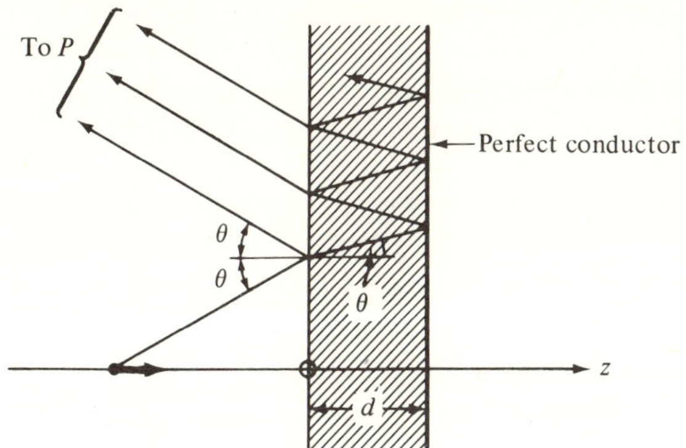

FIG. 5.6.2 Ray-optical interpretation of the reflection coefficient.

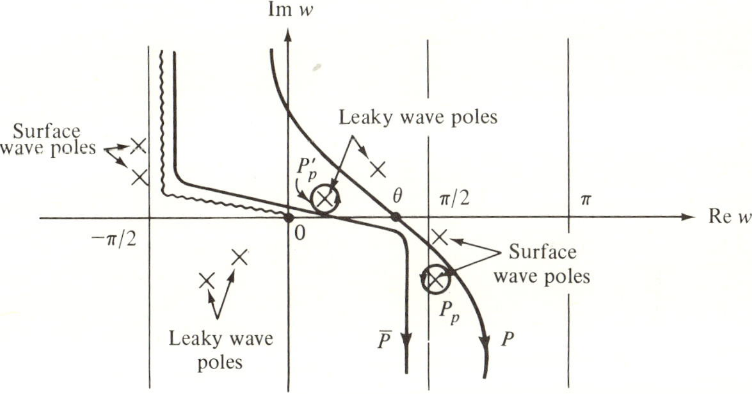

FIG. 5.6.3 Contours of integration and singularities in the w plane.

the relevant surface-wave and leaky-wave poles in the complex w plane are situated as in Fig. 5.6.3; in the lossless case, the surface-wave poles are located on the vertical lines $\operatorname{Re} w_\alpha = \pm\pi/2$. One notes that in contrast to the direct-wave and reflected-wave contributions which possess a $1/k_1\hat{r}$ or $1/k_1 r$ variation characteristic of spherically diverging waves, the surface-wave and leaky-wave fields behave like cylindrical waves with a $1/\sqrt{k_1 r \sin\theta} = 1/\sqrt{k_1\rho}$ dependence. Thus, from purely geometrical considerations, the decay of the fields in a surface wave or leaky wave is slower than that in the geometric-optical field contributions. However, the behavior of the amplitude factors and the exponential terms must likewise be considered in assessing the importance of a particular residue term.

For real ϵ greater than 1, Eqs. (7) can be satisfied for imaginary values of $q = i|q|$ and real values of p. These solutions, finite in number, characterize surface waves with the exponential dependence

$$e^{ik_1 r \cos(w_v - \theta)} = e^{-|q||z+z'|/d} \exp[ik_1\sqrt{1 + (|q|/k_1 d)^2}\,\rho], \tag{9}$$

undamped in the coordinate ρ parallel to the slab surface, but appreciable only when both the source and observation points are situated near the surface. The name "surface wave" derives from the field being confined near the interface in this wave species. In the assumed absence of dissipation, if $|z + z'| \to 0$, the surface waves in the dielectric constitute the dominant contribution to the field because of their above-mentioned cylindrical wave character, as opposed to the spherical wave behavior of the geometric optical field. In fact, in the range $\theta \to \pi/2$ where the surface-wave contribution is most pronounced, one notes that $\Gamma(k_1 \sin \theta) \to 1$, $r \to \hat{r} \to \rho$, so the geometric optical field vanishes to $O(1/k_1\rho)$. The surface-wave variation $O(1/\sqrt{k_1\rho})$ is therefore to be compared with a contribution of $O(1/k_1^2\rho^2)$ arising from a second-order evaluation of the steepest-descent integral in Eq. (5.3.16a). It is noted from Eq. (9) that the undamped propagation constant of the surface wave in the ρ direction, κ_ρ, is given by $\kappa_\rho = k_1\sqrt{1 + (|q|/k_1 d)^2} > k_1$, so its phase velocity $v = \omega/\kappa_\rho$ is less than the phase velocity $v_f = \omega/k_1$ of a plane wave in free space. Hence, these surface waves are also referred to as "slow" waves (see Sec. 5.4d).

Concerning the leaky-wave contributions in Eq. (6), corresponding to complex roots of Eqs. (7), we note from Fig. 5.6.3 that, in their domains of existence, $(w_{\mu r} - \theta) < 0$, $w_{\mu i} > 0$, where $w_{\mu r}$ and $w_{\mu i}$ are the real and imaginary parts of w_μ, respectively. Since

$$e^{ik_1 r \cos(w_\mu - \theta)} = \exp\{ik_1 r[\cos(w_{\mu r} - \theta)\cosh w_{\mu i} - i \sin(w_{\mu r} - \theta)\sinh w_{\mu i}]\}, \tag{10a}$$

$$= \exp\{ik_1[\gamma|z + z'| + \beta\rho]\}, \tag{10b}$$

where

$$\gamma = \cos w_{\mu r} \cosh w_{\mu i} - i \sin w_{\mu r} \sinh w_{\mu i},$$
$$\beta = \sin w_{\mu r} \cosh w_{\mu i} + i \cos w_{\mu r} \sinh w_{\mu i}, \tag{10c}$$

a typical leaky-wave contribution *decreases* exponentially in *all* directions $\theta > \theta_\mu$. These contributions to the far field are therefore negligible unless $w_{\mu i} \to 0$ (i.e., for leaky-wave poles situated very near the real w axis). As noted in Sec. 5.3e, although the leaky-wave fields increase exponentially with $|z + z'|$ [see Eqs. (10b) and (10c)], their restricted domain of existence $\theta > \theta_\mu$ limits the maximum allowable value of $|z + z'|$ (for a given ρ) in such a manner that the decay along ρ overcomes the growth along $|z|$ (see Fig. 5.3.8). In view of the decay with ρ, the wave leaks energy continually into the direction perpendicular to the surface.

In reference to the asymptotic reduction of Eqs. (5.3.12) and (5.3.14), which represent the generic form of Eq. (4), the following observation is relevant. In Eq. (5.3.15), the restricted domain of existence of a typical residue contribution appears as the result of the transformation of an exact integral representation along path $\bar{P}$ into another along path P. Since residues represent surface waves or leaky waves, while the steepest-descent integral gives rise (asymptotically) to a space wave or radiation field, it is suggestive to conclude that surface or leaky waves exist only when $\bar{\alpha} > \bar{\alpha}_p$ in Eq. (5.3.15) or in its asymptotic approximation. The latter supposition does not follow, since the asymptotic approximation, unlike the exact representation, contains an exponentially small error against which exponentially small residue terms must be compared [see remarks following Eqs. (4.2.18)]. Only upon specification of the error term, which depends on the radius of convergence of the power-series expansion of the integrand in Eq. (4) about the saddle point, may Eq. (6) be made consistent. It then follows that the absence of the surface-wave residue terms when $\theta < \theta_v$ need not imply the nonexistence of these waves but merely that their amplitude is so small as to be beyond the accuracy of the approximation.

If $\epsilon_2 < \epsilon_1$ (i.e., $\epsilon < 1$), the region $z < 0$ is occupied by the denser medium, so the configuration in Fig. 5.6.1 can be viewed as a dielectric "gap." In this instance, Eqs. (7) do not admit of surface-wave solutions for which p is real and $q = i|q|$; hence, the leaky-wave contributions may be important for the evaluation of the fields in the region $z < 0$. To illustrate this possibility, let us consider the case $0 < \epsilon \ll 1$. Since the gap region $0 < z < d$ appears to be a waveguide bounded by highly reflecting walls at $z = 0, d$, it is suggestive to look for solutions of the resonance equations (7) which differ from the resonant solutions $p_0 = \kappa_{20}d = m\pi$, $m = 0, 1, 2, \ldots$, appropriate to $\epsilon = 0$, by the small quantity δ:

$$p = m\pi + \delta. \tag{11}$$

Upon substituting Eq. (11) into Eqs. (7) and retaining terms to $O(\delta)$ only, one obtains as a first approximation,

$$\delta \cong \frac{-i\epsilon}{m\pi}\sqrt{(m\pi)^2 + (k_1 d)^2(1 - \epsilon)}, \qquad m = 1, 2, \ldots, \tag{12a}$$

valid when $0 < \epsilon \ll 1$, $\epsilon\sqrt{1 + (1 - \epsilon)(k_1 d/m\pi)^2} \ll 1$. For $m = 0$, one must retain terms to $O(\delta^2)$ and finds

$$\delta = e^{-i\pi/4}\sqrt{\epsilon a}, \quad a = k_1 d\sqrt{1 - \epsilon}, \qquad \epsilon a \ll 1, \quad \epsilon \ll a. \tag{12b}$$

For Eq. (12a), $\text{Im } q = -\sin w_{\mu r} \sinh w_{\mu i} = -\epsilon$, while for Eq. (12b), $\text{Im } q = -\epsilon/2$. Since $\epsilon \ll 1$, one finds that the pertinent leaky-wave poles are situated near the real w axis and the attenuation of the leaky waves noted in the preceding paragraph can be small; for δ given in Eq. (12b), one finds $w_\mu \approx \sqrt{\epsilon}$ $\cdot\sqrt{1 + (i/k_1 d)}$. Thus, their contribution for observation points with large $k_1 r$ but small $k_1 r \sin(\theta - w_{\mu r}) \sinh w_{\mu i}$ may be appreciable.[5]

Analytical details

The integral representation for G'_s in Eq. (4) is derived by considerations directly analogous to those utilized in connection with Eqs. (5.5.11), except that for the modal Green's function, the boundary condition $dg_{z2}/dz = 0$ at $z = d$ must be added [see Eq. (2d)]. The corresponding modal network problem is sketched in Fig. 5.6.4, and g_z is proportional to the current in the network. The solution for $z < 0$ is given in Eq. (5.3.1), where $Z_1 = \kappa_1/\omega\epsilon_1$ (E modes)

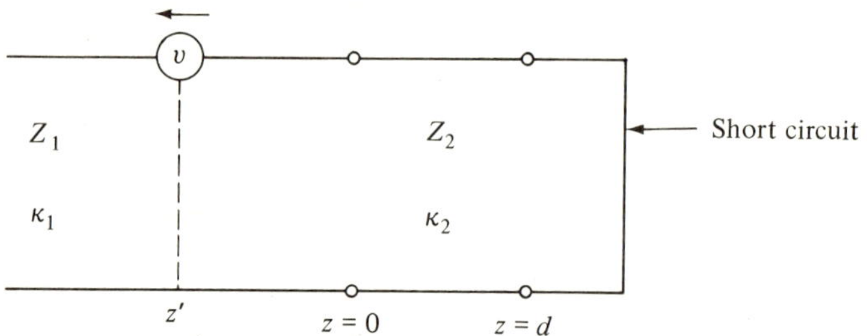

FIG. 5.6.4   Equivalent modal network (*E* modes).

and $\overrightarrow{Z}(0)$, the input impedance of the short-circuited transmission line seen from $z = 0$, is given via Eq. (2.4.24a) as

$$\overrightarrow{Z}(0) = jZ_2 \tan \kappa_2 d, \quad Z_2 = \frac{\kappa_2}{\omega\epsilon_2}, \quad \kappa_{1,2} = \sqrt{k_{1,2}^2 - \xi^2}. \tag{13}$$

Equation (5.3.1a) with $\xi = k_1 \sin w$ thus yields the reflection coefficient in Eq. (4a), and Eqs. (3) and (4) follow from the general integral representation in Eq. (5.2.11), with $j \to -i$ to account for the difference in time dependence [see also Eq. (5.4.7a)]. When the source point z' is located inside the slab region $0 < z \leq d$, the modal Green's function g_z has the more general form given in Eqs. (2.4.28) or (2.4. 29) (see also Sec. 3.4a).

The solution for observation points in the slab region $0 < z \leq d$ may be derived by inspection from that for g_{z1} in the exterior region $z < 0$ [Eq. (5.3.1)]. Since $dg_{z2}/dz = 0$ at $z = d$, g_{z2} must be proportional to $\cos \kappa_2(d - z)$. Then from the continuity $g_{z1} = g_{z2}$ at $z = 0$, one has

$$g_{z2}(z, z') = g_{z2}(0, z') \frac{\cos \kappa_2(d - z)}{\cos \kappa_2 d} = g_{z1}(0, z') \frac{\cos \kappa_2(d - z)}{\cos \kappa_2 d}, \tag{14}$$

thereby justifying Eq. (5).

The asymptotic evaluation of the integral in Eq. (4) proceeds as in Sec. 5.5a, except that in the function $f(w)$ in Eq. (5.5.14), the reflection coefficient is taken from Eq. (4a). A first-order saddle-point evaluation yields the second term on the right-hand side of Eq. (6), and higher-order terms, denoted by the symbol $O(1/k_1 r)$, may be derived by the procedure in Sec. 4.2b.

Concerning the singularities of the integrand in Eq. (4) in the complex w plane, we note first the branch points at $\sin w = 0$ arising from the argument of the Hankel function. The pertinent branch point at $w = 0$, and the associated branch cut, are shown in Fig. 5.6.3. No branch points are located at $w_b = \pm\sin^{-1}\sqrt{\epsilon}$, or correspondingly at $\xi_b = \pm k_2$, since $\Gamma(k_1 \sin w)$ is an even function of $\sqrt{\epsilon - \sin^2 w}$. This is in accord with the general observation made in Sec. 5.3a because region 2 is terminated at $z = d$ (i.e., $z = +\infty$ is inaccessible). The branch-point pair at $z_b = \pm k_1$ in the complex wavenumber plane has been removed by the transformation $\xi = k_1 \sin w$. However, $\Gamma(k_1 \sin w)$ has pole singularities at the zeros w_α of the denominator on the right-hand side of Eq. (4a). Upon introducing the new parameters p and q in Eq. (7a), one obtains the simultaneous equations (7) defining the location of the pole singularities. These equations can also be derived from an application of the transverse resonance argument in Eq. (2.4.36). As seen from the $z = 0$ terminal in Fig. 5.6.4, $\overrightarrow{Z}(0)$ is given in Eq. (13), $\overleftarrow{Z}(0) = Z_1$, whence the tranvserse resonance relation $\overleftarrow{Z}(0) + \overrightarrow{Z}(0) = 0$ yields Eqs. (7) (with $j \rightarrow -i$).

For non-dissipative media, ϵ and l^2 are positive; Eqs. (7) can then be satisfied for imaginary values of $q = i|q|$ and real values of p. Simple poles corresponding to these values of q are finite in number and are located at $w_{vr} = \pi/2$, $w_{vi} < 0$.† From Eq. (4), residue terms arising from such poles have a z dependence $\exp[i|z + z'| \cos w_v] = \exp[-|q||z + z'|/d]$ [see Eq. (9)] and a p dependence $H_0^{(1)}(k_1 \rho \sin w_v)$, the latter behaving like $(k_1\rho)^{-1/2} \exp(ik_1\rho \sin w_v)$ for large $k_1\rho \sin w_v$. These residues represent the surface waves discussed previously. They constitute proper solutions of the transverse resonance equations (7), carry no real power in the direction perpendicular to the surface, and play an important role in the determination of the spectrum of waves that can be guided along the dielectric surface (see Sec. 3.4a). A "proper" modal solution satisfies the source-free Maxwell field equations, the boundary conditions on the perturbing surface, and it decays at infinity.

One observes from Eqs. (7) that for $q = i|q|$ and p real, $|q| = -k_1 d \sinh w_{vi}$ must lie between the limits $0 \leq |q| < k_1 d\sqrt{\epsilon - 1}$. Equations (7) can be solved graphically by a construction analogous to that in Fig. 3.4.3, yielding the same general picture except that the non-circular contours in Fig. 3.4.3 are displaced to the left through an interval $\pi/2$. It is then evident that for real ϵ and k_1, and $\epsilon > 1$, N solutions exist in the range $(N-1)\pi < k_1 d\sqrt{\epsilon - 1} < N\pi$, $N = 1, 2, \ldots$. In particular, one solution exists for arbitrarily small $k_1 d\sqrt{\epsilon - 1}$ (i.e., there is no low-frequency cutoff). For ϵ positive real and less than unity, and k_1 real, *no* solutions are possible for real p and positive imaginary q.

In addition to the surface-wave roots, Eqs. (7) possess an infinity of other complex roots[22] in the half-strip $0 < \text{Re } w < \pi/2$, $\text{Im } w > 0$, to mention only those pertinent to the present problem. For these "leaky-wave" roots $w_\alpha = w_\mu$, $|\exp[i|z + z'|k_1 \cos w_\mu]| = \exp[k_1|z + z'| \sin w_{\mu r} \sinh w_{\mu i}]$ increases with distance from the surface of the dielectric. The field variation along the slab is given by $H_0^{(1)}(k_1 \rho \sin w_\mu)$, thereby leading for large $k_1\rho \sin w_\mu$ to the dependence noted in Eqs. (10). The leaky waves cannot represent a field solution everywhere in space because of their unlimited growth at $z \rightarrow -\infty$. Therefore, they do not belong to the proper spectrum of waves (modes) that can be guided along the dielectric slab. These "non-modal" waves may arise, however, in a field representation for a limited region of space wherein they decay [region $\theta > \theta_\mu$ in Eq. (6); see also Fig. 5.3.8].

---

†It is evident from Eqs. (7) that if w_α is a solution of these equations, then $-w_\alpha$ is also a solution. Moreover, when ϵ is real, one notes upon taking the complex conjugate that Eqs. (7) are also satisfied by $-q^*$, implying for real k_1 that $\pi - w_\alpha^*$ is also a solution.

To ascertain the disposition of the surface-wave poles with respect to the path of integration $\bar{P}$ in Fig. 5.6.3, we assume for the moment that ϵ_1 and ϵ_2 are slightly complex but that $\arg \epsilon_2 = \arg \epsilon_1$, so ϵ is real. Since $p = \kappa_2 d$, $q = \kappa_1 d$, and $\operatorname{Im} \kappa_{1,2} > 0$ for the assumed $\exp(-i\omega t)$ time dependence, it follows that $\operatorname{Im} p > 0$, $\operatorname{Im} q > 0$. From Eq. (7), one then has $\operatorname{Re} q < 0$. From Eqs. (5.3.5a) and (5.3.8), it is noted that the requirement $\operatorname{Re} q < 0$, $\operatorname{Im} q > 0$, is satisfied in the strip $\pi/2 < \operatorname{Re} w < \pi$, $\operatorname{Im} w < 0$. This establishes the location of the pertinent surface-wave poles and path of integration as shown in Fig. 5.6.3. Typical leaky-wave poles located in the strip $-\pi/2 < \operatorname{Re} w < \pi/2$ are also shown. In the limit $\arg \epsilon_1 \to 0$, it is evident that the path $\bar{P}$ is to be indented to the left around the surface-wave poles situated on the half-line $\operatorname{Re} w = \pi/2$, $\operatorname{Im} w < 0$. During the deformation of the path $\bar{P}$ into the steepest-descent path P through the saddle point at $w = \theta$, surface-wave or leaky-wave poles may be intercepted, and these give rise to the residue terms expressed by the sums in Eq. (6). As noted above, the residue contributions can be obtained *exactly* by retaining the Hankel function in Eq. (4). The form in Eq. (6) has been simplified by employing the condition $k_1 \rho \gg 1$ for the Hankel function [see Eq. (5.3.13)], a condition imposed on the evaluation of the saddle-point integral. The residue at a typical pole w_α is proportional to [see Eqs. (4) and (5.3.15)]

$$R_\alpha = \sqrt{\sin w_\alpha}\, e^{ik_1 r \cos(w_\alpha - \theta)} B_\alpha, \tag{15}$$

where

$$B_\alpha = [(w - w_\alpha)\Gamma(k_1 \sin w)]_{w=w_\alpha} = \frac{2p \sin p}{\left[\dfrac{d}{dw}(p \sin p + i\epsilon q \cos p)\right]_{w=w_\alpha}}. \tag{15a}$$

After carrying out the differentiation and recalling Eqs. (7), one obtains, for B_α,

$$B_\alpha = \frac{i}{\epsilon_1 A_\alpha^2 k_1 \sin w_\alpha}, \tag{16}$$

with A_α^2 given in Eq. (6b). For the surface waves, $q = i|q|$, p real; $A_\alpha^2 \equiv A_v^2$ in Eq. (6b) is then identical with the normalization term for the surface-wave spectrum in Eq. (3.4.26c), save for the replacement $\epsilon_1 \leftrightarrow \epsilon_2$ arising because of the different medium designations in Figs. 5.6.1 and 3.4.5.

It should be emphasized that the result in Eq. (6) is valid when $\theta \not\approx \theta_{v,\mu}$ (i.e., when the steepest-descent path does not pass near a pole singularity). To effect a continuous transition in Eq. (6) as the steepest-descent path crosses a pole, or when a pole is near the saddle point, one employs the modified saddle-point procedures in Sec. 4.4a.

Alternative representation (radial transmission formulation)

As noted in the preceding section, near the surface of the dielectric slab configuration of Fig. 5.6.1, with $\epsilon_2 > \epsilon_1$, the dominant contribution to the far electromagnetic fields due to a longitudinal electric current situated near the slab arises from surface waves which propagate in the p direction like cylindrical waves and decay exponentially away from the interface. The amplitudes of excitation of these waves given in Eq. (6) were not obtained directly from the z-transmission formulation leading to Eq. (4), but emerged only after an asymptotic evaluation of the far fields observed near the interface. Because the surface waves represent fields that propagate *along* the interface, it is suggestive that a radial transmission formulation (i.e., a representation in terms of the mode spectrum in the z domain) of the scalar E-mode Green's function G' yields their excitation amplitudes in a more direct manner.

A p-transmission representation for G' can be written down at once from Eq. (3.3.39b), wherein $u \equiv \rho$, $v \equiv \phi$. The orthonormal eigenfunctions for the ϕ and z domains were obtained in Sec. 3.4 and are repeated below for convenience. In the z domain, the scalar eigenfunctions $\Phi_r(z)$ must satisfy boundary conditions appropriate to E modes along z in the configuration of Fig. 5.6.1. This configuration is identical with that in Fig. 3.4.5, provided that $x, x', \epsilon_1, \epsilon_2$ in the latter are replaced by $-z, -z', \epsilon_2, \epsilon_1$, respectively. As noted from Eqs. (3.4.25)–(3.4.27), the mode spectrum has a discrete (surface wave) and a continuous part; we list only the functions required for a field representation in the region $z < 0$, with $z' < 0$.

z domain

Discrete part:

$$\Phi_r(z) \equiv \Phi_v(z) = \frac{e^{|q_v||z|/d}}{\sqrt{\epsilon_1 A_v}}, \tag{17a}$$

where $q_v = i|q_v|$ is a solution of the transverse resonance equations (7), and A_v is given by Eq. (6b).

Continuous part:

$$\Phi_r(z) \equiv \Phi(\zeta, z) = \frac{1}{\sqrt{2\pi}}[e^{i\zeta z} - \overrightarrow{\Gamma}(\zeta)e^{-i\zeta z}], \qquad 0 < \zeta < \infty, \tag{17b}$$

where

$$\overrightarrow{\Gamma}(\zeta) \equiv \Gamma(\zeta) = \frac{-i\xi_1 \tan \xi_1 d - \zeta\epsilon}{-i\xi_1 \tan \xi_1 d + \zeta\epsilon}, \qquad \xi_1 = \sqrt{k_1^2(\epsilon - 1) + \zeta^2}. \tag{17c}$$

The replacement $j \to -i$, appropriate to an assumed $\exp(-i\omega t)$ time dependence, has been made above. Also, the variable ξ in Eqs. (3.4.25)–(3.4.27) has been replaced by ζ to avoid confusion with the notation in the present chapter.

φ domain

The appropriate eigenfunctions are listed in Eq. (3.2.51b):

$$\Phi_m(\phi) = \frac{1}{\sqrt{2\pi}}e^{im\phi}, \qquad m = 0, \pm 1, \pm 2, \ldots. \tag{18}$$

The radial characteristic Green's function $g'_\rho(\rho, \rho'; \lambda_\rho, \lambda_{\phi m})$ is identical with that listed in Eq. (3.4.93):

$$g'_\rho(\rho, \rho'; \lambda_\rho, \lambda_{\phi m}) = \frac{\pi i}{2} J_m(\eta \rho_<) H_m^{(1)}(\eta \rho_>), \qquad \eta = \sqrt{\lambda_\rho} = \sqrt{k_1^2 - \lambda_z}. \quad (19)$$

The relation $\lambda_\rho = k_1^2 - \lambda_z$ follows from Eq. (3.3.38b); λ_{zr} is defined as

$$[-1/\Phi_r(z)][d^2\Phi_r(z)/dz^2].$$

Upon substituting Eqs. (17) and (18) into Eq. (3.3.39b), one obtains for the radial transmission representation of $G'(\mathbf{r}, \mathbf{r}')$ [$\rho' = 0$ in Fig. 5.6.1, hence only the $m = 0$ term in Eq. (18) contributes since $J_m(0) = 0$ for $m \neq 0$]:

$$G'(\mathbf{r}, \mathbf{r}') = \frac{1}{2\pi} \frac{\pi i}{2} \sum_v \frac{e^{|q_v|(z+z')/d}}{\epsilon_1 A_v^2} H_0^{(1)}(\sqrt{k_1^2 + |q_v|^2/d^2}\,\rho)$$

$$+ \frac{1}{2\pi} \frac{\pi i}{2} \int_0^\infty \Phi(\zeta, z)\Phi^*(\zeta, z') H_0^{(1)}(\sqrt{k_1^2 - \zeta^2}\,\rho)\, d\zeta, \quad (20)$$

where $\sqrt{k_1^2 - \zeta^2}$ is positive when real, and equal to $i\sqrt{\zeta^2 - k_1^2}$ when $|\zeta| > k_1$. The sum in Eq. (20) extends over *all* solutions $q_v = i|q_v|$, p_v real, of the transverse resonance equations (7), and represents the spectrum of surface waves excited by the source in the presence of the dielectric slab. Upon approximating the Hankel function by its asymptotic form in Eq. (5.3.13) and recalling Eq. (7a), one verifies that the surface-wave amplitudes in Eq. (20) are identical with those in Eq. (6). Hence, as anticipated, the excitation amplitudes of the surface-wave spectrum are determined directly from the radial transmission representation. However, to assess the influence of the continuous spectrum, one must still estimate the contribution from the integral in Eq. (20).

It is instructive to trace in detail the transition from the z-transmission formulation to the ρ-transmission representation in Eq. (20). It is convenient to begin in the ξ plane, wherein G' is represented via Eqs. (5.2.11) and (5.3.1) for $z, z' < 0$ and for a time dependence $\exp(-i\omega t)$ as

$$G'(\mathbf{r}, \mathbf{r}') = \frac{i}{8\pi} \int_{\infty e^{i\pi}}^\infty \xi H_0^{(1)}(\xi \rho)[e^{i\xi|z-z'|} - \Gamma(\zeta)e^{-i\xi(z+z')}]\frac{d\xi}{\zeta}, \qquad \zeta = \sqrt{k_1^2 - \xi^2},$$

$$(21)$$

where $\Gamma(\zeta)$ is the reflection coefficient given in Eq. (4a). The transformation $\xi = k_1 \sin w$ in Eq. (21) then yields the formulations in Eqs. (3) and (4). Equation (21) involves a *modal* representation in the radial domain and therefore constitutes a *transmission* formulation along z. The path of integration proceeds as in Fig. 5.3.6(a) relative to the branch-point singularities in the complex ξ plane; no branch points exist at $\xi = \pm k_2$. The ρ-transmission formulation is to be obtained from Eq. (21) by a deformation of the integration path into the upper half of the ξ plane around the singularities of $g_z(z, z')$.† Since this deformation is possible for arbitrary $\mathbf{r}$ and $\mathbf{r}'$ only in those regions wherein Im $\zeta >$

---

†There exists an intimate relation between the procedure carried out here in the ξ plane, and the characteristic Green's-function technique described in Sec. 3.3c. In the notation of Eqs. (3.3.37) and (3.3.38), $\xi \equiv \sqrt{\lambda_u}$ and $\lambda_z = k_1^2 - \lambda_u \equiv \zeta^2$.

0, it is convenient to choose branch cuts emanating from the branch points $\xi = \pm k_1$ as in the second of Figs. 5.3.3(a), so $\text{Im}\,\zeta > 0$ on the entire top sheet of the two-sheeted ξ plane. In the discussion carried out in connection with Fig 5.6.3, it is noted that the singularities of $\Gamma(\zeta)$ are either surface-wave poles ξ_ν for which $\text{Im}\,\zeta_\nu > 0$, $\text{Re}\,\zeta_\nu < 0$, or leaky-wave poles ξ_μ for which $\text{Im}\,\zeta_\mu < 0$. Thus, only the surface-wave poles appear on the top sheet of the two-sheeted ξ-plane, as shown in Fig. 5.6.5. To highlight the progress of the contour of

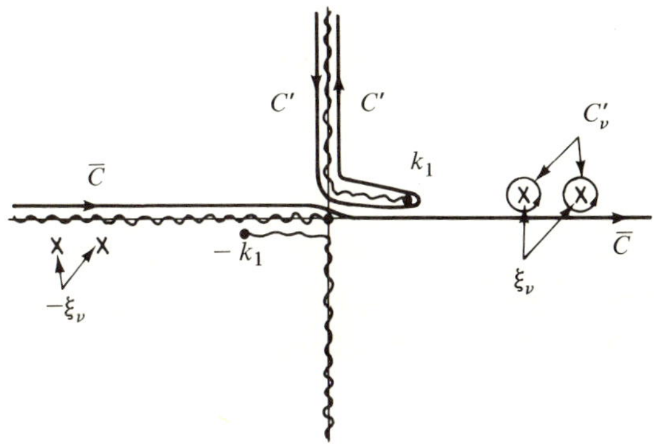

FIG. 5.6.5   Contours of integration and singularities in the ξ plane.

integration $\bar{C}$, the various singularities at $\xi = \pm k_1$ and $\xi = \xi_\nu$ have been given small complex parts.

 A deformation of the contour $\bar{C}$ about the singularities of the integrand in Eq. (21) in the upper half of the ξ plane can now be carried out directly. By virtue of Eq. (5.3.13), the Hankel function vanishes exponentially at $|\xi| \to \infty$, $\text{Im}\,\xi > 0$. Since $\text{Im}\,\zeta > 0$ on the entire top sheet, the terms inside the brackets of Eq. (21) vanish exponentially at $|\xi| \to \infty$. Thus, there is no contribution to the integral from a contour at $|\xi| \to \infty$, $0 < \arg \xi < \pi$; the integral taken over the contour $\bar{C}$ is by Cauchy's theorem equal to the integral taken over the contour C' surrounding the branch cut plus the integrals taken over the contours C'_ν surrounding the surface-wave poles of $\Gamma(\zeta)$. The latter formulation constitutes the desired radial transmission representation since, as in Eq. (20), it involves the mode spectrum in the z domain. The discrete spectrum arises from the residue contributions at the surface-wave poles and yields identically the series portion of Eq. (20). To show that the contribution from the continuous spectrum in Eq. (20) is the same as that arising from the branch-cut integral taken over the contour C', one introduces ζ as a new variable in Eq. (21) [see Eq. (5.4.10) et seq.] and simplifies, noting that $\Gamma(-\zeta) = \Gamma^*(\zeta)$ when ζ is real.

 In the w plane, the contour $\bar{C}$ in Fig. 5.6.5 transforms into the contour $\bar{P}$ in Fig. 5.6.3, whereas the contour C' maps into a path P' which runs along the imaginary axis from $w = i\infty$ to $w = 0$, along the real axis from $w = 0$ to $w =$

π, and then along the vertical line from $w = \pi$ to $w = \pi - i\infty$. In a far-field asymptotic calculation of the contribution to Eq. (20) from the continuous spectrum as represented on the infinite contour P', one deforms P' into the steepest-descent path P in Fig. 5.6.3 and evaluates the steepest-descent integral as in Eq. (6). Unless $\theta = \pi/2$, there exists the possibility of crossing during the path deformation one or more of those surface-wave poles which lie closest to the real w axis of Fig 5.6.3. The resulting residue contributions cancel corresponding terms in the sum on the right-hand side of Eq. (20), so not all of the possible surface waves are included in the asymptotic representation of the fields observed above the dielectric interface. Moreover, one or more of the leaky-wave poles may have to be taken into account. The final result then agrees with that in Eq. (6), which is derived from an initial z-transmission formulation.

Modifications for an ungrounded slab

When the source is located exterior to an ungrounded slab as in Fig 5.6.6(a), the preceding analysis must be modified only through the insertion of the appropriate modal Green's function g_z whose evaluation is schematized in network form in Fig. 5.6.6(b). Instead of treating the problem in Fig. 5.6.6 directly, it

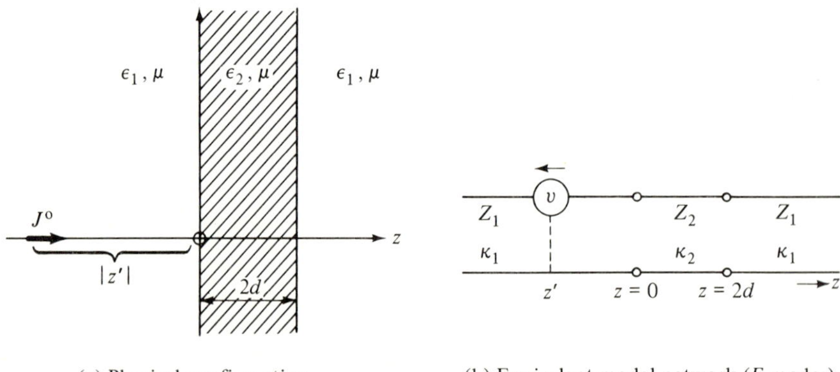

(a) Physical configuration   (b) Equivalent modal network (E modes)

FIG. 5.6.6   Longitudinal current element and ungrounded dielectric slab.

is convenient, because of the network symmetry about the plane $z = d$, to consider two auxiliary problems arising from symmetric and antisymmetric voltage excitations, as shown in Figs 5.6.7(b) and 5.6.7(a), respectively. The corresponding electromagnetic field problems are also shown.

From the network picture one notes that the antisymmetric voltage excitation gives rise to a voltage null (short circuit) at the symmetry plane $z = d$, while the symmetric voltage excitation gives rise to a current null (open circuit) at $z = d$. Correspondingly, the tangential electric and magnetic fields vanish

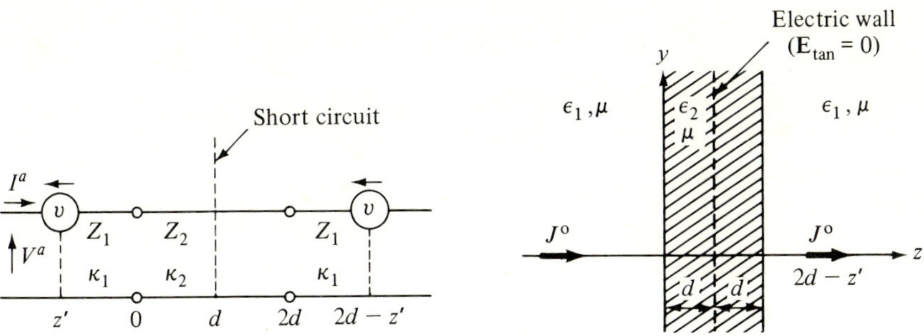

(a) Antisymmetric excitation

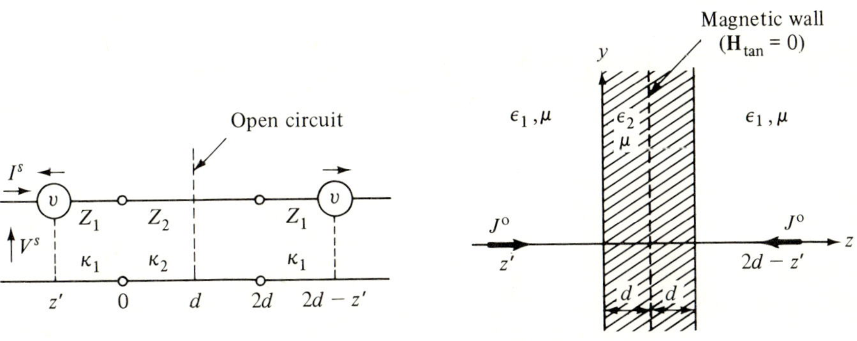

(b) Symmetric excitation

FIG. 5.6.7 Auxiliary field and network problems.

at the symmetry plane $z = d$ for the antisymmetric and symmetric longitudinal electric current excitations, respectively. It is evident that a superposition of field and network responses for the configurations in Fig. 5.6.7(a) and (b) yields twice the response for the structure in Fig. 5.6.6. Let $\mathbf{E}^a$, $\mathbf{H}^a$, and $\mathbf{E}^s$, $\mathbf{H}^s$ denote at any point the electromagnetic fields arising from the excitation by the anti-symmetric and symmetric current elements, respectively. Then the electromagnetic fields in Fig. 5.6.6(a) are given by

$$\mathbf{E} = \tfrac{1}{2}(\mathbf{E}^s + \mathbf{E}^a), \qquad \mathbf{H} = \tfrac{1}{2}(\mathbf{H}^s + \mathbf{H}^a). \tag{22}$$

The electric and magnetic walls in Fig. 5.6.7(a) and (b), respectively, isolate the region $z < d$ from the region $z > d$, so it suffices to treat only the bisected problems in the region $z < d$.

The antisymmetric excitation, leading to the configuration in Fig. 5.6.1, has already been treated. For the symmetric excitation, the requirement $g_z^s = 0$ at $z = d$ leads via Eq. (2.4.24b) to the replacement of $\overrightarrow{Z}(0)$ in Eq. (13) by [exp $(j\omega t)$ dependence]

$$\vec{Z}^s(0) = -jZ_2 \cot \kappa_2 d, \qquad Z_2 = \frac{\kappa_2}{\omega\epsilon_2}; \tag{23}$$

the corresponding reflection coefficient Γ^s is

$$\Gamma^s(\xi) = \frac{j\kappa_2 \cot \kappa_2 d + \epsilon\kappa_1}{j\kappa_2 \cot \kappa_2 d - \epsilon\kappa_1}, \qquad \kappa_{1,2} = \sqrt{k_{1,2}^2 - \xi^2}, \tag{23a}$$

and the three-dimensional Green's function $G'(\mathbf{r}, \mathbf{r}')$ is readily constructed therefrom. For a field evaluation in the slab region, g_{z2} in Eq. (14) is replaced by [see Eq. (5)]

$$g_{z2}^s(z, z') = g_{z2}(0, z')\frac{\sin \kappa_2(d - z)}{\sin \kappa_2 d} = g_{z1}(0, z')\frac{\sin \kappa_2(d - z)}{\sin \kappa_2 d}. \tag{24}$$

It may be noted that while the asymptotic field solution for the open-circuit bisection (symmetric excitation) takes the same form as in Eq.(6), the transcendental equations defining the pole singularities of $\Gamma(k_1 \sin w)$ are now, instead of Eqs. (7),

$$\epsilon q = -ip \cot p, \qquad p^2 - q^2 = (k_1 d)^2(\epsilon - 1), \tag{25}$$

with p and q defined as in Eq. (7a). Surface-wave solutions of these equations for real $\epsilon > 1$, with $q = i|q|$, p real, may be found by a graphical construction as in Fig. 3.4.3; one observes that N solutions exist in the range $(N + \frac{1}{2})\pi < k_1 d\sqrt{\epsilon - 1} < (N + \frac{3}{2})\pi$, $N = 0, 1, 2, \ldots$. No solution is found when $k_1 d\sqrt{\epsilon - 1} < \pi/2$, so the surface waves are now subject to a low-frequency cutoff, in contrast to the short-circuit bisection case; $\epsilon < 1$ likewise does not yield surface-wave solutions.

It may be mentioned that in view of Fig 5.6.7, a knowledge of the field solutions for the bisected structures in the region $z < 0$ implies a knowledge of the solution for the configuration in Fig. 5.6.6 both for $z < 0$ and $z > 2d$. With reference to Figs. 5.6.6 and 5.6.7 and Eq. (22), the field on the side $z < 0$ containing the source is obtained by adding the contributions due to *identically oriented* sources, whereas the transmitted field on the side $z > 2d$ is obtained by adding the contributions due to *oppositely oriented* sources, in the bisected arrangement.

5.6b   Other Source Configurations

Transverse electric current element

$$\hat{\mathbf{J}}(\mathbf{r}, t) = Il\delta(\boldsymbol{\rho})\delta(z - z')e^{-i\omega t}\mathbf{x}_0. \tag{26}$$

If the current element in Fig. 5.6.1 or 5.6.6(a) is directed along the x axis, the equivalent excitation in the modal network of Fig. 5.6.4 or 5.6.6(b) is a shunt current generator located at z' as in Fig. 5.5.6, and both E and H modes must be considered. The analysis for observation points $z < 0$ proceeds exactly as in Sec. 5.5b, wherein the transverse current element is situated in front of a semiinfinite dielectric medium. The only difference is that $\Gamma'(\xi)$ and $\Gamma''(\xi)$ in Eqs. (5.5.31) must be replaced by the E- and H-mode reflection coefficients, respectively, seen looking to the right at the $z = 0$ plane in Fig. 5.6.1 or 5.6.7(b). As in Sec. 5.6a, the fields radiated in the presence of the dielectric slab can be inferred by superposition from those in the two auxiliary configurations wherein an electric and magnetic wall, respectively, is placed along the $z = 0$ plane as in Fig. 5.6.7. The electric wall bisection arises from an excitation by the x-directed current element at z' and an oppositely directed element of the same strength at $2d - z'$, while the magnetic wall bisection occurs when the direction of both current elements is the same. The E-mode reflection coefficients for the bisected configurations seen looking to the right at $z = 0$ are given in Eqs. (4a) and (23a). For the H modes, Eqs. (13) and (23) for $\vec{Z}(0)$ remain valid except that the formulas for the characteristic impedances are now $Z''_{1,2} = \omega\mu/\kappa''_{1,2}$. It may also be recalled that in view of the unbounded cross section transverse to z, $\kappa'_{1,2} = \kappa''_{1,2} = \sqrt{k^2_{1,2} - \xi^2}$. Thus, the modal reflection coefficient for the short-circuit bisection is

$$\Gamma''(\xi) = \frac{j\kappa_1 \tan \kappa_2 d - \kappa_2}{j\kappa_1 \tan \kappa_2 d + \kappa_2}, \qquad (27a)$$

whereas for the open-circuit bisection,

$$\Gamma''^s(\xi) = \frac{j\kappa_1 \cot \kappa_2 d + \kappa_2}{j\kappa_1 \cot \kappa_2 d - \kappa_2}. \qquad (27b)$$

When these formulas (with $j \to -i$) are inserted into Eqs. (5.5.31), they yield integral representations for the functions $V'_t \mathscr{S}'_1$ and $V'_t \mathscr{S}''_1$ for both the open- and short-circuit bisections, from which the electromagnetic fields in the region $z < 0$ may be derived via Eqs. (5.2.1). For the region $0 < z \leq d$, modifications analogous to those in Eqs. (5) and (24) are required.

An asymptotic evaluation in the region $z < 0$ in Figs. 5.6.1 and 5.6.7 or $z > 2d$ in Fig. 5.6.6 may be carried out as before and leads to expressions analogous to those in Eq. (6). It may be noted that the H-mode resonance conditions are quite similar to those for the E modes in Eqs. (7) and (25). In particular, the short-circuit bisection in Eq. (27a) yields pole singularities of $\Gamma''(\xi)$; these are specified by Eq. (25) provided that the term ϵq is replaced by q, and the same replacement makes Eq. (7) applicable to the open-circuit bisection case in Eq. (27b).

Transverse electric line current

$$\hat{\mathbf{J}}(\mathbf{r}, t) = I\delta(\hat{\boldsymbol{\rho}} - \hat{\boldsymbol{\rho}}')e^{-i\omega t}\mathbf{x}_0. \qquad (28)$$

This problem is directly analogous to the line-source excitation of a semi-infinite medium (Sec. 5.5d), provided that the half-space region $z > 0$ is replaced by the slab configurations in Fig. 5.6.1 or 5.6.7. In the analysis, it is implied that the input impedances $\vec{Z}(0)$ and reflection coefficients Γ are replaced by their counterparts for the slab region, as specified in the preceding sections. The asymptotic evaluation of the resulting integrals, which have the same form as in Eq. (5.5.54b), is almost identical with that for the point-source problem since the Hankel function in Eq. (3b) is replaced by its asymptotic approximation in Eq. (5.3.13). The derivation of explicit results is left as an exercise for the reader.

5.7 TIME-HARMONIC SOURCES IN THE PRESENCE OF A CONSTANT-IMPEDANCE SURFACE

For a half-space or layered region $z > 0$, it is possible to define in certain parameter ranges a surface impedance Z_s that characterizes in an *approximate* fashion the relation between the tangential electric and magnetic fields at the $z = 0$ plane:

$$\mathbf{E}_t(\boldsymbol{\rho}, 0) = Z_s \mathbf{H}_t(\boldsymbol{\rho}, 0) \times \mathbf{z}_0. \tag{1}$$

When the surface impedance concept can be employed, it simplifies substantially the solution of an electromagnetic boundary-value problem in the region $z < 0$ since it eliminates the need for a detailed exploration of the fields in the region $z > 0$. In effect, the surface impedance on the $z = 0$ plane terminates the region $z < 0$ [see Figs. 5.6.1 and 5.3.1 and Eq. (1.5.40)].

5.7a   Longitudinal Electric Current Element

$$\hat{\mathbf{J}}(\mathbf{r}, t) = Il\delta(\boldsymbol{\rho})\delta(z - z')e^{-i\omega t}\mathbf{z}_0. \tag{2}$$

We consider the physical configuration in Fig. 5.7.1, wherein a longitudinal electric current element of strength $J^o = Il$, I being the current in the element

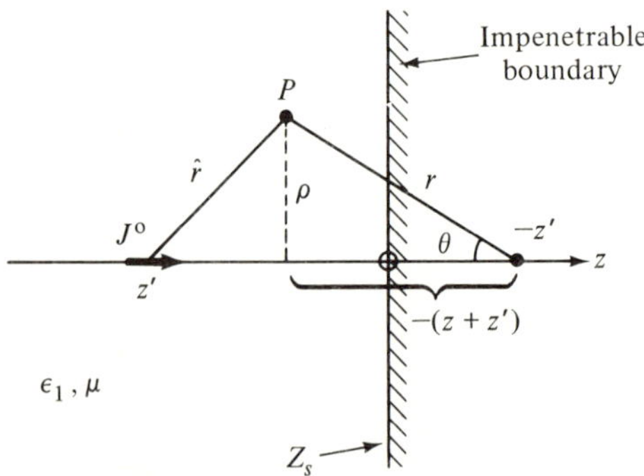

FIG. 5.7.1   Longitudinal current element and constant-impedance surface.

and l its infinitesimal length, is situated at the point $z' < 0$ on the z axis in the presence of a plane surface at $z = 0$ having a constant surface impedance Z_s (Fig. 5.7.1). The electromagnetic fields may be derived via Eqs. (5.2.1) and (5.2.4c) from a scalar E-mode Green's function $G'(\mathbf{r}, \mathbf{r}')$ which satisfies the differential equation

$$(\nabla^2 + k_1^2)G'(\mathbf{r}, \mathbf{r}') = -\delta(\mathbf{r} - \mathbf{r}'), \qquad z, z' < 0, \tag{3}$$

subject to a radiation condition at infinity, and to the following boundary condition [obtained by utilizing Eqs. (1) and (5.2.1)]:

$$\frac{\partial G'}{\partial z} = i\omega\epsilon_1 Z_s G' = ik_1 \bar{Z}_s G' \qquad \text{at } z = 0. \tag{3a}$$

$\bar{Z}_s = Z_s/\sqrt{\mu/\epsilon_1}$ is the surface impedance normalized to the wave impedance in the unbounded medium described by the constitutive parameters ϵ_1, μ, and $k_1 = \omega\sqrt{\mu\epsilon_1}$ is the wavenumber. In a cylindrical coordinate description of the cross section transverse to z, G' may be represented as follows:

$$G'(\mathbf{r}, \mathbf{r}') = G_{f1}(\mathbf{r}, \mathbf{r}') + G'_s(\mathbf{r}, \mathbf{r}'), \tag{4}$$

where $G_{f1} = (1/4\pi\hat{r})\exp(ik_1\hat{r})$ is the free-space Green's function for the unbounded region [Eq. (5.4.2b)], and G'_s, which arises from the presence of the boundary at $z = 0$, has the integral representation:

$$G'_s = \frac{-ik_1}{8\pi} \int_{\bar{P}} \sin w \, H_0^{(1)}(k_1 r \sin\theta \sin w)e^{ik_1 r \cos\theta \cos w}\Gamma(k_1 \sin w) \, dw, \tag{5}$$

where

$$\Gamma(k_1 \sin w) = \frac{\bar{Z}_s - \cos w}{\bar{Z}_s + \cos w}. \tag{5a}$$

The path of integration is given in Fig. (5.3.6b) and the coordinates $\hat{r}$, r, and θ are defined in Fig. 5.7.1.

For large values of $k_1 r \sin\theta$, the integral in Eq. (5) may be evaluated asymptotically and yields a saddle-point contribution as well as a possible residue contribution arising from the pole w_p of the reflection coefficient in Eq. (5a):

$$G' \sim \frac{e^{ik_1\hat{r}}}{4\pi\hat{r}} - \Gamma(k_1 \sin\theta)\frac{e^{ik_1 r}}{4\pi r}\left[1 + O\left(\frac{1}{k_1 r}\right)\right]$$

$$+ \frac{e^{-i\pi/4}}{\sqrt{\sin w_p}}\sqrt{\frac{k_1}{2\pi r \sin\theta}}\cos w_p \, e^{ik_1 r \cos(w_p - \theta)} U(\theta - \theta_p)\left[1 + O\left(\frac{1}{k_1 r}\right)\right], \tag{6}$$

where

$$\cos w_p = -\bar{Z}_s, \tag{6a}$$

and θ_p, the observation angle for which the steepest-descent path P in Fig. 5.3.6(b) crosses the pole w_p, is defined as in Eq. (5.6.8). $U(\theta - \theta_p)$ is the Heaviside unit function defined in Eq. (5.6.6), and its contributing range is limited to inductive surface impedances for which $\text{Im } \bar{Z}_s < 0$ [$\exp(-i\omega t)$ dependence].

Discussion

The physical interpretation of the above solution in terms of geometric-optical and surface-wave contributions is identical with that for Eq. (5.6.6). The solution is valid for those observation points $\theta > 0$ for which the steepest-descent path P does not pass near the pole $w_p(\theta \napprox \theta_p)$. If the pole is located near the steepest-descent path, in particular near the saddle point, the asymptotic evaluation must be carried out by the modified saddle-point procedure described in Sec. 4.4a. For the case where the surface impedance Z_s is equal to $\sqrt{\mu/\epsilon_2}$ representative of a highly lossy medium with complex dielectric constant ϵ_2 occupying the half-space region $z > 0$, one has $\bar{Z}_s \approx 1/\sqrt{\epsilon}$, $\epsilon = \epsilon_2/\epsilon_1$. For corresponding ranges of source and observation point coordinates, the asymptotic field solution then becomes identical with that in Eq. (5.5.7) which is derived from an exact analysis.

In view of Eq. (6a), the propagation characteristics of the residue contribution in Eq. (6) can be expressed simply in terms of the surface impedance $\bar{Z}_s$:

$$\frac{e^{ik_1 r \cos(w_p - \theta)}}{\sqrt{r \sin \theta}} = \frac{1}{\sqrt{\rho}} \exp\left(-ik_1 \bar{Z}_s |z + z'|\right) \exp\left(ik_1 \sqrt{1 - \bar{Z}_s^2}\, \rho\right). \tag{7}$$

Thus, one notes that the residue field contribution decays exponentially away from the surface $z = 0$ if $\mathrm{Im}\, \bar{Z}_s < 0$ and constitutes a proper solution of the Maxwell field equations in this case where $\mathrm{Im}\, \sqrt{1 - \bar{Z}_s^2} > 0$. By an extension of the notion of a surface wave on a lossless structure ($\bar{Z}_s = -i|\bar{Z}_s|$), the wave on a dissipative surface with $\mathrm{Im}\, \bar{Z}_s < 0$ might be termed a "lossy" surface wave. For the highly lossy case, $\bar{Z}_s \approx \epsilon^{-1/2}$, $\sin w_p \approx 1 + i/2|\epsilon|$, a comparison at $z = z' = 0$ of the amplitude of excitation of the surface wave in Eq. (6) with that given by the second term in Eq. (5.5.8b) yields a value for the latter half as large as for the former. This discrepancy occurs since, for $\theta = \pi/2$, the pole is located *on* the steepest-descent path near the saddle point, and the path of integration is indented around the pole in the form of a semicircle. Hence only a half-residue contributes, instead of the full residue given in Eq. (6) (see Sec. 4.4a).

Analytical details

The integral representation in Eq. (5) is derived in a manner directly analogous to that employed in Sec. 5.6. The schematic representation of the modal Green's function problem of determining $g_{zi}(z, z')$ in Eq. (5.3.1) is given in Fig. 5.7.2, with the surface impedance Z_s representing the termination on the modal

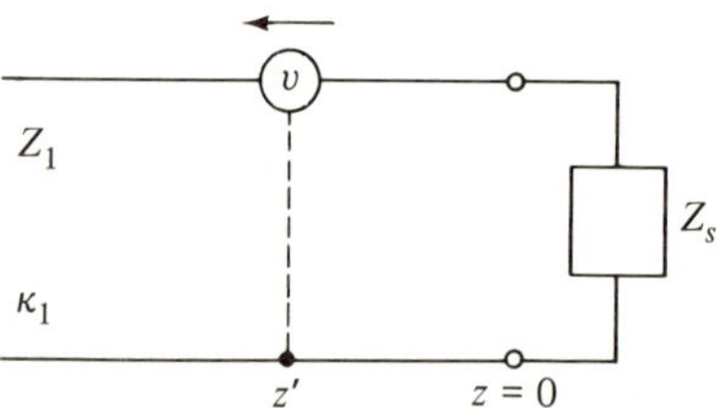

FIG. 5.7.2   Equivalent modal network (E modes).

transmission line. To deduce the above result one employs the field in Eq. (5.2.11) (with $j \to -i$) and Eq. (3a) to derive the modal boundary condition $dg'_{zi}/dz = i\omega\epsilon_1 Z_s g'_{zi}$, and thence, via Eqs. (5.2.6a) and (2.2.15), the network in Fig. 5.7.2. It is recalled that $Y'_i(z, z')$ and therefore g'_{zi} represents a normalized current; from the E-mode transmission-line equations (2.2.15), $(1/i\omega\epsilon_1)(dg'_{zi}/dz)$

is the corresponding voltage, and the ratio of the voltage to the current at $z = 0$ yields the surface impedance Z_s. The formula for the reflection coefficient in Eq. (5a) follows from Eq. (5.3.1a), with $\overrightarrow{Z_i}(0) = Z_s$ and $k_{ti} = \xi = k_1 \sin w$.

The asymptotic evaluation of the integral in Eq. (5) proceeds as in Sec. 5.6. Instead of the infinite set of leaky-wave poles and the finite set of surface-wave poles encountered in the slab problem, there exists now only a single relevant pole at

$$\cos w_p = -\bar{Z}_s, \quad \text{i.e., } w_p = \cos^{-1}(-\bar{Z}_s) \pm 2m\pi, \qquad m = 0, 1, 2, \ldots . \quad (8a)$$

For passive surface impedances, which absorb a finite (or zero) amount of power, the requirement of real power flow into the surface [i.e., $\text{Re} \int_S \mathbf{E}_t \times \mathbf{H}_t^* \cdot \mathbf{z}_0 \, dS \geq 0$] implies via Eq. (1) that $\text{Re} \, \bar{Z}_s \geq 0$. It then follows from Eq. (8a) that, at the pertinent pole singularity,

$$\frac{3\pi}{2} > \text{Re} \, w_p \geq \frac{\pi}{2}; \qquad \text{Im} \, w_p \gtrless 0 \quad \text{if Im } \bar{Z}_s \gtrless 0. \qquad (8b)$$

For lossless surfaces ($\text{Re} \, \bar{Z}_s = 0$), $\text{Re} \, w_p = \pi/2$. From Eqs. (8) it is evident that the pole approaches the line $\text{Re} \, w_p = \pi/2$ from the right as $\text{Re} \, \bar{Z}_s \to 0$. If $\text{Im} \, \bar{Z}_s < 0$, $\text{Re} \, \bar{Z}_s = 0$, it then follows that the path of integration is indented to the left around the pole situated on the half-line $\text{Re} \, w_p = \pi/2$, $\text{Im} \, w_p < 0$. As noted from the discussion in Secs. 5.6 and 5.3e, this latter pole is identified as a "surface-wave pole" since its residue contribution to the integral in Eq. (5) gives rise to a surface wave [see Eq. (7)]. The requirement for the existence of a surface wave on a lossless (reactive) surface is, therefore, $\bar{Z}_s = -i|\bar{Z}_s|$ [i.e., the surface is inductive; one recalls that the assumed time dependence is $\exp(-i\omega t)$]. This behavior follows at once from a transverse resonance argument which requires via Eq. (2.4.36) that $\overrightarrow{Z_i}(0) + \overleftarrow{Z_i}(0) = 0$ at a pole. In the present case, wherein only E modes are involved, $Z_s + Z_i' = 0$ from Fig. 5.7.2. In a surface wave, the fields decay exponentially away from the surface implying that the associated mode is non-propagating. Since, for a non-propagating E mode, $Z_i' = i|Z_i'|$ is capacitive, the resonance equation can be satisfied only if $\bar{Z}_s = -i|\bar{Z}_s|$. The pole is not intercepted when $\text{Im} \, \bar{Z}_s > 0$.

An alternative representation in terms of the waves *guided along the surface* may be carried out as in Sec. 5.6 (see also Sec. 5.7b).

An image formulation

It is of interest to note a formulation[23] in which the perturbing properties of the constant impedance surface on fields radiated by a given source distribution can be taken into account rigorously by a suitably defined distribution of image sources. For excitation by a longitudinal electric current element, the effect of the impedance surface on the field is contained in the scattered contribution G_s to the scalar Green's function G. For the present discussion it is convenient to employ the integral representation in the ξ plane as in Eq. (5.5.3a), with $\Gamma(\xi)$ defined in Eq. (5a) (note: $\xi = k_1 \sin w$). Upon substituting

$$\Gamma(\xi) = \frac{k_1\bar{Z}_s - \kappa(\xi)}{k_1\bar{Z}_s + \kappa(\xi)} = -1 + \frac{2k_1\bar{Z}_s}{k_1\bar{Z}_s + \kappa(\xi)}, \qquad \kappa(\xi) = \sqrt{k_1^2 - \xi^2}, \qquad (9)$$

into Eq. (5.5.3a), one obtains

$$G'_s = G'_{s1} + G'_{s2}, \tag{10}$$

$$G'_{s1} = \frac{i}{8\pi}\int_{\infty e^{i\pi}}^{\infty} \xi H_0^{(1)}(\xi\rho)\frac{e^{-i\kappa(z+z')}}{\kappa}\,d\xi = \frac{\exp\left[ik_1\sqrt{\rho^2 + (z+z')^2}\,\right]}{4\pi\sqrt{\rho^2 + (z+z')^2}}, \tag{10a}$$

$$G'_{s2} = -\frac{ik_1\bar{Z}_s}{4\pi}\int_{\infty e^{i\pi}}^{\infty} \xi H_0^{(1)}(\xi\rho)\frac{e^{-i\kappa(z+z')}}{\kappa}\frac{1}{k_1\bar{Z}_s + \kappa}\,d\xi. \tag{10b}$$

The closed-form evaluation of G'_{s1}, carried out via Eq. (5.4.12c), is interpretable as a contribution from an image source of strength $(+1)$ located at $z = -z'$ as in Fig. 5.7.1. To reformulate G'_{s2}, one employs the identity

$$\frac{1}{k_1\bar{Z}_2 + \kappa} = \int_0^{\infty} e^{-(k_1\bar{Z}_s + \kappa)\gamma}\,d\gamma, \tag{11}$$

which is valid when $\mathrm{Re}\,\bar{Z}_s > 0$; k_1 is assumed to be real, and $\mathrm{Re}\,\kappa \geq 0$ for all ξ on the path of integration in Eq. (10b). Upon inserting Eq. (11) into Eq. (10b) and performing a permissible interchange of the orders of integration, one obtains

$$G_{s2} = -2k_1\bar{Z}_s\int_0^{\infty} d\gamma\, e^{-k_1 Z_s\gamma}\frac{i}{8\pi}\int_{\infty e^{i\pi}}^{\infty} \xi H_0^{(1)}(\xi\rho)\frac{e^{i\kappa[(-z+z')+i\gamma]}}{\kappa}\,d\xi, \tag{12a}$$

$$= \int_0^{\infty}\left[-2k_1\bar{Z}_s e^{-k_1 Z_s\gamma}\right]\frac{e^{ik_1\psi}}{4\pi\psi}\,d\gamma, \qquad \psi = \sqrt{\rho^2 + [-(z+z')+i\gamma]^2}. \tag{12b}$$

The transition from Eq. (12a) to (12b) is accomplished via Eq. (10a), wherein $\sqrt{\rho^2 + \chi^2}$ has been continued analytically from positive real values into a permitted domain of complex values of $\chi = -(z+z') + i\gamma$, $(z+z') < 0$, $\gamma \geq 0$, for which $\mathrm{Im}\,\sqrt{\rho^2 + \chi^2} > 0$. One notes from Fig. 5.3.3(a) (upon substituting $\xi = i\chi$) that the latter condition is satisfied in the first and third quadrants of the complex χ plane, thus resulting in the analytic continuation of the right-hand side of Eq. (10a) from real values of χ to $\chi = -(z+z') + i\gamma$. Since $\mathrm{Re}\,\kappa \geq 0$ along the path of integration, this continuation is also valid for the integral on the left-hand side of Eq. (10a). Equation (12b) can be interpreted as arising from a continuous distribution of image point sources situated in an infinite medium at the complex location $z = -z' + i\gamma$ and having an exponentially decaying strength denoted by the factor inside the brackets in the integrand.

For the special case $\mathrm{Im}\,\bar{Z}_s > 0$, the image formulation above can be made physical by locating the image sources at *real* values of the coordinates. We note that $\mathrm{Im}\,\psi > 0$ throughout the fourth quadrant of the complex γ plane [branch points are located at $\gamma = \pm\rho - i(z+z')$, $(z+z') < 0$], so the $\exp(ik_1\psi)$ term in the integrand of Eq. (12b) vanishes at $|\gamma| \to \infty$ in the fourth quadrant of the γ plane. The same is true for exp $(-k_1\bar{Z}_s\gamma)$ provided that Im $\bar{Z}_s > 0$. Thus, there is no contribution to the integral from a quarter-circle of infinite radius in the fourth quadrant, and the contour of integration can be deformed into the negative imaginary axis of the γ plane. Upon introducing the change of variable $\gamma = i\mu$, one obtains

$$G_{s2} = \int_0^{-\infty} [-2ik_1\bar{Z}_s e^{-ik_1\bar{Z}_s\mu}] \frac{\exp\{ik_1\sqrt{\rho^2 + [z + (z' + \mu)]^2}\}}{4\pi\sqrt{\rho^2 + [z + (z' + \mu)]^2}}\,d\mu,$$

$$\text{Im } \bar{Z}_s > 0, \qquad (13)$$

which now comprises the contributions from a continuous set of image point sources situated along the segment $z' \leq z < \infty$ of the positive z axis (Fig. 5.7.3). It is noted that the condition Im $\bar{Z}_s > 0$, which admits of an exact

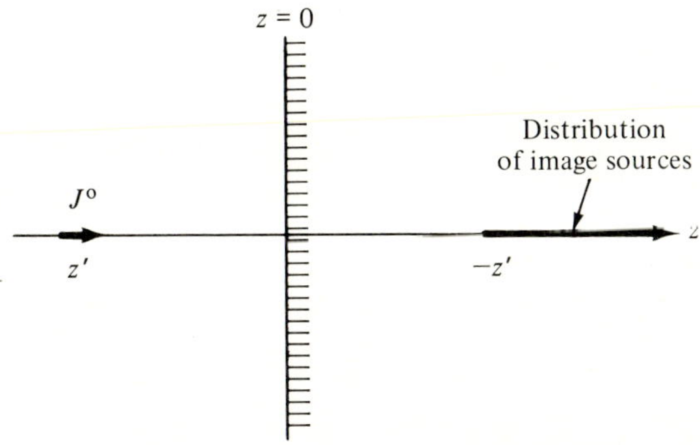

FIG. 5.7.3 Image formulation when Im $Z_s > 0$.

representation of the perturbing effect of the constant impedance surface in terms of physical image sources, excludes the case wherein a surface wave can be propagated along the structure. This implies that an image source distribution of the type noted above cannot produce in the region $z < 0$ a field that behaves in accordance with Eq. (7).

5.7b Transverse Magnetic Line Current

$$\hat{\mathbf{M}}(\mathbf{r}, t) = V\delta(\hat{\boldsymbol{\rho}} - \hat{\boldsymbol{\rho}}')e^{-i\omega t}\mathbf{x}_0. \qquad (14)$$

When the dipole in Fig. 5.7.1 is replaced by a magnetic line current of strength V parallel to the x axis, the electromagnetic field has the non-vanishing components H_x, E_y, E_z [see the dual of Eqs. (5.5.56) or (5.4.31)],

$$H_x = i\omega\epsilon_1 V\bar{G}', \qquad E_y = -V\frac{\partial\bar{G}'}{\partial z}, \qquad E_z = V\frac{\partial\bar{G}'}{\partial y}, \qquad (15)$$

which are derivable from a two-dimensional scalar E-mode Green's function $\bar{G}'$ satisfying the differential equation

$$\left(\frac{\partial^2}{\partial y^2} + \frac{\partial^2}{\partial z^2} + k_1^2\right)\bar{G}'(\hat{\boldsymbol{\rho}}, \hat{\boldsymbol{\rho}}') = -\delta(\hat{\boldsymbol{\rho}} - \hat{\boldsymbol{\rho}}'), \qquad \hat{\boldsymbol{\rho}} = (y, z) \qquad (16)$$

subject to a radiation condition at infinity, and to the boundary condition

$$\frac{\partial \bar{G}'}{\partial z} = i\omega\epsilon_1 Z_s \bar{G}' = ik_1 \bar{Z}_s \bar{G}' \qquad \text{at } z = 0, \qquad (16a)$$

with k_1 the wavenumber in the medium and $\bar{Z}_s = Z_s/\sqrt{\mu/\epsilon_1}$. The solution may be written as

$$\bar{G}'(\hat{\boldsymbol{\rho}}, \hat{\boldsymbol{\rho}}') = \bar{G}_f(\hat{\boldsymbol{\rho}}, \hat{\boldsymbol{\rho}}') + \bar{G}'_s(\hat{\boldsymbol{\rho}}, \hat{\boldsymbol{\rho}}'), \qquad (17)$$

where $\bar{G}_f$ is the free-space Green's function [Eq. (5.4.25)],

$$\bar{G}_f = \frac{i}{4} H_0^{(1)}(k_1|\hat{\boldsymbol{\rho}} - \hat{\boldsymbol{\rho}}'|), \qquad (17a)$$

while $\bar{G}'_s$ accounts for the presence of the impedance sheet at $z = 0$ and may be written in a z-transmission modal representation as

$$\bar{G}'_s(\hat{\boldsymbol{\rho}}, \hat{\boldsymbol{\rho}}') = \begin{cases} -\dfrac{i}{4\pi}\displaystyle\int_{-\infty}^{\infty} \dfrac{e^{i\eta(y-y')}\exp\left[-i\sqrt{k_1^2 - \eta^2}(z + z')\right]}{\sqrt{k_1^2 - \eta^2}}\Gamma(\eta)\,d\eta, & (17b) \\[2.5em] -\dfrac{i}{4\pi}\displaystyle\int_{\bar{P}} e^{ik_1 R \cos(w-\varphi)}\Gamma(k_1 \sin w)\,dw. & (17c) \end{cases}$$

$\Gamma(k_1 \sin w)$ is given in Eq. (5a) and the path $\bar{P}$ is defined in Fig. 5.3.6b. R and φ are polar coordinates measured from the image point, whence $|z + z'| = R \cos\varphi$, $y - y' = R \sin\varphi$.

For large values of $k_1 R$, the integral in Eq. (17c) may be evaluated asymptotically to yield a reflected-wave and a surface-wave contribution as in Eq. (6):

$$\bar{G}'_s \sim -\frac{1}{4\pi}\sqrt{\frac{2\pi}{k_1 R}}\, e^{i(k_1 R + \pi/4)}\Gamma(k_1 \sin\varphi)\left[1 + O\left(\frac{1}{k_1 R}\right)\right]$$
$$+ e^{ik_1 R \cos(w_p - \varphi)}\cot w_p\, U(\varphi - \varphi_p), \qquad (18)$$

where $\cos w_p = -\bar{Z}_s$, $U(\varphi - \varphi_p)$ is the Heaviside unit function, φ_p is defined as in Eq. (5.6.8), and Eq. (7) applies with $\theta \to \varphi$, $y \to \rho$. When $k_1 \hat{R}$ is also large, where $\hat{R} = |\hat{\boldsymbol{\rho}} - \hat{\boldsymbol{\rho}}'|$ is the distance from the source to the observation point, the Hankel function in Eq. (17a) may be approximated as in Eq. (5.3.13) and one finds that

$$\bar{G}' \sim \frac{e^{i\pi/4}}{2\sqrt{2\pi}}\left[\frac{e^{ik_1\hat{R}}}{\sqrt{k_1\hat{R}}} - \Gamma(k_1 \sin\varphi)\frac{e^{ik_1 R}}{\sqrt{k_1 R}}\right] + e^{ik_1 R \cos(w_p - \varphi)}\cot w_p\, U(\varphi - \varphi_p).$$
$$(19)$$

The interpretation of this result is directly analogous to that for the point-source excitation in Sec. 5.7a except that the geometric-optical fields, the first two terms in Eq. (19), are now in the form of cylindrical waves emanating from $\hat{R} = 0$ and $R = 0$, respectively, while the surface-wave contribution is in the form of a *plane wave*. Evidently, when a surface wave is excited (Im $\bar{Z}_s < 0$)

and when the surface impedance is purely reactive, the surface-wave contribution dominates near the boundary plane $z = 0$.

Equation (17) follows from Eqs. (5.2.13a) (with $\alpha = 0$), (5.4.33), and from the considerations in Sec. 5.7a (see also Fig. 5.7.2), so no additional details need be given.

Alternative representation

As in Sec. 5.6, Eq. (5.6.20), it is possible to reformulate Eqs. (17) in such a manner that they emphasize the guiding properties of the surface along the y direction. This is accomplished by deforming the integration contour in the complex η plane about the singularities of g_{zi} in the integral representation for $\bar{G}'$ [see Eq. (5.2.13a), with $j \to -i$]. To this end, one requires in addition to Eq. (17b) the integral representation in Eq. (5.4.36b) for the free-space field $\bar{G}_f$. We note first that the composite integral representation is insensitive to the algebraic sign of $y - y'$ (as verified by letting $\eta \to -\eta$), so $y - y'$ can be replaced by $|y - y'|$. The singularities of $\Gamma(\eta)$ in Eq. (17b) are simple poles at

$$\sqrt{k_1^2 - \eta_p^2} = -k_1 \bar{Z}_s, \qquad \text{i.e., } \eta_p = \pm k_1 \sqrt{1 - \bar{Z}_s^2}, \tag{20}$$

and branch points at $\eta = \pm k_1$. The branch cuts are chosen as in Fig. 5.6.5 to assure $\operatorname{Im} \sqrt{k^2 - \eta^2} > 0$ on the entire top sheet of the two-sheeted Riemann surface in the complex η plane. One notes from Eq. (20) that the poles η_p lie on the top sheet only when $\operatorname{Im} \bar{Z}_s < 0$, in which case they are located in the first and third quadrants. A deformation of the contour of integration about the singularities of the integrand in the upper half of the η plane can now be carried out by considerations analogous to those following Eq. (5.6.21) and yields the desired reformulation

$$\bar{G}'(\hat{\boldsymbol{\rho}}, \hat{\boldsymbol{\rho}}') = \frac{-\bar{Z}_s}{\sqrt{1 - \bar{Z}_s^2}} e^{ik_1 \sqrt{1 - \bar{Z}_s^2}\,|y - y'|} e^{ik_1 Z_s(z + z')} U(-\operatorname{Im} \bar{Z}_s)$$

$$+ \frac{i}{4\pi} \int_{-\infty}^{\infty} \frac{e^{i\sqrt{k_1^2 - \zeta^2}\,|y - y'|}}{\sqrt{k_1^2 - \zeta^2}} \left[e^{i\zeta(z - z')} - \frac{k_1 \bar{Z}_s - \zeta}{k_1 \bar{Z}_s + \zeta} e^{-i\zeta(z + z')} \right] d\zeta, \tag{21}$$

where $\operatorname{Im} \sqrt{k_1^2 - \zeta^2} \geq 0$. The replacement of $|z - z'|$ by $(z - z')$ in the integrand of Eq. (21) is justified by noting that the portion of the integral containing the $\exp[i\zeta(z - z')]$ term is insensitive to the algebraic sign of $(z - z')$ (let $\zeta \to -\zeta$ in the integrand). From Eq. (5.4.36b), the contribution to the integral from the $\exp[i\zeta(z - z')]$ term is equal to the free-space Green's function

$$(i/4) H_0^{(1)}[k_1 \sqrt{(y - y')^2 + (z - z')^2}].$$

The first term on the right-hand side of Eq. (21) represents a surface wave (discrete spectrum) propagating in the y direction [see Eq. (19)], while the remaining integral represents the continuous spectrum of reflected waves. For imaginary values of $\bar{Z}_s$, the continuous spectrum contribution can be written in a form analogous to that in Eq. (5.6.20).

5.7c   Other Elementary Source Configurations

If the constant impedance surface in Fig. 5.7.1 is excited by other funda-mental source distributions, such as a transverse current element, line currents, or ring currents, the solution for the radiated fields is obtained by a modal analysis and synthesis procedure as in the corresponding treatments in Secs. 5.5 and 5.6, provided only that the modal reflection coefficients defined therein are replaced by those appropriate to the present problem. The E-mode reflection coefficient has already been stated in Eq. (5a). For the H modes, the modal characteristic impedance is $Z_i'' = \omega\mu/\kappa_i''$ [see Eq. (5.3.1c)], so that, from Eq. (5.3.1a),

$$\vec{\Gamma}_i''(0) = \frac{\kappa_i'' - k_1 \bar{Y}_s}{\kappa_i'' + k_1 \bar{Y}_s}, \qquad \bar{Y}_s = \frac{1}{\bar{Z}_s}, \tag{22a}$$

which becomes for $\kappa_i'' = \sqrt{k_1^2 - \xi^2} = k_1 \cos w$,

$$\Gamma''(k_1 \sin w) = \frac{\cos w - \bar{Y}_s}{\cos w + \bar{Y}_s}. \tag{22b}$$

Asymptotic field calculations are carried out as in Secs. 5.7a and 5.7b.

5.7d   Continuous Distribution of Transverse Magnetic Line Currents

Excitation of surface waves by an aperture

Consider the configuration in Fig. 5.7.4(a), wherein a space bounded by the plane $z = 0$ with surface impedance Z_s is excited by electromagnetic fields

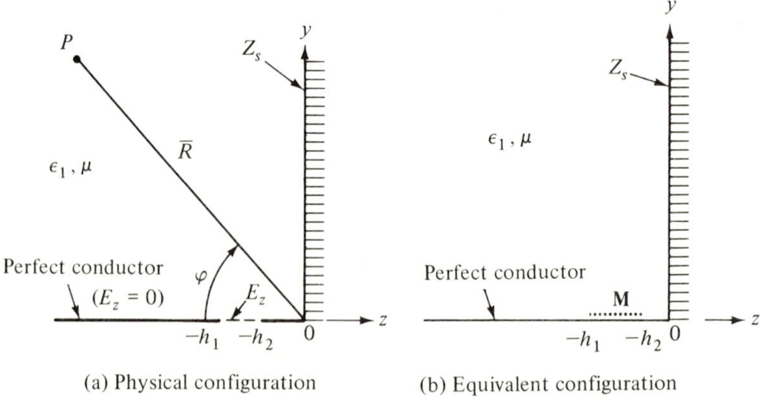

FIG. 5.7.4   Excitation of surface waves by an aperture.

in an aperture perforating a perfectly conducting plane located at $y = 0$.[24] The surrounding medium is characterized by the material constants ϵ_1 and μ. The aperture is in the form of a slit of width $h = h_1 - h_2$ and extends to infinity in the x direction; the aperture field is assumed to be independent of x, with a single magnetic-field component $\mathbf{H} = \mathbf{x}_0 H$ and a tangential electric-field component E_z. This type of aperture field arises, for example, if the aperture is excited from the region $y < 0$ by a parallel-plate waveguide operating in the TEM mode, with the waveguide walls extending along the planes $z = -h_1$ and $z = -h_2$. By the equivalence relation in Eq. (1.5.33a), the effect of the aperture field can be represented in terms of an equivalent magnetic current distribution $\mathbf{M} = \mathbf{E} \times \mathbf{y}_0 = -\mathbf{x}_0 E_z$ flowing on a perfectly conducting surface as shown in Fig. 5.7.4(b). From the considerations in Sec. 1.5b, the electromagnetic fields everywhere in the region $z < 0$, $y > 0$, are determined uniquely from a knowledge of $M_x = -E_z$. In view of the invariance of the configuration with respect to the x coordinate, it follows that everywhere in the region $z < 0$, $y > 0$, the magnetic field has only an x component, $H_x(\hat{\rho})$, in terms of which the electric-field components E_y and E_z are calculable as in Eq. (15). From Eq. (15), one obtains the fields due to a line source of magnetic currents having a strength V, so that by superposition, the magnetic field due to the aperture source distribution is given by

$$H_x(\hat{\rho}) = 2i\omega\epsilon_1 \int_{-h_1}^{-h_2} V(z')\bar{G}'(\hat{\rho}; 0, z')\, dz', \qquad \hat{\rho} = (y, z), \qquad (23)$$

where $\bar{G}'$ is the Green's function in Eq. (17), and the factor 2 arises from the imaging effect of the perfect conductor along the $y = 0$ plane in Fig. 5.7.4(b).

For an exact calculation, it is convenient to employ the y-guided-wave representation in Eq. (21) since the dependence on z' is continuous, thereby facilitating the integration required in Eq. (23). This is not the case for the z-transmission formulation [see Eq. (5.4.36b)]. At large distances from the aperture, one may insert the asymptotic approximation for $\bar{G}'$ which takes the form shown in Eq. (19). If $R \gg h_1$, where h_1 is the coordinate defining the upper edge of the aperture (Fig. 5.7.4), one may approximate $\hat{R}$ and R as follows:

$$\hat{R} = \bar{R} + z' \cos\varphi, \qquad R = \bar{R} - z' \cos\varphi, \qquad (24)$$

where $\bar{R} = (y^2 + z^2)^{1/2}$ is the distance from the origin to the observation point (y, z). Thus,

$$H_x(\hat{\rho}) \sim \frac{-\omega\epsilon}{\sqrt{2\pi k_1 \bar{R}}} e^{i(k_1\bar{R} - \pi/4)} [A(\cos\varphi) - \Gamma(k_1 \sin\varphi)A(-\cos\varphi)]$$

$$- 2i\omega\epsilon_1 \frac{A(\bar{Z}_s)\bar{Z}_s}{\sqrt{1 - \bar{Z}_s^2}} [\exp(ik_1\sqrt{1 - \bar{Z}_s^2}\,y + ik_1\bar{Z}_s z)]U(\varphi - \varphi_p), \qquad (25)$$

where

$$A(\gamma) = \int_{-h_1}^{-h_2} V(z')e^{ik_1 z'\gamma}\, dz'. \qquad (25a)$$

In the limit $\varphi \to \pi/2$, $\Gamma(k_1 \sin\varphi) \to 1$ and the contributions from the direct and reflected waves cancel to $O(1/\sqrt{k_1\bar{R}})$. Hence, for surface impedance values such that $\varphi_p < \pi/2$, the field near the surface is given by the last term in Eq. (25), which represents the surface wave propagating along the sheet.

In a true diffraction problem wherein the aperture is excited by an appropriate source distribution located in the region exterior to the quarter space shown in Fig. 5.7.4(a), the equivalent aperture current $M_x = V$ is unknown and must be determined from a self-consistent solution of the *total* electromagnetic boundary-value problem. Such a solution is usually difficult to attain. In many instances, however, the feeding mechanism is such that one may use a suitable approximation for M_x. A frequent technique, described variously as the "Kirchhoff" or "physical optics" procedure, is suitable for aperture widths large compared with the free-space wavelength; it involves the approximation of the aperture field by the value of the incident field in the aperture domain. For example, if the aperture is excited by a parallel-plate waveguide, the incident field (TEM mode) in the waveguide is constant in the $y = 0$ plane. Hence, as a first approximation, M_x would be a constant, and the integrals in Eq. (25) can be evaluated at once. While the physical optics approximation is usually poor for near-field evaluations (it is definitely wrong near the aperture edges where the fields change rather violently), it generally yields a good approximation of dominant effects in the far field provided that the aperture width is large.

Alternatively, when the slot is very narrow, one may write

$$-A(\gamma) \cong e^{-ikh\gamma}V, \qquad h_{1,2} = h \pm \delta, \quad 0 < \delta \ll \lambda, \tag{26a}$$

where h is the coordinate at the slot center and V is the voltage across the slot,

$$V = \int_{-h-\delta}^{-h+\delta} V(z')\,dz' = \int_{-h-\delta}^{-h+\delta} E_z(z')\,dz'. \tag{26b}$$

A comparison of Eqs. (25) and (26) with Eqs. (15) and (19) establishes the equivalence between the radiation from a voltage V excited narrow slot centered at $z = -h$ in a perfectly conducting plane and that from a magnetic current source of strength V located at $z = -h$ on a similar plane.

Radiation from a terminated reactive surface—comparison of various approximations

In surface-wave antenna applications, one is frequently interested in the radiation from a terminated surface along which a surface wave can be propagated.[25] The physical configuration is shown in Fig. 5.7.5(a), wherein a reactive surface with constant normalized surface impedance $\bar{Z}_s = -i\bar{X}_s$, $\bar{X}_s > 0$ (i.e., an inductance for the implied time dependence $\exp(-i\omega t)$) occupies the half-plane $z = 0$, $y < 0$, while the half-plane $z = 0$, $y > 0$, is a perfect conductor ($\bar{Z}_s = 0$). We assume that a plane surface wave, whose magnetic field vector is directed parallel to the x axis, propagates along the reactive surface in the $+y$ direction:

$$H_{\text{inc}}(\hat{\rho}) = H_0 e^{i\kappa y}e^{k_1\bar{X}_s z}, \quad \kappa = k_1\sqrt{1 + \bar{X}_s^2}, \quad \bar{X}_s > 0, \quad z < 0, \tag{27}$$

where the subscript $_{\text{inc}}$ denotes the incident field, and H_0 is a constant amplitude factor. One verifies at once that H_{inc} in Eq. (27) satisfies the homogeneous wave equation $(\nabla^2 + k^2)H_{\text{inc}} = 0$, the impedance boundary condition

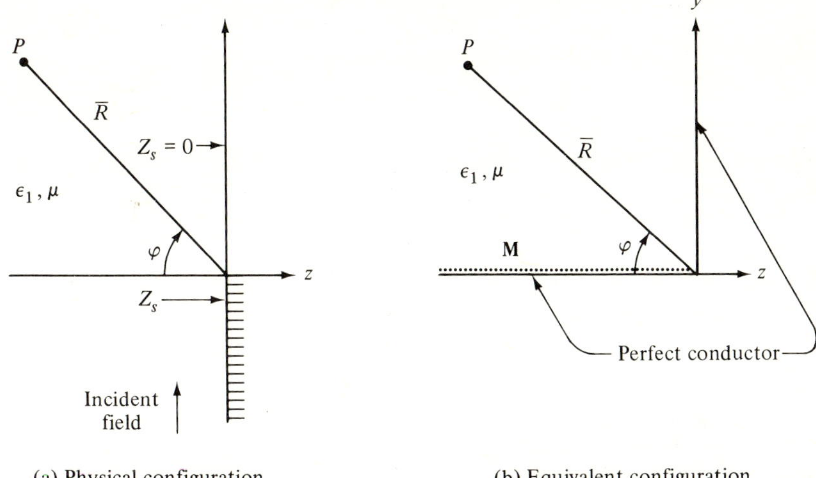

(a) Physical configuration   (b) Equivalent configuration

FIG. 5.7.5   Terminated reactive surface.

$(E_{\text{inc}})_y = (i/\omega\epsilon_1)\partial H_{\text{inc}}/\partial z = -Z_s H_{\text{inc}}$ on the reactive surface, and posseses at $y = -\infty$ the ingoing behavior characteristic of an idealized plane-wave excitation. The required functional dependence in a surface wave is inferred directly from the first term on the right-hand side of Eq. (21).

As in the problem sketched in Fig. 5.7.4, the field in the quarter-space region $y > 0$, $z < 0$, of Fig. 5.7.5(a) can be viewed as arising from an equivalent magnetic current distribution $M_x = -E_z$ flowing on a perfectly conducting plane at $y = 0$ [Fig. 5.7.5(b)]. Since the "aperture" in the present problem occupies the entire half-plane $z < 0$, $y = 0$, the equivalent magnetic current distribution extends from $z = 0$ to $z = -\infty$. The magnetic field H_x in the region $z < 0$, $y > 0$ [Fig. 5.7.5(b)] can then be calculated as in Eq. (23), provided that $h_1 = \infty$, $h_2 = 0$, and $\overline{G}'$ represents the Green's function for the half-space region bounded by a perfect conductor at $z = 0$. The result for the latter is obtained at once from either of Eqs. (17) or (21) by letting $\overline{Z}_s = 0$, as appropriate to a perfect conductor. The far field at $\overline{R} \to \infty$, $\varphi > 0$, in Fig. 5.7.5(b) is then given via Eq. (25) by

$$H_x(\hat{\boldsymbol{\rho}}) \sim \frac{-2\omega\epsilon_1}{\sqrt{2\pi k_1 \overline{R}}} e^{i(k_1\overline{R} - \pi/4)} \int_{-\infty}^{0} V(z') \cos\left(k_1 z' \cos\varphi\right) dz', \tag{28}$$

a result that demands a knowledge of $M_x(z') = V(z')$, or equivalently of E_z, in the aperture plane $y = 0$.

If the surface reactance $\overline{X}_s$ is small, more precisely, if $k_1 \overline{X}_s$ is small, the surface wave is "loosely bound" [i.e., the field in Eq. (27) extends appreciably away from the reactive surface]. The discontinuity in reactance at the junction of the two surfaces in Fig. 5.7.5(a) is likewise small, and it is reasonable to assume as a first approximation that the actual aperture field is approximated by the incident field

$$(E_{\text{inc}})_z = \frac{i}{\omega\epsilon_1}\frac{\partial H_{\text{inc}}}{\partial y} = \frac{-\kappa}{\omega\epsilon_1}H_{\text{inc}}, \tag{29a}$$

whence

$$V(z') \simeq \frac{\kappa}{\omega\epsilon_1}H_0 e^{k_1\bar{X}_s z'}. \tag{29b}$$

One then obtains the following result for the radiated far field, on performing the elementary integration required in Eq. (28):

$$H_x(\hat{\boldsymbol{\rho}}) \sim \frac{e^{i(k_1\bar{R}-\pi/4)}}{\sqrt{2\pi k_1\bar{R}}} H_0 \frac{2\bar{X}_s\sqrt{1+\bar{X}_s^2}}{\bar{X}_s^2+\cos^2\varphi}. \tag{30}$$

It is of interest to compare the approximate result in Eq. (30), whose validity is expected to be greatest for very small values of $\bar{X}_s$, with an exact formula[25,26] obtained via an integral equation procedure. The exact expression for the square of the radiated far-field magnitude is

$$|H_x|^2 \sim \frac{|H_0|^2}{\pi k_1\bar{R}}\frac{\bar{X}_s^2 v\cos^2\varphi}{(1+v)(1-\sin\varphi)(\bar{X}_s^2+\cos^2\varphi)(v-\sin\varphi)}, \qquad v = \sqrt{1+\bar{X}_s^2}, \tag{31a}$$

while the approximate expression in Eq. (30) yields

$$|H_x|^2 \sim \frac{|H_0|^2}{\pi k_1\bar{R}}\frac{2\bar{X}_s^2 v^2}{(\bar{X}_s^2+\cos^2\varphi)^2}. \tag{31b}$$

A variety of additional approximations can be built about the procedure wherein an induced field in an infinite aperture is approximated by the value of the incident field. Consider the region S in Fig. 5.7.5(a), wherein $y>0$, $z<0$, and apply Green's theorem to the two functions $\bar{G}'$ and H_x which, together with their first and second derivatives, are defined within S and on its boundary s:

$$\int_S [\bar{G}'(\hat{\boldsymbol{\rho}},\hat{\boldsymbol{\rho}}')\nabla'^2 H_x(\hat{\boldsymbol{\rho}}') - H_x(\hat{\boldsymbol{\rho}}')\nabla'^2\bar{G}'(\hat{\boldsymbol{\rho}},\hat{\boldsymbol{\rho}}')]\,dS'$$

$$= \oint_s \left[\bar{G}'(\hat{\boldsymbol{\rho}},\hat{\boldsymbol{\rho}}')\frac{\partial}{\partial n'}H_x(\hat{\boldsymbol{\rho}}') - H_x(\hat{\boldsymbol{\rho}}')\frac{\partial}{\partial n'}\bar{G}'(\hat{\boldsymbol{\rho}},\hat{\boldsymbol{\rho}}')\right]ds', \tag{32}$$

where n is normal to s and increases in the direction out of S. H_x is the magnetic field which satisfies in S the homogeneous wave equation

$$(\nabla^2+k^2)H_x(\hat{\boldsymbol{\rho}}) = 0, \qquad \hat{\boldsymbol{\rho}} = (y,z). \tag{33a}$$

On the perfectly conducting boundary $y>0$, $z=0$, denoted by s_1, it is required that $E_y=0$, whence

$$\frac{\partial H_x}{\partial z} = 0 \text{ on } s_1. \tag{33b}$$

On the quarter circle at $\bar{R}\to\infty$, denoted by s_2, H_x is to satisfy a radiation condition [see Eq. (1.5.34c)] since all sources of the field are contained in the region $y<0$. Correspondingly, it is convenient to choose $\bar{G}'$ as a Green's function satisfying the inhomogeneous wave equation

$$(\nabla^2 + k_1^2)\bar{G}'(\hat{\boldsymbol{\rho}}, \hat{\boldsymbol{\rho}}') = -\delta(\hat{\boldsymbol{\rho}} - \hat{\boldsymbol{\rho}}'), \qquad \hat{\boldsymbol{\rho}} \text{ and } \hat{\boldsymbol{\rho}}' \text{ in } S, \qquad (34a)$$

the boundary condition

$$\frac{\partial \bar{G}'}{\partial z} = 0 \text{ on } s_1, \qquad (34b)$$

and a radiation condition on s_2. Its behavior in the aperture plane $y = 0$, required for uniqueness, is left open for the moment. Upon substituting Eqs. (33) and (34) into Eq. (32), one obtains an expression for H_x in S in terms of its value and that of its derivative in the aperture plane:

$$H_x(\hat{\boldsymbol{\rho}}) = -\int_{-\infty}^{0} \left[\bar{G}'(\hat{\boldsymbol{\rho}}, \hat{\boldsymbol{\rho}}') \frac{\partial}{\partial y'} H_x(\hat{\boldsymbol{\rho}}') - H_x(\hat{\boldsymbol{\rho}}') \frac{\partial}{\partial y'} \bar{G}'(\hat{\boldsymbol{\rho}}, \hat{\boldsymbol{\rho}}') \right]_{y'=0} dz'. \qquad (35)$$

The contributions to the boundary integral arising from the contours s_1 and s_2 vanish in view of Eqs. (33b), (34b), and the radiation condition, respectively.

Alternative formulations for H_x can now be obtained from Eq. (35) by specifying the behavior of $\bar{G}'$ in the aperture plane, thus completing the definition of $\bar{G}'$ and permitting its evaluation. Three convenient choices are the following:

$$\bar{G}' = 0 \text{ at } y' = 0, \qquad (36a)$$

$$\frac{\partial \bar{G}'}{\partial y'} = 0 \text{ at } y' = 0, \qquad (36b)$$

$$\bar{G}' = \bar{G}'_h. \qquad (36c)$$

Use of the first two functions eliminates the first and second terms, respectively, in the integrand of Eq. (35). The third choice, the half-space Green's function $\bar{G}'_h$, is that appropriate to an unlimited y domain and satisfies a radiation condition at $\bar{R} \to \infty$ in the entire half-space region $z < 0$, as well as condition (34b) on the entire $z = 0$ plane; its use retains both terms in the integrand of Eq. (35). If values of H_x and $\partial H_x/\partial y$ in the aperture plane are known *exactly* from a rigorous solution of the given diffraction problem, then H_x is uniquely determined in S and its calculation via Eq. (35) with any of the above-mentioned Green's functions leads to the same result. On the other hand, if *approximate* values are employed for H_x and $\partial H_x/\partial y$ in the aperture plane, the fields in S calculated via different Green's functions are generally different.

To illustrate these remarks we consider once again the problem sketched in Fig. 5.7.5(a), with the assumption that H_x and (or) $\partial H_x/\partial y$ in the aperture plane are approximated by the value of H_{inc} and (or) $\partial H_{\text{inc}}/\partial y$ in Eq. (27), respectively. The Green's function satisfying Eqs. (34) and (36b) is evidently the one described in Fig. 5.7.5(b), so the resulting expression for H_x in Eq. (35) is identical with that in Eq. (28), with the recognition that $-V(z') = E_z = (1/i\omega\epsilon_1)\partial H_x/\partial y$. Thus, the choice of $\bar{G}'$ as in Eq. (36b) leads to the far-magnetic-field formula in Eq. (30).

For a choice of $\bar{G}'$ in accord with Eq. (36a), we note by inspection that the required boundary condition at $y' = 0$ can be met by imaging two half-space Green's functions $\bar{G}'_h$ in the $y = 0$ plane. Thus,

$$H_x(\hat{\boldsymbol{\rho}}) = \int_{-\infty}^{0} H_x(\hat{\boldsymbol{\rho}}') \left[\frac{\partial}{\partial y'} [\bar{G}_h'(\hat{\boldsymbol{\rho}}; y', z') - \bar{G}_h'(\hat{\boldsymbol{\rho}}; -y', z')] \right]_{y'=0} dz'. \tag{37}$$

It might be pointed out that Eq. (37) can be interpreted as representing an equivalent configuration wherein the aperture plane has been replaced by a magnetic wall carrying an equivalent electric current distribution $J_z \propto H_x$. Upon performing the derivative operation on the asymptotic formula for $\bar{G}_h'$, substituting into Eq. (37), and employing the value H_{inc} from Eq. (27) for H_x, one obtains

$$H_x(\hat{\boldsymbol{\rho}}) \sim \frac{e^{i(k_1\bar{R}-\pi/4)}}{\sqrt{2\pi k_1 \bar{R}}} H_0 \frac{2\bar{X}_s \sin \varphi}{\bar{X}_s^2 + \cos^2 \varphi}. \tag{38}$$

Finally, we employ the half-space Green's function $\bar{G}_h'$ directly. From simple image considerations, $\bar{G}_h'(\hat{\boldsymbol{\rho}}; 0, z')$ is equal to one half the Green's function satisfying condition (36b), while $(\partial \bar{G}_h'/\partial y')_{y'=0}$ is equal to one half the y' derivative of the Green's function satisfying condition (36a), all functions being

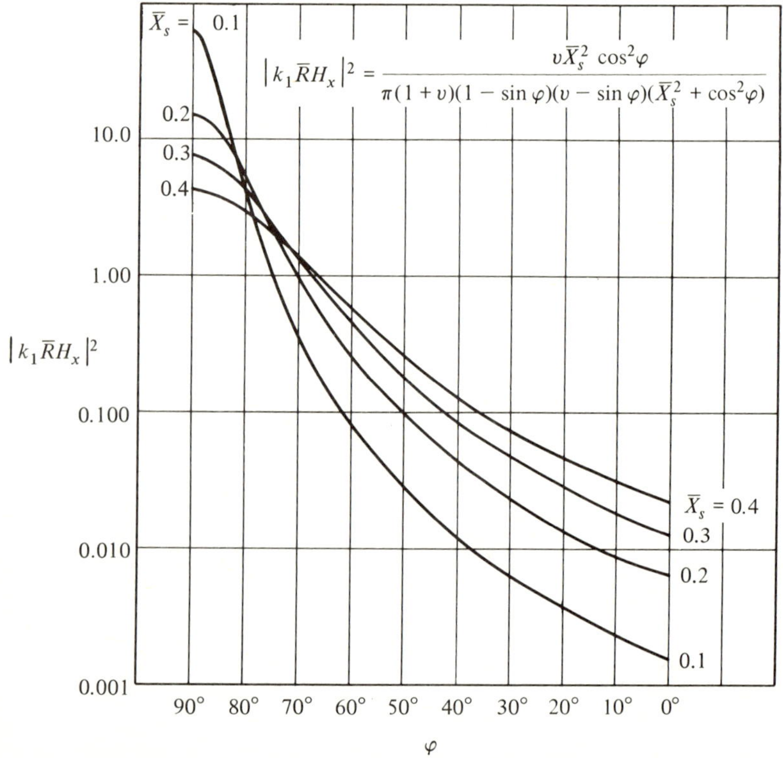

FIG. 5.7.6(a) Radiation pattern of a terminated surface-wave antenna.

evaluated at $y' = 0$. Thus, an approximate evaluation, utilizing for H_x and $\partial H_x/\partial y$ in the $y = 0$ plane the values H_{inc} and $\partial H_{\text{inc}}/\partial y$, respectively, yields a result for H_x in S which is the average of that obtained through use of the two other Green's functions in Eqs. (36a) and (36b). The far field is in this case

$$H_x(\hat{\boldsymbol{\rho}}) \sim \frac{e^{i\,(k_1\bar{R}-\pi/4)}}{\sqrt{2\pi k_1\bar{R}}}\, H_0\, \frac{\bar{X}_s(\sin\varphi + \sqrt{1 + \bar{X}_s^2})}{\bar{X}_s^2 + \cos^2\varphi}. \tag{39}$$

The far-field power patterns $|k_1\bar{R}H_x|^2$ calculated from Eqs. (31a), (31b), (38), and (39) are plotted in Figs. 5.7.6(a)–(d) for values of $\bar{X}_s = 0.1, 0.2, 0.3, 0.4$. All patterns are peaked in the endfire direction $\varphi = 90°$, as expected from a surface-wave illumination of the aperture plane. The fields obtained from the approximate solution in Eq. (39) and the exact solution in Eq. (31a) agree almost perfectly in the entire quarter-space region $0 \le \varphi \le 90°$. The results from Eqs. (31b) and (38) are larger and smaller, respectively, than the correct values, with the greater deviations occurring at larger angles of elevation from

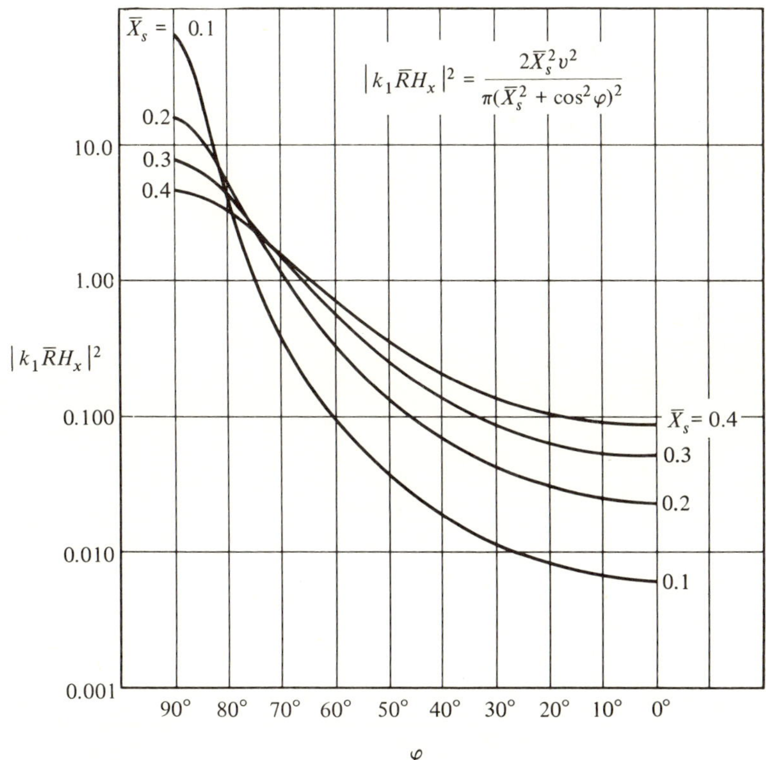

FIG. 5.7.6(b) Radiation pattern of a terminated surface-wave antenna.

the surface. For $\bar{X}_s = 0.1$, the field amplitudes decay sharply as φ decreases from 90°, and any of the above formulations yield acceptable results down to $\varphi \approx 55°$, at which point $|H_x|^2$ decreases to approximately $\frac{1}{1000}$ of its maximum value. The deviations for smaller values of φ are exaggerated on the logarithmic scale but are associated with negligibly small field amplitudes. This agreement will improve further for $\bar{X}_s < 0.1$, thereby justifying any of the above approximation procedures for very small values of surface reactance. For larger values of $\bar{X}_s$, Eq. (39) is to be preferred.

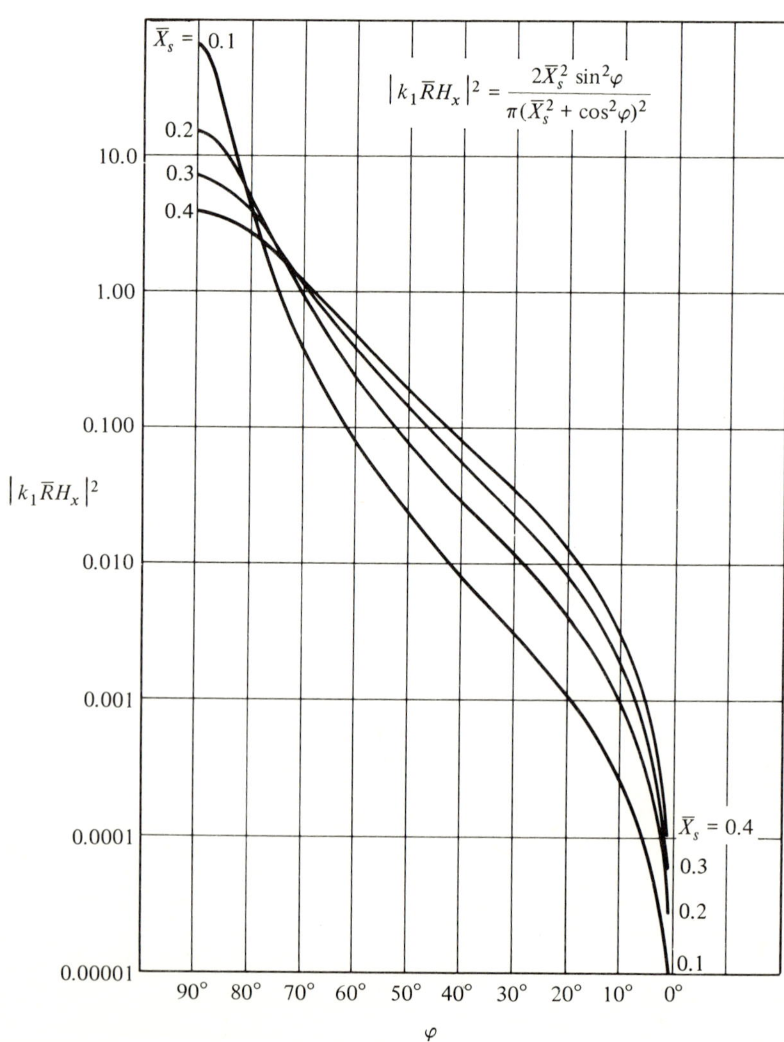

FIG. 5.7.6(c) Radiation pattern of a terminated surface-wave antenna.

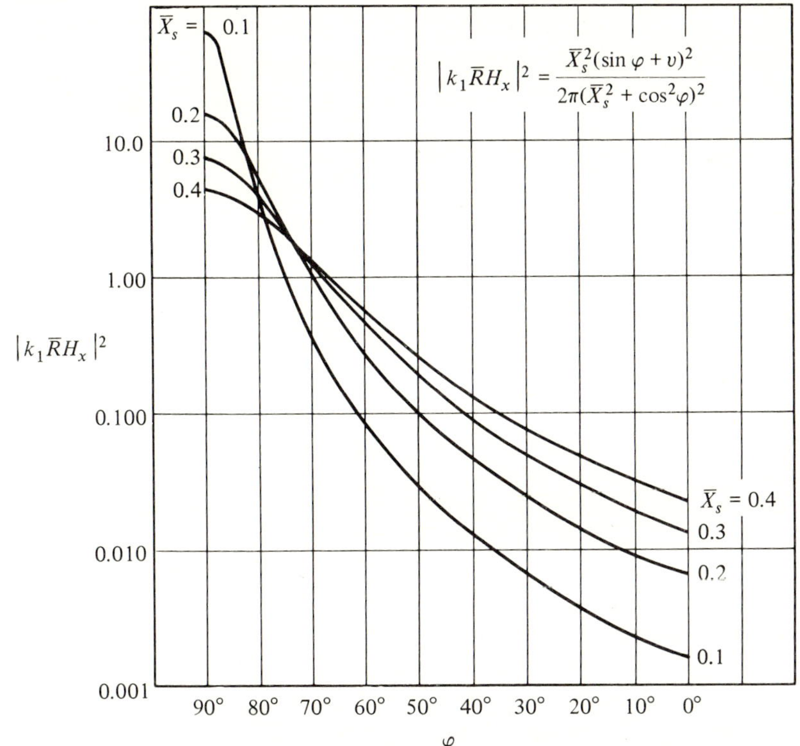

FIG. 5.7.6(d) Radiation pattern of a terminated surface-wave antenna.

5.8 SOURCES IN THE PRESENCE OF MEDIA WITH CONTINUOUS PLANAR STRATIFICATION—ARBITRARY PROFILES

5.8a General Field Properties

The radiation problems in the preceding sections have dealt with regions whose material properties along a rectilinear z coordinate may change discontinuously, as for example, at an interface between two different constant or homogeneous dielectric media. In this section and in Sec. 5.9 we consider the effects of a continuously varying medium on fields radiated by prescribed source distributions. The terms "constant" and "continuously varying" apply only to the macroscopic properties of the medium and represent averages of microscopic effects arising from the more detailed physical structure. The characterization of macroscopically homogeneous isotropic media by a constant permittivity and permeability requires sufficiently long wavelengths. However, many important electromagnetic propagation problems in nature take place under conditions where the macroscopic properties vary in a continuous manner, even for microwave, or lower, frequencies. This is true of such gaseous media as the earth's atmosphere, or for the "electron plasma gas" constituting the ionosphere, where variations in particle density give rise to macroscopic inhomogeneities. Artificial inhomogeneous media also find application, for example, in microwave lenses.

Exact solutions for fields radiated by prescribed sources in an inhomogeneous medium can be obtained only for those special variations in medium constants that lead to differential equations whose solutions are known (see Sec. 5.9 for illustrative examples). However, if the properties of the medium vary slowly over an interval of length equal to the local wavelength, it is possible to derive approximate expressions for the fields that apply for any slow functional variation of permittivity $\epsilon(\mathbf{r})$ and (or) permeability $\mu(\mathbf{r})$ or, to use the terminology of geometrical optics, of the refractive index $n(\mathbf{r}) = [\mu(\mathbf{r})\epsilon(\mathbf{r})/ \mu_0\epsilon_0]^{1/2}$, ϵ_0 and μ_0 being the permittivity and permeability of free space. These approximate expressions are shown to be equal to those obtained from geometric-optical considerations. The corresponding propagation phenomena can be interpreted in terms of real geometric-optical rays that proceed along curved paths determined by the detailed structure of the refractive index; a geometric-optical ray through a point is tangent to the direction of energy flow at that point. Because of the curvature of the rays, there may exist regions of space that are not penetrated by any real rays (see Fig. 5.8.4); in these "shadow" regions, the usual geometric-optical description must be augmented by an alternative analysis. Also, geometrical optics fails in the vicinity of caustics and focal points, and results in these transition regions must be derived from a more rigorous approach. These general observations have been made in Secs. 1.6 and 1.7.

To fit within the framework of plane-stratified regions, the discussion is confined to media whose refractive index varies along the z direction only [i.e., $n(\mathbf{r}) = n(z)$]. Consequently, the electromagnetic fields may be derived from two scalar potential functions as in Sec. 5.2; however, the differential equations satisfied by these potentials are modified by the continuous stratification. The required formulation was given in Sec. 2.3e and is summarized below. For slow variations of medium properties, the differential equations for the potential functions are solved directly by asymptotic techniques that yield geometrical optics as the first approximation exterior to the above-mentioned transition regions (see also Sec. 1.7). Thereafter, a solution is obtained by applying the WKB approximation to rigorous integral representations for the potentials, and it is shown that the result justifies the direct use of geometrical optics in certain regions of space; but, as before, a modification is required in transition regions where rapid field variations take place. These general conclusions are verified in Sec. 5.9, where exact solutions for certain special profiles $n(z)$ are examined.

5.8b Derivation of the Time-Harmonic Field from Scalar Potentials

It is shown in Sec. 2.3e that the electromagnetic fields excited by time-harmonic electric point currents $\hat{\mathbf{J}}(\mathbf{r}, t) = \mathbf{J}^o\delta(\mathbf{r} - \mathbf{r}') \exp{(j\omega t)}$ and magnetic point currents $\hat{\mathbf{M}}(\mathbf{r}, t) = \mathbf{M}^o \delta(\mathbf{r} - \mathbf{r}') \exp(j\omega t)$ in a region with continuously varying permittivity $\epsilon(z)$ and permeability $\mu(z)$ can be represented as [see Eqs. (2.3.26) and (2.3.40)]

$$\mathbf{E}(\mathbf{r}, \mathbf{r}') = \frac{\epsilon(z')}{\epsilon(z)} \nabla \times \nabla \times \mathbf{z}_0 \Pi'(\mathbf{r}, \mathbf{r}') - j\omega\mu(z')\nabla \times \mathbf{z}_0 \Pi''(\mathbf{r}, \mathbf{r}'), \quad (1a)$$

$$\mathbf{H}(\mathbf{r}, \mathbf{r}') = j\omega\epsilon(z')\nabla \times \mathbf{z}_0 \Pi'(\mathbf{r}, \mathbf{r}') + \frac{\mu(z')}{\mu(z)} \nabla \times \nabla \times \mathbf{z}_0 \Pi''(\mathbf{r}, \mathbf{r}'), \quad (1b)$$

where the *E*- and *H*-mode Hertz potentials Π' and Π'', respectively, are related via Eqs. (2.3.39) to the scalar functions $\mathcal{S}'_d$ and $\mathcal{S}''_d$:

$$j\omega\epsilon(z')\Pi'(\mathbf{r}, \mathbf{r}') = \frac{1}{j\omega\epsilon(z')} \mathbf{J}^o \cdot \nabla' \times \nabla' \times \mathbf{z}_0 \mathcal{S}'_d(\mathbf{r}, \mathbf{r}') - \mathbf{M}^o \cdot \nabla' \times \mathbf{z}_0 \mathcal{S}'_d(\mathbf{r}, \mathbf{r}'),$$

$$(1c)$$

$$j\omega\mu(z')\Pi''(\mathbf{r}, \mathbf{r}') = \mathbf{J}^0 \cdot \nabla' \times \mathbf{z}_0 \mathcal{S}''_d(\mathbf{r}, \mathbf{r}') + \frac{1}{j\omega\mu(z')} \mathbf{M}^o \cdot \nabla' \times \nabla' \times \mathbf{z}_0 \mathcal{S}''_d(\mathbf{r}, \mathbf{r}'),$$

$$(1d)$$

with the vector operators reducible as in Eqs. (5.2.1e) and (5.2.1f). The functions $\mathcal{S}'_d$ and $\mathcal{S}''_d$ are related as follows to the scalar Green's functions G' and G'' [see footnote to Eq. (5.2.3a)],

$$-\nabla_t^2 \mathcal{S}'_d(\mathbf{r}, \mathbf{r}') = j\omega\epsilon(z')G'(\mathbf{r}, \mathbf{r}'), \quad -\nabla_t^2 \mathcal{S}''_d(\mathbf{r}, \mathbf{r}') = j\omega\mu(z')G''(\mathbf{r}, \mathbf{r}'), \quad (2)$$

which, subject to appropriate boundary conditions, satisfy the equations

$$[\mathcal{D}_\epsilon^2(z) + \nabla_t^2 + k^2(z)]G'(\mathbf{r}, \mathbf{r}') = -\delta(\mathbf{r} - \mathbf{r}'), \quad k^2(z) = \omega^2\mu(z)\epsilon(z), \quad (3a)$$

$$[\mathcal{D}_\mu^2(z) + \nabla_t^2 + k^2(z)]G''(\mathbf{r}, \mathbf{r}') = -\delta(\mathbf{r} - \mathbf{r}'), \quad (3b)$$

with

$$\mathcal{D}_\alpha^2(z) \equiv \alpha(z)\frac{\partial}{\partial z}\frac{1}{\alpha(z)}\frac{\partial}{\partial z}. \quad (3c)$$

By inserting Eqs. (2) into Eqs. (3), one obtains the differential equations satisfied by $\mathcal{S}'_d$ and $\mathcal{S}''_d$ (see Eqs. 5.2.2 for a homogeneous medium). It may be verified that Eqs. (3a) and (3b) can also be written in the alternative form

$$[\nabla^2 + \bar{k}_\epsilon^2(z)]\frac{G'(\mathbf{r}, \mathbf{r}')}{\sqrt{\epsilon(z)}} = -\frac{\delta(\mathbf{r} - \mathbf{r}')}{\sqrt{\epsilon(z')}}, \quad (4a)$$

$$[\nabla^2 + \bar{k}_\mu^2(z)]\frac{G''(\mathbf{r}, \mathbf{r}')}{\sqrt{\mu(z)}} = -\frac{\delta(\mathbf{r} - \mathbf{r}')}{\sqrt{\mu(z')}}, \quad (4b)$$

where $\bar{k}_\alpha(z)$ is the modified wavenumber

$$\bar{k}_\alpha(z) = \left[k^2(z) - \sqrt{\alpha(z)}\frac{d^2}{dz^2}\frac{1}{\sqrt{\alpha(z)}}\right]^{1/2}. \quad (4c)$$

In a *z*-transmission modal formulation, the solutions for the scalar functions may be expressed in the following manner:

$$\mathcal{S}'_d(\mathbf{r}, \mathbf{r}') = \sum_i \frac{\Phi_i(\boldsymbol{\rho})\Phi_i^*(\boldsymbol{\rho}')}{k_{ti}'^2} Y_i'(z, z'), \quad k_{ti}' \neq 0, \quad (5a)$$

$$\mathscr{S}_d''(\mathbf{r}, \mathbf{r}') = \sum_i \frac{\psi_i(\boldsymbol{\rho})\,\psi_i^*(\boldsymbol{\rho}')}{k_{ti}''^2} Z_i''(z, z'), \qquad k_{ti}'' \neq 0, \tag{5b}$$

and

$$G'(\mathbf{r}, \mathbf{r}') = \frac{1}{j\omega\epsilon(z')} \sum_i \Phi_i(\boldsymbol{\rho})\Phi_i^*(\boldsymbol{\rho}')\,Y_i'(z, z'), \tag{6a}$$

$$G''(\mathbf{r}, \mathbf{r}') = \frac{1}{j\omega\mu(z')} \sum_i \psi_i(\boldsymbol{\rho})\psi_i^*(\boldsymbol{\rho}')Z_i''(z, z'). \tag{6b}$$

$\Phi_i(\boldsymbol{\rho})$ and $\psi_i(\boldsymbol{\rho})$ are the scalar eigenfunctions in the cross sections transverse to z and were discussed in Chapter 3. $Y_i'(z, z')$ and $Z_i''(z, z')$ are, respectively, the E-mode current excited by a unit voltage generator and the H-mode voltage excited by a unit current generator on a nonuniform transmission line with propagation constant $\kappa_i(z)$ and characteristic impedance $Z_i(z)$ (see Figs. 2.4.6 and 2.4.7). The comments pertaining to Eqs. (5.2.4) apply also in the present case, and the entire formulation is in fact analogous to that in Sec. 5.2a, to which it reduces when ϵ and μ are piecewise constant. In view of the z-dependent functions $\epsilon(z)$ and $\mu(z)$, it is convenient to deal with the functions $\mathscr{S}_d'$ and $\mathscr{S}_d''$ rather than with their counterparts $\mathscr{S}'$ and $\mathscr{S}''$ in Eqs. (5.2.5).

The functions $Y_i'(z, z')$ and $Z_i''(z, z')$ may be related as in Eq. (5.2.6a) to the one-dimensional modal Green's functions

$$g_{zi}'(z, z') = \frac{1}{j\omega\epsilon(z')}\,Y_i'(z, z'), \qquad g_{zi}''(z, z') = \frac{1}{j\omega\mu(z')}Z_i''(z, z'), \tag{7}$$

which satisfy the differential equations

$$[D_\epsilon^2(z) + \kappa_i'^2(z)]g_{zi}'(z, z') = -\delta(z - z'), \tag{7a}$$

$$[D_\mu^2(z) + \kappa_i''^2(z)]g_{zi}''(z, z') = -\delta(z - z'), \tag{7b}$$

where $D_a^2(z)$ is defined as in Eq. (3c), with $\partial/\partial z \to d/dz$, and

$$\kappa_i^2(z) = k^2(z) - k_{ti}^2. \tag{7c}$$

General methods for solving these equations subject to appropriate boundary conditions are discussed in Chapter 3. If ϵ and (or) μ are discontinuous at a plane $z = z_1$, the required continuity of the voltage and current across the junction point in the equivalent modal network implies via Eqs. (2.3.10) the continuity of the following quantities:

$$g_{zi}', \quad \frac{1}{\epsilon(z)}\frac{d}{dz}g_{zi}'; \quad g_{zi}'', \quad \frac{1}{\mu(z)}\frac{d}{dz}g_{zi}''. \tag{8}$$

In view of Eqs. (5) and (6), continuity conditions of the same type apply to the three-dimensional Green's functions G' and G'', and to the functions $\mathscr{S}_d'$ and $\mathscr{S}_d''$.

Since the transverse eigenfunctions for media with piecewise constant or with continuously variable characteristics along z are identical, the treatment in Sec. 5.2b pertaining to eigenfunction representations in transversely unbounded regions remains valid provided that the modal Green's functions g_{zi} are determined from Eqs. (7). A basic feature employed in Eqs. (5.2.8) and subsequently, namely the evenness of g_{zi} with respect to the transverse wavenumber $k_{ti} = \xi$, still obtains as one may realize by considering the continuously varying medium with permittivity $\epsilon(z)$ and permeability $\mu(z)$ to be approximated by a series of thin layers. A typical layer extends from $(z_j - \delta)$ to $(z_j + \delta)$ and has a constant permittivity $\epsilon(z_j)$ and permeability $\mu(z_j)$. $g_{zi}(z, z')$ evaluated as in Sec. 5.3a is now a function of the various $\kappa_j = [\omega^2 \mu(z_j)\epsilon(z_j) - \xi^2]^{1/2}$ (i.e., an even function of ξ). This property continues to hold in the limiting case where the number of layers increases indefinitely and each layer width 2δ approaches 0, in which instance one obtains the original continuous medium. From this consideration, it may also be anticipated that when the wavenumber $k(z)$ varies continuously and approaches a *finite* value k_N at $z = \infty$, the function $g_{zi}(z, z')$ has branch points at $k_{ti} = \pm k_N$ (see Fig. 5.3.1), while a finite value k_1 of $k(z)$ at $z = -\infty$ implies existence of branch points at $k_{ti} = \pm k_1$.

5.8c   Direct Ray-Optical Solution in a Slowly Varying Medium

Although it is not possible to solve Eqs. (3) or the corresponding equations for $\mathscr{S}'_d$, $\mathscr{S}''_d$ for arbitrarily prescribed variations in $k(z)$, approximate methods are applicable when the relative variation is small over a length interval equal to the local wavelength $2\pi/k(z)$. It is convenient to characterize this condition in terms of the refractive index $n(z)$ as

$$\left| \frac{dn/dz}{k_0 n^2} \right| \ll 1, \tag{9}$$

where k_0 is the wavenumber in vacuum and $k(z) = k_0 n(z)$, so that this slow variability criterion for a specified $n(z)$ may be met by choosing k_0 large enough (short wavelength limit). Since k_0 is a large parameter, it is suggestive to seek an approximate representation for the field or the scalar potentials in a series involving inverse powers of k_0. This development, which may be carried out directly from the differential equations, has been described in Sec. 1.7b; to a lowest order of approximation, it yields a solution that can be interpreted in geometric-optical terms. The general discussion in Sec. 1.7b is now specialized to the case of planar stratification. We seek field or potential solutions of the form

$$u(\mathbf{r}) \sim u_0(\mathbf{r})e^{ik_0\psi(\mathbf{r})}, \tag{10}$$

where u_0 and ψ are k_0-independent amplitude and phase functions, respectively [however, see remarks following Eq. (1.7.22a)]. Basic to the evaluation of these functions is the determination of the ray trajectories.

Ray trajectories

Ray trajectories are specified by the ray equation [see Eq. (1.7.23)]

$$\frac{d}{ds} n \frac{d\mathbf{R}}{ds} = \nabla n, \qquad \text{with} \left(\frac{d\mathbf{R}}{ds}\right)^2 = 1, \tag{11}$$

where s is the distance coordinate along the ray and $\mathbf{R} = \mathbf{x}_0 x + \mathbf{y}_0 y + \mathbf{z}_0 z$ is the radius vector from an arbitrarily chosen coordinate center to a point on the ray (coordinates along a ray are denoted by $\mathbf{R}$ instead of $\mathbf{r}$ as in Sec. 1.7b). When n varies only with z, this equation can be written in component form as follows:

$$\frac{d}{ds}\,n\,\frac{dx}{ds} = 0 = \frac{d}{ds}\,n\,\frac{dy}{ds}, \tag{12a}$$

$$\frac{d}{ds}\,n\,\frac{dz}{ds} = \frac{dn}{dz}. \tag{12b}$$

From Eq. (12a), $n(dx/ds)$ and $n(dy/ds)$ are constant along a ray, so their ratio, and therefore dy/dx, are constant also. The constancy of dy/dx implies that the projection of a ray curve upon the xy plane is a straight line, whence each ray is a curve confined entirely to a plane perpendicular to the xy plane. Without loss of generality, the coordinate system may be chosen so that the ray under consideration is confined to the plane $x = 0$. With reference to Fig. 5.8.1(a), one notes that

$$\frac{dy}{ds} = \sin\theta, \qquad \frac{dz}{ds} = \cos\theta, \tag{13}$$

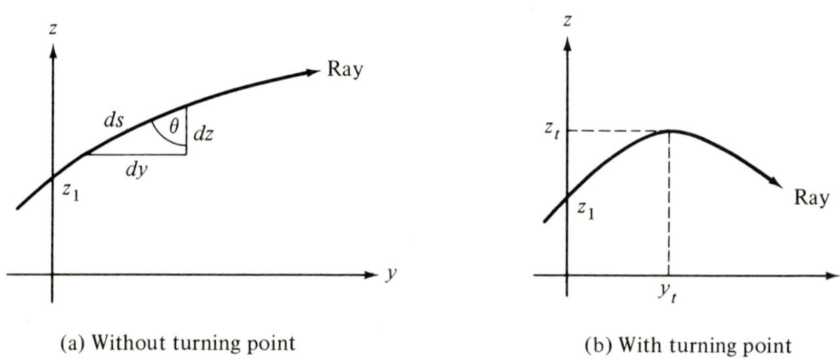

(a) Without turning point
(b) With turning point

FIG. 5.8.1   Ray trajectories.

where θ is the angle between the ray vector $\mathbf{s}$ and the positive z axis. Hence, from Eq. (12a),

$$n \sin\theta = a = \text{constant along a ray}, \tag{14}$$

which statement is known as Snell's law. From Eq. (12b),

$$\frac{dn}{dz} = \frac{d}{ds}(n\cos\theta) = \cos\theta\,\frac{dn}{dz}\frac{dz}{ds} - n\sin\theta\,\frac{d\theta}{ds}, \tag{15}$$

whence

$$\frac{d\theta}{ds} = -\frac{\sin\theta}{n}\frac{dn}{dz}. \tag{16}$$

Thus, $d\theta/ds < 0$ if $dn/dz > 0$, whence the ray inclination angle decreases with increasing s; the ray bends toward the z axis when the propagation vector has a component in the direction of increasing n.

From Eqs. (13) and (14),

$$\frac{dy}{dz} = \tan\theta = \frac{a}{\sqrt{n^2 - a^2}}, \tag{17}$$

and the ray curve $y = y(z)$ is therefore given by

$$y(z) = a \int_{z_1}^{z} \frac{dz}{\sqrt{n^2(z) - a^2}}, \tag{18}$$

with the arbitrary reference $y(z_1) = 0$. If a ranges over a certain set of real numbers for which the square root is real, Eq. (18) defines a corresponding family of rays, all passing through the common point $(0, z_1)$. The square root $\sqrt{n^2 - a^2}$ is conveniently defined to be positive when real and if $a > 0$, one observes from Eq. (17) that $dy/dz > 0$ and the ray in the first quadrant of Fig. 5.8.1(a) is upgoing with respect to the reference point z_1. The solution for a downgoing ray corresponding to the same value of a is obtained upon interchanging the integration limits in Eq. (18), thereby yielding $dy/dz < 0$. With this definition of the square root, θ is restricted to $0 \le \theta \le \pi/2$ and measures the inclination with respect to the negative z direction if the ray is downgoing. It is noted that $\tan\theta = \infty$ at a level z_t defined by

$$n(z_t) = a, \tag{19}$$

so the ray becomes horizontal at z_t [see Fig. 5.8.1(b)]. This may occur when the refractive index decreases along the direction of propagation of the incident ray. Since Eq. (18) has real solutions only when $n^2(z) \ge a^2$, the ray does not penetrate the region where $n^2(z) < a^2$ but bends back toward the direction of incidence after reaching the level $z = z_t$, the "turning point" in the z-coordinate space. If a turning point is present, y is a double-valued function of z, and separate equations apply to the upgoing and downcoming parts of the ray. Along the upgoing part, y is given by Eq. (17); for the downgoing part, z and z_1 in Eq. (18) are replaced by z_t and z, respectively, and the value $y(z_t) \equiv y_t$ from Eq. (18) is added to the right-hand side [see Eq. (28)].

Phase change along a ray

With the ray trajectories determined, one may evaluate the phase change in the field as it advances a specified distance along a ray. From Eq. (1.7.28), the phase difference between two points $\mathbf{R}_1$ and $\mathbf{R}$ on a ray is given by

$$\psi(\mathbf{R}) - \psi(\mathbf{R}_1) = \int_{\mathbf{R}_1}^{\mathbf{R}} n\,ds \equiv \psi - \psi_1, \tag{20}$$

which becomes, for the present problem,

$$\psi - \psi_1 = \int_{\mathbf{R}_1}^{\mathbf{R}} n[(\mathbf{y}_0 \cdot \mathbf{s})\,dy + (\mathbf{z}_0 \cdot \mathbf{s})\,dz] = \int_{\mathbf{R}_1}^{\mathbf{R}} n(\sin\theta\,dy + \cos\theta\,dz), \tag{20a}$$

where **s** is a unit vector along the ray, and the integration path proceeds along the ray. In view of Eq. (13), one has, for an upgoing ray,

$$\psi - \psi_1 = a(y - y_1) + \int_{z_1}^{z} \sqrt{n^2(z) - a^2}\, dz, \tag{20b}$$

with ψ_1 denoting the phase at the point (y_1, z_1) on the ray. One observes that the phase function ψ is real only at those points for which $n^2(z) \geq a^2$; if $n^2(z) < a^2$, ψ is imaginary and yields via Eq. (10) an exponentially decaying ("non-propagating") solution. The concept of "complex rays" has been introduced to deal with complex solutions of the ray equations, but the physical attributes of complex ray fields (energy transport, etc.) remain to be clarified.[27] If there is a turning point, the calculation of the phase is carried out separately for the upward and downward portions along the ray as above. For an upgoing ray and $\sqrt{n^2 - a^2}$ positive, one obtains from Eq. (20b) the required phase increase with z. For a downgoing ray, z has to appear in the lower integration limit in Eq. (20b) in order to yield the required phase increase with decreasing z [see Eq. (29)].

In order to determine the amplitude function u_0 in Eq. (10), details of the source configuration must be given. Two examples, excitation by a line source and a point source, are treated for illustration.

Excitation by a transverse electric line current

As an illustration of how to determine the amplitude of the fields excited by a confined source distribution, consider a line source of electric currents directed parallel to the x axis and located at the point $(0, z')$ in a medium with variable permittivity $\epsilon(z)$ and constant permeability μ_0, as shown in Fig. 5.8.2.[27,28]

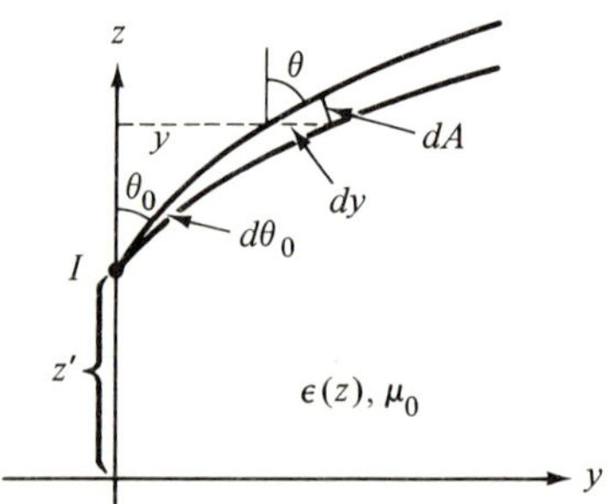

FIG. 5.8.2   Radiation from a line source: $\hat{\mathbf{J}}(\mathbf{r}.\,t) = I\delta(\hat{\boldsymbol{\rho}} - \hat{\boldsymbol{\rho}}')e^{-i\omega t}\mathbf{x}_0.$

This source generates an electric field which has only an x component, as may be verified from Eqs. (1), with $\partial/\partial x \equiv 0$ [see also Eqs. (5.2.4) and (5.4.31)]. All rays will be plane curves parallel to the yz plane, and each ray may be characterized conveniently by the angle, θ_0, at which it emerges from the source. To find the ray amplitude from energy conservation in a ray tube [see Eq. (1.7.34)], it is necessary to evaluate the cross-sectional area dA of the tube bounded by rays leaving the source at angles θ_0 and $\theta_0 + d\theta_0$; a length dimension dx parallel to the x axis is suppressed. From Fig. 5.8.2,

$$dA = dy \cos\theta = \frac{\partial y}{\partial\theta_0} d\theta_0 \cos\theta, \tag{21}$$

where dy, at a fixed value of z, has been related to θ_0 via the parametric form $y = y(\theta_0)$, $z = z(\theta_0)$, of the ray equation. If $P(\theta_0)$ is the angular power density (power per radian) emitted by the source in the direction θ_0, the power into the ray tube, $P(\theta_0)\,d\theta_0$, remains constant inside the tube and the intensity (power density) at any point along the ray tube is given by

$$\bar{S} = \frac{P(\theta_0)\,d\theta_0}{dA} = \frac{P(\theta_0)}{(\partial y/\partial\theta_0)\cos\theta}. \tag{22}$$

The partial derivative $\partial y/\partial\theta$, for a fixed value of z, can be determined from Eq. (18) and the relation $a = n(z)\sin\theta$, whence

$$\frac{\partial y}{\partial\theta_0} = \frac{\partial y}{\partial a}\frac{\partial a}{\partial\theta_0} = n(z')\cos\theta_0 \int_{z'}^{z} \frac{n^2 dz}{(n^2 - a^2)^{3/2}}, \tag{23}$$

whence, from Eq. (1.7.39),

$$E_0(y, z) = \sqrt{\frac{\zeta\bar{S}(y, z)}{n(z)}} = \left[\frac{\zeta P(\theta_0)}{n(z)\cos\theta\, n(z')\cos\theta_0 \int_{z'}^{z} [(n^2\,dz)/(n^2 - a^2)^{3/2}]} \right]^{1/2},$$

$$\zeta = \sqrt{\frac{\mu_0}{\epsilon_0}}. \tag{24}$$

This expression, valid for $z > z'$, applies also to $z < z'$ provided that one exchanges z and z' in the integration limits and restricts $0 \le \theta \le \pi/2$ as noted previously. The resulting symmetry in the source-point and observation-point coordinates is in accord with the reciprocity principle.

To calculate $P(\theta_0)$ it is necessary to consider the line-source field in the vicinity of the source point $(0, z')$. Since $k_0 \gg 1$ and the relative medium parameters change only slightly over a distance interval equal to the local wavelength, it is possible to choose an observation point $\mathbf{r} = (y, z)$ near enough to the source point $\mathbf{r}'$ to have $n(z) \approx n(z')$ and yet far enough to yield

$$k_0 n(z')\,|\mathbf{r} - \mathbf{r}'| \gg 1.$$

Thus, the pertinent local power pattern is that produced by the source when embedded in a medium with constant refractive index $n(z')$. If the source strength is chosen as $I = (ik_0\zeta)^{-1}$, the x component of the electric field, $\mathbf{E}_f = \mathbf{x}_0 E_f$, is given by the free-space Green's function [see Eq. (5.4.25)]:

$$E_f = \frac{i}{4} H_0^{(1)}[k_0 n(z')|\mathbf{r} - \mathbf{r}'|] \sim \frac{1}{4}\sqrt{\frac{2}{\pi k_0 n(z')|\mathbf{r} - \mathbf{r}'|}}\, e^{ik_0 n(z')\,|\mathbf{r}-\mathbf{r}'|+i\pi/4}. \tag{25}$$

Thus, the radiation pattern is independent of angle, and $P(\theta_0) = P = \text{constant}$ is given by

$$P = \frac{\text{total power}}{2\pi} = \frac{n(z')|E_f|^2}{2\pi\zeta} 2\pi|\mathbf{r} - \mathbf{r}'| = \frac{1}{8\pi k_0\zeta}. \tag{26}$$

Substitution of Eq. (26) into Eq. (24) yields an expression for the magnitude E_0 of the electric field, and the actual field $\mathbf{E} = \mathbf{x}_0 E$ is then obtained from $E \sim E_0 \exp(ik_0\psi)$, with the phase function ψ defined in Eq. (20b). The value of the initial phase ψ_1 depends on the phase of the source function and has to be calculated from the requirement that $E \to E_f$ as $\mathbf{r} \to \mathbf{r}'$. Since $(y_1, z_1) \equiv (y', z')$, this comparison yields $k_0\psi_1 = \pi/4$, whence the final result for the line-source Green's function along a ray in a plane-stratified medium is, for $z > z'$,[27]

$E(\mathbf{R}, \mathbf{R}')$

$$\sim \frac{1}{2\sqrt{2\pi k_0}} \frac{\exp\left[ik_0 ay + ik_0\int_{z'}^{z}\sqrt{n^2(z) - a^2}\,dz + i\pi/4\right]}{[n^2(z) - a^2]^{1/4}[n^2(z') - a^2]^{1/4}\left[\int_{z'}^{z} [n^2(z)\,dz]/[n^2(z) - a^2]^{3/2}\right]^{1/2}} \tag{27}$$

where the identity $n(z)\cos\theta = [n^2(z) - a^2]^{1/2}$ has been utilized, with $a = n(z')\sin\theta_0$ along the ray that emerges from the source at the angle θ_0. For $z < z'$, one interchanges z and z' in the integration limits. If $n(z) \to 1$ as $z \to \infty$, the constant a can likewise be regarded as $\sin\theta_\infty$, where θ_∞ is the angle of emergence of the ray at $z \to \infty$. The result in Eq. (27) is then identical with that in Eq. (46), which is obtained from an asymptotic analysis of a rigorous solution. It is recalled that $\sqrt{n^2 - a^2}$ is defined to be positive when real; its analytic continuation to the range $n^2 < a^2$ is achieved by the definition $\text{Im}\sqrt{n^2 - a^2} > 0$. In Eq. (27), the field is expressed in terms of the ray parameter a, which is constant on the ray curve of Eq. (18). Alternatively, for prescribed (y, z), Eq. (18) defines $a = a(y, z)$ and hence permits the elimination of a from Eq. (27) provided that one may solve for $a(y, z)$ explicitly [this is possible only for special profiles $n(z)$].

If $n^2(z)$ decreases as z increases and $n^2(z) = a^2$ at $z = z_t$, a ray that leaves the source at an angle θ_0 propagates up to z_t and is turned back into the region $z < z_t$. The field along the ray up to and including the turning point (y_t, z_t) in Fig. 5.8.1(b) can be calculated from Eq. (27). For the refracted part of the ray in the region $y \geq y_t$, $z \leq z_t$, the ray equation (18) is replaced by

$$y(z) = a\int_{z}^{z_t} \frac{dz}{\sqrt{n^2(z) - a^2}} + y_t, \qquad y_t = y(z_t) = a\int_{z'}^{z_t}\frac{dz}{\sqrt{n^2(z) - a^2}} \tag{28}$$

where y_t is the value of y at the turning point, and the remaining integral traces the ray curve beyond the turning point. Similarly, the phase function ψ is given by

$$\psi = ay + \int_{z'}^{z_t}\sqrt{n^2 - a^2}\,dz + \int_{z}^{z_t}\sqrt{n^2 - a^2}\,dz. \tag{29}$$

The field evaluation along the ray past the turning point can now be carried out as above except that Eq. (28) must be used in the calculation of the cross-section element in Eq. (21), and the phase function taken as that in Eq. (29).

The fields so determined are matched to those obtained from Eq. (27) at the turning point (y_t, z_t). Details of this evaluation are described in connection with the rigorous analysis in Sec. 5.8d.

Excitation by a longitudinal electric current element

A procedure similar to the above may be employed for the point-source problem. If we are dealing with a z-directed electric or magnetic current element in the region of Fig. 5.8.2, the electromagnetic fields can be determined from a scalar Green's function that satisfies the equation $(\nabla^2 + k_0^2 n^2)G = -\delta(\mathbf{r} - \mathbf{r}')$ [see Eqs. (5.2.4c), (1) and (3)].† G is angularly symmetric about the z axis, and rays leaving the source at the angle θ_0 all lie on a surface of revolution defined by Eq. (18) provided that y is replaced by the radial cylindrical variable ρ. The power contained between two such ray surfaces remains constant, and if $P(\theta_0)$ denotes the power per unit solid angle radiated by the source in the direction θ_0, then the power flow into the two neighboring surfaces bounded by rays with $\theta = \theta_0$ and $\theta_0 + d\theta_0$ is given by $2\pi P(\theta_0) \sin \theta_0 d\theta_0$. Thus, the intensity at any point along a ray leaving the source at angle θ_0 is

$$\bar{S} = \frac{2\pi P(\theta_0) \sin \theta_0 \, d\theta_0}{dA} = \frac{P(\theta_0) \sin \theta_0}{\rho \cos \theta [\partial \rho / \partial \theta_0]}, \tag{30}$$

where

$$dA = 2\pi \rho d\rho \cos \theta.$$

All other considerations are now directly analogous to those employed for the line-source problem, with y replaced by ρ. The behavior of G near $\mathbf{R}'$ is given by $(1/4\pi|\mathbf{R} - \mathbf{R}'|) \exp [ik_0 n(z')|\mathbf{R} - \mathbf{R}'|]$ [see Eq. (5.4.2b)], whence for arbitrary $\mathbf{R}$ along the ray characterized by $a = n(z') \sin \theta_0$:

$$G \sim \frac{\sqrt{a} \, \exp \left(ik_0 a\rho + ik_0 \int_{z'}^{z} \sqrt{n^2 - a^2} \, dz\right)}{4\pi\sqrt{\rho} \, [n^2(z) - a^2]^{1/4}[n^2(z') - a^2]^{1/4}\left\{\int_{z'}^{z} n^2 \, dz/[n^2 - a^2]^{3/2}\right\}^{1/2}}. \tag{31}$$

The modifications when a refracted ray has passed through a turning point are again analogous to those for the line-source problem.

Excitation by an incident plane wave

The limiting case of an incident plane wave, polarized with its electric vector perpendicular to z, may be obtained from Eq. (27) upon letting $y' \to -\infty$, $z' \to -\infty$, with $y' = z' \tan \theta_\infty$. To generalize the equation so as to permit arbitrary values of the source-point coordinate y', we replace y by $y - y'$. If it is assumed that the refractive index approaches the constant value n_1 as $z \to -\infty$, the integral in the denominator of Eq. (27) can be evaluated as

---

†The differential operator in Eqs. (3a) and (3b) also contains a term of the form $(1/n)(dn/dz)\partial G/\partial z \sim k_0 G(dn/dz)$, which is neglected in comparison with $k_0^2 n^2 G$, in view of Eq. (9).

$n_1^2[n_1^2 - a^2]^{-3/2}(-z')$ in the limit $z' \to -\infty$. The phase integral cannot be similarly approximated because negligible phase terms must be small compared to unity, rather than small compared to $|z'|$. Since $a = n(z) \sin \theta(z) = n_1 \sin \theta_\infty$, where θ_∞ is the incident ray angle at $z \to -\infty$, $y \to -\infty$, one may write Eq. (27) as

$$E \sim A \frac{\sqrt{n_1 \cos \theta_\infty}}{[n^2(z) - n_1^2 \sin^2 \theta_\infty]^{1/4}} \exp\left\{ ik_0 n_1 [y \sin \theta_\infty + z' \cos \theta_\infty] \right.$$

$$\left. + ik_0 \int_{z'}^{z} \sqrt{n^2(\eta) - n_1^2 \sin^2 \theta_\infty}\, d\eta \right\}, \quad (32)$$

where

$$A = \frac{1}{2\sqrt{2\pi k_0 n_1 \hat{\rho}'}}\, e^{ik_0 n_1 \hat{\rho}' + i\pi/4}, \qquad \hat{\rho}' = \sqrt{y'^2 + z'^2}. \quad (32\text{a})$$

The factor A is set equal to unity if the incident plane wave is to have unit amplitude [see Eq. (5.4.30b)]. Upon rearranging the exponent so that the term $n_1 z' \cos \theta_\infty$ is absorbed into the integrand, one may pass to the limit $z' \to -\infty$ to obtain the expression for the field due to a plane wave of unit amplitude, incident from $z = -\infty$ at an angle θ_∞ with the positive z axis,

$$E \sim \frac{\sqrt{n_1 \cos \theta_\infty}}{[n^2(z) - n_1^2 \sin^2 \theta_\infty]^{1/4}} \exp\left\{ ik_0 n_1 [y \sin \theta_\infty + z \cos \theta_\infty] \right.$$

$$\left. + ik_0 \int_{-\infty}^{z} [\sqrt{n^2(\eta) - n_1^2 \sin^2 \theta_\infty} - n_1 \cos \theta_\infty]\, d\eta \right\} \quad (33)$$

This expression applies when $n(z) > n_1 \sin \theta_\infty$ for all z.

If there is a turning point z_t, Eq. (33) applies up to z_t; the corresponding expression for the refracted-ray field requires consideration of a "canonical" problem and can be deduced from the limit of Eq. (55) as $z' \to -\infty$. The result for $z \to \infty$ is given in Eq. (57), and the formula for $z' \to \infty$ is obtained therefrom by interchanging z and z' since $\bar{G}_2''$ is symmetrical in z and z'. In Eq. (57), it is assumed that $z, z' > z_t$ and that $n(\infty) = 1$. Adapting the result to the present case where $z' \to -\infty$, $y' \to -\infty$, $z_t > z, z'$, $n(-\infty) = n_1$, and using Eq. (32a), one finds, for the field along the refracted ray,

$$E_r \sim e^{-i\pi/2} \frac{\sqrt{n_1 \cos \theta_\infty}}{[n^2(z) - n_1^2 \sin^2 \theta_\infty]^{1/4}} e^{ik_0[n_1 y \sin \theta_\infty - n_1 z \cos \theta_\infty] + ik_0 \psi_r}, \quad (34)$$

where

$$\psi_r = 2\int_{-\infty}^{z_t} \left[\sqrt{n^2(\eta) - n_1^2 \sin^2 \theta_\infty} - n_1 \cos \theta_\infty \right] d\eta$$

$$- \int_{-\infty}^{z} \left[\sqrt{n^2(\eta) - n_1^2 \sin^2 \theta_\infty} - n_1 \cos \theta_\infty \right] d\eta + 2z_t n_1 \cos \theta_\infty. \quad (34\text{a})$$

The ray envelope, the caustic, is in this case the straight line $z = z_t$ in Fig. 5.8.3, and the factor $\exp(-i\pi/2)$ represents the phase shift introduced when a ray passes through the caustic. If a physical reflecting plane is located at $z = z_0$, with $n(z_0) > n_1 \sin \theta_\infty$, the reflected field is also given by Eq. (34),

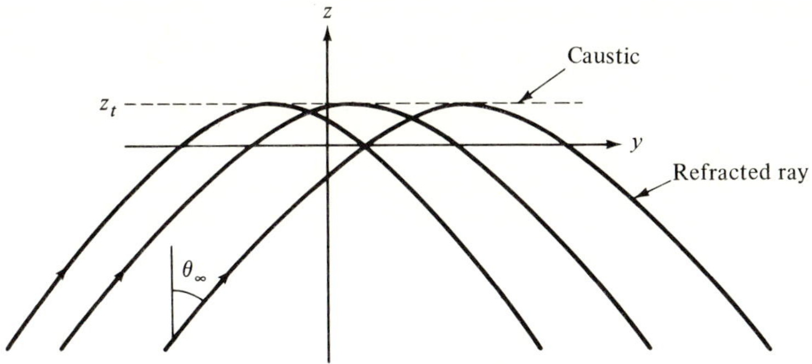

FIG. 5.8.3   Incident plane wave and turning point.

except that z_t is replaced by z_0 and the reflection coefficient $\exp(-i\pi/2)$ is replaced by the one for the plane surface. Since the ray is reflected at the incident angle, the ray trajectories are symmetrical about the reflection point.

5.8d   Asymptotic Evaluation of a Typical Radiation Integral for a Medium with Monotonic Variation

The electromagnetic fields excited by arbitrary source distributions can be evaluated in terms of the scalar functions $\mathscr{S}'_d$ and $\mathscr{S}''_d$, or for special source configurations, in terms of the scalar Green's functions G' or G'' [see Eqs. (1) and (2)]. General modal expansions for these functions, derived from a z-transmission analysis, are given in Eqs. (5) and (6); when the excitation is in the form of point or line currents in a transversely unbounded region, they reduce to the integral representations in Eqs. (5.2.11)–(5.2.14), with g_{zi} specified by Eqs. (7). The nonuniform transmission line equations for g_{zi} can be solved in terms of known functions only for special functional variations $k(z)$. However, for a slowly varying medium satisfying condition (9) (it is convenient to characterize this relation in terms of the large wavenumber k_0), one may approximate $g_{zi}(z, z')$ without specifying its detailed functional form. The nature of the approximation depends on the analytic properties of $[\kappa_i(z)/k_0]^2 = [n^2(z) - (k_{ti}/k_0)^2]$. In Sec. 3.5, this aspect has been pursued in regard to the functional dependence on z and z', with k_{ti} taken as a fixed parameter, whereas for the present application, the functional dependence on k_{ti} remains to be explored for any specific choice of the space coordinates z and z'. Since the wavenumber variable k_{ti} ranges over an infinite interval, $(\kappa_i/k_0)^2$ may take on positive and negative values (when n^2 is real), and the differential Eqs. (7a) and (7b) possess turning points where κ_i vanishes. The asymptotic forms of the solutions of Eqs. (7a) and (7b) may then be given in terms of Airy functions and remain valid over the entire range of variation of κ_i. Comparison of Eqs. (7) and Eq. (3.5.34) shows that $x \to z$, $\Omega \to k_0$, $\alpha(x, \Omega) = \alpha_0(x) \to [\kappa_i(z)/k_0]^2$, and with these identifications, the relevant asymptotic approximations may be synthesized from Eqs. (3.5.45)–(3.5.48). Over most of the range of interest where $\kappa_i \neq 0$, one may employ the WKB approximations for the Airy functions as given in Eq. (3.5.47).

Since $-\infty < k_{ti} < \infty$, different k_{ti}-intervals correspond to positive or negative values of $(\kappa_i/k_0)^2$, and these intervals change with the location of the source and observation points z' and z, respectively. The detailed construction of the asymptotic form for $g_{zi}(z, z')$ depends on the algebraic sign of $(\kappa_i/k_0)^2$ as well as on the boundary conditions at the endpoints of the z-domain, and must in general be carried out separately in each of the above-mentioned ranges of k_{ti}. For a subsequent asymptotic evaluation of the integrals representing the scalar potential functions G', G'' or $\mathscr{S}'_d$, $\mathscr{S}''_d$ in the limit of large k_0, only the wave functions corresponding to real κ_i will be considered since they alone represent propagating wave processes; regions wherein κ_i^2 is negative (Im $\kappa_i > 0$) yield exponentially decaying contributions.

To illustrate these remarks, we take a refractive index profile that increases monotonically from a positive value n_1 at $z = -\infty$ to a positive value n_2 at $z = +\infty$. When $k_{ti}^2 < k_0^2 n_1^2$, κ_i is real for all values of z and waves may propagate toward infinity in both directions. The corresponding one-dimensional Green's function g_{zi} must satisfy the radiation condition at $|z| \to \infty$ and is given in Eq. (3.5.39b):

$$g_{zi}(z, z') \sim \frac{\exp\left\{ik_0 \int_{z_<}^{z_>} [\kappa_i(\zeta)/k_0]d\zeta\right\}}{-2ik_0\left[\dfrac{\kappa_i(z)}{k_0}\dfrac{\kappa_i(z')}{k_0}\right]^{1/2}}, \qquad k_{ti}^2 < k_0^2 n_1^2. \tag{35a}$$

When $k_0^2 n_1^2 < k_{ti}^2 < k_0^2 n_2^2$, the interval contains turning points z_i at which $\kappa_i(z_i) = 0$: $k_{ti} = \pm k_0 n(z_i)$. Since $n_1 < n(z_i)$, κ_i^2 is negative when $z < z_i$ so that no propagation toward $z = -\infty$ is possible. To assure that the solution remains bounded at $z = -\infty$, a decaying exponential must be chosen, with the consequent appearance of an additional reflected wave in the region $z > z_i$ when $z' > z_i$ [see Eq. (3.5.39e)]:

$$g_{zi}(z, z') \sim \frac{\exp\left\{ik_0\int_{z_<}^{z_>} [\kappa_i(\zeta)/k_0]d\zeta\right\}}{-2ik_0\left[\dfrac{\kappa_i(z)}{k_0}\dfrac{\kappa_i(z')}{k_0}\right]^{1/2}} - i\,\frac{\exp\left\{ik_0\left(\int_{z_i}^{z} + \int_{z_i}^{z'}\right)[\kappa_i(\zeta)/k_0]d\zeta\right\}}{-2ik_0\left[\dfrac{\kappa_i(z)}{k_0}\dfrac{\kappa_i(z')}{k_0}\right]^{1/2}}.$$

$$\tag{35b}$$

No propagating solutions exist in the remaining interval $k_{ti}^2 > k_0^2 n_2^2$ wherein g_{zi} decays exponentially. The factor $-i$ multiplying the second term in Eq. (35b) represents the "reflection coefficient" at the caustic $z = z_i$ of the continuously refracted ray system [see Fig. 5.8.3, with $z \to -z$, and remarks following Eq. (58)]; if a physical reflecting boundary is located at the fixed coordinate $z = z_0$, $z_0 > z_i$, Eq. (35b) applies as well provided that $z_i \to z_0$ and the appropriate reflection coefficient Γ at the boundary is inserted instead of $-i$. While the expressions in Eqs. (35) have been deduced for k_{ti}-intervals on the real axis, they remain valid for neighboring complex values of k_{ti} and therefore define g_{zi} in the corresponding portions of the complex k_{ti}-plane.

An example: Excitation by an electric line current

When the solutions in Eqs. (35) are inserted into Eqs. (5.2.11)–(5.2.14), one obtains explicit integral representations for the scalar functions from which the electromagnetic fields can be calculated. Further reduction of these integrals may be achieved by saddle point techniques since the integrands contain the large parameter k_0. Although expressions for g_{zi} have been given only in the range κ_i real, they suffice for the location of real stationary points which contribute propagating fields. No such contribution arises from the remaining ranges of k_{ti}, and no further attention is given to the specific structure of the integrand therein. The asymptotic procedure will be illustrated for the specific example of a line source of electric current directed parallel to x and located at $z = z'$ (see Fig. 5.8.2) in a region wherein the refractive index decreases monotonically from $n_2^2 = 1$ at $z = \infty$ to $n_1^2 = -\infty$ at $z = 0$, with $n^2(z_0) = 0$, $z_0 > 0$. The results may then be compared with the exact solution in Sec. 5.9a for a special profile of this type. The fields can in this instance be derived from the scalar H-mode Green's function $\bar{G}''(\hat{\boldsymbol{\rho}}, \hat{\boldsymbol{\rho}}')$, $\hat{\boldsymbol{\rho}} = (y, z)$ [see Eq. (5.2.13a), with $j \to -i$ to account for the $\exp(-i\omega t)$ dependence, and Eq. (5.4.31)]:

$$\bar{G}''(\hat{\boldsymbol{\rho}}, \hat{\boldsymbol{\rho}}') = \frac{1}{2\pi} \int_{-\infty}^{\infty} e^{i\eta(y-y')} g_{zi}(z, z')d\eta, \qquad k_{ti} = \eta, \tag{36}$$

where g_{zi} is given in Eq. (35b) for the significant range of the integration interval.

As in Sec. 5.3c, it is convenient to introduce the complex angle variable w via the transformation $\eta = k_0 \sin w$, with the contributions arising from the first and second terms in Eq. (35b) denoted by $\bar{G}_1''$ and $\bar{G}_2''$, respectively;

$$\bar{G}''(\hat{\boldsymbol{\rho}}, \hat{\boldsymbol{\rho}}') \sim \bar{G}_1''(\hat{\boldsymbol{\rho}}, \hat{\boldsymbol{\rho}}') + \bar{G}_2''(\hat{\boldsymbol{\rho}}, \hat{\boldsymbol{\rho}}'), \tag{37}$$

where, with $y' = 0$,

$$\bar{G}_{1,2}''(\hat{\boldsymbol{\rho}}, \hat{\boldsymbol{\rho}}') = \frac{i}{4\pi} \int_{\bar{P}} f_{1,2}(w) e^{ik_0 q_{1,2}(w)} dw, \tag{38}$$

$$f_1(w) = \frac{\cos w}{\{[n^2(z) - \sin^2 w][n^2(z') - \sin^2 w]\}^{1/4}}, \qquad f_2(w) = -if_1(w), \tag{38a}$$

$$q_1(w) = y \sin w + \int_{z_<}^{z_>} \sqrt{n^2(\zeta) - \sin^2 w} \, d\zeta, \tag{39a}$$

$$q_2(w) = y \sin w + \int_{z_w}^{z} \sqrt{n^2(\zeta) - \sin^2 w} \, d\zeta + \int_{z_w}^{z'} \sqrt{n^2(\zeta) - \sin^2 w} \, d\zeta, \tag{39b}$$

and $n^2(z_w) = \sin^2 w$ defines z_w. The asymptotic approximation introduces spurious branch points w_b at $\sin^2 w_b = n^2(z)$ or $n^2(z')$; the exact solution only has branch points at $\sin^2 w_b = 1$, corresponding to $\eta_b = k_0 n_2 = k_0$ at $z = \infty$ [see the remarks following Eq. (8)]. The path $\bar{P}$ is shown in Fig. 5.3.6(b). Asymptotic approximations for $\bar{G}_1''$ and $\bar{G}_2''$ are derived below and lead to the results in Eqs. (46) and (55), respectively, which represent the geometric-optical field in the illuminated region. Near the caustic separating the illuminated and shadow regions (Fig. 5.8.4) or in the shadow region, Eqs. (60) and (61) are relevant.

Asymptotic evaluation

The saddle points w_s of $q_1(w)$ are determined by

$$0 = \frac{dq_1(w_s)}{dw_s'} = \cos w_s \left[y - \sin w_s \int_{z_<}^{z_>} \frac{d\zeta}{\sqrt{n^2(\zeta) - \sin^2 w_s}} \right]. \tag{40}$$

The solution $\cos w_s = 0$ is excluded because the resulting $\sin^2 w_s = 1$ is greater than $n^2(z)$ in the medium, thereby falling outside the range of propagating solutions. Hence, the relevant saddle points of $q_1(w)$ are specified implicitly by the equation

$$y = \sin w_s \int_{z_<}^{z_>} \frac{d\zeta}{\sqrt{n^2(\zeta) - \sin^2 w_s}}. \tag{41}$$

Since $n^2(z)$ varies monotonically between z and z', this equation admits of real solutions for w_s provided that $\sin w_s$ does not exceed the smaller of the values $n(z)$ or $n(z')$. Conversely, for each $\sin w_s < n(z)$ or $n(z')$, Eq. (41) defines a real curve in the yz domain which is precisely the geometric-optical ray trajectory [see Eq. (18)]. Thus, the real saddle points $0 \le w_s \le \sin^{-1} n(z)$ or $\sin^{-1} n(z')$ can be associated with real ray solutions, which define the trajectories of energy flow from the source point z' to the observation point z. Since $n(z) \to 1$ as $z \to \infty$, one has, from Eq. (41),

$$y \to z_> \tan w_s, \qquad \text{as} \quad z_> \to \infty. \tag{42}$$

If the source is located at some finite point z', then $z_> = z$, and w_s can be identified as the angle between the emerging ray at $z = \infty$ and the z axis (see Fig. 5.8.4). Conversely, if the source is located at $z' = \infty$, w_s represents the inclination of an incident ray. Each ray can thus be characterized by a particular value of w_s. For a source location $z' < \infty$, the maximum value of w_s is given by

$$w_{sm} = \sin^{-1} n(z'), \tag{43}$$

whence the region illuminated by real rays is confined at $z = \infty$ to the wedge $|\varphi| < w_{sm}$, with the observation angle φ measured from the positive z axis. For the refractive index in Eq. (5.9.19a), the rays form a family of hyperbolas as shown in Fig. 5.9.5. If $z < z'$, Eq. (41) is valid up to the point z_t which satisfies the equation

$$n(z_t) = \sin w_s. \tag{44}$$

When $z = z_t$, $dy/dz = \infty$ from Eq. (41), whence the "turning point" $z = z_t$ (starting point of refracted ray portion in Fig. 5.8.4) defines the level at which

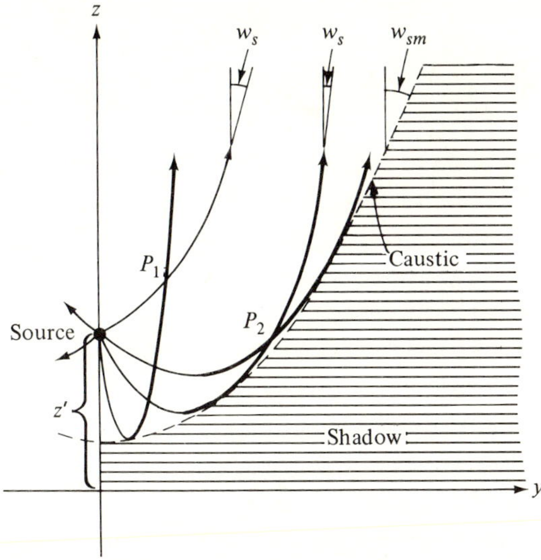

FIG. 5.8.4 Graphical interpretation of saddle–point condition. Direct rays [Eq. (41)] shown light; refracted rays [Eq. (52)] shown heavy.

the ray is parallel to the y axis and is bent back toward the positive z direction.

The first-order asymptotic approximation to $\bar{G}_1''$ can now be inferred from Eqs. (4.2.1). Utilizing the result,

$$\left.\frac{d^2q_1}{dw^2}\right|_{w_s} \equiv q_1''(w_s) = -\cos^2 w_s \int_{z<}^{z>} \frac{n^2(\zeta)d\zeta}{[n^2(\zeta) - \sin^2 w_s]^{3/2}}, \tag{45}$$

one obtains[27]

$$\bar{G}_1'' \sim \frac{1}{2\sqrt{2\pi k_0}}$$

$$\times \frac{\exp\left\{ik_0\left[y\sin w_s + \int_{z<}^{z>}\sqrt{n^2(\zeta) - \sin^2 w_s}\,d\zeta\right] + i/\pi 4\right\}}{\{[n^2(z)-\sin^2 w_s][n^2(z')-\sin^2 w_s]\}^{1/4}\left[\int_{z<}^{z>}\left\{[n^2(\zeta)\,d\zeta]/[n^2(\zeta)-\sin^2 w_s]^{3/2}\right\}\right]^{1/2}},$$

$$\tag{46}$$

with w_s determined from the solution of the ray equation (41) and all roots taken positive. This expression can be interpreted as describing the geometric-optical field along a "direct" ray that proceeds from the source to the observation point without having been turned by refraction; it agrees with the result in Eq. (27) derived from purely geometric-optical considerations. When $z \to \infty$, with z' finite, the integral in the denominator of Eq. (46) can be approximated by $z/\cos^3 w_s$. Since $n(z) \to 1$ as $z \to \infty$, and $z/\cos w_s = \hat{\rho}$, with w_s representing the ray inclination at $z = \infty$ and $\hat{\rho}$ denoting the radius vector [see Fig. 5.9.3(a)], Eq. (46) reduces to

$$\bar{G}_1'' \sim \frac{\sqrt{\cos w_s}}{2\sqrt{2\pi k_0 \hat{\rho}}} \frac{\exp\left\{ik_0\left[y \sin w_s + \int_{z'}^{z} \sqrt{n^2(\zeta) - \sin^2 w_s}\, d\zeta \right] + i\pi/4\right\}}{[n^2(z') - \sin^2 w_s]^{1/4}},$$

$$z \to \infty, \qquad (47)$$

which result agrees with that derived for the special case in Eq. (5.9.15a). For another check, assume a homogeneous medium with $n(z) = 1$, in which instance one obtains, from Eq. (46),

$$\bar{G}_1'' \sim \frac{e^{ik_0[y \sin w_s + |z - z'| \cos w_s] + i\pi/4}}{2\sqrt{2\pi k_0 \hat{\rho}}}, \qquad n(z) = 1, \qquad (48)$$

the correct expression for the free-space Green's function.

It is of interest to note that Eq. (46) tends to a finite limiting value at the turning points $n(z_t) = \sin w_s$ or $n(z_t') = \sin w_s$, although the asymptotic formulas (35) are not valid there. A power-series expansion of $n^2(z)$ and $n^2(\zeta)$ about the point z_t yields the limiting expression

$$\lim_{z \to z_t} [n^2(z) - \sin^2 w_s]^{1/4} \left\{ \int_z^{z'} \frac{n^2(\zeta)\, d\zeta}{[n^2(\zeta) - \sin^2 w_s]^{3/2}} \right\}^{1/2} \to \left[\frac{n(z_t)}{(d/dz_t)n(z_t)} \right]^{1/2}, \qquad (49)$$

whence

$$\bar{G}_1'' \sim \frac{1}{2\sqrt{2\pi k_0}} \frac{\exp\left\{ik_0\left[y \sin w_s + \int_{z_t}^{z'} \sqrt{n^2(\zeta) - \sin^2 w_s}\, d\zeta \right] + (i\pi/4)\right\}}{[n^2(z') - \sin^2 w_s]^{1/4}(n(z_t)/\{dn(z_t)/dz_t\})^{1/2}}, \qquad z = z_t.$$

$$(50)$$

The asymptotic evaluation of $\bar{G}_2''$ in Eq. (38) can be carried out in a similar manner and yields the field along a "refracted" ray. Since

$$\frac{d}{dw} \int_{z_w}^{z} \sqrt{n^2(\zeta) - \sin^2 w}\, d\zeta$$

$$= \int_{z_w}^{z} \frac{d}{dw} \sqrt{n^2 - \sin^2 w}\, d\zeta - [n^2(z_w) - \sin^2 w]^{1/2} \frac{dz_w}{dw}, \qquad (51)$$

and $n^2(z_w) = \sin^2 w$, one obtains the following equation for the saddle points w_s of $q_2(w)$:

$$F(y, z, w_s) = 0, \qquad (52a)$$

where $z_{w_s} \equiv z_t$, with $n(z_t) = \sin w_s$, and

$$F(y, z, w) = y - \sin w\left[\int_{z_w}^{z} \frac{d\zeta}{\sqrt{n^2(\zeta) - \sin^2 w}} + \int_{z_w}^{z'} \frac{d\zeta}{\sqrt{n^2(\zeta) - \sin^2 w}} \right]. \qquad (52b)$$

Equation (52a) is precisely the equation for a refracted geometric-optical ray that has passed through the turning point z_t and emerges at $z \to \infty$ with an inclination given by the angle w_s [see Eq. (28) and darkly drawn rays in Fig. 5.8.4]. In the calculation of

$$q_2''(w_s) = \cos w_s \frac{d}{dw_s} F(y, z, w_s), \tag{53}$$

required for the asymptotic evaluation from Eq. (4.2.1b), the operation d/dw_s cannot be carried out directly on Eq. (52b) since the contribution from differentiation of the integrand and the lower integration limit diverges. However, after integration by parts,

$$\int_{z_t}^{z} \frac{d\zeta}{\sqrt{n^2 - \sin^2 w_s}} = \frac{\sqrt{n^2(z) - \sin^2 w_s}}{n(z)n'(z)} - \int_{z_t}^{z} \sqrt{n^2 - \sin^2 w_s} \frac{d}{d\zeta}\left(\frac{1}{nn'}\right) d\zeta, \tag{54}$$

with a similar result for the z'-dependent integral, the ensuing expression can be differentiated, and yields

$$\sec w_s \frac{d}{dw_s} F(y, z, w_s)$$

$$= \sin^2 w_s \left[\frac{1}{n(z)n'(z)\sqrt{n^2(z) - \sin^2 w_s}} + \frac{1}{n(z')n'(z')\sqrt{n^2(z') - \sin^2 w_s}} \right]$$

$$- \left[\int_{z_t}^{z} + \int_{z_t}^{z'} \right] \frac{1 + \sin^2 w_s (d/d\xi)[n'(\xi)n(\xi)]^{-1}}{\sqrt{n^2(\xi) - \sin^2 w_s}} d\xi, \tag{54a}$$

where $n'(\alpha) \equiv dn/d\alpha$. The asymptotic approximation of $\bar{G}_2''$ is then found from Eqs. (4.2.1b) or (4.2.1c):[27]

$$\bar{G}_2'' \sim \frac{1}{2\sqrt{2\pi k_0}} \times$$

$$\frac{\exp\left\{ ik_0\left[y \sin w_s + \int_{z_t}^{z} \sqrt{n^2(\zeta) - \sin^2 w_s}\, d\zeta + \int_{z_t}^{z'} \sqrt{n^2(\zeta) - \sin^2 w_s}\, d\zeta \right] \mp i\pi/4 \right\}}{\{[n^2(z) - \sin^2 w_s][n^2(z') - \sin^2 w_s]\}^{1/4}\sqrt{\sec w_s}\, |dF(y, z, w_s)/dw_s|}, \tag{55}$$

which formula could again have been derived from geometrical optics (Sec. 5.8c). w_s is defined in Eq. (52a), and the upper and lower signs in Eq. (55) apply when $dF/dw_s < 0$ and $dF/dw_s > 0$, respectively. All roots are taken as positive.

As $z \to z_t$, one finds, from Eqs. (44) and (54a),

$$\sec w_s \frac{dF}{dw_s} \to \frac{n(z_t)}{n'(z_t)} \frac{1}{\sqrt{n^2(z) - \sin^2 w_s}}, \tag{56}$$

so Eqs. (55) and (50) agree at the turning point z_t. As $z \to \infty$, $dF/dw_s \to -z/\cos^2 w_s = -\hat{\rho}/\cos w_s$,[†] and

---

†This result follows at once from Eq. (52b), which reduces to $F = y - z \tan w + $ constant as $z \to \infty$. After some manipulation it can also be derived from the limiting form of Eq. (54a) if one assumes that $1 - n^2(z) \sim bz^{-m}[1 + O(z^{-1})]$, where m and b are positive constants.

$$\bar{G}_2'' \sim \frac{\sqrt{\cos w_s}}{2\sqrt{2\pi k_0 \hat{\rho}}} e^{-i\pi/2} \times$$

$$\frac{\exp\left[ik_0\left(y \sin w_s + \int_{z_t}^{z} \sqrt{n^2 - \sin^2 w_s}\, d\zeta + \int_{z_t}^{z'} \sqrt{n^2 - \sin^2 w_s}\, d\zeta\right) + i\pi/4\right]}{[n^2(z') - \sin^2 w_s]^{1/4}},$$

$$(57)$$

which result agrees with that derived for the special case in Eq. (5.9.15). Since for a monotonically increasing refractive index profile with $n'(z) > 0$, one has dF/dw_s positive when $z \approx z_t$ and negative when $z \to \infty$, dF/dw_s has a zero at an intermediate coordinate location. The elimination of w_s from the simultaneous equations

$$F(y, z, w_s) = 0, \qquad \frac{d}{dw_s} F(y, z, w_s) = 0, \qquad (58)$$

yields a curve, the caustic, which forms the envelope of the refracted-ray family (Fig. 5.8.4). A ray characterized by a given w_s touches the caustic at the coordinates $(\bar{y}, \bar{z})$ for which dF/dw_s vanishes. It is noted from Eq. (55) that a phase change of $\pi/2$ is introduced into the ray field after emerging from the caustic.

It has been mentioned that the solution of Eqs. (41) and (52) for real saddle points w_s is easily visualized from the ray diagram in Fig. 5.8.4 that represents a plot of these equations, with each ray curve characterized by a certain w_s. One finds that every observation point (y, z) in the illuminated region is reached by two rays; the values of w_s belonging to these rays represent the saddle points. In the vicinity of the source (point P_1 in Fig. 5.8.4), one of the rays is a direct ray arising from Eq. (41) while the other is refracted and corresponds to a saddle point of Eq. (52). However, if the observation point (y, z) lies near the caustic (point P_2 in Fig. 5.8.4), both rays are refracted and arise from saddle points of Eq. (52); Eq. (41) has no pertinent solution. $\bar{G}''$ is then given by the sum of two expressions of the form in Eq. (55), with w_s replaced by the two solutions w_1 and w_2 of Eq. (52). It is also noted from Fig. 5.8.4 that of the two rays passing through a point near the caustic, one ray has already touched the caustic while the other has not; if the saddle points corresponding to the former and the latter are denoted by w_1 and w_2 respectively, then $w_1 < w_2$ since it is recalled that $w_{1,2}$ represent the angles of emergence of the two rays at $z \to \infty$. Moreover, from the preceding discussion, $(dF/dw)_{w_2} > 0$ and $(dF/dw)_{w_1} < 0$. As the observation point approaches the caustic, $w_1 \to w_2$ and $(dF/dw)_{w_{1,2}} \to 0$, whence a point on the caustic gives rise to a second-order saddle point of $q_2(w)$.

Evaluation near the caustic

The asymptotic formula (55) fails for observation points on the caustic since $dF/dw_s = 0$. The required modification of the asymptotic expression is obtained from Eq. (4.5.7) and is valid for observation points on or near the caustic [for a uniform approximation at arbitrary observation points, see Eq. (4.5.2)]:

$$\bar{G}_2'' \big|_{w_{1,2}} \sim \pm \tfrac{1}{4} f_2(w_0) \left[\frac{2}{k_0 q_2^{(3)}(w_0)} \right]^{1/3} e^{ik_0[q_2(w_1) + q_2(w_2)]/2}\{\mathrm{Bi}(x) \pm i\mathrm{Ai}(x)\}, \qquad (59)$$

where the upper and lower signs are appropriate to contributions from path segments leading over the saddle points w_1 and w_2, respectively, with $q''(w_1) < 0$ and $q''(w_2) > 0$. $\mathrm{Ai}(x)$ and $\mathrm{Bi}(x)$ are the Airy functions defined in Sec. 4.2e, and

$$x = k_0^{2/3} X, \qquad (59a)$$

where

$$\tfrac{4}{3} X^{3/2} = i[q_2(w_1) - q_2(w_2)], \qquad X^{1/2} \cong [\tfrac{1}{2} q_2^{(3)}(w_0)]^{1/3} \left(\frac{w_1 - w_2}{2} \right) e^{i\pi/2}, \qquad (59b)$$

with $q^{(3)}(w_0) > 0$ since $w_1 < w_2$ and $q_2''(w_{1,2}) \lessgtr 0$. In these equations, $w_1 \approx w_2$, $w_0 \approx (w_1 + w_2)/2$, $q_2''(w_0) = 0$, whence $q_2(w_1) > q_2(w_2)$, and

$$q_2'(w_0) \approx -\frac{q^{(3)}(w_0)}{8}(w_1 - w_2)^2, \qquad q^{(3)} \equiv \frac{d^3 q}{dw^3}. \qquad (59c)$$

For large values of $|x|$ [since $k_0 \gg 1$, this condition can be satisfied for very small $(w_1 - w_2)$], use of the asymptotic approximation for the Airy functions in Eqs (4.2.51) reduces Eq. (59) to the expression given in Eq. (55), subject to $f_2(w_{1,2}) \approx f_2(w_0)$. On the other hand, when $x \to 0$, formula (59) must be retained and yields for the sum of the contributions from w_1 and w_2:

$$\bar{G}'' \sim \bar{G}_2'' \sim \frac{i}{2} f(w_0) \left[\frac{2}{k_0 q_2^{(3)}(w_0)} \right]^{1/3} e^{ik_0[q_2(w_1) + q_2(w_2)]/2} \mathrm{Ai}(x). \qquad (60a)$$

In particular, when $x = 0$,

$$\bar{G}'' \sim \frac{i}{2} f(w_0) \left[\frac{2}{k_0 q_2^{(3)}(w_0)} \right]^{1/3} e^{ik_0 q_2(w_0)} \mathrm{Ai}(0), \qquad (60b)$$

whence the dependence on $k_0^{-1/3}$ here, as compared with the smaller $k_0^{-1/2}$ in Eq. (55) at an ordinary point along a ray, is indicative of the field enhancement on the caustic.

To determine the field behavior on the dark side of the caustic, we consider the real function $F(y, z, w_0) = q_2'(w_0) \sec w_0$ in Eq. (52b), which vanishes for observation points y and z on the caustic. If the observation point moves along the line $z = $ constant, $y > 0$, the algebraic sign of F changes as the caustic is traversed, and since $q^{(3)}(w_0)$ is positive, one notes from Eq. (59c) that $q_2'(w_0)$ is negative on the illuminated side where w_1 and w_2 are real. Consequently, $q_2'(w_0) > 0$ on the dark side [this is also evident from Eq. (52b) since F is surely positive for sufficiently large y], whence $(w_1 - w_2)^2 < 0$. The parameter X in Eq. (59b) is therefore positive, and the field decays in view of the behavior of $\mathrm{Ai}(x)$ for large positive x [see Eq. (4.2.42a)]:

$$\mathrm{Ai}(x) \sim \frac{1}{2\sqrt{\pi}\, x^{1/4}} e^{-(2/3)x^{3/2}}, \qquad x \gg 1. \qquad (61)$$

Because X is positive, one notes from Eq. (59b) that $\arg(w_1 - w_2) = -\pi/2$, so w_2 and w_1 move into the upper and lower halves of the complex w plane, respectively. The asymptotic formula resulting from Eqs. (60a) and (61) has the same characteristics as the analytic continuation of $\bar{G}_2''|_{w_2}$ in Eq. (55) from real to complex values of w_2.

It is of interest to observe the similarity between the description of the field near the caustic in the present example, which involves curved ray trajectories and a curved ray envelope, and the problem in Sec. 7.5e, wherein the rays are straight lines (see also References 28–30).

5.8e Propagation in Ducts (Guided Modes)

If the refractive index profile is not a monotonic function of z but has a maximum n_m at some finite value z_m as shown in Fig. 5.8.5(a), then a ray characterized by the parameter a [see Eq. (18)] lying in the range $n_\beta < a < n_m$ has two turning points z_1 and z_2 since $n(z_{1,2}) = a$. For propagation, $n(z) \geq a$, whence the ray in question is guided or "trapped" in the region $z_1 \leq z \leq z_2$,

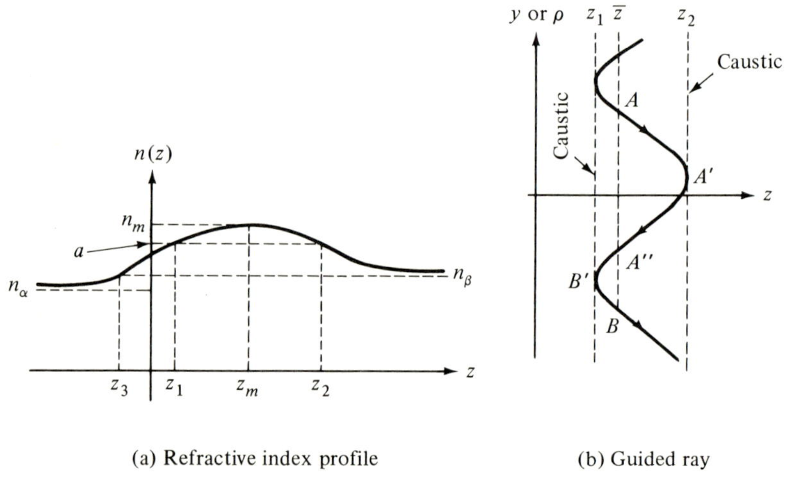

(a) Refractive index profile   (b) Guided ray

FIG. 5.8.5   Duct propagation.

which forms an effective waveguide or duct, as shown in Fig. 5.8.5(b). Outside the duct, $n(z) < a$ and the associated fields are exponentially damped. The duct width changes with a; for the profile in Fig. 5.8.5(a), ray trapping occurs only when $n_\beta < a < n_m$. Rays with $n_\alpha < a < n_\beta$ are turned once at some point z_t, where $-\infty < z_t < z_3$, while those with $a < n_\alpha$ are not turned at all and progress from $-\infty$ to $+\infty$. All rays characterized by the parameter $a = n(z_{1,2})$ touch the lines $z = z_1$ and $z = z_2$, which therefore form the caustics or envelopes of the ray system.

If the field associated with the ray in Fig. 5.8.5(b) is to be a guided or trapped mode that maintains itself indefinitely along the transverse y or ρ direction, certain demands of self-consistency must be met. These can be phrased as the requirement that the magnitude and phase of the field at B is equal to that at A, where the points A and B are separated by a single spatial period.

Let the mode in question be characterized by the transverse (to z) wavenumber $k_0 a$; the longitudinal variation is then defined via Eqs. (7), with solutions given in Sec. 3.3b. If the field amplitude at A is arbitrarily set equal to unity, the reflected wave field at A'' in Fig. 5.8.5(b) has an amplitude $\vec{\Gamma}(\bar{z})$, where $\vec{\Gamma}(\bar{z})$ is the modal reflection coefficient seen from the point $z = \bar{z}$ when looking in the direction of increasing z. In passing from A'' to B, the reflected wavefield with amplitude $\vec{\Gamma}(\bar{z})$ is incident upon the boundary $z = z_1$ and emerges at B with an amplitude $\vec{\Gamma}(\bar{z}) \overleftarrow{\Gamma}(\bar{z})$, where $\overleftarrow{\Gamma}(\bar{z})$ is the reflection coefficient observed at $\bar{z}$ when looking in the direction of decreasing z. Imposition of the above-mentioned self-consistency requirement then leads to the condition

$$\vec{\Gamma}(\bar{z})\overleftarrow{\Gamma}(\bar{z}) = 1, \tag{62}$$

which must be satisfied if the field in question is to represent a guided mode. Equation (62) is simply an alternative statement of the transverse resonance relation

$$\overleftarrow{Z}(\bar{z}) + \vec{Z}(\bar{z}) = 0, \tag{63}$$

phrased in terms of the impedances seen by the mode at $z = \bar{z}$ [see Eq. (3.3.30)]. Equations (62) or (63) can generally be satisfied only for a certain set of discrete values of a, the eigenvalues for the modes in question.

If the medium is slowly varying, the reflection coefficients can be determined from the WKB approximation or from geometrical optics [see Eqs. (35b) or (20b)]. The z-dependent part of the phase change along a ray traveling from A to A' in Fig. 5.8.5(b) is then given by the phase integral $k_0 \int_{\bar{z}}^{z_2} \sqrt{n^2(\eta) - a^2}\,d\eta$. At the boundary z_2, the wave amplitude is changed by the reflection coefficient $\vec{\Gamma}(z_2)$†; in passing from A' to B' and from B' to B, the appropriate phase increments are added, as well as the reflection coefficient $\overleftarrow{\Gamma}(z_1)$ arising from the boundary at z_1. Thus, Eq. (62) can be written as

$$\vec{\Gamma}(z_2)\overleftarrow{\Gamma}(z_1) \exp\left(i2k_0 \int_{z_1}^{z_2} \sqrt{n^2(\eta) - a^2}\,d\eta\right) = 1. \tag{64}$$

In the assumed absence of dissipation, the reflection coefficient magnitudes are equal to unity and may be expressed as $\Gamma(z_{1,2}) = \exp(ic_{1,2})$, where c_1 and c_2 are real constants. If losses are present, these considerations still apply, with complex values for $c_{1,2}$.[31,32] Thus, the transverse resonance condition becomes

$$k_0 \int_{z_1}^{z_2} \sqrt{n^2(\eta) - a_m^2}\,d\eta = m\pi - \frac{c_1 + c_2}{2}, \tag{65}$$

where the values of a corresponding to different m have been denoted by a_m, the eigenvalues for the mth mode. Since the left-hand side of the equation is positive, m takes on those integer values (or zero) that render the right-hand

---

†This discussion applies to the general case where plane reflecting boundaries are present at z_1 and z_2; if the boundary is formed by a caustic, the phase of the ray is changed by $-\pi/2$, thereby giving rise to an effective reflection coefficient $\exp(-i\pi/2)$ [see Eq. (35b) and remarks following Eq. (58)].

side positive. For the case where physical boundaries are located at z_1 and z_2 and $a_m^2 < n^2(z_{1,2})$, the points $z_{1,2}$ remain fixed; if the ray turns before reaching the physical boundary (if any), z_1 and (or) z_2 are defined by $n^2(z_{1,2}) = a_m^2$, and c_1 and (or) c_2 are equal to $-\pi/2$, accounting for the phase change at the caustic. The distance L between points A and B along the ray path can be evaluated from the ray equation (28),

$$ L_m = 2 \int_{z_1}^{z_2} \frac{a_m \, dz}{\sqrt{n^2(z) - a_m^2}}, \tag{66} $$

where $a_m = n(z) \sin \theta(z)$ is a constant along the ray.

From the preceding remarks, showing the connection between the guided modes in a variable medium and the transverse resonance relations (62) or (63), it is apparent that the guiding effect of the inhomogeneity is observed most directly if a radiation problem is analyzed in terms of transmission along the transverse (y or ρ) coordinate. The transformation of the z-transmission representations in Eqs. (5), (6), or (36) into an alternative form highlighting propagation transverse to z has been discussed in general terms in Sec. 3.3c and has been illustrated in detail for the dielectric slab problem in Sec. 5.6. Analytically, the transformation involves the deformation of the integration contour away from the real axis to enclose the singularities of the longitudinal characteristic Green's function $g_z(z, z'; \lambda_z)$. If the z domain is unbounded, the singularities comprise branch points, but poles may also arise if the transverse resonance equations have discrete solutions that satisfy the radiation condition at infinity. The associated spectrum of modes capable of propagating in the transverse direction then has both a continuous and a discrete part. In the presence of impenetrable physical boundaries at $z_{1,2}$, one has a conventional, although inhomogeneously filled, waveguide, and the spectrum is usually purely discrete. If trapped modes can exist and the source and observation points are located in or near the duct region, the contribution from the continuous spectrum is usually negligible and the fields are well represented in terms of the guided modes alone [see remarks following Eq. (5.6.9)]. In the problem of radiation from a longitudinally directed electric current element, the resulting formulation is directly analogous to that for the homogeneous dielectric slab problem [Eq. (5.6.20)]. If an exact solution of the eigenvalue problem cannot be found, the specific form of the eigenfunctions $\Phi_r(z)$ can be ascertained from the WKB approximation provided that the duct is wide and the medium is slowly varying. The utility of the various representations for the evaluation of the fields in the duct region is discussed in connection with a specific example in Sec. 5. 9b.

5.9   SOURCES IN THE PRESENCE OF MEDIA WITH CONTINUOUS PLANAR STRATIFICATION—SPECIAL PROFILES

While the formal presentation in Sec. 5.8a is appropriate to radiation problems in continuously varying media with arbitrary refractive index profiles, explicit solutions (albeit in the form of representations) can be constructed only for special variations of $n^2(z)$ for which Eqs. (5.8.7a) and (5.8.7b) can be solved in terms of known functions. From the by-no-means-exhaustive sampling of profiles in Sec. 3.6, the inverse-square dependence has been selected for detailed study since it incorporates many features associated with a general category of monotonic variation. Another example, the Epstein profile descriptive of a continuous transition, is discussed in a more condensed fashion.

5.9a Inverse-square Profile

Properties of the medium

We begin our discussion of the radiation characteristics of sources in the presence of special inhomogeneous media by considering a region whose permittivity varies from a finite, constant value ϵ_0 at $z \to \infty$ in a monotonic manner characterized by the equation

$$\epsilon(z) = \epsilon_0\left(1 - \frac{p^2}{k_0^2 z^2}\right), \qquad k_0^2 = \omega^2 \mu_0 \epsilon_0. \tag{1}$$

p is an arbitrary constant and the permeability μ_0 is assumed to be constant throughout the region. As will be seen below, the nonuniform transmission-line equations (5.8.7a) and (5.8.7b) have a particularly simple solution in this case. For p^2 positive real and independent of ω, the permittivity $\epsilon(z)$ in Eq. (1) may represent approximately the effect of a lossless, cold, isotropic, electron plasma medium whose electron density $N(z)$ [or equivalently the plasma frequency $\omega_p(z)$ as in Eq. (1.1.60)] is given by

$$N(z) = \frac{p^2 m}{\mu_0 e^2}\frac{1}{z^2} = \frac{\omega_p^2(z) m \epsilon_0}{e^2}, \tag{1a}$$

where m and e signify the electronic mass and charge, respectively. A sketch of both $\epsilon(z)$ and $N(z)$ for real values of p is shown in Fig. 5.9.1.

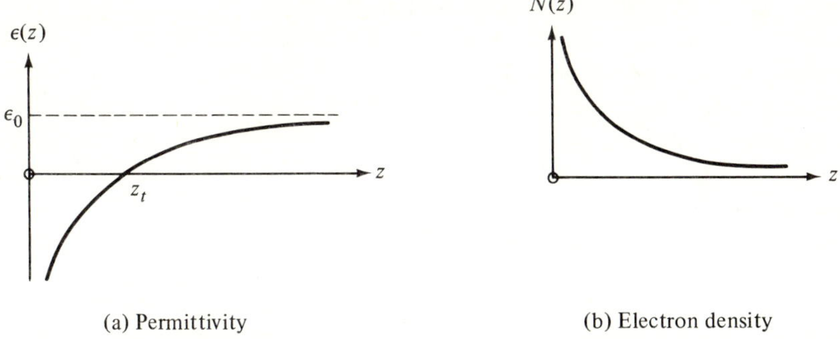

(a) Permittivity   (b) Electron density

FIG. 5.9.1 Inverse-square profile (lossless plasma).

For real p, the permittivity passes through zero at the turning point $z_t = p/k_0$. For $z > z_t$, $k(z) = \omega\sqrt{\mu_0 \epsilon(z)}$ is real, while for $z < z_t$, $k(z)$ is imaginary [Im $k \gtrless 0$ for $\exp\left(\genfrac{}{}{0pt}{}{-i\omega t}{+j\omega t}\right)$ time dependence]. If the continuously variable medium is approximated by a series of thin layers, with a typical layer between the planes $z_i - \delta$ and $z_i + \delta$ characterized by the constant permittivity $\epsilon(z_i)$, then for the $\exp(j\omega t)$ dependence, a plane wave propagating in the $\pm$ direction in the ith layer has a variation $\exp(\mp jk_i z)$, where $k_i = \omega\sqrt{\mu_0 \epsilon(z_i)}$. Hence, the wave propagates when $z_i > z_t$, and attenuates when $z_i < z_t$. Although the electron density in the above model increases indefinitely as $z \to 0$, thus invalidating the physical conditions required for a collision-free (i. e., lossless) plasma, this mathematical singularity has little effect on the radiation characteristics of a source located at finite z values in this medium provided that z_t is reasonably large. Since the region $|z| < z_t$ is essentially impenetrable to electromagnetic waves, the detailed nature of the medium for $|z| \ll z_t$ is immaterial. A physically more realistic density profile, as in Fig. 5.9.2, wherein $N(z)$ in-

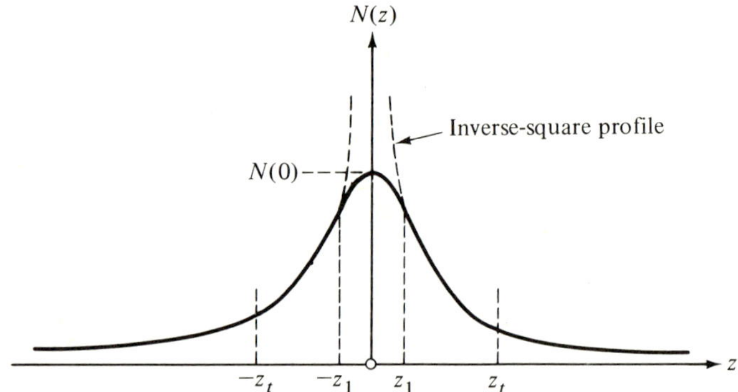

FIG. 5.9.2   Physical profile with monotonically increasing electron density.

creases monotonically to a finite maximum value $N(0)$, can be well approximated by the inverse-square profile provided that the deviation of $N(z)$ from the inverse-square variation occurs only in the region $|z| < z_1$, where $z_1 \ll z_t$, and that the source is located at $z' > z_1$. To demonstrate the validity of these remarks concerning the lack of sensitivity of the radiated field on the detailed medium properties in the region $z \ll z_t$, we shall consider the problem wherein a perfectly conducting plane is located at $z = d$, and will show that the influence of this perturbing structure is negligible when $d \ll z_t$. In the preceding discussion it has been assumed that $\epsilon(z)$ is real. The effect of losses in the medium can be taken into account at any fixed frequency ω by an appropriate complex choice of p in Eq. (1).

Although the inverse-square medium is a special case, it is to be expected that the radiation characteristics of sources therein are representative of those encountered in other media with monotonic variations of $\epsilon(z)$ that pass from a constant value at $|z| \to \infty$ to negative values at $z \to 0$. The present model has the virtue of simplicity.

***Solution for excitation by a longitudinal magnetic dipole or by a
transverse electric line current***

Consider the problem of radiation from either a time-harmonic longitudinal
magnetic current element

$$\mathbf{M(r)} = M^\circ \delta(\mathbf{r} - \mathbf{r}')\mathbf{z}_0, \tag{2a}$$

or from a transverse electric line current,

$$\mathbf{J(r)} = I\delta(\hat{\mathbf{\rho}} - \hat{\mathbf{\rho}}')\mathbf{x}_0, \qquad \hat{\mathbf{\rho}} = (y, z), \tag{2b}$$

located in the variable medium shown in Fig. 5.9.3(a), with a perfectly conducting plane at $z = d$. From Eqs. (5.8.1), (5.2.4), (5.4.31a), and (5.4.31b) the
solutions for the electromagnetic fields can be inferred from the scalar H-mode

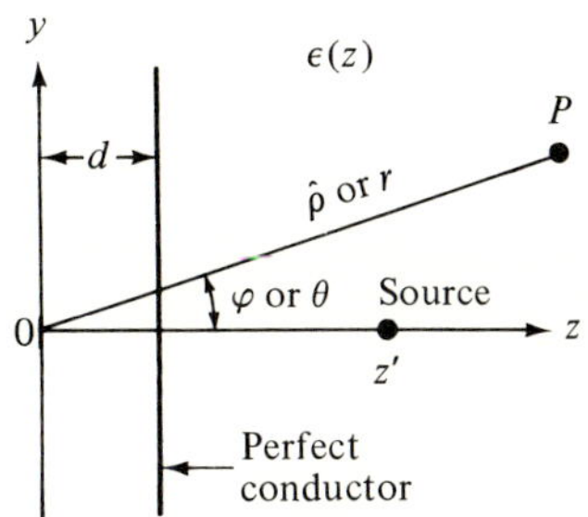

(a) Physical configuration

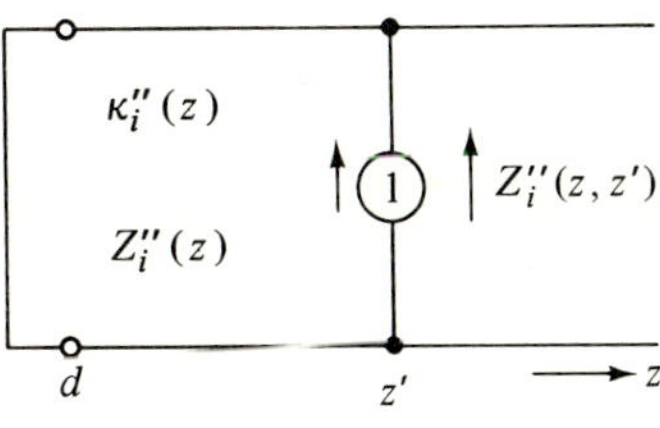

(b) Equivalent modal network (H modes)

FIG. 5.9.3 Physical configuration and network equivalent in the
presence of a perfectly conducting plane.

Green's functions $G''(\mathbf{r}, \mathbf{r}')$ and $\bar{G}''(\hat{\mathbf{\rho}}, \hat{\mathbf{\rho}}')$, respectively, which satisfy the differential equation (5.8.3b) (with $\partial/\partial x \equiv 0$ for the line source) subject to a radia-
tion condition at infinity, and to G'', $\bar{G}'' = 0$ at $z = d$. The solution may be
obtained in a z-transmission representation as in Eq. (5.8.6b) and therefore requires a knowledge of the modal Green's function $g''_{zi}(z, z') = (j\omega\mu_0)^{-1}Z''_i(z, z')$.
g''_{zi} satisfies the differential equation (5.8.7b) subject to vanishing at $z = d$ and
to a radiation condition at $z \to \infty$; its determination is equivalent to finding
the voltage $Z''_i(z, z')$ excited in the modal network of Fig. 5.9.3(b). As noted
in conjunction with Eqs. (5.8.7), the characteristic impedance $Z''_i(z)$ and prop-
agation constant $\kappa''_i(z)$ are given by $Z''_i(z) = \omega\mu_0/\kappa''_i(z)$, $\kappa''_i(z) = [\omega^2\mu_0\epsilon(z) -
k''^2_{ti}]^{1/2}$.

Substitution of the expression for $\epsilon(z)$ in Eq. (1) into Eq. (5.8.7b) yields the
differential equation for the modal Green's function

$$\left[\frac{d^2}{dz^2} + k_0^2\left(1 - \frac{p^2}{k_0^2z^2}\right) - k_t^2\right]g''_{zi}(z, z') = -\delta(z - z'). \tag{3}$$

This equation is identical with Eq. (3.3.3); its solution is given in Eq. (3.3.14a),
with Eq. (3.3.4), in terms of homogeneous solutions $\overrightarrow{V}_i(z)$ and $\overleftarrow{V}_i(z)$ below,

which satisfy the boundary conditions at the upper and lower z termini, respectively. The homogeneous equation (3) is solved by the spherical Bessel functions

$$c_\nu(\kappa z) \equiv \sqrt{\frac{\pi \kappa z}{2}}\, C_{\nu+1/2}(\kappa z), \quad \kappa = \sqrt{k_0^2 - k_t^2}, \quad p^2 = \nu(\nu+1), \qquad (3a)$$

where $C_\mu(z)$ represents any linear combination of the cylinder functions $J_\mu(z)$, $N_\mu(z)$, $H_\mu^{(1)}(z)$, $H_\mu^{(2)}(z)$. In contrast to the z-dependent $\kappa(z)$, κ denotes the value of $\kappa(z)$ at $z \to \infty$. From the network picture in Fig. 5.9.3(b) it is recognized that the solution must be outgoing at $z = \infty$, whence one selects via Eqs. (5.3.13) [for an $\exp(j\omega t)$ dependence],

$$\overrightarrow{V}_i(z) = h_\nu^{(2)}(\kappa z). \qquad (4a)$$

At the lower endpoint $z = d$, the voltage vanishes, so

$$\overleftarrow{V}_i(z) = j_\nu(\kappa z) - \frac{j_\nu(\kappa d)}{h_\nu^{(2)}(\kappa d)}\, h_\nu^{(2)}(\kappa z). \qquad (4b)$$

The particular combination of functions chosen in Eq. (4b) simplifies the determination of the Wronskian required in Eq. (3.3.14a). Since

$$J_\mu(z)\frac{d}{dz} N_\mu(z) - N_\mu(z)\frac{d}{dz} J_\mu(z) = \frac{2}{\pi z}, \qquad (5a)$$

it follows from Eq. (3) that

$$j_\nu(z)\frac{d}{dz} n_\nu(z) - n_\nu(z)\frac{d}{dz} j_\nu(z) = 1, \qquad (5b)$$

and the Wronskian of $\overleftarrow{V}$ and $\overrightarrow{V}$ is evaluated as

$$W(\overleftarrow{V}, \overrightarrow{V}) = \overleftarrow{V}\frac{d\overrightarrow{V}}{dz} - \overrightarrow{V}\frac{d\overleftarrow{V}}{dz} = -j\kappa. \qquad (5c)$$

Thus, the solution is given by [see also Eq. (2.7.12a) for a related problem in spherical regions],

$$g_{zi}''(z, z') = -\frac{j}{\kappa}\left[j_\nu(\kappa z_<) - \frac{j_\nu(\kappa d)}{h_\nu^{(2)}(\kappa d)}\, h_\nu^{(2)}(\kappa z_<)\right] h_\nu^{(2)}(\kappa z_>). \qquad (6)$$

Regarding the definition of $\kappa = \sqrt{k_0^2 - k_t^2}$, the requirements $\kappa = k_0$ when $k_t = 0$, and $\mathrm{Im}\,\kappa < 0$ when $k_t > k_0$, are imposed.

A check of Eq. (6) is possible for the limiting case $p = 0$ (i. e., $\nu = 0$), for which the medium in Eq. (1) reduces to the vacuum $\epsilon(z) = \epsilon_0$. Since

$$j_0(x) = \sin x, \quad h_0^{(1,2)}(x) = \mp j e^{\pm jx}, \qquad (7)$$

one finds, from Eq. (6),

$$g_{zi}''(z, z') = \frac{1}{\kappa}\left[\sin \kappa z_< - \sin \kappa d\, e^{-j\kappa(z_<-d)}\right] e^{-j\kappa z_>}, \qquad (8)$$

which agrees with the result in Eqs. (3.4.4) and (3.4.5), with Eqs. (3.4.21), upon letting $\epsilon_1 = \epsilon_2 = \epsilon_0$, and replacing d by $-d$ as appropriate to the geometry in Fig. 5.9.3(a).

Substitution of Eq. (6) into Eq. (5.2.11) or (5.2.13a) yields the formal solution for the scalar Green's functions appropriate to problems of radiation from a longitudinal magnetic dipole and a transverse electric line current, respectively. For the former case, $\kappa = (k_0^2 - \xi^2)^{1/2}$, while for the latter case, $\kappa = (k_0^2 - \eta^2)^{1/2}$. Pertinent singularities of $g_{zi}''(z, z')$ near the real axis (*on the real axis for k_0 real*) in the complex ξ or η plane are the branch points located at $\kappa = 0$ (i.e., $\xi = \pm k_0$ and $\eta = \pm k_0$, respectively), in agreement with the general observations at the end of Sec. 5.8b. The path of integration avoids these singularities as in Fig. 5.3.5(a). Additional singularities exist at the zeros κ_p determined by the equation $h_\nu^{(2)}(\kappa_p d) = 0$. The location of these complex zeros will be discussed below.

As in previous examples, subsequent considerations will be carried out for an $\exp(-i\omega t)$ dependence. In this instance,

$$g_{zi}''(z, z') = \frac{i}{\kappa}\left[j_\nu(\kappa z_<) - \frac{j_\nu(\kappa d)}{h_\nu^{(1)}(\kappa d)} h_\nu^{(1)}(\kappa z_<) \right] h_\nu^{(1)}(\kappa z_>)\,, \tag{9a}$$

$$= \frac{i}{2\kappa}\left[h_\nu^{(2)}(\kappa z_<) - \frac{h_\nu^{(2)}(\kappa d)}{h_\nu^{(1)}(\kappa d)} h_\nu^{(1)}(\kappa z_<) \right] h_\nu^{(1)}(\kappa z_>)\,. \tag{9b}$$

The requirement on κ is now $\operatorname{Im} \kappa \geq 0$. The frequently more useful form in Eq. (9b) follows from Eq. (9a) upon writing $j_\nu = \frac{1}{2}(h_\nu^{(1)} + h_\nu^{(2)})$. The preceding remarks concerning integration paths and singularities now refer to Fig. 5.3.6a. In particular, one obtains for the two-dimensional Green's function $\bar{G}''(\hat{\boldsymbol{\rho}}, \hat{\boldsymbol{\rho}}')$ in Eq. (5.2.13a), upon letting $j \to -i$ to account for the $\exp(-i\omega t)$ dependence and introducing the change of variable $\eta = k_0 \sin w$,

$$\bar{G}''(\hat{\boldsymbol{\rho}}, \hat{\boldsymbol{\rho}}') = \frac{i}{4\pi} \int_{\bar{P}} e^{ik_0 y \sin w} h_\nu^{(1)}(k_0 z_> \cos w) A(z_<, w)\, dw\,, \tag{10}$$

$$A(z_<, w) = h_\nu^{(2)}(k_0 z_< \cos w) - \frac{h_\nu^{(2)}(k_0 d \cos w)}{h_\nu^{(1)}(k_0 d \cos w)} h_\nu^{(1)}(k_0 z_< \cos w)\,, \tag{10a}$$

$$= 2\left[j_\nu(k_0 z_< \cos w) - \frac{j_\nu(k_0 d \cos w)}{h_\nu^{(1)}(k_0 d \cos w)} h_\nu^{(1)}(k_0 z_< \cos w) \right]\,. \tag{10b}$$

The path of integration in the complex w plane is shown in Fig. 5.3.6(b). Instead of the branch point at $w = 0$, there exists in the present case a branch point, and associated branch cut, at $w = \pi/2$. The analogous representation for the three-dimensional Green's function $G''(\mathbf{r}, \mathbf{r}')$ may be derived from Eq. (5.2.11).

Asymptotic evaluation as $z_> \to \infty$

Since the integral in Eq. (10) cannot be evaluated in closed form, we consider an approximate calculation of the far field. It is convenient to introduce polar coordinates via $z = \hat{\rho} \cos \varphi$, $y = \hat{\rho} \sin \varphi$ as shown in Fig. 5.9.3(a), and assume that $k_0 z_> \equiv k_0 z \to \infty$. As long as the integration path avoids the vicinity of $w = \pi/2$, one has $|k_0 z \cos w| \gg |v|$, so the asymptotic formula in Eq. (5.3.13) may be employed for the representation of $h_\nu^{(1)}(k_0 z \cos w)$, whence

$$\bar{G}''(\hat{\boldsymbol{\rho}}, \hat{\boldsymbol{\rho}}') \sim \frac{1}{4\pi} e^{-iv\pi/2} \int_{\bar{P}} e^{ik_0\hat{\rho}\cos(w-\varphi)} A(z', w)\,dw \ . \tag{11}$$

In the asymptotic evaluation of this integral, $A(z', w)$ is considered a slowly varying amplitude function as compared to $\exp[ik_0\hat{\rho}\cos(w - \varphi)]$. Hence, the saddle point, determined by the exponential function, is located at $w = \varphi$, and the steepest descent path P through the saddle point proceeds as shown in Fig. 5.3.6(b). In the deformation of $\bar{P}$ into P, the only pertinent singularities of $A(z', w)$ are poles located at the zeros w_p of $h_\nu^{(1)}(k_0 d \cos w)$. These poles are complex[33]; those relevant for the present problem are finite in number and lie in the half-strip $0 < \mathrm{Re}\, w < \pi/2$, $\mathrm{Im}\, w > 0$. The pole singularities belong to the "leaky-wave" category (see Sec. 5.3e) and their contribution to the far field is exponentially small since $\mathrm{Im}\, w_p > 0$. Even for $k_0 d \gg 1$, when $w_p \to \pi/2$,[†] their effect is unimportant, since the range $w \approx \pi/2$ (i.e., $\varphi \approx \pi/2$) is excluded from consideration in view of the restriction previously imposed on the integration path. Hence, the contribution from the leaky-wave poles can be neglected, and the asymptotic evaluation of the integral in Eq. (11) yields the first-order result [Eq. (5.3.16a)]

$$\bar{G}'' \sim \bar{G}_f B(z', \varphi)\,, \qquad z \to \infty\,, \quad \varphi \not\approx \pi/2\,, \tag{12}$$

where $\bar{G}_f$ is the free-space field due to a line source at $\hat{\rho} = 0$ [Eqs. (5.4.25) et seq.],

$$\bar{G}_f = \frac{i}{4}\sqrt{\frac{2}{\pi k_0 \hat{\rho}}}\, e^{i(k_0\hat{\rho}-\pi/4)}\,, \tag{12a}$$

while $B(z', \varphi)$ expresses the distortion of the free-space pattern

$$B(z', \varphi) = -ie^{-iv\pi/2} A(z', \varphi)\,. \tag{13}$$

The expression for $A(z', \varphi)$ is given in Eqs. (10a) or (10b).

For the point-source Green's function $G''(\mathbf{r}, \mathbf{r}')$ [see Eq. (5.2.11)], substitution of Eq. (9b) yields an integral representation as in Eq. (10) provided that $\exp(ik_0 y \sin w)$ is replaced by $\frac{1}{2}k_0 \sin w\, H_0^{(1)}(k_0\rho \sin w)$. If $|k_0\rho \sin w| \gg 1$, $|k_0 z \cos w| \gg |v|$ (i. e., $k_0\rho \gg 1$, $k_0 z \gg |v|$, $w \not\approx 0$, $\pi/2$), the Hankel functions $H_0^{(1)}(k_0\rho \sin w)$ and $h_\nu^{(1)}(k_0 z \cos w)$ can be replaced by their asymptotic approximations. The resulting integrand has the same form as that in Eq. (11), save for a factor $\sqrt{\sin w}$, and the integral is evaluated asymptotically as before, yielding the result

$$G''(\mathbf{r}, \mathbf{r}') \sim G_f B(z', \theta), \qquad r \to \infty, \quad \theta \not\approx 0, \ \frac{\pi}{2}\,, \tag{14}$$

[†] Since the zeros α_p of $h_\nu^{(1)}(\alpha)$ are located at finite values of $|\alpha|$ (see Reference 33), $\cos w_p = \alpha_p/k_0 d \to 0$ when $k_0 d \gg 1$.

where $r = \sqrt{\rho^2 + z^2}$, $z = r \cos \theta$ [see Fig. 5.9.3(a)], and G_f is the free-space Green's function

$$G_f = \frac{e^{ik_0 r}}{4\pi r}. \tag{14a}$$

In Eqs. (11)–(14) it has been assumed that the source is located at finite z' while the observation point moves to $z \to \infty$. Manifestly, via the replacement $\mathbf{r} \leftrightarrow \mathbf{r}'$, the above results remain valid for the reciprocal situation wherein the source point moves to infinity and the observation point is situated arbitrarily.

Ray-optical interpretation

While the pattern function $B(z', \varphi)$, and hence the electromagnetic fields derived from the asymptotic forms of the Green's functions $\bar{G}''$ and G'', can be calculated from available numerical tables of the cylinder functions, it is desirable for an interpretation utilizing the concepts of geometrical optics to approximate the cylinder functions comprised in $B(z', \varphi)$ by their representations in Eqs. (5.4.77a). If v is assumed to be large, one may use the approximation $p^2 \approx (v + \frac{1}{2})^2$ in Eq. (3a) and represent the Hankel function as in Eq. (19). Equation (12) can then be written as

$$\bar{G}''(\hat{\boldsymbol{\rho}}, \hat{\boldsymbol{\rho}}') \sim \bar{G}_1''(\hat{\boldsymbol{\rho}}, \hat{\boldsymbol{\rho}}') + \bar{G}_2''(\hat{\boldsymbol{\rho}}, \hat{\boldsymbol{\rho}}'), \tag{15}$$

where

$$\bar{G}_1''(\hat{\boldsymbol{\rho}}, \hat{\boldsymbol{\rho}}') = \frac{\sqrt{\cos \varphi} \, \exp\left[ik_0 y \sin \varphi + ik_0 \int_{z'}^{z} \sqrt{n^2(\zeta) - \sin^2 \varphi} \, d\zeta + i\pi/4 \right]}{2 \sqrt[4]{2\pi k_0 \hat{\rho}} \, \sqrt[4]{n^2(z') - \sin^2 \varphi}}, \tag{15a}$$

$$\bar{G}_2''(\hat{\boldsymbol{\rho}}, \hat{\boldsymbol{\rho}}') =$$
$$- \frac{\sqrt{\cos \varphi} \, \exp\left[ik_0 y \sin \varphi + ik_0 \left[\int_{d}^{z'} + \int_{d}^{z} \right] \sqrt{n^2(\zeta) - \sin^2 \varphi} \, d\zeta + (i\pi/4) \right]}{2\sqrt{2\pi k_0 \hat{\rho}} \, \sqrt[4]{n^2(z') - \sin^2 \varphi}}. \tag{15b}$$

Equation (15a) applies in the range $\sin \varphi \leq \sin \varphi_c \equiv n(z')$, while Eq. (15b) is limited to observation angles $\sin \varphi \leq \sin \varphi_0 \equiv n(d)$. If $\varphi_0 < \varphi < \varphi_c$, the argument of the Hankel functions $H_p^{(1,2)}(k_0 d \cos \varphi)$ is smaller than the order p and $H_p^{(1)} \sim -H_p^{(2)}$ [see Eq. (5.4.77b)]. In this case, $\bar{G}_2''$ is still found to be given by Eq. (15b) provided that the lower integration limits are changed from d to z_t, where z_t is defined by the equation

$$n(z_t) = \sin \varphi, \tag{16}$$

and that the multiplicative constant -1 is replaced by $\exp(-i\pi/2)$. When $\varphi > \varphi_c$, the arguments of all the Hankel functions appearing in $A(z', \varphi)$ are smaller than their order p, whence $A(z', \varphi)$ becomes a decaying function with increasing φ. The corresponding implication in Eqs. (15) is $[n^2(\zeta) - \sin^2 \varphi]^{1/2} = i[\sin^2 \varphi - n^2(\zeta)]^{1/2}$ when $\sin \varphi > n(\zeta)$. Near the transition regions $\varphi \approx \varphi_0$

or $\varphi \approx \varphi_c$, one must employ the Airy-function approximations in Eq. (5.4.77c) [see also Eq. (5.8.60)].

Equations (15a) and (15b) distinguish three different ranges of observation angles φ when the observation point P is located at infinity. In all cases, the radiation field is described locally by plane waves propagating at the angle φ, since $n(\zeta) \to 1$ as $\zeta \to \infty$, so the essential behavior of the exponentials is represented by $\exp[ik_0 y \sin\varphi + ik_0 z \cos\varphi] = \exp(ik_0\hat{\rho})$ [see also Eq. (12)]. However, Eqs. (15) have been expressed in a manner that permits a study of the

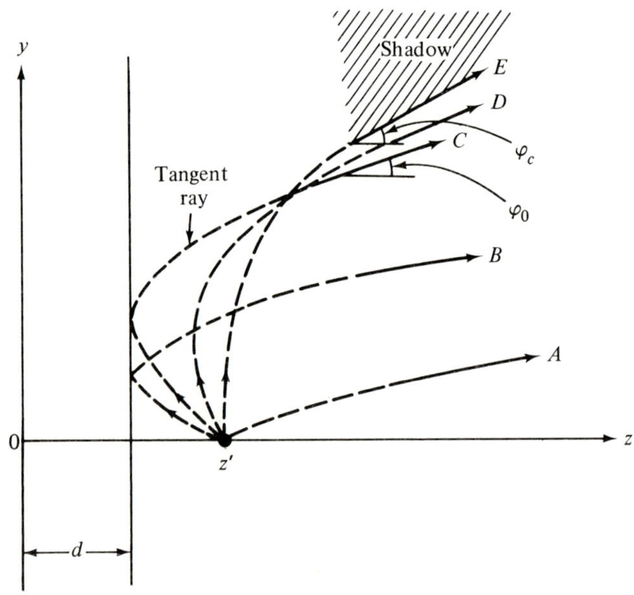

FIG. 5.9.4 Ray picture.

progress of each wave contribution from the source point to $z \to \infty$. As shown in Sec. 5.8d [see Eq. (5.8.47) and (5.8.57)], $\bar{G}_1''$ and $\bar{G}_2''$ represent precisely the fields along geometric-optical rays following curved trajectories in the variable medium, thereby lending significance to what might otherwise appear as an arbitrary rearrangement of Eq. (12). The complete ray trajectories are shown dashed in Fig. 5.9.4 (see Fig. 5.8.4), but their asymptotic direction, to which Eqs. (15) apply, is indicated by the solid arrows. Ray A reaches the observation point P directly from the source while ray B first travels to the plane at $z = d$, is reflected, and then proceeds to P. Comparison of the phase behavior of $\bar{G}_1''$ (ray A) and $\bar{G}_2''$ (ray B) lends direct support to this interpretation; one also observes in the reflected ray term the additional factor -1, which represents the reflection coefficient of the perfectly conducting plane; for a plane having another value of surface impedance, -1 would be replaced by the appropriate plane-wave reflection coefficient. The domain of existence of ray B

is bounded by the tangent ray C, which just grazes the surface and emerges with the angle φ_0.[†]

When $\varphi_0 < \varphi < \varphi_c$, Eq. (15a) yields the direct ray A as before, but the modified Eq. (15b) now describes the field along ray D which is refracted back toward the positive z axis before reaching the boundary. Its maximum penetration into the medium is given by the coordinate z_t defined in Eq. (16). The factor $\exp(-i\pi/2)$ replacing the reflection coefficient (-1) arises from the phase change experienced by this ray when touching the caustic, which forms the envelope of the refracted ray system [see Fig. 5.8.4 and Eq. (5.8.55)]. Ray E, whose initial direction at the source point z' is parallel to the y axis, emerges at $z \to \infty$ at the angle $\varphi_c = \sin^{-1} n(z')$, which also defines the asymptote of the caustic curve. This exhausts the class of propagating geometric-optical rays; observation points specified by angles $\varphi > \varphi_c$ lie in the geometric-optical shadow region and can be reached only by diffraction. The fields in the shadow region can be calculated from Eq. (12) or, equivalently, from the analytic continuation of the appropriately modified ray formulas (15). The general conclusions arrived at in Sec. 5.8d are therefore verified.

Implied in the preceding remarks is the assumption that $n^2(z') > 0$, $n^2(d) > 0$, so both the source point and the bounding plane are located in parts of the medium which can support propagating waves. If $n^2(d) < 0$, $n^2(z') > 0$, the boundary cannot be reached by propagating waves (real rays) and the ray picture comprises only types A and D; the ray traveling along the negative z axis shows the deepest penetration to the point z_0 where $n(z) = 0$ (Fig. 5.8.4). The deeper the plane is embedded in the nonpropagating portion of the medium, the smaller is its effect on the refracted wave field. An estimate can be obtained from $A(z', \varphi)$ in Eq. (10b), which contains a factor of the form

$$Q = H_p^{(2)}(k_0 z' \cos \varphi) + (1 - \Delta)H_p^{(1)}(k_0 z' \cos \varphi), \qquad \Delta = \frac{2J_p(k_0 d \cos \varphi)}{H_p^{(1)}(k_0 d \cos \varphi)}.$$

$$(17)$$

When $n^2(d) < 0$, one has $k_0 d \cos \varphi < p$ and Δ is a small quantity for any φ; when $n^2(d) > 0$, these remarks apply to the rays of type D in Fig. 5.9.4. If $n^2(z') < 0$, the source point lies in the nonpropagating portion; no real geometric-optical rays exist in this case, $H_p^{(1)} \sim -H_p^{(2)}$, and the factor Q is small for all values of φ. If $n^2(z') > 0$, these remarks apply to observation points in the shadow region $(k_0 z' \cos \varphi) < p$. Since the $H_p^{(1)}$ term in Eq. (17) is associated with a refracted ray when $\varphi_c > \varphi > \varphi_0$, one may interpret Δ as the correction to the refracted-wave amplitude introduced by the plane boundary. The magnitude of $1/\Delta$ increases with the ratio z_t/d, where $z_t = p/(k_0 \cos \varphi)$ is the z coordinate at the turning point (i. e., at the point of maximum penetration)

[†]It has been postulated[27] that the tangent ray C excites at its point of contact with the surface a diffracted ray which travels along the surface into the shadow region and sheds energy continually as it progresses. Since the emerging diffracted ray is congruent to ray C and its amplitude decays exponentially during the part of its travel along the boundary, its contribution is negligible at $\hat{\rho} \to \infty$.

along the ray; the farther the turning point is located from the plane, the smaller is the correction of the refracted-wave field. The correction can be interpreted as arising from a decaying wave that has entered the refraction shadow, has been reflected by the plane, and has emerged again into the propagating wave domain.

The far-field calculation above has been carried out under the assumption $z_> \to \infty$, thereby justifying the use of the simple asymptotic formula (5.3.13) for $h_\nu^{(1)}(k_0 z \cos w)$ in the integrand of Eq. (10). A better approximation is obtained via the Debye formula (5.4.77a). If k_0 is assumed to be the large parameter, a short-wavelength representation of the radiated field can be derived by employing this expression for all the cylinder functions appearing in the integrand of Eq. (10). This formulation is then valid for arbitrary observation points provided that they are located many wavelengths away from the source. In order to phrase the result in a manner which highlights its validity for general refractive index variations, rather than only the special one in Eq. (1), it is pertinent to write Eq. (5.4.77a) in the WKB form [see Eq. (3.5.37)]:

$$H_s^{(1,2)}(k_0 z \cos w)$$

$$\sim \left(\frac{2}{\pi k_0 z \cos w \cos \beta} \right)^{1/2} e^{\pm i k_0 z \cos w [\cos \beta + (\beta - \pi/2) \sin \beta] \mp i\pi/4}, \tag{18}$$

$$\sim \left(\frac{2}{\pi k_0 z \sqrt{n^2(z) - \sin^2 w}} \right)^{1/2} \exp\left[\pm i k_0 \int_{z_w}^{z} \sqrt{n^2(\zeta) - \sin^2 w}\, d\zeta \mp i\pi/4 \right], \tag{19}$$

where

$$n^2(z) = 1 - \frac{s^2 - 1/4}{k_0^2 z^2}, \quad s^2 = p^2 + \tfrac{1}{4}; \quad z_w = \frac{s}{k_0 \cos w}, \quad n(z_w) = \sin w, \tag{19a}$$

and

$$\sin \beta \equiv \frac{s}{k_0 z \cos w} \approx \sqrt{1 - n^2(z \cos w)}, \qquad \cos \beta \approx n(z \cos w). \tag{19b}$$

The asymptotic representation in Eq. (18) can be employed along the entire integration path provided that the path avoids the vicinity of $w = \pi/2$ (see Sec. 6.A1). The approximation in Eq. (19b) is valid when s is reasonably large, so $s^2 - 1/4 \approx s^2$. This assumption, together with the formula

$$\int_{z_w}^{z} \sqrt{1 - \frac{s^2}{k_0^2 \zeta^2} - \sin^2 w}\, d\zeta = z \cos w \left[\cos \beta + \left(\beta - \frac{\pi}{2} \right) \sin \beta \right], \tag{20}$$

has been employed in going from Eq. (18) to (19). Via Eq. (19), the solution (10a) of the differential equation (3) can be expressed in a form valid (asymptotically, for large k_0) for *any* variation of refractive index that increases monotonically to the value of unity at $z = \infty$ and contains no turning points or singularities in the interval in question. This follows from a comparison with the more generally derived Eq. (5.8.35b), which corresponds to the special case $d = 0$. The subsequent asymptotic evaluation may then be performed as

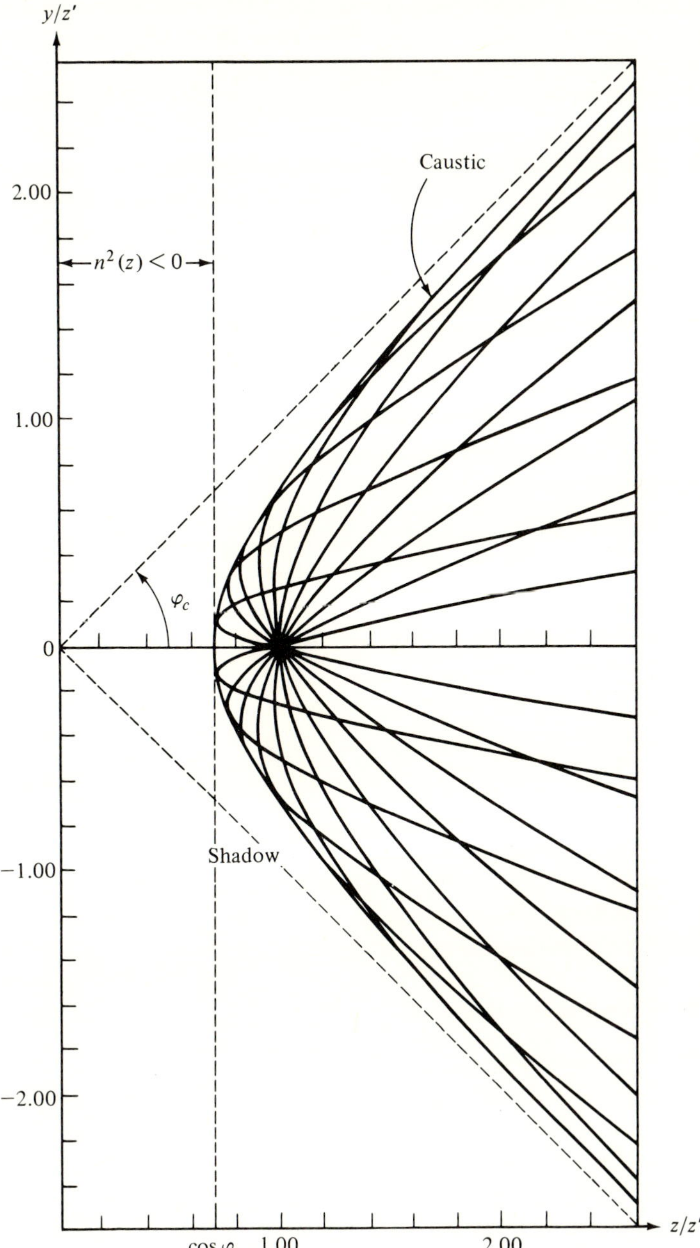

FIG. 5.9.5 Ray configuration and caustic: $[\varphi_c = \cos^{-1}(p/k_0 z') = \pi/4]$.

in Sec. 5.8d and yields a ray-optical interpretation valid everywhere along the trajectories in Fig. 5.9.4.

The geometric-optical ray configuration

We consider now the geometric-optical ray family, defined by the ray equations (5.8.18) and (5.8.28), in the inverse-square profile whose refractive index $n(z) = \sqrt{\epsilon(z)/\epsilon_0}$ is given in Eq. (1). The integration may be performed explicitly and one finds that the rays form a family of hyperbolas described by the equation

$$(z'^2 - z^2)\sin^2\varphi_c + y^2(\cos^2\varphi_c + \cot^2\alpha) + 2z'y\sin^2\varphi_c\cot\alpha = 0, \qquad (21)$$

where $\varphi_c = \sin^{-1}[n(z')]$ defines the extent of the illuminated region at infinity (see Fig. 5.9.4) and α or $\pi + \alpha$ is the angle with the positive z axis at which the ray leaves the line source at $(0, z')$. The relation between α and the angle of emergence $\varphi = w_s$ of the ray at infinity is obtained from Eq. (5.8.18) [see also Eq. (5.8.41)]:

$$\left.\frac{dy}{dz}\right|_{(0,z')} = \frac{\sin\varphi}{\sqrt{\cos^2\varphi - (p^2/k_0^2 z'^2)}} = \tan\alpha. \qquad (22)$$

The caustic is also found to be a hyperbola whose equation is determined by eliminating the parameter $\cot\alpha$ between Eq. (21) and its derivative with respect to $\cot\alpha$:[34]

$$\frac{z^2}{z'^2\cos^2\varphi_c} - \frac{y^2}{z'^2\sin^2\varphi_c} = 1. \qquad (23)$$

The caustic is tangent to the line $z = z'\cos\varphi_c = p/k_0$ on which the refractive index $n(z) = 0$. A plot of the ray configuration for $\varphi_c = \pi/4$ is given in Fig. 5.9.5. If a plane surface is inserted as in Fig. 5.9.4, rays in Fig. 5.9.5 that are intercepted by the surface give rise to reflected rays. The illuminated region is then bounded by the tangent ray C in Fig. 5.9.4 up to its point of tangency with the caustic, and by the caustic thereafter.

5.9b   Radiation in a Duct

The guided-mode spectrum

To illustrate the general procedure concerning duct propagation as discussed in Sec. 5.8e, we consider a region bounded at $z = d$ by a perfectly conducting infinite plane surface; in the space $z \leq d$, the dielectric constant is assumed to vary as in Eq. (1), with p^2 large. The refractive index profile is shown in Fig. 5.9.6(a) and passes through zero at z_0. Thus, no wave propagation is possible when $z < z_0$, and the effective maximum waveguide height is $d - z_0$. The rays are again the hyperbolas described at the end of Sec. 5.9a; however, because of the presence of the wall, reflections take place as shown in Fig. 5.9.6(b). Since a reflected ray emerges at the angle of incidence, it traces out identical trajectories between reflections. In terms of the picture in Fig. 5.9.6(b) and

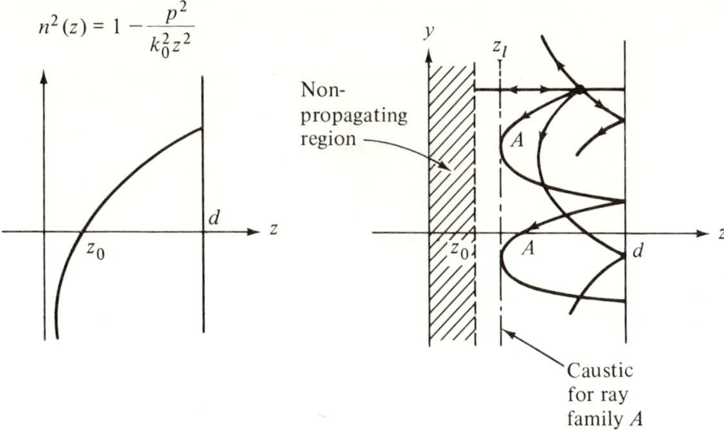

(a) Refractive index variation   (b) Ray picture

FIG. 5.9.6 Inhomogeneously filled waveguide.

with reference to Eq. (5.8.62), $\overrightarrow{\Gamma}(z_2) \equiv \Gamma(d) = -1$, while $\overleftarrow{\Gamma}(z_1) = \exp(-i\pi/2)$ since z_1 represents the caustic for the mode in question. Thus, the transverse resonance equation is

$$k_0 \int_{z_t}^{d} \sqrt{n^2(\eta) - a_m^2}\, d\eta = (m + \tfrac{3}{4})\pi, \qquad m = 0, 1, 2, \ldots \tag{24}$$

with $n^2(z_t) = a_m^2$. The integral has been evaluated in Eq. (20) and yields the implicit relation

$$k_0 d \cos\theta_m \left[\cos\beta_m + \left(\beta_m - \frac{\pi}{2}\right)\sin\beta_m \right] = \left(m + \frac{3}{4}\right)\pi, \tag{24a}$$

where

$$a_m = \sin\theta_m, \quad \beta_m = \sin^{-1}\left(\frac{p}{k_0 d \cos\theta_m}\right).$$

The implications of this equation are studied by comparison with the exact result. For two-dimensional H-mode propagation with respect to z (i. e., $\mathbf{E} = E_x \mathbf{x}_0$ and $\partial/\partial x \equiv 0$), we seek solutions $Q(y, z)$ of the homogeneous equation (5.8.3b), $[\nabla^2 + k^2(z)]Q(y, z) = 0$, that remain bounded at $z = 0$ and vanish at $z = d$. Via Eq. (3a), such solutions are

$$Q(y, z) = \exp(i\eta_l y)\sqrt{z}\, J_s(\sqrt{k_0^2 - \eta_l^2}\, z), \qquad s = \sqrt{p^2 + \tfrac{1}{4}} \approx p, \tag{25}$$

provided that the η_l satisfy the condition

$$J_s(\sqrt{k_0^2 - \eta_l^2}\, d) = 0. \tag{25a}$$

If χ_l denotes the lth non-vanishing solution of $J_s(\chi) = 0$, then

$$\eta_l = \sqrt{k_0^2 - \frac{\chi_l^2}{d^2}}. \tag{26}$$

The pertinent roots χ_l are positive real and form an increasing sequence whence the wavenumber η_l along y is real only as long as $\chi_l < k_0 d$; for $\chi_l > k_0 d$, η_l is imaginary and this implies that the wave does not propagate along the y direction. The modes in question are seen to be quite similar to those in a perfectly conducting circular waveguide filled with a homogeneous dielectric ϵ_0; the refractive index singularity at $x = 0$ produces a virtual boundary, thereby making the region effectively a closed waveguide. The Bessel function in Eq. (25) can be rewritten as $J_s(\chi_l z/d)$. Since $J_s(\chi)$ is a decaying function when $s \gtrsim \chi$, the lth-mode pattern in the z direction oscillates when $z > sd/\chi_l$ and decays when $z < sd/\chi_l$. The effective height of the duct for the lth mode is given by $d - z_l$, where

$$z_l = \frac{s}{\chi_l} d = z_0 \frac{k_0 d}{\chi_l}, \tag{27}$$

and the last equality results from the definition $n^2(z_0) = 0$. Thus, the duct height increases with increasing l. The wavefunction Q is unchanged when $\eta_l y$ changes by multiples of 2π; the modal wavelength λ_l is given by

$$\lambda_l = \frac{2\pi}{\eta_l}, \tag{28}$$

and increases with increasing l.

The zeros χ_l may be evaluated approximately if $J_s(\chi)$ is represented by its asymptotic representation in Eq. (5.4.77a), valid when s is large and $[1 - (s/\chi)] \gg \chi^{-2/3}$. The resulting condition is

$$\cos\left\{\chi\left[\cos\beta + \left(\beta - \frac{\pi}{2}\right)\sin\beta\right] - \frac{\pi}{4}\right\} = 0, \qquad \sin\beta = \frac{s}{\chi} \approx \frac{p}{\chi}. \tag{29}$$

If one puts $\eta = k_0 \sin\theta$, $\chi = k_0 d \cos\theta$, Eq. (29) yields the same result as Eq. (24), whence the geometric-optical formula is certainly adequate to approximate the higher-order zeros of $J_s(\chi)$.

Radiation from a line source

If an electric line current is located in the region $0 < z < d$ in Fig. 5.9.3(a) [see also Fig. 5.9.6(b)], the modal network problem in Fig. 5.9.3(b) must be

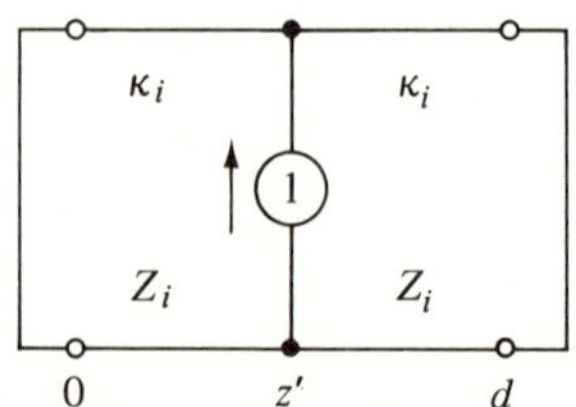

FIG. 5.9.7   Modal network problem.

modified and replaced by the one in Fig. 5.9.7. $\overleftarrow{V}_i$ and $\overrightarrow{V}_i$ in Eqs. (4) are now taken as

$$\overleftarrow{V}_i(z) = j_v(\kappa z), \qquad \overrightarrow{V}_i(z) = h_v^{(1)}(\kappa z) - \frac{h_v^{(1)}(\kappa d)}{h_v^{(2)}(\kappa d)} h_v^{(2)}(\kappa z), \qquad (30)$$

in order to satisfy the boundary conditions $\overleftarrow{V}_i(0) = 0 = \overrightarrow{V}_i(d)$. From Eq. (3.3.14a), the modal Green's function is then found to be [exp $(-i\omega t)$ dependence]

$$g_{zi}''(z, z') = \frac{i\pi}{4} \sqrt{zz'}[H_s^{(1)}(\kappa z_>)H_s^{(2)}(\kappa d) - H_s^{(2)}(\kappa z_>)H_s^{(1)}(\kappa d)]\frac{J_s(\kappa z_<)}{J_s(\kappa d)}, \qquad (31)$$

where $s = (v + \frac{1}{2})$ and $\kappa = \sqrt{k_0^2 - \eta^2}$. Substitution into Eq. (5.2.13a) (with $j \to -i$) or into Eq. (5.8.36) completes the formal solution of the problem in a z-transmission representation, and convergence of the integral is assured by the choice Im $\kappa \geq 0$.

The disposition of the integration path with respect to possible singularities of g_{zi}'' on the real η axis remains to be clarified. Upon use of the circuit relations for the cylinder functions,

$$J_s(\chi e^{i\pi}) = e^{is\pi}J_s(\chi), \qquad H_s^{(1)}(\chi e^{i\pi}) = -e^{-is\pi}H_s^{(2)}(\chi), \qquad (32a)$$

$$H_s^{(2)}(\chi e^{i\pi}) = e^{is\pi}H_s^{(1)}(\chi) + \frac{\sin 2s\pi}{\sin s\pi} H_s^{(2)}(\chi), \qquad \chi = \kappa d, \qquad (32b)$$

one verifies that g_{zi}'' is an even function of κ, whence $\kappa = 0$ is a regular point in the complex η plane (see Sec. 5.3a). Thus, the only singularities are the simple poles η_l arising from the simple zeros of $J_s(\kappa d)$,

$$\eta_l = \pm\sqrt{k_0^2 - \kappa_l^2}, \qquad l = 1, 2, \ldots, \qquad (33a)$$

where $\kappa_l d$ is the lth non-vanishing zero defined by

$$J_s(\kappa_l d) = 0. \qquad (33b)$$

The absence of branch-point singularities in g_{zi}'' implies that the mode spectrum in the z domain is wholly discrete. If s is assumed positive real, the κ_l constitute an infinite discrete set of real numbers [one notes from Eq. (32a) that $-\kappa_l$ also satisfies Eq. (33b)]. The corresponding η_l are real if $\kappa_l^2 < k_0^2$ and imaginary if $\kappa_l^2 > k_0^2$, as noted in connection with Eq. (26). If slight dissipation is assumed

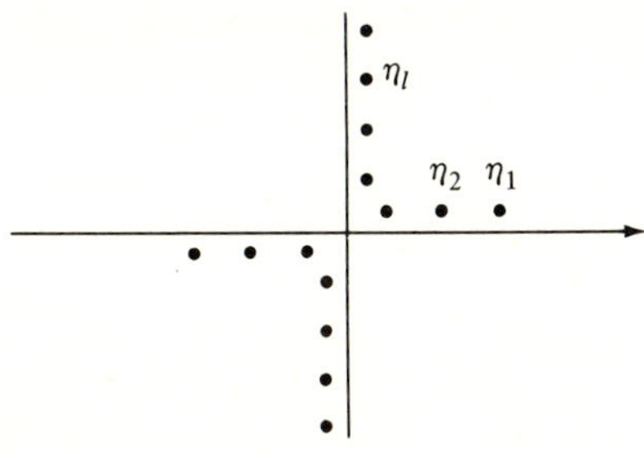

FIG. 5.9.8 Singularities in the complex η plane.

by giving k_0^2 a small positive imaginary part, then the poles are displaced into the complex plane as shown in Fig. 5.9.8, and their location relative to the integration path is unambiguous as Im $k_0 \to 0$.

The guided-mode expansion

A guided-wave representation utilizing y in Fig. 5.9.6(b) as the propagation direction is achieved upon deforming the contour of integration about the pole singularities in the upper or lower halves of the complex η plane.[28] Since g''_{zi} is an even function of η, the integrand in Eq. (5.8.36) remains unchanged if $y - y'$ is replaced by $|y - y'|$ [see Eqs. (5.4.11) and (5.4.12d)]. After verifying from the asymptotic formula for the Hankel function that g''_{zi} decays exponentially as $|\eta| \to \infty$, one may deform the integration path around the poles η_l in the upper half of the η plane to obtain, after a residue evaluation,

$$
\bar{G}'' = \begin{cases}
-\dfrac{\pi\sqrt{zz'}}{2d} \displaystyle\sum_{l=1}^{\infty} \dfrac{\kappa_l}{\eta_l} \dfrac{H_s^{(1)}(\kappa_l d)}{J_s'(\kappa_l d)} J_s(\kappa_l z) J_s(\kappa_l z') e^{i\eta_l |y - y'|}, & (34a) \\[3ex]
\dfrac{i\sqrt{zz'}}{d^2} \displaystyle\sum_{l=1}^{\infty} \dfrac{1}{\eta_l [J_s'(\kappa_l d)]^2} J_s(\kappa_l z) J_s(\kappa_l z') e^{i\eta_l |y - y'|}, & (34b)
\end{cases}
$$

with Im $\eta_l \geq 0$. These alternative formulations follow from the expressions

$$
H_s^{(1)}(\kappa_l d) = -H_s^{(2)}(\kappa_l d) = iN_s(\kappa_l d) = \frac{-2i}{J_s'(\kappa_l d)\pi\kappa_l d}, \tag{34c}
$$

which are a direct consequence of Eq. (33b) when applied to Eq. (5a) and to the definition $H_s^{(1,2)} = J_s \pm iN_s$. Since only a finite number of the η_l are real, the field far from the source plane $y = y'$ is given by a finite number of the guided modes. The series is particularly well suited for direct calculation of the field if only a small number of propagating guided modes exists.

The geometric-optical series

In order to derive a representation that is easily identifiable in terms of the geometric-optical rays sketched in Fig. 5.9.6(b), it is desirable to express g''_{zi} as a superposition of terms containing only products of traveling wave (i.e., Hankel) functions. Such a formulation leads to integrals similar to those in Eqs. (5.8.38), whose stationary-phase approximation is directly interpretable as the geometric-optical solution. The Bessel function in the numerator of Eq. (31) is easily decomposed to yield two additive terms involving the constituent Hankel functions. To achieve a similar additive decomposition of $J_s(\kappa d)$ in the denominator, we use the power-series expansion

$$
\frac{1}{2J_s(\kappa d)} = \frac{1}{H_s^{(2)}(\kappa d)}[1 + t(\kappa d)]^{-1} = \frac{1}{H_s^{(2)}(\kappa d)} \sum_{m=0}^{\infty} (-t)^m, \quad t = \frac{H_s^{(1)}(\kappa d)}{H_s^{(2)}(\kappa d)}, \tag{35}
$$

which converges if $|t| < 1$. The integration path can be slightly distorted so that this condition is satisfied, and substitution into Eqs. (31) and (5.8.36) yields the series

$$\bar{G}'' = \sum_{m=0}^{\infty} G_m, \tag{36}$$

where G_m comprises the four integrals

$$G_m = \sum_{i=1}^{4} g_i, \tag{37}$$

$$g_{1,2} = (-1)^m \frac{i\sqrt{zz'}}{8} \int_{-\infty}^{\infty} H_s^{(2,1)}(\kappa z_<) H_s^{(1)}(\kappa z_>) t^m e^{i\eta|y-y'|} d\eta, \tag{37a}$$

$$g_{3,4} = (-1)^{m+1} \frac{i\sqrt{zz'}}{8} \int_{-\infty}^{\infty} H_s^{(2,1)}(\kappa z_<) H_s^{(2)}(\kappa z_>) t^{m+1} e^{i\eta|y-y'|} d\eta. \tag{37b}$$

Since $z, z' < d$ and $\operatorname{Im} \kappa \geq 0$, one verifies upon use of the asymptotic formula (5.3.13) that the integrals are convergent. However, each term in the series representation for g''_{zi} is no longer an even function of κ, so branch points exist at $\kappa = 0$ and we define $\operatorname{Im} \kappa \geq 0$ on the path of integration.

Substitution of the WKB formula (19) for the various Hankel functions casts these expressions into a general form that remains valid even for slow refractive index variations other than the special one in Fig. 5.9.6(a). The asymptotic evaluation of each integral, after introducing the transformation $\eta = k_0 \sin w$, proceeds as in Sec. 5.8d and will not be repeated here. It is of interest, however, to examine the various saddle-point conditions and their relation to the ray diagram sketched in Fig. 5.9.6(b). The saddle-point conditions are (for convenience, $y' = 0$)

$y =$

$$\begin{cases}
\sin w_s \left[\int_{z<}^{z>} \dfrac{d\zeta}{\sqrt{n^2(\zeta) - \sin^2 w_s}} + 2m \int_{z_t}^{d} \dfrac{d\zeta}{\sqrt{n^2(\zeta) - \sin^2 w_s}} \right], & \text{for } g_1, \quad (38a) \\[3ex]
\sin w_s \left[\int_{z_t}^{z} + \int_{z_t}^{z'} + 2m \int_{z_t}^{d} \right], & \text{for } g_2, \quad (38b) \\[3ex]
\sin w_s \left[\int_{z}^{z_t} + \int_{z'}^{z_t} + (2m+2) \int_{z_t}^{d} \right] = \sin w_s \left[\int_{z'}^{d} + \int_{z}^{d} + 2m \int_{z_t}^{d} \right], & \text{for } g_3, \quad (38c) \\[3ex]
\sin w_s \left[\int_{z>}^{z<} + (2m+2) \int_{z_t}^{d} \right] = \sin w_s \left[\int_{z>}^{d} + \int_{z_t}^{z<} + \int_{z_t}^{d} + 2m \int_{z_t}^{d} \right], & \text{for } g_4, \\[3ex]
& \hspace{4em} (38d)
\end{cases}$$

where all the integrands have the same form as in Eq. (38a), and $n(z_t) = \sin w_s$.

For an interpretation of these geometric-optical ray equations, let us consider the curves described when w_s is a fixed positive number lying in the range $0 < \sin w_s < n(d)$, where $n(d)$ is the refractive index value at the conducting plane $z = d$. If the source point F at $(0, z')$ is located so that $n(z') > 0$, and if $\sin w_s < n(z')$, then Eqs. (38) define the rays shown in Fig. 5.9.9. As noted in Sec. 5.8d, the ray trajectories are tangent to the line $z = z_t$ and would emerge at $z \to \infty$ with an angle of inclination w_s. Two separate ray systems exist: the first is associated with a ray that travels directly from the source F

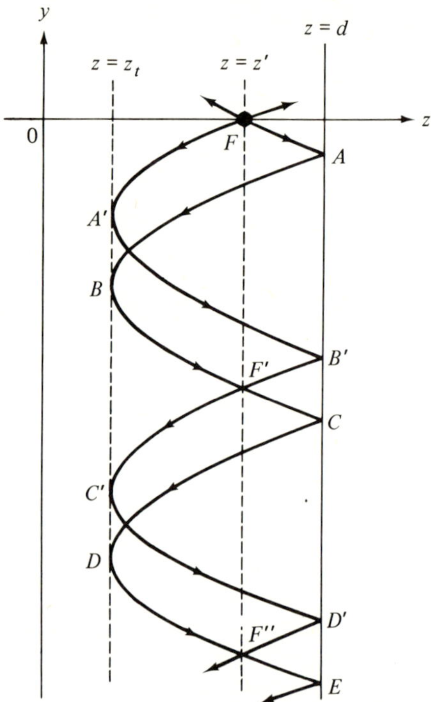

FIG. 5.9.9 Ray trajectories for specified w_s.

to the boundary point A, while the second arises from a ray departing in the opposite direction and being turned back at A'. It is recalled that a ray travels to the right or left when the observation-point coordinate z appears in the upper or lower integration limit, respectively. Equation (38a), with $m = 0$, describes the direct ray segments FA' and FA when $z < z'$ and $z > z'$, respectively; when $m = 1$, the corresponding segments are $F'C'$ and $F'C$ on the first repetition of the cycle; when $m = 2$, one has $F''E'$ and $F''E$, etc. Equation (38b) accounts for the refracted ray $A'B'$ when $m = 0$, ray $C'D'$ when $m = 1$, etc. Equation (38c) describes the reflected ray AB when $m = 0$, CD when $m = 1$, etc. Finally, Eq. (38d), with $m = 0$, yields the ray segments $B'F'$ and BF' when $z > z'$ and $z < z'$, respectively, and correspondingly for $m \geq 1$. Thus, the integrals g_i in Eq. (37) account separately for the fields associated with the various portions of the ray trajectories listed above. The factors $(-1)^m$ arise from the mth reflection of a ray at the plane $z = d$ (reflection coefficient $= -1$), while the factors $[\exp(-i\pi/2)]^m$ entering into the asymptotic approximation of t in Eq. (35) characterize the m contacts of the pertinent ray with the caustic surfaces. The geometric-optical series is convenient for field calculations in a wide duct if the observation point P is located not too far from the source point F, since the effect of the multiply reflected rays, whose amplitude decreases with the path length between F and P [see Eq. (5.8.55)], is small in this case.

While either the guided mode or the geometric-optical expansion can be derived from the contour integral representation (5.8.36) as above, it is of interest to note that one series can also be transformed into the other via the Poisson transformation.[35]

5.9c An Equivalence Relation for Fields in a Homogeneous and an Inverse-Square Medium

An interesting equivalence may be established between a class of two-dimensional (x-independent) field problems in the inhomogeneous medium described by Eq. (1), and a class of field problems with certain rotational symmetry in vacuum.[36] In consequence of the orthogonality of the exponential functions in the interval $0 \leq \phi \leq 2\pi$, the scalar Green's function $\mathring{G}_f$ for a ring source with a progressive phase variation $\exp(i\hat{p}\phi)$, $\hat{p} = 0, \pm 1, \pm 2..$ [see Eq. (5.4.72a)] excites in vacuum a field having this ϕ dependence everywhere. Thus, one may write [Eqs. (5.4.78)]

$$\mathring{G}(\mathbf{r};\, \rho', z') = e^{i\hat{p}\phi}\mathring{G}(\hat{\boldsymbol{\rho}}, \hat{\boldsymbol{\rho}}'), \qquad \hat{\boldsymbol{\rho}} = (y, \rho), \tag{39}$$

where $\mathring{G}$ is the two-dimensional Green's function defined by

$$\left(\hat{\nabla}^2 + k_0^2 - \frac{\hat{p}^2}{\rho^2}\right)\mathring{G}(\hat{\boldsymbol{\rho}}, \hat{\boldsymbol{\rho}}') = -\delta(\rho - \rho')\delta(y - y'), \quad \hat{\nabla}^2 = \nabla^2 - \frac{1}{\rho^2}\frac{\partial^2}{\partial\phi^2}. \tag{39a}$$

To facilitate subsequent interpretation, the axis of the ring source has been chosen as y, with $(x, z) \to (\rho, \phi)$ in the transverse plane. If $(k_0^2 - \hat{p}^2/\rho^2)^{1/2}$ is viewed as an effective wavenumber in a variable medium, the structure of Eq. (39a) is the same as for Green's-function problems involving the inverse-square profile with ρ regarded as a *rectilinear* coordinate, except that the operator $\hat{\nabla}^2 = (1/\rho)(\partial/\partial\rho)(\rho\partial/\partial\rho) + (\partial^2/\partial y^2)$ is not in the rectilinear form $(\partial^2/\partial\rho^2) + (\partial^2/\partial y^2)$. The latter deficiency may be removed by the transformation

$$\mathring{G}(\hat{\boldsymbol{\rho}}, \hat{\boldsymbol{\rho}}') = \sqrt{\frac{\rho'}{\rho}}\; \bar{G}(\hat{\boldsymbol{\rho}}, \hat{\boldsymbol{\rho}}'), \tag{40}$$

which yields the following defining equation for $\bar{G}$:

$$\left[\frac{\partial^2}{\partial\rho^2} + \frac{\partial^2}{\partial y^2} + k^2(\rho)\right]\bar{G}(\hat{\boldsymbol{\rho}}, \hat{\boldsymbol{\rho}}') = -\delta(\rho - \rho')\delta(y - y'), \tag{41}$$

with

$$k^2(\rho) = k_0^2\left(1 - \frac{\hat{p}^2 - \frac{1}{4}}{k_0^2\rho^2}\right). \tag{41a}$$

If ρ is now interpreted as the rectilinear coordinate z and $p^2 = \hat{p}^2 - \frac{1}{4}$, then $\bar{G}(y, z; y', z')$ is exactly the two-dimensional H-mode Green's function for the variable medium in Eq. (1).

The equivalence applies also in the presence of rotationally symmetric obstacles with a contour S described by the equation $f(y, \rho) = 0$, on which $\mathring{G}$ satisfies the boundary condition

$$\mathring{G} = \gamma \frac{\partial \mathring{G}}{\partial n} \qquad \text{on} \quad S, \tag{42a}$$

where n is in the direction of the normal to S and γ is a constant. After transforming according to Eq. (40), one observes the relation

$$\bar{G} = \gamma \left(\frac{\partial}{\partial n} - \frac{1}{2\rho} \frac{\partial \rho}{\partial n} \right) \bar{G} \qquad \text{on} \quad S, \tag{42b}$$

which implies that when $\gamma = 0$, the boundary condition $\mathring{G} = 0$ on S also yields $\bar{G} = 0$ on S. Since $\bar{G}(y, z; y', z')$ is proportional to the electric field E_x in the problems considered [see Eqs. (5.8.1) and (5.4.31)], the latter condition is relevant for perfectly conducting obstacles that are embedded in the inhomogeneous medium and are described by the equation $f(y, z) = 0$. This class of cylindrical diffraction problems in the line-source-excited inhomogeneous medium is therefore equivalent to a corresponding class of problems in vacuum wherein the obstacle is rotationally symmetric and the excitation is due to a phased ring source (see Fig. 5.9.10). Since the refractive index $n^2(0) = -\infty$,

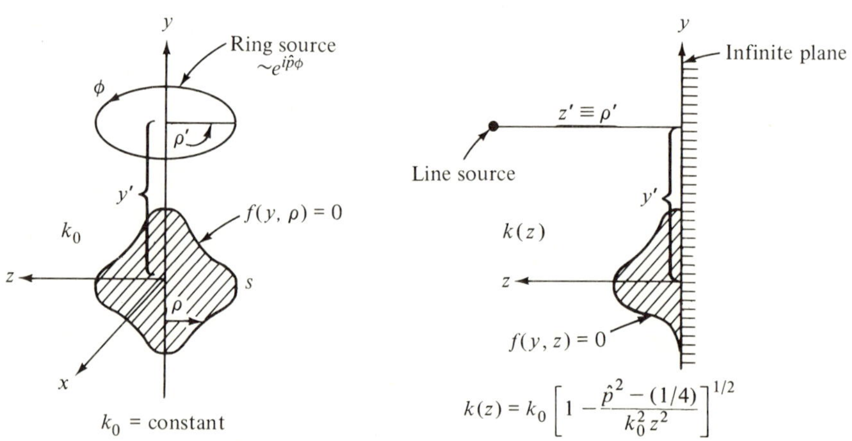

FIG. 5.9.10 Related diffraction problems.

the field vanishes on the plane $z = 0$ and is therefore not disturbed by the insertion of a conducting plane. Various specific configurations are shown in Fig. 5.9.11. It is worth noting that the refractive index varies in a relatively arbitrary (though monotonic) manner over the surfaces shown in Fig. 5.9.11(c), (d), and (e), for which exact solutions of the three-dimensional analogue are known. This refractive index behavior in the presence of an obstacle is not usually found in problems studied in the literature, and the solutions may therefore serve as a valuable check on approximate methods that have been proposed for the analysis of diffraction problems under more general conditions.

The far-field formulas for the ring source [Eq. (5.4.74)] and for the line source in the variable medium [Eq. (12), with $A(z', \varphi) = 2j_{p+1/2}(k_0 z' \cos \varphi)$] may be compared as a simple illustration. After allowing for the different coordinate designation in Eq. (5.4.74) (see also Fig. 5.4.14), one finds readily that the two results are related via Eqs. (39) and (40).

The above-described equivalence may also be employed for the derivation of geometric-optical ray characteristics in one problem in terms of those in another. The ray structure in the plane of the ring source has already been described in Fig. 5.4.14 and is found to be confined to a region of space exterior

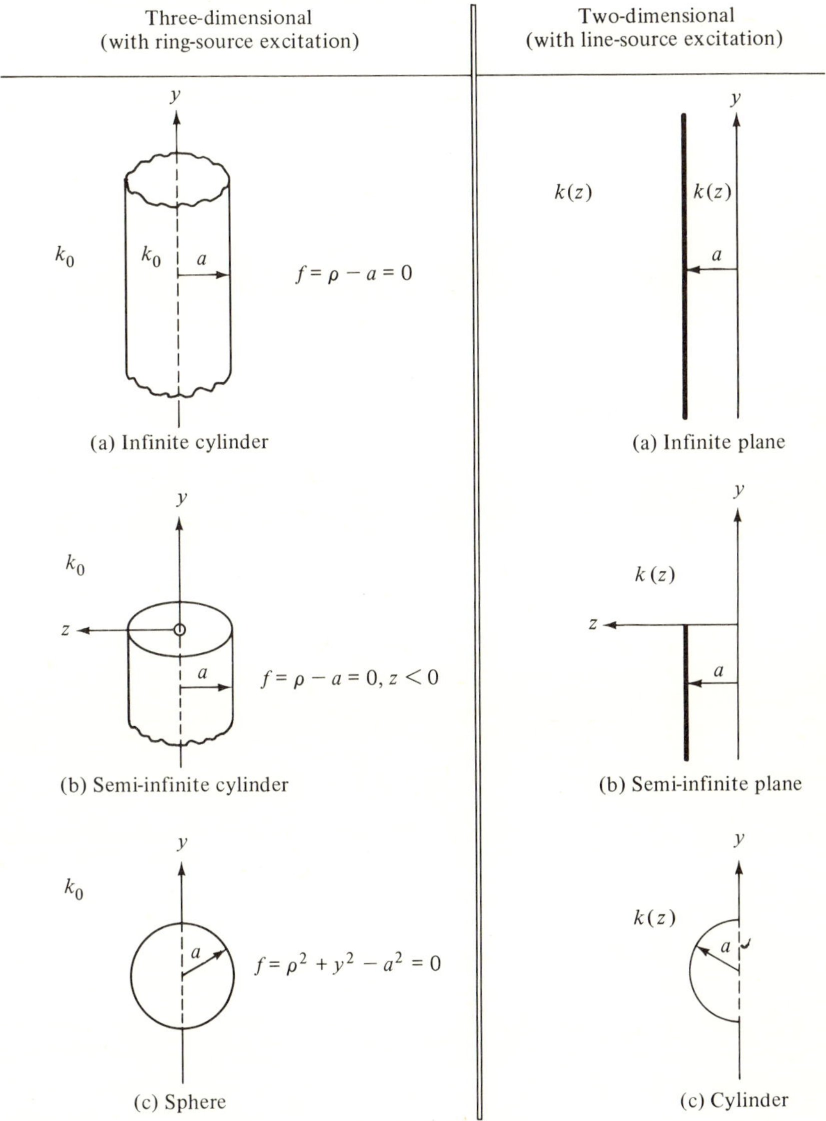

FIG. 5.9.11   Some equivalent configurations.

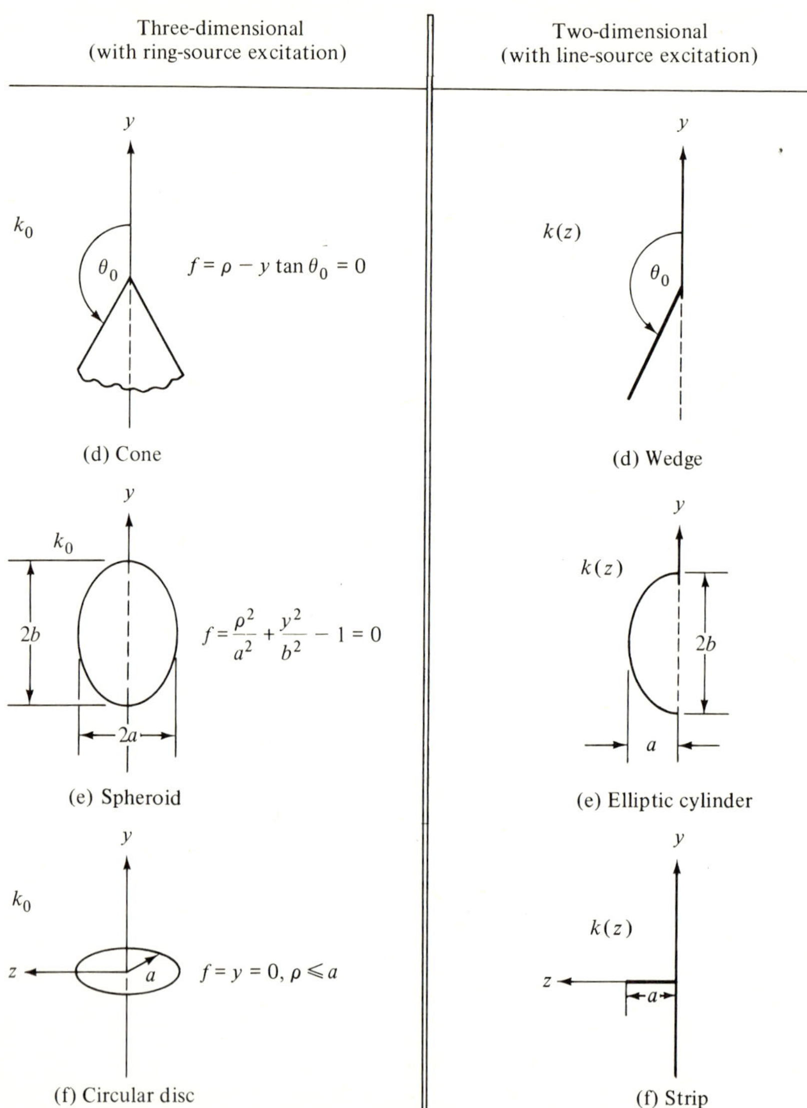

FIG. 5.9.11   Some equivalent configurations (*cont.*)

to a centered circular shadow zone. To determine the ray picture everywhere, it is convenient to go back to Eq. (5.4.72b) and to perform a first-order asymptotic evaluation of this integral when k_0 is large and $\hat{p}/k_0$ is finite. The stationary points ϕ_s, located where $(\partial/\partial\phi')[|\mathbf{r} - \mathbf{r}'| + (m/k_0)\phi'] = 0$, are found to be specified implicitly by

$$\frac{\rho \sin (\phi - \phi_s)}{\sqrt{\rho^2 + \rho'^2 + y^2 - 2\rho\rho' \cos (\phi - \phi_s)}} = \cos \varphi_c, \tag{43}$$

or, upon solving for $\cos(\phi - \phi_s)$, with $0 < \phi - \phi_s < \pi$,

$$\cos(\phi - \phi_s) = \frac{\cos\varphi_c}{\rho}\left[\rho'\cos\varphi_c \pm \sin\varphi_c\sqrt{\frac{\rho^2}{\cos^2\varphi_c} - \frac{y^2}{\sin^2\varphi_c} - \rho'^2}\,\right], \quad (44)$$

where it is recalled that the y and z coordinates in Eq. (5.4.72b) are to be interchanged (with $z' = 0$), and

$$\cos\varphi_c = \frac{p}{k_0\rho'} < 1, \qquad p \approx \hat{p}. \quad (45)$$

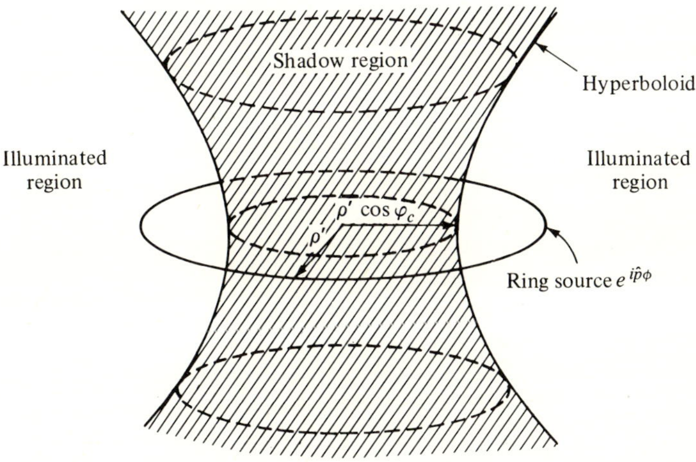

(a) Ray-optical domains

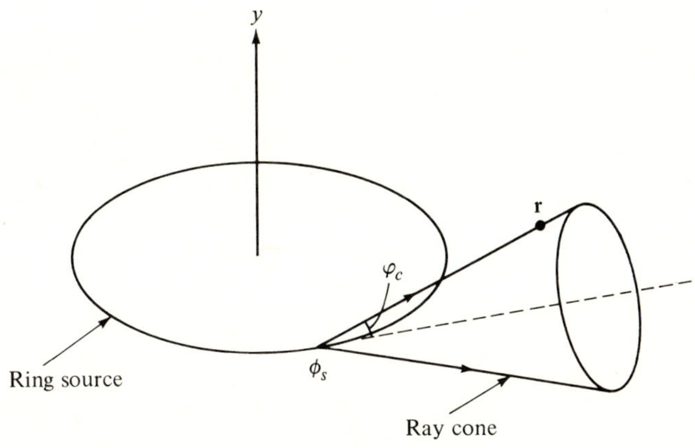

(b) Ray cone

FIG. 5.9.12   Ray structure for phased ring source.

One observes that there are two real stationary points when $(\rho/\cos\varphi_c)^2 - (y/\sin\varphi_c)^2 > \rho'^2$, and no real stationary points when the inequality is reversed. Since real ϕ_s gives rise to undamped propagating fields, the region illuminated by the ring source is bounded by the surface

$$\frac{\rho^2}{\rho'^2\cos^2\varphi_c} - \frac{y^2}{\rho'^2\sin^2\varphi_c} = 1, \tag{46}$$

a hyperboloid as shown in Fig. 5.9.12(a). All field points (ρ, y) corresponding to the same value of ϕ_s in Eq. (43) lie on a right circular cone of total angle $2\varphi_c$, with its apex situated at $\phi' = \phi_s$ and with its axis tangent to the ring, as shown in Fig. 5.9.12(b). This cone is formed by the geometric-optical rays emanating from the ring-source element at $\phi' = \phi_s$, as seen by combining Figs. 5.4.10 and 5.4.14 (with $\psi \to \varphi_c$). The caustic in Eq. (46) is generated by revolving the ray cone in Fig. 5.9.12(b) about the ring axis y.

The ray trajectories associated with the line source in the inhomogeneous medium may be obtained from those above after recalling that the phase function in the former is derived from the latter by suppressing the ϕ variable [see Eqs. (39) and (40)], substituting z for ρ, and leaving y invariant. Interpreted in geometrical terms, a ray element at (ρ, ϕ, y) in the ring-source problem maps into an element at $(\rho = z, y)$, with the z and y components in the transformed configuration equal to the original ρ and y components, respectively. A two-dimensional ray is therefore constructed as in Fig. 5.9.13 by rotating the points

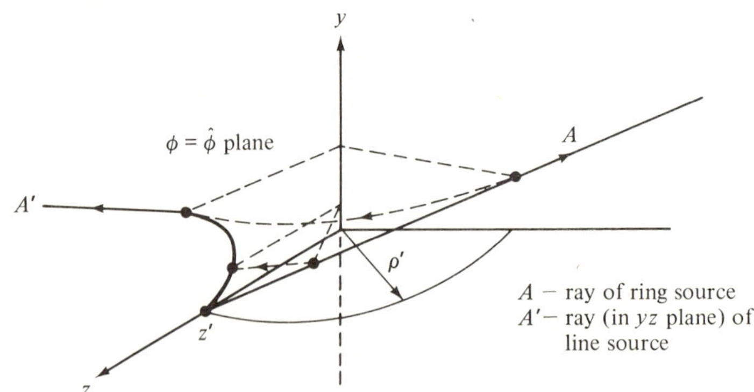

FIG. 5.9.13   Mapping of rays.

on a three-dimensional ray about the y axis into the radial plane $\phi = \hat{\phi}$, which intersects the ring at the point of emergence of the ray. This construction makes evident that the turning point on a ray in the variable medium corresponds in the ring-source problem to the point of nearest approach to the y axis; that a direct and a refracted ray pass through each point (z_1, y_1) in the illuminated region (the plane $y = y_1$ cuts the ray cone in a hyperbolic trace which is intersected at two points by the cylinder $\rho = z_1$; that the pertinent rays are the mapping of those which pass through the points of intersection);

and that the illuminated region is bounded by the hyperbolic caustic in Eq. (23).

To determine the equation of a ray that leaves the source in the inhomogeneous medium at an angle α with the positive z axis, one must select the appropriate ray on the three-dimensional cone. This may be accomplished by considering a plane surface through the cone axis which must intersect the ray cone at the appropriate angle. Since the cone axis is given by $\rho \cos(\phi - \hat{\phi}) = \rho'$, the equation of a plane inclined at the angle β with the $y = 0$ plane is

$$\rho \cos(\phi - \hat{\phi}) - \rho' = y \cot \beta. \tag{47}$$

According to the mapping prescription, a ray contained in this plane has an initial slope equal to $\tan \beta$, whence the desired angle is $\beta = \alpha$. The corresponding ray, found upon eliminating $\cos(\phi - \hat{\phi})$ between Eqs. (47) and (43) (with $\hat{\phi} = \phi_s$), is then transformed into the $\phi = \hat{\phi}$ plane by letting $\rho = z$, $\rho' = z'$. The resulting equation is that of a hyperbola and may be reduced to Eq. (21).

5.9d Continuous Transition (Epstein Profile)

When the permittivity in the medium varies continuously and monotonically from a constant value $\epsilon(-\infty) = (1 + v)\epsilon_0$ at $z = -\infty$ to $\epsilon(+\infty) = \epsilon_0$ at $z = +\infty$, the following functional dependence of $\epsilon(z)$ may be employed to represent such a variation:

$$\epsilon(z) = \left(1 + \frac{v}{1 + e^{\tau z}}\right)\epsilon_0, \tag{48}$$

where v and τ are positive constants (Fig. 5.9.14). The resulting differential equation for the H-mode Green's function G'' [Eq. (5.8.3b)] can be solved in this instance; only this function is required when the excitation is in the form

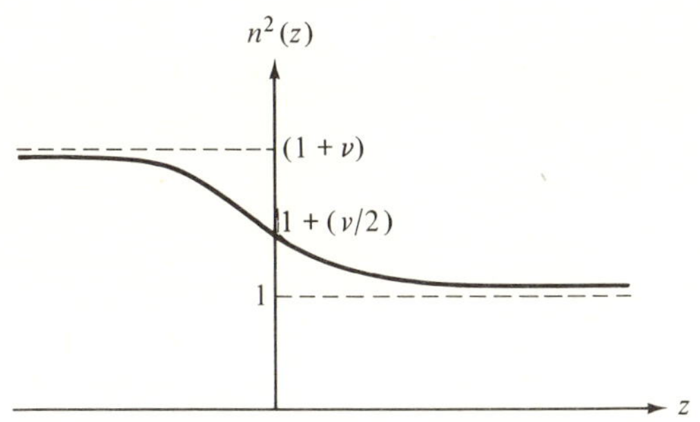

FIG. 5.9.14 Continuous transition.

of a longitudinal magnetic dipole or a transverse electric line current as in Eqs. (2). The resulting z-transmission modal representation [see Eqs. (5.2.11) and (5.2.13)] requires a knowledge of the modal Green's function g''_{zi} which satisfies the differential equation (5.8.7b) with $\mu = \mu_0$, $k^2(z) = k_0^2 \epsilon(z)$, and $k_0^2 = \omega^2 \mu_0 \epsilon_0$. The solution can be expressed in terms of hypergeometric functions as shown in Sec. 3.6b.[37-39] In particular, from Eqs. (3.5.16) and the discussion preceding Eq. (3.6.43),

$$g_{zi}''(z, z') = \frac{\hat{V}_5(z_<)\hat{V}_3(z_>)}{W[\hat{V}_3, \hat{V}_5]}, \tag{49}$$

where the expressions for $\hat{V}_3$, $\hat{V}_5$, and W are obtained from Eqs. (3.6.18)–(3.6.23), (3.6.37), and (3.6.44). The explicit functional form of $\hat{V}_3$ and $\hat{V}_5$ depends on the relative location of the source- and observation-point coordinates z and z', and we shall consider here only the case $z < 0$, $z' < 0$, where both the source and observation points are situated in the optically denser medium. Modifications for other cases, including those where the medium in Eq. (48) is joined at some point z_0 to one with constant $\epsilon(z_0)$, have been considered in Reference 40. From the above-mentioned reference to Sec. 3.6b, one infers the following representations for $\hat{V}_3$ and $\hat{V}_5$, suitable when $z < 0$, $z' < 0$†:

$$\hat{V}_5(z) = \zeta^{(1-\gamma)/2}(1 - \zeta)F(\alpha - \gamma + 1, \beta - \gamma + 1; 2 - \gamma; \zeta), \tag{50a}$$

$$\hat{V}_3(z) = \zeta^{(\gamma-1)/2}(1 - \zeta)\left[\frac{\Gamma(1 - \gamma)\Gamma(\alpha + 1 - \beta)}{\Gamma(1 - \beta)\Gamma(\alpha + 1 - \gamma)}F(\alpha, \beta; \gamma; \zeta)\right.$$

$$+ \frac{\Gamma(\gamma - 1)\Gamma(\alpha + 1 - \beta)e^{i\pi(\gamma-1)}}{\Gamma(\gamma - \beta)\Gamma(\alpha)}\zeta^{(1-\gamma)}$$

$$\left. \cdot F(\alpha - \gamma + 1, \beta - \gamma + 1; 2 - \gamma; \zeta)\right], \tag{50b}$$

where $\zeta = -e^{\tau z}$, and α, β, and γ are defined in Eqs. (3.6.37). Substitution into Eq. (49) yields

$$g_{zi}''(z, z') = g_1(z, z') + g_2(z, z'), \tag{51}$$

where

$$g_1(z, z') = \frac{e^{i\kappa_1|z-z'|}}{-2i\kappa_1}[(1 + e^{\tau z})(1 + e^{\tau z'})F(\alpha, \beta; \gamma; -e^{\tau z_>})$$

$$\times F(\alpha - \gamma + 1, \beta - \gamma + 1; 2 - \gamma; -e^{\tau z_<})], \tag{51a}$$

$$g_2(z, z') = A\frac{e^{-i\kappa_1(z+z')}}{-2i\kappa_1}[(1 + e^{\tau z})(1 + e^{\tau z'})$$

$$\times F(\alpha - \gamma + 1; \beta - \gamma + 1; 2 - \gamma; -e^{\tau z})$$

$$\times F(\alpha - \gamma + 1; \beta - \gamma + 1; 2 - \gamma; -e^{\tau z'})], \tag{51b}$$

$$A = \frac{\Gamma(1 - \beta)\Gamma(\alpha + 1 - \gamma)\Gamma(\gamma - 1)}{\Gamma(\gamma - \beta)\Gamma(\alpha)\Gamma(1 - \gamma)}, \qquad \kappa_1 = \sqrt{k_0^2(1 + v) - \xi^2}. \tag{51c}$$

κ_1 is the propagation constant in the medium to the left of the transition region whose effective width is defined by τ. If $|z|$ and $|z'|$ are large enough so that $\exp(\tau z)$ and $\exp(\tau z')$ are very small, the factors inside the brackets in Eqs. (51a) and (51b) are slowly varying and almost equal to unity [see Eq. (3.6.21c)]

---

†In this section, $\Gamma(w)$ denotes the gamma function and should not be confused with the same symbol used elsewhere for the reflection coefficient.

and may therefore be regarded as distortions of the incident- and reflected-wave fields, respectively.

As a check on Eqs. (51) we examine the limiting cases $v = 0$ (homogeneous medium from $z = -\infty$ to $z = +\infty$) and $\tau = \infty$ (abrupt change of dielectric constant at $z = 0$). When $v = 0$, one has $\alpha = 1$, $\beta = 1 + (2i\kappa_2/\tau) = \gamma$, $\kappa_2 = \sqrt{k_0^2 - \xi^2}$; since $\Gamma(0) = \infty$, one finds that $A = 0$ in Eq. (51c). Moreover,

$$F(\alpha, \beta; \beta; \zeta) = (1 - \zeta)^{-\alpha}, \qquad F(\alpha, \beta; \alpha; \zeta) = (1 - \zeta)^{-\beta}, \qquad (52)$$

so the expression inside the brackets in Eq. (51a) equals unity, thereby yielding the correct value for the free-space Green's function. When $\tau = \infty$, the expressions inside the brackets in Eqs. (51a) and (51b) again have the value unity. In this case, $\alpha = \beta = \gamma \to 1$, or, more precisely,

$$1 - \beta = -\frac{i}{\tau}(\kappa_1 + \kappa_2), \quad 1 - \gamma = -\frac{2i}{\tau}\kappa_1, \quad \gamma - \beta = \frac{i}{\tau}(\kappa_1 - \kappa_2), \quad (53a)$$

which, with the formula

$$\Gamma(w) = \frac{1}{w} + O(1), \qquad \text{as } w \to 0, \qquad (53b)$$

yields the following result for the reflection coefficient A [see also Eq. (3.6.30b)]:

$$A = \frac{\kappa_1 - \kappa_2}{\kappa_1 + \kappa_2}, \quad \kappa_1 = \sqrt{k_0^2(1 + v) - \xi^2}, \quad \kappa_2 = \sqrt{k_0^2 - \xi^2}, \qquad (54)$$

the correct expression for an abrupt transition as given in Eq. (5.5.54c). The modal Green's function then reduces correctly to that in Sec. 5.5d.

Substitution of Eq. (49) into Eqs. (5.2.11) or (5.2.13) yields the Green's functions required for the solution of the problem of radiation from a longitudinal magnetic current element or from a transverse electric line current, respectively, in the medium characterized by the spatially varying permittivity in Eq. (48). To render the integrand unique along the integration contour that extends from $\xi = -\infty$ to $\xi = \infty$ along the real ξ axis (or similarly along the real η axis), it is necessary to investigate the singularities of $g''_{zi}(z, z')$. We recall first [see Eqs. (3.6.37)] that the square roots κ_1 and κ_2 are defined to have positive imaginary parts when the radicands are non-positive. One may then easily verify that the integrals converge exponentially as $|\xi| \to \infty$. $g''_{zi}(z, z')$ has first-order branch-point singularities at $\kappa_{1,2} = 0$, i.e., at

$$\xi = \pm k_0, \qquad \xi = \pm k_0\sqrt{1 + v}. \qquad (55)$$

As regards pole singularities, we note that $\Gamma(w)$ has simple poles at $w = -n = 0, -1, -2, \ldots$, that $F(\alpha, \beta; \gamma; \zeta)$ has simple poles at the poles of $\Gamma(\gamma)$, and that, in particular,[4]

$$\lim_{\gamma \to -n} \frac{F(\alpha, \beta; \gamma; \zeta)}{\Gamma(\gamma)} = \frac{\alpha(\alpha + 1) \cdots (\alpha + n)\beta(\beta + 1) \cdots (\beta + n)}{(n + 1)!}$$
$$\times \zeta^{n+1} F(\alpha + n + 1, \beta + n + 1; n + 2; \zeta). \qquad (56)$$

Moreover, $\Gamma(w)$ has no zeros. From these remarks it appears that both g_1 and g_2 have simple pole singularities at $\gamma = -n$; however, detailed study utilizing Eq. (56) shows that the sum $(g_1 + g_2)$ is regular at $\gamma = -n$. g_1 and (or) g_2 also have poles at points in the complex ξ plane at which

$$2 - \gamma = -n, \quad 1 - \beta = -n, \quad \alpha + 1 - \gamma = -n, \quad n = 0, 1, 2, \ldots. \tag{57}$$

Equation (57) can be satisfied only if γ, β, and $\alpha - \gamma$ are real (i.e., if κ_1 and κ_2 are imaginary). If there are solutions of this equation with $\kappa_{1,2} = +i|\kappa_{1,2}|$, they represent surface waves which individually satisfy the radiation condition at $z \to \pm \infty$; they comprise discrete components in the spectrum of waves that can be guided along the inhomogeneity in the direction transverse to z. Examination of Eqs. (57) and (3.6.37) reveals, however, that such solutions do not exist. If branch cuts are chosen so that $\mathrm{Im}\,\kappa_{1,2} > 0$ on the top sheet of the four-sheeted Riemann surface representing the complex ξ plane, then no pole singularities are present on the top sheet, and the integration path proceeds as in Fig. 5.3.6(a).

An asymptotic evaluation of the integrals in Eqs. (5.2.11) or (5.2.13) for arbitrary z and z' is quite difficult. However, if $|z|$ and $|z'|$ are large enough so that the factors inside the braces in Eqs. (51a) and (51b) may be considered as slowly varying, the integrals can be evaluated asymptotically as in Sec. 5.3d. The saddle points corresponding to the g_1 and g_2 portions of the integrands are located at $\xi_s = k_0(1 + v) \sin \theta$, where θ is, respectively, the angle between the z axis and the radius vector from the source point (ρ', z') and the image point $(\rho', -z')$ to the observation point (ρ, z). The resulting first-order asymptotic approximation (in which branch-cut integral contributions yielding lateral-wave effects are neglected) then looks like that in Eqs. (5.5.5) or (5.5.58) except that the distortion factors inside the brackets in Eqs. (51a) and (51b) are included in the incident- and reflected-field contributions, respectively, and that the reflection coefficient is given by A in Eq. (51c), with ξ replaced by ξ_s. The behavior of the reflection coefficient was discussed in connection with Eq. (3.6.39).

PROBLEMS

1. An azimuthally directed electric current element

$$\hat{\mathbf{J}}(\mathbf{r}, t) = Il\delta(\boldsymbol{\rho} - \boldsymbol{\rho}')\delta(z - z')e^{-i\omega t}\boldsymbol{\phi}_0', \qquad \boldsymbol{\phi}_0' = \mathbf{y}_0 \cos \phi' - \mathbf{x}_0 \sin \phi', \tag{1}$$

where I is the current in the element and l is its length, is located at (ρ', ϕ', z') inside a perfectly conducting circular waveguide with radius b.

(a) Referring to Sec. 5.2a, show that the longitudinal electric and magnetic fields excited by the current element are given by:

$$E_z(\mathbf{r}, \mathbf{r}') = \frac{Il}{-i\omega\epsilon} \frac{\partial}{\rho'\partial\phi'} \frac{\partial}{\partial z'} \sum_i \Phi_i(\boldsymbol{\rho})\Phi_i^*(\boldsymbol{\rho}') \frac{e^{i\kappa_i'|z-z'|}}{-2i\kappa_i'}, \tag{2a}$$

$$H_z(\mathbf{r}, \mathbf{r}') = -Il \frac{\partial}{\partial\rho'} \sum_i \psi_i(\boldsymbol{\rho})\psi_i^*(\boldsymbol{\rho}') \frac{e^{i\kappa_i''|z-z'|}}{-2i\kappa_i''}, \tag{2b}$$

where Φ_i and ψ_i are the scalar mode functions in Eqs. (3.2.75) and (3.2.76).

(b) An electric ring current with radius ρ' and azimuthal variation $\exp(iv\phi')$, $v =$ integer, can be synthesized by multiplying Eq. (1) by $\exp(iv\phi')$ and integrating over ϕ' between the limits 0 and 2π. Show that the longitudinal electric fields $\overset{\circ}{E}_z$ and $\overset{\circ}{H}_z$ excited by the ring current are given by:

$$\overset{\circ}{E}_z(\mathbf{r}; \rho', z') = \frac{Il \, \mathrm{sgn}\,(z - z')ve^{iv\phi}}{\omega\epsilon\rho'b^2} \sum_n \frac{J_v(\beta_n\rho)J_v(\beta_n\rho')e^{i\sqrt{k^2-\beta_n^2}|z-z'|}}{J_v'^2(\beta_n b)} \qquad (3a)$$

and

$$\overset{\circ}{H}_z(\mathbf{r}; \rho', z') = \frac{iIle^{iv\phi}}{b^2} \sum_n \frac{\beta_n'}{J_v(\beta_n'b)J_v''(\beta_n'b)} \frac{J_v(\beta_n'\rho)J_v'(\beta_n'\rho')}{\sqrt{k^2-\beta_n'^2}} e^{i\sqrt{k^2-\beta_n'^2}|z-z'|},$$

$$\qquad (3b)$$

where $\beta_n = x_{nm}/b$, $\beta_n' = x_{nm}'/b$, with $J_m(x_{nm}) = J_m'(x_{nm}') = 0$.

(c) Repeat the calculations in parts (a) and (b) for excitation by a magnetic current element and magnetic current ring source, respectively.

2. Referring to Fig. 5.3.4, discuss the behavior of $\kappa(\xi) = \sqrt{k^2 - \xi^2}$ on a two-sheeted Riemann surface cut along the straight line segment connecting $-k$ and $+k$. Also discuss the corresponding mapping to the w plane via the transformation $\xi = k \sin w$ and compare with the discussion in Sec. 5.3c, which pertains to an alternative choice of branch cuts.

3. Show that when f in the integrand of Eq. (5.2.21) has pole singularities located at $w = \pm jw_1$, w_1 positive real, the time function $A(\tau)$ contains in addition to the result in Eq. (5.2.23) the contribution:

$$A_1(\tau) = \mp\pi j\{(w + jw_1)[f(\bar{\alpha} + w) - f(\bar{\alpha} - w)]\}_{w=-jw_1}\delta\left(\tau - \frac{L}{c}\cosh w_1\right), \quad (4)$$

where the upper and lower signs apply when the integration path around the poles is indented into the half-planes $\mathrm{Re}\, w < 0$ and $\mathrm{Re}\, w > 0$, respectively. Note that $A_1(\tau) = 0$ when f is an even function of w.

4. Derive the two-dimensional free-space Green's function representation in Eq. (5.4.36b) by integrating over x' the cylindrical waveguide representation (5.4.10) of the three-dimensional Green's function. *Hint*: Show first that for $\alpha > 0$, $s > 0$ [cf. Eq. (5.4.12d)]

$$\frac{e^{-j\alpha s}}{s} = \frac{-j}{2}\int_{-j\infty}^{j\infty} H_0^{(2)}(\sqrt{\alpha^2 - \zeta^2}\,s)\,d\zeta = \frac{1}{2}\int_{-\infty}^{\infty} H_0^{(2)}(\sqrt{\alpha^2 + \mu^2}\,s)\,d\mu. \quad (5)$$

5. The cylindrical waveguide representation for the free-space Green's function G_f in Eq. (5.4.10) may by written alternatively as

$$G_f = \frac{i}{4\pi}\int_0^\infty \xi J_0(\xi\rho)\frac{e^{i\sqrt{k^2-\xi^2}|z|}}{\sqrt{k^2-\xi^2}}\,d\xi. \qquad (6)$$

By employing the steepest-descent procedure to evaluate the integral, show that on the axis $\rho = 0$, G_f is given by

$$G_f \sim \frac{e^{ik|z|}}{4\pi|z|}, \qquad (7)$$

thereby verifying that the asymptotic result in Eq. (5.4.9) applies also at $\theta = 0$.

6. Electric line currents flowing parallel to z are distributed over a cylindrical surface $\rho = \rho'$; the current distribution has an azimuthal phase variation $\exp(im\,\phi)$,

m = integer. The electric field E_z generated by this source configuration is proportional to a scalar potential $u(\rho, \phi; \rho')$ defined by

$$\left(\frac{1}{\rho}\frac{\partial}{\partial\rho}\rho\frac{\partial}{\partial\rho} + \frac{1}{\rho^2}\frac{1}{\partial\phi^2} + k^2\right)u(\rho, \phi; \rho') = -\frac{\delta(\rho - \rho')}{\rho'}e^{im\phi}, \tag{8}$$

subject to a radiation condition at infinity. A time-dependence $\exp(-i\omega t)$ is implied.

Show that u is given by

$$u(\rho, \phi; \rho') = \frac{\pi i}{2}J_m(k\rho_<)H_m^{(1)}(k\rho_>)e^{im\phi}. \tag{9}$$

Assuming $k\rho' \gg 1$ and $(m/k\rho') < 1$, use the Debye asymptotic formulas in Eqs. (5.4.77) to derive an asymptotic approximation for u. Show that this approximation agrees with the ray-optical result in Problem 29 of Chapter 1.

7. Use the asymptotic expansion of the time-harmonic, two-dimensional free space Green's function,

$$\bar{G}_f(\hat{\boldsymbol{\rho}}) = \frac{i}{4}H_0^{(1)}(k\hat{\rho}) = \frac{e^{i[k\hat{\rho} + (\pi/4)]}}{2\sqrt{2\pi k\hat{\rho}}}\sum_{m=0}^{\infty}\frac{(0, m)}{(2ik\hat{\rho})^m}, \qquad k = \frac{\omega}{c} \tag{10}$$

where

$$(0, m) \equiv \frac{(1)(3^2)(5^2)\ldots(2m - 1)^2}{2^{2m}m!}, \qquad (0, 0) \equiv 1, \tag{10a}$$

to construct via Eqs. (1.7.80) and (1.7.81) the behavior of the time-dependent Green's function $\hat{\bar{G}}_f$ near the wavefront. Verify the validity of this result by expanding the exact solution in Eq. (5.4.42), $\hat{\bar{G}}_f = \{4\pi^2[t^2 - (\hat{\rho}/\bar{c})^2]\}^{-1/2}U(\bar{c}t - \hat{\rho})$, in a series about $\bar{c}t = \hat{\rho}$.

8. An x-directed electric line current with impulsive behavior $\hat{\mathbf{J}}(\mathbf{r}, t) = I\delta(t)\cdot\delta(\hat{\boldsymbol{\rho}} - \hat{\boldsymbol{\rho}}')\mathbf{x}_0$ is situated at $\hat{\boldsymbol{\rho}}' = (0, z')$, $z' < 0$, in the presence of a non-dispersive lossless dielectric half-space (see Fig. 5.5.7). The electromagnetic fields can be derived from the two-dimensional, time-dependent Green's functions $\hat{\bar{G}}_{1,2}$ defined in Eqs. (5.5.65). When $\epsilon_2 > \epsilon_1$, where ϵ_1 and ϵ_2 are the dielectric constants for the regions $z < 0$ and $z > 0$, respectively, the result for $\hat{\bar{G}}_1$ in $z < 0$ is given in Eqs. (5.5.66)-(5.5.67).

When $\epsilon_1 > \epsilon_2$, show the result in Eqs. (5.5.66)-(5.5.67) remains valid for $\varphi < \sin^{-1}\sqrt{\epsilon}$, $\epsilon = \epsilon_2/\epsilon_1$, where φ is the observation angle measured from the image point as in Fig. 5.5.7. Show that for $\varphi > \sin^{-1}\sqrt{\epsilon}$ in the half-space $z < 0$, one must add to this result a contribution $\hat{\bar{G}}_b$ whose time-harmonic form [for an $\exp(-i\omega t)$ dependence] is as follows:

$$\bar{G}_b = \frac{i}{4\pi}\int_{P_1}e^{-s(R/\bar{c}_1)\cos(w - \varphi)}\Gamma(k_1\sin w)\,dw, \tag{11}$$

where R is the distance from the image point, $\bar{c}_1$ is the propagation speed in the region $z < 0$, and

$$\Gamma(k_1\sin w) = \frac{\cos w - \sqrt{\epsilon - \sin^2 w}}{\cos w + \sqrt{\epsilon - \sin^2 w}}, \tag{11a}$$

and the path P_1 runs from $w = \varphi + i0$ to $w = \varphi - i0$ around the branch point at $w_b = \sin^{-1}\sqrt{\epsilon}$, with the branch cut chosen along the contour $\mathrm{Re}\sqrt{\epsilon - \sin^2 w} = 0$. Use the procedure in Eqs. (5.2.19)-(5.2.23) to show that the time-dependent result is as follows:

$$\hat{\hat{G}}_b = -Im\left(\frac{\cos\beta + i\sqrt{\sin^2\beta - \epsilon}}{\cos\beta - i\sqrt{\sin^2\beta - \epsilon}}\right)$$

$$\times \frac{U(\varphi - \sin^{-1}\sqrt{\epsilon})U\{t - [(R/\bar{c}_1)\cos(w_b - \varphi)]\}U[(R/\bar{c}_1) - t]}{2\pi\sqrt{(R/\bar{c}_1)^2 - t^2}}, \quad (12)$$

where $\beta = \varphi + \cos^{-1}(\bar{c}_1 t/R)$ and $U(\alpha) = 1$ or 0 for $\alpha > 0$ or $\alpha < 0$, respectively. Show that Eq. (12) defines a lateral wave whose relation to the direct and reflected waves is as shown in Fig. 5.5.8 when $z' = 0$. Sketch the incident, reflected, and lateral wavefronts when $z' \neq 0$ and indicate on this sketch the domain of existence of the lateral wave contribution.

9. Assume the semiinfinite dielectric medium in Fig. 5.5.7 to be excited by a source distribution that gives rise only to H modes with respect to the z direction and is confined to the interface (i.e., $z' = 0$).

(a) Show that for observation point locations on the interface ($z = 0$), the voltage in a typical mode is given by:

$$V = -\frac{2}{k_1^2 - k_2^2}\left[\frac{d^2}{dz^2}V_{f1} - \frac{d^2}{dz^2}V_{f2}\right]_{z=z'=0}, \quad (13)$$

where $k_\alpha^2 = \omega^2\mu_0\epsilon_\alpha$, $\alpha = 1, 2$ and $V_{f\alpha}$ is the voltage on an *infinite* transmission line characteristic of medium α.

(b) Since only H modes are excited, the electromagnetic fields can be inferred from an appropriate scalar Green's function, to be called $G(\mathbf{r}, \mathbf{r}')$. Show via modal synthesis that for $z = z' = 0$, G can be expressed in terms of the elementary Green's functions $G_{f\alpha}$ for an *infinite* medium having a dielectric constant ϵ_α:

$$G(\mathbf{r}, \mathbf{r}')\Big|_{z=z'=0} = -\frac{2}{k_1^2 - k_2^2}\left[\frac{\partial^2}{\partial z^2}G_{f1}(\mathbf{r}, \mathbf{r}') - \frac{\partial^2}{\partial z^2}G_{f2}(\mathbf{r}, \mathbf{r}')\right]_{z=z'=0}. \quad (14)$$

(c) For the case of an electric line source as in Fig. 5.5.7, show that Eq. (14) can be written for an $\exp(-i\omega t)$ time dependence as:

$$G\Big|_{z=z'=0} = \frac{i}{2(k_1^2 - k_2^2)}\left[\left(\frac{d^2}{dy^2} + k_1^2\right)H_0^{(1)}(k_1 y) - \left(\frac{d^2}{dy^2} + k_2^2\right)H_0^{(1)}(k_2 y)\right]$$

$$= \frac{i}{2(h_1^2 - h_2^2)}[h_1 H_1^{(1)}(h_1) - h_2 H_1^{(1)}(h_2)], \quad h_\alpha = k_\alpha y, \quad y > 0. \quad (15)$$

For $y < 0$, replace y by $|y|$. It is noted that the far field ($k_\alpha y \gg 1$) varies like $y^{-3/2}$; compare this with the results in Eq. (5.5.58).

(d) For the case of a longitudinal magnetic current element located at $\mathbf{r}' = 0$, show that Eq. (14) can be written for an $\exp(-i\omega t)$ time dependence as:

$$G\Big|_{z=z'=0} = -\frac{1}{2\pi(g_1^2 - g_2^2)}\left[(ig_1 - 1)\frac{e^{ig_1}}{\rho} - (ig_2 - 1)\frac{e^{ig_2}}{\rho}\right], \quad g_\alpha = k_\alpha\rho, \quad (16)$$

where ρ is the radial distance from the source to the observation point. Verify that $G \to 1/4\pi\rho$ as $\rho \to 0$.

10. A source distribution with assumed time-dependence $\exp(j\omega t)$ is located in the presence of a highly lossy dielectric half-space as shown in Fig. P5.1a. Referring

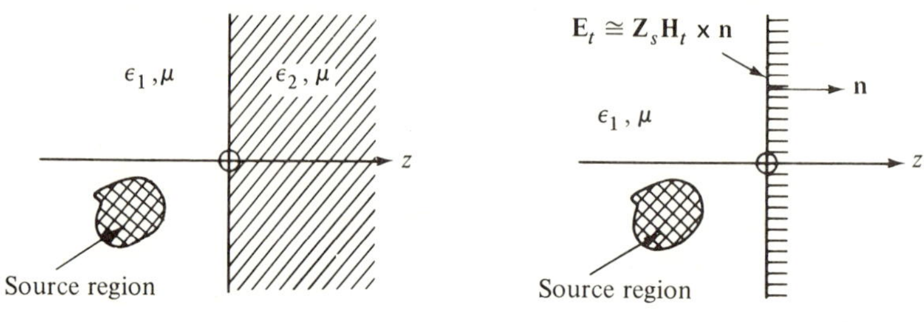

(a) Actual configuration
$|\epsilon_2|\gg\epsilon_1$

(b) Approximate equivalent

FIG. P5.1   Approximately equivalent configurations.

to the equivalent modal network problems in Fig. 5.5.3 or 5.5.6 and assuming that $\epsilon_2 = \epsilon_{2r} - j\sigma/\omega \approx -j\sigma/\omega$, where ϵ_{2r} is the real part of the dielectric constant in the lossy medium and $\sigma \gg \omega\epsilon_{2r}$ is its conductivity, show that for modes with $k_{ti}^2 \ll |k_1^2\epsilon|$, the modal impedance Z_{2i} for the ith E or H mode is given approximately by:

$$Z_{2i} \approx \sqrt{\frac{\mu}{\epsilon_2}} = \frac{\omega\mu e^{j\pi/4}}{\sqrt{2}}\delta, \qquad \delta = \sqrt{\frac{2}{\omega\mu\sigma}}, \tag{17}$$

where k_{ti} is the transverse wavenumber for the ith mode, $k_{1,2}^2 = \omega^2\mu\epsilon_{1,2}$, $\epsilon = \epsilon_2/\epsilon_1$, and δ is the skin depth. When the source region is not too near the interface (give a criterion!), show that the approximation in Eq. (17) can be used for *all* modes whose contribution to the reflected field is not negligible, and that the equivalent network problem reduces to that in Fig. 5.7.2 or its H-mode analogue, with $Z_s = \sqrt{\mu/\epsilon_2}$ denoting the surface impedance [see also Eq. (5.7.1) and Fig P5.1b]. By applying the concept of a locally plane boundary, show that the surface impedance approximation remains valid also for a curved interface provided that the skin depth δ and the wavelength inside the lossy medium are much smaller than (1) the wavelength $2\pi/k_1$ in the exterior medium, (2) the smallest radius of curvature of the boundary surface, and (3) the distance from the source region to the observation point.† Show that when $\epsilon_2 \gg \epsilon_1$, with ϵ_2 real (lossless medium), the surface impedance approximation remains valid for the plane interface but not, in general, for a curved interface if the latter gives rise to multiple internal reflections.

11. A source distribution is located exterior to the grounded dielectric slab shown in Fig. 5.6.1. If the slab width d is so small that $|k_2 d| \ll 1$, with k_2 either real or complex, and if the source configuration excites only H modes with respect to the z direction (perpendicular to the slab), show that under restrictions analogous to those in Problem 10, the grounded slab may be replaced by the surface impedance

---

†M. A. Leontovich, *Investigations of Propagation of Radio Waves*, Part II, Moscow (1948). Also T. B. A. Senior, "Impedance boundary conditions for imperfectly conducting surfaces," *Appl. Sci. Res.* **8**, Sec. B.

$$Z_s = j\sqrt{\frac{\mu}{\epsilon_1}}\,k_1 d \qquad \text{(for } H \text{ modes)}. \tag{18}$$

Show also that no such replacement is possible for the E-mode case.

12. For the grounded dielectric slab in Fig. 5.6.1 calculate the far fields when the source and/or the observation point is located inside the slab region. Obtain and interpret a radial transmission representation analogous to that discussed in Sec. 5.6a.

13. The ray-optical approximation of the field radiated by a time-harmonic line source of electric currents in a plane stratified medium with refractive index $n = n(z)$ is given in Eq. (5.8.27). This result is applicable for observation points on rays emanating from the source point, before such rays reach a turning point z_t (if any) where $n(z_t) = a$, a being the ray parameter. Use the refracted ray equation (5.8.28) to derive a corresponding expression for the field valid on ray segments after turning. Show that the result agrees with the asymptotic evaluation of the exact solution in Eq. (5.8.55).

REFERENCES

1. STRATTON, J. A., *Electromagnetic Theory*, Secs. 6.9-6.11. New York: McGraw-Hill, 1941.

2. MORSE, P. M. and H. FESHBACH, *Methods of Theoretical Physics*, Sec. 4.8. New York: McGraw-Hill, 1953.

3. KNOPP, K., *Theory of Functions*, Part II, Sec. II. New York: Dover Publications, 1947.

4. MAGNUS, W. and F. OBERHETTINGER, *Formulas and Theorems for the Special Functions of Mathematical Physics*, p. 22. New York: Chelsea Publishing Co., 1954.

5. TAMIR, T. and A. A. OLINER, "Guided complex waves. I. Fields at an interface. II. Relation to radiation patterns," *Proc. IEE (London)* **110** (1963), pp. 310-334.

6. CLEMMOW, P. C., "The resolution of a dipole field into transverse electric and transverse magnetic waves," *Proc. IEE (London)* **110** (1963), p. 107.

7. SOMMERFELD, A. N., *Partial Differential Equations in Physics*, Sec. 19. New York: Academic Press, 1949.

8. FRANK, I. M. and I. G. TAMM, *Dokl. Akad. Nauk. (USSR)* **14** (1937), p. 109.

9. PANOFSKY, W. K. H. and M. PHILLIPS, *Classical Electricity and Magnetism*, Chapter 19. Reading, Mass.: Addison-Wesley Publishing Co., 1962.

10. SOMMERFELD, A. N., *Partial Differential Equations in Physics*, Chapter 6; in particular, pp. 256-257. New York: Academic Press, 1949.

11. BREKHOVSKIKH, L. M. *Waves in Layered Media*, Secs. 21-23. New York: Academic Press, 1960.

12. BAÑOS, A., *Dipole Radiation in the Presence of a Conducting Half Space*, Sec. 4.10. New York: Pergamon Press, 1966.

13. STRATTON, J. A., *Electromagnetic Theory*, Secs. 9.5-9.6. New York: McGraw-Hill, 1941.

14. WAIT, J. R., *Electromagnetic Waves in Stratified Media*, Chapter 2. New York: Macmillan, 1962.

15. CLEMMOW, P. C., *The Plane Wave Spectrum Representation of Electromagnetic Fields*, Chapter 5. New York: Pergamon Press, 1966.

16. FOCK, V. A., *Electromagnetic Diffraction and Propagation Problems*, Sec. 11.1. New York: Pergamon Press, 1965.

17. SOMMERFELD, A. N., *Partial Differential Equations in Physics*, Sec. 33. New York: Academic Press, 1949.

18. BANOS, A., *Dipole Radiation in the Presence of a Conducting Half Space*, Chapter 1. New York: Pergamon Press, 1966.

19. deHOOP, A. T. and H. J. FRANKENA, "Radiation of pulses generated by a vertical electric dipole above a plane, non-conducting earth," *Appl. Sci. Res.*, Sec. B 8 (1960), p. 369.

20. JONES, D. S., *The Theory of Electromagnetism*, Sec. 10.1. New York: Macmillan, 1964.

21. SITENKO, A. G. and V. S. TALICH, "Cerenkov effect in the motion of a charge above a boundary between two media," *J. Tech. Phys.* (*Russian*) **29** (1959), pp. 1074-1085.

22. COLLIN, R. E., *Field Theory of Guided Waves*, Chapter 11. New York: McGraw-Hill, 1960.

23. SOMMERFELD, A. N., *Partial Differential Equations in Physics*, p. 250. New York: Academic Press, 1949.

24. CULLEN, A. L., "The excitation of plane surface waves," *Proc. IEE* (*London*) **101**, Part IV, 1954; "A note on the excitation of surface waves," ibid., **104**, Part C, 1957.

25. BARLOW, H. E. M. and J. BROWN, *Radio Surface Waves*, Chapter 12. London: Oxford University Press, 1962.

26. KAY, A. F., "Scattering of a surface wave by a discontinuity in reactance," *IRE Trans. on Antennas and Propagation* **AP-7** (Jan. 1959).

27. SECKLER, B. D. and J. B. KELLER, "Geometrical theory of diffraction in inhomogeneous media," *J. Acoust. Soc.* **31**(1959), pp. 192-205. Also B. D. SECKLER and J. B. KELLER, "Asymptotic theory of diffraction in inhomogeneous media," *J. Acoust. Soc.* **31** (1959), pp. 206-216.

28. BREKHOVSKIKH, L. M., *Waves in Layered Media.*, Sec. 38. New York: Academic Press, 1960.

29. LUDWIG, D., "Uniform asymptotic expansions at a caustic," *Comm. Pure and Appl. Math.* **19** (1966), pp. 215-260.

30. KRAVTSOV, Yu. A., "A modification of the geometrical optics method," *Radiofizika* **7** (Russian) (1964), pp. 664–673.

31. BREKHOVSKIKH, L. M., *Waves in Layered Media.*, Sec. 27. New York: Academic Press, 1960.

32. BUDDEN, K. G., *The Waveguide Mode Theory of Wave Propagation*, Chapter 8. Englewood Cliffs: Prentice-Hall, 1961.

33. WATSON, G. N., *A Treatise on the Theory of Bessel Functions*, Chapter 15. Cambridge, England: Cambridge University Press, 1944.

34. COURANT, R., *Differential and Integral Calculus*, Vol. II, Sec. 3.5. New York: Interscience, 1947.

35. BREMMER, H., *Terrestrial Radio Waves*, Chapters 8 and 9. New York: Elsevier Publishing Co., 1949. See also *Handbuch der Physik* **16**, p. 350.

36. FELSEN, L. B. and L. LEVEY, "A relation between a class of boundary value problems in a homogeneous and an inhomogeneous region," *IEEE Trans. on Antennas and Propagation* **AP-14** (1966), pp. 308–317.

37. EPSTEIN, P. S., "Reflection of waves in an inhomogeneous absorbing medium," *Proc. Natl. Acad. Sci.* (USA) **16** (1930), p. 627.

38. ECKART, C., *Phys. Rev.* **35** (1930), p. 1303.

39. RAWER, K., *Ann. d.* Physik **35** (1939), p. 385.

40. ECKART, G., "Wellenoptische Behandlung der Strahlung eines magnetischen Dipols in einem eben geschichteten Medium mittels der Methode von Epstein," *Bayerische Akad. d. Wissenschaften, Mathematisch-Naturwissenschaftliche Klasse*, Munich, Germany, December 1960.

6. Fields in Cylindrical and Spherical Regions

6.1 DISTINCTIVE FIELD CHARACTERISTICS

In contrast to the unbounded cross sections in Chapter 5, this chapter is concerned with partially bounded waveguide cross sections. In particular, we investigate in detail the effect of waveguide walls whose location is describable simply in terms of a circular-cylindrical or spherical coordinate system; included therein are the important configurations of the circular cylinder, the wedge, the sphere and the cone. Fields in cylindrical configurations are analyzed in detail (Secs. 6.1–6.7), but only a brief summary of the analogous spherical configurations is given in Sec. 6.8. The generic waveguide boundary in cylindrical coordinates is the tipped wedge configuration in Fig. 6.1.1, comprising a cylinder

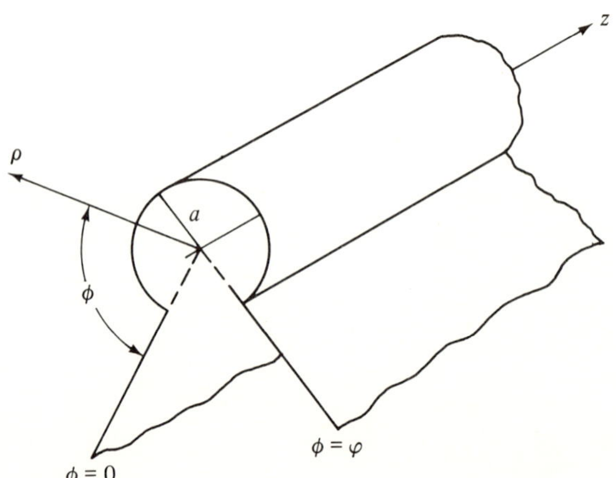

FIG. 6.1.1 Tipped wedge configuration.

with radius a superposed upon a wedge formed by two intersecting half-planes; the axis of the cylinder coincides with the wedge apex (z axis), and the sides of the wedge are defined by the angular locations $\phi = 0$ and $\phi = \varphi$ in a cylindrical coordinate system. We shall be concerned with the exterior region $0 \le \phi \le \varphi$, $a \le \rho < \infty$. When $a = 0$, one recovers the wedge, and for the special case $a = 0$, $\varphi = 2\pi$, one obtains the configuration of a half-plane. The case of a cylinder with radius a cannot be obtained directly from the structure in Fig 6.1.1 since even for $\varphi = 2\pi$, the two wedge faces coalesce into a single septum. However, the difficulty may be easily circumvented by a modification of the angular boundary conditions to furnish the required 2π periodicity of the fields. In spherical coordinates, analogous considerations apply to the generic region depicted in Fig. 6.8.1.

Formal solutions for radiation problems in the region of Fig. 6.1.1 can be written down from the formulas in Sec. 5.2, provided only that one inserts the mode functions or characteristic Green's functions appropriate to the new waveguide cross section (see Sec. 2.2). The real task, as in Chapter 5, is the reduction of these formal representations to obtain the approximate evaluation of radiated fields. The presence of waveguide boundaries gives rise to effects not encountered in the study of totally unbounded cross sections. For example, because of the presence of the waveguide walls, there may exist one or more geometrical regions from which the source configuration is not visible. In the terminology of geometrical optics, these are the "shadow" or "dark" regions; at high frequencies (i.e., in the quasi-optic range) the fields in these dark regions are weak compared to those in the "illuminated" regions, from which the source configuration can be seen directly. The evaluation and ray-optical interpretation of the time-harmonic high-frequency fields in the illuminated and shadow regions, and in the transition regions surrounding the light-shadow boundaries (particularly for the wedge), is of major concern in this chapter. Field solutions in the spherical region depicted in Fig. 6.8.1 can be constructed from the modal representations in Sec. 2.6, with the eigenfunctions taken from the appropriate sections of Chapter 3. The presence of these boundaries gives rise to quasi-optic phenomena analogous to those described for cylindrical regions.

Examination of Fig. 6.1.1 shows that when a source is located at the point $\mathbf{r}' = (\rho', \phi', z')$, the regions of illumination and shadow are separated by a plane boundary which is tangent to the obstacle and passing through $\mathbf{r}'$; see Figs. 6.3.1 and 6.7.1 for the special cases of the wedge and cylinder and note that the latter configuration admits two boundary planes. Evidently, the description of shadow boundary planes does not involve the axial (z) coordinate. It may be anticipated that similar quasi-optic phenomena are associated with a point source at $\mathbf{r}'$, a line source at $\boldsymbol{\rho}' = (\rho', \phi')$, and a plane wave whose propagation vector lies in the boundary plane. This aspect is exploited in the analysis where field solutions for the plane wave or the point source are derived from two-dimensional line-source results. It is also suggestive, in view of the dominant role played by the angular coordinates in specifying the shadow boundary, that field penetration by diffraction into the shadow region is described best in terms of angularly propagating waves. The validity of this conjecture is confirmed by the analysis in Secs. 6.3–6.7, which shows that an angular transmission representation for the fields yields a solution wherein the quasi-optic (geometric-optical and diffracted) constituents appear explicitly and can be approximated readily in the limit of high frequencies. Alternatively, no well-defined illuminated and shadow regions may be distinguished in the low-frequency regime, and for approximate evalution of the field in this parameter range, a radial transmission representation is found to be the more rapidly convergent. Although the z-transmission formulation is not especially useful for field calculations in the presence of an obstacle as in Fig. 6.1.1, it forms a convenient starting point for deriving alternative representations for the scalar functions $\mathscr{S}'(\mathbf{r}, \mathbf{r}')$ and $\mathscr{S}''(\mathbf{r}, \mathbf{r}')$, and thence for the general vector electromagnetic field. These general considerations are pursued in Sec. 6.2 for the cylindrical geometry and in Sec. 6.8b for the spherical geometry.

In the high-frequency range, the fields in the illuminated and shadow regions may be described almost everywhere in terms of rays that propagate locally like plane waves and are capable of accounting for the effects of diffraction as well as those of geometrical optics. This description fails in the vicinity of boundaries that delimit the domain of existence of a particular ray species. Examples are provided by the light-shadow boundary which confines the incident ray species, and by an analogous boundary for the geometrically reflected rays. In these transition regions, the field changes so rapidly as to invalidate a local plane-wave characterization which, in the equiphase plane, requires slow variation over a wavelength interval. Although these regions are of narrow angular extent (see Figs. 6.4.4 and 6.7.1 for cylindrical regions), they play an important role since they must provide a smooth transformation from one ray-optical field type to another. Transition functions for the wedge configuration are studied in detail in the following sections. Because of the observations made previously concerning the similarity of shadow formation for point-source, line-source, or plane-wave excitation, the transition functions exhibit a corresponding behavior.

The preceding ray and wavefront considerations apply also to excitation by impulsive sources since there exists an intimate connection between the time-harmonic high-frequency field and the transient field near the time of arrival of the various wavefronts (see Sec. 1.6c). For various wedge configurations, it is found, however, that the transient response at *all* observation times can be evaluated in closed form, thereby providing an insight into the field behavior at observation times immediately and long after the passage of a wavefront. Special attention is given to the perfectly conducting half-plane, which constitutes one of the classical diffraction problems both for impulsive and time-harmonic excitation.

Since the angular transmission approach is fundamental for the quasi-optic formulation of the fields, it is natural to explore initially those configurations analyzed most simply in terms of angularly propagating waves. As in regions describable in terms of propagation along a rectilinear coordinate, bilaterally matched (reflectionless) structures fall into this category. In angular coordinates, this condition is satisfied by "perfectly absorbing" boundaries at $\phi = 0, \varphi$ and $\theta = \theta_{1,2}$ in cylindrical and spherical regions, respectively (see Fig. 3.4.11). The matched boundary condition is equivalent to regarding the angular space as being infinitely extended so that a traveling wave experiences no reflection. This concept of an infinite angular space, introduced in Sec. 3.4b, is exploited in the analysis. From the resulting solutions for the perfectly absorbing case, one may synthesize by an image construction effects either of reflecting boundaries at $\phi = 0, \varphi$ or $\theta = \theta_{1,2}$, or of periodicity when no physical boundaries are present in the angular domain (see Sec. 3.4b). The image procedure grants insight into the different phenomena encountered in regions with rectilinear and curved transmission coordinates. In the former, all images are visible from the source point and therefore contribute to the geometric-optical field, whereas in the latter, the more distant images are obscured by space curvature and contribute only to diffraction.

6.2 GREEN'S FUNCTION REPRESENTATIONS IN CYLINDRICAL REGIONS

6.2a Derivation of the Field from Scalar Potentials

The electromagnetic fields radiated by an arbitrarily oriented time-harmonic electric or magnetic current element in the presence of the perfectly conducting configuration in Fig. 6.1.1 can be derived as in Sec. 5.2a from the vector Hertz potentials $\mathbf{z}_0\Pi'(\mathbf{r}, \mathbf{r}')$ and $\mathbf{z}_0\Pi''(\mathbf{r}, \mathbf{r}')$ expressive, respectively, of the E-mode and H-mode contributions with respect to the axial coordinate z. When the source direction is parallel to z, the Hertz potentials are proportional to the scalar Green's functions $G'(\mathbf{r}, \mathbf{r}')$ or $G''(\mathbf{r}, \mathbf{r}')$ [Eqs. (5.2.4c)], whereas the response to sources directed transverse to z requires a knowledge of the potential functions $\mathscr{S}'(\mathbf{r}, \mathbf{r}')$ and $\mathscr{S}''(\mathbf{r}, \mathbf{r}')$ [Eqs. (5.2.4a) and (5.2.4b)]. The functions $\mathscr{S}'$, $\mathscr{S}''$ and G', G'' satisfy the differential equations (5.2.2) and (5.2.3), respectively; in a z-transmission representation, their solutions are given by Eqs. (5.2.5), with the scalar-mode functions $\Phi_i(\boldsymbol{\rho})$ and $\psi_i(\boldsymbol{\rho})$ chosen to satisfy the required boundary conditions on the waveguide boundary sketched in Fig. 6.1.1. When the source behavior is impulsive in time, these considerations are modified as in Sec. 5.2c (see also Sec. 1.6).

While the separability of the vector electric and magnetic fields into E and H modes requires the choice of the preferred axial coordinate z, the scalar Green's functions G', G'' and the potential functions $\mathscr{S}'$, $\mathscr{S}''$ may be represented in any convenient alternative form corresponding to transmission along one of the transverse coordinates. As mentioned in Sec. 6.1, the class of problems involving a boundary of the type shown in Fig. 6.1.1 is analyzed most conveniently in terms of angular transmission. This applies to the three-dimensional fields resulting from point-source excitation as well as to the two-dimensional fields excited by a line source extending parallel to the z axis; since the physical configuration has z-invariant properties (see Fig. 6.1.1), an axial line source of constant strength will excite z-independent fields. The two-dimensional z-independent Green's function $\bar{G}(\mathbf{\rho}, \mathbf{\rho}')$ may be synthesized from the three-dimensional function $G(\mathbf{r}, \mathbf{r}')$ by integrating over z' between $-\infty$ and $+\infty$. When this operation is performed on Eqs. (5.2.3), the resulting function satisfies the two-dimensional wave equation (3) below, so $\bar{G}(\mathbf{\rho}, \mathbf{\rho}')$ may also be derived directly from this equation.

It may be noted from Eqs. (5.2.1) and (5.2.4c) (see also Sec. 5.4c) that a line source of electric current

$$\hat{\mathbf{J}}(\mathbf{r}, t) = \mathbf{z}_0 I \delta(\mathbf{\rho} - \mathbf{\rho}')e^{-i\omega t}, \qquad \mathbf{\rho} = (\rho, \phi), \tag{1a}$$

generates an electromagnetic field whose components are given by

$$E_z = \frac{I}{i\omega\epsilon} [\nabla_t^2 \bar{G}' + \delta(\mathbf{\rho} - \mathbf{\rho}')] = i\omega\mu I \bar{G}', \quad H_\rho = I\frac{\partial \bar{G}'}{\rho\,\partial\phi}, \quad H_\phi = -I\frac{\partial \bar{G}'}{\partial\rho}, \tag{1b}$$

$$E_\rho = E_\phi = H_z = 0,$$

whereas a line source of magnetic current,

$$\hat{\mathbf{M}}(\mathbf{r}, t) = \mathbf{z}_0 V \delta(\mathbf{\rho} - \mathbf{\rho}')e^{-i\omega t} \tag{2a}$$

generates an electromagnetic field whose components are given by

$$H_z = i\omega\epsilon V \bar{G}'', \quad E_\rho = -V\frac{\partial \bar{G}''}{\rho\,\partial\phi}, \quad E_\phi = V\frac{\partial \bar{G}''}{\partial\rho}, \quad H_\rho = H_\phi = E_z = 0. \tag{2b}$$

In these equations, $\bar{G}'$ and $\bar{G}''$ still denote E- and H-mode solutions with respect to the z axis which satisfy the equations

$$(\nabla_t^2 + k^2)\frac{\bar{G}'(\mathbf{\rho}, \mathbf{\rho}')}{\bar{G}''(\mathbf{\rho}, \mathbf{\rho}')} = -\delta(\mathbf{\rho} - \mathbf{\rho}') = -\frac{\delta(\rho - \rho')}{\rho'}\delta(\phi - \phi'),$$

$$\nabla_t^2 = \frac{1}{\rho}\frac{\partial}{\partial\rho}\rho\frac{\partial}{\partial\rho} + \frac{1}{\rho^2}\frac{\partial^2}{\partial\phi^2}, \tag{3}$$

subject to the following conditions on the perfectly conducting boundary s in Fig. 6.1.1:

$$\bar{G}' = 0, \quad \frac{\partial \bar{G}''}{\partial n} = 0 \quad \text{on } s, \tag{3a}$$

where $k = \omega\sqrt{\mu\epsilon}$ is the (constant) wavenumber in the medium and n is in the direction normal to s. While the two-dimensional Green's function $\bar{G}(\mathbf{\rho}, \mathbf{\rho}')$ may be constructed from the three-dimensional Green's function $G(\mathbf{r}, \mathbf{r}')$ by integration over z' as noted above, it is important to recognize that the converse is also true. To demonstrate this fact, we note that from Eq. (5.4.52), the two-dimensional Green's function $\bar{G}(\mathbf{r}, \mathbf{\rho}')$ corresponding to a z-directed line source with linearly progressing phase $\exp(i\zeta z')$ is given by

$$\bar{G}(\mathbf{r}, \mathbf{\rho}') = \int_{-\infty}^{\infty} e^{i\zeta z'} G(\mathbf{r}, \mathbf{r}')\,dz' = e^{i\zeta z}\bar{G}_\zeta(\mathbf{\rho}, \mathbf{\rho}'), \tag{3b}$$

where $\bar{G}_\zeta(\boldsymbol{\rho},\boldsymbol{\rho}')$ satisfies Eq. (3) with k^2 replaced by $k^2 - \zeta^2$ [see also Eqs. (5.4.46)]. By Fourier inversion of Eq. (3b),

$$G(\mathbf{r},\mathbf{r}') = \frac{1}{2\pi}\int_{-\infty}^{\infty} e^{-i\zeta z'}\bar{G}(\mathbf{r},\boldsymbol{\rho}')\,d\zeta = \frac{1}{2\pi}\int_{-\infty}^{\infty} e^{i\zeta(z-z')}\bar{G}_\zeta(\boldsymbol{\rho},\boldsymbol{\rho}')\,d\zeta, \qquad (3c)$$

whence the three-dimensional Green's function $G(\mathbf{r},\mathbf{r}')$ can be recovered from the two-dimensional Green's function $\bar{G}(\boldsymbol{\rho},\boldsymbol{\rho}')$ on replacing k by $(k^2 - \zeta^2)^{1/2}$ and performing the operation $(1/2\pi)\int_{-\infty}^{\infty} d\zeta \exp[i\zeta(z-z')]$.

It is of interest to observe that no field components in addition to those in Eqs. (1b) or (2b) are required even for other z-invariant boundary shapes, or for penetrable boundaries. Regions interior to s filled with a z-independent but otherwise arbitrary dielectric material are also included herein provided that the boundary conditions are modified accordingly. The reader may verify that the components in Eqs. (1b) or (2b) suffice for enforcement of the continuity of tangential electric and magnetic fields.

6.2b   Angular Transmission Representation

In an angular transmission formulation, the scalar Green's functions G', G'' or $\bar{G}'$, $\bar{G}''$ are represented in terms of eigenfunctions in the (ρ, z) domain and one-dimensional Green's functions along the angular coordinate ϕ. In view of the translational invariance of the configuration in Fig. 6.1.1 with respect to z, the three-dimensional solutions G', G'' are related to the two-dimensional solutions $\bar{G}'$, $\bar{G}''$ by the transformation noted in Eq. (3c). Owing to greater simplicity, the analysis of various diffraction problems will be carried out primarily for two-dimensional cases, and three-dimensional results pertaining to point-source excitation will be deduced therefrom.

For two-dimensional (z-independent), time-harmonic source arrangements, the completeness relation for eigenfunctions that algebraize the radial operator

$$\frac{1}{\rho}\frac{\partial}{\partial\rho}\rho\frac{\partial}{\partial\rho} + k^2$$

in Eq. (3) is given in the generic form [see Eqs. (3.4.91) and (3.4.94e)]

$$\rho'\delta(\rho - \rho') = \sum_p \Phi_p(k\rho)\bar{\Phi}_p(k\rho'), \qquad (4)$$

where $\Phi_p(k\rho)$, the eigenfunction for the radial domain, and $\bar{\Phi}_p(k\rho)$, the adjoint eigenfunction, are listed in Sec. 3.4c for various boundary conditions. With z in the interval $-\infty < z < \infty$ for the three-dimensional time-harmonic case, the algebraization of the (ρ, z) dependent part of the $\nabla^2 + k^2$ operator,

$$\frac{1}{\rho}\frac{\partial}{\partial\rho}\rho\frac{\partial}{\partial\rho} + \frac{\partial^2}{\partial z^2} + k^2,$$

is achieved by the two-dimensional eigenfunctions characterized by the completeness relation

$$\rho'\delta(\rho - \rho')\delta(z - z') = \frac{1}{2\pi} \int_{-\infty}^{\infty} d\zeta \sum_{p} e^{i\zeta(z-z')}\Phi_p(\sqrt{k^2 - \zeta^2}\rho)\bar{\Phi}_p(\sqrt{k^2 - \zeta^2}\rho'),$$

$$(5)$$

which is obtained from the two-dimensional result in Eq. (4) by the rule stated in Eq. (3c): k in the former is replaced by $\sqrt{k^2 - \zeta^2}$ and the operation

$$\frac{1}{2\pi} \int_{-\infty}^{\infty} d\zeta \exp\left[i\zeta(z - z')\right]$$

is performed subsequently. For impulsive excitation, k^2 on the left-hand side of Eq. (3) or its three-dimensional counterpart is replaced by the temporal operator $-\partial^2/\bar{c}^2\partial t^2$, whose algebraization is achieved by eigenfunctions satisfying the completeness relation

$$\delta(t - t') = \frac{1}{2\pi} \int_{-\infty}^{\infty} e^{-i\omega(t-t')}\, d\omega, \qquad \omega = k\bar{c}. \tag{6}$$

The relevant representation theorem (completeness relation) for

$$\rho'\delta(\rho - \rho')\delta(t - t') \quad \text{and} \quad \rho'\delta(\rho - \rho')\delta(z - z')\delta(t - t')$$

is then obtained by applying the integral operator in Eq. (6) to the time-harmonic formulation on the right-hand sides of Eqs. (4) and (5), respectively (see also Sec. 5.2c).

In the angular transmission representation, the dependence on the ϕ coordinate is given in terms of the angular Green's function $g_{\phi p}(\phi, \phi')$, which satisfies the differential equation

$$\left(\frac{d^2}{d\phi^2} + p^2\right)g_{\phi p}(\phi, \phi') = -\delta(\phi - \phi'), \tag{7}$$

subject to appropriate boundary conditions at the endpoints of the ϕ domain. Various solutions are given in Sec. 3.4b, and their availability permits direct construction of formal solutions for scalar Green's functions by the method described in Sec. 3.3c.

Time-harmonic line source

The two-dimensional Green's functions $\bar{G}'$ and $\bar{G}''$ satisfy Eq. (3) and, via Eqs. (4) and (7), have the formal solution

$$\bar{G}(\boldsymbol{\rho}, \boldsymbol{\rho}') = \sum_{p} \Phi_p(k\rho)\bar{\Phi}_p(k\rho')g_{\phi p}(\phi, \phi'). \tag{8}$$

The appropriate form of the functions Φ_p and $g_{\phi p}$ for the E- or H-mode cases depends on the specific boundary shapes comprised under the wedge or cylinder configurations sketched in Fig. 6.1.1.

Time-harmonic point source

The three-dimensional Green's functions satisfy the differential equations†

---

†The separable cylindrical coordinate representation of the delta function follows from the requirement that $\int_V dV\delta(\mathbf{r} - \mathbf{r}') = 1$, $\mathbf{r}'$ in V, with the volume element represented as $dV = \rho\, d\rho\, d\phi\, dz$.

$$\left(\frac{1}{\rho}\frac{\partial}{\partial\rho}\rho\frac{\partial}{\partial\rho} + \frac{1}{\rho^2}\frac{\partial^2}{\partial\phi^2} + \frac{\partial^2}{\partial z^2} + k^2\right)\begin{matrix}G'(\mathbf{r},\mathbf{r}')\\ G''(\mathbf{r},\mathbf{r}')\end{matrix}$$

$$= -\frac{\delta(\rho-\rho')}{\rho'}\delta(\phi-\phi')\delta(z-z'), \tag{9}$$

subject to the boundary conditions [see Eq. (3a)]

$$G' = 0, \quad \frac{\partial G''}{\partial n} = 0 \text{ on } s. \tag{9a}$$

The formal solution is given by

$$G(\mathbf{r},\mathbf{r}') = \frac{1}{2\pi}\int_{-\infty}^{\infty} d\zeta \sum_p e^{i\zeta(z-z')}\Phi_p(\sqrt{k^2-\zeta^2}\rho)\bar{\Phi}_p(\sqrt{k^2-\zeta^2}\rho')g_{\phi p}(\phi,\phi'). \tag{10}$$

Impulsive line source

The two-dimensional time-dependent Green's functions $\hat{\bar{G}}'(\boldsymbol{\rho},\boldsymbol{\rho}';t,t')$ and $\hat{\bar{G}}''(\boldsymbol{\rho},\boldsymbol{\rho}';t,t')$ satisfy the wave equation

$$\left(\frac{1}{\rho}\frac{\partial}{\partial\rho}\rho\frac{\partial}{\partial\rho} + \frac{1}{\rho^2}\frac{\partial}{\partial\phi^2} - \frac{1}{\bar{c}^2}\frac{\partial^2}{\partial t^2}\right)\hat{\bar{G}} = -\frac{\delta(\rho-\rho')}{\rho'}\delta(\phi-\phi')\delta(t-t'), \tag{11}$$

subject to appropriate spatial boundary conditions and to quiesence for $t < t'$. The formal solution is given by

$$\hat{\bar{G}} = \frac{1}{2\pi}\int_{-\infty}^{\infty} d\omega\, e^{-i\omega(t-t')} \sum_p \Phi_p(k\rho)\bar{\Phi}_p(k\rho')g_{\phi p}(\phi,\phi'), \qquad \omega = k\bar{c}. \tag{12}$$

Impulsive point source

The three-dimensional time-dependent Green's functions $\hat{G}'(\mathbf{r},\mathbf{r}';t,t')$ and $\hat{G}''(\mathbf{r},\mathbf{r}';t,t')$ satisfy the wave equation

$$\left(\frac{1}{\rho}\frac{\partial}{\partial\rho}\rho\frac{\partial}{\partial\rho} + \frac{1}{\rho^2}\frac{\partial^2}{\partial\phi^2} + \frac{\partial^2}{\partial z^2} - \frac{1}{\bar{c}^2}\frac{\partial^2}{\partial t^2}\right)\hat{G}$$

$$= -\frac{\delta(\rho-\rho')}{\rho'}\delta(\phi-\phi')\delta(z-z')\delta(t-t'), \tag{13}$$

subject to appropriate spatial boundary conditions and to quiescence for $t < t'$. The formal solution is given by

$$\hat{G} = \frac{1}{(2\pi)^2}\int_{-\infty}^{\infty} d\omega\, e^{-i\omega(t-t')}$$

$$\times \int_{-\infty}^{\infty} d\zeta \sum_p e^{i\zeta(z-z')}\Phi_p(\sqrt{k^2-\zeta^2}\rho)\bar{\Phi}_p(\sqrt{k^2-\zeta^2}\rho')g_{\phi p}(\phi,\phi'). \tag{14}$$

Plane-wave incidence

By letting ρ' and z' tend to infinity in the preceding results and employing the normalizations in Eqs. (5.4.6c) or (5.4.30b) (with $\hat{\rho}' \to \rho'$), one may derive the wavefunctions corresponding to plane-wave incidence. These wavefunctions satisfy in the finite domain the homogeneous wave equations obtained by equating to zero the right-hand sides in Eqs. (3), (9), (11), and (13).

The rule for deducing three-dimensional from two-dimensional solutions, stated in connection with Eq. (5), may also be employed in the present case. Let $\bar{u}(\boldsymbol{\rho}, \phi'; k)$ denote the two-dimensional wave function derived from $\bar{G}(\boldsymbol{\rho}, \boldsymbol{\rho}'; k)$ in Eq. (3) by letting $\rho' \to \infty$ along the angular direction ϕ'. For reasons to become evident, the dependence on k has been shown explicitly. The function $\bar{u}$ satisfies the source-free equation (3) subject to appropriate boundary conditions and corresponds to an incident plane wave

$$\bar{u}_{\text{inc}}(\boldsymbol{\rho}, \phi'; k) = \exp\left[-ik\rho \cos(\phi - \phi')\right]$$

propagating perpendicular to the z axis. The three-dimensional wavefunction $u(\mathbf{r}; \theta', \phi'; k)$ corresponding to a plane wave

$$u_{\text{inc}}(\mathbf{r}; \theta', \phi'; k) = \exp\left[-ik\rho \sin\theta' \cos(\phi - \phi') - ikz \cos\theta'\right], \tag{15}$$

incident obliquely along the direction θ', ϕ', where $\theta' = \tan^{-1}(\rho'/z')$, satisfies the wave equation

$$\left(\frac{1}{\rho}\frac{\partial}{\partial\rho}\rho\frac{\partial}{\partial\rho} + \frac{1}{\rho^2}\frac{\partial^2}{\partial\phi^2} + \frac{\partial^2}{\partial z^2} + k^2\right)u(\mathbf{r}; \theta', \phi'; k) = 0, \tag{16}$$

subject to appropriate boundary conditions. Evidently,

$$u_{\text{inc}}(\mathbf{r}; \theta', \phi'; k) = \bar{u}_{\text{inc}}(\boldsymbol{\rho}, \phi'; k \sin\theta')e^{-ikz \cos\theta'}, \tag{17}$$

and since the obstacle configuration in Fig. 6.1.1 is invariant with respect to z, the wavefunction u has the same z dependence. Thus, the z-independent part of u satisfies the two-dimensional wave equation with k replaced by $k \sin\theta'$, so

$$u(\mathbf{r}; \theta', \phi'; k) = \bar{u}(\boldsymbol{\rho}, \phi'; k \sin\theta')e^{-ikz \cos\theta'}. \tag{18}$$

In connection with the remarks following Eq. (5), ζ takes on the fixed value $k \cos\theta'$, thereby making the ζ integration unnecessary, and $\sqrt{k^2 - \zeta^2} \to k \sin\theta'$.

The preceding considerations cannot be applied directly to the scalar functions $\mathscr{S}'$ and $\mathscr{S}''$ since the transverse part of the differential operator $(\nabla_t^2 + \partial^2/\partial z^2 + k^2)\nabla_t^2$ in Eq. (5.2.2) is algebraized readily in a z-transmission, but not in a ϕ-transmission, representation [see Eqs. (5.2.5) and note that $\nabla_t^2 = -k_t^2$]. It is then best to proceed via the contour-integral representation deduced from Eqs. (5.2.5) (Sec. 3.3c) and to derive the angular transmission formulation from the z transmission formulation by deformation of contours in the complex k_t plane. The results resemble those obtained for G' and G'' except for the presence of a factor corresponding to $1/k_{ti}^2$ in Eqs. (5.2.5a) and (5.2.5b), and also for a possible residue contribution arising from the additional pole at $k_t = 0$ in the complex k_t plane.

Before proceeding to the application of these results, it is well to recall that the eigenfunction expansions in Eqs. (8), (10), (12), and (14) are to be considered as formal and that they may be employed only for a class of "representable" functions (i.e., functions for which the representations are convergent). In view of the somewhat anomalous behavior of the radial eigenfunctions $\Phi_p(k\rho)$ [see remarks following Eqs. (6.3.1)], the expansions are found to converge only for restricted locations of source and observation points. A modified procedure [see Eq. (6.3.8)] must be employed to make arbitrary locations accessible.

6.3 WEDGE-TYPE PROBLEMS—INTEGRATION TECHNIQUES

The first class of problems to be discussed involves geometries that result from Fig. 6.1.1 when the cylinder radius $a \to 0$. The boundary conditions in the radial domain $0 < \rho < \infty$ require a radiation condition at $\rho \to \infty$, and boundedness at $\rho = 0$ to satisfy the "edge condition," which delimits the permissible growth of the fields near the wedge apex (see Sec. 1.5b). The relevant completeness relation is given by the Lebedev–Kontorovitch transform theorem [see Eq. (3.4.94b)]:

$$\rho' \delta(\rho - \rho') = \frac{1}{2} \int_{-i\infty}^{i\infty} \mu J_\mu(k\rho) H_\mu^{(1)}(k\rho') \, d\mu = \frac{1}{4} \int_{-i\infty}^{i\infty} \mu H_\mu^{(1)}(k\rho) H_\mu^{(1)}(k\rho') \, d\mu, \tag{1a}$$

$$= \frac{1}{4} \int_0^{i\infty} \mu(1 - e^{i2\mu\pi}) H_\mu^{(1)}(k\rho) H_\mu^{(1)}(k\rho') \, d\mu. \tag{1b}$$

Comparison with the generic form in Eq. (6.2.4) permits identification of the eigenfunctions $\Phi_p(k\rho)$ and $\bar{\Phi}_p(k\rho)$, with the continuous mode index p denoted by μ.† Substitution of Eqs. (1) into Eqs. (6.2.8), (6.2.10), (6.2.12), or (6.2.14) yields the modal representation of the two- or three-dimensional Green's functions in terms of a modal Green's function $g_{\phi p}(\phi, \phi') \to g(\phi, \phi'; \mu)$ whose explicit form depends on the assumed boundary conditions on the wedge faces at $\phi = 0$ and $\phi = \varphi$ (the subscript ϕ on g_ϕ will be suppressed for convenience). From Eq. (6.A19a) it is found that since $H_\mu^{(1)}(k\rho)$ grows like $\exp(|\mu| \pi/2)$ as $\mu \to i\infty$, the radial eigenfunction (angular transmission) representation can be employed only when $g(\phi, \phi'; \mu)$ decays sufficiently rapidly to overcome the $\exp(|\mu| \pi)$ behavior arising from the integrand in Eq. (1b). Since g exhibits the exponential behavior [see Eqs. (6.4.1) and (6.5.2)]

$$g(\phi, \phi'; \mu) \propto \exp(-|\mu| |\phi - \phi'|), \qquad \text{as } \mu \to i\infty, \tag{2}$$

the representation applies only when $|\phi - \phi'| > \pi$ (i.e., in the geometric-optical shadow region).

The consequences of these anomalous convergence characteristics are illustrated in detail for the two-dimensional Green's function descriptive of excitation by a time-harmonic line source.

6.3a Time-harmonic Line-source Excitation

Solution in integral form

From Eqs. (6.2.8) and (1), one obtains the representation for the two-dimensional time-harmonic Green's function

$$\bar{G} = \tfrac{1}{4} \int_0^{i\infty} \mu(1 - e^{i2\mu\pi}) H_\mu^{(1)}(k\rho) H_\mu^{(1)}(k\rho') g(\phi, \phi'; \mu) \, d\mu, \tag{3}$$

which, as noted above, converges only in the angular domain $|\phi - \phi'| > \pi$ coincident with the geometric-optical shadow (Fig. 6.3.1). In the shadow region,

---

†In the remainder of this chapter, μ denotes a separation parameter and is not to be confused with the same symbol used elsewhere for the permeability.

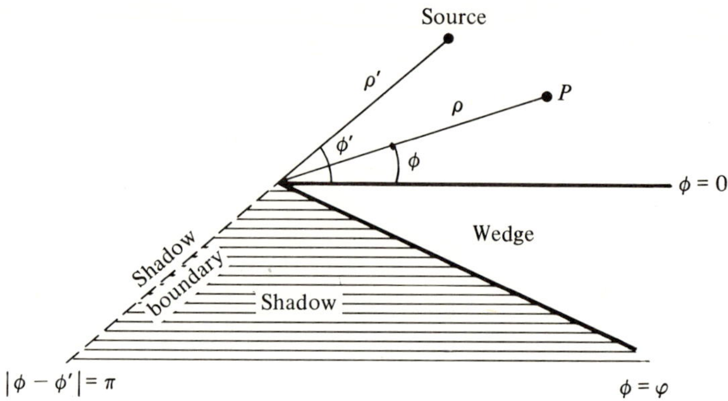

FIG. 6.3.1 Illuminated and shadow regions in a wedge diffraction problem.

the field comprises only outgoing waves, whereas in the illuminated region, incoming waves exist as well. Since the radial dependence in Eq. (3) involves only outgoing-wave Hankel functions [the time dependence is $\exp(-i\omega t)$], the incoming field is not directly expressible in this form. This deficiency is repaired below by an alternative representation valid for all $|\phi - \phi'|$. Because the shadow boundary $|\phi - \phi'| = \pi$ delimits the domain of Eq. (3), this representation is expected to be appropriate to a study of the quasi-optic properties of the diffraction field.

To obtain a formulation valid for all $|\phi - \phi'|$, we replace the Hankel functions of imaginary order μ by the expression [see Eq. (6.B3)]

$$H_\mu^{(1)}(k\rho)H_\mu^{(1)}(k\rho') = \frac{1}{\pi}\int_{i\infty}^{-i\infty} H_0^{(1)}(k\sqrt{\rho^2 + \rho'^2 + 2\rho\rho'\cos w})e^{i\mu(w-\pi)}\,dw, \quad (4)$$

and invert the orders of integration (permissible for $|\phi - \phi'| > \pi$) to obtain

$$\bar{G}(\boldsymbol{\rho}, \boldsymbol{\rho}') = \frac{1}{8\pi}\int_{i\infty}^{-i\infty} H_0^{(1)}(k\chi)A(\phi, \phi'; w)\,dw, \quad \chi = \sqrt{\rho^2 + \rho'^2 + 2\rho\rho'\cos w}, \quad (5)$$

where

$$A(\phi, \phi'; w) = 2\int_0^{i\infty} \mu(1 - e^{i2\mu\pi})e^{i\mu(w-\pi)}g(\phi, \phi'; \mu)\,d\mu, \qquad |\phi - \phi'| > \pi. \quad (6)$$

While the integral representation on the right-hand side of Eq. (6) converges only when $|\phi - \phi'| > \pi$ [see Eq. (2)], it will be found possible to evaluate the integral in closed form so that the resulting *function* $A(\phi, \phi'; w)$ is defined even when $|\phi - \phi'| < \pi$. To proceed further, the integrand in Eq. (5) must be examined in the entire complex w plane. The analytic properties of the Hankel function are discussed in connection with Fig. 6.3.2; for the type of problem considered, $A(\phi, \phi'; w)$ will generally have pole singularities in the complex w plane. In particular, one pole, located at $w_p = \pi - |\phi - \phi'|$, moves across the integration path in Eq. (5) as $|\phi - \phi'|$ decreases through π, and near this pole, A behaves like

$$A(\phi, \phi'; w) = \frac{-1}{w - w_p} + D(w), \qquad w_p = \pi - |\phi - \phi'|, \qquad (7)$$

where $D(w)$ is regular near w_p. Hence, when $|\phi - \phi'| < \pi$, the value of $\bar{G}$ differs from the expression in Eq. (5) by the residue at w_p, and one obtains, for *all* $|\phi - \phi'|$,

$$\bar{G}(\boldsymbol{\rho}, \boldsymbol{\rho}') = \frac{i}{4} H_0^{(1)}(k|\boldsymbol{\rho} - \boldsymbol{\rho}'|)U(\pi - |\phi - \phi'|) + \frac{1}{8\pi}\int_{i\infty}^{-i\infty} H_0^{(1)}(k\chi)A(\phi, \phi'; w)\, dw,$$

$$|\boldsymbol{\rho} - \boldsymbol{\rho}'| = \sqrt{\rho^2 + \rho'^2 - 2\rho\rho' \cos(\phi - \phi')}, \qquad (8)$$

where $U(x) = 1$ when $x > 0$, $U(x) = 0$ when $x < 0$, and the closed-form result is employed for A. The first term on the right-hand side of Eq. (8), the free-space Green's function [see Eq. (5.4.25)], exists only in the "illuminated" region $|\phi - \phi'| < \pi$ in Fig. 6.3.1, from which the source is visible. The integral represents a correction to the free-space field, and therefore displays directly the diffraction effects introduced by the wedge. It is noted that the functional form of the diffraction field in its dependence on $(\boldsymbol{\rho}, \boldsymbol{\rho}')$ is the same in the illuminated and shadow regions. When $A(\phi, \phi'; w)$ has other pole singularities, which also move across the integration path for certain ranges of ϕ, ϕ', their residue contributions must be included in a similar manner. Such additional poles furnish the reflected field of geometrical optics (see Sec. 6.5).

Asymptotic approximation

To effect an asymptotic evaluation of the diffraction integral in Eq. (8) via the steepest-descent method, it is necessary to investigate the analytic properties of the square-root function χ in Eq. (5) throughout the complex w plane. χ has first-order branch points at

$$w = w_b = \pm\left[n\pi \pm i\cosh^{-1}\left(\frac{a^2}{2\rho\rho'}\right)\right], \qquad n = 1, 3, 5, \ldots, \qquad (9)$$

where $a^2 = \rho^2 + \rho'^2 > 2\rho\rho'$. If branch cuts are drawn along the curves $\mathrm{Re}\,\chi = 0$ as in Fig. 6.3.2, the algebraic sign of $\mathrm{Re}\,\chi$ can change only upon passing through a cut. On the top sheet of the Riemann surface, we define

$$\chi = (\rho + \rho') > 0 \text{ when } w = 0, \qquad (10)$$

whence $\mathrm{Re}\,\chi > 0$ on the entire top sheet. $\mathrm{Im}\,\chi$ changes sign upon crossing the curves $\mathrm{Im}\,\chi = 0$, shown dashed in Fig. 6.3.2. To determine the algebraic sign of $\mathrm{Im}\,\chi$ in the various regions, note that near a branch point

$$\frac{\chi}{\sqrt{2\rho\rho'}} \approx \sqrt{i(w - w_b)\sinh w_{bi}}, \qquad w \approx w_b, \qquad (11)$$

with $\sqrt{w - w_b}$ defined so as to yield $\mathrm{Re}\,\chi > 0$. For $w_b = \pi + i|w_{bi}|$, for example, one has $\sqrt{w - w_b} = \sqrt{|w - w_b|}\, e^{i\alpha/2}$, $\pi/2 > \alpha > -3\pi/2$, whence $\arg \chi = (\alpha/2) + (\pi/4)$. Thus, $\mathrm{Im}\,\chi > 0$ for $-\pi/2 < \alpha < \pi/2$, and $\mathrm{Im}\,\chi < 0$

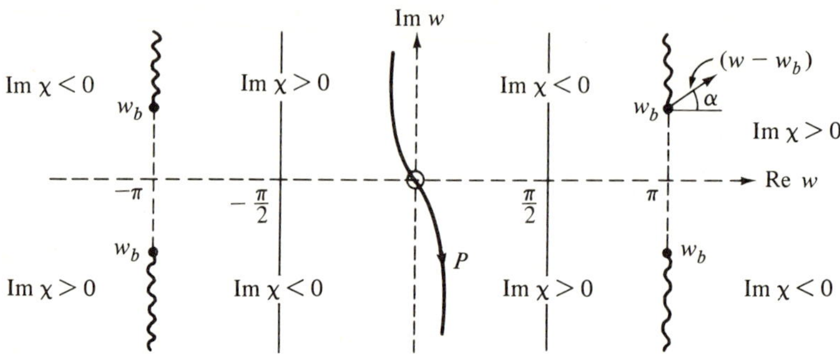

FIG. 6.3.2 Behavior of $\chi = \sqrt{a^2 + 2\rho\rho' \cos w}$, $a^2 > 2\rho\rho' > 0$, in the complex w plane.

for $-3\pi/2 < \alpha < -\pi/2$. Similar considerations apply to the other branch points and lead to a behavior of Im χ shown in Fig. 6.3.2. From these considerations and the asymptotic behavior $H_0^{(1)}(x) \propto \exp(ix)$ as $x \to \infty$, it is verified that the contour of integration in Eq. (8) can be deformed away from the imaginary axis onto a path P as in Fig. 6.3.2, on which Im $\chi > 0$.

If $k\rho$ or $k\rho'$ in Eq. (8) becomes very large, χ likewise becomes large and the Hankel function may be replaced by its asymptotic approximation in Eq. (5.3.13b),

$$H_\mu^{(1)}(k\chi) \sim \sqrt{\frac{2}{\pi k\chi}}\, e^{i\,(k\chi - \pi/4 - \mu\pi/2)}, \quad |k\chi| \gg |\mu|. \tag{12}$$

The integrand then contains an exponential factor $e^{ik\chi\,(w)}$, with k playing the role of the large parameter, whence a pertinent saddle point of $\chi(w)$ is located at $w = 0$; the functions A to be encountered do not behave exponentially in w. Along the steepest-descent path SDP at $w = 0$, one has, from Eq. (4.2.5),

$$\arg dw = \arg \sqrt{\frac{-2}{i\chi''(0)}} = -\frac{\pi}{4}, \tag{13}$$

so the SDP crosses the saddle point at an angle of $-45°$. In general, $\chi(w) = \chi(0) + is^2$, $s^2 > 0$, along the SDP, i.e.,

$$\operatorname{Re} \chi(w) = \operatorname{Re} \chi(0) = \rho + \rho'. \tag{14}$$

As $w_i \to i\infty$, $\cos w \to (\cosh w_i)e^{-iw_r}$, and $\chi(w) \to [2\rho\rho' \cos w]^{1/2}$. Re $\chi(w)$ must be finite [see Eq. (14)], so $w_r \to \mp\pi$ as $w_i \to \pm\infty$. Since there are no singularities between the imaginary axis and the SDP, the original contour can be deformed into the steepest-descent path and the integral evaluated according to Eq. (4.2.1b). Thus, to $O(1/\sqrt{k\rho})$ or $O(1/\sqrt{k\rho'})$, for $|\phi - \phi'| \not\approx \pi$ (i.e., w_p not near the saddle point at $w = 0$),

$$\bar{G}(\boldsymbol{\rho}, \boldsymbol{\rho}') \sim \frac{i}{4}H_0^{(1)}(k|\boldsymbol{\rho} - \boldsymbol{\rho}'|)U(\pi - |\phi - \phi'|) - 2A(\phi, \phi'; 0)C(k\rho)C(k\rho'), \tag{15}$$

where $A(\phi, \phi'; 0)$ is the closed-form expression for the integral on the right-hand side of Eq. (6), with $w = 0$, and

$$C(\alpha) = \frac{e^{i(\alpha + \pi/4)}}{2\sqrt{2\pi\alpha}}. \tag{15a}$$

This result has a simple physical interpretation in terms of a geometric-optical and diffraction field, as will be emphasized in connection with the various problems treated in Secs. 6.4–6.6.

Higher-order terms in the asymptotic expansion of the diffraction integral may be derived directly from Eq. (3) by using the method described in Eqs. (6.4.8)–(6.4.12).

Transition effects (uniform asymptotic formulation)

In view of Eq. (7) the simple formula in Eq. (15) fails in the vicinity of the shadow boundary $|\phi - \phi'| = \pi$ and must be augmented by a transition term derived by accounting explicitly for the presence of the pole at $w_p = \pi - |\phi - \phi'| \approx 0$ near the saddle point of the integrand in Eq. (8). Upon employing Eq. (12) for $H_0^{(1)}(k\chi)$, utilizing Eq. (7), and recalling the fact that the major contribution to the integral arises from the vicinity of $w = 0$, one may remove from the integrand all slowly varying factors and write

$$\frac{1}{8\pi} \int_{i\infty}^{-i\infty} H_0^{(1)}(k\chi) A(\phi, \phi'; w)\, dw \sim -\frac{e^{-i\pi/4}}{4\pi\sqrt{2\pi k\chi(0)}}(I + I'), \tag{16}$$

where

$$I = \int_{\mathrm{SDP}} \frac{e^{ik\chi(w)}}{w - w_p}\, dw, \qquad \chi(w) = \sqrt{\rho + \rho'^2 + 2\rho\rho' \cos w}, \quad w_p \approx 0, \tag{16a}$$

and I' is given by the same integral with $(w_p - w)^{-1}$ replaced by $D(w)$ in Eq. (7). Since $D(w)$ is regular near $w = 0$, the asymptotic evaluation of I' is straightforward and leads to a result as in Eq. (17a), with $1/w_p$ replaced by $D(0)$.

While the asymptotic evaluation of I, with $\chi(w)$ as indicated, can be carried out via the procedure in Sec. 4.4a, it is simpler and sufficiently accurate for the present purpose to approximate $\chi(w)$ by

$$\chi(w) = \sqrt{(\rho + \rho')^2 + 2\rho\rho'(\cos w - 1)} \approx (\rho + \rho') + \frac{\rho\rho'}{\rho + \rho'}(\cos w - 1), \tag{16b}$$

since the major contribution to I arises from the vicinity of the saddle point $w = 0$. The resulting integral in Eq. (16a) is then identical in form with $I_1(\Omega, \alpha, \beta)$ in Eq. (4.2.23), so the asymptotic expression (4.4.34) can be employed directly. It is useful to group the terms as follows:

$$I \equiv \int_{\mathrm{SDP}} \frac{e^{ik\tau \cos w}}{w - w_p}\, dw \sim I^d + I^t, \tag{17}$$

where τ stands for $\rho\rho'/(\rho + \rho')$,

$$I^d = \sqrt{\frac{2\pi}{k\tau}}\, \frac{e^{i(k\tau - \pi/4)}}{-w_p}, \qquad\qquad w_p = \pi - |\phi - \phi'|, \tag{17a}$$

$$\bar{I}^t = i\pi(\mathrm{sgn}\, w_p)e^{ik\tau}\left[F(\xi) - \frac{e^{i\pi/4}}{\sqrt{2\pi}\,\xi}\right], \qquad \xi = \sqrt{k\tau}\left|\sin\frac{w_p}{2}\right|, \tag{17b}$$

and $\mathrm{sgn}\,(x) = \pm 1$, $x \gtrless 0$. The function F is defined as

$$F(\xi) = \frac{2}{\sqrt{\pi}}e^{-i2\xi^2}\int_{(1-i)\xi}^{\infty} e^{-y^2}\,dy, \tag{18}$$

and has the asymptotic approximation

$$F(\xi) \sim \frac{e^{i\pi/4}}{\sqrt{2\pi}\,\xi} + O\left(\frac{1}{\xi^3}\right), \qquad \xi \gg 1, \tag{18a}$$

so that $\bar{I}^t$ contributes only when ξ is small. Since $F(\xi)$ remains bounded and the singularity in the second term of Eq. (17b) cancels that due to $\bar{I}^d$, the asymptotic solution for $\bar{I}$ remains valid as $w_p \to 0$. For sufficiently large ξ where $\bar{I}^t$ is negligible, the result $\bar{I} \sim \bar{I}^d$ represents the usual asymptotic approximation obtained by the ordinary saddle-point technique. $\bar{I}^t$ may therefore be regarded as a correction term that must be added to the simple asymptotic solution in the transition region surrounding the shadow boundary, wherein ξ is small. Suppose that $\bar{I}^t$ is negligible when $\xi \geq \xi_m$; then the transition region is bounded approximately by the curve

$$\xi_m\sqrt{k\tau} = k\tau\left|\sin\frac{w_p}{2}\right| \approx \frac{k|x|}{2}, \tag{19}$$

where $x = \tau w_p$ is the distance from the curve, a parabola, to the shadow boundary and, for ρ' very large, τ is essentially the distance coordinate along the shadow boundary (see Fig. 6.4.4). From Fig. 4.4.3(a) it is noted that the asymptotic approximation in Eq. (18a) holds with good accuracy when $\xi \geq 3$, so we may put $\xi_m = 3$ in Eq. (19); for greater accuracy, a larger value of ξ_m may be taken.

Upon combining Eqs. (16) and (17), one may write the following correction $\bar{G}^t$ to $\bar{G}$ which must be employed in the transition region $|\phi - \phi'| \approx \pi$:

$$\bar{G}^t(\boldsymbol{\rho}, \boldsymbol{\rho}') = -\tfrac{1}{2}C(k\rho + k\rho')\,\mathrm{sgn}\,(\pi - |\phi - \phi'|)\left[F(\xi) - \frac{e^{i\pi/4}}{\xi\sqrt{2\pi}}\right], \tag{20}$$

where ξ and $C(\alpha)$ are defined in Eqs. (17b) and (15a), respectively, and $\tau = \rho\rho'/(\rho + \rho')$. When this expression is added to the result in Eq. (15), the composite formula remains uniformly valid for arbitrary observation points. It is of interest to note that as $|\phi - \phi'| \to \pi$, $F(\xi) \to 1$, and the sum of Eqs. (15) and (20) may be written as

$$\bar{G}(\boldsymbol{\rho}, \boldsymbol{\rho}') \sim \tfrac{1}{2}\bar{G}_{\mathrm{inc}} + O\left(\frac{1}{k\sqrt{\rho\rho'}}\right), \qquad |\phi - \phi'| = \pi, \tag{21}$$

where $\bar{G}_{\mathrm{inc}}$ is the incident wavefunction

$$\bar{G}_{\mathrm{inc}} = \frac{i}{4}H_0^{(1)}(k|\boldsymbol{\rho} - \boldsymbol{\rho}'|) \sim C(k\rho + k\rho'), \tag{21a}$$

and it has been recognized that $|\boldsymbol{\rho} - \boldsymbol{\rho}'| \to (\rho + \rho')$ when $|\phi - \phi'| = \pi$. Thus, for large $k\rho$ and $k\rho'$, the field on the shadow boundary has a value equal to $\frac{1}{2}$ that of the incident field.

The present discussion deals only with the pole singularity at $w_p = \pi - |\phi - \phi'|$; when A in Eq. (7) has other relevant poles descriptive of reflected waves, each pole gives rise to similar transition phenomena near the appropriate angular coordinates.

6.3b Time-harmonic Plane-wave and Point-source Excitations

Solutions in integral form

If the source point ρ' moves to infinity along the angle ϕ', one obtains in the limit the result for an incident plane wave. In this instance, one of the Hankel functions in Eq. (3) is replaced by the asymptotic form in Eq. (12). Although μ in Eq. (3) also covers an infinite range, the integrand decreases exponentially with increasing μ, and the error made by employing Eq. (12) for all μ can be shown to be proportional to $\exp(-\alpha N)$, where $\alpha = |\phi - \phi'| - \pi$ and $1 \ll N \ll k\rho'$. N is a positive number such that Eq. (12) can be employed in the range $|\mu| \leq N$. As $\rho' \to \infty$, N can likewise be made arbitrarily large and the error term goes to zero. Substitution of the contour integral representation for $H_\mu^{(1)}(k\rho)$ for imaginary μ,

$$H_\mu^{(1)}(k\rho) = \frac{1}{\pi}\int_{i\infty}^{-i\infty} e^{ik\rho\,\cos w + i\mu\,(w-\pi/2)}\,dw, \tag{22}$$

then leads to an expression as in Eq. (8), provided that one replaces the Hankel function

$$\frac{i}{4}H_0^{(1)}(k\sqrt{\rho^2 + \rho'^2 + 2\rho\rho'\cos\beta}) \qquad \text{by } C(k\rho')e^{ik\rho\,\cos\beta}, \tag{23}$$

where $C(k\rho')$ is defined in Eq. (15a). For an incident plane wave of unit amplitude, $C \equiv 1$ [see Eq. (5.4.30b)] and one obtains the wavefunction

$$\bar{u}(\boldsymbol{\rho}; \phi') = e^{-ik\rho\,\cos(\phi-\phi')}U(\pi - |\phi - \phi'|) - \frac{i}{2\pi}\int_{i\infty}^{-i\infty} e^{ik\rho\,\cos w}A(\phi, \phi'; w)\,dw, \tag{24}$$

where A is the closed form for the integral on the right-hand side of Eq. (6). The first term on the right-hand side of Eq. (24) represents the incident plane-wave field in the illuminated region, whereas the second term expresses the diffraction field [see remarks following Eq. (8) for possible reflected wave contributions]. The result for an obliquely incident plane wave is deduced from Eq. (6.2.18).

The three-dimensional scalar Green's function $G(\mathbf{r}, \mathbf{r}')$ appropriate to excitation by a point source located at $\mathbf{r}' = (\rho', \phi', z')$ can be obtained from Eq. (8) by application of the rule stated after Eq. (6.2.5). Since k appears only in the argument of the Hankel functions, application of Eq. (5.4.12d),

$$\int_{-\infty}^{\infty} H_0^{(1)}(\sqrt{k^2 - \zeta^2}\,q)e^{i\zeta p}\,d\zeta = -2i\frac{e^{ik\sqrt{q^2+p^2}}}{\sqrt{q^2+p^2}}, \tag{25}$$

leads at once to the result

$$G(\mathbf{r}, \mathbf{r}') = \frac{e^{ik|\mathbf{r}-\mathbf{r}'|}}{4\pi|\mathbf{r} - \mathbf{r}'|} U(\pi - |\phi - \phi'|) - \frac{i}{8\pi^2} \int_{i\infty}^{-i\infty} \frac{e^{iky}}{\gamma} A(\phi, \phi'; w) \, dw, \quad (26)$$

where [see Eq. (5)]

$$\gamma = [\rho^2 + \rho'^2 + (z - z')^2 + 2\rho\rho' \cos w]^{1/2},$$

$$|\mathbf{r} - \mathbf{r}'| = \sqrt{|\mathbf{\rho} - \mathbf{\rho}'|^2 + (z - z')^2}. \tag{26a}$$

Thus, the three-dimensional Green's function $G(\mathbf{r}, \mathbf{r}')$ results from the two-dimensional $\bar{G}(\mathbf{\rho}, \mathbf{\rho}')$ in Eq. (8) upon replacement of

$$\frac{i}{4} H_0^{(1)}(k\sqrt{\rho^2 + \rho'^2 + 2\rho\rho' \cos \beta}) \text{ by}$$

$$\frac{\exp\left[ik\sqrt{\rho^2 + \rho'^2 + 2\rho\rho' \cos \beta + (z - z')^2}\right]}{4\pi\sqrt{\rho^2 + \rho'^2 + 2\rho\rho' \cos \beta + (z - z')^2}}. \tag{27}$$

As before, the first term on the right-hand side of Eq. (26) furnishes the incident point-source field in the illuminated region while the second term yields the diffraction effect [see remarks following Eq. (8) for possible reflected-wave contributions].

Asymptotic evaluation

The diffraction integral for the plane-wave scattering problem in Eq. (24) is already in the standard form, so an asymptotic evaluation for large values of $k\rho$ may be carried out directly by the methods leading to Eqs. (15) and (20). The result to $O(1/\sqrt{k\rho})$, uniformly in $|\phi - \phi'|$, is found to be

$$\bar{u}(\mathbf{\rho}; \phi') \sim \bar{u}^0 + \bar{u}^d + \bar{u}^t, \tag{28}$$

where

$$\bar{u}^o = e^{-ik\rho \cos(\phi-\phi')} U(\pi - |\phi - \phi'|), \tag{28a}$$

$$\bar{u}^d = -\frac{e^{ik\rho + i\pi/4}}{\sqrt{2\pi k\rho}} A(\phi, \phi'; 0), \tag{28b}$$

$$\bar{u}^t = -e^{ik\rho} \operatorname{sgn}(\pi - |\phi - \phi'|) \left[\frac{F(\xi)}{2} - \frac{e^{i\pi/4}}{2\sqrt{2\pi} \, \xi}\right], \tag{28c}$$

with

$$\xi = \sqrt{k\rho} \left|\sin \frac{\pi - |\phi - \phi'|}{2}\right|. \tag{28d}$$

$F(\xi)$ is defined in Eq. (18). As in the result for the line source, the transition term $\bar{u}^t$ must be included only when $|\phi - \phi'| \approx \pi$; near other angular directions where $A(\phi, \phi'; 0)$ may diverge [see remarks following Eq. (8)], analogous transition functions must be employed. When $|\phi - \phi'| = \pi$, one has

$$\bar{u}(\mathbf{\rho}; \phi') \sim \frac{e^{ik\rho}}{2} + O\left(\frac{1}{\sqrt{k\rho}}\right), \tag{29}$$

thereby confirming again that the field on the shadow boundary is asymptotically equal to one half the incident field.

Higher-order terms in the asymptotic expansion of $\bar{u}^d$ may be derived as in Eqs. (6.4.9)–(6.4.12), subject to the modifications in Sec. 6.4e.

The γ-dependent diffraction integral for the point source in Eq. (26) may be evaluated asymptotically by the same procedure as in Sec. 6.3a, owing to its similarity with the integral in Eq. (8) with $H_0^{(1)}(k\chi) \sim e^{ik\chi}/\sqrt{k\chi}$. Since $\gamma = [\chi^2 + (z - z')^2]^{1/2}$ from Eq. (26a), many of the expressions encountered in the line-source problem can be taken over for the point-source case provided that $\rho^2 + \rho'^2$ is replaced by $\rho^2 + \rho'^2 + (z - z')^2$. Thus, one finds for large $k\rho$, $k\rho'$, and uniformly in $|\phi - \phi'|$:

$$G(\mathbf{r}, \mathbf{r}') \sim G^o + G^d + G^t, \tag{30}$$

where

$$G^o = \frac{e^{ik|\mathbf{r}-\mathbf{r}'|}}{4\pi|\mathbf{r} - \mathbf{r}'|}\, U(\pi - |\phi - \phi'|), \tag{30a}$$

$$G^d = -\frac{e^{ikl+i\pi/4}}{4\pi\sqrt{2\pi k\rho\rho'l}}A(\phi, \phi'; 0), \qquad l = \sqrt{(\rho + \rho')^2 + (z - z')^2}, \tag{30b}$$

$$G^t = -\frac{e^{ikl}}{8\pi l}\,\mathrm{sgn}\,(\pi - |\phi - \phi'|)\left[F(\xi) - \frac{e^{i\pi/4}}{\xi\sqrt{2\pi}}\right], \tag{30c}$$

with

$$\xi = \sqrt{\frac{k\rho\rho'}{l}}\left|\sin\frac{\pi - |\phi - \phi'|}{2}\right|. \tag{30d}$$

$F(\xi)$ is defined in Eq. (18). Transition functions analogous to G^t in Eq. (30c) must be included near other angles, where $A(\phi, \phi'; 0)$ may have singularities [see remarks following Eq. (8)]. On the shadow boundary $|\phi - \phi'| = \pi$,

$$G(\mathbf{r}, \mathbf{r}') \sim \frac{e^{ikl}}{8\pi l} + O\left(\frac{1}{\sqrt{k\rho\rho'l}}\right), \tag{31}$$

in accord with a similar result in Eq. (21) (see Fig. 6.4.3 for a geometrical interpretation).

6.3c. Pulsed-source Configurations

When the source configurations in the preceding sections have an impulsive behavior characterized by the delta function $\delta(t - t')$ as in Eq. (5.2.15), the transient solutions may be obtained from the time-harmonic Green's functions by performing an integration as in Eq. (6.2.12) et seq. As noted in Sec. 5.2c, the explicit recovery of the time-dependent result is simplified substantially when the time-harmonic solution has the form given in Eqs. (1.6.37) or (1.6.38). This is indeed the case for the expressions in Eqs. (24) and (26), so the transient response to plane-wave or point-source excitation may be recovered by direct application of Eq. (1.6.39) (see also Sec. 5.2c; A in Sec 5.2c and here denotes different quantities).

Since the time-harmonic plane wave $\exp[-ik\rho\cos(\phi-\phi')]$ corresponds to the plane-wave pulse $\delta[t-t'+(\rho/\bar{c})\cos(\phi-\phi')]$, the temporal response in the presence of the wedge configuration is given via Eqs. (24) and (1.6.39) for arbitrary ρ, ρ', t, t' by

$$\hat{u}(\boldsymbol{\rho},\phi';t) = \delta\left(t-t'+\frac{\rho}{\bar{c}}\cos(\phi-\phi')\right)U(\pi-|\phi-\phi'|)$$

$$-\frac{1}{\pi}\frac{\operatorname{Re}A[\phi,\phi';i\cosh^{-1}(\bar{c}(t-t')/\rho)]}{\sqrt{(t-t')^2-(\rho/\bar{c})^2}}U\left(t-t'-\frac{\rho}{\bar{c}}\right), \qquad (32)$$

where $\bar{c}$ is the propagation speed in the medium and $A(\phi,\phi';w)$ is the closed form of the function defined in Eq. (6). The first term on the right-hand side of Eq. (32) represents the incident plane-wave pulse in the illuminated region of Fig. 6.3.1, and the second term yields the diffraction field. Since the incident pulse does not reach the edge $\rho=0$ until time $t=t'$, no diffraction takes place until $t>t'$. The diffracted pulse spreads cylindrically outward from the edge and has a behavior characteristic of an equivalent line source at the edge [see Eq. (5.4.42)] modified by the amplitude factor $\operatorname{Re}A$. Additional physical implications are discussed in connection with examples in Secs. 6.4 and 6.5 (see Fig. 6.4.5). When $A(\phi,\phi';w)$ has pole singularities in addition to the one at $w=\pi-|\phi-\phi'|$, other plane-wave pulse contributions, descriptive of the reflected field of geometrical optics, arise [see remarks following Eq. (8) and Sec. 6.5].

For excitation by an impulsive point source, the Green's function $\hat{G}(\mathbf{r},\mathbf{r}';t,t')$, which satisfies Eq. (6.2.13) for arbitrary $\mathbf{r}$, $\mathbf{r}'$, t, t', is obtained via Eq. (26) and the procedure following Eq. (1.6.37):

$$\hat{G}(\mathbf{r},\mathbf{r}';t,t') = \frac{\delta(t-t'-|\mathbf{r}-\mathbf{r}'|/\bar{c})}{4\pi|\mathbf{r}-\mathbf{r}'|}U(\pi-|\phi-\phi'|)$$

$$+\frac{\bar{c}}{4\pi}\frac{\operatorname{Re}A(\phi,\phi';i\beta)}{\rho\rho'\sinh\beta}U\left(t-t'-\frac{l}{\bar{c}}\right), \qquad (33)$$

where $\beta=\cosh^{-1}\{[\bar{c}^2(t-t')^2-\rho^2-\rho'^2-(z-z')^2]/2\rho\rho'\}$, and l is defined in Eq. (30b). The interpretation of this result is analogous to the above [see Fig. 6.4.3 and Eq. (5.4.14b), and also the remarks concerning additional contributions arising from A].

The two-dimensional Green's function in Eq. (8) does not exhibit the structure specified in Eq. (1.6.37), so the formulas in Sec 5.2c cannot be applied directly (note the different meaning of A in Sec.5.2c). It is possible, however, to achieve the desired format after certain preliminary manipulations. Upon letting $k\to is/\bar{c}$, where s is positive and $\bar{c}$ is the speed of light in the exterior medium, and recalling that

$$H_\nu^{(1)}(iz) = \frac{2}{\pi i}e^{-i\nu\pi/2}K_\nu(z), \qquad (34)$$

where $K_\nu(z)$ is the modified Bessel function, one obtains

$$\bar{G}(\boldsymbol{\rho}, \boldsymbol{\rho}') = \frac{1}{2\pi} K_0\left(s\frac{|\boldsymbol{\rho} - \boldsymbol{\rho}'|}{\bar{c}}\right) U(\pi - |\phi - \phi'|) + I(\boldsymbol{\rho}, \boldsymbol{\rho}'; s), \qquad (35)$$

with

$$I(\boldsymbol{\rho}, \boldsymbol{\rho}'; s) = \frac{i}{4\pi^2} \int_{-i\infty}^{i\infty} K_0\left(\frac{s}{\bar{c}}\sqrt{\rho^2 + \rho'^2 + 2\rho\rho'\cos w}\right) A(\phi, \phi'; w)\, dw. \qquad (36)$$

The Laplace inversion of the first term on the right-hand side of Eq. (35) is given in Eq. (5.4.42). To transform the second term into a representation as in Eq. (1.6.34), introduce

$$K_0(x) = \int_1^{\infty} \frac{e^{-xt}}{\sqrt{t^2 - 1}}\, dt, \qquad x > 0, \qquad (37)$$

to obtain

$$I(\boldsymbol{\rho}, \boldsymbol{\rho}'; s) = \frac{i}{4\pi^2} \int_{-i\infty}^{i\infty} dw\, A(\phi, \phi'; w) \int_{f/\bar{c}}^{\infty} \frac{e^{-s\tau}}{\sqrt{\tau^2 - f^2/\bar{c}^2}}\, d\tau, \qquad (38)$$

where

$$f \equiv f(\beta) = \sqrt{\rho^2 + \rho'^2 + 2\rho\rho'\cosh\beta} > 0, \qquad \beta = -iw. \qquad (38a)$$

The desired formulation results upon interchange of the order of integrations which, in Eq. (38), cover the range $f(\beta) < \bar{c}\tau < \infty$ and $-\infty < \beta < \infty$. If the β integration is performed first, one has $-\psi(\tau) < \beta < \psi(\tau)$, $\psi(\tau) = \cosh^{-1}[\bar{c}^2\tau^2 - \rho^2 - \rho'^2)/2\rho\rho']$, while $f(0) < \bar{c}\tau < \infty$. Thus

$$I(\boldsymbol{\rho}, \boldsymbol{\rho}'; s) = \int_0^{\infty} e^{-s\tau} Q(\tau)\, d\tau, \qquad (39)$$

where

$$Q(\tau) = \begin{cases} 0, & \bar{c}\tau < (\rho + \rho'), \quad (40a) \\[2ex] -\dfrac{1}{4\pi^2} \displaystyle\int_{-\psi(\tau)}^{\psi(\tau)} \frac{A(\phi, \phi'; i\beta)}{\sqrt{\tau^2 - f^2(\beta)/\bar{c}^2}}\, d\beta, & \bar{c}\tau > (\rho + \rho'). \quad (40b) \end{cases}$$

The line-source Green's function satisfying Eq. (6.2.11) is therefore given for arbitrary $\boldsymbol{\rho}, \boldsymbol{\rho}'\, t, t'$ by

$$\hat{G}(\boldsymbol{\rho}, \boldsymbol{\rho}'; t, t') = \frac{1}{2\pi\sqrt{(t - t')^2 - |\boldsymbol{\rho} - \boldsymbol{\rho}'|^2/\bar{c}^2}} U(\pi - |\phi - \phi'|)$$

$$\times U\left(t - t' - \frac{|\boldsymbol{\rho} - \boldsymbol{\rho}'|}{\bar{c}}\right)$$

$$- \left[\frac{1}{4\pi^2} \int_{-\psi(t-t')}^{\psi(t-t')} \frac{\mathrm{Re}\, A(\phi, \phi'; i\beta)}{\sqrt{(t - t')^2 - f^2(\beta)/\bar{c}^2}} d\beta\right] U\left(t - t' - \frac{\rho + \rho'}{\bar{c}}\right). \qquad (41)$$

A physical interpretation of this result, whose form is more complicated than that for the plane-wave or point-source excitation, is provided in connection with Fig. 6.4.2. It should again be recalled that additional contributions representative of reflected field constituents may be present in Eqs. (32), (33) and (41), as noted already in connection with Eq. (32).

The preceding results simplify substantially for observation times near the time of arrival of the first response [i.e., for values of $(t - t')$ which annul the arguments of the various Heaviside unit functions in Eqs. (32), (33), or (41)]. One observes by inspection that the values of $\hat{G}$ in Eqs. (32) and (33) near the time of arrival of the diffracted wavefront are ascertained directly upon replacement of $A[\phi, \phi'; \Omega]$ by $A[\phi, \phi'; 0]$, whereas the corresponding result in Eq. (41) requires a reduction of the integral as $\psi(t - t') \to 0$. The reader may verify that the expressions obtained in this manner agree with those predicted from the time-harmonic high-frequency asymptotic formulas in Eqs. (15), (24) and (26), after relations (1.6.45) and (1.6.47) have been invoked.

6.4 PERFECTLY ABSORBING WEDGE

A wedge configuration analyzed most simply by the angular transmission analysis is the "perfectly absorbing" wedge defined by the condition that all angularly propagating waves are absorbed completely at the wedge faces; the "black screen" investigated by Sommerfeld[1] belongs to this category, although his analysis differs from the one carried out here. While this boundary condition is not easily phrased as a relation between the *total* electric and magnetic fields at the reflectionless wedge surfaces $\phi = 0$ and $\phi = \varphi$, its specification in terms of a *mode* propagating along the ϕ direction is elementary: $g(\phi, \phi'; \mu)$ must comprise only outgoing waves from the source location $\phi = \phi'$. Thus, the angular Green's function defined in Eq. (6.2.7) is given by Eq. (3.4.55); the reflectionless condition on the wedge faces is the same for E modes and H modes, so no distinction need be made. With $g_\phi^\infty \equiv g_\infty$, one has

$$g_\infty(\phi, \phi'; \mu) = \frac{e^{i\mu|\phi - \phi'|}}{-2i\mu}, \qquad \mathrm{Im}\ \mu > 0. \tag{1}$$

It is observed from the structure of the Green's function that the condition of perfect absorption (matched condition) at $\phi = 0, \varphi$, is equivalent to regarding the ϕ space as being infinitely extended (hence the subscript ∞), and to imposing a radiation condition at $|\phi| = \infty$. For the analogous result pertaining to bilaterally unbounded transmission along a rectilinear coordinate, see Eq. (5.4.7), with $j \to -i$. The behavior of g_∞ in Eq. (1) evidently conforms with that in Eq. (6.3.2), and substitution into Eq. (6.3.6) permits the closed-form evaluation of $A(\phi, \phi'; w)$:

$$A(\phi, \phi'; w) = i\int_0^{i\infty} (1 - e^{i2\mu\pi})e^{i\mu(w - \pi + |\phi - \phi'|)}\, d\mu \tag{2a}$$

$$= \frac{1}{\pi - |\phi - \phi'| - w} + \frac{1}{\pi + |\phi - \phi'| + w}. \tag{2b}$$

Unlike the integral representation in Eq. (2a), the result in Eq. (2b) is defined for all values of $|\phi - \phi'|$, and its analytic structure in w reveals the presence of a simple pole at $w_p = \pi - |\phi - \phi'|$ as specified in Eq. (6.3.7). The second pole at $w = -(\pi + |\phi - \phi'|)$ never crosses the imaginary axis in the complex w plane, so the solutions given in Sec.6.3 apply without modification.

6.4a Time-harmonic Line-source Excitation

When the perfectly absorbing wedge is excited by a line source of electric current (Fig. 6.4.1) having a density prescribed by

$$\hat{\mathbf{J}}(\mathbf{r}, t) = I\delta(\boldsymbol{\rho} - \boldsymbol{\rho}')e^{-i\omega t}\mathbf{z}_0, \tag{3a}$$

or by a line source of magnetic current with

$$\hat{\mathbf{M}}(\mathbf{r}, t) = V\delta(\boldsymbol{\rho} - \boldsymbol{\rho}')e^{-i\omega t}\mathbf{z}_0, \tag{3b}$$

the electromagnetic fields E_z, H_ρ, H_ϕ, and H_z, E_ρ, E_ϕ excited by the electric and magnetic currents, respectively, can be derived from a two-dimensional scalar Green's function [see Eqs. (6.2.1) and (6.2.2)]. Since the boundary conditions on the perfectly absorbing wedge faces are the same in the two cases,

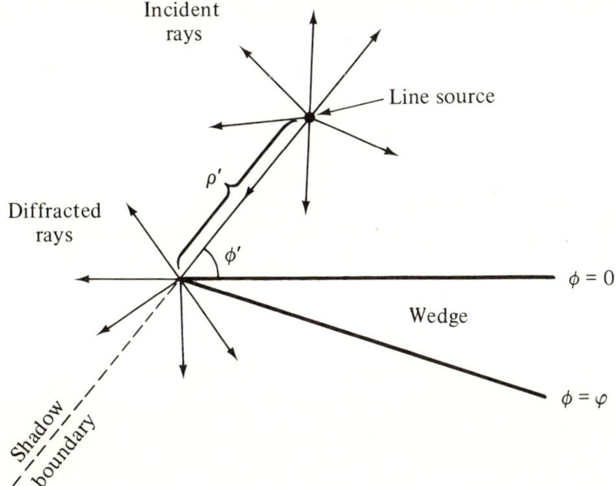

FIG. 6.4.1. Line-source excitation of a perfectly absorbing wedge.

the E- mode and H- mode Green's functions are identical and shall be denoted by $\bar{G}_\infty(\boldsymbol{\rho}, \boldsymbol{\rho}')$, which satisfies the inhomogeneous wave equation

$$\left(\frac{1}{\rho}\frac{\partial}{\partial\rho}\rho\frac{\partial}{\partial\rho} + \frac{1}{\rho^2}\frac{\partial^2}{\partial\phi^2} + k^2\right)\bar{G}_\infty(\boldsymbol{\rho}, \boldsymbol{\rho}') = -\frac{\delta(\rho - \rho')}{\rho'}\delta(\phi - \phi') \tag{4}$$

in the domain $0 < (\rho, \rho') < \infty$, $-\infty < (\phi, \phi') < \infty$, subject to boundedness at $\rho = 0$ and a radiation condition at $\rho \to \infty$, $|\phi| \to \infty$. The subscript ∞ denotes the solution for the angularly unbounded domain $-\infty < \phi < \infty$. The solution in the geometric-optical shadow region $|\phi - \phi'| > \pi$ is given via Eqs. (6.3.3) and (1) by

$$\bar{G}_{\infty}(\boldsymbol{\rho}, \boldsymbol{\rho}') = \frac{i}{8} \int_0^{i\infty} (1 - e^{i2\mu\pi}) e^{i\mu|\phi - \phi'|} H_{\mu}^{(1)}(k\rho) H_{\mu}^{(1)}(k\rho') \, d\mu, \tag{5}$$

whereas the result obtained from Eqs. (6.3.8) and (2) is valid for *all* $|\phi - \phi'|$. For $k\rho \gg 1$, $k\rho' \gg 1$, but with $|\phi - \phi'| \not\approx \pi$, one has the asymptotic approximation given in Eq. (6.3.15),

$$\bar{G}_{\infty} \sim \bar{G}_{\infty}^0 + \bar{G}_{\infty}^d, \tag{6}$$

$$\bar{G}_{\infty}^0 = \frac{i}{4} H_0^{(1)}(k|\boldsymbol{\rho} - \boldsymbol{\rho}'|) U(\pi - |\phi - \phi'|), \tag{6a}$$

$$\bar{G}_{\infty}^d = -2A(\phi, \phi'; 0) C(k\rho) C(k\rho') = -iA(\phi, \phi'; 0)\frac{e^{ik(\rho+\rho')}}{4\pi k\sqrt{\rho\rho'}}, \tag{6b}$$

$$A(\phi, \phi'; 0) = \frac{1}{\pi - |\phi - \phi'|} + \frac{1}{\pi + |\phi - \phi'|}, \tag{6c}$$

whereas for *arbitrary* $|\phi - \phi'|$, one must add to Eq. (6) the transition function from Eq. (6.3.20):

$$\bar{G}_{\infty}^t(\boldsymbol{\rho}, \boldsymbol{\rho}') = -\frac{e^{i[k(\rho+\rho')+\pi/4]}}{4\sqrt{2\pi k(\rho + \rho')}} \left[F(\xi) - \frac{e^{i\pi/4}}{\xi\sqrt{2\pi}} \right] \text{sgn}\,(\pi - |\phi - \phi'|), \tag{7}$$

where

$$\xi = \sqrt{\frac{k\rho\rho'}{\rho + \rho'}} \left| \cos \frac{\phi - \phi'}{2} \right|. \tag{7a}$$

and $F(\xi)$ is defined in Eq. (6.3.18). $U(\alpha)$ denotes the Heaviside unit function and $C(\alpha)$ is defined in Eq. (6.3.15a).

Discussion

The $\bar{G}_{\infty}^0$ term represents the incident field in the illuminated region $|\phi - \phi'| < \pi$, whereas the term $\bar{G}_{\infty}^d$ furnishes to $O(1/\sqrt{k\rho})$, $O(1/\sqrt{k\rho'})$ the diffraction field exterior to the transition region surrounding the light-shadow boundary [see Eq. (9) for a calculation of higher-order terms]. The diffracted contribution associated with the factor $C(k\rho)$ can be interpreted as an outgoing cylindrical wave that emanates from the edge and has an angular distribution given by $-2A(\phi, \phi'; 0)$. This wave is excited by the cylindrical wave incident from the source and the factor $C(k\rho')$ yields precisely the strength of the incident wave at the wedge apex. The process can be schematized clearly through the use of diffracted rays (see Sec 1.7d). In Fig. 6.4.1, the incident field is represented by rays emanating radially from the source, with the strength of the field along an incident ray given by

$$\frac{i}{4} H_0^{(1)}(k|\boldsymbol{\rho} - \boldsymbol{\rho}'|) \sim C(k|\boldsymbol{\rho} - \boldsymbol{\rho}'|) = \frac{e^{i(k|\boldsymbol{\rho} - \boldsymbol{\rho}'|+\pi/4)}}{2\sqrt{2\pi k|\boldsymbol{\rho} - \boldsymbol{\rho}'|}}.$$

The incident rays exist everywhere in the illuminated region $|\phi - \phi'| < \pi$. Upon striking the wedge surface, an incident ray is completely absorbed, so no reflected rays exist. The ray striking the edge, however, is scattered in all directions and gives rise to the spectrum of diffracted rays. The field amplitude along a diffracted ray excited by an incident ray (local plane wave) of unit amplitude is given by $-2A(\phi, \phi'; 0)C(k\rho)$ as noted from Eq. (6.3.28b). In the present case, the field along the incident ray striking the edge has an amplitude $C(k\rho')$ whence this factor appears in Eq. (6b). This simple ray-optical interpretation breaks down in the transition region $|\phi - \phi'| \approx \pi$, where the amplitude factor of the local plane wave varies rapidly, and the more complicated description involving $\bar{G}_{\infty}^{t}$ must be employed [see Fig. 6.4.4 and Eq. (6.3.19) concerning the extent of the transition region].

Attention may again be called to the above-noted absence of a geometrically reflected wave contribution and also to the lack of dependence of $\bar{G}_{\infty}$ on the wedge angle $\phi = \varphi$ in Fig. 6.4.1. This is a direct consequence of the wedge boundary condition that all "angularly propagating waves" are completely absorbed. Because of this condition of angular matching, the location of the wedge faces is of no importance; the matched condition is disturbed only near the apex $\rho = 0$, whence diffraction phenomena arise from this region. While the physical realizability of this type of absorber is questionable, it provides a useful and simple mathematical model for the study of diffraction by absorbing structures and can give physical insight into diffraction effects encountered in physically more meaningful, but mathematically more complicated configurations.

Higher-order terms in the asymptotic expansion

The result given in Eq. (6b) represents to $O(1/\sqrt{k\rho})$, $O(1/\sqrt{k\rho'})$ the asymptotic solution for $\bar{G}_{\infty} \sim \bar{G}_{\infty}^{d}$ in the shadow region. To obtain higher-order terms in the asymptotic representation of the diffraction field, it is convenient to proceed directly from Eq. (5) and to substitute for the Hankel functions the complete expansions

$$H_{\mu}^{(1)}(k\rho) \sim \sqrt{\frac{2}{\pi k\rho}}\, e^{i[k\rho - \mu\pi/2 - \pi/4]} \sum_{m=0}^{\infty} \frac{(\mu, m)}{(-2ik\rho)^m}, \qquad k\rho \gg |\mu|, \quad (8a)$$

$$(\mu, m) = \frac{(4\mu^2 - 1^2)(4\mu^2 - 3^2)\ldots[4\mu^2 - (2m - 1)^2]}{2^{2m}m!}, \quad (\mu, 0) \equiv 1, \quad (8b)$$

and similarly for $H_{\mu}^{(1)}(k\rho')$. Inversion of the orders of summation and integration yields the following asymptotic series for $\bar{G}_{\infty}^{d}$:

$$\bar{G}_{\infty}^{d}(\boldsymbol{\rho}, \boldsymbol{\rho}') \sim C(k\rho)C(k\rho') \sum_{m=0}^{\infty} \sum_{n=0}^{\infty} \frac{I_{mn}(\phi, \phi')}{(-2ik\rho)^m(-2ik\rho')^n}, \qquad (9)$$

where

$$I_{mn}(\phi, \phi') = -2i\int_{0}^{\infty} (e^{-i\mu\pi} - e^{i\mu\pi})e^{i\mu|\phi - \phi'|}(\mu, m)(\mu, n)\, d\mu, \quad |\phi - \phi'| > \pi \quad (9a)$$

and $C(\alpha) = (8\pi\alpha)^{-1/2} \exp(i\alpha + i\pi/4)$. Use of the asymptotic approximation (8) for all values of μ in the integral of Eq. (5) involves an exponentially small error that can be neglected [see discussion preceding Eq. (6.3.22)], and termwise integration of the resulting asymptotic series is permitted.

Since

$$(\mu, m) = \frac{1}{m}[\mu^2 - \frac{1}{4} - m(m-1)](\mu, m-1), \tag{10}$$

and a multiplicative factor μ^2 in the integrand of Eq. (9a) can be replaced by the differentiation $-\partial^2/\partial\phi^2$ when $\phi \neq \phi'$, there exists among the diffraction coefficients I_{mn} the recursion relation

$$I_{mn} = L_m(\phi)I_{m-1,n}, \qquad m = 1, 2, \ldots, \tag{11a}$$

where

$$L_m(\phi) = -\frac{1}{m}\left[\frac{\partial^2}{\partial\phi^2} + m(m-1) + \tfrac{1}{4}\right]. \tag{11b}$$

Application of the operator $L_n(\phi)$ on I_{mn} increases by unity the index n. Thus, the higher-order diffraction coefficients can be derived from I_{00} by successive application of $L_m(\phi)$ and $L_n(\phi)$, and, by comparing Eqs. (2a) and (9a) with $m = n = 0$, one observes that

$$I_{00}(\phi, \phi') = -2A(\phi, \phi'; 0) = \frac{-2}{\pi - |\phi - \phi'|} - \frac{2}{\pi + |\phi - \phi'|}. \tag{12}$$

While the above derivation is subject to the restriction $|\phi - \phi'| > \pi$, it is noted from Eq. (6.3.8) that the diffraction integral exhibits the same functional dependence $A(\phi, \phi'; w)$ for *all* values of $|\phi - \phi'|$. Hence, the asymptotic expansion (9) for $\bar{G}_\infty^d$ is actually valid for *all* ϕ, ϕ' outside the transition region $|\phi - \phi'| \approx \pi$, provided that the I_{mn} are represented in closed form via Eqs. (11) and (12), and not in terms of the integral formula (9a), which converges only when $|\phi - \phi'| > \pi$. Evidently, the lowest-order result $C(k\rho)C(k\rho')I_{00}(\phi, \phi')$ agrees with the one in Eq. (6b) derived by an alternative method.

6.4b Impulsive Line-source Excitation

When the line source in Fig. 6.4.1 has an impulsive time behavior characterized by $\delta(t - t')$, the fields are derivable from the Green's function $\hat{\bar{G}}_\infty(\boldsymbol{\rho}, \boldsymbol{\rho}'; t, t')$, which is defined by

$$\left(\frac{1}{\rho}\frac{\partial}{\partial\rho}\rho\frac{\partial}{\partial\rho} + \frac{1}{\rho^2}\frac{\partial^2}{\partial\phi^2} - \frac{1}{\bar{c}^2}\frac{\partial^2}{\partial t^2}\right)\hat{\bar{G}}_\infty(\boldsymbol{\rho}, \boldsymbol{\rho}'; t, t')$$

$$= -\frac{\delta(\rho - \rho')}{\rho'}\delta(\phi - \phi')\delta(t - t'), \tag{13}$$

subject to boundedness at $\rho = 0$, and quiescence for $t < t'$. The solution is given in Eq. (6.3.41) upon substitution of $A(\phi, \phi'; i\beta)$ from Eq. (2b). For observation times $t \approx t' + (\rho + \rho')/\bar{c}$, the second term in Eq. (6.3.41), to be denoted by $\hat{\bar{G}}_\infty^d$, may be simplified as follows [see also remarks following Eq. (6.3.41)]:

$$\hat{\bar{G}}_\infty^d \approx -\frac{A(\phi, \phi'; 0)\bar{c}}{4\pi\sqrt{\rho\rho'}}U\left(t - t' - \frac{\rho + \rho'}{\bar{c}}\right), \tag{14a}$$

and furnishes the behavior of the diffraction field in the vicinity of the diffracted wavefront. In contrast, near the incident wavefront where $t \approx t' + |\mathbf{\rho} - \mathbf{\rho}'|/\bar{c}$, the first term in Eq. (6.3.41) yields

$$\hat{\bar{G}}^0_\infty \approx \frac{\bar{c}}{2\pi\sqrt{2|\mathbf{\rho} - \mathbf{\rho}'|}} \frac{1}{\sqrt{\bar{c}(t - t') - |\mathbf{\rho} - \mathbf{\rho}'|}}$$

$$\times\, U(\pi - |\phi - \phi'|)U\left(t - t' - \frac{|\mathbf{\rho} - \mathbf{\rho}'|}{\bar{c}}\right), \qquad (14b)$$

whence it is observed that the field discontinuity across a wavefront is weakened by diffraction.

As noted already, the first term on the right-hand side of Eq. (6.3.41) represents the response in the absence of the wedge and exists only outside the shadow region of Fig. 6.4.1; it constitutes the geometric-optical contribution. The second term represents a cylindrically spreading diffracted pulse which exists in the entire region exterior to the wedge and reaches the observation point $P(\rho, \phi)$ at time $t - t' = (\rho + \rho')/\bar{c}$, i.e., after a time interval required to travel both the distance ρ' from the source point to the edge and the distance ρ from the edge to the observation point. In view of the "absorbing" wedge faces, there is no reflected-pulse contribution. The incident and diffracted wave fronts are sketched in Fig. 6.4.2. The normals to the wavefronts, the rays, are equivalent to those sketched in Fig. 6.4.1.

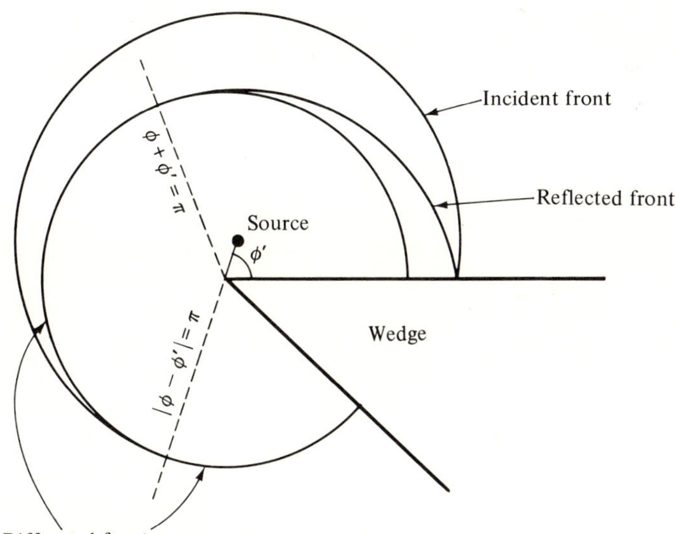

FIG. 6.4.2 Diffraction of cylindrical pulse by a wedge (omit reflected front for perfectly absorbing case). The incident front is centered at the source, the reflected front at the source image with respect to the horizontal wedge face, and the diffracted front is centered at the edge.

6.4c   Time-harmonic Point-source Excitation

The electromagnetic fields radiated by a z-directed element of electric current

$$\hat{\mathbf{J}}(\mathbf{r}, t) = J^{o}\delta(\mathbf{r} - \mathbf{r}')e^{-i\omega t}\mathbf{z}_0. \tag{15a}$$

or magnetic current

$$\hat{\mathbf{M}}(\mathbf{r}, t) = M^{o}\delta(\mathbf{r} - \mathbf{r}')e^{-i\omega t}\mathbf{z}_0 \tag{15b}$$

in the presence of a perfectly absorbing wedge can be ascertained via the discussion in Sec. 6.2a from a scalar Green's function that satisfies the equation

$$\left(\frac{1}{\rho}\frac{\partial}{\partial\rho}\rho\frac{\partial}{\partial\rho} + \frac{1}{\rho^2}\frac{\partial^2}{\partial\phi^2} + \frac{\partial^2}{\partial z^2} + k^2\right)G_\infty(\mathbf{r}, \mathbf{r}') = -\frac{\delta(\rho - \rho')}{\rho'}\delta(\phi - \phi')\delta(z - z')$$

$$\tag{16}$$

in the domain $0 < (\rho, \rho') < \infty$, $-\infty < (\phi, \phi') < \infty$, $-\infty < (z, z') < \infty$, subject to the boundary conditions

$$G_\infty \text{ finite at } \rho = 0; \quad \text{radiation condition at } \rho \to \infty, \quad \phi \to \pm\infty, \quad z \to \pm\infty. \tag{16a}$$

From Eqs. (6.2.10), (6.3.1), and (6.4.1), the following representation applies in the geometric-optical shadow region $|\phi - \phi'| > \pi$ (see Fig. 6.3.1):

$$G_\infty(\mathbf{r}, \mathbf{r}') = \frac{1}{8\pi}\int_{-\infty}^{\infty} d\zeta \int_{0}^{i\infty} \mu(1 - e^{i2\mu\pi})H_\mu^{(1)}(\sqrt{k^2 - \zeta^2}\rho)H_\mu^{(1)}(\sqrt{k^2 - \zeta^2}\rho')$$

$$\times \frac{e^{i\mu|\phi - \phi'|}}{-2i\mu}e^{i\zeta(z - z')}d\mu, \tag{17}$$

whereas for *arbitrary* observation points, the alternative formulation in Eq. (6.3.26) is appropriate, with $A(\phi, \phi'; w)$ taken from Eq. (2b).

For large values of $k\rho$ and $k\rho'$ the asymptotic approximation for $G_\infty(\mathbf{r}, \mathbf{r}')$ is given in Eqs. (6.3.30), with $A(\phi, \phi'; 0)$ substituted from Eq. (2b). The three constituents in Eq. (6.3.30) have direct physical interpretations. G_∞^0 in Eq. (6.3.30a) represents the incident spherical wave in the illuminated region (i.e., the region wherein the source point $\mathbf{r}'$ is visible from the observation point $\mathbf{r}$); since the perfectly absorbing wedge faces do not give rise to a reflected field, G_∞^0 furnishes the entire geometric-optical contribution. G_∞^d in Eq. (6.3.30b) provides the diffraction field exterior to the transition region $|\phi - \phi'| \approx \pi$ surrounding the light-shadow boundary. Viewed in terms of geometrical rays, an incident ray striking the edge gives rise to a cone of diffracted rays, with the cone angle α equal to the angle between the incident ray and the edge as shown in Fig. 6.4.3. The length parameter l appearing in the exponential $\exp(ikl)$ expresses precisely the phase increment for a local plane-wave field following the trajectory $(l_1 + l_2)$ between the source and observation points, in accord with the ray-optical interpretation. For observation points lying near the shadow boundary $|\phi - \phi'| = \pi$, the transition function G_∞^t in Eq. (6.3.30c) is effective and assures the continuity of the total field G_∞.

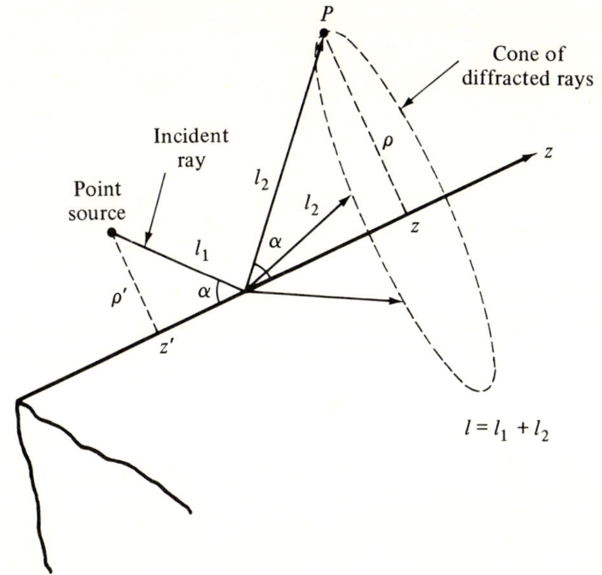

FIG. 6.4.3   Diffracted rays caused by an incident spherical wave.

6.4d   Impulsive Point-Source Excitation

When the point source has the impulsive temporal behavior $\delta(t - t')$, the relevant scalar Green's function is defined by the equation

$$\left(\frac{1}{\rho}\frac{\partial}{\partial\rho}\rho\frac{\partial}{\partial\rho} + \frac{1}{\rho^2}\frac{\partial^2}{\partial\phi^2} + \frac{\partial^2}{\partial z^2} - \frac{1}{\bar{c}^2}\frac{\partial^2}{\partial t^2}\right)\hat{G}_\infty(\mathbf{r}, \mathbf{r}'; t, t')$$

$$= -\frac{\delta(\rho - \rho')}{\rho'}\delta(\phi - \phi')\delta(z - z')\delta(t - t'), \qquad (18)$$

in the domain $0 < (\rho, \rho') < \infty$, $-\infty < (\phi, \phi') < \infty$, $-\infty < (z, z') < \infty$, $-\infty < (t, t') < \infty$, subject to finiteness at $\rho = 0$, and to quiescence when $t < t'$. In an angular transmission representation, the solution for $\hat{G}_\infty$ in the shadow region $|\phi - \phi'| > \pi$ is given by $(2\pi)^{-1}\int_{-\infty}^{\infty} G_\infty(\mathbf{r}, \mathbf{r}') \exp[-i\omega(t - t')]d\omega$, with the time-harmonic Green's function G_∞ taken from Eq. (17) [see Eq. (6.2.14)]. Alternatively, and more directly, one has the closed form solution for arbitrary $\mathbf{r}, \mathbf{r}', t, t'$ in Eq. (6.3.33), the first term of which represents the incident spherical pulse in the illuminated region while the second term [with use of Eq. (2b)] represents the diffraction field $\hat{G}_\infty^d$. In this instance, the diffracted wavefront reaches the observation point along a ray path l_1 that strikes the edge and is reflected therefrom along l_2 at the incidence angle. The diffracted rays, perpendicular to the wavefront, lie on a cone as shown in Fig. 6.4.3.

6.4e   Time-harmonic Plane-wave Excitation

By moving the line source in Fig. 6.4.1 to infinity along the direction ϕ', one may derive the result for plane-wave incidence, with the incident wave, when properly normalized, given by exp $[-ik\rho \cos(\phi - \phi') - i\omega t]$. The corresponding wave function $\bar{u}_\infty(\rho, \phi')$ for the perfectly absorbing wedge satisfies the homogeneous wave equation

$$\left(\frac{1}{\rho}\frac{\partial}{\partial\rho}\rho\frac{\partial}{\partial\rho} + \frac{1}{\rho^2}\frac{\partial^2}{\partial\phi^2} + k^2\right)\bar{u}_\infty(\rho, \phi') = 0 \tag{19}$$

in the domain $0 < \rho < \infty$, $-\infty < (\phi, \phi') < \infty$, subject to boundedness at $\rho = 0$, and a radiation condition on the scattered portion $\bar{u}_\infty^d(\rho, \phi')$ at $(\rho, |\phi|) \to \infty$. In the shadow region $|\phi - \phi'| > \pi$, the solution may be derived from Eq. (5) via the replacement of $H_\mu^{(1)}(k\rho')$ by the first term of its asymptotic expansion in Eq. (8a) and use of the normalization condition in Eq. (5.4.30b),

$$C(k\rho') = \frac{1}{2\sqrt{2\pi k\rho'}} e^{i(k\rho' + \pi/4)} \to 1, \tag{20}$$

to furnish an incident plane wave of unit amplitude. Thus,

$$\bar{u}_\infty(\rho, \phi') = \frac{i}{2}\int_0^{i\infty} (1 - e^{i2\mu\pi})e^{i\mu[|\phi - \phi'| - \pi/2]}H_\mu^{(1)}(k\rho)\,d\mu, \qquad |\phi - \phi'| > \pi. \tag{21}$$

Alternatively, the representation in Eq. (6.3.24) may be employed for *arbitrary* ρ, ϕ, ϕ', with $A(\phi, \phi'; w)$ substituted from Eq. (2b).

For $k\rho \gg 1$, an asymptotic form of the solution, $\bar{u}_\infty \sim \bar{u}_\infty^0 + \bar{u}_\infty^d + \bar{u}_\infty^t$, is given in Eq. (6.3.28), with $A(\phi, \phi'; 0)$ taken from Eq. (2b). $\bar{u}_\infty^0$ in Eq. (6.3.28a) represents the plane wave incident along the direction ϕ'; because of the presence of the wedge (Fig. 6.4.4), its domain of existence is the region $|\phi - \phi'| < \pi$, the illuminated region deduced from simple geometric-optical considerations. $\bar{u}_\infty^d$ in Eq. (6.3.28b) can be interpreted as an outgoing cylindrical wave that appears to emanate from the edge $\rho = 0$; because of the cylindrical "spreading coefficient," the cylindrical wave is smaller then the incident-wave contribution by a factor $O(1/\sqrt{k\rho})$. The cylindrical wave contributes both in the illuminated and shadow regions and represents the diffraction effects almost everywhere; this simple description of the diffraction field fails in the vicinity of the shadow boundary $|\phi - \phi'| = \pi$, where the cylindrical-wave amplitude diverges like $1/(\pi - |\phi - \phi'|)$. In this geometric-optical transition region, whose width may be characterized as in Eq. (6.3.19), the transition function $\bar{u}_\infty^t$ in Eq. (6.3.28c) assumes importance. These considerations are schematized in Fig. 6.4.4.

Higher-order terms in the asymptotic expansion for the diffracted wave $\bar{u}_\infty^d$ may be derived as for the line-source problem. In fact, it is found that $\bar{u}_\infty^d$ is given by the right-hand side of Eq. (9) provided that $C(k\rho') = 1$ and only the terms I_{mo}, $m = 0, 1, 2, \cdots$, are retained. Closed-form expressions for I_{mo} are derived via Eqs. (11) and (12), and the resulting form of $\bar{u}_\infty^d$ is valid for all $|\phi - \phi'| \not\approx \pi$.

Results for oblique incidence with respect to the z axis follow directly from Eq. (6.2.18).

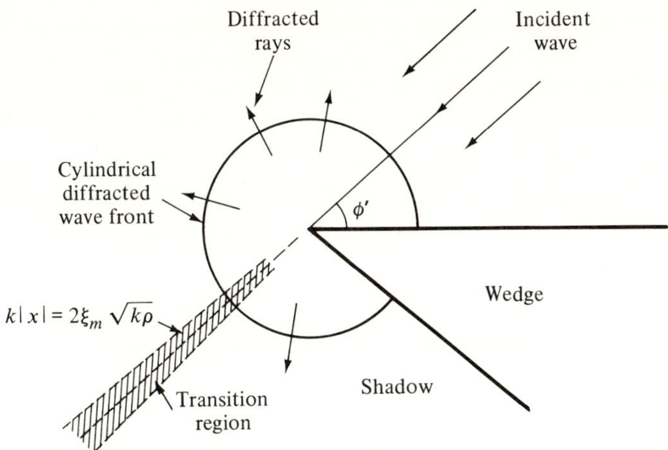

FIG. 6.4.4 Geometrical interpretation of asymptotic field solution for an incident plane wave.

6.4f Impulsive Plane-wave Excitation

When the incident field has the form of a plane-wave pulse $\delta[t - t' + (\rho/\bar{c}) \cos(\phi - \phi')]$, the wave function $\hat{u}_\infty(\mathbf{\rho}, \phi'; t, t')$ for the perfectly absorbing wedge satisfies the time-dependent wave equation,

$$\left(\frac{1}{\rho}\frac{\partial}{\partial\rho}\rho\frac{\partial}{\partial\rho} + \frac{1}{\rho^2}\frac{\partial^2}{\partial\phi^2} - \frac{1}{\bar{c}^2}\frac{\partial^2}{\partial t^2}\right)\hat{u}_\infty(\mathbf{\rho}, \phi'; t, t') = 0, \tag{22}$$

in the domain $0 < \rho < \infty$, $-\infty < (\phi, \phi') < \infty$, $-\infty < (t, t') < \infty$, subject to boundedness at $\rho = 0$, and quiescence for $t < t'$. $\bar{c}$ is the propagation speed in the medium.

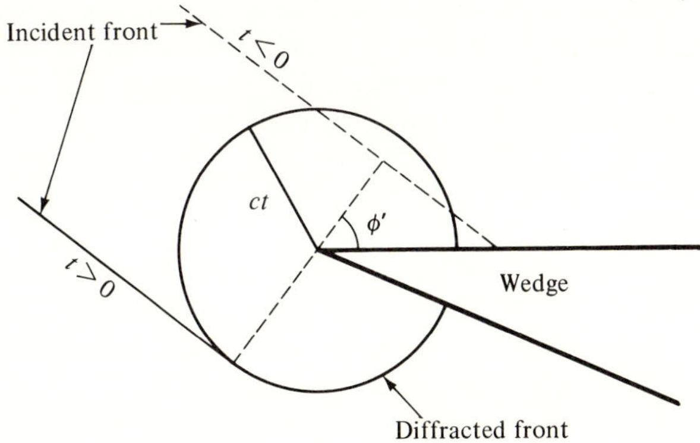

FIG. 6.4.5 Diffraction of plane-wave-pulse by a perfectly absorbing wedge.

The solution for arbitrary $(\mathbf{\rho}, t)$ is given in Eq. (6.3.32), with $A(\phi, \phi' ; w)$ taken from Eq. (2b), and its physical interpretation for $t' = 0$ is seen from Fig. 6.4.5. Diffraction does not take place until observation times $t > 0$ since the incident pulse does not reach the edge until $t = 0$. Absence of a reflected pulse is attributed to the perfectly absorbing wedge faces.

6.5 PERFECTLY CONDUCTING WEDGE AND HALF-PLANE

6.5a Angular Transmission Representation

When the wedge faces at $\phi = 0$, φ in Fig. 6.4.1 are perfectly conducting, it is necessary to distinguish between the boundary conditions satisfied by the E and H mode potential fuunctions in Secs. 5.2 and 6.2 (see Reference 2 for a listing of various results and literature citations). For the E modes with respect to z (i.e., $H_z \equiv 0$), the potential functions vanish on the wedge faces, whereas for the H modes (with $E_z \equiv 0$), the normal derivatives vanish [see Eqs. (2.3. 36a) and (2.3.37a)]. In terms of the one-dimensional angular Green's functions satisfying Eq. (6.2.7), these requirements imply that

$$g'(\phi, \phi' ; \mu) = 0 \text{ at } \phi = 0, \varphi, \qquad \text{for } E\text{-mode case,} \qquad (1a)$$

$$\frac{d}{d\phi} g''(\phi, \phi' ; \mu) = 0 \text{ at } \phi = 0, \varphi, \qquad \text{for } H\text{-mode case,} \qquad (1b)$$

where the $'$ and $''$ superscripts have been introduced to distinguish the two situations. In Eq. (6.2.7), p has been replaced by μ as in Sec. 6.3, and the subscript ϕ has been deleted. Solutions may be given in the closed forms of Eqs. (3.4. 51),

$$g'(\phi, \phi' ; \mu) = \frac{\sin \mu\phi_< \sin \mu(\varphi - \phi_>)}{\mu \sin \mu\varphi}, \qquad (2a)$$

$$g''(\phi, \phi' ; \mu) = \frac{\cos \mu\phi_< \cos \mu(\varphi - \phi_>)}{- \mu \sin \mu\varphi}, \qquad (2b)$$

where $\phi_<$ and $\phi_>$ denote the lesser and greater, respectively, of the angle coordinates ϕ and ϕ'. Alternatively, in terms of a series representation involving images in an infinitely extended ϕ space [Eq. (3.4.59)],

$$\begin{matrix} g'(\phi, \phi' ; \mu) \\ g''(\phi, \phi' ; \mu) \end{matrix} = \sum_{n=-\infty}^{\infty} g_\infty(\phi, 2n\varphi + \phi' ; \mu) \mp \sum_{n=-\infty}^{\infty} g_\infty(\phi, 2n\varphi - \phi' ; \mu) \qquad (3)$$

where g_∞ denotes the Green's function in Eq. (6.4.1). Substitution of these expressions into Eq. (6.3.3), or its counterparts derived from Eqs. (6.2.10), (6.2. 12), and (6.2.14), provides the solution for the multidimensional scalar Green's functions in the geometrical-optical shadow region.

To obtain an angular transmission representation for the Green's functions valid everywhere, it is necessary to evaluate the function $A(\phi, \phi' ; w)$ defined in Eq. (6.3.6). Substitution of Eq. (3) into Eq. (6.3.6), interchange of the order of summation and integration, and use of Eq. (6.4.2b) yields

$$A(\phi, \phi' \,; w) = B(\phi, \phi' \,; w) \mp B(\phi, -\phi' \,; w), \qquad \text{for} \quad \begin{matrix} E \text{ modes,} \\ H \text{ modes,} \end{matrix} \tag{4}$$

where $B = B_1 + B_2$ with

$$B_{1,2}(\phi, \phi' \,; w) = \sum_{n=-\infty}^{\infty} \frac{1}{\pi \mp [w - (\phi - \phi') + 2n\varphi]}, \qquad \phi < \phi'. \tag{5}$$

This series can be summed into a closed form via the representation for the cotangent function,

$$\cot x = \frac{1}{x} + 2x \sum_{n=1}^{\infty} \frac{1}{x^2 - (n\pi)^2} = \sum_{n=-\infty}^{\infty} \frac{1}{x - n\pi}. \tag{6}$$

The last expression in Eq. (6) has the same form as the series in Eq. (5); while it diverges as written there, the series obtained by grouping together terms with $\pm n$, i.e., $[(x - n\pi)^{-1} + (x + n\pi)^{-1}]$, does converge and yields the first equality in Eq. (6). The series in Eq. (5) is to be understood in this sense, and it is assumed that the images have been grouped so as to yield a convergent result. Thus,

$$B_{1,2} = \frac{\pi}{2\varphi} \cot \left[\frac{\pi}{2\varphi} [\pi \mp w \pm (\phi - \phi')] \right], \tag{7}$$

and B can be expressed compactly by use of the formula

$$\cot (\alpha + \beta) + \cot (\alpha - \beta) = \frac{2 \sin 2\alpha}{\cos 2\beta - \cos 2\alpha}. \tag{8}$$

An analogous expression may be derived for $\phi > \phi'$.

When these results are substituted into Eqs. (6.3.8), (6.3.24), or (6.3.26), one obtains expressions for the various Green's functions at arbitrary observation points. In addition to the pole singularity of $A(\phi, \phi' \,; w)$ at $\pi - |\phi - \phi'|$, which has been accounted for explicitly in Sec. 6.3, attention must be given to possible residue contributions from other poles, to be considered subsequently [see remarks following Eq. (6.3.8)]. Since the functions multiplying $A(\phi, \phi' \,; w)$ in the diffraction integrals are even functions of w and the integration interval is symmetrical about $w = 0$, these integrals are affected only by the even part (in w) of A or B. The result is therefore unchanged if w in B_1 and (or) B_2 is replaced by $-w$, and one may construct a variety of functions $B(\phi, \phi' \,; w)$, giving rise to the same diffraction integral (i.e., having the same even part).† Several representations derived in the literature by alternative methods of analysis are given below. From Eqs. (7) and (8), one obtains directly

$$B(\phi, \phi' \,; w) = \frac{\pi}{\varphi} \sin \frac{\pi^2}{\varphi} \frac{1}{\cos [(\pi/\varphi)(w - \phi + \phi')] - \cos (\pi^2/\varphi)}, \tag{9a}$$

while replacement of w by $-w$ in B_2 yields[1]

$$B(\phi, \phi' \,; w) = \frac{\pi}{\varphi} \frac{\sin [(\pi/\varphi)(\pi - w)]}{\cos [(\pi/\varphi)(\phi - \phi')] - \cos [(\pi/\varphi)(\pi - w)]}. \tag{9b}$$

Or, upon writing $2B_1(w) = B_1(w) + B_1(-w)$,

---

† These remarks apply also to Sec. 6.4.

$$B_1 = \frac{\pi}{2\varphi} \frac{\sin\left[(\pi/\varphi)(\pi + \phi - \phi')\right]}{\cos(\pi w/\varphi) - \cos\left[(\pi/\varphi)(\pi + \phi - \phi')\right]},\tag{9c}$$

whence[3]

$$B(\phi, \phi'\,;w) = \frac{\pi}{2\varphi}\left[\frac{\sin\left[(\pi/\varphi)(\pi + \phi - \phi')\right]}{\cos(\pi w/\varphi) - \cos\left[(\pi/\varphi)(\pi + \phi - \phi')\right]}\right.$$

$$\left. + \frac{\sin\left[(\pi/\varphi)(\pi - \phi + \phi')\right]}{\cos(\pi w/\varphi) - \cos\left[(\pi/\varphi)(\pi - \phi + \phi')\right]}\right].\tag{9d}$$

These representations for B are, of course, equal only in connection with the diffraction integrals. It is noted that the contributing part of B is an even function of $\phi - \phi'$, so this quantity can be replaced by $|\phi - \phi'|$.

The pole singularities of $A(\phi, \phi'\,;w)$ in the complex w plane are made evident by the series representation in Eq. (5). When $|\phi - \phi'| > \pi$ and $0 \leq (\phi, \phi') \leq \varphi$, none of the poles crosses the integration path along the imaginary axis in the complex w plane, so no residue contributions appear; as in Eq. (6.3.5), this condition corresponds to observation points located in the geometrical shadow region of Fig. 6.4.1 and can be satisfied only for wedge angles $\varphi > \pi$. However, when the angular interval between the observation point coordinate ϕ and the coordinate ϕ' of the source point or ϕ'_n, $\hat{\phi}'_n$ of an image point is less than $\pm\pi$, a relevant pole singularity traverses the integration path and gives rise to a residue term. Thus, the first term on the right-hand side of Eq. (6.3.8) is now replaced by the finite sums

$$\bar{G}^0(\boldsymbol{\rho}, \boldsymbol{\rho}') = \sum_n \frac{i}{4} H_0^{(1)}[kR(\phi'_n)]U(\pi - |\phi - \phi'_n|)$$

$$\mp \sum_n \frac{i}{4} H_0^{(1)}[kR(\hat{\phi}'_n)]U(\pi - |\phi - \hat{\phi}'_n|),\tag{10}$$

where $n = 0, \pm 1, \ldots, 0 \leq (\phi, \phi') \leq \varphi$. The upper and lower sign in Eq. (10) applies to the E modes and H modes, respectively, and

$$\phi'_n = 2n\varphi + \phi', \ \hat{\phi}'_n = 2n\varphi - \phi', \ R(\alpha) = \sqrt{\rho^2 + \rho'^2 - 2\rho\rho'\cos(\phi - \alpha)}.$$
$$\tag{10a}$$

$R(\alpha)$ represents the distance from the observation point (ρ, ϕ) to a (real or image) source point at (ρ', α). A similar modification occurs in Eqs. (6.3.24) and (6.3.26) pertaining to plane-wave and point-source excitation, respectively. The term corresponding to $\phi'_0 = \phi'$ in Eq. (10) represents the incident wave in the illuminated region, whereas the other terms represent exactly the singly and multiply reflected waves predicted on the basis of geometrical optics. It is evident that many reflections are possible for small wedge angles φ, whereas, when $\varphi > \pi$, a maximum of three terms remains in the series: ϕ'_0, $\hat{\phi}'_0$, ϕ'_1. These conclusions are schematized in Fig. 6.5.1, from which it is observed that an observation point P in a trough region $\varphi < \pi$ may be reached by the direct ray a, two singly reflected rays $b_{1,2}$, doubly reflected rays c, etc., each of which is represented by a term in Eq. (10), with the $\mp$ signs accounting for the reflec-

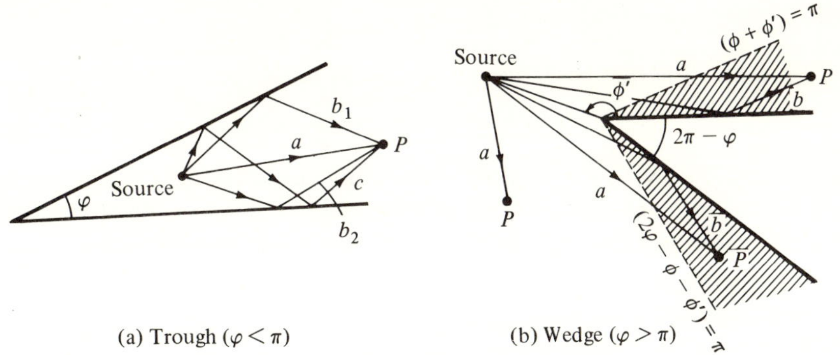

(a) Trough ($\varphi < \pi$)   (b) Wedge ($\varphi > \pi$)

FIG. 6.5.1 Geometric-optical contributions.

tion coefficients -1 and $+1$ for E modes and H modes, respectively, at a perfect conductor. In the wedge-shaped region $\varphi > \pi$, on the other hand, there is at most one singly reflected ray that contributes only when the observation point lies in the shaded region, whereas no reflected contribution arises for other observation-point locations. The Heaviside unit functions in Eq. (10) provide for these limited domains of existence of a particular geometric-optical ray species.

One observes from Eq. (9a) that $B \equiv 0$ when $\varphi = \pi/n$, where n is any positive integer. Hence, the diffraction integrals vanish identically, and the line-source excited field is representable *exactly* in closed form by its image contributions in Eq. (10), with analogous conclusions applicable to plane-wave or point-source excitation.

6.5b Radial Transmission Representation

While the angular transmission representation discussed in Sec. 6.5a is useful for the calculation of the quasi-optic field in the range of large $k\rho$ and $k\rho'$, it is inappropriate when either the source point or the observation point lies near the edge ($k\rho'$ or $k\rho$ small). In this parameter range, it is useful to employ a radial transmission formulation, which involves a one-dimensional Green's function $g_\rho(\rho, \rho'; \mu)$ in the radial domain, and eigenfunctions $\Phi_\mu(\phi)$ in the angular domain. The two-dimensional Green's function $\bar{G}(\boldsymbol{\rho}, \boldsymbol{\rho}')$ corresponding to excitation by a time-harmonic line source is now represented by (see Sec. 3.3c)

$$\bar{G}(\boldsymbol{\rho}, \boldsymbol{\rho}') = \sum_\mu \Phi_\mu(\phi)\Phi_\mu(\phi')g_\rho(\rho, \rho'; \mu), \tag{11}$$

which formulation is alternative to the one given in Eq. (6.2.8), and may be derived therefrom by passing to the complex μ plane and deforming contours as in Sec. 3.3c. The orthonormal angular eigenfunctions are listed in Eqs. (3.2.47) and (3.2.48), and the radial Green's function g_ρ is given in Eq. (3.4.93). Thus,[1]

$$\bar{G}'(\boldsymbol{\rho}, \boldsymbol{\rho}') = \frac{\pi i}{\varphi} \sum_{m=1}^{\infty} J_\mu(k\rho_<)H_\mu^{(1)}(k\rho_>) \sin \mu\phi \sin \mu\phi', \qquad \mu = \frac{m\pi}{\varphi}, \tag{12}$$

and

$$\bar{G}''(\boldsymbol{\rho}, \boldsymbol{\rho}') = \frac{\pi i}{2\varphi} J_0(k\rho_<)H_0^{(1)}(k\rho_>) + \frac{\pi i}{\varphi} \sum_{m=1}^{\infty} J_\mu(k\rho_<)H_\mu^{(1)}(k\rho_>) \cos \mu\phi \cos \mu\phi'.$$

(13)

Since $J_\mu(k\rho_<) \sim (k\rho_</\mu)^\mu$, $\mu \gg k\rho_<$, the series converges rapidly and can be used for numerical evaluation when either the source point or the observation point is located near the edge. In particular, the behavior of $\bar{G}'$ and $\bar{G}''$ near the edge is given by

$$\bar{G}' \propto (k\rho)^{\pi/\alpha}, \quad \bar{G}'' \propto \text{constant} + O[(k\rho)^{\pi/\alpha}], \qquad \text{as } \rho \to 0. \qquad (14)$$

One may verify that the electromagnetic fields derived from Eqs. (12) and (13) satisfy the edge conditions in Eq. (1.5.37).

In view of the remarks made in connection with Eq. (6.2.5), one may derive the corresponding representations for the three-dimensional Green's functions $G'(\mathbf{r}, \mathbf{r}')$ and $G''(\mathbf{r}, \mathbf{r}')$ from Eqs. (12) and (13) upon

replacing k by $\sqrt{k^2 - \zeta^2}$ and performing the operation $(1/2\pi) \int_{-\infty}^{\infty} d\zeta e^{i\zeta(z-z')}$,

(15)

with $\text{Im} \sqrt{k^2 - \zeta^2} \geq 0$.

By letting $\rho' \to \infty$, one may derive from Eqs. (12) and (13) the result for excitation by an incident plane wave $\exp[-ik\rho \cos(\phi - \phi')]$. Since $J_\mu(k\rho_<)$ decays when $\mu > k\rho_<$, the contributing range of the series is bounded in μ, and negligible error is introduced through use of the asymptotic approximation in Eq. (6.3.12) for all relevant μ as $k\rho_> \to \infty$. Imposition of the normalization condition Eq. (5.4.30b) yields the following expressions for the wavefunctions:

$$\bar{u}'(\boldsymbol{\rho}, \phi') = \frac{4\pi}{\varphi} \sum_{m=1}^{\infty} J_\mu(k\rho) \sin \mu\phi \sin \mu\phi' e^{-i\mu\pi/2}, \qquad \mu = \frac{m\pi}{\varphi}, \qquad (16a)$$

$$\bar{u}''(\boldsymbol{\rho}, \phi') = \frac{2\pi}{\varphi} \left\{ J_0(k\rho) + 2 \sum_{m=1}^{\infty} J_\mu(k\rho) \cos \mu\phi \cos \mu\phi' e^{-i\mu\pi/2} \right\}. \qquad (16b)$$

Results for oblique incidence are obtained at once via Eq. (6.2.18).

6.5c   Time-harmonic Line-source Excitation

The electromagnetic fields excited by a line source of electric or magnetic currents may be inferred from the scalar E-and H-mode Green's functions $\bar{G}'(\boldsymbol{\rho}, \boldsymbol{\rho}')$ and $\bar{G}''(\boldsymbol{\rho}, \boldsymbol{\rho}')$, respectively, as in Eqs (6.2.1b) and (6.2.2b).[2] These Green's functions satisfy in the domain $0 < (\rho, \rho') < \infty$, $0 \leq (\phi, \phi') \leq \varphi$ the wave equation (6.2.3), subject to the boundary conditions

$$\bar{G}' = 0, \qquad \frac{\partial \bar{G}''}{\partial \phi} = 0 \quad \text{at } \phi = 0, \varphi, \qquad (17)$$

boundedness at $\rho = 0$, and a radiation condition at $\rho \to \infty$. For arbitrary values of $(\boldsymbol{\rho}, \boldsymbol{\rho}')$, the solution is given in a radial transmission representation by Eqs. (12) and (13), and in an angular transmission representation by Eq. (6.3.8), subject to the modifications in Eqs. (4) and (10). When the wedge angle $\varphi > \pi$, only three terms contribute to the geometric-optical series in Eq. (10), and one obtains the result

$$\bar{G}(\boldsymbol{\rho}, \boldsymbol{\rho}') = \bar{G}^0(\boldsymbol{\rho}, \boldsymbol{\rho}') + \bar{G}_1(\boldsymbol{\rho}, \boldsymbol{\rho}'), \tag{18}$$

where

$$\bar{G}^0(\boldsymbol{\rho}, \boldsymbol{\rho}') = \frac{i}{4} H_0^{(1)}[kR(\phi')](U(\pi - |\phi - \phi'|)$$

$$\mp \frac{i}{4} H_0^{(1)}[kR(-\phi')]U(\pi - \phi - \phi')$$

$$\mp \frac{i}{4} H_0^{(1)}[kR(2\varphi - \phi')]U[\pi - (2\varphi - \phi - \phi')]. \tag{18a}$$

$U(x)$ is the Heaviside unit function defined in Eq. (6.3.8) and $\bar{G}_1$ is the diffraction integral

$$\bar{G}_1(\boldsymbol{\rho}, \boldsymbol{\rho}') = \frac{1}{8\pi} \int_{i\infty}^{-i\infty} H_0^{(1)}(k\chi)[B(\phi, \phi'; w) \mp B(\phi, -\phi'; w)]dw. \tag{18b}$$

The upper sign is appropriate to the E-mode Green's function $\bar{G}'$, while the lower sign yields the H-mode Green's function $\bar{G}''$. R is defined in Eq. (10a), various forms for B are given in Eqs. (9), and

$$\chi = (\rho^2 + \rho'^2 + 2\rho\rho' \cos w)^{1/2}.$$

When $k\rho \gg 1$, $k\rho' \gg 1$, but with $|\phi - \phi'|$, $(\phi + \phi')$, $(2\varphi - \phi - \phi') \napprox \pi$, one has the asymptotic approximation

$$\bar{G} \sim \bar{G}^0 + \bar{G}^d, \tag{19}$$

with $\bar{G}^0$ taken from Eq. (18a) [which may be simplified for large arguments by use of Eq. (6.3.12)], and $\bar{G}^d$ obtained as in Eq. (6.3.15):

$$\bar{G}^d = -2C(k\rho)C(k\rho')[B(\phi, \phi'; 0) \pm B(\phi, -\phi'; 0)]$$

$$= \frac{-ie^{ik(\rho+\rho')}}{4k\sqrt{\rho\rho'}\varphi} \left(\sin \frac{\pi^2}{\varphi}\right) \left[\frac{1}{\cos[(\pi/\varphi)(\phi - \phi')] - \cos(\pi^2/\varphi)}\right.$$

$$\left.\mp \frac{1}{\cos[(\pi/\varphi)(\phi + \phi')] - \cos(\pi^2/\varphi)}\right], \quad \text{for } \frac{E \text{ modes,}}{H \text{ modes.}} \tag{19a}$$

For *arbitrary* ϕ and ϕ', the following transition functions must be added to Eq. (19):

$$\bar{G}^t(\boldsymbol{\rho}, \boldsymbol{\rho}') = \bar{G}_\infty^t(\boldsymbol{\rho}; \rho', \phi') \mp \bar{G}_\infty^t(\boldsymbol{\rho}; \rho', -\phi') \mp \bar{G}_\infty^t(\boldsymbol{\rho}; \rho', 2\varphi - \phi'),$$

$$\text{for } \frac{E \text{ modes,}}{H \text{ modes,}} \tag{20}$$

where $\bar{G}_\infty^t(\boldsymbol{\rho}; \rho', \phi')$ is given in Eq. (6.4.7), it being recalled that $\bar{G}_\infty^t(\boldsymbol{\rho}; \rho', \phi')$ is to be so used that $\xi \to 0$ implies that $\pi - |\phi - \phi'| \to 0$.

Discussion

The physical interpretation of the result in Eq. (19) in terms of geometric-optical and diffraction effects proceeds as for the perfectly absorbing wedge in

Sec. 6.4a. As in Fig. 6.4.1, the diffracted contribution $\bar{G}^d$ is in the form of a cylindrical wave emanating from the edge and penetrating all the various geometric-optical domains. Owing to the perfect reflectivity of the wedge faces in the present case, the geometric-optical field $\bar{G}^0$ includes not only the incident field in the illuminated region but also the fields reflected from the sides of the wedge, which latter contributions have already been analyzed in connection with Fig. 6.5.1. The diffracted field amplitude in Eq. (19a) diverges on the shadow boundary $|\phi - \phi'| = \pi$ and also on the lines $\phi = \pi - \phi'$, $\phi = (2\varphi - \phi' - \pi)$ which bound the domain of existence of the waves reflected according to geometrical optics from the wedge faces at $\phi = 0$ and $\phi = \varphi$, respectively [see Fig. 7.4.5(a) for a plot of the diffracted field variation]. Continuity of $\bar{G}$ through these transition regions is assured by use of the transition functions in Eq. (20), only one of which is relevant at a time since these functions are negligibly small outside their particular transition domain. As noted from Eq. (5), the behavior of $B(\phi, \phi'; w)$ near $w_p = \pi - |\phi - \phi'_n|$ or $w_p = \pi - |\phi - \hat{\phi}'_n|$, where ϕ'_n and $\hat{\phi}'_n$ are the image locations in Eq. (10a), is of the same form as for the previously investigated case $w_p = \pi - |\phi - \phi'|$. The relevant transition functions may therefore be inferred directly from the infinite angular space problem in Sec. 6.4, with one transition function corresponding to each image solution. For a restricted range of observation angles $0 \leq \phi \leq \varphi$, only a limited number of these may become contributory [see Eq. (10)]. In connection with the field values *on* the boundaries of the various geometric-optical domains, see Eq. (6.3.21) and its analogue for the reflected waves.

Higher-order terms in the asymptotic expansion for $\bar{G}^d$ may be derived by considerations analogous to those in Eqs. (6.4.9) et seq., starting from Eqs. (2) and (6.3.3).

6.5d   Impulsive Line-source Excitation

When the line source has the impulsive time dependence $\delta(t - t')$, the fields are derivable from the Green's function $\hat{\bar{G}}(\rho, \rho'; t, t')$, which satisfies the differential equation (6.4.13), subject to the stated boundary conditions, with the modification for the ϕ domain given Eq. (17).[3] The result for arbitrary ρ, ρ', t, t', may be taken directly from Eq. (6.3.41), with $A(\phi, \phi'; i\beta)$ inserted from Eqs. (4) and (9b), and additional terms included to account for the reflected fields corresponding to Eq. (10):

$$\hat{\bar{G}}(\rho, \rho'; t, t') = \sum_n \left\{ \frac{U(\pi - |\phi - \phi'_n|)U(t - t' - [R(\phi'_n)/\bar{c}])}{2\pi\sqrt{(t - t')^2 - [R^2(\phi'_n)/\bar{c}^2]}} \right.$$

$$\mp \frac{U(\pi - |\phi - \hat{\phi}'_n|)U(t - t' - [R(\hat{\phi}'_n)/\bar{c}])}{2\pi\sqrt{(t - t')^2 - [R^2(\hat{\phi}'_n)/\bar{c}^2]}} \right\}$$

$$- \left\{ \frac{1}{4\pi^2} \int_{-\cosh^{-1}M}^{\cosh^{-1}M} \frac{d\beta}{\sqrt{(t - t')^2 - [f^2(\beta)/\bar{c}^2]}} \right.$$

$$\times [\operatorname{Re} B(\phi, \phi'; i\beta) \mp \operatorname{Re} B(\phi, -\phi'; i\beta)] \right\} U\!\left(t - t' - \frac{\rho + \rho'}{\bar{c}}\right), \qquad (21)$$

where the upper and lower signs refer to the E- and H-mode cases, respectively;

the various definitions in Eq. (10a) apply, $f(\beta) = (\rho^2 + \rho'^2 + 2\rho\rho' \cosh \beta)^{1/2}$, $M = [\bar{c}^2(t - t')^2 - \rho^2 - \rho'^2]/2\rho\rho'$, and, from Eqs. (9), with $\psi = (\phi - \phi')$,

$$\text{Re } B(\phi, \phi'; i\beta) =$$

$$\frac{(\pi/\varphi)\,[\cos\,(\pi\psi/\varphi)\,\cosh\,(\pi\beta/\varphi) - \cos\,(\pi^2/\varphi)]\,\sin\,(\pi^2/\varphi)}{\cosh^2\,(\pi\beta/\varphi) + \cos^2\,(\pi\psi/\varphi) - \sin^2\,(\pi^2/\varphi) - 2\cos^2(\pi^2/\varphi)\cos\,(\pi\psi/\varphi)\,\cosh\,(\pi\beta/\varphi)}$$

$$(21a)$$

When the wedge angle $\varphi > \pi$, only the terms with $\hat{\phi}'_0$, $\hat{\phi}'_0$, $\hat{\phi}'_1$ contribute. The physical interpretation of this result is discussed in connection with Fig. 6. 4. 2. For angles $\varphi = (\pi/n)$, $n = 1, 2, \ldots$, one has $B \equiv 0$, and the geometrical optics portion of the solution is exact.

6.5e   Time-harmonic Point-source Excitation

The fields excited by a longitudinal electric or magnetic dipole as in Eqs. (6.4.15) may be inferred from the scalar E-mode Green's function $G'(\mathbf{r}, \mathbf{r}')$ and the scalar H-mode Green's function $G''(\mathbf{r}, \mathbf{r}')$, respectively, which satisfy Eq. (6.4.16) in the domain $0 < (\rho, \rho') < \infty$, $0 \leq (\phi, \phi') \leq \varphi$, $-\infty < (z, z') < \infty$, subject to the boundary conditions

$$G' = 0, \qquad \frac{\partial G''}{\partial \phi} = 0 \quad \text{at } \phi = 0, \varphi, \tag{22}$$

boundedness at $\rho = 0$, and a radiation condition at $r \to \infty$.[2,3]

For arbitrary values of $\mathbf{r}$ and $\mathbf{r}'$, solutions are given in a radial transmission representation by [see Eqs (12), (13), and (15)]

$$G(\mathbf{r}, \mathbf{r}') = \frac{i}{4\varphi} \int_{-\infty}^{\infty} d\zeta \sum_{m=0}^{\infty} \epsilon_m J_\mu(\sqrt{k^2 - \zeta^2}\,\rho_<)$$

$$\times H_\mu^{(1)}(\sqrt{k^2 - \zeta^2}\,\rho_>)e^{i\zeta\,(z-z')} \begin{Bmatrix} \sin \mu\phi \sin \mu\phi' \\ \cos \mu\phi \cos \mu\phi' \end{Bmatrix}, \qquad \text{for} \begin{array}{l} E \text{ modes,} \\ H \text{ modes,} \end{array} \tag{23}$$

where $\epsilon_m = 1$, $m = 0$, while $\epsilon_m = 2$, $m \geq 1$, and $\mu = m\pi/\varphi$. The integration path proceeds as in Fig. 5.3.6a, with $\text{Im }\sqrt{k^2 - \zeta^2} \geq 0$. In an angular transmission representation, the solution for arbitrary $\mathbf{r}$ and $\mathbf{r}'$ is given by Eq. (6.3. 26), provided that $A(\phi, \phi'\,;w)$ is taken from Eqs. (4) and (9), and that additional geometric-optical constituents analogous to those in Eq. (10) are included. Thus,

$$G(\mathbf{r}, \mathbf{r}') = G^0(\mathbf{r}, \mathbf{r}') + G_1(\mathbf{r}, \mathbf{r}'), \tag{24}$$

where G^0 is the geometric-optical contribution,

$$G^o(\mathbf{r}, \mathbf{r}') = \sum_n \left\{ \frac{e^{ik|\mathbf{r}-\mathbf{r}'_n|}}{4\pi|\mathbf{r} - \mathbf{r}'_n|} U(\pi - |\phi - \phi'_n|) \mp \frac{e^{ik|\mathbf{r}-\hat{\mathbf{r}}'_n|}}{4\pi|\mathbf{r} - \hat{\mathbf{r}}'_n|} U(\pi - |\phi - \hat{\phi}'_n|)\right\},$$

$$\text{for}\quad \begin{array}{l} E \text{ modes,} \\ H \text{ modes,} \end{array} \tag{24a}$$

and $|\mathbf{r} - \mathbf{r}'_n| = \sqrt{R^2(\phi'_n) + (z - z')^2}$, $|\mathbf{r} - \hat{\mathbf{r}}'_n| = \sqrt{R^2(\hat{\phi}'_n) + (z - z')^2}$, with R defined in Eq. (10a). The function G_1 is given in the integral form

$$G_1(\mathbf{r}, \mathbf{r}') = -\frac{i}{8\pi^2} \int_{i\infty}^{-i\infty} \frac{e^{ik\gamma}}{\gamma} [B(\phi, \phi' ; w) \mp B(\phi, -\phi' ; w)]dw, \qquad \text{for} \quad \begin{array}{l} E \text{ modes,} \\ H \text{ modes,} \end{array}$$

$$\text{(24b)}$$

where $\gamma = [\rho^2 + \rho'^2 + (z - z')^2 + 2\rho\rho' \cos w]^{1/2}$.

When $k\rho$ and $k\rho'$ are large, $G(\mathbf{r}, \mathbf{r}')$ has the asymptotic approximation

$$G(\mathbf{r}, \mathbf{r}') \sim G^o(\mathbf{r}, \mathbf{r}') + G^d(\mathbf{r}, \mathbf{r}') + G^t(\mathbf{r}, \mathbf{r}'), \qquad (25)$$

where G^o is given in Eq. (24a), G^d is the diffracted contribution in Eq. (6.3.30b),

$$G^d = -\frac{e^{i(kl+\pi/4)}}{4\varphi\sqrt{2\pi k\rho\rho'l}} \sin \frac{\pi^2}{\varphi}$$

$$\times \left[\frac{1}{\cos\left[(\pi/\varphi)(\phi - \phi')\right] - \cos(\pi^2/\varphi)} \mp \frac{1}{\cos\left[(\pi/\varphi)(\phi + \phi')\right] - \cos(\pi^2/\varphi)} \right],$$

$$\text{for} \quad \begin{array}{l} E \text{ modes,} \\ H \text{ modes,} \end{array} \qquad\qquad\qquad (25a)$$

and G^t is the transition function,

$$G^t(\mathbf{r} ; \rho', \phi', z') = \sum_n [G^t_\infty(\mathbf{r} ; \rho', \phi'_n, z') \mp G^t_\infty(\mathbf{r} ; \rho', \hat{\phi}'_n, z')], \qquad \text{for} \quad \begin{array}{l} E \text{ modes,} \\ H \text{ modes.} \end{array}$$

$$\text{(25b)}$$

$G^t_\infty(\mathbf{r} ; \rho', \phi', z')$ is given in Eq. (6.3.30c), and l is defined in Eq. (6.3.30b) (see also Fig. 6.4.3 for a physical interpretation). For wedge angles $\varphi > \pi$, only the contributions from $\phi'_0 = \phi'$, $\hat{\phi}'_0 = -\phi'$ and $\hat{\phi}'_1 = 2\varphi - \phi'$ are relevant in Eqs. (24a) and (25b). G^t is negligible except for observation points in transition regions defined by $|\phi - \phi'_n| \approx \pi$ or $|\phi - \hat{\phi}'_n| \approx \pi$.

6.5f   Impulsive Point-source Excitation

When the point source has the temporal behavior $\delta(t - t')$, the fields can be derived from the scalar Green's functions $\hat{G}'(\mathbf{r}, \mathbf{r}' ; t, t')$ and $\hat{G}''(\mathbf{r}, \mathbf{r}' ; t, t')$ corresponding to the E and H-mode cases, respectively. They satisfy Eq. (6.4.18) in the domain $0 < (\rho, \rho') < \infty$, $0 \le (\phi, \phi') \le \varphi$, $-\infty < (z, z') < \infty$, $-\infty < (t, t') < \infty$, subject to the stated boundary conditions, with the modification for the ϕ domain as in Eq. (22). The solution for arbitrary $\mathbf{r}, \mathbf{r}', t, t'$ is taken from Eq. (6.3.33), with the inclusion of reflected geometric-optical contributions:

$$\hat{G}(\mathbf{r}, \mathbf{r}' ; t, t') = \sum_n \left\{ \frac{\delta(t - t' - |\mathbf{r} - \mathbf{r}'_n|/\bar{c})}{4\pi|\mathbf{r} - \mathbf{r}'_n|} U(\pi - |\phi - \phi'_n|) \right.$$

$$\mp \frac{\delta(t - t' - |\mathbf{r} - \hat{\mathbf{r}}'_n|/\bar{c})}{4\pi|\mathbf{r} - \hat{\mathbf{r}}'_n|} U(\pi - |\phi - \hat{\phi}'_n|) \bigg\}$$

$$+ \frac{\bar{c}}{4\pi} \frac{[\operatorname{Re} B(\phi, \phi' ; i\beta) \mp \operatorname{Re} B(\phi, -\phi' ; i\beta)]}{\rho\rho' \sinh \beta} U\left(t - t' - \frac{l}{\bar{c}}\right),$$

$$\text{for} \quad \begin{array}{l} E \text{ modes,} \\ H \text{ modes,} \end{array} \qquad\qquad (26)$$

where $\mathbf{r}'_n$, $\hat{\mathbf{r}}'_n$, l, β are defined in Eqs. (24a), (6.3.30b), and (6.3.33), respectively, and Re B is taken from Eq. (21a).

6.5g   Time-harmonic Plane-wave Excitation

When a plane wave $\exp\left[-ik\rho\cos(\phi-\phi')-i\omega t\right]$ is incident on the perfectly conducting wedge from the direction ϕ', the corresponding wave function $\bar{u}(\boldsymbol{\rho}, \phi')$ satisfies Eq. (6.4.19) in the domain $0 < \rho < \infty$, $0 \leq (\phi, \phi') \leq \varphi$, subject to boundedness at $\rho = 0$, a radiation condition on the scattered portion at $\rho \to \infty$, and to

$$\bar{u}'(\boldsymbol{\rho}, \phi') = 0, \qquad \frac{\partial \bar{u}''(\boldsymbol{\rho}, \phi')}{\partial \phi} = 0 \quad \text{at } \phi = 0, \varphi. \tag{27}$$

The E-mode function $\bar{u}'$ is relevant when the electric vector in the incident field is parallel to the z axis, whereas the H-mode function $\bar{u}''$ applies to magnetic-field polarization paralled to z.[2,3]

In a radial transmission representation, the solution is given by Eqs. (16), whereas in an angular transmission representation, the solution follows from Eq. (6. 3. 24) provided that one includes geometric-optical reflected contributions as in Eq. (10) and inserts $A(\phi, \phi'; w)$ from Eqs. (4) and (9):

$$\bar{u}(\boldsymbol{\rho}, \phi') = \bar{u}^{\circ}(\boldsymbol{\rho}, \phi') + \bar{u}_1(\boldsymbol{\rho}, \phi'), \tag{28}$$

where $\bar{u}^{\circ}$ is the geometric-optical field

$$\bar{u}^{\circ}(\boldsymbol{\rho}, \phi') = e^{-ik\rho\cos(\phi-\phi')} U(\pi - |\phi - \phi'|) \mp e^{-ik\rho\cos(\phi+\phi')} U(\pi - \phi - \phi')$$
$$\mp\, e^{-ik\rho\cos(2\varphi-\phi-\phi')} U[\pi - (2\varphi - \phi - \phi')], \tag{28a}$$

while $\bar{u}_1$ is the diffraction integral

$$\bar{u}_1(\boldsymbol{\rho}, \phi') = \frac{-i}{2\pi} \int_{i\infty}^{-i\infty} e^{ik\rho\cos w}[B(\phi, \phi'; w) \mp B(\phi, -\phi'; w]\, dw. \tag{28b}$$

The upper and lower signs correspond to the E- and H-mode wavefunctions, respectively, and the formula in Eq. (28a) is valid when $\varphi > \pi$; for arbitrary values of φ, additional terms as in Eq. (10) are required.

For large values of $k\rho$, the following asymptotic approximation applies:

$$\bar{u}(\boldsymbol{\rho}, \phi') \sim \bar{u}^o(\boldsymbol{\rho}, \phi') + \bar{u}^d(\boldsymbol{\rho}, \phi') + \bar{u}^i(\boldsymbol{\rho}, \phi'), \tag{29}$$

where, for $\varphi > \pi$, $\bar{u}^o$ is given in Eq. (28a), while the diffracted field $\bar{u}^d$ is obtained directly from Eq. (19a) subject to the normalization in Eq. (6.4.20) [see also Eqs. (6.3.28)]:

$$\bar{u}^d(\boldsymbol{\rho}, \phi') = -\frac{e^{i(k\rho+\pi/4)}\sqrt{\pi}}{\varphi\sqrt{2k\rho}} \sin\frac{\pi^2}{\varphi}$$
$$\times \left[\frac{1}{\cos[(\pi/\varphi)(\phi-\phi')] - \cos(\pi^2/\varphi)} \mp \frac{1}{\cos[(\pi/\varphi)(\phi+\phi')] - \cos(\pi^2/\varphi)}\right]. \tag{29a}$$

In the transition regions surrounding the shadow boundary $|\phi - \phi'| = \pi$ or the reflected wave boundaries $\phi = \pi - \phi', \phi = 2\varphi - \phi' - \pi$, one must employ the appropriate transition function contained in

$$\bar{u}^t(\mathbf{\rho}, \phi') = \bar{u}^t_\infty(\mathbf{\rho}, \phi') \mp \bar{u}^t_\infty(\mathbf{\rho}, -\phi') \mp \bar{u}^t_\infty(\mathbf{\rho}, 2\varphi - \phi'), \quad \text{for} \quad \begin{matrix} E \text{ modes,} \\ H \text{ modes,} \end{matrix} \qquad (29\text{b})$$

with $\bar{u}^t_\infty(\mathbf{\rho}, \phi')$ given in Eq. (6.3.28c).

The physical interpretation of the asymptotic result is analogous to that in Fig. 6.4.4, provided that the reflected-wave contribution in $\bar{u}^o$ is included.

For oblique incidence, the solution follows from an application of Eq. (6.2.18).

6.5h   Impulsive Plane-wave Excitation

When the incident field is in the form of the plane-wave pulse $\delta[t - t' + (\rho/\bar{c}) \cos (\phi - \phi')]$, the wavefunctions $\hat{u}'(\mathbf{\rho}, \phi'; t, t')$ and $\hat{u}''(\mathbf{\rho}, \phi'; t, t')$ corresponding to the E- and H-mode cases, respectively, satisfy the wave equation (6.4.22) in the domain $0 < \rho < \infty$, $0 \le (\phi, \phi') \le \varphi$, $-\infty < (t, t') < \infty$, subject to the stated boundary conditions, with modifications for the ϕ domain as in Eq. (27). The solution for arbitrary $\mathbf{\rho}, \phi', t, t'$ is given by Eqs. (6.3.32) and (4), provided that geometric-optical reflected wave contributions are included:

$$\hat{u}(\mathbf{\rho}, \phi'; t, t') = \delta\left[t - t' + \frac{\rho}{\bar{c}} \cos (\phi - \phi')\right] U(\pi - |\phi - \phi'|)$$

$$\mp \delta\left[t - t' + \frac{\rho}{\bar{c}} \cos (\phi + \phi')\right] U(\pi - \phi - \phi')$$

$$\mp \delta\left[t - t' + \frac{\rho}{\bar{c}} \cos (2\varphi - \phi - \phi')\right] U[\pi - (2\varphi - \phi - \phi')]$$

$$- \frac{1}{\pi} \frac{U(t - t' - \rho/\bar{c})}{\sqrt{(t - t')^2 - (\rho/\bar{c})^2}} [\operatorname{Re} B(\phi, \phi'; i\beta) \mp \operatorname{Re} B(\phi, -\phi'; i\beta)],$$

$$\text{for} \quad \begin{matrix} E \text{ modes,} \\ H \text{ modes,} \end{matrix} \qquad (30)$$

where $\beta = \cosh^{-1}[\bar{c}(t - t')/\rho]$, $\operatorname{Re} B$ is given in Eq. (21a), and it has been assumed that $\varphi > \pi$. The physical interpretation of this result is the same as in Fig. 6.4.5, with inclusion of the reflected-wave fronts.

6.5i   Special Case: the Half-plane

Although the formulas in the preceding sections apply for all wedge angles $0 < \varphi \le 2\pi$, simplifications occur for certain special values of φ. It has already been noted that when $\varphi = \pi/n$, $n = 1, 2\ldots$, the geometrical optics field derived by the theory of images provides the exact solution. Another important special case is provided by $\varphi = 2\pi$, in which instance the wedge degenerates into a half-plane. While this obstacle configuration still gives rise to diffraction, many of the previously derived solutions can be reduced to a simpler form.[3,4]

Time-harmonic line-source excitation

When $\varphi = 2\pi$, the diffraction integral in Eq. (18b) can be reduced to a more elegant form derived first by MacDonald.[5] Instead of dealing directly with Eq. (18b), it is preferable to start from the pulse solution in Eq. (39) and to recover therefrom the time-harmonic result upon multiplication by $\exp(-i\omega t')$ and integration over t' from $-\infty$ to $+\infty$. Since

$$Q = \int_{-\infty}^{\infty} U[\hat{t} - \beta] \frac{e^{-i\omega t'}}{\sqrt{(\bar{c}\hat{t})^2 - R^2(\phi')}}\, dt'$$

$$= \int_{-\infty}^{t-\beta} \frac{e^{-i\omega t'}}{\sqrt{(\bar{c}\hat{t})^2 - R^2(\phi')}}\, dt', \qquad \beta \text{ real}, \tag{31}$$

where $\hat{t} = t - t'$, the change of variable $\bar{c}\hat{t} = R(\phi')\cosh x$, with $R(\phi')$ defined in Eq. (10a), yields

$$Q = \frac{1}{\bar{c}} e^{-i\omega t} \int_{\xi'}^{\infty} e^{ikR(\phi')\cosh x}\, dx, \quad \xi' = \cosh^{-1}\left[\frac{\bar{c}\beta}{R(\phi')}\right], \qquad k = \frac{\omega}{\bar{c}}. \tag{31a}$$

If $\beta = R(\phi')/\bar{c}$, one has $\xi' = 0$, and the resulting integral is recognized as [see Eq. (6.3.22), with $\mu = 0$]

$$\int_0^{\infty} e^{ikR(\phi')\cosh x}\, dx = \frac{\pi i}{2} H_0^{(1)}[kR(\phi')]. \tag{32}$$

Upon omitting the common time factor $\exp(-i\omega t)$, one finds the desired expression for the steady-state Green's function ($\phi' < \pi$):

$$\bar{G}(\boldsymbol{\rho}, \boldsymbol{\rho}') = \bar{I}(\boldsymbol{\rho}; \rho', \phi') \mp \bar{I}(\boldsymbol{\rho}; \rho', -\phi'), \tag{33}$$

where

$$\bar{I}(\boldsymbol{\rho}; \rho', \phi') = \frac{i}{4} H_0^{(1)}[kR(\phi')] U(\pi - |\phi - \phi'|)$$

$$- \frac{1}{4\pi} \operatorname{sgn}(\pi - |\phi - \phi'|) \int_{\xi}^{\infty} e^{ikR(\phi')\cosh x}\, dx,$$

$$\xi = \cosh^{-1}\left[\frac{\rho + \rho'}{R(\phi')}\right]. \tag{33a}$$

The upper sign in Eq. (33) refers to the E-mode case while the lower sign refers to the H-mode case. In view of Eq. (32), Eq. (33a) can also be written as

$$\bar{I}(\boldsymbol{\rho}; \rho', \phi') = \frac{1}{4\pi} \int_{-\xi}^{\infty} e^{ikR(\phi')\cosh x}\, dx, \qquad |\phi - \phi'| < \pi, \tag{34}$$

so the result for $\bar{G}(\boldsymbol{\rho}, \boldsymbol{\rho}')$ can be expressed solely in terms of the integral of $\exp[ikR(\phi')\cosh x]$.

A series representation of the solution may be obtained from Eqs. (12) and (13) with $\mu = m/2$.

Impulsive line-source excitation

When $\varphi = 2\pi$, the integral appearing in Eq. (21) can be expressed in closed form[3] because of the simplification

$$\mathrm{Re}\, B(\phi, \phi'; i\beta) = \frac{\cos(\psi/2)\cosh(\beta/2)}{\cosh\beta + \cos\psi}, \qquad \varphi = 2\pi. \tag{35}$$

Introduction into the integral in Eq. (21) of the successive changes of variable

$$\cosh\beta = 1 + v^2, \qquad \cosh\frac{\beta}{2}\,d\beta = \sqrt{2}\,dv, \tag{36a}$$

and

$$v = b\sin\gamma, \qquad b = \sqrt{\frac{(\bar{c}\hat{t})^2 - (\rho + \rho')^2}{2\rho\rho'}}, \tag{36b}$$

yields, with $\hat{t} = t - t'$,

$$D(\phi, \phi'; \hat{t}) = -\frac{\bar{c}}{2\pi^2}\frac{\cos(\psi/2)}{\sqrt{\rho\rho'}}\int_0^{\pi/2}\frac{d\gamma}{1 + \cos\psi + b^2\sin^2\gamma}, \tag{37}$$

where $D(\phi, \phi'; \hat{t})$ denotes the contribution to the integral in Eq. (21) arising from the $\mathrm{Re}\, B(\phi, \phi'; i\beta)$ term. The integral in Eq. (37) can be evaluated in closed form,

$$D(\phi, \phi'; \hat{t}) = -\frac{\cos(\psi/2)}{\sqrt{\rho\rho'}}\frac{\bar{c}}{4\pi}\frac{1}{\sqrt{2\cos^2(\psi/2)}\sqrt{1 + \cos\psi + b^2}}$$

$$= -\frac{\bar{c}}{4\pi\sqrt{(\bar{c}\hat{t})^2 - R^2(\phi')}}\,\mathrm{sgn}\cos\frac{\psi}{2}, \tag{38}$$

where $\mathrm{sgn}\, x = \pm 1$, $x \gtrless 0$, and $R(\phi')$ is defined in Eq. (10a). Thus, Eq. (21) furnishes the following form for the half-plane Green's function,

$$\left.\begin{array}{c}\hat{G}'(\boldsymbol{\rho}, \boldsymbol{\rho}'; t) \\ \hat{G}''(\boldsymbol{\rho}, \boldsymbol{\rho}'; t)\end{array}\right\} = \frac{\bar{c}}{2\pi}U[\pi - |\phi - \phi'|]U\!\left[\hat{t} - \frac{R(\phi')}{\bar{c}}\right]\frac{1}{\sqrt{(\bar{c}\hat{t})^2 - R^2(\phi')}}$$

$$\mp \frac{\bar{c}}{2\pi}U[\pi - (\phi + \phi')]U\!\left[\hat{t} - \frac{R(-\phi')}{\bar{c}}\right]\frac{1}{\sqrt{(\bar{c}\hat{t})^2 - R^2(-\phi')}}$$

$$- \frac{\bar{c}}{4\pi}U\!\left[\hat{t} - \frac{(\rho + \rho')}{\bar{c}}\right]\left[\frac{\mathrm{sgn}(\pi - |\phi - \phi'|)}{\sqrt{(\bar{c}\hat{t})^2 - R^2(\phi')}} \mp \frac{\mathrm{sgn}(\pi - \phi - \phi')}{\sqrt{(\bar{c}\hat{t})^2 - R^2(-\phi')}}\right], \tag{39}$$

with $\hat{t} = t - t'$. In view of the symmetry of the half-plane configuration, it has been assumed without loss of generality that $\phi' < \pi$. The various geometric optical regions entering into Eq. (39) are illustrated in Fig. 6.4.2. The result in Eq. (39) demonstrates again how diffraction weakens the singularity that exists across the incident wavefront; the diffracted field is, in fact, finite, except along the geometric-optical boundary lines.

Time-harmonic point-source excitation

The diffraction integral in Eq. (24b) can be reduced to a simpler form involving a finite integration of the Hankel function which replaces the exponential function in Eq. (31a).[4,5] In the series representation in Eq. (23), put $\mu = m/2$.

Impulsive point-source excitation

The result is given in Eq. (26), with the simplified form from Eq. (35), and with only the $\phi'_n = \phi'$, $\hat{\phi}'_n = -\phi'$ terms included (for $\phi' < \pi$).

Time-harmonic plane-wave excitation

When $\varphi = 2\pi$, the diffraction integral in Eq. (28b) can be reduced to the Fresnel-integral type of function $F(\xi)$ defined in Eq. (6.3.18).[1,4] One notes from Eq. (9b) that

$$B(\phi, \phi'; w) = \frac{\cos (w/2)}{2\left(\cos \dfrac{\phi - \phi'}{2} - \sin \dfrac{w}{2}\right)}, \qquad \varphi = 2\pi, \qquad (40)$$

whence the resulting integral in Eq. (28b) can be simplified by the change of variable $\cos w = 1 + is^2$, $-\infty < s < \infty$, i.e.,

$$s = \sqrt{2}\, e^{i\pi/4} \sin \frac{w}{2}, \qquad \frac{dw}{ds} = \sqrt{2}\, e^{-i\pi/4} \sec \frac{w}{2}. \qquad (41)$$

Thus,

$$\bar{I}(\boldsymbol{\rho}, \phi') = \frac{-i}{2\pi} \int_{i\infty}^{-i\infty} e^{ik\rho \cos w}\, B(\phi, \phi'; w)\, dw$$

$$= -\frac{i}{8\pi} e^{ik\rho} \int_{-\infty}^{\infty} e^{-k\rho s^2} \frac{ds}{s - b}, \qquad b = \sqrt{2}\, e^{i\pi/4} \cos \frac{\phi - \phi'}{2}, \qquad (42)$$

which can be expressed alternatively in terms of the error-function complement [see Eqs. (4.4.5a) and (4.4.14); the fraction $1/(s - b)$ in the integrand of Eq. (42) can be replaced by $b/(s^2 - b^2)$ because of the symmetrical integration interval]. Hence, for $\phi' < \pi$ but for arbitrary $\boldsymbol{\rho}$, one finds that the wavefunction in Eq. (33) reduces to

$$\bar{u}(\boldsymbol{\rho}, \phi') = e^{-ik\rho \cos (\phi - \phi')} U(\pi - |\phi - \phi'|) \mp e^{-ik\rho \cos (\phi + \phi')} U(\pi - \phi - \phi')$$

$$- e^{ik\rho} \frac{F(\xi_1)}{2} \operatorname{sgn}(\pi - |\phi - \phi'|) \pm e^{ik\rho} \frac{F(\xi_2)}{2} \operatorname{sgn}(\pi - \phi - \phi'), \qquad (43)$$

for $\dfrac{E}{H}$ modes, where

$$F(\xi) = \frac{2}{\sqrt{\pi}} e^{-i2\xi^2} \int_{(1-i)\xi}^{\infty} e^{-x^2}\, dx, \qquad \xi_{1,2} = \sqrt{k\rho}\left|\cos \frac{\phi \mp \phi'}{2}\right|. \qquad (43a)$$

By an asymptotic evaluation of the function $F(\xi)$ as in Eq. (6.3.18a), one may derive the results in Eq. (29), with $\varphi = 2\pi$.

A harmonic-series form of the solution is given in Eqs. (16), with $\mu = m/2$.

Impulsive plane-wave excitation

The solution is given by Eq. (30), simplified through use of Eq. (35).

6.6 WEDGE WITH VARIABLE IMPEDANCE WALLS

In an angular transmission representation, the description of the perfectly absorbing and perfectly conducting wedge configurations has given rise, respectively, to the simple boundary conditions in Eqs. (6.4.1) and (6.5.1), imposed on the angular characteristic Green's function $g(\phi, \phi' ; \mu)$ in Eq. (6.2.7). More general than either of these is the linear homogeneous boundary condition

$$\frac{dg}{d\phi} = \mp ia_{1,2}g \quad \text{at } \phi = \begin{matrix} 0 \\ \varphi \end{matrix}, \tag{1}$$

where a_1 and a_2 are arbitrary constants. The perfectly absorbing wedge result follows from the identification $a_{1,2} = \mu$, whereas in the perfectly conducting case, $a_{1,2} = \infty$ and $a_{1,2} = 0$ for the E-mode and H-mode problems, respectively.†

If the one-dimensional characteristic Green's function $g(\phi, \phi' ; \mu)$ is to be utilized to synthesize a multidimensional Green's function, it follows that the latter must satisfy on the wedge faces $\phi = 0, \varphi$ boundary conditions of the same type as in Eq.(1). This requirement can be interpreted in terms of a surface impedance Z_s, as will be demonstrated for the time-harmonic line source excitations in Eqs. (6.2.1a) or (6.2.2a). For the electric line current in Eq. (6.2.1a), the electromagnetic fields are derived as in Eq. (6.2.1b) from a scalar Green's function $\bar{G}(\rho, \rho')$ which satisfies the two-dimensional wave equation (6.2.3). The boundary condition

$$\frac{\partial \bar{G}}{\partial \phi} = \mp ia_{1,2}\bar{G} \quad \text{at } \phi = \begin{matrix} 0 \\ \varphi \end{matrix}, \tag{2}$$

can thus be rephrased as

$$E_z = \mp Z_{s1,2}H_\rho \quad \text{at} \quad \phi = \begin{matrix} 0 \\ \varphi \end{matrix}, \qquad (E\text{-mode case}), \tag{3}$$

where

$$Z_{s1,2} = \zeta \frac{k\rho}{a_{1,2}} \tag{3a}$$

are the surface impedances on the wedge faces at $\phi = 0, \varphi$, respectively; ζ is the wave impedance in the unbounded exterior medium, and k is the wavenumber. Alternatively, for the magnetic current excitation in Eq. (6.2.2a), the electromagnetic fields may be derived from the same scalar Green's function, provided that in view of [see Eq. (6.2.2b)]

$$E_\rho = \pm Z_{s1,2}H_z \quad \text{at} \quad \phi = \begin{matrix} 0 \\ \varphi \end{matrix}, \qquad (H\text{-mode case}), \tag{4}$$

the surface impedances are related to $a_{1,2}$ as follows:

$$Z_{s1,2} = \zeta \frac{a_{1,2}}{k\rho}. \tag{4a}$$

---

† It is recalled that E and H modes distinguish solutions with $H_z \equiv 0$ and $E_z \equiv 0$, respectively. Also, μ denotes a separation parameter and is not to be confused with the same symbol employed elsewhere for the permeability. The time dependence is $\exp(-i\omega t)$.

Thus, to ensure constancy of the parameters $a_{1,2}$ required for separability of the boundary condition in Eq. (2), the surface impedances *increase linearly with p* for the *E*-mode problem [Eq. (3a)] but *decrease inversely with p* for the *H*-mode problem [Eq. (4a)], with Re $a_{1,2} > 0$ for a passive surface impedance. It is of interest to observe that in view of Eqs. (2)–(4), diffraction by a constant impedance wedge cannot be solved by the separation of variables procedure and therefore requires more sophisticated mathematical techniques than the variable impedance problems described above.

While the angular characteristic Green's function g satisfying Eqs. (1) and (6.2.7) (with $p \to \mu$) can readily be constructed for arbitrary values of $a_{1,2}$ [see Eq. (3.4.51)], salient features of the electromagnetic field due to the presence of a variable surface impedance may be demonstrated by prescribing this boundary condition on one surface only. The other wedge face is assumed to be either perfectly absorbing (Sec. 6.6a) or perfectly conducting (Sec. 6.6b).

6.6a   One Perfectly Absorbing and One Variable Impedance Wall

Representation emphasizing quasi-optic properties

When the wedge face at $\phi = 0$ has a surface impedance Z_s (*E*-mode problem) or admittance $1/Z_s$ (*H*-mode problem) which increases linearly away from the edge $p = 0$, while the wedge face at $\phi = \varphi$ is perfectly absorbing, the relevant angular characteristic Green's function satisfies the differential equation (6.2.7) (with $p \equiv \mu$), the boundary condition (1) at $\phi = 0$, and a "no reflection" condition at $\phi = \varphi$; the latter is equivalent to extending the ϕ domain to $+\infty$ and requiring a radiation condition. The solution is given by [see Eq. (3.4.51), with $\vec{\Gamma} = 0$]

$$g(\phi, \phi' ; \mu) = g_\infty(\phi, \phi' ; \mu) + \overleftarrow{\Gamma}(\mu)g_\infty(\phi, -\phi' ; \mu), \qquad (5)$$

with

$$g_\infty(\phi, \phi' ; \mu) = \frac{e^{\pm i\mu|\phi - \phi'|}}{\mp 2i\mu}, \quad \text{Im } \mu \gtrless 0,$$

where g_∞ in Eq. (6.4.1) is the Green's function for the bilaterally unbounded angular space, while

$$\overleftarrow{\Gamma}(\mu) = \frac{\pm\mu - a_1}{\pm\mu + a_1}, \qquad \text{Im } \mu \gtrless 0, \qquad (5a)$$

is the reflection coefficient at $\phi = 0$ [note that c_1 in Eq. (3.4.53) corresponds to $-a_1$]. This result can be interpreted in terms of the response to the source at ϕ' and a single weighted image at $-\phi'$ on a bilaterally unbounded angular transmission line.

The two-dimensional Green's function corresponding to excitation by a line source as in Fig. 6.3.1 may be obtained by substitution of Eq. (5) into Eq. (6.3.3):

$$\bar{G}(\boldsymbol{\rho}, \boldsymbol{\rho}') = \bar{G}_\infty(\boldsymbol{\rho}, \boldsymbol{\rho}')$$

$$+ \frac{1}{4} \int_0^{i\infty} \mu(1 - e^{i2\mu\pi}) H_\mu^{(1)}(k\rho) H_\mu^{(1)}(k\rho') \overleftarrow{\Gamma}(\mu) g_\infty(\phi, -\phi'\,;\mu)\, d\mu. \qquad (6)$$

When Eq. (6.3.8), with $A(\phi, \phi'\,;w)$ taken from Eq. (6.4.2b), is employed for the perfectly absorbing wedge Green's function $\bar{G}_\infty(\boldsymbol{\rho}, \boldsymbol{\rho}')$, this result is valid for arbitrary $|\phi - \phi'|$, but the integral in Eq. (6) converges only for $\phi + \phi' > \pi$ [see discussion following Eq. (6.3.3)]. To obtain a representation valid at all observation angles, we proceed as in Sec. 6.3 and utilize Eq. (6.3.4) to find for the second term on the right-hand side of Eq. (6)[6]:

$$W_1 = \frac{1}{4\pi} \int_{i\infty}^{-i\infty} dw\, H_0^{(1)}(k\chi) \int_0^{i\infty} \mu e^{i\mu(w-\pi)} (1 - e^{i2\mu\pi}) \overleftarrow{\Gamma}(\mu) g_\infty(\phi, -\phi'\,;\mu)\, d\mu, \qquad (7)$$

which, upon use of the identity

$$\overleftarrow{\Gamma}(\mu) = \frac{\mu - a_1}{\mu + a_1} = 1 - \frac{2a_1}{\mu + a_1}, \qquad (8)$$

can be expressed as

$$W_1 = \bar{G}_\infty(\boldsymbol{\rho}\,;\rho', -\phi') + \int_{i\infty}^{-i\infty} H_0^{(1)}(k\chi) W_2\, dw, \qquad (9)$$

where

$$W_2 = \frac{-ia_1}{4\pi} \int_0^{i\infty} e^{i\mu(w-\pi)} (1 - e^{i2\mu\pi}) e^{i\mu(\phi+\phi')} \frac{d\mu}{\mu + a_1} \qquad (9a)$$

$$= \frac{-ia_1}{4\pi} [N(w + \phi + \phi' - \pi, a_1) - N(w + \phi + \phi' + \pi, a_1)]. \qquad (9b)$$

N is defined as

$$N(\alpha, a_1) = \int_0^{i\infty} e^{i\mu\alpha} \frac{d\mu}{\mu + a_1} = \int_0^\infty \frac{e^{-x\alpha}}{x - ia_1}\, dx \qquad (10a)$$

$$= e^{-i\alpha a_1} E_1(-i\alpha a_1), \qquad (10b)$$

where E_1 is the exponential integral

$$E_1(y) = \int_y^\infty \frac{e^{-x}}{x}\, dx. \qquad (10c)$$

While the integral representation in Eq. (10a) requires $\mathrm{Re}\,\alpha > 0$ and ia_1 nonpositive, the exponential integral formulation in Eq. (10b) is also valid when $\mathrm{Re}\,\alpha < 0$, since $E_1(y)$ is a regular function of y in the complex y plane cut along the negative real axis; $E_1(y) \sim -\ln y$ as $y \to 0$, so a branch-point singularity is located at $y = 0$. The integral W_2 in Eq. (9b) has branch-point singularities at $w_{b1} = \pi - (\phi + \phi')$ and at $w_{b2} = \pi + (\phi + \phi')$. Therefore, as $\phi + \phi'$ decreases through π, the branch point at w_{b1} moves across the integration path, and the value of W_2 differs from that given in Eq. (9b) by the branch-cut contribution, as shown in Fig. 6.6.1. Thus, for arbitrary values of ϕ, ϕ' one has:

$$\bar{G}(\boldsymbol{\rho}, \boldsymbol{\rho}') = \bar{G}_{\infty}(\boldsymbol{\rho}, \boldsymbol{\rho}') + \bar{G}_{\infty}(\boldsymbol{\rho}; \rho', -\phi')$$

$$- \frac{ia_1}{4\pi} \int_{i\infty}^{-i\infty} H_0^{(1)}(k\chi)[N(w + \phi + \phi' - \pi, a_1) - N(w + \phi + \phi' + \pi, a_1)]dw$$

$$- U(\pi - \phi - \phi')\frac{ia_1}{4\pi} \int_{P_b} H_0^{(1)}(k\chi)N(w + \phi + \phi' - \pi, a_1)dw, \qquad (11)$$

where the representations (6.3.8), with Eq. (6.4.2b), are to be employed for $\bar{G}_{\infty}$, and where N is expressed as in Eq. (10b). The first two terms on the right-hand side of Eq. (11) represent the Green's function for a perfectly reflecting surface at $\phi = 0$, while the remaining contributions constitute correction terms

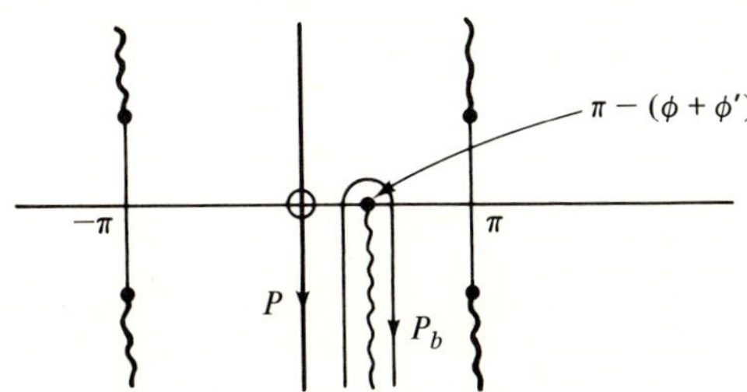

FIG. 6.6.1 Contours of integration in the complex w plane when $\phi + \phi' < \pi$.

that account for the presence of the surface impedance Z_s. The branch cut surrounded by P_b in Fig. 6.6.1 has been chosen so that $\text{Im}(k\chi) > 0$ along P_b (see Fig. 6.3.2). The above solution, derived from an angular transmission formulation, is directly suited to a study of the field behavior when k is large.

Asymptotic evaluation

To effect an asymptotic evaluation of the integrals in Eq. (11) for large values of k, we proceed as in Sec. 6.3 and replace $H_0^{(1)}(k\chi)$ by its asymptotic expression (6.3.12). The first integral has the same form as in Eq. (6.3.8) and yields by comparison with Eq. (6.3.15) the asymptotic result

$$4ia_1[N(\phi + \phi' - \pi, a_1) - N(\phi + \phi' + \pi, a_1)]C(k\rho)C(k\rho'), \quad C(x) = \frac{e^{i(x+\pi/4)}}{2\sqrt{2\pi x}}, \tag{12a}$$

with N defined in Eq. (10b). Since $\exp(ik\chi)$ decays along P_b, the major contribution to the second integral arises from the vicinity of the branch point $w_b = \pi - (\phi + \phi')$. Upon approximating $\chi(w)$ by $R + (\rho\rho'/R)(\cos w - \cos w_b)$, $R \equiv \chi(w_b)$ [see Eq. (6.3.16b)], one finds

$$Q_1 \equiv \frac{i}{4}\int_{P_b} H_0^{(1)}(k\chi)N\,dw \sim C(kR)e^{-ik(\rho\rho'/R)\cos w_b}\int_{P_b} e^{ik(\rho\rho'/R)\cos w}N\,dw. \tag{12b}$$

To simplify the integral in Eq. (12b), it is convenient to introduce the change of variable $w = w_b - iz$, whence

$$Q_2 \equiv \int_{P_b} e^{i\alpha \cos w} N(w - w_b, a_1)\, dw = i \int_{P_b'} e^{i\alpha \cos (w_b - iz)} N(-iz, a_1)\, dz, \tag{13}$$

where the contour P_b' encloses in the positive sense the branch cut along the positive real z axis. Since the exponential integral possesses the series representation[7]

$$E_1(y) = -\gamma - \ln y - \sum_{n=1}^{\infty} \frac{(-y)^n}{n!\, n}, \qquad \gamma = 0.5772, \tag{14}$$

its multivaluedness is contained in the $\ln y$ term, which alone contributes to the integral over the contour P_b'. Upon recalling that $\ln ze^{i2\pi} = \ln z + i2\pi$, one may reduce Eq. (13) to

$$Q_2 = 2\pi \int_0^{\infty} e^{i\alpha \cos (w_b - iz) - a_1 z}\, dz, \qquad w_b = \pi - (\phi + \phi') > 0. \tag{15}$$

An approximate evaluation of Q_2 for large positive α can be carried out by approximating $\cos(w_b - iz) \approx \cos w_b + iz \sin w_b$, whence

$$Q_2 \sim \frac{2\pi}{\alpha \sin w_b + a_1} e^{i\alpha \cos w_b} \tag{16}$$

Thus,

$$Q_1 \sim C(kR) \frac{2\pi}{(k\rho\rho'/R)\sin(\phi + \phi') + a_1}, \qquad R = \sqrt{\rho^2 + \rho'^2 - 2\rho\rho'\cos(\phi + \phi')}. \tag{17}$$

Upon substituting these results into Eq. (11) and employing Eq. (6.4.6) for the asymptotic representation of $\bar{G}_{\infty}$, one obtains[6]

$$\bar{G}(\boldsymbol{\rho}, \boldsymbol{\rho}') \sim \bar{G}^0(\boldsymbol{\rho}, \boldsymbol{\rho}') + \bar{G}^d(\boldsymbol{\rho}, \boldsymbol{\rho}'), \tag{18}$$

where the geometric-optical field is given by

$$\bar{G}^o = \frac{i}{4} H_0^{(1)}(k|\boldsymbol{\rho} - \boldsymbol{\rho}'|) U(\pi - |\phi - \phi'|)$$

$$+ C(kR)\overleftarrow{\Gamma}\left[\frac{k\rho\rho'}{R}\sin(\phi + \phi')\right] U(\pi - \phi - \phi'), \tag{18a}$$

with $\overleftarrow{\Gamma}(\mu)$ defined in Eq. (5a) and $C(x) = (8\pi x)^{-1/2} \exp(ix + i\pi/4)$. The diffracted field is

$$\bar{G}^d = C(k\rho)C(k\rho')\{-2A(\phi, \phi'; 0) - 2A(\phi, -\phi'; 0)$$

$$+ 4ia_1[N(\phi + \phi' - \pi, a_1) - N(\phi + \phi' + \pi, a_1)]\}, \tag{18b}$$

with

$$A(\phi, \phi'; 0) = \frac{1}{\pi - |\phi - \phi'|} + \frac{1}{\pi + |\phi - \phi'|}, \tag{18c}$$

and N taken from Eq. (10b). When $a_1 = 0$, then $\overleftarrow{\Gamma} = 1$, and $a_1 N(\alpha, a_1) \to 0$, so $\bar{G}$ reduces to

$$\bar{G}(\boldsymbol{\rho}, \boldsymbol{\rho}') = \bar{G}_{\infty}(\boldsymbol{\rho}; \rho', \phi') + \bar{G}_{\infty}(\boldsymbol{\rho}; \rho', -\phi'), \tag{19}$$

the correct solution for the perfectly conducting (zero impedance) case corresponding to magnetic current excitation [see Eq. (4a)]. When $a_1 \rightarrow \infty$, then $\overleftarrow{\Gamma} \rightarrow -1$ and[7]

$$E_1(y) \sim \frac{e^{-y}}{y} \sum_{m=0}^{\infty} \frac{m!}{(-y)^m}, \qquad |y| \rightarrow \infty, \, |\arg y| < \frac{3\pi}{2}. \qquad (20)$$

Thus,

$$4ia_1[N(\alpha, a_1) - N(\beta, a_1)] = -4\left[\frac{1}{\alpha} - \frac{1}{\beta}\right] + O\left(\frac{1}{a_1}\right), \qquad |a_1| \rightarrow \infty, \quad (21)$$

and $\bar{G}$ reduces to the expression in Eq. (19), with the plus sign on the right-hand side replaced by minus. The resulting formula accounts correctly for a perfect conductor in the case of electric current excitation [see Eq. (3a)]. The exponential integral in Eq. (10b) permits the detailed calculation of the diffraction effect for any finite surface impedance of the form specified in Eqs. (3a) or (4a). The physical interpretation of the diffracted-wave contribution $\bar{G}^d$ is as shown in Fig. 6.4.1. It may be noted that the quasi-optic field solution in Eq. (18) does not exhibit explicitly a term that can be identified as a surface wave, the existence of which on the variable impedance boundary is discussed in connection with Eq. (28). In contrast to a surface wave on a constant impedance plane (Sec. 5.7), its contribution in the present instance is contained in the diffraction field function $\bar{G}^d$.

When a_1 is very small, the surface impedance changes rapidly from a value of zero at $\rho = 0$ to large values at small but finite distances from the edge [see Eq. (3a)]. Since the last term inside the braces in Eq. (18b) can be neglected in this case, the diffraction field given by the first two terms is the same as for an infinite impedance wedge. This result emphasizes the fact that the diffraction effects are influenced not merely by the surface properties *at* the edge but also in its vicinity, and that the geometric-optical concepts of localized reflection or diffraction can be employed only when the surface properties do not vary rapidly over a distance equal to the local wavelength. The case of slow variation obtains when a_1 is large, and in this instance (as noted above) the diffraction field is essentially the same as for a zero impedance wedge, thereby justifying the use of the local reflection principle.

The field contribution $\bar{G}^o$ comprises the direct wave and a reflected wave that appears to originate from an image source located at $(\rho', -\phi')$. Hence, the reflected wave reaches the observation point via a ray path leaving the surface at the incident angle δ as shown in Fig. 6.6.2. The reflected-wave amplitude is given by

$$\overleftarrow{\Gamma}[\xi \sin(\phi + \phi')] = \frac{\xi \sin(\phi + \phi') - a_1}{\xi \sin(\phi + \phi') + a_1}, \qquad \xi = \frac{k\rho\rho'}{R}. \qquad (22)$$

From geometrical optics one expects an incident ray to be reflected at the specular angle, with the reflected-ray amplitude given by the plane-wave reflection coefficient determined from the properties of the surface *at* the point of reflection. For the present case of a variable surface impedance $Z_s(\rho)$, the geometric-

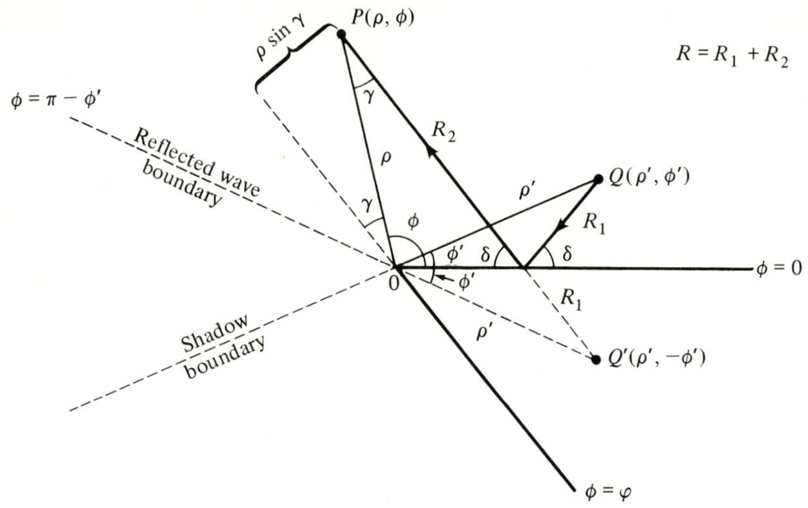

FIG. 6.6.2 Reflected rays and geometric-optical domains.

optical reflection coefficient $\overleftarrow{\Gamma}{}^{o}$ at a point $\boldsymbol{\rho} = (x, 0)$ is determined via the expression [see Eqs. (2.4.3), (2.2.15c), and (2.2.15d)]

$$\overleftarrow{\Gamma}{}^{o} = \begin{cases} -\dfrac{Z_s(x) - \zeta \sin \delta}{Z_s(x) + \zeta \sin \delta}, & H\text{-mode case}, & (23a) \\[3mm] \dfrac{Z_s(x) - \zeta/\sin \delta}{Z_s(x) + \zeta/\sin \delta}, & E\text{-mode case}, & (23b) \end{cases}$$

where δ is the angle of incidence measured from the surface $\phi = 0$, and ζ is the wave impedance in the medium.

To show that $\overleftarrow{\Gamma}$ in Eq. (22) is identical with $\overleftarrow{\Gamma}{}^{o}$, we note first that $\overleftarrow{\Gamma}$ does not vary along a reflected ray, i.e.,

$$\frac{k\rho\rho'}{R} \sin (\phi + \phi') = \text{constant along a reflected ray.} \qquad (24)$$

This follows from the equation of a reflected ray, $\rho \sin \gamma = \text{const.}$ (see Fig. 6.6.2), and from the law of sines applied to the triangle OPQ': $\rho' \sin (\phi + \phi') = R \sin \gamma$. From the law of sines applied to the triangle OQS, where S is the point of reflection $R_2 = 0$, one has, furthermore: $R_1 \sin (\pi - \delta) = \rho' \sin \phi'$, so

$$\frac{k\rho\rho'}{R} \sin (\phi + \phi') = k\rho \sin \delta \quad \text{at} \quad \phi = 0. \qquad (24a)$$

Thus, Eq. (22) can be rewritten as

$$\overleftarrow{\Gamma} = \frac{kx \sin \delta - a_1}{kx \sin \delta + a_1}, \qquad (24b)$$

which reduces via Eqs. (3a) and (4a) to Eqs. (23a) and (23b). Because of the dependence of $\overleftarrow{\Gamma}$ on x, a group of parallel reflected rays ($\delta = $ constant) does not define a uniform plane-wave front.

Equations (18) are not applicable in the transition regions surrounding the shadow boundary $|\phi - \phi'| = \pi$ and the reflected-wave boundary $(\phi + \phi') = \pi$. Since the dependence on $|\phi - \phi'|$ is contained entirely in the term $\bar{G}_\infty(\boldsymbol{\rho}, \boldsymbol{\rho}')$ in Eq. (11), the transition across the shadow boundary is described also in this case by the function $\bar{G}^t_\infty$ in Eq. (6.4.7). This leads to the important conclusion, already verified in Secs. 6.4 and 6.5, that the light–shadow transition behavior is independent of the physical composition of the wedge, and that to a lowest order in the asymptotic representation, the strength of the field on the shadow boundary is one half that of the incident field. The behavior across the reflected-wave boundary does, however, depend on the physical properties of the reflecting surface. Although we do not discuss here the reflected-wave transition function for the variable impedance surface with finite a_1, the limiting cases $a_1 = 0, \infty$, are covered via the results in Sec. 6.5.

From the similarity of the formal solutions in Eqs. (6) and (6.3.3), it is noted that the procedure described in conjunction with Eq. (6.4.9) can also be employed in the present case to derive higher-order terms in the asymptotic expansion of $\bar{G}^d$.[6] The higher-order coefficients $I_{mn}(\phi, \phi')$ are derivable by differentiation from the lowest-order function $I_{00}(\phi, \phi')$ in Eq. (18b). In a consistent evaluation of $\bar{G}$ to a higher order in $1/k$ ($k \gg 1$) the branch-cut integral in Eq. (11) must be evaluated more carefully, since diffraction effects in the illuminated region arise not only from the edge but also from local variations of the surface impedance.

Representation emphasizing guided-wave properties: surface wave

In the analysis of radiation in the presence of an infinite plane surface with constant surface impedance Z_s (see Sec. 5.7), it is noted that an electric (magnetic) line current parallel to the boundary will excite a surface wave if Z_s is capacitive (inductive). The guiding properties of the surface become evident in a most direct manner if one chooses a modal formulation in which the transmission axis is parallel to the surface. To explore these properties of the variable impedance surface in the wedge configuration, it is to be expected that a radial transmission representation will prove most fruitful. Such a representation, involving explicitly the radial characteristic Green's function g_ρ in Eq. (3.4.93) with $\tau = k^2$, is given formally in Eq. (6.5.11), and can be derived from the angular transmission formulation (6.3.3) by deforming the integration path around the singularities of $g(\phi, \phi'; \mu)$. As noted in Eq. (3.4.55), g_∞ is represented discontinuously across the positive real μ axis, on which lies the continuous spectrum of eigenvalues associated with the infinitely extended angular domain. Since the initial integration path can be chosen to run along the entire imaginary μ axis [see Eq. (6.3.1)], all other singularities of $g(\phi, \phi'; \mu)$ located in the right half of the complex μ plane contribute to the representation in terms of the angular mode spectrum; hence it is pertinent to examine the analytic properties of the reflection coefficient $\overleftarrow{\Gamma}(\mu)$ in Eq. (5). Since from Eq. (5a),

$$\overleftarrow{\Gamma}(\mu) = \frac{\mu - a_1}{\mu + a_1} \qquad \text{when Im } \mu > 0, \tag{25a}$$

and

$$\overleftarrow{\Gamma}(\mu) = \frac{-\mu - a_1}{-\mu + a_1} \qquad \text{when Im } \mu < 0, \tag{25b}$$

and Re $a_1 > 0$ for a passive surface impedance [see Eqs. (3a) and (4a)], one notes that $\overleftarrow{\Gamma}(\mu)$ has no singularities in the first quadrant of the complex μ plane. However, a simple pole singularity exists in the fourth quadrant at

$$\mu_p = a_1, \qquad \text{provided that Im } a_1 < 0. \tag{26}$$

The restriction Im $a_1 < 0$ implies that Im $Z_s > 0$ (capacitive) for the electric current excitation and Im $Z_s < 0$ (inductive) for the magnetic current excitation [note the time dependence $\exp(-i\omega t)$]. Thus, the pole singularity of $\overleftarrow{\Gamma}(\mu)$ contributes what will be identified as a surface wave under conditions analogous to those encountered with a constant impedance surface.

The desired representation is now obtained directly, via Eqs. (3.4.93) (with $\tau = k^2$, $\lambda = \mu^2$), (6.3.1a), and (6.2.8),

$$\bar{G}(\boldsymbol{\rho}, \boldsymbol{\rho}') = \frac{1}{2} \int_{-i\infty}^{i\infty} \mu J_\mu(k\rho_<) H_\mu^{(1)}(k\rho_>) g(\phi, \phi'; \mu)\, d\mu, \tag{27}$$

with $g(\phi, \phi'; \mu)$ inserted from Eq. (5). The integration path can be closed by addition of quarter-circles at $|\mu| \to \infty$ in the first and fourth quadrants because of the decaying behavior of $J_\mu(k\rho_<) H_\mu^{(1)}(k\rho_>)$ therein [see Eqs. (6.A12) and (6.A16)]. Application of Cauchy's theorem transforms Eq. (27) into[8]

$$\bar{G}(\boldsymbol{\rho}, \boldsymbol{\rho}') = -\pi a_1 e^{-i a_1 (\phi + \phi')} J_{a_1}(k\rho_<) H_{a_1}^{(1)}(k\rho_>) U(-\text{Im } a_1) + \bar{G}_1(\boldsymbol{\rho}, \boldsymbol{\rho}'), \tag{28}$$

where

$$\bar{G}_1 = \frac{i}{4} \int_0^\infty d\mu \left[e^{i\mu(\phi - \phi')} + e^{-i\mu(\phi - \phi')} + \frac{\mu - a_1}{\mu + a_1} e^{i\mu(\phi + \phi')} \right.$$

$$\left. + \frac{\mu + a_1}{\mu - a_1} e^{-i\mu(\phi + \phi')} \right] J_\mu(k\rho_<) H_\mu^{(1)}(k\rho_>)$$

$$= \frac{i}{4} \int_{-\infty}^\infty d\mu \left[e^{i\mu(\phi - \phi')} + \frac{\mu - a_1}{\mu + a_1} e^{i\mu(\phi + \phi')} \right] J_{|\mu|}(k\rho_<) H_{|\mu|}^{(1)}(k\rho_>). \tag{28a}$$

While Eq. (27) is subject to the restrictions $|\phi - \phi'| > \pi$, $(\phi + \phi') > \pi$, these conditions can be removed after contour deformation. $\bar{G}_1$ contains the contribution from the continuous spectrum in the ϕ domain, while the first term in Eq. (28) arises from a discrete spectral component representative of a surface wave.

The surface wave on this variable impedance boundary possesses certain interesting properties that distinguish it from its counterpart on a constant impedance surface. Since Im $a_1 < 0$, the field decays exponentially away from the $\phi = 0$ plane along any circular arc centered at the apex. As $\rho \to \infty$, the Hankel function can be represented by its asymptotic approximation, whence the surface-wave contribution $\bar{G}_s$ varies like $(1/\sqrt{k\rho}) \exp (ik\rho - ia_1\phi)$ [i.e., an outgoing cylindrical wave whose angular intensity decays like $\exp (-|\operatorname{Im} a_1| \phi)$]. Thus, energy is radiated into the surrounding medium—a consequence of the non-constancy of the surface impedance. Because its far-field dependence is $O(1/\sqrt{k\rho})$, the surface-wave contribution is not readily distinguished from the ordinary radiation field, which shows the same radial decay [Eq. (18b)]. However, the surface-wave field is orthogonal to the field in the continuous spectrum, as one may verify on multiplying $\bar{G}_1$ by $\exp (-a_1\phi)$ and integrating over ϕ between $\phi = 0$ and $\phi = \infty$. Hence, the surface wave can be excited in pure form by a source distribution that matches its angular variation $\exp (-ia_1\phi)$. This property is of interest in connection with radiation from variable impedance surface-wave antennas.[9] While the orthogonality statement requires for its exact validity an infinite angular interval, the exponential decay of the surface-wave field permits the effective truncation of the interval at some finite $\phi = \varphi$.

As $\rho \to 0$, one finds $\bar{G}_s \sim (k\rho)^{a_1} = (k\rho)^{\operatorname{Re} a_1} \exp [i(\operatorname{Im} a_1 \ln k\rho)]$. Thus, $|\bar{G}_s| \to 0$, and the associated electromagnetic fields satisfy the edge condition in Eq. (1.5.37). The phase is seen to fluctuate violently near the edge.

6.6b Two Variable-Impedance Walls

If the boundary conditions at the wedge faces $\phi = 0$ and $\phi = \varphi$ are specified by a surface impedance of the type indicated in Eqs. (3a) or (4a), the two-dimensional Green's function $\bar{G}(\boldsymbol{\rho}, \boldsymbol{\rho}')$ satisfies the corresponding boundary condition (2). The associated angular characteristic Green's function $g(\phi, \phi'; \mu)$ is given in closed form in Eq. (3.4.51), or in terms of an image representation in Eq. (3.4.57). Substitution of Eq. (3.4.57) into Eq. (6.3.3) yields an integral representation for $\bar{G}(\boldsymbol{\rho}, \boldsymbol{\rho}')$ valid in the geometrical shadow region $|\phi - \phi'| > \pi$. To obtain a representation valid everywhere, it is necessary to treat separately those image contributions that characterize geometrical reflections from the wedge faces (see also Sec. 6.5). If $\varphi > \pi$, only two image terms fall into this category: the image located at $-\phi'$ which has already been treated in Eq. (11), and the image at $2\varphi - \phi'$ for which an analogous representation is employed. The response from the source in the infinitely extended angular region is represented as in Eq. (6.3.8), with Eq. (6.4.2b), and the resulting formulation then applies to all observation points.

In the asymptotic evaluation for large values of k, one obtains a geometric-optical field contribution as in Eq. (18a), with the addition of a similar term to account for possible reflections from the wedge face at $\phi = \varphi$; the physical interpretation is shown in Fig. 6.6.2. The diffracted-wave contribution arising from the true source and from each of the image sources in the infinite angular region has been shown to have the same functional form, as a function of ϕ

and ϕ', for arbitrary locations of source and observations points [see Eq. (18b)]. An integral representation of the angular diffraction function in the shadow region $|\phi - \phi'| > \pi$ is obtained upon substituting Eq. (3.4.51) (with $j \to -i$) into Eq. (6.3.3) and following the procedure leading to Eq. (6.4.9), whence[6]

$$\bar{G}^d(\boldsymbol{\rho}, \boldsymbol{\rho}') \sim C(k\rho)C(k\rho')I_{00}(\phi, \phi'), \tag{29}$$

where $C(x) = (8\pi x)^{-1/2} \exp{(ix + i\pi/4)}$, and

$$I_{00}(\phi, \phi') = -4 \int_0^{i\infty} (e^{-i\mu\pi} - e^{i\mu\pi})\mu g(\phi, \phi'; \mu)\, d\mu, \qquad |\phi - \phi'| > \pi. \tag{29a}$$

If the integral in Eq. (29a) can be evaluated in closed form, the resulting function $I_{00}(\phi, \phi')$ is valid for all ϕ, ϕ'. Higher-order terms in the asymptotic expansion can be constructed from I_{00} as in Eq. (6.4.9).

A representation emphasizing the guiding properties of the wedge surface can be constructed by considerations analogous to those leading to Eq. (28). Since $g(\phi, \phi'; \mu)$ in Eq. (3.4.51) is an even function of μ, its only singularities in the complex μ plane are poles, whence the angular spectrum is discrete. From the integral representation (27), one may express $\bar{G}$ as a series of residues arising from the poles of g, after closing the contour at infinity in the right half of the complex μ plane. The result is listed below for the special case $a_2 = 0$, $a_1 = i\,|a_1|$ (i.e., the wedge face at $\phi = \varphi$ is perfectly reflecting while the surface impedance at $\phi = 0$ is purely imaginary). Under these circumstances, the pertinent pole singularities separate into two sets: a discretely infinite number on the positive real μ axis, and a single pole on the negative imaginary μ axis; the latter pole approaches the imaginary axis from the fourth quadrant and is therefore included in the residue evaluation. The result is found to be[8,10] [see Eqs. (3.4.62)]

$$\bar{G}(\boldsymbol{\rho}, \boldsymbol{\rho}') = \frac{\pi i}{\varphi} \frac{\cosh \eta(\varphi - \phi)\, \cosh \eta(\varphi - \phi')}{1 + \dfrac{1}{\varphi|a_1|}\sinh^2 \eta\varphi}\, J_{-i\eta}(k\rho_<)H^{(1)}_{-i\eta}(k\rho_>)$$

$$+ \frac{\pi i}{\varphi} \sum_\xi \frac{\cos \xi(\varphi - \phi)\, \cos \xi(\varphi - \phi')}{1 - \dfrac{1}{\varphi|a_1|}\sin^2 \xi\varphi}\, J_\xi(k\rho_<)H^{(1)}_\xi(k\rho_>), \tag{30}$$

where the eigenvalues ξ and η are the positive solutions of the transcendental equations

$$\cot{(\xi\varphi)} = -\frac{\xi}{|a_1|}, \qquad \coth{(\eta\varphi)} = \frac{\eta}{|a_1|}. \tag{30a}$$

The series representation in Eq. (30) is rapidly convergent when either the source point or the observation point is located near the apex.

When $|a_1| \to 0$, then $\eta \to 0$ and the ξ equation in (30a) reduces to $\sin \xi\varphi = 0$, so $\xi = m\pi/\varphi$, $m = 1, 2, \ldots$; moreover, one can show from Eq. (30a) that $\eta^2\varphi/|a_1| \to 1$, $(1/|a_1|)\sin^2 \xi\varphi \to 0$. Thus, $\bar{G}(\boldsymbol{\rho}, \boldsymbol{\rho}')$ reduces properly to the expression in Eq. (6.5.13).

The first term in Eq. (30) represents a surface wave in which the field intensity decays away from the surface $\phi = 0$. This becomes particularly evident when $|a_1|$ is large so that $\eta \approx |a_1|$. The surface-wave contribution then is approximately equal to that in Eq. (28), the interpretation of which was discussed in detail.

6.7 DIFFRACTION BY A CIRCULAR CYLINDER

6.7a Line-source Excitation

The circular cylinder results from the configuration in Fig. 6.1.1 on elimination of the angular boundaries at $\phi = 0, \varphi$. For the z-independent problem of excitation by a line source parallel to the cylinder axis, the region exterior to the cylinder may be viewed as a radial waveguide with propagation along the ρ coordinate, or as an angular waveguide with propagation along the ϕ coordinate (see Sec. 3.3c). While both formulations are treated in this section, special attention is given to the angular transmission representation which, as noted in Sec. 6.1, is well suited to the study of high-frequency phenomena. The various geometric optical domains depicted in Fig. 6.7.1 are characterized by different behavior of the high-frequency field. We shall explore in detail the asymptotic form of the field in the illuminated and shadow zones, and interpret the solution in ray-optical terms. Not discussed herein are the more complicated transition phenomena occurring in the shaded regions surrounding the shadow boundaries in Fig. 6.7.1 (see, however, References 11–14; Reference 14(a) has a representative bibliography).

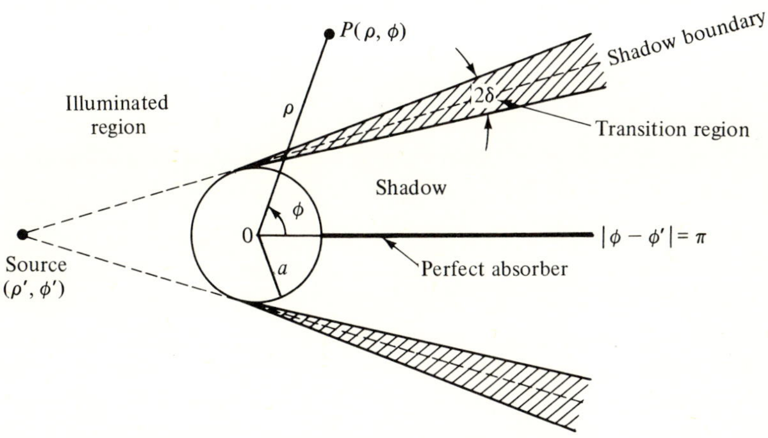

FIG. 6.7.1 Geometric-optical domains in cylinder diffraction problem.

Alternative representations of the formal solution

Diffraction problems in the region exterior to the impenetrable cylinder in Fig. 6.7.1 can be analyzed in the same manner as problems of diffraction by a wedge. In the angular transmission analysis, the only formal difference stems from the change in the angular waveguide cross section of the wedge from $0 < \rho < \infty$ to $a \le \rho < \infty$, with appropriate boundary conditions satisfied at $\rho = a$. As for the wedge, the simplest angular transmission problem is one wherein the angular region is infinitely extended; this artificial configuration can be made "physical" by placing at a convenient plane an "angularly matched" surface which absorbs completely all angularly propagating modes. If the surface is located at $|\phi - \phi'| = \pi$ as in Fig. 6.7.1 (i.e., deep in the geometrical shadow), the difference between this auxiliary field solution and the true solution in a periodic ϕ domain (or, equivalently, in a region with a perfectly reflecting termination at $|\phi - \phi'| = \pi$, as in Figs. 3.4.12 and 3.4.13) arises from multiple reflections. In the quasi-optic range $ka \gg 1$, where the cylinder radius is large compared to the wavelength of the incident radiation, the field in the shadow region is small, and reflections from a perturbing surface deep in the shadow have only a small effect. Hence, one expects the infinite angular space solution to yield a good approximation to the diffraction field at high frequencies—a conjecture substantiated by detailed analysis. If the cylinder radius is small compared to the wavelength ($ka \ll 1$), there is no well-defined shadow region, and the angular transmission approach offers no advantage. Instead, a radial transmission analysis is fruitful, with the cylinder representing a reflecting surface for radially propagating waves. We present here primarily the formal analysis, which permits the construction of the desired solution with a minimal effort; when appropriate, reference is made to the technical literature for additional details.

The above-mentioned alternative representations of the solution for the two-dimensional line-source Green's function $\bar{G}(\boldsymbol{\rho}, \boldsymbol{\rho}')$, can be constructed directly from Sec. 3.3c. The Green's function satisfies the equation

$$\left(\frac{1}{\rho}\frac{\partial}{\partial\rho}\,\rho\,\frac{\partial}{\partial\rho} + \frac{1}{\rho^2}\frac{\partial^2}{\partial\phi^2} + k^2\right)\bar{G}(\boldsymbol{\rho}, \boldsymbol{\rho}') = -\frac{\delta(\rho - \rho')}{\rho'}\delta(\phi - \phi') \qquad (1)$$

in the domain $a < (\rho, \rho') < \infty$, $0 \le (\phi, \phi') \le 2\pi$, subject to a radiation condition at $\rho \to \infty$, and to the following condition on the cylinder surface:

$$\frac{\partial\bar{G}}{\partial\rho} = -ik\bar{C}\bar{G} \quad \text{at } \rho = a, \qquad \bar{C} = \text{constant.} \qquad (2)$$

This condition is interpretable in terms of a surface impedance Z_s at $\rho = a$. For excitation by a line source of electric currents (E-mode case), one has, from Eq. (6.2.1b) [for an $\exp(-i\omega t)$ variation],

$$E \equiv E_z = i\omega\mu I\bar{G}, \qquad H_\phi = -I\frac{\partial\bar{G}}{\partial\rho}, \qquad (3a)$$

whence, at $\rho = a$,

$$E_z = Z'_s H_\phi, \qquad Z'_s = \frac{\zeta}{\bar{C}}, \qquad \zeta = \left(\frac{\mu}{\epsilon}\right)^{1/2}. \qquad (3b)$$

For a magnetic line current (H-mode case), by duality,

$$E_\phi = -Z_s'' H_z, \qquad Z_s'' = \zeta \bar{C}. \tag{4}$$

Passivity requirements are met if Re $\bar{C} \geq 0$. The special case of a perfectly conducting cylinder is obtained when $Z_s \to 0$, i.e., $\bar{C} \to 0$ or ∞ for the H- and E-mode cases, respectively.

The radial characteristic Green's function and completeness relation are given in Eqs. (3.4.97) and (3.4.98):

$$g_\rho(\rho, \rho', \lambda) = \frac{\pi i}{2}\Big[J_\mu(k\rho_<) + \bar{f}(\mu) H_\mu^{(1)}(k\rho_<)\Big] H_\mu^{(1)}(k\rho_>), \qquad \mu = \sqrt{\lambda}, \tag{5}$$

where $\bar{f}(\mu)$ is a radial reflection coefficient,

$$\bar{f}(\mu) = -\frac{b(\mu)}{d(\mu)},$$

$$b(\mu) = J_\mu'(ka) + i\bar{C}J_\mu(ka), \qquad d(\mu) = H_\mu'^{(1)}(ka) + i\bar{C}H_\mu^{(1)}(ka), \tag{5a}$$

and the prime on J_μ and $H_\mu^{(1)}$ denotes the derivative with respect to the argument. Also,

$$\rho'\delta(\rho-\rho') = -\pi i \sum_{\mu_p} \frac{\mu_p b(\mu_p)}{\left[\dfrac{\partial}{\partial\mu}d(\mu)\right]_{\mu_p}} H_{\mu_p}^{(1)}(k\rho) H_{\mu_p}^{(1)}(k\rho'), \tag{6}$$

where the μ_p are defined by

$$d(\mu_p) = 0. \tag{6a}$$

For the angular domain, one has, from Eq. (3.4.51c), with the subscript on g_ϕ suppressed,

$$g(\phi, \phi'; \hat{\lambda}) = \begin{cases} \dfrac{-\cos\hat{\mu}[\pi - |\phi - \phi'|]}{2\hat{\mu}\sin\hat{\mu}\pi}, & \tag{7a} \\[2mm] \displaystyle\sum_{n=-\infty}^{\infty} g_\infty(\phi, 2n\pi + \phi'), & \tag{7b} \end{cases}$$

where $\hat{\mu} = \sqrt{\hat{\lambda}}$ and $g_\infty(\phi, \phi'; \hat{\lambda})$ is the characteristic Green's function for the infinite angular space:

$$g_\infty(\phi, \phi'; \hat{\lambda}) = \frac{e^{i\hat{\mu}|\phi - \phi'|}}{-2i\hat{\mu}}, \qquad \text{Im } \hat{\mu} > 0. \tag{7c}$$

The completeness relation is [from Eq. (3.2.50c)]

$$\delta(\phi - \phi') = \frac{1}{2\pi}\sum_{m=-\infty}^{\infty} e^{im(\phi - \phi')}. \tag{8}$$

Alternative modal representations for the two-dimensional Green's function $\bar{G}(\boldsymbol{\rho}, \boldsymbol{\rho}')$ are constructed directly from the z-independent version of Eqs. (3.3.37) and (3.3.38b). In particular, one has the contour-integral representation

$$\bar{G}(\boldsymbol{\rho}, \boldsymbol{\rho}') = -\frac{1}{\pi i}\int_{C_3 + C_4} g_\rho(\rho, \rho'; \lambda) g(\phi, \phi'; \lambda)\mu\, d\mu, \tag{9}$$

where the contours C_3 and C_4 are shown in Fig. 6.7.2. The contour remains open about the pole at $\mu = 0$, whence the principal value of the integral is taken at that point. Since the integrand in Eq. (9) is an odd function of μ, the contour C_4 can be reflected about the origin into the contour C_3', and $\int_{C_3+C_4} = \int_{C_3+C_3'}$. In terms of the image representation in Eq. (7b), Eq.(9) becomes

$$\bar{G}(\boldsymbol{\rho}, \boldsymbol{\rho}') = \sum_{n=-\infty}^{\infty} \bar{G}_\infty(\boldsymbol{\rho}, \boldsymbol{\rho}_n'), \qquad \boldsymbol{\rho}_n' = (\rho', \phi_n'), \qquad \phi_n' = \phi' + 2n\pi, \quad (10a)$$

where, in view of Eq. (7c),

$$\bar{G}_\infty(\boldsymbol{\rho}, \boldsymbol{\rho}_n') = \frac{1}{2\pi} \int_{-\infty}^{\infty} g_\rho(\rho, \rho'; \lambda) e^{i\mu|\phi - \phi_n'|} d\mu. \quad (10b)$$

Since the integrand in Eq. (10b) has no singularities on the real μ axis, the integration can be taken from $\mu = -\infty$ to $\mu = \infty$. Because g_ρ is an even function of μ, replacement of μ by $(-\mu)$ yields the same integrand except for the

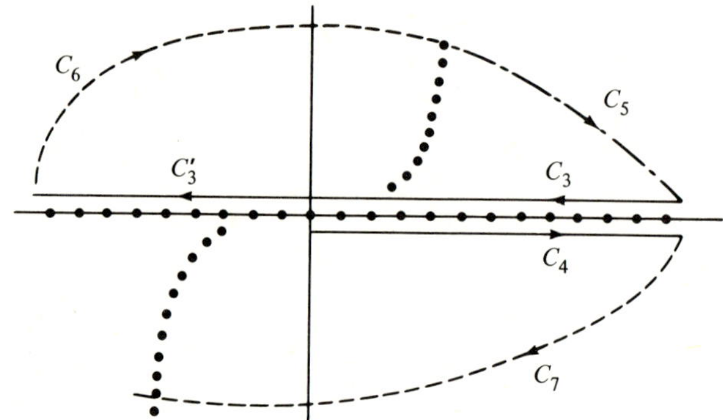

FIG. 6.7.2.   Paths of integration and singularities in the complex μ plane.

occurrence of $-|\phi - \phi'|$. Hence, the absolute value is of no consequence and $|\phi - \phi'|$ can be replaced by $\phi - \phi'$. $\bar{G}_\infty$ and g_ρ can therefore be interpreted as Fourier transform pairs in the infinitely extended angular space. The $n = 0$ term in Eq. (10a) represents the Green's function for the configuration in Fig. 6.7.1 and yields the dominant contribution in the quasi-optic range $ka \gg 1$, as shown below. When $n \neq 0$, a rapidly convergent representation for large ka is obtained by closing the integration path by an infinite semicircle in the upper half of the μ plane and evaluating the integral in terms of the residues at the poles of g_ρ. Thus, the following representation, rapidly convergent for large ka as shown below, but valid for any cylinder radius, is suitable for all values of $|\phi - \phi'| \leq \pi$:

$$\bar{G}(\boldsymbol{\rho}, \boldsymbol{\rho}') = \frac{1}{2\pi} \int_{-\infty}^{\infty} g_\rho(\rho, \rho'; \lambda) e^{i|\mu||\phi - \phi'|} d\mu$$

$$+ \frac{\pi}{2} \sum_{n=-\infty}^{\infty}{}' \sum_{\mu_p} \frac{b(\mu_p)}{\left.\dfrac{\partial}{\partial\mu} d(\mu)\right|_{\mu_p}} H_{\mu_p}^{(1)}(k\rho) H_{\mu_p}^{(1)}(k\rho') e^{i\mu_p|\phi - \phi'_n|}, \tag{11}$$

where $\mu = \sqrt{\lambda}$; the prime on the sum over n denotes omission of the $n = 0$ term, and the sum represents the contribution, from the images in the infinite angular space, which restores the required periodicity along the boundary $|\phi - \phi'| = \pi$.

Employing the angular eigenfunctions of Eq. (8), one obtains the radial transmission representation

$$\bar{G}(\boldsymbol{\rho}, \boldsymbol{\rho}') = \frac{i}{4} \sum_{n=-\infty}^{\infty} e^{in(\phi - \phi')} \left[J_n(k\rho_<) - \frac{b(n)}{d(n)} H_n^{(1)}(k\rho_<) \right] H_n^{(1)}(k\rho_>), \tag{12}$$

$$= \frac{i}{4} H_0^{(1)}(k|\boldsymbol{\rho} - \boldsymbol{\rho}'|) + \bar{G}_s(\boldsymbol{\rho}, \boldsymbol{\rho}'), \tag{13a}$$

$$\bar{G}_s(\boldsymbol{\rho}, \boldsymbol{\rho}') = -\frac{i}{4} \sum_{n=-\infty}^{\infty} e^{in(\phi - \phi')} H_n^{(1)}(k\rho) H_n^{(1)}(k\rho') \frac{b(n)}{d(n)}. \tag{13b}$$

Since $b(n)/d(n) \to 0$ as $a \to 0$, the contribution from the $J_n(k\rho_<)$ term in the series is recognized as the free-space Green's function $\bar{G}_f$ whose closed-form result is given in Eq. (5.4.25). The remaining series representation for the scattered field $\bar{G}_s$ converges rapidly when ka is small and can therefore be used for numerical evaluation of the diffraction effects of a *small cylinder*.

The preceding remarks concerning the convergence properties of the image representation in Eq. (11) will now be justified. Although the discussion can be carried out for arbitrary values of the surface impedance Z_s,[15] we assume for convenience that $\bar{C} = \infty$ in Eq. (5a), so

$$b(\mu) \to J_\mu(ka) i\bar{C}, \qquad d(\mu) \to H_\mu^{(1)}(ka) i\bar{C}. \tag{14}$$

Convergence of the integral representation for $\bar{G}_\infty$ in Eq. (10b) is established upon noting from Eqs. (6.A12) and (6.A16) that g_ρ behaves for $\mu \to \infty$ like

$$g_\rho \sim \frac{1}{\mu} \left[\left(\frac{\rho_<}{\rho_>} \right)^\mu + \text{const.} \left(\frac{a^2}{\rho\rho'} \right)^\mu \right]. \tag{15}$$

Since both $\rho_</\rho_>$ and $a^2/\rho\rho'$ are less than unity, g_ρ tends to zero exponentially as $\mu \to \infty$. [Note: $A^\mu = \exp(\mu \ln A)$.] The same applies as $\mu \to -\infty$, since g_ρ is an even function of μ. Equation (15) actually remains valid along an infinite semicircle in the μ plane extending to the right of the lines on which the zeros of $H_\mu^{(1)}(ka)$ are located [see Fig. 6.7.2; also curves $C_{1,2}$ in Fig. 6.A1(a)].

If the integration path is closed by the addition of the path segments C_5 and C_6 in Fig. 6.7.2, one assesses the behavior of the integrand in Eq. (10b) on C_5 directly from Eq. (15). Since Re $\mu > 0$, Im $\mu > 0$ on C_5, both g_ρ and $\exp(i\mu|\phi - \phi'_n|)$ converge exponentially as $|\mu| \to \infty$, and there is no contribution to the integral unless $\rho = \rho'$, $\phi = \phi'_n$. The contribution from C_6 is examined conveniently by introducing a change of variable from μ to $-\mu$, thereby furnishing an integral over the contour C_7. While the angularly dependent term $\exp\left(-i\mu|\phi - \phi_n'|\right)$ still converges exponentially when $\phi \neq \phi_n'$, g_ρ in Eq. (15) converges exponentially when Re $\mu > 0$, $\rho \neq \rho'$, but diverges over the path segment lying to the left of the negative imaginary μ axis on C_7. However, if $\mu = |\mu|e^{i\psi}$, and $\psi = -(\pi/2) - \epsilon$, $0 < \epsilon \ll 1$, each term in the integrand of Eq. (10b) has a magnitude characterized by

$$|A^\mu e^{-i\mu\varphi}| = e^{|\mu|[\epsilon|\ln A| - \varphi]}, \qquad 0 < A < 1, \tag{15a}$$

where, from Eq. (6.A21), $\epsilon \leq \pi[2 \ln(2|\mu|/eka)]^{-1}$ (see Sec. 6.A5 for a discussion of the behavior of μ_p). Hence, if $\varphi \neq 0$, the expression in Eq. (15) decays as $|\mu| \to \infty$, and there is no contribution to the integral from the path segment C_6 provided that $\phi \neq \phi_n'$. The latter restriction affects only the $n = 0$ term and eliminates observation along the source location angle $\phi = \phi'$.

One notes from the above discussion that, for $\phi \neq \phi'$, the contour deformation can be carried out for all the $\bar{G}_\infty(\mathbf{\rho}, \mathbf{\rho}_n')$ terms, and the resulting integral can be evaluated in terms of the residues at the complex zeros μ_p of $H_\mu^{(1)}(ka)$ in the first quadrant of the μ plane. Hence, one may write an expression for $\bar{G}(\mathbf{\rho}, \mathbf{\rho}')$ as in Eq. (11), wherein the integral is omitted and the series over n includes the $n = 0$ term. While the resulting series converges for all $|\phi - \phi_n'| \neq 0$, the μ_p series associated with the $n = 0$ term converges rapidly only when $|\phi - \phi'| > \hat{\phi}$, where the angle $\hat{\phi} = \gamma_1 + \gamma_2$ is as determined in Eq. (18). When $|\phi - \phi'| < \hat{\phi}$, initial terms in the series increase exponentially, and the series is therefore unsuitable for numerical evaluation.

We verify first the convergence of the "residue series" over μ_p for arbitrary n. From Eqs. (6.A22)–(6.A26), as $|\mu_p| \to \infty$,

$$J_{\mu_p}(ka) \equiv \frac{1}{2} H_{\mu_p}^{(2)}(ka) \sim O\left(\frac{1}{\sqrt{\zeta}}\right), \qquad \zeta = |\mu_p|, \dagger \tag{16a}$$

$$\frac{\partial}{\partial\mu} H_\mu^{(1)}(ka)\bigg|_{\mu_p} \sim O\left(\frac{\ln(2\zeta/eka)}{\sqrt{\zeta}}\right), \tag{16b}$$

$$H_{\mu_p}^{(1)}(k\rho) \sim O\left\{\frac{1}{\sqrt{\zeta}} \exp\left[\frac{\zeta\pi}{2}\frac{\ln(\rho/a)}{\ln(2\zeta/eka)}\right]\right\}. \tag{16c}$$

While $H_{\mu_p}^{(1)}(k\rho)$, $H_{\mu_p}^{(1)}(k\rho')$ diverge as $\zeta \to \infty$, the decay of $\exp\left(-\zeta|\phi - \phi_n'|\right)$ assures convergence for all n as long as $\phi \neq \phi_n'$ (i.e., $\phi \neq \phi'$). To assess the rapidity of convergence of the residue series, we examine the initial terms in the expansion (for a more detailed discussion of the convergence properties, see Reference 14). Since $H_{\mu_p}^{(2)}(ka)$ and $(\partial/\partial\mu_p)H_{\mu_p}^{(1)}(ka)$ do not have an exponential dependence [see Eqs. (6.A24) and (6.A26)], it suffices to focus attention on the remaining terms in Eq. (11). For large ka, the lowest values of μ_p are $O[ka + \alpha(ka)^{1/3}]$, $\alpha = $ constant, Im $\alpha > 0$ [see Eq. (6.A35)]; if $k\rho, k\rho' > O[ka + \alpha(ka)^{1/3}]$, the Hankel functions of argument $k\rho, k\rho'$ can be approximated by

---

$\dagger$ Here, ζ should not be confused with the same symbol used elsewhere for wave impedance.

their Debye asymptotic representations in Eq. (6.A1). The exponential dependence of the summand in Eq. (11) is, therefore,

$$H^{(1)}_{\mu_p}(k\rho)H^{(1)}_{\mu_p}(k\rho')e^{i\mu_p|\phi-\phi'_n|} \sim \exp\{i[\sqrt{(k\rho)^2 - \mu_p^2} + \sqrt{(k\rho')^2 - \mu_p^2}]\}$$

$$\times \exp\left\{i\mu_p\left[|\phi - \phi'_n| - \cos^{-1}\frac{\mu_p}{k\rho} - \cos^{-1}\frac{\mu_p}{k\rho'}\right]\right\}. \qquad (17)$$

Since $\mu_p = ka(1 + \Delta)$, with $\mathrm{Im}\,\Delta > 0$ and increasing for successive terms in the series, one may verify that the exponential decays if

$$|\phi - \phi'_n| > \gamma_1 + \gamma_2, \qquad \gamma_1 = \cos^{-1}\frac{a}{\rho'}, \qquad \gamma_2 = \cos^{-1}\frac{a}{\rho}. \qquad (18)$$

The residue series—physical interpretation

The condition (18) has a simple physical interpretation, as seen from Fig. 6.7.3. For $n = 0$, the condition requires that the source point Q be invisible

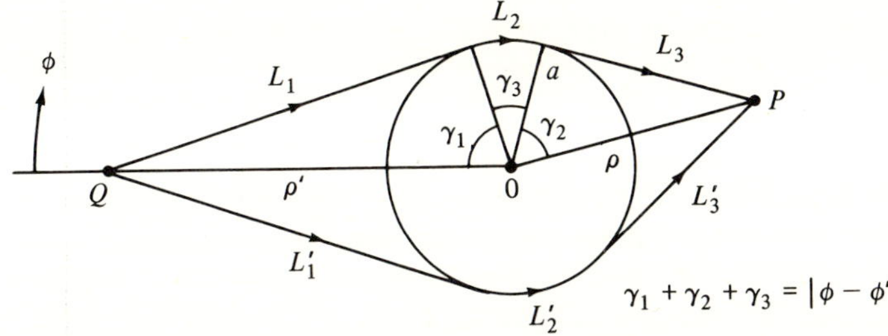

FIG. 6.7.3   Geometrical interpretation of field in shadow region.

from the observation point P (i.e., the observation point lies in the geometrical shadow region behind the cylinder). The decay in Eq. (17) is proportional to $\exp[-(\mathrm{Im}\,\mu_p)\gamma_3]$, where γ_3 is the angle subtended at the origin by the points of tangency of the lines L_1 and L_3 from the source and observation points to the cylinder surface. Equation (17) can be written as

$$e^{ik(L_1+L_2+L_3)}e^{-\chi_p L_2}, \qquad \chi_p = \frac{\mathrm{Im}\,\mu_p}{a}, \qquad (19)$$

so each term in the residue series for $n = 0$ may be interpreted as being associated with an incident wave which strikes the cylinder at a glancing angle, travels along the cylinder surface, and sheds energy continually during its progress; due to the leakage of energy, the wave amplitude decays according to the factor $\exp(-\chi_p L_2)$. The associated interpretation in terms of geometrical rays is also evident from Fig. 6.7.3. Because the wave "creeps" along the cylinder surface, it has been termed "creeping wave,"[16] and its ray description "creeping ray."[17] The creeping rays, launched by a glancing incident ray, represent the diffraction effects and provide the means of guiding energy into the geometrical shadow region. Each μ_p has its own creeping ray; however, because of the increasing imaginary part of successive μ_p, only the lowest are of importance. Since the creeping-ray amplitude decays exponentially, whereas that on a diffracted ray due to an edge decays algebraically [see Eq. (6.3.15)], the shadow behind a gently curved object is darker than that behind a pointed structure.

Each of the image terms $n \neq 0$ in the series solution (11) is also easily interpreted via the creeping-ray picture. Suppose that $0 < (\phi - \phi') < \pi$. Since $\phi'_n = \phi' + 2n\pi$, one has $|\phi - \phi'_n| = 2|n|\pi + (\phi - \phi')$ or $2n\pi - (\phi - \phi')$ when $n < 0$ and $n > 0$, respectively. Thus, an $n < 0$ image term represents a ray that is launched by the glancing ray L_1 and reaches the observation point P after having encircled the cylinder n times. The $n > 0$ image terms are launched by the alternative glancing ray L'_1. The $n = 1$ term describes the ray traveling along path $(L'_1 + L'_2 + L'_3)$; in general, the nth term describes a ray that is launched by an incident ray along L'_1 and reaches the observation point P after having encircled the cylinder $(n - 1)$ times. Because of the exponential decay, proportional to the length of travel of a ray on the cylinder surface, only the nearest image terms are of importance. Since $|\phi - \phi'_n| \geq (2|n| - 1)\pi$ for all $|\phi - \phi'| < \pi$ when $n \neq 0$, and $(\gamma_1 + \gamma_2) < \pi$ in Fig. 6.7.3, one notes that the residue-series representation is rapidly convergent and that the image terms in the infinitely extended angular space contribute only to the diffraction effects of the cylinder.

If $|\phi - \phi'| < \gamma_1 + \gamma_2$, the observation point P is located in the illuminated region and the creeping-ray contribution arising from the $n = 0$ term disappears. The residue series corresponding to the $n = 0$ term is no longer rapidly convergent, and it is more appropriate to represent the contribution $\bar{G}_\infty(\rho, \rho')$ in terms of the integral in Eq. (11). From a saddle-point evaluation (carried out below) it is found that the $\bar{G}_\infty(\rho, \rho')$ term furnishes in the illuminated region precisely the geometric optical (incident wave and reflected wave) field. In the transition region between the illuminated and shadow regions (Fig. 6.7.1), a more detailed treatment of the integral is required (see remarks below). It is to be emphasized, however, that all geometric-optical, transition, and dominant diffraction effects are contained in the $n = 0$ term; in the illuminated or shadow regions the image terms contribute only the higher-order diffraction effects. For example, if $0 < \phi - \phi' < \gamma_1 + \gamma_2$, the $n = 1$ image term yields the creeping ray launched along L'_1 and emerging along L'_3 in Fig. 6.7.4, while the $n = -1$ image term yields the creeping ray launched along $\hat{L}_1$ and emerging along $\hat{L}_3$.

Equation (11), derived directly via the image concept in an infinitely extended angular domain, may be compared with an alternative representation[16] that utilizes Eq. (9) and also aims at a formulation useful in the illuminated or shadow regions. A residue series for the shadow region can be obtained without difficulty from Eq. (9) by contour deformation and residue evaluation. Equation (9) is not directly suitable, however, for application of the saddle-point method to field evaluation in the illuminated region; a difficulty arises

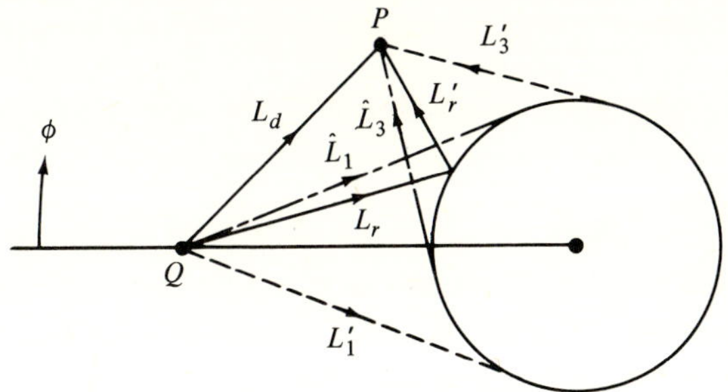

FIG. 6.7.4   Geometrical interpretation of field in illuminated region.

because the resulting saddle points lie on the real μ axis, where the singularities of g at the zeros of $\sin \mu\pi$ introduce rapid fluctuations. To cope with this complication, one may introduce the decomposition

$$\cos \{\mu [\pi - |\phi - \phi'|]\} = e^{\pm i\mu\pi} \cos \mu(\phi - \phi') \mp i e^{i\mu|\phi-\phi'|} \sin \mu\pi, \qquad \text{Im } \mu \gtrless 0, \tag{20}$$

which, when substituted into Eq. (9), permits the contribution from the first term to be evaluated in terms of a rapidly convergent residue series for all $0 < |\phi - \phi'| \leq \pi$. The second term contributes precisely the integral in Eq. (11), whose integrand has no singularities on the real axis and can therefore be evaluated accurately by the saddle-point method, as shown below. The image approach, leading directly to Eq. (11), avoids the necessity for the two separate representations mentioned above.

Illuminated region—geometric-optical field

As noted above, to evaluate the contribution from $\bar{G}_\infty(\boldsymbol{\rho}, \boldsymbol{\rho}')$ when the observation point lies in the illuminated region, it is convenient to retain the integral representation in Eq. (11) and to seek an asymptotic evaluation by the saddle-point method. For simplicity, it is assumed again that $\bar{C} = \infty$ [see Eq. (14)], so $\bar{f}(\mu) = -J_\mu(ka)/H_\mu^{(1)}(ka)$ in Eq. (5); the calculation for other values of $\bar{C}$ proceeds in a similar manner. Since ka, $k\rho$, and $k\rho'$ are large quantities, it is suggestive to represent the cylinder functions in the integrand by their Debye asymptotic approximations in Eqs. (6.A1)–(6.A7) and to look for possible stationary points in the resulting exponent. If real saddle points exist, the corresponding asymptotic approximation will yield a field that propagates without exponential decay; admissible complex saddle points lead to attenuating solutions that characterize the fields in the shadow region where the rapidly convergent residue series provides an alternative method of calculation. Real saddle points should therefore occur only when the observation point lies in the illuminated region; this conjecture is confirmed by the analysis.

It is preferable to write the integral in the form[16]

$$\bar{G}_\infty(\boldsymbol{\rho}, \boldsymbol{\rho}') = \bar{G}_\infty^{(1)}(\boldsymbol{\rho}, \boldsymbol{\rho}') + \bar{G}_\infty^{(2)}(\boldsymbol{\rho}, \boldsymbol{\rho}'), \tag{21}$$

where

$$\bar{G}_{\infty}^{(1)} = \frac{i}{8} \int_{C} e^{i\mu|\phi - \phi'|} H_{\mu}^{(2)}(k\rho_{<})H_{\mu}^{(1)}(k\rho_{>})\, d\mu, \tag{21a}$$

$$\bar{G}_{\infty}^{(2)} = -\frac{i}{8} \int_{C} e^{i\mu|\phi - \phi'|} H_{\mu}^{(1)}(k\rho)H_{\mu}^{(1)}(k\rho')\frac{H_{\mu}^{(2)}(ka)}{H_{\mu}^{(1)}(ka)}\, d\mu, \tag{21b}$$

where the integration path C proceeds as shown in Fig. 6.7.5. As noted pre-

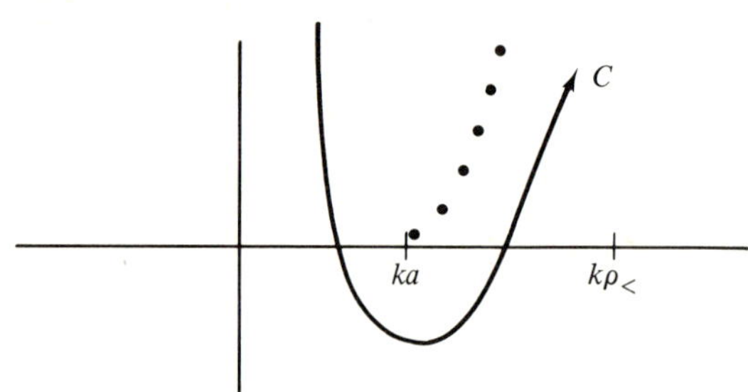

FIG. **6.7.5** Integration path for saddle-point evaluation in the complex μ plane.

viously, this path deformation is permissible and the combined integrand of $\bar{G}_{\infty}$ decays in the upper half of the complex μ plane. Attention is therefore centered on the path segment in the vicinity of the real axis where the Debye approximation for the cylinder functions may be used [Eqs. (6.A1)–(6.A7)]. The resulting expression for $\bar{G}_{\infty}^{(1)}$ is then as follows:

$$\bar{G}_{\infty}^{(1)} \sim \frac{i}{4\pi} \int \frac{1}{[k\rho(\sin \beta_{1})k\rho'(\sin \beta_{2})]^{1/2}} e^{i\psi_{1}}\, d\mu, \tag{22}$$

where

$$\psi_{1} = \mu|\phi - \phi'| + k\rho_{>}(\sin \beta_{1} - \beta_{1}\cos \beta_{1}) - k\rho_{<}(\sin \beta_{2} - \beta_{2}\cos \beta_{2}), \tag{22a}$$

$$\beta_{1} = \cos^{-1}\frac{\mu}{k\rho_{>}}, \qquad \beta_{2} = \cos^{-1}\frac{\mu}{k\rho_{<}}, \tag{22b}$$

and the restriction $0 < \mathrm{Re}\,\beta_{1,2} < \pi$ must be imposed [Eq. (6.A4b)]. The saddle points μ_{s} of $\psi_{1}(\mu)$ are determined from $d\psi_{1}/d\mu = 0$:

$$\cos^{-1}\left(\frac{\mu_{s}}{k\rho_{<}}\right) - \cos^{-1}\left(\frac{\mu_{s}}{k\rho_{>}}\right) + |\phi - \phi'| = 0; \tag{23}$$

also, one finds that

$$\frac{d^{2}\psi_{1}}{d\mu^{2}}\bigg|_{\mu_{s}} = \frac{1}{k\rho_{>}\sin \beta_{1s}} - \frac{1}{k\rho_{<}\sin \beta_{2s}} < 0, \tag{24}$$

where β_{1s} and β_{2s} denote the values of β_{1} and β_{2} corresponding to μ_{s}.

The saddle-point condition (23) admits of the physical interpretation shown in Fig. 6.7.6(a). Since β_{1s} and β_{2s} must be positive in view of the restriction

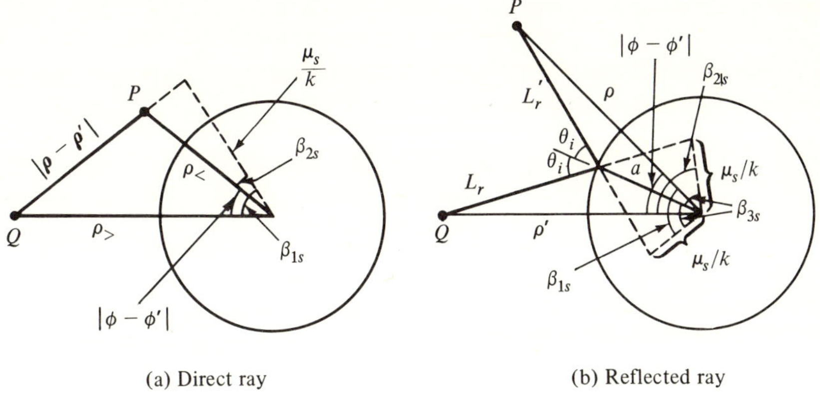

<table>
<tr><td>(a) Direct ray</td><td>(b) Reflected ray</td></tr>
</table>

FIG. 6.7.6   Physical interpretation of saddle-point condition.

following Eq. (22b), one observes that a real solution is possible only when $|\phi - \phi'| < \pi/2$. The integration path is readily deformed through the saddle point at $\mu_s = k\rho_< \cos \beta_{2s} = k\rho_> \cos \beta_{1s}$ and leaves the saddle point at a $-45°$ angle [see Eq. (24), which follows from $\beta_{1s} > \beta_{2s}$]; it may be noted that the poles in Fig. 6.7.5 do not occur in the integrand of Eq. (21a). Use of Eq. (4.2.1a) and reference to Fig. 6.7.6(a) then leads directly to the expression

$$\bar{G}_{\infty}^{(1)} \sim \frac{e^{ik|\mathbf{\rho}-\mathbf{\rho}'|+i\pi/4}}{2\sqrt{2\pi}\sqrt{k|\mathbf{\rho}-\mathbf{\rho}'|}}, \qquad \text{when } |\phi - \phi'| < \frac{\pi}{2}, \qquad (25)$$

which result represents the asymptotic form of the free-space Green's function [see Eqs. (5.4.36) and (5.4.37)]. The integral $\bar{G}_{\infty}^{(1)}$ therefore furnishes the incident (direct) field in the region $|\phi - \phi'| < \pi/2$ and yields a negligibly small contribution when $|\phi - \phi'| > \pi/2$.

The integral $\bar{G}_{\infty}^{(2)}$ in Eq. (21b) is treated in the same manner, subject to the recognition that the asymptotic approximation for $H_{\mu}^{(1,2)}(ka)$ is dependent on whether $\mu < ka$ or $\mu > ka$. In the latter instance, $H_{\mu}^{(1)}(ka) \sim -H_{\mu}^{(2)}(ka)$ [see Eqs. (6.A1) and (6.A6)], and the resulting integrand resembles that in Eq. (21a). By proceeding as before, one finds that the saddle-point condition is now given by

$$\beta_{1s} + \beta_{2s} = |\phi - \phi'|, \qquad \mu_s = k\rho_> \cos \beta_{1s} = k\rho_< \cos \beta_{2s} > ka, \qquad (26)$$

and admits of a graphical interpretation as in Fig. 6.7.6(a) provided that $|\phi - \phi'| > \pi/2$. It is also noted that the restriction $\mu_s/k > a$ defines those rays which connect P and Q without touching the cylinder, thereby restricting the validity of Eq. (26) (for real μ_s) to the illuminated region. The second derivative of the phase function $\psi(\mu_s)$ is now positive, so the path through the saddle point $\mu_s > ka$ proceeds as in the right-hand portion of Fig. 6.7.5. The contribution of this saddle point to $\bar{G}_{\infty}^{(2)}$ is found to be the same as in Eq. (25), with $|\phi - \phi'| > \pi/2$, so the two results together furnish the incident field in the entire illuminated region; the analysis actually excludes information along the line $|\phi - \phi'| = \pi/2$, although the solution remains valid there.

When $\mu < ka$, the Debye approximation may be employed for all the Hankel functions in the integrand of Eq. (21b):

$$\bar{G}_{\infty}^{(2)}\Big|_{\mu<ka} \sim \frac{i}{4\pi} \int \frac{1}{[k\rho(\sin\beta_1)k\rho'(\sin\beta_2)]^{1/2}} e^{i\psi_2}\, d\mu, \tag{27}$$

where

$$\psi_2 = \mu|\phi - \phi'| + k\rho(\sin\beta_1 - \beta_1\cos\beta_1) + k\rho'(\sin\beta_2 - \beta_2\cos\beta_2)$$
$$-2ka(\sin\beta_3 - \beta_3\cos\beta_3), \tag{27a}$$

$$\mu = k\rho\cos\beta_1 = k\rho'\cos\beta_2 = ka\cos\beta_3.$$

The saddle-point condition $d\psi_2/d\mu = 0$ now reads

$$\beta_{1s} + \beta_{2s} - 2\beta_{3s} = |\phi - \phi'|, \tag{28}$$

and may be interpreted graphically as in Fig. 6.7.6(b). $[d^2\psi_2/d\mu^2]_{\mu_s}$ is found to be negative, so the steepest-descent path traverses the saddle point at $\mu_s < ka$ in the manner shown in the left-hand portion of Fig. 6.7.5. Equation (4.2.1b), together with parameters defined in Fig. 6.7.6(b), then yields the reflected-wave contribution,

$$\bar{G}_{\infty}^{(2)}\Big|_{\mu_s<ka} \sim \frac{e^{ik(L_r+L_r')+i\pi/4}}{2\sqrt{2\pi k}} \sqrt{\frac{a\cos\theta_i}{2L_rL_r' + (L_r + L_r')a\cos\theta_i}}, \tag{29}$$

where L_r and L_r' are the ray trajectories in Fig. 6.7.6(b), θ_i is the angle of incidence, and it has been recognized that

$$kL_r = k\rho'\sin\beta_{2s} - ka\sin\beta_{3s}, \qquad kL_r' = k\rho\sin\beta_{1s} - ka\sin\beta_{3s},$$
$$\sin\beta_{3s} = \cos\theta_i. \tag{29a}$$

Equation (29) may be shown to represent exactly the reflected field predicted from geometrical optics, so we verify again the validity of the geometrical-optics method in the range of large k. For finite values of $\bar{C}$, Eq. (29) must be multiplied by the reflection coefficient of a plane interface appropriate to a plane wave incident at the angle θ_i.

These simple expressions become invalid when the observation point approaches the light–shadow boundary, in which instance $\mu_s \to ka$. One observes from Fig. 6.7.5 that this condition corresponds to a coalescence of two saddle points near the sequence of poles, therefore requiring a more sophisticated asymptotic evaluation of the integral. Moreover, it is then no longer possible to approximate $H_\mu^{(1,2)}(ka)$ by the Debye formulas which are valid when $|\mu_s - ka| > O[(ka)^{1/3}]$; one requires instead the uniform approximation in terms of Airy functions [Eq. (6.A29)]. Details of the asymptotic field evaluation in this instance may be found in the literature.[11-14] However, some simple observations may be made concerning the angular extent of the transition zone separating the illuminated and shadow regions. Since $\mu_s = ka\cos\beta_{3s}$, the condition $|\mu_s - ka| > O[(ka)^{1/3}]$ may be restated as $(1 - \cos\beta_{3s}) = (1 - \sin\theta_i) > O[(ka)^{2/3}]$, or $\theta_i < (\pi/2) - O[(ka)^{-1/3}]$. The geometric-optical formula comprising the direct field and the reflected field in Eq. (29) is therefore valid only when $\theta_i = (\pi/2) - \delta$, where δ is an angle of $O[(ka)^{-1/3}]$. Similarly, by examining the exponent in Eq. (19), one finds that the residue series terms in the shadow region decay according to $\exp(-\gamma_3 \,\mathrm{Im}\,\mu_p)$, where $\mathrm{Im}\,\mu_p$ is proportional to $(ka)^{1/3}$. The series therefore converges rapidly only when $\gamma_3 > \delta$ and then permits the physical interpretation of the diffracted field in the shadow in terms of the rays sketched in Fig. 6.7.3.A simple ray-optical description of the field in the illuminated and shadow regions, respectively, applies therefore only exterior to the transition zones shown shaded in Fig. 6.7.1.

6.7b Point-source Excitation

The line-source Green's functions obtained in Sec. 6.7a may be employed directly for the construction of the three-dimensional Green's functions descriptive of the fields excited by a longitudinal electric or magnetic current element exterior to a perfectly conducting cylinder [see Fig. (6.7.7)]. As described in

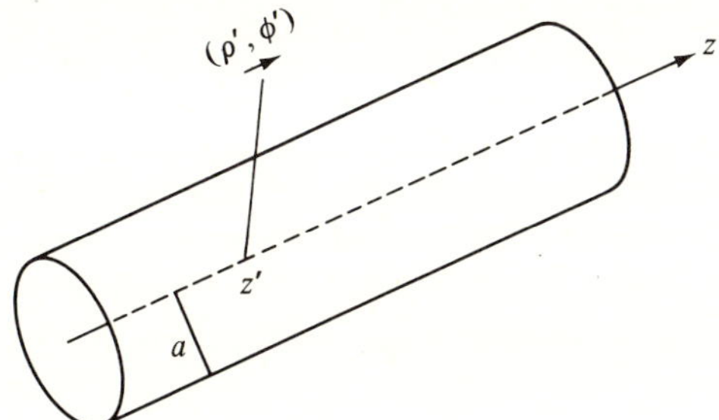

FIG. 6.7.7 Point-source excitation.

Sec. 6.2, one replaces k by $\sqrt{k^2 - \zeta^2}$ in the line source result, and then performs the operation $(1/2\pi)\int_{-\infty}^{\infty} d\zeta \exp[i\zeta(z - z')]$. From these considerations, we construct at once the following alternative representations corresponding to the radial transmission, angular transmission, and general contour integral representations discussed in Sec. 3.3c[14]:

$$G(\mathbf{r}, \mathbf{r}') = \begin{cases} \dfrac{i}{8\pi} \displaystyle\sum_{n=-\infty}^{\infty} \int_{-\infty}^{\infty} d\zeta\, e^{in(\phi-\phi')} e^{i\zeta(z-z')} \left[J_n(k_\zeta \rho_<) - \dfrac{b(n)}{d(n)} H_n^{(1)}(k_\zeta \rho_<) \right] \\ \qquad\qquad \times\, H_n^{(1)}(k_\zeta \rho_>), \hfill (30a) \\[2em] \dfrac{1}{4} \displaystyle\sum_{\mu_p} \int_{-\infty}^{\infty} d\zeta \sum_{n=-\infty}^{\infty} e^{i\zeta(z-z')}\, \left.\dfrac{b(\mu_p)}{\dfrac{\partial}{\partial\mu}d(\mu)}\right|_{\mu_p} H_{\mu_p}^{(1)}(k_\zeta \rho) H_{\mu_p}^{(1)}(k_\zeta \rho')\, e^{i\mu_p|\phi-\phi'_n|}, \\ \hfill (30b) \\[2em] \dfrac{1}{8\pi} \displaystyle\int_{C_3+C_4} d\mu \int_{-\infty}^{\infty} d\zeta\, e^{i\zeta(z-z')} \left[J_\mu(k_\zeta \rho_<) - \dfrac{b(\mu)}{d(\mu)} H_\mu^{(1)}(k_\zeta \rho_<) \right] \\ \qquad\qquad \times\, H_\mu^{(1)}(k_\zeta \rho_>) \dfrac{\cos\mu[\pi - |\phi - \phi'|]}{\sin\mu\pi}, \hfill (30c) \end{cases}$$

where $k_\zeta = \sqrt{k^2 - \zeta^2}$, with Im $k_\zeta \geq 0$, and ϕ'_n is defined in Eq. (10a). Since the cylinder is assumed to be perfectly conducting, one puts $\bar{C} = \infty$ in Eq. (5a) for the E-mode case with respect to z (electric current element) and $\bar{C} = 0$ for the H-mode case (magnetic current element). It is to be emphasized that the replacement $k \to k_\zeta$ is also to be made in the functions $b(\mu)$ and $d(\mu)$.

Other representations may be derived from the above by deforming the integration path in the complex ζ plane about the branch-point singularities of k_ζ. The results then correspond to a z-transmission formulation, one of which is given by

$$G(\mathbf{r}, \mathbf{r}') = \frac{i}{8\pi} \sum_{n=-\infty}^{\infty} e^{in(\phi-\phi')}$$

$$\times \int_{-\infty}^{\infty} d\eta\, \eta \left[J_n(\eta\rho') - \frac{b(n)}{d(n)} H_n^{(1)}(\eta\rho') \right] H_n^{(1)}(\eta\rho) \frac{e^{ik_\eta|z-z'|}}{k_\eta}, \tag{31}$$

where $k_\eta = \sqrt{k^2 - \eta^2}$, with Im $k_\eta \geq 0$. The derivation is left as an exercise for the reader.

Equation (31) is useful for the derivation of results pertaining to vector point sources with *arbitrary orientation*. We recall from Sec. 2.3c [see also Eqs. (5.2.1)] that the fields may in this case be derived from the potential functions $\mathscr{S}'(\mathbf{r}, \mathbf{r}')$ and $\mathscr{S}''(\mathbf{r}, \mathbf{r}')$, which, in a z-transmission representation, differ from $G'(\mathbf{r}, \mathbf{r}')$ and $G''(\mathbf{r}, \mathbf{r}')$ only through the presence of the factors $1/k_{ti}'^2$ and $1/k_{ti}''^2$, respectively. In the cylindrical waveguide, $k_{ti}' = k_{ti}'' = \eta$ [see Eq. (3.2.46b)], so the formula for $\mathscr{S}$ is obtained by dividing the integrand in Eq. (31) by η^2. For a proper interpretation of this result, the reader should refer to the remarks following Eq. (5.2.10).

6.8 FIELDS IN SPHERICAL REGIONS

6.8a Introduction

The generic configuration for separable field representations in a spherical coordinate system comprises a combination of spheres, circular cones, and planes as shown in Fig. 6.8.1 (see also Fig. 2.7.1). Although only single spherical and conical surfaces are shown, regions contained between two concentric spheres and two coaxial cones can also be accommodated. Certain physical attributes of fields in the presence of opaque spherical boundaries (e.g., shadowing effects) are analogous to those discussed in Secs. 6.1 and 6.7 for cylindrical scatterers, while other characteristics are strongly affected by the different properties of the cylindrical and spherical coordinates (e.g., scattering by an edge at $\rho = 0$ and by a tip at $r = 0$). The intent in this section is to obtain alternative representations for fields in spherical regions and to contrast them with analogous results for cylindrical regions. Because of the formal similarity of field representations utilizing radially or angularly guided waves in cylindrical and spherical coordinates, alternative modal expansions of fields or their potentials can be written down at once (Sec. 6.8b). To illustrate some

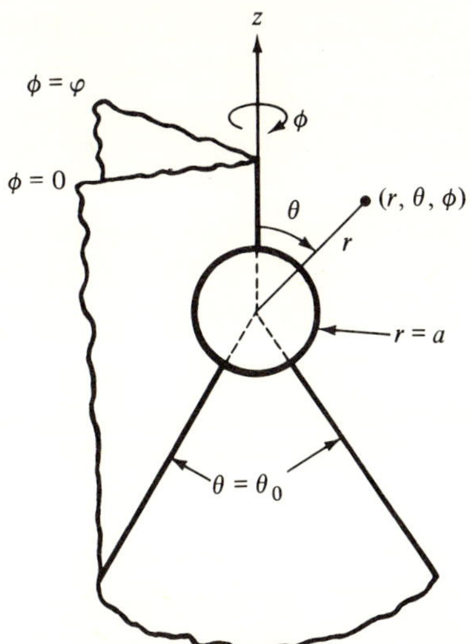

FIG. 6.8.1 Configuration accommodated by a spherical coordinate system.

fundamental differences between scattering in two and three dimensions, field solutions for a perfectly conducting cone are evaluated asymptotically in the high-frequency limit to yield the field diffracted by a conical tip (Sec. 6.8c). For asymptotic results on the thoroughly investigated problem of scattering by a sphere (see Reference 11) and for additional results on the cone, as well as for a representative bibliography, the reader may wish to consult References 18–20.

6.8b Alternative Field Representations

Free space

The fields radiated by arbitrarily prescribed source distributions in an unbounded homogeneous medium may be derived by differentiation from the scalar Green's function $G_f = [4\pi|\mathbf{r} - \mathbf{r}'|]^{-1} \exp[-jk|\mathbf{r} - \mathbf{r}'|]$ [see Eq. (5.4.2b)].[†] A radial transmission representation for G_f follows from Eqs. (2.6.11) with $\zeta Y_i'(r, r') = Z_i''(r, r')/\zeta$ taken from Eq. (2.7.11), and the mode functions inserted from Eqs. (3.4.79a) and (3.2.50):

$$G_f(\mathbf{r}, \mathbf{r}') = \frac{e^{-jk|\mathbf{r}-\mathbf{r}'|}}{4\pi|\mathbf{r} - \mathbf{r}'|}$$

$$= \frac{1}{j4\pi krr'} \sum_{m=0}^{\infty} \sum_{n=m}^{\infty} \epsilon_m \frac{(2n + 1)(n - m)!}{(n + m)!}$$

$$\times \cos m(\phi - \phi')P_n^m(\cos \theta)P_n^m(\cos \theta')j_n(kr_<)h_n^{(2)}(kr_>), \qquad (1)$$

[†] In this section, the time dependence is $\exp(+j\omega t)$. For $\exp(-i\omega t)$ dependence, replace j by $-i$ and $h_\nu^{(2)}$ by $h_\nu^{(1)}$.

where $\epsilon_m = 1$, $m = 0$, and $\epsilon_m = 2$, $m \geq 1$. When the source is located on the z axis, $\theta' = 0$, and since $P_n(1) = 1$, only the $m = 0$ term contributes. Thus,

$$G_f(\mathbf{r}, \mathbf{r}') = \frac{1}{j4\pi krr'} \sum_{n=0}^{\infty} (2n + 1)P_n(\cos\theta)j_n(kr_<)h_n^{(2)}(kr_>), \qquad \theta' = 0, \qquad (2a)$$

and when the source point is at the origin, only the $n = 0$ term remains [see Eq. (2.7.4c)]:

$$G_f(\mathbf{r}, \mathbf{r}') = \frac{1}{j4\pi r} h_0^{(2)}(kr), \qquad r' = 0. \qquad (2b)$$

The spherical Bessel functions of integral order n may be expressed in terms of a finite number of trigonometric functions:

$$j_0(x) = \sin x, \quad j_1(x) = \frac{\sin x}{x} - \cos x, \quad j_2(x) = \left(\frac{3}{x^2} - 1\right)\sin x - \frac{3}{x}\cos x, \quad (3a)$$

$$n_0(x) = -\cos x, \quad n_1(x) = -\sin x - \frac{\cos x}{x},$$

$$n_2(x) = \left(1 - \frac{3}{x^2}\right)\cos x - \frac{3}{x}\sin x. \qquad (3b)$$

Thus, $h_0^{(2)}(x) = j\exp(-jx)$, so Eq. (2b) yields directly the closed-form expression for G_f. One confirms the earlier observation (Sec. 2.7a) that the spherical (radial) transmission-line analysis leads to a rapidly convergent representation for the radiated fields when the sources are confined to a small region about the origin. In particular, a scalar point source at the origin excites only a single spherical mode.

While the fields radiated by an electric or magnetic current element may be calculated by differentiating the closed-form expression for G_f, it is instructive to derive the result directly from the radial transmission representation. If the dipole is chosen to lie on the z axis, the fields are azimuthally symmetrical and Eq. (2a) is applicable. With J_r^0 and M_r^0 denoting the respective dipole strengths, the azimuthal field components, E_ϕ'' for a magnetic and H_ϕ' for an electric dipole, may then be derived from Eqs. (2.6.4) and (2.6.11) as

$$\left.\begin{array}{c} \dfrac{H_\phi'}{-J_r^0} \\[2mm] \dfrac{E_\phi''}{M_r^0} \end{array}\right\} = \frac{1}{r'}\frac{\partial G}{\partial\theta} = \frac{1}{j4\pi krr'^2} \sum_{n=1}^{\infty} (2n + 1)P_n^1(\cos\theta)j_n(kr_<)h_n^{(2)}(kr_>), \qquad (4)$$

since $P_n^1 = (d/d\theta)P_n$ and $P_0^1 = 0$. As $r' \to 0$, only the $n = 1$ term contributes [see Eq. (2.7.4c)] and the result is

$$\frac{1}{r'}\frac{\partial G}{\partial\theta}\bigg|_{r'=0} = -\frac{k\sin\theta}{j4\pi r}h_1^{(2)}(kr), \qquad h_1^{(2)}(kr) = -e^{-jkr} + j\frac{e^{-jkr}}{kr}. \qquad (4a)$$

Thus, while the $n = 0$ mode is the lowest mode in the scalar problem, the vector dependence of the source changes this to $n = 1$.

Expressions for $\mathscr{S}'/rr'$ and $\mathscr{S}''/rr'$ in Eqs. (2.6.9) are obtained by including the factor $[n(n+1)]^{-1}$ in the summand of Eq. (1) and excluding the term $n = 0$.

The sphere

When a source distribution is located exterior to a perfectly conducting sphere, the electromagnetic field may be derived from the scalar functions G', G'' or $\mathscr{S}'$, $\mathscr{S}''$, whose radial transmission representations are obtained from those in free space on replacing the modal Green's function in Eq. (2.7.11) by those in Eqs. (2.7.12). For example, if the excitation is from a vertical (radial) electric dipole and the coordinate system is oriented so that the source lies on the z axis, then only G' is required for the calculation of the field, and is given via Eqs. (2.7.12), (3.4.67), (3.3.43), (3.4.101), and (2a), in the following alternative forms, with $h_t^{\prime(2)}(ka) = 0$: [18]

$$G'(\mathbf{r}, \mathbf{r}') = \begin{cases} G_f(\mathbf{r}, \mathbf{r}') - \dfrac{1}{j4\pi krr'} \sum_{n=0}^{\infty} (2n+1)P_n(\cos\theta)h_n^{(2)}(kr)h_n^{(2)}(kr')\dfrac{j_n'(ka)}{h_n^{\prime(2)}(ka)}, & \text{(5a)} \\[3ex] \dfrac{1}{4j\pi krr'} \sum_t \dfrac{(2t+1)\,j_t'(ka)}{\sin t\pi(\partial/\partial t)h_t^{\prime(2)}(ka)}P_t(-\cos\theta)h_t^{(2)}(kr)h_t^{(2)}(kr') & \text{(5b)} \end{cases}$$

with an intermediate contour integral representation also possible [see Eq. (3.3.43c)]. The conversion of Eq. (5a) into Eq. (5b) is known as the Watson transformation;[21] the original treatment did not, however, utilize the characteristic Green's function concept.

Equation (5a) expresses the Green's function for the sphere as a correction on the free-space result, and is rapidly convergent when ka is small. Alternatively, for large ka, the "residue series" (5b) is suitable for a field calculation in the geometrical shadow region. Since $t = O(ka)$ (see Sec. 6.A5), one may employ the asymptotic approximation (3.4.66b) for the Legendre function provided that $\theta \not\approx 0, \pi$. The resulting series then has the same form as in the problem of scattering by a circular cylinder (Sec. 6.7a), and the same analysis may be employed to deduce the fields in various geometric-optical domains. The physical interpretation of the behavior of the various field constituents is also directly analogous, due cognizance being given to the three-dimensional, but rotationally symmetric, character of the ray structure in the present instance (see Figs. 6.7.3 and 6.7.4). Simple ray pictures fail along the axis $\theta = 0, \pi$ where $P_t(-\cos\theta)$ cannot be approximated in terms of exponentials; the failure may be explained physically since the creeping rays all intersect along this axis, thereby forming a caustic of the creeping-ray system. As in the cylinder problem, it is convenient to extend the θ domain to $-\infty$ and $+\infty$, and to represent the angular characteristic Green's functions g_θ' and g_θ'' in terms of images in this infinitely extended space. This representation is easily achieved when the Legendre functions may be approximated by their trigonometric forms (i.e., for sufficiently large ν), but may also be phrased for arbitrary ν by introducing the "traveling-wave" Legendre functions in Eq. (3.4.71).

If the electric current element on the z axis is transverse (for example, $\mathbf{J}^0 = \boldsymbol{\theta}_0 J^0$, with $\theta' = \phi' = 0$) then the Hertz potentials $\boldsymbol{\Pi}'$ and $\boldsymbol{\Pi}''$ may be derived from $\mathscr{S}'$ and $\mathscr{S}''$ in Eqs. (2.6.9) as follows:

$$j\omega\epsilon r\boldsymbol{\Pi}'(\mathbf{r}, \mathbf{r}') = \frac{J^0}{r'}\frac{\partial^2}{\partial\theta'\,\partial r'}\,\mathscr{S}'(\mathbf{r}, \mathbf{r}'), \tag{6a}$$

$$r\boldsymbol{\Pi}''(\mathbf{r}, \mathbf{r}') = \frac{J^0}{r'\sin\theta'}\frac{\partial}{\partial\phi'}\,\mathscr{S}''(\mathbf{r}, \mathbf{r}'), \tag{6b}$$

with the limiting operation θ', $\phi' \to 0$ to be carried out after the differentiations have been performed. The radial transmission representation for $\mathscr{S}'/rr'$ has the same form as the right-hand side of Eq. (1), provided that one replaces the radial function in Eq. (2.7.11) by that in Eq. (2.7.12b), includes a factor $(n+1)^{-1} \cdot n^{-1}$ in the summand, and excludes the $n = 0$ term in the series. In view of the relations listed in Eq. (3.4.79d), only the $m = 1$ term remains after the limit $\theta' = 0$ has been invoked. Thus,

$$j\omega\epsilon r\boldsymbol{\Pi}'(\mathbf{r}, \mathbf{r}') = \frac{J^0}{r'}\frac{\partial}{\partial r'}\Bigg\{\frac{-\cos\phi}{j4\pi k}\sum_{n=1}^{\infty}\frac{(2n+1)}{n(n+1)}P_n^1(\cos\theta)$$
$$\times\left[j_n(kr_<) - \frac{j_n'(ka)}{h_n'^{(2)}(ka)}h_n^{(2)}(kr_<)\right]h_n^{(2)}(kr_>)\Bigg\}, \tag{7a}$$

and similarly for $\boldsymbol{\Pi}''$,

$$r\boldsymbol{\Pi}''(\mathbf{r}, \mathbf{r}') = \frac{J^0}{r'}\Bigg\{\frac{-\sin\phi}{j4\pi k}\sum_{n=1}^{\infty}\frac{(2n+1)}{n(n+1)}P_n^1(\cos\theta)$$
$$\times\left[j_n(kr_<) - \frac{j_n(ka)}{h_n^{(2)}(ka)}h_n^{(2)}(kr_<)\right]h_n^{(2)}(kr_>)\Bigg\}, \tag{7b}$$

which expressions are now inserted into Eq. (2.6.4). As before, the result separates into the free-space field, which may be evaluated in closed form, and a contribution accounting for the effect of the sphere, with the latter rapidly convergent when ka is small.

Alternative representations for the functions inside the braces in Eqs. (7) may again be obtained by utilizing the characteristic Green's functions. By referring to Eq. (3.3.43a) with $q \equiv m = 1$, one may write

$$\mathscr{S}'(\mathbf{r}, \mathbf{r}') = \frac{-\cos(\phi - \phi')}{j2\pi^2}\int_{C_\theta} g_\theta(\theta, \theta'; 1; \lambda_\theta)g_r'(r, r'; \lambda_\theta)\frac{d\lambda_\theta}{\lambda_\theta}, \tag{8}$$

where C_θ surrounds the singularities of $g_\theta = g_\theta^\circ$ [Eq. (3.4.67)] on the positive real axis in the positive sense, but excludes the singularities of g_r' [Eq. (2.7.12b)], and the pole at the origin. This contour may now be deformed about the poles of g_r', thereby yielding a residue-series representation which converges rapidly in the shadow region of the sphere when $ka \gg 1$. Analogous considerations may be applied to $\mathscr{S}''$.

By moving the transverse dipole to infinity along the z axis, one may derive the result for an incident plane wave with electric field $\mathbf{E}_{\text{inc}} = \mathbf{x}_0 \exp(jkz)$. This is achieved by letting $r' \to \infty$, replacing $h^{(2)}(kr')$ by its asymptotic form, and setting

$$\frac{-j\omega\mu J^0 e^{-jkr'}}{4\pi r'} = 1.$$

Results for a dielectric sphere may be obtained by inserting the appropriate radial characteristic Green's functions (see Fig. 2.7.4).

The cone

When the scatterer is a perfectly conducting semiinfinite cone as in Fig. 6.8.2, the angular functions in Eqs (3.4.68), (3.4.69), (3.4.80)–(3.4.83), and the radial

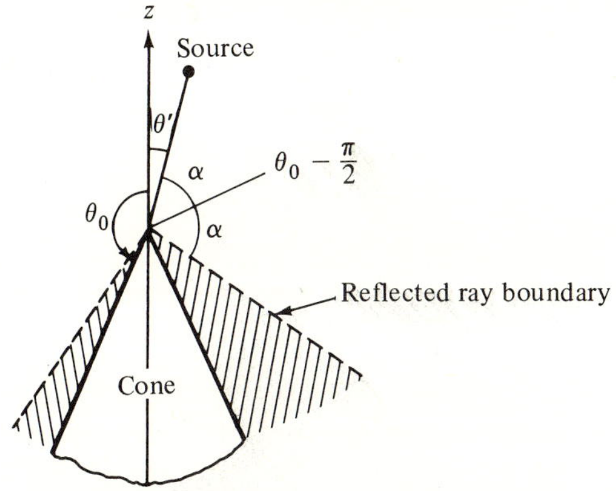

FIG. 6.8.2   Conical obstacle $(\alpha = \theta_0 - \pi/2 - \theta')$.

functions in Eqs (2.7.11), and (3.4.100) are appropriate. The Green's function G' may thus be written in the following alternative forms [see Eqs. (3.3.43)][19]:

$$rr'G'(\mathbf{r}, \mathbf{r}') = \begin{cases}
\begin{aligned}
&\frac{j}{4k}\sum_{m=0}^{\infty}\epsilon_m \cos m(\phi - \phi')\sum_p(2p+1)\frac{\Gamma(p+m+1)}{\Gamma(p-m+1)}\\
&\quad\times\frac{P_p^{-m}(-\cos\theta_0)P_p^{-m}(\cos\theta')}{[\sin(p-m)\pi](\partial/\partial p)P_p^{-m}(\cos\theta_0)}\\
&\quad\times j_p(kr_<)h_p^{(2)}(kr_>)P_p^{-m}(\cos\theta),
\end{aligned} & \text{(9a)}\\[2em]
\begin{aligned}
&rr'G_f(\mathbf{r},\mathbf{r}') + \frac{1}{4\pi^2 k}\sum_{m=0}^{\infty}\epsilon_m\cos m(\phi-\phi')\\
&\quad\times\oint_{C_\theta} j_\nu(kr_<)h_\nu^{(2)}(kr_>)g'_{\theta s}(\theta,\theta';m^2;\lambda)\,d\lambda,
\end{aligned} & \text{(9b)}\\[2em]
\begin{aligned}
&rr'G_f(\mathbf{r},\mathbf{r}') - \frac{1}{16\pi k}\sum_{m=0}^{\infty}\epsilon_m\cos m(\phi-\phi')\int_{-1/2-j\infty}^{-1/2+j\infty}(2\nu+1)\\
&\quad\times\frac{\Gamma(\nu+m+1)}{\Gamma(\nu-m+1)\sin(\nu-m)\pi}\frac{P_\nu^{-m}(-\cos\theta_0)}{P_\nu^{-m}(\cos\theta_0)}\\
&\quad\times h_\nu^{(2)}(kr)h_\nu^{(2)}(kr')P_\nu^{-m}(\cos\theta)P_\nu^{-m}(\cos\theta')\,d\nu,
\end{aligned} & \text{(9c)}
\end{cases}$$

where $\nu = p$ are the positive zeros of $P_\nu^{-m}(\cos\theta_0)$, $\lambda = \nu(\nu+1)$ (i.e., $\nu = \sqrt{\lambda+1/4}-1/2$), $\epsilon_m = 1, m = 0$, and $\epsilon_m = 2, m \geq 1$. G_f represents the free-space Green's function [Eq. (1)] and $g'_{\theta s}$ denotes the second term in Eq. (3.4.68), which expresses the perturbation introduced by the conical surface. The contour C_θ in the complex λ plane, which surrounds the real zeros $\lambda_p = p(p + 1)$ of g'_θ in the positive sense, has been deformed about the singularities of g_r (branch point at $\lambda = -1/4$, with branch cut drawn along negative real axis) to achieve the θ-transmission representation in Eq. (9c) [see also Eq. (6.3.3); in the final reduction, we have utilized the fact that $h_\nu^{(1)} h_\nu^{(2)}$ and $g'_{\theta s}$ are even functions of $\nu + 1/2$]. Analogous representations may be written for the H-mode Green's function G'' on replacing $P_\nu^{-m}(\pm \cos \theta_0)$ by $(d/d\theta_0)P_\nu^{-m}(\pm \cos \theta_0)$,[22] and for $\mathscr{S}'$ and $\mathscr{S}''$ on including the term $1/p(p + 1)$ in Eq. (9a). In the latter instance, the angular transmission representation is found to be[23]

$$\mathscr{S}'_s(\mathbf{r}, \mathbf{r}') = Q'(\mathbf{r}, \mathbf{r}') + \bar{\mathscr{S}}'_s(\mathbf{r}, \mathbf{r}'), \tag{10}$$

where $\mathscr{S}'_s \equiv \mathscr{S}' - \mathscr{S}_f$ and

$$Q'(\mathbf{r}, \mathbf{r}') = \frac{1}{4\pi^2 k} \sum_{m=0}^{\infty} \epsilon_m \cos m(\phi - \phi') \, [2\pi j \, (\text{Residue at } \lambda = 0)], \tag{10a}$$

$$\bar{\mathscr{S}}'_s(\mathbf{r}, \mathbf{r}') = \frac{-1}{16\pi k} \sum_{m=0}^{\infty} \epsilon_m \cos m(\phi - \phi') \int_{-(1/2)-j\infty}^{-(1/2)+j\infty} \frac{d\nu}{\nu(\nu + 1)} [\quad], \tag{10b}$$

[ ] denoting the integrand in Eq. (9c). The corresponding expression for $\mathscr{S}''_s$ differs only in that $P_\nu^{-m}(\pm \cos \theta_0)$ is replaced by $(d/d\theta_0)P_\nu^{-m}(\pm \cos \theta_0)$. One may show via Eqs. (3.4.65) and (3.4.66a) that

$$Q' = -Q'' = \frac{j}{2\pi k} j_0(kr_<)h_0^{(2)}(kr_>) \sum_{m=1}^{\infty} \frac{\cos m(\phi - \phi')}{m}$$

$$\times \left[\tan \frac{\theta}{2} \tan \frac{\theta'}{2} \tan^2 \left(\frac{\pi - \theta_0}{2} \right) \right]^m + (m = 0) \text{ term}. \tag{10c}$$

It may also be verified from substitution into Eqs. (2.6.10) that the $m = 0$ term in Eq. (10c) does not contribute to the electromagnetic fields calculated from $\mathscr{S}'$ and $\mathscr{S}''$, so it need be of no further concern.

By employing the large-p approximations for the functions in Eq. (9a) [see Eqs. (2.7.4) and (3.4.66)], one may show that the radial transmission representation converges everywhere, but *converges rapidly* only when either the source or the observation point is located near the cone tip (kr or kr' small; see Sec. 6.5c). This formulation is useful to check that the "tip condition" [Eq. (1.5.39)] delimiting the permissible growth of the fields near the tip singularity is satisfied. The series may be employed, for example, for the calculation of the currents induced near the cone tip by an incident plane wave (one replaces $h_p^{(2)}(kr_>)$ by its asymptotic approximation for $r' \to \infty$, and renormalizes) or for the evaluation of the radiation pattern due to sources placed near the tip, subject however to the availability of the eigenvalues p. When kr and kr' are large, series representations of the type (9a) are inconvenient for calculation since the terms decay in magnitude only when p is larger than $kr_<$. It is then more suitable to employ the integral representation in Eq. (9c) which exhibits the perturbation effect of the cone explicitly. Since the representation theorem involves the

Kontorovich–Lebedev transform, which applies only to a limited class of functions, it is necessary to impose restrictions that assure the convergence of the integral. The considerations are directly analogous to those encountered in the wedge problem [see the discussion following Eq. (6.3.3)], and furnish the conclusion that the contour deformation leading from Eq. (9b) to Eq. (9c) is permissible provided that $\theta + \theta' < 2\theta_0 - \pi$. As in the case of the wedge, the angles θ satisfying this restriction define a region that excludes rays reflected specularly at the cone surface (i.e., the domain $\theta > \theta' + 2\alpha$ in Fig. 6.8.2). In its domain of validity, the second term in Eq. (9c) may therefore be expected to represent the diffraction effect that accounts for the deviation of the high-frequency field from that predicted by geometrical optics.

6.8c The Cone—Diffracted Field at High Frequencies

Asymptotic expansion

In the high-frequency range where kr and kr' are large, the integral in Eq. (9c) may be evaluated asymptotically by the procedure described in Sec. 6.4a. and leads to the following result for $G'_s = G' - G_f$ when $(\theta + \theta') < 2\theta_0 - \pi$:[23]†

$$G'_s(\mathbf{r}, \mathbf{r}') \sim \frac{i}{4\pi^2 k} \frac{e^{ik\,(r+r')}}{rr'} \sum_{n=0}^{\infty} \frac{A'_n(\theta, \theta'; \phi, \phi')}{(-2ikr)^n}, \tag{11}$$

where the coefficients A'_n are given by

$$A'_n = \sum_{m=0}^{\infty} \varepsilon_m \cos m(\phi - \phi') \int_{-\infty}^{\infty} x e^{x\pi}(ix, n) g'_{\theta s}\, dx, \tag{11a}$$

$$g'_{\theta s} = \frac{\pi}{2} \frac{K_x^{-m}(\cos \theta) K_x^{-m}(\cos \theta') \Gamma(ix + m + \tfrac{1}{2})}{(-1)^{m+1} \cosh \pi x\, \Gamma(ix - m + \tfrac{1}{2})} \frac{K_x^{-m}(-\cos \theta_0)}{K_x^{-m}(\cos \theta_0)}. \tag{11b}$$

In these equations, the variable $x = -i(\nu + 1/2)$ has been introduced, as well as Mehler's notation

$$K_x^{\mu}(\cos \theta) \equiv P_{-1/2+ix}^{\mu}(\cos \theta) \tag{11c}$$

for the Legendre function of order $ix - \tfrac{1}{2}$, not to be confused with the modified Bessel function, denoted elsewhere by K_α. Also,

$$(ix, n) \equiv \frac{(-1)^n}{n!}(x^2 + \tfrac{1}{4})(x^2 + \tfrac{1}{4} + 2)\cdots[x^2 + \tfrac{1}{4} + n(n-1)]$$

$$= \frac{-1}{n}[x^2 + \tfrac{1}{4} + n(n-1)](ix, n-1), \qquad n = 1, 2, \ldots, \tag{11d}$$

with $(ix, 0) \equiv 1$. While it is straightforward to write an expansion for large but arbitrary kr and kr' as in Sec. 6.4a, Eq. (11) is restricted to $kr' \gg kr$; i.e., the asymptotic series in Eq. (6.4.8a) has been employed for $h_\nu^{(1)}(kr)$, whereas $h_\nu^{(1)}(kr')$ has been represented by its leading term [see Eqs. (5.4.6) for passing to the limit of plane-wave incidence]. The dominant contribution,

$$G'_s \sim \frac{ie^{ik\,(r+r')}}{4\pi^2 krr'} A'_0(\theta, \theta'; \phi, \phi'), \tag{12}$$

---

†For the high-frequency calculation in this section the time dependence is $\exp(-i\omega t)$.

is, however, correct to $O(1/kr)$ and $O(1/kr')$ regardless of whether $r > r'$ or $r < r'$.

In the analogous expression G_s'' for the H-mode case, the last factor in Eq. (11b) involves the derivatives of the Mehler function with respect to θ_0, and the corresponding expansion coefficients are $A_n''(\theta, \theta'; \phi, \phi')$. Similarly, for the functions $\mathscr{S}_s'$ and $\mathscr{S}_s''$, the integral in Eq. (10b) may be expanded asymptotically and one finds

$$\bar{\mathscr{S}}_s'(\mathbf{r}, \mathbf{r}') \sim \frac{-i}{4\pi^2 k} e^{ik(r+r')} \sum_{n=0}^{\infty} \frac{B_n'(\theta, \theta'; \phi, \phi')}{(-i2kr)^n}, \tag{13}$$

where

$$B_n' = \sum_{m=0}^{\infty} \epsilon_m \cos m(\phi - \phi') \int_{-\infty}^{\infty} x e^{x\pi} \frac{(ix, n)}{x^2 + \frac{1}{4}} g_{\theta_s}' \, dx. \tag{13a}$$

The analogous result for $\mathscr{S}_s''$ has expansion coefficients B_n'' that contain g_{θ_s}'' instead of g_{θ_s}'. In view of Eq. (11d) and the differential equation for the Mehler functions,

$$\left[L(\theta) - \frac{m^2}{\sin^2 \theta} - (\tfrac{1}{4} + x^2)\right] K_x^{-m}(\cos \theta) = 0, \quad L(\theta) = \frac{1}{\sin \theta} \frac{d}{d\theta} \sin \theta \frac{d}{d\theta}, \tag{14}$$

one may verify that

$$C_n = -\frac{1}{n} [r^2 \nabla_t \cdot {}_t\nabla + n(n-1)] C_{n-1}, \quad n = 1, 2, \ldots, \tag{15a}$$

where C_n stands for any of the coefficients A_n', A_n'', B_n', B_n'', and $r^2 \nabla_t \cdot {}_t\nabla$ is defined in Eq. (2.6.2). Thus, all the coefficients may be obtained recursively from a knowledge of the $n = 0$ term. Moreover, in view of Eq. (14), the A_n may be derived from the B_n via

$$r^2 \nabla_t \cdot {}_t\nabla B_n = A_n, \quad n = 0, 1, 2, \ldots. \tag{15b}$$

The field derived from Eq. (12) for a radial electric dipole, or from Eqs. (10) and (13) (and their equivalent for $\mathscr{S}''$) for an arbitrarily oriented source, has a simple physical interpretation. The phase at the observation point P corresponds to a wave that has traveled from the source to the cone tip along the radius vector $\mathbf{r}'$ and then from the tip to P along the radius vector $\mathbf{r}$. The dependence on r and r' in Eq. (12) shows that both the incoming and the outgoing waves are spherical waves, that the incoming wave is centered at r' while the scattered wave is centered at the tip, and that the scattered wave has an angular dependence specified by the "diffraction coefficient" A_0. This process admits of a pictorial representation in terms of diffracted rays as shown in Fig. 6.8.3, the total field comprising free space and diffracted contributions. These observations apply in the region $\theta < \theta_c$ that excludes the geometric-optical rays reflected from the side of the cone. The formulation of the problem in the region $\theta_0 \geq \theta > \theta_c$ is considerably more involved, since the integrals in Eqs. (9c) and (10b) are then no longer convergent. As in the similar analysis of the wedge problem (Secs. 6.3a and 6.5c), it is necessary to subtract the geometrically re-

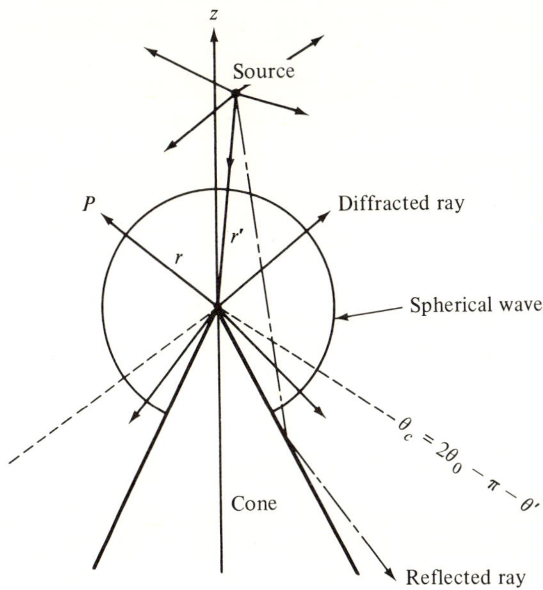

FIG. 6.8.3   Interpretation of high-frequency diffracted field.

flected fields before a convergent integral representation for the remainder can be obtained. Although this procedure is relatively straightforward for the wedge where the plane sides give rise to a reflected field which is of the same type as the incident field, the curved boundary of the cone destroys this simple feature and complicates the analysis. A study of this has been carried out for fields with simple azimuthal behavior[19,24] [excited by a ring source centered on the axis and phased according to exp $(im\phi)$], and it is found that the *functional form* of the diffraction coefficients A'_n, A''_n or B'_n, B''_n is the same throughout the entire region $0 < \theta \leq \theta_0$, although their above-described integral representation is valid only when $\theta < \theta_c$. Thus, if it is possible to evaluate the integral, the resulting function of $\theta, \theta', \phi, \phi'$ is expected to apply everywhere. The extraction of the geometric-optical contribution is facilitated by formulating the angular transmission line problem in terms of images in an infinitely extended θ space as in Eq. (3.4.74). It should be emphasized that the simple characterization of the high-frequency field in terms of contributions from direct, reflected, and diffracted rays is invalid in transition regions bounding the domain of existence of the reflected rays.

Approximation for small cone angles

While the evaluation of the diffraction coefficients must generally be carried out numerically,[19,25] approximate results may be derived for cones having small apex angles. In this range, $\theta_0 \approx \pi$, and the Legendre functions involving θ_0 may be approximated by elementary forms. Since $F(a, b; c; 0) = 1$, one finds from Eqs. (3.4.65) that for $\varphi \to 0$,

$$P_{\nu}^{-m}(\cos \varphi) \sim \frac{1}{\Gamma(1+m)}\left(\frac{\varphi}{2}\right)^{m}, \qquad m \geq 0, \tag{16a}$$

$$P_{\nu}^{-m}(-\cos \varphi) \sim \frac{\Gamma(m)}{\Gamma(1+m+\nu)\Gamma(m-\nu)}\left(\frac{2}{\varphi}\right)^{m}, \qquad m \geq 1, \tag{16b}$$

whereas[26]

$$P_{\nu}(-\cos \varphi) \sim \frac{2 \sin \nu\pi}{\pi} \ln \frac{\varphi}{2}, \qquad (m = 0). \tag{16c}$$

Also,

$$\frac{d}{d\varphi} P_{\nu}^{-m}(\cos \varphi) \sim \frac{1}{2\Gamma(m)}\left(\frac{\varphi}{2}\right)^{m-1}, \qquad m \geq 1, \tag{17a}$$

$$\frac{d}{d\varphi} P_{\nu}^{-m}(-\cos \varphi) \sim \frac{-\Gamma(1+m)}{2\Gamma(1+m+\nu)\Gamma(m-\nu)}\left(\frac{2}{\varphi}\right)^{m+1}, \qquad m \geq 1, \tag{17b}$$

and the expressions for $m = 0$ may be obtained from Eqs. (16a) and (16b) via the formula

$$\frac{d}{d\varphi} P_{\nu}(\cos \varphi) = P_{\nu}^{1}(\cos \varphi) = -\nu(\nu+1)P_{\nu}^{-1}(\cos \varphi). \tag{17c}$$

Thus, as $\theta_0 \to \pi$,

$$\frac{P_{\nu}(-\cos \theta_0)}{P_{\nu}(\cos \theta_0)} \sim \frac{\pi \csc \nu\pi}{2 \, ln \, [(\pi - \theta_0)/2]},$$

$$\frac{P_{\nu}^{-m}(-\cos \theta_0)}{P_{\nu}^{-m}(\cos \theta_0)} \sim O[(\pi - \theta_0)^{2m}], \qquad m \geq 1, \tag{18a}$$

and

$$\frac{\dfrac{d}{d\theta_0} P_{\nu}^{-m}(-\cos \theta_0)}{\dfrac{d}{d\theta_0} P_{\nu}^{-m}(\cos \theta_0)} \sim \begin{cases} \dfrac{\Gamma(m-\nu)\Gamma(1+m+\nu)}{-\Gamma(m)\Gamma(1+m)}\left(\dfrac{\pi-\theta_0}{2}\right)^{2m}, & m \geq 1, \tag{18b} \\[2ex] -\dfrac{\pi\nu(\nu+1)}{\sin \nu\pi}\left(\dfrac{\pi-\theta_0}{2}\right)^{2}, & m = 0. \tag{18c} \end{cases}$$

When the $m = 0$ term contributes to the field, a lowest-order approximation in $\pi - \theta_0$ therefore involves only the $m = 0$ term for the E modes, but the $m = 0$ and $m = 1$ terms for the H modes.

After these approximations have been substituted into Eqs. (11a), (13), and their counterparts for A_n'', B_n'', there remain certain integrals that may be evaluated in closed form;[19,27]

$$I_1 = \int_0^{\infty} x \frac{\tanh \pi x}{\cosh \pi x} K_x(\cos \theta)K_x(\cos \theta')\, dx = \frac{1}{\pi(\cos \theta + \cos \theta')}, \tag{19a}$$

$$I_2 = \int_0^{\infty} x \frac{\tanh \pi x}{\cosh \pi x} K_x^1(\cos \theta)K_x^1(\cos \theta')\, dx = \frac{2}{\pi} \frac{\sin \theta \sin \theta'}{(\cos \theta + \cos \theta')^3}, \tag{19b}$$

$$I_3 = \int_0^{\infty} x \frac{\tanh \pi x}{\cosh \pi x} \frac{K_x^1(\cos \theta)K_x^1(\cos \theta')}{(x^2 + 1/4)}\, dx = \frac{\tan (\theta/2) \tan (\theta'/2)}{\pi(\cos \theta + \cos \theta')}, \tag{19c}$$

$$I_4 = \int_0^\infty x \frac{\tanh \pi x}{\cosh \pi x}(1/4 + x^2)K_x(\cos \theta)K_x(\cos \theta')\,dx = \frac{2}{\pi}\frac{1 + \cos\theta\cos\theta'}{(\cos\theta + \cos\theta')^3},$$

$$(19d)$$

all results being subject to the restriction $\theta + \theta' < \pi$. It is then found that the $n = 0$ diffraction coefficients are given by

$$A_0'(\theta, \theta'; \phi, \phi') \approx \frac{-\pi}{2(\cos\theta + \cos\theta')\ln[(\pi - \theta_0)/2]}[1 + o(1)], \qquad (20a)$$

$$A_0''(\theta, \theta'; \phi, \phi') \approx 2\pi\left(\frac{\pi - \theta_0}{2}\right)^2\left[\frac{1 + \cos\theta\cos\theta' + 2\sin\theta\sin\theta'\cos(\phi - \phi')}{(\cos\theta + \cos\theta')^3}\right]$$

$$\times [1 + o(1)], \qquad (20b)$$

where $o(1)$ denotes terms which vanish as $\theta_0 \to \pi$. Thus, one has for the scattered portion of the scalar Green's functions G_s' and G_s'' when $\theta_0 \approx \pi$ and $\theta + \theta' < 2\theta_0 - \pi,$[23]

$$G_s'(\mathbf{r}, \mathbf{r}') \sim -\frac{ie^{ik(r+r')}}{(4\pi r')kr}\frac{1}{2(\cos\theta + \cos\theta')\ln[(\pi - \theta_0)/2]}, \qquad (21a)$$

$$G_s''(\mathbf{r}, \mathbf{r}') \sim \frac{ie^{ik(r+r')}}{(4\pi r')kr}\frac{2}{(\cos\theta + \cos\theta')^3}$$

$$\times [1 + \cos\theta\cos\theta' + 2\sin\theta\sin\theta'\cos(\phi - \phi')]\left(\frac{\pi - \theta_0}{2}\right)^2, \qquad (21b)$$

which results may be employed for the calculation of the fields due to a radial electric or magnetic dipole. The restriction $\theta + \theta' < 2\theta_0 - \pi$ is retained in view of the observations following Eqs. (10), although the *approximation* $\theta_0 \approx \pi$ has relaxed the convergence condition to $(\theta + \theta') < \pi$.

When the source is a transverse electric dipole, for example, $\mathbf{J}^0 = \boldsymbol{\phi}^0 J_\phi^0$, the electric field may be determined from Eqs. (2.6.4) and (2.6.9) as follows:

$$\frac{r'\mathbf{E}}{J_\phi^0} = \mathbf{r}_0\left(\frac{\partial^2}{\partial r^2} + k^2\right)\frac{1}{\sin\theta'}\frac{\partial^2}{\partial r'\,\partial\phi'}\frac{\mathscr{S}'}{-i\omega\epsilon}$$

$$+ \boldsymbol{\theta}_0\left(\frac{\partial^2}{r\,\partial r\,\partial\theta}\frac{\partial^2}{\sin\theta'\,\partial r'\,\partial\phi'}\frac{\mathscr{S}'}{-i\omega\epsilon} - \frac{i\omega\mu}{r\sin\theta}\frac{\partial^2}{\partial\phi\,\partial\theta'}\mathscr{S}''\right)$$

$$+ \boldsymbol{\phi}_0\left(\frac{1}{r\sin\theta}\frac{\partial^2}{\partial r\,\partial\phi}\frac{\partial^2}{\sin\theta'\,\partial r'\,\partial\phi'}\frac{\mathscr{S}'}{-i\omega\epsilon} + \frac{i\omega\mu}{r}\frac{\partial^2}{\partial\theta\,\partial\theta'}\mathscr{S}''\right). \qquad (22)$$

Since $\partial/\partial\phi'$ appears in all terms involving $\mathscr{S}'$, the ϕ'-independent $m = 0$ term in Eq. (13) does not contribute. The dominant contribution to $\mathscr{S}_s'$ for a small-angle cone is then found from the $m = 1$ term in the diffraction coefficient B_0',

$$B_0'(\theta, \theta'; \phi, \phi')\,|_{m=1} \approx -2\pi\left(\frac{\pi - \theta_0}{2}\right)^2\frac{\tan(\theta/2)\tan(\theta'/2)}{\cos\theta + \cos\theta'}\cos(\phi - \phi'), \quad (23a)$$

whereas to the same order of approximation in $\pi - \theta_0$, both the $m = 0$ and $m = 1$ terms contribute to B_0'' and yield

$$B_0''(\theta, \theta'; \phi, \phi') \approx \pi\left(\frac{\pi - \theta_0}{2}\right)^2\frac{1}{\cos\theta + \cos\theta'} - B_0'(\theta, \theta'; \phi, \phi')\,|_{m=1}. \quad (23b)$$

The distant scattered field may now be obtained by substituting Eqs. (23) into Eq. (13) and its counterpart for $\mathscr{S}''$ and recalling Eq. (10). The result simplifies substantially when the source lies on the axis, so $\theta' = 0$. By moving the dipole to infinity and normalizing as in the paragraph following Eq. (8), one finds for the total electric field in the region $\theta < (2\theta_0 - \pi)$ due to a plane wave incident parallel to the cone axis

$$\mathbf{E} \approx \mathbf{x}_0 e^{-ikz} + (\boldsymbol{\theta}_0 \cos\phi - \boldsymbol{\phi}_0 \sin\phi)\frac{e^{ikr}}{ikr}\left(\frac{\pi - \theta_0}{2}\right)^2 \sec^4\frac{\theta}{2}, \qquad (24)$$

which result is valid in the region $\theta < 2\theta_0 - \pi$. While the exact diffraction coefficient diverges on the reflected wave boundary $\theta = 2\theta_0 - \pi$, the approximate evaluation for $\theta_0 \approx \pi$ has shifted the divergence to $\theta = \pi$. Along the cone axis $\theta = 0$, Eq. (24) reduces to

$$\mathbf{E} \approx \mathbf{x}_0\left[e^{-ikz} - i\left(\frac{\pi - \theta_0}{2}\right)^2\frac{e^{ikz}}{kz}\right]. \qquad (24a)$$

For arbitrary cone angles in the range $\pi/2 < \theta_0 < \pi$, the integrals in Eqs. (11a), (13a), etc., cannot be evaluated in closed form; numerically computed curves are available and have been presented in References 19 and 25. It is found that the fields calculated from these curves are well approximated by the physical optics expression; moreover, for wide-angle cones $\theta_0 \approx \pi/2$, one may derive approximate closed-form results. These aspects are illustrated in the Problems section at the end of this chapter.

APPENDIX 6A. ASYMPTOTIC FORMULAS FOR $H_\nu^{(1)}(z)$ AND $H_\nu^{(2)}(z)$

For many of the problems discussed in the text, it has been necessary to utilize asymptotic expressions for Bessel functions of large complex argument z or order ν. The pertinent formulas, listed in this appendix, comprise two different sets: (1) for the range $|\nu/z| \gg 1$ or $|\nu/z| \ll 1$, and (2) for the range $|\nu/z| \approx 1$. Expressions for (1) were derived by Debye[28]; those for (2) are due primarily to Langer[29] and Olver[30] (see also Reference 31).

6A.1 Large, Unequal Order and Argument

If the argument and order are not approximately equal [more precisely, if $|\nu - z| > O(|\nu|^{1/3})$], three different asymptotic representations for $H_\nu^{(1),(2)}(z)$ suffice to cover the entire complex plane divided as in Fig. 6A.1.

For $H_\nu^{(1)}(z)$, one has

Region I

$$H_\nu^{(1)}(z) \sim \sqrt{\frac{2}{\pi^2 z \sin\gamma}}\, e^{-i\pi/4} e^{iz(\sin\gamma - \gamma\cos\gamma)} \sum_{n=0}^{\infty} \frac{A_n \Gamma[n + (1/2)] e^{-in\pi/2}}{[(z\sin\gamma)/2]^n}. \qquad (A1)$$

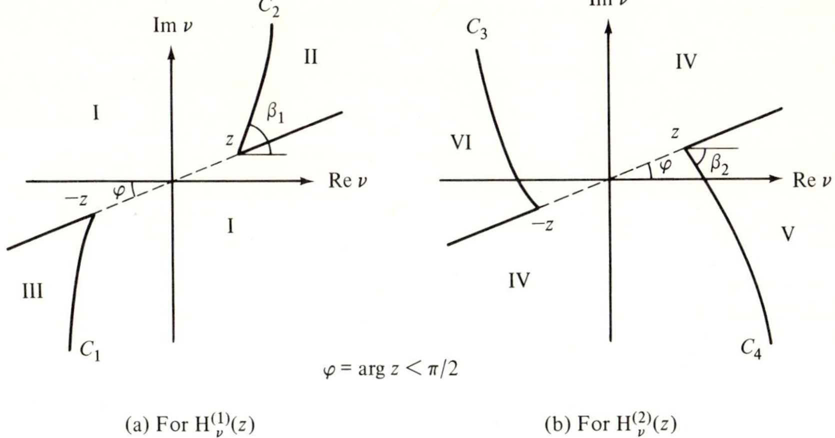

FIG. 6A.1   Various regions in the complex ν plane.

Region II

$$H_\nu^{(1)}(z) \sim -\sqrt{\frac{2}{\pi^2 z \sin \gamma}}\, e^{i\pi/4} e^{-iz\,(\sin \gamma - \gamma \cos \gamma)} \sum_{n=0}^{\infty} \frac{A_n \Gamma[(n + (1/2)] e^{in\pi/2}}{[(z \sin \gamma)/2]^n}. \tag{A2}$$

Region III

$$H_\nu^{(1)}(z) \sim -\sqrt{\frac{2}{\pi^2 z \sin \gamma}}\, e^{i\pi/4} e^{-iz[\sin \gamma + (2\pi - \gamma) \cos \gamma]} \sum_{n=0}^{\infty} \frac{A_n \Gamma[n + (1/2)] e^{in\pi/2}}{[(z \sin \gamma)/2]^n}. \tag{A3}$$

In these formulas, $\Gamma(x)$ is the gamma function, with

$$\cos \gamma = \frac{\nu}{z}, \qquad |\arg z| < \pi/2, \tag{A4a}$$

and γ is restricted so that

$$0 < \operatorname{Re} \gamma < \pi. \tag{A4b}$$

The first few of the expansion coefficients A_n are

$$A_0 = 1, \quad A_1 = \tfrac{1}{8} + \tfrac{5}{24} \cot^2\gamma, \quad A_2 = \tfrac{3}{128} + \tfrac{77}{576} \cot^2\gamma + \tfrac{385}{3456} \cot^4\gamma. \tag{A4c}$$

For $H_\nu^{(2)}(z)$, one has

Region IV

$$H_\nu^{(2)}(z) \sim \sqrt{\frac{2}{\pi^2 z \sin \gamma}}\, e^{i\pi/4} e^{-iz\,(\sin \gamma - \gamma \cos \gamma)} \sum_{n=0}^{\infty} \frac{A_n e^{in\pi/2} \Gamma[n + (1/2)]}{[(z \sin \gamma)/2]^n}. \tag{A5}$$

Region V

$$H_\nu^{(2)}(z) \sim -\sqrt{\frac{2}{\pi^2 z \sin \gamma}}\, e^{-i\pi/4} e^{iz\,(\sin \gamma - \gamma \cos \gamma)} \sum_{n=0}^{\infty} \frac{A_n \Gamma[n + (1/2)] e^{-in\pi/2}}{[(z \sin \gamma)/2]^n}. \tag{A6}$$

Region VI

$$H_\nu^{(2)}(z) \sim -\sqrt{\frac{2}{\pi^2 z \sin \gamma}}\, e^{-i\pi/4} e^{iz[\sin \gamma + (2\pi - \gamma)\cos \gamma]} \sum_{n=0}^{\infty} \frac{A_n \Gamma[n + (1/2)] e^{-in\pi/2}}{[(z \sin \gamma)/2]^n}. \quad (A7)$$

Equations (A4) apply here as well. The zeros of $H_\nu^{(1)}(z)$ for large ν, z lie on the curves C_1 and C_2 which form a partial boundary of regions III and II, respectively. To obtain an expression for $H_\nu^{(1)}(z)$ near or on C_1, one adds Eqs. (A1) and (A3), while the sum of Eqs. (A1) and (A2) is appropriate for a representation of $H_\nu^{(1)}(z)$ near or on C_2. This addition is required since in the above formulas, in which the exponential terms assume large magnitude, a further contribution of exponentially small magnitude has been omitted. Near C_1 or C_2, however, the magnitude of the exponential terms approaches unity, and both contributions must be taken into account. Similar considerations apply to the representation of $H_\nu^{(2)}(z)$ in the vicinity of the contours C_3 and C_4 in Fig. 6.A.1(b). The slope of $C_{1,2}$ at $\nu = \mp z$ is defined by the angle $\beta_1 = (\pi + \arg z)/3$, while the corresponding angle for $C_{3,4}$ is $\beta_2 = (\pi - \arg z)/3$ [see Eq. (A35)]. As $|\nu| \to \infty$ along the curves, both $|\mathrm{Re}\, \nu|$ and $|\mathrm{Im}\, \nu|$ increase but the growth of $|\mathrm{Im}\, \nu|$ is more rapid than that of $|\mathrm{Re}\, \nu|$, whence $|\arg \nu| \to \pi/2$ [see Eqs. (A21) and the discussion following Eq. (A44) for the behavior along C_2].

On the line $\arg (\nu/z) = 0$, $|\nu| > |z|$, formulas (A1) and (A6) apply for $H_\nu^{(1)}$ and $H_\nu^{(2)}$, respectively, with $\gamma = i\delta, \delta > 0$. On the line $|\arg (\nu/z)| = \pi$, $|\nu| > |z|$, the appropriate formulas are (3) and (5), with $\gamma = \pi + i\delta, \delta > 0$.

Formulas for

$$J_\nu(z) = \tfrac{1}{2}[H_\nu^{(1)}(z) + H_\nu^{(2)}(z)] \quad (A8)$$

can be obtained by addition from the above, except in those cases when $H_\nu^{(1)}(z) \sim -H_\nu^{(2)}(z)$, in which instance the Bessel function has a behavior characterized by an exponentially small contribution. Thus, when ν and z are large and positive and $\nu > z$, $J_\nu(z)$ is represented asymptotically by the real function

$$J_\nu(z) \sim \tfrac{1}{2} S(-i\delta), \qquad \delta > 0, \quad (A9)$$

where $S(\gamma)$ denotes the expression on the right-hand side of Eq. (A1).

6A.2 Large Argument

When z is large and $|z| \gg |\nu|$ (i.e. $\gamma \to \pi/2$), formulas (1) and (5) can be replaced by the simpler expressions

$$H_\nu^{(1)}(z) \sim \sqrt{\frac{2}{\pi z}}\, e^{i(z - \nu\pi/2 - \pi/4)} \sum_{n=0}^{\infty} \frac{(\nu, n)}{(-2iz)^n}, \qquad -\pi < \arg z < 2\pi, \quad (A10a)$$

$$H_\nu^{(2)}(z) \sim \sqrt{\frac{2}{\pi z}}\, e^{-i(z - \nu\pi/2 - \pi/4)} \sum_{n=0}^{\infty} \frac{(\nu, n)}{(2iz)^n}, \qquad -2\pi < \arg z < \pi, \quad (A10b)$$

where

$$(\nu, n) \equiv \frac{[4\nu^2 - 1^2][4\nu^2 - 3^2] \cdots [4\nu^2 - (2n - 1)^2]}{2^{2n} n!}, \qquad (\nu, 0) \equiv 1. \quad (A10c)$$

A formula for $J_\nu(z)$ is obtained from Eq. (A8).

6A.3 Large Order

When ν is large and $|\nu| \gg |z|$, the results in Sec. A1 may likewise be simplified; $\cos \gamma$ is now very large and can be approximated by its dominant exponential. Suppose that $\gamma = i\xi$, $\operatorname{Re}\xi > 0$, so that $2\nu/z = 2\cos\gamma \approx \exp(\xi) \approx -2i\sin\gamma$. Then, from the $n = 0$ term in Eq. (A1),

$$H_\nu^{(1)}(z) \sim \sqrt{\frac{2}{\pi z(i\nu/z)}}\, e^{-i\pi/4} e^{-\nu + \nu\xi} = -i\sqrt{\frac{2}{\pi\nu}}\left(\frac{2\nu}{ez}\right)^\nu. \tag{A11}$$

Since $\gamma = i\ln(2\nu/z) = i\ln|2\nu/z| - \arg(\nu/z)$, one notes from Eq. (A4b) that the validity of Eq. (A11) is restricted to that portion of region I in Fig. A1(a), in which $-\pi < \arg(\nu/z) < 0$. Expressions appropriate to other portions of the complex ν plane, obtained in a similar manner, are summarized below, and apply in the range $|\arg(\nu/z)| < \pi$:

$$H_\nu^{(1)}(z) \sim \begin{cases} e^{i(\pm 1 - 1)\pi/4}\sqrt{\dfrac{2}{\pi\nu}}\left(\dfrac{2\nu}{ez}\right)^{\mp\nu}, & \arg\left(\dfrac{\nu}{z}\right) \gtrless 0, & \text{in region I,} & \text{(A12a)} \\[3ex] -i\sqrt{\dfrac{2}{\pi\nu}}\left(\dfrac{2\nu}{ez}\right)^\nu, & & \text{in region II,} & \text{(A12b)} \\[3ex] -\sqrt{\dfrac{2}{\pi\nu}}\,e^{-i2\pi\nu}\left(\dfrac{2\nu}{ez}\right)^{-\nu}, & & \text{in region III.} & \text{(A12c)} \end{cases}$$

$$H_\nu^{(2)}(z) \sim \begin{cases} e^{i(\pm 1 + 1)\pi/4}\sqrt{\dfrac{2}{\pi\nu}}\left(\dfrac{2\nu}{ez}\right)^{\pm\nu}, & \arg\left(\dfrac{\nu}{z}\right) \gtrless 0, & \text{in region IV,} & \text{(A13a)} \\[3ex] i\sqrt{\dfrac{2}{\pi\nu}}\left(\dfrac{2\nu}{ez}\right)^\nu, & & \text{in region V,} & \text{(A13b)} \\[3ex] -\sqrt{\dfrac{2}{\pi\nu}}\,e^{i2\pi\nu}\left(\dfrac{2\nu}{ez}\right)^{-\nu}, & & \text{in region VI.} & \text{(A13c)} \end{cases}$$

These formulas are continuous across the lines $\arg(\nu/z) = 0$ and $|\arg(\nu/z)| = \pi$, and may therefore be employed also on these boundaries. The continuity across the line $\arg(\nu/z) = 0$ is evident. To show the equality of Eqs. (A12a) and (A12c) on the line $\arg(\nu/z) = \pm\pi$, it is noted that the values of ν in regions I and III are related there via $\nu_{\text{I}} = \nu_{\text{III}}e^{i2\pi}$. Similar considerations apply to Eqs. (A13a) and (A13c). The factors $(2\nu/ez)^{\pm\nu}$ approach unit magnitude near the curves C_2 and C_4, while $|(2\nu e^{\pm i\pi}/ez)^\nu| \to 1$ near the curves $C_{1,3}$. In the vicinity of any of these contours, the pertinent expressions for $H_\nu^{(1),\,(2)}$ are given by the formulas valid on either side.

The asymptotic formula for the Bessel function is obtained directly from the power-series expansion

$$J_\nu(z) = \sum_{n=0}^{\infty} \frac{(-1)^n (z/2)^{\nu + 2n}}{n!\,\Gamma(\nu + n + 1)}, \tag{A14}$$

upon substitution of the large-argument expression for the gamma function,

$$\Gamma(\nu + \alpha) \sim \sqrt{\frac{2\pi}{\nu}}\left(\frac{\nu}{e}\right)^\nu \nu^\alpha, \qquad |\nu| \to \infty, \qquad |\arg\nu| < \pi, \quad \alpha > 0. \tag{A15}$$

Thus,

$$J_\nu(z) \sim \frac{1}{\sqrt{2\pi\nu}} \left(\frac{2\nu}{ez}\right)^{-\nu}, \qquad |\arg \nu| < \pi. \qquad (A16)$$

Equations (A12) and (A13) could also have been derived from Eq. (A16) and the formula

$$H_\nu^{(1,2)}(z) = \frac{J_{-\nu}(z) - e^{\mp i\nu\pi} J_\nu(z)}{\pm i \sin \nu\pi}. \qquad (A17)$$

To assess the magnitude of the factor $(2\nu/ez)^{\pm\nu}$, we let

$$\nu = \zeta e^{i\psi}, \qquad \zeta = |\nu| \to \infty, \qquad (A18a)$$

to obtain

$$\left|\left(\frac{2\nu}{ez}\right)^{\pm\nu}\right| = \exp\left\{\pm\zeta\left[\cos\psi \ln\frac{2\zeta}{e|z|} - \sin\psi\,(\psi - \arg z)\right]\right\}. \qquad (A18b)$$

This term grows faster than exponentially whenever $\pm\cos\psi > 0$, decays whenever $\pm\cos\psi < 0$, and assumes unit magnitude only when $|\psi| \to \pi/2$ [see Eq. (A21)]. Hence, the Hankel functions grow faster than exponentially as $|\nu| \to \infty$ everywhere except when $|\psi| \to \pi/2$. In the latter range (i.e., near the curves $C_{1,2,3,4}$) the simple expressions in Eqs. (A12) and (A13) must be modified as mentioned above, and a study of the resulting formulas reveals that for $|\nu| \to \infty$, $H_\nu^{(1)}(z)$ is small in the region between the curves C_1 and C_4 but large outside this region, while $H_\nu^{(2)}(z)$ is small in the region between the curves C_2 and C_3 but large outside this region. $J_\nu(z)$ in Eq. (A16) decays to the right of the curves C_2 and C_4 but increases elsewhere. On the imaginary axis, for z positive real,

$$H_\nu^{(1)}(z),\, J_\nu(z) \sim O[\zeta^{-1/2}e^{\zeta\pi/2}]; \quad H_\nu^{(2)}(z) \sim O[\zeta^{-1/2}e^{-\zeta\pi/2}], \qquad \psi = \frac{\pi}{2}, \qquad (A19a)$$

$$H_\nu^{(1)}(z) \sim O[\zeta^{-1/2}e^{-\zeta\pi/2}]; \quad H_\nu^{(2)}(z),\, J_\nu(z) \sim O[\zeta^{-1/2}e^{\zeta\pi/2}], \qquad \psi = -\frac{\pi}{2}. \qquad (A19b)$$

Because of its bearing on the cylinder diffraction problem in the text, we now consider behavior of $H_\nu^{(1,2)}(z)$, $(\partial/\partial z)H_\nu^{(1)}(z)$, $(\partial/\partial\nu)H_\nu^{(1)}(z)$, and $J_\nu(z)$ on the curve C_2, as well as that of $H_\nu^{(1)}(y)$, $y \neq z$. In the vicinity of C_2, the asymptotic approximation of $H_\nu^{(1)}(z)$ comprises the sum of Eqs. (A12a) and (A12b):

$$H_\nu^{(1)}(z) \sim \sqrt{\frac{2}{\pi\nu}}\left[\left(\frac{2\nu}{ez}\right)^{-\nu} - i\left(\frac{2\nu}{ez}\right)^{\nu}\right]. \qquad (A20)$$

Each term inside the brackets takes on unit magnitude when

$$\psi = \frac{\pi}{2} - \delta, \qquad \delta \simeq \frac{\pi}{2\ln(2\zeta/ez)} \quad \text{or} \quad \zeta = \frac{ez}{2}e^{\pi/2\delta}, \qquad (A21)$$

where z is assumed to be positive real for convenience. Equation (A21) actually defines the distant portions of the curve C_2 on which the zeros of $H_\nu^{(1)}(z)$ are located since Eq. (A20) can vanish only when each of the exponential terms has the same magnitude. One observes from a comparison of Eqs. (A19) and (A21) that all functions vary extremely rapidly as $|\psi| \to \pi/2$.

Since the distant zeros of $H_\nu^{(1)}(y)$, y positive and unequal to z, also have $|\psi| \to \pi/2$, the representation of $H_\nu^{(1)}(y)$ near C_2 likewise comprises the two terms in Eq.(A20). On C_2,

$$\delta \ln \frac{2\zeta}{ey} = \delta \ln\left(\frac{z}{y}e^{\pi/2\delta}\right) = \frac{\pi \ln (z/y)}{2 \ln (2\zeta/ez)} + \frac{\pi}{2}. \qquad (A22a)$$

so

$$\left|\left(\frac{2\nu}{ey}\right)^{\pm\nu}\right| = \exp\left[\pm\frac{\zeta\pi}{2}\frac{\ln (z/y)}{\ln (2\zeta/ez)}\right]. \qquad (A22b)$$

Thus, one or the other of the terms in Eq. (A20) (with z replaced by y) dominates when $z > y$ or $z < y$, respectively, and one finds for any positive y,

$$|H_\nu^{(1)}(y)| \sim \sqrt{\frac{2}{\pi\zeta}} \exp\left[\frac{\zeta\pi}{2}\frac{|\ln (z/y)|}{\ln (2\zeta/ez)}\right] \quad \text{on } C_2, \qquad \text{as } \zeta \to \infty. \qquad (A23)$$

Upon differentiating Eq. (A20) with respect to ν, noting that $|(2\nu/ez)^{\pm\nu}|$ is $O(1)$ on C_2 and that the terms in the brackets of Eq. (A20) must cancel at the zeros of $H_\nu^{(1)}(z)$, one finds that

$$\frac{\partial}{\partial\nu} H_\nu^{(1)}(z) \sim O\left(\frac{\ln (2\zeta/ez)}{\sqrt{\zeta}}\right), \qquad (A24)$$

along C_2 and at the zeros of $H_\nu^{(1)}(z)$. Similarly,

$$\frac{\partial}{\partial z}H_\nu^{(1)}(z) \sim O(\sqrt{\zeta}), \qquad (A25)$$

and, also,

$$J_\nu(z) \sim O\left(\frac{1}{\sqrt{\zeta}}\right), \quad H_\nu^{(2)}(z) \sim O\left(\frac{1}{\sqrt{\zeta}}\right), \qquad \zeta \to \infty, \qquad (A26)$$

along C_2 and at the zeros of $H_\nu^{(1)}(z)$.

Since

$$H_{-\nu}^{(1,2)}(z) = e^{\pm i\nu\pi}H_\nu^{(1,2)}(z), \qquad (A27)$$

the zeros of $H_\nu^{(1,2)}(z)$ are disposed symmetrically with respect to the origin in the complex ν plane as indicated by the curves C_1, C_2 and C_3, C_4. One may also verify that the locus of the zeros of $(d/dz)H_\nu^{(1)}(z)$ behaves in a manner similar to $C_{1,2}$ as $\nu \to \infty$.

6A.4 Large and Almost Equal Order and Argument

When $\nu \approx z$, then $\gamma \to 0$ in the Debye expansions in Eqs. (A1)–(A7) and the resulting expressions become invalid. In this range the Hankel functions can be approximated by Airy functions that provide the means of passing smoothly from the region $|\nu - z| > O(|\nu|^{1/3})$, wherein the Debye formulas are valid, to the region $|\nu - z| \le O(|\nu|^{1/3})$. The expressions below have been obtained by Olver.[30]

If an arbitrary complex number τ is defined by the relation

$$z = v + \tau v^{1/3}, \tag{A28}$$

then $H_v^{(1,2)}(z)$ has the asymptotic expansion as $v \to \infty$,

$$H_v^{(1,2)}(z) \sim \frac{2^{1/3}}{v^{1/3}} [\mathrm{Ai}(-2^{1/3}\tau) \mp i\,\mathrm{Bi}(-2^{1/3}\tau)] \sum_{n=0}^{\infty} \frac{A_n(\tau)}{v^{2n/3}}$$

$$+ \frac{2^{1/3}}{v^{1/3}} [\mathrm{Ai}'(-2^{1/3}\tau) \mp i\,\mathrm{Bi}'(-2^{1/3}\tau)] \sum_{n=0}^{\infty} \frac{B_n(\tau)}{v^{2n/3}}, \tag{A29}$$

where the first few coefficients are

$$A_0 = 1, \quad A_1 = -\frac{\tau}{5}, \quad A_2 = -\frac{9}{100}\tau^5 + \frac{3}{35}\tau^2,$$

$$B_0 = 0, \quad B_1 = \frac{3}{10}\tau^2, \quad B_2 = -\frac{17}{3}\tau^3 + \frac{1}{70}. \tag{A29a}$$

The expansion for $H_v^{(1)}(z)$ is valid when $-\pi/2 < \arg v < 3\pi/2$, while that for $H_v^{(2)}(z)$ applies when $-3\pi/2 < \arg v < \pi/2$. For values of v outside these ranges, one may employ Eq. (A27). $\mathrm{Ai}(x)$, $\mathrm{Bi}(x)$ and $\mathrm{Ai}'(x)$, $\mathrm{Bi}'(x)$, denote the Airy functions and their x derivatives, respectively (see Chapter 4, Appendix B). To identify the $n = 0$ term in Eq. (A29) with the similar result in Eq. (4.5.33), one observes from Eq. (4.5.35a) that η in these equations is given approximately by

$$\eta \approx 2^{1/3}\tau z^{-2/3}, \qquad z \approx v. \tag{A30}$$

Expressions for the derivatives $(d/dz)H_v^{(1,2)}(z) \equiv H_v'^{(1,2)}(z)$ are obtained by differentiation of the asymptotic series for $H_v^{(1,2)}(z)$. The Airy functions and their zeros are tabulated in Reference 32.

6A.5 The Zeros of $H_v^{(1)}(z)$, $H_v'^{(1)}(z)$, and Related Results

From the results in the preceding section, it is noted that the zeros of $H_v^{(1,2)}(z)$ or $H_v'^{(1,2)}(z)$ when $v \approx z$, $|z|$ large, are given in first approximation by the zeros of the appropriate Airy function combinations or their derivatives.

Since the Airy differential equation

$$\left(\frac{d^2}{d\sigma^2} + \sigma\right)\frac{\mathrm{Ai}(-\sigma)}{\mathrm{Bi}(-\sigma)} = 0 \tag{A31}$$

is also solved by $\sqrt{\sigma}\,Z_{1/3}(\tfrac{2}{3}\sigma^{3/2})$, where $Z_{1/3}(x)$ is any Bessel function of order $\frac{1}{3}$ and argument x, one finds from a comparison of the large-σ asymptotic solution for the function $A_1(-\sigma)$ [see Eqs. (4.2.51)] and the Hankel function of order $1/3$ that

$$A_1(-\sigma) \equiv \mathrm{Ai}(-\sigma) - i\,\mathrm{Bi}(-\sigma) = e^{i\pi/6}\sqrt{\frac{\sigma}{3}}\,H_{1/3}^{(1)}(\tfrac{2}{3}\sigma^{3/2})$$

$$= -2e^{-i2\pi/3}\mathrm{Ai}(-\sigma e^{-i2\pi/3}). \tag{A32}$$

Although it is more convenient to deal directly with the Airy function formulation [see Eqs. (A39)], many investigations in the literature have utilized the $\frac{1}{3}$-order Bessel functions; for this reason, both treatments are included. Upon introducing the change of variable

$$\tfrac{2}{3}\sigma^{3/2} = \xi e^{-i\pi},\qquad\qquad (A33)$$

and employing Eq. (A17), one may show that the zeros $\bar{\xi}_p$ of $H_{1/3}^{(1)}(\xi e^{-i\pi})$ are identical with those of the equation

$$J_{-1/3}(\bar{\xi}_p) + J_{1/3}(\bar{\xi}_p) = 0,\qquad\qquad (A34)$$

which are real. With $\sigma = 2^{1/3}\tau$, $\tau = (z-\nu)\nu^{-1/3}$, and z fixed, the zeros $\pm\xi_p$ of $H_\nu^{(1)}(z)$ are given via Eqs. (A29), (A33), and (A34) by

$$\xi_p \sim z + 2^{-1/3}(\tfrac{3}{2}\bar{\xi}_p)^{2/3}e^{i\pi/3}z^{1/3},\quad H_{\xi_p}^{(1)}(z) = 0,\qquad \bar{\xi}_p > 0.\qquad (A35)$$

To a first approximation, the zeros of $H_\nu'^{(1)}(z)$ coincide with those of $A_1'(-2^{1/3}\tau) \equiv A_1'(-\sigma)$, which, in view of the formula

$$\frac{d}{d\xi}\,[\xi^\mu H_\mu^{(1)}(\xi e^{-i\pi})] = \xi^\mu H_{\mu-1}^{(1)}(\xi e^{-i\pi})\qquad\qquad (A36)$$

applied to $\mu = \tfrac{1}{3}$, and Eqs. (A17), (A32), and (A33), can be seen to be derivable from the purely real zeros $\bar{\eta}_p$ of the equation

$$J_{2/3}(\bar{\eta}_p) - J_{-2/3}(\bar{\eta}_p) = 0.\qquad\qquad (A37)$$

If $\pm\eta_p$ denotes the zeros of $H_\nu'^{(1)}(z)$ in the complex ν plane, then

$$\eta_p \sim z + 2^{-1/3}(\tfrac{3}{2}\bar{\eta}_p)^{2/3}e^{i\pi/3}z^{1/3},\quad H_{\eta_p}'^{(1)}(z) = 0,\qquad \bar{\eta}_p > 0.\qquad (A38)$$

If z is positive real, the zeros of $H_\nu^{(2)}(z)$ and $H_\nu'^{(2)}(z)$ are given by the complex conjugates of ξ_p and η_p, respectively.

The zeros $\bar{\xi}_p$ and $\bar{\eta}_p$ can also be expressed as zeros of the Airy function $\text{Ai}(-\alpha)$ or of its derivative. From Eqs. (4.2.32) and (4.2.34) one may verify that

$$A_1(-\alpha e^{-i2\pi/3}) = -2e^{i2\pi/3}\,\text{Ai}(-\alpha),\qquad \alpha = (\tfrac{3}{2}\xi)^{2/3},\qquad (A39a)$$

$$A_2(-\alpha e^{-i2\pi/3}) = e^{i2\pi/3}A_1(-\alpha),\qquad\qquad A_{2,1}(x) \equiv \text{Ai}(x) \pm i\,\text{Bi}(x),\quad (A39b)$$

whence the $\bar{\xi}_p$ also satisfy the equation

$$\text{Ai}(-\bar{\alpha}_p) = 0,\qquad \bar{\alpha}_p = (\tfrac{3}{2}\bar{\xi}_p)^{2/3},\qquad\qquad (A40)$$

while for the $\bar{\eta}_p$,

$$\text{Ai}'(-\bar{\beta}_p) = 0,\qquad \bar{\beta}_p = (\tfrac{3}{2}\bar{\eta}_p)^{2/3}.\qquad\qquad (A41)$$

$\bar{\alpha}_p$ and $\bar{\beta}_p$ are tabulated in reference 32. The first few zeros are

$$\begin{aligned}
\bar{\alpha}_1 &= 2.3381, \quad \bar{\alpha}_2 = 4.0879, \quad \bar{\alpha}_3 = 5.5205,\\
\bar{\beta}_1 &= 1.0188, \quad \bar{\beta}_2 = 3.2482, \quad \bar{\beta}_3 = 4.8201.
\end{aligned}\qquad (A41a)$$

One also verifies from Eqs. (A29) that

$$\frac{\partial H_\nu^{(1,2)}(z)}{\partial\nu} \sim -H_\nu'^{(1,2)}(z) \sim \left(\frac{2}{\nu}\right)^{2/3}A_1'(-2^{1/3}\tau),\qquad (A42a)$$

$$\frac{\partial H_\nu'^{(1,2)}(z)}{\partial \nu} \sim -\frac{d^2 H_\nu^{(1,2)}(z)}{dz^2} = \left(1 - \frac{\nu^2}{z^2}\right) H_\nu^{(1,2)}(z) + \frac{1}{z} H_\nu'^{(1,2)}(z), \quad \text{(A42b)}$$

with the last equality resulting from the Bessel differential equation.

The preceding expressions are the lowest-order asymptotic approximations and can be improved by developing the pertinent quantities in asymptotic series involving inverse powers of z. Such expansions have been obtained by a number of investigators; including the second terms in the expansions, one finds[14,33]

$$\xi_p \sim z + \left(\frac{z}{2}\right)^{1/3}\left[t_p + \left(\frac{2}{z}\right)^{2/3}\frac{t_p^2}{60} + \cdots\right], \qquad t_p = \bar{\alpha}_p e^{i\pi/3}, \quad \text{(A43a)}$$

$$\frac{1}{[\partial H_\nu^{(1)}(z)/\partial \nu]_{\nu=\xi_p}} \sim \frac{1}{2}\left(\frac{z}{2}\right)^{2/3}\frac{e^{-i\pi/3}}{\text{Ai}'(-\bar{\alpha}_p)}\left[1 + \left(\frac{2}{z}\right)^{2/3}\frac{t_p}{10} + \cdots\right], \quad \text{(A43b)}$$

$$\eta_p \sim z + \left(\frac{z}{2}\right)^{1/3}\left[q_p + \left(\frac{2}{z}\right)^{2/3}\frac{q_p^2}{60}\left(1 + \frac{8}{3\bar{\eta}_p^2}\right) + \cdots\right], \quad q_p = \bar{\beta}_p e^{i\pi/3},$$

$$\text{(A44a)}$$

$$\frac{1}{[\partial H_\nu'^{(1)}(z)/\partial \nu]_{\nu=\eta_p}} \sim \frac{1}{[1 - (\eta_p/z)^2]H_{\eta_p}^{(1)}(z)}\left[1 + \frac{1}{6}q_p\left(\frac{2}{z}\right)^{2/3} + \cdots\right], \quad \text{(A44b)}$$

where for the first few values of p,

$$\text{Ai}(-\bar{\beta}_1) = +0.5356, \quad \text{Ai}(-\bar{\beta}_2) = -0.4190, \quad \text{Ai}(-\bar{\beta}_3) = +0.3804,$$

$$\text{Ai}'(-\bar{\alpha}_1) = +0.7012, \quad \text{Ai}'(-\bar{\alpha}_2) = -0.8031, \quad \text{Ai}'(-\bar{\alpha}_3) = +0.8652.$$

$$\text{(A44c)}$$

The preceding discussion has dealt with the zeros of $H_\nu^{(1,2)}(z)$ and $H_\nu'^{(1,2)}(z)$ in the complex ν plane when the large parameter z is specified and $\nu \approx z$. It can be shown, however, that the magnitude of the zeros is always greater than z, so ξ_p and η_p in Eqs. (A35) and (A38) do indeed constitute the first zeros of $H_\nu^{(1)}(z)$ and $H_\nu'^{(1)}(z)$, respectively. These zeros lie on a straight line which, for real z, departs into the first quadrant at an angle of $60°$ from the point $\nu = z$ (see Fig. 6.A.1). They are of the first order and infinite in number, but their value as $|\xi_p| \to \infty$ or $|\eta_p| \to \infty$ is no longer given by the equations above; instead, they are derived from the representations in Eqs. (A12), which are appropriate to the range $|\nu| \to \infty$. One finds that $\arg \eta_p \to \pi/2$ [see Eq. (A21)], whence, as indicated in Fig. 6A.1, the locus of zeros bends toward the vertical direction as $|\eta_p| \to \infty$.[34]

APPENDIX 6B.
MISCELLANEOUS FORMULAS INVOLVING CYLINDER FUNCTIONS

The product of two modified Hankel functions of imaginary argument can be represented as[3]

$$K_{i\nu}(a)K_{i\nu}(b) = \frac{1}{2}\int_{-\infty}^{\infty} K_0(\sqrt{a^2 + b^2 + 2ab\cosh x})e^{i x\nu}\, dx, \qquad \nu \text{ real}, \quad \text{(B1)}$$

where a and b are assumed to be positive and the square root is positive. Because of the asymptotic behavior of $K_0(|y|) \sim |y|^{-1/2} \exp(-|y|)$, $|y| \to \infty$, the integrand decays faster than exponentially as $|x| \to \infty$. The integral remains convergent when a and b are imaginary since $|H_0^{(1)}(\sqrt{\alpha + \beta \cosh x})|$ $\sim \exp(-|x|/4)$ as $|x| \to \infty$, with α, β real. Upon employing the formula

$$K_{i\nu}(-i\xi) \equiv \frac{\pi i}{2} e^{-\nu\pi/2} H_{i\nu}^{(1)}(\xi), \tag{B2}$$

one obtains, from Eq. (B1),

$$H_{i\nu}^{(1)}(\xi)H_{i\nu}^{(1)}(\eta) = \frac{1}{\pi i} \int_{-\infty}^{\infty} H_0^{(1)}(\sqrt{\xi^2 + \eta^2 + 2\xi\eta \cosh x})e^{(ix+\pi)\nu} \, dx, \tag{B3}$$

where ξ and η and the square root are assumed positive.

The Hankel function of arbitrary order μ and argument x can be represented by the Sommerfeld integral,

$$H_\mu^{(1)}(x) = \frac{1}{\pi} \int_{i\infty - \epsilon}^{-i\infty + \epsilon} e^{ix \cos w} e^{i\mu(w - \pi/2)} \, dw, \tag{B4}$$

with ϵ chosen so that the integral converges. For positive real x and arbitrary μ, one has $0 < \epsilon < \pi$. If μ is imaginary, one may verify that

$$\lim_{a \to \infty} \int_{ia}^{ia-\epsilon} e^{ix \cos w + i\mu(w - \pi/2)} \, dw = 0, \tag{B5}$$

so the integration path may be deformed into the imaginary axis. Thus, for real ν and positive x,

$$H_{i\nu}^{(1)}(x) = \frac{1}{\pi} \int_{i\infty}^{-i\infty} e^{ix \cos w} e^{-\nu(w - \pi/2)} \, dw. \tag{B6}$$

PROBLEMS

1. Show that the high-frequency asymptotic solution for the time-harmonic point source Green's function for a wedge, comprising the geometric optical contribution in Eq. (6.3.30a) and the diffracted contribution in Eq. (6.3.30b), can be constructed from the corresponding plane wave results in Eqs. (6.3.28a) and (6.3.28b) on application of the ray optical method discussed in Sec. 1.7d.

2. Derive the time-harmonic electromagnetic fields described by the scalar line source Green's functions for a perfectly conducting wedge [see Eqs. (6.5.12) and (6.5.13)] and show that the fields satisfy the edge condition of Eq. (1.5.37). Repeat the calculation for the fields excited by an axial electric or magnetic dipole source [see Eq. (6.5.23)].

3. Expand the time-dependent point source Green's functions for a perfectly conducting wedge, as given in Eq. (6.5.26), in powers of $(t - t') - (L/c)$, where $t = t' + L/c$ denotes the time of arrival of the wavefronts and L is an appropriate distance parameter, to obtain series of the form (1.7.81). Show that on use in Eq. (1.7.80) of the leading expansion coefficients, one may construct the time-harmonic high-frequency asymptotic result in Eqs. (6.5.24a) and (6.5.25a).

4. Use the procedure detailed in connection with Eq. (6.4.9) to derive higher order terms in the asymptotic expansion of the diffracted part of the Green's function for a wedge with a linearly varying surface impedance [see Eqs. (6.6.6) and (6.6.18b)].

5. A distribution of magnetic currents

$$\mathbf{M}(\rho) = \mathbf{z}_0 M_0 e^{i\xi\rho} e^{-i\omega t}, \qquad l_1 < \rho < l_2, \tag{1}$$

with $M_0 = $ constant, is located on the $\phi = 0$ face of a perfectly conducting wedge and simulates radiation from an array of axial line sources progressively phased in the ρ direction; the second wedge face is located at $\phi = \varphi$. The radiated magnetic field $H = H_z$ is parallel to the edge (z axis) and is given by:

$$H_z(\mathbf{\rho}) = ik\sqrt{\frac{\epsilon}{\mu}} \int_{l_1}^{l_2} [\bar{G}''(\mathbf{\rho}, \mathbf{\rho}')M(\rho')]_{\phi'=0} \, d\rho', \tag{2}$$

where $\bar{G}''(\mathbf{\rho}, \mathbf{\rho}')$ is the H-mode Green's function in Sec. 6.5c, whose asymptotic form for large ρ is given in Eqs. (6.5.19) and (6.5.20). Show that for $\rho \to \infty$, $kl_1 \gg 1$, and $\varphi > \pi$, the magnetic field in Eq. (2) is given for arbitrary ϕ by:

$$H_z(\mathbf{\rho}) \sim ikl_1\sqrt{\frac{\epsilon}{\mu}} M_0 \frac{e^{i[k\rho + (\pi/4)]}}{\sqrt{2\pi k\rho}}[A(\xi, k, \phi) + B(\xi, k, \phi) + C(\xi, k, \phi)], \tag{3}$$

where

$$A(\xi, k, \phi) = \frac{e^{i2(\gamma_2{}^2 - \delta_2{}^2)} - e^{i2(\gamma_1{}^2 - \delta_1{}^2)}}{i2(\gamma_1^2 - \delta_1^2)} U(\pi - \phi)$$

$$= e^{i(\xi - k\cos\phi)(l_1+l)} \frac{2\sin[(\xi - k\cos\phi)l]}{(\xi - k\cos\phi)l} U(\pi - \phi), \qquad 2l = l_2 - l_1, \tag{3a}$$

$$B(\xi, k, \phi) = \frac{i\pi}{2\varphi} \sin\frac{\pi^2}{\varphi} \frac{1}{[\cos(\pi^2/\varphi) - \cos(\pi\phi/\varphi)]} \frac{1}{\gamma_1\sqrt{kl_1}}$$

$$\times [e^{i2\gamma_1{}^2}F(\gamma_1) - e^{i2\gamma_2{}^2}F(\gamma_2)], \tag{3b}$$

$$C(\xi, k, \phi) = \text{sgn}(\pi - \phi)\frac{i}{4(\gamma_1^2 - \delta_1^2)}\left\{e^{i2\gamma_2{}^2}\left[F(\delta_2) - \frac{\gamma_2}{\delta_2}F(\gamma_2)\right]\right.$$

$$\left. - e^{i2\gamma_1{}^2}\left[F(\delta_1) - \frac{\gamma_1}{\delta_1}F(\gamma_1)\right]\right\}, \tag{3c}$$

$$\gamma_{1,2} = \sqrt{\left(\frac{k+\xi}{2}\right)l_{1,2}}, \qquad \delta_{1,2} = \sqrt{kl_{1,2}}\left|\cos\frac{\phi}{2}\right|,$$

$$\gamma_2^2 - \delta_2^2 = \frac{l_2}{2}(\xi - k\cos\phi) = \frac{l_2}{l_1}(\gamma_1^2 - \delta_1^2),$$

$$F(y) = \frac{2}{\sqrt{\pi}}e^{-i2y^2}\int_{(1-i)y}^{\infty} e^{-x^2}\, dx. \tag{3d}$$

Show that $A(\xi, k, \phi)$ represents the angular pattern function when the source distribution is contained in an infinite plane, that the effect of the edge outside the transition region $\phi \approx \pi$ is given by $B(\xi, k, \phi)$, and that the entire expression (3) is required to yield the field behavior in the transition domain. More precisely, note that the exterior of the transition region is defined by $\delta_1 \gg 1$ and that since C is $0(\delta_1^{-2})$, it can be neglected in this range. Verify the evaluation of the following integrals, required for calculation of the result in Eq. (3):

$$I_1 = \int_a^b \frac{e^{i\alpha\rho}}{\sqrt{\rho}}\, d\rho = e^{i\pi/4}\sqrt{\frac{\pi}{\alpha}}\left[e^{i\alpha a}F\left(\sqrt{\frac{\alpha a}{2}}\right) - e^{i\alpha b}F\left(\sqrt{\frac{\alpha b}{2}}\right)\right]. \qquad (4)$$

$$I_2 = \int_a^b d\rho\, e^{i\alpha\rho}F(\beta\sqrt{\rho}) = I_2' + I_2'', \qquad (5)$$

where

$$I_2' = \frac{2}{\sqrt{\pi}}\int_a^b d\rho\, e^{i\gamma\rho}\int_{(1-i)\beta\sqrt{b}}^\infty dx\, e^{-x^2} = \frac{e^{i\gamma b} - e^{i\gamma a}}{i\gamma}\frac{2}{\sqrt{\pi}}S(\beta\sqrt{b}), \qquad (5a)$$

$$I_2'' = \frac{2}{\sqrt{\pi}}\int_a^b d\rho\, e^{i\gamma\rho}\int_{(1-i)\beta\sqrt{\rho}}^{(1-i)\beta\sqrt{b}} dx\, e^{-x^2}, \qquad (5b)$$

with

$$S(y) = \int_{(1-i)y}^\infty dx\, e^{-x^2}, \qquad \gamma = \alpha - 2\beta^2. \qquad (6)$$

For the evaluation of I_2'', interchange the order of integration and obtain:

$$I_2'' = \frac{2}{\sqrt{\pi}}\int_{(1-i)\beta\sqrt{a}}^{(1-i)\beta\sqrt{b}} dx\, e^{-x^2}\int_a^{(ix^2/2\beta^2)} d\rho\, e^{i\gamma\rho}$$

$$= -\frac{2}{\sqrt{\pi}}\frac{1}{i\gamma}e^{i\gamma a}[S(\beta\sqrt{a}) - S(\beta\sqrt{b})]$$

$$+ \frac{2}{\sqrt{\pi}}\frac{1}{i\gamma\sqrt{\delta}}[S(\beta\sqrt{\delta a}) - S(\beta\sqrt{\delta b})], \qquad \delta = 1 + \frac{\gamma}{2\beta^2} = \frac{\alpha}{2\beta^2}, \qquad (7)$$

whence

$$I_2 = \frac{1}{i\gamma}\left\{e^{i\alpha b}\left[F(\beta\sqrt{b}) - \frac{1}{\sqrt{\delta}}F\left(\sqrt{\frac{\alpha b}{2}}\right)\right] - e^{i\alpha a}\left[F(\beta\sqrt{a}) - \frac{1}{\sqrt{\delta}}F\left(\sqrt{\frac{\alpha a}{2}}\right)\right]\right\}$$

$$(8)$$

6. A perfectly conducting half-plane occupying the region $y < 0$ in the $z = 0$ plane is illuminated by a plane wave incident at an angle θ_0 with respect to the z axis (Fig. P6.1). The incident wave is polarized so that its electric vector is parallel to the edge whence, due to translational symmetry along x, the total electric field E has only an x component.

(a) Show that on use of Eq. (5.4.26a) and the equivalence theorem leading to Eq. (1.5.33), the secondary field E_s produced by the induced currents on the half-plane can be expressed for a time-dependence $\exp(-i\omega t)$ as

$$E_s(\hat{\rho}) = -\frac{\omega\mu}{4}\int_{-\infty}^0 H_0^{(1)}(k|\hat{\rho} - \hat{\rho}'|)J(y')dy', \qquad \hat{\rho} = (y, z) \qquad (9)$$

where the x-directed currents $J(y')$ flow on both sides of the screen. Show that the requirement $E = 0$ on the screen, where

$$E = E_i + E_s, \qquad E_i = e^{ik(z\cos\theta_0 + y\sin\theta_0)}, \qquad (10)$$

is the total electric field, also satisfies the edge condition (1.5.37), and derive on substitution of Eq. (9) an integral equation for the induced currents $J(y)$ (see also Problem 11 of Chapter 1). While the integral equation can be solved by the Wiener–Hopf technique,† utilize the physical optics approximation for the evaluation of the scattered field, and assume that the induced currents are the same as for an infinite plane at $z = 0$. Thus, on the illuminated side $z = -0$,

$$J(y) \approx 2H_{yi}|_{z=0} = \frac{2k \cos \theta_0}{\mu\omega} e^{iky \sin \theta_0}, \tag{11}$$

while on the dark side $z = +0$, one takes $J(y) \approx 0$. For k large, and observation points $\hat{\boldsymbol{\rho}}$ sufficiently far from the screen, use the asymptotic form for $H_0^{(1)}$ [see Eq. (5.3.13b)] to show that

$$E_s \approx -\frac{\sqrt{k} \cos \theta_0 e^{-i\pi/4}}{\sqrt{2\pi}} \int_{-\infty}^{0} \frac{\exp\{ik[|\hat{\boldsymbol{\rho}} - \hat{\boldsymbol{\rho}}'| + y' \sin \theta_0]\}}{|\hat{\boldsymbol{\rho}} - \hat{\boldsymbol{\rho}}'|^{1/2}} \, dy'. \tag{12}$$

(b) Evaluate the integral in Eq. (12) by the saddle-point method. Show that there is a saddle point y_s' located at

$$y_s' = y - |z| \tan \theta_0 \tag{13}$$

Interpret the saddle-point condition as shown in Fig. P6.1 and relate to the

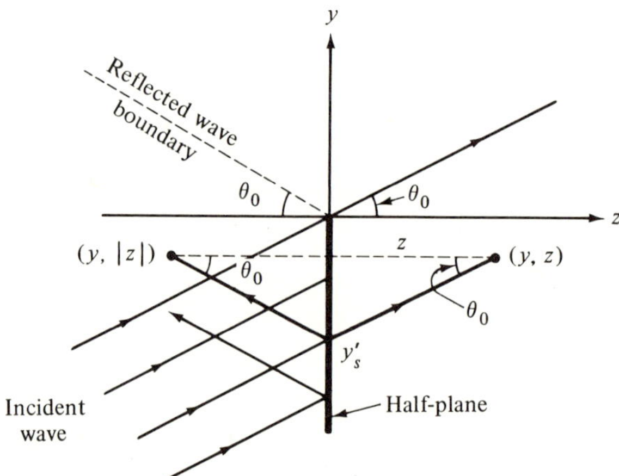

FIG. P6.1 Physical configuration and interpretation of saddle point condition.

geometric optical properties of the field; note in particular the role played by $|z|$ in Eq. (13) (the induced current distribution radiates symmetrically). Verify that the integration path can be deformed through the saddle point as shown in Fig. P6.2, provided that $-\infty < y_s' < 0$. Use the asymptotic formula in Eq. (4.2.20) to obtain the approximate value of the integral in Eq. (12), comprising contributions from the saddle point, when relevant, and from the endpoint $y' = 0$. Show that the saddle-point contribution E_{s1} is given by

$$E_{s1} \sim -e^{ik[|z| \cos \theta_0 + y \sin \theta_0]} \left[1 + O\left(\frac{1}{k}\right)\right], \tag{14a}$$

---

†B. Noble, *Methods Based on the Wiener–Hopf Technique for the Solution of Partial Differential Equations*, Pergamon Press, New York (1958), Chapter 2.

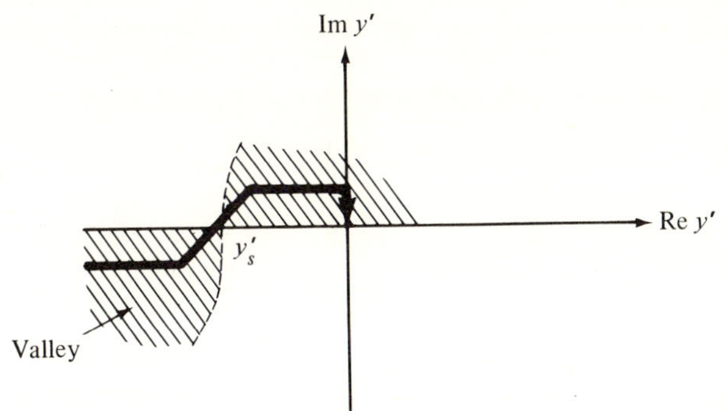

FIG. P6.2   Deformed integration path in y'-plane.

while the endpoint contribution E_{s2} is

$$E_{s2} \sim \frac{-\cos\theta_0}{\sin\theta - \sin\theta_0} \frac{e^{i[k\hat{\rho}+(\pi/4)]}}{\sqrt{2\pi k\hat{\rho}}} \left[1 + O\left(\frac{1}{k}\right)\right], \qquad \sin\theta = \frac{y}{\hat{\rho}} = \frac{y}{\sqrt{y^2+z^2}}$$

$$(14b)$$

Add the incident field E_i to the secondary field $E_{s1} + E_{s2}$ and interpret the result in ray-optical terms. Comment particularly on the role of the saddle point y'_s and the endpoint $y' = 0$ in selecting those portions of the induced current distribution that are principally responsible for establishing the field at a given observation point (y, z). Compare with the ray-optical construction in Fig. 1.7.8, with $n_1 = n_2$ [the factor $\exp(ik_0 n R_8)/R_8$ in Eq. (1.7.64d) should be set equal to unity for an incident plane wave]. When $y'_s \to 0$, improve the asymptotic evaluation by use of the uniform approximation in Sec. 4.6a.

(c) Compare the results of part (b) with the asymptotic approximation of the exact solution [Eq. (6.5.29), with $\varphi = 2\pi$]. Comment on why the geometric-optical portion of the field in part (b) is given correctly whereas the diffracted portion is not. Show from the exact field solution in Eq. (6.5.43) that the induced currents are given exactly by :

$$J(y') = \frac{k}{\omega\mu} \left\{ 2\cos\theta_0\, e^{iky'\sin\theta_0} + 2\sqrt{\frac{2}{-\pi ky'}}\, e^{-i[ky'-\pi/4]} \right.$$

$$\left. \cdot \sin\frac{\phi'}{2}[1 - 2\sqrt{2}\,\xi e^{-i[2\xi^2+(\pi/4)]}Q[(1-i)\xi]\right\}$$

$$(15)$$

where $Q[(1-i)\xi] \equiv S(\xi)$ in Eq. (6) and

$$\xi = \sqrt{-ky'}\cos\frac{\phi'}{2}, \qquad \phi' = \frac{\pi}{2} - \theta_0.$$

$$(15a)$$

The first term in Eq. (15) is the physical optics approximation in Eq. (11); the second represents a rigorous correction and contains the contributions from the illuminated and the dark sides of the half-plane. Show that the correction is appreciable only near the edge [see Fig. 4.4.3(a)], and relate this observation to the validity, or not, of the results in Eqs. (14a) and (14b).

(d) Obtain an improved expression for the diffracted field E_{s2} by substituting Eq. (15) into Eq. (9) and using the asymptotic approximation for the Hankel function as before. Note that the endpoint contribution due to the physical optics term in Eq. (15) yields Eq. (14). Show that the correction E_s' to E_s arising from the first term inside the large square brackets in Eq. (15) is given by

$$E_s' \sim \frac{-e^{[ik\hat{\rho}+(\pi/4)]}}{\sqrt{2\pi k \hat{\rho}}} \frac{\sqrt{2}\,\sin(\phi'/2)}{\sqrt{1+\sin\theta}}, \qquad \sqrt{2}\,\sin\frac{\phi'}{2} = \sqrt{1-\sin\theta_0}, \qquad (16)$$

where it has been noted that $|\hat{\boldsymbol{\rho}} - \hat{\boldsymbol{\rho}}'| = \hat{\rho} - y' \sin\theta + O(y'^2)$ near $y' = 0$, whence the calculation involves evaluation of the integral

$$I_1 = \int_{-\infty}^{0} \frac{e^{-ik(1+\sin\theta)y'}}{\sqrt{-y'}}\,dy' = \sqrt{\frac{\pi}{-ik(1+\sin\theta)}}. \qquad (16a)$$

Note that calculation of the correction E_s'' due to the last term in the large square brackets in Eq. (15) requires consideration of the integral

$$I_2 = \int^{0} \exp\left[-ik(1+\sin\theta)y' - i2\xi^2\right] Q\,dy' = \int^{0} \exp[ik(\sin\theta_0 - \sin\theta)y']Q\,dy'$$

$$(17a)$$

$$= \frac{1}{ik(\sin\theta_0 - \sin\theta)} \int^{0} Q\frac{d}{dy'}\exp[ik(\sin\theta_0 - \sin\theta)y']\,dy'. \qquad (17b)$$

Use integration by parts, and $Q(0) = \sqrt{\pi}/2$, $dQ/d\xi = -(1-i)\exp(2i\xi^2)$, to obtain

$$I_2 = \frac{1}{ik(\sin\theta_0 - \sin\theta)}\left\{\frac{\sqrt{\pi}}{2} - \frac{e^{-i\pi/4}\sqrt{k}}{\sqrt{2}}\cos\frac{\phi'}{2}\int^{0}\frac{e^{-ik(1+\sin\theta)y'}}{\sqrt{-y'}}\,dy'\right\}, \qquad (17c)$$

and evaluate via Eq. (16a). Show that the resulting contribution E_s'' is obtained as

$$E_s'' \sim \frac{-e^{i[k\hat{\rho}+(\pi/4)]}}{\sqrt{2\pi k\hat{\rho}}} \frac{\cos\theta_0}{(\sin\theta_0 - \sin\theta)}\left[1 - \frac{\sqrt{1+\sin\theta_0}}{\sqrt{1+\sin\theta}}\right]. \qquad (18)$$

The total diffracted field E_{s2} is comprised of the sum of E_s', E_s'', and the physical optics formula in Eq. (14b). Show that the resulting expression agrees with the exact formula in Eq. (6.5.29), with $\varphi = 2\pi$.

7. (a) A perfectly conducting wedge formed of the two intersecting half-planes $\phi = 0$ and $\phi = \varphi$ is a "separable" configuration in either a cylindrical or spherical coordinate system. When the wedge is excited by a current element that lies on a straight line passing through the apex at point A, show that the analysis is simplified in spherical coordinates since the element is then radial with respect to an origin chosen at A. [Note that fields due to radial current elements can be derived from a single spherical potential function as in Eqs. (2.6.11.).] Discuss how this simplification may be utilized for non-axially oriented tangential magnetic current elements located on one of the wedge faces, and for excitation by small slot radiators represented by such elements.

(b) Using spherical coordinates, show that the H mode Green's function $G''(\mathbf{r}, \mathbf{r}')$ is given in a radial transmission representation by [$\exp(-i\omega t)$ dependence]:

$$G''(\mathbf{r}, \mathbf{r}') = \frac{i}{2k\varphi rr'} \sum_{m=0}^{\infty} \epsilon_m \cos p\phi \cos p\phi' f_p(r, r'; \theta, \theta') \qquad (19a)$$

where

$$f_p(r, r'; \theta, \theta') = \sum_{n=0}^{\infty} [2(n + p) + 1] \frac{\Gamma(n + 2p + 1)}{n!} P_{n+p}^{-p}(\cos \theta)$$

$$\cdot P_{n+p}^{-p}(\cos \theta') j_{n+p}(kr_<) h_{n+p}^{(1)}(kr_>) ; \tag{19b}$$

$p = (m\pi/\varphi)$, $\epsilon_0 = 1$, and $\epsilon_m = 2$, $m \geq 1$. Use this Green's function to derive the fields radiated by a magnetic current element located on one of the wedge faces. The orientation of the element is arbitrary as long as it is not parallel to the edge. Show that the edge condition is satisfied [see Eq. (1.5.39)]. Repeat for the E-mode Green's function $G'(\mathbf{r}, \mathbf{r}')$.

(c) Derive from Eqs. (19) the result for a plane wave of unit amplitude incident from (θ', ϕ') (Let $r' \to \infty$, use the asymptotic form for $h_{n+p}^{(1)}(kr')$, and replace $[\exp (ikr')]/4\pi r'$ by unity);

$$u''(\mathbf{r}; \theta', \phi') = \frac{2\pi}{\varphi} \sum_{m=0}^{\infty} \epsilon_m \cos p\phi \cos p\phi' \, e^{-ip\pi/2} \bar{f}_p(r, \theta; \theta') \tag{20a}$$

where

$$\bar{f}_p(r, \theta; \theta') = \sum_{n=0}^{\infty} [2(n + p) + 1] \frac{\Gamma(n + 2p + 1)}{n!(kr)} P_{n+p}^{-p}(\cos \theta)$$

$$\cdot P_{n+p}^{-p}(\cos \theta') j_{n+p}(kr) e^{-in\pi/2}. \tag{20b}$$

By using an analysis in cylindrical coordinates, show that u'' is also given by [see Eqs. (6.5.16b) and (6.2.18)] :

$$u''(\mathbf{r}; \theta', \phi') = \frac{2\pi}{\varphi} e^{-ikr \cos \theta \cos \theta'} \sum_{m=0}^{\infty} \epsilon_m \cos p\phi \cos p\phi' e^{-ip\pi/2} J_p(kr \sin \theta \sin \theta').$$

$$\tag{21}$$

Compare Eqs. (20) and (21) to deduce the addition theorem

$$J_p(kr \sin \theta \sin \theta') e^{-ikr \cos \theta \cos \theta'} = \bar{f}_p(r, \theta; \theta'). \tag{22}$$

8. Consider the series

$$S = \sum_{p=1}^{\infty} f(\nu_p), \tag{23}$$

where the index ν_p is assumed to be given by the simple zeros of a certain transcendental function $g(\nu)$; i.e.,

$$g(\nu_p) = 0, \qquad p = 1, 2, 3, \ldots. \tag{24}$$

For example, if $\nu_p = 1, 2, 3, \ldots$, the appropriate function is $g(\nu) = \sin \nu\pi$. Assume that $f(\nu)$ and $g(\nu)$ are analytic functions almost everywhere in the complex ν plane.

(a) Show that S can be expressed in terms of the contour integral

$$S = \frac{1}{2\pi i} \int_C f(\nu) \frac{dg(\nu)/d\nu}{g(\nu)} d\nu, \tag{25}$$

where C is a contour which encloses in the positive sense the zeros ν_p, $p = 1, 2, \ldots$, of $g(\nu)$, but no other singularities of the integrand.

(b) Apply Eqs. (23) and (25) to the cylinder Green's function representation in Eq. (6.7.12) and derive a contour integral representation. Show that the integrand so obtained does not decay as $\nu \to \infty$ in the complex plane, thereby preventing path deformations leading to an alternative representation. Modify the integrand by adding a function $\psi(v)$ which has no singularities inside C (i.e., it does not alter the value of S) but is so chosen that the resulting integrand decays at infinity. Carrying out a path deformation about the singularities of $f(v)$, derive the residue series representation in Eq. (6.7.11). Compare this approach with the one in the text utilizing the characteristic Green's functions. Which is more direct? (The above-described procedure is usually referred to as the *Watson transformation*, after G. N. Watson† who used this technique in connection with the problem of diffraction by a sphere.)

9. Derive the three-dimensional cylinder Green's function in Eq. (6.7.31) directly by treating the configuration in Fig. 6.7.7 as a uniform waveguide along z, with the transverse mode functions given in Eqs. (3.4.99).

10. Multiply Eqs. (6.7.30a) and (6.7.31) by $\exp(im\phi')$, m = integer, and integrate over ϕ', to derive alternative representations for the Green's function for a cylinder excited by a ring source with radius $\rho' > a$ and having a phase variation $\exp(im\phi)$. Referring to Sec. 5.9c, compare the result with the two-dimensional Green's function for the variable medium described in Eq. (5.9.1) [see Eq. (5.9.9), etc.].

11. In the configuration in Fig. 6.7.7. assume that either the source point or the observation point is located far from the cylinder surface.

(a) Evaluate the integral in Eq. (6.7.30a) asymptotically by the method of saddle points to derive the E-mode result (for $k\rho_> \sin \theta_> \gg 1$ and an $e^{-i\omega t}$ dependence):

$$G'(\mathbf{r}, \mathbf{r}') \sim \frac{e^{ikr_>}}{4\pi r_>} \sum_{n=0}^{\infty} \epsilon_n \cos n(\phi - \phi')e^{-in\pi/2}e^{-ikz_< \cos \theta_>}$$

$$\cdot [J_n(kr_< \sin \theta \sin \theta') - \frac{J_n(ka \sin \theta_>)}{H_n^{(1)}(ka \sin \theta_>)} H_n^{(1)}(kr_< \sin \theta \sin \theta')], \quad (26)$$

where $z_>, r_>, \theta_>$ stand for z, r, θ when $\rho > \rho'$, and for z', r', θ' when $\rho < \rho'$. The converse holds for $z_<, r_<, \theta_<$. r, θ are spherical polar coordinates defined as follows: $r_> \cos \theta_> = z_>$, $r_> \sin \theta_> = \rho_>$, etc., and $\epsilon_n = 1$, $n = 0$, $\epsilon_n = 2$, $n \neq 0$. Derive an analogous expression for the H-mode Green's function $G''(\mathbf{r}, \mathbf{r}')$. Evaluate the electromagnetic field components when the source is a longitudinal electric or magnetic dipole [see Eqs. (5.2.1)].

(b) By moving the source to infinity, specialize Eq. (26) to the case of plane-wave incidence and show that if the incident field is derived from the wave function

$$u'_{\text{inc}} = \exp[-ikr \sin \theta \sin \theta' \cos(\phi - \phi') - ikr \cos \theta \cos \theta'], \quad (27)$$

the corresponding expression for the total wave function is $u' = u'_{\text{inc}} + u'_s$, where

$$u'_s = -e^{-ikr \cos \theta \cos \theta'} \sum_{n=0}^{\infty} \epsilon_n \cos n(\phi - \phi'') e^{-in\pi/2} \frac{J_n(ka \sin \theta')}{H_n^{(1)}(ka \sin \theta')}$$

$$\cdot H_n^{(1)}(kr \sin \theta \sin \theta'). \quad (28)$$

Derive the corresponding result for the H-mode wave function $u''(\mathbf{r}, \mathbf{r}')$. By a

---

†G. N. Watson, "Diffraction of electric waves by the earth," *Proc. Roy. Soc. (London)*, **A95** (1919), pp. 83–99. See also H. Bremmer, *Terrestrial Radio Waves*, Elsevier Publishing Co., New York (1949), Chapter. 3.

suitable superposition of *E*- and *H*-mode constituents, derive the fields due to an arbitrarily polarized incident plane wave.

(c) Specialize the *H*-mode analogue of Eq. (26) to the case where the source is located on the cylinder and show that the *H*-mode radiated far field may be derived from

$$G''(\mathbf{r}, \mathbf{r}') = \frac{e^{ikr}}{4\pi r} \frac{2ie^{-ikz'\cos\theta}}{\pi ka \sin\theta} \sum_{n=0}^{\infty} \epsilon_n \cos n(\phi - \phi') \frac{e^{-in\pi/2}}{H_n'^{(1)}(ka \sin\theta)}. \tag{29}$$

Use this result to derive expressions for the fields radiated by an axial slot in the cylinder, assuming that the electric field in the slot has only an E_ϕ component which is specified (this excitation is equivalent to a distribution of axial magnetic currents $\mathbf{M} = \mathbf{E}_\phi \times \boldsymbol{\rho}_0$ on the smooth cylinder surface).

12. Repeat Problem 11, starting from the representation for $G(\mathbf{r}, \mathbf{r}')$ in Eq. (6.7.30b). Show that an alternative expression for $G''(\mathbf{r}, \mathbf{r}')$ in Eq. (29) is

$$G''(\mathbf{r}, \mathbf{r}') \sim \frac{e^{ikr}}{4r} e^{-ikz'\cos\theta}$$

$$\cdot \sum_{\mu_p} \frac{H_{\mu_p}^{(1)}(ka \sin\theta)\, H_{\mu_p}'^{(2)}(ka \sin\theta)\cos[\mu_p(\pi - |\phi - \phi'|)]e^{-i\mu_p\pi/2}}{(\partial/\partial\mu)H_\mu'^{(1)}(ka \sin\theta)|_{\mu_p}\sin\mu_p\pi}, \tag{30}$$

where $H_{\mu_p}'^{(1)}(ka \sin\theta) = 0$. Discuss the convergence properties of this series. Show that for $ka \sin\theta \gg 1$, the series converges rapidly in the shadow region $|\phi - \phi'| > \pi/2$; give a simplified form of the summand in this instance. Show that the field in the shadow region may be interpreted as arising from a creeping ray that leaves the source point along a helical trajectory on the cylinder surface and then departs tangentially toward the distant observation point *P* (see Fig. P6.3).

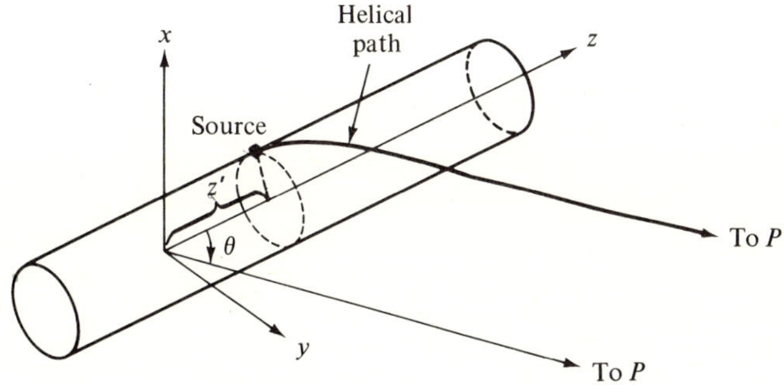

FIG. P6.3 Creeping rays on a cylinder (oblique propagation).

13. It was shown that for line-source excitation, the residue series in Eq. (6.7.11) resulting from the angular transmission analysis converges provided that $\phi \neq \phi'$ [include the integral in Eq. (6.7.11) as the $n = 0$ term in the sum; see remarks preceding Eq. (6.7.16)]. Derive the result for plane-wave incidence by letting $\rho' \to \infty$, thereby replacing $H_{\mu_p}^{(1)}(k\rho')$ by its asymptotic form [see Eq. (5.3.13)]. Show that the resulting series now converges only in the shadow region.

14. A line source of magnetic currents is located at (ρ', ϕ') in a cylindrical coordi-

nate system centered at the apex of a perfectly conducting wedge. The wedge faces are described by the half-planes $\phi = 0$ and $\phi = \varphi$. From Eqs. (6.3.3) and (6.5.2b), the magnetic field H_z is proportional to the scalar Green's function ($e^{-i\omega t}$ dependence)

$$\bar{G}(\mathbf{\rho}, \mathbf{\rho}') = \frac{1}{2} P \int_{-i\infty}^{i\infty} \mu J_\mu(k\rho') H_\mu^{(1)}(k\rho) g_\phi(\phi, \phi'; \mu)\, d\mu, \tag{31}$$

where ρ and ρ' may be interchanged,

$$g_\phi(\phi, \phi'; \mu) = -\frac{\cos \mu\phi_< \cos \mu(\varphi - \phi_>)}{\mu \sin \mu\varphi}, \tag{31a}$$

and "P" denotes that the principal value of the integral is taken at $\mu = 0$.

(a) Assume that an aperture of infinite length parallel to the edge is located on the wedge face $\phi = 0$, and that the electric field in the aperture has only a component E_ρ which is specified. Utilizing Eq. (31), derive an expression for the magnetic field in the shadow region. (Note that the aperture field is given equivalently in terms of a magnetic current distribution $\mathbf{M} = \mathbf{E} \times \mathbf{\phi}_0$ on the unperforated wedge.)

(b) If the aperture occupies the entire wedge face and the electric field is assumed to have the progressive phase dependence

$$E_\rho = e^{i\xi\rho}, \qquad \operatorname{Im} \xi \geq 0, \tag{32}$$

use the formulas†

$$\int_0^\infty H_\mu^{(1)}(k\rho') e^{i\xi\rho'}\, d\rho' = \frac{2e^{-i\mu\pi/2} \sin \mu\,[(\pi/2) - \gamma]}{k \cos \gamma \sin \mu\pi}, \qquad |\operatorname{Re} \mu| < 1, \tag{33a}$$

$$\int_0^\infty J_\mu(k\rho') e^{i\xi\rho'}\, d\rho' = \frac{e^{i\mu\gamma}}{k \cos \gamma}, \qquad \operatorname{Re} \mu > -1, \tag{33b}$$

with $\gamma = \sin^{-1}(\xi/k)$, to show that the magnetic field H_z is now proportional to

$$S(\mathbf{\rho}) = \frac{-1}{k \cos \gamma} P \int_{-i\infty}^{i\infty} J_\mu(k\rho) \frac{e^{-i\mu\pi/2} \sin \mu[(\pi/2) - \gamma] \cos \mu(\varphi - \phi)}{\sin \mu\pi \sin \mu\varphi}\, d\mu \tag{34a}$$

$$= -\frac{1}{2k \cos \gamma} P \int_{-i\infty}^{i\infty} H_\mu^{(1)}(k\rho) \frac{e^{i\mu\gamma} \cos \mu(\varphi - \phi)}{\sin \mu\varphi}\, d\mu. \tag{34b}$$

Show that these integrals converge in the region $\phi > (\pi/2) - \operatorname{Re} \gamma$ so that Eqs. (34) are valid in this extended domain. For real γ, interpret this region as the geometric-optical shadow region for the source distribution in Eq. (32) extending over a semi-infinite interval (cf. Problem 28 of Chapter 1 for the radiation characteristics of a progressively phased sheet current).

(c) Show that the integral in Eq. (34a) may be evaluated in terms of the residues at the poles of the integrand in the right half of the μ plane, leading to the result‡, §

---

†W. Magnus and F. Oberhettinger, *Formulas and Theorems for the Special Functions of Mathematical Physics*, Chelsea Pub. Co., New York (1954), pp. 131, 133.

‡G. D. Malyuzinec, "Radiation of sound by vibrating boundaries of an arbitrary wedge," *Acoustical Journal (USSR)*, **1** (1955), p. 144.

§L. B. Felsen, "Radiation from source distributions covering one face of a perfectly conducting wedge," *IEEE Trans. on Antennas and Propagation* **AP-12** (1964), p. 653.

$$S(\boldsymbol{\rho}) = \frac{\pi i}{2k\varphi} J_0(k\rho) + \frac{\pi i}{k\varphi} \sum_{n=1}^{\infty} \cos v\phi \, J_v(k\rho) \frac{e^{-iv\pi/2}}{\cos(v\pi/2)}$$

$$- \frac{2}{k} \sum_{n=1}^{\infty} \frac{\cos(2n-1)(\varphi-\phi)}{\sin(2n-1)\varphi} J_{2n-1}(k\rho), \qquad v = \frac{n\pi}{\varphi}. \tag{35}$$

This formula is useful for the calculation of the field near the origin.

(d) By using the integral representation (5.4.36c) for $H_\mu^{(1)}(k\rho)$ and following the procedure leading to Eq. (6.3.8), show that Eq. (34b) can be transformed into (for $\varphi > \pi$)

$$S(\boldsymbol{\rho}) = \frac{i}{k \cos \gamma} e^{ik\rho \cos [(\pi/2)-\gamma-\phi]} U\left[\frac{\pi}{2} - \gamma - \phi\right]$$

$$+ \frac{1}{2\varphi k \cos \gamma} \int_{i\infty}^{-i\infty} e^{ik\rho \cos w} \frac{\sin[(\pi/\varphi)(w + \gamma - \pi/2)]}{\cos[(\pi/\varphi)(w + \gamma - \pi/2)] - \cos(\pi\phi/\varphi)} dw, \tag{36}$$

where $U(x)$ is the Heaviside unit function. This result is valid for all angles in the interval $0 < \phi < \varphi$. Identify the first term as the geometric-optical field and the second as the diffraction field. Calculate the radiation field by evaluating the diffraction integral asymptotically for large values of $k\rho$ and for all values of ϕ. Interpret the asymptotic field formula in ray-optical terms.

15. This problem deals with alternative representations for the z-independent two-dimensional Green's functions and for the three-dimensional Green's functions $G'(\mathbf{r}, \mathbf{r}')$, $G''(\mathbf{r}, \mathbf{r}')$, $\mathscr{S}'(\mathbf{r}, \mathbf{r}')$, $\mathscr{S}''(\mathbf{r}, \mathbf{r}')$ for the (perfectly conducting) tipped wedge configuration shown in Fig. 6.1.1.

(a) Since the radial characteristic Green's function in Eq. (3.4.97) consists of two parts, one of which represents the result in the absence of the cylindrical boundary while the other accounts for the presence of the boundary, show that the total Green's functions obtained by the radial transmission formulation may be separated as follows: the Green's function for the untipped wedge plus a correction term due to the cylinder. Show that this representation is useful when the cylinder radius is small.

(b) Repeat the preceding considerations for the angular transmission formulation by resorting to the image superposition in the infinite angular space (see Eq. (6.7.10), modified in accord with Eq. (3.4.57)). Show that the resulting representation comprises the Green's function for the cylinder configuration in Fig. 6.7.1, plus correction terms due to the presence of the wedge faces at $\phi = 0$, φ. Show that this representation is useful when the cylinder radius is large and both wedge faces are invisible from the source. Derive an asymptotic approximation for the fields, following the procedure in Sec. 6.7.

(c) When the cylinder radius is large and one of the wedge faces is visible from the source, show from an asymptotic evaluation by the saddle-point method that the dominant contribution to the scattered field in the illuminated region may be interpreted as corresponding to geometric-optical rays arising from reflection at the cylinder only, from reflection at the wedge face only, and from multiple reflections as shown in Fig. P6.4. Construct the field amplitude and phase along a multiply reflected ray directly from geometrical optics and compare with the asymptotic solution.

(d) Repeat the analysis in (c) when both wedge faces are visible from the source.

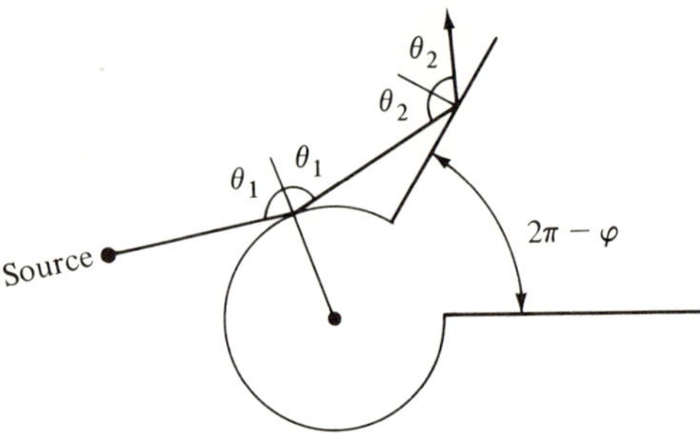

FIG. P6.4   Multiply reflected rays on a cylindrically tipped wedge.

16. A cone defined by the surface $\theta = \theta_0$ is tipped by a sphere at $r = a$, all surfaces being perfectly conducting. Derive alternative representations for the Green's functions for this configuration. In particular, obtain a radial transmission formulation wherein the cone Green's function appears explicitly, with the effect of the sphere occurring as a perturbation, and an angular transmission formulation wherein the sphere Green's function appears explicitly, with the conical portions of the surface contained as a perturbation. Discuss the convergence properties of these representations (see also Problem 15 for the analogous configuration in cylindrical geometry) and show that the radial transmission result is convenient for small spherical tips and arbitrary cones while the angular transmission result is useful for large tips and narrow angle cones.

17. A cylinder with radius a is excited by a line source of magnetic currents of unit strength directed parallel to the cylinder axis at a location $\phi = \phi'$, $\rho = \rho' > a$. The boundary conditions on the cylinder are specified in terms of the angularly dependent surface impedance

$$\bar{Z}(\phi) = -\frac{E_\phi}{H_z} = Z\sqrt{\frac{\mu}{\epsilon}}[1 + \alpha \cos p(\phi - \phi_0)] \qquad \text{at } \rho = a. \qquad (37)$$

Z denotes a normalized reference impedance, $\sqrt{\mu/\epsilon}$ is the impedance of free space, α is a real constant descriptive of the modulation amplitude, $p = 2\pi a/L$ is an integer, with L denoting the spatial period of the impedance modulation, and ϕ_0 is a phase reference. For a passive structure, it is required that $\text{Re } \bar{Z} \geq 0$, and an $\exp(-i\omega t)$ dependence is implied.

(a) Show that the only component of magnetic field is given by $H_z = i\omega\epsilon\bar{G}$ where the scalar Green's function $\bar{G}(\boldsymbol{\rho}, \boldsymbol{\rho}')$ satisfies the inhomogeneous wave equation

$$(\nabla^2 + k^2)\bar{G}(\boldsymbol{\rho}, \boldsymbol{\rho}') = -\delta(\boldsymbol{\rho} - \boldsymbol{\rho}'), \qquad (38)$$

subject to

$$\frac{\partial \bar{G}}{\partial \rho} = -ikZ[1 + \alpha \cos p(\phi - \phi_0)]\bar{G} \qquad \text{at } \rho = a, \qquad (38a)$$

and a radiation condition at $\rho \to \infty$. By assuming a representation in terms of angular eigenfunctions,

$$\bar{G}(\boldsymbol{\rho}, \boldsymbol{\rho}') = \frac{1}{2\pi} \sum_{n=-\infty}^{\infty} e^{-in(\phi-\phi')} h(\rho, \rho'; n) \tag{39}$$

$$h(\rho, \rho'; n) = \frac{\pi i}{2} [J_n(k\rho_<) + A(n)H_n^{(1)}(k\rho_<)]H_n^{(1)}(k\rho_>), \tag{39a}$$

show that this series represents the formal solution provided that the coefficients $A(n)$ satisfy the inhomogeneous second-order difference equation†

$$\frac{iZ\alpha}{2}[c_1(n, p)A(n + p) + c_1(n, -p)A(n - p)] + d(n)A(n) = c_2(n), \tag{40}$$

where

$$c_1(n, p) = e^{ip(\phi'-\phi_0)} \frac{H_{n+p}^{(1)}(ka)H_{n+p}^{(1)}(k\rho')}{H_n^{(1)}(k\rho')} \tag{40a}$$

$$d(n) = H_n'^{(1)}(ka) + iZH_n^{(1)}(ka), \tag{40b}$$

$$c_2(n) = -b(n) - \frac{iZ\alpha}{2}\left[\frac{J_{n+p}(ka)H_{n+p}^{(1)}(k\rho')}{H_n^{(1)}(k\rho')} e^{ip(\phi'-\phi_0)}\right.$$

$$\left. + \frac{J_{n-p}(ka)H_{n-p}^{(1)}(k\rho')}{H_n^{(1)}(k\rho')} e^{-ip(\phi'-\phi_0)}\right] \tag{40c}$$

$$b(n) = J_n'(ka) + iZJ_n(ka). \tag{40d}$$

(b) To solve Eq. (40), assume that $A(n)$ has the power series expansion

$$A(n) = \sum_{m=0}^{\infty} f(n, m)\alpha^m, \tag{41}$$

where α is a small parameter. Show that

$$f(n, 0) = -\frac{b(n)}{d(n)}, \tag{41a}$$

$$f(n, 1) = \frac{Z/(\pi ka)}{d(n)H_n^{(1)}(k\rho')}\left[\frac{e^{ip(\phi'-\phi_0)} H_{n+p}^{(1)}(k\rho')}{d(n + p)} + \frac{e^{-ip(\phi'-\phi_0)} H_{n-p}^{(1)}(k\rho')}{d(n - p)}\right], \tag{41b}$$

and that the coefficients $f(n, m)$, $m \geq 2$, may be derived recursively from a knowledge of $f(n \pm p, m - 1)$.

(c) To $O(\alpha)$, derive an alternative expression for $\bar{G}(\boldsymbol{\rho}, \boldsymbol{\rho}')$ analogous to the one in Eq. (6.7.11). Discuss its convergence properties. Simplify the resulting formulas by passing to the limit of plane-wave incidence ($\rho' \to \infty$). Show that the dominant characteristics of the high frequency field ($ka \gg 1$) are determined for plane wave incidence by the integrals

$$\bar{G}(\boldsymbol{\rho}, \boldsymbol{\rho}') = \bar{G}_0(\boldsymbol{\rho}, \boldsymbol{\rho}') + \alpha\bar{G}_1(\boldsymbol{\rho}, \boldsymbol{\rho}') + O(\alpha^2), \tag{42}$$

where

$$\bar{G}_0 = \int_{-\infty}^{\infty} e^{i\mu|\phi-\phi'|-i\mu\pi/2}\left[J_\mu(k\rho) - \frac{b(\mu)}{d(\mu)} H_\mu^{(1)}(k\rho)\right] d\mu \tag{42a}$$

pertains to the constant impedance cylinder, and

---

†L. B. Felsen and C. J. Marcinkowski, "Diffraction by a cylinder with a variable surface impedance, "*Proc. Roy. Soc. (London)*, **A267**(1962), pp. 329–350.

$$\bar{G}_1 = \hat{G}_1(p) + \hat{G}_1(-p), \tag{42b}$$

$$\hat{G}_1(p) = \frac{Z}{\pi k a} e^{i p [\phi' - \phi_0 - (\pi/2)]} \int_{-\infty}^{\infty} \frac{e^{i \mu [\phi' - \phi - (\pi/2)]} H_\mu^{(1)}(k\rho)}{d(\mu) d(\mu + p)} \, d\mu. \tag{42c}$$

18. This problem is concerned with the asymptotic evaluation and physical interpretation of the Green's function for a cylinder with periodically varying surface impedance.

(a) From a saddle-point evaluation of the integrals in Eqs. (42a) and (42c), show that the field in the illuminated region of the cylinder is given by the asymptotic approximation:†

$$\bar{G} = e^{-ik\rho \cos(\phi - \phi')} + \bar{G}_r, \tag{43}$$

$$\begin{aligned} \bar{G}_r \sim{}& A_0(0) \Gamma_0 D(0) e^{iks} \\ &+ \alpha [A_0(p) \Gamma_1(p) D(p) e^{iks(p)} + A_0(-p) \Gamma_1(-pD(-p) e^{iks(-p)}] + O(\alpha^2), \end{aligned} \tag{44}$$

where for $n = 0, \pm 1$,

$$A_0(np) = e^{-ika \cos[\theta_i(np)]}, \tag{44a}$$

$$D(np) = \frac{1}{\sqrt{1 + [s(np)]/[r(np)]}}, \qquad r(np) = a \frac{\cos[\theta_r(np)]}{1 + [d\theta_r(np)/d\theta_i(np)]}, \tag{44b}$$

$$\Gamma_0 = \frac{\cos\theta_i - Z}{\cos\theta_i + Z}; \qquad \Gamma_1(p) = \frac{-Z \cos\theta_i(p) e^{ip(\phi' \pm \theta_i)}}{\{Z + \cos[\theta_i(p)]\}\{Z + \cos[\theta_r(p)]\}}, \qquad \theta_n \gtrless 0. \tag{44c}$$

The relation between $\theta_i(np)$ and $\theta_r(np)$ is given by the "grating law"

$$\sin[\theta_r(np)] = \sin[\theta_i(np)] \pm \frac{np}{ka}, \qquad \theta_\eta \gtrless 0. \tag{44d}$$

The parameters θ_i, θ_r, and s denote the angles of incidence, reflection, and the distance along a reflected ray from the cylinder surface to the observation point, respectively, as shown in Fig. P6.5; depending on the incident ray in question, one defines $\theta_n = \pm \theta_i$. Interpret the first term in Eq. (44) as a specularly reflected ray ($\theta_r = \theta_i$) on a cylinder with constant surface impedance Z [cf. Eq. (6.7.29), with $L_r \to \infty$ and subsequent normalization to plane-wave incidence], and the second and third terms as first-order reflected "grating rays." Show that the divergence coefficient $D(np)$ follows from a conservation-of-energy argument applied to an incident- and reflected-ray tube for the general case where the angles of incidence and reflection are not equal (cf. Problems 30 and 32 of Chapter 1). Discuss the symmetry properties of the field in Eq. (44) for $\phi' = \phi_0$ and $\phi' \neq \phi_0$. Discuss the behavior of the reflected grating rays and delimit their domains of existence relative to the cylindrical boundary.

(b) When the surface impedance varies slowly [$(p/ka) \ll 1$], show that the three reflected rays corresponding to $n = 0, \pm 1$ in Eq. (44d) have only slightly different trajectories which intersect at the observation point P and originate from three closely adjacent points on the cylinder surface, two of these being equidistant from the specular reflection point. Show that the contribution from the

---

†C. J. Marcinkowski and L. B. Felsen, "On the geometrical optics of curved surfaces with periodic impedance properties," *J. Res. NBS*, **66D** (1962), pp. 699–705.

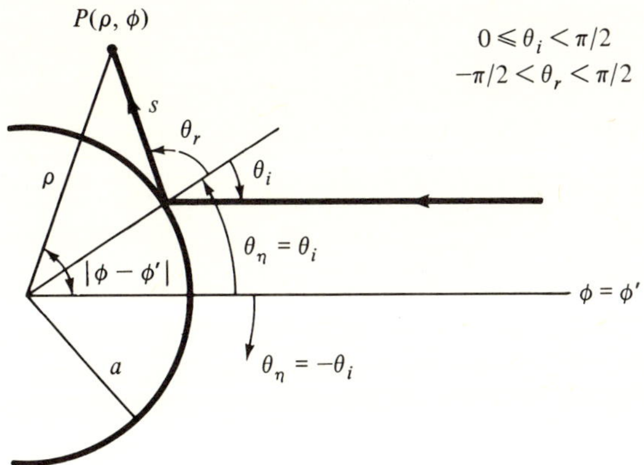

FIG. P6.5 Ray structure for a cylinder with variable surface impedance.

three reflected rays may be combined into a single specularly reflected ray as in the first term in Eq. (44) provided that Γ_0 is replaced by the *local* reflection coefficient Γ_0' at the point of reflection $\hat{\phi}$:

$$\Gamma_0' = \frac{\cos\theta_i - \bar{Z}(\hat{\phi})/\sqrt{\mu/\epsilon}}{\cos\theta_i + \bar{Z}(\hat{\phi})/\sqrt{\mu/\epsilon}}, \tag{44e}$$

with $\bar{Z}(\phi)$ given in Eq. (37). Interpret this result as validating the simple geometric-optical construction of the field in terms of the local parameters on the surface near the reflection point [cf. Eqs. (6.6.23)].

19. A line source of electric current is located exterior to a perfectly conducting cylinder whose surface is described by the equation $\rho = a + b\cos p\phi$, where p is an integer.† If b is a small parameter, show that the solution to this problem can be obtained as a special case of the one in Problem 17.

20. Starting from the relations in Eq. (6.8.16), and from‡

$$\frac{d}{d\theta_0} P_\nu^{-q}(\cos\theta_0) = \cos(\nu - q)\pi \frac{d}{d\theta_0} P_\nu^{-q}(-\cos\theta_0)$$

$$+ \frac{\sin(\nu - q)\pi}{\sin q\pi}\left[\cos\nu\pi \frac{d}{d\theta_0} P_\nu^{-q}(-\cos\theta_0) - \frac{\Gamma(\nu - q + 1)}{\Gamma(\nu + q + 1)} \frac{d}{d\theta_0} P_\nu^{q}(-\cos\theta_0)\right], \tag{45}$$

show that when $\theta_0 \approx \pi$, the solutions of $(d/d\theta_0)P_p^{-q}(\cos\theta_0) = 0$ are given approximately by:

$$p \approx q + k - \frac{\Gamma(2q + n + 1)}{\Gamma(1 + q)\Gamma(1 + n)\Gamma(q)}\left(\frac{\pi - \theta_0}{2}\right)^{2q}, \tag{46}$$

where $n = 0, 1, 2, \ldots$ and $\Gamma(x)$ is the gamma function.

---

†P. C. Clemmow and V. H. Weston, "Diffraction of a plane wave by an almost circular cylinder," *Proc. Roy. Soc. (London)*, **A264** (1961), p. 246.

‡E. W. Hobson, *The Theory of Spherical and Ellipsoidal Harmonics*, Cambridge Univ. Press (1931), p. 407.

Repeat to obtain the solutions of $P_p^{-q}(\cos\theta_0) = 0$.

When $\theta_0 \approx \pi/2$, obtain approximate values of the zeros from the asymptotic formula in Eq. (3.4.66b).

21. Construct the H-mode Green's function $G''(\mathbf{r}, \mathbf{r}')$ for a perfectly conducting semiinfinite cone in a representation corresponding to Eq. (6.8.9a), and pass to the special case of axial plane-wave back-scattering ($r' \to \infty$ with subsequent normalization, and $\theta, \theta' \to 0$) [use an exp $(-i\omega t)$ dependence]. Replace the function $j_p(kr)$ by its Sommerfeld integral

$$j_p(kr) = \frac{1}{2\pi}\sqrt{\frac{\pi kr}{2}} \int_{\hat{P}} \exp\{ikr\cos w + i[p + (1/2)][w - (\pi/2)\} \, dw, \quad (47)$$

where $\hat{P}$ is the path shown in Fig. P6.6. Interchange the order of summation

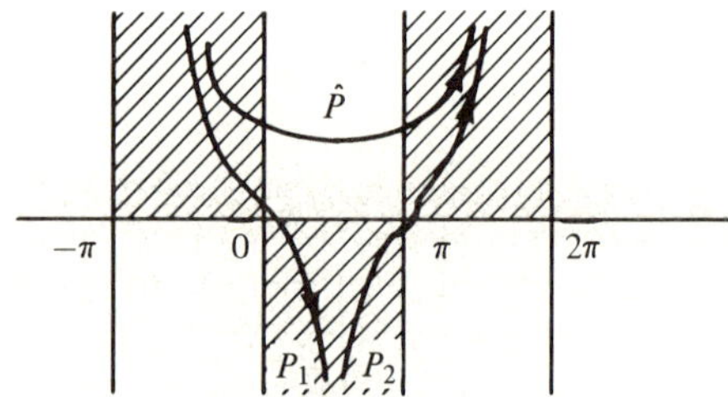

FIG. P6.6 Integration paths.

and integration (justify this) to obtain the representation[†]

$$G'' = e^{i\pi/4}\sqrt{\frac{\pi}{2kz}} \int_{\hat{P}} e^{ikz\cos w} T(w)dw \,, \quad (48)$$

where

$$T(w) = \frac{1}{2\pi}\csc^2\frac{\theta_0}{2}e^{i(w-\pi)/2}$$

$$- \sum_p \frac{(d/d\theta_0)P_p(-\cos\theta_0)[p + (1/2)]}{(\sin p\pi)(\partial^2/\partial p \partial\theta_0)P_p(\cos\theta_0)} e^{i[p + (1/2)](w-\pi)} \,, \quad (48a)$$

and $(d/d\theta_0)P_p(\cos\theta_0) = 0$,

For an approximate evaluation when $\theta_0 \approx \pi/2$, use the asymptotic formula (3.4.66b) to determine the zeros p and also to replace the Legendre functions, to show that

$$T(w) \approx \frac{1}{\theta_0}\sum_{n=0}^{\infty}\left(p + \frac{1}{2}\right)e^{i[p + (1/2)](w-\pi)} \,, \qquad p \approx \left(n + \frac{1}{4}\right)\frac{\pi}{\theta_0} - \frac{1}{2}, \quad (49)$$

and write for this geometric series its closed form exhibiting double poles at $w = \pi$ and $w = \pi - 2\theta_0$. Upon substituting the closed-form expression for $T(w)$ into Eq. (48), show that the contour can be deformed into the steepest-descent paths P_1 and P_2 through the saddle points at $w = 0$ and $w = \pi$, respectively. From an asymptotic evaluation for large kr (see Sec. 4.2d), show that

†L. B. Felsen, "Back-scattering from wide-angle and narrow-angle cones," *J. Applied Physics* **26** (1955), p. 138.

$$G'' \sim e^{-ikz} - \frac{1}{2}\sqrt{\frac{\pi}{2}} e^{i(kz+\pi/4)} \frac{dF}{d\xi}, \qquad \xi \text{ small}, \tag{50a}$$

$$\sim e^{-ikz} + \frac{i}{4[\theta_0 - (\pi/2)]^2} \frac{e^{ikz}}{kz}, \qquad \xi \text{ large}, \tag{50b}$$

where $\xi = \sqrt{kz}|\cos\theta_0|$, $\theta_0 \approx \pi/2$, and $F(\xi) = (2/\sqrt{\pi})e^{-i2\xi^2}\int_{(1-i)\xi}^{\infty} e^{-y^2}dy$.

22. Repeat the calculation in Problem 21 for an x-directed electric current element on the z axis which is then moved to infinity. Show that the total electric field on the z axis is given by

$$E_x \sim e^{-ikz} + \frac{1}{2}\sqrt{\frac{\pi}{2}} e^{i[kz+(\pi/4)]} \frac{dF}{d\xi}, \tag{51}$$

where ξ and $F(\xi)$ are defined in Problem 21.

23. The discussion in Problems 21 and 22 has dealt with the case $\theta_0 = (\pi/2) + \delta$, where δ is small and positive. How must the result in Eqs. (50) be modified when $\delta < 0$? [Note the location of the poles of $T(w)$.] Explain in terms of ray focusing along the cone axis.

24. A uniform plane wave with electric field $E = \exp(-ikz)$ is incident along the axis of a perfectly conducting semiinfinite cone described by the equation $\theta = \theta_0 > \pi/2$. Approximate the induced current distribution on the cone by the physical optics value $\mathbf{J} = 2\mathbf{H}_{\text{inc}} \times \boldsymbol{\theta}_0$, where $\mathbf{H}_{\text{inc}}$ is the incident magnetic field and $\boldsymbol{\theta}_0$ is the normal unit vector directed into the cone surface. Calculate the electric field E_s due to this induced current distribution at a distant observation point on the z axis and show that it yields

$$E_s \sim -\frac{ie^{ikz}}{4kz} \tan^2(\pi - \theta_0). \tag{52}$$

Since the result agrees well with that obtained by a more accurate calculation when $\theta_0 \approx \pi/2$ (see Problems 20 and 21) or $\theta_0 \approx \pi$ [see Eq. (6.8.24a)], is the use of the physical optics approximation justified in the present instance? Explain, paying attention to the presence of the cone tip.

25. A semiinfinite cone is attached to a semiinfinite plane in the manner shown in Fig. P6.7, and all surfaces are assumed to be perfectly conducting. This configuration is excited by a ring source centered on the z axis. If the intensity variation along the ring source is proportional to $\sin(\phi/2)$, show that the E mode Green's function $G'(\mathbf{r}, \mathbf{r}')$ is given by:

$$G'(\mathbf{r}, \mathbf{r}') = \sqrt{\frac{r'\sin\theta'}{r\sin\theta}} \sin\frac{\phi}{2} \, \bar{G}'(\boldsymbol{\rho}, \boldsymbol{\rho}'), \qquad \boldsymbol{\rho} = (r, \theta), \tag{53}$$

where $\bar{G}'(\boldsymbol{\rho}, \boldsymbol{\rho}')$ is the two-dimensional wedge Green's function in Eqs. (6.5.12) or (6.5.18), with (r, θ) interpreted as cylindrical coordinates and the wedge faces situated at $\theta = 0, \theta_0$.

(*Note*: $P_\nu^{-1/2}(\cos\theta) = \{\sqrt{2}\sin[(\nu + 1/2)\theta]\}/\{[\nu + (1/2)]\sqrt{\pi\sin\theta}\}$.)

26. (a) Employ the formulas [see Part (b) and Problem 5 of Chapter 4]

$$\lim_{\substack{\nu\to\infty \\ \theta\to 0}} \nu^q P_\nu^{-q}(\cos\theta) = J_q(\nu\theta), \tag{54a}$$

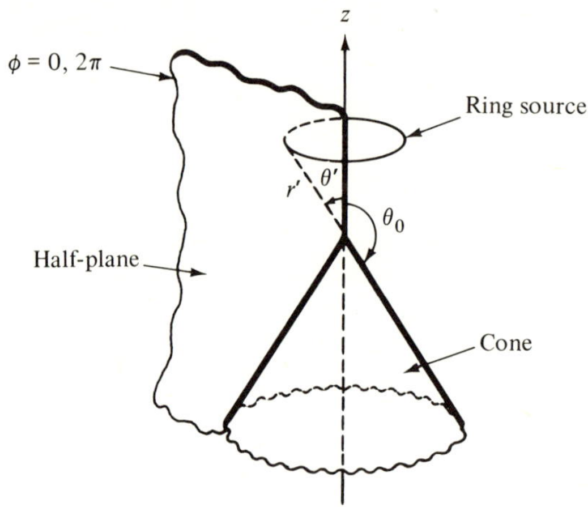

FIG. P6.7   Cone-tipped half-plane.

and

$$\lim_{\substack{\nu \to \infty \\ \theta \to 0}} \frac{\nu^q P_\nu^{-q}(-\cos\theta)}{\sin\nu\pi}\bigg|_{\arg\nu\gtrless 0} = \mp i e^{\pm iq\pi} H_q^{(1,2)}(\nu\theta) \tag{54b}$$

in Eq. (3.4.71) to show that

$$\lim_{\substack{|\xi| \to \infty \\ \theta \to 0}} E_q^{(1,2)}(\xi, \theta) = \xi^{-q} H_q^{(1,2)}(\xi\theta), \qquad \arg\xi \neq 0. \tag{54c}$$

Combine this result with Eqs. (3.4.72) to obtain an asymptotic approximation valid for small and large values of θ:

$$E_q^{(1,2)}(\xi, \theta) \sim \xi^{-q} \sqrt{\frac{\theta}{\sin\theta}} \, H_q^{(1,2)}(\xi\theta), \qquad |\xi| \to \infty. \tag{55}$$

(b) Use Eqs. (54a), (6.A17), and the formula

$$P_\nu^{-q}(-\cos\theta) = \frac{\sin\nu\pi}{\sin q\pi} P_\nu^{-q}(\cos\theta) - \frac{\sin(\nu - q)\pi}{\sin q\pi} \frac{\Gamma(\nu - q + 1)}{\Gamma(\nu + q + 1)} P_\nu^q(\cos\theta), \tag{56}$$

to derive Eq. (54b).

27. When the cone angle $\theta_0 \approx \pi$ and source and observation points are located far from the tip and near the cone surface, a conical obstacle appears locally like the surface of a cylinder (see Fig. P6.8). Show that in order to explore the transition from the scattering problem for the cone to that for the circular cylinder, one may imagine the origin to be pulled to infinity, with some selected point, say P, serving as a fulcrum. Define the following limiting quantities:

$$\lim_{\substack{r \to \infty \\ \theta \to \pi}} r(\pi - \theta) = \rho = \lim_{\substack{r' \to \infty \\ \theta \to \pi}} r'(\pi - \theta) \tag{57}$$

and similarly for ρ' (with $\theta \to \theta'$). Show that the radius a of the resulting cylinder is obtained by letting θ or θ' equal θ_0, and $(r_> - r_<) \to |z - z'| = |\bar{z} - \bar{z}'|$, where $\bar{z}$ is measured from a new origin located near $\rho = 0$ or $\rho' = 0$ on the z axis.

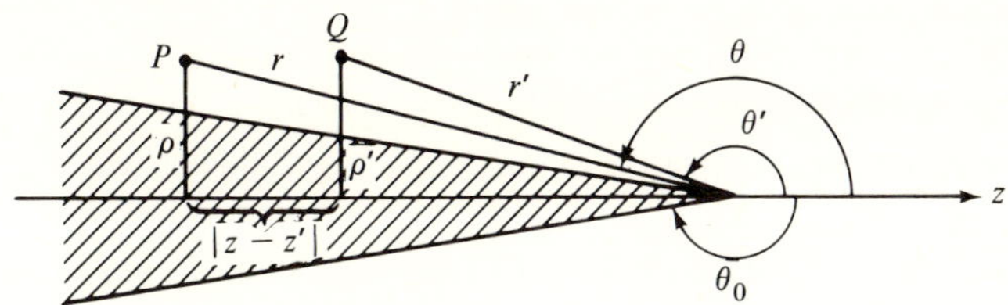

FIG. P6.8   Transition from cone to cylinder.

Show that in consequence of the above transition, the radial characteristic Green's function in Eq. (2.7.11) is replaced by that in Eq. (2.7.10) [with the switch to an $\exp(-i\omega t)$ dependence],

$$g_r \to \frac{i}{2k} h_\nu^{(2)}(kr_<) h_\nu^{(1)}(kr_>) , \qquad (58)$$

since no reflections occur from the distant cone tip (assume a slightly dissipative medium). Since this modified g_r is an even function of $\nu + (1/2)$, as is g'_θ in Eq. (3.4.68), show that the lower portion of the integration path in Eq. (6.8.9b) (with g'_θ inserted instead of $g'_{\theta s}$, and the G_f term omitted) may be reflected into the third quadrant of the complex ν plane, whence the resulting contour integral representation involves a path passing above the entire real ν-axis.

Use the asymptotic formulas for the cylinder functions (Sec. 6.A1) to show that

$$g_r \to \frac{i}{2} \frac{\exp\left(i\sqrt{k^2 - \xi^2}\,|z - z'|\right)}{\sqrt{k^2 - \xi^2}} , \qquad \mathrm{Im}\,\sqrt{k^2 - \xi^2} > 0 , \qquad (59)$$

where $\xi = t/a$, and the change of variable

$$t = \frac{\nu + (1/2)}{\alpha} , \qquad \alpha = \csc\theta_0 , \qquad \theta_0 \to \pi , \qquad (59a)$$

has been introduced. Also show via Eqs. (54) and (57), etc., that

$$g'_\theta[\theta, \theta'; m^2; \nu(\nu + 1)] \to \frac{i\pi}{2} \left[J_m(\xi\rho_<) H_m^{(1)}(\xi\rho_>) - H_m^{(1)}(\xi\rho) H_m^{(1)}(\xi\rho') \frac{J_m(\xi a)}{H_m^{(1)}(\xi a)} \right] ,$$
$$\mathrm{Im}\,\xi > 0 , \qquad (60)$$

and derive a similar result for the H mode case, g''_θ. Compare with the cylinder Green's function in Eq. (6.7.30a).

28. Show that the electromagnetic fields derived from the scalar functions in Eqs. (6.8.9) and (6.8.10) satisfy the "tip condition" as $r \to 0$ [see Eq. (1.5.39)].

REFERENCES

1. SOMMERFELD, A., "Mathematische Theorie der Diffraktion," *Math. Ann.* **47** (1896). See also Chapter 20, Vol. 2 in P. FRANK und R. V. MISES, *Die Differential- und Integralgleichungen der Mechanik und Physik*, Braunschweig, Germany: Vieweg and Son, 1961.

2. BOWMAN, J. J., T. B. A. SENIOR, and P. L. E. USLENGHI (Eds.), *Electromagnetic and Acoustic Scattering by Simple Shapes*, Chapter 6. Amsterdam: North Holland Publishing Co., 1969. Further references may be found here.

3. OBERHETTINGER, F., "On the diffraction and reflection of waves and pulses by wedges and corners," *J. Res. (NBS)* **61** (1958), pp. 343–365.

4. BOWMANN, J. J., et al., *Electromagnetic and Acoustic Scattering by Simple Shapes*, Chapter 8. Amsterdam: North Holland Publishing Co., 1969.

5. MCDONALD, H. M., "A class of diffraction problems," *Proc. London Math. Soc.* **14** (1915), pp. 410–427.

6. FELSEN, L. B., "High-frequency diffraction by a wedge with a linearly varying surface impedance." *Appl. Sci. Res.* **B9** (1961), pp. 170–188.

7. ERDELYI, A., et al., *Higher Transcendental Functions*, Bateman Manuscript Project, Vol. 2, pp. 143–147. New York: McGraw-Hill, 1953.

8. FELSEN, L. B., "Electromagnetic properties of wedge and cone surfaces with a linearly varying surface impedance," *IRE Trans. on Antennas and Propagation*, **AP-7** (1959), pp. S231–S243.

9. FELSEN, L, B., "Radiation from a tapered surface wave antenna," *IRE Trans. on Antennas and Propagation*, **AP-8** (1960), pp. 577–586.

10. TALANOV, V. I., "On surface electromagnetic waves in systems with non-uniform impedance," *Radiofisika (Russian)* **2** (1959), pp. 132–133.

11. FOCK, V. A., *Electromagnetic Diffraction and Propagation Problems*, Chapters 2, 9, 13. New York: Pergamon Press, 1965.

12. KING, R. W. P. and T. T. WU, *The Scattering and Diffraction of Waves*. Cambridge, Mass.: Harvard University Press, 1959.

13. JONES, D. S., *The Theory of Electromagnetism*, Secs. 8.6–8.8. New York: Macmillan, 1964.

14(a). BOWMAN, J, J., et al., *Electromagnetic and Acoustic Scattering by Simple Shapes*, Secs. I.2.13.5–I.2.13.6 and Chapter 2. Amsterdam: North Holland Publishing Co., 1969.

14(b). WAIT, J. R. *Electromagnetic Radiation from Cylindrical Structures*. New York: Pergamon Press, 1959.

15. WAIT, J. R., "Pattern of an antenna on a curved lossy surface," *IRE Trans. on Antennas and Propagation*, **AP-6** (1958), pp. 348–359.

16. FRANZ, W., *Theorie der Beugung Elektromagnetischer Wellen*, Secs. 11–16. Berlin: Springer, 1957.

17. KELLER, J. B., "Diffraction by a convex cylinder," *IRE Trans. on Antennas and Propagation*, **AP-4** (1956), pp. 312–321.

18. BOWMAN, J. J., et al., *Electromagnetic and Acoustic Scattering by Simple Shapes*, Chapter 10. Amsterdam: North Holland Publishing Co., 1969.

19. BOWMAN, J. J., et al., *Electromagnetic and Acoustic Scattering by Simple Shapes*, Chapter 18. Amsterdam: North Holland Publishing Co., 1969.

20. RUCK, G, T., D, E. BARRICK, W. D. Stuart, and C. K. Krichbaum, *Radar Cross Section Handbook*, Chapters 3 and 6. New York: Plenum Press, 1970.

21. WATSON, G. N., "The diffraction of electric waves by the earth," *Proc. Roy. Soc. (London)*, **A95** (1918), pp. 83-99.

22. CARSLAW, H. S., "Scattering of sound waves by a cone," *Math. Ann.* **75** (1914), p. 133.

23. FELSEN, L. B., "Plane wave scattering by small angle cones," *IRE Trans. on Antennas and Propagation* **AP-5** (1957), p. 404. Also, "Asymptotic expansions of the diffracted wave for a semi-infinite cone," *IRE Trans. on Antennas and Propagation*, **AP-5**, 1957, p. 121, and "Alternative field representations in regions bounded by spheres, cones and planes," *IRE Trans. on Antennas and Propagation*, **AP-5**, 1957, p. 109.

24. FELSEN, L. B., "Radiation from ring sources in the presence of a semi-infinite cone," *IRE Trans. on Antennas and Propagation* **AP-7** (1959), p. 168.

25. GORYANOV, A. S., "Diffraction of a plane electromagnetic wave propagated along the axis of a cone," *Radio Engrg. and Electronics* (English translation from Russian) **6** (1961), pp. 47-57.

26. HOBSON, E. W., *The Theory of Spherical and Ellipsoidal Harmonics*, p. 225. Cambridge, England: Cambridge University Press, 1931.

27. FELSEN, L. B., "Some definite integrals involving conical functions," *J. Math. Phys.* **35** (1956), p. 127.

28. DEBYE, P., "Näherungsformeln für Zylinderfunktionen für grosse Werte des Arguments und unbeschränkt veränderliche Werte des Index," *Math. Ann.* **67** (1909), pp. 535-558.

29. LANGER, R. E., "On the asymptotic solutions of ordinary differential equations with application to Bessel functions of large order," *Trans. Amer. Math. Soc.* **33** (1931), pp. 23-64,

30. OLVER, F. W. J., "Some new asymptotic expansions for Bessel functions of large orders," *Proc. Camb. Phil. Soc.* **48** (1952), pp. 414-427.

31. WATSON, G. N., *A Treatise on the Theory of Bessel Functions*, Sec. 8.61. Cambridge, England: Cambridge University Press, 1944.

32. MILLER, J. C. P., *The Airy Integral*, Cambridge, England: Cambridge University Press, 1946.

33. FRANZ, W. and R. GALLE, "Semiasymptotische Reihen für die Beugung einer ebenen Welle am Zylinder," *Z. f. Naturforschung* **10a** (1955), pp. 374-378.

34. MAGNUS, W. and L. KOTIN, "The zeros of the Hankel function as a function of its order," *Numerische Mathematik* **2** (1960), pp. 228-244. See also J. B. KELLER, S. I. RUBINOW, and M. GOLDSTEIN, "Zeros of Hankel functions and poles of scattering amplitudes," *J. of Math. Phys.* **4**, No. 6 (June 1963).

7. Fields in Uniaxially Anisotropic Regions

7.1 INTRODUCTION

The present and following chapters deal with the effects of anisotropy in a medium. Under suitable restrictions, macroscopic electromagnetic properties of physical media—including solid-state crystals, magnetized ferrites, magneto-plasmas, and artificial dielectrics—may be described in terms of a dyadic permittivity ϵ and (or) permeability μ. Plane waves in such media have been studied for some time in connection with the propagation of visible light in crystals,[1,2] the propagation of radio waves in the ionosphere,[3,4] and the propagation of microwaves in ferrite-loaded waveguides.[5] More recently, problems of radiation from stationary or moving sources, and of diffraction by objects, in an anisotropic medium have gained in importance. To understand electromagnetic propagation in a medium whose characteristics are a function of the propagation direction, it is useful to consider first the simplest type of anisotropy—the uniaxial. In this case, the constitutive parameters may be described in an appropriate coordinate system by a diagonal dyadic with two identical (transverse) elements that differ from the remaining (longitudinal) element. Physical media representable in this manner include uniaxial crystals, and plasmas or ferrites in a strong external magnetic field. The formulation and solution of radiation and diffraction problems in uniaxial regions, and the physical interpretation of the results, constitutes the substance of the present chapter. Problems in such media may sometimes be reduced to equivalent ones in isotropic regions, thereby facilitating the analysis. In the more complicated case of gyrotropic media, characterized by non-diagonal ϵ and (or) μ dyadics, no such simplification is generally possible; gyrotropic effects are treated in Chapter 8.

As in the isotropic case, propagation in anisotropic media can be described in terms of guided waves or modes, but there are characteristic differences. To elucidate these differences, let us first consider unbounded homogeneous media. In an unbounded isotropic non-spatially-dispersive† medium, modes with transverse field dependence $\exp\left[i(\mathbf{k}_t \cdot \boldsymbol{\rho} - \omega t)\right]$ propagate along the guiding direction z with wavenumbers $\pm\kappa(\mathbf{k}_t, \omega)$; for given $\mathbf{k}_t, \omega$, there are two such mode types (E and H), yielding four waves with the same isotropic wavenumbers $\pm\kappa, \pm\kappa$ independently of the choice of guiding direction (see Sec. 1.4 and Sec. 7.2). In an unbounded anisotropic, non-spatially-dispersive medium, guided waves with the same transverse field dependence likewise exist; there are again four possible waves for each $\mathbf{k}_t, \omega$, but the wavenumbers along the guiding direction are not in general negatives of one another. However, for the special case of a uniaxial medium, wherein the guiding axis z is chosen along the uniaxial direction, the four waves for given $\mathbf{k}_t, \omega$ separate into two types with wavenumbers $\pm\kappa', \pm\kappa''$ $(\kappa' \neq \kappa'')$. Furthermore, the characteristic impedance properties of these wave types are similar to those for an isotropic medium and lead to similar uncoupled transmission line descriptions. Modes remain uncoupled even in the presence of uniaxially anisotropic, planar stratification, or boundaries along z provided that all regions have z as the uniaxial direction. However, mode coupling is introduced in general by boundaries or stratification oblique to the uniaxial direction.

To illustrate these remarks, we consider the dispersion equations relating the longitudinal and transverse wavenumbers κ and $\mathbf{k}_t$ of a plane wave in a homogeneous anisotropic medium. If the electric field is taken as $\mathbf{E}(\mathbf{r}) = \bar{\mathbf{E}} \exp(i\mathbf{k} \cdot \mathbf{r} - i\omega t)$, with $\bar{\mathbf{E}}$ denoting a constant amplitude and polarization, then on eliminating $\mathbf{H}(\mathbf{r})$ from the source-free form of Eqs. (1.5.4la), one has

$$\left[\mathbf{k} \times (\mathbf{k} \times \mathbf{1}) + \frac{\omega^2}{c^2}\boldsymbol{\epsilon}'\right] \cdot \bar{\mathbf{E}} = 0, \qquad \boldsymbol{\epsilon}' = \frac{\boldsymbol{\epsilon}}{\epsilon_0}, \qquad c^2 = \frac{1}{\mu_0 \epsilon_0}, \tag{1}$$

in a medium with scalar permeability μ_0 and dyadic permittivity $\boldsymbol{\epsilon}$. In a uniaxial medium, $\boldsymbol{\epsilon}$ has the form

$$\boldsymbol{\epsilon} = \mathbf{1}_t \epsilon_t + \mathbf{z}_0 \mathbf{z}_0 \epsilon_z, \qquad \mathbf{1}_t = \mathbf{1} - \mathbf{z}_0 \mathbf{z}_0. \tag{2}$$

The parameters $\epsilon_t = \epsilon_0$ and $\epsilon_z = \epsilon_0(1 - \omega_p^2/\omega^2)$ are representative of a cold plasma under the influence of an infinitely strong axial magnetic field [see Eqs. (1.5.20), with $\omega_c = \infty$]. Substituting Eq. (2) into Eq. (1) and taking the dot product with $\mathbf{z}_0$, $\mathbf{k}_t$, and $\mathbf{z}_0 \times \mathbf{k}_t$, respectively, where $\mathbf{k}_t = \mathbf{k} - \mathbf{z}_0\kappa$, one obtains the three equations:

$$\left(\frac{\omega^2}{c^2}\epsilon_z' - k_t^2\right)\bar{E}_z + \kappa(\bar{\mathbf{E}}_t \cdot \mathbf{k}_t) = 0, \tag{3a}$$

$$k_t^2 \kappa \bar{E}_z + \left(\frac{\omega^2}{c^2}\epsilon_t' - \kappa^2\right)(\bar{\mathbf{E}}_t \cdot \mathbf{k}_t) = 0, \tag{3b}$$

$$\left(\frac{\omega^2}{c^2}\epsilon_t' - k^2\right)(\bar{\mathbf{E}}_t \cdot \mathbf{z}_0 \times \mathbf{k}_t) = 0. \tag{3c}$$

---

† In a non-spatially-dispersive medium, the dielectric constant and permeability are independent of the spatial derivative operator ∇, or in a $(\mathbf{k}, \omega)$ basis, of the wavevector $\mathbf{k}$.

When $\bar{E}_z = 0$, Eqs. (3a) and (3b) yield $\bar{\mathbf{E}}_t \cdot \mathbf{k}_t = 0$, whence, from Eq. (3c), the only non-vanishing electric-field component is parallel to $\mathbf{z}_0 \times \mathbf{k}_t$ provided that the plane-wave field obeys the dispersion equation

$$k^2 = k_0^2 \epsilon_t', \qquad k_0^2 = \frac{\omega^2}{c^2}. \tag{4}$$

Thus, the uniaxial medium can support plane waves with the electric field polarized perpendicular both to the z axis and to the wavevector $\mathbf{k}$ along which the equiphase surfaces advance (H modes with respect to z); one notes from Eq. (4) that these waves are propagated as in an *isotropic* medium with effective permittivity ϵ_t. The latter behavior follows from the transverse isotropy of the $\boldsymbol{\epsilon}$ dyadic in Eq. (2) and from the ineffectiveness of the longitudinal dependence of $\boldsymbol{\epsilon}$ for transverse-electric waves.

When $\bar{E}_z \neq 0$, $\bar{\mathbf{E}}_t \cdot \mathbf{k}_t \neq 0$ but $\bar{\mathbf{E}}_t \cdot (\mathbf{z}_0 \times \mathbf{k}_t) = 0$, Eqs. (3a) and (3b) describe waves with electric-field polarization orthogonal to that discussed above provided that κ and k_t satisfy the dispersion equation

$$\kappa^2 + \frac{k_t^2}{\epsilon} = k_0^2 \epsilon_t', \qquad \epsilon = \frac{\epsilon_z}{\epsilon_t}. \tag{5}$$

One may verify from the relation $\mathbf{H} = (i\omega\mu)^{-1} \nabla \times \mathbf{E}$ that $H_z = 0$, whence these waves are E modes with respect to z. Since from Eq. (5), $\kappa^2 + k_t^2 = k^2$ is not constant, E-mode waves display anisotropic behavior. The dependence of κ on $\mathbf{k}_t$, expressed in Eqs. (4) and (5),

$$\kappa \equiv \kappa' = \sqrt{k_0^2 \epsilon_t' - \left(\frac{k_t^2}{\epsilon}\right)} \qquad \text{for } E \text{ modes}, \tag{6a}$$

$$\kappa \equiv \kappa'' = \sqrt{k_0^2 \epsilon_t' - k_t^2} \qquad \text{for } H \text{ modes}, \tag{6b}$$

indicates that, in contrast with isotropic media, E and H modes in a uniaxially anisotropic region are propagated with different phase speeds unless $k_t = 0$.

Further insight into propagation characteristics in a uniaxial medium is gained from a study of the wavenumber surfaces traced out, for real κ and k_t, by the endpoints of the wavevector $\mathbf{k}$ in Eqs. (4) and (5). If the medium is a plasma with $\epsilon_t' = 1$, the wavenumber surface for H modes is a sphere with radius k_0 as in vacuum. For the E modes, two distinct situations arise, depending on whether in Eq. (5), $\epsilon = 1 - (\omega_p^2/\omega^2)$ is positive or negative. As shown in Fig. 7.1.1, when $\epsilon > 0$, the endpoints of the vector $\mathbf{k}$ lie on an ellipsoid, and when $\epsilon < 0$ on a hyperboloid, of revolution about the z axis. As noted in Sec. 1.6, the direction of energy transport [i.e., the direction of the group velocity vector $\mathbf{v}_g$, or of the time-averaged Poynting vector $\bar{\mathbf{S}} = \mathrm{Re}\,(\mathbf{E} \times \mathbf{H}^*)$] in a plane wave with wavevector $\mathbf{k}$, coincides with that of the normal to the surface at the point $\mathbf{k}$; to ascertain the directions of $\mathbf{k}$, $\mathbf{v}_g$ or $\bar{\mathbf{S}}$ in the (ρ, z) coordinate space, the ρ and z axes are superposed on the k_t and κ axes, respectively. While $\mathbf{k}$ and $\mathbf{v}_g$ are parallel for H modes, propagating as in vacuum, their directions differ for the E modes; this situation is depicted in Fig. 7.1.1, where $\bar{\theta}$ and θ are the angles between the positive z axis and $\mathbf{k}$ and

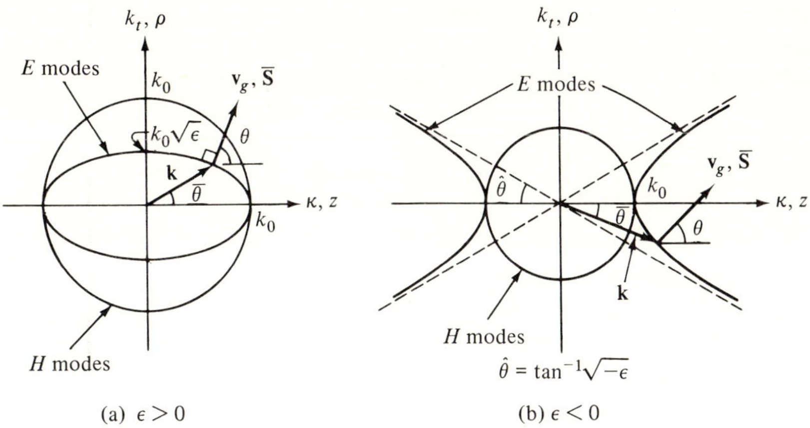

FIG. 7.1.1 Wavenumber surfaces for a uniaxially anisotropic plasma.

$\mathbf{v}_g$, respectively. One observes that *E*- mode plane waves propagate in all space directions when $\epsilon > 0$, but when $\epsilon < 0$ their group velocity (or ray) vectors are confined to the interior of the conical regions $|\tan \theta| < \tan \theta_c = 1/\sqrt{|\epsilon|}$. The limiting angle θ_c is defined by the perpendicular to the asymptote for the wavenumber surface in Fig. 7.1.1(b), and since the asymptote makes an angle $\hat{\theta} = \tan^{-1} \sqrt{|\epsilon|}$ with the *z* axis, the above expression for θ_c follows. For ray directions $\theta_c < \theta < \pi - \theta_c$, **k** is complex and the wave is exponentially damped. Thus, plane-wave propagation for $\epsilon < 0$ is confined to certain angular domains in space, and if the medium is excited by a point source, it is to be expected that the energy is confined to regions supporting "real rays." This conclusion is confirmed by the analysis in Sec. 7.3. For $\epsilon > 0$ in Fig. 7.1.1(a), the *p* and *z* components of *E*- mode **k** and $\mathbf{v}_g$ vectors have the same algebraic sign, and hence describe a propagation mechanism similar to that in an isotropic medium. This is not true for the strongly anisotropic, open-branched *E*-mode surface in Fig. 7.1.1(b). Opposite signs of corresponding wavevector and group velocity components imply "backward" wave characteristics (i.e., oppositely directed phase and energy propagation).

Because of the simple form of the dispersion equation, the relation between the angles θ and $\bar{\theta}$ is determined readily. Since the slope of the *k* versus $\bar{\theta}$ plot with respect to the κ axis is given by $dk_t/d\kappa$, the slope of the normal to the curve is $-d\kappa/dk_t$, which is evaluated from Eq. (5) as

$$\tan \theta = -\frac{d\kappa}{dk_t} = \frac{1}{\epsilon}\frac{k_t}{\kappa} = \frac{1}{\epsilon} \tan \bar{\theta}. \tag{7}$$

When $\epsilon < 0, \theta$ and $\bar{\theta}$ are measured from opposite sides of the *z* axis as in Fig. 7.1.1(b).

One observes from Fig. 7.1.1 that because of the rotational symmetry of the wavenumber surfaces with respect to the *z* axis, the two values of κ' or κ''

corresponding to given $\mathbf{k}_t$, ω are negatives of one another, with positive and negative κ distinguishing waves carrying energy along the $+z$ and $-z$ directions, respectively. As shown in Sec. 7.2. this feature permits the network representation of the propagation process in a manner similar to that employed for isotropic media, even in the presence of perfectly conducting waveguide boundaries parallel to z and of stratification along z. One finds that the H-mode portion of the field is the same as in an isotropic region while the E-mode portion can be deduced from the isotropic solution by a simple scaling of coordinates. These procedures are illustrated in Sec. 7.3, wherein are constructed explicit results for various sources in an unbounded uniaxially anisotropic region. In the discussion, emphasis is placed on the physical interpretation of field behavior, especially on the use of ray techniques and on the resolution of the "infinity catastrophe" for radiation from point sources in a medium with an open-branched wavenumber surface.

When, as is sometimes necessary, the $(\boldsymbol{\rho}, z)$ coordinate system is rotated with respect to the principal axes of the E-mode dispersion surfaces, the two solutions $\kappa_1'(\mathbf{k}_t, \omega)$ and $\kappa_2'(\mathbf{k}_t, \omega)$ are no longer related by $\kappa_1' = -\kappa_2'$ (see Fig. 7.5.2). This lack of reflection symmetry with respect to the new z direction precludes a network formulation of the field problem since a conventional transmission-line representation requires identical propagation constants for waves traveling in opposite directions. Rotated coordinates are introduced to describe planar boundaries or interfaces obliquely oriented with respect to the uniaxial direction. Such boundaries introduce coupling between E and H modes, and the associated field problem must be attacked by the more general techniques presented in Chapter 8 for arbitrary anisotropy. Simplifications occur for two-dimensional problems wherein the excitation is a magnetic line current perpendicular to the uniaxial direction but parallel to planar or cylindrical boundary surfaces; only E modes are required in this case. When the boundaries are perfectly conducting, or have a special surface impedance, coordinate scaling may be used to relate the uniaxial medium solutions to corresponding isotropic ones. As shown in Sec. 7.4, uniaxial media exhibit peculiarly anisotropic phenomena such as non-specular reflection of rays, producing distorted imaging of a line source field in a perfectly reflecting plane. Diffraction by a half-plane is an example exhibiting the influence of anisotropy on edge diffraction patterns.

When an oblique interface separates a uniaxial from an isotropic medium, coordinate scaling does not simplify the two-dimensional field problem in the manner noted above. It is then necessary to employ a representation in terms of non-reflection symmetric plane-wave modes $\kappa_1' \neq -\kappa_2'$. Section 7.5 contains a formulation of this problem, and a detailed asymptotic evaluation of the radiation field that exhibits effects of anisotropy on reflected, refracted, and diffracted (lateral) wave constituents in both the uniaxial and isotropic medium. Although the media in the present chapter are uniaxial, an interpretation involving rays and wavenumber surfaces permits inferences to be drawn for more general anisotropy.

7.2 NETWORK FORMULATION OF FIELD PROBLEM

7.2a Derivation of the Transmission-Line Equations

Consider a uniform waveguide region (Fig. 7.2.1) bounded (if at all) by a perfectly conducting surface whose generators are parallel to the z-axis of a cylindrical coordinate system. The medium filling the region is characterized

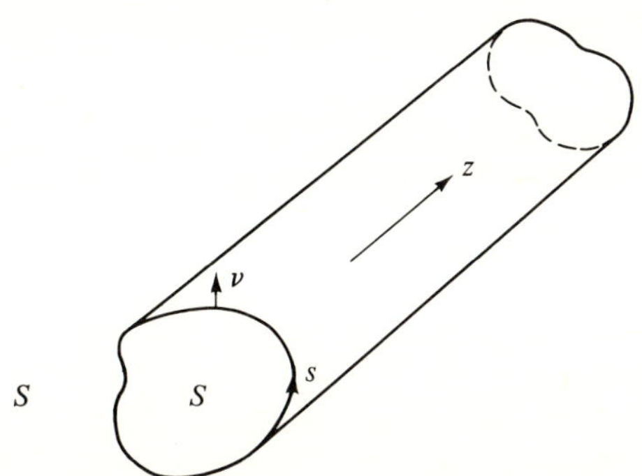

FIG. 7.2.1   Optic axis and waveguide boundaries.

by a permittivity dyadic $\boldsymbol{\epsilon} = \boldsymbol{\epsilon}(z)$ and a permeability dyadic $\boldsymbol{\mu} = \boldsymbol{\mu}(z)$ (i.e., the medium properties may vary along the z direction; if the region is comprised of homogeneous layers, the z axis is oriented perpendicular to the layer interfaces). The dyadics are assumed to have the special form appropriate to uniaxial anisotropy,[1,2] with the "optic axis" parallel to z:

$$\boldsymbol{\alpha}(z) = \mathbf{1}_t \alpha_t(z) + \mathbf{z}_0 \mathbf{z}_0 \alpha_z(z), \tag{1}$$

where $\mathbf{1}_t$ is the unit dyadic in the transverse (to z) vector space, $\mathbf{z}_0$ is a unit vector along the positive z axis, and $\boldsymbol{\alpha}$ represents either $\boldsymbol{\epsilon}$ or $\boldsymbol{\mu}$.

Upon following the same procedure as for the isotropic problem in Sec. 2.2a, one may reduce the Maxwell field equations [with an $\exp(-i\omega t)$ dependence],

$$\nabla \times \mathbf{E}(\mathbf{r}) = i\omega\boldsymbol{\mu}(z) \cdot \mathbf{H}(\mathbf{r}) - \mathbf{M}(\mathbf{r}), \tag{2a}$$

$$\nabla \times \mathbf{H}(\mathbf{r}) = -i\omega\boldsymbol{\epsilon}(z) \cdot \mathbf{E}(\mathbf{r}) + \mathbf{J}(\mathbf{r}), \tag{2b}$$

to their equivalent form for the transverse (to z) components $\mathbf{E}_t$ and $\mathbf{H}_t$,

$$-\frac{\partial \mathbf{E}_t}{\partial z} = -i\omega\mu_t \left[\mathbf{1}_t + \frac{\nabla_t \nabla_t}{k_1^2} \right] \cdot (\mathbf{H}_t \times \mathbf{z}_0) + \mathbf{M}_{te} \times \mathbf{z}_0, \qquad k_1^2 = \omega^2 \mu_t \epsilon_z, \tag{3a}$$

$$-\frac{\partial \mathbf{H}_t}{\partial z} = -i\omega\epsilon_t \left[\mathbf{1}_t + \frac{\nabla_t \nabla_t}{k_2^2} \right] \cdot (\mathbf{z}_0 \times \mathbf{E}_t) + \mathbf{z}_0 \times \mathbf{J}_{te}, \qquad k_2^2 = \omega^2 \mu_z \epsilon_t. \tag{3b}$$

In these equations, $\mathbf{E}, \mathbf{H}, \mathbf{J},$ and $\mathbf{M}$ denote the vector electric field, magnetic field, applied electric current density, and applied magnetic current density, respectively. The equivalent transverse source currents $\mathbf{M}_{te}$ and $\mathbf{J}_{te}$ are given by

$$\mathbf{M}_{te} = \mathbf{M}_t + \frac{\boldsymbol{\nabla}_t \times \mathbf{z}_0 J_z}{i\omega\epsilon_z}, \qquad \mathbf{J}_{te} = \mathbf{J}_t - \frac{\boldsymbol{\nabla}_t \times \mathbf{z}_0 M_z}{i\omega\mu_z}, \tag{3c}$$

where the subscripts t and z distinguish transverse and longitudinal components. From a knowledge of $\mathbf{E}_t$ and $\mathbf{H}_t$, one determines E_z and H_z via the relations

$$E_z = \frac{1}{-i\omega\epsilon_z}[\boldsymbol{\nabla}_t \cdot (\mathbf{H}_t \times \mathbf{z}_0) - J_z], \qquad H_z = \frac{1}{-i\omega\mu_z}[\boldsymbol{\nabla}_t \cdot (\mathbf{z}_0 \times \mathbf{E}_t) - M_z].$$

$$\tag{4}$$

On the perfectly conducting transverse boundaries s of the region, the tangential components of the total electric field are required to vanish, whence, if J_z is assumed to vanish on s,

$$\mathbf{v} \times \mathbf{E}_t = 0 = \boldsymbol{\nabla}_t \cdot (\mathbf{H}_t \times \mathbf{z}_0) \quad \text{on } s, \tag{5}$$

where $\mathbf{v}$ is a unit vector normal to s. If the region is unbounded in the transverse domain, a radiation condition is imposed at infinity. The boundary conditions at the z termini of the region are left unspecified for the moment.

As in Chapter 2, to satisfy Eqs. (3)–(5), one seeks a representation for $\mathbf{E}_t$ and $\mathbf{H}_t$ in terms of solutions of the source-free field equations in the given region. By virtue of the symmetry of the region with respect to the z coordinate, the transverse part of a source-free electric and magnetic field solution can be expressed in the form $V_i(z)\,\mathbf{e}_i(\boldsymbol{\rho})$ and $I_i(z)\mathbf{h}_i(\boldsymbol{\rho})$, respectively, where $\mathbf{e}_i$ and $\mathbf{h}_i$ are transverse vector functions of the transverse vector coordinate $\boldsymbol{\rho}$ only. In view of the homogeneity of the configuration in the cross-sectional domain and the simple form of Eqs. (3a) and (3b), the transverse eigenvalue problem (characterized by the operator $\boldsymbol{\nabla}_t\boldsymbol{\nabla}_t$) is the same as for the corresponding isotropic case, whence the set of eigenfunctions can be similarly decomposed into E and H modes relative to the z direction (see Sec. 8.2 for formulation of the eigenvalue problem). Hence [see Eqs. (2.2.8)]

$$\mathbf{E}_t(\mathbf{r}) = \sum_i V_i'(z)\,\mathbf{e}_i'(\boldsymbol{\rho}) + \sum_i V_i''(z)\,\mathbf{e}_i''(\boldsymbol{\rho}), \tag{6a}$$

$$\mathbf{H}_t(\mathbf{r}) = \sum_i I_i'(z)\,\mathbf{h}_i'(\boldsymbol{\rho}) + \sum_i I_i''(z)\,\mathbf{h}_i''(\boldsymbol{\rho}), \qquad \mathbf{h}_i = \mathbf{z}_0 \times \mathbf{e}_i, \tag{6b}$$

$$\mathbf{J}_{te}(\mathbf{r}) = \sum_i i_i'(z)\,\mathbf{e}_i'(\boldsymbol{\rho}) + \sum_i i_i''(z)\,\mathbf{e}_i''(\boldsymbol{\rho}), \tag{6c}$$

$$\mathbf{M}_{te}(\mathbf{r}) = \sum_i v_i'(z)\,\mathbf{h}_i'(\boldsymbol{\rho}) + \sum_i v_i''(z)\,\mathbf{h}_i''(\boldsymbol{\rho}), \tag{6d}$$

where the single and double primes denote E-mode and H-mode quantities, respectively. The vector-mode functions $\mathbf{e}_i'$, $\mathbf{e}_i''$, $\mathbf{h}_i'$, $\mathbf{h}_i''$, identical with those for isotropic waveguide regions, satisfy Eqs. (2.2.10) and the orthogonality conditions (2.2.11b). The source "voltages" v_i and "currents" i_i being known, it is desired to find the modal "voltages" V_i and "currents" I_i.

Substitution of Eqs. (6a–6d) into Eqs. (3a) and (3b) yields via Eqs. (2.2.10) the transmission line equations for the modal amplitudes V_i and I_i (note: $\boldsymbol{\nabla}_t \cdot \mathbf{e}_i'' = \boldsymbol{\nabla}_t \cdot \mathbf{h}_i' = 0$ in S):

$$-\frac{dV_i(z)}{dz} = -i\kappa_i(z)Z_i(z)I_i(z) + v_i(z), \tag{7a}$$

$$-\frac{dI_i(z)}{dz} = -i\kappa_i(z)Y_i(z)V_i(z) + i_i(z), \qquad Y_i(z) = \frac{1}{Z_i(z)}, \tag{7b}$$

where, for the E modes, the propagation constant κ_i and the characteristic impedance Z_i are defined as

$$Z_i'(z) = \frac{1}{Y_i'(z)} = \frac{\kappa_i'(z)}{\omega\epsilon_t(z)}, \qquad \kappa_i'(z) = \sqrt{\frac{\epsilon_t(z)}{\epsilon_z(z)}}\sqrt{k_1^2(z) - k_{ti}'^2}, \tag{8a}$$

while, for the H modes,

$$Z_i''(z) = \frac{1}{Y_i''(z)} = \frac{\omega\mu_t(z)}{\kappa_i''(z)}, \qquad \kappa_i''(z) = \sqrt{\frac{\mu_t(z)}{\mu_z(z)}}\sqrt{k_2^2(z) - k_{ti}''^2}. \tag{8b}$$

It should be noted that the characteristic impedances for the uniaxial and isotropic cases are different, and more importantly, the propagation wavenumbers $\kappa_i' \neq \kappa_i''$ are anisotropic (see Sec. 7.1). The voltage and current source terms v_i and i_i are given in terms of the specified applied sources $\mathbf{J}$ and $\mathbf{M}$ as [see Eqs. (2.2.14)]

$$v_i = \int_S \mathbf{M}_{te} \cdot \mathbf{h}_i^* \, dS = \int_S \mathbf{M} \cdot \mathbf{h}_i^* \, dS + Z_i^* \int_S \mathbf{J} \cdot \mathbf{e}_{zi}^* \, dS, \tag{9a}$$

$$i_i = \int_S \mathbf{J}_{te} \cdot \mathbf{e}_i^* \, dS = \int_S \mathbf{J} \cdot \mathbf{e}_i^* \, dS + Y_i^* \int_S \mathbf{M} \cdot \mathbf{h}_{zi}^* \, dS, \tag{9b}$$

where

$$Y_i'' \mathbf{h}_{zi}'' = \mathbf{z}_0 \frac{\nabla_t \cdot \mathbf{h}_i''}{-i\omega\mu_z}, \qquad \mathbf{h}_{zi}' \equiv 0, \tag{9c}$$

$$Z_i' \mathbf{e}_{zi}' = \mathbf{z}_0 \frac{\nabla_t \cdot \mathbf{e}_i'}{-i\omega\epsilon_z}, \qquad \mathbf{e}_{zi}'' \equiv 0. \tag{9d}$$

From a network viewpoint, V_i and I_i are the voltage and current on a source-excited transmission line, and their solution can be inferred by network techniques.

If $\boldsymbol{\epsilon}$ and $\boldsymbol{\mu}$ are discontinuous at $z = z_1$, the required continuity of $\mathbf{E}_t$ and $\mathbf{H}_t$ in Eqs. (6a) and (6b) implies the continuity of V_i and I_i, in view of the invariability along z of the transverse vector-mode functions. Hence, the modal network representation for this discontinuous case comprises a simple junction of the two transmission lines representative of the regions $z < z_1$ and $z > z_1$, respectively.

The real power carried in the ith mode along the positive z direction is given by

$$\bar{S}_{zi} = \text{Re} \int_S (V_i \mathbf{e}_i) \times (I_i \mathbf{h}_i)^* \cdot \mathbf{z}_0 \, dS = \text{Re}(V_i I_i^*), \tag{10a}$$

where all quantities are root mean square and use has been made of the orthogonality condition (2.2.11b). If the region is homogeneous, lossless, extends

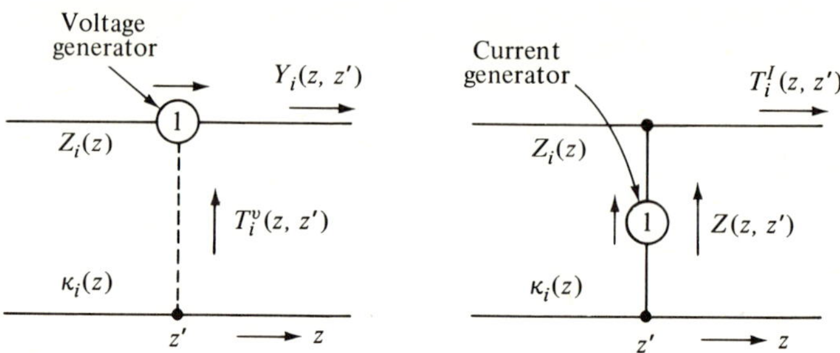

FIG. 7.2.2 Network problems for the determination of the modal Green's functions.

to $z = \infty$, and if all sources are contained in the space $z < z_0$, each modal field in $z > z_0$ comprises a traveling wave $\propto \exp{(i\kappa_i z)}$ which propagates without reflection toward $z = \infty$. In this case, the transmission-line equations, or the network representation in Fig. 7.2.2, yield $I_i = Y_i V_i$, whence

$$\bar{S}_{zi} = Y_i|V_i|^2, \qquad z > z_0, \quad \kappa_i \text{ real.} \tag{10b}$$

Equation (10b) is relevant only for the propagating modes (κ_i real) since $\bar{S}_{zi} = 0$ when κ_i is imaginary. In view of the assumption that all sources are confined in $z < z_0$, power must be carried *toward* $z = \infty$ and $\bar{S}_{zi}$ must be positive. This *radiation condition* implies that

$$Y_i > 0 \qquad \text{when } \kappa_i \text{ is real and positive,} \tag{10c}$$

and serves to specify the multivalued propagation constants κ_i' and κ_i'' (similar conclusions are reached for waves carrying power to $z = -\infty$). Thus, for a propagating mode,

$$\kappa_i' \gtrless 0 \text{ if } \epsilon_t \gtrless 0 \qquad \text{and} \qquad \kappa_i'' \gtrless 0 \text{ if } \mu_t \gtrless 0. \tag{10d}$$

Negative values of ϵ_t or μ_t may still admit propagating modes. If (ϵ_t/ϵ_z) and k_1^2 are both negative, κ_i' is real for all values of k_{ti}' and there is no E-mode cutoff; dual considerations apply to H modes. It should be noted that in contrast to wave propagation in isotropic media, it is possible to have a phase variation $\exp{(-i|\kappa_i|z)}$ (i.e., phase progressing in the $-z$ direction) associated with power flow in the $+z$ direction. This observation emphasizes the importance of phrasing the radiation condition in anisotropic media in terms of outward power flow rather than in terms of "outgoing waves". The latter condition, implying phase propagation away from the source region, is equivalent to the (energy) radiation condition in isotropic, but not in anisotropic, media. In this connection, it may be noted that because of the mode orthogonality properties (2.2.11b), the total real power $\bar{S}_z$ in a lossless medium is given by $\bar{S}_z = \sum_i \bar{S}_{zi}$, with only the propagating modes included. If each of the $\bar{S}_{zi}$ satisfies the radiation condition, so does the total power $\bar{S}_z$. If κ_i is imaginary (non-propagating wave), the requirement of boundedness at infinity, in conjunction with the assumed dependence $\exp(i\kappa_i z)$, implies that $\mathrm{Im}\ \kappa_i > 0$, thereby specifying the multivalued functions in Eqs. (8) for all values of the transverse wavenumbers k_{ti}.

If the inhomogeneous medium is thought to be comprised of a series of thin homogeneous layers, the above conditions must be satisfied for each constituent traveling wave carrying power in the positive z direction and are therefore still applicable.

7.2b Formulation in Terms of Potential Functions

Instead of evaluating the electromagnetic fields via Eqs. (6a) and (6b) and (4), it is frequently convenient to deal with scalar potential functions from which the vector fields are derived by differentiation. The desired formulation is achieved upon noting as in Eqs. (2.3.1) that the vector-mode functions can be derived from scalar-mode functions Φ_i and ψ_i, or, more conveniently, from the scalar potential functions $I'(\mathbf{r})$ and $V''(\mathbf{r})$ of Eqs. (2.3.4). After proceeding as in Sec. 2.3, one derives the analogue of Eqs. (2.3.38):[6]

$$\mathbf{E}(\mathbf{r}) = \frac{1}{-i\omega\epsilon_t(z)}\mathbf{Q}_\epsilon I'(\mathbf{r}) - \mathbf{L}_1 V''(\mathbf{r}), \tag{11a}$$

$$\mathbf{H}(\mathbf{r}) = \mathbf{L}_1 I'(\mathbf{r}) - \frac{1}{i\omega\mu_t(z)}\mathbf{Q}_\mu V''(\mathbf{r}), \tag{11b}$$

where $\mathbf{L}_1$ and $\mathbf{Q}_\alpha$ are the vector differential operators

$$\mathbf{Q}_\alpha = \nabla_t \frac{\partial}{\partial z} - \mathbf{z}_0 \frac{\alpha_t(z)}{\alpha_z(z)}\nabla_t^2 = \nabla \times \nabla \times \mathbf{z}_0 - \left[\frac{\alpha_t(z)}{\alpha_z(z)} - 1\right]\nabla_t^2 \mathbf{z}_0$$

$$\mathbf{L}_1 = \nabla \times \mathbf{z}_0 \equiv -\mathbf{z}_0 \times \nabla_t. \tag{11c}$$

In Eq. (11c), α stands for either ϵ or μ. Equations (11a) and (11b) are valid at any *source-free* point z [i.e., any point at which $v_i(z) = i_i(z) = 0$]. One verifies that the E-mode potential function $I'(\mathbf{r})$ satisfies at any source-free point z the wave equation [see Eqs. (2.3.44)]:

$$\left[\frac{\epsilon_z(z)}{\epsilon_t(z)}\mathscr{D}_\epsilon^2 + \nabla_t^2 + k_1^2(z)\right]I'(\mathbf{r}) = 0, \qquad k_1^2(z) = \omega^2\mu_t(z)\epsilon_z(z), \tag{12a}$$

$$\mathscr{D}_\epsilon^2 \equiv \epsilon_t(z)\frac{\partial}{\partial z}\frac{1}{\epsilon_t(z)}\frac{\partial}{\partial z}, \tag{12b}$$

while, for the H-mode potential function $V''(\mathbf{r})$,

$$\left[\frac{\mu_z(z)}{\mu_t(z)}\mathscr{D}_\mu^2 + \nabla_t^2 + k_2^2(z)\right]V''(\mathbf{r}) = 0, \qquad k_2^2(z) = \omega^2\mu_z(z)\epsilon_t(z). \tag{12c}$$

One notes from Eqs. (2.3.36a) and (2.3.37a) that on the perfectly conducting transverse boundaries s of the region, I' and V'' satisfy the boundary conditions

$$I'(\mathbf{r}) = 0, \qquad \frac{\partial V''(\mathbf{r})}{\partial v} = 0 \quad \text{on } s. \tag{12d}$$

7.2c   The Dyadic Green's Functions

General case

For waveguide regions filled with the uniaxially anisotropic medium specified in Eq. (1), the dyadic Green's functions are derived by following the steps leading to Eqs. (2.3.40). Let $Y_i(z, z')$ and $T_i^V(z, z')$ denote the current and voltage, respectively, observed at z due to excitation at z' by a unit strength voltage generator $v_i(z') \equiv -1$, and let $Z_i''(z, z')$ and $T_i^I(z, z')$ denote the voltage and current, respectively, observed at z due to excitation at z' by a unit-strength current generator $i_i(z') \equiv -1$ (see Fig. 7.2.2). Then the E-mode current $I_i'(z, z')$ and H-mode voltage $V_i''(z, z')$ due to the simultaneous presence of $v_i(z')$ and $i_i(z')$ are given by Eqs. (2.3.19) and (2.3.20) provided that ϵ and μ are replaced by $\epsilon_t(z')$ and $\mu_t(z')$, respectively, and that $j \to -i$ to account for the time dependence $\exp(-i\omega t)$ in the present discussion. Instead of Eq. (2.3.39), one now obtains

$$I'(\mathbf{r}, \mathbf{r}') = -\mathbf{L}_1' \mathcal{S}_d'(\mathbf{r}, \mathbf{r}') \cdot \mathbf{M}^\circ - \frac{1}{i\omega\epsilon_t(z')} \mathbf{Q}_\epsilon' \mathcal{S}_d'(\mathbf{r}, \mathbf{r}') \cdot \mathbf{J}^\circ, \tag{13a}$$

$$V''(\mathbf{r}, \mathbf{r}') = \mathbf{L}_1' \mathcal{S}_d''(\mathbf{r}, \mathbf{r}') \cdot \mathbf{J}^\circ - \frac{1}{i\omega\mu_t(z')} \mathbf{Q}_\mu' \mathcal{S}_d''(\mathbf{r}, \mathbf{r}') \cdot \mathbf{M}^\circ, \tag{13b}$$

where the scalar functions $\mathcal{S}_d'$ and $\mathcal{S}_d''$ have the modal representations

$$\mathcal{S}_d'(\mathbf{r}, \mathbf{r}') = \sum_i \frac{1}{k_{ti}'^2} Y_i'(z, z')\, \Phi_i(\boldsymbol{\rho})\, \Phi_i^*(\boldsymbol{\rho}'), \tag{13c}$$

$$\mathcal{S}_d''(\mathbf{r}, \mathbf{r}') = \sum_i \frac{1}{k_{ti}''^2} Z_i''(z, z')\, \psi_i(\boldsymbol{\rho})\psi_i^*(\boldsymbol{\rho}'). \tag{13d}$$

The prime on the operators $\mathbf{L}_1'$ and $\mathbf{Q}_\alpha'$ [see Eq. (11c)] denotes that all variables are replaced by their primed values. In deriving Eqs. (13a) and (13b) it is assumed that differentiation and summation operations can be freely interchanged, leading to the appearance of the factor $1/k_{ti}^2$ in the summands of $\mathcal{S}_d'$ and $\mathcal{S}_d''$. In this connection, the remarks made in connection with Eq. (2.3.24) are pertinent.

The dyadic Green's functions defined in Eqs. (2.3.26) are then found to be given

$$\mathcal{X}(\mathbf{r}, \mathbf{r}') = -\frac{\mathbf{Q}_\epsilon}{i\omega\epsilon_t(z)} \frac{\mathbf{Q}_\epsilon'}{i\omega\epsilon_t(z')} \mathcal{S}_d'(\mathbf{r}, \mathbf{r}') + \mathbf{L}_1 \mathbf{L}_1' \mathcal{S}_d''(\mathbf{r}, \mathbf{r}'), \tag{14a}$$

$$\mathcal{Y}(\mathbf{r}, \mathbf{r}') = \mathbf{L}_1 \mathbf{L}_1' \mathcal{S}_d'(\mathbf{r}, \mathbf{r}') - \frac{\mathbf{Q}_\mu}{i\omega\mu_t(z)} \frac{\mathbf{Q}_\mu'}{i\omega\mu_t(z')} \mathcal{S}_d''(\mathbf{r}, \mathbf{r}'), \tag{14b}$$

$$\mathcal{T}_e(\mathbf{r}, \mathbf{r}') = -\frac{\mathbf{Q}_\epsilon}{i\omega\epsilon_t(z)} \mathbf{L}_1' \mathcal{S}_d'(\mathbf{r}, \mathbf{r}') - \mathbf{L}_1 \frac{\mathbf{Q}_\mu'}{i\omega\mu_t(z')} \mathcal{S}_d''(\mathbf{r}, \mathbf{r}'), \tag{14c}$$

$$\mathcal{T}_m(\mathbf{r}, \mathbf{r}') = \mathbf{L}_1 \frac{\mathbf{Q}_\epsilon'}{i\omega\epsilon_t(z')} \mathcal{S}_d'(\mathbf{r}, \mathbf{r}') + \frac{\mathbf{Q}_\mu}{i\omega\mu_t(z)} \mathbf{L}_1' \mathcal{S}_d''(\mathbf{r}, \mathbf{r}'). \tag{14d}$$

$\mathcal{S}_d'(\mathbf{r}, \mathbf{r}') = \mathcal{S}_d'(\mathbf{r}', \mathbf{r})$ and $\mathcal{S}_d''(\mathbf{r}, \mathbf{r}') = \mathcal{S}_d''(\mathbf{r}', \mathbf{r})$, so the dyadic Green's functions satisfy the usual reciprocity conditions (2.3.28); this is a consequence of the symmetrical tensors $\boldsymbol{\epsilon} = \tilde{\boldsymbol{\epsilon}}$ and $\boldsymbol{\mu} = \tilde{\boldsymbol{\mu}}$, which characterize the medium [see Eq. (1) and Sec. 1.5b]. The modal Green's functions $Y_i'(z, z')$ and $Z_i''(z, z')$ satisfy via Eqs. (7a) and (7b) the second-order differential equations

$$[D_\epsilon^2 + \kappa_i'^2(z)]\, Y_i'(z, z') = i\omega\epsilon_t(z')\delta(z - z'), \tag{15a}$$

$$[D_\mu^2 + \kappa_i''^2(z)]\, Z_i''(z, z') = i\omega\mu_t(z')\delta(z - z'), \tag{15b}$$

where the differential operator D_α^2 is defined in Eq. (12b), with the partial derivatives replaced by total derivatives. Hence, one verifies from Eqs. (13c), (13d), and (2.3.43) that

$$\nabla_t'^2\, \mathscr{S}_d'(\mathbf{r}, \mathbf{r}') = i\omega\epsilon_z(z')G'(\mathbf{r}, \mathbf{r}'), \qquad \nabla_t'^2\, \mathscr{S}_d''(\mathbf{r}, \mathbf{r}') = i\omega\mu_z(z')G''(\mathbf{r}, \mathbf{r}'), \tag{16}$$

where G' and G'' are scalar three-dimensional Green's functions that solve the inhomogeneous wave equation

$$\left[\frac{\epsilon_z(z)}{\epsilon_t(z)}\mathscr{D}_\epsilon^2 + \nabla_t^2 + k_1^2(z)\right]G'(\mathbf{r}, \mathbf{r}') = -\delta(\mathbf{r} - \mathbf{r}'), \qquad k_1^2 = \omega^2\mu_t\epsilon_z, \tag{17a}$$

$$\left[\frac{\mu_z(z)}{\mu_t(z)}\mathscr{D}_\mu^2 + \nabla_t^2 + k_2^2(z)\right]G''(\mathbf{r}, \mathbf{r}') = -\delta(\mathbf{r} - \mathbf{r}'), \qquad k_2^2 = \omega^2\mu_z\epsilon_t. \tag{17b}$$

From the modal representations [see Eqs. (13c), (13d), and (16)]

$$G'(\mathbf{r}, \mathbf{r}') = \frac{-1}{i\omega\epsilon_z(z')}\sum_i Y_i'(z, z')\,\Phi_i(\boldsymbol{\rho})\Phi_i^*(\boldsymbol{\rho}'), \tag{18a}$$

$$G''(\mathbf{r}, \mathbf{r}') = \frac{-1}{i\omega\mu_z(z')}\sum_i Z_i''(z, z')\psi_i(\boldsymbol{\rho})\psi_i^*(\boldsymbol{\rho}'). \tag{18b}$$

one recognizes that G' and G'' satisfy on the transverse boundaries s of the region the boundary conditions

$$G' = 0, \qquad \partial G''/\partial\nu = 0 \quad \text{on } s. \tag{19}$$

Concerning the boundary conditions in the z domain, a radiation condition is imposed on $Y_i'(z, z')$ and $Z_i''(z, z')$ if the region extends to infinity along z. If two media characterized by $\epsilon_1(z)$, $\boldsymbol{\mu}_1(z)$ and $\epsilon_2(z)$, $\boldsymbol{\mu}_2(z)$, respectively, are joined at the plane $z = z_1$, the required continuity of $\mathbf{E}_t$ and $\mathbf{H}_t$ implies that of the modal voltages and currents. Since $Y_i'(z, z')$ represents a current, and $(1/\kappa_i'\, Y_i')(d/dz)Y_i'(z, z') = (1/\omega\epsilon_t)(d/dz)Y_i'(z, z')$ represents a voltage,† one notes from Eq. (18a) that, for $z_1 \neq z'$,

$$G' \text{ and } \frac{1}{\epsilon_t(z)}\frac{\partial G'}{\partial z} \qquad \text{are continuous at } \quad z = z_1, \tag{20a}$$

while, from dual considerations,

$$G'' \text{ and } \frac{1}{\mu_t(z)}\frac{\partial G''}{\partial z} \qquad \text{are continuous at } \quad z = z_1. \tag{20b}$$

---

†We recall the meaning of $Y_i'(z)$ and $Y_i'(z, z')$, which denote, respectively, the modal characteristic admittance and the current response to unit voltage excitation.

If the region is terminated at $z = z_2$ in a surface with constant impedance $\mathscr{Z}$, i.e., $\mathbf{E}_t = \mathscr{Z}\mathbf{H}_t \times \mathbf{z}_0$ at $z = z_2$, one has, from Eqs. (6a) and (6b), $V_i(z_2) = \mathscr{Z}I_i(z_2)$, whence from Eqs. (18a), (18b), (7a) and (7b),

$$\frac{\partial G'}{\partial z} = i\omega\epsilon_t(z)\mathscr{Z}G',$$

$$\text{at } z = z_2. \tag{20c}$$

$$\frac{\partial G''}{\partial z} = \frac{i\omega\mu_t(z)}{\mathscr{Z}}G'',$$

For the above-stated boundary conditions, G' and G'' in the waveguide region are uniquely determined by Eqs. (17), (19), and (20); Eqs. (18) constitute their solution in terms of a z-transmission modal representation. These remarks apply also to $\mathscr{S}'_d$ and $\mathscr{S}''_d$. It is noted that the E and H modes are not coupled by these boundary conditions.

Longitudinal sources

If the current elements $\mathbf{J}^\circ$ and $\mathbf{M}^\circ$ in Eqs. (2.3.26) are longitudinal, i.e., $\mathbf{J}^\circ = \mathbf{z}_0 J^\circ$, $\mathbf{M}^\circ = \mathbf{z}_0 M^\circ$, substantial simplification results in Eqs. (14) since $\mathbf{z}_0 \cdot \mathbf{L}'_1 = 0$ and $\mathbf{z}_0 \cdot \mathbf{Q}'_\alpha = -[\alpha_t(z')/\alpha_z(z')]\,\mathbf{V}'^2_t$. Hence, in view of Eq. (16),

$$\mathbf{E}(\mathbf{r}, \mathbf{r}') = -\frac{J^\circ}{i\omega\epsilon_t(z)}\mathbf{Q}_\epsilon G'(\mathbf{r}, \mathbf{r}') - M^\circ \mathbf{L}_1 G''(\mathbf{r}, \mathbf{r}'), \tag{21a}$$

$$\mathbf{H}(\mathbf{r}, \mathbf{r}') = J^\circ \mathbf{L}_1 G'(\mathbf{r}, \mathbf{r}') - \frac{M^\circ}{i\omega\mu_t(z)}\mathbf{Q}_\mu G''(\mathbf{r}, \mathbf{r}'). \tag{21b}$$

Piecewise constant media

If $\epsilon_t, \epsilon_z, \mu_t$, and μ_z are constant, then $\mathscr{D}^2_\alpha \to \partial^2/\partial z^2$. Upon introducing the change of variable

$$\zeta = \sqrt{\frac{\epsilon_t}{\epsilon_z}}\, z, \tag{22}$$

one may write the expression inside the brackets in Eq. (17a) as $\nabla^2_\zeta + k^2_1$, where $\nabla^2_\zeta = \nabla^2_t + \partial^2/\partial\zeta^2$. If (ϵ_t/ϵ_z) is positive real, ζ can be interpreted as a real coordinate, whence the modified Eq. (17a) represents a Green's function problem in a homogeneous isotropic $(\boldsymbol{\rho}, \zeta)$ space characterized by a wavenumber $k_1 = \omega\sqrt{\mu_t\epsilon_z}$.[6,7] Consider the isotropic Green's function defined by the equation

$$(\nabla^2_\zeta + k^2_1)G'_\zeta(\boldsymbol{\rho}, \zeta; \boldsymbol{\rho}', \zeta') = -\delta(\boldsymbol{\rho} - \boldsymbol{\rho}')\delta(\zeta - \zeta'); \tag{23}$$

then the anisotropic Green's function G' is given in terms of G'_ζ as [note: $\delta(\alpha x) = (1/|\alpha|)\delta(x)$]

$$G'(\boldsymbol{\rho}, z; \boldsymbol{\rho}', z') = \sqrt{\frac{\epsilon_t}{\epsilon_z}}\, G'_\zeta\left(\boldsymbol{\rho}, \sqrt{\frac{\epsilon_t}{\epsilon_z}}z; \boldsymbol{\rho}', \sqrt{\frac{\epsilon_t}{\epsilon_z}}\, z'\right). \tag{24}$$

The boundary conditions on G'_ζ are inferred from those on G' in Eqs. (19) and (20). If ϵ_t/ϵ_z is not positive real, one may still derive G' from the isotropic Green's function G'_ζ if the resulting function can be continued analytically into the appropriate domain of complex ϵ_t/ϵ_z.

Dual considerations apply to G'' in Eq. (17b).

Isotropic media

For isotropic media with $\epsilon_t(z) = \epsilon_z(z) \equiv \epsilon(z)$, $\mu_t(z) = \mu_z(z) \equiv \mu(z)$, the operator $\mathbf{Q}_\alpha$ in Eq. (11c) reduces to $\nabla \times \nabla \times \mathbf{z}_0$, and the results become identical with those in Sec. 2.3d.

7.3 SOURCES IN UNBOUNDED MEDIA

The results in Sec. 7.2 are now applied to radiation problems in an unbounded homogeneous medium characterized by a uniaxially anisotropic permittivity $\boldsymbol{\epsilon}$ and a scalar permeability μ_0:

$$\boldsymbol{\epsilon} = \epsilon_0(\mathbf{1}_t + \mathbf{z}_0\mathbf{z}_0\epsilon) \rightarrow \epsilon_0 \begin{pmatrix} 1 & 0 & 0 \\ 0 & 1 & 0 \\ 0 & 0 & \epsilon \end{pmatrix}, \qquad \boldsymbol{\mu} = \mathbf{1}\mu_0, \tag{1}$$

where $\mathbf{1}$ and $\mathbf{1}_t = \mathbf{1} - \mathbf{z}_0\mathbf{z}_0$ are the unit dyadic and the dyadic in the domain transverse to the optic (z) axis, respectively; $\mathbf{z}_0$ is a unit vector along z, and ϵ_0 and μ_0 are the constitutive parameters in vacuum. Because H-mode fields in this medium behave as in vacuum and E-mode fields can be derived from vacuum solutions by coordinate scaling (see Sec. 7.2), the results for fields and potentials in the uniaxial medium can be written down directly from corresponding isotropic solutions in Sec. 5.4. The discussion in this section is therefore concerned with the interpretation of such scaled solutions, with emphasis on features of the radiation process typical of an anisotropic environment. In this connection, rays and wavenumber surfaces are utilized for relating the uniaxial results to radiation characteristics observed in more general anisotropic media.

When $\epsilon > 0$, the wavenumber surface is ellipsoidal but resembles the spherical surface for isotropic media. The propagation mechanism in this case may be regarded as undergoing a continuous distortion in the transition from an isotropic to an anisotropic regime. This is not true for $\epsilon < 0$, when the wavenumber surface is open-branched and thus admits propagating wave solutions for arbitrarily large transverse and longitudinal wavenumbers. The partial differential equations descriptive of wave behavior along the open branches are hyperbolic, thus permitting the existence of field discontinuities and singularities on "characteristic cones" (shadow boundaries) similar to those encountered in time-dependent, impulse-excited fields in an isotropic medium. For *point-* and *line-*source excitation, the singularities on the shadow boundaries are so strong as to render the total radiated power infinite, an anomaly referred to in the literature as the "infinity catastrophe." While the infinity indicates the inadequacy of the equivalent dielectric description of a cold, lossless plasma as in Eq. (1), it is of interest to explore the cause of the infinity and methods for its removal. This is done in detail in Secs. 7.3a and 7.3e.

The presentation begins in Sec. 7.3a with radiation from a time-harmonic current element oriented along the optic axis. After a discussion of the closed-form solution, obtained by coordinate scaling of the vacuum fields, attention is given to a modal (integral) representation that must be employed under more general conditions, when results cannot be expressed in closed form. Analytic properties of the integrand, arising especially from saddle points and singularities, are related to features observed in the closed-form expressions, thereby providing further insight into the radiation mechanism. Sections 7.3b, 7.3c, and 7.3e deal with other time-harmonic source configurations and Sec. 7.3d explores the radiation properties of a uniformly moving point charge.

7.3a Dipoles Oriented Along the Optic Axis

Time-harmonic electric source current density

$$\hat{\mathbf{J}}(\mathbf{r}, t) = Il\delta(\mathbf{r})e^{-i\omega t}\mathbf{z}_0. \tag{2}$$

The electromagnetic fields in the uniaxial medium described by Eq. (1) can be derived from the scalar E-mode Green's function $G'_f(\mathbf{r}, \mathbf{r}')$ defined by [see Eq. (7.2.17a)]

$$\left[\epsilon\left(\frac{\partial^2}{\partial z^2} + k_0^2\right) + \nabla_t^2\right] G'_f(\mathbf{r}, \mathbf{r}') = -\delta(\mathbf{r} - \mathbf{r}'), \tag{3a}$$

subject to an energy radiation condition at ∞. The solution is given by

$$G'_f(\mathbf{r}, \mathbf{r}') = \frac{e^{ik_0 N(\theta)|\mathbf{r}-\mathbf{r}'|}}{4\pi N(\theta)|\mathbf{r} - \mathbf{r}'|}, \tag{3b}$$

where

$$k_0 = \omega\sqrt{\mu_0\epsilon_0} \qquad \text{and} \qquad N(\theta) = \sqrt{\cos^2\theta + \epsilon\sin^2\theta}, \tag{3c}$$

with $N(\theta)$ denoting the ray refractive index. Field components are derived via Eqs. (7.2.21) and yield for $r' = 0$, in spherical coordinates,[6]

$$E_r = \sqrt{\frac{\mu_0}{\epsilon_0}}\frac{Ik_0 l\cos\theta\, e^{ik_0 rN(\theta)}}{2\pi k_0 r^2 N^2(\theta)}\left(1 + \frac{i}{k_0 rN(\theta)}\right), \tag{4a}$$

$$E_\theta = \sqrt{\frac{\mu_0}{\epsilon_0}}\frac{-iIk_0 l\epsilon\sin\theta\, e^{ik_0 rN(\theta)}}{4\pi rN^3(\theta)}$$

$$\times\left[1 + \left(1 + 2\frac{\epsilon - 1}{\epsilon}\cos^2\theta\right)\left(\frac{i}{k_0 rN(\theta)} - \frac{1}{k_0^2 r^2 N^2(\theta)}\right)\right] \tag{4b}$$

$$H_\phi = \frac{-iIk_0 l\epsilon\sin\theta\, e^{ik_0 rN(\theta)}}{4\pi rN^2(\theta)}\left(1 + \frac{i}{k_0 rN(\theta)}\right), \tag{4c}$$

$$H_r = H_\theta = E_\phi = 0, \tag{4d}$$

where I is the current in the element and l is an infinitesimal length. For arbitrary $\mathbf{r}'$, one replaces r by $|\mathbf{r} - \mathbf{r}'|$. The coordinate system, with $\rho = r\sin\theta$, $z - z' = r\cos\theta$, is the same as for the dipole in an isotropic medium (Fig. 5.4.1). Provided that one chooses Im $N(\theta) \geq 0$, the results are valid for $0 \leq \arg\epsilon \leq \pi$ (i.e., for the dissipative case Im $\epsilon > 0$ and also for the "hyperbolic" case $\epsilon < 0$, Im $\epsilon = 0$). The radiated power density $\bar{\mathbf{S}}$ in a lossless

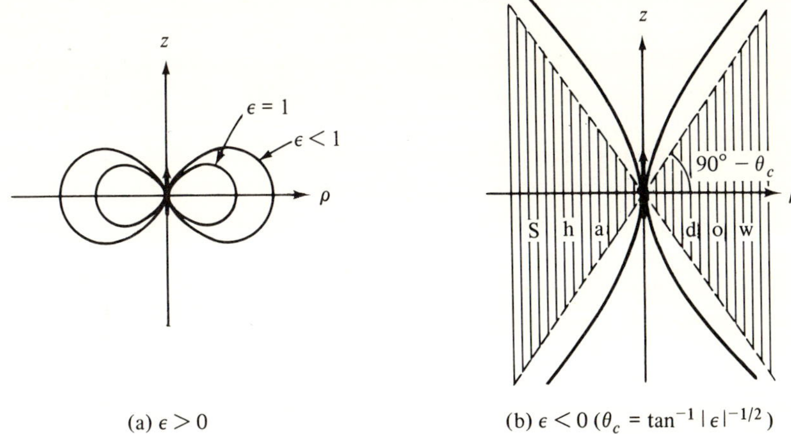

FIG. 7.3.1 Pattern functions: $A(\theta) = \sin^2\theta/N^5(\theta)$.

medium is given via Eqs. (4) by

$$\bar{S}_r = \begin{cases} \operatorname{Re}(E_\theta H_\phi^*) = \sqrt{\dfrac{\mu_0}{\epsilon_0}}\dfrac{\epsilon^2|Ik_0l|^2\sin^2\theta}{(4\pi)^2 r^2 N^5(\theta)}, & N \text{ real} \qquad (5a) \\[4mm] 0, & N \text{ imaginary} \quad (5b) \end{cases}$$

$$\bar{S}_\theta = \bar{S}_\phi = 0, \qquad N \text{ real or imaginary} \tag{5c}$$

Typical shapes of the radiation pattern function $A(\theta) = \sin^2\theta/N^5(\theta)$ are shown in Fig. 7.3.1. The total radiated power is

$$P = 2\pi r^2 \int \bar{S}_r \sin\theta\, d\theta = \begin{cases} \text{finite,} & \epsilon > 0, \qquad (6a) \\[2mm] \infty, & \epsilon < 0. \qquad (6b) \end{cases}$$

Discussion

The solution in Eq. (3b) is derived by a direct application of the coordinate scaling in Eq. (7.2.22) to the isotropic free-space Green's function in Eq. (5.4.2b) when arg $\epsilon = 0$, with subsequent analytic continuation into the range $0 < \arg \epsilon \le \pi$ [see also Eq. (9) et seq.]. When $\epsilon = 1$, Eqs. (4) reduce properly to the vacuum fields in Eqs. (5.4.3).

The simple closed-form solutions in Eqs. (4) and (5) reveal interesting features that are characteristic of radiation phenomena in anisotropic regions. When $\epsilon > 0$, the effect of anisotropy is primarily to distort, but not to alter fundamentally, the equiphase surfaces, radiation patterns, etc., encountered in an isotropic environment, whereas negative real values of ϵ yield novel and anomalous characteristics. This may be anticipated from the partial differential equation (3a), which is elliptic when $\epsilon > 0$ but hyperbolic when $\epsilon < 0$, with the consequent appearance of (characteristic) surfaces of singularity in the latter instance. Thus, it is no longer sufficient to define the far zone as $k_0 r \gg 1$ as in the isotropic case but instead one must have $k_0 r N(\theta) \gg 1$. When $\epsilon > 0$, a far zone exists at sufficiently great distance from the source, and the corresponding radiation field, given by Eqs. (4) with the bracketed expressions equated to unity, varies like $1/r$ and is transverse to r. This is not true when $\epsilon < 0$, since $N(\theta)$ then possesses a zero on the cone $\theta_c = \tan^{-1}(|\epsilon|^{-1/2})$, on which all the field components have singularities regardless of the value of r. To make matters worse, the corresponding radiated power density in Eq. (5a) behaves near θ_c like $(\theta_c - \theta)^{-5/2}$, which singularity is non-integrable near θ_c and leads to the result in Eq. (6b). Phrased in terms of the radiation resistance $R = P/|I|^2$, one concludes that $R = \infty$ when $\epsilon < 0$, a behavior found to be exhibited also by dipoles in a more general gyrotropic medium with open-branched refractive index surfaces (hyperbolic regime; see Sec. 8.3). It must be kept in mind, however, that the infinities disappear when a loss mechanism is included (ϵ complex), and that the entire description of a plasma medium by the simple tensor in Eq. (1), with $\epsilon = [1 - (\omega_p^2/\omega^2)]$, is valid only in the small signal approximation, which is clearly violated near θ_c. It is worth noting that even within the framework of the present idealized model, the field singularities for $\epsilon < 0$ disappear if the source is distributed and hence characterized by a more regular spatial function than the delta function $\delta(\mathbf{r} - \mathbf{r}')$ descriptive of the Hertzian dipole element.[8,9] This aspect is considered further in Sec. 7.3e.

One observes from Eq. (5a) that energy flows *radially outward* from the source, thereby confirming the physical mechanism of energy transport along rays discussed in Sec. 1.6. Propagating rays exist only in the illuminated region wherein $N(\theta) > 0$, and no energy is carried into the shadow region $N(\theta) = i|N(\theta)|$, where the fields are evanescent. The zones of illumination, $|\tan \theta| < \tan\theta_c = |\epsilon|^{-1/2}$, and shadow, $|\tan \theta| > \tan \theta_c$, are in evidence in Fig. 7.3.1(b). In view of the anisotropy, the rays are not normal to the equiphase surfaces defined by the equation $rN(\theta) = $ constant, or

$$\frac{z^2}{|\epsilon|} \pm \rho^2 = \frac{B^2}{|\epsilon|}, \qquad \epsilon \gtrless 0,$$

where $z = r \cos \theta$, $\rho = r \sin \theta$, and B is a positive constant. Evidently, the surfaces are elliptical when $\epsilon > 0$ and hyperbolic when $\epsilon < 0$, as shown in Fig. 7.3.2(a). It is instructive to compare these plots with the wavenumber diagrams in Fig. 7.3.2(b), from which one may also deduce directly certain salient features of the radiation field (see Sec. 1.6). For example, from the configuration of normals (rays) that can be drawn to the surfaces in Fig. 7.3.2(b), it follows that all space is accessible to radiation when $\epsilon > 0$, whereas only the region $|\tan \theta| < |\epsilon|^{-1/2}$ is illuminated when $\epsilon < 0$.

Modal procedure

While the point-source radiation problem in the infinite, homogeneous, uniaxially anisotropic medium can be solved in closed form as in Eq. (3b), no such simple solution is possible for partially bounded regions. Instead, one

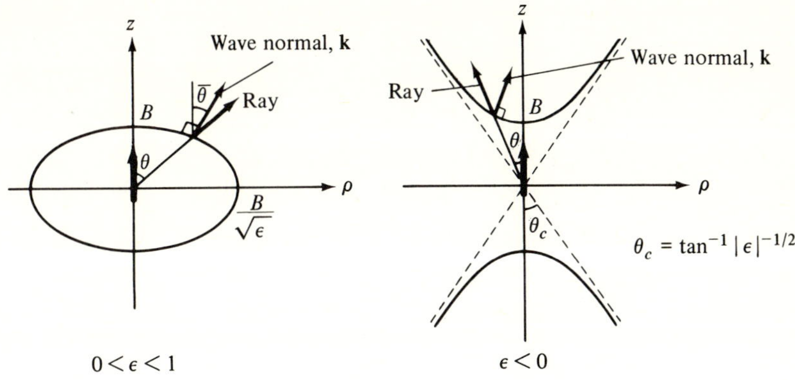

(a) Equiphase surfaces, wave normals, and rays

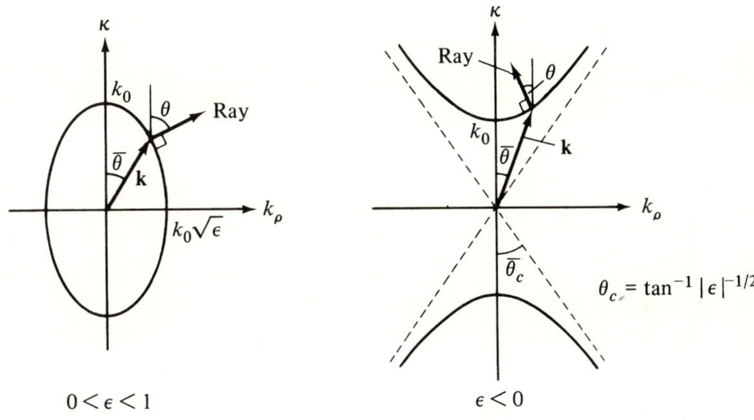

(b) Wavenumber surfaces

FIG. 7.3.2　Equiphase and wavenumber surfaces.

must resort to a modal representation of the fields and then attempt an approximate evaluation by asymptotic methods. It is instructive to apply this procedure to an infinite medium problem in order to gain insight into the difficulties encountered in the asymptotic evaluation and to establish the accuracy of the approximate solution by comparison with the exact one. Because of the symmetry of the fields about the z-axis, a circular waveguide description of the transverse cross section is appropriate, and the corresponding mode functions are given in Eq. (3.2.78b):

$$\Phi_i(\boldsymbol{\rho}) \equiv \phi_\xi(\rho) = \sqrt{\frac{\xi}{2\pi}}\, J_0(\xi\rho), \quad 0 < \xi < \infty; \qquad k_{ti} \equiv \xi; \quad \sum_i \to \int_0^\infty d\xi. \quad (7a)$$

Since the region is unbounded along z, one has from a simple network calculation,

$$Y_i'(z, z') = \frac{1}{2Z_i'} e^{i\kappa_i'|z-z'|}, \tag{7b}$$

with Z_i' and κ_i' specified in Eq. (7.2.8a). Upon substituting these results into Eq. (7.2.18a) and extending the ξ integration interval to run from $-\infty$ to $+\infty$, one obtains the desired representation [see Eq. (5.4.7a) for the analogous formulation in an isotropic medium; note the different time dependence $\exp(j\omega t)$]:†

$$G_f'(\mathbf{r}, \mathbf{r}') = \frac{i}{8\pi\epsilon} \int_{\infty e^{i\pi}}^{\infty} \xi H_0^{(1)}(\xi\rho) \frac{e^{i\kappa(\xi)|z-z'|}}{\kappa(\xi)} d\xi, \qquad \kappa(\xi) = \sqrt{k_0^2 - \frac{\xi^2}{\epsilon}}. \tag{8}$$

The radiation condition, as in Eq. (7.2.10d), requires that $\kappa > 0$ when real, and $\kappa = i|\kappa|$ when imaginary, whence the integration path is the same as in Fig. 5.3.6; the branch points are located at $\xi_b = \pm k_0 \sqrt{\epsilon}$, and the choice of branch cuts assures that $\text{Im } \kappa > 0$ on the entire top Riemann sheet. When $\epsilon = 1$, Eq. (8) reduces to the isotropic Green's-function representation in Eq. (5.4.7a).

It is of interest to point out that the radiation condition on the integrand in Eq. (8) may also be imposed from a study of the refractive index curves in Fig. 7.3.2(b) (with $k_\rho \equiv \xi$). The cylindrical waves in the integral representation degenerate into plane waves far from the source, and when $z > z'$, these waves must carry energy toward $z = +\infty$. Hence the vectors $\mathbf{v}_g$ or $\bar{\mathbf{S}}$ associated with these waves must have a component along the $+z$ direction. From general considerations [see Eq. (1.7.53a)] it is known that the angle between the $\mathbf{k}$ and $\mathbf{v}_g$ vectors is less than $90°$, so the required portions of the dispersion curves in Fig. 7.3.2(b) are those for which $\kappa \geq 0$. This alternative method of imposing the radiation condition through use of the refractive index plots is frequently to be preferred in more complicated situations (gyrotropic media) where the analytical form of the characteristic admittances Y_i is much more involved than in the present case.

When ϵ is positive real, the integral in Eq. (8) evidently converges uniformly and absolutely since $H_0^{(1)}(\xi\rho) \sim (\xi\rho)^{-1/2} \exp(i\xi\rho)$, $-\pi < \arg \xi < 2\pi$, and $\text{Im } \kappa > 0$ as $\xi \to \infty$. The case of negative real ϵ is approached unambiguously by assuming losses in the medium, whence $0 < \arg \epsilon < \pi$. One observes from Fig. 5.3.3(b) applied to $\text{Im } k > 0$ that $\arg \sqrt{k_0^2\epsilon - \xi^2}$ varies from $(\arg \epsilon)/2$ to $\pi/2$ as ξ moves from zero to infinity along the real axis. From $\kappa = \epsilon^{-1/2}(k_0^2\epsilon - \xi^2)^{1/2}$, one has

$$0 \leq \arg \kappa \leq \frac{\pi - \arg \epsilon}{2} \qquad \text{for real } \xi, \tag{9}$$

with the lower and upper limits corresponding to $\xi = 0$ and $\xi = \pm\infty$, respectively. Thus, for $0 \leq \arg \epsilon < \pi$, $\text{Im } \kappa > 0$ as $|\xi| \to \infty$ along the real axis

---

†For the calculation of radiated power, a plane-wave representation as in Sec. 1.2c may be more convenient than the guided-wave representation employed here.

and the convergence of the integral remains unaffected. In consequence, the formula in Eq. (8) may be employed over the extended range Im $\epsilon > 0$.

When arg $\epsilon = \pi$, κ is positive real along the entire real ξ axis and the integral in Eq. (8) is no longer absolutely, but instead conditionally, convergent. Since the derivation of the electromagnetic fields requires the repeated differentiation of G', it is desirable to obtain a representation that retains the exponential convergence as $\xi \to \infty$. Evidently, such a formulation involves the deformation of the integration path into the complex plane. If branch cuts are introduced along the lines Re $\kappa = 0$ (see Fig. 7.3.3), then Re $\kappa > 0$ on the entire top sheet of the two-sheeted ξ plane. Thus, as $\xi \to \infty$, $\kappa \sim \pm \xi \sqrt{|\epsilon|}$ for Re $\xi \gtrless 0$, so that in the right half-plane, the exponential dependence in the integrand is exp $(i\Psi)$, with $\Psi = \xi\rho + \kappa|z - z'| \sim \xi(\rho + |z - z'|/\sqrt{|\epsilon|})$, while in the left half-plane, $\Psi \sim \xi(\rho - |z - z'|/\sqrt{|\epsilon|})$. Thus, Im $\Psi > 0$ when

$$\text{Im } \xi > 0, \qquad \text{as } |\xi| \to \infty \text{ with Re } \xi > 0, \tag{10a}$$

$$\left. \begin{array}{ll} \text{Im } \xi > 0, & |\tan \theta| > \tan \theta_c \\ \text{Im } \xi < 0, & |\tan \theta| < \tan \theta_c \end{array} \right\} \text{ as } |\xi| \to \infty \text{ with Re } \xi < 0, \tag{10b}$$

where the observation angle θ and the limiting angle θ_c are defined in Eqs. (4) and in Fig. 7.3.2(b), respectively. A distortion of the endpoints of the integration path away from the real ξ axis according to Eqs. (10) then assures the exponential convergence of the integral in Eq. (8) when arg $\epsilon = \pi$. The different convergence requirements in Eq. (10b) for $|\tan \theta| \lessgtr \tan \theta_c$ (i.e., for observation points in the illuminated or shadow regions) are already indicative of the critical role played by the shadow boundary cone $|\tan \theta| = \tan \theta_c$. For observation points on the shadow boundary, $\Psi \sim 0$, the integrand behaves asymptotically like $\xi^{-1/2}$, and the integral diverges.

An asymptotic evaluation of the integral in Eq. (8) can be carried out as in Sec. 5.3, except that we prefer to remain in the complex ξ plane for the present analysis, in view of the simple interpretation of the saddle-point condition via the refractive index plots. The Hankel function is replaced by its large argument approximation in Eq. (5.3.13b), and the resulting exponential in the integrand has the previously noted form exp $(i\Psi) = \exp [ir(\xi \sin \theta + \kappa|\cos \theta|)]$. For large values of r, the dominant contribution arises from the vicinity of the saddle point ξ_s, which is obtained from a solution of the equation $(d/d\xi)[\xi \sin \theta + \kappa|\cos \theta|] = 0$,

$$\left. \frac{d\kappa(\xi)}{d\xi} \right|_{\xi_s} = -\frac{\xi_s}{\epsilon \kappa(\xi_s)} = -|\tan \theta|. \tag{11a}$$

This relation is easily solved for ξ_s,

$$\xi_s = \frac{k_0 \epsilon \sin \theta}{N(\theta)}, \qquad \kappa(\xi_s) = \frac{k_0|\cos \theta|}{N(\theta)}, \tag{11b}$$

with $N(\theta)$ so defined that arg $N(\theta) = -\text{arg } \kappa(\xi_s)$. Equation (11a) is identical with Eq. (7.1.7) if $k_t \equiv \xi$, and therefore has a simple graphical interpretation: the real values of ξ_s, $\kappa(\xi_s)$, which define a propagating wave, correspond to points on the k versus $\bar{\theta}$ plot having a normal inclined to the positive z axis at an angle θ (see Fig. 7.3.2). Since the normal to the surface coincides with the ray direction, one verifies from the saddle-point condition that the rays proceed radially from the source to the observation point. Conversely, the k versus $\bar{\theta}$ plots may be employed for the location of the saddle point(s): the saddle points ξ_s correspond to those surface points that have a normal $\mathbf{v}$ directed along θ, with $\mathbf{k} \cdot \mathbf{v} \geq 0$. As noted in Sec. 1.6, this fundamental condition remains valid for propagation problems in more general media where $\kappa(\xi)$ is no longer the simple function specified in Eq. (8).

The saddle point ξ_s lies on the positive real ξ axis when ϵ is positive, on the negative real ξ axis when $\epsilon < 0$ and $N(\theta) > 0$, and in the upper half of the complex ξ plane when $\epsilon < 0$ and $N(\theta)$ is positive imaginary. The two first-mentioned conditions could also have been predicted from a study of Figs. 7.1.1(a) and 7.1.1(b). A closer examination of the expression for ξ_s in Eq. (11b) for a slightly dissipative medium with $0 < \text{Im } \epsilon \ll 1$ reveals that the saddle point lies slightly above the real ξ axis when $N(\theta)$ is essentially real, and lies slightly to the left of the imaginary axis, in the interval $|\xi_s| > k_0\sqrt{|\epsilon|}$, when $N(\theta)$ is essentially imaginary. For positive real ϵ, Eq. (8) is very similar to Eq. (5.4.7a), and the asymptotic evaluation of the integral along the steepest-descent path through the saddle point proceeds as for a source in vacuum. For negative real ϵ, with $|\tan \theta| < \tan \theta_c$, the endpoints of the integration path are deformed into the first and third quadrants of the ξ plane, as noted in Eqs. (10). Since

$$\left. \frac{d^2\kappa}{d\xi^2} \right|_{\xi_s} = \frac{-N^3(\theta)}{k_0 \epsilon |\cos^3 \theta|}, \tag{12}$$

this quantity is positive in the present instance and represents a measure of the curvature K of the refractive index plots via the relation $K = (d^2\kappa/d\xi^2)_{\xi_s} \cdot |\cos^3 \theta|$; both its algebraic sign, and an estimate of its magnitude, may be inferred directly from an inspection of the curves in Fig. 7.1.1. Because $d^2\kappa/d\xi_s^2 > 0$, the steepest-descent path (SDP) through the saddle point is inclined at a $+45°$ angle with the positive ξ direction [see Eq. (4.2.5)]. The SDP is defined by the equation $\text{Re } \Psi(\xi) = \text{Re } \Psi(\xi_s)$ (see Sec. 4.1b), so

$$\text{Re } (\kappa|\cos \theta| + \xi \sin \theta) = k_0 N(\theta), \qquad \kappa = \sqrt{k_0^2 + \frac{\xi^2}{|\epsilon|}}. \tag{13a}$$

If the branch cuts from $\xi_b = \pm ik_0\sqrt{|\epsilon|}$ are chosen along the straight lines $\text{Re } \kappa = 0$, then $\text{Re } \kappa > 0$ on the entire top sheet of the two-sheeted κ surface, and, as noted earlier, $\kappa \sim \pm\xi/\sqrt{|\epsilon|}$ as $\xi \to \infty$ in the half-planes $\text{Re } \xi \gtrless 0$. Consequently, the SDP is asymptotic to the straight lines

$$\text{Re } \xi^\pm \equiv \xi_r^\pm = \pm\sqrt{|\epsilon|}k_0 \left[|\cos \theta| \mp \sqrt{|\epsilon|} \sin \theta\right]^{1/2}; \tag{13b}$$

one easily verifies that $\xi_r^- < \xi_s$ and that the path crosses the imaginary ξ axis below the branch point $\xi_b = ik_0\sqrt{|\epsilon|}$. The SDP therefore takes the form shown in Fig. 7.3.3(a); it passes through the branch cut arising from the Hankel function in the integrand of Eq. (8). Since the integrand decays exponentially in the first and third quadrants of the complex ξ plane, the original integration path may be deformed into the SDP without intercepting any singularities.

The resulting first-order asymptotic approximation of the integral for $\epsilon \gtrless 0$, $N(\theta) > 0$, may be obtained from Eq. (4.2.1b) and yields [see also Eq. (1.6.21)]

$$G'_f(\mathbf{r}, \mathbf{r}') \sim \frac{e^{ik_0 r N(\theta)}}{4\pi r N(\theta)}, \qquad r \gg 1, \quad N(\theta) \text{ real}, \tag{14}$$

the same expression as the exact result in Eq. (3). Although Eq. (14) happens to agree with the exact formula, it was derived subject to the assumptions that the r-independent factors in the integrand of Eq. (8) are slowly varying and that $\kappa''(\xi_s) \not\approx 0$, $\xi_s \not\approx 0$; the condition $\xi_s \not\approx 0$ arises from the replacement of the Hankel function by its asymptotic form, valid only when $|\xi \rho| \gg 1$ along the entire path. One observes from Eq. (12) that $\kappa''(\xi_s) \to 0$ as $N(\theta) \to 0$, so the validity of the asymptotic expression breaks down as the observation point approaches the shadow boundary. In this region, the saddle point moves to $\xi_s = -\infty$, where the curvature K of the refractive index diagram tends to zero—an indication that difficulties may arise in the asymptotic evaluation. From the exact field expressions in Eqs. (4), it is noted that the asymptotic result to $O(1/r)$ is meaningful only when $r \gg 1$ implies also $k_0 r N(\theta) \gg 1$, and this requirement must be imposed here to delimit the range of validity of the first-order asypmptotic formula. Since $\xi_s \propto k_0/N(\theta)$, the condition

$$\frac{k_0^2 r}{|\xi_s|} \gg 1 \tag{14a}$$

serves to restrict the admissible range of saddle-point locations. These observa-

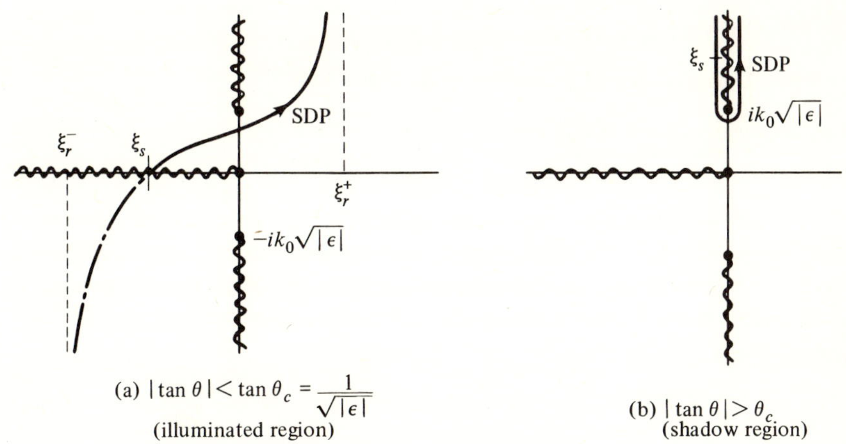

(a) $|\tan \theta| < \tan \theta_c = \dfrac{1}{\sqrt{|\epsilon|}}$
(illuminated region)

(b) $|\tan \theta| > \theta_c$
(shadow region)

FIG. 7.3.3 Steepest-descent paths in the complex ξ plane.

tions will also apply to asymptotic calculations in gyrotropic media (see Sec. 8.3c) if the refractive index surface contains an open branch.

If $|\tan \theta| > \tan \theta_c$, the saddle point lies slightly to the left of the imaginary axis and above the branch point at $\xi_b = ik_0\sqrt{|\epsilon|}$. In the interval $\xi = i|\xi|$, $|\xi| > k_0\sqrt{|\epsilon|}$, the propagation constant κ is imaginary, and if the branch cuts are drawn as before, $\kappa = i|\kappa|$ when $\arg(\xi - \xi_b) = \pi/2$ while $\kappa = -i|\kappa|$ when $\arg(\xi - \xi_b) = -3\pi/2$. Thus, Eq. (13a), with the right-hand side set equal to zero, is satisfied along a path SDP that surrounds the branch cut [Fig. 7.3.3(b)]. In view of Eqs. (10), the exponential function in the integrand of Eq. (8) decays in the upper half of the ξ plane, so the original integration path may be deformed into the SDP without encountering any singularities. The resulting first-order asymptotic evaluation yields

$$G'_f(\mathbf{r}, \mathbf{r}') \sim \frac{e^{-k_0 r |N(\theta)|}}{4\pi i r |N(\theta)|}, \qquad k_0 r |N(\theta)| \gg 1, \tag{15}$$

an expression that represents the analytic continuation of Eq. (14) to positive imaginary values of $N(\theta)$. The asymptotic evaluation therefore confirms the previously observed division into illuminated and shadow regions when $\epsilon < 0$, but gives no information about the field behavior in the immediate vicinity of the shadow boundary.

For the dissipative case with Im $\epsilon > 0$, the saddle point is located in the upper half of the ξ plane and the resulting asymptotic results are found to be the same as in Eq. (14), with Im $N(\theta) > 0$.

Time-harmonic magnetic source current density

$$\hat{\mathbf{M}}(\mathbf{r}, t) = Vl\delta(\mathbf{r})e^{-i\omega t}\mathbf{z}_0. \tag{16}$$

The relevant scalar H-mode Green's function $G''_f(\mathbf{r}, \mathbf{r}')$ is, in view of Eqs. (1) and (7.2.17b), identical with the one in vacuum [see Eq. (3), with $N(\theta) = 1$] and the fields derived therefrom are also the same as in vacuum [see Eq. (5.4.16) et seq.].

7.3b   Dipoles Oriented Transverse to the Optic Axis

Time-harmonic electric source current density

$$\hat{\mathbf{J}}(\mathbf{r}, t) = Il\delta(\mathbf{r})e^{-i\omega t}\mathbf{x}_0. \tag{17}$$

A dipole oriented transverse to the optic (z) axis in the uniaxially anisotropic medium of Eq. (1) excites both E- and H-mode fields which are derivable via Eqs. (7.2.14) from the scalar potential functions $\mathscr{S}'_d(\mathbf{r}, \mathbf{r}')$ and $\mathscr{S}''_d(\mathbf{r}, \mathbf{r}')$, respectively; these functions satisfy the differential equations [see Eqs. (7.2.16) and (7.2.17)]

$$\left[\epsilon\left(\frac{\partial^2}{\partial z^2} + k_0^2\right) + \nabla_t^2\right]\nabla_t^2\, \mathscr{S}'_f(\mathbf{r}, \mathbf{r}') = \epsilon\delta(\mathbf{r} - \mathbf{r}'), \qquad -\mathscr{S}'_f = \frac{\mathscr{S}'_d}{i\omega\epsilon_0}, \tag{18a}$$

$$(\nabla^2 + k_0^2)\nabla_t^2\, \mathscr{S}''_f(\mathbf{r}, \mathbf{r}') = \delta(\mathbf{r} - \mathbf{r}'), \qquad -\mathscr{S}''_f = \frac{\mathscr{S}''_d}{i\omega\mu_0}, \tag{18b}$$

subject to radiation conditions at infinity, with $k_0 = \omega\sqrt{\mu_0\epsilon_0}$. The *H*-mode potential function in Eq. (18b) is identical with the function $\mathscr{S}_f(\mathbf{r}, \mathbf{r}')$ defined in Eq. (5.4.18) for an unbounded isotropic space with *wavenumber* $k = k_0$. The *E*-mode potential function may be related to $\mathscr{S}_f$ via the coordinate scale transformation $\zeta = z\epsilon^{-1/2}$ when ϵ is positive real, since from

$$\left(\frac{\partial^2}{\partial\zeta^2} + \nabla_t^2 + k_0^2\epsilon\right)\nabla_t^2\mathscr{S}'_f = \sqrt{\epsilon}\;\delta(\boldsymbol{\rho} - \boldsymbol{\rho}')\delta(\zeta - \zeta'), \tag{19}$$

it is recognized that

$$\mathscr{S}'_f(\mathbf{r}, \mathbf{r}') = \sqrt{\epsilon}\;\mathscr{S}_f\left(\boldsymbol{\rho}, \frac{z}{\sqrt{\epsilon}};\; \boldsymbol{\rho}', \frac{z'}{\sqrt{\epsilon}}\right), \tag{20}$$

provided that $\mathscr{S}_f$ is evaluated in a medium with wavenumber $k = k_0\sqrt{\epsilon}$. As in Sec. 7.3a, analytic continuation may be invoked to validate the resulting function for $0 < \arg\epsilon \leq \pi$.

The electromagnetic fields may now be deduced directly from the isotropic medium formulas in Sec. 5.4b. The *H*-mode constituents are identical with those in Sec. 5.4b, whereas the *E*-mode constituents can be determined from their isotropic medium counterparts by applying the above-mentioned scale transformation.[7] The details are left as an exercise for the reader.

Time-harmonic magnetic source current density

$$\hat{\mathbf{M}}(\mathbf{r}, t) = Vl\delta(\mathbf{r})e^{-i\omega t}\mathbf{x}_0. \tag{21}$$

The fields are derivable via Eqs. (7.2.14) from the scalar functions $\mathscr{S}'_f$ and $\mathscr{S}''_f$ in Eqs. (18).

7.3c Linearly Phased Line Currents Oriented Along the Optic Axis

Time-harmonic electric source current density

$$\hat{\mathbf{J}}(\mathbf{r}, t) = Ie^{i\alpha z}\delta(\boldsymbol{\rho})e^{-i\omega t}\mathbf{z}_0. \tag{22}$$

Since the current distribution in Fig. 7.3.4 is directed along the optic axis, the electromagnetic fields can be derived from a two-dimensional scalar *E*-mode Green's function (see Sec. 7.2)

$$[\nabla_t^2 + (k_0^2 - \alpha^2)\epsilon]\,\bar{G}'_f(\boldsymbol{\rho}, \boldsymbol{\rho}') = -\delta(\boldsymbol{\rho} - \boldsymbol{\rho}'), \qquad \nabla_t^2 = \nabla^2 - \frac{\partial^2}{\partial z^2}, \tag{23}$$

subject to an energy radiation condition at infinity. $\boldsymbol{\rho}$ is the vector coordinate transverse to z and $k_0 = \omega\sqrt{\mu_0\epsilon_0}$. All field components vary like $\exp(i\alpha z)$ whence $\partial^2/\partial z^2 \to -\alpha^2$. The solution is given by[6]

$$\bar{G}'_f(\boldsymbol{\rho}, \boldsymbol{\rho}') = \begin{cases} \dfrac{i}{4}\,H_0^{(1)}(\sqrt{\epsilon}\,\sqrt{k_0^2 - \alpha^2}\,\rho) & \text{when } \epsilon > 0, \tag{24a} \\[2em] -\dfrac{i}{4}\,H_0^{(2)}(\sqrt{|\epsilon|}\sqrt{\alpha^2 - k_0^2}\,\rho) & \text{when } \epsilon < 0, \tag{24b} \end{cases}$$

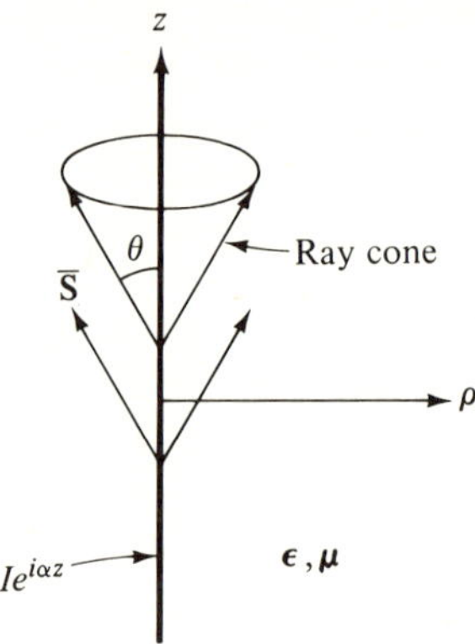

FIG. 7.3.4 Progressively phased line current, $Ie^{i\alpha z}$, $\alpha > 0$;
$\epsilon = \epsilon_0(\mathbf{1}_t + \mathbf{z}_0\mathbf{z}_0\epsilon)$, $\mu = \mu_0\mathbf{1}$.

with the square roots defined as follows:

$$\sqrt{k_0^2 - \alpha^2} > 0 \quad \text{when } k_0 > \alpha, \qquad \text{Im } \sqrt{k_0^2 - \alpha^2} > 0 \quad \text{when } k_0 < \alpha,$$

$$\sqrt{\alpha^2 - k_0^2} > 0 \quad \text{when } k_0 < \alpha, \qquad \text{Im } \sqrt{\alpha^2 - k_0^2} < 0 \quad \text{when } k_0 > \alpha. \tag{24c}$$

The fields are derived from $\bar{G}'_f$ via Eqs. (7.2.14):

$$E_\rho = \frac{i\alpha I\sqrt{\epsilon}\,\hat{\kappa}}{4\omega\epsilon_0} e^{i\alpha z} H_1^{(1)}(\sqrt{\epsilon}\,\hat{\kappa}\rho), \tag{25a}$$

$$E_z = -\frac{\hat{\kappa}^2 I}{4\omega\epsilon_0} e^{i\alpha z} H_0^{(1)}(\sqrt{\epsilon}\,\hat{\kappa}\rho), \tag{25b}$$

$$H_\phi = \frac{iI\sqrt{\epsilon}\,\hat{\kappa}}{4} e^{i\alpha z} H_1^{(1)}(\sqrt{\epsilon}\,\hat{\kappa}\rho), \tag{25c}$$

$$E_\phi = H_\rho = H_z = 0, \tag{25d}$$

where $\hat{\kappa} = \sqrt{k_0^2 - \alpha^2}$. These results apply when $0 \leq \arg \epsilon \leq \pi$, provided that Im $\hat{\kappa} > 0$ when $\hat{\kappa}$ is non-real; for negative real ϵ, one may employ the alternative formulation in Eq. (24b).

The average radiated power density is given for real ϵ by

$$\bar{S}_\rho = \begin{cases} -\text{Re}(E_z H_\phi^*) = \dfrac{|\hat{\kappa}|^2}{8\pi k_0 \rho}\sqrt{\dfrac{\mu_0}{\epsilon_0}}\,|I|^2, & \sqrt{\epsilon}\,\hat{\kappa} \text{ real,} \tag{26a} \\[2ex] 0, & \sqrt{\epsilon}\,\hat{\kappa} \text{ imaginary,} \tag{26b} \end{cases}$$

$$\bar{S}_z = \text{Re}(E_\rho H_\phi^*) = \frac{\alpha|\sqrt{\epsilon}\,\hat{\kappa}|}{8\pi k_0 \rho}\sqrt{\frac{\mu_0}{\epsilon_0}}\,|I|^2, \qquad \sqrt{\epsilon}\,\hat{\kappa} \text{ real or imaginary.} \tag{26c}$$

The radiated power flow direction θ and the wavevector direction $\bar{\theta}$ with respect to the z axis are defined by

$$\tan \theta = \frac{\bar{S}_\rho}{\bar{S}_z} = \frac{|\hat{\kappa}|}{\alpha|\sqrt{\epsilon}|} = \frac{1}{\epsilon}\tan\bar{\theta}, \tag{27}$$

where $\bar{\theta}$ is the direction of the wavevector $\mathbf{k} = \mathbf{z}_0\alpha + \boldsymbol{\rho}_0\sqrt{\epsilon}\,\hat{\kappa}$.

For large values of $|\sqrt{\epsilon}\,\hat{\kappa}|\rho$, Eqs. (25) may be simplified through use of the asymptotic formula $H_n^{(1)}(w) \sim (2/\pi w)^{1/2} \exp\left[i(w - \pi/4 - n\pi/2)\right]$.

Discussion

The solution in Eq. (24a) follows directly from the analogous isotropic medium result in Eq. (5.4.47), and the formula in Eq. (24b) is deduced therefrom by analytic continuation via Eq. (24c) and $H_0^{(1)}(we^{i\pi}) = -H_0^{(2)}(w)$.

Since the Hankel functions denote propagating wave solutions in a lossless medium only when the argument is real, one observes that the source distribution in Fig. 7.3.4 radiates when

$$\text{(a)}\quad \epsilon > 0,\ k_0 > \alpha \qquad \text{and} \qquad \text{(b)}\quad \epsilon < 0,\ k_0 < \alpha. \tag{28}$$

These restrictions for cases (a) and (b) imply that the phase speeds along the source are larger and smaller, respectively, than the speed of light in vacuum; thus, different excitation functions are required to produce radiation when $\epsilon > 0$ and $\epsilon < 0$. While, from Eq. (27), the energy propagation vector always has an outward radial component in accord with the energy radiation condition, the corresponding phase fronts move outward when $\epsilon > 0$ but toward the source axis (backward wave) when $\epsilon < 0$, as is observed either from Eq. (27) or from Eqs. (24a) and (24b).

These conclusions may also be arrived at directly from the wavenumber surfaces. In view of the translational invariance of the physical configuration with respect to the z axis, the radiated fields (i.e., the local plane-wave fields at a large distance from the axis) are characterized by the dependence exp $(i\alpha z)$ impressed along the source. α therefore assumes for the relevant wave constituents the role of the longitudinal wavenumber κ, and one observes from Fig. 7.3.5 that propagation (i.e., intersection of the plane $\kappa = \alpha$ and the wavenumber surface) occurs for $\alpha < k_0$ when $\epsilon > 0$ and for $\alpha > k_0$ when $\epsilon < 0$.

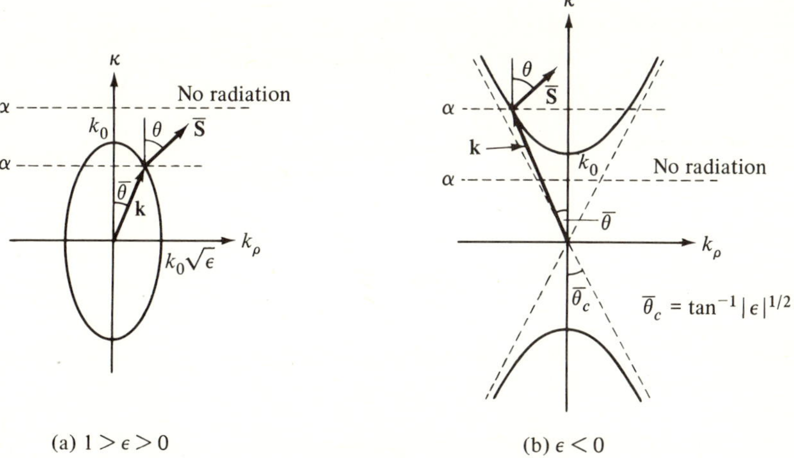

FIG. 7.3.5 Interpretation via wavenumber surfaces.

Evidently, when $\bar{S}_\rho > 0$, as required by the radiation condition, the transverse wavenumber $k_\rho > 0$ for $\epsilon > 0$, but $k_\rho < 0$ for $\epsilon < 0$, thereby confirming the conclusions reached earlier; in discussing the local plane-wave behavior of the far field, k_ρ may be regarded as a *rectilinear* wavenumber, thereby making negative values of k_ρ meaningful. Since the wavenumber surfaces are symmetric about the source axis, the rays along $\bar{S}$ defined by the intersection with the $\kappa = \alpha$ plane emerge from the source at the constant angle $\theta = \tan^{-1}[|\hat{\kappa}|/(\alpha\sqrt{|\epsilon|})]$, so the ray configuration forms a right circular cone (Fig. 7.3.4). This symmetry is destroyed when the source axis is inclined with respect to the optic axis in the medium.

Time-harmonic magnetic source current density

$$\hat{\mathbf{M}}(\mathbf{r}, t) = V e^{i\alpha z}\delta(\boldsymbol{\rho})e^{-i\omega t}\mathbf{z}_0. \tag{29}$$

The source configuration excites only H modes, so the fields are identical with those in vacuum (Sec. 5.4d).

7.3d Point Charge in Uniform Straight Motion Along the Optic Axis

The source current density is given by

$$\hat{\mathbf{J}}(\mathbf{r}, t) = qv\delta(z - vt)\delta(\boldsymbol{\rho})\mathbf{z}_0, \tag{30}$$

where q is the charge and $v = $ constant is the particle speed (see Fig. 7.3.6). On introducing the Fourier transform

$$Q(\mathbf{r}, \omega) = \int_{-\infty}^{\infty} e^{i\omega t}\hat{Q}(\mathbf{r}, t)\, dt, \tag{31a}$$

one finds for the source function

$$\mathbf{J}(\mathbf{r}, \omega) = q e^{i(\omega/v)z}\delta(\boldsymbol{\rho})\mathbf{z}_0, \tag{31b}$$

which is the same harmonic source as in Eq. (22), with $I = q, \alpha = \omega/v$. With this identification, the fields follow from Eqs. (25), and the radial harmonic power density is given from Eq. (26a) by

$$\bar{S}_\rho(\mathbf{r}, \omega) = \frac{\left(\dfrac{1}{\beta^2} - 1\right)k_0\sqrt{\mu_0/\epsilon_0}\, q^2}{8\pi\rho}, \qquad \beta < 1, \tag{32}$$

where $\beta = v/c$ and $k_0 = \omega/c$, with c denoting the speed of light in vacuum.

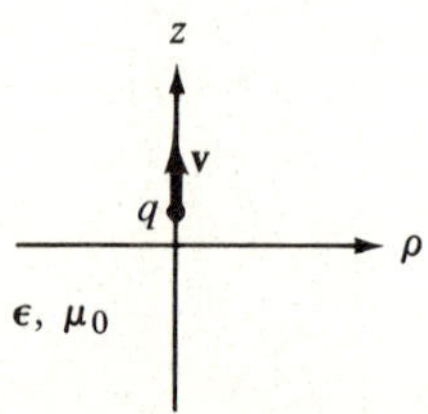

FIG. 7.3.6 Point charge moving along optic axis. Constitutive parameters: $\boldsymbol{\epsilon} = \epsilon_0(\mathbf{1}_t + \mathbf{z}_0\mathbf{z}_0\epsilon), \boldsymbol{\mu} = \mathbf{1}\mu_0$.

Since $\beta < 1$, it follows that $k_0 < \alpha \, (= k_0/\beta)$ and radiation takes place only for negative values of ϵ [see Eq. (28)]. The total energy radiated per unit length along the trajectory is obtained from Eq. (5.2.30) as

$$W_1 = \frac{1}{\pi} \int_0^\infty \bar{S}_\rho(\mathbf{r}, \omega) \cdot 2\pi\rho \, d\omega. \tag{33}$$

In general, this result cannot be evaluated unless the frequency dependence of ϵ is stated. If, as is the case for certain plasmas, $\epsilon = 1 - \omega_p^2/\omega^2$ is negative only for $\omega < \omega_p$, the upper integration limit is ω_p, whence[7]

$$W = \frac{q^2 \mu_0 \omega_p^2}{8\pi} \left(\frac{1}{\beta^2} - 1 \right). \tag{34}$$

This result implies somewhat surprisingly that slower moving particles radiate more strongly than those in rapid motion, a behavior quite different from that encountered in isotropic dielectrics (Sec. 5.4e). It must, of course, be kept in mind that the particle speed is assumed uniform in the above calculation, a condition incompatible with the non-negligible deceleration that accompanies substantial radiation.

The basic features of the radiation process may again be inferred directly from the wavenumber plots in Fig. 7.3.5. From Eq. (31b), the relevant wave constituents excited by the particle have $\alpha = \omega/v = k_0/\beta > k_0$, so no radiation occurs for $\epsilon > 0$ in Fig. 7.3.5(a). When $\epsilon < 0$, the sketch in Fig. 7.3.5(b) implies that radiation is of the backward-wave type, with phase progression toward the particle trajectory. The graph illustrates also the basic "Cerenkov coherence condition," which states that in a small frequency interval about ω, the particle excites those waves whose wavevector component along the trajectory matches the spectral component ω/v.

7.3e   Line Currents Oriented Perpendicular to the Optic Axis

Magnetic source current density

$$\hat{\mathbf{M}}(\mathbf{r}, t) = V\delta(\hat{\boldsymbol{\rho}} - \hat{\boldsymbol{\rho}}')e^{-i\omega t}\mathbf{x}_0. \tag{35}$$

With the line current oriented as in Fig. 7.3.7, the electromagnetic fields can be derived from the two-dimensional scalar E-mode Green's function $\bar{G}'_f(\hat{\boldsymbol{\rho}}, \hat{\boldsymbol{\rho}}')$ that satisfies the differential equation

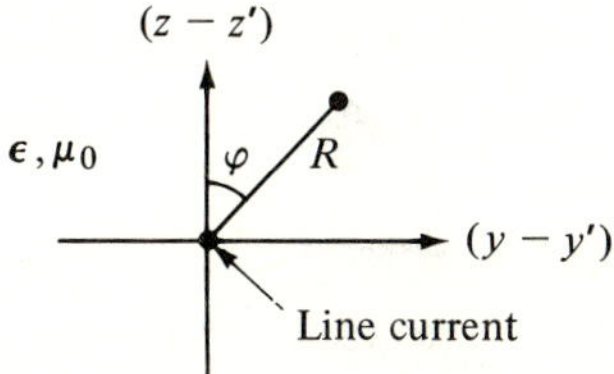

FIG. 7.3.7   Line current transverse to optic axis;
$\epsilon = \epsilon_0(\mathbf{1}_t + \mathbf{z}_0\mathbf{z}_0\epsilon)$, $\boldsymbol{\mu} = \mu_0\mathbf{1}$.

$$\left(\frac{\partial^2}{\partial z^2} + \frac{1}{\epsilon}\frac{\partial^2}{\partial y^2} + k_0^2\right)\bar{G}'_f(\hat{\boldsymbol{\rho}}, \hat{\boldsymbol{\rho}}') = -\delta(\hat{\boldsymbol{\rho}} - \hat{\boldsymbol{\rho}}'), \qquad \hat{\boldsymbol{\rho}} = (y, z), \qquad (36)$$

subject to an energy radiation condition at infinity. The solution for $0 \leq \arg \epsilon \leq \pi$ is given by

$$\bar{G}'_f = \frac{i}{4}\sqrt{\epsilon}\, H_0^{(1)}[k_0 R N(\varphi)], \qquad N(\varphi) = \sqrt{\cos^2\varphi + \epsilon \sin^2\varphi}, \qquad (37)$$

where $k_0 = \omega\sqrt{\mu_0\epsilon_0}$, $R = \sqrt{(y - y')^2 + (z - z')^2}$, and $\mathrm{Im}\ N(\varphi) \geq 0$. The field components are derived from $\bar{G}'_f$ as follows:

$$H_x \equiv H = i\omega\epsilon_0 V \bar{G}'_f = -\frac{\omega\epsilon_0\sqrt{\epsilon}\,V}{4} H_0^{(1)}[k_0 R N(\varphi)], \qquad (38a)$$

$$E_y = -V\frac{\partial \bar{G}'_f}{\partial z} = \frac{ik_0\sqrt{\epsilon}\,V}{4}\frac{\cos\varphi}{N(\varphi)} H_1^{(1)}[k_0 R N(\varphi)], \qquad (38b)$$

$$E_z = \frac{V}{\epsilon}\frac{\partial \bar{G}'_f}{\partial y} = -\frac{ik_0\sqrt{\epsilon}\,V}{4}\frac{\sin\varphi}{N(\varphi)} H_1^{(1)}[k_0 R N(\varphi)], \qquad (38c)$$

$$H_y = H_z = E_x = 0. \qquad (38d)$$

In a cylindrical coordinate system, Eqs. (38b) and (38c) may be combined to yield for the field components,

$$E_R = E_z\cos\varphi + E_y\sin\varphi = 0, \qquad (38e)$$

$$E_\varphi = E_y\cos\varphi - E_z\sin\varphi = \frac{ik_0\sqrt{\epsilon}\,V}{4N(\varphi)} H_1^{(1)}[k_0 R N(\varphi)]. \qquad (38f)$$

The average radiated power density for real ϵ is given by

$$\bar{S}_R = -\mathrm{Re}(E_\varphi H_x^*) = \begin{cases} \sqrt{\dfrac{\epsilon_0}{\mu_0}}\dfrac{|\epsilon|k_0}{8\pi R N^2(\varphi)}|V|^2, & N \text{ real,} & (39a) \\[2ex] 0, & N \text{ imaginary,} & (39b) \end{cases}$$

$$\bar{S}_\varphi = \bar{S}_x = 0, \qquad (39c)$$

and the total radiated power is

$$P = R\int_0^{2\pi}\bar{S}_R\, d\varphi = \begin{cases} \text{finite,} & \epsilon > 0, & (40a) \\[1ex] \infty, & \epsilon < 0. & (40b) \end{cases}$$

Asymptotic expressions for the field components are found from Eqs. (38) by replacing the Hankel functions by their large-argument approximation,

$$H_n^{(1)}(w) \sim (2/\pi w)^{1/2}\exp\left[i(w - \pi/4 - n\pi/2)\right].$$

Discussion

The formulation of the field in terms of the Green's function G'_f as in Eqs. (38) is accomplished by proceeding directly from the Maxwell field equations subject to the constraint $\partial/\partial x = 0$. Alternatively, the results may be deduced by simplification of Eqs. (7.2.14), (7.2.16), and (7.2.17) [see also the analogous treatment for the isotropic case, Eqs. (5.4.31)]. In the derivation of Eqs. (39), use is made of the Wronskian for the cylinder functions; it has also been recognized for imaginary N, that $H_\nu^{(1)}(iw) = (2/\pi i) \exp(-\nu\pi i/2)K_\nu(w)$, where the modified Hankel function $K_\nu(w)$ is real when ν and w are real. One observes from Eq. (39a) that the source illuminates all regions of space when $\epsilon > 0$, while the region $|\tan\varphi| > |\epsilon|^{-1/2}$ constitutes for $\epsilon < 0$ a shadow region wherein the fields are evanescent (N imaginary). On the shadow boundary $|\tan\varphi| = \tan\varphi_c = |\epsilon|^{-1/2}$, where $N = 0$, the fields and the radiated power density are infinite. The infinities are weaker than those for point-source excitation [see Eqs. (5a) and (39a)] but still strong enough to render the total radiated power, and therefore the radiation conductance $\bar{S}/|V|^2$, infinite.

The interpretation of these results follows also from the wavenumber surfaces. Since the fields are independent of x, the constituent plane waves in the far zone are characterized by a vanishing wavenumber ξ (or k_x) along the x direction. The only relevant portions of the wavenumber surfaces are therefore the curves obtained by intersection with the $k_x = 0$ plane, and these curves are the same as in Fig. 7.3.2(b) if k_ρ is interpreted as the wavenumber $k_y \equiv \eta$ along the y direction. Since the normals to the wavenumber surface at $k_x = 0$ lie in the $\eta\kappa$ plane, the rays (trajectories of energy flow) are perpendicular to the x axis and emanate radially from the source, as noted also from Eqs. (39). Various aspects discussed in connection with Fig. 7.3.2 therefore remain applicable, subject to the identification $k_\rho \to \eta$, $\rho \to y$.

Removal of the infinity in the radiated power

It has been found that the fields radiated either by a point source or a line source in a uniaxially anisotropic medium characterized by an open-branched wavenumber surface exhibit infinities along the shadow boundary θ_c or $\varphi_c = \tan^{-1}|\epsilon|^{-1/2}$, and that the *total* radiated power is also infinite. While we have noted in Sec. 7.3a that these singularities disappear for a more realistic description of the physical medium, it will now be shown that even within the confines of the lossless uniaxially anisotropic model, finite fields and power are obtained for a *distributed* source configuration.[6,8,9] Although the calculation is performed for the two-dimensional case, analogous conclusions apply also to the three-dimensional field.

It may be anticipated on physical grounds that a distributed source softens the singular point-source field behavior since the presence of neighboring elements in the source distribution tends to diffuse the shadow boundary. This effect may be observed from a comparison of the point-source and line-source results in Eqs. (4), (5), and (38), (39), respectively, since the line-source field exhibits a weaker singularity than the point-source field.

It is convenient to employ an integral representation for $\bar{G}_f'$ analogous to that in Eq. (8). The appropriate modal basis involves the plane waves $\Phi_\eta(y) = (2\pi)^{-1/2} \exp(i\eta y)$, $-\infty < \eta < \infty$, and utilization of Eq. (7.2.18a) yields, for the magnetic field,

$$H = C \int_{-\infty}^{\infty} \frac{e^{i\eta y + i\kappa |z-z'|}}{\kappa} \, d\eta, \qquad \kappa(\eta) = \sqrt{k_0^2 - \frac{\eta^2}{\epsilon}}, \tag{41}$$

where C includes various constants. The convergence properties of this integral are essentially the same as in Eq. (8), and the previous observations apply here as well. If the source is distributed in the plane $z = z'$ with an amplitude variation given by $f(y')$, the corresponding magnetic field $\hat{H}$ is given by

$$\hat{H} = C \int_{-\infty}^{\infty} \frac{e^{i\eta y + i\kappa |z-z'|}}{\kappa} \, F(\eta) \, d\eta, \tag{42}$$

where $F(\eta)$ is the Fourier transform of the source function $f(y')$,

$$F(\eta) = \int f(y') e^{-i\eta y'} \, dy', \tag{42a}$$

and the integration extends over the source region. For a line source, $f(y') = \delta(y')$, $F(\eta) = 1$, and one recovers Eq. (41). A more regular, distributed source has a spectrum $F(\eta)$ that decays as $\eta \to \infty$ and therefore deemphasizes the contribution from these regions to the integral in Eq. (42). If the integral is then absolutely convergent, although κ is real over the entire integration interval, the previously noted divergence in the fields (as $\varphi \to \varphi_c$) disappears. If $F(\eta_s)$, where η_s denotes the saddle point as in Eqs. (11), drops off sufficiently rapidly, an asymptotic evaluation of the integral for $\epsilon < 0$ will yield a vanishing result as $\varphi \to \varphi_c$ since the decay of $F(\eta_s)$ overcomes the growth of the earlier asymptotic approximation as $\eta_s \to \infty$.

These considerations may also be applied to the total radiated power. The y component of electric field is calculated from Eq. (38b):

$$\hat{E}_y = \frac{-C}{\omega\epsilon_0} \int_{-\infty}^{\infty} e^{i\eta y + i\kappa(z-z')} F(\eta) \, d\eta, \qquad z > z', \tag{43}$$

and the total power flow in the $+z$ direction is then given by

$$\bar{S}_z = \frac{|C|^2}{\omega\epsilon_0} \text{Re} \left[\int_{-\infty}^{\infty} dy \int_{-\infty}^{\infty} d\eta \int_{-\infty}^{\infty} d\eta' \frac{e^{i(\eta'-\eta)y + i(\kappa'-\kappa)(z-z')}}{\kappa} F^*(\eta')F(\eta) \right] \tag{44}$$

$$= \frac{2\pi|C|^2}{\omega\epsilon_0} \text{Re} \int_{-\infty}^{\infty} \frac{|F(\eta)|^2}{\sqrt{k_0^2 - (\eta^2/\epsilon)}} \, d\eta. \tag{45}$$

In Eq. (44), $\kappa' \equiv \kappa(\eta')$, and evaluation of the y integration in terms of the delta function $\delta(\eta - \eta')$ leads to Eq. (45), which could also have been written down directly from Eq. (7.2.10b). If $\epsilon > 0$, the integration extends only over the range $|\eta| \leq k_0\sqrt{\epsilon}$ and the integral is finite for all reasonably behaved $F(\eta)$. If $\epsilon < 0$, the integration interval is infinite and $\bar{S}_z$ is finite only when

$$\int^{\infty} \frac{|F(\eta)|^2}{\eta} \, d\eta < \infty, \tag{46}$$

a mild restriction satisfied by most source distributions. For example, if

$$f(y') = \begin{cases} A = \text{constant} & \text{when} \quad -a < y' < a, \\ 0 & \text{when} \quad |y'| > a, \end{cases} \tag{47}$$

then

$$F(\eta) = \frac{2A \sin \eta a}{\eta}, \tag{48}$$

and the inequality (46) certainly applies. The total radiated power P is equal to $2\bar{S}_z$ and is calculated by integrating the average power flow density $\bar{S}$ over a surface completely enclosing the source. In the present instance, the surface comprises two infinite parallel planes at $z = z' \pm \Delta$, $\Delta > 0$, and two segments of width 2Δ located at $y = \pm\infty$. Since the far field decays like $1/\sqrt{R}$ [see asymptotic form of Eqs. (38)], the only non-vanishing contribution arises from the planes at $z = z' \pm \Delta$, whence $P = 2\bar{S}_z$.

Thus, finiteness of radiated power in a lossless anisotropic region with an open-branched wavenumber surface is achieved when the *spatial transform of the source distribution decays sufficiently rapidly at infinity*, and this general conclusion remains valid even for gyrotropic media and for three-dimensional configurations although the *specific* requirements may vary for different situations.

Electric line source current density

$$\hat{\mathbf{J}}(\mathbf{r}, t) = I\delta(\hat{\boldsymbol{\rho}} - \hat{\boldsymbol{\rho}}')e^{-i\omega t}\mathbf{x}_0. \tag{49}$$

This source configuration excites only H modes with respect to the optic axis, and the resulting fields are the same as in vacuum (see Sec. 5.4c).

Electric dipolar source current density

$$\hat{\mathbf{J}}(\mathbf{r}, t) = A\delta(\hat{\boldsymbol{\rho}} - \hat{\boldsymbol{\rho}}')e^{-i\omega t}(\mathbf{y}_0 \cos \alpha + \mathbf{z}_0 \sin \alpha). \tag{50}$$

This source arrangement consists of a line distribution of electric current elements oriented perpendicular to the line axis z, at an angle α with respect to the y axis (Fig. 7.3.8).

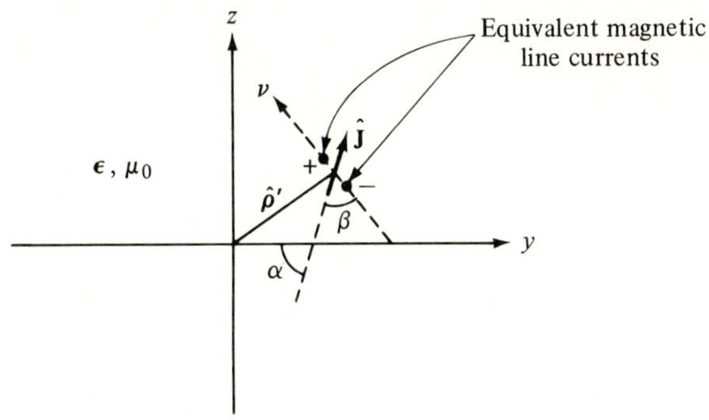

FIG. 7.3.8 Electric dipole line current; $\epsilon = \epsilon_0(\mathbf{1}_t + \mathbf{z}_0\mathbf{z}_0\epsilon)$, $\mu = \mu_0\mathbf{1}$.

Since $\partial/\partial x \equiv 0$, the non-vanishing field components are easily shown to be $H_x \equiv H$, E_y, E_z, and the Maxwell field equations reduce to

$$\frac{\partial E_z}{\partial y} - \frac{\partial E_y}{\partial z} = i\omega\mu_0 H, \tag{51a}$$

$$\frac{\partial H}{\partial z} = -i\omega\epsilon_0 E_y + J_y, \qquad \frac{\partial H}{\partial y} = i\omega\epsilon_0\epsilon E_z - J_z. \tag{51b}$$

Upon substituting for $\partial E_z/\partial y$ and $\partial E_y/\partial z$ from Eq. (51b) into Eq. (51a), one obtains the inhomogeneous wave equation

$$\left(\frac{\partial^2}{\partial z^2} + \frac{1}{\epsilon}\frac{\partial^2}{\partial y^2} + k_y^2\right) H = -A\left(\frac{1}{\epsilon}\sin\alpha\frac{\partial}{\partial y} - \cos\alpha\frac{\partial}{\partial z}\right)\delta(y - y')\delta(z - z'), \tag{52}$$

which can be solved in terms of the two-dimensional Green's function $\bar{G}'_f$ in Eq. (37):

$$H = -A\left(\frac{1}{\epsilon}\sin\alpha\frac{\partial}{\partial y'} - \cos\alpha\frac{\partial}{\partial z'}\right)\bar{G}'_f, \tag{53}$$

with the recognition that $\partial/\partial y = -\partial/\partial y'$, $\partial/\partial z = -\partial/\partial z'$, in view of the dependence of $\bar{G}'_f$ on $y - y'$ and $z - z'$. The operator inside the brackets in Eq. (53) may be interpreted as $\mathbf{v} \cdot \nabla' = v(\partial/\partial v)$, where $\mathbf{v}$ is a vector with components $v_y = (1/\epsilon)\sin\alpha$, $v_z = -\cos\alpha$. Consequently,

$$\mathbf{v} \cdot (\mathbf{y}_0\cos\alpha + \mathbf{z}_0\sin\alpha) = [(1/\epsilon) - 1]\sin\alpha\cos\alpha = v\cos\beta,$$

$$v = \sqrt{\cos^2\alpha + (1/\epsilon^2)\sin^2\alpha}, \tag{54}$$

where β is the angle between $\mathbf{v}$ and the source direction $\hat{\mathbf{J}}$.

For non-dissipative media with real ϵ, β is a real angle, and in view of Eq. (37), the result in Eq. (53) may be interpreted as the field due to two closely spaced, oppositely directed magnetic currents—a magnetic "line dipole" directed along $\mathbf{v}$. In the isotropic case $\epsilon = 1$, $\mathbf{v}$ and $\hat{\mathbf{J}}$ are mutually perpendicular, and Eq. (53) reduces to Eq. (5.4.39a). In the uniaxial medium, $\mathbf{v} \cdot \hat{\mathbf{J}} = 0$ only when the electric dipole direction is parallel or perpendicular to the optic axis ($\alpha = 0, \pi/2$). A source distribution of the type considered here is induced, for example, when a narrow conducting strip centered at $\hat{\mathbf{p}}'$ and oriented along α is excited by an incident field with $H \equiv H_x$.

Highly directive, distributed magnetic current source:

$$\hat{\mathbf{M}}(\mathbf{r}, t) = \begin{cases} \cos\dfrac{\pi u}{a}\, e^{iu\zeta}\delta(v)e^{-i\omega t}\mathbf{x}_0, & -a \leq u \leq a, \tag{55a}\\[2mm] 0, & \text{elsewhere,} \tag{55b} \end{cases}$$

with the coordinates u and v defined in Fig. 7.3.9. This functional dependence of $\hat{\mathbf{M}}$ simulates the excitation due to a progressively phased antenna with a tapered amplitude distribution, and is certainly smooth enough to assure avoidance of the shadow boundary singularities that arise for the isolated line

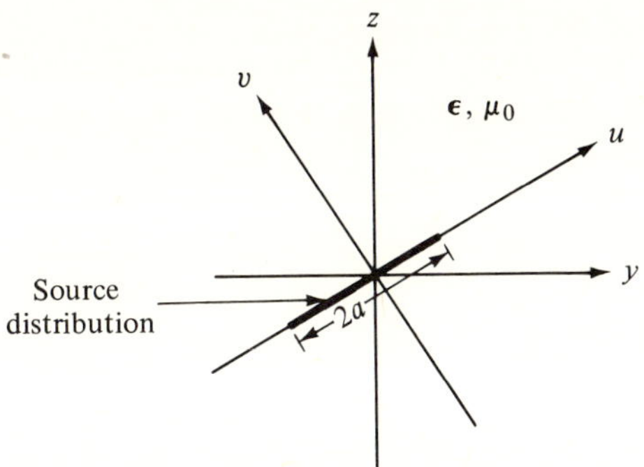

FIG. 7.3.9 Physical configuration; $\epsilon = \epsilon_0(\mathbf{1}_t + \mathbf{z}_0\mathbf{z}_0\epsilon)$, $\boldsymbol{\mu} = \mu_0\mathbf{1}$.

source when $\epsilon < 0$ (see earlier discussion in this section). When the antenna extends over many wavelengths ($k_0 a \gg 1$), the radiated field is highly directive. Such sources find important application in antenna engineering and may be realized either by a series of closely spaced individual elements or by a continuous distribution, as considered here.

The radiation pattern of this antenna is obtained by integrating the far-field form of Eq. (38a),

$$H \sim \frac{-\omega\epsilon_0\sqrt{\epsilon}\, e^{-i\pi/4}}{2\sqrt{2\pi}\sqrt{k_0 RN(\varphi)}}\, e^{ik_0 RN(\varphi) - i\mathbf{k}\,\cdot\, v_0 u}, \tag{56}$$

over the source region. In this equation, $R = \sqrt{y^2 + z^2} \to \infty$, φ is the angle between R and the positive z axis, $\mathbf{u}_0$ is a unit vector along u, and $\mathbf{k}$ is the phase propagation vector corresponding to the ray progressing in the φ direction. The resulting magnetic field H_T is then given by

$$H_T \sim H|_{u=0}F, \tag{57}$$

where F is the array factor

$$F = \pi a \frac{\cos\left[a(\zeta - \mathbf{k}\cdot\mathbf{u}_0)\right]}{(\pi/2)^2 - \left[a(\zeta - \mathbf{k}\cdot\mathbf{u}_0)\right]^2}. \tag{58}$$

Since $\mathbf{k}\cdot\mathbf{u}_0 a = (\mathbf{z}_0\kappa + \mathbf{y}_0\eta)\cdot\mathbf{u}_0 a$ is proportional to $1/N(\varphi)$, the ratio $F/\sqrt{N(\varphi)}$ vanishes as $N(\varphi) \to 0$, so one verifies the regularity of the field near the shadow boundary when $\epsilon < 0$. It is recalled from Eq. (11b) that $k_0 RN(\varphi) = \eta y + \kappa z = \mathbf{k}\cdot\mathbf{R}$ since $\eta = k_0\epsilon \sin\varphi/N(\varphi)$, $\kappa = k_0\cos\varphi/N(\varphi)$; the relation between $\mathbf{k}$ and the radius vector $\mathbf{R}$ (which is parallel to the group velocity vector $\mathbf{v}_g$) may also be inferred from Fig. 7.3.2.

If $k_0 a \gg 1$, the array factor has sharp maxima at those angular locations for which

$$\zeta = \mathbf{k}\cdot\mathbf{u}_0. \tag{59}$$

The corresponding ray angles define the major lobes in the radiation pattern, and they may be easily inferred from the $\mathbf{k}$ versus φ plot. Evidently, condition

(59) selects those values of **k** whose projection onto the positive u axis has a length equal to ζ, the wavenumber of the progressively phased source distribution in Eq. (55a). The procedure is schematized in Fig. 7.3.10 for the case

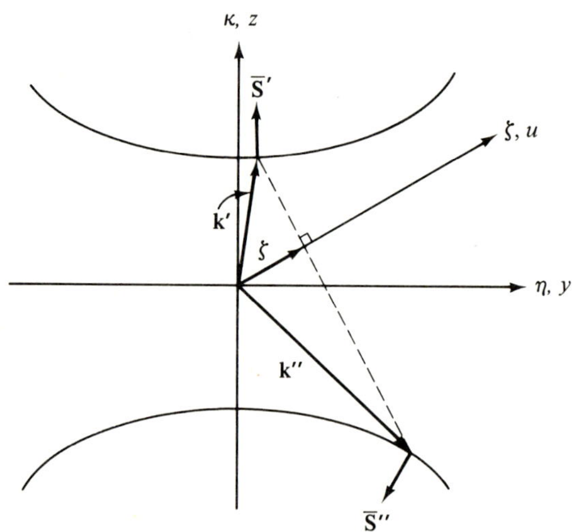

FIG. 7.3.10   Directive, distributed source: directions of pattern maxima ($\epsilon > 0$).

$\epsilon < 0$, $\zeta > 0$, and one observes that two values **k**$'$ and **k**$''$ satisfy Eq. (59). The corresponding rays $\overline{\mathbf{S}}'$ and $\overline{\mathbf{S}}''$ point essentially in the directions of the major lobes in the radiation pattern since, except near the shadow boundary, the angularly dependent factor $1/\sqrt{N(\varphi)}$ in $H|_{u=0}$ of Eq. (57) is slowly varying over the angular width of a narrow beam and introduces only a small shift in the location of the pattern maxima determined from these considerations. Owing to the anisotropy, the radiation pattern is asymmetric about the source axis u, and the lobe corresponding to $\overline{\mathbf{S}}''$ actually points in the backward direction, although the phase of the antenna progresses along $+u$. The construction in Fig. 7.3.10 may also be employed for the inverse problem of determining the values of ζ and the directions $\mathbf{u}_0$ that yield a pattern maximum along a specified angle.

The above procedure for determining the locations of the pattern maxima due to highly directive sources in an anisotropic medium is also useful under conditions when the wavenumber surface has a more complicated shape or has additional branches. In essence, one looks for the directions of the rays whose wavenormal projections on the antenna axis match the wavenumber of the phase progression along the source.

A numerical example (Fig. 7.3.11) illustrates the validity of the preceding remarks for a special case in which $\mathbf{u}_0 = \mathbf{z}_0$, $\epsilon = -1$, $a = 4\lambda_0$, and ζ takes on

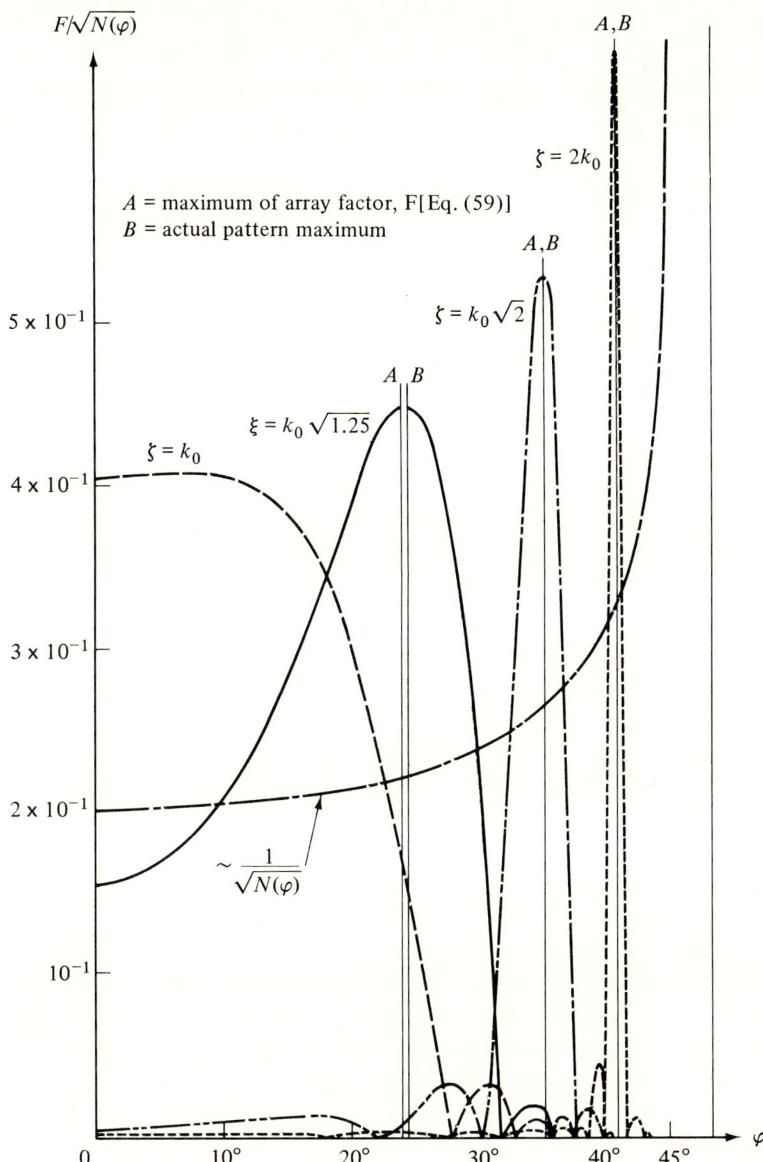

FIG. 7.3.11 Radiation pattern of a highly directive antenna.

the values $\zeta/k_0 = 1, \sqrt{1.25}, \sqrt{2}, 2$. A comparison between the actual magnetic field patterns calculated from Eq. (57) and the location of the pattern maxima predicted from Eq. (59) shows good agreement when the pattern is sufficiently narrow.

7.4 DIFFRACTION BY STRUCTURES EMBEDDED IN AN INFINITE, HOMOGENEOUS PLASMA

The calculation of the effect of a perturbing structure on radiation from a source distribution is complicated substantially if the surrounding medium is anisotropic. Simplifications arise under special conditions; in uniaxially anisotropic regions, one may distinguish two special classes of diffraction problems that are easily reduced via scaling to equivalent problems in an isotropic space.

7.4a Optic Axis Parallel to Axis of a Perfectly Conducting Cylindrical Obstacle

In the first class of scattering problems, the scattering object is a perfectly conducting cylinder of *arbitrary* cross section (e.g., a circular, elliptic, or parabolic cylinder; a wedge, half-plane, or strip; etc.) excited by an *arbitrary* source configuration, with the cylinder axis oriented parallel to the optic axis of the medium. This cylindrical scatterer forms the boundaries of a uniform waveguide region, and the resulting radiation or diffraction problem is then solved by the procedure described in Sec. 7.2. From Eqs. (7.2.24), the solutions may be expressed directly in terms of those for the isotropic case, whence one may construct results for a whole class of uniaxial diffraction problems by a simple scaling of the axial (z) coordinate in the conventional isotropic expressions. An analysis of these results in the limit of short and long wavelengths, valuable in providing an insight into some of the diffraction effects encountered in anisotropic regions, is left as an exercise for the reader. It may be recalled that the H-mode constituents of the field are *identical* with those in the isotropic case, and that the coordinate scaling need be applied only to the E-mode constituents.

7.4b Optic Axis Perpendicular to Axis of a Perfectly Conducting Cylindrical Obstacle

Formulation and reduction of the boundary-value problem

The second class of problems, involving cylindrical scatterers of arbitrary cross section oriented perpendicular to the optic axis in the medium, is also reducible to equivalent isotropic problems provided that the source distribution does not vary along the direction of the cylinder axis. The solutions may incorporate effects caused by a continuous change in direction of the optic (z) axis over the obstacle boundary when the latter is curved, whereas in the problems of Sec. 7.4a, the optic axis is always parallel to the boundary.

Let the cylinder axis be parallel to x and let the excitation be comprised of a line source of magnetic currents

$$\hat{\mathbf{M}}(\mathbf{r}, t) = V\delta(\hat{\mathbf{\rho}} - \hat{\mathbf{\rho}}')e^{-i\omega t}\mathbf{x}_0, \qquad \hat{\mathbf{\rho}} = (y, z). \tag{1}$$

Parenthetically, we note that a line source of electric currents excites only H

modes in the direction perpendicular to x; these are not affected by the presence of the uniaxially anisotropic medium. The medium is described again by the constitutive parameters

$$\epsilon = \epsilon_0(\mathbf{1}_t + \mathbf{z}_0\mathbf{z}_0\epsilon), \qquad \mu = \mu_0\mathbf{1}. \tag{2}$$

As in the absence of the scatterer, the non-vanishing field components are $H_x \equiv H, E_y, E_z$. On the obstacle surface A specified by the equation $A(y, z) = 0$, the impedance boundary condition

$$E_{\text{tan}} = ZH \tag{3a}$$

is assumed, where $Z(y, z)$ denotes the surface impedance and E_{tan} is the electric field component tangential to A. If n represents the direction normal to A and α is the angle between the tangent to A and the positive y axis [see Fig. 7.4.1(a)], then $E_{\text{tan}} = (E_z \sin\alpha + E_y \cos\alpha)$ and the relations $E_y =$

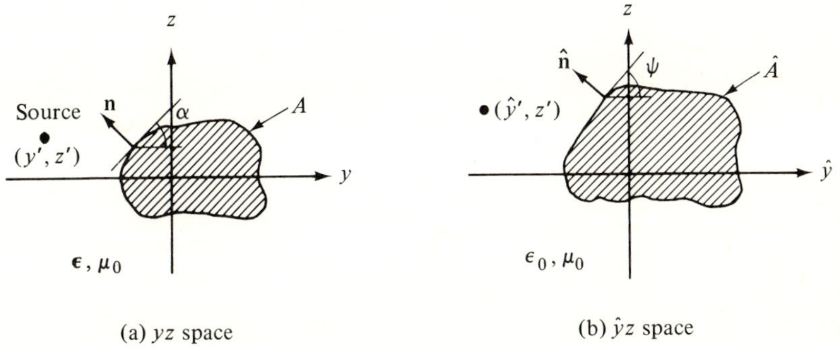

(a) *yz* space   (b) *ŷz* space

FIG. 7.4.1   Equivalent configurations.

$-(i\omega\epsilon_0)^{-1}\partial H/\partial z$, $E_z = (i\omega\epsilon_z)^{-1}\partial H/\partial y$, $\epsilon_z = \epsilon\epsilon_0$, may be used to rewrite Eq. (3a) as [see also Eqs. (7.3.38)]

$$\frac{\partial H}{\partial y} - \epsilon \cot\alpha \frac{\partial H}{\partial z} = i\omega\epsilon\epsilon_0 \frac{Z}{\sin\alpha} H \qquad \text{on } A. \tag{3b}$$

In an isotropic medium ($\epsilon = 1$), the expression on the left-hand side of the equation is equal to $-\csc\alpha\, \partial H/\partial n$, and the impedance boundary condition then takes the familiar form $\partial H/\partial n \propto ZH$ on A. This condition does not hold in the anisotropic medium, where the normal derivative of H is replaced by an "oblique derivative" whose direction may be inferred from considerations analogous to those following Eq. (7.3.53).

The boundary-value problem can now be stated as follows. As in Eqs. (7.3.36) and (7.3.38a), the magnetic field H is defined by the differential equation

$$\left(\frac{\partial^2}{\partial z^2} + \frac{1}{\epsilon}\frac{\partial^2}{\partial y^2} + k_0^2\right) H(\hat{\boldsymbol{\rho}}, \hat{\boldsymbol{\rho}}') = -i\omega\epsilon_0 V\delta(\hat{\boldsymbol{\rho}} - \hat{\boldsymbol{\rho}}'), \tag{4}$$

subject to an energy radiation condition at infinity, and to the boundary condition (3b) on the obstacle surface A.

To reduce this boundary-value problem to one in an isotropic medium, it is again suggestive to introduce a coordinate scaling transformation that yields a Laplacian instead of a differential operator as in Eq. (4). Evidently, either the z or y coordinates may be scaled. The former has already been utilized (Sec. 7.2) and involves an equivalent isotropic medium with wavenumber $k_0\sqrt{\epsilon}$. For the present application, the alternative procedure is employed via the definition

$$\hat{y} = \sqrt{\epsilon}\, y, \tag{5}$$

which furnishes the Laplacian in $\hat{y}z$ space, with the medium properties described by the wavenumber k_0. The boundary condition in Eq. (3b) becomes, accordingly,

$$\frac{\partial \hat{H}}{\partial \hat{y}} - \cot\psi\, \frac{\partial \hat{H}}{\partial z} = i\omega\epsilon_0 \frac{\sqrt{\epsilon}}{\sin\alpha} Z\hat{H} \qquad \text{on } \hat{A}, \tag{6a}$$

where $\hat{H}$ and $\hat{G}$ are, respectively, the equivalent isotropic field solution and Green's function

$$\hat{H}(\hat{y}, z; \hat{y}', z') = i\omega\epsilon_0\sqrt{\epsilon}\, V\hat{G}(\hat{y}, z; \hat{y}', z') = H\!\left(\frac{\hat{y}}{\sqrt{\epsilon}}, z; \frac{\hat{y}'}{\sqrt{\epsilon}}, z'\right), \tag{6b}$$

with

$$\cot\psi = \sqrt{\epsilon}\, \cot\alpha, \tag{6c}$$

and $\hat{A}$ defining the surface

$$\hat{A}(\hat{y}, z) = A\!\left(\frac{\hat{y}}{\sqrt{\epsilon}}, z\right) = 0. \tag{6d}$$

ψ is the angle between the tangent to the transformed surface $\hat{A}$ and the positive $\hat{y}$ axis [see Fig. 7.4.1(b)], and if $\hat{n}$ denotes the direction of the outward normal to $\hat{A}$, Eq. (6a) may be written as

$$\frac{\partial \hat{H}}{\partial \hat{n}} = -i\omega\epsilon_0 \hat{Z}\hat{H} \qquad \text{on } \hat{A}, \tag{7a}$$

where $\hat{Z}$ is the equivalent surface impedance

$$\hat{Z} = \sqrt{\epsilon}\, \frac{\sin\psi}{\sin\alpha}\, Z = \left(\frac{\epsilon}{\sin^2\alpha + \epsilon\cos^2\alpha}\right)^{1/2} Z. \tag{7b}$$

Thus, it is found that the boundary conditions on $\hat{H}$ in the equivalent isotropic problem involving the medium ϵ_0 and a scatterer $\hat{A}$ are of the conventional impedance type, so solutions of diffraction problems in the anisotropic medium can be constructed by applying to the customary isotropic Green's function the transformations (5), (6d), and (7b).[6]

In view of Eqs. (6d) and (7b), the auxiliary isotropic problem generally involves both a different obstacle shape† and surface impedance. The obstacle shape is stretched according to Eq. (5), whence a circular cylinder with radius a (i.e., $A = y^2 + z^2 - a^2 = 0$) transforms into a elliptic cylinder, $\hat{A} = (\hat{y}^2/\epsilon) + z^2 - a^2 = 0$; a wedge formed by the two half-planes $A_1 = y - z \cot \alpha_1 = 0$, $A_2 = y - z \cot \alpha_2 = 0$, $z \geq 0$, transforms into an equivalent wedge, $\hat{A}_1 = \hat{y} - z \cot \psi_1 = 0$, $\hat{A}_2 = \hat{y} - z \cot \psi_2 = 0$, $z \geq 0$; etc. A constant surface impedance Z in the yz space generally requires a non-constant surface impedance $\hat{Z}$ in the isotropic space since α is variable over a curved obstacle contour A. The only exceptions occur when

$$\text{(a)} \quad Z = 0 \quad \text{or} \quad \text{(b)} \quad \alpha = \text{piecewise constant on } A; \tag{8}$$

the former condition for a perfectly conducting surface in the yz space transforms unchanged into the $\hat{y}z$ space since $\hat{Z} = 0$, while the latter condition applies to an obstacle contour comprised of plane segments (e.g., an infinite plane, half-plane, wedge, or polygon). The validity of the preceding considerations is apparent only when ϵ is positive real since the geometrical equivalence between the problems sketched in Figs. 7.4.1(a) and (b) does not apply when $\sqrt{\epsilon}$ has an imaginary part. However, if a solution constructed for positive real ϵ can be continued analytically into $0 \leq \arg \epsilon \leq \pi$, the result will be valid throughout this extended range.

This procedure is now applied to various examples which illustrate some of the diffraction phenomena encountered in anisotropic media.

7.4c Half-Space Bounded by a Perfect Conductor

Consider a line source of magnetic current located at $(0, z')$ in the presence of an infinite, perfectly conducting plane inclined at an angle α with the $z = 0$ plane as shown in Fig. 7.4.2(a). This boundary is described by the equation $u = 0$, where the uv coordinate system is rotated through the angle α with respect to y and z. The auxiliary problem in the isotropic $\hat{y}z$ space, involving a perfectly conducting plane inclined at the angle $\psi = \cot^{-1}(\sqrt{\epsilon} \cot \alpha)$, with the source location unaltered, may evidently be solved by the method of images. The location $(\hat{y}', \hat{z}')$ of the image point is obtained by reflection,

$$\hat{y}' = z' \sin 2\psi, \qquad \hat{z}' = -z' \cos 2\psi, \tag{9}$$

and the resulting magnetic field in the half-space $u > 0$ is given by

$$H(y, z; 0, z') = H_i(y, z; 0, z') + H_i(y, z; \frac{\hat{y}'}{\sqrt{\epsilon}}, \hat{z}'), \tag{10}$$

where H_i denotes the infinite space solution in Eq. (7.3.38a).

---

†The transformation $\hat{z} = z/\sqrt{\epsilon}$ in Sec. 7.2 applied to a cylindrical obstacle oriented *parallel* to the z axis leaves the obstacle shape unchanged. This invariance is not maintained by scaling either z or y in the present configuration oriented perpendicular to the z axis.

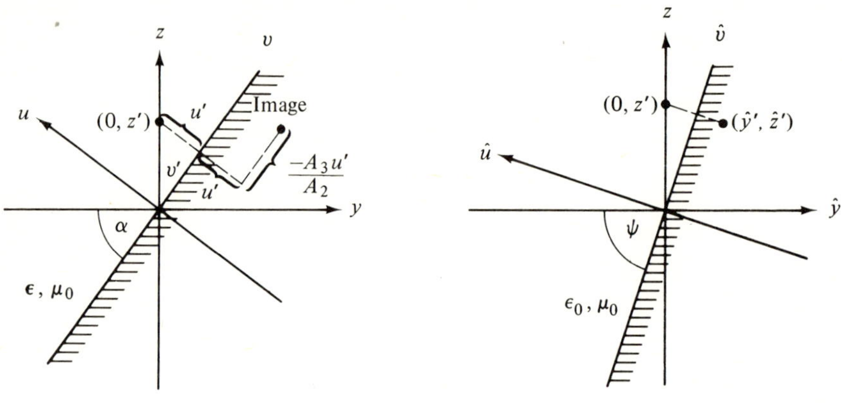

(a) Physical configuration   (b) Auxiliary problem

FIG. 7.4.2   Half-space bounded by a perfect conductor $(0 < \epsilon < 1)$.

It is of interest to determine the image location in the uv coordinate space which is natural to the description of the bounding plane $u = 0$. From Eqs. (5) and (6c), the image location $(\bar{y}', \bar{z}')$ in the yz coordinate system is

$$\bar{y}' = 2z' \frac{\sin \alpha \cos \alpha}{\epsilon A_2}, \qquad \bar{z}' = \hat{z}' = z' \frac{\sin^2 \alpha - \epsilon \cos^2 \alpha}{\epsilon A_2},$$

$$\epsilon A_2 = \sin^2 \alpha + \epsilon \cos^2 \alpha; \tag{11}$$

the corresponding point $(\bar{u}', \bar{v}')$ relative to the uv coordinates is obtained via the relations

$$\bar{u}' = \bar{z}' \cos \alpha - \bar{y}' \sin \alpha, \qquad \bar{v}' = \bar{y}' \cos \alpha + \bar{z}' \sin \alpha, \tag{12}$$

and yields

$$\bar{u}' = -u', \quad \bar{v}' = v' - \frac{A_3 u'}{A_2}, \quad \epsilon A_3 = 2(\epsilon - 1) \sin \alpha \cos \alpha. \tag{13}$$

Thus, the perturbation fields due to the presence of the perfectly reflecting plane in Fig. 7.4.2(a) may be rigorously accounted for in terms of an image source that is, however, displaced from the mirror-image location $(-u', v')$ by the distance $-A_3 u'/A_2$ measured parallel to plane.[6] The image location coincides with the conventional one only when $\epsilon = 1$ (isotropic case) or when $\alpha = 0, \pi/2$ (optic axis parallel or perpendicular to boundary); in the latter instance, the analysis in Sec. 7.4a shows that the mirror-image construction holds for arbitrary source distributions, as in isotropic regions. Although the solution (10) was obtained on the assumption $\epsilon > 0$, it may be continued analytically into the range $0 \leq \arg \epsilon \leq \pi$ and holds in particular when $\epsilon < 0$.

The image displacement has interesting consequences which arise also in media with more general anisotropic properties: it implies the occurrence of non-specular reflections at a plane boundary, as schematized in Fig. 7.4.3 for

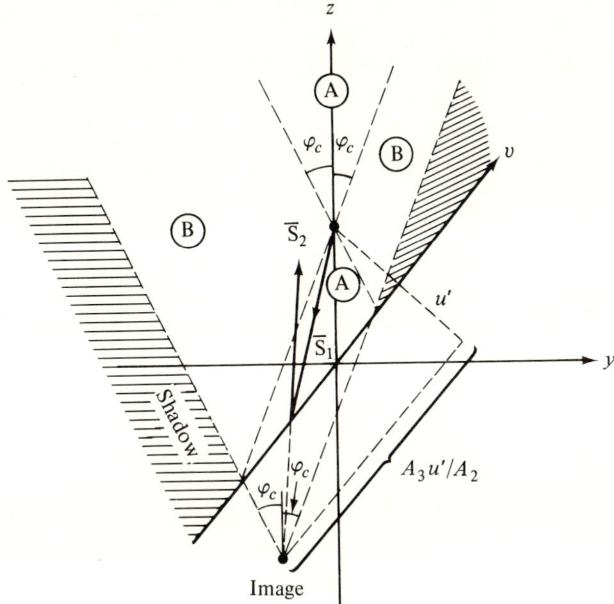

FIG. 7.4.3 Incident and reflected rays when $\epsilon < 0$.

the case $\epsilon < 0$, with $A_2 > 0$. Since $\epsilon < 0$, the propagating plane-wave spectrum issuing from the source and its image is restricted to ray angles φ which satisfy the condition $|\tan \varphi| < \tan \varphi_c = 1/\sqrt{|\epsilon|}$, where φ is measured from the positive z axis (see Sec. 7.3e). Consequently, the direct radiation $\bar{\mathbf{S}}_1$ from the source illuminates only region A in Fig. 7.4.3, while propagating rays $\bar{\mathbf{S}}_2$ from the image source (i.e., the rays reflected at the boundary) penetrate both regions A and B. The shaded area constitutes the shadow zone in which the fields decay, and one observes that the presence of a reflecting boundary enlarges the illuminated region. The non-specular character of the reflection is evident from the construction and, for the case considered, both the incident and reflected rays lie on the same side of the normal to the surface—a phenomenon not encountered in isotropic media.

The directions of the incident and reflected rays may also be inferred from the refractive index diagrams in Fig. 7.1.1, an aspect that is explored further in Sec. 7.5b.

7.4d Half-space Bounded By a Reactive Surface

We repeat the analysis in the preceding section for the case where the plane boundary at $u = 0$ is assumed to have a constant surface impedance $Z \neq 0$.[6] The transformed isotropic problem then involves the plane surface in Fig.

7.4.2(b), with a constant surface impedance $\hat{Z}$ inferred from Eq. (7b). The solution for $\hat{H}$ can no longer be obtained in closed form by the method of images but involves the integral representations in Eqs. (5.7.17) or (5.7.21). For distant observation points, the integrals may be evaluated asymptotically and yield the result in Eq. (5.7.19) which contains three contributions: the infinite space field in the absence of the boundary; the reflected field which appears to arise from the mirror-image point, with an angularly dependent amplitude factor equal to the plane-wave reflection coefficient of the surface; and a surface-wave field that contributes only when the surface impedance is inductive (Im $\hat{Z} < 0$) and when the observation point lies sufficiently close to the boundary. Upon transforming these field solutions into the anisotropic y, z space (with $\epsilon > 0$), one obtains a far field comprising the above-mentioned three constituents. The first two may be interpreted exactly as in Fig. 7.4.3, or rather its analogue for $\epsilon > 0$, and yield the incident and reflected rays, with the latter weighted by the transformed reflection coefficient.

The surface-wave contribution merits further attention. If $\hat{u}$ and $\hat{v}$ denote the coordinates parallel and perpendicular, respectively, to the surface in the equivalent isotropic problem [see Fig. 7.4.2(b)], then the magnetic field $\hat{H}_s$ in the surface wave has the behavior [see Eq. (5.7.21)]

$$\hat{H}_s = \exp\left(ik_y\sqrt{1 - \hat{Z}^2}\,\hat{u} - ik_0\hat{Z}\hat{v}\right), \tag{14}$$

where Im $\hat{Z} < 0$, Re $\hat{Z} \geq 0$, and $\hat{v} \geq 0$. This solution may be transformed into the anisotropic space by noting that $\hat{u} = \hat{y}\cos\psi + z\sin\psi$, $\hat{v} = z\cos\psi - \hat{y}\sin\psi$, and applying Eq. (5). The resulting expression is then written in terms of the u, v coordinates via $y = v\cos\alpha - u\sin\alpha$, $z = v\sin\alpha + u\cos\alpha$, and yields the following variation of the surface-wave field H_s:

$$H_s = e^{iqu + i\beta v}, \tag{15}$$

where q and β are the wavenumbers along the u and v directions, respectively,

$$q = -\beta\frac{A_3}{2A_2} - k_0\frac{Z'}{A_2}, \quad \beta = k_0\sqrt{\epsilon(A_2 - Z'^2)}, \quad Z' = \frac{Z}{\sqrt{\mu/\epsilon_0}}, \tag{15a}$$

with ϵ assumed positive, and Z and A_2 defined in Eqs. (7b) and (11). This relation between q and β is a direct consequence of the dispersion equation (7.1.5) when expressed in terms of the rotated u, v coordinates [see Eq. (7.5.5)]. If the surface is purely reactive, then $Z' = -i|Z'|$, and the field in both Eqs. (14) and (15) decays away from the boundary. However, in the isotropic case, the wavenumber in the perpendicular $\hat{u}$ direction is purely imaginary, whereas it is complex in the anisotropic problem. The constant-phase planes in the anisotropic medium are therefore inclined with respect to the v direction along the boundary.

If the field H_s does indeed define a surface wave, it must carry energy in the v direction parallel to the surface. Since

$$\frac{\partial}{\partial y} = \cos\alpha\frac{\partial}{\partial v} - \sin\alpha\frac{\partial}{\partial u}, \qquad \frac{\partial}{\partial z} = \sin\alpha\frac{\partial}{\partial v} + \cos\alpha\frac{\partial}{\partial u}, \tag{16}$$

and $E_u = E_z \cos \alpha - E_y \sin \alpha$, $E_v = E_z \sin \alpha + E_y \cos \alpha$, one obtains from the electric-field expressions in Eq. (7.3.38),

$$E_u = \frac{1}{i\omega\epsilon_0} \left(A_1 \frac{\partial}{\partial v} + \frac{A_3}{2} \frac{\partial}{\partial u} \right) H, \qquad E_v = -\frac{1}{i\omega\epsilon_0} \left(\frac{A_3}{2} \frac{\partial}{\partial v} + A_2 \frac{\partial}{\partial u} \right) H, \qquad (17)$$

where A_2 and A_3 are defined in Eqs. (11) and (13), respectively, and

$$A_1 = \sin^2 \alpha + \frac{1}{\epsilon} \cos^2 \alpha \equiv \frac{1}{\epsilon} N^2(\alpha). \qquad (17a)$$

A calculation of the time-averaged Poynting vector for $Z' = -i|Z'|$ then shows that

$$\bar{S}_u = -\mathrm{Re}\,(E_{vs} H_s^*) = 0, \qquad (18a)$$

$$\bar{S}_v = \mathrm{Re}\,(E_{us} H_s^*) = \frac{\beta}{\omega\epsilon A_2} |H_s|^2, \qquad (18b)$$

thereby confirming the surface-wave character of the field in Eq. (15).

7.4e   Wedge and Half-plane

We consider briefly the problem of diffraction by a perfectly conducting wedge, the isotropic version of which, treated in detail in Chapter 6, may be transformed according to Eq. (6) to yield a result in the anisotropic space.[6,10] Any of the various representations listed in Sec. 6.5 may be employed in appropriate parameter ranges. For source or observation points located near the edge, the series in Eq. (6.5.13) is convenient and rapidly convergent, while the far field excited by a distant source is calculated best from the asymptotic formulas in Eqs. (6.5.19).

The far-field approximation in the isotropic medium comprises three parts, each of which possesses a simple physical interpretation in terms of geometric-optical rays: (a) the infinite space field in the absence of the obstacle; (b) the reflected field comprising all rays that are reflected specularly from the sides of the wedge (Fig. 6.5.1); and (c) the diffracted field which takes the form of an inhomogeneous cylindrical wave (radially diverging rays) emanating from the wedge apex (see Fig. 6.3.1). This simple description fails in the vicinity of the geometrical shadow and reflected-wave boundaries and must be augmented by a transition solution, as discussed in Chapter 6. No difficulty arises in transforming these solutions into the anisotropic medium when $\epsilon > 0$; the wedge angle is changed thereby according to the remarks made in the paragraph following Eq. (7), and the source location is also affected when $y' \neq 0$. The resulting asymptotic field comprises the same contributions as above, and has an analogous physical interpretation. The primary field (a) is as in Eq. (7.3.37) and exists only in the illuminated region; the shadow boundary is defined by the ray that just grazes the apex, as illustrated for the half-plane in Fig. 7.4.4. As in Fig. 7.4.2, the reflected wave field (b) may be viewed as arising from a displaced image source, one for each of the singly and multiply re-

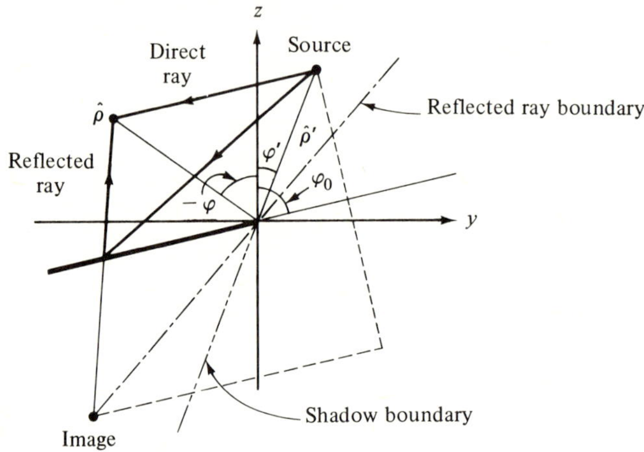

FIG. 7.4.4   Domains of existence of direct and reflected rays.

flected rays, and its domain of existence is bounded by the ray from the image that just grazes the edge. The diffracted field (c) appears to emanate from a line source at the edge and reaches all observation points; it is excited by the incident ray that grazes the edge, its angular intensity being dependent upon the transformed diffraction coefficient as well as on the $1/\sqrt{N(\varphi)}$ term arising from the asymptotic approximation to Eq. (7.3.37) [see Eq. (19)]. Most important in this quasi-optic interpretation of the scattered field is the recognition that the *rays*, and not the wavenormals, enter into the identification of the various field constituents—a property that continues to apply in more general types of anisotropic media. These simple physical ray concepts, in conjunction with the image construction, or more generally the refractive index curves (see Sec. 1.6), may be employed to derive detailed information about the nature of the high-frequency field.

It is also of interest to take note of the edge condition in the anisotropic case. In the isotropic problem, the electric field at a distance $\hat{\rho}$ from the edge diverges like $\hat{\rho}^{(\pi/\delta)-1}$ as $\hat{\rho} \to 0$, where δ is the exterior angle subtended by the wedge. Through the transformation (5), $\hat{\rho} \to \rho N(\varphi)$, $\rho = (y^2 + z^2)^{1/2}$, and

$$\delta = \cot^{-1}(\sqrt{\epsilon}\,\cot\varphi_2) - \cot^{-1}(\sqrt{\epsilon}\,\cot\varphi_1), \qquad \epsilon > 0,$$

with the angles $\varphi_{1,2}$ specifying the location of the wedge faces in the y, z space. The resulting behavior of the electric fields near the edge, $O[\rho^{(\pi/\delta)-1}]$, may be more violent than that observed when the identical structure is embedded in an isotropic medium. For example, if $\varphi_2 = 3\pi/2$ and $-\pi/2 < \varphi_1 < 0$, then $\delta = (3\pi/2) - \cot^{-1}(\sqrt{\epsilon}\,\cot\varphi_1)$. For $0 < \epsilon < 1$, $\cot^{-1}(\sqrt{\epsilon}\,\cot\varphi_1) < \varphi_1$, so $\delta > \delta_0 = 3\pi/2 + |\varphi_1|$, with δ_0 denoting the exterior wedge angle. In an isotropic medium the electric field near the edge behaves like $\rho^{(\pi/\delta_0)-1}$ and therefore grows less rapidly than $\rho^{(\pi/\delta)-1}$.

When ϵ is complex or negative real, the solutions are obtained from those for positive ϵ by analytic continuation. Since the scaling procedure [Eq. (6c)] then transforms real into complex angles, questions of interpretation arise which have been clarified for the special case of a perfectly conducting half plane.[10] It is found that for $\epsilon \gtrless 0$, the incident and reflected fields in their domains of existence are the same as for an infinitely extended plane, with the reflected field derivable from an image source as in Fig. 7.4.3. The boundary of the domain of existence of the incident field (shadow boundary) is still given by the ray that grazes the edge regardless of whether the field along this ray is propagating or evanescent; a similar statement applies to the boundary of the domain of existence of the reflected rays (Fig. 7.4.4). The diffracted field H_d obtained by application of coordinate scaling to the isotropic-medium formula in Eq. (6.5.19a) yields the asymptotic result:

$$H_d \sim [VQ(\hat{\boldsymbol{\rho}}')][f(\varphi, \varphi', \varphi_0)Q(\hat{\boldsymbol{\rho}})], \qquad \hat{\boldsymbol{\rho}} = (\hat{\rho}, \varphi), \quad \hat{\boldsymbol{\rho}}' = (\hat{\rho}', \varphi'), \qquad (19)$$

where V is the strength of the magnetic line current [Eq. (1)], while

$$Q(\hat{\boldsymbol{\rho}}') = -\frac{\omega\epsilon_0\sqrt{\epsilon}}{4}\sqrt{\frac{2}{\pi k_0 \hat{\rho}' N(\varphi')}} e^{ik_0\hat{\rho}'N(\varphi')-i\pi/4} \qquad (19a)$$

is the magnetic field at the edge of the half-plane, generated by a line current of unit strength at $\hat{\boldsymbol{\rho}}'$ [asymptotic form of Eq. (7.3.38)]; $Q(\hat{\boldsymbol{\rho}})$ is the field at the observation point $\hat{\boldsymbol{\rho}}$ generated by a virtual line current of unit strength at the edge, and $f(\varphi, \varphi', \varphi_0)$ is the pattern function associated with the edge-diffracted field:

$$f(\varphi, \varphi', \varphi_0) = \frac{1}{2}\frac{f_1(\varphi, \varphi_0)f_1(\varphi', \varphi_0)}{f_2(\varphi, \varphi_0) + f_2(\varphi', \varphi_0)}, \qquad (19b)$$

where

$$f_1(\varphi, \varphi_0) = \left[1 - \frac{\epsilon \sin \varphi_0 \sin \varphi + \cos \varphi_0 \cos \varphi}{N(\varphi_0)N(\varphi)}\right]^{1/2},$$

$$f_2(\varphi, \varphi_0) = \frac{\epsilon \sin \varphi_0 \sin \varphi + \cos \varphi_0 \cos \varphi}{N(\varphi_0)N(\varphi)}, \qquad (19c)$$

and $N(\varphi) = \cos^2 \varphi + \epsilon \sin^2 \varphi$. In these formulas, which remain valid for positive or negative ϵ provided that $\text{Im } N \geq 0$, the angles φ, φ', and φ_0 measure, respectively, the angular locations of the observation point, source point, and half-plane with respect to the optic (z) axis (Fig. 7.4.4). Near the geometrical shadow and reflected ray boundaries where $f_2(\varphi, \varphi_0) \to -f_2(\varphi', \varphi_0)$, one must employ a modified description in terms of Fresnel integrals [see Eq. (6.5.20)]. When $\epsilon < 0$, the formula fails along the shadow boundaries $N(\varphi_c') = 0$ and $N(\varphi_c) = 0$ associated with the anisotropy in the medium.

The factors inside the brackets in Eq. (19) have a direct physical interpretation; $fQ(\hat{\boldsymbol{\rho}})$ represents the diffracted field due to an incident field having unit amplitude at the edge while $VQ(\hat{\boldsymbol{\rho}}')$ accounts for deviations of the incident field strength from unity. It follows from these considerations that an incident *plane*

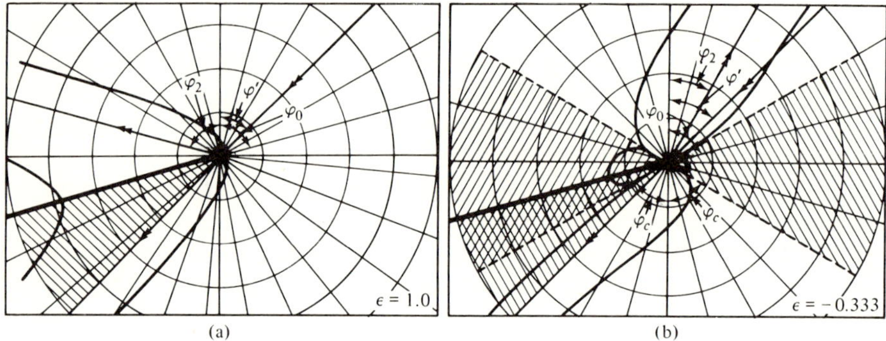

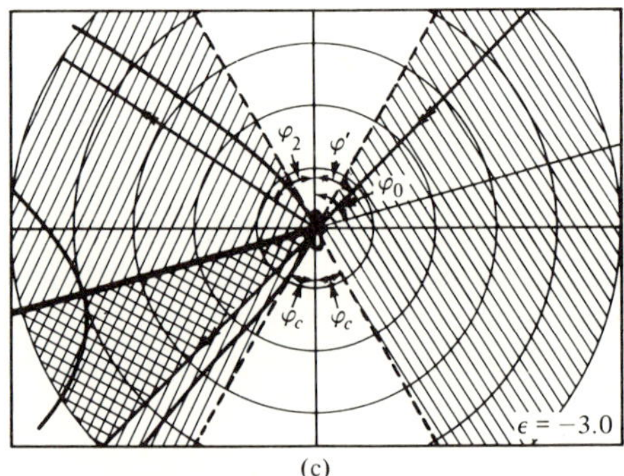

FIG. 7.4.5 Plot of the pattern function $f(\varphi, \varphi', \varphi_0)$ for $\varphi' = 45°$, $\varphi_0 = 75°$, and various ϵ.

wave of the form $\exp[-i(\eta_i y + \kappa_i z)]$ carrying *energy* along φ' gives rise to an edge diffracted field $fQ(\hat{\rho})$. The influence of an anisotropic environment on the pattern function may be assessed from Fig. 7.4.5, which contains a plot of f for (a) $\epsilon = 1$, (b) $\epsilon = -\frac{1}{3}$, (c) $\epsilon = -3$, with the incidence and half-plane angles remaining fixed at $\varphi' = 45°$ and $\varphi_0 = 75°$, respectively. φ_2 locates the boundary of the reflected waves. Case (a) corresponds to an isotropic medium; since $\epsilon < 0$ in cases (b) and (c), the medium supports propagating fields only in the restricted angular domain $|\tan \varphi| < \tan \varphi_c = |\epsilon|^{-1/2}$, whence the incident field is propagating in case (b) but evanescent in case (c). Two shadow zones, shown shaded in Fig. 7.4.5, must be distinguished: the first is due to the geometrical blocking effect of the half-plane, whereas the second, the double-wedge-shaped region $|\tan \varphi| > |\epsilon|^{-1/2}$, arises from anisotropy in the medium. Inside the latter region, f describes the angular variation of the evanescent field. Evidently, the presence of anisotropy may introduce substantial distortion of the isotropic medium pattern which is symmetrical about the half-plane. It

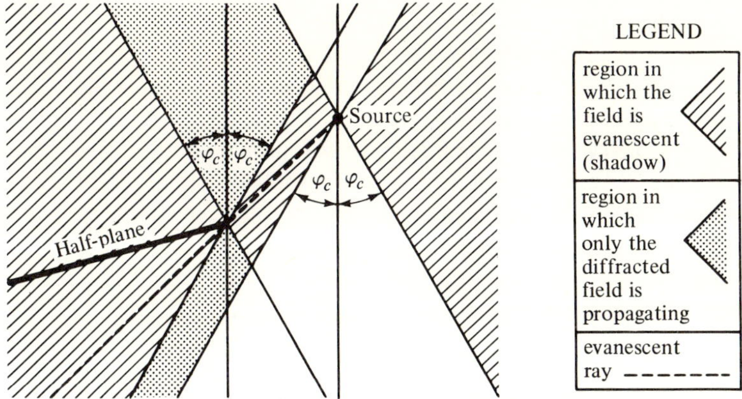

FIG. 7.4.6   Various wave domains when half-plane is in the shadow region of the source in a medium with $\epsilon = \epsilon_0(\mathbf{1}_t + \mathbf{z}_0\mathbf{z}_0\epsilon), \epsilon < 0$.

is of interest to observe that although the incident field may be evanescent when $\epsilon < 0$ [$N(\varphi')$ imaginary], the diffraction field due to the edge couples to propagating waves in the region $|\tan \varphi| < |\epsilon|^{-1/2}$ wherein $N(\varphi)$ is real. Thus, diffraction may extract real energy from an incident evanescent field, by a mechanism somewhat reminiscent of that encountered in the operation of a microwave cutoff attenuator or the tunnel effect in quantum mechanics. Figure 7.4.6 contains a respresentative sketch, wherein the regions illuminated by the source and by diffraction are delineated.

7.5 RADIATION FROM A HOMOGENEOUS PLASMA HALF-SPACE

To observe radiation phenomena associated with an interface bounding a uniaxially anisotropic region, we consider a source embedded in a plasma half-space. If the interface is perpendicular to the z axis, the fields excited by arbitrary source distributions may be constructed via the simple procedure in Sec. 7.2; the equivalent modal network problem is directly analogous to that for the isotropic half-space, shown in Figs. 5.5.3 or 5.5.6. The modal description is in terms of transmission along the z direction, and because of the reflection symmetry with respect to the optic axis, longitudinal wavenumbers $\pm\kappa$ exist for the same transverse wavenumber k_t or $-k_t$ [see Eq. (7.2.8)]. Interface effects are similar to those in isotropic regions, and such peculiarly anisotropic phenomena as non-specular reflections do not occur. The latter are present when the optic axis is inclined with respect to the interface; hence it is instructive to examine a problem of this kind since it is not amenable to a simple network treatment.

Even for the two-dimensional case corresponding to excitation by a magnetic line source as in Sec. 7.4, when the optic axis is not perpendicular to the interface, coordinate scaling does not facilitate the solution since the originally isotropic medium becomes anisotropic in the scaled coordinate frame. Thus, it is necessary to resort to a representation in terms of plane-wave modes propagating along the direction perpendicular to the interface; owing to the obliquity of the optic axis in the anisotropic half-spce, these modes do not possess reflection symmetry. After formulation of this problem in Sec. 7.5a, attention is given in Sec. 7.5b to the reflection and transmission properties of the plane-wave modes, and to the interpretation of nonspecular reflection, etc., through use of the wavenumber surfaces for the two media. The solution for the line-source field, obtained by modal synthesis in Sec. 7.5c, is simplified subsequently for distant observation points. In the plasma half-space, the asymptotic evaluation of the radiation integral by the steepest-descent method yields saddle-point and branch-point contributions that can be identified as geometric-optical and diffracted (lateral-wave) fields, respectively (Sec. 7.5d). The geometric-optical fields are similar to those obtained in Sec. 7.4c for a perfectly conducting boundary. As in the isotropic problem discussed in Sec. 5.5, the lateral-wave fields are excited by a ray incident at the critical angle. However, their behavior is modified substantially by anisotropy, especially when the uniaxial wavenumber surface is open branched; in this event, lateral rays are refracted backward with respect to the incident critical ray. The simple asymptotic calculation providing a ray-optical interpretation must be modified for observation points in the transition region surrounding the angle of total reflection. This parameter regime is characterized by proximity of a saddle point and branch point in the representation integral and thus requires use of the uniform asymptotic procedure discussed in Sec. 4.4c.

Fields in the vacuum half-space are evaluated asymptotically in Sec. 7.5e, and are found to propagate along refracted rays. When the uniaxial wavenumber surface is open branched, the family of refracted rays is found to form a caustic indicative of a kind of field enhancement or focusing not exhibited in an isotropic medium. For observation points near the caustic, modified asymptotic procedures (Secs. 4.5a and 4.5b) must be employed to account for the proximity of two or three saddle points in the refracted field integral. The chapter concludes in Sec. 7.5f with a brief treatment of radiation from a transverse electric dipole source.

7.5a Formulation of the Problem (Line-source Excitation)

The physical configuration is shown in Fig. 7.5.1, where the uniaxially anisotropic half-space occupies the region $u < 0$, and the region $u > 0$ is filled with an isotropic dielectric of permittivity ϵ_0. The uv coordinate system, natural to the description of this geometrical configuration, is inclined at an angle α with respect to the yz axes along which the medium anisotropy is described most directly. On suppressing the $\exp(-i\omega t)$ factor and normalizing the source strength, the single component of magnetic field satisfies in the anisotropic

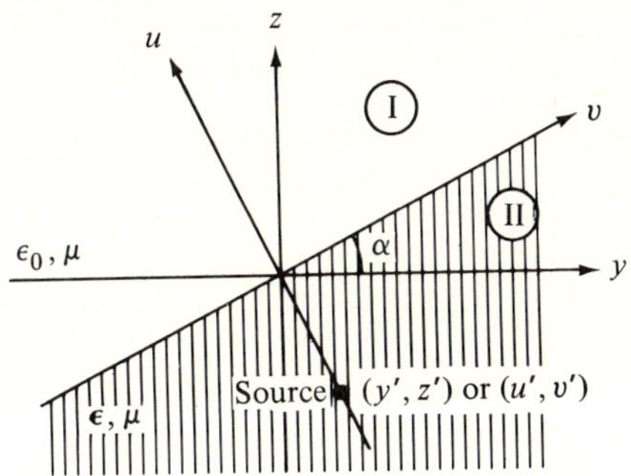

FIG. 7.5.1   Physical configuration.

region II the inhomogeneous wave equation (7.3.36) which, when transformed into the uv frame, reads (with ϵ_0 replaced by ϵ_y for generality)[11]

$$\left(A_1 \frac{\partial^2}{\partial v^2} + A_2 \frac{\partial^2}{\partial u^2} + A_3 \frac{\partial^2}{\partial u\,\partial v} + k_y^2\right) H_2 = -i\omega\epsilon_y\,\delta(u - u')\,\delta(v - v'), \quad u < 0,$$

$$(1)$$

and the electric-field components E_{u2} and E_{v2} are then determined from Eq. (7.4.17). The constants $A_{1,2,3}$ were defined in Eqs. (7.4.11), (7.4.13), and (7.4.17a). In the isotropic region I, characterized by the scalar dielectric constant ϵ_0 and permeability μ, the magnetic field H_1 satisfies the homogeneous equation

$$\left(\frac{\partial^2}{\partial u^2} + \frac{\partial^2}{\partial v^2} + k_0^2\right) H_1 = 0, \qquad u > 0,$$

$$(2)$$

and the corresponding electric fields are

$$E_{u1} = \frac{1}{i\omega\epsilon_0} \frac{\partial H_1}{\partial v}, \qquad E_{v1} = -\frac{1}{i\omega\epsilon_0} \frac{\partial H_1}{\partial u}.$$

$$(2a)$$

The solutions in regions I and II are connected by the continuity conditions $H_1 = H_2$, $E_{v1} = E_{v2}$ at $u = 0$, and imposition of a radiation condition at infinity completes the unique specification of H.

7.5b   Reflection and Transmission of Plane Waves, and the Radiation Condition

The radiation problem specified in the preceding section is solved most naturally by viewing the configuration as a waveguide whose axis is parallel to u, with the mode functions taken as plane waves with cross-sectional dependence $\exp(i\beta v)$, $-\infty < \beta < \infty$. Thus, we seek a representation for H_2 in terms of a continuous spectrum of plane waves of the form

$$\exp(i\mathbf{k} \cdot \mathbf{R}) = \exp[iq(\beta)u + i\beta v].$$

$$(3)$$

Because of their importance in the analysis, these wave solution are subjected to further study. It will be found that the longitudinal wavenumbers $q_1(\beta)$ and $q_2(\beta)$ corresponding to a plane wave that carries energy along the $+u$ and $-u$ directions, respectively, are not connected by the simple relation $q_1 = -q_2$ which applies in regions with reflection symmetry. Thus, as noted in Sec. 7.1, it is no longer possible to schematize propagation phenomena by a bilateral transmission line with a given propagation constant q; instead, unidirectional transmission lines with different properties are required, one for each forward and backward propagating wave. Because of this complication, which is generally characteristic of wave propagation in anisotropic media, the traveling-, and not the standing-, wave description is the more fundamental; in consequence, a network representation in terms of transmission lines loses much of the appeal it possesses in reflection symmetric problems.

The plane wave in Eq. (3) satisfies the homogeneous wave equation (1), and hence the wavenumbers q and β are connected by the dispersion equation

$$A_2 q^2 + A_1 \beta^2 + A_3 q\beta = k_y^2 = k_0^2 \epsilon_y', \qquad \epsilon_y' = \frac{\epsilon_y}{\epsilon_0}, \tag{4}$$

which has the two solutions

$$q_{1,2}(\beta) = \frac{-\beta A_3 \pm 2\sqrt{k_y^2 A_2 - (\beta^2/\epsilon)}}{2A_2}, \qquad \epsilon = \frac{\epsilon_z}{\epsilon_y}. \tag{5}$$

Equation (4) contains the same information as Eq. (7.1.5), but the latter has a simpler form since its wavenumbers $k_t = \eta$ and κ are expressed with respect to the principal axes of the wavenumber surface. A plot of the real solutions of Eq. (5) therefore reproduces the curves in Fig. 7.1.1, with q and β measured in the rotated coordinate system shown in Fig. 7.5.2. If the square root in Eq.

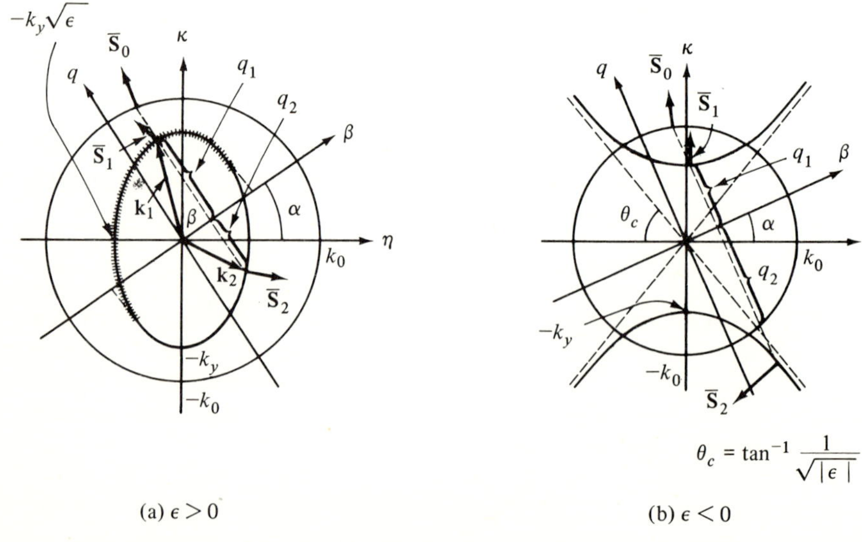

FIG. 7.5.2 Wavenumber surfaces.

(5) is defined to be positive when real, one observes from Fig. 7.5.2(a) that the solution designated as q_1 defines the shaded portion of the ellipse while q_2 specifies the unshaded portion. Since the outward normals on the q_1 branch have a component parallel to the $+q$ axis while those on the q_2 branch have a component parallel to the $-q$ axis, plane waves with wave vector $\mathbf{k}_1 = q_1\mathbf{u}_0 + \beta\mathbf{v}_0$ carry energy along the $+u$ direction, and those with wave vector $\mathbf{k}_2 = q_2\mathbf{u}_0 + \beta\mathbf{v}_0$ carry energy along the $-u$ direction. Evidently, $q_1 \neq -q_2$, and q_1 may be negative or q_2 positive—another example of a "backward wave" with respect to the u axis. Analogous considerations apply when $\epsilon < 0$.

In the presence of an interface parallel to the v axis, the continuity of the tangential field components requires that the incident, reflected, and transmitted waves all have the same dependence $\exp(i\beta v)$. If the wave is incident from the anisotropic half-space $u < 0$, then

$$\bar{H}_2 = g(\beta)e^{i\beta v}[e^{iq_1u} - \Gamma(\beta)e^{iq_2u}], \qquad u < 0, \tag{6a}$$

and

$$\bar{H}_1 = f(\beta)e^{i\beta v}e^{iq_0u}, \qquad q_0 = \sqrt{k_0^2 - \beta^2}, \quad u > 0, \tag{6b}$$

where $g(\beta)$, $g\Gamma(\beta)$, and $f(\beta)$ represent the amplitudes of the incident, reflected, and transmitted (refracted) waves, respectively; q_0, which is positive when real, is the conventional longitudinal propagation constant in an isotropic waveguide. Upon imposing the boundary conditions $H_1 = H_2$ and $E_{v1} = E_{v2}$ at $u = 0$, one obtains the expressions for the reflection and transmission coefficients,

$$\Gamma = 1 - T = \frac{2\epsilon_y'q_0 + A_2(q_2 - q_1)}{2\epsilon_y'q_0 - A_2(q_2 - q_1)}, \qquad T = \frac{f}{g}. \tag{7}$$

In view of the preceding remarks, the plane-wave solutions in Eqs. (6) satisfy the energy radiation condition when q_1, q_2, or q_0 are real. If these quantities are complex, the required decay of the field solutions is assured if

$$\text{Im } q_1 > 0, \quad \text{Im } q_2 < 0, \quad \text{Im } q_0 > 0, \tag{8}$$

thereby completing the specification of the multivalued propagation constants q. If $\epsilon = 1$, the resulting expression for Γ agrees with that listed in Eq. (5.5.3d) for the isotropic half-space problem.

The ray directions, along which energy is transported in the plane waves characterized by the wave vectors $\mathbf{k}_1$, $\mathbf{k}_2$, and $\mathbf{k}_0 = q_0\mathbf{u}_0 + \beta\mathbf{v}_0$, may be determined either analytically or from an inspection of the wavenumber curves in Fig. 7.5.2. The values of q_1 and q_2 belonging to a specified value of β are obtained directly from the graphs, and the corresponding incident- and reflected-ray directions $\bar{\mathbf{S}}_1$ and $\bar{\mathbf{S}}_2$ in the uv coordinate space are determined by constructing the normals to the curves at the points (q_1, β) and (q_2, β), respectively; it is recalled from Eq. (1.7.53a) that the angle between $\mathbf{k}$ and $\bar{\mathbf{S}}$ is less than 90°. Since the dispersion curve for the vacuum region II is the circle described by the equation $q_0^2 + \beta^2 = k_0^2$, the transmitted ray $\bar{\mathbf{S}}_0$ follows the direction of the radius vector at the point on the circle having an abscissa equal to β. This "ray-tracing" procedure applies also when the medium has more general anisotropic properties; it is particularly useful in inhomogeneous regions that may be approximated by piece-wise constant layers having different refractive index diagrams (see Fig. 1.6.7).

An analytical expression relating the ray angles φ_1, φ_2, and φ_0, which are measured from the positive u axis toward the positive v axis and specify the directions on the incident, reflected, and refracted rays, respectively, is obtained from the analogue of Eq. (7.1.7):

$$\tan \varphi = -\frac{dq}{d\beta}. \tag{9a}$$

Evaluation of the derivative from Eq. (5) and use of $\beta = k_0 \sin \varphi_0$ yields

$$\tan \varphi_{1,2} = \frac{A_3}{2A_2} \pm \frac{\sin \varphi_0}{\epsilon A_2 \sqrt{\epsilon_y' A_2 - (\sin^2 \varphi_0)/\epsilon}}, \tag{9b}$$

the generalization of the Snell reflection and refraction law for the rays at a uniaxially anisotropic interface. Unless $A_3 = 0$ (i.e., $\epsilon = 1$, or $\alpha = 0, \pi/2$), $\varphi_2 \neq \pi - \varphi_1$ so reflection at the interface is generally non-specular. This phenomenon has already been encountered in the problem of reflection from a perfectly conducting plane, analyzed by a different method in Sec. 7.4c, wherein the relation between the incident- and reflected-ray directions has been schematized by an image construction. The present discussion also applies to an anisotropic half-space bounded by a perfectly conducting plane if the current reflection coefficient $-\Gamma$ is set equal to unity, thereby assuring that $E_v = 0$ at $u = 0$.

When $\epsilon < 0$, the algebraic signs of φ_1 and φ_0 may be opposite, and the associated phenomenon is one of backward refraction of the *rays* with respect to the direction perpendicular to the interface (i.e., the refracted and incident rays both lie on the same side of the surface normal). This does not apply to the wavevectors, since $\mathbf{k}_1 = q_1 \mathbf{u}_0 + \beta \mathbf{v}_0$ and $\mathbf{k}_0 = q_0 \mathbf{u}_0 + \beta \mathbf{v}_0$ both have the same tangential component $\beta \mathbf{v}_0$. Moreover, if $\sqrt{\epsilon_y' A_2 - (1/\epsilon)}$ is real, total reflection is possible in the plasma half-space. The angles of total reflection are obtained by putting $\varphi_0 = \pm \pi/2$ in the expression for φ_1 in Eq. (9b).

It is relevant to ascertain possible pole singularities of Γ which, if they exist, may give rise to the appearance of surface-wave or leaky-wave contributions in the radiated field. While the discussion so far has been concerned with an anisotropic medium specified by the permittivity parameters ϵ_y and ϵ_z, and an exterior isotropic medium characterized by a different permittivity ϵ_0, it will now be assumed that the anisotropic medium is a plasma and the exterior region is vacuum. In this instance, $\epsilon_y' = 1$. The denominator of the expression for Γ in Eq. (7) then has a simple zero β_p when

$$\sqrt{k_0^2 - \beta_p^2} = -\sqrt{k_0^2 A_2 - (\beta_p^2/\epsilon)}. \tag{10}$$

Squaring and solving for β_p, one finds $\beta_p = \pm k_0 \sin \alpha$, so $\sqrt{k_0^2 - \beta_p^2}$ is positive real. Since the square root on the right-hand side is also defined to be positive when real (on the top sheet of the pertinent Riemann surface), β_p cannot be a solution of Eq. (10) but annuls instead the numerator in Eq. (7). Thus, β_p is a zero of the reflection coefficient and poles do not occur. One observes from Fig. 7.5.2 with $k_y = k_0$ that a wave with wavenumber β_p propagates along the direction of the optic (z) axis in the medium, and experiences no reflection at the interface since $\Gamma = 0$. This result may be understood by noting that when the transverse permittivities in both media are the same, the region appears homogeneous to a wave propagating along the z axis since its electric field has only an E_y component.

7.5c   Modal Representation of the Solution

To utilize the plane-wave solutions $\bar{H}_1$ and $\bar{H}_2$ in the synthesis of the line source fields H_1 and H_2, we assume the representations

$$H_1 = \int_{-\infty}^{\infty} e^{i\beta v + iq_0 u} f(\beta) \, d\beta, \qquad\qquad u > 0, \qquad\qquad (11)$$

$$H_2 = \begin{cases} \int_{-\infty}^{\infty} g(\beta) e^{i\beta v} [e^{iq_1 u} - \Gamma(\beta) e^{iq_2 u}] \, d\beta, & 0 > u > u', & (12a) \\[2ex] \int_{-\infty}^{\infty} g(\beta) e^{i\beta v} [h(\beta) e^{iq_2 u} - \Gamma(\beta) e^{iq_2 u}] \, d\beta, & u < u' < 0. & (12b) \end{cases}$$

The radiation condition has already been satisfied in these formulations since the constituent plane waves describing the primary (i.e., $\Gamma = 0$) field in Eqs. (12) carry energy away from the source region while the reflected and transmitted waves carry energy away from the interface. Equation (7) gives the expression for Γ and the relation between f and g. One may determine the incident-wave amplitudes by writing the primary field as

$$H_{2i} = \int_{-\infty}^{\infty} e^{i\beta v} Q(u, \beta) \, d\beta, \qquad\qquad (13)$$

substituting this Fourier integral representation into Eq. (1), recalling that $2\pi\delta(v - v') = \int_{-\infty}^{\infty} e^{i\beta(v-v')} \, d\beta$, and then deriving the differential equation for the transform $Q(u, \beta)$:

$$\left(A_2 \frac{d^2}{du^2} + i\beta A_3 \frac{d}{du} - A_1 \beta^2 + k_y^2 \right) Q = \frac{-i\omega\epsilon_y}{2\pi} e^{-i\beta v'} \delta(u - u'). \qquad (14)$$

A solution of this equation that satisfies continuity at $u = u'$ and the radiation condition is given by (for an alternative procedure utilizing first-order transmission-line equations, see Sec. 8.2g)

$$Q = \begin{cases} ae^{i(q_1 u + q_2 u')}, & u > u', \\[1ex] ae^{i(q_2 u + q_1 u')}, & u < u', \end{cases} \qquad\qquad (15)$$

with the constant a evaluated from the jump condition on the derivative,

$$A_2 \frac{dQ}{du}\bigg|_{u'-0}^{u'+0} = -\frac{i\omega\epsilon_y}{2\pi} e^{-i\beta v'}. \tag{16}$$

Consequently,

$$a = \frac{-\omega\epsilon_y e^{-i(q_1 u' + q_2 u' + \beta v')}}{2\pi A_2 (q_1 - q_2)}, \tag{17}$$

and a comparison of Eqs. (12), (13), (15), (17) leads to the identification,

$$g(\beta) = \frac{-\omega\epsilon_y e^{-i(q_1 u' + \beta v')}}{2\pi A_2 (q_1 - q_2)}, \tag{18a}$$

$$h(\beta) = e^{i(q_1 - q_2)u'}, \tag{18b}$$

thereby completing the integral representation for the primary field whose closed-form solution is given in Eq. (7.3.38a). Since each of the constituent plane waves in Eqs. (12a) and (11) satisfies the boundary conditions at the interface, so does the total magnetic field, whence the integral representations in Eqs. (11) and (12) constitute the desired solution; the convergence properties of the integrals when $\epsilon < 0$ are discussed in connection with Eq. (7.3.10). The multivalued functions $q_{0,1,2}$ are defined uniquely on the integration path, in view of the specifications given in Sec. 7.5b.

7.5d   Asymptotic Evaluation in the Plasma Half-space

While the infinite space contribution arising from the first term in the integrand of Eq. (12a) can be written in the closed form in Eq. (7.3.38a), no such simple evaluation is possible for the remaining term. Hence, we employ asymptotic methods to derive simple expressions for the fields observed at great distances from the source. As noted previously, it will be assumed that $k_y = k_0$; in addition, A_2 in Eq. (7.4.11) is restricted to be positive. This evidently places no constraint on α or ϵ when $\epsilon > 0$. For $\epsilon < 0$, however, it is implied that $\tan \alpha < \sqrt{|\epsilon|}$ [i.e., the q axis in Fig. 7.5.2(b) intersects the branches of the hyperbolic dispersion curve]. If $A_2 > 0$, the requirements Im $q_1 > 0$, Im $q_2 < 0$, when q is complex, are met by specifying that Im $\sqrt{k_0^2 A_2 - (\beta^2/\epsilon)} > 0$ when the radicand is negative; the square root is positive when real. The $\epsilon > 0$ case is similar to the previously treated problem of radiation from an isotropic half-space (Sec. 5.5) and is therefore not considered in detail. When $\epsilon < 0$, however, the analysis contains distinctive features that merit further attention. In this instance, the branch points at $\beta_b = \pm k_0 \sqrt{\epsilon A_2}$ lie on the imaginary axis; the integration path in the complex β plane appears as in Fig. 7.5.3(a), with the branch cuts drawn so that the various square roots have positive imaginary parts on the entire top sheet of the four-sheeted β surface.

As in the isotropic problems in Sec. 5.5, it is convenient to introduce a new variable w via

$$\beta = k \sin w, \qquad k = k_0 \sqrt{\epsilon A_2} > 0, \tag{19}$$

so that we have

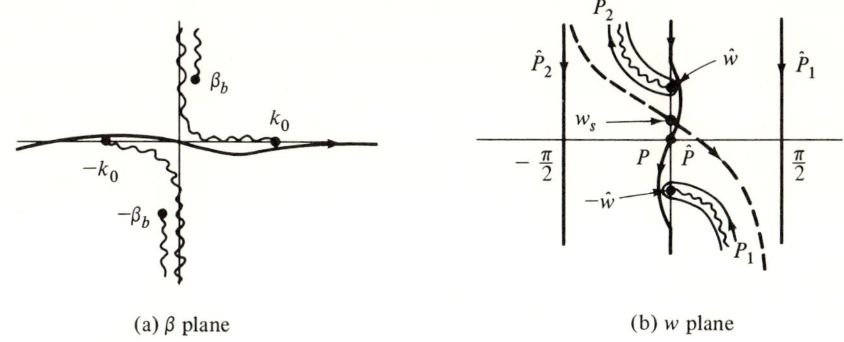

(a) β plane   (b) w plane

FIG. 7.5.3 Integration paths and singularities in the complex β and w planes ($\epsilon < 0$).

$$q_{1,2} = \frac{-kA_3 \sin w \pm 2(k/\sqrt{\epsilon})\cos w}{2A_2}. \tag{20}$$

For positive ϵ, the integration path in the w plane is then essentially the same as $\bar{P}$ in Fig. 5.3.6(b), while P in Fig. 7.5.3(b) is the corresponding path when ϵ is negative, with $\sqrt{\epsilon} = i\sqrt{|\epsilon|}$. In order to cast the reflected-field contribution in Eq. (12) into a form that exhibits an exponential dependence $\exp\left[i\gamma\cos(w - w_s)\right]$, where γ and w_s do not depend on w, it is useful to define the parameter w_s via

$$\tan w_s = \frac{(v - v') - (A_3/2A_2)(u - u')}{|u + u'|/\sqrt{\epsilon A_2}}. \tag{21}$$

The integral representation for the reflected field in region II may then be written as

$$H_{2r} = -\frac{\omega\epsilon_0\sqrt{\epsilon}}{4\pi}\int_P \Gamma(k\sin w)e^{i\gamma\cos(w-w_s)}\,dw, \qquad \gamma = \frac{k_0}{\sqrt{A_2}}\frac{|u + u'|}{\cos w_s}, \tag{22}$$

which is in the familiar form suitable for asymptotic evaluation when γ is large.

The only pertinent singularities in the w plane are branch points located at $\hat{w} = \pm\sin^{-1}(k_0/k)$. $w_s \pm n\pi$, $n = 0, 1, 2, \ldots$, specifies the saddle points of the integrand. For positive ϵ, w_s is real and the pertinent saddle point is located in the interval $-\pi/2 < w_s < \pi/2$, while the pertinent branch points $\hat{w}$ are complex and lie on the lines $|\text{Re } w| = \pi/2$. The deformation of P into the steepest-descent path through w_s, and the subsequent asymptotic evaluation of the integral, proceeds therefore as in Sec. 5.3. When $\epsilon < 0$, w_s is complex and it is convenient to introduce a real parameter δ,

$$w_s = i\tanh^{-1}\delta, \qquad \delta = \frac{(v - v') - (A_3/2A_2)(u - u')}{|u + u'|/\sqrt{|\epsilon|A_2}}. \tag{23}$$

If $-1 < \delta < 1$, then $w_s = i\phi_s$, where ϕ_s is real and varies between $-\infty$ and $+\infty$; if $\pm\delta > 1$, then $w_s = \pm\pi/2 + i\coth^{-1}\delta$. Since w_s represents the saddle point of the integrand in Eq. (22), one must distinguish between the cases $|\delta| < 1$ and $|\delta| > 1$ in the asymptotic evaluation. For $|\delta| < 1$, γ is real, and Im cos $(w - w_s) = \sin w_r \sinh (\phi_s - w_i)$ is positive either when $\sin w_r > 0$, $w_i < \phi_s$, or when $\sin w_r < 0$, $w_i > \phi_s$, where w_r and w_i denote the real and imaginary parts of w, respectively. Hence, the integral converges exponentially along a path such as $\hat{P}$ in Fig. 7.5.3(b). For $\pm\delta > 1$, $\gamma = i|\gamma|$, whence Im $[\gamma \cos (w - w_s)] > 0$ when $\sin w_r \gtrless 0$, and the integral converges exponentially along paths $\hat{P}_{1,2}$ when $\pm\delta > 1$.

For an asymptotic evaluation when $|\gamma|$ is large, the integration path P is deformed into the steepest-descent path (SDP) through the appropriate saddle points. When $|\delta| < 1$, the SDP is defined by the requirement Re cos $(w - w_s)$ $= 1$, which yields the contour $\hat{P}$ in Fig. 7.5.3(b). P can be deformed into $\hat{P}$ by the addition of path segments P_1 and (or) P_2 around the branch-point singularities (the branch cuts have been drawn conveniently as shown) and at $|w_i| = \infty$. In view of the preceding discussion concerning the exponential decay of the integrand, no contribution accrues from the latter. Thus,[11]

$$H_{2r} = -\frac{i\omega\epsilon_0\sqrt{|\epsilon|}}{4\pi} [I + U(|\hat{w}| + \phi_s)I_1 + U(|\hat{w}| - \phi_s)I_2], \qquad |\delta| < 1, \qquad (24)$$

where

$$I = \int_{\hat{P}} F\,dw, \quad I_1 = \int_{P_1} F\,dw, \quad I_2 = \int_{P_2} F\,dw, \qquad (24a)$$

with F representing the integrand in Eq. (22). $U(x) = 1$, $x > 0$, and $U(x) = 0$, $x < 0$.

When $|\delta| > 1$, γ is imaginary and the saddle point w_s lies on the line $w_r = \pi/2$ for $\delta > 1$, and $w_r = -\pi/2$ for $\delta < -1$. The steepest-descent path for $\delta > 1$ is defined by the equation Im cos $(w - w_s) = 0$, which determines the line $w_r = \pi/2$. Since the integrand in Eq. (22) converges exponentially in the strip $\sin w_r > 0$, the contour P can be deformed into the path $\hat{P}_1$ provided that the branch-cut integral over P_1 is included. Thus, for $\delta > 1$,

$$H_{2r} = \frac{-i\omega\epsilon_0\sqrt{|\epsilon|}}{4\pi} [\hat{I}_1 + I_1], \qquad \hat{I}_1 = \int_{\hat{P}_1} F\,dw, \qquad (25a)$$

while, from analogous considerations for $\delta < -1$,

$$H_{2r} = \frac{-i\omega\epsilon_0\sqrt{|\epsilon|}}{4\pi} [\hat{I}_2 + I_2], \qquad \hat{I}_2 = \int_{\hat{P}_2} F\,dw. \qquad (25b)$$

The geometric-optical field

The asymptotic evaluation of the steepest-descent path integral can now be carried out directly, and from the formula in Eq. (4.2.7), one obtains the lowest-order approximation,

$$H_{2r}\big|_{\text{SDP}} \sim -\frac{\omega\epsilon_0\sqrt{\epsilon}}{2\sqrt{2\pi}} \Gamma(k \sin w_s) \frac{e^{i(\gamma - \pi/4)}}{\sqrt{\gamma}} + O(\gamma^{-3/2}), \qquad (26)$$

valid for positive or negative ϵ; if γ is imaginary, one takes $\gamma = i|\gamma|$. This contribution is subsequently interpreted as the geometric-optical part of the reflected field, and with the asymptotic approximation of the primary field in Eq. (7.3.38a), the total geometric-optical field H_{2g} in the plasma half-space may be written as

$$H_{2g} \sim \frac{-\omega\epsilon_0\sqrt{\epsilon}\,e^{-i\pi/4}}{2\sqrt{2\pi}}\left\{\frac{e^{i\chi}}{\sqrt{\chi}}\left[1 + O\!\left(\frac{1}{\chi}\right)\right] + \Gamma(k\sin w_s)\frac{e^{i\gamma}}{\sqrt{\gamma}}\left[1 + O\!\left(\frac{1}{\gamma}\right)\right]\right\},$$

$$(27)$$

valid when $|\chi|$ and $|\gamma|$ are large. The higher-order terms, not shown explicitly, may be evaluated by the procedure in Sec. 4.2. Here

$$\chi = k_0 R N(\varphi), \qquad N(\varphi) = \sqrt{\cos^2(\varphi - \alpha) + \epsilon\sin^2(\varphi - \alpha)}, \qquad (27a)$$

while alternative expressions for γ are

$$\gamma = \frac{k_0}{\sqrt{A_2}}|u + u'|\sqrt{1 + \tan^2 w_s}$$

$$= k_0\sqrt{\epsilon A_2}\left\{\frac{|u + u'|^2}{\epsilon A_2^2} + \left[(v - v') - \frac{A_3}{2A_2}(u - u')\right]^2\right\}^{1/2} = L_1 + L_2, \qquad (27b)$$

$$L_1 = k_0 R_1 N(\varphi_1) = \beta_s \bar{v} + q_1(\beta_s)|u'|, \qquad (27c)$$

$$L_2 = k_0 R_2 N(\varphi_2) = \beta_s(v - \bar{v}) - q_2(\beta_s)|u|. \qquad (27d)$$

R is the distance from the source to the observation point, φ is the angle between R and the positive u axis, and $N(\varphi)$ is the ray refractive index [see Eq. (7.3.37), where φ is measured from the z axis]. L_1 represents the phase change along a ray $\bar{S}_1$ which leaves the source at an angle φ_1 and travels the distance R_1 to the interface, while L_2 is the phase change along a reflected ray $\bar{S}_2$ which travels the distance R_2 from the interface to the observation point P, with an inclination specified by the angle φ_2 (see Fig. 7.5.4).

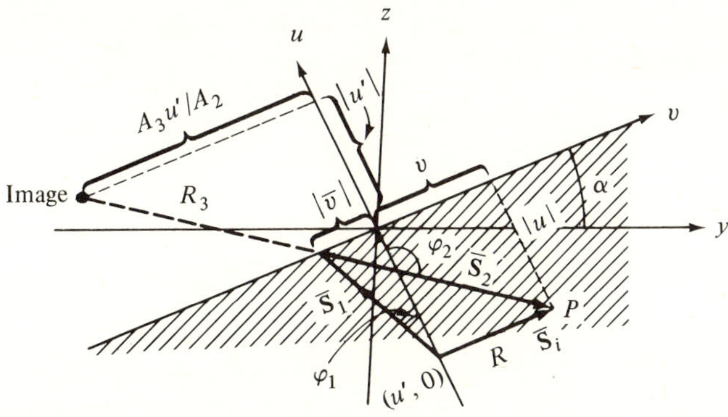

FIG. 7.5.4 Physical interpretation of far fields in the plasma ($\epsilon > 0$): geometrical-optical contribution.

When $\epsilon < 0$, both the primary and reflected fields are confined to illuminated regions in which χ and γ, respectively, are real; the disposition of these regions is the same as in the problem of reflection from an infinite, perfectly conducting plane (Fig. 7.4.3). Equation (27) fails in the vicinity of the incident or reflected field shadow boundaries on which χ or γ vanishes. The transition through the former is accomplished via the exact expression in Eq. (7.3.38a). Since the major contribution to the reflected field integral in the vicinity of its shadow boundary arises from a saddle point at very large distance from the origin, and since $\Gamma(k \sin w)$ is a well-behaved function which may be approximated by $\Gamma(k \sin w_s)$ and removed from the integrand in Eq. (22), the remaining integral may be evaluated exactly in terms of $H_0^{(1)}(\gamma)$. This latter expression may be employed in the transition region where $\gamma \to 0$ and exhibits a singularity similar to that in the primary field.

Since φ_1 and φ_2 are precisely the angles required to satisfy the reflection law in Eq. (9b), the asymptotic field solution has a simple physical interpretation: the field energy leaves the source along radial rays that, many wavelengths aways from the source, behave locally like plane waves. These incident rays are reflected at the interface, the amplitude and direction of the emerging ray being the same as for the corresponding plane wave. The ray amplitudes decay in accord with the divergence coefficients $|\chi|^{-1/2}$ or $|\gamma|^{-1/2}$ that are also explained in terms of the energy conserved in a ray tube (see Sec. 1.6b), and the factor outside the braces in Eq. (27) has to do with the normalization of the primary field. Thus, the first-order asymptotic solution confirms the predictions of geometric-optical "ray tracing" in an anisotropic medium. To substantiate this interpretation, it is convenient to return to Eqs. (12) and to examine the saddle-point condition for the reflected field in the β plane. If $v' = 0$ for simplicity, the pertinent exponential function is $\exp[i\beta v + iq_2 u - iq_1 u']$, and the saddle points β_s are defined by

$$-\frac{dq_2}{d\beta} + \frac{dq_1}{d\beta}\frac{|u'|}{|u|} + \frac{v}{|u|} = 0, \qquad \beta = \beta_s. \tag{28}$$

In view of relation (9a) between the ray angle and the derivative of q, one may write, instead of Eq. (28),

$$\tan \varphi_2 - \frac{|u'|}{|u|}\tan \varphi_1 + \frac{v}{|u|} = 0, \tag{29}$$

where

$$\tan \varphi_1 = -\frac{dq_1}{d\beta}\bigg|_{\beta_s}, \qquad \tan \varphi_2 = -\frac{dq_2}{d\beta}\bigg|_{\beta_s}. \tag{29a}$$

Equation (29) may be interpreted graphically as in Fig. 7.5.4, with $\tan \varphi_1 = \bar{v}/|u|$, $\tan \varphi_2 = -(v - \bar{v})/|u|$. Since β remains fixed at β_s, Eq. (29a) is evidently equivalent to Eq. (9b). Thus, the saddle-point condition (28) selects that value of β_s for which the incident and reflected ray directions are such that the reflected ray $\bar{S}_2$ passes through the observation point P. The phase variation of the reflected field corresponding to the saddle point β_s, $\gamma = \beta_s v + q_1(\beta_s)|u'| - q_2(\beta_s)|u|$, is equivalent to Eq. (27b).

It is of interest to observe that the image construction for the reflected field, rigorously applicable when reflection takes place from a perfect conductor (Sec. 7.4c), is also valid for the asymptotic field solution in Eq. (27). For verification, it suffices to show that the phase path L_1 from the source to the interface along R_1 in Fig. 7.5.4 is the same as the phase path L_1' from the image to the interface along R_3, provided that the anisotropic medium fills all of space. Now, L_1 is given in Eq. (27c), while

$$L_1' = \beta_s\left(\frac{A_3}{A_2} u' + \bar{v}\right) - q_2(\beta_s)|u'| = k_0 R_3 N(\varphi_2). \tag{30}$$

But since $q_1 + q_2 = -\beta A_3/A_2$ [see Eq. (5)], one confirms easily that $L_1' = L_1$. In view of the image construction, the region illuminated by the reflected field, and the extent of the shadow region, may also be inferred as in Fig. 7.4.3 when $\epsilon < 0$. The reader may verify that the directions of incident–reflected ray pairs described analytically in Eq. (29) may also be inferred from the wave-number surfaces in Fig. 7.5.2.

The lateral waves

The branch-cut integrals I_1 and I_2 in Eqs. (24) and (25) yield field contributions that can be identified as lateral waves. When $1 > \epsilon > 0$, the branch points $\hat{w} = \pm\sin^{-1}(1/\sqrt{\epsilon A_2}) = \pm\sin^{-1}(1/\sqrt{\sin^2\alpha + \epsilon\cos^2\alpha})$ are complex and unless both the source and observation points lie on the interface, the resulting branch-cut integral is exponentially small, as in the analogous isotropic problem discussed in Sec. 5.5e. When $\epsilon < 0$, however, the branch points lie on the imaginary axis in Fig. 7.5.3(b) where the exponential term in the integrand of Eq. (22) is undamped, and their contribution to the field in the plasma half-space may be important. In fact, the branch-cut integral contributions are found to be dominant in the shadow regions wherein both χ and γ in Eq. (27) are imaginary; their role is more significant here than in the analogous isotropic case (Fig. 5.5.2), where the geometric-optical fields are always stronger (it is recalled that the branch-cut integral is $O[(\text{distance})^{-3/2}]$; however, when losses are present, the geometric-optical fields are exponentially damped and may then be smaller than the lateral-wave fields). Thus, the branch-cut integrals for $\epsilon < 0$ merit detailed investigation. Their subsequent physical interpretation in terms of lateral waves exhibits features quite different from those encountered in the isotropic problem.[11]

The asymptotic evaluation of the branch-cut integral I_1 in Eq. (24a) may be carried out exactly as in Eqs. (5.5.17)–(5.5.22) provided that the saddle point does not lie near the branch point (i.e., $-\phi_s \not\approx |\hat{w}|$). For $|\delta| < 1$, one introduces the change of variable

$$\cos(w - w_s) = \cos(w + \hat{w}) + is^2, \qquad -\infty < s < \infty, \tag{31}$$

where P_1 in Fig. 7.5.3(b) denotes the resulting path on which s is real and increases from $-\infty$ to $+\infty$ along the direction indicated by the arrow. Near $s = 0$, from which region the major contribution to the integral originates when $\gamma \gg 1$,

$$\frac{dw}{ds} = \frac{2s}{\sinh(\phi_s + |\hat{w}|)} + O(s^3), \tag{32a}$$

$$s \cong \sqrt{\sinh(\phi_s + |\hat{w}|)}\sqrt{w + \hat{w}}, \qquad 0 < \arg(w + \hat{w}) < 2\pi, \tag{32b}$$

$$\Gamma(k \sin w) = 1 + \frac{2e^{-i\pi/4}}{\cosh|\hat{w}|} \sqrt{\frac{|\epsilon|\sinh 2|\hat{w}|}{\sinh(\phi_s + |\hat{w}|)}}\, s + O(s^2). \tag{32c}$$

The integral can then be written as

$$I_1 = e^{i\gamma \cosh(\phi_s + |\hat{w}|)} \int_{-\infty}^{\infty} \left(\frac{dw}{ds}\Gamma\right) e^{-\gamma s^2}\, ds, \tag{33}$$

and the lowest-order term in the asymptotic approximation for large γ arises from the s^2 term in the power-series expansion of $\Gamma\, dw/ds$. The resulting formula is [see Eq. (4.2.17)]

$$I_1 \sim \frac{2\sqrt{\pi}\, e^{-i\pi/4} e^{i\gamma \cosh(\phi_s + |\hat{w}|)}}{\cosh|\hat{w}|\, [\gamma \sinh(\phi_s + |\hat{w}|)]^{3/2}} \sqrt{|\epsilon|\sinh 2|\hat{w}|} + O(\gamma^{-5/2}), \tag{34}$$

and the analogous result for I_2 is the same except for the replacement of ϕ_s by $-\phi_s$. With $w_s = i \tanh^{-1}\delta$, these expressions may be shown to apply as well when $|\delta| > 1$.

For subsequent interpretation,[11] it is useful to employ alternative expressions that are derived most simply in terms of the original β variable. Since the branch points at $w = \pm i|\hat{w}|$ correspond to $\beta = \pm k_0$, one finds that, for $v' = 0$,

$$\gamma \cosh(|\hat{w}| \pm \phi_s) = \pm k_0 v + q_1(\pm k_0)|u'| - q_2(\pm k_0)|u|; \tag{35}$$

it may also be shown that

$$\frac{|u + u'|}{\cosh\phi_s}\sinh(|\hat{w}| \pm \phi_s) = \pm\sqrt{|\epsilon|}A_2 \cosh|\hat{w}| \left[v - |u|\frac{dq_2}{d\beta}\bigg|_{\pm k_0} + |u'|\frac{dq_1}{d\beta}\bigg|_{\pm k_0}\right]. \tag{36}$$

In each of these branch-cut contributions, the transverse wavenumber β remains *fixed* at $+k_0$ or at $-k_0$, and the various phase increments for different observation points are therefore not achieved by varying the ray directions as in the geometric-optical field. Because $\beta = \pm k_0$ corresponds to a ray that travels parallel to the interface in the vacuum region, the phenomenon of critical refraction may be expected to play a role. It is suggestive to define phase increments along the critically refracted ray paths and so let

$$L_3(k_0) \equiv L_3 = [k_0 \tan\varphi_3 + q_1(k_0)]|u'|, \qquad \tan\varphi_3 = -\frac{dq_1}{d\beta}\bigg|_{k_0}, \tag{37a}$$

$$L_4(k_0) \equiv L_4 = k_0(v - |u'|\tan\varphi_3 + |u|\tan\varphi_5), \qquad \tan\varphi_5 = -\frac{dq_2}{d\beta}\bigg|_{k_0}, \tag{37b}$$

$$L_5(k_0) \equiv L_5 = -[k_0 \tan\varphi_5 + q_2(k_0)]|u|. \tag{37c}$$

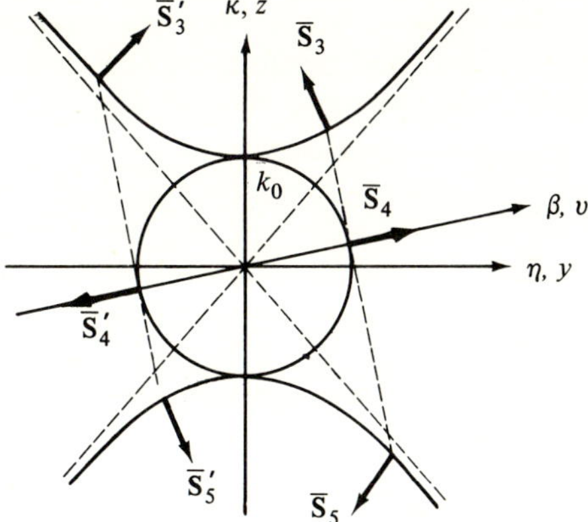

(a) Dispersion curves

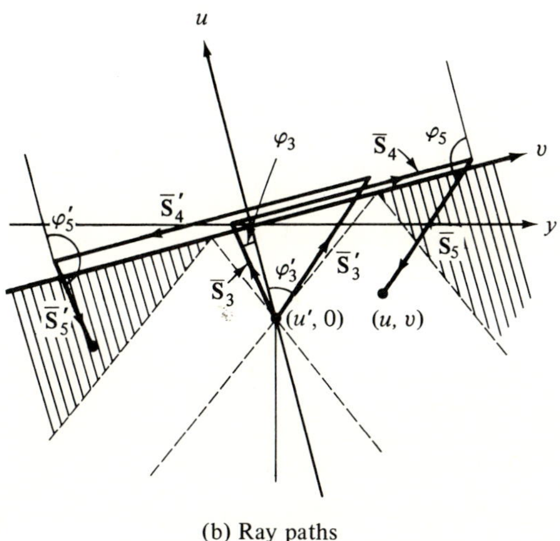

(b) Ray paths

FIG. 7.5.5 Lateral rays ($\epsilon < 0$).

L_3 is the phase change along ray $\bar{\mathbf{S}}_3$ in Fig. 7.5.5(b) which travels from the source to the interface at the first critical angle φ_3 [see Eq. (27c), with $\beta_s = k_0$], L_4 is the phase change along ray $\bar{\mathbf{S}}_4$ which travels in the *vacuum* region parallel to the interface, and L_5 is the phase change along ray $\bar{\mathbf{S}}_5$ which travels from the interface to the observation point (u, v) in the plasma at the second critical angle φ_5 [see Eq. (27d), with $\beta_s = k_0$ and $\bar{v}$ equal to the value of v at the starting point of $\bar{\mathbf{S}}_5$]. The directions of $\bar{\mathbf{S}}_3$, $\bar{\mathbf{S}}_4$, and $\bar{\mathbf{S}}_5$ may also be inferred from the wavenumber diagram in Fig. 7.5.5(a); note the backward refraction since $\epsilon < 0$.

Along the ray path $(L_3 + L_4 + L_5)$, it is possible to reach different observation points by changing the length, but not the direction, of L_4 or L_5, and keeping L_3 fixed. The entire trajectory then corresponds to the constant value $\beta_s = k_0$ as required, and defines a "lateral wave" on an anisotropic interface (see Fig. 5.5.2 for the isotropic case).

With these definitions, the branch-cut integral contributions H_{2b} to the reflected field in the plasma half-space may be written as

$$H_{2b} \sim -\frac{\omega \epsilon_0 e^{i\pi/4}}{\sqrt{2\pi}} \left(\frac{|\epsilon|}{1 + |\epsilon| A_2} \right) \left[\frac{e^{i(L_3 + L_4 + L_5)}}{L_4^{3/2}} U(L_4) + \frac{e^{i(L_3' + L_4' + L_5')}}{L_4'^{3/2}} U(L_4') \right], \qquad (38)$$

where $U(x)$ is the Heaviside unit function, and

$$L_3' = L_3(-k_0), \qquad L_4' = L_4(-k_0), \qquad L_5' = L_5(-k_0), \qquad (38a)$$

are the phase increments along the lateral ray path corresponding to $\beta_s = -k_0$. Thus, the branch-cut integrals may be interpreted as representing the fields along lateral rays that propagate parallel to the interface in the vacuum region, are excited by a ray incident at the critical angle, and shed energy back into the plasma by refraction. In contrast to lateral waves on an isotropic interface, it is observed that backward refraction obtains, as a result of which a certain range of observation points may be reached by two lateral rays. Moreover, the lateral rays penetrate the shadow zone, one ray for each of the areas shown shaded in Fig. 7.5.5(b), thereby providing a means of transferring energy into an otherwise inaccessible region; in this zone, the geometric optical field H_{2g} in Eq. (27) is exponentially small. The magnitude of the lateral ray field varies like $L_4^{-3/2}$ or $L_4'^{-3/2}$, and for distant observation points in the illuminated region, this decay with distance is stronger than that of the geometric optical field. An exception occurs near the angles of total reflection defined by $L_4 = 0$ or $L_4' = 0$. In their vicinity, Eq. (38) becomes invalid and must be replaced by the more precise formula in Eq. (45), which permits the field calculation in or near this transition region.

Fields in the vicinity of the angle of total reflection

For an asymptotic evaluation of the integral I_1 in the vicinity of the angle of total reflection where $\phi_s \rightarrow -|\hat{w}|$, the change of variable introduced in Eq. (31) is inappropriate since dw/ds in Eq. (32a) is then not slowly varying near $s = 0$. The difficulty arises since the branch point is located near the saddle point. It is convenient to employ as a new variable the function†

$$\tau = \sqrt{1 + \frac{\sin^2 w}{\sinh^2 |\hat{w}|}}, \qquad \sinh |\hat{w}| = \frac{1}{\sqrt{|\epsilon| A_2}}, \qquad \epsilon < 0, \qquad (39)$$

so the branch point $w = -\hat{w}$ maps into $\tau = 0$. The mapping derivative,

$$\frac{dw}{d\tau} = \frac{\tau}{(\sin w \cos w)|\epsilon| A_2}, \qquad (40)$$

---

†The integration variable τ here is not to be confused with the function $\tau(s) = q(z)$ employed in Chapter 4.

is now well behaved near $\tau = 0$, and, for $w \approx -i|\hat{w}|$,

$$\cos w = \cosh|\hat{w}| \sqrt{1 - \frac{\tau^2}{1 + |\epsilon|A_2}}$$

$$\approx \cosh|\hat{w}| \left[1 - \frac{\tau^2}{2(1 + |\epsilon|A_2)} - \frac{\tau^4}{8(1 + |\epsilon|A_2)^2} - \cdots \right], \tag{41a}$$

$$\sin w = -i(\sinh|\hat{w}|)\sqrt{1 - \tau^2} \approx -i\sinh|\hat{w}|\left(1 - \frac{\tau^2}{2} - \frac{\tau^4}{8} - \cdots\right), \tag{41b}$$

$$\frac{dw}{d\tau}\Gamma = i\left(1 - \frac{2\tau}{\sqrt{A_2}\cosh|\hat{w}|}\right)\tau\tanh|\hat{w}| + O(\tau^3), \tag{41c}$$

with the square roots defined to be positive when the radicands are positive.

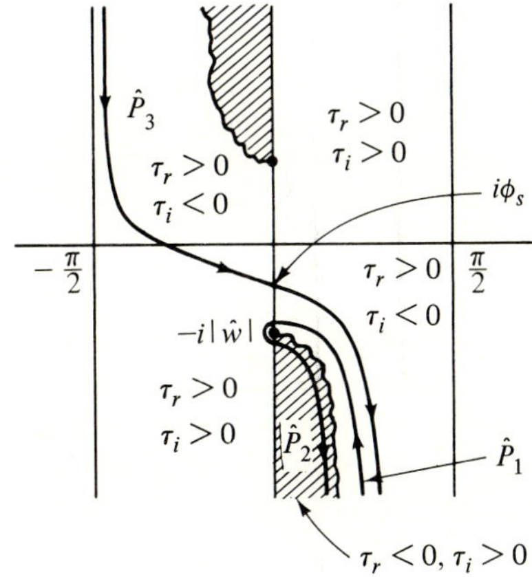

(a) w plane

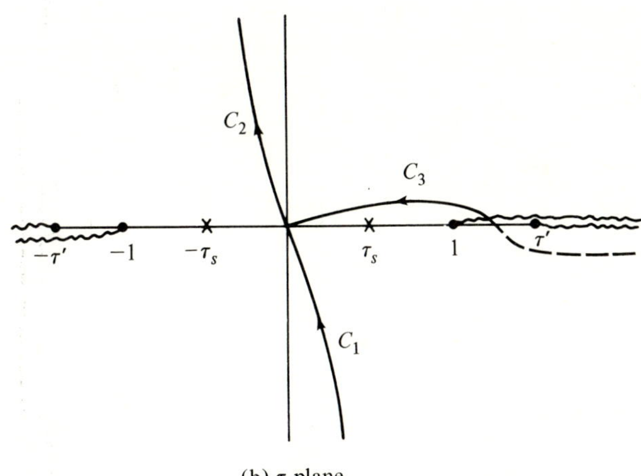

(b) τ plane

FIG. 7.5.6   Contours and singularities in the w and τ planes.

The behavior of $\tau_r = \text{Re}\,\tau$ and $\tau_i = \text{Im}\,\tau$ in the various sections of the complex w plane is shown in Fig. 7.5.6(a), whence one infers that the path $\hat{P}_1 + \hat{P}_2$ maps into the contour $C_1 + C_2$ in Fig. 7.5.6(b). Since

$$\cos (w - w_s) = \sinh|\hat{w}|\,(\cosh \phi_s \sqrt{1 + |\epsilon|A_2 - \tau^2} + \sinh \phi_s \sqrt{1 - \tau^2}),$$

$$(42)$$

the integrand of I_1 in Eq. (24a) has saddle points at

$$\tau = 0, \pm\tau_s, \qquad \text{with } \tau_s = \sqrt{1 - (\sinh \phi_s / \sinh |\hat{w}|)^2} \geq 0, \qquad (43)$$

so there exists on the real τ axis a confluence of three equally spaced, first-order saddle points as $\phi_s \to -|\hat{w}|$. The integrand, which decays away from $\tau = 0$ along $C_{1,2}$, also possesses branch points at $\tau = \pm 1$ and $\pm\tau' = \pm \coth |\hat{w}|$ [see Fig. 7.5.6(b)] which are, however, far removed from the saddle-point region $\tau \approx 0$. The problem of performing an integration when a saddle point and a branch point tend to coalesce has thus been transformed into one of integration in the presence of three adjacent saddle points (see Sec. 4.4c). With the indicated choice of branch cuts, $\sqrt{1 - \tau^2}$ and $\sqrt{\coth^2 |\hat{w}| - \tau^2}$ have positive real parts on the entire top sheet of the four-sheeted Riemann surface, and positive (negative) imaginary parts in the second and fourth (first and third) quadrants.

For large γ, the dominant contribution to the integral I_1 arises from the vicinity of $\tau = 0$, and since the saddle-point configuration is similar to the one occurring in the integral representation for the parabolic cylinder function, I_1 may be approximated in terms of this function (for related problems in isotropic media, see Reference 12). The required analysis has been presented in Sec. 4.5b, and in terms of the notation employed there, $\tau \to z, \Omega = \gamma, q(\tau) = i \cos (w - w_s)$. Since the integration path $C_1 + C_2$ is symmetrical with respect to $\tau = 0$, and $\Gamma\, dw/d\tau \sim \tau$ as $\tau \to 0$ [see Eq. (41c)], the lowest-order contribution arises from the τ^2 term in the power-series expansion of $\Gamma\, dw/d\tau$, and the corresponding parabolic cylinder function is of order $-\frac{3}{2}$. From Eqs. (42) and (43),

$$q(\tau_s) - q(0) = -2i \sinh^2 \left(\frac{\phi_s + |\hat{w}|}{2}\right), \qquad (44a)$$

$$q(\tau_s) + q(0) = 2i \cosh^2 \left(\frac{\phi_s + |\hat{w}|}{2}\right), \qquad (44b)$$

so, from an application of Eq. (4.5.44)[11],

$$I_1 \sim \overset{\circ}{I}_1 e^{-i3\pi/8 - i\zeta^2/4} \zeta^{3/2} D_{-3/2}(\zeta e^{-i\pi/4}), \qquad \zeta = 2\sqrt{\gamma}\, \sinh \left(\frac{\phi_s + |\hat{w}|}{2}\right) \geq 0, \qquad (45)$$

where $\overset{\circ}{I}_1$ is the simple asymptotic formula in Eq. (34). The terms multiplying $\overset{\circ}{I}_1$ provide a correction that approaches unity when $\zeta \gg 1$ [see Eq. (66a)], and compensate for the singularity in $\overset{\circ}{I}_1$ when $\zeta \to 0$. Because γ is assumed to be large, it is possible to have $\zeta \gg 1$, although $\phi_s + |\hat{w}|$ is small; thus, Eq. (45) is required only when $\phi_s + |\hat{w}| = O(\gamma^{-1/2})$, and Eq. (34) suffices for larger values of $\phi_s + |\hat{w}|$. On the boundary confining the domain of existence of the totally reflected waves, $\zeta = 0$, and use of the formula for $D_{-3/2}(0)$ [Eq. (4.5.41)] yields the result

$$I_1 \sim \left(\frac{2}{\gamma}\right)^{3/4} \frac{4\pi\sqrt{|\epsilon|}\tanh|\hat{w}|\,e^{-i5\pi/8}\,e^{i\gamma}}{\Gamma(\tfrac{1}{4})}. \tag{46}$$

No confusion should arise between the gamma function $\Gamma(\tfrac{1}{4})$ and the same symbol Γ employed elsewhere for the reflection coefficient. The distance dependence in the branch-cut integral contribution is seen to increase from $\gamma^{-3/2}$ far from the angle of total reflection to $\gamma^{-3/4}$ at this angle.

While the asymptotic evaluation of the integral I, in Eq. (24a), along the steepest-descent path $\hat{P}$ yields a lowest-order contribution [Eq. (26)] that is not affected as $\phi_s \to -|\hat{w}|$, the higher-order terms are influenced by the proximity of the branch point near the saddle point. As $\phi_s \to -|\hat{w}|$, one has, with reference to Fig. 7.5.6,

$$I = \begin{cases} \left(\int_{\hat{P}_3} - \int_{\hat{P}_1}\right)F\,dw = \left(\int_{C_3} - \int_{C_1}\right)F\frac{dw}{d\tau}\,d\tau, & -\phi_s < |\hat{w}|, \quad \text{(47a)} \\[2ex] \int_{\hat{P}_3 + \hat{P}_2} F\,dw = \int_{C_3 + C_2} F\frac{dw}{d\tau}\,d\tau, & -\phi_s > |\hat{w}|, \quad \text{(47b)} \end{cases}$$

where $F = \Gamma \exp\left[i\gamma\cos(w - w_s)\right]$. It is convenient to write $\Gamma = 1 + (\Gamma - 1)$, since the contribution from the first term can then be evaluated exactly in terms of the Hankel function. For the remaining term, follow the procedure above to express the integral over paths $C_{1,2}$ in terms of $D_{-3/2}[\zeta\exp(-i\pi/4)]$; the one over path C_3 involves $D_{-3/2}[-\zeta\exp(-i\pi/4)]$. Thus,[11]

$$I = I' + I'' \tag{48}$$

$$I' = \int_{\hat{P}} e^{i\gamma\cos(w-w_s)}\,dw \sim \sqrt{\frac{2\pi}{\gamma}}\,e^{i(\gamma-\pi/4)} + O(\gamma^{-3/2}), \tag{48a}$$

$$I'' = \int_{\hat{P}} (\Gamma - 1)e^{i\gamma\cos(w-w_s)}\,dw \sim \frac{-iI_1}{2D_{-3/2}(\zeta e^{-i\pi/4})}$$

$$\times\,[D_{-3/2}(-\zeta e^{-i\pi/4}) \mp iD_{-3/2}(\zeta e^{-i\pi/4})], \qquad -\phi_s \lesseqgtr |\hat{w}|, \tag{48b}$$

with I_1 given in Eq. (45).

Although I'' is discontinuous across the total reflection boundary $\phi_s = -|\hat{w}|$, this is not true of the sum $\bar{I} = I'' + I_1 U(|\hat{w}| + \phi_s)$ which occurs in Eq. (24). Via the relation

$$D_{1/2}(z) = 2^{-3/2}[e^{i\pi/4}D_{-3/2}(iz) + e^{-i\pi/4}D_{-3/2}(-iz)], \tag{49}$$

one may write

$$\bar{I} \sim \mathring{I}_1 e^{-i5\pi/8 - i\zeta^2/4}\zeta^{3/2}\sqrt{2}\,D_{1/2}(-\zeta e^{i\pi/4}), \tag{50}$$

which expression remains valid for $-\phi_s \gtrsim |\hat{w}|$. For $-\zeta \gg 1$, Eq. (50) may be reduced via the asymptotic formula in Eq. (66a), while for $\zeta \gg 1$, one uses Eq. (66b). When applied to the above expressions for $\bar{I} + I'$, these asymptotic results reduce to those in Eqs. (27) and (38) in the vicinity of $-\phi_s \approx |\hat{w}|$.

7.5e Asymptotic Evaluation in the Vacuum Half-Space

Ray interpretation of the saddle-point condition—caustic and cusp

The field solution in the vacuum region $u > 0$ is given in Eq. (11):

$$H_1 = -\frac{\omega\epsilon_0}{\pi} \int_{-\infty}^{\infty} F(\beta)e^{i\psi(\beta)}\, d\beta, \tag{51}$$

where

$$F(\beta) = \frac{1}{2q_0 - A_2(q_2 - q_1)}, \tag{51a}$$

and

$$\psi(\beta) = \beta(v - v') + q_0 u + q_1|u'|. \tag{51b}$$

Instead of transforming to the w plane as before, it is preferable to carry out the present analysis in the β plane. Upon introducing a change of scale via $\beta = k_0\beta'$, one may write $\psi(\beta) = k_0\hat{\psi}(\beta')$ and, if k_0 is large, the integrand is in a form suitable for the application of asymptotic methods. The change to the variable β' will not be carried out but it will be kept in mind that $\psi(\beta)$ contains a large parameter k_0. Thus, the stationary points β_s, obtained from $d\psi/d\beta = 0$, are expected to play an important role in the asymptotic approximation:

$$v = u \tan \varphi_0 + |u'| \tan \varphi_1, \qquad \text{when } \beta = \beta_s, \tag{52}$$

where $\tan \varphi_1 \equiv -dq_1/d\beta$ and $\tan \varphi_0 \equiv -dq_0/d\beta = \beta(k_0^2 - \beta^2)^{-1/2}$. The solution of the latter equation yields $\beta_s = k_0 \sin \varphi_0$ and the associated value of φ_1 is then inferred from Eq. (9b). Subsequent discussion is restricted to the espe-

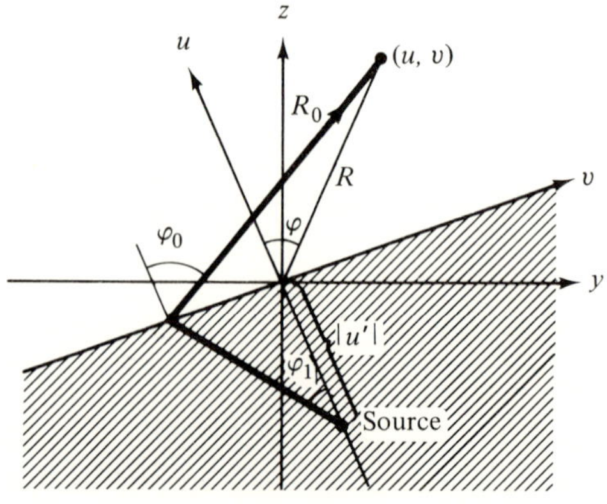

FIG. 7.5.7 Refracted ray when $\epsilon < 0$.

cially interesting case $\epsilon < 0$, with $A_2 > 0$. Also, for simplicity and without loss of generality, v' has been set equal to zero.

The important saddle points are those for which $\psi(\beta_s)$ is real since the associated field contribution $\sim \exp[i\psi(\beta_s)]$ is then undamped. These real solutions correspond to real values of φ_1 and φ_0, which represent the angles of incidence and refraction, respectively, of a ray that emerges from the plasma into the vacuum region (see Sec. 7.5b). The equation of the refracted ray is precisely the one given in Eq. (52), and the pertinent ray picture is shown in Fig. 7.5.7. One observes from Eq. (9b) (with $\epsilon < 0$ and $A_2 > 0$) that φ_0 and φ_1 may have opposite signs, thereby resulting in backward refraction; an increase in φ_0 causes φ_1 to decrease (i.e., $|\varphi_1|$ increases since $\varphi_1 < 0$ when $\varphi_0 > 0$), thus leading to a crossing of successive rays. When $\varphi_0 = \pm\pi/2$, the corresponding values of φ_1, the angles of critical refraction, are the ones required to launch the lateral waves. It is shown below (see Fig. 7.5.8) that because of the occurrence of

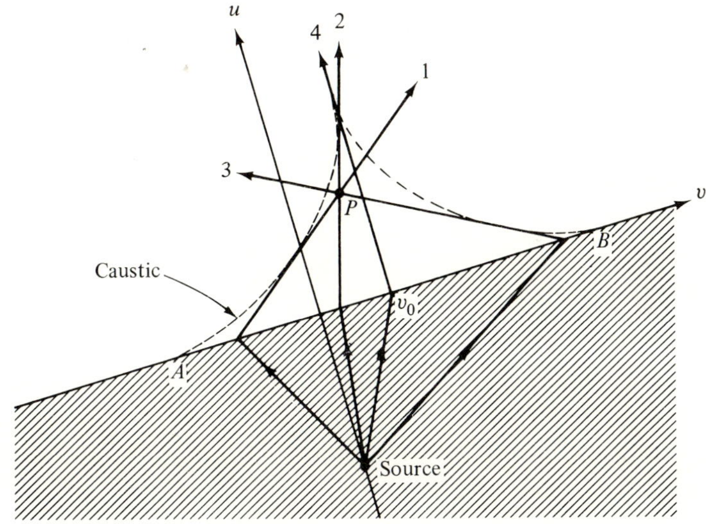

FIG. 7.5.8   Refracted ray configuration when $\epsilon < 0$.

backward refraction, a certain region in the vacuum half-space is reached by three propagating rays. This region is bounded by a caustic possessing a cusp at some distance above the interface. Thus, the *straight* interface on the *homogeneous* anisotropic half-space acts like an optical lens that focuses the fields radiated into the vacuum[13]—a phenomenon quite distinct from focusing in isotropic media where inhomogeneities or curved boundaries are required.

The real saddle-point solutions $\beta_s = k_0 \sin\varphi_0$ of Eq. (52) are best visualized from a plot of the ray configuration shown in Fig. 7.5.8. One notes that an observation point $P(u, v)$ near the interface may be reached by three rays (e.g., by 1, 2, and 3) but that more distant points are met by only a single ray. In the former case, the two rays 1 and 2 are refracted on the same side of the normal ray 4 and have positive angles φ_{01} and φ_{02}, while ray 3 is refracted on the opposite side and has a negative angle φ_{03}. The picture is reversed when the observation point lies on the other side of the normal ray 4 whose equation is given by $\varphi_0 = 0$, or $v_0 = |u'|A_3/2A_2$. The ray configuration in the vacuum region but not in the plasma is, in fact, symmetrical about the line $v = v_0$. A caustic (ray envelope), whose shape is found to be a hypocycloid, separates the region of three ray crossings from the exterior region wherein the rays diverge. The contact points A and B between the caustic and the interface correspond to $\varphi_0 = \pm\pi/2$ and are located at

$$v = v_0 \mp \frac{|u'|}{|\epsilon|A_2\sqrt{A_2 + 1/|\epsilon|}}, \qquad v_0 = \frac{|u'|A_3}{2A_2}; \qquad (53a)$$

the intercept of a ray with the line $v = v_0$ is given by

$$u = \frac{|u'|\cos\varphi_0}{|\epsilon|A_2\sqrt{A_2 + (\sin^2\varphi_0)/|\epsilon|}},$$

and the cusp C corresponds to $\varphi_0 = 0$, whence

$$u = \frac{|u'|}{|\epsilon|A_2^{3/2}}, \quad v = v_0, \quad \text{at the cusp.} \qquad (53b)$$

To determine the equation of the caustic, the parameter φ_0 is eliminated between Eq. (52) and the derivative of Eq. (52) with respect to φ_0:

$$\frac{u}{|u'|} = \frac{\sqrt{|\epsilon|}\cos^3\varphi_0}{(|\epsilon|A_2 + \sin^2\varphi_0)^{3/2}}. \qquad (54a)$$

After a bit of manipulation, one obtains for $\varphi_0 > 0$ the equation of one branch of a hypocycloid,[13]

$$\frac{(v_0 - v)^{2/3}}{[a(A_2\sqrt{A_2 + 1/|\epsilon|})^{-1}]^{2/3}} + \frac{u^{2/3}}{A_2^{-1}a^{2/3}} = 1, \qquad a = \frac{|u'|}{|\epsilon|}; \qquad (54b)$$

for the other branch, corresponding to $\varphi_0 < 0$, $v_0 - v$ is replaced by $v - v_0$. It is evident that Eq. (54b) yields correctly the endpoints in Eqs. (53a) and (53b).

Asymptotic field evaluation

In the evaluation of the integral in Eq. (51) by the saddle-point method, the path of integration is deformed into steepest-descent paths traversing the various contributing saddle points. The division of the complex β plane into "mountain" and "valley" regions in the vicinity of each saddle point depends on the second derivative of $\psi(\beta)$:

$$\psi''(\beta) = q_0''(\beta)u + q_1''(\beta)|u'|, \qquad (55)$$

where

$$q_0'' = \frac{-k_0^2}{(k_0^2 - \beta^2)^{3/2}}, \quad q_1'' = \frac{k_0^2}{|\epsilon|(k_0^2 A_2 + \beta^2/|\epsilon|)^{3/2}}, \quad \beta = k_0\sin\varphi_0. \qquad (55a)$$

Since q_0'' and q_1'' have opposite signs in the important region $|\beta| < k_0$, where both quantities are real, $\psi''(\beta)$ may be positive, negative, or zero. If one considers a ray with a specified angle of refraction, φ_0, then as the observation point moves along the ray, $\psi''(\beta)$ is positive at the starting point $u = 0$ since $q_1''(\beta) > 0$, but is negative at sufficiently large values of u where the $q_0''u$ term dominates. The change in sign occurs at the point of tangency with the caustic where $\psi''(\beta) = 0$. An examination of the ray picture in Fig. 7.5.8 reveals that, of the three rays reaching a given observation point inside the caustic, two have touched the caustic but one has not, and that the angle φ_0 for the latter is intermediate between the other two. Thus, at the three corresponding saddle points $\beta_s = \beta_{1,2,3}$, one has $\psi''(\beta_1) < 0, \psi''(\beta_2) > 0, \psi''(\beta_3) < 0$, while at the single real saddle-point exterior to the caustic, $\psi''(\beta_s) < 0$. The resulting steepest-descent paths in the vicinity of the saddle points then have the direction along the $45°$ lines shown in Fig. 7.5.9, and, on the remainder of the path,

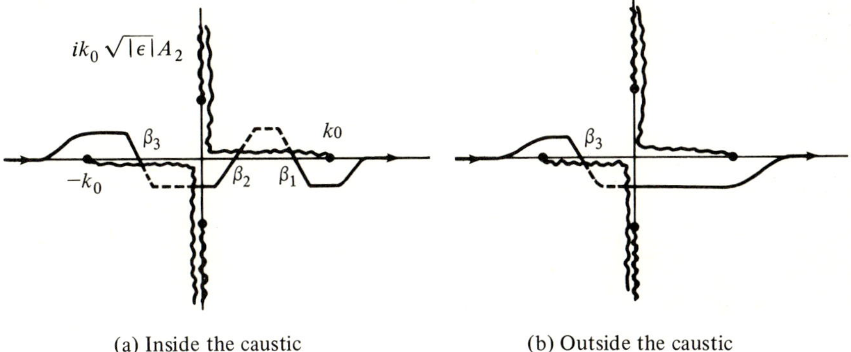

(a) Inside the caustic   (b) Outside the caustic

FIG. 7.5.9   Integration paths for different observation points.

the magnitude of the integrand in Eq. (51) may be shown to be exponentially smaller than at the saddle points. Consequently, the integral may be approximated in terms of the contributions arising from the vicinity of the saddle points only. The branch cuts associated with $\beta = \pm k_0$ have been so chosen that $\text{Im } q_0 > 0$ on the entire top sheet of the four-sheeted β plane. The branch cuts arising from q_1 in Eq. (5) have been drawn so that $\text{Re }\sqrt{} > 0$ on the top sheet.

Figure 7.5.9(a) pertains to an observation point P as shown in Fig. 7.5.8, while Fig. 7.5.9(b) is relevant when the observation point lies along ray 3 but outside the caustic. If the observer moves along ray 3 (i.e., β_3 is fixed), then β_1 and β_2 approach one another and coalesce into a double saddle point on the caustic; exterior to the caustic, the saddle points β_1 and β_2 are complex and the dominant contribution to the integral arises from the vicinity of β_3 alone. While the explicit solution of Eq. (52) for the saddle points β_s is somewhat involved when u and v are arbitrary, it is easily obtained when the observation point lies along the normal ray $v = v_0$. In this instance,

$$\beta_2 = 0, \quad \beta_1 = -\beta_3 = k_0\left(\frac{1 - A_2 b^2}{1 + b^2/|\epsilon|}\right)^{1/2}, \qquad b = \frac{u|\epsilon|A_2}{|u'|}, \tag{56}$$

and one observes that β_1 and β_3 are imaginary when $u > \hat{u}$, where $\hat{u} = |u'|[[|\epsilon|A_2^{3/2}]^{-1}$ is the ordinate at the cusp. Since $\beta_1 = \beta_2 = \beta_3 = 0$ at the cusp, one has $\psi'(0) = \psi''(0) = \psi'''(0) = 0$, so a third-order saddle-point evaluation is required for the asymptotic approximation.

The asymptotic formulas for the magnetic field in Eq. (51) may now be written down directly.[13] By the ordinary saddle-point procedure [Eq. (4.2.1b)], one obtains for observation points inside the caustic,

$$H \sim -\omega\epsilon_0\sqrt{\frac{2}{\pi}} \sum_{i=1}^{3} \frac{F(\beta_i)}{\sqrt{|\psi''(\beta_i)|}} \, e^{i\psi(\beta_i) + i(\pi/4)\,\mathrm{sgn}\,\psi''(\beta_i)}, \tag{57}$$

while outside the caustic,

$$H \sim -\omega\epsilon_0\sqrt{\frac{2}{\pi}} \frac{F(\beta_s)}{\sqrt{|\psi''(\beta_s)|}} \, e^{i\psi(\beta_s) - i\pi/4}, \tag{58}$$

where $\beta_s = k_0 \sin \varphi_0$. Near the caustic, $\beta_1 \approx \beta_2$ or $\beta_3 \approx \beta_2$, and the required formula must account explicitly for the proximity of two saddle points. When $\beta_1 \approx \beta_2$ [see Eq. (4.5.7)]

$$H \sim -\omega\epsilon_0\sqrt{\frac{2}{\pi}} \frac{F(\beta_3)}{\sqrt{|\psi''(\beta_3)|}} \, e^{i\psi(\beta_3) - i\pi/4}$$
$$- 2\omega\epsilon_0\left[\frac{2}{-\psi'''(\beta_0)}\right]^{1/3} e^{i[\psi(\beta_1) + \psi(\beta_2)]/2} \, \mathrm{Ai}(\xi)F(\beta_0), \tag{59}$$

where $\beta_0 = (\beta_1 + \beta_2)/2$, with $\psi''(\beta_0) = 0$, $\mathrm{Ai}(\xi)$ is the Airy function

$$\mathrm{Ai}(\xi) = \frac{1}{2\pi i} \int_{\infty e^{-i2\pi/3}}^{\infty e^{i2\pi/3}} e^{\xi t - (t^3/3)} \, dt, \qquad \mathrm{Ai}(0) = \frac{1}{3^{2/3}\Gamma(\frac{2}{3})}, \tag{59a}$$

and

$$\xi = \{-\tfrac{3}{4} i[\psi(\beta_1) - \psi(\beta_2)]\}^{2/3}, \tag{59b}$$

with

$$\xi = |\xi|e^{+i\pi}, \qquad \text{when} \quad \psi(\beta_1) - \psi(\beta_2) > 0. \tag{59c}$$

The last relation specifies the choice of the cube root. When $-\xi \gg 1$, one may employ the asymptotic formula in Eq. (4.2.51),

$$\mathrm{Ai}(\xi) \sim \frac{e^{-i\pi/4}}{2\sqrt{\pi}(-\xi)^{1/4}} [e^{i(2/3)(-\xi)^{3/2}} + ie^{-i(2/3)(-\xi)^{3/2}}], \tag{60}$$

to simplify the second expression in Eq. (59). With the relations

$$-\xi \approx \frac{1}{4}\left[-\frac{\psi'''(\beta_0)}{2}\right]^{2/3} (\beta_1 - \beta_2)^2, \quad \psi'(\beta_0) \approx \left[-\frac{\psi'''(\beta_0)}{8}\right](\beta_1 - \beta_2)^2, \tag{61a}$$

and

$$i\left[-\frac{\psi'''(\beta_0)}{2}\right]^{-1/3} \approx \left[\frac{-2\xi^{1/2}}{i|\psi''(\beta_{1,2})|}\right]^{1/2}, \tag{61b}$$

the large-ξ approximation of Eq. (59) reduces to Eq. (57), thereby permitting the smooth calculation of H as the observation point approaches the left branch of the caustic. Analogous considerations apply to the other branch. It is implied in these formulas that the region of overlap between Eqs. (57) and (59) occurs when $\beta_1 \approx \beta_2 \approx \beta_0$, so $F(\beta_1) \approx F(\beta_2) \approx F(\beta_0)$ and $\psi'''(\beta_1) \approx \psi'''(\beta_2) \approx \psi'''(\beta_0) < 0$. From Eq. (61a), this restriction is compatible with the requirement $-\xi \gg 1$ provided that $-\psi'''(\beta_0) \gg 1$ (i.e., the wavenumber k_0 or the coordinates u and $|u'|$ must be sufficiently large). For a uniform approximation not subject to $\beta_1 \approx \beta_2 \approx \beta_0$, one employs the more general result in Eq. (4.5.2).

The transition through a point on the left branch of the caustic for which $\psi'(\beta_0) = \psi''(\beta_0) = 0$ is followed most directly if the observation point approaches along the line

$$u = |u'|\frac{q_1''(\beta_0)}{-q_0''(\beta_0)}, \tag{62}$$

on which $\psi''(\beta_0) = 0$ [see Eq. (55)]. On this line, $\psi'(\beta_0) = v + q_0'(\beta_0)u + q_1'(\beta_0)|u'|$ is positive when the observation point lies to the right of the caustic, $\psi'(\beta_0) = 0$ on the caustic, and $\psi'(\beta_0) < 0$ after the observation point has passed through the caustic. From Eq. (61a), $(\beta_1 - \beta_2)^2 < 0$ when $\psi'(\beta_0) < 0$, so ξ is positive. For $\xi \gg 1$, the asymptotic approximation in Eq. (4.2.42a)

$$\text{Ai}(\xi) \sim \frac{1}{2\sqrt{\pi}\,\xi^{1/4}}e^{-(2/3)\xi^{3/2}}, \tag{63}$$

shows that the field decays exponentially on the "dark" side of the caustic, so the resulting contribution is negligible. Thus, Eq. (59) reduces correctly to Eq. (58) since, by the same considerations as above, one may find an overlapping region for these formulas.

The simplest description of the field near the cusp is obtained when the observation point approaches along the line $v = v_0$. From Eq. (56), the saddle points β_1 and β_3 are then located symmetrically with respect to $\beta_2 = 0$, either on the real or on the imaginary axis. For real $\beta_{1,3}$, the integration path proceeds as in Fig. 7.5.9(a) and the three saddle points contribute, while for imaginary $\beta_{1,3}$, only the saddle point $\beta_2 = 0$ is significant and the path is similar to the one in Fig. 7.5.9(b), with β_3 replaced by β_2 at the origin. When $u < \hat{u}$ and not near the cusp, the asymptotic field solution is the same as in Eq. (57), with rays 2 and 4 in Fig. 7.5.8 coincident. Since $F(\beta)$, $\psi(\beta)$, and $\psi''(\beta)$ are even functions of β on the line $v = v_0$, one may rewrite Eq. (57) as

$$H \sim -\omega\epsilon_0\sqrt{\frac{2}{\pi}}\left\{\frac{F(0)}{\sqrt{\psi''(0)}}e^{i[\psi(0)+\pi/4]} + \frac{2F(\beta_1)}{\sqrt{|\psi''(\beta_1)|}}e^{i[\psi(\beta_1)-\pi/4]}\right\}, \quad u < \hat{u}, \quad v = v_0. \tag{64}$$

For u sufficiently larger than $\hat{u}$, Eq. (58) applies with $\beta_s = 0$ whence the field along the ray $v = v_0$ experiences a phase retardation of $90°$ upon passing through the cusp at $u = \hat{u}$. These formulas fail as $u \to \hat{u}$ since $\psi''(0) \to 0$. $\psi'(0) = 0$ on $v = v_0$ and $\psi'''(0)$ vanishes identically so that the first non-vanishing derivative is $\psi''''(0)$. Hence, the cusp (focal) region is characterized by a coalescence of three saddle points.

The formula required for the detailed transition through the focal region is

$$H \sim \frac{i\omega\epsilon_0 e^{+i3\pi/8}}{-2^{1/4}\sqrt{\pi}} e^{i[\psi(0)+\psi(\beta_1)]/2} D_{-1/2}\{e^{-i(3\pi/4)}\sqrt{2[\psi(\beta_1) - \psi(0)]}\}\left[\frac{24}{-\psi''''(0)}\right]^{1/4} F(0),$$

$$(65)$$

where $D_{-1/2}(z)$ is the parabolic cylinder function of order $-\frac{1}{2}$, $[-\psi''''(0)]$ is positive, and $\arg[\psi(\beta_1) - \psi(0)]$ equals zero or π when $u < \hat{u}$ and $u > \hat{u}$, respectively. Use of the large-argument approximations,

$$D_{-\nu}(z) \sim \frac{e^{-z^2/4}}{z^\nu}\left[1 + O\left(\frac{1}{z^2}\right)\right], \qquad |\arg z| < \frac{3\pi}{4}, \qquad (66a)$$

$$D_{-\nu}(z) \sim \frac{e^{-z^2/4}}{z^\nu} - \frac{\sqrt{2\pi}}{\Gamma(\nu)}\frac{e^{z^2/4+i\nu\pi}}{z^{1-\nu}} + O[z^{\nu-3}, z^{-(2+\nu)}], \qquad -\frac{\pi}{4} > \arg z > -\frac{5\pi}{4}, \qquad (66b)$$

reduces Eq. (65) to Eqs. (58) and (64), respectively, thereby permitting the complete transition through the cusp. In this reduction, the following approximate identities are helpful:

$$\psi''(\beta_1) \approx -2\psi''(0), \qquad \left[\frac{24}{-\psi''''(0)}\right]^{1/4} \approx -\frac{i2\sqrt{2}\,[\psi(\beta_1) - \psi(0)]^{1/4}}{\sqrt{\psi''(\beta_1)}}, \qquad (67)$$

and it is assumed that the argument of the parabolic cylinder function may be large although β_1 does not differ greatly from zero (i.e., k_0 or u and $|u'|$ are large). One observes that the field far from the caustic behaves like $|\psi''|^{-1/2}$, near the caustic like $|\psi'''|^{-1/3}$, and near the cusp like $|\psi''''|^{-1/4}$. Since ψ contains a large parameter, these variations illustrate the successive field-strength enhancement as the observer approaches the caustic and the focal region.

The above-described focusing phenomena occur within a confined region near the interface and are not in evidence at great distances from the boundary. This follows from the ray picture in Fig. 7.5.8 which shows that observation points outside the caustic are reached by a single ray only; the corresponding far-field solution is given by the single term in Eq. (58). It may also be emphasized that analogous considerations are applicable for media with more general anisotropy (e.g., gyrotropic), in which the refractive index curves have a more complicated shape than that shown in Fig. 7.5.2 (see Sec. 8.3b). The transmitted field is generally given by integrals of the type shown in Eq. (51), but the functional dependence of the wavenumber q_1 on β is then specified by a dispersion relation that is usually much more involved than the one in Eq. (4). Focusing effects may arise whenever $\psi''(\beta) \to 0$ [see Eq. (55)] (i.e., when the real quantities q_0'' and q_1'' have opposite algebraic signs). The sign of q'' is related to the curvature of the $q(\beta)$ versus β curves, so an inspection of the refractive index plots reveals the possible existence, or not, of focal regions. From these considerations, focusing effects are absent when the uniaxially anisotropic half-space has a refractive index plot as in Fig. 7.5.2(a) since q_0'' and q_1'' then have the same algebraic sign.

7.5f Radiation from a Transverse Electric Dipole

The preceding sections have dealt with the radiation from a line source of magnetic current embedded in a uniaxially anisotropic plasma half-space. In view of the symmetries inherent in the physical structure for this special form of excitation, the electromagnetic field problem has been reducible to a single scalar boundary value problem, and a mathematical formulation has been possible in terms of E modes only. Since the plasma medium generally has a different effect upon the E and H modes, as noted from the two different refractive index diagrams in Figs. 7.1.1, it is instructive to treat a configuration in which both mode types must be considered.[13] A transversely directed dipole source belongs in this category, and incorporates, furthermore, certain three-dimensional aspects that are absent in the line-source excitation.

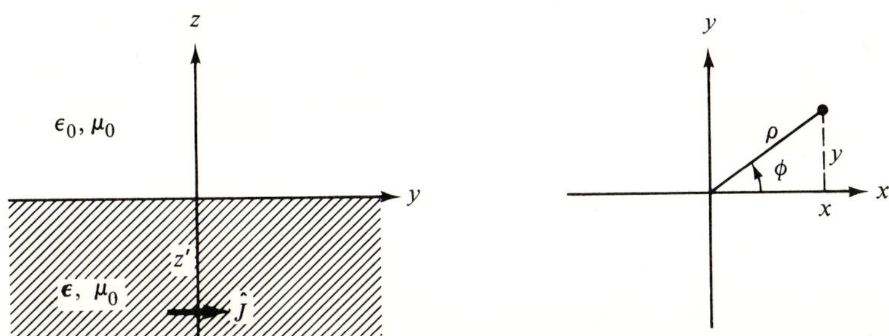

FIG. 7.5.10 Three-dimensional problem.

The physical arrangement is shown in Fig. 7.5.10, where the source distribution is taken as an electric current element oriented along the y direction,

$$\hat{\mathbf{J}}(\mathbf{r}, t) = \mathbf{y}_0 \delta(\mathbf{r} - \mathbf{r}')e^{-i\omega t}, \quad \mathbf{r}' \equiv (x', y', z') = (0, 0, z'), \quad z' < 0. \qquad (68)$$

To simplify the analysis, the interface is now assumed to be coincident with the $z = 0$ plane, corresponding to $\alpha = 0$ in Fig. 7.5.1. Alternative representations for the electromagnetic fields may be written down directly from the general results in Sec. 7.2. It is convenient to calculate first the voltage and current potentials $V''(\mathbf{r})$ and $I'(\mathbf{r})$ in Eqs. (7.2.13a) and (7.2.13b) and then to compute the fields via Eqs. (7.2.11a) and (7.2.11b). To utilize the symmetries inherent in excitation by a point source, it is convenient, as in the isotropic problem in Sec. 5.5b, to regard the infinite cross section transverse to z as a cylindrical waveguide, with the scalar mode functions Φ_i and ψ_i taken from Eqs. (3.2.78b). Since the dipole is located on the waveguide axis $\rho' = 0$, the resulting integral representations for the functions $\nabla'_t \mathscr{S}'_d$ and $\nabla'_t \mathscr{S}''_d$ may be simplified by considerations directly analogous to those employed in Sec. 5.5b, and one may easily derive the expressions for the potential functions. For the E modes,

$$I'(\mathbf{r}) = \frac{i}{4\pi} \sin \phi \frac{\partial}{\partial z'} \int_0^\infty J_1(\beta\rho) \frac{h(\beta)}{\kappa'_1} \, d\beta, \tag{69}$$

where

$$h(\beta) = \begin{cases} e^{i\kappa'_1 |z-z'|} - \Gamma'(\beta)e^{-i\kappa'_1(z+z')}, & z < 0, \\ [1 - \Gamma'(\beta)]e^{-i\kappa'_1 z'} e^{i\kappa_2 z}, & z > 0, \end{cases} \tag{69a}$$

and

$$\Gamma'(\beta) = \frac{\kappa'_2 - \kappa'_1}{\kappa'_2 + \kappa'_1}, \quad \kappa'_1 = \sqrt{k_0^2 - \frac{\beta^2}{\epsilon}}, \quad \kappa'_2 = \sqrt{k_0^2 - \beta^2}, \quad \epsilon = \frac{\epsilon_z}{\epsilon_0}, \tag{69b}$$

with the square roots defined to be positive when real, and positive imaginary otherwise. The longitudinal dependence in Eq. (69a) is the same as in Eqs. (6a) and (6b) if one recalls that $\alpha = 0$ in the present problem, so $q_1 = \kappa'_1$, $q_2 = -\kappa'_1$, $q_0 = \kappa'_2$, $u = z$, $u' = z'$. The modal network problem is the same as in Fig. 5.5.6, with the characteristic impedance and propagation constant taken from Eq. (7.2.8a).

For the H modes, the modal parameters κ''_i and Z''_i are identical inside and outside the plasma medium [see Eq. (7.2.8b), with $\mu_t = \mu_z = \mu_0$, $\epsilon_t = \epsilon_0$], so the corresponding network problem is the same as for an infinite vacuum. The potential function V'' is then given by

$$V''(\mathbf{r}) = -\frac{\omega\mu}{4\pi} \cos \phi \int_0^\infty J_1(\beta\rho) \frac{e^{i\kappa''_2 |z-z'|}}{\kappa''_2} \, d\beta, \qquad \kappa''_2 \equiv \kappa'_2, \tag{70}$$

which expression may be evaluated in closed form[7] via the formula [see Eq. (5.4.19) and the first term of Eq. (5.5.31)]

$$\int_0^\infty J_1(\beta\rho) \frac{e^{i\kappa''_2 \zeta}}{\kappa''_2} \, d\beta = \frac{1}{k_0\rho} [\exp(ik_0|\zeta|) - \exp(ik_0\sqrt{\rho^2 + \zeta^2})]. \tag{70a}$$

Such a simple reduction is not possible for the E-mode integral in Eq. (69). The different effect of the plasma medium on the E-mode and H-mode constituents is clearly evident since Eq. (70), in contrast to Eq. (69), shows no dependence on the boundary at $z = 0$.

The electromagnetic fields may now be determined from Eqs. (7.2.11a) and (7.2.11b). Some interesting differences between two-dimensional and three-dimensional field solutions, especially in the vicinity of caustics and cusps, may be ascertained from an asymptotic analysis in the vacuum half-space, with the choice $\epsilon < 0$. To simplify the discussion, we shall consider the magnetic field in the plane $x = 0$ which then has only a single component, $H \equiv H_x$; the field components at other points in space exhibit quite a similar behavior, modified by the azimuthal variation $\sin \phi$ or $\cos \phi$. In the plane $x = 0$, with $z > 0$, the E-mode contribution H' is found to be

$$H' = \frac{1}{2\pi} \int_0^\infty \frac{\kappa_1'}{\kappa_1' + \kappa_2'} \frac{d}{d\rho} J_1(\beta\rho) e^{i(\kappa_2'z + \kappa_1'|z'|)} \, d\beta, \tag{71}$$

while the closed-form expression for the H-mode portion of the field is[7]

$$H'' = \frac{1}{4\pi y^2} \left[e^{ik_0(z + |z'|)} - \frac{z + |z'|}{r} e^{ik_0 r} \right], \qquad r = \sqrt{y^2 + (z + |z'|)^2}. \tag{72}$$

Since $x = 0$, one has $\rho = \sqrt{x^2 + y^2} = y$. The total magnetic field in the $x = 0$ plane is given by $H = H' + H''$.

To facilitate an asymptotic evaluation of the integral in Eq. (71), it is convenient to separate observation points on the z axis ($\rho = 0$) from those at large distances from this axis. For the former, use of

$$\frac{d}{d\rho} J_1(\beta\rho)\Big|_{\rho=0} = \frac{\beta}{2}, \tag{73}$$

leads to

$$H' = \frac{1}{4\pi} \int_0^\infty \frac{\kappa_1'}{\kappa_1' + \kappa_2'} \beta e^{i(\kappa_2'z + \kappa_1'|z'|)} \, d\beta, \qquad \rho = 0, \tag{74}$$

while for the latter, an infinite integral representation is useful [see Eqs. (3.2.68) et seq.]:

$$H' = \frac{1}{4\pi} \frac{\partial}{\partial\rho} \int_{-\infty}^\infty \frac{\kappa_1'}{\kappa_1' + \kappa_2'} H_1^{(1)}(\beta\rho) e^{i(\kappa_2'z + \kappa_1'|z'|)} \, d\beta. \tag{75}$$

In the last expression, the integration path is indented into the upper half of the complex β plane to avoid the branch-point singularity at $\beta = 0$. For an asymptotic procedure accommodating arbitrary values of ρ, see Reference 14.

The asymptotic calculation of H' in Eq. (75) may now proceed in the usual manner with a replacement of the Hankel function by its large-argument approximation in Eq. (5.3.13b), provided that the integration path is deformed away from the region $\beta = 0$. The resulting integrand is identical in form with that in Eq. (51), except for a differently defined amplitude function $F(\beta)$, whence the saddle-point configuration is the same as in Sec. 7.5e provided that the rectilinear variable v is identified with the radial variable ρ. A geometrical interpretation of the saddle-point condition leads to the ray diagram in Fig. 7.5.8, modified so that $\alpha = 0$, with observation points lying in the region $v > 0$ since the radial variable ρ is always positive. In view of the rotational symmetry of the ray system about the $\rho = 0$ axis, the caustic in Fig. 7.5.8 is now a surface of revolution and the (line) cusp at C in the two-dimensional problem becomes a point cusp. Upon noting the angles of inclination φ_0 of the refracted rays that pass through an observation point with $\rho > 0$ (i.e., $v > 0$ in Fig. 7.5.8), and recalling the saddle-point condition $\beta_s = k_0 \sin \varphi_0$, one may verify that the integration path in Eq. (75) may be deformed into the steepest-descent paths through the various saddle points without encountering any of the branch-point singularities. The asymptotic approximation of the E-mode contribution to the field is therefore analogous to that in Eqs. (57) and (58) for observation points off the caustic, and to that in Eq. (59) for observation points near or on the caustic. Since the previously mentioned Hankel function approximation contains a factor $\rho^{-1/2}$, the three-dimensional field decays properly according to (distance)$^{-1}$, in contrast to the (distance)$^{-1/2}$ dependence of the two-dimensional field in the line-source problem.

In the preceding considerations, valid for observation points "far" from the axis $\rho = 0$, the contributing saddle points have not been located near the origin of the complex β plane. If $\rho = 0$, however, the integrand of Eq. (74) has the saddle points

$$\beta_1 = k_0 \left(\frac{1 - b^2}{1 + b^2/|\epsilon|} \right)^{1/2}, \quad \beta_2 = 0, \quad b = \frac{z|\epsilon|}{|z'|}, \tag{76}$$

as noted from Eq. (56), with $\alpha = 0$. The asymptotic evaluation of the contribution H_1' from β_1 when $0 < z < \hat{z}$, where $\hat{z} = |z'|/|\epsilon|$ is the coordinate at the cusp, proceeds as in the line-source problem and yields

$$H_1' \sim \frac{1}{4\pi} \sqrt{\frac{2\pi}{\psi''(\beta_1)}} \frac{\beta_1 \kappa_1'(\beta_1)}{\kappa_1'(\beta_1) + \kappa_2'(\beta_1)} e^{i\psi(\beta_1) - i\pi/4}, \quad \psi(\beta) = \kappa_2' z + \kappa_1' |z'|, \tag{77}$$

thereby exhibiting a dependence $\sim$ (distance)$^{-1/2}$. This behavior is explained by the observation that points on the axis of the rotationally symmetric ray configuration are reached by refracted rays arriving from all angles $0 < \phi \leq 2\pi$, so the z axis is a caustic of the refracted ray system. Hence, there exists on the axis a field enhancement not evident at off-axis points.

The saddle point $\beta_2 = 0$ describes a ray that progresses along the z axis and represents a conventional field contribution uninfluenced by the axial focusing described above. Its decay should therefore be described by the ordinary inverse distance dependence. In an asymptotic evaluation of the integral in Eq. (74) for this case, one must take account of the vanishing of the amplitude function $g(\beta) = \beta \kappa_1' [\kappa_1' + \kappa_2']^{-1}$ at $\beta = 0$. On application of the procedure in Sec. 4.2b to the semiinfinite interval $0 \leq \beta < \infty$, the saddle point at $\beta = 0$ is transformed via $\psi(\beta) = \psi(0) + is^2$ into a saddle point at $s = 0$, and the amplitude function $G(s) = g(\beta) \, d\beta/ds$ is then expanded in a power series about $s = 0$. Since $G(0) = 0$, the first non-vanishing term arises from $G'(0)s$, and the corresponding integral to be evaluated is $\int_0^\infty s e^{-s^2} \, ds = \frac{1}{2}$. The first-order asymptotic approximation H_2' arising from the saddle point at $\beta_2 = 0$ is then found to be

$$H_2' \sim \frac{1}{4\pi} \frac{i}{2\psi''(0)} e^{i\psi(0)}, \tag{78}$$

and the total E-mode contribution to the magnetic field on the z axis is $H' \sim H_1' + H_2'$. The difference between the three-dimensional field behavior here and the two-dimensional one described previously is to be noted and has been explained in physical terms. Since $\psi''(0) \gtrless 0$ when $z \lessgtr \hat{z}$, one notes a phase retardation of π along a ray passing through a point cusp.

Formulas (77) and (78) break down when $z \approx \hat{z}$ because then $\beta_1 \to 0$ and $\psi''(0) \to 0$. In this case, one may employ the more accurate formula derived in Sec. 4.5b, which accounts explicitly for the simultaneous presence of three closely spaced saddle points. The resulting expression for H' is[13]

$$H' \sim \frac{1}{4\pi} \frac{e^{i[\psi(0)+\psi(\beta_1)]/2-i\pi/4}}{4\sqrt{2}} \sqrt{\frac{24}{-\psi''''(0)}} \, D_{-1}\{e^{-i3\pi/4}\sqrt{2[\psi(\beta_1) - \psi(0)]}\}, \qquad (79)$$

with the square root in the argument of the parabolic cylinder function defined as in Eq. (65). By considerations analogous to those carried out for Eq. (65), this formula may be reduced to the approximations in Eqs. (77) and (78), provided that $\beta_1 \approx 0$, $\sqrt{\psi(\beta_1) - \psi(0)} \gg 1$, and the combined expressions therefore permit the calculation of H' along the entire positive z axis.

The E-mode reflected field in the plasma half-space may be evaluated in a similar manner, and exhibits a behavior analogous to that in the line-source problem if proper account is taken of the equivalence between v and ρ.

PROBLEMS

1. From the E-mode dispersion equation (7.1.5), derive the refractive index $n(\bar{\theta})$, and therefore the ray refractive index $N(\theta)$ in Eq. (7.3.3c), by use of the relations $N(\theta) = n(\bar{\theta}) \cos(\theta - \bar{\theta})$, $\tan(\theta - \bar{\theta}) = (1/n)(\partial n/\partial\bar{\theta})$ (see Fig. 7.1.1 and Problem 33 of Chapter 1).

2. Apply the coordinate scaling transformation in Eq. (7.2.22) to the integral representation in Eq. (5.4.12c) for the isotropic free-space Green's function to deduce the Green's function for the unbounded uniaxially anisotropic medium. Compare the result with the integral formulation in Eq. (7.3.8) and show that the two expressions are equivalent when $\epsilon > 0$. Explain why the expression in Eq. (7.3.8) remains valid when $0 < \arg \epsilon \leq \pi$ whereas the integral expression obtained by coordinate scaling does not.

3. Show that the Green's function in Eq. (7.3.14) for an unbounded uniaxially anisotropic medium can be expressed in the invariant form of Eq. (1.7.56), which involves the geometrical properties of the dispersion surface shown in Fig. 7.1.1.

4. A perfectly conducting cylinder with radius $\rho = a$ is embedded in an infinite homogeneous, uniaxial plasma whose optic axis is parallel to that of the cylinder. Time-harmonic excitation is provided by an axial electric dipole located at a distance $\rho > a$.
   (a) Referring to results in Sec. 6.7 for the cylinder diffraction problem in an isotropic medium, and employing coordinate scaling [see Eq. (7.2.22)], derive expressions for the total electromagnetic fields in the uniaxial environment.
   (b) For $\omega_p < \omega$ and $\omega_p > \omega$, derive approximate field solutions for very small

cylinder radius, and compare the results with those for the isotropic case. Discuss the physical properties of the scattered field, especially for $\omega < \omega_p$ ($\epsilon < 0$), utilizing the concept of an equivalent line source (cf. Sec. 7.3c) to represent the effect of the induced currents on the cylinder.

5. A line source of unit strength magnetic currents in a uniaxially anisotropic dielectric ϵ is located at Q in the presence of a smoothly curved, perfectly conducting cylindrical surface as shown in Fig. P7.1. The optic axis extends along the

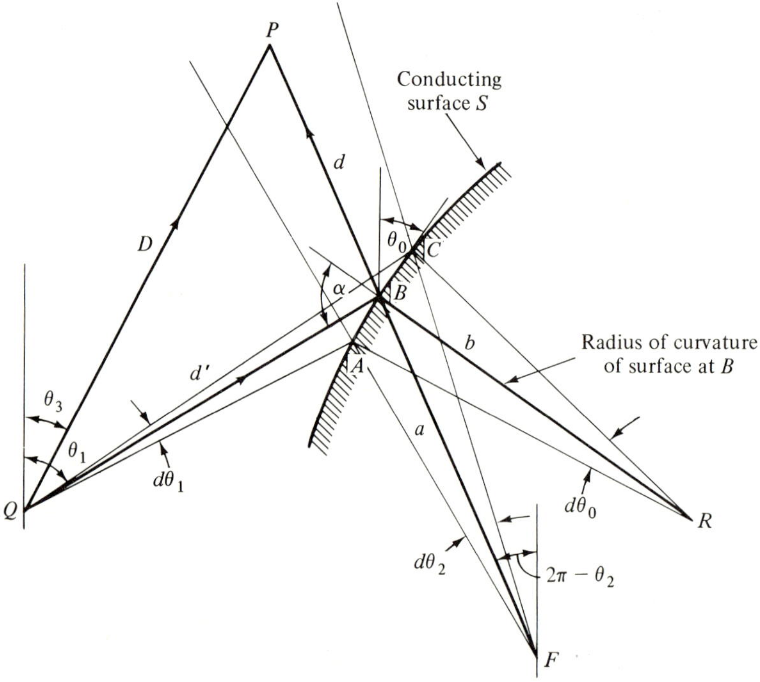

FIG. P7.1   Curved scatterer in anisotropic dielectric.

y direction ($\theta = 0$). Using the methods of geometrical optics (Sec. 1.7) and referring to Fig. P7.1 for definition of distance and angle variables, show that the magnetic field at the observation point P is given by:[10]

$$H = H_i + H_r \tag{1}$$

$$H_i = -\frac{\omega\epsilon_0\sqrt{\epsilon}}{2\sqrt{2\pi k_0\,DN(\theta_3)}}\,e^{ik_0 DN(\theta_3)-i\pi/4} \tag{1a}$$

$$H_r = \frac{-\omega\epsilon_0\sqrt{\epsilon}}{2\sqrt{2\pi}}\,\sqrt{\frac{N(\theta_2)\sin(\theta_1-\theta_0)}{N^2(\theta_1)\sin(\theta_2-\theta_0)}}$$

$$\times\frac{\exp\{ik_0[d'\,N(\theta_1) + d\,N(\theta_2)] - (i\pi/4)\}}{\sqrt{k_0\{d' + [d\,m\sin(\theta_1-\theta_0)/\sin(\theta_0-\theta_2)] + (d'\,d\,n/b)[1/\sin(\theta_0-\theta_2)]\}}} \tag{1b}$$

where $\boldsymbol{\epsilon} = \mathbf{y}_0\mathbf{y}_0\epsilon_0 + \mathbf{z}_0\mathbf{z}_0\epsilon_0\epsilon$, $N(\theta) = \sqrt{\cos^2\theta + \epsilon\sin^2\theta}$, $m = (\partial\theta_2/\partial\theta_1)$, and $n = (\partial\theta_2/\partial\theta_0)$; the latter quantities may be evaluated from the ray reflection law in Eq. (7.5.9b) [with $\varphi \to \theta + \alpha$, $\theta_0 = (\pi/2) - \alpha$].

Show that this result reduces correctly to the one in Problem 30 of Chapter 1 for a cylindrical surface in an isotropic medium when $\epsilon = 1$, and to the infinite plane result in Sec. 7.4c when $b \to \infty$.

6. Utilizing the respective optical path lengths L_{12} and L_{23} from a source point 1 to an observation point 3 along incident and reflected trajectories, with point 2 lying on a plane boundary, show that the ray reflection law in Eq. (7.5.9b) can be derived by applying Fermat's principle

$$\delta(L_{12} + L_{23}) = 0, \qquad L_{ab} = \int_a^b N\,ds, \tag{2}$$

requiring the optical length to be an extremum. δ denotes the variational derivative and N the ray refractive index.

7. A uniform waveguide region is bounded transversely (if at all) by a perfectly conducting surface and is filled with an anisotropic dielectric having $\epsilon_t = $ constant, $\epsilon_z(\boldsymbol{\rho}, z) = $ arbitrary. The permeability $\mu_t = \mu_z = \mu$ is constant and isotropic.
(a) Show that, in this region, H modes can exist and propagate as in an isotropic homogeneous medium with wavenumer $k = \omega\sqrt{\mu\epsilon_t}$.
(b) Show that E modes exist also and specify the corresponding field problem.

8. Consider the configuration shown in Fig. P7.2, where a line source of magnetic currents is located in the vacuum half-space $z > 0$ and the field is observed in

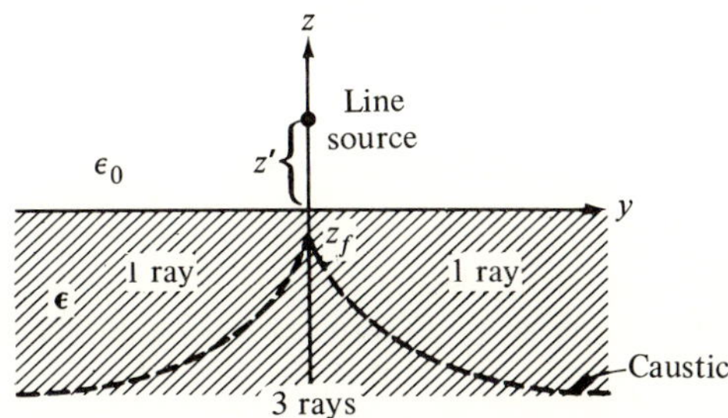

FIG. P7.2   Caustic for source exterior to plasma half-space ($|z_f| = z'|\epsilon|$).

the uniaxially anisotropic plasma half-space $z < 0$ described by the dielectric tensor $\boldsymbol{\epsilon} = \mathbf{y}_0\mathbf{y}_0\epsilon_0 + \mathbf{z}_0\mathbf{z}_0\epsilon_0\epsilon$.
(a) Show that for $\epsilon_z < 0$, the refracted ray system forms a caustic given by the equation

$$\left|\frac{z\sqrt{|\epsilon|}}{z'}\right|^{2/3} - \left|\frac{y\sqrt{1 + |\epsilon|}}{z'|\epsilon|}\right|^{2/3} = |\epsilon|.$$

(b) Determine the asymptotic field behavior (for large k_0) everywhere in the plasma half-space.

REFERENCES

1. BORN, M. and E. WOLF, *Principles of Optics*, Chapter XIV. New York: Pergamon Press, 1959.

2. LANDAU, L. D. and E. M. LIFSHITZ, *Electrodynamics of Continuous Media*, Chapter II. New York: Pergamon Press, 1960.

3. BUDDEN, K. G., *Radio Waves in the Ionosphere.* Cambridge, England: Cambridge University Press, 1961.

4. GINZBURG, V. L., *The Propagation of Electromagnetic Waves in Plasmas.* New York: Pergamon Press, 1964.

5. LAX, B. and J. BUTTON, *Microwave Ferrites and Ferrimagnetics.* New York: McGraw-Hill, 1962.

6. FELSEN, L. B., "Propagation and diffraction in uniaxially anisotropic regions. I—Theory. II—Applications," *Proc. IEE (London)* 111 (1964), pp. 445–464.

7. CLEMMOW, P. C., *The Plane Wave Spectrum Representation of Electromagnetic Fields*, Chapter 8. New York: Pergamon Press, 1966.

8. ARBEL, E. and L. B. FELSEN, "Theory of radiation from sources in anisotropic media. Part I—General sources in stratified media. Part II—Point source in an infinite homogeneous medium," in *Electromagnetic Theory and Antennas* (E. C. Jordan, Ed.), pp. 391–459. New York: Pergamon Press, 1963.

9. STARAS, H., "The impedance of an electric dipole in a magneto-ionic medium" *IEEE Trans. on Antennas and Propagation* **AP-12** (1964), pp. 695–702.

10. RULF, B. and L. B. FELSEN, "Diffraction by objects in anisotropic media," *Proc. of Symposium on "Quasi-Optics,"* **XIV**, pp. 107–147, Brooklyn, New York: Polytechnic Press, 1964.

11. FELSEN, L. B., "Radiation from a uniaxially anisotropic plasma half space," *IEEE Trans. on Antennas and Propagation* **AP-11** (1963), pp. 469–484.

12. BREKHOVSKIKH, L. M., *Waves in Layered Media*, Sec. 22. New York: Academic Press, 1960.

13. FELSEN, L. B., "Focusing by an anisotropic plasma interface," *IEEE Trans. on Antennas and Propagation* **AP-12** (1964), pp. 624–635.

14. BAÑOS, A., *Dipole Radiation in the Presence of a Conducting Half Space*, Chapter 3. New York: Pergamon Press, 1966.

8. Fields in Anisotropic Regions

8.1 INTRODUCTION

The descriptive simplicity of wave excitation and propagation in uniaxial media no longer obtains in the case of general anisotropic media. As noted in Chapter 7, the reduced electromagnetic field equations in a uniaxial medium are characterized by a permittivity dyadic ϵ and permeability dyadic μ that are diagonal in appropriate real coordinate frames. In arbitrary anisotropic media, on the other hand, these dyadics cannot be so diagonalized and in a real coordinate basis, all matrix (tensor) elements of ϵ and μ are in general non-vanishing. However, in gyrotropic regions, which possess a symmetry axis along, say, the $\mathbf{b}_0$ direction, the ϵ and μ dyadics assume the form

$$\epsilon = \epsilon_t + \epsilon_z \mathbf{b}_0 \mathbf{b}_0, \qquad \mu = \mu_t + \mu_z \mathbf{b}_0 \mathbf{b}_0, \tag{1}$$

where ϵ_t and μ_t are in general non-diagonal dyadics transverse to $\mathbf{b}_0$; for a uniaxial medium, ϵ_t and μ_t are of course diagonal.

In the present chapter we shall consider guided-wave descriptions in anisotropic media, with special emphasis on electromagnetic fields in gyrotropic regions. Descriptions of general linear fields in anisotropic media can be effected either in terms of first-order field equations or, on elimination of non-electromagnetic field variables, can be reduced to higher-order equations for just the electromagnetic variables. As noted in Sec. 1.5, this reduction process implies the introduction of equivalent permittivity and permeability dyadics that in general are spatially and temporally dispersive operators containing ∇ and $\partial/\partial t$. The reduced field formulation frequently leads to analytical complexities in identification of energy expressions, reciprocity properties, eigenmodes, etc., and may even omit non-electromagnetic types of wave phenomena. The first-order formulation, in which electromagnetic and non-electromagnetic field variables are given equal status, avoids many of these difficulties. Nevertheless, because of the widespread use of reduced descriptions in the literature, both first-order and reduced descriptions will be employed in the following. In certain cases, for example in the guided-wave descriptions of a cold magneto-plasma with non-spatially dispersive parameters, the reduced description is adequate and does not give rise to the complexities noted above.

In the guided-wave description of a homogeneous linear field developed in Sec. 1.4, the overall field is represented as a superposition of eigenmodes of the form $\Psi_\alpha \exp(i\kappa_\alpha z)$. The variable z distinguishes the guiding or symmetry axis, κ_α denotes the eigenvalue or mode wavenumber, and the field eigenvector Ψ_α depends on the spatial variable $\boldsymbol{\rho}$ transverse to z and on the time t. Explicitly, the overall field representation takes the form (at source-free points)

$$\Psi(\mathbf{r}, t) = \sum_\alpha a_\alpha(0)\Psi_\alpha(\boldsymbol{\rho}, t)e^{i\kappa_\alpha z} = \sum_\alpha a_\alpha(z)\Psi_\alpha(\boldsymbol{\rho}, t), \tag{2}$$

where $a_\alpha(z)$ distinguishes the amplitude of the αth mode at z. For time harmonicity and unbounded cross sections, $\Psi_\alpha(\boldsymbol{\rho}, t) = \Psi_\alpha \exp[i(\mathbf{k}_t \cdot \boldsymbol{\rho} - \omega t)]$, while for bounded cross sections, $\Psi_\alpha(\boldsymbol{\rho}, t) = \Psi_\alpha(\boldsymbol{\rho}) \exp(-i\omega t)$; in either event, if the Ψ_α and their orthogonality properties are known, it is a simple matter to evaluate the amplitude coefficients $a_\alpha(z)$ from a knowledge of the total field Ψ at any reference plane z. In the following sections, the requisite information about the mode functions Ψ_α and wavenumbers κ_α will be ascertained for anisotropic media (Sec. 8.2) and gyrotropic media (Secs. 8.2 and 8.3).

Differences in the nature of wave propagation in isotropic and anisotropic regions are revealed by contrasting the transmission line behavior of mode amplitudes $a_\alpha(z)$ in such media. In all cases, assuming uniformity along the direction z, the z dependence of the a_α at source-free points is determined by the equations

$$\frac{d}{dz}a_\alpha = i\kappa_\alpha a_\alpha. \tag{3}$$

Only for regions reflection symmetric with respect to z, wherein there exists for each wave a_α a reflected wave with appropriate field symmetry and propagation wavenumber $\kappa_{-\alpha} = -\kappa_\alpha$, is it possible to develop a conventional transmission line description in terms of voltage and current amplitudes as we have done for isotropic and uniaxial regions. As noted in Sec. 7.1, in the guided-wave description of propagation in isotropic (unbounded) regions, one introduces a $\mathbf{k}_t, \omega$ basis of modes with an $\exp[i(\mathbf{k}_t \cdot \boldsymbol{\rho} - \omega t)]$ dependence and thereby reduces the overall field problem to two independent E- and H-mode transmission-line problems with identical propagation wavenumbers $\pm\kappa'_\alpha = \pm\kappa''_\alpha = \pm\kappa_\alpha$ but unequal characteristic impedances $\pm Z'_\alpha \neq \pm Z''_\alpha$. In a uniaxial region with transmission direction chosen along the uniaxial axis (i.e., $\mathbf{b}_0$ parallel to $\mathbf{z}_0$), the field problem is still reducible (Sec. 7.2) to conventional E- and H-mode transmission-line problems, but with the parameters $\pm\kappa'_\alpha \neq \pm\kappa''_\alpha$ and $\pm Z'_\alpha \neq \pm Z''_\alpha$ characterizing the direct and reflected waves. However, non-conventional transmission-line descriptions (see Sec. 8.2.h) are possible for gyrotropic regions, or more generally for any region admitting $\pm\kappa_\alpha$ waves.

For a gyrotropic region with transmission along the gyrotropic axis, one finds direct and reflected waves with $\pm\kappa'_\alpha \neq \pm\kappa''_\alpha$. Moreover, the corresponding characteristic impedances are not $\pm$ of one another but rather one has Z'_α and $-Z'^*_\alpha$; also, one has Z''_x and $-Z''^*_\alpha$, which are not the same as the corresponding primed quantities. For a gyrotropic region with the direction of transmission chosen perpendicular to the gyrotropic axis (see Sec. 8.4b), one finds similar direct and reflected waves, but the characteristic impedances of these waves are not simply related. In the case of a general anisotropic region, one has $\kappa_x \neq \kappa_\beta$ (i.e., it is not generally possible to find modes with equal but opposite wavenumbers). A schematization of these line descriptions in an unbounded electromagnetic region is displayed in Fig. 8.1.1.

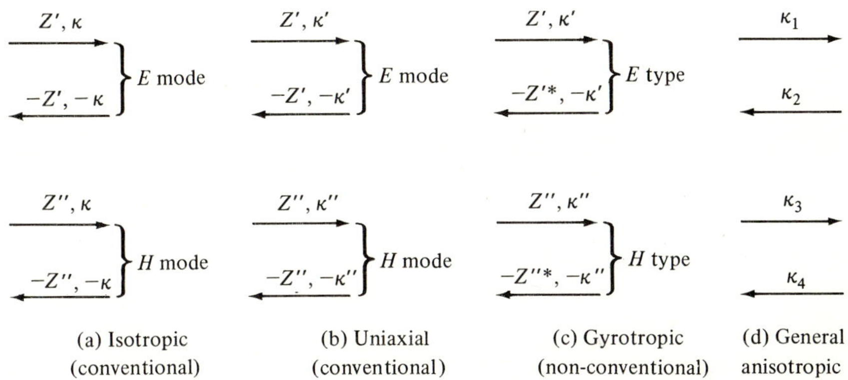

FIG. 8.1.1 Transmission-line descriptions in unbounded, homogeneous media (transmission direction along symmetry direction, if any).

8.2 GUIDED-WAVE REPRESENTATION IN ANISOTROPIC MEDIA (REDUCED FORMULATION)

8.2a Formulation for Arbitrary Media

As noted in Sec. 8.1, a reduced description of electromagnetic radiation and scattering in non-spatially dispersive media involves only the electromagnetic fields **E** and **H** and employs an equivalent permittivity dyadic $\boldsymbol{\epsilon}$ and permeability dyadic $\boldsymbol{\mu}$. For generality, both the permittivity $\boldsymbol{\epsilon} = \boldsymbol{\epsilon}(\mathbf{r}, \omega)$ and permeability $\boldsymbol{\mu} = \boldsymbol{\mu}(\mathbf{r}, \omega)$ are assumed $\mathbf{r}$ and ω dependent, thereby making the discussion applicable to inhomogeneous anisotropic media under steady-state conditions. This reduced formulation will be seen as a generalization of procedures utilized previously in Chapters 5 and 7 for isotropic and uniaxial media.

In a reduced formulation, the steady-state Maxwell equations in a source-excited, inhomogeneous, anisotropic medium can be written in the operator form [omitting the time dependence $\exp(-i\omega t)$]

$$L\Psi(\mathbf{r}) = -\Phi(\mathbf{r}), \tag{1a}$$

where, as in Sec. 1.4,

$$L \to -i\begin{bmatrix} \omega\boldsymbol{\epsilon} & -i\boldsymbol{\nabla}\times\mathbf{1} \\ i\boldsymbol{\nabla}\times\mathbf{1} & \omega\boldsymbol{\mu} \end{bmatrix}, \quad \Psi \to \begin{bmatrix} \mathbf{E} \\ \mathbf{H} \end{bmatrix}, \quad \Phi \to \begin{bmatrix} \mathbf{J} \\ \mathbf{M} \end{bmatrix}, \tag{1b}$$

$\mathbf{J}$ and $\mathbf{M}$ being the equivalent steady-state electric and magnetic current densities that appear in a reduced description on eliminating dynamical variables.

To effect a guided-wave representation with z as the transmission direction, one decomposes the field operator L into components that depend on either $\boldsymbol{\nabla}_t$ or $\partial/\partial z$ $[\boldsymbol{\nabla} = \boldsymbol{\nabla}_t + \mathbf{z}_0(\partial/\partial z)]$,

$$L = -i\left(K - \frac{\Gamma}{i}\frac{\partial}{\partial z}\right), \tag{2a}$$

where

$$K \to \begin{bmatrix} \omega\boldsymbol{\epsilon} & -i\boldsymbol{\nabla}_t\times\mathbf{1} \\ i\boldsymbol{\nabla}_t\times\mathbf{1} & \omega\boldsymbol{\mu} \end{bmatrix} \quad \text{and} \quad \Gamma \to \begin{bmatrix} 0 & -\mathbf{z}_0\times\mathbf{1} \\ \mathbf{z}_0\times\mathbf{1} & 0 \end{bmatrix}. \tag{2b}$$

In a z-oriented vector basis the dyadics $\boldsymbol{\epsilon}$ and $\boldsymbol{\mu}$ may be represented for a general gyroelectric and gyromagnetic medium as

$$\boldsymbol{\epsilon} = \boldsymbol{\epsilon}_t + \mathbf{z}_0\mathbf{z}_0\epsilon_z + \mathbf{z}_0\bar{\boldsymbol{\epsilon}}_{zt} + \bar{\boldsymbol{\epsilon}}_{tz}\mathbf{z}_0,$$
$$\boldsymbol{\mu} = \boldsymbol{\mu}_t + \mathbf{z}_0\mathbf{z}_0\mu_z + \mathbf{z}_0\bar{\boldsymbol{\mu}}_{zt} + \bar{\boldsymbol{\mu}}_{tz}\mathbf{z}_0, \tag{2c}$$

where $\mathbf{u}_t$ is a transverse dyadic such that $\mathbf{u}_t\cdot\mathbf{z}_0 = 0 = \mathbf{z}_0\cdot\mathbf{u}_t$ and $\bar{\mathbf{u}}_{tz}, \bar{\mathbf{u}}_{zt}$ are vectors transverse to $\mathbf{z}_0$, $\mathbf{u}$ representing either $\boldsymbol{\epsilon}$ or $\boldsymbol{\mu}$. For a gyrotropic medium $\bar{\boldsymbol{\epsilon}}_{zt} = 0 = \bar{\boldsymbol{\epsilon}}_{tz}$ and $\bar{\boldsymbol{\mu}}_{zt} = 0 = \bar{\boldsymbol{\mu}}_{tz}$; this permits a simplification in the analysis, but for the time being we shall continue to treat the general case.

As in the isotropic case, the longitudinal components E_z and H_z of electric and magnetic fields are derivable from the transverse components $\mathbf{E}_t$ and $\mathbf{H}_t$. Thus from Eqs. (1) and (2) one derives

$$\begin{bmatrix} E_z \\ H_z \end{bmatrix} = \begin{bmatrix} -\epsilon_z^{-1}\bar{\boldsymbol{\epsilon}}_{zt} & (i\omega\epsilon_z)^{-1}\boldsymbol{\nabla}_t\times\mathbf{z}_0 \\ -(i\omega\mu_z)^{-1}\boldsymbol{\nabla}_t\times\mathbf{z}_0 & -\mu_z^{-1}\bar{\boldsymbol{\mu}}_{zt} \end{bmatrix}\cdot\begin{bmatrix} \mathbf{E}_t \\ \mathbf{H}_t \end{bmatrix} + \begin{bmatrix} \dfrac{J_z}{i\omega\epsilon_z} \\ \dfrac{M_z}{i\omega\mu_z} \end{bmatrix}$$
$$\to \mathcal{N}\bar{\Psi} + \theta_z, \tag{3a}$$

where it is recalled that $\boldsymbol{\nabla}_t\times\mathbf{z}_0\cdot\mathbf{A}_t \equiv \boldsymbol{\nabla}_t\cdot(\mathbf{z}_0\times\mathbf{A}_t)$ and

$$\bar{\Psi} \to \begin{bmatrix} \mathbf{E}_t \\ \mathbf{H}_t \end{bmatrix}, \quad \theta_z \to \begin{bmatrix} \dfrac{J_z}{i\omega\epsilon_z} \\ \dfrac{M_z}{i\omega\mu_z} \end{bmatrix}. \tag{3b}$$

On elimination of the longitudinal field components from Eqs. (1) and (2) via

Eqs. (3a), one can obtain defining equations for the transverse wavevector $\bar{\Psi}(\mathbf{r})$. Because of the relative complexity of the resulting transverse equations in the general anisotropic case, we shall not exhibit them explicitly (however, see the source-free transverse modal equations (6) below).

To obtain the mode functions characteristic of guided waves propagating along z, we seek source-free ($\Phi = 0$) solutions of Eq. (1a) with the form

$$\Psi_\alpha(\mathbf{r}) = \Psi_\alpha(\boldsymbol{\rho})e^{i\kappa_\alpha z}, \tag{4a}$$

where the $\Psi_\alpha(\boldsymbol{\rho})$ and κ_α are eigenfunctions and eigenvalues, respectively, and $\mathbf{r} = \boldsymbol{\rho} + z\mathbf{z}_0$. The presence of the $\exp(i\kappa_\alpha z)$ factor implies that medium properties are independent of z, a restriction subsequently removed. As in Sec. 1.4, substitution of Eq. (4a) into the source-free equation (1a) yields the eigenvalue equation

$$K\Psi_\alpha = \kappa_\alpha \Gamma\Psi_\alpha \tag{4b}$$

subject on the transverse boundary s, if any, to the dyadic impedance condition

$$\mathbf{v} \times \mathbf{E}_\alpha = \mathscr{L} \cdot \mathbf{H}_\alpha \qquad \text{on } s, \tag{4c}$$

or in a $\mathbf{s}_0$, $\mathbf{z}_0$ basis,

$$-\mathbf{s}_0 \cdot \mathbf{E}_\alpha = \mathscr{L}_{zs}\mathbf{H}_\alpha \cdot \mathbf{s}_0 + \mathscr{L}_{zz}\mathbf{H}_\alpha \cdot \mathbf{z}_0,$$
$$\mathbf{z}_0 \cdot \mathbf{E}_\alpha = \mathscr{L}_{ss}\mathbf{H}_\alpha \cdot \mathbf{s}_0 + \mathscr{L}_{sz}\mathbf{H}_\alpha \cdot \mathbf{z}_0, \tag{4d}$$

where the tangential and normal unit vectors $\mathbf{s}_0$ and $\mathbf{v}$, respectively, are related by $\mathbf{v} \times \mathbf{z}_0 = \mathbf{s}_0$. The transverse nature of the electromagnetic field implies that the z components of the mode fields are derivable from the transverse components. If one decomposes the mode fields as

$$\mathbf{E}_\alpha = \mathbf{E}_{t\alpha} + E_{z\alpha}\mathbf{z}_0, \qquad \mathbf{H}_\alpha = \mathbf{H}_{t\alpha} + H_{z\alpha}\mathbf{z}_0, \tag{5}$$

and eliminates the longitudinal z components via [see Eq. (3a)],

$$-i\omega\mu_z H_{z\alpha} - i\omega\bar{\boldsymbol{\mu}}_{zt} \cdot \mathbf{H}_{t\alpha} = \nabla_t \cdot (\mathbf{z}_0 \times \mathbf{E}_{t\alpha}),$$
$$-i\omega\epsilon_z E_{z\alpha} - i\omega\bar{\boldsymbol{\epsilon}}_{zt} \cdot \mathbf{E}_{t\alpha} = \nabla_t \cdot (\mathbf{H}_{t\alpha} \times \mathbf{z}_0), \tag{5a}$$

the eigenvalue equation (4) can be expressed solely in terms of transverse components as (see Sec. 2.2a)

$$\omega\left[\boldsymbol{\epsilon}_t - \frac{\bar{\boldsymbol{\epsilon}}_{tz}\bar{\boldsymbol{\epsilon}}_{zt}}{\epsilon_z} - \frac{1}{\omega^2}\mathbf{z}_0 \times \nabla_t \frac{1}{\mu_z}\nabla_t \times \mathbf{z}_0\right] \cdot \mathbf{E}_{t\alpha}$$

$$+ i\left[\frac{\bar{\boldsymbol{\epsilon}}_{tz}}{\epsilon_z}\mathbf{z}_0 \times \nabla_t - \mathbf{z}_0 \times \nabla_t \frac{\bar{\boldsymbol{\mu}}_{zt}}{\mu_z}\right] \cdot \mathbf{H}_{t\alpha} = \kappa_\alpha \mathbf{H}_{t\alpha} \times \mathbf{z}_0,$$

$$\omega\left[\boldsymbol{\mu}_t - \frac{\bar{\boldsymbol{\mu}}_{tz}\bar{\boldsymbol{\mu}}_{zt}}{\mu_z} - \frac{1}{\omega^2}\mathbf{z}_0 \times \nabla_t \frac{1}{\epsilon_z}\nabla_t \times \mathbf{z}_0\right] \cdot \mathbf{H}_{t\alpha}$$

$$- i\left[\frac{\bar{\boldsymbol{\mu}}_{tz}}{\mu_z}\mathbf{z}_0 \times \nabla_t - \mathbf{z}_0 \times \nabla_t \frac{\bar{\boldsymbol{\epsilon}}_{zt}}{\epsilon_z}\right] \cdot \mathbf{E}_{t\alpha} = \kappa_\alpha \mathbf{z}_0 \times \mathbf{E}_{t\alpha}, \tag{6}$$

which evidently assume a much simpler form in the gyrotropic case $\bar{\boldsymbol{\epsilon}}_{zt} = \bar{\boldsymbol{\epsilon}}_{tz} = \bar{\boldsymbol{\mu}}_{zt} = \bar{\boldsymbol{\mu}}_{tz} = 0$.

To ascertain orthogonality properties of the mode functions $\boldsymbol{\Psi}_\alpha$, one considers the eigenvalue problem adjoint to Eq. (4),

$$K^+ \boldsymbol{\Psi}_\alpha^+ = \kappa_\alpha^* \boldsymbol{\Gamma}^+ \boldsymbol{\Psi}_\alpha^+, \tag{7a}$$

where the adjoint operators and eigenfunctions are

$$K^+ \rightarrow \begin{bmatrix} \omega\tilde{\boldsymbol{\epsilon}}^* & -i\boldsymbol{\nabla}_t \times \mathbf{1} \\ i\boldsymbol{\nabla}_t \times \mathbf{1} & \omega\tilde{\boldsymbol{\mu}}^* \end{bmatrix}, \quad \boldsymbol{\Gamma}^+ = \boldsymbol{\Gamma}, \quad \boldsymbol{\Psi}_\alpha^+ \rightarrow \begin{bmatrix} \mathbf{E}_\alpha^+ \\ \mathbf{H}_\alpha^+ \end{bmatrix} \tag{7b}$$

subject to the condition on the boundary (if any)

$$\mathbf{v} \times \mathbf{E}_\alpha^+ = \tilde{\mathscr{L}}^* \cdot \mathbf{H}_\alpha^+. \tag{7c}$$

The adjoint fields $\mathbf{E}_\alpha^+$ and $\mathbf{H}_\alpha^+$ satisfy the source-free equations in a transposed-conjugate (Hermitian adjoint) medium with parameters $\tilde{\boldsymbol{\epsilon}}^*$ and $\tilde{\boldsymbol{\mu}}^*$ obtained from the original medium by the replacements ($\sim$ denotes the transpose and $*$ the complex-conjugate operation):

$$\boldsymbol{\epsilon}_t \rightarrow \tilde{\boldsymbol{\epsilon}}_t^*, \, \bar{\boldsymbol{\epsilon}}_{tz} \leftrightarrow \bar{\boldsymbol{\epsilon}}_{zt}^*, \, \boldsymbol{\mu}_t \rightarrow \tilde{\boldsymbol{\mu}}_t^*, \, \bar{\boldsymbol{\mu}}_{tz} \leftrightarrow \bar{\boldsymbol{\mu}}_{zt}^*, \, \mathscr{L} \rightarrow \tilde{\mathscr{L}}^*. \tag{7d}$$

In a conventional manner (see Sec. 1.4) one infers from the eigenvalue equations (4) and (7) the orthogonality properties

$$(\boldsymbol{\Psi}_\alpha^+, \boldsymbol{\Gamma}\boldsymbol{\Psi}_\beta) = 2N_\alpha \delta_{\alpha\beta} = (\bar{\boldsymbol{\Psi}}_\alpha^+, \boldsymbol{\Gamma}\bar{\boldsymbol{\Psi}}_\beta), \tag{8a}$$

where $\delta_{\alpha\beta}$ is the Kronecker symbol that equals 1 or 0 depending on whether or not κ_α is equal to κ_β; N_α is the αth-mode normalization constant, and the $\bar{\boldsymbol{\Psi}}_\alpha$ and $\bar{\boldsymbol{\Psi}}_\alpha^+$ are transverse eigenvectors with components $\mathbf{E}_{t\alpha}, \mathbf{H}_{t\alpha}$ and $\mathbf{E}_{t\alpha}^+, \mathbf{H}_{t\alpha}^+$, respectively. More specifically, the orthogonality properties (8a) are expressed as

$$\iint\limits_S [\mathbf{E}_{t\alpha}^{+*} \cdot \mathbf{H}_{t\beta} \times \mathbf{z}_0 + \mathbf{H}_{t\alpha}^{+*} \cdot \mathbf{z}_0 \times \mathbf{E}_{t\beta}] \, dS = 2N_\alpha \delta_{\alpha\beta}, \tag{8b}$$

where the integral is extended over the cross section S transverse to z. As emphasized in Sec. 8.1, in the same waveguide there generally do not exist mode solutions with eigenvalues $+\kappa_\alpha$ and $-\kappa_\alpha$, thereby eliminating the possibility of a standing-wave transmission-line description in terms of interfering $+\kappa_\alpha$ and $-\kappa_\alpha$ waves. The latter obtains only for reflection symmetric regions.

The preceding description, valid for a general uniform electromagnetic waveguide, simplifies substantially in special situations.

8.2b Lossless Regions

If the anisotropic medium and waveguide boundary (if any) are lossless (i.e., reactive), then the constitutive medium parameters are Hermitian:

$$\boldsymbol{\epsilon} = \tilde{\boldsymbol{\epsilon}}^*, \qquad \boldsymbol{\mu} = \tilde{\boldsymbol{\mu}}^*, \qquad (i\mathscr{L}) = (i\tilde{\mathscr{L}})^*. \tag{9}$$

Under these conditions, it follows from a comparison of Eqs. (2b), (4), and (7) that

$$K = K^+ \qquad \text{and} \qquad \Psi_\alpha^+ = \Psi_{\alpha*}, \tag{10a}$$

where $\Psi_{\alpha*}$ is an eigenvector of the original waveguide with eigenvalue κ_α^*. It should be noted that eigenvalue problems of the type (4) do not in general possess real κ_α eigenvalues even in the Hermitian case, and furthermore, if κ_α is an eigenvalue, so also is its conjugate κ_α^*. For the special case of real eigenvalues κ_α one has $\Psi_\alpha^+ = \Psi_\alpha$, while for imaginary κ_α, one has $\Psi_\alpha^+ = -\Psi_\alpha$. In view of Eq. (10a), the orthogonality relation (8b) for a lossless region takes the form

$$\iint_S [\mathbf{E}_{t\alpha*}^* \cdot \mathbf{H}_{t\beta} \times \mathbf{z}_0 + \mathbf{H}_{t\alpha*}^* \cdot \mathbf{z}_0 \times \mathbf{E}_{t\beta}]dS = 0 \qquad \text{if } \kappa_\alpha \neq \kappa_\beta. \tag{10b}$$

The appearance of κ_α in the orthogonality statement is due to the fact that κ_α, rather than κ_α^* of Eq. 7(a), distinguishes the eigenvector $\Psi_\alpha^{+*} = \Psi_{\alpha*}^*$ appearing in Eq. (10b).

8.2c Lossy (Symmetric) Regions

If the anisotropic medium and waveguide boundary (if any) are dissipative but symmetric, then the constitutive parameters are described by symmetric dyadics:

$$\boldsymbol{\epsilon} = \tilde{\boldsymbol{\epsilon}}, \qquad \boldsymbol{\mu} = \tilde{\boldsymbol{\mu}}, \qquad \mathscr{L} = \tilde{\mathscr{L}}. \tag{11}$$

For these conditions, comparison of Eqs. (2b), (4), and (7) reveals that

$$K^+ = AK^*A \qquad \text{and} \qquad \Psi_\alpha^+ = A\Psi_{-\alpha}^*, \tag{12a}$$

where

$$A = A^{-1} \to \begin{bmatrix} 1 & 0 \\ 0 & -1 \end{bmatrix}, \qquad A\Gamma = -\Gamma A.$$

$A\Psi_{-\alpha}^*$ is an eigenvector of the original waveguide with eigenvalue $-\kappa_\alpha$; it is derivable from Ψ_α on replacing κ_α by $-\kappa_\alpha$, taking the complex conjugate of the field components, and reversing the sign of H_α. The orthogonality relation (8b) now becomes

$$\iint_S [\mathbf{E}_{t,-\alpha} \cdot \mathbf{H}_{t\beta} \times \mathbf{z}_0 - H_{t,-\alpha} \cdot \mathbf{z}_0 \times \mathbf{E}_{t\beta}]\, dS = 0 \qquad \text{if } -\kappa_\alpha \neq \kappa_\beta. \tag{12b}$$

It should be observed in this case that if κ_α is an eigenvalue, so also is $-\kappa_\alpha$. However, a conventional transmission-line description is generally not possible since via Eq. (12a), Ψ_α and $\Psi_{-\alpha}$ are not related as in the conventional reflection symmetric case; differences arise because the characteristic wave impedances are complex quantities in the dissipative case.

8.2d Transverse Anisotropy (Reflection Symmetry)

If the medium is gyrotropic, so that the anisotropy is confined to a plane transverse to z, then

$$\bar{\boldsymbol{\epsilon}}_{tz} = \bar{\boldsymbol{\epsilon}}_{zt} = \bar{\boldsymbol{\mu}}_{tz} = \bar{\boldsymbol{\mu}}_{zt} = \mathscr{L}_{zs} = \mathscr{L}_{sz} = 0, \tag{13}$$

and the region is said to be reflection symmetric with respect to a plane transverse to z. To examine the symmetry properties of Ψ vectors in such regions, it is convenient to introduce the reflection operator R defined by[†]

$$R \rightarrow \begin{bmatrix} \mathbf{1}_t - \mathbf{z}_0 \mathbf{z}_0 & 0 \\ 0 & -\mathbf{1}_t + \mathbf{z}_0 \mathbf{z}_0 \end{bmatrix} \quad \text{whence } R\Psi \rightarrow \begin{bmatrix} \mathbf{E}_t - E_z \mathbf{z}_0 \\ -\mathbf{H}_t + H_z \mathbf{z}_0 \end{bmatrix}. \tag{13a}$$

If Eq. (13) is satisfied, then $RK = KR$, $R\Gamma = -\Gamma R$, and hence if Ψ_α is a mode solution of Eq. (4) corresponding to κ_α, then $R\Psi_\alpha$ is a mode solution with eigenvalue $-\kappa_\alpha$. Similarly, if Ψ_α^+ has eigenvalue κ_α^*, then $R\Psi_\alpha^+$ has eigenvalue $-\kappa_\alpha^*$. Thus, one infers from Eq. (8a) the orthogonality statements

$$(\Psi_\alpha^+, \Gamma\Psi_\beta) = 0 \qquad\qquad \text{if } \kappa_\alpha \neq \kappa_\beta,$$

$$(\Psi_\alpha^+, \Gamma R\Psi_\beta) = 0 = (R\Psi_\alpha^+, \Gamma\Psi_\beta) \qquad \text{if } \kappa_\alpha \neq -\kappa_\beta,$$

and therefore, on addition,

$$(\Psi_\alpha^+, \Gamma(\Psi_\beta + R\Psi_\beta)) = 0 = (\Psi_\alpha^+ + R\Psi_\alpha^+, \Gamma\Psi_\beta) \qquad \text{if } \kappa_\alpha^2 \neq \kappa_\beta^2, \tag{14a}$$

which reads, in component form,

$$\iint_S \mathbf{H}_{t\alpha}^{+*} \cdot \mathbf{z}_0 \times \mathbf{E}_{t\beta} \, dS = 0 = \iint_S \mathbf{E}_{t\alpha}^{+*} \cdot \mathbf{H}_{t\beta} \times \mathbf{z}_0 \, dS \qquad \text{if } \kappa_\alpha^2 \neq \kappa_\beta^2. \tag{14b}$$

It should be noted that the orthogonality statements (14) do not distinguish between $\pm\kappa$ modes.

For the lossless (Hermitian) case, in view of Eq. (10a), Eqs. (14b) become

$$\iint_S \mathbf{H}_{t\alpha}^* \cdot \mathbf{z}_0 \times \mathbf{E}_{t\beta} \, dS = 0 \qquad \text{if } \kappa_\alpha^2 \neq \kappa_\beta^2. \tag{15}$$

For the lossy symmetric case, in view of Eq. (12a), Eqs. (14b) take the form

$$\iint_S \mathbf{H}_{t\alpha} \cdot \mathbf{z}_0 \times \mathbf{E}_{t\beta} \, dS = 0 \qquad \text{if } \kappa_\alpha^2 \neq \kappa_\beta^2. \tag{16}$$

8.2e Isotropic Regions

In the case of an isotropic medium and waveguide boundary (if any),

$$\boldsymbol{\epsilon} = \mathbf{1}\epsilon, \qquad \boldsymbol{\mu} = \mathbf{1}\mu, \qquad \mathscr{L} = \mathscr{L} \, (\mathbf{z}_0 \mathbf{z}_0 + \mathbf{s}_0 \mathbf{s}_0), \tag{17}$$

where ϵ, μ, and $\mathscr{L}$ are scalars, and $\mathbf{s}_0$ is defined in Eq. (4d). In the lossless case, wherein ϵ, μ, and $i\mathscr{L}$ are real, the orthogonality statement (15) again obtains, whereas in the lossy case, wherein these parameters are complex, Eq. (16) is applicable.

[†]It is also possible to define a reflection operator that is the negative of the one in Eq. (13a).

8.2f   Regions with *E*- and *H*-mode Decompositions

The total fields corresponding to the transverse eigenfunctions $\mathbf{E}_{t\alpha}$ and $\mathbf{H}_{t\alpha}$ generally have both longitudinal components $E_{z\alpha}$ and $H_{z\alpha}$. To explore the possibility of separating such eigenfields into *E* modes and *H* modes with respect to the *z* direction, one may investigate whether Eqs. (6) can be satisfied when either $H_{z\alpha} = 0$ or $E_{z\alpha} = 0$, respectively. The *z* components of the field are specified in terms of $\mathbf{E}_{t\alpha}$ and $\mathbf{H}_{t\alpha}$ by Eqs. (5a); these equations, together with Eqs. (6), hold little promise for achieving the indicated simplification in the general case. In an isotropically filled waveguide, however, Eqs. (6) become

$$\kappa_\alpha \mathbf{E}_{t\alpha} = \omega \left[\mathbf{1}_t \mu + \frac{1}{\omega^2} \nabla_t \frac{1}{\epsilon} \nabla_t \right] \cdot (\mathbf{H}_{t\alpha} \times \mathbf{z}_0), \tag{18a}$$

$$\kappa_\alpha \mathbf{H}_{t\alpha} = \omega \left[\mathbf{1}_t \epsilon + \frac{1}{\omega^2} \nabla_t \frac{1}{\mu} \nabla_t \right] \cdot (\mathbf{z}_0 \times \mathbf{E}_{t\alpha}), \tag{18b}$$

where on the isotropic guide wall *s* of impedance $\mathscr{Z}$ [see Eq. (17)] and unit normal $\mathbf{v}$,

$$\mathbf{v} \times \mathbf{E}_{t\alpha} = \mathscr{Z} \mathbf{H}_{t\alpha} \qquad \text{on } s. \tag{18c}$$

For an *E* mode, Eq. (5a) implies $\nabla_t \cdot (\mathbf{z}_0 \times \mathbf{E}_{t\alpha}) = 0$, and Eq. (18b) yields

$$\kappa_\alpha \mathbf{H}_{t\alpha} = \omega\epsilon \mathbf{z}_0 \times \mathbf{E}_{t\alpha}, \tag{19a}$$

whereas for an *H* mode, with $\nabla_t \cdot (\mathbf{H}_{t\alpha} \times \mathbf{z}_0) = 0$,

$$\kappa_\alpha \mathbf{E}_{t\alpha} = \omega\mu \mathbf{H}_{t\alpha} \times \mathbf{z}_0. \tag{19b}$$

Thus, the requirement $E_{z\alpha} = 0$ or $H_{z\alpha} = 0$ forces the vectors $\mathbf{E}_{t\alpha}$ and $\mathbf{H}_{t\alpha}$ to be mutually perpendicular.

It is convenient to consider how boundary impedance and medium inhomogeneity individually affect separability of the modal field into *E* and *H* modes. The impedance boundary condition (18c) implies that

$$-\mathbf{s}_0 \cdot \mathbf{E}_{t\alpha} = \mathscr{Z} H_{z\alpha}, \qquad \mathscr{Z} \mathbf{s}_0 \cdot \mathbf{H}_{t\alpha} = E_{z\alpha} \quad \text{on } s, \tag{20}$$

so a tangential component of $\mathbf{H}_{t\alpha}$ on the boundary generates a longitudinal component $E_{z\alpha}$, while a tangential component of $\mathbf{E}_{t\alpha}$ on the boundary generates a longitudinal component $H_{z\alpha}$. Thus, even if an *E*-mode field is capable of existing in the waveguide medium, its transverse tangential electric component on the boundary couples to the longitudinal magnetic field, thereby destroying the *E*-mode character unless $\mathscr{Z} = 0$ or ∞. Analogous remarks apply to the *H* modes. A waveguide bounded by a wall with finite surface impedance therefore does not generally admit separate *E*- and *H*-mode fields, although such a decomposition may be possible for special symmetries. For example, in the presence of *E*-mode excitation which does not generate transverse electric field components parallel to the boundary, no *H*-mode coupling is required.

A transversely inhomogeneous medium filling the waveguide cross section also prohibits in general the separate existence of *E* and *H* modes. This may be verified by examination of the boundary conditions across an interface $\tilde{s}$ which separates two cross-sectional regions characterized by constant constitutive parameters μ_1, ϵ_1, and μ_2, ϵ_2, respectively. If $\tilde{\mathbf{s}}_0$ and $\tilde{\mathbf{v}}_0$ denote transverse unit vectors parallel and perpendicular to $\tilde{s}$, respectively, with $\tilde{\mathbf{v}}_0 \times \tilde{\mathbf{z}}_0 = \tilde{\mathbf{s}}_0$, then the boundary conditions on $\tilde{s}$ require that

$$\tilde{\mathbf{v}}_0 \times \mathbf{E}_{\alpha 1} = \tilde{\mathbf{v}}_0 \times \mathbf{E}_{\alpha 2}, \qquad \epsilon_1 \tilde{\mathbf{v}}_0 \cdot \mathbf{E}_{\alpha 1} = \epsilon_2 \tilde{\mathbf{v}}_0 \cdot \mathbf{E}_{\alpha 2}, \tag{21a}$$

$$\tilde{\mathbf{v}}_0 \times \mathbf{H}_{\alpha 1} = \tilde{\mathbf{v}}_0 \times \mathbf{H}_{\alpha 2}, \qquad \mu_1 \tilde{\mathbf{v}}_0 \cdot \mathbf{H}_{\alpha 1} = \mu_2 \tilde{\mathbf{v}}_0 \cdot \mathbf{H}_{\alpha 2}. \tag{21b}$$

Imposition of the E-mode condition $\nabla_t \cdot \mathbf{z}_0 \times \mathbf{E}_{\alpha 1} = 0$, with $\kappa_\alpha \mathbf{H}_{t\alpha} = \omega \epsilon \mathbf{z}_0 \times \mathbf{E}_{t\alpha}$, in regions 1 and 2 permits the last of Eqs. (21b) to be written as

$$\mu_1 \epsilon_1 \tilde{\mathbf{s}}_0 \cdot \mathbf{E}_{t\alpha 1} = \mu_2 \epsilon_2 \tilde{\mathbf{s}}_0 \cdot \mathbf{E}_{t\alpha 2}, \tag{22}$$

which is incompatible with the first of Eqs. (21a) unless $\mu_1 \epsilon_1 = \mu_2 \epsilon_2$ or $\tilde{\mathbf{s}}_0 \cdot \mathbf{E}_{t\alpha} = 0$. An analogous result is obtained for the H modes. Thus, a decomposition into E and H modes is generally impossible when the local propagation speed $c = 1/\sqrt{\mu\epsilon}$ has a transverse variation, unless special symmetries eliminate the existence of tangential tansverse field components.

It is evident from the preceding discussion that E and H modes *may* exist individually in a homogeneously filled cross section bounded by perfectly conducting walls. Upon introducing the characteristic admittance by renormalizing $\mathbf{H}_{t\alpha}$ according to $\mathbf{H}_{t\alpha} = Y_\alpha \mathbf{h}_\alpha$ and setting $\mathbf{E}_{t\alpha} = \mathbf{e}_\alpha$, one obtains for the E modes [see Eqs. (18) and (19)],

$$\mathbf{h}_\alpha = \mathbf{z}_0 \times \mathbf{e}_\alpha, \quad Y_\alpha = \frac{\omega\epsilon}{\kappa_\alpha}, \quad \nabla_t \nabla_t \cdot \mathbf{e}_\alpha = -(k^2 - \kappa_\alpha^2)\mathbf{e}_\alpha, \quad k^2 = \omega^2 \mu\epsilon, \tag{23a}$$

and, for the H modes,

$$\mathbf{e}_\alpha = \mathbf{h}_\alpha \times \mathbf{z}_0, \quad Y_\alpha = \frac{\kappa_\alpha}{\omega\mu}, \quad \nabla_t \nabla_t \cdot \mathbf{h}_\alpha = -(k^2 - \kappa_\alpha^2)\mathbf{h}_\alpha. \tag{23b}$$

The normalized mode set $\mathbf{e}_\alpha, \mathbf{h}_\alpha$ is identical with that defined in Eqs. (2.2.10).

In a uniaxially anisotropic medium with

$$\boldsymbol{\epsilon} = \mathbf{1}_t \epsilon_t + \mathbf{z}_0 \mathbf{z}_0 \epsilon_z, \qquad \boldsymbol{\mu} = \mathbf{1}_t \mu_t + \mathbf{z}_0 \mathbf{z}_0 \mu_z, \tag{24}$$

the eigenvalue problem in Eqs. (6) reduces to

$$\kappa_\alpha Z_\alpha \mathbf{e}_\alpha = \omega \left[\mathbf{1}_t \mu_t + \frac{1}{\omega^2} \nabla_t \frac{1}{\epsilon_z} \nabla_t \right] \cdot \mathbf{h}_\alpha \times \mathbf{z}_0, \tag{25a}$$

$$\kappa_\alpha Y_\alpha \mathbf{h}_\alpha = \omega \left[\mathbf{1}_t \epsilon_t + \frac{1}{\omega^2} \nabla_t \frac{1}{\mu_z} \nabla_t \right] \cdot \mathbf{z}_0 \times \mathbf{e}_\alpha. \tag{25b}$$

These equations are almost identical with Eqs. (18a) and (18c) and the same considerations lead to the conclusion that a field decomposition into E and H modes is possible when the cross section is filled homogeneously and is bounded by a perfectly conducting wall. The resulting normalized mode set has been employed in Sec. 7.2.

8.2g Modal Representations for the Reduced Electromagnetic Field

The mode functions described in Secs. 8.2a through 8.2f may be employed to represent the electromagnetic field excited by arbitrary source configurations in various types of uniform waveguides. In abstract notation, the steady-state problem posed in Eqs. (1) and (2) may be solved for the transverse field $\bar{\Psi}(\mathbf{r})$ of Eq. (3b) by utilizing the complete set of transverse modes $\bar{\Psi}_\alpha(\boldsymbol{\rho})$ defined in Eq. (8a) [see also Eqs. (4b) and (7a)],

$$\bar{\Psi}(\mathbf{r}) = \sum_\alpha a_\alpha(z)\bar{\Psi}_\alpha(\boldsymbol{\rho}), \tag{26a}$$

where the summation† is to be extended over all α, which distinguish the eigenvalues κ_α of both the forward and backward travelling modes of amplitudes $a_\alpha(z)$. An $\exp(-i\omega t)$ dependence is implied. On adding to the transverse field representation (26a) the associated longitudinal field components, defined in terms of the transverse fields $\bar{\Psi}$ and $\bar{\Psi}_\alpha$ by Eqs. (3) and (5a), one obtains even at source points the total field representation

$$\Psi(\mathbf{r}) = \sum_\alpha a_\alpha(z)\Psi_\alpha(\boldsymbol{\rho}) - \theta_z(\mathbf{r}), \tag{26b}$$

where Ψ and Ψ_α are complete wavevectors of the excited and modal fields, respectively, and θ_z, defined in Eq. (3b), is related to the source wavevector. To determine the modal amplitudes $a_\alpha(z)$, we first observe from the orthogonality property (8b) and from (26b) that $a_\alpha = (2N_\alpha)^{-1}(\Psi_\alpha^+, \Gamma\Psi)$, a result that is independent of θ_z. Transformation of the field equations (1a) and (2a) by multiplication with the adjoint mode function $\Psi_\alpha^+(\boldsymbol{\rho})$ and integration over the transverse waveguide cross section then yields as the defining equation for the $a_\alpha(z)$ [we recall from Eq. (7a) that $(\Psi_\alpha^+, K\Psi) = (K^+\Psi_\alpha^+, \Psi) = (\kappa_\alpha^*\Gamma^+\Psi_\alpha^+, \Psi) = 2N_\alpha\kappa_\alpha a_\alpha$]:

$$\left(\frac{d}{dz} - i\kappa_\alpha\right)a_\alpha(z) = -b_\alpha(z), \tag{27}$$

where the source amplitude b_α is given by‡

$$b_\alpha = (2N_\alpha)^{-1}(\Psi_\alpha^+, \boldsymbol{\Phi}), \tag{28a}$$

$$= (2N_\alpha)^{-1}\iint [\mathbf{J}_t \cdot \mathbf{E}_\alpha^{+*} + J_z E_{z\alpha}^{+*} + \mathbf{M}_t \cdot \mathbf{H}_\alpha^{+*} + M_z H_{z\alpha}^{+*}]\, dS \tag{28b}$$

On integration of the second and fourth terms in Eq. (28b) by parts (assuming that J_z vanishes on the guide walls), one can write

$$b_\alpha = (2N_\alpha)^{-1}\iint [\mathbf{J}_{te} \cdot \mathbf{E}_\alpha^{+*} + \mathbf{M}_{te} \cdot \mathbf{H}_\alpha^{+*}]\, dS, \tag{29a}$$

---

†In the present abstract notation, $\displaystyle\sum_\alpha$ implies both an integration over $\mathbf{k}_t$ and a sum over eigenvalues for each $\mathbf{k}_t$, ω [see the time-harmonic analogue of Eq. (31)].

‡Treatment of excitation by arbitrary sources may be simplified by use of Green's functions. The transmission equations then assume the form given in Eq. (1.4.15).

where

$$\mathbf{J}_{te} = \mathbf{J}_t - \frac{1}{i\omega} \nabla_t \times \frac{\mathbf{M}_z}{\mu_z} - \frac{\bar{\epsilon}_{zt} J_z}{\epsilon_z}, \qquad \mathbf{M}_{te} = \mathbf{M}_t + \frac{1}{i\omega} \nabla_t \times \frac{\mathbf{J}_z}{\epsilon_z} - \frac{\bar{\mu}_{zt} M_z}{\mu_z};$$

$$(29\mathrm{b})$$

$\mathbf{J}_{te}$ and $\mathbf{M}_{te}$ are equivalent transverse current densities characteristic of a field description in terms of transverse fields only.

Alternative to the abstract procedure in Eqs. (26) and (27), one may obtain a field representation by first eliminating the longitudinal field components from Eqs. (1) and (2), and thereby derive explicit equations for the transverse field components (see Sec. 2.2a). The resulting transverse field equations resemble Eqs. (6), with $i\kappa_\alpha$ replaced by d/dz, and contain the equivalent transverse currents $\mathbf{J}_{te}$ and $\mathbf{M}_{te}$ defined in Eq. (29b) [see Eqs. (2.2.4) and (2.2.5) for the special case of homogeneous isotropic regions bounded by lossless walls]. The explicit modal representation of the transverse fields now becomes

$$\mathbf{E}_t(\mathbf{r}) = \sum_\alpha a_\alpha(z)\mathbf{E}_{t\alpha}(\boldsymbol{\rho}), \qquad \mathbf{H}_t(\mathbf{r}) = \sum_\alpha a_\alpha(z)\mathbf{H}_{t\alpha}(\boldsymbol{\rho}). \qquad (30)$$

Substitution of these expansions into the transverse fields equations, recalling Eqs. (6) and the orthogonality property (10b), then yields the above transmission-line equation (27) for the modal amplitudes $a_\alpha(z)$. Since there are generally no solutions admitting both κ_α and $-\kappa_\alpha$ in the *same* region, the transmission lines representative of Eq. (27) are unilateral and propagate waves in one direction only. Because of this restriction, the conventional transmission-line procedure is of limited value in this general formulation.

As noted in Eq. (26b), to represent the *total* fields in a region containing sources, it is necessary to stipulate at each point $\mathbf{r}$ not only the total modal fields $\mathbf{E}_\alpha = \mathbf{E}_{t\alpha} + E_{z\alpha}\mathbf{z}_0$ and $\mathbf{H}_\alpha = \mathbf{H}_{t\alpha} + H_{z\alpha}\mathbf{z}_0$ but also the longitudinal source currents J_z and M_z in order to obtain a complete representation. This lack of vector completeness for the total fields is to be anticipated since the eigenvalue problem refers only to the transverse space.

8.2h   Non-Conventional Transmission-Line Descriptions

The radiation of energy from a source in an unbounded, homogeneous, and stationary medium may be described in terms of a set of guided modes carrying energy away from the source. A complete set of such modes (see Sec. 1.4) comprises waves carrying energy both away from and toward the source, but only the former are excited in an unbounded medium. A boundary surface scatters the radiated waves, and the description of this scattering process requires in general the sophisticated analytical techniques of diffraction theory. However, if the boundary surface and the medium are appropriately simple, the scattering description may be effected by relatively simple transmission-line methods. These methods have been discussed in Sec. 2.4 for the simple case of an isotropic medium. In the present section, we develop similar but less conventional transmission-line methods for certain choices of transmission direction in appropriate gyrotropic media and thereby introduce a more general form of transmission-line theory.

Representation of a linear field in terms of a complete set of characteristic guided waves reduces an overall field description to a simple determination of wave amplitudes. In spatially homogeneous and stationary regions, each amplitude is determined by an ordinary differential equation and may be shown (see Sec. 1.4) to have a simple exponential dependence on distance along the guide direction. In unbounded regions, each wave amplitude is uncoupled from the other amplitudes and hence separately evaluable; the amplitude of a wave at any point is finite if the wave in question carries energy away from a source, or zero if the energy transport is toward a source. The presence of a boundary surface generally introduces coupling among the waves traveling toward or from sources and thereby makes any one amplitude dependent on the amplitudes of other waves. For boundary surfaces of arbitrary shape, all waves are coupled at the boundary; consequently, the evaluation of wave amplitudes may be prohibitively difficult, thus vitiating the utility of a guided-mode representation. However, for regions with media and boundaries of suitable symmetry, the wave coupling introduced by boundary surfaces may be simple and readily taken into account. In the present section we shall analyze mode coupling caused by plane-parallel boundaries in stratified gyrotropic media. This analysis can always be performed in terms of the traveling waves defined in Sec. 1.4. For guiding structures containing isotropic media and guide walls, not only traveling-wave (scattering) but also the standing-wave (impedance) descriptions of conventional transmission-line analysis can be employed, as has been indicated in Sec. 2.4. In more general anisotropic structures admitting $\pm\kappa$ waves, a simple but less conventional transmission-line analysis may still be possible.

For simplicity, we shall limit the discussion to reduced electromagnetic field descriptions in homogeneous media with non spatially dispersive ϵ and μ parameters. As shown in Sec. 1.4, a general field in a transversely unbounded region may be represented as

$$\Psi(\mathbf{r}, t) = \iiint \sum_\alpha a_\alpha(\mathbf{k}_t, \omega; z)\Psi_\alpha e^{i(\mathbf{k}_t\cdot\mathbf{\rho}-\omega t)} \frac{d\mathbf{k}_t\,d\omega}{(2\pi)^3}, \tag{31}$$

where for the special case of a reduced electromagnetic description

$$\Psi(\mathbf{r}, t) \rightarrow \begin{bmatrix} \mathbf{E}(\mathbf{r}, t) \\ \mathbf{H}(\mathbf{r}, t) \end{bmatrix}.$$

If the medium is Hermitian, i.e., the field operator $K = K^+$ (see Sec. 8.2b), then the eigenvectors Ψ_α and their adjoints Ψ_α^+ are representable as

$$\Psi_\alpha \rightarrow \begin{bmatrix} Z_\alpha \mathbf{e}_\alpha + \mathbf{e}_{z\alpha} \\ \mathbf{h}_\alpha + \mathbf{h}_{z\alpha} \end{bmatrix}, \qquad \Psi_\alpha^+ = \Psi_{\alpha*} \rightarrow \begin{bmatrix} Z_{\alpha*}\mathbf{e}_{\alpha*} + \mathbf{e}_{z\alpha*} \\ \mathbf{h}_{\alpha*} + \mathbf{h}_{z\alpha*} \end{bmatrix}, \tag{32}$$

where $\mathbf{e}_\alpha$, $\mathbf{h}_\alpha$ and $\mathbf{e}_{z\alpha}$, $\mathbf{h}_{z\alpha}$ are mode vectors transverse and parallel to z for the eigenvalue κ_α, while the adjoint eigenvectors $\mathbf{e}_{\alpha*}$, $\mathbf{h}_{\alpha*}$ and $\mathbf{e}_{z\alpha*}$, $\mathbf{h}_{z\alpha*}$ are the corresponding mode vectors with eigenvalue κ_α^*. The orthogonality property (1.4.4c) of the Ψ_α implies that in polarization space,

$$(\Gamma\Psi_\alpha^+, \Psi_\beta) = Z_\beta \mathbf{e}_\beta \cdot \mathbf{h}_{\alpha*}^* \times \mathbf{z}_0 + Z_{\alpha*}^* \mathbf{h}_\beta \cdot \mathbf{z}_0 \times \mathbf{e}_{\alpha*}^* = 2N_\alpha \delta_{\alpha\beta}. \tag{33}$$

Since the normalization of Ψ_α and $\Psi_{\alpha*}$ is arbitrary, it is convenient to choose the mode parameters Z_α and $Z_{\alpha*}$ so as to effect

$$\mathbf{e}_\alpha \cdot \mathbf{h}_{\alpha*}^* \times \mathbf{z}_0 = 1 \qquad \text{and} \qquad \mathbf{h}_\alpha \cdot \mathbf{z}_0 \times \mathbf{e}_{\alpha*}^* = 1 \tag{34}$$

whence the normalization constant is

$$2N_\alpha = Z_\alpha + Z_{\alpha*}^* \tag{35}$$

In view of Eq. (35) and the orthogonality properties of the complete wave-vector $\Psi_\alpha \exp[i(\mathbf{k}_t \cdot \boldsymbol{\rho} - \omega t)]$, one infers, just as in Eq. (1.4.6b), that the guided-wave amplitude $a_\alpha(\mathbf{k}_t, \omega; z)$ is related to the fields $\mathbf{E}(\mathbf{r}, t)$ and $\mathbf{H}(\mathbf{r}, t)$ on the plane z by

$$a_\alpha(\mathbf{k}_t, \omega; z) = \frac{1}{2N_\alpha} \iiint [\mathbf{E} \cdot \mathbf{h}_{\alpha*}^* \times \mathbf{z}_0 + Z_{\alpha*}^* \mathbf{H} \cdot \mathbf{z}_0 \times \mathbf{e}_{\alpha*}^*]e^{-i(\mathbf{k}_t \cdot \boldsymbol{\rho} - \omega t)} \, d\boldsymbol{\rho} \, dt, \tag{36}$$

where, for prescribed $\mathbf{k}_t, \omega$ and non-spatially dispersive medium parameters, the polarization index α distinguishes only four waves,† conforming to the four possible transverse (to z) components of the two vectors $\mathbf{E}$ and $\mathbf{H}$. In source-free regions, the z dependence of the a_α follows from Eq. (27) as

$$\left(\frac{d}{dz} - i\kappa_\alpha\right)a_\alpha(\mathbf{k}_t, \omega; z) = 0. \tag{37}$$

In a homogeneous isotropic or anisotropic Hermitian medium, the relation (36) between the total fields $\mathbf{E}$ and $\mathbf{H}$ and the wave amplitude $a_\alpha(\mathbf{k}_t, \omega; z)$ is valid at any plane z. Let us consider the coupling effects introduced by the presence of a planar boundary surface at z perpendicular to the guiding direction $\mathbf{z}_0$. If the planar boundary is spatially homogeneous and non-time-varying, only modes of the same $\mathbf{k}_t, \omega$ will be coupled at this boundary surface to the wave of amplitude $a_\alpha(\mathbf{k}_t, \omega; z)$. As noted above, there are four such modal waves with given $\mathbf{k}_t, \omega$. Hence, if $\Psi(z) \equiv \Psi(\mathbf{k}_t, \omega; z)$ denotes the $\mathbf{k}_t, \omega$ component of the total field $\Psi(\mathbf{r}, t)$ at the plane z, then on omission of the $\mathbf{k}_t, \omega$ arguments, Eq. (31) implies that

$$\Psi(z) = \sum_\alpha a_\alpha(z)\Psi_\alpha, \tag{38}$$

where for the reduced electromagnetic field description (in non spatially dispersive media), α distinguishes four possible waves. A basic analytical problem is that of ascertaining the interdependence among the $a_\alpha(z)$ caused by a planar boundary surface at some plane z_0.

In particular, we examine the coupling among the wave amplitudes a_α for Hermitian media that admit waves with wavenumbers (eigenvalues) $\pm\kappa_\alpha$ and transverse-mode vectors $\mathbf{e}_\alpha = \mathbf{e}_{-\alpha}, \mathbf{h}_\alpha = \mathbf{h}_{-\alpha}$. Such media are more general than

---

†The secular determinant for $\kappa_\alpha = \kappa_\alpha(\mathbf{k}_t, \omega)$ is of fourth degree for non spatially dispersive ϵ and μ.

the reflection symmetric regions described in Sec. 8.2d. For prescribed $\mathbf{k}_t, \omega$, one then finds two distinctive pairs of waves with wavenumbers† $\pm \kappa'_\alpha$ and $\pm \kappa''_\alpha$. Accordingly, for each $\mathbf{k}_t, \omega$, one can represent the transverse fields $\mathbf{E}_t(z), \mathbf{H}_t(z)$ at z by Eqs. (32) and (38) as

$$\mathbf{E}_t(z) = V'_\alpha(z)\mathbf{e}'_\alpha + V''_\alpha(z)\mathbf{e}''_\alpha,$$
$$\mathbf{H}_t(z) = I'_\alpha(z)\mathbf{h}'_\alpha + I''_\alpha(z)\mathbf{h}''_\alpha, \tag{39a}$$

where for both the $'$ and $''$ waves, V_α and I_α are given by

$$V_\alpha(z) = Z_\alpha a_\alpha(z) + Z_{-\alpha} a_{-\alpha}(z),$$
$$I_\alpha(z) = a_\alpha(z) + a_{-\alpha}(z). \tag{39b}$$

In general, the "characteristic impedance" parameters Z_α and $Z_{-\alpha}$ are not simply related (see Sec. 8.4b), whereas in conventional transmission-line descriptions, one recalls that $Z_{-\alpha} = -Z_\alpha$. Since it is assumed that $\mathbf{e}_\alpha = \mathbf{e}_{-\alpha}$, $\mathbf{h}_\alpha = \mathbf{h}_{-\alpha}$, one can derive from the orthogonality property (33) and the normalization in Eq. (34) the additional orthogonality properties (in polarization space)

$$\mathbf{e}_\beta \cdot \mathbf{h}^*_{\alpha*} \times \mathbf{z}_0 = \delta_{\alpha\beta}, \qquad \mathbf{h}_\beta \cdot \mathbf{z}_0 \times \mathbf{e}^*_{\alpha*} = \delta_{\alpha\beta}, \tag{40}$$

for mode vectors $\mathbf{e}_\alpha, \mathbf{h}_\alpha$ distinguished by κ^2_α. It should be noted, as in Sec. 2.2, that the transverse-vector-mode functions $\mathbf{e}_\alpha \exp[i(\mathbf{k}_t \cdot \boldsymbol{\rho} - \omega t)]$, $\mathbf{h}_\alpha \exp[i(\mathbf{k}_t \cdot \boldsymbol{\rho} - \omega t)]$ determine a complete set of eigenvectors in the $\boldsymbol{\rho}, t$ and transverse vector (polarization) space. These vectors are distinguished by eigenvalues κ^2_α and are to be contrasted with the eigenvectors $\boldsymbol{\Psi}_\alpha \exp[i(\mathbf{k}_t \cdot \boldsymbol{\rho} - \omega t)]$ whose eigenvalues are κ_α. Accordingly, the mode summation for $\mathbf{e}_\alpha, \mathbf{h}_\alpha$ mode vectors, as employed in Chapters 2 and 3, ranges only over positive values of the index α, whereas for the $\boldsymbol{\Psi}_\alpha$-mode vectors the mode summation comprises $\pm\alpha$ values.

One can infer from Eqs. (36) and (40) the following relation between the "traveling-wave" amplitudes a_α and the "voltage" and "current" amplitudes V_α and I_α at z:

$$a_\alpha(z) = \frac{1}{2N_\alpha}[V_\alpha(z) + Z^*_{\alpha*}I_\alpha(z)],$$

$$a_{-\alpha}(z) = \frac{1}{2N_{-\alpha}}[V_\alpha(z) + Z^*_{-\alpha*}I_\alpha(z)], \tag{41}$$

where, since the voltage and current amplitudes are associated with κ^2_α, there is no distinction between $\pm\alpha$ subscripts for these quantities. In transmission-line analyses one customarily introduces, looking in the direction of increasing z, a (voltage) "reflection coefficient" $\Gamma_\alpha(z)$ and a "terminal impedance" $Z_\alpha(z)$ defined by

$$\Gamma_\alpha(z) = \frac{a_{-\alpha}(z)N_{-\alpha}}{a_\alpha(z)N_\alpha}, \qquad Z_\alpha(z) = \frac{V_\alpha(z)}{I_\alpha(z)}, \tag{42}$$

---

†Note that the notation $\kappa'_\alpha = \kappa_\alpha$, $\kappa''_\alpha = \kappa_\beta$ could also be used to distinguish these modes.

to characterize coupling introduced between the $\pm\alpha$ waves. From Eqs. (41), one then derives the relation

$$\Gamma_\alpha(z) = \frac{Z_\alpha(z) + Z^*_{-\alpha*}}{Z_\alpha(z) + Z^*_{\alpha*}} \tag{43a}$$

and its converse,

$$Z_\alpha(z) = \frac{-Z^*_{-\alpha*} + Z^*_{\alpha*}\Gamma_\alpha(z)}{1 - \Gamma_\alpha(z)}. \tag{43b}$$

Equations (43) are to be contrasted with the conventional transmission-line relations

$$\Gamma_\alpha(z) = \frac{Z_\alpha(z) - Z_\alpha}{Z_\alpha(z) + Z_\alpha}, \qquad \frac{Z_\alpha(z)}{Z_\alpha} = \frac{1 + \Gamma_\alpha(z)}{1 - \Gamma_\alpha(z)}, \tag{44}$$

as derived in Eqs. (2.4.12) for a medium wherein $-N_{-\alpha} = N_\alpha$ and $-Z^*_{-\alpha*} = Z^*_{\alpha*} = Z_\alpha$.

If the impedance $Z_\alpha(z')$ is known at some terminal plane $z = z'$, as occurs if boundary conditions on this plane are given, then in order to calculate the field at some other plane z, it is necessary to ascertain the reflection coefficient $\Gamma_\alpha(z)$ or $Z_\alpha(z)$ thereon. To this end, one first evaluates $\Gamma_\alpha(z')$ from the known $Z_\alpha(z')$ by Eq. (43a) applied at $z = z'$. Since, in a source-free region, the traveling-wave amplitudes at z and z' are related in accordance with Eq. (37) by

$$a_\alpha(z) = a_\alpha(z')e^{i\kappa_\alpha(z-z')}$$

then, from Eq. (42), one finds that

$$\Gamma_\alpha(z) = \Gamma_\alpha(z')e^{-2i\kappa_\alpha(z-z')} \tag{45}$$

is the desired reflection coefficient at z. To find the corresponding relation between the impedances at z and z', one substitutes Eqs. (45) and (43a) into Eq. (43b) and obtains

$$Z_\alpha(z) = \frac{-2Z^*_{\alpha*}Z^*_{-\alpha*} - i[(Z^*_{\alpha*} - Z^*_{-\alpha*})\cot\kappa_\alpha(z - z') + i(Z^*_{\alpha*} + Z^*_{-\alpha*})]Z_\alpha(z')}{2Z_\alpha(z') + Z^*_{\alpha*} + Z^*_{-\alpha*} - i(Z^*_{\alpha*} - Z^*_{-\alpha*})\cot\kappa_\alpha(z - z')}. \tag{46}$$

For a conventional transmission line, wherein $-Z^*_{-\alpha*} = Z^*_{\alpha*} = Z_\alpha$, Eq. (46) reduces to [see Eq. (2.4.10) with $j \to -i$]

$$\frac{Z_\alpha(z)}{Z_\alpha} = \frac{Z_\alpha - iZ_\alpha(z')\cot\kappa_\alpha(z - z')}{Z_\alpha(z') - iZ_\alpha\cot\kappa_\alpha(z - z')}. \tag{47}$$

For the non-reflection symmetric media under consideration, Eqs. (45) and (46) provide the basis for a general transmission-line analysis of the coupling of waves by a planar boundary surface at z'.

For prescribed $\mathbf{k}_t, \omega$, the average electromagnetic power density in the positive z direction can be calculated from the real part of the complex Poynting vector,

$$P_z = \operatorname{Re}\left[\mathbf{E}(z) \cdot \mathbf{H}(z)^* \times \mathbf{z}_0\right]. \tag{48}$$

For this purpose one requires, in addition to the field representations (38) and (39a), a representation in terms of adjoint wavevectors (for Hermitian media)

$$\Psi(z) = \sum_\alpha a_\alpha^+(z)\Psi_\alpha^+ = \sum_\alpha a_{\alpha^*}(z)\Psi_{\alpha^*} \tag{49}$$

and correspondingly, by Eq. (32),

$$\begin{aligned}
\mathbf{E}_t(z) &= V'_{\alpha^*}(z)\,\mathbf{e}'_{\alpha^*} + V''_{\alpha^*}(z)\,\mathbf{e}''_{\alpha^*}, \\[4pt]
\mathbf{H}_t(z) &= I'_{\alpha^*}(z)\,\mathbf{h}'_{\alpha^*} + I''_{\alpha^*}(z)\,\mathbf{h}''_{\alpha^*},
\end{aligned} \tag{50a}$$

where, for both the ′ and ″ modes,

$$\begin{aligned}
V_{\alpha^*}(z) &= Z_{\alpha^*}a_{\alpha^*}(z) + Z_{-\alpha^*}a_{-\alpha^*}(z), \\[4pt]
I_{\alpha^*}(z) &= a_{\alpha^*}(z) + a_{-\alpha^*}(z).
\end{aligned} \tag{50b}$$

In view of the orthonormality properties (40), one then finds on substitution of Eqs. (39a) and (50a) into (48) the average power relation

$$P_z = \operatorname{Re}\sum_{\alpha>0} V_\alpha(z)I^*_{\alpha^*}(z) = \operatorname{Re}\sum_{\alpha>0} V^*_{\alpha^*}(z)I_\alpha(z), \tag{51}$$

where, in the impedance description, the mode sum must be only over the $+\alpha$ modes. For the special case of propagating modes in isotropic media, wherein $V_{\alpha^*} = V_\alpha, I_{\alpha^*} = I_\alpha$, Eq. (51) reduces to the conventional relation

$$P_z = \operatorname{Re}\sum_{\alpha>0} V_\alpha(z)I^*_\alpha(z).$$

The transmission-line analysis outlined above is applicable in reflection symmetric Hermitian media (i.e., media admitting $\pm\kappa_\alpha$ waves for which $\mathbf{e}_\alpha = \mathbf{e}_{-\alpha}, \mathbf{h}_\alpha = \mathbf{h}_{-\alpha}, \mathbf{e}_{\alpha^*} = \mathbf{e}_{-\alpha^*}, \mathbf{h}_{\alpha^*} = \mathbf{h}_{-\alpha^*}$). Such conditions are fulfilled in non-dissipative gyrotropic media, but in general $Z^*_{\alpha^*} \neq Z_\alpha$ and $\mathbf{e}_\alpha \neq \mathbf{e}_{\alpha^*}$, $\mathbf{h}_\alpha \neq \mathbf{h}_{\alpha^*}$. In gyrotropic media with the gyrotropic axis parallel to the transmission direction, one finds that $Z^*_{\alpha^*} = Z_\alpha$, but $\mathbf{e}_\alpha \neq \mathbf{e}_{\alpha^*}, \mathbf{h}_\alpha \neq \mathbf{h}_{\alpha^*}$; thus the transmission-line relations (43) and (46) become conventional but the power relation (51) does not (see Sec. 8.3). In gyrotropic media with axis perpendicular to $\mathbf{z}_0$, one can find modes with $\mathbf{e}_\alpha = \mathbf{e}_{\alpha^*}, \mathbf{h}_\alpha = \mathbf{h}_{\alpha^*}$ but $Z^*_{\alpha^*} \neq Z_\alpha$, hence requiring the general transmission-line relations (46).

8.3 GUIDED WAVES IN A COLD MAGNETOPLASMA (GUIDE AXIS PARALLEL TO GYROTROPIC AXIS)

8.3a Evaluation of the Mode Functions

The guided-wave considerations of the preceding section on anisotropic media will now be specialized to the case of a gyrotropic medium wherein the guide and gyrotropic axes are parallel. The steady-state wavevectors Ψ_α descriptive of guided modes in such a medium will be exhibited explicitly, and their application to the theory of radiation from time-harmonic sources will be illustrated. As noted in Sec. 8.2, representation of radiated fields may be phrased either in terms of "first-order" Ψ vectors incorporating all electromagnetic and dynamical field variables, or in terms of "reduced" Ψ vectors displaying only electromagnetic variables.

Guided waves in a homogeneous gyrotropic medium, such as a collisionless cold magnetoplasma, are described by wavevectors of the form $\Psi_a(\mathbf{r}, t) = \Psi_\alpha \exp[j(\omega t - \mathbf{k} \cdot \mathbf{r})]$,[†] where the mode suffix a denotes both the polarization index α and the periodicities, ω, $\mathbf{k}_t$; $\mathbf{k}_t = \mathbf{k} - \mathbf{z}_0 \kappa_\alpha$ is the transverse wavenumber. In polarization space, for prescribed ω, $\mathbf{k}_t$, the Ψ_α satisfy the eigenvalue problem [see Eq. (8.2.2)]

$$L\Psi_\alpha = 0 = j(K - \kappa_\alpha \Gamma)\Psi_\alpha, \tag{1a}$$

where K and Γ are Hermitian operators and κ_α is the eigenvalue. The Ψ_α possess the orthogonality properties [see Eqs. (8.2.8a)]

$$(\Psi_\alpha^+, \Gamma\Psi_\beta) = 2N_\alpha \, \delta_{\alpha\beta} = (\bar{\Psi}_\alpha^+, \Gamma\bar{\Psi}_\beta); \tag{1b}$$

for the non-dissipative case, as observed in Sec. 8.2b, the adjoint wavevector $\Psi_\alpha^+ = \Psi_\alpha$ for $\kappa_\alpha = \kappa_\alpha^*$ and $\Psi_\alpha^+ = \Psi_{\alpha^*}$ for complex κ_α. Although these properties can be inferred a priori from the general form of the operator L indicated in Eq. (1a), one must know the specific structure of L to evaluate explicitly the various components of Ψ_α.

For a magnetoplasma fluid in which only electrons are mobile, the first-order form of L is displayed in Eq. (1.1.68) and the reduced form in Eqs. (8.2.1) and (8.2.2); in the first-order description, the components of Ψ are given by $\Psi \to (\mathbf{E}, \mathbf{H}, p, \mathbf{v})$, whereas in the reduced description, $\Psi \to (\mathbf{E}, \mathbf{H})$. The elimination of the dynamical variables p and $\mathbf{v}$, the electron pressure and average velocity, respectively, is characteristic of the reduced description. If, in addition, the magnetic field $\mathbf{H}$ is eliminated from the source-free plasma field equations, the resulting equation for the electric field $\mathbf{E}(\mathbf{r}, t)$ may be written in the form

$$\mathscr{Y}\left(\nabla, \frac{\partial}{\partial t}\right) \cdot \mathbf{E}(\mathbf{r}, t) = 0; \tag{2}$$

it may be noted that the dyadic "admittance" operator $\mathscr{Y}$ is the inverse of the electric component $\mathscr{G}_{11}$ of the plasma Green's function defined in Eq. (1.1.57).

Although solution of Eq. (2) is desired in a $\mathbf{k}_t$, ω basis, we shall find the solution in a $\mathbf{k}$, ω basis with $\mathbf{k} = \mathbf{k}_t + \kappa \mathbf{z}_0$. In a $\mathbf{k}$, ω basis (wherein $\nabla = -j\mathbf{k}$ and $\partial/\partial t = j\omega$, see Sec. 1.2), the admittance operator $\mathscr{Y}$ for a gyrotropic medium with permittivity $\boldsymbol{\epsilon}$ and permeability $\boldsymbol{\mu} = \mu_0 \mathbf{1}$ is given by

$$\mathscr{Y}(\mathbf{k}, \omega) = j\omega\epsilon_0\left[\boldsymbol{\epsilon} + \frac{\mathbf{k} \times (\mathbf{k} \times \mathbf{1})}{k_0^2}\right], \qquad k_0^2 = \omega^2 \mu_0 \epsilon_0, \tag{3a}$$

where, for the gyrotropic axis in the direction $\mathbf{b}_0$,

$$\frac{\boldsymbol{\epsilon}}{\epsilon_0} = \epsilon_1 \mathbf{1}_\perp - j\epsilon_2 \mathbf{b}_0 \times \mathbf{1}_\perp + \epsilon_3 \mathbf{1}_b. \tag{3b}$$

[†]Note that the time dependence in Sec. 8.3 is chosen to be $\exp(+j\omega t)$.

$\mathbf{1}_b = \mathbf{b}_0 \mathbf{b}_0$ and $\mathbf{1}_\perp = \mathbf{1} - \mathbf{b}_0 \mathbf{b}_0$ are unit dyadics longitudinal and transverse to $\mathbf{b}_0$, the unit vector defining the direction of the gyrotropic axis [in Eqs. (3), primes used elsewhere in this text to denote normalization with respect to ϵ_0 are omitted for simplicity]. For the special case of a cold collisionless magneto-plasma

$$\epsilon_1 = 1 - \frac{\omega_p^2}{\omega^2 - \omega_c^2}, \quad \epsilon_2 = -\frac{\omega_c}{\omega}\frac{\omega_p^2}{\omega^2 - \omega_c^2}, \quad \epsilon_3 = 1 - \frac{\omega_p^2}{\omega^2}, \tag{4}$$

where, as derived in Eqs. (1.5.20)–(1.5.22), the quantities ω_p and ω_c are the electron plasma and cyclotron angular frequencies, respectively.

To determine explicitly the eigenvalues κ_α and eigenvectors $\boldsymbol{\Psi}_\alpha$, it is generally more convenient to solve a reduced equation of the form (2) rather than the original eigenvalue problem (1). As was shown in Sec. 1.4, and noted in connection with Eq. (1a), in a homogeneous transversely unbounded region with guide axis along $\mathbf{z}_0$, there exists a set of guided modes of the form $\boldsymbol{\Psi}_a(\mathbf{r}, t) = \boldsymbol{\Psi}_\alpha \exp[j(\omega t - \mathbf{k}\cdot\mathbf{r})]$, with $\mathbf{k} = \mathbf{k}_t + \kappa_\alpha \mathbf{z}_0$. Accordingly, since $\mathbf{E}_a(\mathbf{r}, t) = \mathbf{E}_\alpha \exp[j(\omega t - \mathbf{k}\cdot\mathbf{r})]$, one infers from Eq. (2) that

$$\mathscr{Y}(\mathbf{k}, \omega) \cdot \mathbf{E}_\alpha = 0, \tag{5a}$$

and hence a non-vanishing mode field $\mathbf{E}_\alpha$ exists for those $\kappa = \kappa_\alpha$ that for prescribed ω, $\mathbf{k}_t$ satisfy the secular equation

$$\det \mathscr{Y}(\mathbf{k}, \omega) = 0. \tag{5b}$$

To within a normalization constant, the mode fields $\mathbf{E}_\alpha$ corresponding to the κ_x are then derivable from Eq. (2). The remaining mode fields, for either the first-order or reduced descriptions, follow from $\mathbf{E}_\alpha$ and the source-free plasma field equations (1.1.54) as

$$\mathbf{H}_\alpha = \frac{\mathbf{k} \times \mathbf{E}_\alpha}{\omega \mu_0},$$

$$\mathbf{v}_\alpha = \frac{1}{j\omega\mu_0 n_0 q}[(k_0^2 - k^2)\mathbf{1} + \mathbf{k}\mathbf{k}] \cdot \mathbf{E}_\alpha, \tag{6}$$

$$p_\alpha = \frac{\gamma p_0}{\omega}\mathbf{k} \cdot \mathbf{v}_\alpha.$$

For a cold plasma wherein $p_0 = 0$, it is evident that $p_\alpha = 0$. Thus, for prescribed ω, $\mathbf{k}_t$, where $\mathbf{k} = \mathbf{k}_t + \kappa\mathbf{z}_0$, the orthogonality property derivable from Eq. (8.2.8b) in a cold plasma can be expected to have the form

$$\mathbf{E}_\alpha^{+*} \cdot \mathbf{H}_\beta \times \mathbf{z}_0 + \mathbf{H}_\alpha^{+*} \cdot \mathbf{z}_0 \times \mathbf{E}_\beta = 2N_\alpha \delta_{\alpha\beta}, \tag{7}$$

with the adjoint components identifiable as $\mathbf{E}_\alpha^+ = \mathbf{E}_\alpha$, $\mathbf{H}_\alpha^+ = \mathbf{H}_\alpha$ for real κ_α, and as $\mathbf{E}_\alpha^+ = \mathbf{E}_{\alpha*}$, $\mathbf{H}_\alpha^+ = \mathbf{H}_{\alpha*}$ for complex κ_α; it should be recalled from Eq. (8.2.10a) that $\mathbf{E}_{\alpha*}$, $\mathbf{H}_{\alpha*}$ are components of the mode with eigenvalue κ_α^*. For modes with different ω, $\mathbf{k}_t$, the orthogonality statement (7) also requires integration over both the cross section and time as in Eq. (1.4.2b).

The considerations in Eqs. (4)–(6), which are independent of the relative orientation of guide and gyrotropic axes, will now be specialized to the case wherein the guide axis is parallel to the static magnetic field impressed on the plasma. As a vector basis we choose the right-handed set of unit vectors $\mathbf{z}_0$, $\mathbf{k}_{t0}$, $\mathbf{z}_0 \times \mathbf{k}_{t0} = \hat{\mathbf{k}}_{t0}$ determined by the guide axis $\mathbf{z}_0$ and the transverse unit vector $\mathbf{k}_{t0}$, the latter defined by the decomposition $\mathbf{k} = \kappa \mathbf{z}_0 + k_t \mathbf{k}_{t0}$. In this basis, one represents the mode electric field as

$$\mathbf{E} = E_z \mathbf{z}_0 + E_{t'} \mathbf{k}_{t0} + E_{t''} \hat{\mathbf{k}}_{t0}, \qquad \hat{\mathbf{k}}_{t0} = \mathbf{z}_0 \times \mathbf{k}_{t0}, \tag{8}$$

and, furthermore,

$$\mathbf{1}_b = \mathbf{z}_0 \mathbf{z}_0,$$

$$\mathbf{k} \times (\mathbf{k} \times \mathbf{1}) = -k_t^2 \mathbf{z}_0 \mathbf{z}_0 - \kappa^2 \mathbf{k}_{t0} \mathbf{k}_{t0} - k^2 \hat{\mathbf{k}}_{t0} \hat{\mathbf{k}}_{t0} + \kappa k_t(\mathbf{z}_0 \mathbf{k}_{t0} + \mathbf{k}_{t0} \mathbf{z}_0).$$

Thus, for the cold plasma defined by the permittivity parameters in Eq. (4), the defining equation (5a) for $\mathbf{E}$ may be represented as

$$j\omega\epsilon_0 \begin{pmatrix} \epsilon_3 - \dfrac{k_t^2}{k_0^2} & \dfrac{\kappa k_t}{k_0^2} & 0 \\[2ex] \dfrac{\kappa k_t}{k_0^2} & \epsilon_1 - \dfrac{\kappa^2}{k_0^2} & j\epsilon_2 \\[2ex] 0 & -j\epsilon_2 & \epsilon_1 - \dfrac{k_t^2 + \kappa^2}{k_0^2} \end{pmatrix} \begin{pmatrix} E_z \\[2ex] E_{t'} \\[2ex] E_{t''} \end{pmatrix} = 0. \tag{9}$$

The corresponding secular equation (5b) then yields for the eigenvalues κ the dispersion equation[†1]

$$\kappa'^4 + [(1 + \epsilon_1')k_t'^2 - 2\epsilon_1']\kappa'^2 + \epsilon_1'(k_t'^2 - 1)(k_t'^2 - \epsilon_\perp') = 0, \tag{10}$$

where we have employed the normalized quantities (for $\epsilon_3 > 0$)[†]

$$\kappa' = \frac{\kappa}{k_0\sqrt{\epsilon_3}}, \quad k_t' = \frac{k_t}{k_0\sqrt{\epsilon_3}}, \quad \epsilon_1' = \frac{\epsilon_1}{\epsilon_3}, \quad \epsilon_\perp' = \frac{\epsilon_1'^2 - \epsilon_2'^2}{\epsilon_1'}, \quad \epsilon_2' = \frac{\epsilon_2}{\epsilon_3}. \tag{10a}$$

κ' versus k_t' plots of the dispersion equation (10) in various ω ranges appear in Fig. 8.3.2.

For prescribed ω and $\mathbf{k}_t$, there are four eigenvalues κ_α, given by the solution of the quartic equation (10) as

$$\kappa_\alpha' = \pm\sqrt{\epsilon_1' - (\epsilon_1' + 1)(k_t'^2/2) \pm \Delta[(\epsilon_1' - 1)/2]} \tag{11}$$

with

$$\Delta = \sqrt{k_t'^4 + (1 - k_t'^2)[2\epsilon_2'/(\epsilon_1' - 1)]^2} \ \operatorname{sgn}(\epsilon_1' - 1). \tag{11a}$$

For spatially dispersive ϵ, additional roots are possible. The sign of the discriminant Δ has been chosen to facilitate subsequent discussion of the analytic properties of the mode functions in the $\mathbf{k}_t$ plane. For each eigenvalue κ_α in Eq. (11), one finds on substitution into Eq. (9) that the corresponding mode electric-field components are related by

---

†When $\epsilon_3 < 0$, the normalization of κ and k_t is with respect to $k_0\sqrt{|\epsilon_3|}$. In Eq. (10), one then replaces κ' by $j\kappa'$ and k_t' by jk_t'.

$$\frac{E_z}{E_{t'}} = \frac{k_t' \kappa_\alpha'}{k_t'^2 - 1}, \qquad \frac{E_{t''}}{E_{t'}} = -j\frac{\epsilon_2'}{k_t'^2 + \kappa_\alpha'^2 - \epsilon_1'}, \tag{12a}$$

whence, from Eq. (6), the corresponding mode magnetic field is given by

$$\mathbf{H} = \frac{k_t}{\omega\mu_0} E_{t''}\mathbf{z}_0 - \frac{E_{t''}}{Z_\alpha''}\mathbf{k}_{t0} + \frac{E_{t'}}{Z_\alpha'}\hat{\mathbf{k}}_{t0}; \tag{12b}$$

the parameters

$$Z_\alpha' = \frac{\omega\mu_0}{\kappa_\alpha}(1 - k_t'^2), \qquad Z_\alpha'' = \frac{\omega\mu_0}{\kappa_\alpha} \tag{12c}$$

reduce to the characteristic E- and H-mode impedances in the isotropic case $\epsilon_2' = 0$, $\epsilon_1' = 1$ [see Eqs. (2.2.15c) and (2.2.15d)]. Disregarding the velocity components of the κ_α mode, which are readily evaluated from Eqs. (6) and (12a), we shall exhibit only reduced wavevectors $\mathbf{\Psi}_\alpha$ defined by

$$\mathbf{\Psi}_\alpha \rightarrow \begin{bmatrix} \mathbf{E}_\alpha \\ \mathbf{H}_\alpha \end{bmatrix} = \begin{bmatrix} Z_\alpha\mathbf{e}_\alpha + \mathbf{e}_{z\alpha} \\ \mathbf{h}_\alpha + \mathbf{h}_{z\alpha} \end{bmatrix}. \tag{13a}$$

From Eqs. (12) and (8), on setting $E_{t'}/(\sqrt{2}\,Z_\alpha')$ equal to unity, one can identify[2]

$$\sqrt{2}\,\mathbf{e}_\alpha = \frac{\Delta - k_t'^2}{\Delta}\mathbf{k}_{t0} - j\frac{\delta}{\Delta}\hat{\mathbf{k}}_{t0}, \qquad \sqrt{2}\,\mathbf{e}_{z\alpha} = -\frac{k_t}{\omega\epsilon_0\epsilon_3}\mathbf{z}_0, \tag{13b}$$

$$\sqrt{2}\,\mathbf{h}_\alpha = j\frac{\Delta + k_t'^2}{\delta}\mathbf{k}_{t0} + \hat{\mathbf{k}}_{t0}, \qquad \sqrt{2}\,\mathbf{h}_{z\alpha} = -j\frac{k_t}{\kappa_\alpha}\frac{\Delta + k_t'^2}{\delta}\mathbf{z}_0,$$

where $\delta = 2\epsilon_2'/(\epsilon_1' - 1)$ and the mode impedance in Eq. (13a), denoted by

$$Z_\alpha = \frac{\omega\mu_0}{\kappa_\alpha}\frac{\Delta(1 - k_t'^2)}{\Delta - k_t'^2} = \frac{\omega\mu_0}{\kappa_\alpha}\frac{\Delta^2 + k_t'^2\Delta}{\delta^2}, \tag{13c}$$

is chosen to secure the normalization in Eq. (15). It should be noted that $\mathbf{e}_\alpha$ and $\mathbf{h}_\alpha$ are fixed for given ω and $\mathbf{k}_t$, while Z_α may assume four distinct values, depending on the choice of sign for the eigenvalue κ_α in Eq. (11) that distinguishes the various $\mathbf{\Psi}_\alpha$ wavevectors. The adjoint mode vectors $\mathbf{e}_{\alpha*}$ and $\mathbf{h}_{\alpha*}$, and adjoint mode impedance $Z_{\alpha*}$, follow from Eqs. (13b) and (13c) on the replacements

$$\Delta \rightarrow \Delta^* \qquad \text{and} \qquad \kappa_\alpha \rightarrow \kappa_\alpha^*; \tag{13d}$$

note that $Z_{\alpha*}^* = Z_\alpha$ but $\mathbf{e}_\alpha \neq \mathbf{e}_{\alpha*}$ and $\mathbf{h}_\alpha \neq \mathbf{h}_{\alpha*}$, so there exist conventional transmission-line relations but the power expression involves the amplitudes of both the orginal and adjoint modes (see Sec. 8.2h).

The derivation of the transverse mode functions in Eq. (13a) can equally well be effected by use of the transverse field equations (8.2.25), which may be written as $(\epsilon_z = \epsilon_3)$

$$\kappa_\alpha Z_\alpha \mathbf{e}_a = \omega\mu_0\left(\mathbf{1}_t + \frac{\nabla_t\nabla_t}{k_0^2\epsilon_z}\right)\cdot\mathbf{h}_a \times \mathbf{z}_0,$$

$$\kappa_\alpha Y_\alpha \mathbf{h}_a = \omega\epsilon_0\left(\boldsymbol{\epsilon}_t + \frac{\nabla_t\nabla_t}{k_0^2}\right)\cdot\mathbf{z}_0 \times \mathbf{e}_a, \tag{14}$$

where $Y_\alpha = 1/Z_\alpha$, $\boldsymbol{\epsilon}_t = \epsilon_1 \mathbf{1}_t - j\epsilon_2 \mathbf{z}_0 \times \mathbf{1}_t$, and the suffix a on $\mathbf{e}_a$, $\mathbf{h}_a$ implies that they include the $\exp[j(\omega t - \mathbf{k}_t \cdot \boldsymbol{\rho})]$ dependence. The indicated derivation via the reduced equation (2) or (5a) for the electric field $\mathbf{E}$ constitutes a more general procedure, applicable as well to spatially dispersive systems.

As anticipated from the reflection symmetry with respect to the gyrotropic axis, $\pm\kappa_\alpha$ eigenvalues are possible. Consequently, the orthogonality properties (1b) or (7) can be simplified by distinguishing modes by κ_α^2 rather than κ_α. Thus, from Eq. (13a), on adding the relations (7) for $+\kappa_\alpha$ and $-\kappa_\alpha$ and choosing $N_\alpha = Z_\alpha$, one infers for prescribed ω, $\mathbf{k}_t$ the orthonormality properties

$$
\begin{aligned}
\mathbf{e}_\alpha \cdot \mathbf{h}_\beta^* \times \mathbf{z}_0 = \delta_{\alpha\beta} &\qquad \text{for real } \kappa_\alpha, \\
\mathbf{e}_{\alpha^*} \cdot \mathbf{h}_\beta^* \times \mathbf{z}_0 = \delta_{\alpha\beta} &\qquad \text{for complex } \kappa_\alpha,
\end{aligned}
\tag{15}
$$

where now $\delta_{\alpha\beta} = 1$ if $\kappa_\alpha^2 = \kappa_\beta^2$ and $\delta_{\alpha\beta} = 0$ if $\kappa_\alpha^2 \neq \kappa_\beta^2$. One notes that mode functions distinguished by suffices α and α^*, respectively, correspond to eigenvalues κ_α and κ_α^* and thus to opposite choices of sign before Δ in Eq. (11).

The above mode functions $\mathbf{e}_\alpha$ and $\mathbf{h}_\alpha$, distinguished by eigenvalues κ_α^2, permit a transmission-line representation of the steady-state transverse electric and magnetic fields and their sources in the cold magnetoplasma as

$$
\begin{aligned}
\mathbf{E}_t(\mathbf{r}) &= \sum_a V_\alpha(z)\mathbf{e}_\alpha e^{-j\mathbf{k}_t \cdot \boldsymbol{\rho}}, \\
\mathbf{H}_t(\mathbf{r}) &= \sum_a I_\alpha(z)\mathbf{h}_\alpha e^{-j\mathbf{k}_t \cdot \boldsymbol{\rho}}, \\
\mathbf{M}_{te} \times \mathbf{z}_0 &= \sum_a v_\alpha(z)\mathbf{e}_\alpha e^{-j\mathbf{k}_t \cdot \boldsymbol{\rho}}, \\
\mathbf{z}_0 \times \mathbf{J}_{te} &= \sum_a i_\alpha(z)\mathbf{h}_\alpha e^{-j\mathbf{k}_t \cdot \boldsymbol{\rho}},
\end{aligned}
\tag{16a}
$$

where the summation $\sum_a \equiv \sum_\alpha \iint d\mathbf{k}_t/(2\pi)^2$, the α subscript distinguishes "ordinary" and "extraordinary" waves in polarization space as will be explained in Sec. 8.3b, and the voltage (V_α) and current (I_α) amplitudes are determined by transmission-line equations with propagation constant κ_α and characteristic impedance $Z_\alpha = 1/Y_\alpha$ given by Eqs. (11) and (13c). The transmission-line equations are derived by following a procedure similar to that in Sec. 2.2 [however, subject to the orthogonality conditions in Eqs. (8.2.15) or (8.2.16)], and take the form

$$
-\frac{dV_\alpha}{dz} = j\kappa_\alpha Z_\alpha I_\alpha + v_\alpha, \qquad -\frac{dI_\alpha}{dz} = j\kappa_\alpha Y_\alpha V_\alpha + i_\alpha,
\tag{16b}
$$

with source voltages and currents

$$
v_\alpha(z) = \int_S \mathbf{M}_{te} \cdot \mathbf{H}_a^{+*}\, dS, \qquad i_\alpha(z) = \int_S \mathbf{J}_{te} \cdot \mathbf{E}_a^{+*}\, dS,
\tag{16c}
$$

where S denotes the waveguide cross section and the equivalent source currents $\mathbf{M}_{te}$ and $\mathbf{J}_{te}$ are defined in Eq. (8.2.29b). The details are left as an exercise for the reader.

Correspondingly, knowledge of the complete set of wavevectors $\boldsymbol{\Psi}_\alpha$, each distinguished by the eigenvalue κ, permits a representation of an arbitrary steady-state field $\Psi(\mathbf{r})$ as shown in Eq. (8.2.26b). In a cold magnetoplasma, using wavevectors defined in Eqs. (13), one thereby obtains in component form the field representation

$$\mathbf{E}(\mathbf{r}) = \sum_\alpha \int a_\alpha(z)(Z_\alpha \mathbf{e}_\alpha + \mathbf{e}_{z\alpha})e^{-j\mathbf{k}_t\cdot\boldsymbol{\rho}} \frac{d\mathbf{k}_t}{(2\pi)^2} + \frac{\mathbf{J}_z}{j\omega\epsilon_0\epsilon_z},$$

$$\mathbf{H}(\mathbf{r}) = \sum_\alpha \int a_\alpha(z)(\mathbf{h}_\alpha + \mathbf{h}_{z\alpha})e^{-j\mathbf{k}_t\cdot\boldsymbol{\rho}} \frac{d\mathbf{k}_t}{(2\pi)^2} + \frac{\mathbf{M}_z}{j\omega\mu_0\mu_z}. \tag{17}$$

Solution of the defining equation (8.2.27) for a_α yields

$$a_\alpha(z) = -\int_{-\infty}^{z} b_\alpha(z')e^{-j\kappa_\alpha(z-z')}\,dz', \qquad \alpha = \alpha_>, \tag{18a}$$

for modes with $\kappa_\alpha = \kappa_{\alpha_>}$ carrying power in the increasing z direction, and

$$a_\alpha(z) = \int_{z}^{\infty} b_\alpha(z')e^{-j\kappa_\alpha(z-z')}\,dz', \qquad \alpha = \alpha_<, \tag{18b}$$

for modes with $\kappa_\alpha = \kappa_{\alpha_<}$ carrying power in the descreasing z direction; the source term b_α follows from Eq. (8.2.28) as (note $N_\alpha = Z_\alpha$)

$$b_\alpha(z) = \frac{1}{2Z_\alpha}\iint [(Z_{\alpha^*}^*\mathbf{e}_{\alpha^*}^* + \mathbf{e}_{z\alpha^*}^*)\cdot\mathbf{J}(\mathbf{r}) + (\mathbf{h}_{\alpha^*}^* + \mathbf{h}_{z\alpha^*}^*)\cdot\mathbf{M}(\mathbf{r})]e^{+j\mathbf{k}_t\cdot\boldsymbol{\rho}}\,dS. \tag{18c}$$

The results in Eqs. (17) and (18) follow equally well from the general Green's function representation in Eq. (1.4.17), the field representation (1.1.73a), and Eqs. (13).

8.3b Wavenumber Surfaces

The real solutions of the dispersion equation (10) define wavenumber (or refractive index) surfaces in (k_t', κ') space whose shape depends strongly on $\epsilon_1, \epsilon_2, \epsilon_3$ via the wave frequency ω, the plasma frequency ω_p, and the cyclotron frequency ω_c [see Eq. (4)]. For various ranges of $X = (\omega_p/\omega)^2$ and $Y = \omega_c/\omega$ bounded by the solid curves in Fig. 8.3.1, the surface shapes may be grouped into the eight categories shown in Fig. 8.3.2; the X and Y notation is customary in ionospheric propagation theory.[3-5] The branches labeled "o" and "e" represent, respectively, the real "ordinary" and "extraordinary" mode solutions; they correspond to the plus and minus signs, respectively, under the radical in Eq. (11), and the corresponding κ_α' values are denoted by $\pm\kappa_o'$ and $\pm\kappa_e'$ [the $\pm$ here refer to the signs outside the radical in Eq. (11)]. In k_t' ranges for which no branch is shown, κ_o' and (or) κ_e' is complex, with Im $\kappa_{o,e}' < 0$ in order to satisfy the radiation condition for traveling modal fields [see Eq. (18a) requiring decay of $\exp(-j\kappa_{o,e}z)$ as $z \to \infty$]. This definition, based on the analytic properties of the multivalued functions $\kappa_{o,e}'$, does not always coincide with an alternative one used in the literature wherein "extraordinary" and "ordinary" distinguishes those wave types whose propagation in the direction transverse to z is, or is not, affected by the presence of the external magnetic field B_0.

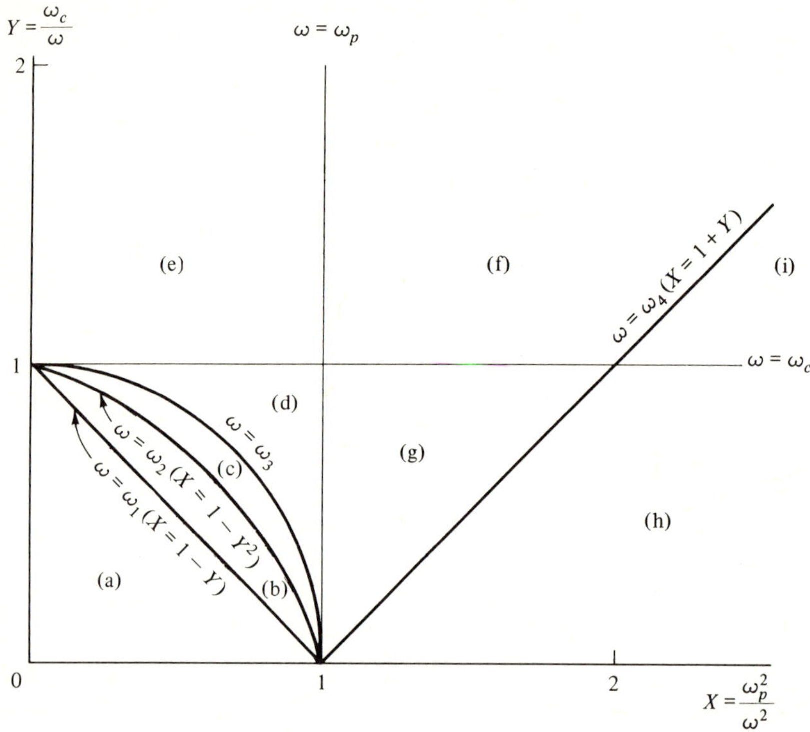

FIG. 8.3.1 Subdivision of the XY plane (the portion shown excludes the low-frequency domain $\omega \to 0$). The frequencies delineating various regions correspond to the following parameter values: $\omega_{1,4}$: $\epsilon_1^2 = \epsilon_2^2$; ω_2: $\epsilon_1 = 0$; ω_3: $2\epsilon_1'^2 = (\epsilon_1' + 1)(\epsilon_1'^2 - \epsilon_2'^2)$.

Transverse propagation corresponds to $\kappa' = 0$, and since $B_0 = 0$ (i.e., $\omega_c = 0$) implies an isotropic medium with wavenumber $k = k_0\sqrt{\epsilon_3}$ [see Eqs. (3b) and (4)], the corresponding k_t' values are $k_t'^2 = 1$ for the ordinary modes and $k_t'^2 \neq 1$ for the extraordinary modes.

The longitudinal wavenumbers $\kappa'_{o,e}$ in Eq. (11) evidently possess branch-point singularities in the complex k_t' plane.[2,6] Most relevant are those occurring for real values of k_t' and signifying the k_t' terminus of the interval of propagation of a non-evanescent mode with real κ'. Such branch points $\bar{k}'_{ti}$ may be located on Fig. 8.3.2 by the condition $d\kappa'/dk_t' = \infty$. $\bar{k}_{t1}'^2 = 1$ and $\bar{k}_{t2}'^2 = (\epsilon_1'^2 - \epsilon_2'^2)/\epsilon_1'$ distinguish k_t' values at which $\kappa'_{o,e} = 0$ while branch points at $\bar{k}'_{t4} = \pm \delta[(\delta/2) - \sqrt{(\delta^2/4) - 1}]$ arise from $\Delta = 0$ in Eq. (11a), and it is recalled that $\delta = 2\epsilon_2'/(\epsilon_1' - 1)$. The algebraic sign of $\kappa'_{o,e}$, when real, must be determined in accord with the radiation condition requiring that the waves with dependence $\exp(-jk\kappa'_{o,e}z)$ carry *energy* in the $+z$ direction. From the discussion in Sec. 1.7c, the average power flow density vector $\bar{\mathbf{S}} = \text{Re}\,(\mathbf{E} \times \mathbf{H}^*)$ is normal to the wavenumber surface and drawn so that $\mathbf{k}' \cdot \bar{\mathbf{S}} > 0$, where $\mathbf{k}' = \mathbf{k}_t' + \mathbf{z}_0\kappa'$ is

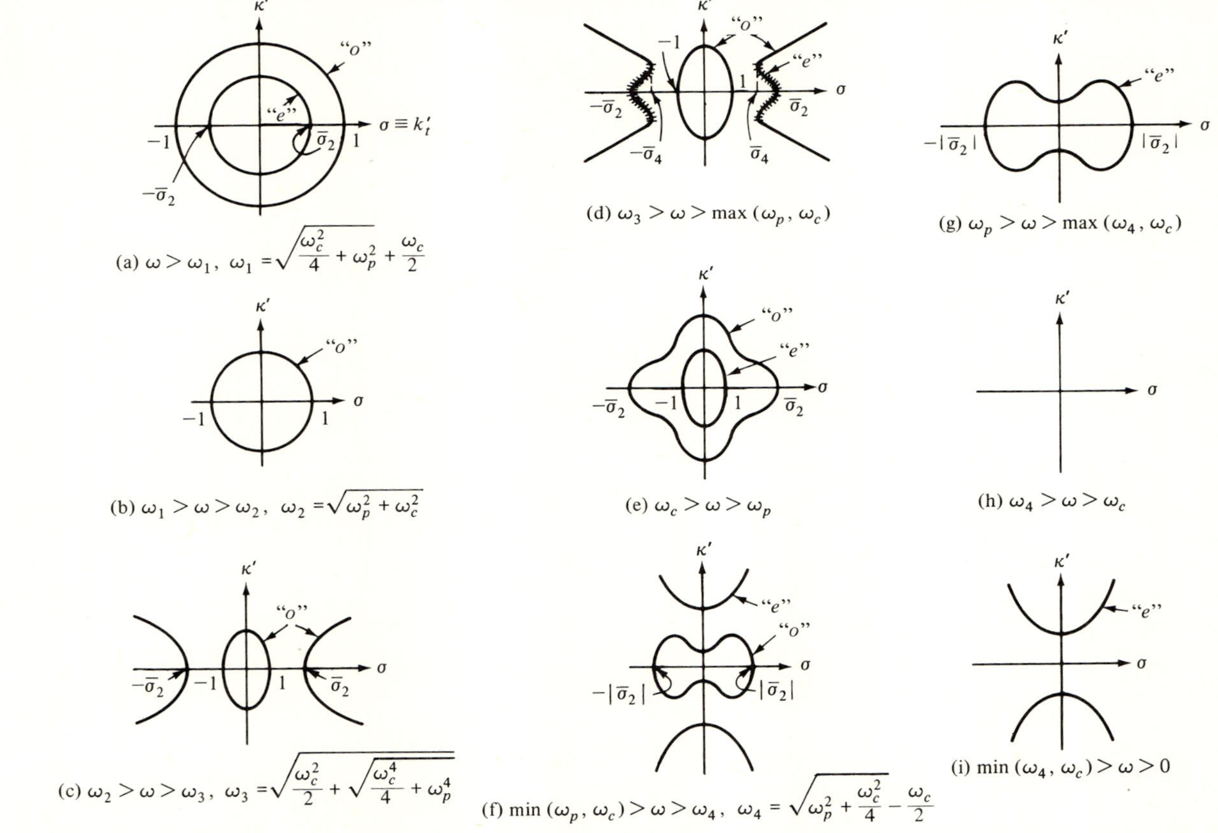

FIG. **8.3.2** Wavenumber surfaces for various ranges of plasma parameters shown in Fig. 8.3.1. $\sigma \equiv k'_t = k_t(k_0\sqrt{\epsilon_3})^{-1}$ and $\kappa' = \kappa(k_0\sqrt{\epsilon_3})^{-1}$ denote the normalized transverse and longitudinal wavenumbers, respectively, while $\bar{\sigma}_i \equiv \bar{k}'_{ti}$ locate branch points of κ'.

the normalized wavevector. Since we are interested in waves with $\bar{\mathbf{S}} \cdot \mathbf{z}_0 > 0$, this implies, for example, that $\kappa'_{o,e} > 0$ for the propagating ranges in Figs. 8.3.2(a) and 8.3.2(e); that $\kappa'_0 > 0$ for $k_t'^2 < 1$ and $\kappa'_0 < 0$ for $\bar{k}_{t2}^2 < k_t'^2 < \infty$ in Fig 8.3.2(c); and that $\kappa'_e > 0$ for $-\infty < k_t'^2 < \infty$ in Fig. 8.3.2(i). When passing from the propagating to the non-propagating range, the analytic continuation of $\kappa'_{o,e}$ must be performed so that $\operatorname{Im} \kappa'_{o,e} < 0$ and thereby specifies the corresponding sign of $\operatorname{Re} \kappa'_{o,e}$.

The explicit form of the mode functions $\mathbf{e}_\alpha = \mathbf{e}_{o,e}$, $\mathbf{h}_\alpha = \mathbf{h}_{o,e}$ and their adjoints for the ordinary and extraordinary modes follows from Eqs. (13) for $\epsilon_3 > 0$ as

$$\sqrt{2}\,\mathbf{e}_{o,e} = \frac{\pm\Delta - k_t'^2}{\pm\Delta}\,\mathbf{k}_{to} \mp j\frac{\delta}{\Delta}\,\hat{\mathbf{k}}_{to}, \quad \sqrt{2}\,\mathbf{e}_{o*,e*} = \frac{\pm\Delta^* - k_t'^2}{\pm\Delta^*}\,\mathbf{k}_{to} \mp j\frac{\delta}{\Delta^*}\,\hat{\mathbf{k}}_{to},$$

$$\tag{19a}$$

$$\sqrt{2}\,\mathbf{h}_{o,e} = j\frac{k_t'^2 \pm \Delta}{\delta}\,\mathbf{k}_{t0} + \hat{\mathbf{k}}_{to}, \quad \sqrt{2}\,\mathbf{h}_{o*,e*} = j\frac{k_t'^2 \pm \Delta^*}{\delta}\,\mathbf{k}_{to} + \hat{\mathbf{k}}_{to}, \tag{19b}$$

with the characteristic impedances $Z_{o,e}$ in Eq. (13c) given correspondingly by

$$Z_{o,e} = \frac{\omega\mu_0}{k\kappa'_{o,e}}\frac{\Delta^2 \pm k_t'^2\Delta}{\delta^2}, \qquad k = k_0\sqrt{\epsilon_3}, \quad \kappa'_{o,e} = \frac{1}{k}\,\kappa_{o,e}. \tag{20}$$

According to Eq. (15), the mode functions in Eqs. (19a, b) satisfy the ortho-normality conditions

$$\mathbf{e}_{e,o} \cdot \mathbf{h}_{o,e}^* \times \mathbf{z}_0 = 0, \qquad \mathbf{e}_{e,o} \cdot \mathbf{h}_{e,o}^* \times \mathbf{z}_0 = 1, \qquad \Delta \text{ real,} \tag{21a}$$

$$\mathbf{e}_{e*,o*} \cdot \mathbf{h}_{o,e}^* \times \mathbf{z}_0 = 0, \qquad \mathbf{e}_{e*,o*} \cdot \mathbf{h}_{e,o}^* \times \mathbf{z}_0 = 1, \qquad \Delta \text{ imaginary,} \tag{21b}$$

with the recognition that $\kappa'_o = -\kappa'^*_e$ when Δ is imaginary (the minus sign being chosen to assure $\operatorname{Im} \kappa'_{o,e} < 0$). It then follows from Eq. (20) that

$$Z_0 = -Z_e^* \qquad \text{when } \Delta \text{ is imaginary.} \tag{21c}$$

When the medium is lossy, the matrix elements in Eq. (4) are complex and the mode orthogonality condition (8.2.8b), or its simplified form (8.2.14b) for the present reflection symmetric case, involves the eigenfunctions for the adjoint medium. It can be shown that results for this case can be obtained by analytic continuation from those for real Δ.

8.3c   Green's Functions for Unbounded Regions

Modal representation

Application of the preceding results is now illustrated for the problem of radiation from a time-harmonic electric current element in an unbounded, cold homogeneous, and lossless magnetoplasma. While radiation in an unbounded medium can also be described in terms of the plane-wave modes of Sec. 1.2,[7,8] we employ the guided-wave formulation[1,2] in anticipation of the plane-stratified medium to be discussed in Sec. 8.3d. Since the gyrotropic axis is parallel to the symmetry axis z, one may employ Eqs. (16) or (17), which represent the electromagnetic fields and their source excitations in terms of the eigenfunctions in Eqs. (19).

We assume the current element to be directed parallel to the z axis:

$$\mathbf{J}(\mathbf{r}) = \mathbf{z}_0 J^0 \delta(\mathbf{r}), \qquad \mathbf{M}(\mathbf{r}) \equiv 0. \tag{22}$$

Employing the field formulation in Eqs. (16), one finds, in view of Eqs. (8.2.29b) that $\mathbf{J}_{te} = 0$, $\mathbf{M}_{te} = (j/\omega\epsilon_3\epsilon_0)(\nabla_t \times \mathbf{J}_z)$, so the current generator $i_\alpha(z) = 0$ while $v_\alpha(z) = v_\alpha \delta(z)$ is given by (note: $\mathbf{H}_\alpha^+ = \mathbf{h}_{\alpha *} + \mathbf{h}_{z\alpha *}$)

$$v_\alpha = \frac{j}{\omega\epsilon_3\epsilon_0} \int\limits_{-\infty}^{\infty}\!\!\int J^0 \delta(\mathbf{\rho})\nabla_t \cdot \mathbf{h}_{a*}^*(\mathbf{\rho}) \times \mathbf{z}_0 \, dx \, dy = -\frac{k_t J^0}{\sqrt{2}\,\omega\epsilon_3\epsilon_0} = e_{z\alpha} J^0, \tag{23}$$

where $\mathbf{h}_a(\mathbf{\rho}) = \mathbf{h}_\alpha \exp{(-j\mathbf{k}_t \cdot \mathbf{\rho})}$ from Eq. (13b), a denotes both α and $\mathbf{k}_t$, and α stands for o or e. Also recall that $\epsilon_z = \epsilon_3\epsilon_0$. The solution of the relevant transmission-line equations (2.2.15a) and (2.2.15b) in the unbounded medium is given for $\epsilon_3 > 0$ by Eqs. (2.4.19):

$$V_\alpha(z) = -(\operatorname{sgn} z)\frac{v_\alpha}{2} e^{-j\kappa_\alpha |z|} = (\operatorname{sgn} z)Z_\alpha I_\alpha(z) \tag{23a}$$

with $Z_\alpha = 1/Y_\alpha$ given in Eq. (20). The total electromagnetic fields may now be written down from Eqs. (16) and (8.2.3).

Alternatively, proceeding from Eqs. (17) and (22), one finds in a source-free region,

$$\mathbf{E}(\mathbf{r}) = \sum_{\alpha\gtrless} \int [b_\alpha(Z_\alpha \mathbf{e}_\alpha + \mathbf{e}_{z\alpha})e^{-j(\mathbf{k}_t\cdot\mathbf{\rho}+\kappa_\alpha z)}] \frac{d\mathbf{k}_t}{(2\pi)^2}, \qquad z \gtrless 0, \tag{24a}$$

$$\mathbf{H}(\mathbf{r}) = \sum_{\alpha\gtrless} \int [b_\alpha(\mathbf{h}_\alpha + \mathbf{h}_{z\alpha})e^{-j(\mathbf{k}_t\cdot\mathbf{\rho}+\kappa_\alpha z)}] \frac{d\mathbf{k}_t}{(2\pi)^2}, \qquad z \gtrless 0. \tag{24b}$$

The suffixes $\alpha_>$ and $\alpha_<$ denote the modes carrying power in the increasing $(z > 0)$ and decreasing $(z < 0)$ directions respectively, and, as indicated, the summation is over one type of mode or the other depending on whether $z \gtrless 0$; furthermore, from Eq. (18c), with $b_\alpha(z) = b_\alpha \delta(z)$,

$$b_\alpha = -(\operatorname{sgn} z)\frac{e_{z\alpha *}^*}{2Z_\alpha} J^0 = \frac{v_\alpha}{2Z_\alpha}, \qquad e_{z\alpha *}^* = e_{z\alpha}. \tag{24c}$$

Because of reflection symmetry, to each value κ_α there corresponds an eigenvalue $-\kappa_\alpha$ (see Sec. 8.2d), so the two ranges $z \gtrless 0$ can be accommodated by the $\alpha_>$ expression provided that in the range $z < 0$, one replaces κ_α by $-\kappa_\alpha$. Equations (24) then yield the same result as that obtained from Eqs. (16) and (8.2.3).

Since the integrands in Eqs. (24a) and (24b) depend only on $\mathbf{k}_t$ and ρ, it is convenient to exploit the azimuthal symmetry and change to a Fourier–Bessel (cylindrical-wave) representation. On noting that $\mathbf{k}_{to} = \mathbf{\rho}_0 \cos\phi - \mathbf{\phi}_0 \sin\phi$, $\hat{\mathbf{k}}_{to} = \mathbf{\rho}_0 \sin\phi + \mathbf{\phi}_0 \cos\phi$, one finds from Eqs. (5.2.7) with $f(k_t) = 1$, $\mathbf{\rho}' = 0$, that

$$\int e^{-j\mathbf{k}_t\cdot\mathbf{\rho}} \, d\mathbf{k}_t = 2\pi \int_0^\infty J_0(k_t\rho)k_t \, dk_t, \qquad d\mathbf{k}_t = k_t \, dk_t \, d\phi, \tag{24d}$$

while from ρ differentiation of the same formula with $f(k_t) = 1/k_t$, $\boldsymbol{\rho}' = 0$,

$$\int_0^\infty dk_t \int_0^{2\pi} d\phi \, k_t \cos\phi \, e^{-jk_t\rho\cos\phi} = -j2\pi \int_0^\infty J_1(k_t\rho)k_t \, dk_t. \tag{24e}$$

Moreover, by direct integration of the ϕ integral,

$$\int_0^\infty dk_t \int_0^{2\pi} d\phi \, k_t \sin\phi \, e^{-jk_t\rho\cos\phi} = 0. \tag{24f}$$

When these results are utilized in Eqs. (24a), (24b), and (13), with $\boldsymbol{\rho}_0$ and $\boldsymbol{\phi}_0$ fixed, one finds on employing the transformation leading from Eq. (3.2.62) to (3.2.69) that for $z \neq 0$ and $\epsilon_3 > 0$,[2]†

$$\mathbf{E}_{t_{o,e}} = jA(\mathrm{sgn}\, z)\int_{-\infty}^\infty \sigma^2\left(\boldsymbol{\rho}_0 \frac{\sigma^2 \mp \Delta}{\pm\Delta} + \boldsymbol{\phi}_0 \frac{j\delta}{\pm\Delta}\right) H_1^{(2)}(k\sigma\rho)e^{-jk\kappa'_{o,e}|z|} \, d\sigma, \tag{25a}$$

$$E_{z_{o,e}} = A\int_{-\infty}^\infty \frac{\sigma^2 \mp \Delta}{\pm\Delta} \frac{\sigma^3\kappa'_{o,e}}{1-\sigma^2} H_0^{(2)}(k\sigma\rho)e^{-jk\kappa'_{o,e}|z|} \, d\sigma, \tag{25b}$$

$$\mathbf{H}_{t_{o,e}} = \frac{A}{\zeta}\int_{-\infty}^\infty \sigma^2\left(\boldsymbol{\rho}_0 \frac{\delta\kappa'_{o,e}}{\pm\Delta} + \boldsymbol{\phi}_0 j\frac{\sigma^2 \mp \Delta}{\pm\Delta} \frac{\kappa'_{o,e}}{1-\sigma^2}\right) H_1^{(2)}(k\sigma\rho)e^{-jk\kappa'_{o,e}|z|} \, d\sigma, \tag{25c}$$

$$H_{z_{o,e}} = -j(\mathrm{sgn}\, z)\frac{A\delta}{\zeta}\int_{-\infty}^\infty \frac{\sigma^3}{\pm\Delta} H_0^{(2)}(k\sigma\rho)e^{-jk\kappa'_{o,e}|z|} \, d\sigma, \qquad \sigma \equiv k'_t, \tag{25d}$$

where the upper and lower signs refer to the ordinary and extraordinary mode contributions, respectively, $A = k^2\zeta J^0/16\pi$ with $\zeta = (\mu_0/\epsilon_0\epsilon_3)^{1/2}$, and $\boldsymbol{\rho}_0$, $\boldsymbol{\phi}_0$, z_0 are unit vectors in a cylindrical coordinate system. For brevity, the notation $\sigma \equiv k'_t$ has been introduced, where k'_t is defined in Eq. (10a). The total fields are then obtained from

$$\mathbf{E} = \mathbf{E}_e + \mathbf{E}_o, \qquad \mathbf{H} = \mathbf{H}_e + \mathbf{H}_0. \tag{26}$$

The integration path is indented into the lower half of the σ plane to avoid the branch point at $\sigma = 0$ arising from the presence of the Hankel functions. The branch-point singularities due to Δ and $\kappa'_{o,e}$ are displaced from the real axis when losses are assumed, and their disposition in the lossless limit is thus made unambiguous; alternatively, one may follow the considerations in Sec. 8.3b for directly determining the path deformation around significant real branch points. These considerations are linked with the radiation condition requiring $\mathrm{Im}\,\kappa'_{o,e} \leq 0$ on the integration path. In the lossless case, where $\kappa'_{o,e}$ may be real over certain ranges of σ, its algebraic sign can be determined as noted in Sec. 8.3b, or by treating the problem in the limit of vanishing dissipation.

    The radiation condition for propagating waves may alternatively be imposed by requiring the mode admittances $Y_{o,e} = 1/Z_{o,e}$ in Eq. (20) to be positive when $\kappa'_{o,e}$ is real. For proof, we begin with the requirement that the total average power flow through a plane $z > 0$ must be non-negative and given by

$$\bar{S}_z = \mathrm{Re}\int\int_{-\infty}^\infty \mathbf{E}_t^* \cdot \mathbf{H}_t \times z_0 \, dx \, dy \geq 0. \tag{27}$$

---

†For $\epsilon_3 < 0$, replace $\sqrt{\epsilon} \to -j\sqrt{|\epsilon|}$, $\sigma \to j\sigma$, $\kappa' \to j\kappa'$.

Upon substituting the modal representations for $\mathbf{E}_t^*$ and $\mathbf{H}_t$ from Eq. (16a), interchanging the orders of the xy and $\mathbf{k}_t$ integrations, and recalling the orthogonality relation (8.2.15),† one finds

$$\bar{S}_z = \mathrm{Re} \iint [V_o^* I_o + V_e^* I_e] \frac{d\mathbf{k}_t'}{(2\pi)^2} \geq 0, \tag{28}$$

with the integrals extending over those portions of the k_t' plane for which $\kappa_{o,e}'^2$ is both real (Δ real) and complex (Δ imaginary), respectively. The relation in Eq. (28) is valid for arbitrary source distributions confined to or below the plane $z = 0$, and the form of $V_{o,e}$ and $I_{o,e}$ for $z > 0$ is then as in Eq. (23a); for the special case of the longitudinal electric dipole, $v_{o,e}$ is specified in Eq. (23). Equation (28) may therefore be written as

$$\bar{S}_z = \mathrm{Re} \iint [Y_0 |V_0|^2 + Y_e |V_e|^2] \frac{d\mathbf{k}_t'}{(2\pi)^2} \geq 0, \tag{29}$$

and since $Y_e = -Y_o^*$ when κ'^2 is complex [see Eq. (21c)], the corresponding integral representative of energy coupling in the non-propagating ordinary and extraordinary modes does not contribute to the real power. When κ' is imaginary with Δ real, $Y_{o,e} = 1/Z_{o,e}$ in Eq. (20) is also imaginary and there is again no contribution to $\bar{S}_z$. The only remaining modes are the propagating modes along z for which $\kappa_{o,e}'$ is real:

$$\bar{S}_z = \iint_{\kappa_o' \text{ real}} Y_0 |V_0|^2 \frac{d\mathbf{k}_t'}{(2\pi)^2} + \iint_{\kappa_e' \text{ real}} Y_e |V_e|^2 \frac{d\mathbf{k}_t'}{(2\pi)^2} \geq 0. \tag{30}$$

Since this inequality must be satisfied for arbitrary source distributions (i.e., arbitrary V_o and V_e) it follows that

$$Y_{o,e} > 0 \qquad \text{when } \kappa'_{o,e} \text{ is real}, \tag{31}$$

which statement implies that each propagating ordinary and extraordinary mode carries energy away from the source region (see also Reference 6).

Asymptotic evaluation of far fields

Since it does not seem possible to express the integrals in Eqs. (25) exactly in terms of known functions (a reduction is possible, however, for the special case of uniaxial anisotropy, $\epsilon_2' = 0$ [see Eq. (7.3.8) et seq.]), it is necessary to resort to asymptotic techniques in order to infer specific information about the behavior of the radiated fields. Generic double Fourier integrals encountered in the theory of radiation by arbitrary sources in anisotropic dispersive media have been discussed in Sec. 1.6b, and an asymptotic evaluation at large distances from the source has been performed by the stationary-phase method. Because of the complexity exhibited by multiple integral representations, the asymptotic

---

†On multiplying $\mathbf{e}_{o,e}$ and $\mathbf{h}_{o,e}$ in Eqs. (19) by $\exp(-j\mathbf{k}_t \cdot \boldsymbol{\rho})$ to synthesize the mode fields $\mathbf{e}_a = \mathbf{E}_{ta}/Z_\alpha$ and $\mathbf{h}_a = \mathbf{H}_{ta}$, one finds from Eq. (8.2.15),

$$\int_{-\infty}^{\infty} \int_{-\infty}^{\infty} \mathbf{e}_a(\boldsymbol{\rho}, \mathbf{k}_{t1}) \cdot \mathbf{h}_{a*}^*(\boldsymbol{\rho}, \mathbf{k}_{t2}) \times \mathbf{z}_0 \, dS = \mathbf{e}_\alpha \cdot \mathbf{h}_{\alpha*}^* \times \mathbf{z}_0 \, \delta(\mathbf{k}_{t1} - \mathbf{k}_{t2})(2\pi)^2.$$

calculation in Sec. 1.6b accounts only for saddle-point contributions without including effects of singularities in the integrands. For the special case under discussion here, the double integrals have been reduced to the single integrals in Eqs. (25), which can be treated more thoroughly by the methods of Chapter 5. For this reason, the asymptotic evaluation is presented again, due regard now being given to the steepest-descent paths through saddle points, and to the manner of dealing with singularities and transition phenomena. Graphical methods for determining saddle points are the same as in Sec. 1.6b and are summarized briefly.

As in the investigations of similar integrals encountered in problems of radiation in isotropic regions (see Sec. 5.3), it is assumed that $|k\sigma\rho| \gg 1$, so the Hankel functions may be replaced by their asymptotic approximation in Eq. (5.3.13). If $k\rho \gg 1$, this condition may be satisfied over the entire integration interval provided that the path is distorted away from the vicinity of $\sigma = 0$. Each of the integrals is then of the form [see Eq. (1.6.22), with $A \equiv L$, $\psi \equiv krM$, $i \to -j$, $k_t \to k\sigma$]

$$J = \int L(\sigma)e^{-jkrM(\sigma)}\, d\sigma, \tag{32}$$

where

$$M(\sigma) = \kappa'(\sigma)|\cos\theta| + \sigma\sin\theta, \tag{32a}$$

and (r, θ) are the spherical coordinates with respect to the source point $\mathbf{r} = 0$,

$$\rho = r\sin\theta, \qquad z = r\cos\theta. \tag{32b}$$

For $kr \gg 1$, the major contribution to the integral arises from the vicinity of the saddle points σ_i at which

$$\frac{d}{d\sigma}M(\sigma) = 0 \qquad \text{or} \qquad \frac{d\kappa'(\sigma)}{d\sigma}\bigg|_{\sigma_i} = -|\tan\theta|. \tag{33}$$

The real solutions $\kappa'(\sigma_i) \equiv \kappa'_i$, σ_i, yield propagating waves in the far field while complex saddle points give rise to exponentially attenuated solutions. The geometrical significance of the saddle-point condition (33) has been discussed in Sec. 1.6b and also in connection with radiation in the uniaxial medium [see Eq. (7.3.11) et seq]: the real saddle points locate those points κ'_i, σ_i on the wavenumber surface for the medium [i.e., the plot of the real κ', $k'_t \equiv \sigma$ solutions of Eq. (11)] at which the normal makes an angle θ with the positive κ'axis. The normal to the wavenumber surface points in the direction of power flow (ray direction), so the contribution from each saddle point represents a wave that, at least in the far field, carries energy radially away from the source region. This radial power flow property does not necessarily apply to the total field contributed by several saddle points, in view of interference that generally occurs between the various field constituents. These remarks are illustrated in general in Fig. 8.3.3 (see Fig. 1.6.5 for further elaboration on the use of the diagram), with Fig. 8.3.2 showing the actual curve shapes corresponding to various parameters in a plasma described by the permittivity tensor in Eq. (3b).

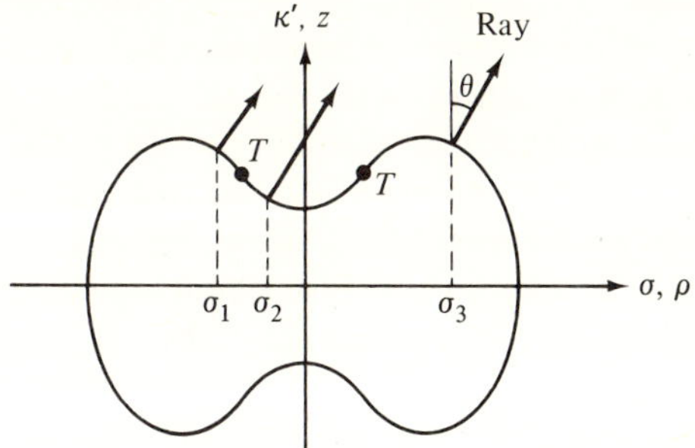

FIG. 8.3.3   Wavenumber surface and saddle points.

To carry out the asymptotic evaluation of the integral in Eq. (32), it remains to be shown that the integration path can be deformed into steepest-descent paths (SDP) through the various saddle points. The steepest-descent contours through the saddle point σ_i are defined by the equation $\operatorname{Re} M(\sigma) = \operatorname{Re} M(\sigma_i)$, but a solution is not simply obtained in view of the complicated structure of $\kappa'(\sigma)$. It suffices, however, to follow the SDP only in the vicinity of σ_i, where its progress is determined readily, provided that the remaining portions C' of the path lie in "valley regions" below the level of the various contributing saddle points. In the valleys of the complex σ plane, one has $\operatorname{Im} M(\sigma) < 0$, whence the contribution from any path segment C' is

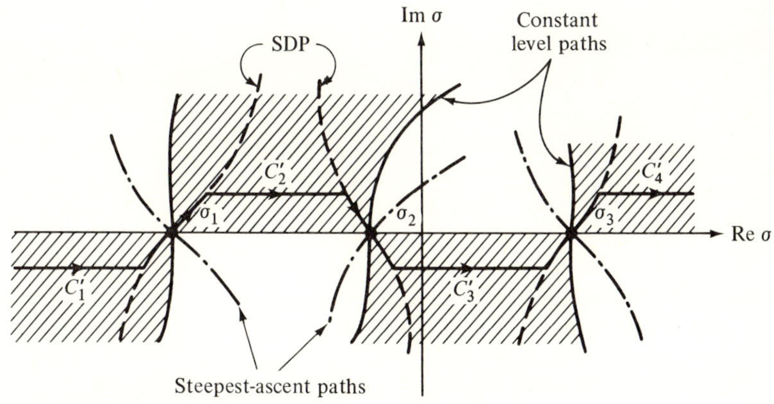

FIG. 8.3.4   Path deformation in the complex σ plane. Valley region $[\operatorname{Im} M(\sigma) < 0]$ is shaded. Note that $[d^2 M(\sigma)/d\sigma^2]_{\sigma_{1,3}} < 0$, $[d^2 M(\sigma)/d\sigma^2]_{\sigma_2} > 0$.

$O[\exp(-akr)]$, where $a > 0$ is the smallest value of $|\operatorname{Im} M(\sigma)|$ along C'. This exponentially small term can be neglected in comparison with the saddle-point approximation, for which $\operatorname{Im} M(\sigma_i) = 0.$[†] For the configuration in Fig. 8.3.3, the disposition of the complex σ plane in the vicinity of the saddle points is shown in Fig. 8.3.4, and one observes that the path may be deformed from the real axis so as to proceed entirely in valley regions when $\sigma \neq \sigma_i$. To achieve this path distortion it is necessary that the algebraic sign of $d^2M(\sigma_i)/d\sigma_i^2$ alternate from one saddle point to the other since the argument of the vector element $\boldsymbol{\sigma} - \boldsymbol{\sigma}_i$ leading away from the saddle point along the SDP is $\mp\pi/4$ when $d^2M(\sigma_i)/d\sigma_i^2 \gtrless 0$ [see Eqs. (4.2.1) and the similar analysis in Sec. 7.5e]. This behavior is met as long as $\kappa'(\sigma)$ is a regular function of σ over the relevant interval on the real axis. One may also verify that none of the branch points in the integrand is crossed during the path deformation so that the asymptotic approximation of the integral in Eq. (32) arises from the saddle points only, with the lowest-order contribution from each taking the form given in Eq. (4.2.1b) (see Sec. 8.3b for the analytic properties of $\kappa'_{o,e}$).

After these preliminaries, the asymptotic solution for the field components in Eqs. (25) at the distant observation point (r, θ) may be written as follows[2]:

$$\boldsymbol{\rho}_0 E_\rho + \mathbf{z}_0 E_z \sim 2jA \sum_i \beta_i \sqrt{R_i} G_i \frac{e^{-jkrN_i(\theta)}}{kr} (\mathbf{q}'_i \times \boldsymbol{\phi}_0), \tag{34a}$$

$$E_\phi \sim 2A \sum_i \beta_i \sqrt{R_i} F_i \frac{e^{-jkrN_i(\theta)}}{kr}, \qquad A = \frac{k^2 \zeta J^0}{16\pi}, \tag{34b}$$

$$\boldsymbol{\rho}_0 H_\rho + \mathbf{z}_0 H_z \sim -\frac{2A}{\zeta} \sum_i \beta_i \sqrt{R_i}\, F_i \frac{e^{-jkrN_i(\theta)}}{kr} (\mathbf{p}'_i \times \boldsymbol{\phi}_0), \tag{34c}$$

$$H_\phi \sim -\frac{2jA}{\zeta} \sum_i \beta_i \sqrt{R_i} G_i \frac{e^{-jkrN_i(\theta)}}{kr}, \qquad \zeta = \sqrt{\frac{\mu_0}{\epsilon_0 \epsilon_3}}, \tag{34d}$$

where the summation extends over all contributing saddle points and $N_i(\theta) \equiv M(\sigma_i)$ is the ray refractive index. R_i is the radius of curvature of the wave-number surface at the point $\kappa'_i, \sigma_i,$[‡]

$$\frac{1}{R_i} \equiv \frac{|d^2\kappa'_i/d\sigma_i^2|}{[1 + (d\kappa'_i/d\sigma_i)^2]^{3/2}} = \left|\frac{d^2\kappa'_i}{d\sigma_i^2}\right| |\cos^3 \theta| = \frac{d^2M(\sigma_i)}{d\sigma_i^2} \cos^2 \theta, \tag{35a}$$

and F_i, G_i are amplitude functions defined as follows:

$$F_i(\theta) = \sqrt{\frac{\sigma_i}{\sin \theta}} \frac{\sigma_i \delta}{\gamma_i \Delta_i} \cos \theta, \qquad G_i(\theta) = \sqrt{\frac{\sigma_i}{\sin \theta}} \frac{\kappa'_i \sigma_i}{1 - \sigma_i^2} \frac{\sigma_i^2 - \gamma_i \Delta_i}{\gamma_i \Delta_i} \cos \theta, \tag{35b}$$

with $\Delta_i \equiv \Delta(\sigma_i)$, and $\gamma_i = +1$ or -1 if the saddle point arises in the ordinary and extraordinary integral, respectively. Finally, $\beta_i = +1$ if $d^2M(\sigma_i)/d\sigma_i^2 < 0$, $\beta_i = -j$ if $d^2M(\sigma_i)/d\sigma_i^2 > 0$, and $\mathbf{p}'_i$, $\mathbf{q}'_i$ are the polarization vectors

---

[†]This estimate of the exponentially small error made by neglecting the contribution from C' is used in addition to the error estimate of the saddle-point evaluation in Sec. 4.2b based on the distance from the saddle point to the nearest singularity in the integrand.

[‡]The symbol $dA(\sigma_i)/d\sigma_i$ denotes evaluation of $dA(\sigma)/d\sigma$ at σ_i, and similarly for higher derivatives.

$$\mathbf{p}'_i = \boldsymbol{\rho}_0\sigma_i + \mathbf{z}_0\kappa'_i, \qquad \mathbf{q}'_i = \boldsymbol{\rho}_0\sigma_i + \mathbf{z}_0\frac{1-\sigma_i^2}{\kappa_i'^2}, \tag{35c}$$

where $k\mathbf{p}'_i\cdot\mathbf{r} \equiv krN_i(\theta)$, so $N_i(\theta) = \mathbf{p}'_i\cdot\mathbf{r}_0$. These results, valid when $\epsilon_3 > 0$, also apply to $\epsilon_3 < 0$ with the substitutions listed in the footnote to Eqs. (25). Further investigation of the formulas in Eqs. (34) is required when the radius of curvature $R_i \to \infty$, and the pertinent considerations are given below. The form of Eqs. (34) is the same as in Eq. (1.6.25) if one observes that in view of the rotational symmetry of the wavenumber surface about the κ' axis, one of the principal radii of curvature is constant.

The structure of the asymptotic formulas substantiates the remarks made earlier about the use of the refractive index surfaces in predicting the number of propagating ray contributions in the far field. As noted in Sec. 1.6b, certain information regarding the ray amplitudes is also obtainable in view of the presence of the radius of curvature R_i in the result; field enhancement is evidently associated with rays corresponding to segments having small curvature. If the surface has points of inflection or open branches, certain rays will contribute only over a limited range of observation angles; in the vicinity of the boundaries of the domains of existence of these rays, $R_i \to \infty$ and a more careful evaluation of the original integral is required. The restricted domain of existence of certain rays associated with a wavenumber plot having points of inflection may be visualized from Fig. 8.3.3, which shows that the saddle points σ_1 and σ_2 yield propagating waves only when $0 \le \theta < \theta_c$, where θ_c is the angle between the positive κ' (or z) axis and the normal at the inflection point T. For $\theta_c < \theta < \pi/2$, σ_1 and σ_2 are complex and the associated ray contributions are evanescent; this range of observation angles therefore defines the "shadow region" for these rays. These considerations do not affect the ray corresponding to σ_3 which contributes at all observation points. Analogous remarks have been made in Sec. 7.3 to explain phenomena associated with an open-branched surface.

Transition region*: *coalescence of two saddle points

As noted earlier, the asymptotic formulas in Eqs. (34) become invalid when $R_i \to \infty$, as, for example, at a point of inflection on the dispersion curve (see points T in Fig. 8.3.3). As the observation angle θ in the far field approaches the value θ_c defining the direction of the normal at T, the two first-order saddle points at σ_1 and σ_2 approach one another and coalesce into a second-order saddle point on the shadow boundary $\theta = \theta_c$. The asymptotic evaluation of the integral must then be carried out in terms of Airy functions (see Sec. 4.5a), and the result is directly analogous to that in Sec. 7.5e [see the portion of Eq. (7.5.59) relating to the saddle points β_1 and β_2]. On the dark side ($\theta > \theta_c$ for the diagram in Fig. 8.3.3), the contributions from the two rays corresponding to $\sigma_{1,2}$ are exponentially small and therefore negligible [the ray refractive index $N(\theta)$ is no longer real], whereas on the shadow boundary, the variation is $\sim (kr)^{-5/6} \exp\left[-jkrN(\theta_c)\right]$. Thus, the fields on the boundary are stronger than at other observation angles where the distance dependence is according to the conventional $(kr)^{-1}$. Larger field concentrations may therefore be expected in the transition region where the two ray contributions from σ_1 and σ_2 have almost identical properties and interact strongly.

Transition region: saddle point moves to infinity

The radius of curvature of the dispersion surface also grows without limit on the remote segments of open branches. A simple but typical example has already been encountered in the uniaxially anisotropic medium in Chapter 7 (see Fig. 7.3.2); the difficulty in the present asymptotic evaluation arises when $\theta \to \theta_c (\sigma_i \to \infty)$, where θ_c relates to the normal to the asymptote of the surface. Open branches of the dispersion curve may arise when $\epsilon_1' < 0$, as found from a study of the real solutions of Eq. (10) when $\kappa', \sigma \to \infty$; the equation for the asymptotes is easily shown to be $\theta = \hat{\theta} = \pm \tan^{-1} \sqrt{-1/\epsilon_1'}$, it being recalled that $\sigma \equiv k_t'$.

To assess the modification of the asymptotic procedure required when $\sigma_i \to \infty$, it suffices to consider the integrands in Eqs. (25) in the range of large σ. From Eq. (11) and its counterpart when $\epsilon_3 < 0$, one finds, for $|\sigma| \gg 1$,

$$\kappa_0' \approx -\sqrt{\epsilon_1' + \frac{\epsilon_2'^2}{\epsilon_1' - 1} - \epsilon_1' \sigma^2}\,, \qquad \epsilon_3 > 0\,, \quad \epsilon_1' < 0\,, \tag{36a}$$

$$\kappa_e' \approx \sqrt{-\epsilon_1' + \frac{\epsilon_2'^2}{1 - \epsilon_1'} - \epsilon_1' \sigma^2}\,, \qquad \epsilon_3 < 0\,, \quad \epsilon_1' < 0\,, \tag{36b}$$

where it has been recognized that $\kappa_0' < 0$ and $\kappa_e' > 0$ (see Fig. 8.3.2), and that the open branch is associated with the ordinary and extraordinary modes when $\epsilon_3 > 0$ and $\epsilon_3 < 0$, respectively. To approximate the factors multiplying the exponential and Hankel function in the integrands of Eqs. (25), we may put $\kappa \infty \sigma, \Delta \approx -\sigma^2$, thereby obtaining simple powers of σ which may be reexpressed as appropriate spatial derivatives of the remaining integrand. The resulting fields may thus be formulated as spatial derivatives of an integral that has the same form as the one in Eq. (7.3.8), and the latter may be evaluated in the closed form listed in Eq. (7.3.3b).† Since the potential function in Eq. (7.3.3b) behaves like $[rN(\theta)]^{-1} \exp[-jkrN(\theta)]$, one observes that a true field singularity exists as $\theta \to \theta_c$ [i.e., $N(\theta) \to 0$] and that the field behavior near this type of shadow boundary in the gyrotropic case is the same as in the simpler uniaxially anisotropic medium.[2] The reader is therefore referred to the detailed discussion of the latter problem in Secs. 7.3a and 7.3e. It is perhaps worth emphasizing again that the strong field and power flow singularities at $\theta = \theta_c$ disappear when losses are present, and also when a smooth extended source distribution is employed in place of the point dipole current.

---

†Allowance must be made for the different time dependence $\exp(j\omega t)$ employed in this chapter. While Eq. (7.3.8) represents directly the form corresponding to Eq. (36b), one may show that the integral corresponding to Eq. (36a) leads to the same type of result.

8.3d   Green's Functions for Plane-Stratified Regions

Representation in terms of ordinary and extraordinary modes

In a medium consisting of piecewise constant layers whose interfaces are perpendicular to the rectilinear coordinate z, the electromagnetic boundary conditions across each interface may be satisfied in a systematic manner in a representation comprised of the guided-wave eigenfunctions.[2,9] If the gyrotropic axis is inclined arbitrarily with respect to the z axis, the general mode functions in Sec. 8. 2 must be employed in each region; since there exists in general no mode pair $\exp(-j\kappa_\alpha z)$ and $\exp(+j\kappa_\alpha z)$ descriptive of waves propagating along the $+z$ and $-z$ directions with the same propagation constant κ_α, the analysis in layered media must be effected in a traveling-wave, rather than a standing-wave, description. However, for the special case of longitudinal orientation of the gyrotropic axis as in this section, the region becomes reflection symmetric and a standing-wave approach may be used to advantage (see Sec. 8.2d). A number of pertinent considerations are presented below.

If the region in question consists of layers, each of which is characterized by a dielectric tensor of the form shown in Eq. (3b) with $\mathbf{b}_0 \equiv \mathbf{z}_0$, the transverse electric and magnetic fields in the νth layer are represented in the form [see Eq. (16)]

$$\mathbf{E}_{t\nu}(\mathbf{r}) = \iint_{-\infty}^{\infty} [V_{o\nu}(z, \mathbf{k}_t)\mathbf{e}_{o\nu}(\mathbf{k}_t) + V_{e\nu}(z, \mathbf{k}_t)\mathbf{e}_{e\nu}(\mathbf{k}_t)] \exp(-j\mathbf{k}_t \cdot \boldsymbol{\rho}) \frac{d\mathbf{k}_t}{(2\pi)^2}, \quad (37a)$$

$$\mathbf{H}_{t\nu}(\mathbf{r}) = \iint_{-\infty}^{\infty} [I_{o\nu}(z, \mathbf{k}_t)\mathbf{h}_{o\nu}(\mathbf{k}_t) + I_{e\nu}(z, \mathbf{k}_t)\mathbf{h}_{e\nu}(\mathbf{k}_t)] \exp(-j\mathbf{k}_t \cdot \boldsymbol{\rho}) \frac{d\mathbf{k}_t}{(2\pi)^2}, \quad (37b)$$

where $\mathbf{e}_{\alpha\nu}$ and $\mathbf{h}_{\alpha\nu}$, $(\alpha = o, e)$, are the eigenfunctions in Eqs. (19) and $V_{\alpha\nu}$, $I_{\alpha\nu}$ are the modal voltage and current which satisfy the transmission-line equations (16a). Let us suppose that the νth layer is separated from the μth layer by a plane interface at $z = z_\nu$; then the required continuity of $\mathbf{E}_t$ and $\mathbf{H}_t$ across z_ν may be satisfied if

$$V_{o\nu}\mathbf{e}_{o\nu} + V_{e\nu}\mathbf{e}_{e\nu} = V_{o\mu}\mathbf{e}_{o\mu} + V_{e\mu}\mathbf{e}_{e\mu}, \quad (38a)$$

$$I_{o\nu}\mathbf{h}_{o\nu} + I_{e\nu}\mathbf{h}_{e\nu} = I_{o\mu}\mathbf{h}_{o\mu} + I_{e\mu}\mathbf{h}_{e\mu}. \quad (38b)$$

In view of the orthogonality relation in Eqs. (21), Eqs. (38) may be reduced to the matrix form

$$\begin{pmatrix} V_\nu \\ I_\nu \end{pmatrix} = T_z \begin{pmatrix} V_\mu \\ I_\mu \end{pmatrix}, \quad V_\nu \rightarrow \begin{pmatrix} V_{o\nu} \\ V_{e\nu} \end{pmatrix}, \quad \text{etc.}, \quad (39)$$

where V_i, I_i, $(i = \nu, \mu)$, are column vectors while T_z is the impedance transfer matrix

$$T_z \rightarrow \begin{pmatrix} t_{\nu\mu} & 0 \\ 0 & \hat{t}_{\nu\mu} \end{pmatrix}, \quad (39a)$$

whose elements $t_{\nu\mu}$, $\hat{t}_{\nu\mu}$ are themselves 2×2 matrices:

$$t_{\nu\mu} \rightarrow \begin{pmatrix} \mathbf{h}^*_{o*\nu} \times \mathbf{z}_0 \cdot \mathbf{e}_{o\mu} & \mathbf{h}^*_{o*\mu} \times \mathbf{z}_0 \cdot \mathbf{e}_{e\mu} \\ \mathbf{h}^*_{e*\nu} \times \mathbf{z}_0 \cdot \mathbf{e}_{o\mu} & \mathbf{h}^*_{e*\mu} \times \mathbf{z}_0 \cdot \mathbf{e}_{e\mu} \end{pmatrix}, \tag{39b}$$

$$\hat{t}_{\nu\mu} \rightarrow \begin{pmatrix} \mathbf{e}^*_{o*\nu} \cdot \mathbf{h}_{o\mu} \times \mathbf{z}_0 & \mathbf{e}^*_{o*\nu} \cdot \mathbf{h}_{e\mu} \times \mathbf{z}_0 \\ \mathbf{e}^*_{e*\nu} \cdot \mathbf{h}_{o\mu} \times \mathbf{z}_0 & \mathbf{e}^*_{e*\nu} \cdot \mathbf{h}_{e\mu} \times \mathbf{z}_0 \end{pmatrix}. \tag{39c}$$

Since the submatrices $t_{\nu\mu}$ and $\hat{t}_{\nu\mu}$ are generally not diagonal, the ordinary and extraordinary modes are coupled at the interface.

In the interior of each slab region, the modal network comprises two uncoupled transmission lines representative of the o and e modes, respectively (see Fig. 8.3.5). The impedance transfer matrix $\hat{T}_z$ for a slab of length d is readily constructed from the transmission-line solution in Sec. 2.4:

$$\hat{T}_z \rightarrow \begin{pmatrix} \cos(k\kappa'd) & jZ\sin(k\kappa'd) \\ jY\sin(k\kappa'd) & \cos(k\kappa'd) \end{pmatrix}, \tag{40}$$

where κ' and $Z = Y^{-1}$ represent the diagonal 2×2 matrices

$$\kappa' \rightarrow \begin{pmatrix} \kappa'_o & 0 \\ 0 & \kappa'_e \end{pmatrix}, \qquad Z \rightarrow \begin{pmatrix} Z_o & 0 \\ 0 & Z_e \end{pmatrix}. \tag{40a}$$

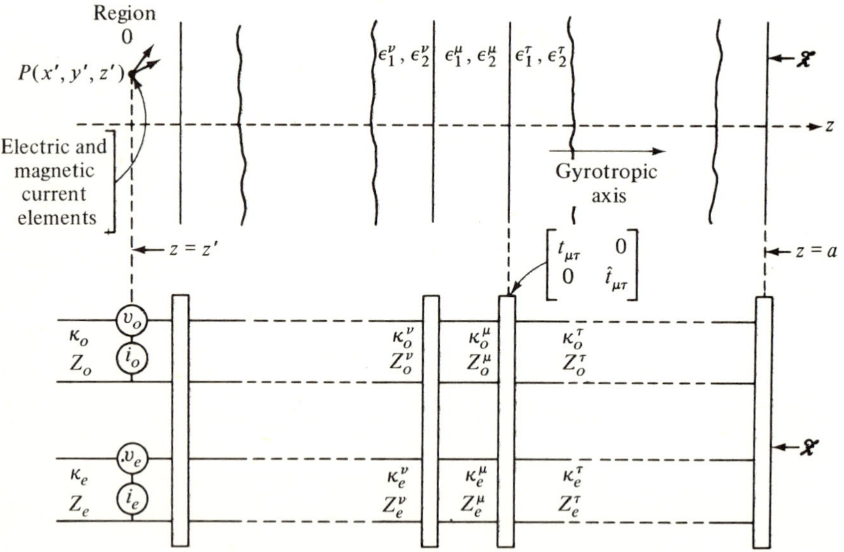

FIG. 8.3.5.   Physical structure and associated network problem.

$\cos(k\kappa'd)$ is a diagonal matrix having as elements $\cos(k\kappa'_o d)$ and $\cos(k\kappa'_e d)$, respectively, with a similar interpretation applicable to $\sin(k\kappa'd)$. The voltage and current vectors V_1, I_1 at the slab face $z = z_1$ are then related to the analogous quantities at $z = z_2$ in the *same* slab via

$$\begin{pmatrix} V_1 \\ I_1 \end{pmatrix} = \hat{T}_z \begin{pmatrix} V_2 \\ I_2 \end{pmatrix}, \qquad z_2 - z_1 = d, \tag{41}$$

and the transition across the interface is accomplished through use of Eq. (39). Repeated application of Eqs. (39) and (41) allows the systematic calculation of the voltage and current at any point in a plane-stratified region and leads to an overall transfer matrix given by the ordered product of the individual T_z and $\hat{T}_z$.

If the region is terminated at $z = a$ by a plane with constant anisotropic surface impedance $\mathcal{Z}$ (i. e., $\mathbf{E}_t = \mathcal{Z} \cdot \mathbf{H}_t \times \mathbf{z}_0$ at $z = a$), then by considerations analogous to those leading to Eqs. (38),

$$V = \hat{\mathcal{Z}} I, \tag{42}$$

where V and I are column vectors as in Eq. (39). and $\hat{\mathcal{Z}}$ is the 2×2 impedance matrix,

$$\hat{\mathcal{Z}} \rightarrow \begin{pmatrix} \mathbf{h}_{o*}^* \times \mathbf{z}_0 \cdot \mathcal{Z} \cdot \mathbf{h}_o \times \mathbf{z}_0 & \mathbf{h}_{o*}^* \times \mathbf{z}_0 \cdot \mathcal{Z} \cdot \mathbf{h}_e \times \mathbf{z}_0 \\ \mathbf{h}_{e*}^* \times \mathbf{z}_0 \cdot \mathcal{Z} \cdot \mathbf{h}_o \times \mathbf{z}_0 & \mathbf{h}_{e*}^* \times \mathbf{z}_0 \cdot \mathcal{Z} \cdot \mathbf{h}_e \times \mathbf{z}_0 \end{pmatrix}, \tag{42a}$$

the mode functions being those for the medium adjoining the boundary. One observes that even a scalar surface impedance $\mathcal{Z}_s$, for which $\mathcal{Z} = 1_t \mathcal{Z}_s$, generally couples the ordinary modes unless $\mathcal{Z} = 0$ (perfect conductor) or $k_t = 0$ (normally incident mode).

Since an abrupt change in the medium properties leads to a coupling between the o and e modes as in Eq. (39), it follows that these modes are coupled continuously in a gyrotropic medium with continuous variation along z. The modal representations in Eqs. (37) may then be employed only locally and the mode functions become z dependent (the discrete index ν goes over into a continuous function of z). This complication disappears in the special case of uniaxial anisotropy (Sec. 7. 2) and, of course, for an isotropic dielectric (Sec. 2.3e). For a treatment of the coupled equations, see Reference 10.

Rapid variations in the medium parameters lead to strong coupling of the o and e modes, thereby making calculation difficult in this representation. It may then be convenient to employ an alternative method wherein the modal equations are uncoupled when $\epsilon_2 = 0$, for an arbitrary z dependence of ϵ_1 and ϵ_3; mode coupling is now introduced by the off-diagonal term in the dielectric tensor in Eq. (3b). The "unperturbed" solution in this instance is evidently either the uniaxial one discussed in Sec. 7.2 ($\epsilon_1 \neq \epsilon_3$) or the isotropic one in Sec. 2. 3e ($\epsilon_1 = \epsilon_3$). In a gyrotropic plasma described by Eqs. (4), $\epsilon_2 = 0$ obtains for $\omega_c = 0$ or ∞, so a perturbation scheme about the isotropic and uniaxial cases is possible for small values of ω_c and $1/\omega_c$, respectively, for relatively arbitrary electron density variations $X(z) = \omega_p^2(z)/\omega^2$.[11]

Asymptotic evaluation of the fields

The reduction of the integral representations for the fields when the observation point is located many wavelengths from the source region may be accomplished by the techniques described in Secs. 5.5 and 7.5, and will be discussed only briefly. For example, the solution of the problem of radiation from a dipole source in a homogeneous, gyrotropic plasma half-space $z < 0$ is given in the form of integrals that, for observation points inside the plasma, include those in Eqs. (25) representing the primary field, as well as a similar set for the perturbation introduced by the interface; the integrands of the latter contain in addition the appropriate modal reflection coefficients derivable from the network considerations earlier in this section. The exponential dependence in the integrands is of the form

$$\exp\left\{-jk[\kappa'_{o,e}(\sigma)\,|z| + \kappa'_{o,e}(\sigma)\,|z'| + \sigma\rho]\right\}, \tag{43}$$

where the subscripts o and e may occur in any combination. An exponent of the same type has been investigated in connection with Eq. (7.5.12), where $q_{1,2}(\beta)$ takes the place of $\kappa'_{o,e}$; the subsequent derivation and interpretation of the saddle-point condition may be utilized also in the present case after it is recalled that the rays have a direction perpendicular to the wavenumber surface. Again, the wavenumber surfaces play an important role in providing a physical picture, as noted in Secs. 1.6 and 1.7d. If the plasma has a wavenumber surface as in diagram (f) of Fig. 8.3.2, for example, then any of the rays A through D in Fig. 8.3.6 correspond to a typical saddle point σ_i as shown. Upon striking the interface, the incident ordinary ray A in Fig 8.3.6(a) excites reflected rays C and D in the ordinary and extraordinary modes, respectively; the latter arise from saddle points in integrands with exponential dependence

$$\exp\left[-j\,|k|\,(\kappa'_o\,|z'| + \kappa'_o\,|z| + \sigma\rho)\right]$$

and

$$\exp\left[-j\,|k|\,(\kappa'_o\,|z'| + \kappa'_e\,|z| + \sigma\rho)\right],$$

respectively. Analogous considerations apply to the incident extraordinary ray B. When the observation point is located in the exterior isotropic half-space $z > 0$, then $k\kappa'_{o,e}(\sigma)$ multiplying $|z|$ in Eq. (43) is replaced by $\kappa = k\kappa' = \sqrt{k_0^2 - k^2\sigma^2}$ for both the E and H modes, and the refracted ray E is perpendicular to the corresponding circular diagram (see Sec. 7.5e, also Fig. 1.7.4). These simple ray pictures, which may also be constructed when the interface is inclined with respect to the z axis, provide a direct insight into the reflection and refraction mechanism pertaining to confined source distributions; the reflected and transmitted ray amplitudes emerge from the amplitude coefficients in the asymptotic approximation. The diagram may also be employed to predict the occurrence of lateral rays that are associated intimately with the phenomenon of total reflection (see Fig. 1.7.7). It is especially interesting to note that in addition to the lateral rays traveling on the vacuum side of the interface [ray path $B_2E_2D_2$ in Fig. 8.3.6(b)], such rays may also occur on the plasma side in view of the mode coupling at the boundary. When σ_i in Fig. 8.3.6(a) moves to the right until it coincides with the σ intercept of the o branch, the incident extraordinary ray B_1 excites an ordinary ray C_1 traveling parallel to the boundary which, in turn, sheds extraordinary rays D_1 back into the medium and refracted rays E_1 into the vacuum region [Fig. 8.3. 6(b)]. It should also be mentioned that the diversity of refractive index profiles in Fig. 8.3.2 admits of focusing phenomena more varied than those described in Sec. 7.5e. In addition to the focusing arising from the e branch in Fig. 8.3.6, a similar effect occurs due to those portions of the o branch having a curvature opposite to that of the circular contour representing the vacuum half-space.

These procedures, applying to an abrupt change in the medium properties, may also be utilized to chart the progress of a ray in a slowly varying medium, as noted in Sec. 1.6. In this instance, one approximates the medium by piecewise constant layers, draws the appropriate wavenumber or refractive index diagrams, and determines the trajectory of the continuously refracted rays as above (internal reflection is neglected in a lowest-order approximation).

Although these simple considerations provide an interpretation of the saddle-point contributions to the integrals, and also of the lateral-wave effects ascribed to branch-point singularities, they do not account for wave types arising from pole singularities that may be crossed during the deformation of the integration path into the steepest-descent contour. Because of the complicated dependence of $\kappa_{o,e}$ on σ, the determination of pole singularities in the modal reflection coefficients poses a formidable task and must generally be accomplished by numerical means (see Secs. 8.4b and 8.4c for a simple example).

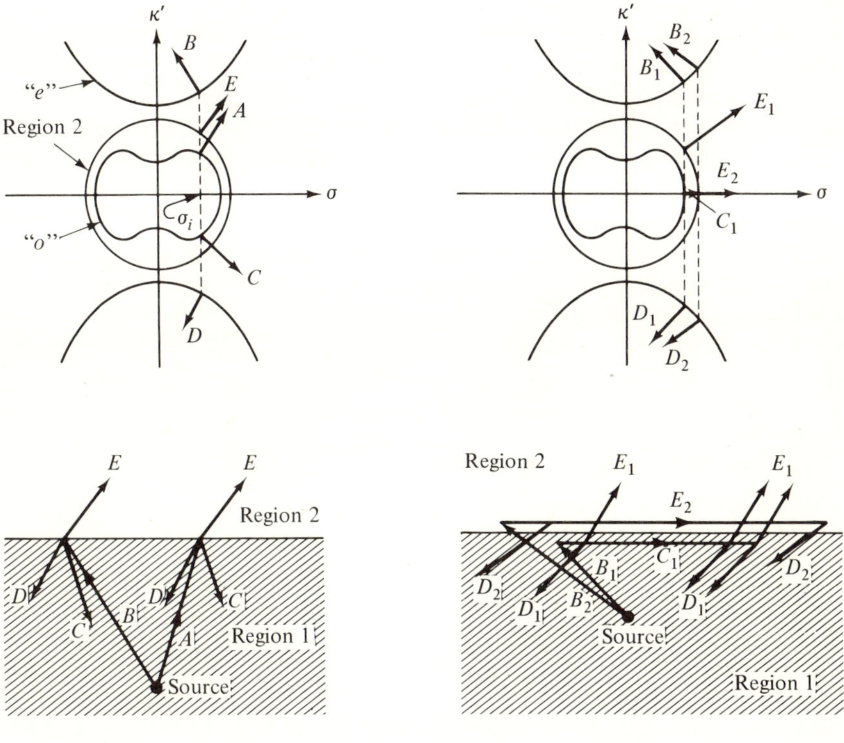

(a) Reflected rays   (b) Lateral rays

FIG. 8.3.6   Ray configurations.

8.4   GUIDED WAVES IN A COLD MAGNETOPLASMA (GUIDE AXIS PERPENDICULAR TO GYROTROPIC AXIS)

As discussed in Sec. 8.2h, the analysis of boundary-value problems in a gyrotropic medium is in general quite involved. However, for special geometries, medium parameters, and excitation, the analysis becomes quite tractable and may be effected by transmission-line methods. For example, it was found in Secs. 7.2b and 7.4b that a certain class of boundary value problems in a uniaxially anisotropic medium can be reduced to equivalent problems in an isotropic environment. Similar reductions occur in a gyrotropic medium if the fields are independent of the coordinate parallel to the gyrotropic axis (i.e., if wave propagation is perpendicular to this axis). To explore this latter class of problems, we shall first ascertain the possible guided waves in a gyrotropic medium with the gyrotropic axis $\mathbf{b}_0$ perpendicular to the guiding direction $\mathbf{z}_0$. In this guide structure the problems in question then correspond to the special case of waves with vector wavenumbers $\mathbf{k}$ in the plane containing $\mathbf{z}_0$ and perpendicular to $\mathbf{b}_0$; however, other waves also exist in this structure, with wavenumbers $\mathbf{k}$ in the plane formed by $\mathbf{b}_0$ and $\mathbf{z}_0$, for instance.

8.4a   Eigenfunctions and Eigenvalues for $\mathbf{b}_0$ Perpendicular to $\mathbf{z}_0$

In a waveguide with $\mathbf{b}_0$ perpendicular to $\mathbf{z}_0$, the guided wavevectors $\mathbf{\Psi}_\alpha$ and eigenvalues κ_α, defined in Secs. 1.4, and 8.2. and 8.3, are strongly dependent on the orientation φ of the gyrotropic axis $\mathbf{b}_0 = \mathbf{k}_{t0} \cos \varphi + \hat{\mathbf{k}}_{t0} \sin \varphi$ in the plane transverse to the guide axis $\mathbf{z}_0$. The notation is the same as that in the reduced electromagnetic description of a gyrotropic medium with $\mathbf{b}_0$ parallel to $\mathbf{z}_0$ and discussed in Sec. 8.3a. The dependence of $\mathbf{\Psi}_\alpha$ and κ_α on φ can be ascertained by solution of the eigenvalue problem (8.3.1a) via the method indicated in Eqs. (8.3.2)–(8.3.6). For $\mathbf{b}_0$ perpendicular to $\mathbf{z}_0$, the E_α component of $\mathbf{\Psi}_\alpha$ can be derived from Eq. (8.3.5a), evaluated in the $\mathbf{z}_0$, $\mathbf{k}_{t0}$, $\hat{\mathbf{k}}_{t0}$ basis defined by Eq. (8.3.8) (assuming $\omega \neq 0$):

$$
\begin{bmatrix}
\epsilon_1 - \dfrac{k_t^2}{k_0^2} & \dfrac{\kappa k_t}{k_0} + j\epsilon_2 \sin \varphi & -j\epsilon_2 \cos \varphi \\[2ex]
\dfrac{\kappa k_t}{k_0^2} - j\epsilon_2 \sin \varphi & \epsilon_1 \sin^2 \varphi + \epsilon_3 \cos^2 \varphi - \dfrac{\kappa^2}{k_0^2} & (\epsilon_1 - \epsilon_3) \sin \varphi \cos \varphi \\[2ex]
j\epsilon_2 \cos \varphi & (\epsilon_1 - \epsilon_3) \sin \varphi \cos \varphi & \epsilon_1 \cos^2 \varphi + \epsilon_3 \sin^2 \varphi - \dfrac{k_t^2 + \kappa^2}{k_0^2}
\end{bmatrix}
$$

$$
\cdot \begin{bmatrix} E_z \\ E_{t'} \\ E_{t''} \end{bmatrix} = 0 \qquad\qquad (1)
$$

with the notation and procedure the same as that employed in Eq. (8.3.9) et seq., for the case $\mathbf{b}_0$ parallel to $\mathbf{z}_0$.

Non-vanishing solutions of Eq. (1) exist for those values $\kappa = \kappa_\alpha$ that satisfy the determinental equation [note that the present normalization $\kappa' = \kappa/k_0$, $k'_t = k_t/k_0$ differs from that in Eq. (8.3.10)]:

$$\kappa'^4 - \left[\epsilon_\perp + \epsilon_3 - 2k'^2_t + \left(1 - \frac{\epsilon_3}{\epsilon_1}\right) k'^2_t \cos^2 \varphi\right]\kappa'^2 + (\epsilon_\perp - k'^2_t)(\epsilon_3 - k'^2_t)$$

$$- \cos^2 \varphi\left[4(\epsilon_1 - \epsilon_3)\sin^2 \varphi - (\epsilon_\perp - \epsilon_3)k'^2_t + \left(1 - \frac{\epsilon_3}{\epsilon_1}\right)k'^4_t\right] = 0, \qquad (2a)$$

which for $\varphi = \pi/2$ becomes

$$[\kappa'^2 - (\epsilon_\perp - k'^2_t)][\kappa'^2 - (\epsilon_3 - k'^2_t)] = 0, \qquad (2b)$$

and for $\varphi = 0$ becomes

$$\kappa'^4 - \left[\epsilon_\perp + \epsilon_3 - \left(1 + \frac{\epsilon_3}{\epsilon_1}\right)k'^2_t\right]\kappa'^2 + \frac{\epsilon_3}{\epsilon_1}\left[\left(\epsilon_1 - k'^2_t\right)^2 - \epsilon_2^2\right] = 0 \qquad (2c)$$

with $\epsilon_\perp = (\epsilon_1^2 - \epsilon_2^2)/\epsilon_1$, and $\epsilon_{1,2,3}$ defined in Eq. (8.3.4).

The wavevectors $\boldsymbol{\Psi}_\alpha$ corresponding to the roots $\kappa = \kappa_\alpha$ of Eq. (2) are somewhat complicated for arbitrary orientation φ of the gyrotropic axis in the transverse plane. For simplicity we consider first the case $\varphi = \pi/2$, wherein the gyrotropic axis is perpendicular to both the guide axis $\mathbf{z}_0$ and the wave direction $\mathbf{k}_0$. For prescribed $\mathbf{k}_t$, ω, one obtains from Eq. (1) and Eq. (8.3.6) wavevectors $\boldsymbol{\Psi}_\alpha$ of the same form as in Eq. (8.3.13a). For one type of mode one finds

$$\mathbf{e}_\alpha = \hat{\mathbf{k}}_{t0}, \qquad \mathbf{e}_{z\alpha} = 0,$$

$$\mathbf{h}_\alpha = -\mathbf{k}_{t0}, \qquad \mathbf{h}_{z\alpha} = \frac{k_t}{\kappa_\alpha}\mathbf{z}_0, \qquad (3a)$$

with the characteristic mode impedance Z_α and wavenumber κ_α given by

$$Z_\alpha = \frac{\omega\mu_0}{\kappa_\alpha}, \qquad \kappa_\alpha = \pm\sqrt{k_0^2\epsilon_3 - k_t^2}. \qquad (3b)$$

This is an ordinary H-type mode [see Eq. (2.2.15d)], with electric field parallel to $\mathbf{b}_0$, appropriate to propagation in a medium with relative dielectric constant ϵ_3, the latter being given in Eq. (8.3.4). Since $\kappa_{\pm\alpha} = \pm\kappa_\alpha$, $\mathbf{e}_\alpha = \mathbf{e}_{-\alpha} = \mathbf{e}_{\alpha^*}$, $\mathbf{h}_\alpha = \mathbf{h}_{-\alpha} = \mathbf{h}_{\alpha^*}$, and $Z^*_{\alpha^*} = Z_\alpha$ for these modes, the transmission-line description is conventional.

There also exists another type of mode for $\varphi = \pi/2$, again obtained from Eqs. (1) and (8.3.6), with

$$\mathbf{e}_\alpha = \mathbf{k}_{t0}, \qquad \mathbf{e}_{z\alpha} = \frac{\kappa_\alpha k_t + jk_0^2\epsilon_2}{k_t^2 - k_0^2\epsilon_1}\mathbf{z}_0,$$

$$\mathbf{h}_\alpha = \mathbf{k}_{t0}, \qquad \mathbf{h}_{z\alpha} = 0, \qquad (4a)$$

and with characteristic impedance and wavenumber given by

$$Z_\alpha = \frac{\kappa_\alpha - jk_t\epsilon_2/\epsilon_1}{\omega\epsilon_0\epsilon_\perp}, \qquad \kappa_\alpha = \pm\sqrt{k_0^2\epsilon_\perp - k_t^2}. \qquad (4b)$$

In this second type of polarization, the "extraordinary" mode, the gyrotropic axis $\mathbf{b}_0$ is parallel to the a.c. magnetic field. Although $\kappa_{\pm\alpha} = \pm\kappa_\alpha$, $\mathbf{e}_\alpha = \mathbf{e}_{-\alpha} = \mathbf{e}_{\alpha^*}$, and $\mathbf{h}_\alpha = \mathbf{h}_{-\alpha} = \mathbf{h}_{\alpha^*}$ for this mode type, the impedance $Z_{\alpha^*}^* \neq Z_\alpha$ and hence the general transmission-line formalism of Eqs. (8.2.43) and (8.2.46) must be employed; in connection with the latter one notes that

$$Z_{\alpha^*}^* = Z_\alpha^*, \qquad Z_{-\alpha^*}^* = -Z_\alpha \qquad \text{for } \kappa_\alpha \text{ real},$$
$$Z_{\alpha^*}^* = -Z_{-\alpha}, \qquad Z_{-\alpha^*}^* = -Z_\alpha \qquad \text{for } \kappa_\alpha \text{ imaginary}.$$

$$(4c)$$

For given $\mathbf{k}_t$, ω, the four $\boldsymbol{\Psi}_\alpha$ wavevectors defined in Eqs. (3a) and (4a) possess the orthonormality properties (8.2.40) and also $\mathbf{e}_\alpha = \mathbf{h}_\alpha \times \mathbf{z}_0$.

For $\mathbf{b}_0 = \mathbf{k}_{t0}$ (i.e., $\varphi = 0$) the mode vectors $\mathbf{e}_\alpha$ and $\mathbf{h}_\alpha$ can again be determined via the general procedure outlined in Sec. 8.3a. However, in this case, $\mathbf{e}_\alpha \neq \mathbf{e}_{-\alpha}$ and $\mathbf{h}_\alpha \neq \mathbf{h}_{-\alpha}$, whence no simple transmission-line (impedance) formalism is apparent under this circumstance. We shall relegate explicit evaluation of these mode vectors to the Problems section.

8.4b   Two-dimensional Boundary-value Problems in Gyrotropic Media

Simplifications in the solution of boundary-value problems in a gyrotropic medium occur for certain choices of excitation and geometry. Such is the case if the fields are independent of the coordinate parallel to the gyrotropic direction $\mathbf{b}_0$ and the medium contains planar or cylindrical scatterers with planes or cylindrical axes parallel to $\mathbf{b}_0$. It is then convenient to regard the system as a waveguide with axis $\mathbf{z}_0$ perpendicular to $\mathbf{b}_0$, whence, as noted in Sec. 8.4a, only the $\varphi = \pi/2$ modes are necessary for the field description. These waves have two distinctive polarizations and can be separately excited, depending on the source orientation.

The ordinary waves, whose wavevectors are indicated in Eq. (3a), have their electric field polarized along the gyrotropic direction $\mathbf{b}_0 \equiv \hat{\mathbf{k}}_{t0}$. The propagation characteristics of these waves are dependent only on ϵ_3 and are independent of the anisotropy transverse to $\mathbf{b}_0$. In the presence of sources, boundaries, or obstacles that do not generate other electric field components, radiation and scattering problems reduce to conventional ones in an isotropic medium with wavenumber $k_0\sqrt{\epsilon_3}$. From Eqs. (3a) and (8.2.28b), it is seen that electric line currents parallel to $\mathbf{b}_0$ excite only these ordinary waves, as do scatterer configurations in the form of arbitrarily shaped perfectly conducting cylinders whose axes are aligned along $\mathbf{b}_0$.

The extraordinary waves described by Eq. (4a) have their a.c. magnetic field polarized along the gyrotropic axis $\mathbf{b}_0 = \hat{\mathbf{k}}_{t0}$. In consequence, wave propagation is now affected by the anisotropy transverse to $\mathbf{b}_0$, but in a sufficiently simple manner to make certain propagation characteristics analogous to isotropic ones, while others exhibit strikingly the influence of anisotropy. Because of the relative simplicity of this class of radiation and scattering problems, having only a single magnetic field component, let us consider scattering of waves from a magnetic line current oriented parallel to $\mathbf{b}_0 = \hat{\mathbf{k}}_{t0}$.

From Eqs. (4a) and (8.2.28b) one verifies that only extraordinary waves are excited in this case; there is no electric field along $\mathbf{b}_0$ and hence only the transverse-to-$\mathbf{b}_0$ part of the dielectric dyadic is effective.

It is instructive to contrast the reduced Maxwell field and transmission-line descriptions for the case of excitation from a magnetic line current.[12] Let us choose a rectangular x, y, z coordinate system oriented such that $\mathbf{x}_0 = \mathbf{k}_{t0}$, $\mathbf{y}_0 = \mathbf{b}_0 = \hat{\mathbf{k}}_{t0}$, with $\mathbf{z}_0$ along the guide axis. Also, let $\mathbf{M} = M\mathbf{y}_0$ be the magnetic line current density, whence $\partial/\partial y \equiv 0$, and denote the non-vanishing field components by $H_{t''} = H_y = H$, by $E_{t'} = E_x$ and by E_z. On elimination of E_z, the reduced Maxwell field equations (8.2.1), with $\boldsymbol{\epsilon}$ given as in Eq. (8.3.3b), can then be written in the form

$$\frac{\partial E_x}{\partial z} = -j\omega\mu_0\left(1 + \frac{1}{k_0^2\epsilon_1}\frac{\partial^2}{\partial x^2}\right)H - j\frac{\epsilon_2}{\epsilon_1}\frac{\partial}{\partial x}E_x - M,$$

$$\frac{\partial H}{\partial z} = j\frac{\epsilon_2}{\epsilon_1}\frac{\partial}{\partial x}H - j\omega\epsilon_0\epsilon_\perp E_x, \tag{5a}$$

and on elimination of E_x as

$$\left(\frac{\partial^2}{\partial x^2} + \frac{\partial^2}{\partial z^2} + k_0^2\epsilon_\perp\right)H = j\omega\epsilon_0\epsilon_\perp M, \tag{5b}$$

where $\epsilon_\perp = (\epsilon_1^2 - \epsilon_2^2)/\epsilon_1$ is the effective relative dielectric constant for the wave propagation. While Eq. (5b) is the same as for an isotropic medium with relative permittivity $\epsilon_\perp$, the effect of anisotropy is evident from the presence of the ϵ_2 dependent terms in Eqs. (5a). It is not difficult to verify from Eq. (8.3.3b) that for a cold electron plasma, $\epsilon_\perp$ has the following behavior as a function of frequency:

$$\begin{aligned} \epsilon_\perp < 0 \qquad &\text{when } \omega < \omega_4 \text{ and } \omega_2 < \omega < \omega_1, \\ \epsilon_\perp > 0 \qquad &\text{where } \omega_4 < \omega < \omega_2 \text{ and } \omega > \omega_1, \end{aligned} \tag{6a}$$

where (see Fig. 8.3.2)

$$\omega_{4,1} = \frac{\sqrt{\omega_c^2 + 4\omega_p^2} \mp \omega_c}{2}, \qquad \omega_2 = \sqrt{\omega_c^2 + \omega_p^2}. \tag{6b}$$

For the above magnetic line current excitation, there also exists a transmission-line description of the z dependence of the extraordinary wave fields. This description is particularly appropriate for wave fields with an x dependence $\exp(-jk_t x)$ and for plane-stratified scatterers with axis of stratification along z. As indicated in the general transmission-line formalism of Sec. 8.2h, one describes the αth-mode behavior in terms of a (voltage) reflection coefficient $\Gamma_\alpha(z)$ which is related to the impedance $Z_\alpha(z)$ at some scattering plane at z by Eq. (8.2.43). For given k_t, ω, the extraordinary modes have characteristic impedances given by Eqs. (4b) and (4c), whence by Eq. (8.2.43) the reflection coefficient, looking in the direction of increasing z, becomes for propagating modes with real κ_α,

$$\Gamma_\alpha(z) = \frac{Z_\alpha(z) - Z_\alpha}{Z_\alpha(z) + Z_\alpha^*}. \tag{7}$$

For example, if there exists a perfectly conducting plane at $z = 0$ so that $Z_\alpha(0) = 0$, then by Eqs. (4b) and (7), the reflection coefficients $\overrightarrow{\Gamma}_\alpha, \overleftarrow{\Gamma}_\alpha$ at $z = 0$, looking in the positive and negative direction, respectively, are

$$\overrightarrow{\Gamma}_\alpha(0) = -\frac{Z_\alpha}{Z_\alpha^*} = -\frac{1 + j(k_t\epsilon_2/\kappa_\alpha\epsilon_1)}{1 - j(k_t\epsilon_2/\kappa_\alpha\epsilon_1)}, \quad \overleftarrow{\Gamma}_\alpha(0) = -\frac{Z_{-\alpha}}{Z_{-\alpha}^*} = -\frac{1 - j(k_t\epsilon_2/\kappa_\alpha\epsilon_1)}{1 + j(k_t\epsilon_2/\kappa_\alpha\epsilon_1)},$$
$$\tag{8}$$

whence by an equation of the form (8.2.45), the reflection coefficient at any plane z can be readily ascertained. The non-symmetric results in Eqs. (8) for the reflection coefficient of a perfectly conducting plane in a gyrotropic medium are to be contrasted for an isotropic medium with the single symmetric reflection coefficient of value -1. Furthermore, the requirement $E_x = 0$ on a perfectly conducting plane at $z = 0$ is seen from Eq. (5a) to imply a relation between $\partial H/\partial z$ and $\partial H/\partial x$ that contrasts with the $\partial H/\partial z = 0$ requirement in an isotropic medium. In this case the effect of medium anisotropy produces non–reciprocal behavior since $\Gamma_\alpha(0)$ is not an even function of k_t (i.e., waves with transverse periodicity described by $-k_t$ are reflected differently from those with $+k_t$).

The above transmission analysis in terms of z-guided waves is no longer useful when the scatterer is a perfectly conducting cylinder of arbitrary cross section, with axis oriented parallel to $\mathbf{b}_0$ and to the magnetic line current direction. In this case, an incident wave with prescribed k_t is then coupled to an infinity of scattered waves with $-\infty < k_t < \infty$. Even if the scattering surface S lies on a coordinate surface $u = $ constant in an orthogonal (u, v) coordinate system that renders the wave equation (5b) separable, and if the incident field is represented in terms of v-dependent eigenfunctions, the boundary condition $E_v = 0$ on S will generally introduce coupling between the modes. An exception occurs for a circular cylinder if the azimuthal eigenfunctions are chosen as $\exp(-jn\phi)$, $n = 0, \pm 1, \pm 2, \ldots$.

8.4c   Radiation from a Magnetic Line Source in the Presence of a Perfectly Conducting Plane

To exhibit the effects of anisotropy alluded to in Sec. 8.4b, we shall consider in more detail the problem of radiation from a y-directed magnetic line source in the presence of a perfectly conducting plane boundary at $z = 0$ (Fig. 8.4.1). The entire configuration is embedded in a cold plasma, with the external magnetic field axis $\mathbf{b}_0$ parallel to y. The transverse electric-field component E_x is obtainable from the single magnetic-field component $H_y \equiv H$ from the second of Eqs. (5a), while the longitudinal component E_z follows in a similar manner from $E_z = (j\omega\epsilon_0\epsilon_\perp)^{-1}[j(\epsilon_2/\epsilon_1)(\partial H/\partial z) + (\partial H/\partial x)]$. H satisfies Eq. (5b) with $M = \delta(x)\delta(z - z')$, subject on $z = 0$ to the boundary condition

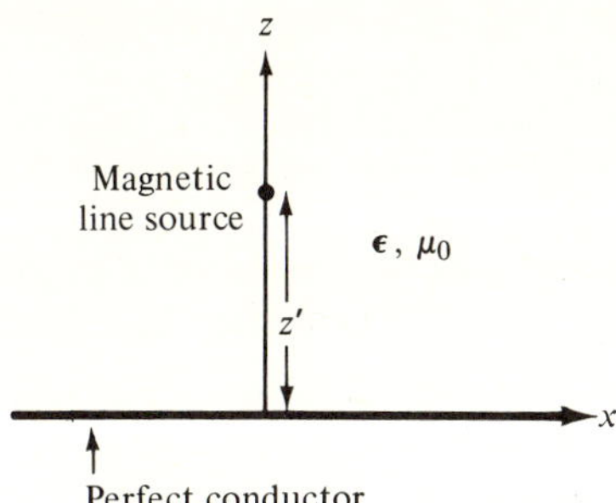

FIG. 8.4.1   Line source and perfectly conducting plane.

$E_x = 0$ [seeEq. (5a)], whence

$$\frac{\partial H}{\partial z} - j\frac{\epsilon_2}{\epsilon_1}\frac{\partial H}{\partial x} = 0 \qquad \text{at } z = 0. \tag{9}$$

In an unbounded medium, the magnetic Green's function solution H_0 is given by [see Eqs. (5.4.26a) and (5.4.40)]

$$H_0(\mathbf{r}, \mathbf{r}') = -\frac{\omega\epsilon_\perp}{4} H_0^{(2)}(k\sqrt{x^2 + (z - z')^2}), \qquad \mathbf{r} = (x, z), \tag{10}$$

with $k = k_0\sqrt{\epsilon_\perp}$. When $\epsilon_\perp < 0$, no radiation occurs. Since propagation is transverse to the direction of the gyrotropic axis (corresponding to $\kappa' = 0$ in Fig. 8.3.2), the wavevector and ray directions coincide in this class of problems. One may also verify that the ray direction is radially outward from the source point $\mathbf{r}'$. It is not difficult to ascertain that the boundary conditions for a perfectly conducting plane are not satisfied by the insertion of an image source at $(-z', 0)$, so it is necessary to employ a modal representation. In view of the translational invariance along x, the transverse dependence of a modal field is $\exp(-j\xi x)$, $-\infty < \xi < \infty$, whence $H(\mathbf{r}, \mathbf{r}')$ can be represented as

$$H(\mathbf{r}, \mathbf{r}') = \int_{-\infty}^{\infty} I(z, z'; \xi) \frac{e^{-j\xi x}}{2\pi} d\xi, \tag{11}$$

where $\xi \equiv k_t$ and the 2π factor has been included for normalization. Insertion into Eq. (5b) and use of the Fourier integral representation for the delta function $\delta(x) = (1/2\pi) \int_{-\infty}^{\infty} \exp(-j\xi x) d\xi$ yields

$$\frac{d^2 I}{dz^2} + (k^2 - \xi^2)I = j\omega\epsilon_\perp\delta(z - z'), \qquad k = k_0\sqrt{\epsilon_\perp}. \tag{12a}$$

subject to the boundary condition

$$\epsilon_2\xi I - \epsilon_1\frac{dI}{dz} = 0 \qquad \text{at } z = 0, \tag{12b}$$

and a radiation condition at $z \to \infty$. With $\kappa = \sqrt{k^2 - \xi^2}$ defined as the modal propagation constant, these equations are in the standard transmission line form. By the traveling-wave techniques described in Chapter 2 [see Eq. (2.4.23b)], the solution for I may be written as

$$I = -j\omega\epsilon_\perp \frac{e^{-j\kappa|z-z'|} - \Gamma e^{-j\kappa(z+z')}}{2j\kappa}, \qquad \kappa = \sqrt{k^2 - \xi^2}, \quad \mathrm{Im}\,\kappa \le 0, \qquad (13)$$

where use has been made of the one-dimensional Green's function for the unbounded z-domain in Eq. (5.4.7). The modal reflection coefficient Γ is given by Eq. (8),

$$\Gamma = -\frac{1 - (\xi\epsilon_2/j\kappa\epsilon_1)}{1 + (\xi\epsilon_2/j\kappa\epsilon_1)}. \qquad (13a)$$

The present problem is similar to that in Sec. 5.7, which deals with the radiation from sources in an isotropic medium in the presence of a plane surface having a constant surface impedance, the difference between the two problems arising from the functional form of Γ in Eq. (13a). Thus, the asymptotic analysis required in the derivation of explicit far-field formulas is essentially the same as in Sec. 5.7b. As noted previously, an important feature is introduced by the behavior $\Gamma(\xi) \ne \Gamma(-\xi)$, implying that waves with identical spatial periodicity, but incident on opposite sides of the surface normal, are affected unequally by the boundary. This non-reciprocal characteristic is ascribable to the gyrotropic properties of the medium.

Unidirectional surface wave

Since $\Gamma(\xi)$ in Eq. (13a) has the form corresponding to an inductive surface impedance in an isotropic medium, one expects from Sec. 5.7 to find a surface wave that decays exponentially with z and propagates undamped in the x direction. Indeed, the reflection coefficient Γ has poles ξ_p when

$$\sqrt{k^2 - \xi_p^2}\,\epsilon_1 = j\xi_p\epsilon_2, \qquad (14)$$

which equation may easily be solved by squaring to yield $\xi_p = \pm k_0\sqrt{\epsilon_1}$. To obtain a surface-wave solution with real ξ_p, ϵ_1 must be positive and in that case, $\xi_p^2 > k^2$, so $\kappa_p = -j|\kappa_p|$ is imaginary. The algebraic sign of ξ_p is therefore identical with that of $-\epsilon_2/\epsilon_1$ and, since $\epsilon_1 > 0$, $\mathrm{sgn}\,\xi_p = -\mathrm{sgn}\,\epsilon_2$; the sign of ϵ_2 may be changed by reversing the direction of the d.c. magnetic field impressed on the plasma [see Eq. (8.3.4)]. Thus,

$$\xi_p = \pm k_0\sqrt{\epsilon_1} \qquad \text{when } \epsilon_2 \lessgtr 0, \quad \epsilon_1 > 0. \qquad (15)$$

If a small amount of dissipation is present, then $\mathrm{Im}\,\sqrt{\epsilon_1} < 0$, thereby fixing the pole position with respect to the integration path. In contrast to the situation in isotropic configurations, only a single pole exists instead of the customary symmetrical pair at $\pm\xi_p$ as in Sec. 5.7. This characteristic leads to unidirectional wave propagation, as noted below.[12]

An examination of the elements of the dielectric tensor shows that $\epsilon_1 = (\omega^2 - \omega_p^2 - \omega_c^2)(\omega^2 - \omega_c^2)^{-1}$ is positive when $\omega > \sqrt{\omega_p^2 + \omega_c^2}$ or $\omega < \omega_c$, where ω_c and ω_p denote the electron cyclotron and plasma frequencies, respectively, and that, in particular,

$$\begin{aligned}
\text{(a)} \quad & 0 < \epsilon_1 < 1 \quad \text{when } \omega > \sqrt{\omega_p^2 + \omega_c^2}; \\
\text{(b)} \quad & \epsilon_1 > 1 \quad\quad\; \text{when } \omega < \omega_c.
\end{aligned} \qquad (16)$$

The surface-wave phase velocity $c/\sqrt{\epsilon_1}$ along the y direction parallel to the plane is therefore greater than the velocity c of light in vacuum (fast wave) in case (a), but less than c (slow wave) in case (b). These observations are of interest in connection with the problem of radiation from a line source embedded in a magneto-plasma sheath of finite thickness, in which instance the perturbed slow wave of case (b) remains a surface wave while the perturbed fast wave corresponding to case (a) becomes a leaky wave that may strongly influence the radiation pattern; a fast wave distribution radiates energy away from the boundary (see Sec. 5.5g). One observes also from Eqs. (5), (14), and (15) that the field components in the surface wave are

$$H \equiv H_y = C \exp\left(\pm jk_0\sqrt{\epsilon_1}\,x - k_0|\epsilon_2|\epsilon_1^{-1/2}z\right), \qquad (17a)$$

$$E_z = \pm \frac{k_0}{\omega\epsilon_0\sqrt{\epsilon_1}}H, \qquad E_x = 0, \qquad (17b)$$

where C is a constant-amplitude factor and the upper and lower signs correspond to $\epsilon_2 > 0$ and $\epsilon_2 < 0$, respectively, with $\epsilon_1 > 0$. Since $E_x = 0$, the surface wave is TEM (transverse electromagnetic) with respect to the x direction. This property could have been deduced directly from the requirement that the surface wave is a solution of the source-free Maxwell field equations, satisfies the boundary conditions at $z = 0$, and has field components that behave according to $\exp\left[-j\xi_p x - |\kappa_p|z\right]$. Because E_x must vanish at $z = 0$, it must therefore vanish everywhere. It is noted that the surface wave propagates (and carries energy) only in the $-x$ direction when ϵ_2 is positive, and only in the $+x$ direction when ϵ_2 is negative, thereby exhibiting the unidirectional characteristic mentioned earlier.

A comparison of Eqs. (6) and (16) shows that the surface waves may propagate even when the medium itself cannot sustain propagation; this condition applies when $\epsilon_\perp < 0$, $\epsilon_1 > 0$ [i.e., $\omega_2 < \omega < \omega_3$ or $\omega < \min(\omega_c, \omega_1)$]. One may then conceive of a physical situation that leads to a "thermodynamic paradox" since energy is apparently absorbed by a lossless termination. Consider the configuration in Fig. 8.4.1, without the line source and with a non-absorbing boundary inserted along the $x = 0$ plane; the medium parameters are chosen so that $\epsilon_\perp < 0$ and a surface wave may propagate along the $+x$ direction. Since no propagation is possible in the medium itself nor along the plane toward $x = -\infty$, there is no mechanism to account for the energy transported toward the boundary by an incident surface wave (the lossless $x = 0$ plane may be chosen so as to be incapable of supporting a surface wave). This difficulty exists only in the absence of dissipation and disappears when slight losses are assumed. The boundary value problem is, in fact, poorly posed in the lossless case but may be treated correctly when the non-dissipative problem is derived from the lossy one in the limit of vanishing conductivity.[13,14]

The far field

For large values of $kR = k_0\sqrt{\epsilon_\perp}R$, where $R = \sqrt{x^2 + (z + z')^2}$ is the distance from the image point $(-z', 0)$ to the observation point (x, z), the scattered portion $H_s(\mathbf{r}, \mathbf{r'})$ of the integral in Eq. (11) may be evaluated by the asymptotic techniques employed elsewhere for analogous problems (see Secs. 5.3d and 7.5d). The unperturbed field H_0 in the unbounded medium is given in closed form in Eq. (10). To evaluate H_s we introduce the polar coordinates (R, θ) via $z + z' = R\cos\theta$, $x = R\sin\theta$, and also transform into the w plane according to $\xi = k\sin w$. When k is real and positive, the integration path maps into the contour $\hat{P}$ shown in Fig. 5.3.5(b), whereas for imaginary $k = -j|k|$, the contour follows the imaginary axis in the w plane (see Fig. 8.4.2). The steepest-descent

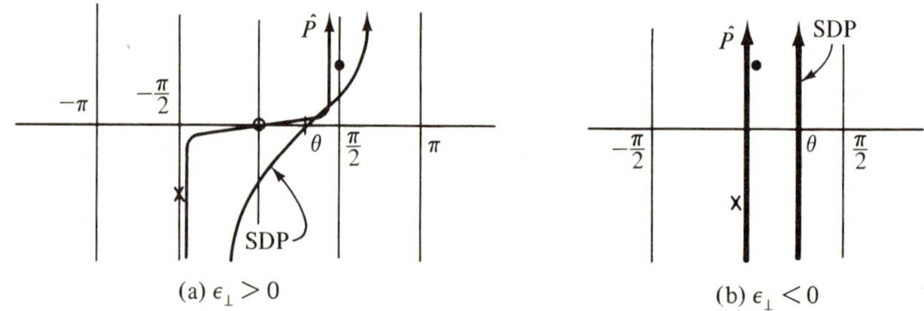

(a) $\epsilon_\perp > 0$ (b) $\epsilon_\perp < 0$

FIG. 8.4.2 Integration paths and pertinent singularities ($\epsilon_1 > 0$)
  •–Pole location when $\epsilon_2 < 0$
  ×–Pole location when $\epsilon_2 > 0$

paths through the saddle point at $w_s = \theta$ are also shown, and one notes that when $\epsilon_2 < 0$, a surface-wave pole may be captured for sufficiently large, positive observation angles θ, but not for $\theta < 0$; the converse is true when $\epsilon_2 > 0$. Since the surface waves have been shown to possess unidirectional characteristics, with the wave going in the $+x$ direction ($\theta > 0$) when $\epsilon_2 < 0$ and in the $-x$ direction ($\theta < 0$) when $\epsilon_2 > 0$, these domains of contribution to the line-source field satisfy the radiation condition requiring the transport of energy away from the source. The asymptotic evaluation of the integral along the steepest-descent path furnishes the reflected field, which appears in this case to originate from the mirror-image point (it is worth emphasizing that the image construction is valid only in the asymptotic sense),

$$H_{SDP} \sim \frac{\omega\epsilon_\perp}{2\sqrt{2\pi}}\,\Gamma(k\sin\theta)\frac{e^{-j(kR-\pi/4)}}{\sqrt{kR}} + O\left(\frac{1}{(kR)^{3/2}}\right), \tag{18}$$

where

$$\Gamma(k\sin\theta) = \frac{\epsilon_2\sin\theta - j\epsilon_1\cos\theta}{\epsilon_2\sin\theta + j\epsilon_1\cos\theta}. \tag{18a}$$

The total far-field approximation is then given by

$$H \sim H_0 + H_{SDP} + U(\theta - \theta_p)\ \text{(surface wave)}, \quad |kR| \gg 1,\ \theta \not\approx \theta_p, \tag{19}$$

where θ_p is the observation angle at which the steepest-descent path crosses the pole and $U(\theta - \theta_p)$ is the Heaviside unit function. This simple formula must be modified in accord with the procedure in Sec. 4.4 if one wishes to include as well observation angles $\theta \approx \theta_p$. When k is imaginary, both the primary and reflected fields are evanescent, but real energy may be extracted from the source by the surface waves.

While attention has already been called to the non-reciprocal behavior exhibited by the surface-wave contribution, the same phenomenon may be observed in connection with the reflected field H_{SDP} since Γ is not an even function of θ. $H(\mathbf{r}, \mathbf{r}')$ is therefore not equal to $H(\mathbf{r}', \mathbf{r})$. An examination of Γ reveals, however, that the simultaneous changes $\theta \to -\theta$, $\epsilon_2 \to -\epsilon_2$, leave the reflection coefficient unchanged; the pole configuration in Fig. 8.4.2 is likewise symmetrized when a reversal in the sign of θ is made jointly with a corresponding modification of ϵ_2. From these considerations, the fields at P in the two configurations depicted in Fig. 8.4.3 are seen to be equivalent, and since a sign change in ϵ_2 produces the transposed tensor $\boldsymbol{\epsilon}$ in Eq. (1.5.20) [see Eq. (8.3.3b)], reciprocity is found to be satisfied if an interchange of source and observation points is accompanied by insertion of the transposed medium. The general conclusions derived in Sec. 1.5b are therefore verified. In the plasma medium under consideration, the transposed medium corresponds to a reversal in the direction of the externally impressed magnetic field.

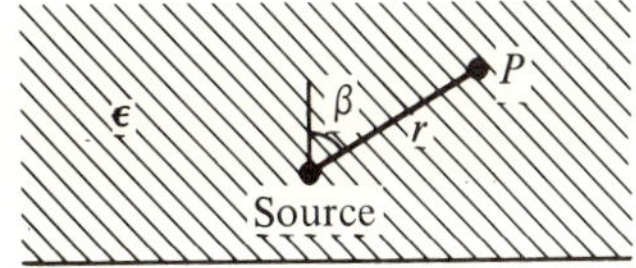

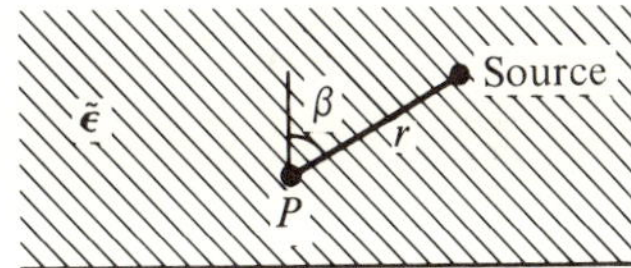

FIG. 8.4.3   Equivalent problems.

The preceding result is relevant for antenna pattern measurements in an anisotropic region. A customary technique in an isotropic environment is to determine the radiation pattern by measuring the energy received by the antenna from plane waves incident at various angles. Use of this procedure in a gyrotropic region requires as well the transposition of the medium parameters.[15]

8.4d   Diffraction by a Half-plane

If the perfectly conducting boundary in Fig. 8.4.1(a) extends only from $x = 0$ to $x = \infty$, with the remainder of space filled with the same anisotropic medium, the resulting boundary-value problem may not be solved by separation of variables but is amenable to treatment by the Wiener–Hopf technique.[16] The details are omitted here and only some physical characteristics of the solution are mentioned. If conditions are adjusted so that a surface wave may propagate along the top face toward $x = -\infty$, then it follows that the bottom face of the screen supports a wave traveling toward $x = +\infty$ (rotate the configuration in Fig. 8.4.1 by 180° about the y axis). These are the only possible wave types and no propagation occurs along the top and bottom faces toward $x = \infty$ and $x = -\infty$, respectively. An incident surface wave on the top face therefore excites a reflected surface wave on the bottom face and a radiation field when $\epsilon_\perp > 0$; for negative $\epsilon_\perp$, no radiation occurs from the edge and the entire incident energy is returned along the bottom side. The process is schematized in Fig. 8.4.4.

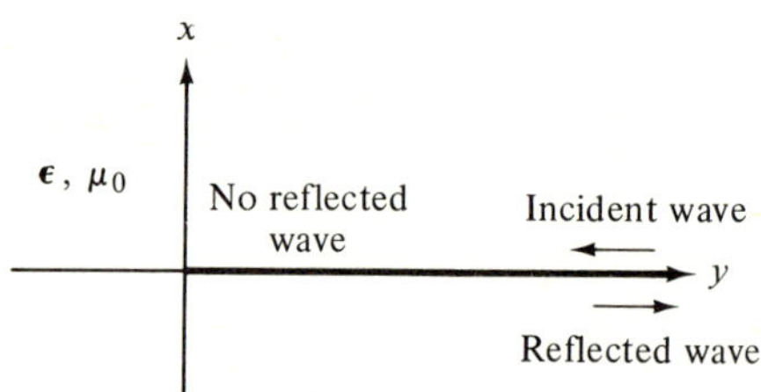

FIG. 8.4.4 Half-plane.

PROBLEMS

1. Show that the transverse field equations for a medium with arbitrary dyadic permittivity $\boldsymbol{\epsilon}(\mathbf{r})$ and permeability $\boldsymbol{\mu}(\mathbf{r})$ are given by (time dependence exp $(j\omega t)$):

$$-\frac{\partial \mathbf{E}_t}{\partial z} = j\omega \left[-\mathbf{z}_0 \times \boldsymbol{\mu}_t \times \mathbf{z}_0 + \frac{1}{\omega^2} \nabla_t \frac{1}{\epsilon_z} \nabla_t + \frac{\mathbf{z}_0 \times \bar{\boldsymbol{\mu}}_{tz}\bar{\boldsymbol{\mu}}_{zt} \times \mathbf{z}_0}{\mu_z} \right] \cdot \mathbf{H}_t \times \mathbf{z}_0$$

$$- \left[\nabla_t \frac{1}{\epsilon_z} \bar{\boldsymbol{\epsilon}}_{zt} \times \mathbf{z}_0 + \frac{\mathbf{z}_0 \times \bar{\boldsymbol{\mu}}_{tz}\nabla_t}{\mu_z} \right] \cdot \mathbf{z}_0 \times \mathbf{E}_t + \mathbf{M}_{te} \times \mathbf{z}_0, \qquad (1a)$$

$$-\frac{\partial \mathbf{H}_t}{\partial z} = j\omega \left[-\mathbf{z}_0 \times \boldsymbol{\epsilon}_t \times \mathbf{z}_0 + \frac{1}{\omega^2} \nabla_t \frac{1}{\mu_z} \nabla_t + \frac{\mathbf{z}_0 \times \bar{\boldsymbol{\epsilon}}_{tz}\bar{\boldsymbol{\epsilon}}_{zt} \times \mathbf{z}_0}{\epsilon_z} \right] \cdot \mathbf{z}_0 \times \mathbf{E}_t$$

$$- \left[\nabla_t \frac{1}{\mu_z} \bar{\boldsymbol{\mu}}_{zt} \times \mathbf{z}_0 + \frac{\mathbf{z}_0 \times \bar{\boldsymbol{\epsilon}}_{tz}\nabla_t}{\epsilon_z} \right] \cdot \mathbf{H}_t \times \mathbf{z}_0 + \mathbf{z}_0 \times \mathbf{J}_{te}, \qquad (1b)$$

where z is the longitudinal direction, with respect to which $\boldsymbol{\epsilon}$ and $\boldsymbol{\mu}$ have the representations in Eq. (8.2.2c), and where the equivalent electric and magnetic source currents $\mathbf{J}_{te}$ and $\mathbf{M}_{te}$ are given in Eq. (8.2.29b) (with $i \to -j$). Show also that the longitudinal field components are obtained from the transverse components as follows:

$$H_z = \frac{1}{j\omega\mu_z} [\nabla_t \cdot (\mathbf{z}_0 \times \mathbf{E}_t) - j\omega\bar{\boldsymbol{\mu}}_{zt} \cdot \mathbf{H}_t - M_z], \qquad (2a)$$

$$E_z = \frac{1}{j\omega\epsilon_z} [\nabla_t \cdot (\mathbf{H}_t \times \mathbf{z}_0) - j\omega\bar{\boldsymbol{\epsilon}}_{zt} \cdot \mathbf{E}_t - J_z]. \qquad (2b)$$

2. While ordinary and extraordinary modes propagate independently in a homogeneous gyrotropic medium, mode coupling occurs in the presence of spatial inhomogeneities. Coupling is weak, and computation thereby facilitated, when

the medium is slowly varying but the anisotropy is arbitrary. Alternatively, in a uniaxially anisotropic medium, inhomogeneities along the gyrotropic axis do not introduce mode coupling (see Sec. 7.2). Weakly coupled modes may therefore be found for a strongly inhomogeneous medium whose anisotropy deviates only slightly from the uniaxial (or from the isotropic, as a special case of the uniaxial medium).[11]

(a) To derive the coupled mode equation, note from Sec. 7.2a that the transverse eigenfunctions in a transversely homogeneous, unbounded, isotropic or uniaxially anisotropic dielectric medium are identical and may be decomposed into E (extraordinary) and H (ordinary) modes with respect to the optic (z) axis. Use this mode set for representation of fields in the gyrotropic plasma described by Eq. (8.3.3b), with $\epsilon_2 \neq 0$. Substitute Eqs. (7.2.6) into the transverse field equations (1), with $\bar{\epsilon}_{zt} = \bar{\epsilon}_{tz} = 0$, $\boldsymbol{\mu} = \mu_0\mathbf{1}$, and follow the procedure in Sec. 7.2a to derive transmission-line equations for the modal voltages and currents. Derive the relations

$$\int_S \mathbf{h}_j'^* \cdot \boldsymbol{\epsilon}_t \cdot \mathbf{h}_i' \, dS = \epsilon_1 \int_S \mathbf{h}_j'^* \cdot \mathbf{h}_i' \, dS - j\epsilon_2 \int_S \mathbf{h}_j'^* \cdot \mathbf{e}_i' \, dS = \epsilon_1 \delta_{ij}$$

$$= \int_S \mathbf{h}_j''^* \cdot \boldsymbol{\epsilon}_t \cdot \mathbf{h}_i'' dS, \tag{3a}$$

$$\int_S \mathbf{h}_j'^* \cdot \boldsymbol{\epsilon}_t \cdot \mathbf{h}_i'' \, dS = \epsilon_1 \int_S \mathbf{h}_j'^* \cdot \mathbf{h}_i'' \, dS + j\epsilon_2 \int_S \mathbf{h}_j'^* \cdot \mathbf{e}_i'' \, dS = -j\epsilon_2 \delta_{ij}$$

$$= \int_S \mathbf{h}_j''^* \cdot \boldsymbol{\epsilon}_t \cdot \mathbf{h}_i' \, dS, \tag{3b}$$

and show therefrom that the E-mode amplitudes satisfy the equations

$$-\frac{dV_i'}{dz} = j\kappa_i' Z_i' I_i' + v_i', \qquad -\frac{dI_i'}{dz} = j\kappa_i' Y_i' V_i' + i_i' + \kappa_i' Y_i' \frac{\epsilon_2}{\epsilon_1} V_i'', \tag{4a}$$

and the H modes satisfy

$$-\frac{dV_i''}{dz} = j\kappa_i'' Z_i'' I_i'' + v_i'', \qquad -\frac{dI_i''}{dz} = j\kappa_i'' Y_i'' V_i'' + i_i'' - \kappa_i' Y_i' \frac{\epsilon_2}{\epsilon_1} V_i', \tag{4b}$$

where κ_i, $Z_i = 1/Y_i$, v_i and i_i are the uniaxial medium quantities in Eqs. (7.2.8) and (7.2.9). Show that the uniaxial equations (7.2.7) (with $\epsilon_t \equiv \epsilon_1 \to \epsilon_0$, $\mu_t = \mu_z = \mu_0$) are recovered when $\epsilon_2 = 0$, whence the presence of a non-vanishing ϵ_2 introduces coupling between E and H modes.

(b) A source of inhomogeneity in an anisotropic plasma [see Eqs. (8.3.3b) and (8.3.4)] is a variation in the plasma density or, equivalently, in the parameter $X \equiv (\omega_p/\omega)^2$, with the external magnetic field H_0, and therefore $Y \equiv \omega_c/\omega$, taken to be constant. In this instance, a gyrotropic deviation from the uniaxial case $Y = \infty$ may be analyzed by representing the voltages and currents as series involving integral powers of the small, constant parameter $(1/Y)$:†

$$V_i = \sum_{n=0} \frac{1}{Y^n} V_i^{(n)}, \qquad I_i = \sum_{n=0} \frac{1}{Y^n} I_i^{(n)}. \tag{5}$$

---

†When Y is small, analogous considerations may be applied to develop an expansion in terms of Y^n. The customary ionospheric notation Y should not be confused with the same symbol used elsewhere for characteristic admittance.

Other Y-dependent quantities are expanded in a similar manner, and like coefficients of $(1/Y)^n$ are equated. Show that the zeroth order approximations $V_i^{(0)}$ and $I_i^{(0)}$ are the solutions of the uniaxial medium equations (7.2.7) (with $\epsilon_1 \to \epsilon_0$), while the higher-order terms $n \geq 1$ are obtained from:

$$-\frac{dV_i'^{(n)}}{dz} = j\kappa_i'^{(0)} Z_i'^{(0)} I_i'^{(n)}, \tag{6a}$$

$$-\frac{dI_i'^{(n)}}{dz} = j\kappa_i'^{(0)} Y_i'^{(0)} V_i'^{(n)} + \omega\epsilon_0 X[V_i''^{(n-1)} + V_i''^{(n-3)} + \ldots$$

$$+ j(V_i'^{(n-2)} + V_i'^{(n-4)} + \ldots)], \tag{6b}$$

$$-\frac{dV_i''^{(n)}}{dz} = j\kappa_i''^{(0)} Z_i''^{(0)} I_i''^{(n)}, \tag{6c}$$

$$-\frac{dI_i''^{(n)}}{dz} = j\kappa_i''^{(0)} Y_i''^{(0)} V_i''^{(n)} - \omega\epsilon_0 X[V_i'^{(n-1)} + V_i'^{(n-3)} + \ldots$$

$$- j(V_i''^{(n-2)} + V_i''^{(n-4)} + \ldots)], \tag{6d}$$

where the quantities with index $^{(0)}$ pertain to $Y = \infty$, and quantities with negative index vanish identically. Observe that the voltages and currents of any order $n \geq 1$ satisfy the same basic (zeroth order) transmission-line equations, with the continuously distributed source terms provided by the solutions of lower order. For example, when $n = 2$, show that the lower-order solutions $V_i''^{(1)}$ and $V_i'^{(0)}$ act as current sources for the E modes while $V_i'^{(1)}$ and $V_i''^{(0)}$ play a similar role with respect to the H modes. Equations (6) therefore represent a recursive system wherein the true sources v_i, i_i excite the uniaxial quantities $V_i^{(0)}, I_i^{(0)}$, which in turn generate $V_i^{(1)}, I_i^{(1)}$, etc.

(c) Obtain a typical solution by finding first the Green's functions $Z_i'^{(0)}(z, z')$, $Z_i''^{(0)}(z, z')$ and $[T_i''^I(z, z')]^{(0)}$, $[T_i''^I(z, z')]^{(0)}$ [see Eq. (7.2.15b) and Fig. 7.2.2], and then performing an integration over the source region distributed continuously along z:

$$V_i'^{(1)}(z) = -\omega\epsilon_0 \int Z_i'^{(0)}(z, z')X(z')V_i''^{(0)}(z')dz', \tag{7a}$$

$$V_i''^{(1)}(z) = \omega\epsilon_0 \int Z_i''^{(0)}(z, z')X(z')V_i'^{(0)}(z')dz', \tag{7b}$$

$$V_i'^{(2)}(z) = -\omega\epsilon_0 \int Z_i'^{(0)}(z, z')X(z')[V_i''^{(1)}(z') + jV_i'^{(0)}(z')]dz'$$

$$= -j\omega\epsilon_0 \int Z_i'^{(0)}(z, z')X(z')V_i'^{(0)}(z')dz'$$

$$- (\omega\epsilon_0)^2 \int dz' Z_i'^{(0)}(z, z')X(z') \int dz'' Z_i''^{(0)}(z', z'')X(z'')V_i'^{(0)}(z''), \tag{7c}$$

etc. The integration interval in Eqs. (7) spans the entire longitudinal extent of the source region. Show that the current Green's function T^I appears in the analogous expressions for $I_i^{(n)}$, and that the nth-order approximation involves an n-fold integration of the zeroth-order solutions. Note that since each integral contains the density function $X(z)$, the nth-order term in Eq. (5) involves the factor $[X/Y]^n$. Discuss the convergence properties of the solution.

3. Starting from the eigenvalue equations (8.2.6), when applied to a gyrotropic plasma described by the dielectric tensor in Eq. (8.3.3b), show that the polarization vectors $\mathbf{e}_\alpha$ and $\mathbf{h}_\alpha$ for modes with transverse spatial and time dependence $\exp\left[-j\mathbf{k}_t\cdot\boldsymbol{\rho} + j\omega t\right]$ satisfy the equations

$$\mathbf{P}\cdot\mathbf{e}_\alpha = \kappa_\alpha'^2\mathbf{e}_\alpha, \qquad \tilde{\mathbf{P}}^*\cdot(\mathbf{h}_\alpha \times \mathbf{z}_0) = \kappa_\alpha'^2(\mathbf{h}_\alpha \times \mathbf{z}_0) \tag{8}$$

where $\mathbf{P} = \mathbf{R}\cdot\mathbf{S}$, and

$$\mathbf{R} = \mathbf{1}_t - \mathbf{k}_t'\mathbf{k}_t', \qquad \mathbf{S} = \boldsymbol{\epsilon}_t' - (\mathbf{z}_0 \times \mathbf{k}_t')(\mathbf{z}_0 \times \mathbf{k}_t'). \tag{8a}$$

The notation is that of Eqs. (8.3.10a) and (8.3.13a). Representing $\mathbf{e}_\alpha$ and $\mathbf{h}_\alpha$ in a $\mathbf{k}_{t0}$, $\hat{\mathbf{k}}_{t0}$ vector basis [see Eq. (8.3.8)], show that Eqs. (8) can be solved to yield the result in Eqs. (8.3.13b).

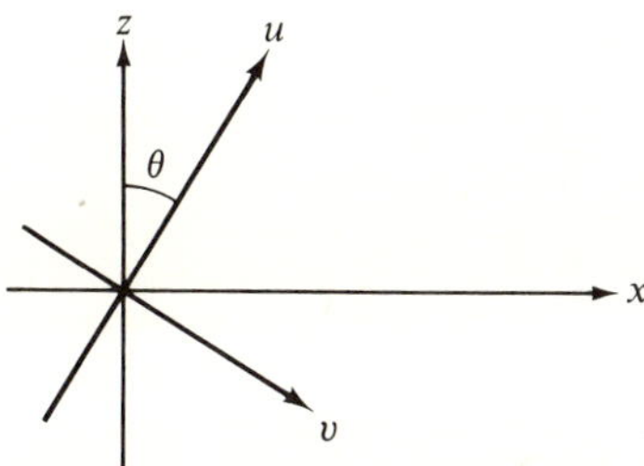

FIG. P8.1   Coordinate systems (y coordinate not shown).

4. In the (v, y, u) coordinate system of Fig. P8.1, a uniaxial medium (with optic axis along u) has the dielectric tensor (normalized to ϵ_0)

$$\boldsymbol{\epsilon} \rightarrow \begin{pmatrix} 1 & 1 & 0 \\ 0 & 1 & 0 \\ 0 & 0 & \epsilon \end{pmatrix}. \tag{9}$$

(a) Show that in the (x, y, z) coordinate frame rotated through an angle θ about the y axis, the $\boldsymbol{\epsilon}$ tensor has the representation:

$$\boldsymbol{\epsilon} \rightarrow \begin{pmatrix} \epsilon_{11} & 0 & \epsilon_{13} \\ 0 & 1 & 0 \\ \epsilon_{13} & 0 & \epsilon_{33} \end{pmatrix} \tag{10}$$

where

$$\epsilon_{11} = \cos^2\theta + \epsilon\sin^2\theta, \qquad \epsilon_{13} = (\epsilon - 1)\sin\theta\cos\theta$$
$$\epsilon_{33} = \epsilon\cos^2\theta + \sin^2\theta. \tag{10a}$$

(b) Show that for a time dependence $\exp(j\omega t)$, x-independent plane-wave solutions of the form

$$\mathbf{E}_a(y, z) = \mathbf{E}_\alpha e^{-jk_0\bar{k}_t y - jk_0\bar{\kappa}_\alpha z}, \qquad \bar{k}_t = k_t/k_0,\ \bar{\kappa}_\alpha = \kappa_\alpha/k_0$$
$$\mathbf{H}_a(y, z) = \mathbf{H}_\alpha e^{-jk_0\bar{k}_t y - jk_0\bar{\kappa}_\alpha z} \tag{11}$$

can exist in this medium provided that $\bar{\kappa}_\alpha$ takes on one of the four values:

$$\bar{\kappa}_\alpha(\bar{k}_t) = \pm \kappa_o, \qquad \kappa_o = \sqrt{1 - \bar{k}_t^2}$$

$$\bar{\kappa}_\alpha(\bar{k}_t) = \pm \kappa_e, \qquad \kappa_e = \frac{1}{\sqrt{\epsilon_{33}}} \sqrt{\epsilon - \bar{k}_t^2}. \tag{11a}$$

κ_o describes ordinary modes which propagate as in vacuum while κ_e describes extraordinary modes. Interpret these $\bar{\kappa}$ versus $\bar{k}_t$ solutions by referring to the wavenumber surfaces as in Fig. 7.1.1, noting that the (x, y, z) coordinate system does not coincide with the principal axes of the medium, and that for the waves considered, one has $k_x \equiv 0$. Show that $\kappa_{o,e} > 0$ and $\kappa_{o,e} < 0$ describe waves carrying energy in the $+z$ and $-z$ directions, respectively.

(c) Show that with $\kappa_{o,e}$ defined as in Eq. (11a), the polarization vectors $\mathbf{E}_\alpha$ and $\mathbf{H}_\alpha$ in Eq. (11) have the form:

$$\mathbf{E}_\alpha \to \mathbf{E}_{o,e} = \mathbf{x}_0 \pm g_{o,e}\mathbf{y}_0 - h_{o,e}\mathbf{z}_0 \tag{12}$$

where

$$g_{o,e} = \frac{\bar{k}_t \kappa_{o,e}(\epsilon_{11} - \bar{k}_t^2 - \kappa_{o,e}^2)}{\epsilon_{13}(1 - \kappa_{o,e}^2)}. \tag{12a}$$

and

$$h_{o,e} = \frac{\epsilon_{11} - \bar{k}_t^2 - \kappa_{o,e}^2}{\epsilon_{13}}. \tag{12b}$$

Also,

$$\mathbf{H}_\alpha \to \mathbf{H}_{o,e} = \frac{k_0}{\omega\mu_0}[-(\kappa_{o,e}g_{o,e} + \bar{k}_t h_{o,e})\mathbf{x}_0 \pm \kappa_{o,e}\mathbf{y}_0 - \bar{k}_t\mathbf{z}_0]. \tag{13}$$

Upper and lower signs correspond to subscripts o and e, respectively. Note that although the eigenvalues $\bar{\kappa}_\alpha$ occur in pairs $\pm\kappa_o$, $\pm\kappa_e$, the transverse (to z) eigenvectors corresponding to $+\kappa_{o,e}$ differ from those corresponding to $-\kappa_{o,e}$.

(d) Show that the mode fields in Eq. (11), with Eqs. (12)-(13), reduce to E modes and H modes with respect to u (i.e., $H_{eu} = 0$ and $E_{ou} = 0$, respectively).

(e) Verify that the modes in Eqs. (12)-(13) satisfy the orthogonality condition in Eq. (8.2.10b).

5. A perfectly conducting cylindrical obstacle embedded in a homogeneous cold magnetoplasma is excited by a line source of magnetic currents of strength M, the arrangement being such that the cylinder axis, gyrotropic axis, and line-source direction are parallel to the z axis. Assuming that in an orthogonal, separable, curvilinear (u, v) coordinate system, the cylinder boundary s in a plane transverse to z is described by the equation $u = $ constant, show that for an $\exp(j\omega t)$ dependence, the electromagnetic fields can be derived from the single component $H_z \equiv H$ satisfying the equation

$$(\nabla_t^2 + k_0^2\epsilon_\perp)H = j\omega\epsilon_0\epsilon_\perp M, \tag{14}$$

subject to a radiation condition at infinity and the boundary condition

$$j\frac{\epsilon_2}{\epsilon_1}\frac{1}{h_v}\frac{\partial H}{\partial v} - \frac{1}{h_u}\frac{\partial H}{\partial u} = 0 \text{ on } s, \tag{14a}$$

where $\epsilon_\perp = (\epsilon_1^2 - \epsilon_2^2)/\epsilon_1$ [see Eqs. (8.4.5b) and (8.3.4)], $\nabla_t = \mathbf{u}_0(\partial/h_u\partial u) + \mathbf{v}_0(\partial/h_v\partial v)$, and h_u, h_v are the metric coefficients in the (u, v) coordinate system. Show that while the method of separation of variables can be employed to solve Eq. (14), separability does not obtain in general for the boundary condition in Eq. (14a) although the boundary coincides with a curve $u = $ constant. However, if the v-dependent orthogonal eigenfunctions $\psi_i(v)$ are such that

$$\frac{1}{h_v}\frac{\partial\psi_i}{\partial v} = \alpha_i\psi_i, \tag{15}$$

where α_i does not depend on v, and if h_u is also independent of v, show that the representation

$$H(u, v) = \sum_i I_i(u)\psi_i(v) \tag{16}$$

leads to a separable form. Note that included among the separable configurations are the plane boundary $u \equiv x = $ constant, whence $h_u \equiv 1$ and $\psi_i(v) \equiv \psi_i(y) \propto \exp(-j\eta y)$, $-\infty < \eta < \infty$ (cf. Sec. 8.4c), and also the circular cylindrical boundary $u \equiv \rho = $ constant, whence $h_u \equiv 1$ and $\psi_i(v) \equiv \psi_i(\phi) \propto \exp(-jn\phi)$, $n = 0, \pm1, \pm2, \ldots$. Construct the solution for the electromagnetic fields in the presence of the circular cylindrical scatterer and compare its form with that for the isotropic medium ($\epsilon_2 = 0$) in Sec. 6.7.

6. Calculate the electromagnetic fields and the power flow properties for the line source in Fig. 8.4.1 in an infinite medium (in the absence of the conducting screen at $z = 0$). Place a source at the image point $(-z', 0)$ and calculate the fields at $z = 0$. Referring to the boundary condition (8.4.9), show that the image source is incapable of accounting properly for the effect of a perfect conductor at $z = 0$. Explain this conclusion in reference to Fig. 8.4.3 and propose an alternative procedure.

7. The sectoral region between two intersecting, perfectly conducting half-planes inclined at an angle $\varphi < \pi$ is filled with a gyrotropic dielectric as in Eq. (8.3.3b), the gyrotropic axis being parallel to the apex. Referring to Eq. (6.5.13) or (6.5.16b) for the formulation in the isotropic case, with $J_\mu(k\rho)$ replaced by the leading terms in its expansion as $\rho \to 0$, construct expressions for the electromagnetic fields near the apex when the medium is gyrotropic and the polarization is such that $\mathbf{H} = \mathbf{z}_0 H$. Calculate the energy stored in the vicinity of the apex (cf. Sec. 1.5b) for the cases of vanishing and non-vanishing dissipation in the medium. To impose a "boundary condition" near the apex, show that the lossless case must be approached as the limiting case of a dissipative medium, and that direct consideration of a purely lossless medium poses an ill-defined problem.[13,14]

REFERENCES

1. CLEMMOW, P. C., *The Plane Wave Spectrum Representation of Electromagnetic Fields*, Sec. 8.2. New York: Pergamon Press, 1966.

2. ARBEL, E. and L. B. FELSEN, "Theory of radiation from sources in anisotropic media. Part I—General sources in stratified media. Part II—Point source in

an infinite homogeneous medium," in *Electromagnetic Theory and Antennas* (E. C. Jordan, Ed.), pp. 391-459. New York: Pergamon Press, 1963.

3. CLEMMOW, P. C. and F. MULLALY, "The dependence of the refractive index in magneto-ionic theory on the direction of the wave normal," in *The Physics of the Ionosphere*, p. 340. London, England: The Physical Society, 1955.

4. ALLIS, W. P., S. J. BUCHSBAUM and A. BERS, *Waves in Anisotropic Plasmas*, Chapter 3. New York: John Wiley Sons. 1962.

5. DESCHAMPS, G. A. and W. L. WEEKS, "Charts for computing the refractive indexes of a magneto-ionic medium," *IRE Trans. on Antennas and Propagation* **AP-10** (1962), p. 305.

6. SESHADRI, S. R. and T. T. WU, "Radiation condition for a magnetoplasma medium, "*Quart. J. of Mech. and Appl. Math.* **23** (1970), Part 2, pp. 285-313.

7. BUNKIN, F. V., "On radiation in anisotropic media," *J. Exp. Theoret. Phys. (USSR)* **32** (1957), p. 338.

8. KOGELNIK, H. and H. MOTZ, "Electromagnetic radiation from sources embedded in an infinite anisotropic medium and the significance of the Poynting vector," in *Electromagnetic Theory and Antennas* (E. C. Jordan, Ed.), pp. 477-493. New York: Pergamon Press, 1963.

9. BARSUKOV, K. A., "Radiation of electromagnetic waves from a point source in a gyrotropic medium with a separation boundary," *Radiotekh. i Elektronika* **4** (1959), p. 1759.

10. BUDDEN, K. G., *Radio Waves in the Ionosphere*, Chapter 18. Cambridge, England: Cambridge University Press, 1961.

11. GROSS, S. and L. B. FELSEN, "Propagation in non-uniform gyrotropic media," *Radio Science (NBS)* **69D** (1965), pp. 333-348.

12. SESHADRI, S. R., "Excitation of surface waves on a perfectly conducting screen covered with anisotropic plasma," *IRE Trans. on Microwave Theory and Techniques* **MTT-10** (1962), pp. 573-578.

13. ISHIMARU, A., "Unidirectional surface waves in anisotropic media," in *Electromagnetic Theory and Antennas* (E. C. Jordan, Ed.), p. 591-601. New York: Pergamon Press, 1963.

14 HURD, R. A., "On the possibility of intrinsic loss occurring at the edges of ferrites," in *Electromagnetic Theory and Antennas* (E. C. Jordan, Ed.), p. 569-571. New York: Pergamon Press, 1963.

15. TAI, C. T., "On the transposed radiating systems in an anisotropic medium," *IRE Trans. on Antennas and Propagation* **AP-9** (1961), p. 502.

16. SESHADRI, S. R., "Scattering of unidirectional surface waves," *IEEE Trans. on Microwave Theory and Techniques* **MTT-11** (1963), p. 238.

Subject Index

Author Index